5th International Conference on Energy, Power and Advanced Thermodynamic Systems (EPATS 2025)

Journal of Physics: Conference Series Volume 3011

Kunming, China
21-23 February 2025

ISBN: 979-8-3313-2152-9
ISSN: 1742-6588

Printed from e-media with permission by:

Curran Associates, Inc.
57 Morehouse Lane
Red Hook, NY 12571

proceedings.com

Some format issues inherent in the e-media version may also appear in this print version.

Institute of Physics
Dirac House, Temple Back
Bristol BS1 6BE UK

Phone: 44 1 17 929 7481
Fax: 44 1 17 920 0979

techtracking@iop.org

Additional copies of this publication are available from:

Curran Associates, Inc.
57 Morehouse Lane
Red Hook, NY 12571 USA
Phone: 845-758-0400
Fax: 845-758-2633
Email: curran@proceedings.com
Web: www.proceedings.com

TABLE OF CONTENTS

Journal of Physics: Conference Series **3011** (2025) 011001 doi:10.1088/1742-6596/3011/1/011001

Preface

The 2025 5th International Conference on Energy, Power and Advanced Thermodynamic Systems (EPATS 2025), held in Kunming, China, from February 21 to 23, 2025. This year's conference, hosted by Tianjin University, China, was conducted entirely online, bringing together a diverse and global community of researchers, academics, and industry professionals. EPATS 2025 served as a platform for the exchange of cutting-edge ideas, innovative research, and practical solutions in the fields of energy, power, and advanced thermodynamic systems. The conference underscored the critical role of interdisciplinary collaboration in addressing the pressing challenges of energy sustainability, efficiency, and resilience in a rapidly evolving world.

The conference featured three distinguished keynote speakers, each of whom delivered thought-provoking presentations that set the tone for the discussions that followed. Assoc. Prof. Muhyiddine Jradi from the University of Southern Denmark, Denmark, opened the keynote sessions with his presentation. His insights into the integration of digital twins in energy management highlighted the transformative potential of digital technologies in optimizing energy systems. Prof. Ting Yang from Tianjin University, China, followed with a presentation, emphasizing the importance of intelligent systems in advancing renewable energy integration. Finally, Prof. Qiang Yang from Zhejiang University, China, concluded the keynote sessions with his talk on "Artificial Intelligence Empowered Power Infrastructures", exploring the pivotal role of AI in enhancing the safety and reliability of power infrastructures. These keynote speeches not only provided a comprehensive overview of the latest advancements but also inspired meaningful discussions among participants.

We extend our heartfelt gratitude to all those who contributed to the success of EPATS 2025. First and foremost, we thank the attendees for their active participation and engagement throughout the conference. We are deeply appreciative of the efforts of the reviewers, whose meticulous evaluation ensured the high quality of the presentations and proceedings. Special thanks go to our keynote speakers, Assoc. Prof. Muhyiddine Jradi, Prof. Ting Yang, and Prof. Qiang Yang, for their invaluable contributions and thought leadership. We also acknowledge the dedication of the authors, whose innovative research formed the backbone of this conference. Lastly, we express our sincere appreciation to the organizing committees, whose tireless work behind the scenes made this event possible.

The Conference General Chairs, Prof. Lingai Luo from the University of Nantes, France, Prof. Jeyraj A/L Selvaraj from Universiti Malaya, Malaysia, and Prof. Charles Chun Yang from Nanyang Technological University, Singapore, played a pivotal role in guiding the conference to its successful conclusion. Their leadership, vision, and commitment were instrumental in shaping EPATS 2025 into a landmark event in the field.

As we look to the future, we are optimistic that the insights and collaborations forged at this conference will pave the way for groundbreaking advancements in energy, power, and thermodynamic systems. We eagerly anticipate the next edition of EPATS, where we will continue to push the boundaries of innovation and sustainability.

EPATS 2025 Conference Committee

Journal of Physics: Conference Series **3011** (2025) 011001 doi:10.1088/1742-6596/3011/1/011001

Committee Member

General Chairs
Prof. Lingai Luo, University of Nantes, France
Prof. Jeyraj A/L Selvaraj, Universiti Malaya (UM), Malaysia
Prof. Charles Chun Yang, Nanyang Technological University, Singapore

Program Committee Chairs
Prof. Santolo Meo, University of Naples Federico II, Italy
Prof. Dimitrios Karamanis, University of Patras, Greece
Prof. Qiang Yang, Zhejiang University, China

Publication Chair
Assoc. Prof. Muhyiddine Jradi, University of Southern Denmark, Denmark

Technical Program Committee
Prof. Fushuan Wen, Zhejiang University, China
Prof. Pierluigi Siano, University of Salerno, Italy
Prof. Ting Yang, Tianjin University, China
Prof. Khalil El-Hami, University of Mohammed 5 of Rabat, Morocco
Prof. I.Neethi Manickam, Renganayagi Varatharaj College of Engineering, India
Assoc. Prof. Muhyiddine Jradi, University of Southern Denmark, Denmark
Assoc. Prof. Ranjith kumar Gatla, Institute of Aeronautical Engineering, India
Assoc. Prof. Xiaomin Kang, University of South China, China
Assoc. Prof. Bowen Zhou, Northeastern University, China
Assoc. Prof. Zafar Said, University of Sharjah, Sharjah, UAE
Senior Researcher Yangyang Zhao, State Grid (Suzhou) City and Energy Research Institute, China
Senior Engineer Yizheng Xu, China Electric Power Research Institute, China
Dr. Gang Huang, Zhejiang University, China
Dr. Qianzhi Zhang, Zhejiang University, China
Dr. Chuangxin Guo, Zhejiang University, China

Journal of Physics: Conference Series **3011** (2025) 011002 doi:10.1088/1742-6596/3011/1/011002

Peer Review Statement

All papers published in this volume have been reviewed through processes administered by the Editors. Reviews were conducted by expert referees to the professional and scientific standards expected of a proceedings journal published by IOP Publishing.

- **Type of peer review:** Single Anonymous
- **Conference submission management system:** Morressier
- **Number of submissions received:** 187
- **Number of submissions sent for review:** 149
- **Number of submissions accepted:** 112
- **Acceptance Rate (Submissions Accepted / Submissions Received × 100):** 59.9
- **Average number of reviews per paper:** 2
- **Total number of reviewers involved:** 20
- **Contact person for queries:**
  Name: Xuexia Ye
  Email: xx.ye@keoaeic.org
  Affiliation: AEIC Academic Exchange Information Centre

Load control technology for cascade optimization operation of comprehensive energy system based on combination variable weight

Ting Yu[1*], Hui Shi[2], Xiaoye Gu[1], Xurong Jin[1], Han Wang[2]

[1]State Grid Ningxia marketing service center (State Grid Ningxia metrology center), Ningxia, Yinchuan, 750001, China;
[2]State Grid NingDong Power Supply Company, Ningxia, Yinchuan, 750411, China.

[*]Corresponding author's e-mail: 15729543036@163.com

Abstract. The load cascade status data of the integrated energy system belongs to multidimensional and multi-source data, and there are uncertain changes. The predicted results have high errors compared to the actual situation, resulting in high network losses and low load balancing rates, which urgently need to be controlled. Combining variable weights to respond to the demand for cascade loads in the integrated energy system and using the variational mode decomposition method to decompose the cascade load sequence data and construct a load forecasting model. Construct the objective function and constraints for load control in the cascade optimization operation of the integrated energy system and use an improved immune ant colony algorithm to solve the objective function of the load prediction model, achieving load control in the cascade optimization operation of the integrated energy system. The experimental results show that the proposed method has a low error between the predicted results and the actual situation of the cascade load in the integrated energy system. Load control can reduce network losses and improve the load balancing rate.

1. Introduction

In a comprehensive energy system, there are multiple forms of energy, such as electricity, heat, gas, etc. The demand for these energy sources varies in time and quantity among user loads. Many users have unreasonable energy load usage, such as some industrial users maintaining the operation of high-power equipment during nonpeak production hours or commercial buildings not wholly shutting down equipment such as air conditioning when there is no one at night. Load control can identify these unreasonable load situations and reduce energy waste by adjusting the operating mode of equipment or shutting down unnecessary equipment. There are supply fluctuations in comprehensive energy supply sources such as power plants, wind farms, and solar power stations. Through load control, some noncritical loads can be reduced to ensure a stable supply of essential loads and improve the reliability of the entire energy supply system [1–2]. Therefore, reference [3] utilizes multi-agent deep reinforcement learning algorithms to control the disturbance load of large-scale integrated energy access to the power grid. The priority sampling mechanism based on the confidence interval upper bound strategy replaces traditional deep Q-learning's uniform random sampling mechanism to solve the load frequency control model. Reference [4] proposes a comprehensive energy system supply and demand dual side load automation control technology based on the PSO-IVACO algorithm. Adopting the stepwise regression method to determine the main influencing factors of energy demand, constructing the objective function and constraint conditions for load control on both sides of supply and demand, and using the PSO-IVACO algorithm to solve the objective function to achieve load automation control. Reference [5] designed a two-stage load optimization method for regional integrated energy system operation that considers the uncertainty of energy fluctuations. Analyze the dynamic changes in the proportion of comprehensive energy and construct a two-stage collaborative optimization model for operation. Implementing load control based on fuzzy chance-constrained planning method.

Based on the above research results, study the load control technology for cascade optimization operation of comprehensive energy systems based on combination variable weight.

2. Comprehensive Energy System Cascade Load Demand Response Matrix Based on Combination Variable Weight

The optimization control of integrated energy system operation load requires a description of the probability distribution of uncertain parameters in response to the above random loads:

$$C\left(\chi(t)\right)=\frac{\chi(t)}{\beta(t)}z\left(\frac{\alpha(t)}{\beta(t)}\right)\exp\left(-\left(\frac{\alpha(t)}{\beta(t)}\right)^{\phi-1}\right) \tag{1}$$

The formula contains the load scale and distribution parameters at the time t; ϕ is the variance; and $z(\cdot)$ is the gamma function.

Under the condition of random energy input [6-7], it is necessary to consider the probability distribution of parameters contained in each type of energy. When controlling the load of cascade operation under different energy input conditions, it is directly used as a fuzzy variable. By considering the time dimension of each parameter within 24 hours, a random parameter probability extraction matrix is established, which can characterize the load demand of cascade operation [8-10]:

$$\left[C\left(\chi(t)\right)\right]=\begin{bmatrix} C_{S_{1,1}} & C_{S_{1,2}} & \cdots & C_{S_{1,f}} \\ C_{S_{2,1}} & C_{S_{2,2}} & \cdots & C_{S_{2,f}} \\ \cdots & \cdots & \cdots & \cdots \\ C_{S_{K,1}} & C_{S_{K,2}} & \cdots & C_{S_{K,f}} \end{bmatrix} \tag{2}$$

In the formula, K is the dimension; f is the number of samples on each gradient; and is the probability of load occurrence.

According to the random probability matrix and the distribution function of various parameters, the response matrix of the cascade operating load demand of the integrated energy system is formed, which is expressed as:

$$L=\begin{bmatrix} w_1V_1C_{S_{1,1}} & w_1V_1C_{S_{1,2}} & \cdots & w_1V_1C_{S_{1,f}} \\ w_2V_2C_{S_{2,1}} & w_2V_2C_{S_{2,2}} & \cdots & w_2V_2C_{S_{2,f}} \\ \cdots & \cdots & \cdots & \cdots \\ w_nV_nC_{S_{K,1}} & w_nV_nC_{S_{K,2}} & \cdots & w_nV_nC_{S_{K,f}} \end{bmatrix} \tag{3}$$

In the formula, w_n is the cumulative distribution combination weight of load probability, and V_1, V_2,..., V_D are the load planning scenarios for cascade operation. By combining random parameters and varying weights, it is possible to respond to different sets of load gradient scenarios corresponding to changes in comprehensive energy system load demand, and based on this, load forecasting can be carried out.

3. Comprehensive energy system load forecasting based on the demand response matrix of cascade operation load

In the load control process in integrated energy systems, it is necessary to predict the future load changes of the integrated energy system based on the cascade operation load demand response matrix as the basis for subsequent load control.

In summary, considering the significant compound fluctuations in the cascade operation load of the integrated energy system, to ensure prediction accuracy, the variational mode decomposition method is adopted to decompose the cascade load sequence data:

$$\delta(L)=\sum_{i=1}^{N} g_s\left(\frac{\varphi a_s(L)}{v_i}\right) \tag{4}$$

In the formula, N represents the number of load data points; g_s represents the Dirac function; a_s represents the modal function; φ represents the decomposition coefficient; v_i represents the load fluctuation state matrix.

Based on the differential load supply and supply decomposition principle of the integrated energy system network, the cascade load sequence is decomposed into the following combinations, namely:

$$W_y = W_s + W_e + W_o \tag{5}$$

In the formula, W_s, W_e, and Wo represent the integrated energy system's power load, generating power load, and interactive power load, respectively.

The load prediction model is constructed with the regional power load curve. The output load prediction results are expressed as follows:

$$Q=\varphi_k \cdot W_y \varsigma\left(-\frac{\delta(L)}{\chi_t}\right)/h_s \tag{6}$$

In the formula, χ_t is the correlation factor for predicting the daily maximum power; h_s represents the prediction parameter, φ_k represents the real-time load data, and ς represents the time class coding function of the load data.

By performing modal decomposition on the integrated energy system's historical load data and constructing a load forecasting model combined with real-time load data, the future load of the integrated energy system is predicted, providing a foundation for further load control.

4. Implementation of load control for cascade optimization operation of integrated energy system based on load forecasting

Based on the results of power load forecasting, the goal is to minimize the difference between load regulation and load demand and establish a load control objective function for the cascade optimization operation of the comprehensive energy system.

$$\min G = \Delta t \sum [Q_{\min,t} - \sum_{a=1}^{N_q} Q_{t,a}] \tag{7}$$

In the formula, Δt represents the time interval; $Q_{min,\,t}$ represents the minimum load prediction value of the integrated energy system in the t period; N_q represents the total number of adjustable loads involved in the cascade optimized operation load control of the integrated energy system; $Q_{t,\,a}$ represents the adjustable load of the integrated energy system source a during the t period.

Constraints for setting the objective function of the integrated energy system: The load operating power constraints are:

$$0 \le Q_{t,a} \le Q_{t,a}^{\max} \tag{8}$$

In Eq, $Q_{t,\,a}^{max}$ represents the maximum value of $Q_{t,\,a}$.

The curve of the power system is affected by the fluctuation of the wind output, so it is necessary to set the load climbing rate constraint conditions:

$$\Delta A_{A,a}^{d} \le Q_{t,a} - Q_{t-1,a} \le \Delta Q_{t,a}^{u} \tag{9}$$

In the formula, $\Delta A_{A,\,a}^{down}$ and $\Delta A_{A,\,a}^{up}$ represent the maximum values corresponding to the upper and lower climbing rate of the energy output a, respectively.

Journal of Physics: Conference Series **3011** (2025) 012001 doi:10.1088/1742-6596/3011/1/012001

Calculating antibody affinity can determine whether it is close to or has reached the optimal solution to the problem. The calculation formula is:

$$\gamma\left(\eta(b)\right)=\frac{f\left(Q_j\right)}{1+\varepsilon\ln\left(1+\eta(b)\right)} \tag{10}$$

In the formula, $f(Q_j)$ represents the fitness of Q_j; ε represents the constant.

If it meets the conditions, stop the algorithm iteration and output the globally optimal solution obtained through optimization, which is the load control scheme for the cascade optimization operation of the integrated energy system.

5. Experiments and Analysis

5.1. Experimental Environment Settings

In the experiment, the KV75mvA-P magnetic balance voltage sensor, ACS712 current sensor, and A44 power sensor were used to collect load data of the integrated energy system. This paper uses the improved ant colony algorithm to solve the control objective function, thus setting the algorithm's parameters, as shown in Table 1.

Table 1. Shows the improved parameter settings of the immune ant colony algorithm.

Algorithm parameters	Magnitude
Ant group scale	50
Iterations	100
Coefficients of contraction expansion	0.5 to 1.0
Quantum potential trap parameters	1.0
Probability of local search	0.2
Inertial weight	0.9 to 0.4
Social learning factors	2.0
Self-cognitive factors	2.0
Ant location range	[0, 1]

5.2. Analysis of the experimental results

Two additional sets of traditional control methods were selected as the control group from literature [3] and literature [4] to verify the study method's validity and applicability. Test the ability of different techniques to predict the cascade load of the integrated energy system, and the prediction results are shown in Figure 1.

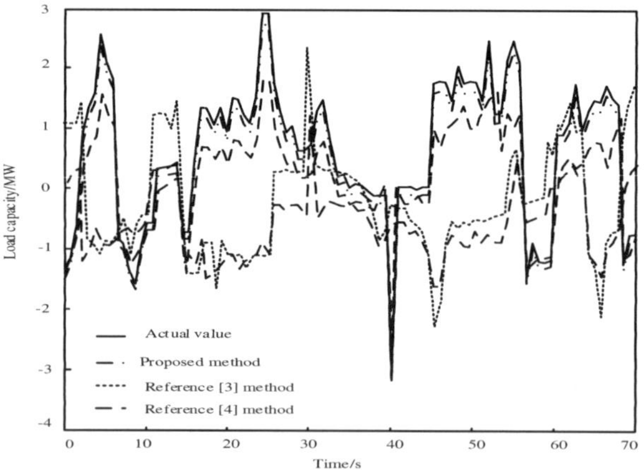

Figure 1. Prediction of Cascade Load of Integrated Energy System under Different Methods Application.

In Figure 1, the method proposed in this paper has a prediction result for the cascade load of the integrated energy system that is closer to its proper situation.

Calculate the hourly network loss of the integrated energy system under different control methods within a day, as shown in Figure 2:

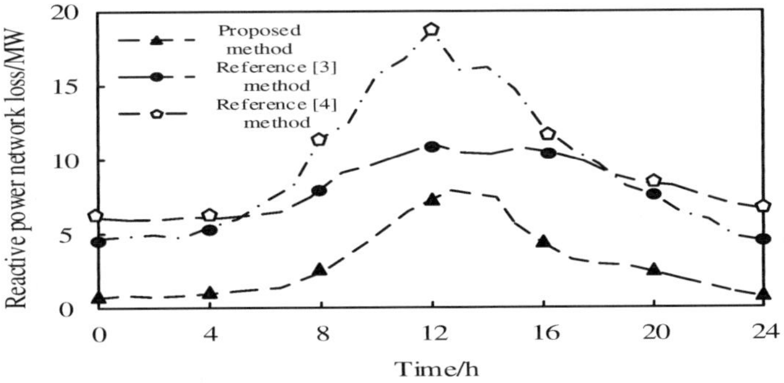

Figure 2. Comparison of network losses in integrated energy systems under different control methods.

From Figure 2, it can be seen that the proposed method has significantly lower network losses than the load control methods proposed in references [3] and [4]. This indicates that the proposed method performs better in reducing network losses, helping to reduce overload risks and improve equipment utilization. In other words, the proposed method can more effectively control the reactive power of the integrated energy system and reduce energy losses.

The higher the load balancing rate, the better the performance of the integrated energy system under the corresponding method control. Calculate the load balancing rate of the integrated energy system under different control methods within one day, as shown in Figure 3:

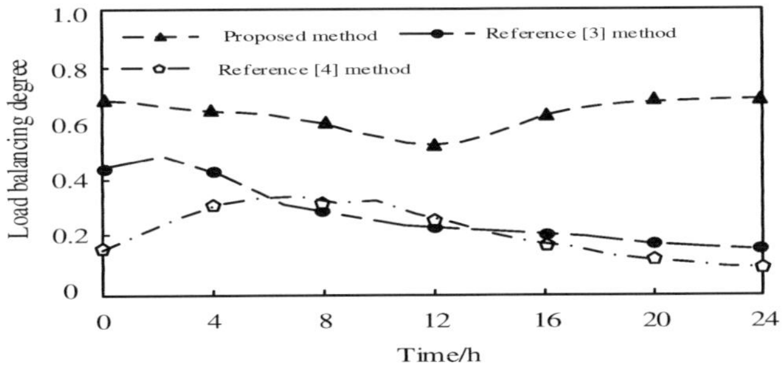

Figure 3. Comparison of load balancing rates of integrated energy systems under different control methods.

Figure 3 shows that the proposed method's load-balancing index is significantly higher than the other two load-balancing control methods. This indicates that the proposed method performs better in load balancing, can more effectively adjust the load of the integrated energy system, and improves the overall operational performance of the integrated energy system.

6. Conclusion

The article studies load control technology for cascade optimization operations of integrated energy systems based on a combination of variable weights. The test results show that the proposed method can accurately predict the cascade load of the integrated energy system based on load monitoring data, optimize load control, effectively reduce network loss, improve load balancing rate, and enhance the operational safety and stability of the integrated energy system.

Journal of Physics: Conference Series **3011** (2025) 012001 doi:10.1088/1742-6596/3011/1/012001

Reference

[1] Hwang H, Yoon A, Moon Y S. Demand response of HVAC systems for hosting capacity improvement in distribution networks: A comprehensive review and case study. Renewable & sustainable energy reviews, 2023, 187(11): 113751.1-113751.21.
[2] Veeramsetty V, Reddy K R, Santhosh M, et al. Short-term electric power load forecasting using random forest and gated recurrent unit. Electrical Engineering, 2022, 104(1):307-329.
[3] Xi L, Wang Y H, Chen S S, et al. Multi-agent collaborative AGC strategy for the integrated energy system. Electric Machines and Control, 2022, 26 (4): 77-88.
[4] Xie X Y, Liu X X, Li C, et al. Automatic Control Technology of Supply and Demand Bilateral Load in Smart City Integrated Energy System. Automation & Instrumentation, 2021, 000 (10): 97-100.
[5] Wang F, Gong D D, Guo H, et al. Two-stage optimization of regional integrated energy system planning-operation with dynamic hydrogen pricing and uncertainties. Power System Protection and Control, 2022, 50 (13): 53-62.
[6] Sekhar C, Dahiya R. Robust framework based on hybrid deep learning approach for short term load forecasting of building electricity demand. Energy, 2023, 268(1): 126660.1-126660.17.
[7] Alizadeh Bidgoli M, Ahmadian A. Multi-stage optimal scheduling of multi-microgrids using deep-learning artificial neural network and cooperative game approach. Energy, 2022, 239(1): 122036.1-122036.15.
[8] NIU Zhi-liang, ZHENG Jian-yong, MEI Fei, et al. United Fast State Estimation Combined Heat and Power Based on Deep Neural Network[J]. Computer Simulation, 2023, 40(11): 58-63.
[9] Harrison, Ambe, Nguimfack-Ndongmo, Jean De Dieu, Alombah, Njimboh Henry, et al. Robust nonlinear MPPT controller for PV energy systems using PSO-based integral backstepping and artificial neural network techniques[J]. International Journal of dynamics and control, 2024, 12(5): 1598-1615.
[10] Ordonez J G., Barco-Jimenez J, Pantoja A., et al. Comprehensive analysis of MPC-based energy management strategies for isolated microgrids empowered by storage units and renewable energy sources[J]. Journal of Energy Storage, 2024, 94(Jul.): 1.1-1.20.

Journal of Physics: Conference Series 3011 (2025) 012002 doi:10.1088/1742-6596/3011/1/012002

A two-stage ultra short-term load forecasting model for comprehensive energy based on FCM

Zhaoye Liu[1], Hongda Bi[1], Sihui Zhu[1], Zheng Liu[2], Botao Zhou[2*]

[1]State Grid Heilongjiang Electric Power Co., Ltd, Harbin, 150000, China
[2]State Grid Electric Power Research Institute Wuhan Efficiency Evaluation Company Limited, Wuhan, 430074, China

[*]Corresponding author's e-mail: 150712822014@163.com

Abstract. The dynamic scheduling requirements of a two-stage integrated energy system with single-layer variable step size are complex and diverse. When conducting ultra short-term load forecasting, it is difficult to confirm the forecasting results of different types of loads, and the degree of fitting with actual values is low. Therefore, a comprehensive energy two-stage ultra short-term load forecasting model based on FCM is proposed. In this study, we use DTW distance and Pearson distance to describe the horizontal and vertical shape similarity between load curves as initial clustering centers, set the number of cluster centers in FCM to match the type of load, iterate the membership degree of load to cluster centers and cluster center matrix using the comprehensive distance between load curves until the result conditions are met, and output the clustering results. Additionally, we calculate the distribution of predicted samples within the corresponding range of each clustering result and output two-stage ultra short-term load forecasting results. In the test results, the designed model achieved a high degree of fit between the predicted electric and hydrogen loads of the two-stage integrated energy system and the actual values, providing assistance in ensuring the smooth operation of the integrated energy system.

1. Introduction

Ultra short-term load forecasting usually refers to predicting the future minute-level power load, which can reduce the generation of redundant backup capacity and enable the production capacity equipment of the integrated energy system to work in a more efficient state [1]. How to better plan primary energy and achieve mutual conversion between various energy sources through ultra-short-term load forecasting is currently a research hotspot.

Martins et al. [2] conducted cluster analysis on historical data sets based on load time series of historical dates and further screened and predicted using temperature time series. However, this method ignores the influence of other real-time changing environmental factors and user subjective factors. Lin W et al. [3] decomposed power load data using EEMD to obtain modal components with different characteristics and optimized the BP neural network using the Adaboost algorithm to improve prediction accuracy. However, BP neural networks are prone to getting stuck in local optima, which increases the complexity of the model. LI Xiao et al. [4] used a temporal attention mechanism to autonomously select historical information of key time points in the GRU network, improving the stability of long-term prediction performance. However, its strong dependence on historical data affects the prediction results. Rafati et al. [5] established a more accurate load estimation model by integrating output data and historical load data of distributed photovoltaics. However, this method did not take into account the randomness and volatility of photovoltaic output, resulting in insufficient accuracy in load forecasting.

In summary, this article involves the design and research of a two-stage ultra-short-term load forecasting model for comprehensive energy based on FCM, aiming to improve the accuracy of ultra-short-term load forecasting.

2. Design of a two-stage ultra short-term load forecasting model for integrated energy

2.1. Selection of load clustering centers for two-stage integrated energy

This article draws on the idea of a density peak clustering algorithm and uses DTW distance to describe the horizontal shape similarity between load curves based on their characteristics. We employ Pearson distance to describe the vertical shape similarity between load curves. Samples with high local density and far distance from high-density samples in all load curves are selected as the initial clustering centers [6-7].

We define the local density ρ_i of each load curve as:

$$\rho_i = \sum_{j \neq i} \exp\left(-\frac{d_{dtw}(x_i, x_j)^2}{d_c} \right) \tag{1}$$

where $d_{dtw}(x_i, x_j)$ is the DTW distance between load curves x_i and x_j, and d_c is the truncation distance. d_c can be determined based on the number of samples. Rodriguez et al. pointed out that the appropriate d_c should be such that the proportion of samples within the range of d_c for each sample point on average accounts for $\tau(1\% - 2.0\%)$ of the entire dataset. In this paper, $\tau = 2.0\%$ is selected.

We define the minimum distance δ_i between each load curve and curves with higher density within a range of d_c:

$$\delta_i = \min_{\rho_i > \rho_j} \left(d_{pea}(x_i, x_j) \right) \tag{2}$$

where $d_{pea}(x_i, x_j)$ is the Pearson distance between load curves x_i and x_j.

Combining Equation (1), for load curve x_i with the highest density globally, there exists $\delta_i = \max_{\rho_i > \rho_j} \left(d_{pea}(x_i, x_j) \right)$.

In order to make the selected initial cluster centers dispersed and have significant morphological differences, sample points with local density ρ_i and distance δ_i from the high-density curve that are both large in all load curves are selected as the initial cluster centers. Therefore, we let $\gamma = \rho_i \delta_i$ calculate the γ values of all curves, sort them in descending order, and select the γ largest load curves as the initial clustering centers for the algorithm.

2.2. FCM-based comprehensive energy two-stage load clustering

This paper performs clustering processing on the original two-stage load of comprehensive energy [8]. Using Equations (1) and (2), we calculate the local density of each load curve and the product of the distance between each curve and its high-density curve. We sort the γ values of all load curves in descending order [9] and select the top K load curves corresponding to the same number of clusters as the initial clustering center of the FCM algorithm. The fusion value of Pearson distance and DTW distance between all load curves is calculated, which is the comprehensive distance $D(x_i, x_j)$ between load curves. This can be expressed as:

$$D(x_i, x_j) = \omega_1 \varphi d_{pea}(x_i, x_j) + \omega_2 d_{dtw}(x_i, x_j) \tag{3}$$

where ω_1 and ω_2 are similarity weights, and φ is the proportionality coefficient, which is used to eliminate the amplitude difference between the two metrics.

The membership degree u_{ij} of the load to the cluster center and the update method of the cluster center matrix v_K are as follows:

$$u_{ij} = \frac{1}{\sum_{K=1}^{c} \left(\frac{D(x_i, x_j)}{D(x_i, v_K)} \right)^{\frac{2}{m-1}}} \tag{4}$$

$$v_i = \frac{\sum_{i=1}^{n} u_{ij} x_i}{\sum_{i=1}^{n} u_{ij}} \tag{5}$$

When $e = \left\| v_i^{k+1} - v_i^{k} \right\|$ is less than the given integer ε in two consecutive clustering outputs, the iteration ends and the corresponding clustering results are output.

2.3. Two-stage ultra short-term load forecasting based on clustering results

Based on the clustering results of load data from a two-stage integrated energy system, the ultra-short-term load forecasting structure based on FCM is shown in Figure 1.

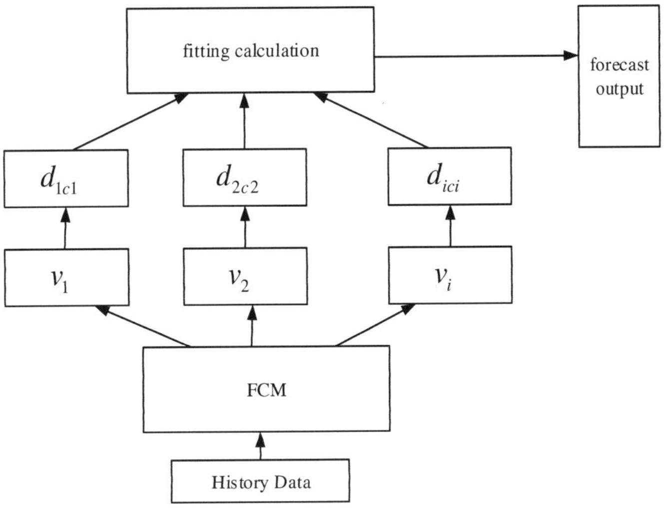

Figure 1. Structure of Ultra Short-Term Load Forecasting Based on FCM.

Based on the FCM-based ultra short-term load forecasting structure shown in Figure 1, for the historical composite data of a two-stage integrated energy system, after determining the training sample set and the prediction sample set, the FCM algorithm is used to cluster the training set and determine the category of the predicted sample. The distribution of the prediction sample x_{ci} within the corresponding range of each cluster center is used to fit and output the final prediction result:

$$y_{ci} = \sum_{i=1}^{n} \frac{(x_{ci} - v_i)}{d_{ici}} v_i \tag{6}$$

where y_{ci} represents the predicted result corresponding to type v_i load data, and d_{ici} represents the distance between predicted samples x_{ci} and v_i.

We calculate the predicted values of each classification load one by one and complete the predictive analysis of the load of the comprehensive energy system that includes two stages. The load forecasting process of this method is shown in Figure 2.

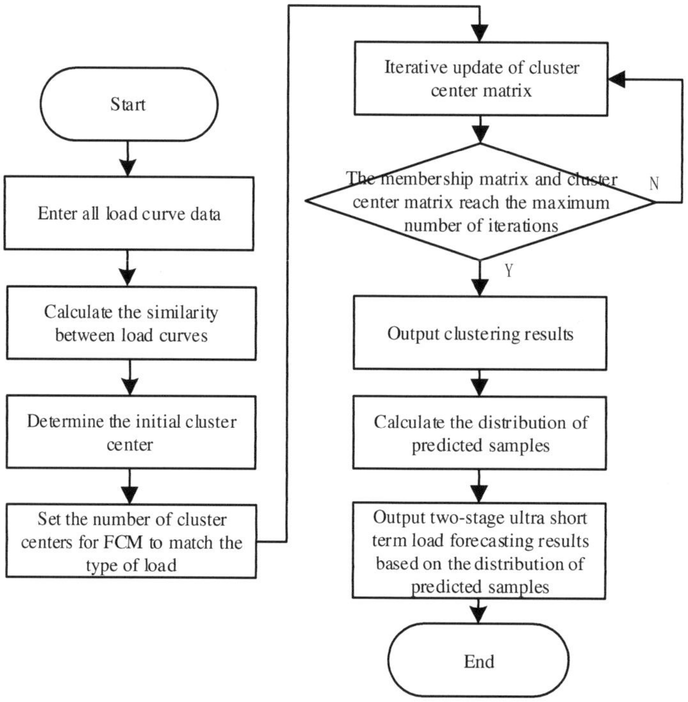

Figure 2. FCM-Based Two-Stage Ultra Short-Term Load Forecasting Process for Comprehensive Energy.

In Figure 2, this article describes the longitudinal shape similarity between load curves using Pearson distance. Based on the comprehensive distance between load curves, samples with high local density and far distance from high-density samples are identified as the initial clustering centers. Using the comprehensive distance between load curves, we iteratively calculate the membership degree of load to cluster centers. The predicted sample distribution that satisfies the clustering results of the iterative process is calculated, and the two-stage ultra short-term load forecasting results are output.

3. Testing and analysis

3.1. Testing environment

The testing phase will use the "Electricity Gas Heat Storage Hydrogen" regional integrated energy system with two-stage P2G as the specific predictive research environment, as shown in Figure 3.

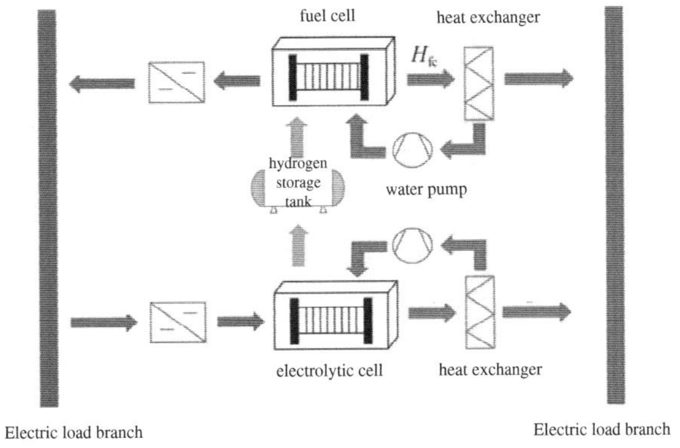

Figure 3. Testing the Comprehensive Energy System.

Statistical analysis was conducted on the typical daily load, daily gas load, daily heat load, and daily hydrogen load data of a two-stage P2G "electricity gas heat storage hydrogen" regional integrated energy system, as shown in Table 1.

Table 1. Typical daily load data of the two-stage P2G “electricity gas heat storage hydrogen” regional integrated energy system.

Time period	Electricity load/kW	Gas load/kW	Heat load/kW	Hydrogen load/kg
0:00-4:00	865.0	773.0	669.0	500.0
4:00-8:00	926.0	916.0	759.0	678.0
8:00-12:00	1,216.0	1,012.0	772.0	1,216.0
12:00-16:00	1,035.0	1,146.0	613.0	1,695.0
16:00-20:00	1,224.0	1,219.0	545.0	1,503.0
20:00-24:00	885.0	874.0	612.0	568.0

We divide the above dataset into a certain proportion of 80% training set and 20% testing set for model training and evaluation. This paper sets up a short-term load forecasting method based on time series clustering and a short-term load forecasting method based on EEMD Adaboost BP as control groups for predicting the specific ultra short-term load situation of the "electricity gas heat storage hydrogen" regional integrated energy system containing two-stage P2G.

3.2. Test results and analysis

Firstly, an analysis was conducted on the electrical load situation corresponding to the first stage of the P2G process and the second stage of the P2G process output by three different prediction methods and models. The data results obtained are shown in Figure 4.

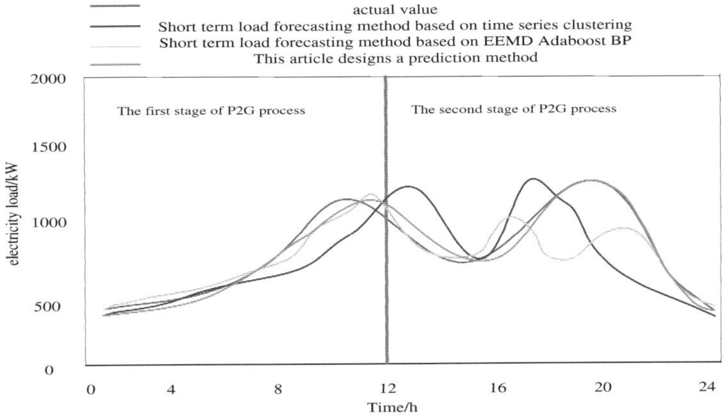

Figure 4. Comparison of Electric Load Forecasting Results.

From Figure 4, it can be seen that the predicted output of the electric load in this article has the highest degree of meshing with the actual load. This is because when clustering the original load data, this article considers the density distribution of the load, which enables the classification results output by FCM to more accurately classify the data types and avoid the interference of noisy data.

Secondly, the analysis of the hydrogen load prediction results and the actual situation resulted in the data shown in Figure 5.

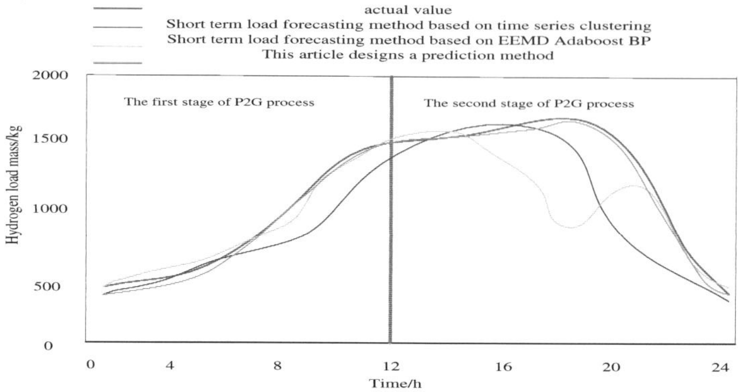

Figure 5. Comparison of Hydrogen Load Prediction Results.

Based on the test results shown in Figure 5, it can be seen that the predicted output of the model designed in this paper is still the optimal value among the three prediction methods and models, which is closest to the actual load parameters. This also provides a more comprehensive verification of the effectiveness of clustering test sample data using FCM in this article.

4. Conclusion

This article proposes a structural design method for ultra-short-term load forecasting based on FCM. The DTW distance and Pearson distance are used to describe the horizontal and vertical shape similarity between load curves, and the samples with high local density and far distance from high-density samples are determined as the initial clustering centers. Based on the distribution of predicted samples within the corresponding range of each cluster center, accurate prediction of ultra-short-term load is achieved.

Reference

[1] Zhang Z B, Gu C H, Wen M. Ultra-Short-Term Load Forecasting Method Based on XGBoost and QRLSTM [J]. Computer Simulation, 2022, 039(001): 90-95,110.

[2] Martins A, Lagarto J, Canacsinh H, et al. Short-term load forecasting using time series clustering[J]. Optimization and Engineering, 2022, 23(4): 2293-2314.

[3] Lin W, Zhang B, Li H L R. Short-term load forecasting based on EEMD-Adaboost-BP[J]. Systems Science & Control Engineering, 2022, 10(1): 846-853.

[4] LI Xiao, LU Xianling. Method for Forecasting Short-Term Power Load Based on Dual-Stage Attention Mechanism and Gated Recurrent Unit Network[J]. Computer Engineering, 2022, 48(2): 291-296,305.

[5] Rafati A, Joorabian M, Mashhour E, et al. Machine learning-based very short-term load forecasting in microgrid environment: evaluating the impact of high penetration of PV systems[J]. Electrical Engineering, 2022, 104(4): 2667-2677.

[6] Yu J, Zhang X, Zhong Q, et al. Investigating Periodic Dependencies to Improve Short-Term Load Forecasting[J]. Energy Engineering, 2024, 121(3): 789-806.

[7] Lee Y G, Oh J Y, Kim D, et al. SHAP Value-Based Feature Importance Analysis for Short-Term Load Forecasting[J]. Journal of Electrical Engineering & Technology, 2023, 18(1): 579-588.

[8] Yang Y, Gao Y, Wang Z, et al. Multiscale-integrated deep learning approaches for short-term load forecasting[J]. International Journal of Machine Learning and Cybernetics, 2024, 15(12): 6061-6076.

[9] Gundu V, Simon S P. Quality analysis of combined similar day and day ahead short-term load forecasting using recurrent neural networks[J]. Sādhanā, 2022, 47(1): 1-11.

Comprehensive evaluation and analysis of anodic gas state parameters of hydrogen recirculation ejector for PEM fuel cells

Jiangkun Zou[1], Jing Li[1*], Zhongjun Hou[2], Keyong Wang[2], Zheng Huang[2] and Huatao Jin[1]

[1]School of Mechanical Engineering, Tongji University, Shanghai, 201804, China
[2]Shanghai Hydrogen Propulsion Technology Co., Ltd., Shanghai, 201804, China

[*]Corresponding author's e-mail: cynthia_li@tongji.edu.cn

Abstract. Hydrogen ejectors in proton exchange membrane fuel cell (PEMFC) systems can reliably recover hydrogen without parasitic power, making them an ideal alternative to pumps. However, improving fuel utilization, enhancing ejector efficiency, and extend the ejector's operating range are important issues currently faced by PEMFC systems. In this study, a comprehensive evaluation system including four indicators (hydrogen entrainment ratio (ω_{H_2}), ejector outlet relative humidity (RH_c), ejector outlet temperature (T_c), and ejector efficiency ($\eta_{ejector}$)) are established based on the operating conditions and structural parameters characteristics of the ejector. The influence of secondary flow relative humidity (RH_s) and secondary flow temperature (T_s) on ejector performance are analyzed. The results show that the hydrogen entrainment ratio decreases as the secondary flow relative humidity or temperature increases. RH_c increases and T_c decreases as the gas mixture reaches the outlet of the ejector. Ts should be controlled below 353K and RH_s below 57% to prevent condensation of water vapor at the ejector outlet, at which point the ejector efficiency is 55.82%. To improve the ejector efficiency, the secondary flow temperature can be further increased, but the secondary flow relative humidity must be controlled accordingly. A theoretical basis for the development of high-temperature PEMFC is also provided.

1. Introduction

The global energy crisis and environmental damage caused by fossil fuels highlight the urgent need for new energy development. Proton exchange membrane fuel cells (PEMFCs) have emerged as a leading clean energy technology, offering significant advantages over conventional power sources. Their electrochemical conversion process produces only water as a byproduct, eliminating harmful emissions and addressing critical environmental concerns. Compared to traditional combustion-based systems, PEMFCs demonstrate superior energy conversion efficiency and power density, making them particularly suitable for applications requiring compact energy solutions, such as automotive propulsion and portable electronics. A key operational benefit lies in their ability to achieve full power output within seconds of activation, overcoming the limitations of many alternative energy systems that require extended warm-up periods [1]. Fuel utilization, a critical factor in PEMFC efficiency, can be improved using hydrogen recirculation systems [2]. Among the three main methods of hydrogen recycling, mechanical pumps consume parasitic power and produce noise [3], while electrochemical hydrogen pumps, though quieter, face challenges in cost and durability [4]. In contrast, hydrogen ejectors offer simplicity, durability, zero parasitic power, low noise, and cost-effectiveness, making them ideal for hydrogen recirculation [5]. However, challenges remain in terms of their narrow operating range and complex internal flow dynamics [6].

At present, most scholars extend the operating range of ejectors by optimizing the structure parameters or innovating the structure. In terms of optimizing the structure parameters, Pei et al. [7] developed an innovative analytical approach to optimize ejector performance under low-load conditions through comprehensive geometric parameter analysis. Their study established a novel pressure drop correlation model that systematically evaluates the interdependent effects of critical dimensions, including the mixing chamber's hydraulic characteristics and nozzle configuration parameters. Through computational simulations and experimental validation, the researchers identified precise dimensional combinations that maximize fluid entrainment efficiency during reduced power operation. Zou et al. [5] additionally analyzed the influence of diffuser divergence angle and length on the entrainment ratio. Liu et al. [2] pioneered the investigation of ejector dynamic responses during system start-up and power transitions, categorizing the transient process into distinct "primary flow impact" and "mixed flow impact" phases. Their work elucidated how velocity differentials influence ejector efficiency. Additionally, they examined key geometric parameters—including nozzle convergence angle, secondary flow tube diameter, mixing tube length, and diffuser angle—identifying optimal configurations such as a mixing tube length of 5.5 times its diameter for peak performance.

In terms of structural innovations, the academic community has made several breakthrough advancements. Researchers led by Brunner et al. [8] proposed an ejector with revolutionary adjustable design that achieves dynamic geometric parameter modulation through precision control of needle valve displacement, significantly improving entrainment ratio performance under low-load conditions. Building upon this, Jenssen's team [9] made further breakthroughs by developing a adjustable ejector for automotive fuel cell systems. These technological advances effectively address the performance degradation challenges of traditional fixed-geometry ejectors under variable operating conditions. These innovative designs represent two important development directions in ejector technology: Precision adjustment capability: Evolution from mechanical to intelligent control System adaptability: Transition from single-condition optimization to full operating range coverage.

Previous studies on the ejector system's boundary conditions are insufficient, particularly regarding their impact on ejector performance. Pei and Lee et al. [7,10] qualitatively examined the effect of relative humidity on entrainment performance but lacked comprehensive analysis of other parameters. Ma et al. [11] quantitatively explored relative humidity's impact on entrainment ratio but did not address ejector efficiency or outlet conditions. As PEMFCs trend toward high-temperature applications, the secondary flow temperature in the anode system becomes critical. This study establishes a comprehensive evaluation index to analyze how secondary flow temperature and relative humidity influence hydrogen entrainment ratio, ejector efficiency, outlet relative humidity, and outlet temperature. A method for maximizing efficiency is proposed, providing theoretical support for high-temperature PEMFC development.

2. Anode recirculation system architecture in PEMFC systems

The hydrogen recirculation subsystem represents a critical component in proton exchange membrane fuel cell (PEMFC) architectures, employing an ejector-based solution for efficient gas management. As depicted in Figure 1(a), the core ejector assembly incorporates several key functional elements: a converging primary nozzle, secondary flow inlet channels, a specially designed suction chamber for flow entrainment, an optimized mixing section, and a pressure recovery diffuser. The schematic in Figure 1(b) demonstrates the complete anode loop configuration where this ejector operates [5]. Within this configuration, high-pressure hydrogen from the storage system first passes through precision pressure regulation components before entering the primary nozzle. The converging geometry induces significant flow acceleration, with the subsequent expansion creating a vacuum effect in the suction chamber through the Venturi principle. This pressure differential actively draws in the secondary flow consisting of unconsumed hydrogen and water vapor from the fuel cell exhaust stream. The two distinct flow paths undergo turbulent mixing in the specially contoured chamber, with the combined flow then directed toward the fuel cell stack's anode inlet. This recirculation approach provides multiple system benefits: it maintains proper humidification of the membrane electrode assembly, improves hydrogen utilization efficiency, and helps manage water content in the anode channels. The continuous flow cycling ensures optimal reactant distribution while preventing gas stagnation that could lead to local fuel starvation. Advanced designs often incorporate additional features such as liquid-vapor separators and humidity control elements to further enhance system performance under varying operating conditions.

The operational effectiveness of fuel cell ejectors is quantitatively assessed through two distinct but interrelated entrainment metrics. These parameters serve as critical indicators of the ejector's gas handling capability and hydrogen recovery efficiency: includes the total entrainment ratio (ω) and the hydrogen entrainment ratio (ω_{H_2}), calculated as follows:

$$\omega = \frac{W_s}{W_p} \tag{1}$$

$$\omega_{H_2} = \frac{y_{s,H_2} \cdot W_s}{W_p} \tag{2}$$

where y_{s,H_2} is the secondary flow hydrogen mass fraction. Before calculating, we make the following assumptions:

1. The molecular gravity of the gas is neglected and the inner wall surfaces of the ejector are considered as adiabatic walls.
2. The working gas is considered to be an ideal gas and is in accordance with the law of ideal gas.
3. The fluid is uniformly distributed radially in temperature and pressure.
4. The radial velocity distribution of secondary flow is not uniform.
5. The fluid flow is in a steady state and water vapor phase changes are ignored.

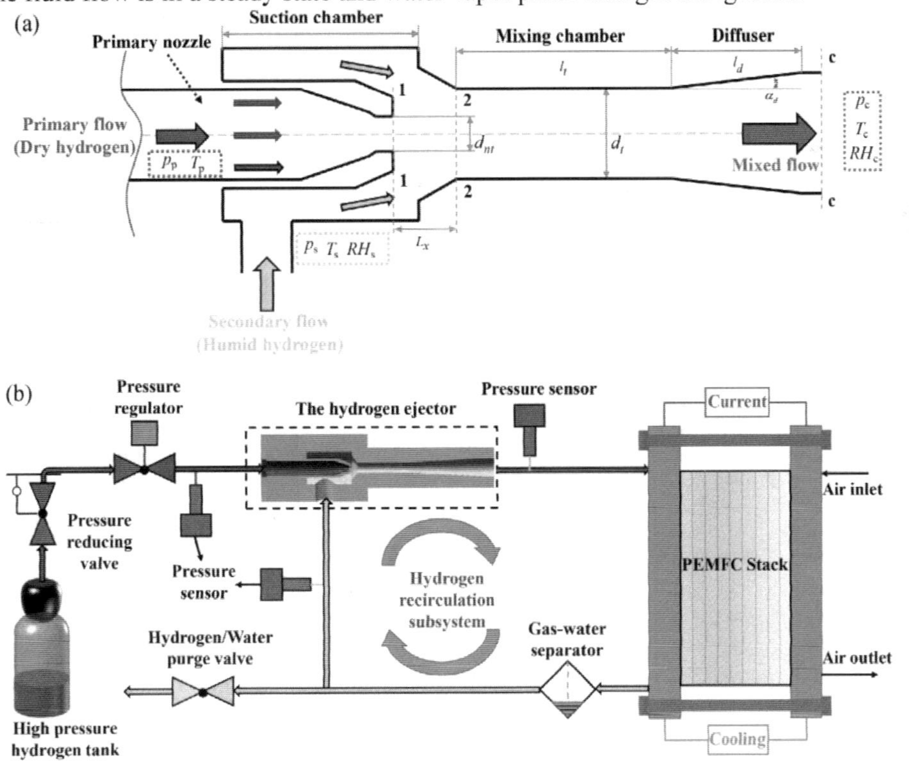

Figure 1. PEMFC systems: (a) ejector (b) anode subsystem schematic.

3. Mathematical model of the comprehensive evaluation system

3.1. Calculation method for mass flow rate

The flow is subsonic when the ratio of p_s / p_p exceeds 0.528. The energy conservation equation and isentropic flow relationship are used to solve the mass flow rate of a primary flow, as well as the Mach number and velocity at the nozzle outlet [6]:

$$W_p = p_p A_t \sqrt{\frac{2\phi_p \kappa \left[\left(p_s/p_p\right)^{2/\kappa} - \left(p_s/p_p\right)^{(1+\kappa)/\kappa} \right]}{(\kappa-1) R_{g,p} T_p}} \tag{3}$$

$$M_t = \sqrt{\frac{2\phi_p \left[1 - \left(p_s / p_p\right)^{(\kappa-1)/\kappa} \right]}{\left(p_s / p_p\right)^{(\kappa-1)/\kappa} (\kappa-1)}} \tag{4}$$

$$V_t = M_t \sqrt{\kappa R_{g,p} T_t} \tag{5}$$

where A_t represents nozzle throat area and ϕ_p is an isentropic coefficient that accounts for flow wear. The primary flow average velocity at 2-2 section at subsonic speed can be obtained as [5]:

$$V_{p,2} = \frac{\int_0^{R'_{p,2}} 2\pi r V_t dr + \int_{R'_{p,2}}^{R_{p,2}} \pi r V_t \left(1 - \cos\left(\frac{R_2 - r}{R_2 - R'_{p,2}} \pi\right)\right) dr}{\pi R_{p,2}^2} \tag{6}$$

According to the assumption, the pressure distribution of the fluid in 2-2 section is uniform. Here, our previous findings are adopted to obtain the formulae for the mass flow rate of the secondary flow at subsonic states by two-dimensional three-segment velocity distributions [5]:

$$W_s = \frac{\pi V_t \rho_s \cdot \left(1 - \cos\left(\frac{R_2 - R_{p,2}}{R_2 - R'_{p,2}} \pi\right)\right) \left(R_2 - R_{p,2}\right)\left(R_2 + R_{p,2} + n_v R_{p,2}\right)}{(n_v + 1)(n_v + 2)} \tag{7}$$

where $R_2 = d_t / 2$, $R_{p,2}$ and $R'_{p,2}$ are determined from literature [5].

In the supersonic flow state, p_s / p_p is less than the critical value 0.528, and the mass flow rate of a primary flow is [6]:

$$W_p = p_p A_t \sqrt{\frac{\phi_p \kappa}{R_{g,p} T_p}} \left(\frac{2}{\kappa+1}\right)^{(\kappa+1)/(2(\kappa-1))} \tag{8}$$

At this time, the nozzle throat Mach number M_t=1. The ambient pressure can then be expressed as the secondary flow pressure as the primary flow expands in the suction chamber. Through the isentropic flow relationship of the primary flow, we can obtain [5,6]:

$$M_{p,2} = \sqrt{\frac{2\left[\left(p_p / p_s\right)^{(\kappa-1)/\kappa} - 1 \right]}{(\kappa-1)}} \tag{9}$$

$$\frac{T_p}{T_{p,2}} = 1 + \frac{1}{2}(\kappa-1) M_{p,2}^2 \tag{10}$$

$$V_{p,2} = M_{p,2} \sqrt{\kappa R_{g,p} T_{p,2}} \tag{11}$$

where $V_{p,2}$ and $M_{p,2}$ are the primary flow average velocity and Mach number at the 2-2 section, respectively. This leads to a method for calculating the mass flow rate of secondary flow at supersonic conditions:

$$W_s = \frac{2\pi\rho_s \cdot V_{p,2}\left(R_2 - R_{p,2}\right)\left(R_2 + R_{p,2} + n_v R_{p,2}\right)}{\left(n_v + 1\right)\left(n_v + 2\right)} \tag{12}$$

3.2. The calculation model of ejector outlet temperature, outlet relative humidity and efficiency

Before solving for W_s, the exponent n_v and the average density ρ_s need to be obtained:

$$n_v = A_1 \exp\left(\lambda_a / 0.05\right) + A_2\lambda_b + A_3\lambda_c^2 + A_4\lambda_c + A_5 \exp\left(-L_x / 2.7512\right) + A_6 L_x + A_7 Z_1 + A_8 \tag{13}$$

$$\rho_s = \frac{p_s}{T_s}\frac{M_s}{G} \tag{14}$$

where λ_a is related to the pressure; λ_b , λ_c are related to the ejector's structural parameters. L_x is the main nozzle distance, A_1–A_8 are the coefficients, and Z_1 depends on the size of the diffuser, which can be found in the literature [5]. M_s is the secondary flow molar mass, which consists mainly of water vapor and hydrogen:

$$M_s = x_{s,va} M_{va} + \left(1 - x_{s,va}\right) M_{H_2} \tag{15}$$

where $x_{s,va}$ is the secondary flow water vapor volume fraction, which can be obtained from the saturated vapor pressure and the relative humidity of the secondary flow.

$$x_{s,va} = \frac{RH_s \cdot p_{s,va}^{sat}}{p_s} \tag{16}$$

$$\log_{10} p_{s,va}^{sat} = 4.6543 - \frac{1435.264}{T_s - 273.15 + 208.302} \tag{17}$$

The equation for the energy balance of the secondary and primary flows from the 2-2 section to the c-c section is as follows [1]:

$$\sum_i W_p \cdot Cp_p^i \cdot T_p + \sum_i W_s \cdot Cp_s^i \cdot T_s = \sum_i \left(W_p + W_s\right) \cdot Cp_c^i \cdot T_c + E_{loss} \tag{18}$$

where E_{loss} is the energy loss during the collision of the secondary and primary flows, which can be expressed as [1]:

$$E_{loss} = 0.5\left(1 - \phi_p\right) W_p \cdot V_{p,2}^2 + 0.5\left(1 - \phi_s\right) W_s \cdot V_{s,2}^2 \tag{19}$$

where $V_{s,2}$ represents average velocity of the secondary flow at 2-2 section, which can be calculated as:

$$V_{s,2} = \frac{W_s}{\rho_s \cdot \pi\left(R_2^2 - R_{p,2}^2\right)} \tag{20}$$

The ejector outlet temperature (T_c) can be calculated from the above energy equation. Then, by definition, the volume fraction of the water vapor at the outlet is given by:

$$x_{c,va} = \frac{y_{c,va} \cdot M_{H_2}}{M_{va} + y_{c,va} \cdot M_{H_2} - y_{c,va} \cdot M_{va}} \tag{21}$$

where $y_{c,va}$ is the mass fraction of the water vapour at the outlet, and the relative humidity at the outlet of the ejector (RH_c) is defined as:

$$RH_c = \frac{x_{c,va} \cdot p_c}{p_{c,va}^{sat}} \tag{22}$$

where $p_{c,va}^{sat}$ is the outlet saturated vapor pressure. The ejector efficiency can be calculated as [1]:

$$\eta_{\text{ejector}} = \omega\left(\frac{T_{\text{p}}}{T_{\text{s}}}\right)\frac{\left(p_{\text{c}}/p_{\text{s}}\right)^{(\kappa_{\text{s}}-1)/\kappa_{\text{s}}}-1}{1-\left(p_{\text{c}}/p_{\text{p}}\right)^{(\kappa-1)/\kappa}} \tag{23}$$

In addition, in order to obtain the primary flow pressure at different electrostack currents, it is necessary to find it indirectly through the equation between current and mass flow rate [1]:

$$W_{\text{p}} = 1.045\times10^{-8}\text{N}_{\text{cell}}\text{I} \tag{24}$$

where N_{cell} is the number of electrical stacks. Combined with equations (4) and (9), the primary flow pressure at subsonic and supersonic speeds can be obtained by current [1].

3.3. Ejector boundary conditions and designed parameters

To study the relationship between anode gas state parameters (including secondary flow relative humidity and temperature) and evaluation indicators, Table 1 provides typical ejector values used in PEMFC systems. Using these data, the variation patterns of four ejector evaluation indices are analyzed under different secondary flow relative humidity and temperature conditions.

Table 1. Typical values for fuel cell systems operation

Parameter condition	Values
Cell active area, A_{cell} (cm^2)	1000
Wear isentropic coefficient, ϕ_{p}	0.95
Wear isentropic coefficient, ϕ_{s}	0.85
Primary flow gas constant, $R_{\text{g,p}}$ (J kg^{-1} K^{-1})	4121
Primary flow temperature, T_{p} (K)	298
Molar mass of hydrogen, M_{H_2} (kg^{-1} mol)	0.002
Molar mass of water vapor, M_{va} (kg^{-1} mol)	0.018
Gas constant of universal G (J mol^{-1} K^{-1})	8.1345
Main nozzle distance, L_x (mm)	6
Diffusion angle, α_d (°)	2.4
Diameter of nozzle, d_{nt} (mm)	2.4
Diffusion length, l_d (mm)	186
Mixing chamber length, l_t (mm)	49
Diameter of mixing chamber, d_t (mm)	8.2
Cell number, N_{cell}	381

Figure. 2 shows the variation of anode pressure drop and inlet and outlet pressures of the ejector with current. In general, the p_{c} is 3bar at high loads and 1.1bar at low loads.

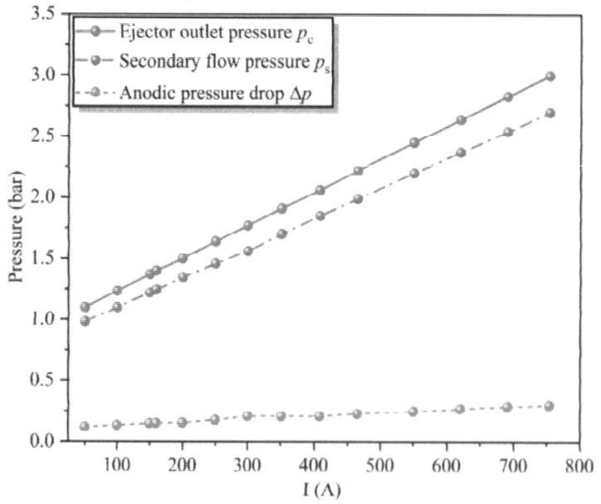

Figure 2. Inlet and outlet boundary conditions for the ejector.

Journal of Physics: Conference Series **3011** (2025) 012003 doi:10.1088/1742-6596/3011/1/012003

4. Results and discussion

Figure. 3 shows the effect of current on entrainment ratio under different RH_s, with T_s set at 353K. It is observable that both the total entrainment ratio and the hydrogen entrainment ratio escalate rapidly as the current rises in the subcritical state, and subsequently decline gradually upon reaching the critical state. Figure. 4 shows that the total entrainment ratio increases almost linearly with increasing RH_s with a maximal value of 7.65 (RH_s =100%, T_s =353K). Whereas, the hydrogen entrainment ratio decreases almost linearly with increasing relative humidity of the secondary flow with a maximal value of 2.03 (RH_s =20%, T_s =353K). The inverse relationship between humidity and hydrogen recovery efficiency can be attributed to competitive molecular displacement in the gas mixture. As the secondary flow's relative humidity rises, water vapor molecules progressively occupy greater volumetric fraction within the gas stream, thereby proportionally reducing hydrogen's molar concentration. Experimental measurements at 353K reveal a significant 26.6% decline in hydrogen entrainment performance (ω_{H_2} = 2.03 → 1.49) when relative humidity escalates from 20% to saturation point (RH_s =100%). This substantial reduction demonstrates the critical trade-off between maintaining adequate system hydration and optimizing fuel recirculation efficiency.

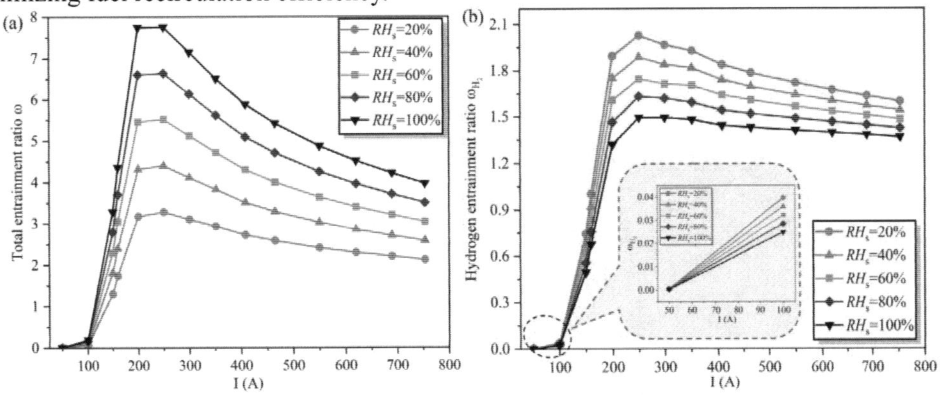

Figure 3. The entrainment ratio changes with current (T_s=353K): (a) total entrainment ratio (b) hydrogen entrainment ratio.

Therefore, in combination with Figure. 3(b), to obtain a higher hydrogen entrainment ratio during the entire operation of the electric stack, the secondary flow relative humidity should be reasonably controlled. Lowering RH_s during operation can enhance the hydrogen entrainment ratio.

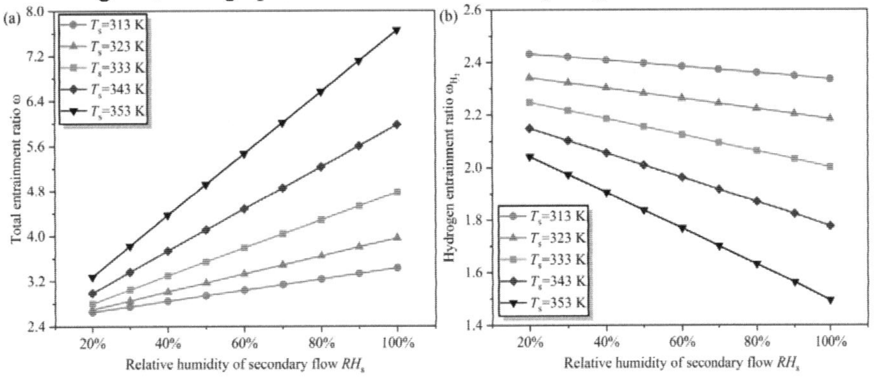

Figure 4. Change of entrainment ratio with RH_s (I=250A): (a) total entrainment ratio (b) hydrogen entrainment ratio.

The results demonstrate significant temperature-dependent characteristics of ejector performance metrics. As revealed in Figure. 5, systematic variations in current produce distinct effects on entrainment behavior across different secondary flow temperature conditions at full relative humidity (RH_s =100%).

The comprehensive data presented in Figure. 6 reveals several important trends: Total entrainment characteristics: Exhibits a nonlinear positive correlation with secondary flow temperature Reaches peak performance (ω =7.65) at T_s =353K with saturated humidity This enhancement stems from improved fluid dynamics at elevated temperatures Hydrogen-specific entrainment behavior: Demonstrates an inverse nonlinear relationship with temperature Achieves maximum hydrogen recovery (ω_{H_2}=2.34) at T_s =313K under saturated conditions Declines by 36.3% across the 313K-353K temperature range The underlying mechanisms involve two primary factors: a) Vapor pressure dynamics: Elevated temperatures increase water's saturated vapor pressure This reduces hydrogen's relative concentration in the secondary flown Directly impacts the hydrogen-specific entrainment metric b) Fluid property variations: Temperature changes alter gas density and viscosity Affects momentum transfer efficiency in the mixing chamber Influences overall entrainment performance These findings, when analyzed alongside Figure. 4's results, clearly establish secondary flow temperature as one of the most influential operational parameters affecting ejector efficiency. The data suggests that maintaining lower secondary flow temperatures (while avoiding condensation issues) represents an effective strategy for optimizing hydrogen recovery rates during fuel cell operation, as particularly evident in Figure. 5 (b)'s comparative analysis.

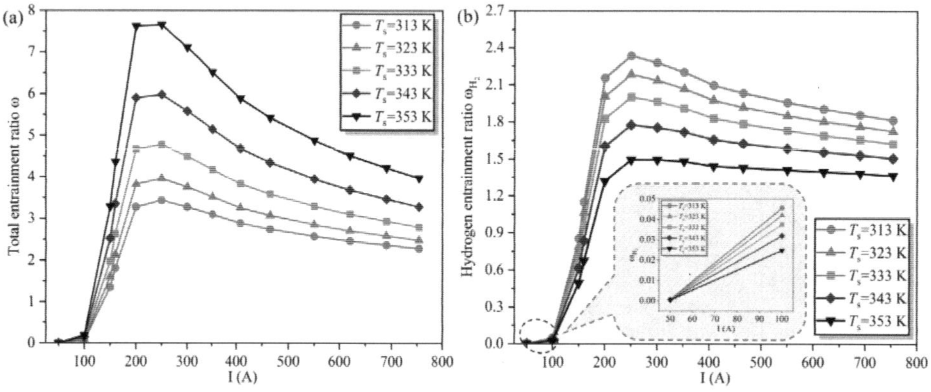

Figure 5. The entrainment ratio changes with current (RH_s=100%): (a) total entrainment ratio (b) hydrogen entrainment ratio.

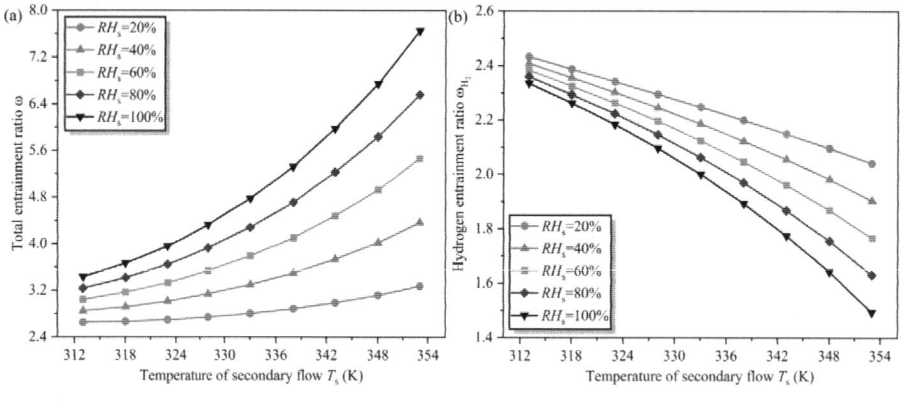

Figure 6. Change of entrainment ratio with T_s (I=250A): (a) total entrainment ratio (b) hydrogen entrainment ratio.

Under the same operating conditions, the influence of secondary flow temperature (T_s) and relative humidity (RH_s) on ejector outlet parameters is analyzed.

Figure. 7(a) shows that outlet relative humidity (RH_c) increases with higher T_s and RH_s due to the rising water vapor concentration in the secondary flow. RH_c is consistently higher than RH_s, indicating that relative humidity rises when the gas mixture reaches the ejector outlet. Since RH_c exceeding 100% can cause water vapor condensation, potentially flooding the stack and affecting fuel supply, RH_c must remain below 100% [2]. For this, as shown in Figure. 7(b), RH_s should be kept below 90% when T_s is 313K and below 57% when T_s is 353K.

Figure. 8 highlights that RH_s has minimal impact on outlet temperature (T_c), while T_s significantly influences it. T_c increases with higher T_s but is always lower than T_s, showing that the gas mixture cools upon exiting the ejector. Combined with Figure. 7, to minimize water vapor condensation, T_s should be maximized without exceeding 353K, while RH_s remains below 57%.

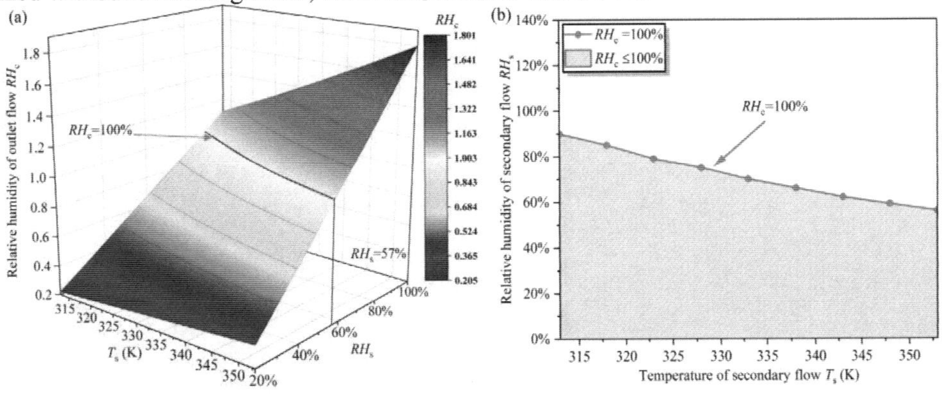

Figure 7. Change law of RH_c: (a) the effect of T_s and RH_s on the RH_c (b) the distribution pattern of T_s and RH_s when RH_c is 100%.

As shown in Figure. 9, the ejector efficiency ($\eta_{ejector}$) increases as the RH_s and T_s increase. From Figure. 10, it can be seen that when T_s rises from 313K to 353K, $\eta_{ejector}$ improves by 2.75% at RH_s =20% and by 37.19% at RH_s =100%. Similarly, as RH_s increases from 20% to 100%, $\eta_{ejector}$ grows by 9.23% at T_s =313K and by 43.66% at T_s =353K. It can be seen that high temperature and high humidity can significantly increase the ejector efficiency. When T_s is 353K and RH_s is 100%, $\eta_{ejector}$ reaches the maximal value of 78.83%. However, in order to satisfy all the evaluation indexes (ω_{H_2}, T_c, RH_c, $\eta_{ejector}$), the secondary flow temperature and relative humidity should be controlled. When T_s is 353K, the RH_s should be controlled to 57%, and the $\eta_{ejector}$ is 55.82%. When T_s is 313K, the RH_s should be controlled to 90%, and the $\eta_{ejector}$ is 40.50%. Therefore, increasing the temperature of the secondary flow and controlling the relative humidity of the secondary flow will be an optimal scheme.

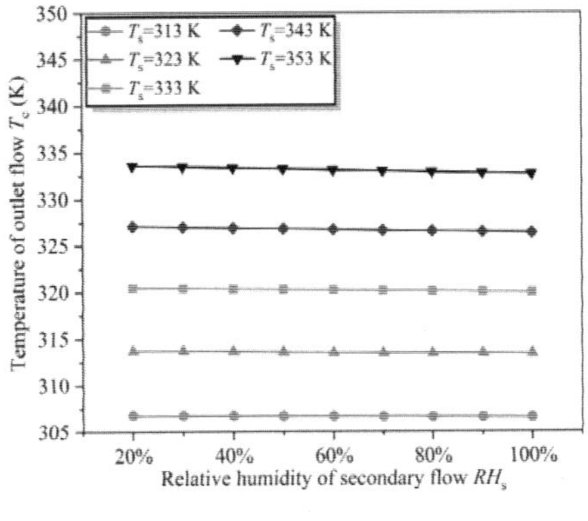

Figure 8. Effect of RH_s and T_s on T_c.

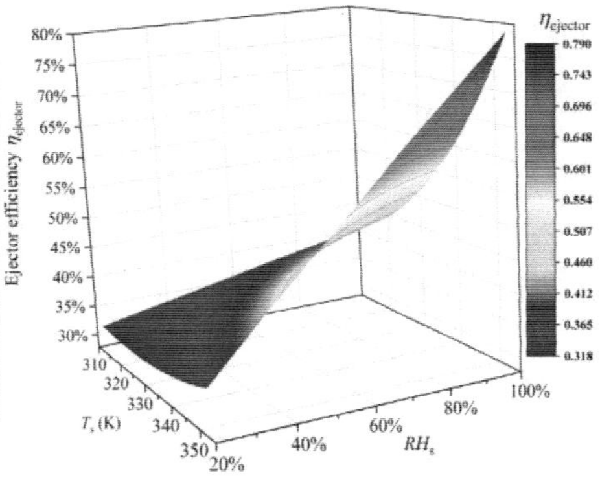

Figure 9. Effect of RH_s and T_s on $\eta_{ejector}$.

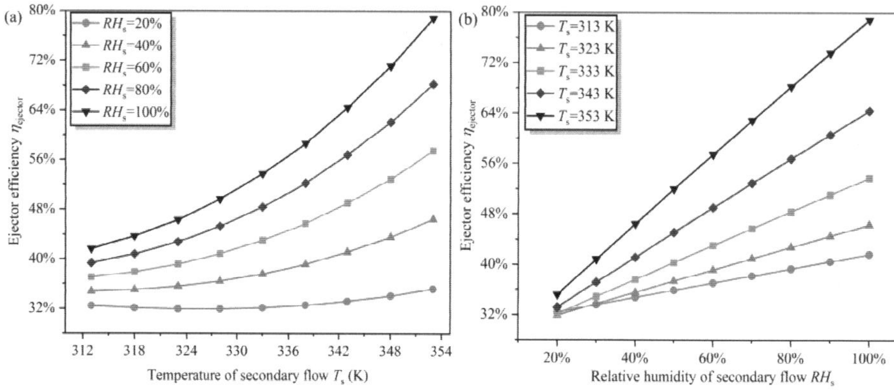

Figure 10. Variation law for the ejector efficiency: (a) effect of T_s on the ejector efficiency (b) Effect of RH_s on the ejector efficiency.

5. Conclusions

This research develops a comprehensive performance assessment framework specifically designed for ejector systems in high-temperature PEMFC applications. The proposed evaluation methodology incorporates multiple quantitative indicators that collectively characterize ejector operation across various working conditions. These metrics serve dual purposes: optimizing current ejector designs while establishing fundamental guidelines for future technological advancements in high-temperature fuel cell systems. These indexes can maximize the efficiency of the ejector. Some conclusions are obtained as follows:

(1) The hydrogen entrainment ratio decreases with increasing secondary flow relative humidity (linearly) and temperature (nonlinearly), with temperature having a stronger impact. At 353 K and 100% relative humidity, the entrainment ratio reaches a minimum of 1.49.Therefore, the secondary flow relative humidity and temperature should be reasonably controlled in the evaluation index.

(2) The outlet relative humidity is always greater than the secondary flow relative humidity, and the outlet temperature is always lower than the secondary flow temperature. When the secondary flow temperature is 313K, the secondary flow relative humidity should be controlled to be within 90%, and when the secondary flow temperature is 353K, the relative humidity of the secondary flow should be controlled to be within 57% to prevent condensation at the outlet.

(3) The ejector efficiency increases as the increase of secondary flow relative humidity and temperature. Under the condition that all evaluation indexes are satisfied, the maximal efficiency of the ejector is 55.82%, the temperature of secondary flow is 353K, and the relative humidity of secondary flow is 57%.

To optimize ejector performance and mitigate water vapor condensation risks, several promising research directions warrant investigation. First, elevating the PEMFC stack operating temperature could be strategically employed to enhance ejector efficiency through improved gas dynamics and reduced humidity sensitivity. However, this approach requires precise thermal management to balance between performance gains and material durability concerns. A complementary strategy involves implementing advanced humidity control systems, where integrating membrane-based dehumidifiers or temperature-swing adsorption units at the stack outlet could maintain optimal secondary flow conditions. Additional considerations for future research should include:

Development of smart control algorithms that synchronize temperature and humidity regulation. Optimize hydrogen recirculation efficiency. Investigation of novel materials for condensation-resistant ejector surfaces. These integrated approaches would address both the fundamental fluid dynamics challenges and practical implementation requirements for next-generation fuel cell systems.

Acknowledgements

This research has been supported by the Explorer Programme of the Shanghai Science and Technology Commission (No. 24TS1402200).

Nomenclature

	Symbols	y	mass fraction
A_t	nozzle throat area (mm^2)		*Subscripts*
Cp	constant-pressure specific heat (J mol^{-1} K^{-1})	c	ejector exit
d_t	diameter of mixing chamber (mm)	H_2	hydrogen
D	diameter (mm)	s	secondary flow
d_{nt}	nozzle throat diameter (mm)	p	primary flow
I	stack current (A)	t	nozzle throat
G	gas constant of universal (J mol^{-1} K^{-1})	sat	saturated
l_d	length of diffusion (mm)	va	steam
l_t	length of mixing chamber (mm)	2	mixing chamber inlet
M	Mach number	1	nozzle exit
M	molar mass (kg mol^{-1})		*Greek letters*
N_{cell}	cell number	α_d	diffusion angle (°)
n_v	exponent of the velocity function	$\eta_{ejector}$	ejector efficiency
p	pressure (bar)	κ_s	specific heat ratio of secondary flow
Δp	pressure drop of anodic ($p_c - p_s$) (bar)	κ	primary flow specific heat
R	radius (mm)	λ_b	diameter ratio, (d_t / d_{nt})
RH	relative humidity (%)	λ_a	pressure ratio, ($p_s^{0.8}$ / $p_p^{1.1}$)
T	temperature (K)	λ_c	dimensionless parameter, (l_t / d_t)
R_g	gas constant (J kg^{-1} K^{-1})	ξ_{exp}	coefficient for friction loss
W	mass flow rate (kg s^{-1})	ρ	density (kg m^{-3})
V	velocity (m s^{-1})	ϕ_p	wear isentropic coefficient
x	mole fraction	ϕ_s	wear isentropic coefficient
L_x	main nozzle distance (mm)	ω	entrainment ratio

Reference:

[1] Dadvar M and Afshari E (2014) Analysis of design parameters in anodic recirculation system based on ejector technology for PEM fuel cells: A new approach in designing. Int J Hydrogen Energ, 39: 12061–12073.
[2] Liu Z, Liu Z, Jiao K, Yang Z, Zhou X and Du Q (2020) Numerical investigation of ejector transient characteristics for a 130-kW PEMFC system. Int J Energ Res, 44: 3697–3710.
[3] Badami M and Mura M (2012) Comparison between 3D and 1D simulations of a regenerative blower for fuel cell applications. Energ Convers Manage, 55: 93–100.
[4] Barbir F and GÖRGÜN H (2007) Electrochemical hydrogen pump for recirculation of hydrogen in a fuel cell stack. J Appl Electrochem, 37: 359–365.
[5] Zou J, Li J, Zhang L and Ming P (2024) Theoretical studies on wide-operating-range ejector based on operating conditions and multi-parameter coupling effect in proton exchange membrane fuel cell system, Appl. Therm. Eng, 251: 123602.
[6] Zhu Y and Li Y (2009) New theoretical model for convergent nozzle ejector in the proton exchange membrane fuel cell system. J Power Sources, 191: 510–519.
[7] Pei P, Ren P, Li Y, Wu Z, Chen D, Huang S and Jia X (2019) Numerical studies on wide-operating-range ejector based on anodic pressure drop characteristics in proton exchange membrane fuel cell system. Appl Energ, 235: 729–738.
[8] Brunner D, Marcks S, Bajpai M, Prasad A and Advani S (2012) Design and characterization of an electronically controlled variable flow rate ejector for fuel cell applications. Int J Hydrogen Energ, 37: 4457–4466.
[9] Jenssen D, Berger O and Krewer U (2017) Improved PEM fuel cell system operation with cascaded stack and ejector-based recirculation. Appl Energ, 195: 324–333.
[10] Lee H, Jegal S and Song S J (2017) Analysis and measurement of relative humidity effects on ejector performance. J Mech Sci Technol, 31: 4237–4244.
[11] Ma T, Cong M, Meng Y, Wang K, Zhu D and Yang Y (2021) Numerical studies on ejector in proton exchange membrane fuel cell system with anodic gas state parameters as design boundary. Int J Hydrogen Energ, 46: 38841–38853.

Influence of surface roughness on vortex shedding characteristics in the flow around a circular cylinder

Xiaoxiang Wang[1], Jiancheng Liu[2,*], Lei Li[2,*], Liangwan Rong[3,*] and Yong Ma[1,4]

[1] School of Ocean Engineering and Technology, Sun Yat-sen University, Zhuhai 519082, China
[2] China Merchants Marine and Offshore Research Institute Co., Ltd., Shenzhen 518067, China
[3] School of Civil Engineering and Transportation, South China University of Technology, Guangzhou 510640, China
[4] Southern Marine Science and Engineering Guangdong Laboratory, Zhuhai 519080, China

*Corresponding author's e-mail: liujiancheng@cmhk.com; lilei@cmhk.com; rongliangwan@163.com

Abstract. The advancement of marine energy resources, including offshore wind power and seabed minerals, has led to the increased construction and utilization of cylindrical structures in marine environments. However, these structures frequently encounter the issue of circumferential flow, causing their surfaces to degrade due to seawater corrosion or the attachment of marine organisms. Leveraging the open-source code *Basilisk*, this study employs a quadtree grid to simulate, numerically and in two dimensions, the flow patterns around a cylinder with varying Reynolds numbers and surface roughness. By examining the vortex features formed around the cylinder, the study determines the distribution of vorticity on its circumferential surface, as well as the Strouhal number and how surface roughness impacts the vortex shedding behavior during flow. The findings indicate that surface roughness significantly affects vortex characteristics and the distribution of vorticity around the cylinder, but has a relatively minor influence on the Strouhal number.

1. Introduction

With the rapid development of the marine energy industry, an increasing number of offshore structures, such as offshore platforms and offshore wind turbines, have emerged for the exploitation and utilization of energy resources like offshore wind power and submarine minerals [1]. Among them, the riser structures of offshore platforms [2–3], pile foundations of offshore wind turbines [4], and mineral lifting pipelines for deep-sea mining [5] are all cylindrical. These cylindrical structures play crucial roles in supporting and bearing various loads, significantly impacting the reliability and safety of offshore structures. Therefore, it is essential to research cylindrical structures in marine energy development.

Many scholars have studied the cylindrical structures in offshore structures, focusing primarily on dynamic response issues [6–9], fatigue damage problems [10–12], and pipeline blockage problems [13–14]. Additionally, these cylindrical structures encounter flow-around issues in practical applications, which affect their hydrodynamic properties and stability. Therefore, besides the aforementioned issues, flow around cylinders is also a problem that cannot be ignored. Regarding the flow around cylinders, current research primarily employs numerical simulations and experiments to study individual cylinders [15–18] or multiple cylinders [19–21] in different scenarios. The research mainly analyzes characteristics such as vorticity distribution, wake characteristics, and the Strouhal number to investigate the flow around cylinders. The findings reveal that, in the case of a solitary cylinder, when the Reynolds number is low, the flow field manifests a relatively straightforward symmetrical configuration, characterized by closely packed and stable streamlines. Conversely, as the Reynolds number escalates, the flow field assumes a more intricate asymmetrical formation, accompanied by the emergence of numerous vortices and turbulent phenomena. For multiple cylinders, the flow and evolution patterns of the wake vary with different arrangements. However, in practical engineering applications, prolonged use will lead to erosion of the cylindrical structures by seawater or biofouling, resulting in a certain roughness on the cylinder surface and altering the flow-around characteristics. Therefore, further research is needed on the flow around cylinders with rough surfaces.

At present, some scholars have studied the flow around a rough cylinder [22–25], but the existing studies have not carried out further studies on the vorticity distribution on the surface of a rough cylinder and the details of the vortex shedding. Therefore, this paper employs the open-source code *Basilisk* to numerically simulate the flow around cylinders with different roughness levels in two dimensions. By using quadtree grids to continuously refine the area around the cylinders, more details of vortex shedding can be captured. This paper undertakes numerical simulations to analyze the flow characteristics surrounding cylinders with differing degrees of roughness. By juxtaposing the computational outcomes derived from these simulations with those pertaining to smooth cylinders, the study ultimately ascertains the influence of roughness on vortex shedding and the distribution of the vorticity field in the flow encircling the cylinders.

The remainder of this paper is organized as outlined below: Section 2 delineates the numerical methodologies employed in this study. Section 3 presents the results and discusses the findings obtained. Section 4 concludes the paper with a summary of the key points.

2. Numerical method

2.1. Governing equation

In the context of flow dynamics around a circular cylinder, the fluid is presumed to be incompressible, hence adhering to the incompressible Navier-Stokes equations. These equations can be solved to compute the fluid behavior as detailed below:

$$\nabla \cdot \boldsymbol{u} = 0 \tag{1}$$

$$\rho(\partial_t \boldsymbol{u} + \boldsymbol{u} \cdot \nabla \boldsymbol{u}) = -\nabla p + \nabla \cdot (2\mu \boldsymbol{D}) \tag{2}$$

where $\boldsymbol{u}$ is the fluid velocity, ρ is the fluid density, μ is the dynamic viscosity, and $\boldsymbol{D} = (\nabla \boldsymbol{u} + \nabla \boldsymbol{u}^T)/2$ is the strain rate tensor.

Equations (1) and (2) are discretized with second-order accuracy in time:

$$\nabla \cdot \boldsymbol{u}_n = 0 \tag{3}$$

$$\rho_{n+\frac{1}{2}}\left(\frac{\boldsymbol{u}_{n+1} - \boldsymbol{u}_n}{\Delta t} + \boldsymbol{u}_{n+\frac{1}{2}} \cdot \nabla \boldsymbol{u}_{n+\frac{1}{2}}\right) = -\nabla p_{n+\frac{1}{2}} + \nabla \cdot \left[\mu_{n+\frac{1}{2}}(\boldsymbol{D}_n + \boldsymbol{D}_{n+1})\right] \tag{4}$$

where n denotes the time step.

Subsequently, utilizing the classical time-splitting projection technique, Equations (3) and (4) can be reformulated as follows:

$$\rho_{n+\frac{1}{2}}\left(\frac{\boldsymbol{u}^* - \boldsymbol{u}_n}{\Delta t} + \boldsymbol{u}_{n+\frac{1}{2}} \cdot \nabla \boldsymbol{u}_{n+\frac{1}{2}}\right) = \nabla \cdot \left[\mu_{n+\frac{1}{2}}(\boldsymbol{D}_n + \boldsymbol{D}^*)\right] \tag{5}$$

$$\boldsymbol{u}_{n+1} = \boldsymbol{u}^* - \frac{\Delta t}{\rho_{n+\frac{1}{2}}} \nabla p_{n+\frac{1}{2}} \tag{6}$$

$$\nabla \cdot \boldsymbol{u}_{n+1} = 0 \tag{7}$$

where $\boldsymbol{u}^*$ is a provisional value. In equation (5), the advection term $\boldsymbol{u}_{n+\frac{1}{2}} \cdot \nabla \boldsymbol{u}_{n+\frac{1}{2}}$ is discretized employing the second-order unsplit upwind method suggested by Bell-Collela-Glaz [26], whereas the diffusion term using the second-order Crank-Nicholson scheme. For the final term, $\boldsymbol{u}^*$ is determined by computing $\boldsymbol{u}_{n+\frac{1}{2}}$, which can be represented through characteristic extrapolation.

Through Equations (6) and (7), the Poisson equation can be obtained:

$$\nabla \cdot \left[\frac{\Delta t}{\rho_{n+\frac{1}{2}}} \nabla p_{n+\frac{1}{2}}\right] = \nabla \cdot \boldsymbol{u}^* \tag{8}$$

The numerical methodology employed in this paper is derived from calculations based on the open-source code *Basilisk* [27-29], in which equation (8) is solved based on the quad-based multilevel solver.

In addition, the vorticity ω is obtained by calculating the curl of the velocity $\boldsymbol{u}$:

$$\omega = \nabla \times \boldsymbol{u} \tag{9}$$

All variables in this paper are represented by dimensionless quantities, as follows:

$$\omega' = \frac{\omega D}{U}, \theta' = \frac{\theta}{\pi}, t' = \frac{tU}{D} \tag{10}$$

where θ represents the parameter in polar coordinates, U denotes the incoming flow velocity, and D signifies the diameter of the cylinder.

Furthermore, this article also utilizes the Reynolds number and the Strouhal number:

$$Re = \frac{\rho UD}{\mu}, St = \frac{fD}{U} \tag{11}$$

where f is the vortex shedding frequency.

2.2. Definition of the computational region and mesh construction

The computational domain employed in this paper is depicted in Figure 1. The diameter of the cylinder, denoted as D, is set to 1 meter. Given the utilization of a quadtree grid for spatial discretization, the two-dimensional computational domain is rectangular. To balance computational accuracy and efficiency, the dimensions of the rectangular domain are chosen as $64D$ in length and $32D$ in width. The origin of the coordinate system is positioned at the center of the cylinder, which is located $16D$ from the left boundary and $16D$ from the lower boundary. The left boundary serves as the velocity inlet, with a uniform incoming flow velocity of $U = 1$ m/s, while the right boundary functions as the pressure outlet. The remaining boundaries are designated as non-slip walls.

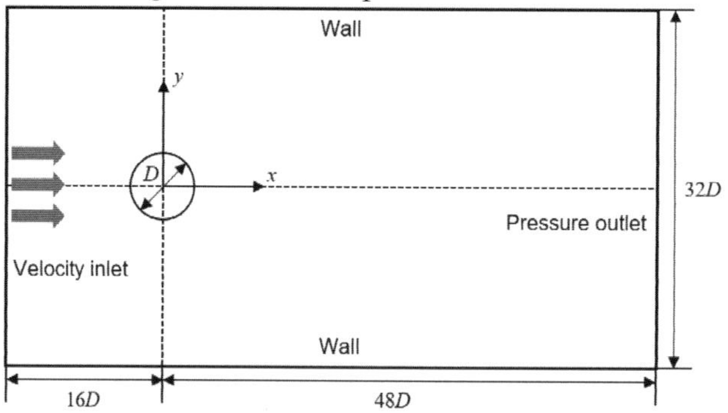

Figure 1. Diagram of the computational domain.

In this paper, quadtree adaptive mesh was used to divide the grid, and the *maxlevel* of the quadtree adaptive mesh was continuously increased to encrypt the grid around the cylinder to capture more flow details. The diagram illustrating the grid segmentation is presented in Figure 2.

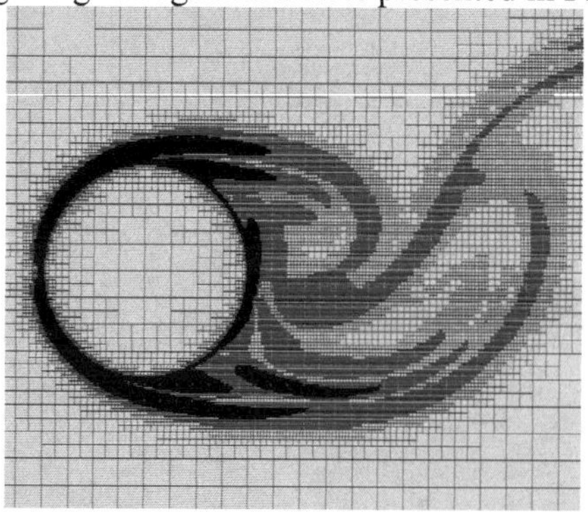

Figure 2. Schematic representation of grid partitioning.

2.3. Setting of Cylinder Surface Roughness

To quantify the roughness of the cylindrical surface to a certain extent, the spatial wave number was defined as β as a parameter to measure the relative density of the rough structure [30], and the roughness of the cylindrical surface was changed by changing β. To make it as simple as possible, the cosine function was used in this paper to deal with the rough cylindrical boundary. The rough boundary equation $\Gamma(\beta,\theta)$ is:

$$\Gamma(\beta,\theta) = 0.5 + 0.025cos(\beta\theta) \tag{12}$$

where θ is the parameter in the polar coordinate system. In this paper, rough cylinders with β of 20, 40, and 60 were mainly calculated, and their surface shapes are shown in Figure 3.

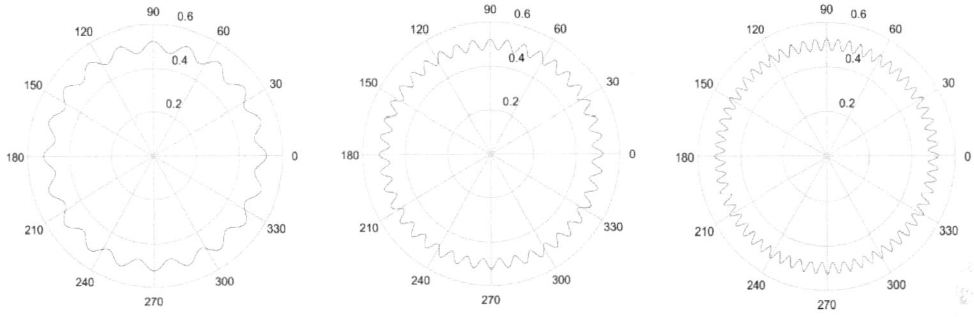

(a) (b) (c)

Figure 3. Surface diagram of cylinder with different roughness. (a) $\beta = 20$; (b) $\beta = 40$;(c) $\beta = 60$.

3. Results and discussion

3.1. Verification of mesh independence

To minimize grid density effects on simulation accuracy, a grid convergence analysis was performed. The vorticity distribution across the rough surface was adopted as the convergence criterion to determine the appropriate *maxlevel* for varying Reynolds numbers and surface roughness conditions.When t' equaled to 2.5, with Reynolds numbers of 50 and 1000, the vorticity distribution on the rough surface along the semicircular perimeter ($\theta' \in [0,1]$) was calculated for different *maxlevel* when β was 10 and 50, respectively. The detailed computational parameters are presented in Table 1.

Table 1. Table of mesh validation condition settings.

Re	β	*maxlevel*
50	10	12,13,14,15,16
	50	
100	10	
	50	

The computational results are depicted in Figure 4, where the rough surface shapes for β values of 10 and 50 are presented. During the numerical computation process, both computational accuracy and duration needed to be taken into consideration. Hence, as illustrated in Figure 4, for a Reynolds number of 50, a *maxlevel* of 13 was deemed suitable for a value of β equal to 10, whereas a *maxlevel* of 15 was considered appropriate for a value of β equal to 50.When the Reynolds number was 1000, a *maxlevel* of 14 and 15 was more suitable for β values of 10 and 50, respectively. Hence, for simulating flow around rough surfaces at a lower Reynolds number, a *maxlevel* of 13 is appropriate for smaller roughness, whereas a *maxlevel* of 15 was suitable for larger roughness. At a higher Reynolds number, a *maxlevel* of 14 was preferable for smaller roughness, and a *maxlevel* of 15 was more fitting for larger roughness.

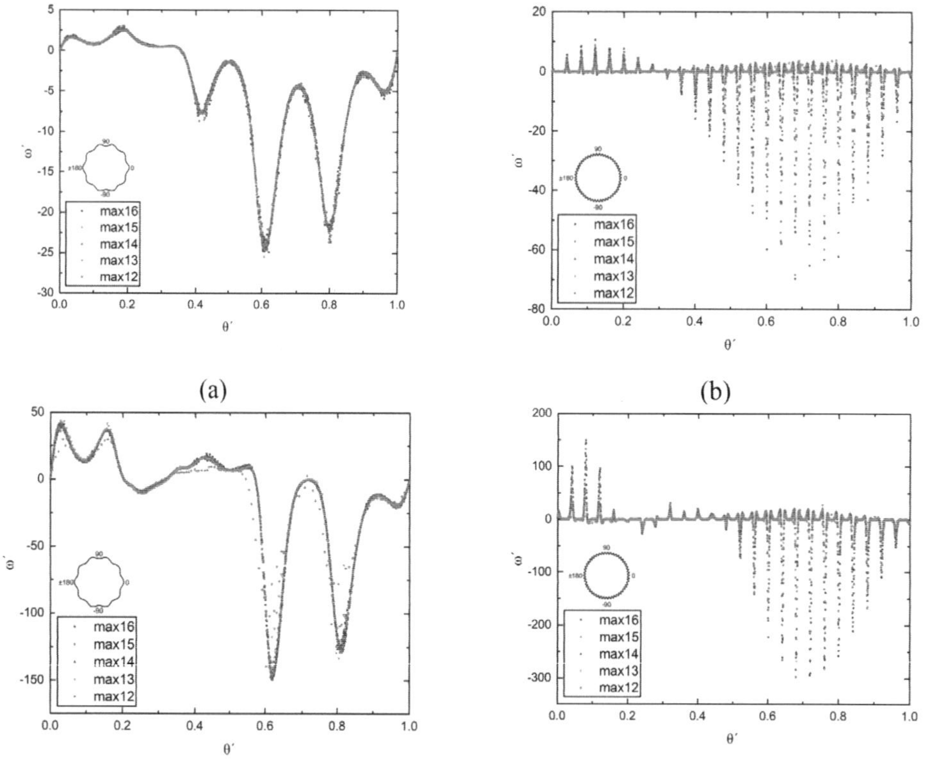

(a) (b)

(c) (d)

Figure 4. Surface vorticity distribution under different Reynolds number and roughness at $t' = 2.5$. (a) $Re = 50, \beta = 10$; (b) $Re = 50, \beta = 50$;(c)$Re = 1000, \beta = 10$;(d)$Re = 1000, \beta = 50$.

3.2. Verification of code accuracy

To ascertain the precision of the computational results presented herein, the vorticity distribution across the surface of a smooth cylinder was computed under two distinct sets of conditions: when Re equaled 1000 at $t' = 0.5$ and 1.5, and when Re equaled 9500 at $t' = 0.9$ and 2.5. The computed results were then compared with those obtained by Mohaghegh et al. [31]. The comparison results are shown in Figures 5 and 6, respectively.

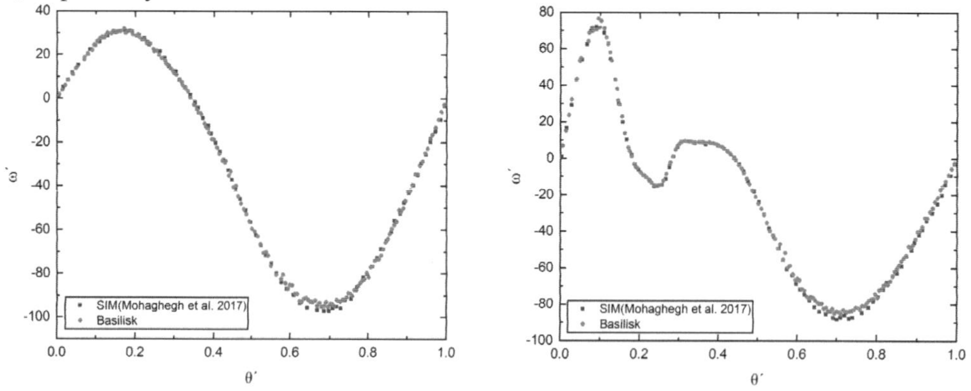

(a) (b)

Figure 5. Vorticity distribution on the surface of a smooth cylinder at different time when $Re = 1000$. (a) $t' = 0.5$; (b) $t' = 1.5$.

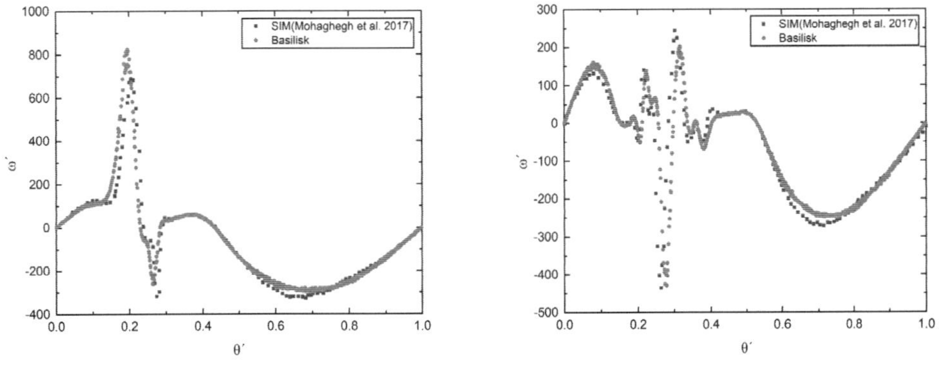

(a) (b)

Figure 6. Vorticity distribution on the surface of a smooth cylinder at different time when $Re = 9500$. (a) $t' = 0.9$; (b) $t' = 2.5$.

The results depicted in Figure 5 reveal that, at a Reynolds number of 1000, the computational outcomes derived from the code utilized in this paper are in close agreement with those documented in the literature. Moreover, as shown in Figure 6, when the Reynolds number reached 9500, although minor discrepancies were observed between the computational results obtained with the code in this paper and those reported in the literature, the overall trends and conclusions remained consistent. Therefore, the computational results achieved using the code in this paper demonstrate a high level of accuracy.

3.3. Vortex characteristics

In the problem of flow around a cylinder, there are often many vortices around the circumference, which will affect the stability and security of the structure, so the characteristics of the vortices are often one of the problems that scholars pay more attention to. To explore the characteristics of vortices around a rough cylinder, this paper selected t' as 200 when the Reynolds number was 500, and calculated the flow around a rough surface with β being 20, 40, and 60, respectively, and gave the vorticity field diagram at the same time, and compared it with a smooth cylindrical surface. Based on the computational results presented in Section 3.1, a *maxlevel* of 15 was chosen, and the corresponding results are illustrated in Figure 7.

The results depicted in Figure 7 indicate that, at a constant Reynolds number, as the surface roughness of the cylinder increased, the length of the trailing vortices generated during the flow around the cylinder augmented. The irregular surface of the cylinder led to the primary formation of vortices at the protruding portions of the surface on the rear side of the cylinder, while the recessed areas exhibited reduced vorticity. On both sides of the cylinder, in the concave part of the surface, there were a few vortices in the opposite direction of the outside. Compared to a smooth cylinder, before flow separation, the vortices on both sides of the rough surface did not adhere closely to the wall but only attached to the wall at each protruding part. This is because, during fluid flow, the protrusions on the rough surface exerted a greater disturbance on the fluid and acted as barriers to the recessed areas behind them.

3.4. Characteristics of vorticity distribution on the circumferential surface

In addition to the vortex characteristics, the vorticity distribution along the circumferential surface is also an important characteristic to be paid attention to in cylindrical flow. To explore the surface vorticity distribution of cylinders with different roughness at different times, in this paper, when the Reynolds number was 500, t' was 5,10,20,50,100, β is 20,40,60, the entire circumferential surface vorticity was calculated, and a comparison of these computational results was made with those obtained for smooth cylinders, as illustrated in Figure 8.

Journal of Physics: Conference Series **3011** (2025) 012004 doi:10.1088/1742-6596/3011/1/012004

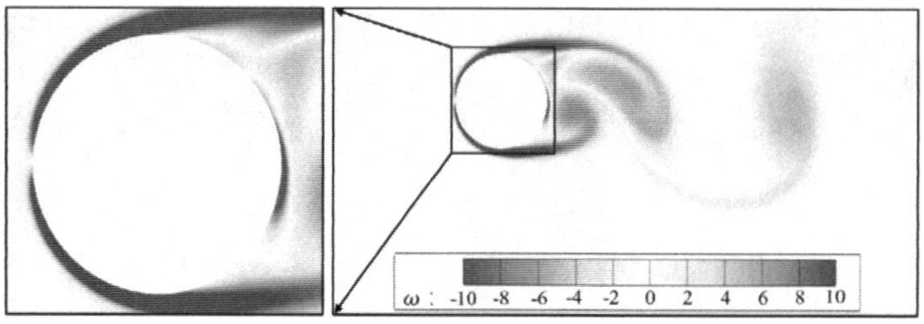

(a)

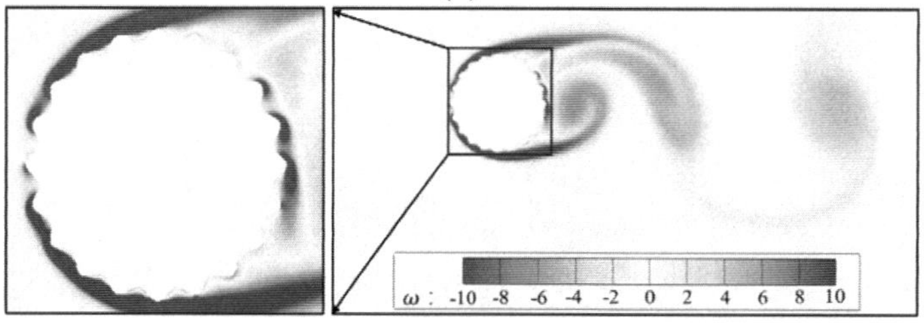

(b)

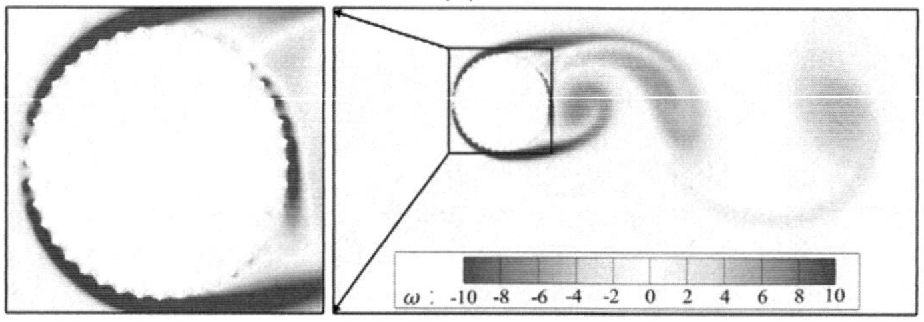

(c)

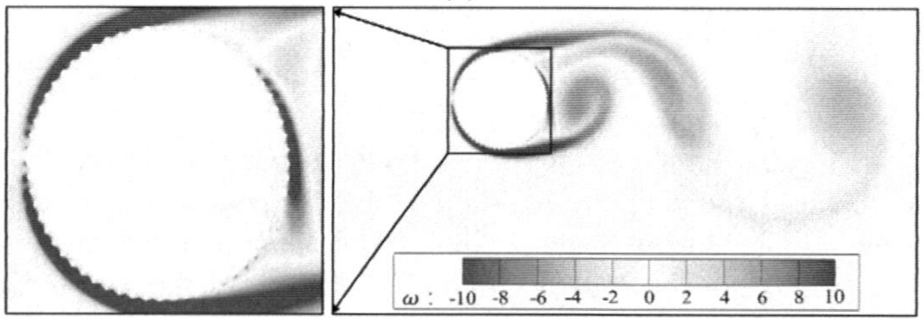

(d)

Figure 7. Vorticity field diagrams for cylinders with varying roughness levels. The left sub-figure presents an enlarged view of the boxed area in the right figure. (a) circle; (b) $\beta = 20$; (c)$\beta = 40$; (d)$\beta = 60$.

As illustrated in Figure 8(a), the vorticity displayed a symmetrical distribution around the entire circumference, with positive and negative values signifying opposing rotational directions of the vortices. Higher vorticity magnitudes were observed on the front side of the cylindrical structure, as opposed to the rear side. Additionally, the enlarged views on both sides revealed that as surface roughness increased, the vorticity profiles on the circumferential surface transformed from smooth curves into curves characterized by spikes. The greater the roughness, the more densely distributed these spikes became, accompanied by larger vorticity values. This result corresponds to Figure 7, where each spike represented the vorticity value at the protrusions on the rough surface, while the vorticity values at the recesses were smaller. Additionally, as seen in Figure 8, as time increased, the rotational directions of the vortices on both sides of the front end of the cylinder remained opposite, while the rotational

Journal of Physics: Conference Series **3011** (2025) 012004 doi:10.1088/1742-6596/3011/1/012004

directions of the vortices at both ends of the aft section of the cylinder changed from opposite to the same.

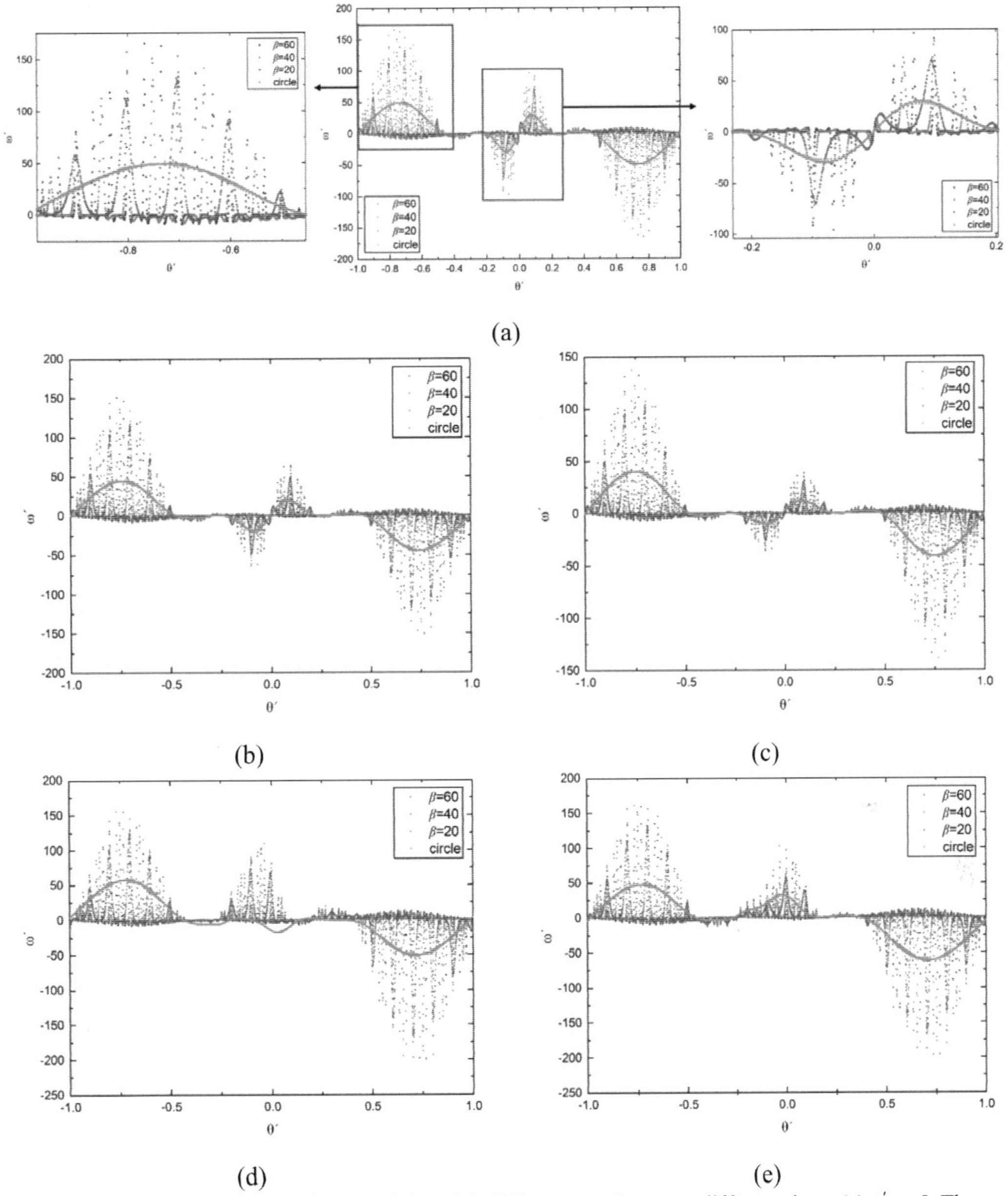

Figure 8. Calculation results of surface vorticity with different roughness at different time. (a) $t' = 5$. The subgraphs on both sides are enlarged images of the box parts; (b) $t' = 10$;(c) $t' = 20$;(d) $t' = 50$; (e) $t' = 100$.

3.5. Strouhal number

When analyzing the flow around cylindrical structures, apart from focusing on the vortex characteristics generated around the cylinder and the vorticity distribution characteristics on its circumferential surface, the resonance phenomenon resulting from vortex shedding is also an issue that cannot be ignored. Therefore, attention must also be given to the frequency of vortex shedding, f. In this paper, the Strouhal number was used to characterize this feature. Calculations of the Strouhal number for surfaces with different roughness levels were conducted at Reynolds numbers of 100, 500, and 1000, respectively, and compared with those for a smooth cylinder. Additionally, the calculation results for the smooth cylinder were validated against the results obtained by Stringer et al. [32]. The findings are depicted in Figure 9.

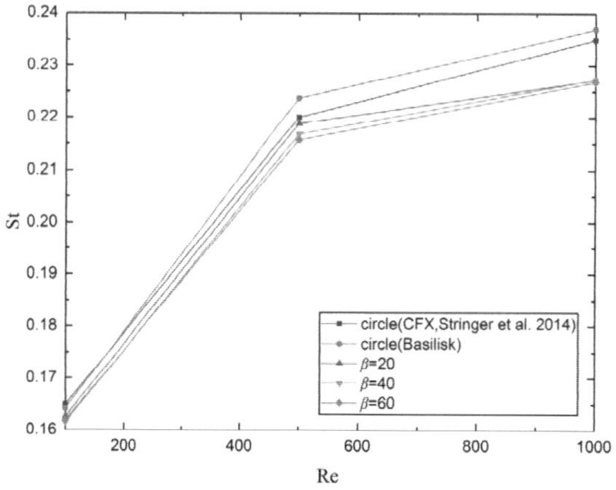

Figure 9. *St* under different roughness and Reynolds number.

As shown in Figure 9, though there existed a slight discrepancy between the computational results attained using the numerical methodology employed in this paper and those documented in the literature, the error was within an acceptable range, thus the accuracy of the calculation results presented herein can be deemed satisfactory. Moreover, an examination of the presented figure elucidates that, within the defined computational parameters of this study, the Strouhal number exhibited a proportional increase commensurate with the augmentation of the Reynolds number. This is because a higher Reynolds number indicates higher fluid viscosity, which leads to faster vortex shedding. However, at the same Reynolds number, as the roughness of the cylindrical surface increased, the Strouhal number decreased, but the difference was not significant compared to a smooth cylindrical structure. Therefore, surface roughness has a relatively minor impact on the Strouhal number during the flow around a cylinder.

4. Conclusion

In this paper, the open-source code *Basilisk* was employed to perform numerical simulations of the flow dynamics around a two-dimensional cylindrical obstacle.Through the systematic variation of the surface texture of the cylinder, characterized by distinct roughness parameters, and the selection of varied Reynolds numbers for the simulation scenarios, a comprehensive comparison was established between the computational outcomes and those pertaining to a baseline smooth cylinder configuration. Finally, the influence of surface roughness on the vortex characteristics around the cylinder, the distribution of vorticity on the circumferential surface, and the Strouhal number were obtained through the analysis of the calculated results. The detailed findings derived from this paper are presented below:

(1) The roughness of the cylindrical surface has a certain impact on the vortex characteristics generated around the cylinder. To commence, with respect to the downstream face of the cylindrical structure, under identical Reynolds number conditions, as the surface roughness increases, the length of the trailing vortices produced during the flow around the cylinder becomes larger. Moreover, due to the uneven surface of the cylinder, the vortices mainly originate from the protruded parts of the surface, while the vorticity in the recessed parts is relatively low. Second, concerning the sides of the cylinder, there are a few reverse vortices in the recessed areas. Unlike a smooth cylinder, on a rough surface, before flow separation, the vortices do not adhere closely to the entire wall surface but only attach tightly to each protrusion.

(2) The roughness of the cylindrical surface has a significant impact on the vorticity distribution on its circumferential surface. Circumferentially, the vorticity distribution exhibits symmetry, characterized by elevated levels of vorticity on the upstream face and diminished vorticity on the downstream face.

An increase in roughness causes the vorticity curve to take a peak-like shape, with the peak distribution density and vorticity values increasing as the roughness increases. The peaks on the curve represent areas of high vorticity corresponding to the protrusions on the circumference, while the troughs represent areas of low vorticity corresponding to the recesses. Over time, the direction of vortex rotation remains opposite at the front end of the cylinder, while the direction of the vortices at the rear side changes from opposite to the same.

(3) The influence exerted by the roughness of the cylindrical surface on the Strouhal number is observed to be relatively insignificant. Within the defined computational scope of the present study, the Strouhal number demonstrates a progressive augmentation with increasing Reynolds number. At a constant Reynolds number, the escalation in surface roughness of the cylinder engenders a marginal diminution in the Strouhal number. However, the disparity in comparison to the smooth cylinder configuration remains statistically non-significant.

Acknowledgments

This work was financially supported by the Guangdong Natural Resources Cooperation ([2024] No. 44).

References

[1] Chen G, Chen S, Qian G, et al. Direct numerical simulations of flows around an array of rough cylinders partially coated with porous media at Re= 3900[J]. Physics of Fluids, 2023, 35(9).
[2] Wu W, Zhao Y, Gou Y, et al. An overview of structural design, analysis and common monitoring technologies for floating platform and flexible cable and riser[J]. China Ocean Engineering, 2022, 36(4): 511-531.
[3] Duan J, Wang X, Zhou J, et al. Effect of internal solitary wave on the dynamic response of a flexible riser[J]. Physics of Fluids, 2023, 35(1).
[4] Zhang J, Wang H. Development of offshore wind power and foundation technology for offshore wind turbines in China[J]. Ocean Engineering, 2022, 266: 113256.
[5] Hu Q, Li Z, Zhai X, et al. Development of hydraulic lifting system of Deep-Sea mineral resources[J]. Minerals, 2022, 12(10): 1319.
[6] Hu B, Wang Z, Du H, et al. Response characteristics of flexible risers in offshore compressed air energy storage systems[J]. Journal of Marine Science and Application, 2019, 18: 353-365.
[7] Zhang D, Zhao B, Bai Y, et al. Dynamic response of DP offshore platform-riser multi-body system based on UKF-PID control[J]. Journal of Marine Science and Engineering, 2022, 10(11): 1596.
[8] Zhang X, Liu C, Wang B, et al. Experimental simulation study on dynamic response of offshore wind power pile foundation under complex marine loadings[J]. Soil Dynamics and Earthquake Engineering, 2022, 156: 107232.
[9] Wu Q, Yang J, Guo X, et al. Experimental study on dynamic responses of a deep-sea mining system[J]. Ocean Engineering, 2022, 248: 110675.
[10] Basack S, Goswami G, Dai Z H, et al. Failure-mechanism and design techniques of offshore wind turbine pile foundation: Review and research directions[J]. Sustainability, 2022, 14(19): 12666.
[11] Gaidai O, Wang F, Yakimov V, et al. Lifetime assessment for riser systems[J]. Green Technology, Resilience, and Sustainability, 2023, 3(1): 4.
[12] Du Y, Kong D, Wang S, et al. Fatigue analysis of jacket foundations for offshore wind turbines[J]. Rock and Soil Mechanics, 2024, 44(12): 8.
[13] Dai Y, Zhang Y, Li X. Numerical and experimental investigations on pipeline internal solid-liquid mixed fluid for deep ocean mining[J]. Ocean Engineering, 2021, 220: 108411.
[14] Liu Y, Zhu X, Song J, et al. Research on bypass pigging in offshore riser system to mitigate severe slugging[J]. Ocean Engineering, 2022, 246: 110606.
[15] Stringer R M, Zang J, Hillis A J.Unsteady RANS computations of flow around a circular cylinder for a wide range of Reynolds numbers[J]. Ocean Engineering, 2014, 87: 1-9.
[16] Parekh C J, Roy A, Harichandan A B. Numerical simulation of incompressible gusty flow past a circular cylinder[J]. Alexandria engineering journal, 2018, 57(4): 3321-3332.
[17] Pereira F S, Eça L, Vaz G, et al. On the simulation of the flow around a circular cylinder at Re= 140,000[J]. International Journal of Heat and Fluid Flow, 2019, 76: 40-56.
[18] Yan J, YANG C. Numerical Simulation of Flow Characteristics of Cylinder Around Flow under Different Reynolds Numbers[J]. Fluid Measurement & Control, 2024,5(04):1-5+14.
[19] Zhu H, Zhong J, Zhou T. Wake structure characteristics of three tandem circular cylinders at a low Reynolds number of 160[J]. Physics of Fluids, 2021, 33(4).
[20] Nguyen V L, Nguyen-Thoi T, Duong V D. Characteristics of the flow around four cylinders of various shapes[J]. Ocean Engineering, 2021, 238: 109690.

[21] Skonecki G M, Buick J M. Numerical Study of Flow around Two Circular Cylinders in Tandem, Side-By-Side and Staggered Arrangements[J]. Fluids, 2023, 8(5): 148.
[22] Han X, Tang Y, Meng Z, et al. Surface roughness effect on cylinder vortex-induced vibration at moderate Re regimes[J]. Ocean Engineering, 2021, 224: 108690.
[23] Michálek P, Procházka P, Uruba V, et al. Influence of surface roughness on the wake structure of a circular cylinder at Reynolds number 5×103 to 12×103[J]. European Journal of Mechanics-B/Fluids, 2022, 96: 15-25.
[24] Sun C, Zhou T, An H, et al. Experimental studies on wake structures of a circular cylinder with different surface roughness heights[J]. International Journal of Heat and Fluid Flow, 2022, 94: 108941.
[25] Sun C, Zhou T, An H, et al. On the study of vortex-induced vibration of circular cylinders covered with different roughness[J]. Applied Ocean Research, 2022, 124: 103215.
[26] Bell J B, Colella P, Glaz H M. A second-order projection method for the incompressible Navier-Stokes equations[J]. Journal of computational physics, 1989, 85(2): 257-283.
[27] Popinet S. Gerris: a tree-based adaptive solver for the incompressible Euler equations in complex geometries[J]. Journal of computational physics, 2003, 190(2): 572-600.
[28] Popinet S, Smith M, Stevens C. Experimental and numerical study of the turbulence characteristics of airflow around a research vessel[J]. Journal of Atmospheric and Oceanic Technology, 2004, 21(10): 1575-1589.
[29] Popinet S. An accurate adaptive solver for surface-tension-driven interfacial flows[J]. Journal of Computational Physics, 2009, 228(16): 5838-5866.
[30] Xu Z, Song S, Xin F, et al. Analytical solution for the viscous flow of small reynolds numbers in rough pipes[J]. Applied Mathematics and Mechanics, 2018,39(02):123-136.
[31] Mohaghegh F, Udaykumar H S. Comparison of sharp and smoothed interface methods for simulation of particulate flows II: Inertial and added mass effects[J]. Computers & Fluids, 2017, 143: 103-119.
[32] Stringer R M, Zang J, Hillis A J.Unsteady RANS computations of flow around a circular cylinder for a wide range of Reynolds numbers[J]. Ocean Engineering, 2014, 87: 1-9.

Design and performance analysis of low flashpoint fuel supply system

Ting Guo[1], Xing Tong[1*], Lesheng Liu[1], Gang Zhao[2], Lijun Shi[2], Xianglei Kong[1]

[1]School of Energy and Power, Jiangsu University of Science and Technology, Zhenjiang, China
[2]Guangdong Guangchuan International Marine Science and Technology Research Institute, Guangzhou Shipyard International Company Limited, Guangzhou, China

[*]Corresponding author's e-mail: 221210802129@stu.just.edu.cn

Abstract: This study investigates the methanol low flash point fuel supply system (LFSS). A simulation model was constructed using HYSYS and EDR software to examine the impact of load, ambient temperature, and operating parameters on system performance. The results indicate that when the load ranges from 25% to 100% and the feed temperature varies between -10°C and 30°C, the pressure fluctuation of the fuel supplied to the engine is 2.8% and 3.37%. These values are superior to the 6.25% pressure fluctuation range required by the engine. Additionally, the fuel temperature stabilized at 36°C-50°C and 38°C-46°C, which falls within the engine's required temperature range of 25°C-50°C. Furthermore, the system's specific power consumption (SPC) remains relatively stable at approximately 0.014 kWh/g when the engine load is between 65% and 100%.

1. Introduction

The rapid expansion of the global economy and the increasing frequency of international trade have elevated maritime transportation to a crucial role in global commerce [1]. While traditional marine fuels power ships, they also contribute significantly to environmental pollution, emitting sulfur oxides, nitrogen oxides, and greenhouse gases [2]. In response to these mounting environmental challenges, the International Maritime Organization (IMO) has implemented a series of regulations aimed at limiting ship emissions. These regulations seek to achieve a 50% reduction in carbon emissions by 2050 and to guide the shipping industry towards a cleaner, more environmentally friendly future [3]. In this context, environmentally friendly alternative fuels and dual-fuel engines with promising potential have become the focus of intense research. A study conducted by the European Union Joint Research Center has identified Liquefied Natural Gas (LNG) and methanol as having greater potential as alternative fuels compared to conventional marine fuels [4].

Current research on fuel supply systems for ships primarily focuses on Liquified Petroleum Gas (LPG) and LNG systems. Seo et al. designed a new LPG supply system that demonstrated a 38% reduction in energy consumption [5]. Hyeon et al. proposed a hybrid gas supply system utilizing both LNG and ammonia as fuel sources, which showed enhanced efficiency and reduced power consumption [6]. Gu et al. designed a low-pressure gas supply system for LNG transportation and vaporization, optimizing the heat transfer area and buffer tank volume [7]. Wang et al. analyzed the economic viability and efficiency of low-pressure fuel gas supply systems in ship operations under various configurations and management schemes [8]. Additionally, Wang et al. examined the transient characteristics of high-pressure gas supply systems using HYSYS software, concluding that operating parameters and buffer tank volume significantly influence system performance [9].

Methanol is a promising marine fuel due to its low carbon emissions and diverse feedstock sources [10]. Table 1 presents the properties of alternative fuels. Methanol is a liquid at room temperature and, as a low-emission, this environmentally friendly fuel has been demonstrated to reduce NOx emissions by 76.78%, SOx emissions by 89%, and PM emissions by 82.56% [11]. Compared to hydrogen, ammonia, and LNG, methanol is more easily storable, with simpler and less expensive fuel tanks that are compatible with most marine engines. Furthermore, in the event of a methanol leak, the fuel is rapidly diluted and has a low flash point, which significantly reduces the risk of a fire caused by a fuel leak, thus improving the safety of ship operations [12]. Methanol has already been adopted as an engine fuel. MAN launched a methanol dual-fuel engine and was scheduled for a sea trial in 2016 [13]. Qu et al. proposed a methanol-fueled integrated hybrid power system. The system demonstrated energy and exergy efficiencies of 55.6% and 57.9% [14]. Li et al. proposed a proton exchange membrane fuel cell/engine-based cogeneration system with methanol fuel. The results indicated that the power generation efficiency of this system was enhanced by 19.55%, and the total life cycle carbon emission was 344.938 g/kWh [15].

Table 1. Alternative fuel properties.

Parameter	LHV (MJ/kg)	Flashpoint (°C)	Emission reduction compared to HFO (%)			
			SOx	Nox	Co_2	PM
Ammonia	18.6	132	100	Compliant with regulation	~90	~90
Methanol	19.9	9	90-97	30-50	11	90
LPG	46	-104	90-100	10-15	13-18	90
LNG	50	--	90-99	20-30	24	90

There is a scarcity of research on the design and optimization of methanol supply systems for ships. During the operation of a ship, the methanol supply system must be capable of dynamically adjusting the fuel flow rate in accordance with the real-time changes in working conditions. This is necessary to adapt to the diverse sailing conditions and engine load demands that may arise. This study innovatively addresses the gap in methanol supply system performance research. The construction of a simulation model and the optimization of the design of a low-flash fuel supply system allow for the improvement of the overall performance and reliability of the system, thereby ensuring the stability and safety of the ship's operation under various working conditions. Furthermore, this research provides a scientific foundation and technical support for the green transformation and sustainable development of the shipbuilding industry.

2. System Description

Figure 1 illustrates the structure of a methanol supply system for a dual-fuel engine. In this system, methanol fuel is initially stored in a fuel cargo tank and connected to a fuel service tank via piping. The methanol in the fuel service tank undergoes pressurization and heating in the fuel supply module to ensure it meets the engine's temperature and pressure requirements. Subsequently, the processed methanol is delivered to the engine through the fuel valve train to meet its operational needs. Notably, excess methanol from the engine is recirculated back to the fuel service tank. Additionally, the system is equipped with a nitrogen tank for inerting, enhancing overall safety.

This study utilized the MAN-6S50ME-C9.6-LGI (M) methanol dual fuel engine model. Table 2 presents the fundamental characteristics of this engine. The engine initially operated in diesel mode and switched to methanol-based operation when the load reached or exceeded 25%.

This study presents a comprehensive analysis and design of equipment parameters and energy consumption for LFSS. Table 3 outlines the primary performance parameters of the LFSS. The study utilizes Aspen HYSYS v11 software for system process simulation [16]. The LFSS employs a two-stage pressurization mode to enhance the stability of methanol fuel pressure supplied to the engine. The system incorporates a low-pressure pump designed to boost methanol fuel pressure to 8 bar. Adhering to the ISO ambient standard temperature of 25°C, this study sets the first-stage heat transfer design temperature to 25°C, minimizing heat loss from the buffer tank. Figure 2 presents a detailed block diagram of the LFSS. Key components include a low-pressure pump, a low-pressure heat exchanger, a buffer tank, a high-pressure pump, and a high-pressure heat exchanger. The methanol fuel is processed in the system as follows: 1) Initially, the substance passes through a filter to remove any impurities; 2) Primary pressurization and heat exchange via a low-pressure pump and low-pressure heat exchanger; 3) Temporary storage in a buffer tank; 4) Secondary pressurization and heat exchange through a circulating pump and high-pressure heat exchanger, ensuring the fuel meets the dual-fuel engine's temperature and pressure requirements; 5) Finally, the fuel is filtered once more to ensure purity, after which it is transferred to the dual-fuel engine. Glycol-water mixtures exhibit low freezing points and high boiling points, with excellent thermophysical properties that ensure stable and efficient heat transfer processes [17]. Consequently, this study employs a 50% water-glycol mixture as the heat transfer medium for methanol fuel. In the system design, the mixture is stored in a GW tank at ambient temperature and pressure. The aqueous glycol solution in the GW tank was transferred to the plate heat exchanger by a GW pump. Subsequently, this solution serves as a heat source to warm the methanol fuel.

Table 2. Parameters of engine.

Parameter	Value
Engine	MAN-6S50ME-C9.6-LGI (M)
NCR	5402 kW×76 rpm
SMCR	6400 kW×80.4 rpm
NMCR	10320 kW×100 rpm
SMCR SPOC	10.83 g/kWh (Tier III)
SMCR SGC	327.4 g/kWh (Tier III)

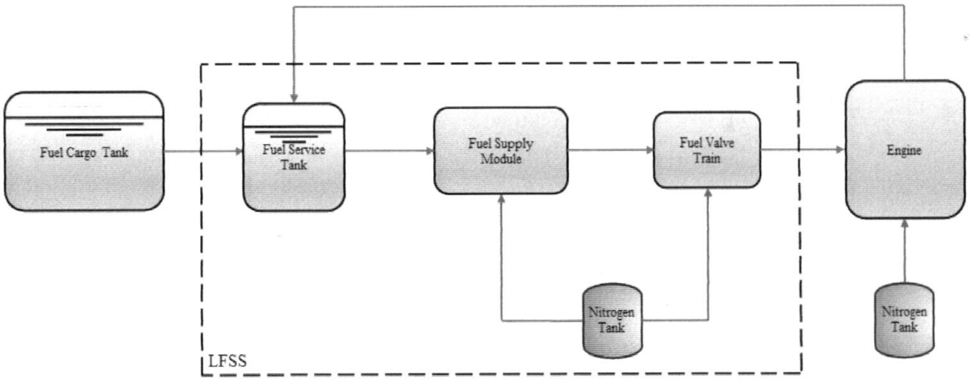

Figure 1. Schematic diagram of the methanol supply system.

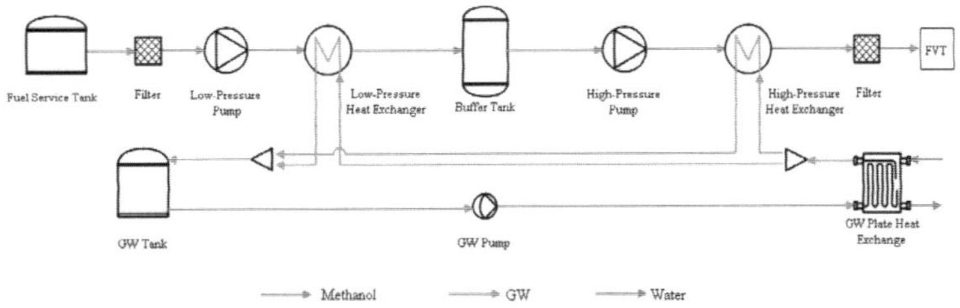

Figure 2. Process flow diagram (PFD) of the low flashpoint methanol fuel supply system.

Table 3. Low flash point fuel supply system parameter table.

Parameter	Value
System design pressure	16 bar
Fuel temperature at the engine inlet	25-50°C
Max. fuel flow	2700 kg/h
Pressure tolerance	±0.5 bar

3. Model Establishment and Numerical Methods

3.1 Model Assumption

To facilitate calculations, this study makes the following assumptions, with specific parameters shown in Table 4: 1) The environmental conditions are taken from the ISO standard, the ambient pressure is 101325 Pa and the temperature is 25°C. 2) Storage pressure in the fuel service tank is 1 bar; methanol is filtered, with a system feed temperature of -10°C and pressure of 20 kPa. 3) Low-pressure pumps are centrifugal type with 70% efficiency. 4) Low-pressure and high-pressure heat exchangers are shell-and-tube types, with glycol-water temperature at 50°C and pressure at 400 kPa. 5) The GW pump is a diaphragm pump with 70% efficiency. 6) The water-glycol heat exchanger is a plate heat exchanger. 7) All flow and heat transfer processes are steady-state; potential and kinetic energy are ignored. 8) It is assumed that there is no pressure drop in the pipeline.

Table 4. Values assumed for process simulation.

Parameter	Value
Ambient conditions	ISO
Methanol temperature	-10°C
Methanol pressure	20 kPa
Glycol-water temperature	50°C
Glycol-water pressure	400 kPa
Low-pressure pump type	Centrifugal
Low-pressure pump efficiency	70%
Low-pressure heat exchanger type	Tubular heat exchanger
High-pressure pump type	Centrifugal
High-pressure pump efficiency	70%
High-pressure heat exchanger type	Tubular heat exchanger
GW pump type	Diaphragm
GW pump efficiency	70%
GW heat exchanger type	Plate heat exchanger

3.2 Mathematical Model

In this paper, we first establish a fuel consumption model for engine load variation and construct an LFSS model. The LFSS model primarily comprises three core components: pump model, heat exchanger model, and tank model.

3.2.1 Fuel Consumption Model. The engine's fuel demand fluctuates with varying loads. Figure 3 illustrates how the engine's fuel demand flow rate changes across different load conditions. The methanol engine fuel flow can be expressed as follows:

$$q = Power \times SGC \tag{1}$$

In the formula, $Power$ is the engine power and SGC is the specific gas consumption.

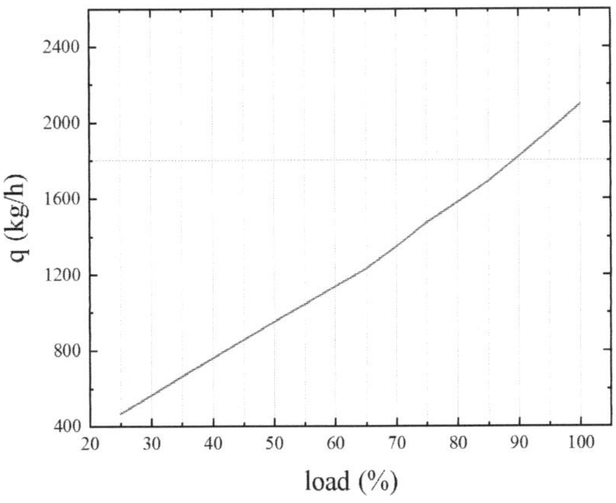

Figure 3. Variation of fuel flow with different loads.

3.2.2 Pump model. Low-pressure pump and high-pressure pump selected centrifugal pumps and the energy balance is calculated as follows:

$$p_{out} = p_{in} + \Delta p \tag{2}$$

$$W = \frac{\Delta p \times q}{3600 \rho \eta} = (H_{in} - H_{out}) \times q \tag{3}$$

In the formula, W and η are the pump power and the pump adiabatic efficiency; q and ρ are the fluid volume flow and the fluid density; H is the fluid specific enthalpy; p is the fluid pressure.

3.2.3 Heat Exchanger Model. The energy balance of the heat exchanger is calculated as follows:

$$\Delta T_{\mathrm{m}} = \frac{(T_{\mathrm{C1}} - T_{\mathrm{H2}}) - (T_{\mathrm{C2}} - T_{\mathrm{H1}})}{\ln \frac{T_{\mathrm{C1}} - T_{\mathrm{H2}}}{T_{\mathrm{C2}} - T_{\mathrm{H1}}}} \tag{4}$$

$$Q = UA\Delta T_{\mathrm{m}} \tag{5}$$

In the formula, U is the total heat transfer coefficient of the heat exchanger, ΔT_{m} is the log mean temperature difference for the heat exchanger, A is the heat transfer area, T_{C1} and T_{C2} are the inlet and outlet temperatures of the cold flow, and T_{H1} and T_{H2} are the inlet and outlet temperatures of the hot flow.

3.2.4 Tank Model. The fluid within the tank is in thermodynamic equilibrium. The conservation of mass and the conservation of energy are as follows:

$$\frac{dm}{dt} = \sum q_{in,i} - \sum q_{out,j} \tag{6}$$

$$\frac{de}{dt} = \sum q_{in,i} h_{in,i} - \sum q_{out,j} h_{out,j} + \vec{Q}_s \tag{7}$$

In the formula, m is the mass of streams in the tank, e is the total energy in the tank, and $\vec{Q}_s$ is the energy input to the tank from the environment. It is assumed to be zero.

3.3 Specific Power Consumption

To facilitate a comparative analysis of the system's energy consumption across various operational scenarios, the SPC metric is employed to evaluate the system's overall performance [18]. The SPC of the system quantifies the energy expended to supply a unit of methanol and is expressed as follows:

$$\mathrm{SPC} = \vec{W} / m \tag{8}$$

In the formula, $\vec{W}$ is the power consumption of the entire system and m is the quality of methanol supplied.

3.4 Model Verification

To verify the model's accuracy, this study compares theoretical calculations with simulation results. As illustrated in Figure 4, there is a high degree of concordance between the two sets of results. Specific analysis shows that the relative errors of the shaft power of the low-pressure pump and the high-pressure pump are 4.49% and 0.58%, respectively, while the relative errors of the heat load of the low-pressure heat exchanger and the high-pressure heat exchanger are 3.87% and 6.02%, respectively. Therefore, the model can meet the theoretical calculation requirements.

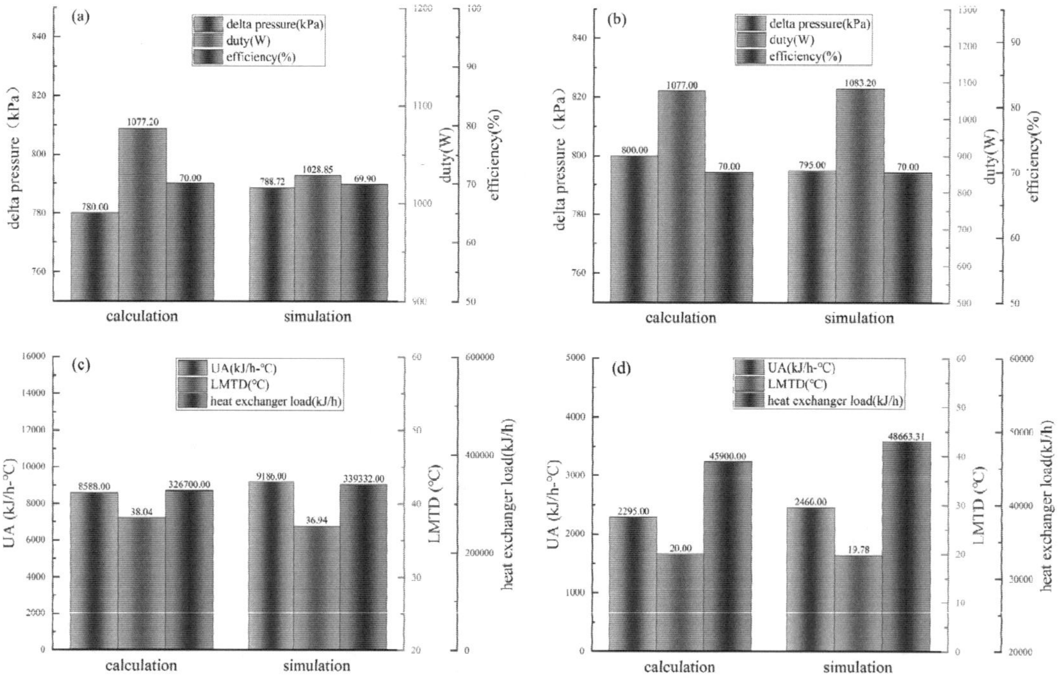

Figure 4. Comparison of the simulation results against the calculated results: (a) low-pressure pump (b) high-pressure pump (c) low-pressure heat exchanger (d) high-pressure heat exchanger.

3.5 Calculation Method

In this study, we derived fuel consumption curves for load variation using linear regression based on engine specification parameters. Subsequently, we designed and modeled the pump and heat exchanger using HYSYS and EDR software, respectively, and validated these equipment models. The study focuses on the effects of load and temperature variations on supply fuel pressure and temperature while analyzing the performance of the pump and heat exchanger. Finally, we used SPC as an evaluation index to assess the performance of the LFSS under varying load conditions.

4. Results and Discussion

4.1 Effect of Load on Supplied Fuel Pressure and Temperature

Assessing the system's response to load variations is crucial for ensuring stable operation. By analyzing the pressure and temperature variations under different load conditions, the fuel system design can be optimized to ensure its reliable operation under different operating conditions.

Figure 5 illustrates the system's performance across a load range of 25% to 100%. The results indicate that the supplied fuel temperature decreases as the load increases, from 48.68°C at 25% load to 35.47°C at 100% load. The pressure variation, however, exhibits a complex trend, initially decreasing, then increasing, and finally decreasing again, but with an overall downward trend. This is primarily due to the increased load resulting in a decreased pump head, which consequently reduces the pressure ratio. Simultaneously, the increased flow rate diminishes the heat exchanger's efficiency. Notably, the pressure shows a slight fluctuation in the 65% to 80% load range, varying from 1625.14 kPa to 1626.86 kPa.

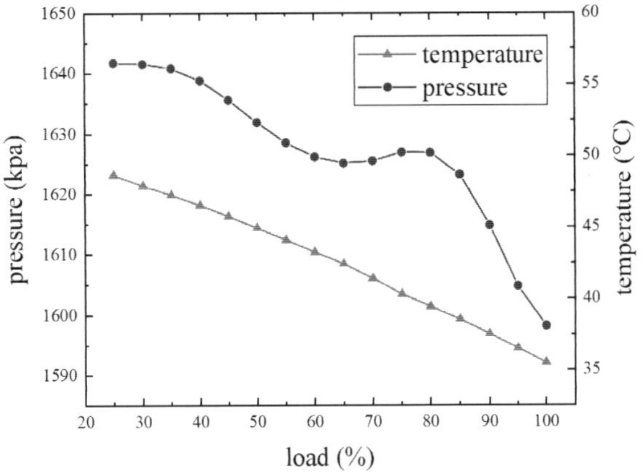

Figure 5. The effect of engine load on fuel temperature and pressure.

This phenomenon can be attributed to two factors: firstly, the decreased head of the low-pressure pump due to increased flow rate, and secondly, the reduced efficiency of the low-pressure heat exchanger, which lowers the methanol fuel temperature and relatively increases its density, thereby enhancing the pressurization effect of the high-pressure pump.

Throughout the load range, the temperature was consistently maintained between 25°C and 50°C, while pressure variations remained within a 50 kPa margin. These data indicate that LFSS meets the operational requirements of the engine and demonstrates good stability and adaptability.

4.2 Effect of Temperature on Supplied Fuel Pressure and Temperature

Marine navigation is characterized by continuous fluctuations in ambient temperature, which can significantly impact the fuel system's feed temperature. This study, therefore, concentrates on examining the effects of methanol feed temperature on the engine fuel system under elevated ambient conditions. To simulate realistic sailing scenarios, the simulation loads were set at 85% of normal sailing loads. All other parameters were maintained in accordance with design specifications to ensure the reliability and practical applicability of the results.

Figure 6 illustrates the relationship between methanol feed temperature and the temperature and pressure supplied to the engine as the methanol feed temperature is increased from -10°C to 30°C. The analysis reveals that the pressure of methanol fuel supplied to the engine demonstrates a gradual declining trend with increasing feed temperature. When the feed temperature reaches 30°C, the fuel pressure diminishes to 1569.36 kPa. This phenomenon can be attributed to the inverse relationship between methanol temperature and density. As the temperature of methanol rises, its density decreases proportionally. Consequently, if the pump head remains constant, the pump's outlet pressure declines, ultimately resulting in a reduction of the methanol fuel pressure supplied to the engine.

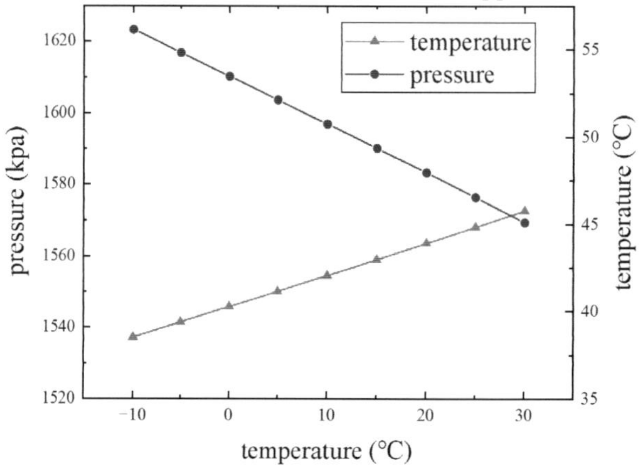

Figure 6. The effect of feed temperature on fuel temperature and pressure.

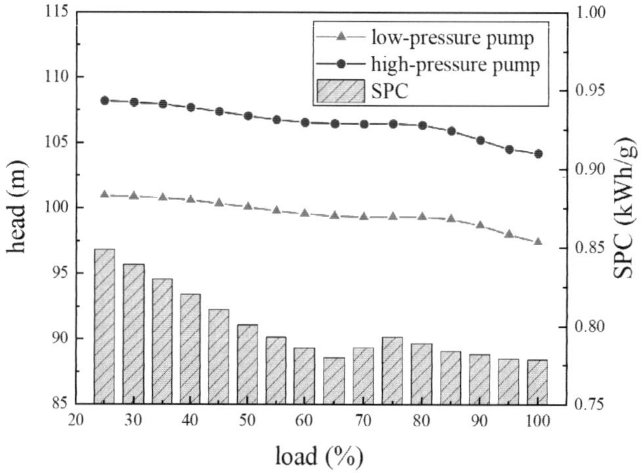

Figure 7. The effect of engine load on system performance.

4.3 Effect of Load Changes on Pump Head and System Power Consumption

Figure 7 illustrates a decreasing trend in the pump head with increasing load. Specifically, the low-pressure pump head decreases from 100.95 m to 97.40 m, while the high-pressure pump head decreases from 108.18 m to 104.22 m. This phenomenon can be attributed to the increased fluid velocity inside the pump due to higher flow rates, which in turn increases friction and turbulence losses. In contrast, the SPC exhibits a complex trend with load. It initially decreases from 0.848 kWh/g to 0.779 kWh/g, then rises to 0.793 kWh/g, and finally decreases to 0.778 kWh/g.

It should be noted that the SPC calculations are based solely on the power consumption of the low-pressure and high-pressure pumps. This is because these pumps pressurize varying flows of methanol to specified pressures, and their power consumption fluctuates with the load. In contrast, the GW pumps deliver the GW to the plate heat exchanger, and the power consumption remains constant.

The observed fluctuations in specific power consumption can be attributed to the system's power consumption being influenced not only by flow rate but also by the low-pressure heat exchanger's performance. As the flow rate increases, the fuel temperature in the buffer tank decreases, leading to increased fuel density. This, in turn, causes a rise in the high-pressure pump's power consumption. The interplay of these factors ultimately results in the observed fluctuations in specific power consumption.

4.4 Low Pressure Pump Performance Analysis
The impact of temperature on transfer pump performance is a crucial factor in determining system efficiency and stability. Therefore, this study analyzes the effect of feed temperature variation on pump performance. Three representative load points were selected for comprehensive analysis: 1) 25% load: This is the lowest load point for switching to methanol fuel mode, which corresponds to a mass flow rate of 466.08 kg/h. This load is critical for understanding system performance at low load operation. 2) 85% load: As a common load for the engine, it has a mass flow rate of 1691.84 kg/h. This load reflects the performance of the system under typical operating conditions. 3) The maximum flow: it corresponds to the maximum design flow of the LFSS of 2700 kg/h. This load is used to evaluate the performance and reliability of the system under extreme conditions.

Figure 8 illustrates the general declining trend in the inlet-outlet differential pressure of the low-pressure pump as the feed temperature increases. Among them, at 466.08 kg/h, the differential pressure of the low-pressure pump decreased from 813.23 kPa to 772.80 kPa with the increase in temperature; at 1691.84 kg/h, the differential pressure decreased from 798.83 kPa to 756.94 kPa; and the most significant change in differential pressure was observed at 2700 kg/h, where the differential pressure decreased from 788.72 kPa to 733.38 kPa. This is due to the fact that as the temperature increases, the density of the methanol fuel decreases, thus weakening the pressurization effect. In addition, the pressurization effect decreases as the fuel flow rate increases from 25% to the maximum fuel flow rate.

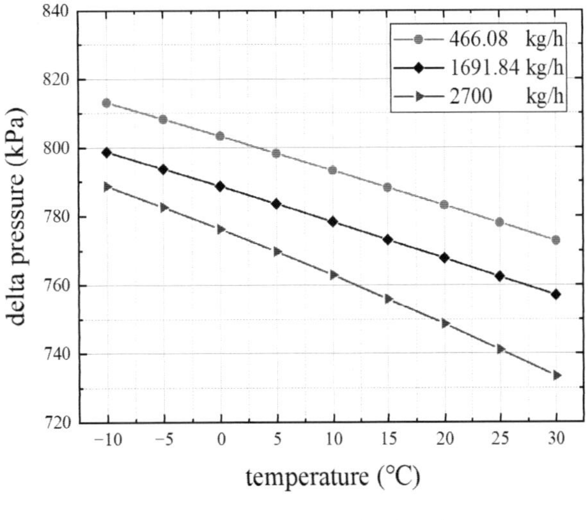

Figure 8. The effect of feed temperature on low-pressure pump head at flow rates of 466.08 kg/h, 1691.84 kg/h, and 2700 kg/h.

4.5 Low Pressure Heat Exchanger Performance Analysis
The performance of low-pressure heat exchangers is susceptible to fluctuations in ambient temperature and varying operating conditions. Consequently, this study examines the impact of different loads, inlet temperatures, and GW temperatures on the efficiency of low-pressure heat exchangers.

Figure 9 shows that as the load increases, the UA value increases from 5420.34 kJ/h-°C to 8955.56 kJ/h-°C, the LMTD increases from 17.24°C to 34.15°C, and the heat load increases from 93426.81 kJ/h to 293610.74 kJ/h. Figure 10 illustrates that as the feed temperature rises, the heat exchanged by the low-pressure heat exchanger decreases, resulting in a reduction of the LMTD value. Figure 11 shows that an increase in GW temperature leads to enhanced heat exchange in the low-pressure heat exchanger, consequently raising the LMTD value.

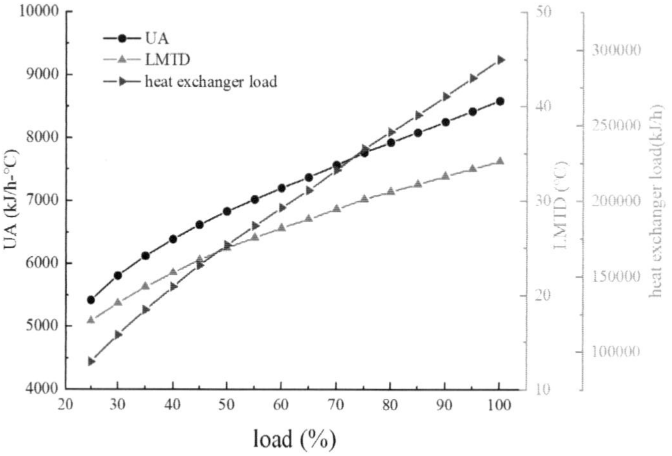

Figure 9. The effect of engine load on low-pressure heat exchanger performance.

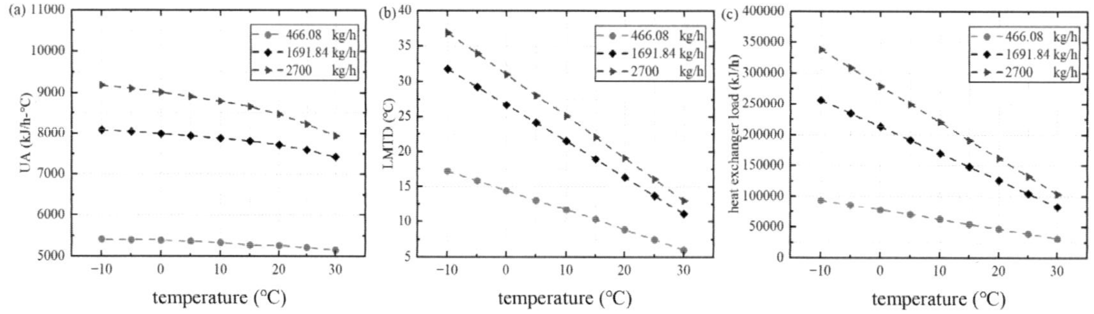

Figure 10 The effect of feed temperature on low-pressure heat exchanger performance at flow rates of 466.08 kg/h, 1691.84 kg/h, and 2700 kg/h.

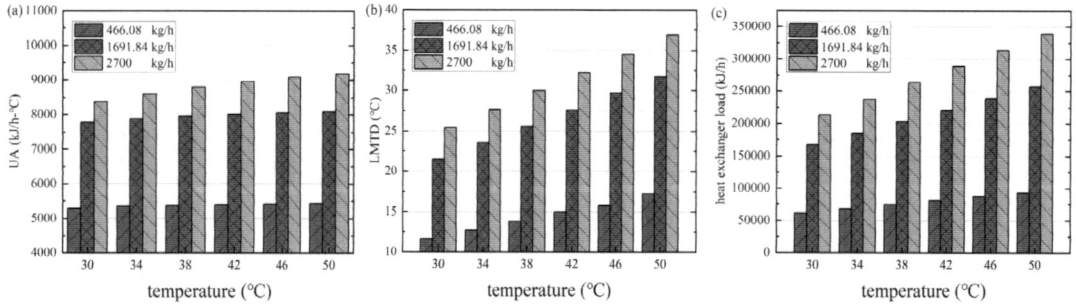

Figure 11. The effect of GW temperatures on low-pressure heat exchanger performance at flow rates of 466.08 kg/h, 1691.84 kg/h, and 2700 kg/h.

5. Conclusions

To meet the requirements for supplying methanol fuel to methanol-fueled engines, this study conducted a series of investigations. First, based on engine fuel requirements, we designed a low flash point fuel supply system for methanol engines and analyzed the system equipment. Subsequently, we focused on the patterns of pressure and temperature changes in the supplied fuel under varying load and ambient temperature conditions. Based on the simulation results, we draw the following conclusions:

(1) With an initial methanol temperature of -10°C, the supplied fuel pressure fluctuation is 2.8% when the engine load varies between 25% and 100%. During normal sailing at 85% load, the supplied fuel pressure fluctuation is 3.37% when the initial methanol temperature ranges from -10°C to 30°C. The supplied fuel pressure variation is significantly less than the 6.25% engine pressure fluctuation requirement, indicating good system pressure stability.

(2) With an initial methanol temperature of -10°C, the supplied fuel temperature remains stable between 36°C and 50°C as the engine load varies from 25% to 100%. During normal navigation at 85% load, the supplied fuel temperature stays between 38°C and 46°C when the initial methanol temperature ranges from -10°C to 30°C. These temperature ranges meet the engine fuel temperature requirement of 25°C to 50°C.

(3) The pump head decreases as the load increases. Simultaneously, the system's specific power consumption shows a fluctuating downward trend with increasing load. In the high load range (65%-100%), the system's specific power consumption is relatively stable (about 0.014 kWh/g), but it fluctuates more significantly (about 0.109 kWh/g) in the low load range (25%-65%).

(4) The differential pressure between the low-pressure pump's inlet and outlet gradually decreases as the feed temperature increases. When the system flow rate is 466.08 kg/h, the inlet-outlet differential pressure decrease is 4.97%, which corresponds to the optimal stability of the low-pressure pump.

The results and methods presented in this paper can guide the development of fuel supply systems for methanol-fueled ships. They provide valuable references for equipment design and operating parameter optimization, ensuring effective heat exchange and pressurization under various working conditions.

Acknowledgment

The paper is supported by the project of the Postgraduate Research & Practice Innovation Program of Jiangsu Province (SJCX23_2216), for which the authors are thankful.

References

[1] T. I. IMO, "Greenhouse gas study," Executive Summary and Final Report, London, 2014.
[2] A. M. Goulielmos, "The Decarbonization of the Shipping Industry and the New Fuel Issue," Modern Economy, vol. 12, no. 12, pp. 1999-2022, 2021.
[3] Z. Y. Tay and D. Konovessis, "Sustainable energy propulsion system for sea transport to achieve United Nations sustainable development goals: a review," Discover Sustainability, vol. 4, no. 1, 2023.
[4] K. Moirangthem, "Alternative fuels for marine and inland waterways: An exploratory study," 2016.
[5] Y. Seo, J. Kim, E. Park, J. Lee, M. Cho, and S. Han, "Analysis of Energy Consumption of Novel Re-Liquefaction System Integrated with Fuel Supply System (FSS) for LPG-Fuelled LPG Carrier to Conventional Systems," Energies, vol. 15, no. 24, 2022.
[6] S. Hyeon, J. Lee, and J. Choi, "Evaluation of Fuel Gas Supply System for Marine Dual-Fuel Propulsion Engines Using LNG and Ammonia Fuel," Energies, vol. 15, no. 17, 2022.
[7] X. Gu, G. Jiang, Z. Guo, and S. Ding, "Design and experiment of low-pressure gas supply system for dual fuel engine," Polish Maritime Research, vol. 27, no. 2, pp. 76-84, 2020.
[8] C. Wang, Y. Ju, and Y. Fu, "Comparative life cycle cost analysis of low pressure fuel gas supply systems for LNG fueled ships," Energy, vol. 218, 2021.
[9] C. Wang, Y. Ju, T. Wang, and S. Zou, "Transient performance study of high pressure fuel gas supply system for LNG fueled ships," Cryogenics, vol. 125, 2022.
[10] S. Brynolf, E. Fridell, and K. Andersson, "Environmental assessment of marine fuels: liquefied natural gas, liquefied biogas, methanol and bio-methanol," Journal of Cleaner Production, vol. 74, pp. 86-95, 2014.
[11] N. R. Ammar, "An environmental and economic analysis of methanol fuel for a cellular container ship," Transportation Research Part D: Transport and Environment, vol. 69, pp. 66-76, 2019.
[12] P. T. Katsumata and W. E. Kastenberg, "Fate and transport of methanol fuel from spills and leaks," Hazardous waste and hazardous materials, vol. 13, no. 4, pp. 485-498, 1996.
[13] S. Mayer, J. Sjöholm, T. Murakami, K. Shimada, N. Kjemtrup, and N. Kjemtrump, "Performance and emission results from the MAN B&W LGI Low-Speed Engine Operating on Methanol," in CIMAC Congress, 2016, pp. 6-10.
[14] J. Qu et al., "Assessment of a methanol-fueled integrated hybrid power system of solid oxide fuel cell and low-speed two-stroke engine for maritime application," Applied Thermal Engineering, vol. 230, 2023.

[15] C. Li et al., “4E analysis of a novel proton exchange membrane fuel cell/engine based cogeneration system with methanol fuel for ship application,” Energy, vol. 282, 2023.
[16] T. Saadi, M.-R. Jeday, and J.-N. Jaubert, “Exergetic analysis of an LPG production plant using HYSYS software,” Energy Procedia, vol. 157, pp. 1385-1390, 2019.
[17] D.-C. Lee, H. Afrianto, H.-S. Chung, and H.-M. Jeong, “Numerical analysis of LNG vaporizer heat transfer characteristic in LNG fuel ship,” Journal of Advanced Marine Engineering and Technology, vol. 37, no. 1, pp. 22-28, 2013.
[18] J. Choi and E.-Y. Park, “Comparative study on fuel gas supply systems for LNG bunkering using carbon dioxide and glycol water,” Journal of Marine Science and Engineering, vol. 7, no. 6, p. 184, 2019.

Design of spectral split PV-TEG hybrid system in outer space

Hongrui Li [1, 2], Shaowei Qing [1, 2, *], Gengxin Xie [3, 4]

[1] Key Laboratory of Low-grade Energy Utilization Technologies and Systems, Ministry of Education, Chongqing University, Chongqing, 400044, China
[2] Department of Energy and Power Engineering, Chongqing University, Chongqing, 400044, China
[3] College of Urban Construction and Environmental Engineering, Chongqing University, Chongqing, 400044, China
[4] Center of Deep Space Exploration, Ministry of Education, Chongqing University, Chongqing, 400044, China

[*] Corresponding author's e-mail: qshaowei@cqu.edu.cn

Abstract. A novel PV-TEG hybrid system applied in outer space is designed, and relevant physical and numerical models are established to study the effects of cut-off wavelength (λ_c) and concentration ratio (*CR*) on the output behaviors of separate PV module, TEG module, and PV-TEG hybrid system. Combining the changing patterns of the output characteristics of PV and TEG modules in terms of λ_c and *CR*, it is concluded that the hybrid system achieves a 20.1% increase in the maximum electrical power density in comparison to a separate PV module. In outer space, the novel PV-TEG hybrid system significantly improves the output behavior compared with a single PV cell and can provide technical support for converting space solar energy into electricity.

1. Introduction

Recently, drastic changes in the global climate and the energy crisis have seriously threatened the survival and development of humankind. Space solar energy is characterized by clean, sustainable, high energy flow density, and abundant energy. Compared with terrestrial solar energy, space solar energy is not affected by factors such as diurnal variation, seasonal variation, climate change, and geographic location [1], which makes it an ideal new-generation energy source. Photovoltaic (PV) cell can convert photon energy into electricity through the photoelectric effect. It is a major way of utilizing solar energy in outer space. However, due to material limitations, PV cells can only convert a portion of the incident space solar energy into electricity, while the rest energy is converted into heat [2]. This part of heat cannot be utilized by the PV cells and also increases the temperature of the cell, weakening its photovoltaic performance [3]. The thermoelectric generator (TEG) is a device that uses the Seebeck effect to convert heat energy directly into electricity. It is commonly used for waste heat recovery [4]. This solid-state and reliable energy technology has been successfully applied for long-duration power supply in deep space exploration [5]. Therefore, PV cell and TEG can form a hybrid system for use in the outer space. The waste heat produced by the PV cell can be used for temperature difference power generation using TEG to realize the utilization of waste heat.

Spectral splitting (SBS) technology is a technology that utilizes a beam splitter to divide the entire solar spectrum into two or more parts at different λ_c. The ultraviolet and visible light portions of the short wavelength band are reflected onto the PV plate, generating electricity through the photoelectric effect. The infrared light is transmitted through the beam splitter to the TEG, where the heat is converted into electricity through the Seebeck effect. Thus, SBS technology enables full solar spectrum utilization.

By using the SBS technology, the PV cell and the TEG will be thermally decoupled and can be operated at different temperatures [6]. Since the efficiency of PV cell strongly depends on the operating temperature, The power conversion efficiency of PV cell can be improved using this decoupling technique by reducing the working temperature of PV cell. In outer space, the PV-TEG hybrid system can be utilized to effectively enhance the output behavior of PV cell and provide technical support for converting space solar energy into electrical energy.

In this paper, a new type of PV-TEG hybrid system based on the SBS technology has been designed for use in outer space. The influence of the *CR* and the λ_c on the output power of individual PV cell module, individual TEG module, and PV-TEG hybrid system are investigated to explore the optimal system working conditions, in order to evaluate the output behavior of the PV-TEG hybrid system in outer space.

2. Designed PV-TEG hybrid system

The main components of spectral split PV-TEG hybrid system are a Fresnel lens, a beam splitter, a PV cell module and a TEG module, as shown in Figure 1. The PV cell module is composed of a GaAs-type PV cell and a heat sink which is attached to the PV cell cold side. The TEG module consists of two parts, a fiber-optic energy transfer part with a planoconvex lens for secondary concentration, and a TE module with an internal reflector, in addition to a heat sink which is attached the TE module cold side. The planoconvex lens can further reduce the spot size of the concentrator and the coupling energy loss with the optical fiber. The internal reflector can reduce the radiation energy loss from the surface of the TEG hot side to outer space. This design can reduce the incident energy loss on the surface of the TEG hot side, which is conducive to improving output behavior of the TEG module.

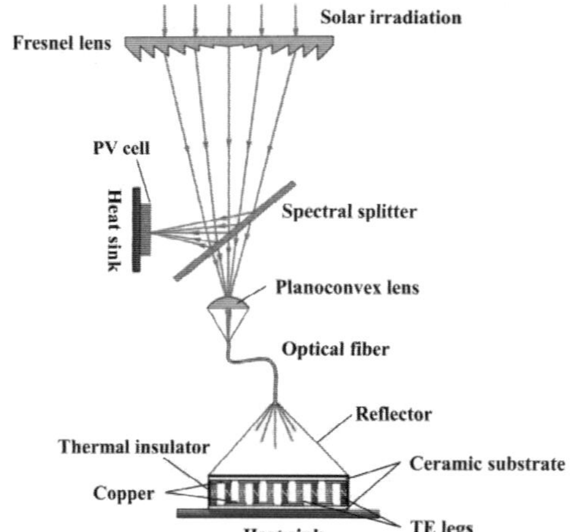

Figure 1. Schematic model of spectral split PV-TEG hybrid system applied in outer space.

In this paper, the space solar irradiance AM0 takes the value of 1353 W/m^2. The total energy Q_{tot} (W) projected onto the Fresnel lens can be calculated as:

$$Q_{tot} = \int_{280}^{4000} A_{Fresnel} G(\lambda) d\lambda \tag{1}$$

where $A_{Fresnel}$ is the area of the Fresnel lens, G(λ) is irradiation intensity of the AM0 spectrum at a specific wavelength.

The energy Q_{PV}(W) reflected to the PV module through the beam splitter can be calculated as:

$$Q_{PV} = \int_{280}^{\lambda} CR A_{PV} \eta_{Fresnel} R_{splitter} G(\lambda) d\lambda \tag{2}$$

where A_{PV} is the area of the PV cell, $\eta_{Fresnel}$ is the concentration efficiency of the Fresnel lens, which is 0.92. $R_{splitter}$ is the reflectance efficiency of the beam splitter, which is 0.95.

The energy Q_{TEG}(W) projected through the beam splitter to the planoconvex lens, converged by the planoconvex lens, and then transmitted by the optical fiber to the surface of the TEG can be calculated as follows:

$$Q_{TEG} = \int_{\lambda}^{4000} CRA_{TEG}\eta_{Fresnel}\eta_{Pl}\tau_{splitter}G(\lambda)d\lambda \tag{3}$$

where A_{TEG} is the area of the TEG, η_{Pl} is the coupling efficiency of the planoconvex lens that collects light and then couples through the optical fiber, and $\tau_{splitter}$ is the transmission efficiency of the beam splitter, which is 0.85.

2.1. *Numerical model of PV*

The numerical model of the PV cell includes thermal and electrical models, which are co-simulated by using MATLAB and COMSOL. The PV cell consists of six layers, which are composed of a glass cover, top EVA, ARC, GaAs, bottom EVA, Tedlar and heat sink, in order from top to bottom. The AM0 spectral energy absorbed by the PV cell is transformed into heat and electricity.

2.1.1. Thermal model. The GaAs layer is capable of converting spectral energy into electricity through the photovoltaic effect, while the remaining part of space solar energy is transformed into heat. The heat conduction equation for each layer can be calculated as:

$$\nabla\cdot(k_i\nabla T_i)+S_i=\rho C_P\nabla T_i, i=1...6 \tag{4}$$

where k_i is the thermal conductivity of each layer, and S_i is the heat of generation (W/m^3) of each layer, which may be computed as:

$$S_i=\frac{(1-\eta_{PVSP})Q_{PV}\alpha_i A_i\tau_i}{V_i A_{PV}} \tag{5}$$

where α_i is the absorptivity of each layer, τ_i is the net transmittance of each layer, and V_i(m^3) is the volume of each layer.

The boundary condition of the PV cell is that the upper surface receives energy and radiates energy outward, and the cold end of the PV cell radiates energy outward through the heat sink plate, surrounded by an adiabatic state. The boundary condition may be computed as:

$$\nabla\cdot(k_i\nabla T_i)+S_i=\rho C_P\nabla T_i+q \tag{6}$$

2.1.2. Electrical model. The electrical model of the PV cell is calculated by using MATLAB and the associated electrical performance properties of GaAs PV cell are shown in Table 1. The electrical performance properties are calculated as follows, with reference temperature T_{ref}=298 K.

The short-circuit current can be computed as:

$$J_{SC}=CR\int_{280}^{\lambda}\eta_{Fresnel}\tau_{PV}\alpha_{PV}R_{splitter}G(\lambda)\frac{e\lambda}{h_P c}EQE(\lambda)d\lambda+\beta_J(T_{PV}-T_{ref}) \tag{7}$$

The open circuit voltage can be computed as:

$$V_{OC}=\frac{h_P c}{\lambda_C}\cdot\frac{V_{OC,1}}{eE_g}+\frac{n_f kT_{PV}}{e}\ln(C)+\beta_v(T_{PV}-T_{ref}) \tag{8}$$

The electrical output power can be computed as:

$$P_{PV}=V_{OC}J_{SC}A_{PV}FF(1-\beta_{ref}(T_{PV}-T_{ref})) \tag{9}$$

The fill factor can be computed as:

$$FF=\frac{v_{OC}-\ln(v_{OC}+0.72)}{1+v_{OC}}\cdot(1-r_s) \tag{10}$$

The normalized open circuit voltage and series resistance are able to be computed as follows:

$$v_{OC}=\frac{eV_{OC}}{kT_{PV}} \tag{11}$$

$$r_s = R_s \frac{J_{SC} A_{PV}}{V_{OC}} \tag{12}$$

Table 1. Electrical properties of GaAs PV cell [7].

Electrical Properties	Value	Units	Electrical Properties	Value	Units
Bandgap of GaAs E_g	1.42	eV	Electron charge e	1.6022×10^{-19}	C
Ideality coefficient of PV cell n_f	1.33	—	Boltzmann constant k	1.38065×10^{-23}	J/K
Open circuit voltage $V_{oc,1}$	0.965	V	Current temperature factor β_J	0.73	A/K/m^2
Series resistance R_s	0.01216	Ω	Voltage temperature factor β_v	-0.0022	mV/K
Planck constant h_p	6.626×10^{-34}	J·s	Efficiency temperature factor β_{ref}	0.002	K^{-1}

2.2. Numerical model of TEG

Afte space solar radiation is focused by the Fresnel lens, it transmits through beam splitter to the surface of the planoconvex lens, concentrated again to reduce the area of the concentrated spot, thereby reducing the coupling energy loss with the fiber surface, and finally transmitted through the optical fiber to the surface of the TEG. The net incident energy on the TEG surface can be calculated as follows:

$$Q_{TEG\_net} = \alpha_{TEG} Q_{TEG} - (1-\alpha_{TEG}\rho_R) A_{TEG} E_b \tag{13}$$

where α_{TEG} is the absorptivity of the TEG surface, which is 0.92; ρ_R is the reflectivity of the reflector; E_b is the spectral radiance of the surface of TEG hot side, which is calculated as follows:

$$E_b = \varepsilon_{TEG}\sigma_B (T_{TEG\_h} - T_0) \tag{14}$$

where ε_{TEG} is the emissivity of the TEG surface, which is 0.92. σ_B is the blackbody radiation constant, which is 5.67×10^{-8} W/(m^2·K^4). T_0 is outer space temperature, which is 2.725 K. T_{TEG_h} is the temperature of the surface of TEG hot side.

The mathematical control equations for the entire simulation process used as a thermoelectric coupling calculation are as follows:

$$\nabla\cdot(\lambda_i \nabla T) + \frac{J^2}{\sigma_i} - \beta_i \vec{J}\cdot\nabla T = 0 \tag{15}$$

$$\nabla\cdot\left(\sigma\left(\nabla\varphi - \alpha\nabla T\right)\right) = 0 \tag{16}$$

Where σ is the electrical conductivity of the material, β is Thompson's coefficient, $\vec{J}$ is the local current density, φ is the electric potential, and $\alpha \nabla T$ is the electromotive force generated by the Seebeck effect.

The electric field and current density can be obtained by the following equations:

$$\vec{E} = -\nabla\varphi + \alpha\nabla T \tag{17}$$

$$\vec{J} = \sigma\vec{E} \tag{18}$$

The boundary conditions for the hot and cold side of TEG module are as follows:

$$-\lambda\frac{\partial T}{\partial z} = \frac{Q_{TEG\_net}}{A_{TEG}} \tag{19}$$

$$-\lambda\frac{\partial T}{\partial z} = \varepsilon_{HS}\sigma_B (T_{HS} - T_0) \tag{20}$$

where ε_{HS} is the emissivity of the TEG cold end heat sink, which is 0.9, and T_{HS} is the heat sink temperature.

3. Results and discussion

In this paper, the effects of different λ_c and CR on the output behavior are analyzed for separate PV and TEG modules, PV-TEG hybrid system at their respective maximum allowable temperatures. In this case, the maximum allowable temperatures of the PV and TEG modules are 623.15 K and 858 K, respectively.

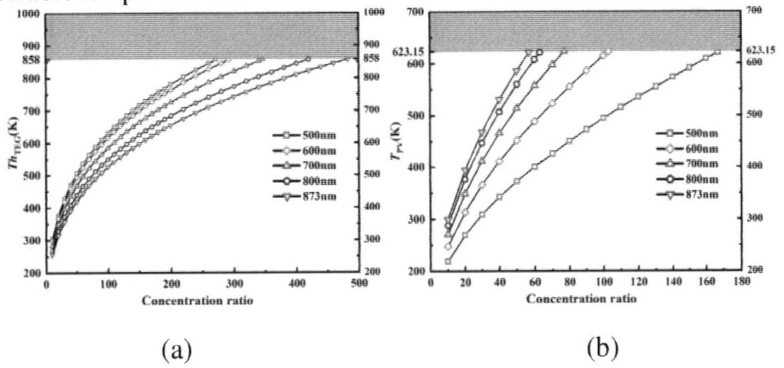

Figure 2. (a) The hot side temperature of TEG and (b) The GaAs layer temperature of PV cell at different λ_c and CR.

As shown in Figure 2, the results show that the larger the λ_c is, the smaller the maximum CR endured by the PV module is, and on the contrary, the larger the CR endured by the TEG module is. The CR of PV and TEG modules shows an opposite trend with the change of the λ_c. Therefore, the determination of the appropriate CR and λ_c is beneficial for the improvement of the output behavior of PV-TEG hybrid system.

3.1. PV module

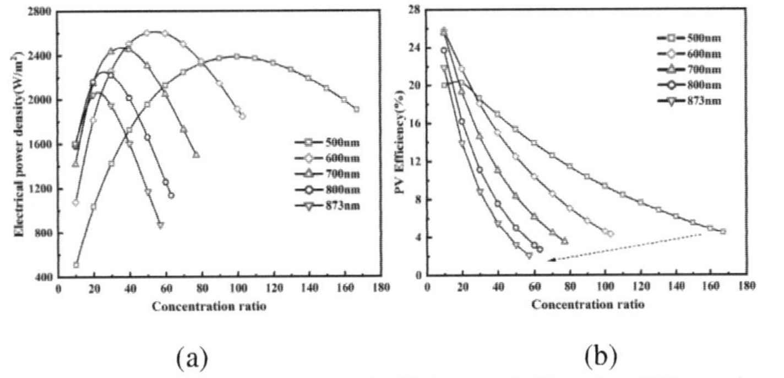

Figure 3. (a) Electrical power density and (b) Electrical efficiency of PV cell at different λ_c and CR.

As shown in Figure 3(a), at a certain λ_c, the electrical power density shows a parabolic variation with increasing and then decreasing as the CR changes. By increasing the λ_c, the maximum electrical power density at this λ_c first increases and then decreases, and when the λ_c is 600 nm, the electrical power density reaches the maximum, which is 2603.2 W/m^2. By increasing the CR, PV cell temperature increases, which makes the electrical efficiency of PV module decrease sharply as shown in Figure 3(b). When the λ_c is small, increasing the same CR increases the energy density less significantly and the temperature of the PV cell rises less. Therefore, the electrical efficiency of the PV module decreases more gently when the phase wavelength is small. By increasing the λ_c, the electrical efficiency corresponding to the maximum CR that can be achieved at a certain determined λ_c shows a gradually decreasing trend.

3.2. TEG module

As shown in Figure 4(a), the electrical power density shows a parabolic variation as the CR increases. The CR corresponding to the point of reaching the maximum electrical power density (about 630 W/m^2) is gradually larger as the λ_c increases. The reason is that the larger the λ_c is, the smaller the energy given to the TEG module by the same *CR* is. Therefore, a larger *CR* is required to reach the temperature corresponding to the optimal ZT value of the TEG. However, *CR* at the maximum allowable temperature is large, resulting in a high energy density into the TEG module and therefore a higher output power compared to the unconcentrated or low concentration.

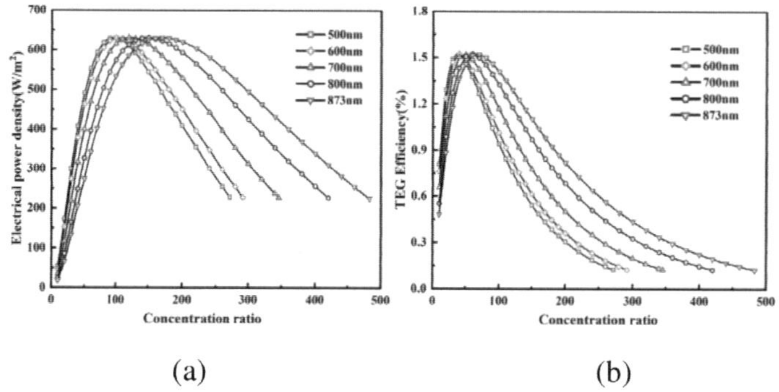

Figure 4. (a) Electrical power density and (b) Electrical efficiency of TEG at different λ_c and *CR*.

As shown in Figure 4(b), the electrical efficiency also shows a parabolic change as the *CR* increases. The *CR* is in the range of 40~70, and the electrical efficiency reaches the maximum value of about 1.5%. TEG electrical efficiency is smaller than the *CR* corresponding to the maximum electrical power density. However, because of the high temperature, TEG electrical efficiency at the maximum temperature is smaller than when it is unconcentrated.Therefore, it can be reasonably selected to make the appropriate *CR* to achieve the maximum electrical power density or electrical efficiency.

3.3. PV-TEG hybrid system

At the same λ_c, the maximum *CR* that PV module can withstand is smaller than the TEG module, so the output behavior of the hybrid system is analyzed in terms of the maximum *CR* that PV module can withstand.

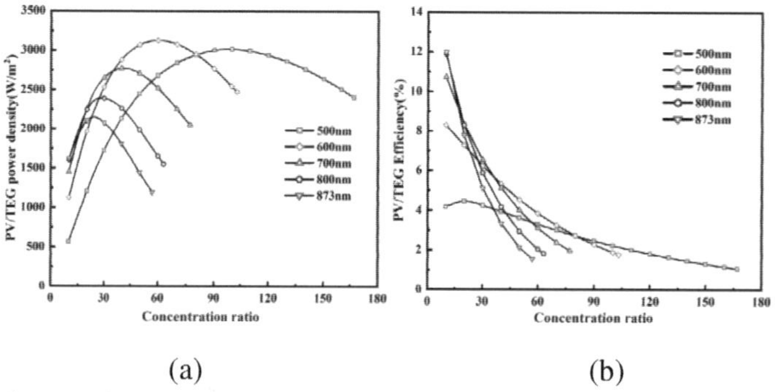

Figure 5. (a) Electrical power density and (b) Electrical efficiency of PV-TEG hybrid system at different λ_c and *CR*.

As shown in Figure 5(a), the trends of PV-TEG hybrid system and PV module with respect to *CR* at different λ_c are basically the same, both showing parabolic changes as the *CR* increases. However, the maximum electrical power density that can be achieved by the hybrid system is enhanced by 20.1% compared with the separate PV module, which significantly improves the output behavior.

The electrical efficiency of PV-TEG hybrid system is shown in Figure 5(b). Electrical efficiency decreases with the increase of the *CR*. The main reason is that incident energy of the PV and TEG modules increase as the *CR* increases, and the temperature of the PV cell and TEG increases. PV cell electrical efficiency shows a negative correlation with the temperature, and TEG electrical efficiency first increases and then decreases as the temperature rises. However, PV module dominates compared to TEG module. Therefore, the electrical efficiency of the PV-TEG hybrid system shows a decreasing trend with an increasing *CR*.

4. Conclusion

In summary, a new PV-TEG hybrid system design applied in outer space is proposed to analyze the rules of output behavior of individual PV module, TEG module, and PV-TEG hybrid system by changing the *CR* and the λ_c.

(1) For PV module, the electrical power density shows a parabolic trend with the *CR*, and the electrical efficiency gradually decreases with the *CR*. λ_c in the interval of 500 nm~873 nm, the range of concentrating ratio where the electrical power density reaches the maximum is 20~100, and with the gradual increasing the λ_c, the *CR* that reaches the maximum power density is gradually smaller.

(2) For TEG module, the electrical power density and efficiency show a parabolic trend with the *CR*, and the *CR* corresponding to the maximum power density is greater than the *CR* that corresponds to the maximum electrical efficiency. In the range of 100~170, the electrical power density reaches a maximum value of about 630 W/m^2. In the range of 40~70, the electrical efficiency reaches a maximum value of about 1.5%.

(3) For PV-TEG hybrid system, at the same λ_c, the maximum *CR* that the PV module can withstand is smaller than TEG module. The trends of the hybrid system and the PV module in terms of *CR* at different λ_c are basically the same, both showing parabolic changes with increasing *CR*. However, the maximum electrical power density that can be achieved by the PV-TEG hybrid system is 20.1% higher than a separate PV module.

References

[1] Fan G H (2022) Heat dissipation analysis and innovative design of sunlight concentration and photoelectric conversion system for space solar power station (in Chinese). Xidian University, Xi'an. [CrossRef]

[2] Sathe T M, Dhoble A S (2017) A review on recent advancements in photovoltaic thermal techniques. J. Renew Sust Energ Rev, 76: 645-672. [CrossRef]

[3] Zhong F H, Liu Z, Zhao S Q, Ai T C, Zou H Y, Qu M, Wei X, Song Y F and Chen H W (2024) A novel concentrated photovoltaic and ionic thermocells hybrid system for full-spectrum solar cascade utilization. J. Appl. Energy, 363: 123033. [CrossRef]

[4] Flores-Hernández D A, Palomino-Resendiz S I, Jost N and Luviano-Juárez A. (2024) Numerical energy evaluation of an optimized hybrid energy harvesting system (CPV-TEG) conendring the effects of solar tracking. J. Appl. Therm. Eng, 247: 123071. [CrossRef]

[5] Chen Z Y, Li M C, Guo R, Wang Y, Zhou D J, Chen Z and Ang R (2023) Performance evolution of thermoelectric modules under constant heat flux. J. Mater. Today Phys., 35: 101136. [CrossRef]

[6] Mojiri A, Taylor R, Thomsen E and Rosengarten G (2013) Spectral beam splitting for efficient conversion of solar energy - a review. J. Renew Sust Energ Rev, 28: 654-663. [CrossRef]

[7] Kandil A A, Awad M M, Sultan I G and Salem S M (2022) Investigating the performance characteristics of low concentrated photovoltaic systems utilizing a beam splitting device under variable cut-off wavelengths. J. Renew. Energy, 196: 375-389. [CrossRef]

Journal of Physics: Conference Series **3011** (2025) 012007 doi:10.1088/1742-6596/3011/1/012007

Calculation and analysis of the temperature field in the tail compartment of civil aircraft based on conjugate heat transfer

Huiqin Yuan*

Shanghai Aircraft Design and Research Institute, Shanghai 201210, China

*Corresponding author's e-mail: yuanhuiqin@comac.cc

Abstract: Accurate temperature field simulation can provide a reference for the formulation and verification of regional temperature requirements of aircraft, and is an important means of forward design of aircraft temperature field. Based on the calculation method of conjugate heat transfer, the temperature field in the tail cabin of a civil aircraft is studied and analyzed in this paper. The main conclusions are as follows: the conjugate heat transfer method is suitable for the simulation of the temperature field in the aircraft area. The results are more accurate than the single fluid calculation. By using thin mesh in the solid domain, the number of mesh can be greatly reduced and the calculation efficiency can be improved. The low-temperature environment of the tail compartment can meet the design requirements. In an extremely high-temperature environment, there is the risk of over-temperature in some areas of the cabin, so it is necessary to adjust the layout of the cabin or change the temperature design requirements. In the fire condition of the APU cabin, the electrical equipment in the cabin near the firewall has the risk of over-temperature, which may cause damage to the equipment.

1. Introduction

With the continuous development of the domestic civil aircraft industry, environmental adaptability has attracted more and more attention. Research shows that when the airborne equipment is working in a temperature environment 20°C higher than its allowable temperature, its failure rate will increase by 8 times [1]. At the same time, the extreme temperature environment will also cause irreversible structural damage and material aging, especially for temperature-sensitive composite materials. Therefore, the impact of aircraft operating environment temperature on airborne equipment and structure has always been an important factor that must be considered in aircraft design and airworthiness certification.

Some models have not carried out the forward design of the temperature field in the aircraft development stage, resulting in a series of thermal design problems such as over-temperature, skin bulge, and uneven temperature distribution in the cockpit during the operation stage, which affects the safety and comfort of aircraft operation. Therefore, it is very necessary to strictly implement the forward design of the temperature field in the whole life cycle of the aircraft. Forward design means that the design process starts from the top-level requirements, defines and decomposes the design tasks from the top down, and then carries out the bottom-up integration and verification after the completion of the product trial production [2, 3]. The complete capture of requirements is the key to forward design. Accurate temperature field simulation can provide an important reference for the formulation and verification of aircraft regional temperature requirements.

Lü et al. [4] conducted simulations on the flow field and temperature field of aircraft electronic pods at different altitudes and speeds and analyzed the heat dissipation performance inside the pods. The conclusion indicates that the higher the flight altitude is and the slower the flight speed is, the better the heat dissipation effect inside the pod is. Zhang et al. [5] numerically simulated the temperature of the electronic equipment compartment of high-speed aircraft, considering the influence of aerodynamic heating on the temperature inside the compartment, and verified the validity of the temperature field simulation results through experiments. Yu et al. [6] used the thermal lattice Boltzmann algorithm and GPU acceleration technology to calculate and analyze the influence of passenger behavior on the flow field and temperature field inside the cabin. The results show that when passengers stand in the aisle, it will change the large-scale circulation, causing heat to accumulate near the aisle. Xue et al. [7] focused on the thermal comfort issue of aircraft cockpits and used CFD to simulate the flow field and temperature field of the cockpit under different supply air temperatures. They also verified the calculation results with experimental data and derived the values of the cabin thermal comfort evaluation parameters PMV and PPD. Romani et al. [8] proposed a dynamic prediction model for the temperature of aircraft cabins and compared the model prediction results with experimental data. The results show that the two are in good agreement. At present, most studies on the temperature field in aircraft cabins focus on the cabin, cockpit, and electronic equipment cabin, while relatively few studies on other high-temperature cabins of aircraft. In addition, most of the published literature simplifies the structure and rarely considers the temperature distribution of real structures such as skin and frame comprehensively.

Taking the tail cabin of a civil aircraft as the research object, this paper studies the temperature distribution in the tail cabin of an aircraft under extremely high temperature, extremely low temperature, and the specific risk scenario of APU cabin fire, considering the influence of structural heat transfer and using the calculation method of conjugate heat transfer.

2. Numerical method

The conjugate heat transfer method treats fluid and solid as a whole. The solution domain includes the fluid domain and the solid domain. Through the whole field, discrete, simultaneous solution, and heat exchange at the flow/solid interface, temperature, and heat balance are achieved through repeated iterations [9].

The fluid domain is to solve the three-dimensional compressible N-S equation in the Cartesian coordinate system as follows:

$$\int_V \frac{\partial U}{\partial t} dV + \oint_{\partial V} \left(F_c - F_v\right) \cdot n dS = 0 \tag{1}$$

where U is the conserved quantity, F_c and F_v are the convective and diffusive fluxes, respectively, S is the source term, V is the control volume, and n is the normal vector pointing outward. In this paper, the finite volume method is used to discretize the control equations. The convective terms are interpolated using the second-order windward format. The separated flow SIMPLE algorithm is chosen to solve the discretized control equations. The turbulence calculation is done using the Reynolds time-averaged method (RANS). The turbulence model is adopted as the Realizable $k-\varepsilon$ model.

The solid domain, i.e., the structural region of the aft module, includes the spherical frame, the frame, the firewall, and so on. In the solid domain, the three-dimensional Fourier heat transfer equation is solved [10], as shown below:

$$\frac{\partial}{\partial t} \int_V U_s dV = \oint_{\partial V} \nabla\left(c_{p,s} T_s\right) \cdot n dS \tag{2}$$

where U_s is the solid domain conserved quantity, which is the product of solid density, specific heat and temperature; $c_{p,s}$ is the constant pressure specific heat capacity of the solid domain. In this paper, the energy transfer control Equation (2) for the solid domain is solved using the separated solid energy model.

The heat exchange at the coupled fluid/solid interface is shown in Figure 1, with the fluid unit on the left and the solid unit on the right. T_f and T_s are respectively the temperatures of the fluid unit and the solid unit. T_m is the temperature of the interface. d_f and d_s are the normal distances of the fluid unit and the solid unit to the interface, respectively. q_f and q_s are respectively the heat fluxes through the interface for the fluid and solid units.

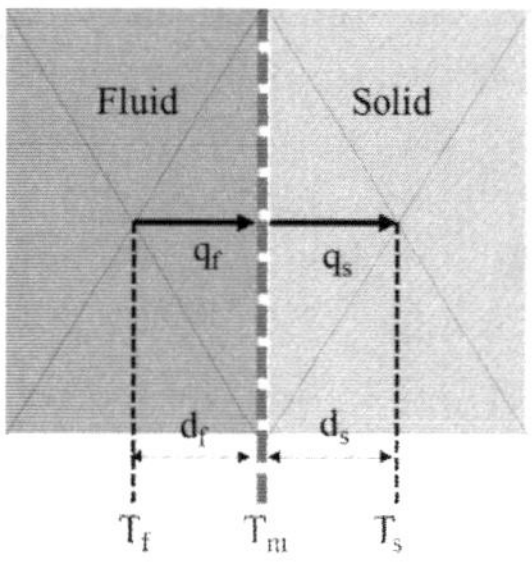

Figure 1. Schematic of the coupled flow/solid interface.

According to the conservation of energy, the heat flux flowing through the interface is the same on both sides.

$$q = -c_{p,f}\frac{T_f - T_m}{d_f} = -c_{p,s}\frac{T_m - T_s}{d_s} \tag{3}$$

where $c_{p,f}$ is the constant pressure specific heat capacity of the fluid domain.

Solving Equation (3) yields the temperature at the coupled flow/solid interface T_m.

$$T_m = \frac{T_f + T_s * \left(c_{p,s}/c_{p,f}\right) * \left(d_f/d_s\right)}{1 + \left(c_{p,s}/c_{p,f}\right) * \left(d_f/d_s\right)} \tag{4}$$

3. Geometric model and mesh

3.1. Geometric model

The tail cabin is located in the rear fuselage of the aircraft. The front end of the tail cabin is isolated from the cabin lavatory and water waste tank by a spherical frame. The rear end is isolated from the APU cabin by a firewall. The cabin is equipped with two electric pumps for the hydraulic system, an accumulator and other equipment and pipelines, an APU exhaust pipe for the air source system, a flight control actuator, and other equipment.

The structure of the tail cabin is relatively complex and there is many of equipment in the cabin. In order to improve the simulation efficiency, on the premise of not affecting the calculation accuracy, the tail cabin model is simplified, ignoring the small structures such as brackets, nuts, and convex that have no influence on heat dissipation, and only retaining the equipment and structures that have a great influence on the temperature field and flow field in the cabin. The simplified model is shown in Figure 2. The solid domain is shown in the green highlighted area in the figure.

If all structures in the tail compartment are meshed as solid domains, the number of meshes will be very large and the calculation amount will be doubled. According to the arrangement scheme of the tail cabin, the structures in the cabin that are susceptible to high temperatures are spherical frames and firewalls. Therefore, this paper only divides the spherical frame and firewall into solid domain grids and conducts conjugate heat transfer analysis. The temperature of the skin can be approximated as the recovery temperature. The frame, equipment frame, and other structures are far away from the heat source and the material can withstand high temperatures, so there is no need to pay attention to their thermal load. They are treated according to the adiabatic wall surface.

Journal of Physics: Conference Series **3011** (2025) 012007 doi:10.1088/1742-6596/3011/1/012007

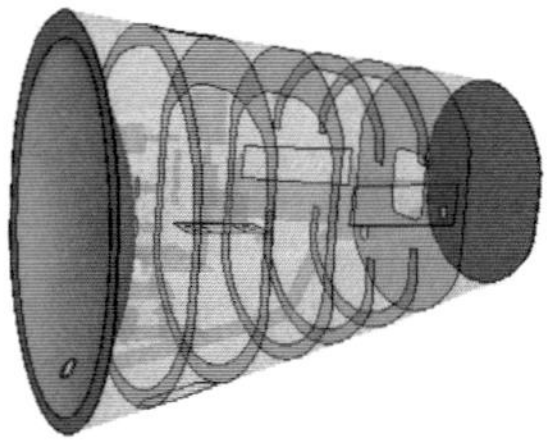

Figure 2. Simplified model of the aft cabin.

3.2. Meshing

The commercial software STARCCM+ was used to mesh polyhedral and prismatic layers of the tail cabin. Since the thickness of the solid domain is only 5~10mm, but the spreading distance is up to 2m, if the thickness size is used as a reference for grid division, the number of grids will be too large. Because the temperature distribution gradient in the spreading direction of the structure is relatively small, this paper adopts a thin-body grid in the solid domain. The number of thin body layers is 3. The solid domain grid is shown in Figure 3. Under the premise of not affecting the calculation accuracy, the number of grids can be greatly reduced.

In order to improve the accuracy of the calculation results, it is also necessary to encrypt and control the inlet and outlet of EMP, APU inlet pipe, and other cabin parts, as shown in Figure 4. The prismatic layer on the wall of the fluid domain is set to 5 layers. The prismatic layer is disabled in the solid domain. The total grid generated is 6.5 million.

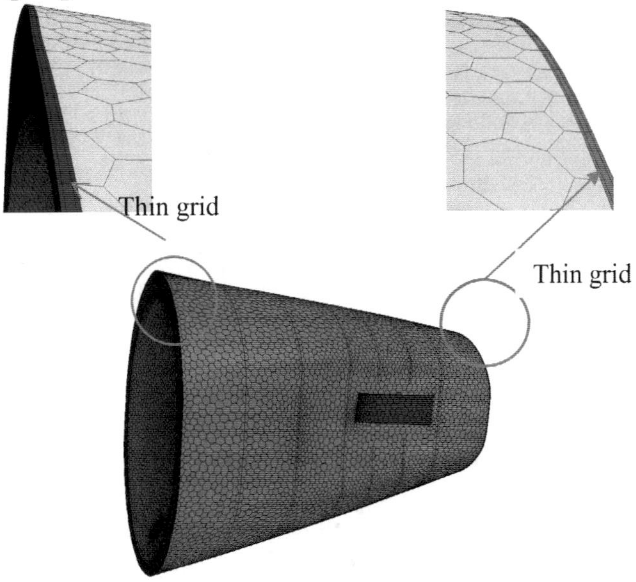

Figure 3. Schematic of the thin-body mesh in the solid domain of the tail cabin.

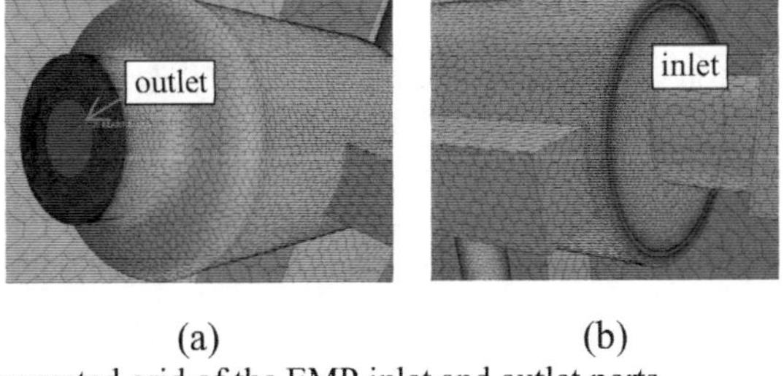

Figure 4. Schematic of the encrypted grid of the EMP inlet and outlet ports.

4. Calculation of working conditions and boundary conditions

4.1. Calculation of working conditions

With reference to the measured data of RTCA/ DO160G [11] and similar models, it is necessary to ensure that the cabin temperature is -55~70°C when the tail cabin is working normally. In addition, according to AC25.869 [12], "the electrical equipment located in the area outside the firewall and in the auxiliary structure of the engine compartment, the materials used and the distance from the fire resistance shall ensure that when the temperature of the firewall near the fire source reaches 1100°C for 15 minutes, these electrical components will not be damaged which may endanger the aircraft". Therefore, in order to obtain the temperature range in the aft cabin during normal operation of the aircraft, as well as to evaluate the effect of APU cabin fire on the aft cabin temperature. This paper will carry out the calculation and analysis of the temperature field in the aft cabin under three working conditions, among which the steady calculation is adopted in Working Conditions 1 and 2, and the unsteady calculation is adopted in Working Condition 3.

According to the operating envelope of aircraft temperature and altitude, combined with the operating scenario of equipment in the tail cabin, it can be seen that the temperature in the tail cabin is the highest when the ground is extremely high. When cruising at extremely low temperatures at high altitudes, the temperature in the aft cabin is the lowest. The specific parameters of calculation working conditions are shown in Table 1.

Table 1. Calculated working conditions.

Working conditions	Descriptive	Outside air temperature (°C)	Height (ft)	Ma
Case1	Extremely high temperature	55	0	0
Case2	Extremely low temperature	-74	39800	0.78
Case3	Ignite	55	0	0

4.2. Boundary conditions

There are two EMPs arranged in the tail cabin, but only EMP3A works in the full range, and EMP 3B only works in the take-off and landing stages. This paper only considers the effect of EMP3A on cabin temperature. The EMP 3A outlet is set as the mass flow inlet boundary, and the EMP 3A inlet is set as the outlet boundary. The APU exhaust air is enabled in the ground stage and closed in the cruise stage, so the APU inlet pipe is set as the temperature boundary in Working Conditions 1 and 3, and the APU inlet pipe is set as the adiabatic wall in Working Condition 2. All other equipment in the cabin is heat source boundaries. The spherical frame and the inside of the firewall are conjugate heat transfer boundaries, except for Working Condition 3. The outer surfaces of the spherical frame and the firewall are set as the convective heat transfer boundary, given the surface heat transfer coefficient and the temperature of the adjacent cabin. In Working Condition 3, the outside of the firewall is set as the temperature boundary, and the temperature is 1100°C.

5. Calculation results and analysis

5.1. Validation of the numerical method

In order to verify the validity of the calculation method, the test data of the tail cabin of similar models are compared with the simulation results. The installation position of the temperature sensor in the tail cabin of similar models is shown in Figure 5. Simulation of cabin temperature field in hot and cold days under the same test conditions was carried out respectively. After convergence of calculation, simulation results at the same position as the temperature sensor were extracted and compared with the measured data, as shown in Table 2.

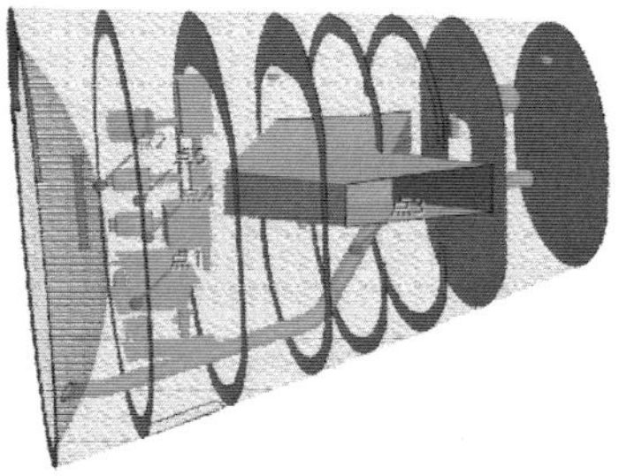

Figure 5. Schematic diagram of sensor position.

Table 2. Comparison of test flight data and simulation results.

Working conditions	Point 1	Point 2	Point 3	Point 4	Point 5
Case1_test	55	69	100	65	57
Case1_cfd	56	70	105	67	59
Case2_test	-29	-2	-27	-24	-28
Case2_cfd	-28	-3	-25	-24	-26

According to Table 2, on a hot day on the ground (Case1), the difference between CFD calculation results and test flight data is at most 5°C, and the average difference is 2.2°C. On cold cruise days (Case2), the maximum difference between CFD calculation results and flight test data is 2°C, and the average difference is 1.2°C. The error is acceptable, indicating that the calculation method is effective.

5.2. Analysis of simulation results for extreme hot and cold days

The key area for temperature assessment in the aft cabin is around the equipment and structure where the EMP high-temperature outlet air flow is directly facing the low-temperature sensitive actuator. Therefore, the temperature distribution cloud map of the EMP center section (Plane 1) and the actuator center section (Plane 2) was extracted for analysis, as shown in Figure 6.

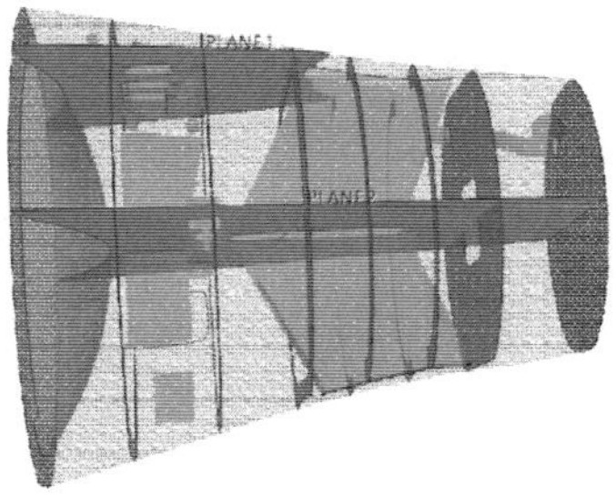

Figure 6. Schematic of cross-section location

According to the simulation results, the temperature in the tail cabin is greatly affected by the external environment and the heating of the equipment in the cabin. On extremely hot days, the average temperature in the tail cabin is 78°C. On extremely cold days, the average temperature in the tail cabin is -46.3°C. The temperature distribution in the cabin is uneven. Under the influence of gravity, the temperature in the upper part of the cabin is significantly higher than that in the lower part. Under the influence of airflow at the high-temperature outlet of EMP3A, the temperature in the front part of the cabin is higher than that in the rear part, as shown in Figure 7 and Figure 8. For the over-temperature region, it is necessary to further evaluate whether the equipment can withstand it. If necessary, it is necessary to change the arrangement plan of the tail cabin or adjust the temperature demand, to increase the high-temperature identification temperature of the equipment in the over-temperature region.

The temperature in the area where the spherical frame is facing the airflow at the outlet of EMP3A is obviously higher. The highest temperature in this area is 82°C on an extremely hot ground day and -34°C on an extremely cold cruise day. In addition, in extremely hot days on the ground, affected by the high-temperature radiation and convective heat transfer of APU exhaust pipes through the frame, the temperature around the APU exhaust pipes on the spherical frame is higher, reaching up to 110°C, as shown in Figure 9.

(a) Case1 (b) Case2

Figure 7. Temperature distribution cloud map of Plane1 for extremely hot/cold days.

(a) Case1 (b) Case2

Figure 8. Temperature distribution cloud map of Plane2 for extremely hot/cold days.

(a) Case1 (b) Case2

Figure 9. Temperature distribution cloud map of spherical frame for extremely hot/cold days.

5.3. Analysis of simulation results for fire condition

According to the arrangement of the tail cabin, the electrical equipment near the firewall has the tail cabin light and fire extinguisher. The specific location is shown in Figure 10. Therefore, this paper mainly analyzes the temperature changes near the tail cabin light and fire extinguisher within 15 minutes after the APU cabin is on fire and the temperature of the firewall is 1100°C.

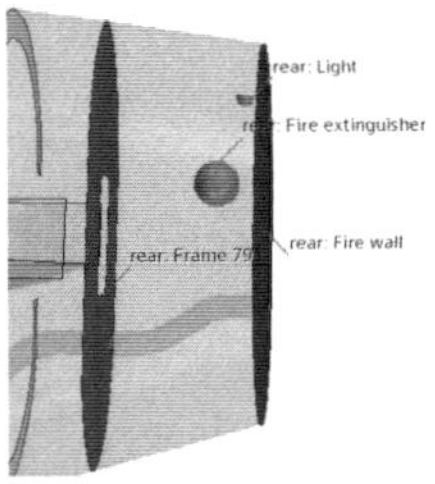

Figure 10. Position of the electrical equipment near the firewall.

Figure 11 shows the temperature curve overtime at the position points near the tail cabin light and the fire extinguisher. It can be seen from the figure that the temperatures near the tail cabin light and the fire extinguisher both rose first and then stabilized with time. After 5 minutes, the temperature of the tail cabin light reached 288°C and that of the fire extinguisher reached 221°C, both exceeding the withstand temperature of the equipment. Among them, the temperature near the tail cabin light was higher because it was closer to the firewall and was more affected by thermal radiation.

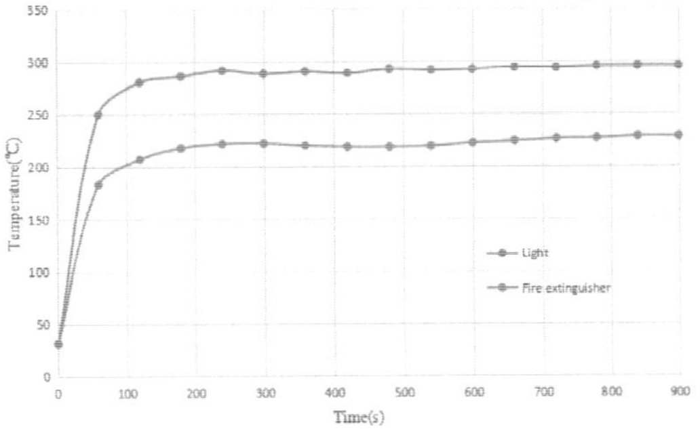

Figure 11. Temperature variation curve of the tail cabin light and the fire extinguisher with time.

In addition, the 79 frame is relatively close to the firewall. Its temperature distribution at 100s, 300s, 500s, and 900s is shown in Figure 12. The maximum and minimum values on the color bar represent the highest and lowest temperatures of the 79 frames respectively. As can be seen from the figure, the area of the high-temperature area of the 79 frame changes more and more over time. The maximum temperature also rises. The temperature distribution of the 79 frames is seriously uneven. The upper temperature is obviously higher than the lower temperature. The maximum temperature difference is about 260°C after 15 minutes. Excessive temperature difference will lead to thermal fatigue of the structure and shorten the life of the structure.

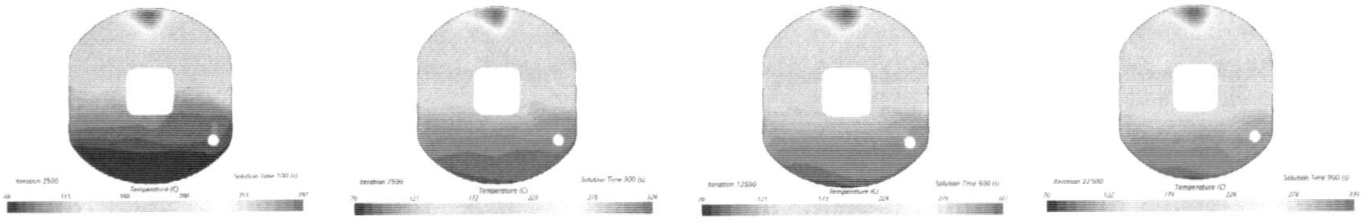

(a) t=100s (b) t=300s (c) t=500s (d) t=900s

Figure 12. Temperature distribution cloud map of the 79 frames at different times.

6. Conclusion

In this paper, the conjugate heat transfer method is used to study the temperature field of aircraft tail cabins. The key issues such as numerical method, model simplification, grid division, and boundary condition setting are explained. The temperature field of the tail cabin under extreme high/low-temperature environments and APU cabin fire conditions is simulated by CFD. The main conclusions are as follows:

(1) The conjugate heat transfer method is suitable for the calculation and analysis of the temperature field in the tail cabin. Compared with the single fluid calculation, it can obtain a more accurate temperature distribution of the structural plane, which is more in line with the actual results, and can provide references for structural design and thermal stress analysis.

(2) Using the conjugate heat transfer method to calculate the temperature field, the thin mesh can be used in the solid domain, which will greatly reduce the number of grids and improve the calculation efficiency without affecting the accuracy of the calculation results.

(3) The temperature distribution in the aft cabin is uneven. The low-temperature environment meets the design requirements. In some areas, there is the risk of over-temperature in the extreme high-temperature environment. It is necessary to evaluate the tolerance of the equipment in the high-temperature area, and adjust the arrangement of the stern cabin or increase the high-temperature identification temperature of the equipment if necessary.

(4) Under the fire condition, the temperature of the electrical equipment in the tail cabin near the firewall increases first with time and then tends to be stable after 5 minutes. The ambient temperature around the tail cabin light and fire extinguisher greatly exceeds the withstand temperature of the equipment, which affects the normal operation of the equipment. In addition, the temperature difference between the upper and lower parts of the 79 frames close to the firewall is too large, reaching up to 260°C, which will reduce the life of the structure.

Acknowledgments

This work was financially supported by the Natural Science Foundation of China (U2333203).

References

[1] Wei Yanjuan, Ming Zhimao. Airborne equipment climate class environment test standard analysis [J]. Journal of environmental technology, 2017, 35 (1): 7. DOI: 10.3969/j.i SSN. 1004-7204.2017.01.016.

[2] Jiao Hongchen, Lei Yong, Zhang Hongyu, et al. Based on the analysis of MBSE spacecraft system modeling and design method [J]. Journal of systems engineering and electronics, 2021, DOI: 10.12305/j.i SSN.1001-506-x.2021.09.19.

[3] Anshi Asia Pacific. Lean R & D 2.0: Promoting Independent Innovation based on Forward Design [J]. Intelligent manufacturing, 2017 (1): 3. DOI: CNKI:SUN:JSFY.0.2017-Z1-046.

[4] Lü Riyi, Yin Dawei, Qian Renjun, et al. Research on Heat Dissipation of Internal Equipment in Electronic Pods [J]. Journal of Projectiles, Rockets, Missiles and Guidance, 2024, 44(1): 69-74.

[5] Zhang Jie, Qin Qiang, Yang Zhibin. Simulation Analysis of the Environmental Temperature within the Electronic Equipment Compartment of High-Speed Aircraft [J]. Aeronautical Computing Technique, 2021, 51(2): 4.

[6] Yu Zhonghao, Zhang Chao, Xiao Guangming, et al. The Impact of Passenger Behaviors on the Flow Field and Temperature Field Distribution in the Cabin [J]. Journal of Engineering Thermophysics, 2024, 45(1): 255 - 262.

[7] Xue Changle, Li Yanna, Zhang Taobo. Research on Thermal Comfort of the Cockpit Based on the PMV Index [J]. 2020.

[8] Romani R, de Goes, Luis Carlos. Cabin Temperature Control Model for Commercial Aircraft[C]//Aiaa Modeling & Simulation Technologies Conference. 2012. DOI: 10.2514/6.2012-4637.

[9] Marineau E C, Schetz J A, Neel R E. Turbulent Navier-Stokes Simulations of Heat Transfer with Complex Wall Temperature Variations [J]. Journal of Thermophysics and Heat Transfer, 2007, 21(3): 525-535.DOI: 10.2514/1.26007.

[10] Li Hongyang, Wang Xiao, Sun Chao. Computational analysis of Temperature Field of unilateral expanded afterbody based on Conjugate Heat Transfer [J]. Journal of Aerodynamics, 2022(008): 037.

[11] RTCA/DO160G, "Environmental Conditions and Test Procedures for Airborne Equipment," RTCA, Incorporated, December 2010.

[12] Federal Aviation Administration (FAA), "Advisory Circular 25.869-1: Electrical System Fire and Smoke Protection," U. S. Department of Transportation, 2004.

Analysis of compressed air energy storage system coupled with thermal storage and photothermal

Zhirong Jiang[1], Shishu Zhang[1], Zhongbo Hu[1], Junjie Di[2], Xu Liu[2], Xiaopeng Wang[1], Renkun Wang[1], Yubin Li[1] and Qing He[2 *]

[1]POWERCHINA Chengdu Engineering Corporation Limited, Chengdu 611100, China
[2]North China Electric Power University, Beijing 102206, China

[*]Corresponding author's e-mail: heq@ncepu.edu.cn

Abstract: With the increasingly prominent issues of environmental pollution and resource depletion, vigorously developing renewable energy has become an important consensus among countries. Addressing the inherent intermittency of renewable energy sources, vigorously developing energy storage has become a top priority. The AA - CAES system coupled with photothermal energy storage can expand the application scenarios of AA - CAES power plants and improve system cycle efficiency. This article conducts research on AA - CAES technology coupled with high-temperature thermal storage and photothermal energy storage (CAES - HTPT) for large-scale CAES power plants above 100 MW. A thermodynamic model for a coupled tower solar collector system is established. The thermodynamic model analyzes the temperature, pressure, and other related thermodynamic parameters of major equipment such as compressors and expanders. By analyzing the operational characteristics and thermodynamic performance of CAES - HTPT, the CAES - HTPT system achieves an energy efficiency of 53.7 %, and the conversion factor of electricity storage is 1.008. Exergy analysis of CAES - HTPT is performed, revealing that the cooling unit experiences the greatest exergy loss, totalling 24234 kW, which represents 53.9 % of the overall exergy loss of the system.

1. Introduction

With the increasingly prominent problems of environmental pollution and resource depletion, large-scale development of renewable energy embodied in wind and solar resources has become a significant consensus for future development strategies of countries around the world. Energy storage has been highly valued as an important means of reducing peak demand and increasing off-peak usage [1]. Compared with energy storage technologies, AA - CAES has relatively relaxed site selection requirements, and it also has the characteristics of high efficiency, large capacity, and mature technology [2]. The integration of solar energy with AA - CAES systems enables the full exploitation of complementary benefits, which subsequently enhances the operational efficiency and performance of AA-CAES systems [3].

At present, academic community has conducted research on the thermodynamic characteristics of using photothermal heat sources as external expansion heat sources for AA - CAES power plants, verifying their feasibility and effectiveness [4]. In a related study, Su [5] proposed an innovative power generation configuration featuring a medium temperature solar field and systematically explored its operational parameters through multi-objective optimization techniques (NSGA - II algorithm). Furthermore, Rahimimotlagh et al. demonstrated a novel integrated energy conversion system combining compressed air energy storage, parabolic trough collectors, and Rankine cycle processes for simultaneous electricity, cooling, and heating production, achieving an outstanding energy round trip efficiency of 62.69 %.

Through the literature research mentioned above, the coupling of CAES and photothermal can improve the output capacity of CAES. The research on the modular configuration technology of power plants with coupled photothermal and thermal storage, and complementary multi-dimensional energy storage, is of great significance for the further promotion and application of CAES technology. This article conducts research on CAES technology coupled with HTPT system for CAES power plants above 100MW and establishes the model of the coupled system. We consider the operational characteristics of CAES - HTPT and apply appropriate performance indicators to evaluate the CAES - HTPT system. We also calculate the operating parameters and performance indicators under standard operating conditions.

2. System principle

Figure 1 shows the flowchart of the CAES - HTPT . The coupling system includes the CAES module, photothermal (PT) module, and high thermal storage (HT) module. During the energy storage process, surplus electricity from the power grid or renewable energy sources drives compressors (CP) to compress air. Coolers are installed at the outlet of each level of the CP. The air at the outlet of cooler 4 is stored in the air storage tank (GS) for the energy release process. Meanwhile, the heat storage medium (HM) in the high temperature cold tank (HTCT) enters the solar thermal field to absorb solar heat and transform into a high-temperature state for storage in the high temperature hot tank (HTHT), ending the energy storage process.

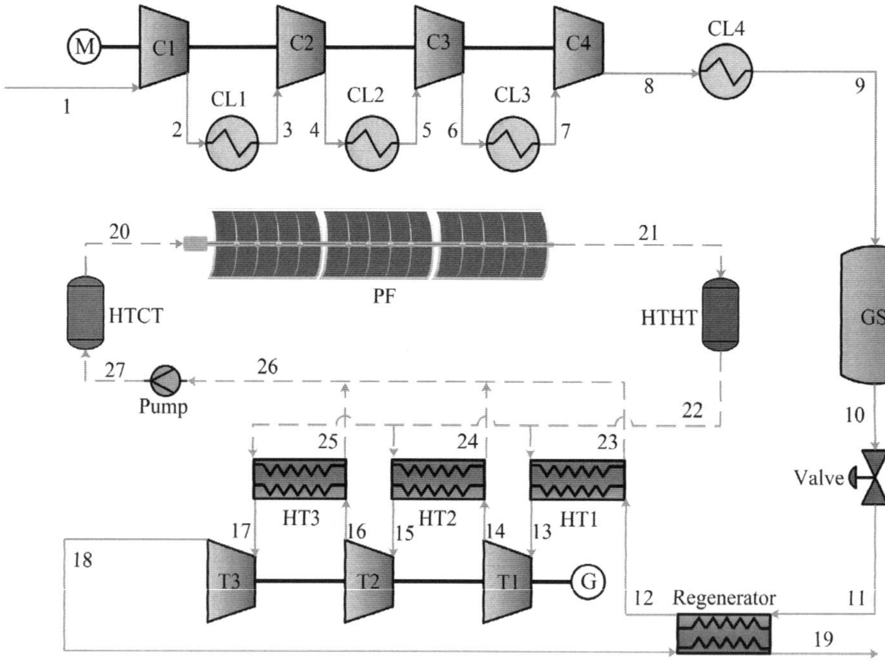

Figure 1. CAES - HTPT system schematic.

The air within GS is first reduced in pressure via a stabilizing valve, lowering it from high to low pressure. The air then passes through the reheater, where it absorbs heat from the exhaust steam of the final stage turbine, effectively preheating the air prior to entering the boiler. The preheated air undergoes a multi-stage expansion process to perform external work and drive the electric motor to output electrical energy. The high temperature HM in the HTHT enters each stage of the heater to heat the air at the inlet of the expansion machine, in order to improve the system output level. The

HM at the outlet of each level of the heater enters the HTCT for storage in the next cycle. At this point, the energy release process comes to an end.

3. Thermodynamic model

The CAES - HTPT system include a CAES module, photothermal module, thermal storage module. The thermodynamic models of modules are established. We consider the operational characteristics of the system and analysis appropriate performance indicators.

3.1. Energy analysis model

Table 1 presents the modeling of the primary component of the CAES - HTPT system [6].

Table 1. Primary component of the CAES-HTPT system.

Component name	Temperature	Power
Compressor	$T_{co}(i)=\{1+[\beta_c^{(\kappa-1)/\kappa}(i)-1]/\eta_c(i)\}T_{ci}(i)$	$P_c=m_c\sum_{i=1}^{N}c_aT_{ci}(i)[\beta_c^{(\kappa-1)/\kappa}(i)-1]/\eta_c(i)$
Turbine	$T_{co}(i)=\beta_e^{(\kappa-1)/\kappa}(i)T_{ei}(i)$	$P_e=m_e\sum_{i=1}^{M}\eta_e(i)c_aT_{ei}(i)[1-\beta_e^{(\kappa-1)/\kappa}(i)]$

The relationship between VGA and the duration of energy storage can be expressed as:

$$V_{GA}=m_{air}t_{ch}/(\rho_{ch}-\rho_{di}) \quad (1)$$

where ρ_{ch} is the density of air under energy storage pressure, kg/m^3; and ρ_{di} is the density of air under energy release pressure, kg/m^3.

The calculation model of the valve is an isenthalpic model, which means:

$$h_{v,in}=h_{v,out} \quad (2)$$

3.2. Performance evaluation indicators

Energy efficiency and energy storage conversion factors as performance indicators of the system are used.

3.2.1. Energy efficiency.

The energy efficiency is:

$$\eta_E=W_{out}/(W_{in}+Q_{solar}) \quad (3)$$

where W_{in} is the system power input, kW; and W_{out} is the system power output, kW.

3.2.2. Energy storage conversion factor.

The energy storage conversion factor is:

$$\eta_{ES}=(W_{out}-W_{solar})/W_{in} \quad (4)$$

$$W_{solar}=Q_{solar}\eta_{solar}/\eta_{collector} \quad (5)$$

where W_{solar} is the converted work of solar energy, kW; η_{solar} is 12.6 %; and $\eta_{collector}$ is 78.3 %.

3.3. Exergy analysis model

The CAES system does not undergo chemical changes. The physical exergy is the exergy value of the fluid, that is:

$$E_{ph}=\dot{m}\left[(h-h_0)-T_0(s-s_0)\right] \quad (6)$$

The exergy loss models of CAES - HTPT are established in Table 2.

Table 2. Exergy loss model of major components.

Component name	Exergy loss	Exergy efficiency
Compressor	$E_{x,c}^{D} = E_{x,c}^{in} - E_{x,c}^{out} + W_c$	$\eta_{cp} = (E_{x,c}^{out} - E_{x,c}^{in}) / W_c$
Turbine	$E_{x,e}^{D} = E_{x,e}^{in} - E_{x,e}^{out} + W_e$	$\eta_e = W_e / (E_{x,e}^{in} - E_{x,e}^{out})$
Cooler	$E_{x,h}^{D} = E_{x,h}^{in} - E_{x,h}^{out} + E_{x,hm}^{in} - E_{x,hm}^{out}$	$\eta_h = (E_{x,hm}^{out} - E_{x,hm}^{in}) / (E_{x,h}^{in} - E_{x,h}^{out})$
Heater	$E_{x,l}^{D} = E_{x,l}^{in} - E_{x,l}^{out} + E_{x,cm}^{in} - E_{x,cm}^{out}$	$\eta_c = (E_{x,1}^{out} - E_{x,1}^{in}) / (E_{x,cm}^{in} - E_{x,cm}^{out})$

4. Energy analysis

This section conducts energy analysis on the system, analyzing the main parameters and performance indicators under basic operating conditions, and studying the trend of the influence of photothermal temperature on the system performance.

4.1. System performance

Table 3 analyzes the performance parameters of the CAES - HTPT. The CAES - HTPT energy efficiency is 53.7 % and the energy storage conversion factor is 1.008, indicating higher energy efficiency and energy storage conversion factor.

Table 3. Performance Parameters of CAES-HTPT.

Item	Designation	Parameter
Coupling mode	Photothermal module	Tower solar collector
	Thermal storage module	Molten salt
Main parameters	Compressor power consumption, kW	82725.0
	Turbine output power, kW	100000.0
	Solar energy input heat, kW	103486.1
	Solar energy conversion power, kW	16652.9
Performance indicators	Energy efficiency, %	53.7
	Energy storage conversion factor	1.008
Design parameters	Volume of high temperature cold tank, m^3	18137.0
	Volume of high temperature heat tank, m^3	16444.7
	Volume of gas storage, m^3	82286.7

4.2. Influence of photothermal temperature

This section explores the trend of the influence of tower solar collector photothermal temperature on CAES - HTPT system performance. Figure 2 shows the effect of the collection temperature of the tower collector on the energy efficiency and the energy storage conversion factor of CAES - HTPT. A higher tower collector temperature is beneficial for improving system performance. Within the research scope, a collection temperature of 565 °C for the tower collector can achieve the highest energy efficiency and energy storage conversion factor for the system.

5. Exergy analysis

Based on the CAES - HTPT design parameters and related calculation results under basic operating conditions, we draw the damage diagrams of each component of the coupled system as shown in Figure 3. From Figure 3, it can be seen that the highest exergy loss (EXL) in the coupled system is the cooling unit, with an exergy loss of 24234 kW, accounting for 53.9 % of the total system EXL. At the same time, the EXL of the valve is 3218 kW, accounting for 7.2 % of the total system EXL. The EXL of the compressor unit is 8916kW, accounting for 19.83 % of the total system EXL. The EXL of the expansion unit is 6768 kW, accounting for 15.06 % of the total system EXL. The sum of the EXL of these four components is 43137 kW, accounting for 95.96 % of the total EXL of the system. It can be seen that the main EXL of the coupled system comes from these four components.

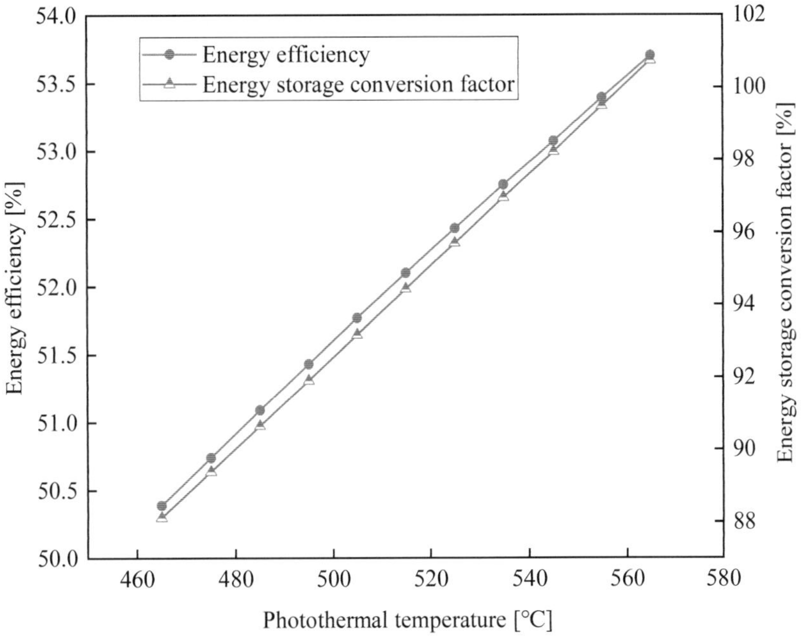

Figure 2. The influence of the collection temperature of tower type collectors on energy efficiency and energy storage conversion factor of CAES - HTPT.

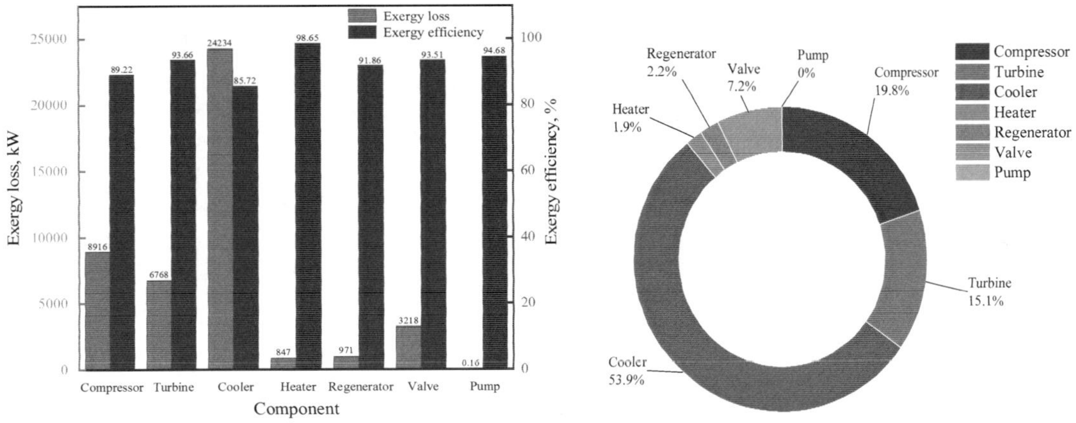

a) Value of exergy loss b) Proportion of exergy loss

Figure 3. CAES-HTPT exergy loss situation.

6. Conclusions

This work conducts research on the modular configuration of coupled solar thermal, thermal storage, and other composite multi-dimensional energy storage complementary power plants. We establish a thermodynamic model for CAES - HTPT. The key findings are summarized as follows: (1) Analyzing the operational characteristics and thermodynamic performance of CAES - HTPT, the energy efficiency of the coupled system is 53.7 %, and the equivalent conversion factor of energy storage of the coupled system is 1.008.

(2) The PT temperature has a significant influence on the performance. The higher tower collector temperature is beneficial for improving system performance. Within the research scope, a tower collector temperature of 565 °C can achieve the highest energy efficiency and energy storage conversion factor for the system.

(3) It is found that the component with the highest EXL in the system is the cooling unit, with EXL of 24234 kW, accounting for 53.9 % of the total system EXL. The sum of the EXL of the valve, compressor unit, expansion unit, and cooler components is 43137 kW, accounting for 95.96 % of the total EXL. The main EXL of the coupled system comes from these four components.

In future research, optimization can be carried out for components with high system exergy losses to reduce their energy losses.

Acknowledgments

This work is supported by the National Natural Science Foundation of China (Grant No. U23B20145) and the Key Research Project of POWERCHINA (No. DJ-HXGG-2023-09).

References

[1] Kılkış, Ş., Krajačić, G., Duić, N., Rosen, M. A., Al-Nimr, M. A. (**2023**) Sustainable development of energy, water and environment systems in the critical decade for climate action. Energy Convers. Manage., 296: 117644.

[2] Chen, L. X., Zhang, L. G., Wang, Y. Z., Xie, M. N., Yang, H. P., Ye, K., Mohtaram, S. (**2023**) Design and performance evaluation of a novel system integrating water-based carbon capture with adiabatic compressed air energy storage. Energy Convers. Manage., 276: 116583.

[3] Han, Z. H., Wang, S., Hu, Z. Q., Zhu, L., Guo, L. D. (**2021**) Analysis on performance and key parameters of AA-CAES + CSP system. Acta Energiae Solaris Sinica, 42 (2): 322-329.

[4] Li, F. Y., Yu, Y. P., Shu, Y., Liu, X. M. (**2023**) Study on characteristics of photovoltaic and photothermal coupling compressed air energy storage system. Process Saf. Environ., 178: 147-155.

[5] Su, D. W. (**2022**) Comprehensive thermodynamic and exergoeconomic analyses and multi-objective optimization of a compressed air energy storage hybridized with a parabolic trough solar collectors. Energy, 244: 122568.

[6] Cui, S. S., Chen, L. J., Chen, S. Y., Sun, Z. T., Mei, S. W. (**2024**) Dynamic modeling and analysis of compressed air energy storage for multi-scenario regulation requirements. J. Energy Storage, 100: 113227.

New energy efficient access timing control based on reduction and weighted optimal hierarchical clustering

Fajun Zhang*, Yuanyuan Zhao, Aoyang Han, Zhi Yin, Zhen Wei

State Grid Qingdao Power Supply Company, Qingdao, 266002, China

*Corresponding author's e-mail: fajunwz@163.com

Abstract. During the development and utilization of new energy, the transient abnormal characteristics of power grid control mode switching are multi-terminal. In order to realize efficient access to new energy, the time sequence control method of efficient access of new energy is studied by combining reduction and weighted optimal hierarchical clustering algorithm with convergence effect as the starting point. Under the coupling structure of reduction and weighted optimal hierarchical clustering, the objective function of new energy efficient access timing control is constructed, the constraint conditions are set, and the timing current of new energy efficient access is calculated. According to the calculation results, the timing control interval of new energy efficient access is divided, and the objective function is solved, thus realizing the timing control of new energy efficient access. The experimental results show that this method has a stable and efficient access timing control effect, good convergence effect, and fast speed.

1. Introduction

The world pays more and more attention to environmental protection and sustainable development, and the development and utilization of new energy has become an important strategic direction for all countries in the world. Renewable energy, such as wind energy and solar energy has become an important option to replace traditional fossil energy because of its clean and renewable characteristics [1-2]. However, new energy power generation is intermittent and uncertain; after large-scale access to the power grid, it brings unprecedented challenges to the stable operation of the power system [3]. How to access new energy efficiently and stably and realize the time sequence control of power systems is a hot issue in current power system research and engineering practice [4]. Efficient time sequence control after new energy is connected to the power grid means that under the premise of ensuring the safe and stable operation of the power grid by optimizing the control strategy, the new energy is used to generate electricity to the maximum extent, the phenomena of wind and light abandonment are reduced, and the energy utilization efficiency is improved. To achieve this goal, it is necessary to comprehensively consider the characteristics of new energy generation, the carrying capacity of the power grid, and the change of load demand and realize the coordination and optimization between new energy generation and power grid operation.

Reference [5] puts forward the effective synchronization control of energy in wireless optical communication systems. The accurate clock drift model is established based on the measurement of real cheap equipment. The design mechanism has been used to evaluate the same random access based on ALOHA, but it is based on time slots, so it increases the maximum achievable throughput compared with traditional access. The throughput and energy efficiency models are established to evaluate the performance of LoRaSync operating network. These models are verified by simulating the simulation environment of large-scale deployment and then used to determine the most energy-saving slot size for any traffic load. As the final verification of the concept, LoRaSync has been implemented and tested on the LoRa test bed to prove the feasibility of the designed solution on real hardware. Reference [6] proposes priority-based control of concurrent and shared resource access mechanisms and proposes a new CAmkES library with a thread model. Reference [7] proposes that transmission timing control between aperiodic and periodic streams in wireless sensor networks can be realized by limiting packet loss and achieving reliable transmission within allowable delay.

On the basis of the above research, this paper proposes a new energy-efficient access timing control based on reduction and weighted optimal hierarchical clustering. In this method, reduction and weighted optimal hierarchical clustering are two important technical means. By reducing redundant information and extracting key features, reduction technology is helpful in simplifying the problem and improving the calculation efficiency. Weighted optimal hierarchical clustering is an effective data grouping method. According to the similarities and differences of data, the data are divided into different categories, and the data in each category has high similarity, while the data in different categories have great differences. In the time sequence control of new energy access to the power grid, weighted optimal hierarchical clustering can be used to classify new energy power generation units and classify units with similar power generation characteristics into one category, thus facilitating the formulation of targeted control strategies and improving the control effect.

2. Design of time sequence control target method for efficient access to new energy

2.1. *Reduction and weighted optimal hierarchical clustering under the new energy efficient access timing control objective function construction*

Weighted optimization means that when dealing with multi-objective or multi-variable optimization problems, the relative importance of each objective or variable is balanced by introducing weighting factors, thus finding a comprehensive optimal solution. In the timing control of new energy access, it is necessary to consider multiple objectives at the same time, such as maximizing the consumption of new energy, minimizing the operating cost of the power grid, and maintaining the stability of the power grid.

The objective function of time sequence control for efficient access to new energy is constructed, and the formula is:

$$\begin{cases} P_{in} = P_{W_{in}} + P_{V_{in}} + P_{A_{in}} \\ P_{out} = P_W + P_V + P_A \end{cases} \tag{1}$$

In the formula, P_{in} and P_{out} respectively represent the objective function of timing control for efficient access of input and output new energy; $P_{W_{in}}$, $P_{V_{in}}$ and $P_{A_{in}}$ respectively represent the input power of wind power, photovoltaic power, and geothermal energy; P_W, P_V and P_A respectively represent the output power of wind power, photovoltaic power, and geothermal energy; Assuming that $\max P$ represents the maximum allowable power fluctuation, when $|P_{in} - P_{out}| \geq \max P$ occurs and multiple energy systems are unbalanced, they need to be actively adjusted and optimally compensated for different load states.

The complexity of new energy generation data and power grid operation status data is reduced by reduction technology. According to the specific needs of new energy access timing control, the targets that need to be optimized are determined, and reasonable weighting factors are assigned to these targets to construct a weighted optimal objective function, which is helpful in reducing data complexity and improving control efficiency to adapt to the uncertainty of new energy generation.

2.2. *Constraint condition setting*

The power flow equation is an important equation used to calculate the voltage and power of each node in the power system in the process of high-efficiency access timing control of new energy. Especially, the power flow equation considers the node voltage and power of the power system and solves the optimal operating point in the time sequence control of high-efficiency access of new electric energy by optimization method, so as to realize the stable operation of the power system. The specific formula for constraint conditions is:

$$\begin{cases} P_j = P_{in} \times U_j \sum_{i=1}^{N} U_g (B_{ij} \cos\theta_{ij} + D_{ij} \sin\theta_{ij}) \\ Q_j = P_{out} \times U_j \sum_{i=1}^{N} U_i (B_{ij} \sin\theta_{ij} - D_{ij} \cos\theta_{ij}) \end{cases} \tag{2}$$

In the formula, P_j represents the active power input to the system node j; Q_j represents the reactive power input to the system node j; U_j represents the voltage value of the node j; N represents all system nodes; B_{ij} represents the real value of the coupling structure of reduction and weighted optimal hierarchical clustering; D_{ij} represents the imaginary part of the coupling structure of reduction and weighted optimal hierarchical clustering; θ_{ij} represents the phase difference between active power and reactive power.

2.3. Objective function solution

2.3.1. Calculation of Time Sequence Current for Efficient Access of New Energy. In the calculation of sequential current for efficient access to new energy, the steady-state value of current is calculated. Under steady-state conditions, the calculation formula of current is shown in Formula (3):

$$I = \frac{\left(U_\alpha \cos\chi - U_\beta \cos\delta\right) \times P_j \times Q_j}{R_d + R_k - R_c} \tag{3}$$

In the formula, U_α represents the DC voltage at the rectifier side at steady state; U_β represents the DC voltage at the inverter side at steady state; χ represents the delay angle; δ represents the turn-off angle of the inverter side; R_d represents the equivalent commutation resistance of the inverter side; R_k represents equivalent commutation resistance of rectifier side; R_c represents equivalent resistance of DC line.

In this process, the current flowing through the equivalent capacitor cannot be ignored. Therefore, the sequential current of efficient access of new energy is calculated by Formula (4):

$$I' = I + I_r \tag{4}$$

In the formula, I_r represents the discharge current of the equivalent capacitor. Thereby complete that calculation of the time sequence current of efficient access to new energy.

2.3.2. Division of Time Sequence Control Interval for Efficient Access to New Energy. The flow rate in the interval with high-efficiency access timing control of new energy will change accordingly. Therefore, dividing the interval of high-efficiency access timing current of new energy can improve the accuracy of control. According to the coverage area of the power generation device, generate the data set of flow velocity interval, as shown in Formula (5):

$$V = \left(v_1, v_2, \ldots, v_n\right) \tag{5}$$

In the formula, $v_1, v_2, \ldots, v_n$ represents the obtained tidal current and velocity data, and each generation node has a corresponding relationship of interval length. Based on this, the number of intervals can be calculated, as shown in Formula (6).

$$M = \frac{\max V - \min V}{S} \tag{6}$$

In the formula, $\max V$ represents the maximum abnormal value in the data set; $\min V$ represents the smallest outlier in the data set; S represents the interval length. In this paper, based on the coupling structure of reduction and weighted optimal hierarchical clustering, the interval segmentation nodes are determined, and the time sequence control interval of efficient access to new energy is divided. The result is shown in Formula (7):

$$Q_k = I' \times F_D \times q_z \times M \times V \tag{7}$$

In the formula, F_D represents the total amount of tidal current power generation in the velocity range; q_z represents the comprehensive weight of output power. According to Formula (7), the time sequence control interval of efficient access of new energy is divided, which lays the foundation for control.

2.3.3. Realization of time sequence control for efficient access to new energy. Through the division results of new energy efficient access timing control interval, it provides support for the realization of new energy efficient access timing control. The time sequence state of efficient access of new energy is regarded as a discrete-time nonlinear dynamic system, and the formula is:

$$\mu(a) = \frac{x(a) \times y(a)}{Q_k} \tag{8}$$

In the formula, $x(a)$ represents the state change value of the energy system; $u(a)$ represents the control input value of the energy system.

Select a positive integer d , and calculate from $d = 1$, where DD represents the number of iterations. After n iterations, the sequence control result of efficient access to new energy is obtained, and the formula is as follows:

$$E(a) = J^d[\mu(a)] + J^{(d+1)}[\mu(a)] + J^{(d+2)}[\mu(a)] + \ldots + J^{(d+n)}[\mu(a)] \tag{9}$$

In the formula, J represents the sequence control cardinal number of efficient access of new energy.

3. Experiment

Taking the time sequence control system of new energy efficient access in a certain area as the experimental object and the time sequence control system of new energy efficient grid connection in a certain area as an example, the system covers two major parts. Among them, part 1 includes 8 power generation units and 23 power demand points; Part 2 includes 12 power generation units and 32 power load points. In order to verify the control efficiency of the proposed method, MATLAB software is used for simulation verification, and the specific configuration parameters of the experiment are shown in Table 1.

Table 1. Table of Experimental Parameters.

Experimental parameters	Numerical value
Line transmission limit of voltage source converter	600MW
Wind speed value	6m/s
starting threshold of win velocity	4.5m/s
Shutdown wind speed	6.2m/s
Sunshine intensity	50000Lux

Working temperature of photovoltaic panel	29℃
System adjustment factor	0.02
sensitivity coefficient	0.5
Iterations	500 times
External temperature	26℃

According to the comparative experiment of experimental parameter setting in Table 1, considering that the new energy access will cause great disturbance to the power system, and then the load stability will be affected, the method of Reference [5] and the method of Reference [6] are used to control the time sequence of new energy efficient access, and the load frequency changes are shown in Figure 1.

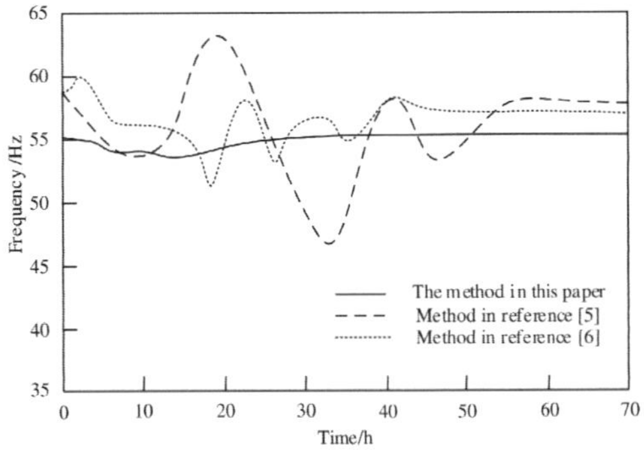

Figure 1. Load frequency change under high-efficiency access timing control of new energy.

As can be seen from Figure 1, the fluctuation amplitude of load frequency is small under the time sequence control of efficient access of new energy in this paper, and the frequency can be controlled to fluctuate up and down at 55 Hz. The load frequency fluctuation range of the methods in Reference [5] and Reference [6] is between 46 Hz and 64 Hz, and the fluctuation fluctuates greatly. This shows that the method in this paper has a stable and efficient timing control effect on new energy access.

In order to further verify the effectiveness of this method, the convergence performance of the three methods is compared, and the experimental results are shown in Figure 2.

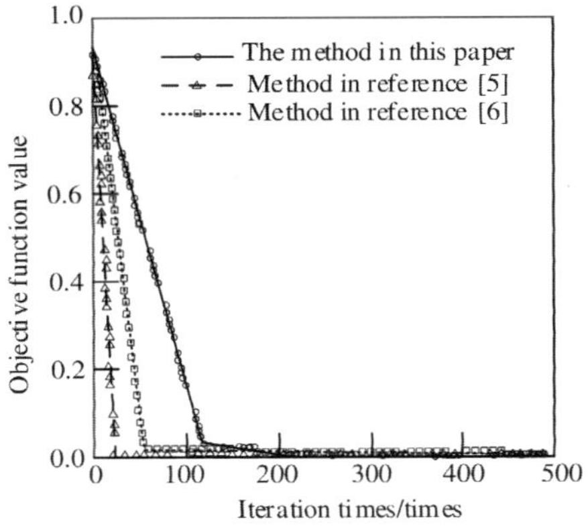

Figure 2. Comparison results of control convergence performance of different methods.

As can be seen from Figure 2, the control convergence effect of this method is better than that of the methods in Reference [5] and Reference [6], and the convergence speed is faster. Thus, this method is more suitable for efficient access timing control of new energy.

4. Conclusion

In this study, aiming at the timing control problem of efficient access of new energy to the power grid, an innovative solution is provided for the coordinated optimization of new energy power generation and power grid operation by introducing reduction technology and weighted optimal hierarchical clustering method. In the research process, the time sequence control objective function of high-efficiency new energy access is constructed, the constraint conditions are set, and the time sequence control objective function is solved. The effectiveness of this method is verified by experiments, which have a stable time sequence control effect of high-efficiency new energy access, good convergence effect, and fast speed.

References

[1] Ren M, Zhou T, Wang C. New energy vehicle innovation network, innovation resources agglomeration externalities and energy efficiency: Navigating industry chain innovation[J]. Technological forecasting and social change, 2024, 200(3): 1-13.
[2] Lin B, Xu C. Reaping green dividend: The effect of China's urban new energy transition strategy on green economic performance[J]. Energy, 2024, 296(1):1-14.
[3] Vereshchagin A A, Volkov A I, Novoselova J V, et al. Harmonizing Energies: The Interplay Between a Nonplanar SalEn - Type Molecule and a TEMPO Moiety in a New Hybrid Energy - Storing Redox - Conducting Polymer[J]. Macromolecular Rapid Communications, 2024, 45(14): 1-10.
[4] Cheng J Z, Feng X Y, Wang C. Dynamic Risk Assessment of Wind Power Generation System Based on Bayesian Network[J]. Computer simulation, 2023, 40(10): 108-112.
[5] Chasserat L, Accettura N, Berthou P. LoRaSync: Energy efficient synchronization for scalable LoRaWAN[J]. Transactions on Emerging Telecommunications Technologies, 2024, 35(2):1-19.
[6] Sudvarg M, Sun Z, Li A, et al. Priority-based concurrency and shared resource access mechanisms for nested intercomponent requests in CAmkES[J]. Real-Time Systems, 2024, 60(1): 76-107.
[7] Koyama A, Tanigawa Y, Tode H. Transmission Timing Control among Both Aperiodic and Periodic Flows for Reliable Transfer by Restricted Packet Loss and within Permissible Delay in Wireless Sensor Networks[J]. IEICE Transactions on Communications, 2023, 106(9): 817-826.

Research on penetrating liquid cooling heat dissipation design and simulation of high heat flux density computing power servers

Ying Wang [*]**, Bin Jiang** [*]**, Chuping Yan, Mingjun Luan, Shuai Qin**

The 15th Research Institute of China Electronics Technology Group Corporation, Beijing 100083, China

[*]Corresponding author's e-mail: dqszljwy@126.com, bj_jiangbin@126.com

Abstract: With the continuous development of technologies such as artificial intelligence, big data, and cloud computing, the demand for high heat flux density computing power servers has witnessed explosive growth. The key technology affecting the stable and reliable operation of these computing power services is thermal design. Traditional air-cooling technology has great difficulty in dissipating heat for them. At present, the research focus of heat dissipation technology for high heat flux density computing power servers is liquid-cooling technology, mainly due to the advantages of liquid-cooling such as strong fluidity and high heat transfer coefficient. In this paper, aiming at the main challenge of thermal design for high heat flux density computing power servers, a through-type liquid-cooling heat dissipation system was constructed for 6U equipment, and a liquid-cooling chassis and coolant flow channels were designed. The design of the through-type liquid-cooling plate is the core of this design. The thermal simulation was carried out, and the results of the thermal simulation show that through-type liquid-cooling can effectively solve the heat dissipation problem of high-power-consumption modules.

1. Introduction

With the continuous development of technologies such as artificial intelligence, big data, and cloud computing, the computing power of high heat flux density computing power servers has increased exponentially, leading to a sharp increase in the heat flux density of electronic devices. More than 55% of the failures of electronic devices are caused by excessively high temperatures. Therefore, it is urgent to solve the bottleneck that restricts the stable and reliable operation of high-power computing platforms and high-density servers – the heat dissipation problem.

The heat dissipation methods used for electronic devices mainly include: natural cooling, forced air cooling, heat pipe phase change cooling, and liquid cooling, etc. In recent years, a large number of research achievements have emerged in the field of liquid cooling. Zhuang et al., Bao et al., and Ke et al. [1-3] focus on the comparison of liquid heat dissipation methods, introducing the limitations of contact (immersion or spray) liquid cooling technology and the advantages of non-contact cold plates. Yang, Ni, Li et al., and Jiang et al. [4-7] attempt to combine thermal simulation analysis methods to conduct structural design of liquid cooling plates to improve the heat dissipation performance. Among them, Li et al. and Jiang et al. also analyze the influence of the welding process on the structure of the liquid cooling plates. Zhu et al., Chen et al., Zhang and Wen [8-10] discuss the influence of the flow rate on heat dissipation performance and the influence of adding a liquid cooling control system on heat dissipation technology. Based on the existing achievements, there are still gaps in the significant improvement of the liquid cooling heat dissipation performance in existing research, and three major problems need to be solved urgently: (1) the improved design of the flow channels of the through-type liquid cooling plates; (2) the influence of the improvement of the welding process of the through-type liquid cooling plates on the flatness, sealing performance, and compression resistance of the liquid cooling plates; (3) the design of the liquid cooling temperature control feedback system and software. In response to the above deficiencies, the innovation of this paper is reflected in: (1) the optimized design of the flow channels of the through-type liquid cooling modules; (2) the improvement of the welding types and process sequences of the through-type liquid cooling modules.

Considering the overall balance design issues in terms of reliability, structural lifespan, performance indicators, and stability [1-3], the research on the through-type liquid cooling and heat dissipation technology can address the heat dissipation problems caused by the continuous increase in heat flux density resulting from the growing high-performance requirements of high-computing-power platforms.

2. Construction of the Penetrating Liquid-Cooling Heat Dissipation System

The construction of a through-type liquid cooling and heat dissipation system generally requires a liquid cooling source, a liquid cooling chassis, a through-type liquid cooling plate, a self-relieving pressure fluid connector, a coolant flow pipeline, etc. The liquid cooling source mainly provides coolant with a certain pressure or flow rate, serving as the power source of the overall system. It also includes a hydraulic pump, a busbar box, a liquid cooling heat exchanger, and related control circuits. The liquid cooling chassis acts as the carrier for all functional boards that require liquid cooling in the entire system. It mainly contains branch and confluence pipelines. These pipelines are used to distribute the coolant input from the liquid cooling source to multiple liquid cooling plates [4] or to collect the coolant flowing out from multiple liquid cooling plates and return it to the recovery tank of the liquid cooling source. Coolant may flow in and out of the through-type liquid cooling plate. Additionally, it is directly attached to high-power thermal devices, and it may absorb the heat generated by these devices. Through the coolant flowing inside it, heat transfer is carried out, thus taking away the heat generated by the high-power thermal devices. Then, the coolant is sent to the heat exchanger by the hydraulic pump. The liquid cooling heat exchanger exchanges heat with the external environment, cools the coolant, and returns it to the liquid cooling chassis to complete a cycle. The schematic diagram of its construction, the diagram of the branch and confluence pipelines, and the diagram of the coolant circulation path are shown in Figure 1 to Figure 3 respectively. Among them, the liquid cooling source, the self-relieving pressure fluid connector, and the coolant flow pipeline are generally regarded as mature off-the-shelf products. This paper does not focus on discussing them but mainly introduces the design of the liquid cooling chassis and the liquid cooling plate in the liquid cooling and heat dissipation system.

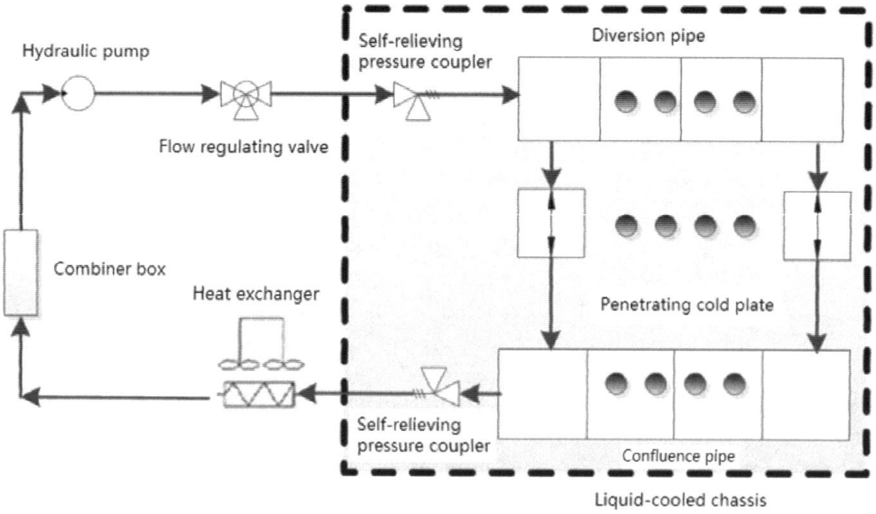

Figure 1. Construction Schematic Diagram of the Penetrating Liquid-Cooling Heat Dissipation System.

Journal of Physics: Conference Series **3011** (2025) 012010 doi:10.1088/1742-6596/3011/1/012010

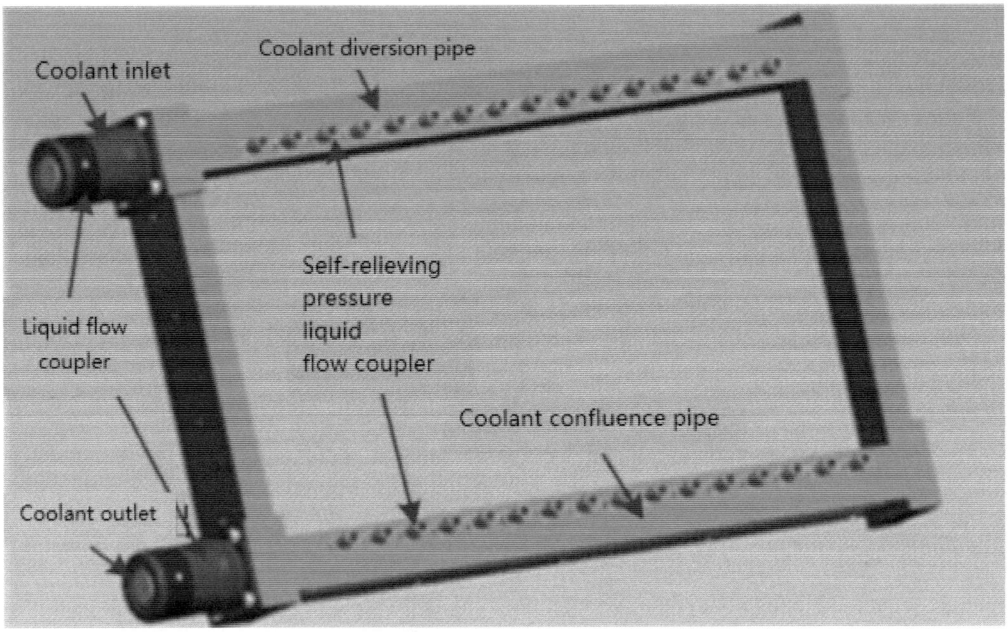

Figure 2. Flow-Dividing and -Combining Pipeline Diagram.

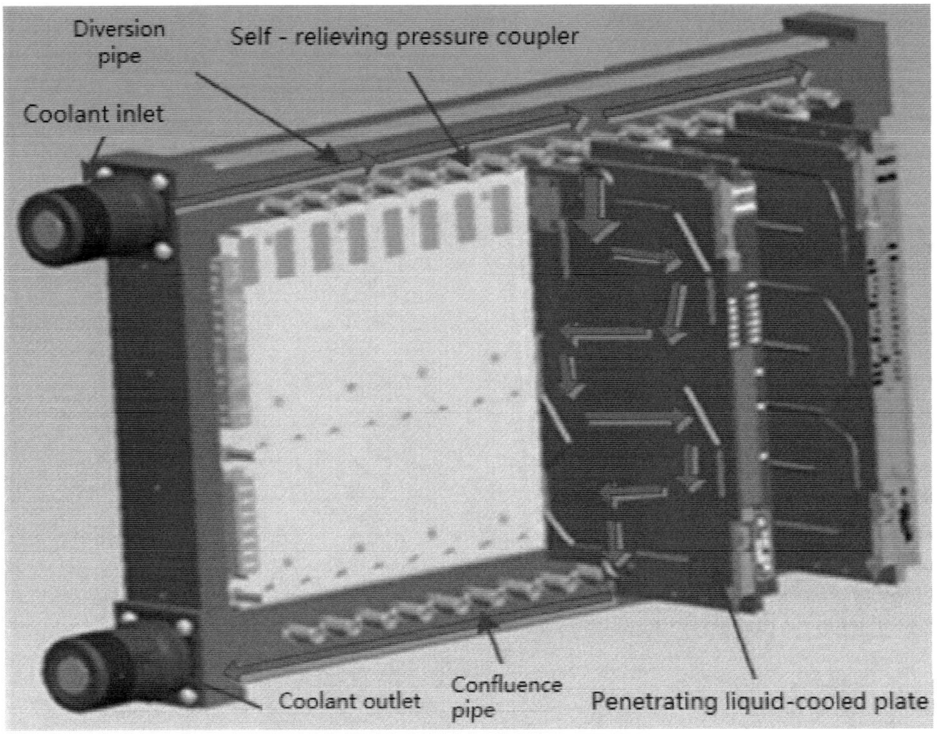

Figure 3. Coolant Circulation Path Diagram.

3. Design of the Liquid-Cooling Chassis

The chassis of the penetrating liquid-cooling heat dissipation system studied in this paper is mainly composed of a chassis, a flow-dividing and -combining pipeline, coolant inlet, and outlet connectors, a

Journal of Physics: Conference Series **3011** (2025) 012010 doi:10.1088/1742-6596/3011/1/012010

penetrating liquid-cooling heat dissipation cold plate, a backplane, and external signal connectors [5]. It can install 16 functional modules, each designed as 5HP, and each functional module adopts separate penetrating liquid cooling. The cold-plate module is used for the mechanical support and convective heat transfer of the thermal power circuit board. The liquid-cooling chassis is the framework foundation of the entire heat dissipation system and also the hub of the flow channels in the entire heat dissipation system. The fluid connector is the interface with the liquid-cooling supply system, playing the role of inlet and outlet of the liquid. The entire liquid-cooling system requires no liquid leakage under a pipeline flow pressure of 1.5 MPa. The overall chassis front view, explosion view, and flow-dividing and -combining pipeline are shown in Figure 4 to Figure 6 respectively.

A front-turning door is designed on the front side of the chassis. The front-turning door is equipped with visible shielding glass. When the front-turning door is closed, the operation of each functional module in the device can be observed through the shielding glass. Handles are designed on the front and rear panels of the device for easy handling. An external interface is designed on the rear side of the device, and a liquid-cooling inlet and outlet socket is set on the left side of the device.

Figure 4. Front View of the Chassis.

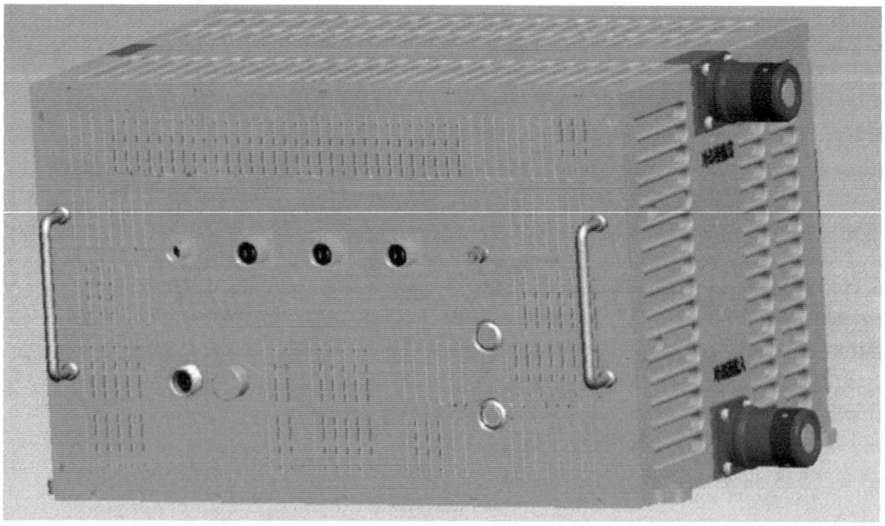

Figure 5. Rear View of the Chassis.

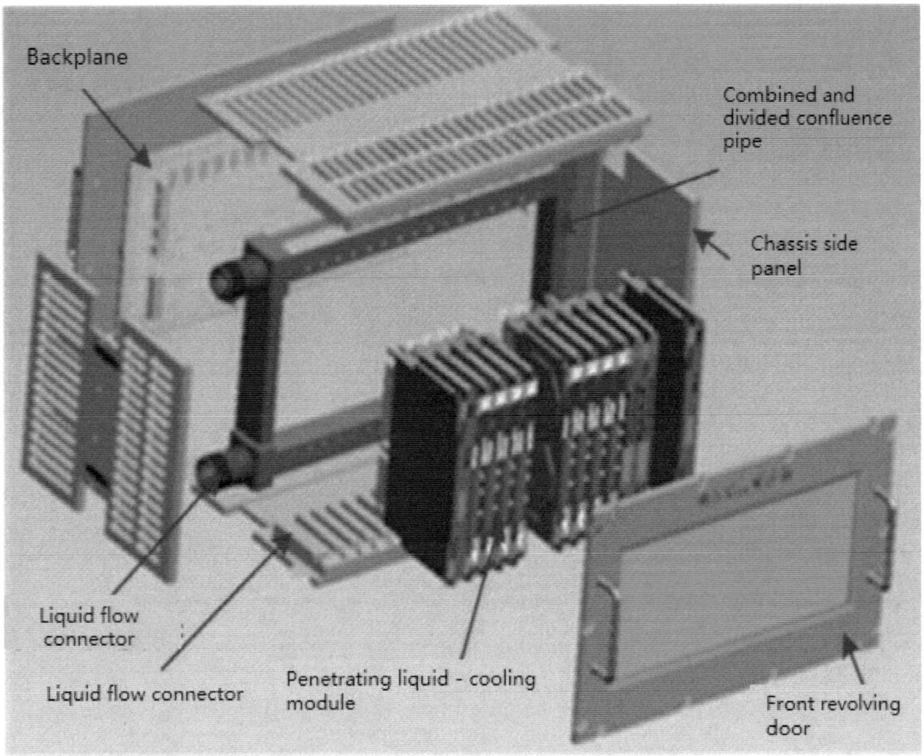

Figure 6. Explosion View of the Chassis.

4. Design of the Coolant Flow Channel

Generally, after the coolant comes out from the hydraulic source, it enters the chassis through the liquid-flow connector on the chassis. The liquid-flow connector is directly connected to the shunt pipe. A self-depressurizing connector is set on the shunt pipe. The coolant enters the penetrating liquid-cooling module through the self-depressurizing connector and then enters the confluence pipe through the self-depressurizing connector on the other side of the penetrating liquid-cooling module. Then, it enters the heat exchanger of the hydraulic source through the liquid-flow connector connected to the confluence pipe, and finally enters the liquid storage tank, thus completing a cycle. The heat absorbed by the penetrating liquid-cooling module is conducted to the heat sink, achieving the cooling effect.

The design of the coolant flow channel mainly involves the liquid-flow connector, the flow-dividing and -combining pipeline, the self-depressurizing connector, and the penetrating liquid-cooling module. By matching the fluid connector of the cold-plate module with the internal fluid connector of the liquid-cooling chassis, the flow channels of the entire heat dissipation system can be connected. At the same time, the signal-transmission connector on the cold-plate module is matched with the signal-transmission connector on the backplane in the liquid-cooling chassis to ensure signal transmission. The design of the coolant flow channel is shown in Figure 7 and Figure 8 respectively.

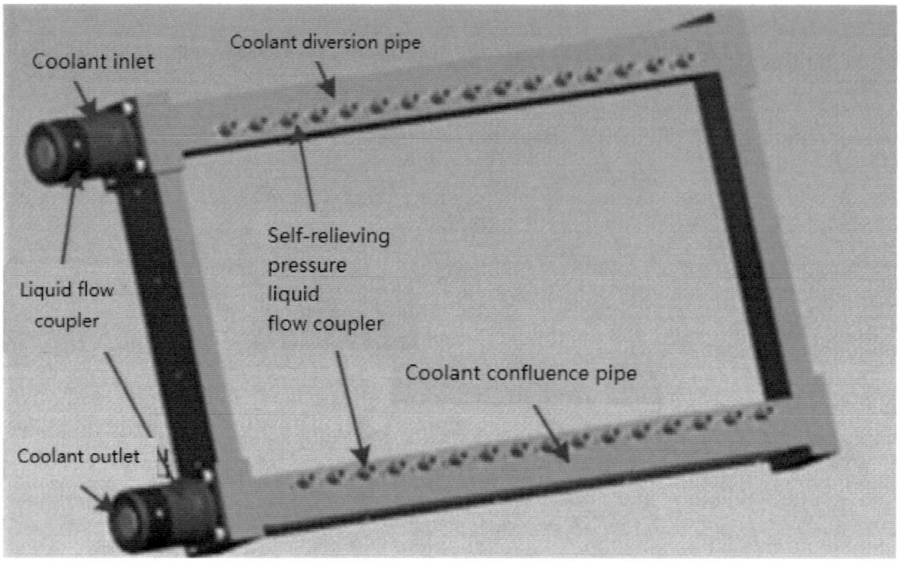

Figure 7. Flow-Dividing and -Combining Pipeline Diagram.

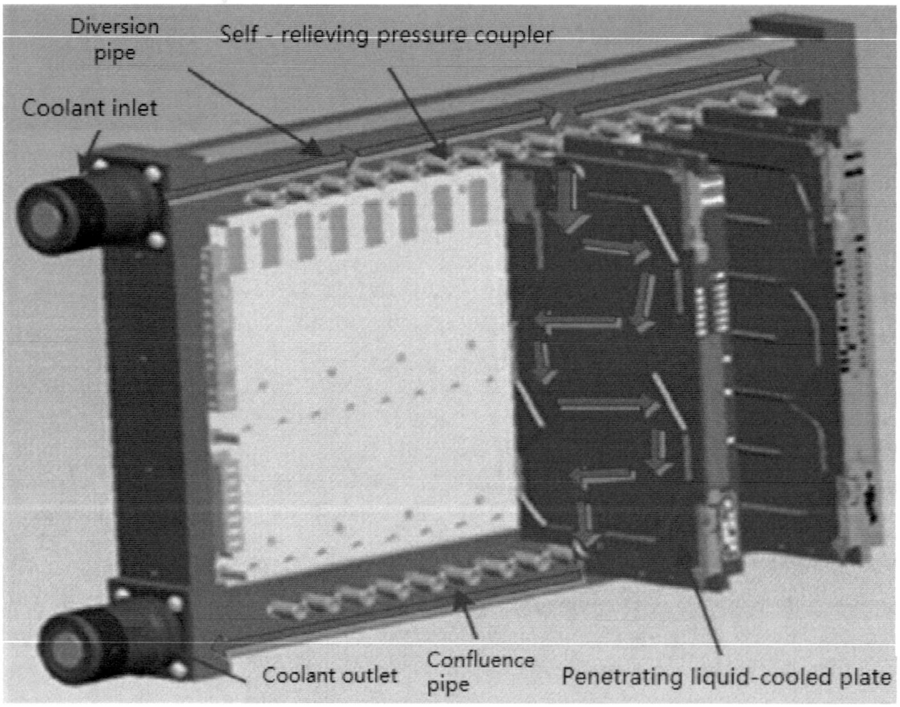

Figure 8. Coolant Circulation Path Diagram.

5. Design of the Penetrating Liquid-Cooling Plate

The through-type liquid-cooling module mainly consists of a cold plate (through which coolant can penetrate), a rear cover plate, a thermal-power circuit board, a wedge-shaped locking strip, a puller, and a self-depressurizing fluid connector. The wedge-shaped strip is composed of a three-section slider, a screw, and a screw sleeve. It is used to lock and fix the cold-plate module to the upper and lower guides of the chassis, facilitating the transfer of the heat dissipated by the thermal-power circuit board to the

Journal of Physics: Conference Series **3011** (2025) 012010 doi:10.1088/1742-6596/3011/1/012010

chassis through the guides. The puller is used for inserting and removing the cold-plate module into and from the chassis slot. The rear cover plate, by cooperating with the cold plate, clamps the thermal-power circuit board and is fixed with screws. It is used to reinforce the thermal-power circuit board, also plays a part in heat conduction and radiation dissipation, and can serve as a shielding plate for shielding purposes. The self-depressurizing fluid connector is used to cooperate with the internal fluid connector in the manifold pipe, thus introducing external fluid into the cold-plate flow channel for circulation, taking away the heat generated by the heating elements of the thermal-power circuit board, and achieving the cooling effect. Regarding the requirements for functional modules, they have relatively high thermal power. The entire device adopts a liquid-cooling module for thermal design. Although the types of functional modules included are different, the design ideas of each module are the same, and basically, the external installation dimensions are consistent. The modules are fixed to the rack-mounting structural components in the cabinet through the puller and the wedge-shaped locking strip.

In the heat dissipation system, the heat source is the thermal power circuit board in the cold plate module. The through-type cold plate, serving as the basis for each heat-consuming unit on the printed circuit board, mainly functions as the installation base for electronic devices. It absorbs the heat generated by the chips on the thermal power circuit board through heat conduction. The absorbed heat can be carried away by the coolant flowing through it, thus achieving the heat dissipation effect.

Heat conduction is the most important way of heat transfer between solids. Heat is transferred from the high-temperature region to the low-temperature region. Thermal resistance is the resistance that heat encounters on the heat flow path, reflecting the heat transfer capacity of the medium or between media. It can help us determine the thermal conduction performance of materials and select appropriate materials. The expression of thermal resistance R_{th} is:

$$R_{th} = \frac{L}{\lambda \cdot A} \tag{1}$$

where L is the length passed in the direction perpendicular to heat conduction, λ is the thermal conductivity of the material, and A is the cross-sectional area passed in the direction perpendicular to heat conduction.

According to Fourier's law, during the heat conduction process, the amount of heat conducted through a given cross-section per unit of time is proportional to the rate of temperature change in the direction perpendicular to the cross-section and the cross-sectional area, and the direction is opposite to the direction of temperature increase.

$$Q_{conduct} = \frac{(T_{heat} - T_{cool})}{R_{th}} \tag{2}$$

When heat is transferred at the interface between two solid media, due to the limitations of machining, the contact surface of the heat sink has roughness, or there are assembly tolerances. As a result, there will always be tiny air gaps between the solid surfaces. Since the thermal resistance of air is very high, the contact thermal resistance has a significant impact on the thermal simulation results. Therefore, thermal conductive grease or thermal conductive pads can be used to fill the contact surfaces between high-power chips and the heat sink (cold plate) to reduce the contact thermal resistance, and heat dissipation is mainly achieved through the conduction method.

As the core of the liquid-cooling module, it is required to have a high thermal conductivity coefficient, low density, good machinability, strong weldability, and a simple and compact structure. Based on these requirements and combined with the calculation of thermal resistance, an aluminum alloy material with good weldability is selected. The following aspects are considered in the design of the cold plate:

(1) Through-type liquid flow channel design: An "S"-shaped flow channel is innovatively designed based on the layout of thermal-power devices. This allows the coolant to pass through the heat-absorbing surface of the cold plate as much as possible, ensuring that the heat generated by thermal-power devices can be promptly absorbed and carried away by the coolant, thus preventing rapid temperature rise.

(2) The flow channel should preferably adopt a wide-length scale. The flow throughout the entire flow channel should be uniform, and it should have good directivity to reduce pressure loss. A 2 mm flow channel [6] should be cut out on a heat sink with a thickness of less than 5 mm (approximately 4.1

mm). Heat exchange ribs should be added inside the flow channel to enhance its overall heat exchange capacity. By milling intermittent grooves on the S-shaped flow channel, on the one hand, the heat exchange area between the fluids is increased, and on the other hand, the collision between the coolant molecules is intensified. When the pressure of the hydraulic pump increases, the coolant begins to gradually exhibit wavy oscillations in the flow channel. The frequency and amplitude of the oscillations increase with the increase in the flow velocity. When the flow velocity increases to a certain extent and the Reynolds number is greater than 2300, many small vortices first appear at the intermittent grooves, which can change the coolant from laminar flow to turbulent flow [7] and enhance the heat exchange capacity.

(3) The design abandons the defect of the large space occupancy of traditional buried-tube liquid cooling. On one side of the liquid-cooling plate, an integrated 360-degree welding process is used to seal the flow channel, which can also withstand a pressure of 1.5 MPa. To prevent the welded cover plate and the entire cold plate from deforming significantly under the action of 1.5 MPa pressure, the welding area needs to be increased and the support spacing shortened during welding.

(4) On the other side of the through-type cold plate, bosses corresponding to the thermal-power chips (devices) need to be milled. A rubber pad with high thermal conductivity (or thermal-conductive grease) can be added at the contact between the bosses and the thermal-power chips (devices). The mating surface between the through-type cold plate and the heat-generating devices should be smooth and flat, and a certain binding and pressing force should be applied to reduce the contact thermal resistance.

(5) Since the processing of the through-type cold plate involves welding [8], to eliminate welding residual stress, relevant low-temperature tempering treatment is required. Additionally, a reasonable design of its processing technology is necessary. It is recommended to first process and weld the flow channel on one side, conduct a compression test, and then perform surface finishing. After low-temperature tempering or aging treatment, the second-stage finishing of the bosses on the lower side is carried out.

(6) After welding, the flow channel of the cold plate needs to be cleaned and dried to ensure cleanliness and pressure testing is carried out. The pressure is 1.5 MPa, and the pressure-holding time is 1 hour. During the test, there should be no leakage in the flow channel, and no abnormalities such as deformation or cracks should occur on the cold plate.

(7) Based on the above requirements, when designing the through-type cold plate, the main materials of the box body and the liquid cooling plate are aluminum alloy, and the grade is 5A06. According to the formula $Q = C\rho V\Delta T$, that is, the heat of the heat-dissipating device, the density of the cooling working medium, the specific heat capacity, and the temperature difference, the liquid flow rate of the working medium can be calculated. The main properties of aluminum are as follows: thermal conductivity $\lambda_{aluminum} = 201$ W/(m·K); specific heat capacity $C_{aluminum} = 871$ J/(kg·K); density $\rho_{aluminum} = 2719$ kg/m³.

Its machinability, weldability, compressive resistance, heat exchange performance of the flow channel, and low thermal resistance for conduction heat dissipation are fully considered. The relevant structural diagrams are shown in Figure 9 to Figure 12 respectively.

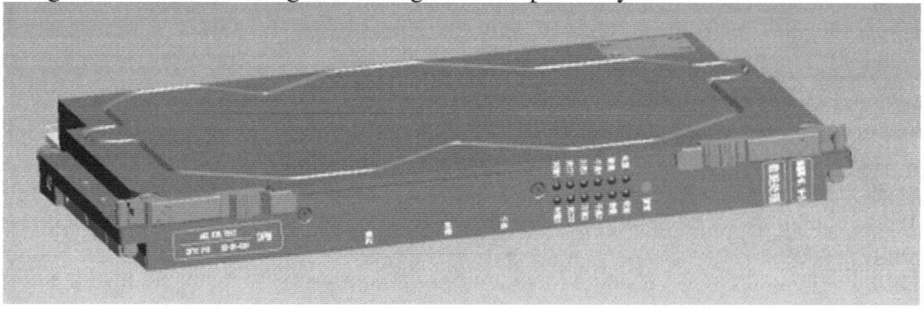

Figure 9. External View of the Penetrating Liquid-Cooling Module.

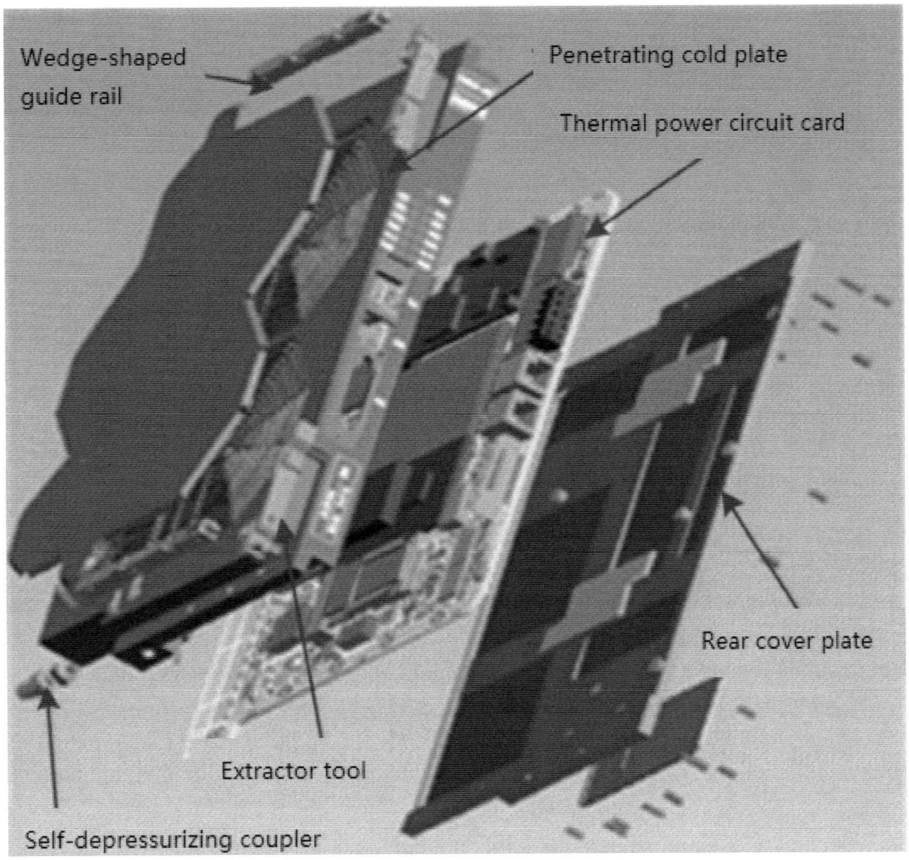

Figure 10. Explosion View of the Penetrating Liquid-Cooling Module.

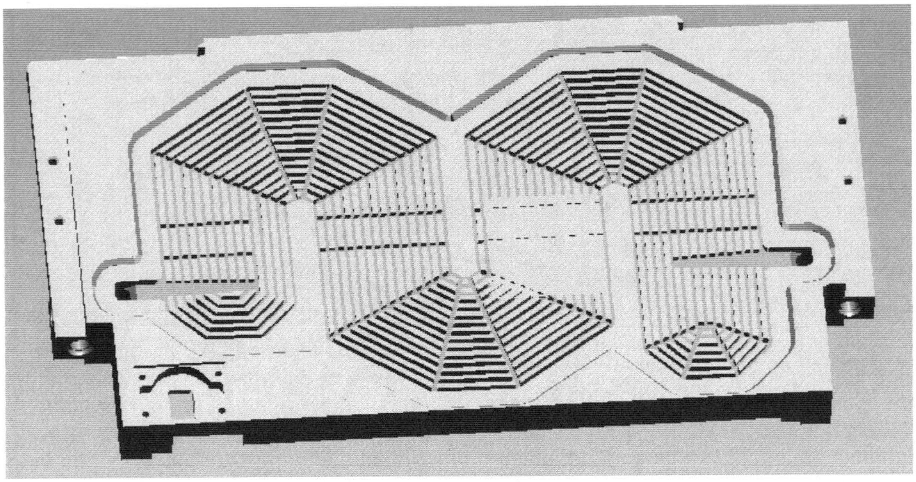

Figure 11. Upper Side View of the Penetrating Liquid-Cooling Plate.

Journal of Physics: Conference Series **3011** (2025) 012010 doi:10.1088/1742-6596/3011/1/012010

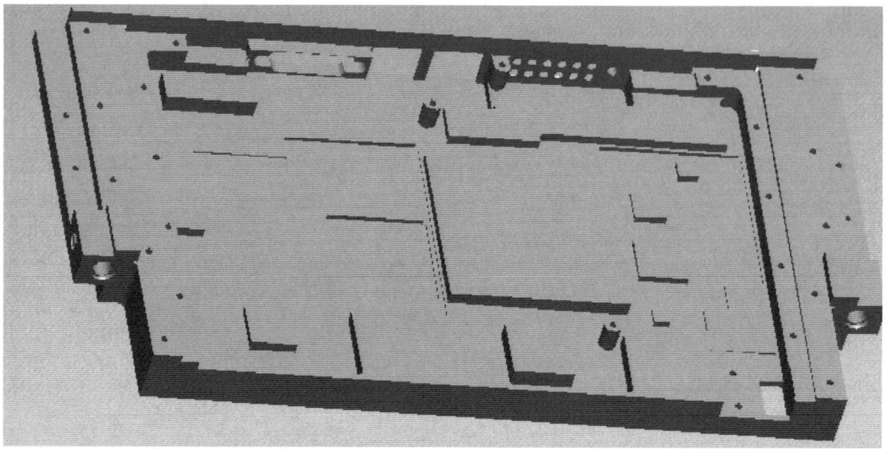

Figure 12. Lower Side View of the Penetrating Liquid-Cooling Plate.

6. Thermal Simulation

Regarding the above-designed penetrating heat-dissipation system, it is necessary to conduct a relevant thermal simulation analysis to determine whether it is feasible and whether it achieves the design purpose. This paper will use relevant fluid simulation software for thermal simulation.

There are a total of 16 penetrating liquid-cooling modules inside this device, and the liquid-cooling flow channels are the same. The data-processing module has the largest thermal power consumption, approximately 237 W. This paper takes it as the simulation object for thermal simulation.

6.1. Initial Conditions

The coolant used is the No. 65 antifreeze coolant. The ambient temperature is considered as the maximum operating temperature of 70 °C. The coolant inlet temperature is 40 °C, the working pressure is 0.8 MPa, the pressure-resistance requirement is 1.5 MPa, and the flow rate is 0.9 LPM (liters per minute). The layout and power values of each thermal power device on its data-processing board are shown in the following figure 13:

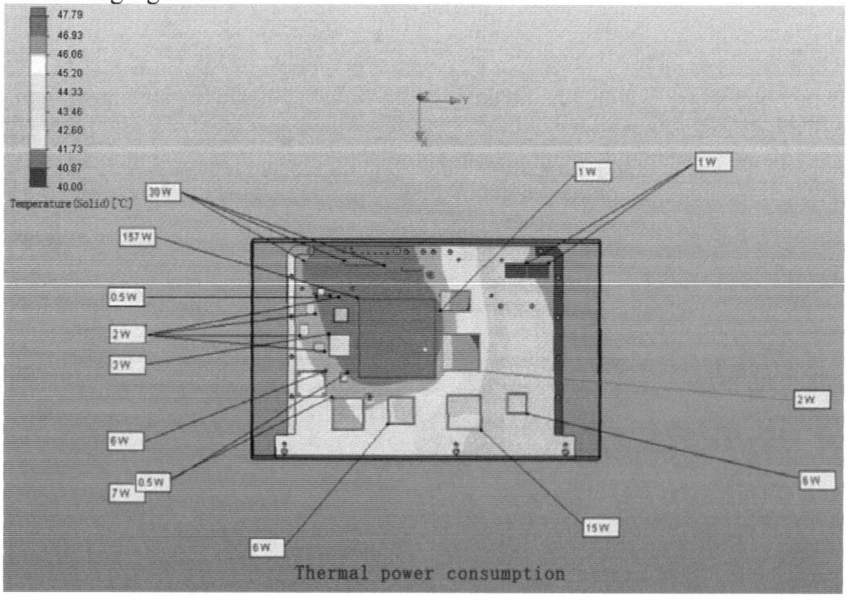

Figure 13. Module Power Distribution Diagram.

6.2. Thermal Simulation Analysis

According to the above thermal design, to know in advance whether the design is feasible, relevant thermal simulation analysis software can be used for thermal simulation. This simulation takes the computing-processing module as an example for simulation analysis. The working fluid for liquid supply is No. 65 coolant, the liquid-supply temperature is 40 °C, the flow resistance of a single module is ≤ 25 KPa and the flow rate of a single module is ≤ 0.9 L/min. According to the relevant boundary conditions, thermal simulation analysis is carried out. The temperature distribution of the entire box, the flow velocity of the coolant, and the pressure simulation diagram of the coolant are shown in Figure 14 to Figure 18 respectively:

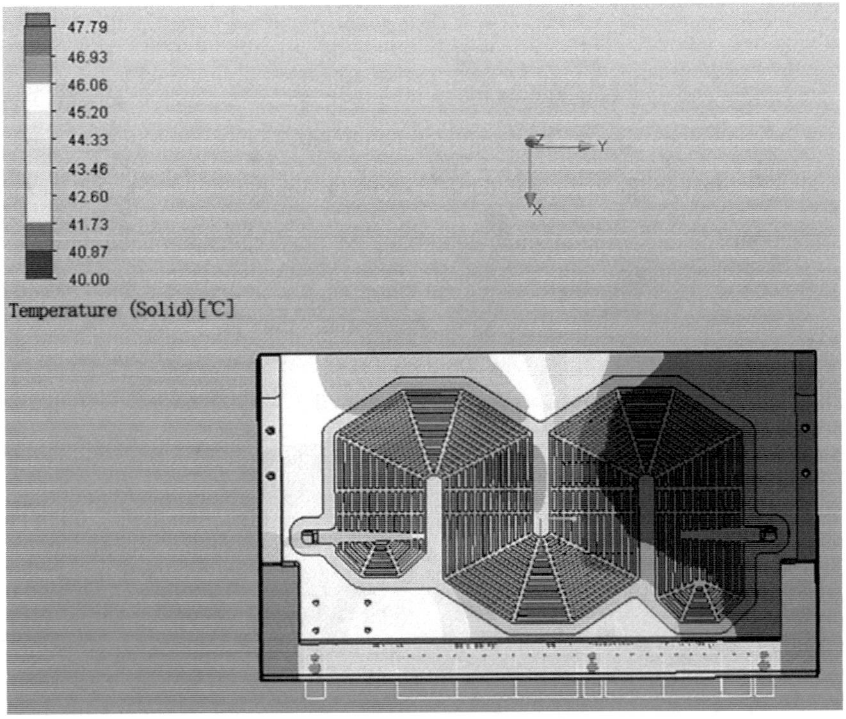

Figure 14. Temperature Cloud Diagram of the Module Surface.

As can be seen from Figure 14, with the increase in the flow path of the coolant flowing through the liquid-cooled plate, the temperature of the entire through-type liquid-cooling module gradually rises from the initial coolant temperature of 40 °C, and only reaches 46.93 °C when flowing out of the liquid-cooled plate, indicating an excellent heat dissipation effect.

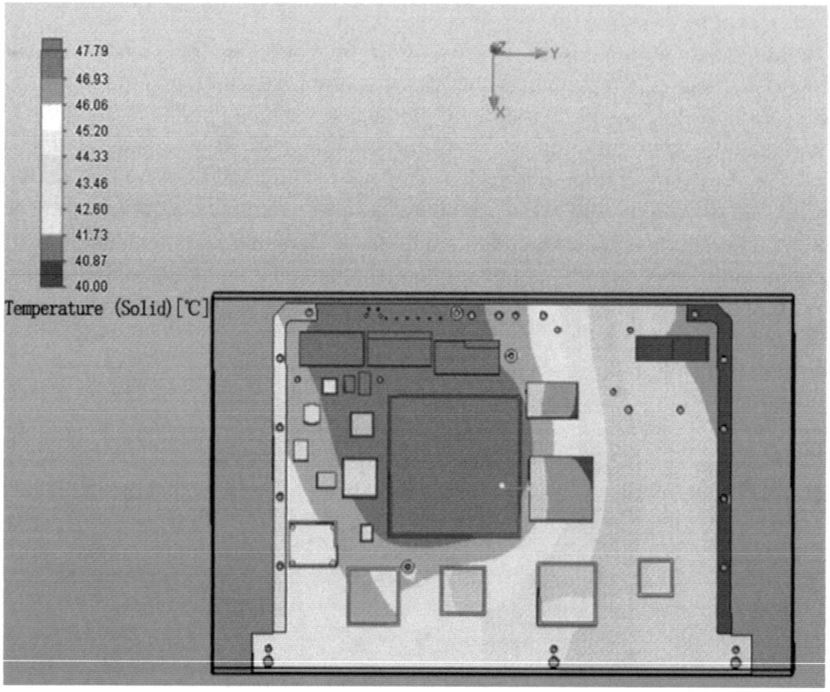

Figure 15. Temperature Cloud Diagram of the Module PCB Board.

Figure 15 shows the operating temperatures and heat distribution areas of each chip on the module's PCB board after the device is operating normally with heat dissipation through the through-type liquid-cooling module. From this figure, it can be seen that the heat is concentrated in the CPU area and the power management area, and the maximum temperature only rises by 7.79 °C, reaching 47.79 °C [9].

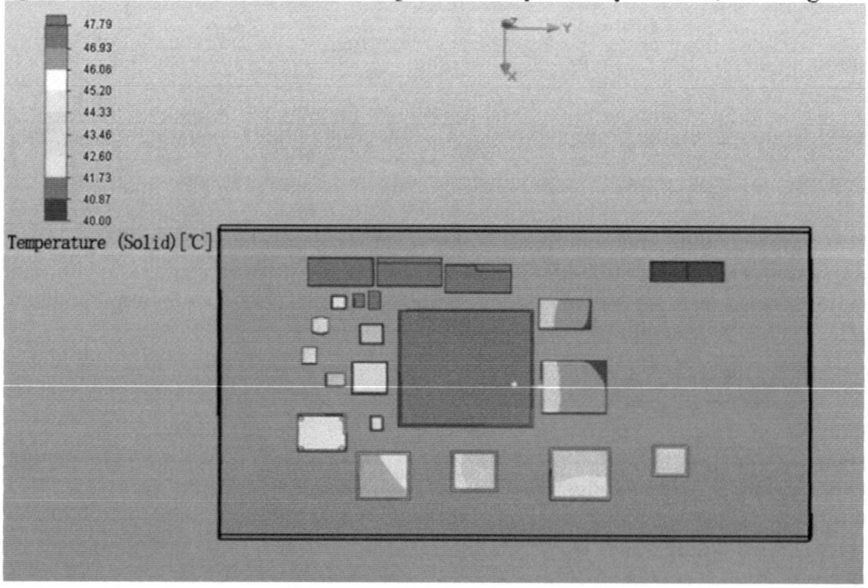

Figure 16. Temperature Cloud Diagram of Each Device on the Module PCB Board.

Figure 16 shows the operating temperature diagram of each chip on the PCB board. It is consistent with the temperature contour map of the PCB board in Figure 15. The heat is concentrated in the CPU area, reaching 47.79 °C.

Journal of Physics: Conference Series **3011** (2025) 012010 doi:10.1088/1742-6596/3011/1/012010

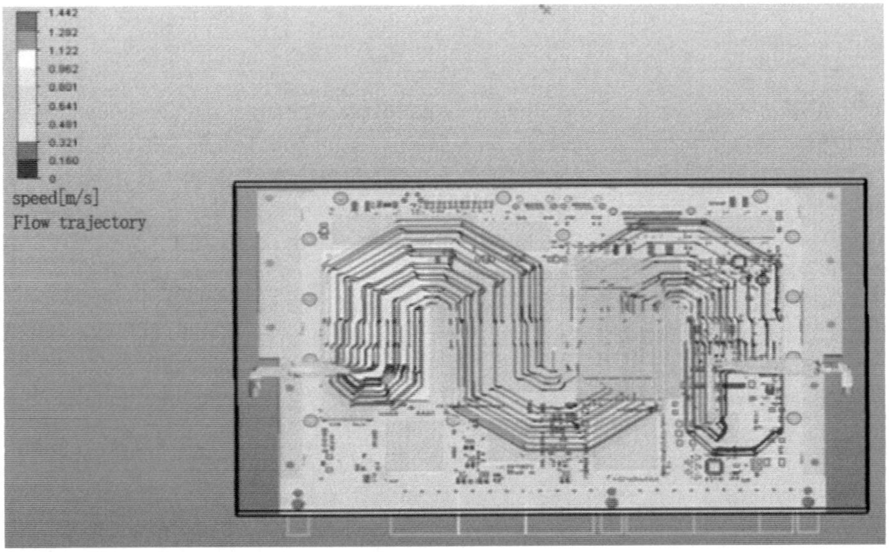

Figure 17. Coolant Flow Velocity Diagram in the Module.

Figure 17 shows the flow traces and flow velocities of the coolant inside the module. The average flow velocity throughout the process is around 0.2 m/s. It can be seen that the coolant flows uniformly throughout the process and has good directivity, enabling efficient heat conduction.

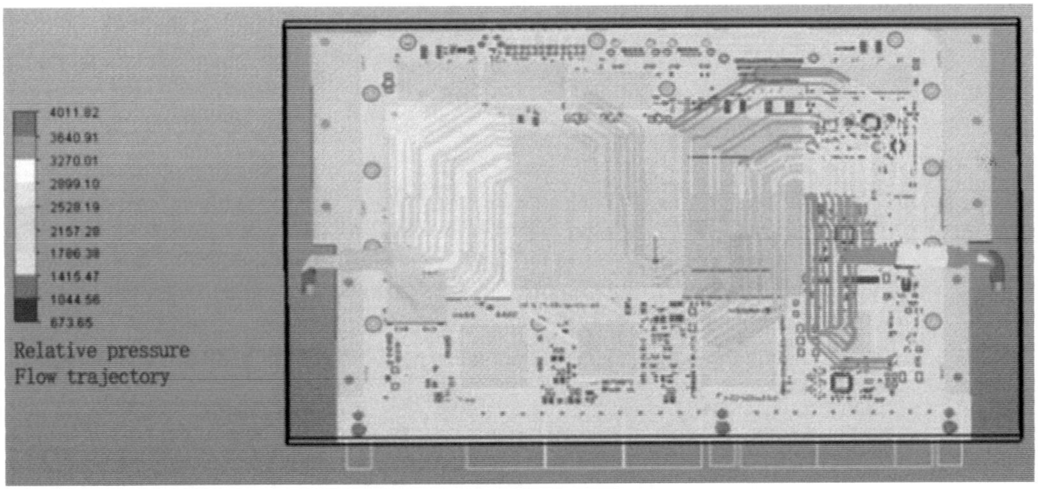

Figure 18. Coolant Pressure Diagram in the Module.

Figure 18 represents the pressure diagram of the coolant inside the module. From the pressure diagram, it can be seen that the pressure of the coolant is slightly higher when it first enters the liquid-cooled plate, about 3500 Pa, which is far lower than the output pressure of the hydraulic pump, 1.5 Mpa. As the flow path increases, the pressure of the coolant stabilizes at around 2200 Pa. With stable pressure, heat can be conducted effectively [10].

Through simulation analysis, when the coolant temperature is 40 °C, the highest temperature in the module is 47.79 °C, mainly concentrated on the CPU chip. From the coolant flow velocity and pressure diagrams of the module, it can be seen that the coolant can evenly sweep through the liquid-cooling flow channels in the module radiator under the drive of the liquid-cooling source. From the simulation results, the liquid-cooling flow channel design, input pressure, and layout of this module are very ideal and can meet the development temperature requirements.

7. Conclusion

Through the optimized design of the through-type liquid-cooled plate, this research has achieved the following: when the ambient temperature is 60 °C, the initial temperature of the coolant is 40 °C, and the maximum temperature of the coolant flowing out of the liquid-cooled plate is 47.49 °C, with a temperature rise of only 7.79 °C. Compared with traditional heat-dissipation solutions such as air-cooling or traditional heat pipes, the heat-dissipation efficiency has been greatly improved, providing an effective and feasible solution for the heat dissipation of high-power chips.

Regarding the current application scenarios, which are mainly divided into fixed and mobile ones, the future research directions can be as follows. For large-computing-power centers such as intelligent computing centers and data centers in fixed command posts, by developing new nanofluids with high thermal conductivity and low corrosion at the material level, the research results can be directly applied to the iteration of the immersion liquid-cooling system for high-computing-power GPU clusters. For mobile high-computing-power servers with limited space and high maintainability requirements, through structural innovation, local micro-channel heat dissipation can be carried out for high-power-consumption chips. For the non-uniform heat source distribution of heterogeneous integrated chips, an adaptive flow-channel structure can be designed. From the perspective of intelligent control, an intelligent flow-rate regulation system based on real-time temperature feedback can be developed. Combining with digital twin technology, the dynamic calibration of the heat-dissipation solution can be realized, which can be extended to the global optimization of the entire liquid-cooling system. The above-mentioned research will contribute to the large-scale application of through-type liquid-cooling technology in high-computing-power and high-density computing platforms, and truly and effectively solve the technical problems of high heat-flux density and high thermal power consumption of high-computing-power servers in the era of intelligent computing.

References

[1] Zhuang Zeyan, Sun Cong, Zhang Qi, Huang Zixuan, Yu Hongyuan. Analysis of the Current Situation of Cold Plate Liquid Cooling Technology and Its Application. Communication Management and Technology, Issue 3, June 2021.
[2] Bao Yunhao, Chen Jianye, Shao Shuangquan. Research Status of High-Efficiency Liquid Cooling Technology in Data Centers. Refrigeration and Air Conditioning, 2023, 23(10): 58-69.
[3] Ke Yuanhua, Cheng Jun, Yang Yingjie, et al. Research on Liquid Cooling Technology in Data Centers [J]. Posts and Telecommunications Design Technology, 2023(12): 35-41.
[4] Yang Weichao. Research on the Through-Type High-Efficiency Heat Dissipation Technology of Reinforced Computing Equipment. Industrial Control Computer, Vol. 34, Issue 11, 2021.
[5] Ni Tongqing. Design, Analysis and Application of the Heat Dissipation Structure of a Liquid Cooling Chassis. Mechanical and Electrical Information, Issue 17, 2023, Total Issue 713.
[6] Li Bang, Chen Qian, Lin Zhi. Analysis of the Influencing Factors of the Single-Phase Immersion Liquid Cooling and Heat Dissipation Performance of Servers. Electronics World, Issue 19, 2021.
[7] Jiang He, Yuan Junfei, Wang Lin, Liang Boyang, Li Shaoqiang. Research on the Coupling Mechanism of Boiling Flow and Heat Transfer Characteristics in Microchannels. Cryogenics and Superconductivity, Vol. 51, Issue 1, 2023.
[8] Zhu Guanghui, Liu Chao, Liu Qianwen, Wang Chao. Structural Optimization of the Vacuum Brazing Fixture for Liquid Cooling Shells. Mechanical Management and Development, Total Issue 227, Issue 3, 2022.
[9] Chen En, Jiang Yongsheng, Wei Jijun, Wang Kun, Zhang Erxin. Design of the Liquid Cooling System for a 1500 kW Electronic Device. Vol. 35, Issue 1, February 2015.
[10] Zhang Ge, Wen Jian. Design and Simulation Analysis of the Microchannel Liquid Cooling Cold Plate. Machinery, Total Issue 663, 2019, 57(11).

Journal of Physics: Conference Series **3011** (2025) 012011 doi:10.1088/1742-6596/3011/1/012011

Low carbon scheduling strategy of electricity-heat integrated energy system considering carbon trading and demand response

Xudong Wang[1], Lin Wang[1], Qiancheng Wang[2*], Ting Yang[2]

[1] State Grid Tianjin Electric Power Company, Tianjin, 300010, China
[2] School of Electrical and Information Engineering, Tianjin University, Tianjin 300072, China

[*]Corresponding author's e-mail: wangqch_cn@163.com

Abstract. Addressing the challenge that the carbon emission trading market is progressively being incorporated into various industries, the existing scheduling strategies are characterized by high carbon quota performance costs and limited carbon reduction potential in demand. This paper studies a low-carbon optimization strategy of an electric-heat integrated energy system (IES), incorporating considerations of carbon quota trading and demand side. Firstly, considering the two trading targets of carbon quota and Chinese certified emission reductions (CCER), a carbon trading model that includes thermal power generation, renewable energy, and load is constructed with daily clearing frequency. Secondly, an electricity-heat IES low-carbon scheduling model is established, including carbon quota trading and demand response, aiming to maximize the system's marginal revenue. Subsequently, an AMP-NSGA-II algorithm is proposed to address the established scheduling method. The simulation outcomes demonstrate that the proposed low-carbon scheduling strategy enhances the system's boundary revenue by 8.7 % and decreases its carbon emissions by 13.2 %.

1. Introduction

Given the growing proportion of renewable energy integrated into the smart grid continues to rise, the trading times of China Emission Allowances (CEAs) and Chinese Certified Emission Reductions (CCERs) in the carbon trading market increasing significantly. Promoting the internalization of carbon emission externalities through flexible trading mechanisms has become an important part of energy protection and emission decrees [1]. Reference [2] used the modeling method of system dynamics to study the mechanism between green certificates and carbon emissions trading. Reference [3] also analyzed the coupling relationship between the green certificate and the electricity market by establishing a causal loop diagram through system dynamics modeling. In reference [4], the uncertainty of carbon trading price is fully considered, and a bi-period randomly programming method is proposed. However, the aforementioned studies have overlooked the significant consequence of the trading system in wind power consumption and failed to implement an equilibrium between low-carbon objectives and economic considerations in the optimization scheduling process. Therefore, a low-carbon optimization strategy for the electricity-heat IES is proposed in this paper to reduce carbon emissions and increase the system revenue of IES.

2. Carbon emissions trading modeling of the integrated energy system

2.1. Carbon trading model of various entities

In this paper, carbon emissions from thermal power plants are determined according to the reference values of the power supply emissions. The carbon emission benefit $C^{\mathrm{CFPU}}(t)$ of thermal power plants is:

$$C^{\text{CFPU}}(t)=C_t^{\text{CE}}-F^{\text{CSD}}E^{\text{CCER}}\lambda^{\text{CFPU}}P^{\text{CFPU}}(t) \tag{1}$$

In Eq. (1), $P^{\text{CFPU}}(t)$ represents the thermal output; λ^{CFPU} is a reference value of carbon emissions from thermal power units, with the unit being t CO_2/(MW·h). E^{CEA} indicates the market price of China Emission Allowances. E_a is the quota applied for following the dispatching plan, with each unit distributed to each dispatching cycle according to the power generation plan. The value of F^{CSD} is drawn from a chi-square distribution, and the value range is [0, 0.05]. E^{CCER} is the market price of CCER.

Distributed photovoltaics, wind farms, and fuel cells sell all CCERs in the carbon emission trading market to obtain profits, and the revenue $C^{\text{RES\_CCER}}(t)$ is:

$$C^{\text{RES\_CCER}}(t)=\left[P^{\text{WT}}(t)+P^{\text{HST}}(t)+P^{\text{ET}}(t)+P^{\text{PV}}(t)\right]\varphi^{\text{TF}}E^{\text{CCER}} \tag{2}$$

where $P^{\text{WT}}(t)$ represents the power of wind power; $P^{\text{PV}}(t)$ represents the photovoltaic power; $P^{\text{HST}}(t)$ and $P^{\text{BT}}(t)$ are the on-grid power of gas energy storage and battery respectively; φ^{TF} represents the marginal emission factor associated with the grid's benchmark scenario.

Both of them earn the difference by selling free quotas to purchase CCERs, and the largest consumer of the load CEA will purchase CEA when the carbon emission exceeds the quota. Exemption from punishment, so the trading strategy is to purchase CCER quotas first, and then purchase CEA based on the emission results to complete the carbon trading clearing. The carbon emission income $C^{\text{LOAD}}(t)$ of the load is:

$$C^{\text{LOAD}}(t)=C_t^{\text{CE}}-F^{\text{CSD}}E_o^{\text{LOAD}}(t)E^{\text{CCER}} \tag{3}$$

where $E_{\text{LOAD}\,o}(t)$ is the free allocation quota assigned to each period according to the scheduling cycle, and o is the number of loads.

2.2. *Reward-penalty ladder carbon trading price model*

For each entity involved in the transaction, carbon emission quotas are determined. The growth interval and rate of carbon transaction costs are then ascertained using historical data [5]. The calculation formula for stepped carbon transaction cost is as follows:

$$C_t^{\text{CE}}=\begin{cases}E^{\text{CEA}}\Delta B_s,\Delta B_s<0\\ E^{\text{CEA}}\Delta B_s,0<\Delta B_s<d\\ E^{\text{CEA}}d+(1+\lambda)E^{\text{CEA}}(\Delta B_s-d),d<\Delta B_s<2d\\ E^{\text{CEA}}d+(1+\lambda)E^{\text{CEA}}d+(1+2\lambda)E^{\text{CEA}}(\Delta B_s-2d),2d<\Delta B_s<3d\\ E^{\text{CEA}}d+(1+\lambda)E^{\text{CEA}}d+(1+2\lambda)E^{\text{CEA}}d+(1+3\lambda)E^{\text{CEA}}(\Delta B_s-3d),\Delta B_s>3d\end{cases} \tag{4}$$

$$\Delta B_s=\begin{cases}\left[E_a-0.98\lambda^{\text{CFPU}}P^{\text{CFPU}}(t)\right]E^{\text{CEA}},s=\text{CFPU}\\ \sum_{o=1}^{M}\left[1.063E_o^{\text{LOAD}}(t)-\delta_o P_o^{\text{LOAD}}(t)\right]E^{\text{CEA}},s=\text{LOAD}\end{cases} \tag{5}$$

where λ represents the carbon emission penalty factor at time t. The variable d represents the growth of the carbon emission penalty range at time t. It is influenced by factors such as the system's production capacity and emission reduction targets.

3. Economic dispatch with low-carbon considerations for electricity-heat IES

The IES comprises a thermal plant, wind farm, fuel cell, photovoltaic power generation, heat pump, boiler, electrolytic cell, and gas storage tank. The heat produced by the thermal power plant and the fuel cell is recovered via the heat pump and subsequently supplied to the load after being heated by the boiler. The electrolytic cell and gas storage tank assist the wind farm and photovoltaic power output device to complete the renewable energy consumption.

3.1. Carbon emission quota modeling considering price demand response

The power price resilient matrix is utilized to depict the propensity of users to shift their load consumption in the context of different electricity price fluctuations.

$$\begin{cases} \tau_{ii} = \dfrac{\partial d_i / d_i}{\partial c_i / c_i} \\ \tau_{ij} = \dfrac{\partial c_i / c_i}{\partial d_j / d_j} \end{cases} \tag{6}$$

where d_i is the electricity quantity, c_i is the electricity price, $\partial d_i/d_i$ is the change rate of electricity quantity, $\partial c_i/c_i$ is the change rate of electricity price, τ_{ii} denotes the inherent elasticity coefficient, which measures the alteration of electricity consumption due to the price change in the same period i. τ_{ij} represents the interaction resilient coefficient.

It can be concluded that the load response model for a certain period can be described as:

$$P_i^{\text{LOAD}} = \begin{bmatrix} P_{1s}^{\text{LOAD}} \\ P_{2s}^{\text{LOAD}} \\ \vdots \\ P_{3s}^{\text{LOAD}} \end{bmatrix} = \begin{bmatrix} P_1^{\text{LOAD}} \cdot (1 + \tau_{11} \dfrac{\Delta c_1}{c}) \\ P_2^{\text{LOAD}} \cdot (1 + \tau_{22} \dfrac{\Delta c_2}{c_2} \\ \vdots \\ P_n^{\text{LOAD}} \cdot (1 + \tau_{nn} \dfrac{\Delta c_n}{c_n} \end{bmatrix} \tag{7}$$

where $P_{\text{LOAD}\,ns}$ represents the load after the adjustment due to the variable electricity rate based on usage time; $P_{\text{LOAD}\,i}$ denotes the initial load; Δc_n represents the resultant price fluctuation from the execution of the variable electricity rate based on usage time.

Drawing on the historical coefficient, the carbon emission quota is defined as follows:

$$E_o^{\text{LOAD}} = \delta_o P_o^{\text{LOAD}} \tag{8}$$

3.2. Objective function

The system's marginal revenue maximization is represented by the following objective function:

$$\max F = C_{\text{f}}^{\text{CFPU}}(t) + C_{\text{f}}^{\text{RES}}(t) + C_{\text{f}}^{\text{LOAD}}(t) \tag{9}$$

where $C_{\text{f}}^{\text{CFPU}}(t)$ represents the revenue from thermal power generation; $C_{\text{f}}^{\text{RES}}(t)$ represents renewable energy revenue; $C_{\text{f}}^{\text{LOAD}}(t)$ is the benefit of load participating in demand response.

$$\begin{cases} C_{\text{f}}^{\text{CFPU}}(t) = P^{\text{CFPU}}(t) d_{\text{u}} + C^{\text{CFPU}}(t) + P_{\text{h}}^{\text{CFPU}}(t) c_{\text{eb}} + C_t^{\text{SER}} d^{\text{SER}} - [a(P^{\text{CFPU}}(t))^2 + bP^{\text{CFPU}}(t) + c)] \\ C_{\text{f}}^{\text{RES}}(t) = P^{\text{WT}}(t) c_{\text{wt}} + P^{\text{HST}}(t) c_{\text{hst}} + P^{\text{PV}}(t) c_{\text{pv}} + C^{\text{RES\_CCER}}(t) + P^{\text{BT}}(t) \eta^{\text{BT}} c_{\text{bt}} - c_{\text{sto}} S_{\text{sto}}(t) \\ C_{\text{f}}^{\text{LOAD}}(t) = \sum_{o=1}^{M} [\lambda_o^{\text{SY}} P_o^{\text{LOAD}} - P_{\text{h},o}^{\text{LOAD}} c_{\text{h}} - C^{\text{LOAD\_CCER}}(t)] \end{cases} \tag{10}$$

where d_{u} represents the thermal power grid connection tariff, $P_{\text{h}}^{\text{CFPU}}(t)$ represents the heating income; and $C_{\text{SER}\,t}$ is the service volume participating in frequency modulation power, reserve capacity, and so on. d^{SER} is the unit price of service quantity. c_{eb} is the price for the acquisition of heat. c_{wt} is the on-grid price of wind power; c_{bt} is the on-grid price of fuel cell; c_{pv} is the photovoltaic grid connection price; η^{BT} represents the conversion efficiency of fuel cell; c_{sto} is the gas storage cost coefficient; $S_{\text{sto}}(t)$ represents the gas storage mass; $P_{\text{h,o}}^{\text{LOAD}}(t)$ represents the thermal energy power consumed by the load, M is the total number, c_{h} is the unit thermal energy sales price, and $\lambda_{\text{o}}^{\text{SY}}$ is the load economic benefit index.

The constraint conditions are categorized into power constraints and heat load constraints. Since the carbon emission trading model is fully linearized, the carbon emission constraint will be transformed into a straightforward output constraint [6].

4. Solving method of the dispatch model based on adaptive multi-population NSGA-II

An adaptive multi-population NSGA-II (AMP-NSGA-II) algorithm is presented in this section. An excellent solution set (EXS) is introduced, which helps to position the individuals in the excellent solution set closer to the Pareto front and thereby enhances the algorithm's astringent characteristics.

(1) Non-dominated sorting: For each individual in the population, the number of individuals dominated by other individuals and the set dominated by other individuals are calculated.

(2) Crowding degree calculation: The crowding degree represents the density of other individuals around the individual. The calculation formula for the crowding degree is:

$$D_i = \sum_{M}^{m=1}\left(h_m(i+1) - h_m(i-1)\right) / \left(h_{m_{\max}} - h_{m_{\min}}\right) \tag{11}$$

where D_i denotes the crowding degree of the i-th individual, M represents the number of objective functions, $h_m(i+1)$ and $h_m(i-1)$ represent the function values of the adjacent individuals to the i-th individual in the m-th objective function, and $h_{m\max}$ and $h_{m\min}$ are the highest and lowest values of the m-th objective function.

(3) Fitness calculation: Fitness is a comprehensive index of individual level and crowding degree. The calculation formula is:

$$F_i = \frac{N - N_i}{N} + \frac{D_i}{D_{\max}} \tag{12}$$

where F_i denotes the fitness of the i-th individual, N is the overall volume of layers, N_i represents the level of the individual, D_i is the crowding degree of the individual, and $D_{\max}$ is the maximum crowding degree.

(4) Multi-population strategy: In AMP-NSGA-II, the population is divided into multiple sub-populations, and each sub-population may apply different crossover and mutation operators.

(5) Dynamic adjustment of sub-groups: The size of each sub-group is dynamically adjusted according to the distribution of individuals in the external archive, thereby improving adaptability to various problems.

5. Case analysis

Taking an industrial thermal-electric integrated energy system as an example [7], the load includes industrial production load and residential load. The system structure includes a 150 MW photovoltaic unit, a 150 MW, a 250 MW unit, a 300 MW thermal power unit, and a 140 MW wind turbine. Day-ahead scheduling is adopted, and the cycle is 24 hours.

Table 1. The proposed strategy and the traditional method optimize the scheduling result

Method	$C_f^{CFPU}(t)$ (yuan ×10^4)	$C_f^{RES}(t)$ (yuan ×10^4)	$C_f^{LOAD}(t)$ (yuan ×10^4)	Carbon emissions (tCO_2)
Traditional	50.78	8.42	47.82	6849.0
The proposed	55.79	9.17	51.43	5944.93

The optimization results of the offered AMP-NSGA-II method and the traditional NSGA-II method are shown in Figure 1. In the 100-iteration process, the objective function of the offered AMP-NSGA-II method converges after 37 iterations, and the calculation time is 13.8 s. This demonstrates the high computational efficiency and accuracy of the AMP-NSGA-II algorithm. As shown in Table 1, the low-carbon optimization method of the IES considering carbon quota trading and demand response and the optimal operation results of the traditional scheduling method are compared. The proposed low-carbon scheduling strategy raises the system's boundary revenue by 8.7 %. The waste heat recovery system, benefiting from the high output of thermal power units, retains a surplus after meeting the residents' heat load. Thermal power and renewable energy make an additional profit of 57,600 yuan, and the residents ' load increases by 7.5 %. Moreover, it lowers the system's carbon emissions by 13.2 % and greatly boosts the energy utilization rate.

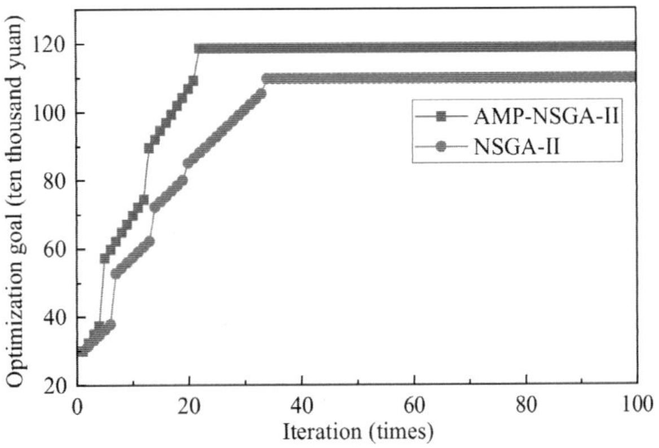

Figure 1. The optimization outcome of the AMP-NSGA-II and the NSGA-II algorithm.

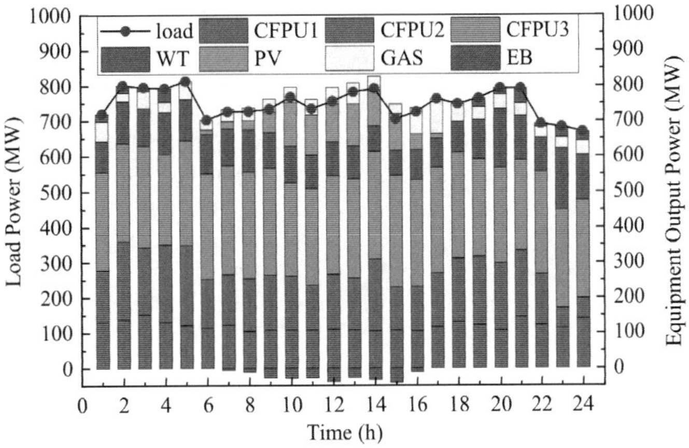

Figure 2. The device output and load fluctuation trajectories of each unit.

Figure 2 illustrates the output and load variation profiles of each plant. It can be seen that thermal power units respond to the reduction of residential load, and wind power and photovoltaic can still maintain a good level of consumption. In systems characterized by poor geographical conditions and a low proportion of new energy, the installation of a waste heat recovery system can be a highly effective approach to energy saving and emission reduction.

To deeply substantiate the efficacy of the proposed strategy, comparative simulation analyses are carried out across three distinct scenarios. Scenario 1: Fixed carbon trading price, only considering CEA trading; Scenario 2: Stepped carbon trading price, only CEA trading is considered; Scenario 3 (the proposed method): Stepped carbon trading price, considering both CEA and CCER transactions. As shown in Figure 3, the optimal scheduling results are presented. Upon comparing scenarios 1 and 2, it is shown that the introduction of a step-by-step carbon trading price results in an 8.37 % decrease in system carbon emissions in scenario 2 compared to scenario 1. This reduction, however, is accompanied by a 3.54 % reduction in system revenue. While the step-by-step carbon trading price employs dynamic pricing to control the carbon production process of IES, it also raises the expenditure of carbon reduction. After the introduction of CCER, the system revenue of Scenario 3 (the proposed method) increased by 4.36 % compared with Scenario 2, and the carbon emissions decreased by 7.57 %. Compared with scenarios 1 and 3, the total revenue of the proposed method increases by 7700 yuan and decreases carbon emissions by 15.3 %. The simulation results presented above validate the economic and low-carbon superiority of this paper over the traditional method.

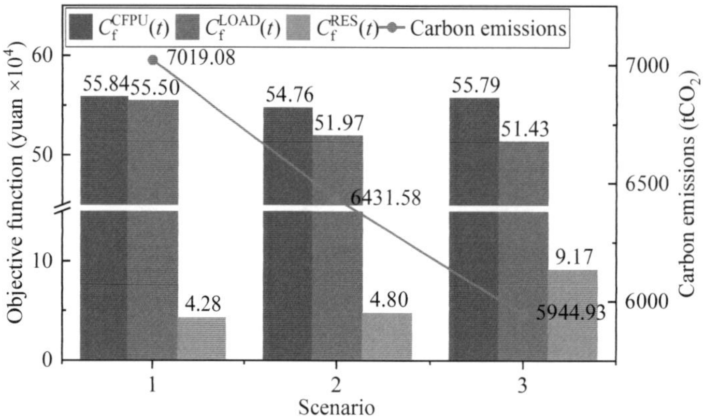

Figure 3. Optimized scheduling results in different scenarios.

6. Conclusion

In this paper, a low-carbon optimization strategy for the electricity-heat IES that considers carbon trading and demand response is proposed. To maximize the system's marginal revenue, a low-carbon optimization strategy incorporating carbon quota trading and the corresponding demand side are established. An AMP-NSGA-II method is introduced to enhance the convergence performance of the low-carbon scheduling model. The consequence indicates that the proposed low-carbon optimization method increases the system's boundary revenue by 8.7 % and reduces carbon emissions by 13.2 %, demonstrating the proposed method can reduce carbon emissions and increase system revenue.

Acknowledgments

Supported by State Grid Corporation of China (Project Code: 5700-202312315A-1-1-ZN, Project Name: Research and application of digital support technology for carbon asset operation service in large cities).

References

[1] Yang P, Jiang H, Liu C, et al. Coordinated optimization scheduling operation of integrated energy system considering demand response and carbon trading mechanism[J]. International Journal of Electrical Power & Energy Systems, 2023, 147: 108902.
[2] Wang L, Ren X, Ma Y, et al. Optimal low-carbon scheduling of integrated energy systems considering stepped carbon trading and source-load side resources[J]. Energy Reports, 2024, 12: 3145-3154.
[3] Ma R, Zhao J, Zhou J, et al. Optimization of Integrated Energy System Scheduling Considering Stepped Carbon Emission Trading Mechanism and Electricity Heat Demand Response[C]//2023 Panda Forum on Power and Energy (PandaFPE). IEEE, 2023: 1932-1938.
[4] Zhang G, Liang J, Li F, et al. Low-carbon scheduling model of electricity-gas-heat integrated energy system considering ladder-type carbon trading mechanism, vehicle charging, and multimode utilization of hydrogen[J]. IEEE Access, 2024.
[5] Zhang Y, Han Y, Liu D, et al. Low-carbon economic dispatch of electricity-heat-gas integrated energy systems based on deep reinforcement learning[J]. Journal of Modern Power Systems and Clean Energy, 2023, 11(6): 1827-1841.
[6] Yi Z, Xu Y, Zhou J, et al. Bi-level programming for optimal operation of an active distribution network with multiple virtual power plants[J]. IEEE transactions on sustainable energy, 2020, 11(4): 2855-2869.
[7] Luo Y, Hao H, Yang D, et al. Multi-objective optimization of integrated energy systems considering ladder-type carbon emission trading and refined load demand response[J]. Journal of Modern Power Systems and Clean Energy, 2023.

A bilateral collaborative optimization strategy considering carbon responsibility allocation

Jiayi Wu[1,*], Xinyu Li[1], Yi Wang[2] and Gang Xu[1]

[1]School of Electrical and Electronic Engineering, North China Electric Power University, Beijing, Beijing, 102206, China
[2]New Energy/Ocean Energy and Power Engineering Institute/Integrated Energy Institute, Kunming Engineering Corporation Limited, Kunming, Yunnan, 650032, China

[*]Corresponding author's e-mail: ydkm666@outlook.com

Abstract. To address renewable energy intermittency in power system stability, this study proposes a carbon-responsibility-embedded coordinated optimization strategy for source-load interactions. The framework systematically coordinates generation flexibility and demand response while integrating carbon responsibility quantification into operational decisions, achieving dual optimization of grid reliability and emission mitigation. Firstly, the framework of source-charge cooperative operation is constructed to solve the power supply and demand problem caused by the volatility of renewable energy output on the source side. Secondly, the carbon responsibility allocation mechanism is introduced to fully consider the carbon emissions on both sides of the source and load. Thirdly, based on the cooperative game theory and taking the lowest total cost of the cooperative operation as the objective function, the optimal scheduling model of the cooperative operation considering carbon responsibility allocation is established. Finally, the feasibility and effectiveness of the proposed source-load cooperative operation strategy are verified by a simulation example.

1. Introduction

As nations accelerate strategic modernization of electricity infrastructure, renewable energy deployment has emerged as a pivotal pathway for achieving grid decarbonization and operational sustainability[1][2]. However, conventional grid operation predominantly depending on generation-side regulation faces prohibitive economic costs amidst increasing renewable penetration and heterogeneous load composition[3]. To address renewable energy intermittency in power system stability, this study proposes a carbon-responsibility-embedded coordinated optimization strategy for source-load interactions. The framework systematically coordinates generation flexibility and demand response while integrating carbon responsibility quantification into operational decisions, achieving dual optimization of grid reliability and emission mitigation.

Current research advances employ hierarchical game-theoretic frameworks for integrated energy system optimization, as demonstrated in literature [4], which establishes dynamic equilibrium between generation portfolios and demand clusters through bidirectional benefit coordination mechanisms. To mitigate wind curtailment challenges stemming from inadequate operational flexibility, literature [5] developed a multi-temporal granularity optimization framework for integrated energy systems, enabling dynamic flexibility reconfiguration across generation-storage-load domains through sequential decision-making architectures. Literature [6] advanced a demand-side flexibility scheduling paradigm that enhances wind power accommodation capacity through adaptive load shaping mechanisms. While existing studies demonstrate economic viability and renewable integration via source-load coordination, their optimization criteria critically overlook carbon externality internalization in operational decision-making processes.

China's power sector has achieved notable progress in carbon-aware operational strategies, particularly through market-driven mechanisms. Literature [7-8] pioneered multi-tiered carbon trading schemes integrated with energy system dispatch, demonstrating 12-18% emission reduction potential through carbon price elasticity. Building upon carbon flow tracing theory, literature [9] established a carbon-intensity-responsive demand response framework that dynamically aligns electricity consumption patterns with grid decarbonization trajectories through time-varying carbon attribution signals, achieving 14.3% emission reduction in distribution networks via price-elastic load shifting. Literature [10] constructs a carbon-led demand response mode, which can change users' energy use behavior and thus reduce carbon emissions. All the above studies have effectively reduced the carbon emissions of the power system, but the above literatures only consider the emission reduction of the source side or the load side. It would be unfair to restrict carbon emissions to only one side [11]. Considering the current situation of the electric-carbon market in China, how to formulate a reasonable carbon responsibility sharing method to promote the synergistic carbon reduction of both sources and charges is a problem worth studying.

Based on the above analysis, based on the cooperative game theory and carbon emission flow theory, this paper proposes a two-sided cooperative optimization strategy considering carbon responsibility allocation.

2. Source load collaborative optimization architecture

The source-load bilateral cooperative operation framework constructed in this paper is shown in Figure 1. The thermal energy emitted is a transmission vessel, and a component of thermal energy and wind energy. The load side is the load aggregator, which is mainly composed of distributed photovoltaic, fixed load, transferable load and energy storage device. Through the introduction of carbon responsibility allocation mechanism, the carbon emission responsibility is allocated to the source side and the load side. On this basis, a cooperative game model between generation scheduler and load aggregator is established, and the cost of source load cooperative alliance is minimized by optimizing the output and load curves of generator sets.

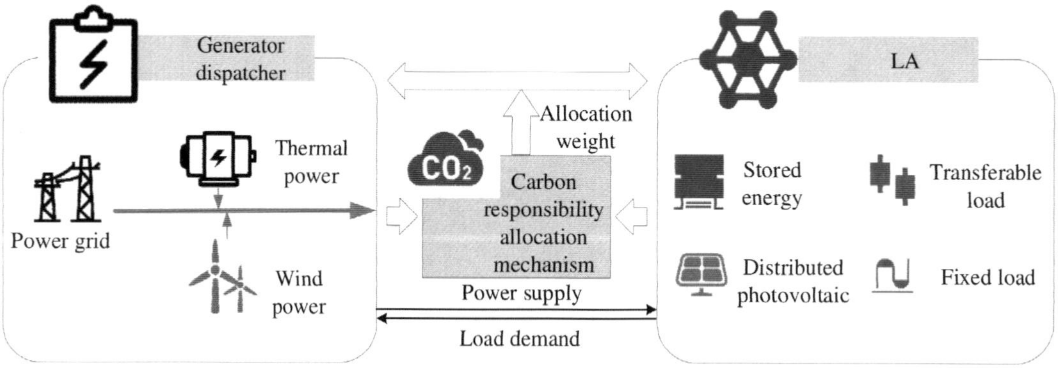

Figure 1. Source load collaborative optimization architecture.

3. Carbon responsibility allocation mechanism

The carbon emission flow tracing framework[12] revolutionizes emission accountability through physical-carbon coupling mechanisms, quantifying nodal carbon flux distribution across energy supply chains while establishing consumption-side liability via nodal carbon intensity indices. Building upon this paradigm, our study pioneers a bilateral carbon accountability allocation model that introduces nodal carbon flux allocation coefficients to dynamically apportion emission responsibilities between generation and demand entities. This methodology concurrently integrates carbon shadow pricing mechanisms[13], transforming emission liabilities into operational cost signals through carbon tax equivalency, thereby establishing techno-economic incentives for source-load collaborative decarbonization.

3.1. Dynamic apportionment of bilateral carbon accountability in generation-demand nexus

In the carbon emission flow tracing framework, the carbon potential of the node is used to evaluate the carbon emission of the source side caused by the node to meet the load demand. If the system has M nodes, its carbon potential is shown in equation (1) :

$$\varepsilon_t^i = \frac{\sum_{n\in I_i} P_t^{n,ap}\rho_t^n + P_t^{i,ap}\delta_i}{\sum_{n\in I_i} P_t^{n,ap} + P_t^{i,ap}}, \quad i = 0,1,2,\ldots,M \tag{1}$$

In the equation: ε_t^i is the carbon potential of node i at time t; I_i is the set of branches of power flow into node i; $P_t^{n,ap}$ is the active power flowing through the branch n at time t; ρ_t^n is the carbon flow density of branch n at time t, and its value is the node carbon potential ε_t^j of source node j of branch n; $P_t^{i,ap}$ is the active power of the power supply connected to node i at time t; δ_i is the carbon emission intensity of the power supply connected to node i, determined by the carbon emission factor and consumption rate of the fuel.

3.2. Weight factor of source-load carbon responsibility

On this basis, it is considered to allocate the carbon emission responsibility to both sides of the source load. The types of units connected to the node include conventional thermal power, renewable energy and energy storage. $\mu \in [0,1]$ is now defined to represent the carbon emission responsibility weight factor of the units and loads connected to the node.

(1) The node is connected to the thermal power unit, and the source load is divided into half of the carbon emission responsibility.

$$\mu_t^F = \mu_t^L = \frac{1}{2} \tag{2}$$

In the equation: μ_t^F and μ_t^L are respectively the carbon responsibility weighting factors of thermal power units and loads at time t.

(2) The node is connected to the new energy unit, and the carbon emission responsibility on both sides of the source load is:

$$\mu_t^R = \begin{cases} \dfrac{P_{M,t} - P_t^{i,ap}}{P_{M,t}}, P_t^{i,ap} < P_t^{i,fh} \\ 0, \varepsilon_t^i = 0\text{或}P_t^{i,ap} \geq P_t^{i,fh} \end{cases} \tag{3}$$

$$\mu_t^L = \begin{cases} 1 - \mu_t^R, e_{i,t}^N \neq 0 \\ 0, \varepsilon_t^i = 0 \end{cases} \tag{4}$$

In the equation: μ_t^R is the carbon responsibility weight factor of new energy units at time t; $P_t^{i,fh}$ is the load of node i at time t; P_M is the sum of the active power of all incoming nodes in the power flow direction.

(3) The node is connected to the energy storage device, and the carbon emission responsibility of both source and load is as follows.

$$\mu_t^E = \begin{cases} \dfrac{P_t^{i,E}}{P_t^{i,E} + P_t^{i,fh}}, P_t^{i,E} < 0 \\ 0, P_t^{i,E} \geq 0 \end{cases} \tag{5}$$

$$\mu_t^L = \begin{cases} 1 - \mu_t^E, \varepsilon_t^i \neq 0 \\ 0, \varepsilon_t^i = 0 \end{cases} \tag{6}$$

In the equation: μ_t^E is the weight factor of carbon responsibility of energy storage equipment at time t.

3.3. Carbon emission cost on both sides of source and load

The expression of carbon emission cost on both sides of source and load is as follows:

$$C_1 = \alpha \sum_{t=1}^{T} \sum_{i \in I_L} \varepsilon_t^i \mu_t^S P_t^{i,ap} \Delta t \tag{7}$$

$$C_2 = \alpha \sum_{t=1}^{T} \sum_{i \in I_S} \varepsilon_t^i \mu_t^L P_t^{i,fh} \Delta t \tag{8}$$

In the equation: C_1 and C_2 represent the carbon emission costs of source and load respectively. α is the unit carbon price; I_L and I_S are load node and power node sets respectively; μ_t^S is the weight factor of carbon responsibility on the source side.

4. A two-sided collaborative optimization model considering carbon responsibility allocation mechanism

The game strategy to reverse the low carbon economy of a bilateral cooperation alliance that takes carbon trading into account aims to reduce the overall cost of a cooperation alliance, which can be expressed as follows:

$$F = \min(F_1 + F_2) \tag{9}$$

In the equation: F_1 is the source side operating cost; F_2 is the load side operation cost.

4.1. The source-side operating cost model

Source side operation cost includes thermal power unit operation cost, carbon emission cost, and wind abandonment penalty cost. The expression is as follows:

$$F_1 = Q_f + Q_w + C_1 \tag{10}$$

In the equation: Q_f is the operating cost of thermal power unit; Q_w is the wind curtailment penalty cost.

(1) Operating costs of thermal power units

$$Q_f = \sum_{t=1}^{T} a(P_t^f)^2 + bP_t^f + c \tag{11}$$

In the equation: T is the scheduling time in one day, and T=24 h; P_t^f is the output power of thermal power unit during t period; a, b and c are operating cost coefficients of thermal power units.

(2) Curtailment penalties cost

$$Q_w = \sum_{t=1}^{T} \gamma_1 (P_t^{w,pre} - P_t^{w,real}) \tag{12}$$

In the equation: γ_1 is the penalty coefficient of wind abandonment; $P_t^{w,pre}$ is the predicted wind power value during the period of t; $P_t^{w,real}$ indicates the actual wind power in period t.

4.2. LA operating cost model

LA operation cost includes LA power purchase cost, light abandonment penalty cost, carbon emission cost, and the expression is:

$$F_2 = Q_L + Q_o + C_2 \tag{13}$$

In the equation: Q_L is the electricity purchase cost of LA; Q_o is the penalty cost of light abandonment.

(1) LA electricity purchase cost.

$$Q_L = \sum_{t=1}^{T} \omega_t (P_{t,1} + P_{t,2} + P_{t,3} + P_{t,4}) \tag{14}$$

In the equation: ω_t is the unit price of electricity purchase in time period t; $P_{t,1}$ is the fixed load during t period; $P_{t,2}$ is the transferable load during the t period; $P_{t,3}$ is the amount of energy storage device charging during t period; $P_{t,4}$ indicates the amount of energy discharged by the energy storage device during the period t.

(2) Penalty cost of light abandonment.

$$Q_o = \sum_{t=1}^{T} \gamma_2 (P_t^{o,pre} - P_t^{o,real}) \tag{15}$$

In the equation: γ_2 is the penalty coefficient of wind abandonment; $P_t^{o,pre}$ is the predicted optical power value of the time period t; $P_t^{o,real}$ indicates the actual optical power in the t period.

4.3. Constraint condition

(1) Constraints of thermal power units.

$$\begin{cases} P_t^{f,\min} \le P_t^f \le P_t^{f,\max} \\ P_t^{f,up} \le P_t^f - P_{t-1}^f \le P_t^{f,down} \end{cases} \tag{16}$$

In the equation: $P_t^{f,\min}$ and $P_t^{f,\max}$ are the upper and lower limits of thermal power output in t period respectively. $P_t^{f,up}$ and $P_t^{f,down}$ are the uphill and downward climb rates of thermal power plants during the time period t, respectively.

(2) Power balance constraints.

$$P_t^f + P_t^w + P_t^o + P_{t,4} = P_{t,3} + P_{t,2} + P_{t,1} \tag{17}$$

(3) Constraints on energy storage devices.

$$\begin{cases} Z_t^L = Z_{t-1}^L(1-\upsilon_L) + \eta_{t,3}P_{t,3} - \dfrac{P_{t,4}}{\eta_{t,4}} \\ Z_t^{L,\min} \le Z_t^L \le Z_t^{L,\max} \\ Z_o^L = Z_a^L \\ 0 \le D_{t,3} + D_{t,4} \le 1 \\ D_{t,3}P_{t,3}^{\min} \le P_{t,3} \le D_{t,3}P_{t,3}^{\max} \\ D_{t,4}P_{t,4}^{\min} \le P_{t,4} \le D_{t,4}P_{t,4}^{\max} \end{cases} \tag{18}$$

In the equation: Z_t^L, Z_o^L and Z_a^L are the state of charge at t time, initial time and final time in the charging and discharging cycle respectively. υ_L is the self-loss rate of electric energy storage; $\eta_{t,3}$ and $\eta_{t,4}$ are the charge and discharge efficiency of t period respectively. $Z_t^{L,\min}$ and $Z_t^{L,\max}$ are the upper and lower limits of the remaining charged energy respectively. $D_{t,3}$ and $D_{t,4}$ are the charge and discharge status markers in the t period, respectively. $P_{t,3}^{\max}$, $P_{t,4}^{\max}$, $P_{t,3}^{\min}$ and $P_{t,4}^{\min}$ are the maximum and minimum values of charge and discharge power.

(4) Transferable load constraints.

$$\begin{cases} -\beta P_{t,1}^{\max} \le P_{t,2} \le \beta P_{t,1}^{\max} \\ \sum_{t=1}^{T} P_{t,2} = 0 \end{cases} \tag{19}$$

In the equation: β is the proportion of transferable load; $P_{t,1}^{\max}$ indicates the maximum fixed electrical load in period t.

(5) Wind and light output constraints.

$$\begin{cases} 0 \le P_t^{w,real} \le P_t^{w,pre} \\ 0 \le P_t^{o,real} \le P_t^{o,pre} \end{cases} \tag{20}$$

4.4. Cooperative game profit distribution strategy

Cooperative alliance means that under certain agreement constraints, all participants form a cooperative alliance, and the overall income when realizing the alliance is greater than the sum of the income when all participants act alone, and the income distributed by all participants after the alliance is greater than the income when they act alone. This paper constructs a two-sided collaborative optimization strategy considering carbon responsibility allocation. The game participants are generation scheduler and load aggregator, aiming at the lowest cost of cooperative alliance, and the cost allocation is carried out on both sides of source and load based on Shapley value allocation method. The expression of income distribution is as follows:

$$\begin{cases} H_i(v) = \sum_{x \in N} K(x)\varphi_i \\ \varphi_i = v(x) - v(\dfrac{x}{i}) \\ K(x) = \dfrac{(|x|-1)!(N-|x|)!}{N!} \end{cases} \tag{21}$$

In the equation: $H_i(v)$ is the income of member i; N is the general alliance; $v(x)$ is the revenue of sub-alliance x; $v(\frac{x}{i})$ is the revenue of the cooperative alliance of other members except member i; $K(x)$ is the distribution coefficient of member i's participation in the cooperative alliance; $(N-|x|)!$ is the arrangement of other members; $v(x)-v(\frac{x}{i})$ is the marginal contribution of member i's participation in different alliances to its own participation in the alliance.

5. Example analysis

5.1. Technical specifications setting

This study employs a regional power grid configuration from literature [14] as the empirical testbed for model validation. The computational implementation utilizes the CPLEX optimization solver to derive equilibrium solutions under constrained operational scenarios. Key system parameters are structured as follows: Table 1 details the technical specifications of thermal generation assets, including minimum stable operating levels and ramping constraints, while Table 2 enumerates network topology parameters and transmission capacity limits. Electricity market dynamics, comprising time-of-use pricing tiers and demand elasticity coefficients, are catalogued in Table 3. Temporal system behaviour is characterized by the 24-hour load profile and wind power forecast pattern illustrated in Figure 2, with renewable curtailment penalty rates set at 450 yuan/(MW·h) to reflect current market conditions.

Table 1. Thermal power unit parameters.

Units	Maximum output /MW	Minimum output /MW	Unit climbing /(MW.h^{-1})	Cost factor			Start-stop cost/yuan
				a/(yuan.MW^{-2})	b/(yuan.MW^{-1})	c/ yuan	
1	500	220	200	0.0031	119.8	6980	31350
2	430	160	200	0.0020	122.1	6550	35500
3	167	48	160	0.0281	136.3	2400	6200
4	125	25	100	0.0137	115.6	4780	3750
5	125	20	80	0.0146	115.7	4670	3800
6	85	20	80	0.0053	153.1	3300	1720

Table 2. Other parameter settings.

Parameters	Numerical value
Unit carbon price α /(yuan /t)	125
Energy storage capacity Z^L /MW	[10,250]
Initial energy storage capacity Z_o^L /MW	30
Energy storage charging and discharging efficiency $\eta_{t,3}$ and $\eta_{t,4}$	0.90, 0.90
Transferable load ratio β	0.1

Table 3. Electricity price parameter.

Time frame	Peak (10:00-13:00, 17:00-22:00)	Flat (7:00-9:00, 14:00-16:00)	Valley (23:00-6:00)
ω_t /yuan	0.75	0.50	0.20

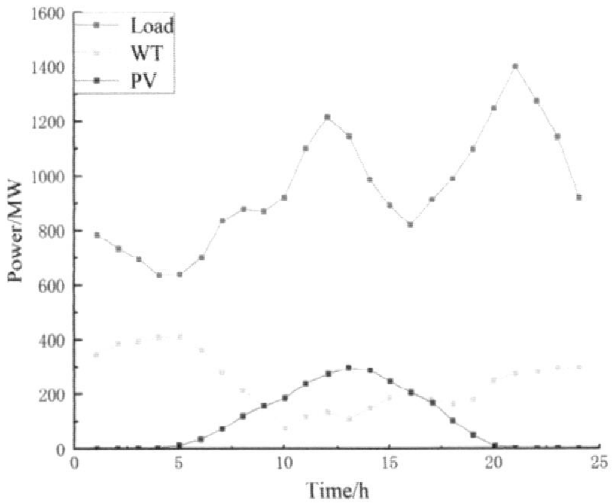

Figure 2. Load curve and wind prediction curve.

5.2. Analysis of cooperative alliance results

In order to verify the validity of the cooperative game model in this paper, two scenarios are set for comparison.

Scenario 1: Two-sided optimization of source load in a non-cooperative alliance;

Scenario 2: Source-load two-sided collaborative optimization in a cooperative alliance.

The unit output curves under the two scenarios are shown in Figure 3 and Figure 4 respectively. The source load cost pairs in each scenario are shown in Table 4.

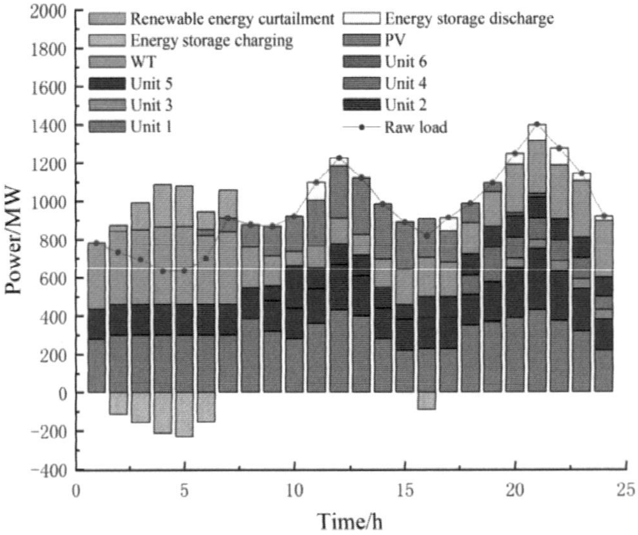

Figure 3. Scenario 1 unit output.

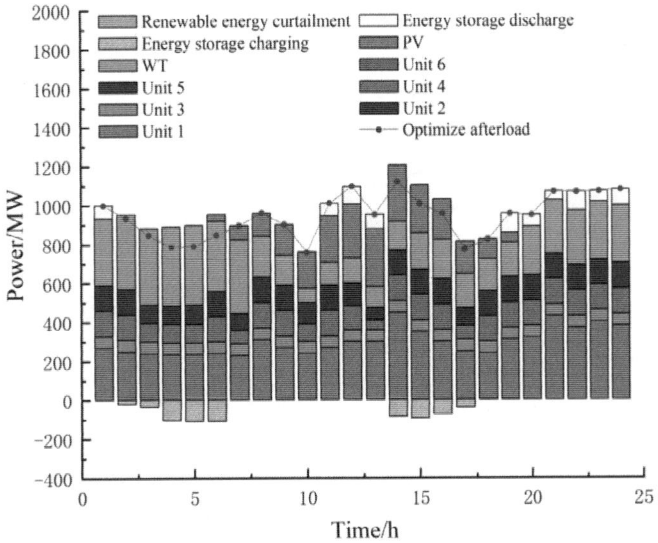

Figure 4. Scenario 2 unit output.

Table 4. Comparison of source load costs in different scenarios.

Cost	Scenario 1	Scenario 2
Power generation cost of thermal power units / 10,000 yuan	273.51	251.07
Start-up and shutdown cost of thermal power units / 10,000 yuan	30.93	0
Transferable load call cost / 10,000 yuan	0	12.92
Cost of purchasing electricity / 10,000 yuan	1401.19	1320.01
Renewable energy power abandonment cost / 10,000 yuan	50.37	0
Total cost / 10,000 yuan	1756.00	1584.00

Figure 3 illustrates the operational challenges of unilateral source-side scheduling, where the absence of load participation forces thermal units to undergo frequent start-stop cycles to address peak-valley load fluctuations, resulting in significant cycling costs and compromised system economy. Concurrently, renewable energy curtailment intensifies during low-demand periods due to insufficient flexibility. In contrast, Figure 4 demonstrates the efficacy of source-load coordination, where optimized dispatch of adjustable loads flattens the load profile and enhances renewable energy utilization. As quantified in Table 4, the cooperative framework reduces total operating costs by 9.8%, achieving complete elimination of thermal unit cycling costs and renewable curtailment expenses. This synergy establishes a cost-efficient paradigm that aligns operational flexibility with decarbonization objectives through institutionalized bilateral coordination.

5.3. Analysis of results of carbon responsibility allocation

In order to verify that the carbon responsibility allocation mechanism introduced in this paper based on weight factors can effectively improve the low carbon and economy of the system, three different mechanisms are set up for comparative analysis.

Mechanism 1: The source side assumes full carbon responsibility under the cooperative alliance;

Mechanism 2: The lower load side of the cooperative alliance bears all the carbon responsibility;

Mechanism 3: The two-sided carbon responsibility allocation mechanism is adopted under the cooperative alliance.

The comparison of the results of different mechanisms is shown in Table 5, and the optimization results of system daily carbon emission under different mechanisms are shown in Figure 5.

Table 5. Comparison of results of different mechanisms.

Mechanism	Total daily carbon emission /t	Total daily carbon emission cost / 10,000 yuan
1	84883.10	529.3861
2	82399.21	510.2439
3	70361.73	446.4023

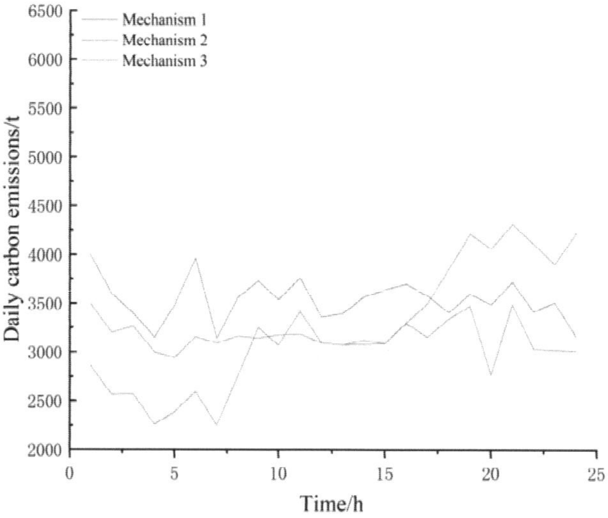

Figure 5. Optimization results of system daily carbon emission under different mechanisms.

As demonstrated in Table 5 and Figure 5, the proposed bilateral carbon responsibility allocation mechanism achieves significant improvements over conventional approaches. Mechanism 1, which exclusively assigns carbon obligations to generation units, results in elevated system emissions and carbon costs due to the lack of demand-side mitigation incentives. Mechanism 2 shifts full responsibility to consumers, reducing emissions marginally but disproportionately increasing user electricity expenditures. In contrast, the weight-factor-based Mechanism 3 equitably distributes carbon accountability between generation and demand sides through physics-informed allocation principles. Compared to Mechanism 2, this framework reduces daily carbon emissions by 14.6% and total decarbonization costs by 12.5% while maintaining balanced economic impacts. The results validate that institutionalizing bilateral carbon responsibility coordination effectively reconciles emission reduction objectives with system cost-efficiency, establishing a paradigm shift from unilateral constraints to collaborative decarbonization governance.

6. Conclusion

This study establishes a novel carbon accountability coordination framework through the synergistic integration of carbon emission flow theory and cooperative game-theoretic principles. The proposed methodology introduces dynamic allocation coefficients that equitably apportion emission responsibilities between generation and demand entities based on their physical contributions to system carbon flux. By embedding these mechanisms within a bilateral optimization architecture, the framework activates collaborative decarbonization behaviors while maintaining economic viability.

Empirical analyses confirm the approach's dual capability to significantly reduce system-wide carbon emissions and mitigation costs compared to conventional unilateral accountability paradigms. The cooperative equilibrium mechanism achieves this through institutionalized flexibility exchange that eliminates renewable curtailment inefficiencies and optimizes operational expenditure, ultimately establishing a replicable model for low-carbon power system governance through source-load symbiosis.

References

[1] Zhou, R., Sun, H., Tang, X., et al. (2018). Low-carbon economic scheduling of wind power-carbon capture virtual power plant under dual carbon constraints. Proceedings of the CSEE, 38(6), 1675-1683, 1904.
[2] Wen, J., Zhou, B., Wei, L. (2022). Preliminary study on the energy storage network of China's future power system. Power System Protection and Control, 50(7), 1-10.
[3] Chen, Z., Li, F., Zhao, X., et al. (2020). Bi-level interactive optimization scheduling considering source-load characteristics. Power System Protection and Control, 48(1), 135-141.
[4] Li, Y., Wang, C., Li, G., et al. (2021). Optimal scheduling of integrated demand response-enabled integrated energy systems with uncertain renewable generations: A Stackelberg game approach. Energy Conversion and Management, 235.
[5] Yang, D., Sheng, H., Hao, Z., et al. (2024). Optimized scheduling of integrated energy systems considering operational flexibility on both source and load sides. Journal of Electric Power, 1-16. https://kns-cnki-net.webvpn.ncepu.edu.cn/kcms/detail/12.1420.TM.20240125.1623.004.html
[6] Ren, D., Wu, J., Mao, X., et al. (2019). Integrated energy system source-load coordination operation optimization considering wind power absorption. Smart Power, 47(9), 37-44.
[7] Cui, Y., Zeng, P., Zhong, W., et al. (2021). Low-carbon economic scheduling of electric-thermal integrated energy system considering ladder-type carbon trading. Electric Power Automation Equipment, 41(3), 10-17.
[8] Qin, T., Liu, H., Wang, J., et al. (2018). Low-carbon economic scheduling of electric-thermal-gas integrated energy systems based on carbon trading. Power System Automation, 42(14), 8-13, 22.
[9] Li, Y., Zhang, N., Du, E., et al. (2022). Research on low-carbon demand response mechanism and benefit analysis of power system based on carbon emission flow. Proceedings of the CSEE, 42(08), 2830-2842.
[10] Cheng, Y. H., Zhang, N., Zhang, B. S., et al. (2020). Low-carbon operation of multiple energy systems based on energy-carbon integrated prices. IEEE Transactions on Smart Grid, 11(2), 1307-1318.
[11] Chang, L., Cao, R., Pang, L., et al. (2024). Research on source-load coordination optimization considering user-side carbon responsibility in the low-carbon context. Power Supply and Distribution, 41(05), 46-55. DOI:10.19421/j.cnki.1006-6357.2024.05.006.
[12] Zhou, T., Kang, C., Xu, Q., et al. (2012). Preliminary study on carbon emission flow analysis theory of power systems. Power System Automation, 36(07), 38-43+85.
[13] Zhou, Q., Feng, D., Xu, C., et al. (2015). Comparative study of direct allocation methods for load-side carbon emission responsibility. Power System Automation, 39(17), 153-159.
[14] Cai, H., Wang, W., Fan, T., et al. (2023). Source-load cooperative operation strategy considering load aggregator. Science Technology and Engineering, 23(04), 1565-1572.

Journal of Physics: Conference Series **3011** (2025) 012013 doi:10.1088/1742-6596/3011/1/012013

Optimization of home load scheduling considering LOLP and user dissatisfaction

Shiwei Qin [1], Hongtao Liu [2*], Yixin Lin [2]

[1] School of Electrical Engineering, Guangxi University, Nanning 530004, China
[2] National Institute of Energy Development Strategy, North China Electric Power University, Beijing 102206, China

[*] Corresponding author's e-mail: liuhongtao@ncepu.edu.cn

Abstract: With the increasing integration ratio of renewable energy sources into the grid, exploring the flexible adjustment capabilities on the demand side has become a critically important task. This paper first establishes a HEMS model from the perspective of household energy use, incorporating three types of flexible loads and electric vehicle home interaction. Next, it proposes an optimization scheduling model for the home energy management system with a comprehensive objective of system load shedding rate and user dissatisfaction. Finally, a case study of a typical household is conducted. The results validate the effectiveness and reasonableness of the proposed method.

1. Introduction

As the costs of photovoltaic and energy storage systems continue to decrease, equipping homes with these systems has become a major direction for the future of the "user-centric" power era [1] and the development of green, low-carbon power systems. Therefore, enhancing the utilization rate of home photovoltaic and energy storage systems has emerged as a hot issue. As a new type of flexibility resource in the power system, flexible loads can be adjusted to some extent from the demand side. Based on the HEMS, appropriate optimization of the scheduling plan for flexible loads is of significant importance for further enhancing the efficiency of photovoltaic and energy storage systems.

HEMS acting as a key carrier linking loads, photovoltaic systems, and energy storage systems, is the central system for optimizing flexible load scheduling [2]. Scholars at home and abroad have conducted extensive research on the optimization scheduling issues of HEMS, most of which aim to minimize electricity costs [3-5]. Jiang et al. [6] proposed a home energy management strategy based on real-time electricity pricing and peak power control, also utilizing the V2G function of electric vehicles to bring good returns to users. However, as user demands for comfort continue to rise, multi-objective optimization that considers both electricity costs and user comfort has increasingly attracted attention. Qu and Han [7] designed a model that minimizes electricity expenses linked to household appliances and a user comfort model, achieving multi-objective optimization. He and Xia [8] pointed out that the home energy management system is a key resource for future virtual power plant aggregations, considering both the economic and comfort aspects of household electricity use, and proposed a multi-objective electricity use model that integrates household user electricity load association, comfort, and satisfaction with electricity payments. A mathematical model for both rigid and flexible loads was established based on user load characteristics. A genetic algorithm was used to solve the model. However, these studies did not address the utilization efficiency of photovoltaic storage systems.

At present, scholars domestically and internationally have also conducted a lot of research in the field of residential flexible loads. Zheng et al. [9] and Liu and Wang [10] respectively established scheduling models for air conditioners and electric vehicles to reduce peak loads and estimated the corresponding peak reduction effects. Higashitani et al. [11] proposed an optimization model considering the complementary scheduling of home photovoltaic storage and electric vehicles. Liu and

Gao [12] established a dual-layer model for scheduling three types of flexible loads in LA. Sun and Liu [13] built a multi-time-scale flexible load cooperative response model, but these models did not consider the satisfaction of flexible loads.

In summary, based on the current research results, this paper first establishes a home energy management system model from the user perspective, including three types of home flexible loads and considering V2H electric vehicles. Next, a multi-objective optimization scheduling model for the home energy management system is proposed, considering the utilization efficiency of the photovoltaic storage system and the dissatisfaction of the three types of flexible loads. Finally, case studies of three different types of households with different energy use habits validate the effectiveness of the proposed model.

2. HEMS and its component models

HEMS serves as an integrated solution for the intelligent interconnection and energy scheduling of household components. Its primary operational logic is predicated on ensuring power balance and optimizing the scheduling of flexible household loads and energy storage strategies based on the input photovoltaic power generation curve. The principal system components of the HEMS include rigid loads, flexible loads, electric vehicles, photovoltaic systems, energy storage systems, and a management terminal. The system diagram is presented as shown in Figure 1.

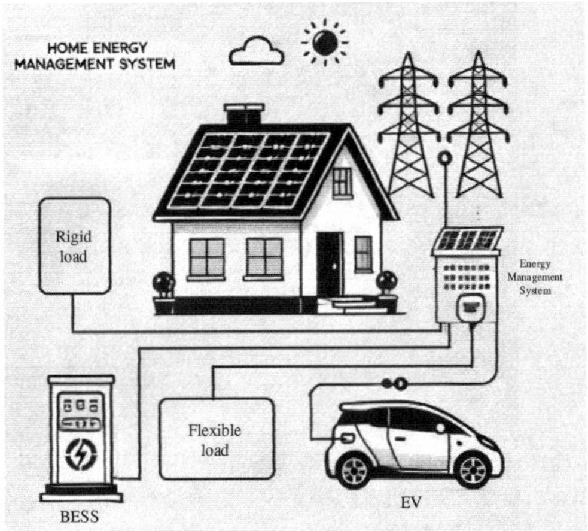

Figure 1. HEMS.

2.1. Pv model

Photovoltaic components utilize the photovoltaic effect of semiconductors to generate direct current (DC) power. The magnitude of the DC power produced is typically influenced by the irradiance, ambient temperature, and the intrinsic characteristics of the components themselves. Moreover, the DC power directly generated by the photovoltaic system requires conversion into alternating current (AC) power through an inverter for utilization by the Home Energy Management System. Consequently, the output model of the photovoltaic components can be represented by the following Equations (1)-(3).

$$P_{\mathrm{PV,DC}} = P_{\mathrm{STC}} k_{\mathrm{pv}} \left(\frac{G(t)}{G_{\mathrm{STC}}} \right) \left[1 + \alpha_{\mathrm{pv}} \left(T(t) - T_{\mathrm{STC}} \right) \right] \tag{1}$$

$$T(t) = T_{\mathrm{a}}(t) + G(t) \left(\frac{T_{\mathrm{oct}} - 20}{0.8} \right) \tag{2}$$

$$P_{\mathrm{pv}}(t) = \eta_{\mathrm{DC,AC}} \cdot P_{\mathrm{pv,DC}}(t) \tag{3}$$

In the equations, $P_{\mathrm{PV,\,DC}}(t)$ denotes DC power generated by the PV in current; $G(t)$ represents the irradiance intensity in current; $T(t)$ indicates the temperature of the photovoltaic components during in current; $T_a(t)$ is the ambient temperature in current; $P_{\mathrm{pv}}(t)$ refers to the actual AC power that can be output. The terms $P_{\mathrm{STC}}, k_{\mathrm{pv}}, G_{\mathrm{STC}}, \alpha_{\mathrm{pv}}, T_{\mathrm{STC}}, T_{\mathrm{oct}},$ and $\eta_{\mathrm{DC,\,AC}}$ respectively correspond to the power generation capacity of the photovoltaic system under rated conditions, the derating factor of the photovoltaic system, the irradiance intensity under rated conditions, the temperature coefficient, the operating temperature under rated conditions, the standard operating temperature of the photovoltaic components, and the efficiency coefficient of the photovoltaic DC inverter.

2.2. Bess model

Bess serves as a valuable tool for promoting the consumption of renewable energy and for peak shaving and valley filling. It also assists in further reducing the comprehensive energy costs for users. On one hand, it can enhance the utilization rate of renewable energy by storing excess electricity produced by the photovoltaic system; on the other hand, it can improve the operational logic of flexible loads, supporting the enhancement of user satisfaction with flexible loads. The amount of electricity stored in the energy storage system can be represented by the following Equation (4).

$$\mathrm{S}_{ess}(t)=\begin{cases}(1-\beta)\mathrm{S}_{ess}(t-1)+P_{\mathrm{soc}}(t)\Delta t\eta_{\mathrm{c}}\\(1-\beta)\mathrm{S}_{ess}(t-1)+P_{\mathrm{soc}}(t)\Delta t\eta_{\mathrm{d}}\end{cases}\tag{4}$$

In the equations, $S_{ess}(t)$ denotes the amount of electricity stored in Bess at the current period; $S_{ess}(t-1)$ refers to the amount of electricity stored in the previous period; $P_{\mathrm{soc}}(t)$ represents the charging or discharging power during the current period; β stands for the self-discharge coefficient of the Bess; η_c and η_d are the charging and discharging loss coefficients of the Bess.

2.3. Flexible load model

Flexible loads are typically classified according to their response characteristics into categories such as shiftable, interruptible, and reducible. However, such classification merely differentiates based on electrical characteristics and does not align well with home living scenarios. Therefore, this paper categorizes flexible loads into three types based on the actual characteristics of household appliances, namely time-sensitive flexible loads, electricity consumption-sensitive flexible loads, and temperature-sensitive flexible loads.

2.3.1. Time-sensitive flexible load. Time-sensitive flexible loads primarily include appliances such as washing machines and rice cookers, which are sensitive to the timing of electricity usage. These loads primarily exhibit shiftable characteristics, meaning their operating times can be adjusted, yet they require continuous power consumption within a certain time frame. The focus for users regarding these loads is on their start or stop times. These loads can be represented by the following Equations (5)-(8).

$$P_{time}(t)=G_{time}(t)\cdot P_{time,st}\tag{5}$$

$$t_{time}^{\min}\le t_{time}^{start}\le(t_{time}^{\max}-n_{time})\tag{6}$$

$$t_{time}^{end}=t_{time}^{start}+n_{time}\tag{7}$$

$$G_{time}(t)=\begin{cases}0,\ \ t\le t_{time}^{start}\bigcup t_{time}^{end}\le t\\1,\ \ t_{time}^{start}\le t\le t_{time}^{end}\end{cases}\tag{8}$$

In the equations, $P_{time}(t)$ represents the power of time-sensitive flexible load in current; $G_{time}(t)$ is the symbol of it; $P_{time,st}$ is its power rating; t_{time}^{start}, t_{time}^{end}, t_{time}^{min}, and t_{time}^{max} represent the start time, shutdown time, earliest start time and latest shutdown time of this type of flexible load, respectively; n_{time} represents the number of hours of work it requires.

2.3.2. Consumption-sensitive flexible load. Electricity consumption-sensitive flexible loads primarily include appliances such as electric bicycles, mobile phones, and portable computers, which are sensitive to the amount of electricity used. These types of flexible loads mainly exhibit interruptible and reducible characteristics, meaning they require a certain amount of electricity within a designated time. The focus for users regarding these loads is on the actual amount of electricity they consume. These loads can be represented by the following Equations (9)-(12).

$$S_{quantity} = \sum \Delta t \cdot P_{quantity}(t), t \in t_{quantity} \tag{9}$$

$$P_{quantity}(t) = 0, t \notin t_{quantity} \tag{10}$$

$$0 \le P_{quantity}(t) \le P_{quantity}^{cst} \tag{11}$$

$$S_{quantity}^{\min} \le S_{quantity} \le S_{quantity}^{\max} \tag{12}$$

In the equations, $S_{quantity}$ represents the total electricity consumption of this flexible load; $P_{quantity}(t)$ represents its power in current; $t_{quantity}$ represents the period of time when electricity is allowed for this type of load; $P_{quantity}^{cst}$ represents the maximum electricity power allowed for this type of load; $S_{quantity}^{min}$ and $S_{quantity}^{max}$ represents the lower limit and upper limit of power allowed for this type of load.

2.3.3. Temperature-sensitive flexible load. Temperature-sensitive flexible loads primarily include appliances such as air conditioners, electric heaters, and water heaters, which are sensitive to temperature variations. These types of flexible loads exhibit interruptible and reducible characteristics, meaning they require a certain amount of electricity within a specified period to maintain a specified temperature range. The focus for users regarding these loads is on the actual temperature maintained. These loads utilize thermodynamic equivalent models, which in this paper can be represented by the following Equations (13)-(15).

$$T_{\text{in}}(t+1) = \mathrm{e}^{-\Delta t/RC} T_{\text{in}}(t) + R(\mathrm{e}^{-\Delta t/RC} - 1)\eta P_{temperary}(t) + (1 - \mathrm{e}^{-\Delta t/RC}) T_{\text{out}}(t) \tag{13}$$

$$T_{set}^{low} \le T_{\text{in}}(t+1) \le T_{set}^{high}, t \in (1,23) \tag{14}$$

$$0 \le P_{temperary}(t) \le P_{temperary}^{\max} \tag{15}$$

In the equations, $T_{\text{in}}(t)$ *and* $T_{\text{out}}(t)$ represent the temperature of indoor and out door; R *and* C are equivalent thermal resistance and equivalent heat capacity; $P_{temperary,i}(t)$ is the power of AC in current; T_{set}^{low}, T_{set}^{high}, and $P_{temperary,i}^{max}$ are the set minimum temperature, maximum temperature and maximum power, respectively.

2.4. EV model

Electric vehicles are currently the most common form of incremental load in China. Due to their equipped energy storage batteries, electric vehicles can not only act as loads when connected to the system but can also serve as storage devices participating in vehicle-to-home and vehicle-to-grid interactions. This paper divides the characteristics of electric vehicles into two phases: During the phase when their battery charge does not meet travel demands, they act as electricity consumption-sensitive flexible loads; once their charge level meets travel demands, they function as part of the energy storage system within the grid. The model for electric vehicles in this paper is presented as Equations (16)-(17).

$$E_{\text{car}}(t) = \begin{cases} (1-\beta)E_{car}(t-1) + G(t)P_{car}(t)\Delta t\eta_1, 0 \le P_{car,i}(t) \\ (1-\beta)E_{car}(t-1) + G(t)P_{car}(t)\Delta t\eta_2, P_{car,i}(t) < 0 \end{cases} \tag{16}$$

$$G_{car}(t) = \begin{cases} 0, & \text{Leave home} \\ 1, & \text{At home} \end{cases} \tag{17}$$

In the equations, $E_{car}(t)$ is the battery status of EV; $P_{car}(t)$ is the charge and discharge power of EV in current; β, η_1, and η_2 represent the self-loss coefficient, charging loss coefficient, and discharge loss coefficient of the EV battery.

3. Operation optimization model

3.1. Evaluation index

The optimization of scheduling for the HEMS needs to be considered from both economic and comfort perspectives. From an economic standpoint, the current residential electricity pricing in China adopts a tiered pricing model. The assessment criteria should emphasize minimizing the net electricity purchased from the grid while maximizing the utilization efficiency of the home photovoltaic and storage system. From the perspective of comfort, the characterization of electrical comfort mainly reflects the deviation between the actual electricity usage plan and the expected plan. The following sections construct the evaluation system based on the load shedding rate of the photovoltaic and storage system and the dissatisfaction of flexible loads.

3.1.1. LOLP. The photovoltaic and storage system must be able to meet household electricity demands as much as possible while maintaining a certain level of reliability and high utilization efficiency. Therefore, the reliability and utilization efficiency of the photovoltaic and storage system can be assessed and characterized by the load-shedding rate. The system load-shedding rate can be calculated using the following Equation (18) [14].

$$F_{lolp} = \frac{\sum_{t=1}^{24}(P_{RL}(t)+P_{FL}(t)+P_{car}(t)+P_{soc}(t)-P_{pv}(t))\cdot\Delta t}{\sum_{t=1}^{24}(P_{RL}(t)+P_{FL}(t))\cdot\Delta t+E_{car}^{set}}\times 100\% \tag{18}$$

In the equations, $P_{RL}(t)$ is the total rigid loads; $P_{FL}(t)$ is the total flexible loads.

3.1.2. Dissatisfaction with flexible load. Flexible loads possess a certain degree of adjustability, yet they still fall within the category of loads; thus, their satisfaction with energy use should not be overlooked. For the three types of flexible loads proposed in this paper, the total dissatisfaction can be calculated using the following Equation (19).

$$F_{fls} = \frac{\sum_{i=1}^{n_{time}}\frac{\left|t_{time,i}^{start}-t_{time,i}^{set}\right|}{t_{time,i}^{\max}-n_i-t_{time,i}^{\min}}}{n_{time}} + \frac{\sum_{i=1}^{n_{quantity}}\frac{\left|S_{quantity,i}-S_{quantity,i}^{set}\right|}{S_{quantity,i}^{\max}-S_{quantity,i}^{\min}}}{n_{quantity}} + \frac{\left|\overline{T}_{in}-T_{in}^{set}\right|}{T_{set}^{high}-T_{set}^{low}} \tag{19}$$

In the equations, n_{time} is the number of time-sensitive load; $n_{quantity}$ is the number of consumption-sensitive load.

3.2. Optimization objective

In the day-ahead scheduling optimization of the Home Energy Management System, it is necessary to consider the impact of different optimization objectives on the optimization results comprehensively. This paper employs a weighted sum method to transform the multi-objective optimization problem into a single-objective optimization problem for solution. Therefore, the objective function of this paper can be seen in the following Equation (20).

$$\min J = \lambda_1 F_{lolp} + \lambda_2 F_{fls} \tag{20}$$

In the equations, λ_1 and λ_2 are coefficient of weight.

3.3. Constraints

3.3.1. Bess constraints. The operation of the energy storage system is limited by the maximum charging and discharging power constraints. Its state of charge should also remain within a certain range. Therefore, the operational constraints of the energy storage system can be expressed as follows Equations (21)-(22).

$$P_{soc,\min} \le P_{soc}(t) \le P_{soc,\max} \tag{21}$$

$$0.2N_{ess} \leq S_{ess}(t) \leq 0.9N_{ess} \tag{22}$$

3.3.2. EV constraints. When electric vehicles transition to function as part of the energy storage system, it is also necessary to set an upper limit for their state of charge and to ensure that the state of charge of the electric vehicles is at a set level upon completion of a cycle. The constraints can be expressed as follows Equations (23)-(24).

$$0.4N_{car} \leq E_{car}(t) \leq 0.8N_{car} \tag{23}$$

$$-3.3kW \leq P_{car}(t) \leq 7kW \tag{24}$$

3.3.3. Power balance constraint

$$\begin{aligned} 0 = P_{\text{pv}}(t) - P_{soc}(t) + P_{grid}(t) - P_{time}(t) \\ - P_{quantity}(t) - P_{temperary}(t) - P_{car}(t) - P_{RL}(t) \end{aligned} \tag{25}$$

3.4. Solution method

In the objective function of this paper, the system load-shedding rate is a nonlinear objective. Therefore, it requires linearization. This paper assumes that the optimization results of flexible load scheduling have minimal impact on electricity consumption; hence, the denominator of the load shedding rate index, which is the total electricity consumption, is replaced with the original electricity plan to linearize the objective function. Finally, the model is solved using the Gurobi commercial solver in Matlab.

4. Case study

4.1. Main parameters

In the case study of this paper, the capacity of the photovoltaic system is set at 6kW, with the energy storage battery having a rated capacity of 3kWh. The charging and discharging efficiency are both set at 0.85, with maximum charging and discharging powers set at 2kW and 1.8kW respectively. The initial charge of the energy storage battery is set at 30%, with the minimum and maximum states of charge set at 20% and 90%, respectively. The electric vehicle battery has a rated capacity of 20kWh, with both charging and discharging efficiencies set at 0.8, and maximum charging and discharging powers set at 7kW and 3.3kW, respectively. The initial charge of the electric vehicle is set at 40%, with permissible maximum and minimum states of charge set at 40% and 80%, respectively. The predicted output power of the photovoltaic source for each time period and the power consumption of the rigid loads in the household for each time period are shown in Figure 2 below. The parameters for flexible loads are shown in the Table1-Table3 below.

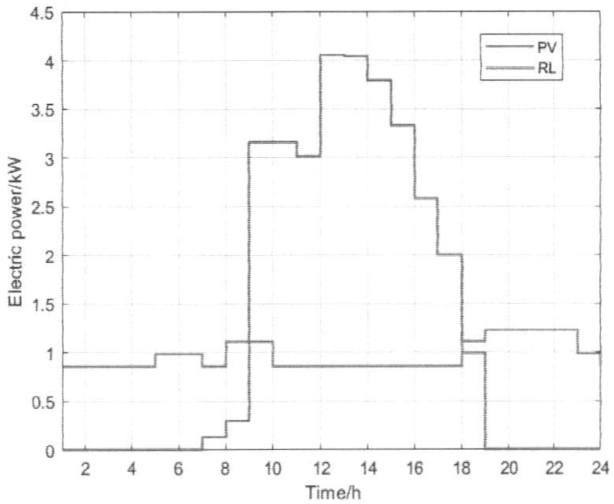

Figure 2. Pv and rigid load power.

Table 1. Time-sensitive flexible load parameters.

Device Name	Rated power rating/kW	Allowed working hours	Duration/hour	Ideal time
Washing machine	0.8	8-17	1	9
Rice cooker	1	9-12, 16-19	1	11, 18
Floor sweeping robot	1.4	8-17	1	9
Dishwasher	0.73	19-22	1	19

Table 2. Consumption-sensitive flexible load parameters.

Device Name	Maximum power/kW	Allowed working hours	Allowable range/kWh	Ideal consumption/kWh
Electric bicycle	0.18	1-7, 11-14, 18-24	0.6-0.96	0.8
Mobile devices	0.1	1-7, 12-14, 18-24	0.2-0.4	0.3
Water heater	1.8	1-24	2-6	4

Table 3. Temperature-sensitive flexible load parameters.

Device Name	Maximum power/kW	Allowed working hours	Allowable temperature/°C	Ideal temperature/°C
Air conditioner	2	1-24	22-27	24

4.2. Results and analysis

Optimization results are presented in Figure 3. Figure 3(a) displays the operational results of the optimized flexible loads. It is evident that the peak electricity demand of flexible loads has been significantly shifted, smoothing the unmanaged peak demand from 18:00 to 20:00 and redistributing it to periods when photovoltaic output is higher at noon and during the low demand periods in the early morning and late night.

Figure 3(b) shows the energy storage behavior of the energy storage system and electric vehicles. It can be observed that the internal storage of the energy storage system and electric vehicles has shifted loads to some extent, enhancing the utilization rate of the electricity generated by the photovoltaics. Additionally, the electric vehicles are set to leave home for short commutes between 7:00-12:00 and 14:00-18:00, with the electricity consumed during commuting being expended in one go upon departure.

Figure 3(c) provides a comparison of household and grid interaction power under three scenarios: without any scheduling optimization, with flexible load scheduling optimization not considering vehicle-to-home (V2H) integration, and optimization according to the method proposed in this paper. It is found that when flexible loads participate in scheduling optimization, they can reduce the household's evening peak electricity power demand by 33.8%, decrease the net electricity purchased from the grid by 6.4 kWh, and lower the peak-to-valley difference in grid interaction power by 18.7%. However, when considering the integration of electric vehicles into the home energy management system as proposed in this paper, the addition of the internal storage system of electric vehicles further shifts the load from the evening peak period to the early morning low demand period. Moreover, the inclusion of electric vehicles also reduces the maximum power fed back to the grid, benefiting grid frequency safety.

Journal of Physics: Conference Series **3011** (2025) 012013 doi:10.1088/1742-6596/3011/1/012013

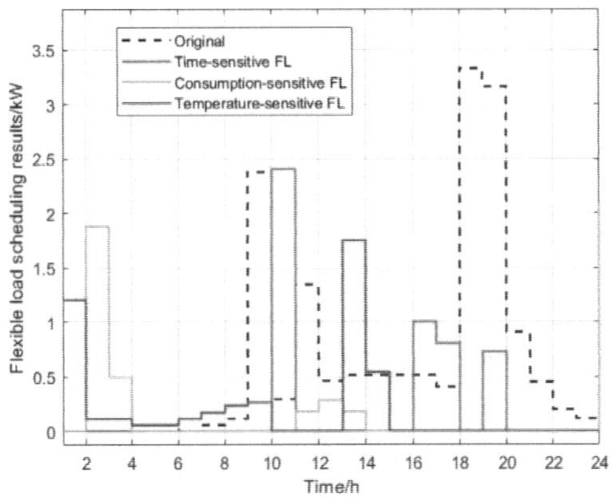

(a)

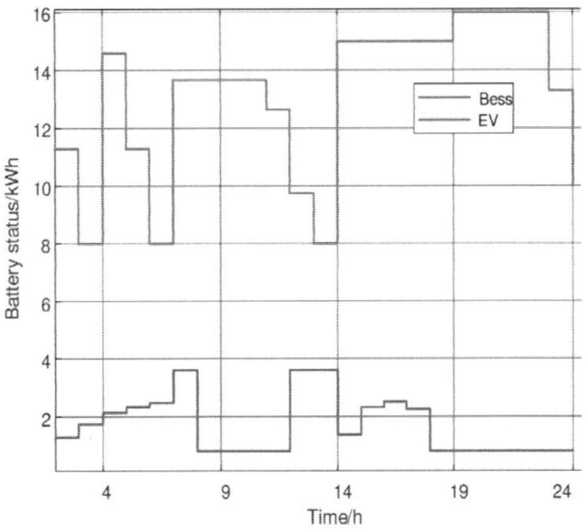

(b)

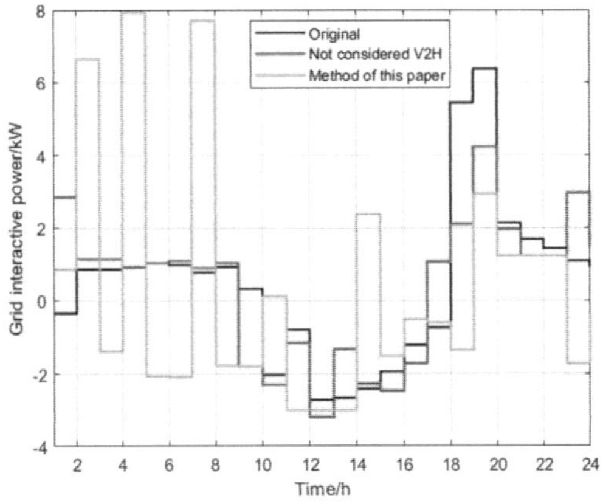

(c)

Figure 3. Scheduling optimization results.

5. Conclusion

This paper establishes a framework under the Home Energy Management System from the perspective of household electricity usage, constructs models for three types of flexible loads, and presents their respective dissatisfaction calculation methods. Based on this, a method for the optimized scheduling of flexible loads and electric vehicle vehicle-to-home (V2H) integration is proposed, considering both system load shedding rate and user dissatisfaction. Case study validation shows that this method can effectively reduce the household load peaks by 33.8%, smooth out fluctuations in electricity power usage, enhance the utilization efficiency of the household photovoltaic system, and reduce the amount of electricity purchased from the grid by 6.4 kWh, effectively alleviating the peak-valley pressures on the grid.

Acknowledgments

This work was financially supported by the National Key Research and Development Program Intergovernmental Projects (2022YFE0129400), the China National Natural Science Foundation (No. 52306007), and the National Grid Central Science and Technology Project (5108-202457041A-1-1-ZN).

Journal of Physics: Conference Series **3011** (2025) 012013 doi:10.1088/1742-6596/3011/1/012013

References

[1] Wang P, Wang D R. (2022) Power Industry Enters "User-centered Era". China Electric power Enterprise management, (07): 46-47.
[2] Zhang Y Y, Zeng P. (2016) Optimal scheduling algorithm for home energy management system in smart grid environment. Power system protection and control, 44(02): 18-26.
[3] Zhang L, Liu Q L. (2023) Household load optimization scheduling strategy based on improved binary particle swarm optimization. China Electric Power, 56(05): 118-128.
[4] Chavali P, Yang P and Nehorai A. (2014) A Distributed Algorithm of Appliance Scheduling for Home Energy Management System. IEEE Transactions on Smart Grid, vol. 5, no. 1, pp: 282-290.
[5] Liu X F, Peng L S. (2024) Based on model predictive control of family energy management optimization scheduling method research. Electric measurement and instrument, 21(10): 26-32.
[6] Jiang X K, Liu C R, Tao Y B. (2024) A home energy management optimization strategy based on user commuter behavior. Journal of Harbin university of science and technology, 29(01): 50-61.
[7] Qu C Y, Han J. (2018) Multi-objective Optimization Model of electricity Consumption Behavior Considering Association and Comfort of Household Appliances. Automation of Power Systems, 42(02): 50-57.
[8] He W G, Xia J R. (2020) Power consumption optimization strategy considering user payment satisfaction and power consumption comfort. Power Demand Side Management, 22(01): 69-74.
[9] Zheng Z Y, Xiong J, Xu Y K. (2024) Consider comfort air conditioning load demand response of cluster control method. Automation and instrumentation, (11): 150-153.
[10] Liu Z M, Wang Y. (2022) Research on Two-stage optimal Scheduling Strategy of peak and valley filling in distribution Network considering EV and BESS. Journal of power supply, http://kns.cnki.net/kcms/detail/12.1420.TM.20220526.1523.004.html.
[11] Higashitani T, Ikegami T, Uemichi A. (2021) Evaluation of residential power supply by photovoltaics and electric vehicles, Renewable Energy, Volume178: 745-756.
[12] Liu X F, Gao B T. (2017) Residential load hierarchical dispatching model based on non-cooperative game. Automation of Electric Power Systems, 41(14): 51-60.
[13] Sun Y, Liu C L. (2019) Multi-time scale demand response cooperation strategy for residential user groups. Power Grid Technology, 43(11): 4170-4177.
[14] Liu H T, Zhai R R. (2023) Multi-objective optimisation of a thermal-storage PV-CSP-wind hybrid power system in three operation modes. Energy, Volume 284.

Journal of Physics: Conference Series **3011** (2025) 012014 doi:10.1088/1742-6596/3011/1/012014

Fast resonant frequency tracking strategy of linear oscillatory motor based on gradient descent method

Wei Huang *, Xiang Li , Baoxu Shi

School of Electrical Engineering, Guangxi University, Nanning, China

*Corresponding author's e-mail: 2212392045@st.gxu.edu.cn

Abstract. Recently, scholars have conducted many studies on the linear oscillatory motor, which is the core driver of linear compressors due to the LOM's simple mechanical structure and high efficiency of energy conversion. According to its special frequency system characteristics, the compressor system can reach maximum efficiency when the motor works at the system's resonant frequency. In this paper, to achieve resonant frequency tracking and ensure a quick response speed, a gradient descent method with discretizing the displacement and speed of the motor is raised. The simulation results verify the rapidity and effectiveness of the proposed method.

1. Introduction

The traditional compressor core driver is driven by the rotating motor, but the piston of the cylinder is reciprocating. To transfer the rotating motion into linear motion, the crankshaft connecting rod is required. Due to the increased mechanical structure, the compressor will cause greater energy loss and noise, and the required space is also larger. The research of linear oscillatory motors [1] has solved the above problems, but for the frequency operating characteristics of the motor, the working efficiency is the highest when the compressor works at a special frequency, which is the compressor resonant frequency. However, as the drive motor of the compressor, the resonant frequency will change with the load change. To guarantee the high efficiency of the motor and compressor, it is greatly necessary to track and make the LOM work at the compressor system resonant frequency.

Given the special frequency characteristics of linear oscillating motors, many scholars have studied the frequency tracking control of LOM. Although the zero-crossing detection [2] of displacement and current signals is easy to implement, this method is sensitive to noise. To avoid phase detection, the average value of the stroke-current product (ASCP) method is also applied in resonant frequency tracking [3]. However, the tracking performance is unstable when the resonant frequency changes, which is caused by the load change. Model reference adaptive system (MRAS) and adaptive full-order state observer (AFO), which are widely used in rotating motors, are used for parameter identification to get the resonant frequency. Besides, a phasor model with orthogonal current [4] is applied in frequency control. However, this method is slow in converging to load changes. In this paper, a gradient descent [5] method that can realize parameter identification and frequency tracking is displayed to solve the problem that the compressor internal mass cannot be accurately obtained and the resonant frequency is constantly changing due to load changes. Simulation can prove the effectiveness and fast convergence of the gradient descent method.

2. LOM model

The LOM model discussed in this paper is consistent with traditional motor structure. The equivalent circuit model and simplified kinematics model are shown in Figure 1.

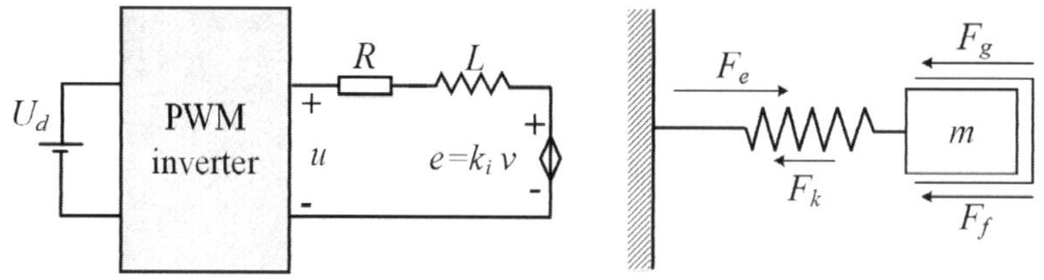

Figure 1. LOM working system. (a) Equivalent circuit. (b) Mechanical model.

$$u = iR + \dot{i}L + e \tag{1}$$

$$F_e = F_f + F_g + F_k + ma \tag{2}$$

where R is the resistance of LOM, L is the inductance, U_d is the DC source of the inverter, $e=k_i v$ is the back electromotive force (EMF), and the LOM can be regarded as a resistance, inductor, and back EMF in series. a is the acceleration of the mover, $F_e=k_i i$ is the thrust generated by the motor, k_i is the electromagnetic thrust coefficient, $F_k=k_s x$ is the spring force of the motor, k_s is the elastic coefficient of the spring, m is the mass of the mover, F_f is the frictional resistance, which can be regarded as $F_f=c_f v$. F_g is the force produced by compressed gas. To simplify the system, the gas force should be linearized. The nonlinear gas force can be displayed as:

$$F_g = k_g x + c_g v \tag{3}$$

where k_g and c_g are the equivalent coefficients of gas elasticity and damping, so Equation (3) can be rewritten as:

$$k_{ei} i = (k_s + k_g)x + (c_f + c_g)v + ma \tag{4}$$

which can be simplified as:

$$k_{ei} i = k_L x + c_L v + ma \tag{5}$$

where $k_L=k_s+k_g$ is the spring coefficient of the compressor and $c_L=c_f+c_g$ is a damping factor of LOM. The equivalent circuit and kinetic can be written in the frequency domain as:

$$\begin{cases} \dot{U} = \dot{I}(R + j\omega L) + k_{ei}\dot{V} \\ \dot{F}_e = k_{ei}\dot{I} = \left[c_L + j\left(\omega L - \dfrac{k_L}{\omega} \right) \right] \dot{V} \end{cases} \tag{6}$$

The power input and output can be written as:

$$\begin{cases} P_{input} = \dot{U}\dot{I} = \left[R + \dfrac{k_{ei} c_L}{c_L^2 - (\omega m - \dfrac{k_L}{\omega})} \right] I^2 \\ P_{output} = \dfrac{k_{ei} c_L}{c_L^2 - (\omega m - \dfrac{k_L}{\omega})} I^2 \end{cases} \tag{7}$$

The compressor working efficiency is:

$$\delta = \frac{P_{input}}{P_{output}} = \frac{k_{ei} c_L}{k_{ei}^2 + R\left[c^2 - \left(\omega m - \dfrac{k_L}{w} \right)^2 \right]} \tag{8}$$

It can be found from Equation (8) that when $\omega m - k_L/m$ is 0 and $\omega^2=k_L/m$, the LOM can get the highest efficiency. Therefore, the resonant frequency tracking method is necessary for LOM.

3. Gradient descent track method

The gradient descent method has a wide range of applications to the solution of the optimal solution of the system, which means that the reference estimate of the model is gradually modified along the negative gradient direction of the objective function until the optimal reference value of the system is reached, and the optimal measure is that the criterion function reaches the minimum value. For the specific system, it can be described as:

$$D(m) = A^T(m)B \tag{9}$$

The parameters are assumed as:

$$\hat{B}(m) = \hat{B}(m-1) + \mu A(m) \tag{10}$$

The proper μ can make $\hat{D}(m)=D(m)$, so that:

$$D(m) = A^T(m)\hat{B}(m) = A^T(m)\hat{B}(m-1) + \mu A^T(m)A(m) \tag{11}$$

The μ can be written as:

$$\mu = \frac{1}{A^T(m)A(m)}\left[D(m) - A^T(m)\hat{B}(m-1)\right] \tag{12}$$

So we can obtain:

$$\hat{B}(m) = \hat{B}(m-1) + \frac{A(m)}{A^T(m)A(m)}\left[D(m) - A^T(m)\hat{B}(m-1)\right] \tag{13}$$

To make sure the denominator is not 0, it can be converted to:

$$\hat{B}(m) = \hat{B}(m-1) + \frac{qA(m)}{p + A^T(m)A(m)}\left[D(m) - A^T(m)\hat{B}(m-1)\right] \tag{14}$$

where $p>0$, $0<q<2$.

In LOM, the parameters m, k_L, and c_L are hard to get directly. These parameters are all in the kinetic model. Besides, the other parameters such as displacement signal x can reached from the sensor. Therefore, to apply the gradient descent in LOM, we could set A $= [k_{ei}i, -x, -v]$, B=$[1/m, k_L/m, c_L/m]^T$, and D=$\dot{v}$. As for $\dot{v}$,

$$\dot{v} = \frac{v(m) - v(m-1)}{\Delta T} \tag{15}$$

where ΔT is the sampling time. The final parameters can be described as:

$$\hat{m} = \frac{1}{B(1)}, k_L = \frac{B(2)}{B(1)}, c_L = \frac{B(3)}{B(1)} \tag{16}$$

The resonant frequency is reached: $f_r = \sqrt{B(2)}/2\pi$. Resonant frequency tracking means parameter identification of k_L/m. By obtaining the value of k_L/m, the resonant frequency f can be obtained.

4. Simulation verification results

To prove the effectiveness and rapidity of the gradient descent frequency track algorithm, the LOM control system is constructed as Figure 2.

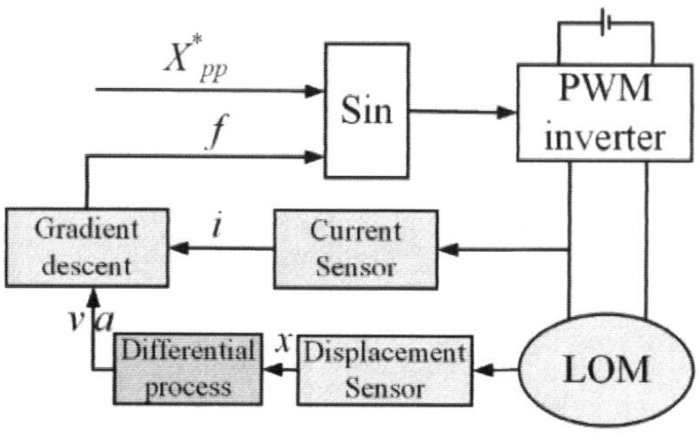

Figure 2. Structure of gradient descent.

The LOM parameters [6] are set as follows:

Table 1. LOM parameters.

Parameter	Value	Unit
m	0.91	kg
L	0.02	mH
R	1	Ω
k_{ei}	18	N/A
k_L	30200	N/m
c_L	10	N/m^2
p	0.01	
q	1.5	
ΔT	1e-6	s

All the parameters in Table 1 of LOM are tested and obtained by static and dynamic experiments. The initial resonant frequency is 29 Hz, which can be obtained from the table and $\frac{\sqrt{k_L/m}}{2\pi}$, to observe the response speed of the frequency tracking method to load changes. At 4 s, a load which is k_g=20000 and c_g=4 is added to the LOM system, which is equivalent to the increase of air pressure when the cylinder is pumped in the compressor. For the gradient descent parameters needed, the motor current and displacement signals can be reached by different sensors.

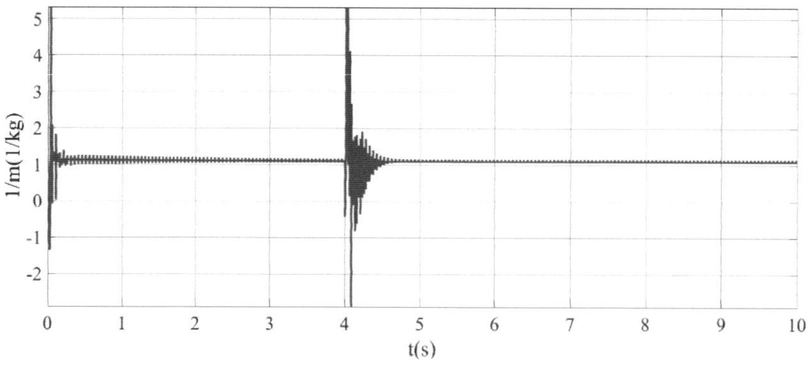

Figure 3. Simulation result of 1/m.

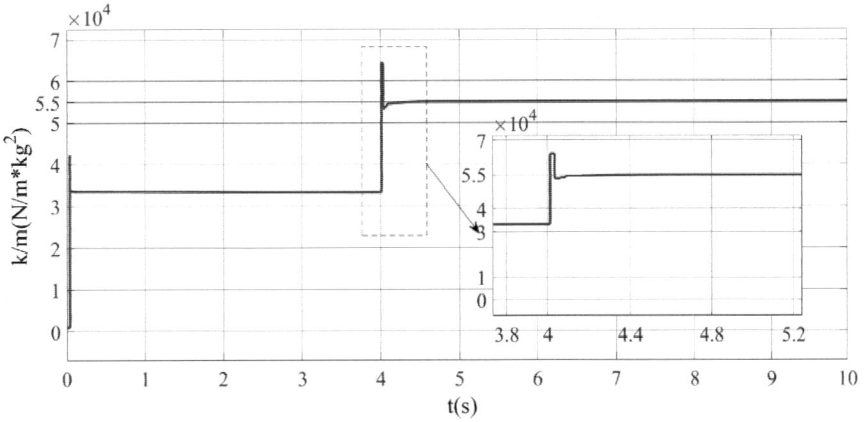

Figure 4. Simulation result of k_L/m.

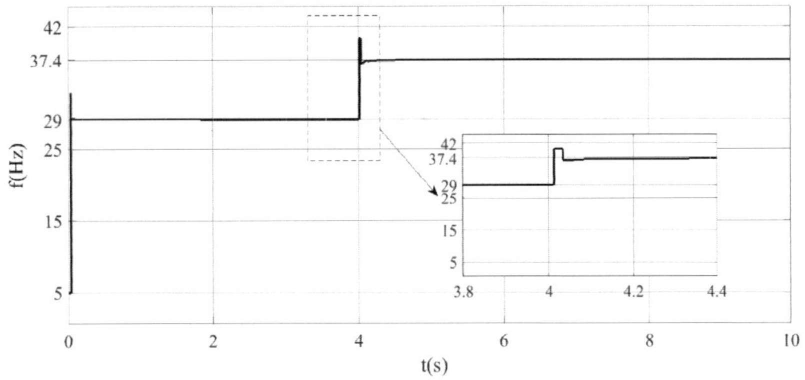

Figure 5. Simulation result of f.

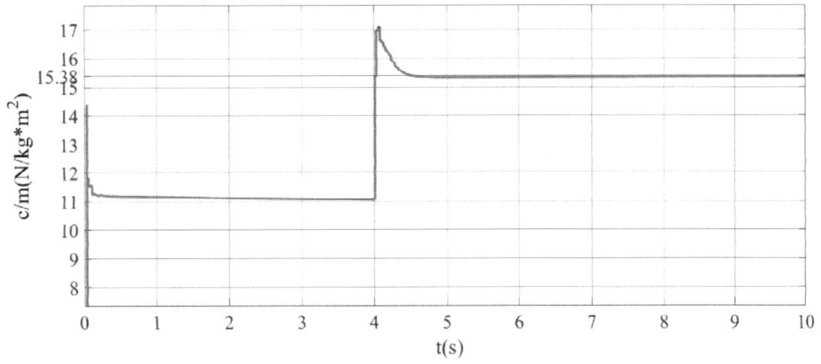

Figure 6. Simulation result of c_L/m.

It can be seen from Figure 4 and Figure 6 that the initial setting parameters can be identified as k_L/m=33200, c_L/m=11. At 4 s, the load with k_g=20000 and c_g=4 is added to the LOM system, and at 0.15 s, k_L and c_L are tracked. In Figure 5, the frequency can be adjusted at the resonant frequency, which is 29 Hz before 4 s and 37.4 Hz after 4 s. In Figure 3, the mover mass remains unchanged, which is the same as the setting. From Figure 4 and Figure 5, parameters k_L/m and f reach stability in 0.1 s and the accuracy of parameters is reliable. The results show that the gradient descent can track the system resonant frequency at a quick response speed.

5. Conclusion

A gradient descent method for tracking the compressor system resonant frequency is presented in this paper. Firstly, the working characteristics of the compressor frequency are discussed and the significance of system resonant frequency is understood. Secondly, the working principle of the gradient descent method is analyzed, and specific transformation is realized according to the LOM equation. At last, the availability and fast convergence of the gradient descent are shown by the results of the simulation.

Acknowledgments

This work was financially supported in part by National Natural Science Foundation of China (No. 52407044), in part by Guangxi Natural Science Foundation (No. 2022GXNSFBA035622), in part by Guangxi Science and Technology Plan Project (No. AD23026124), in part by Guangxi Science and Technology Major Program (No. AA23023012), in part by Guangxi Science and Technology Major Program (No. AA23073019), and in part by Guangxi Key Research and Development Plan (No. AB23049006).

References

[1] H. Lee, S. Ki, J. Kang, et al., "Noise characteristics of linear compressor for refrigerators," in Proceeding of 15th International Conference on Sound and Vibration, Daejeon, Korea, pp. 2714-2719, 2008.
[2] C. Demoulias and K. Gouramanis, "Voltage multiple-zero-crossings at buses feeding large triac-controlled loads," IEEE Trans. Ind. Electron., vol. 54, no. 5, pp. 2853-2863, Oct. 2007.
[3] B. Yang, "Linear compressor control circuit to control frequency based on the piston position of the linear compressor," U.S. Patent 5947693, Sep. 7, 1999.
[4] W. Xu et al., "Fast Efficiency Optimization Control Based on Orthogonal Current Phasor Model for Linear Oscillatory Motor," in IEEE Transactions on Power Electronics, vol. 39, no. 9, pp. 11504-11516, Sept. 2024, doi: 10.1109/TPEL.2024.3412739.
[5] X. Yang and X. Zheng, "Gradient Descent Algorithm-Based Adaptive NN Control Design for an Induction Motor," in IEEE Transactions on Systems, Man, and Cybernetics: Systems, vol. 51, no. 2, pp. 1027-1034, Feb. 2021, doi: 10.1109/TSMC.2019.2894661.
[6] X. Li, M. Jiang, W. Xu, T. Wu and D. Zhang, "Design and test of a less-rare-earth stator-magnet moving-iron transverse-flux linear oscillatory machine for reciprocating drive," IEEE Trans. Ind. Electron., vol. 71, doi: 10.1109/TIE.2023.3269472. no. 3, pp. 2831-2841, March 2024.

Aanalysis of the feasibility of broadband oscillation suppression for PMSG integration into weak grid based on virtual synchronous generator control

Lijian Chen [1], Hongtao Wang [1*], Haobo Wang [1] and Zhen Song [1]

[1] Key Laboratory of Power System Intelligent Dispatch and Control of Ministry of Education, Shandong University, Shandong, Jinan, 250000, China

[*] Corresponding author's e-mail: whtwhm@sdu.edu.cn

Abstract: With the significant growth in installed wind power capacity, the penetration of power electronics has increased substantially. A notable effect of this heightened integration of power electronics into the power system is a decrease in the system's overall inertia, leading to a deterioration in grid stability. Consequently, these alterations pose fresh obstacles to maintaining frequency stability within power systems. However, the ability of wind turbines based on grid-following converters to support weak grids is limited. They fail to meet the operational stability requirements for high shares of renewable energy in the grid. Virtual synchronous generator (VSG)-type converter can enhance system inertia damping while independently regulating AC voltage and frequency, thus providing active support for the system in weak grids with a strong frequency dynamic response capability. In this paper, impedance models of permanent magnet synchronous generators (PMSGs) are developed using an impedance modeling approach. Simulation results validate that the control strategy in this paper offers remarkable stability support for the system.

1. Introduction

Nowadays, with green energy transforming rapidly in the power industry, new energy sources, particularly wind and photovoltaic energy, have experienced rapid development. Wind power and photovoltaic generation have been widely utilized. Up till 2021, the utilized capacity of wind power and photovoltaic power in China has reached 328 GW and 307 GW [1]. China's wind power resources and load centers are geographically mismatched, with wind power generation often located far from high-load areas. As a result, the power generated from renewable energy sources must be transmitted over long distances to meet demand. This long-distance transmission can lead to a weakening of grid intensity after integration into the power grid [2]. Wind turbines, which are attached to the grid with the help of numerous power electronic units, produce intermittent and volatile output. Consequently, large-scale integration of these turbines connecting to the grid can reduce the system's rotational inertia, degrade its anti-interference capability, and even threaten the safety and stability of the power system [3-4].

With the increasing scale of power electronic devices, broadband oscillation happens frequently. It is essential to implement grid-forming wind turbines in weak grids to ensure the safe and stable operation of these grids. Grid-forming turbines offer better adaptability and stability in new energy grids characterized by low grid intensity.

Conventional grid-following wind turbine converters typically operate with maximum power point tracking (MPPT) as the control objective and do not participate in regulating system frequency or voltage. In contrast, grid-forming converters, based on principles similar to synchronous generators, can autonomously generate reference amplitude and phase for the AC voltage [5].

In this paper, a frequency-domain impedance modeling approach for wind turbines is applied. This method divides the new energy grid-connected system into two independently operating subsystems.

Impedance models are developed for each subsystem separately. The wind turbine is modeled as a parallel combination of an ideal current source and an equivalent impedance, while the grid subsystem is modeled as a series combination of an ideal voltage source and an equivalent impedance [6]. This paper focuses on the modeling of the wind turbine section. The equivalent circuit diagram of a wind turbine connected to the grid is constructed using the impedance method, as shown in Figure 1.

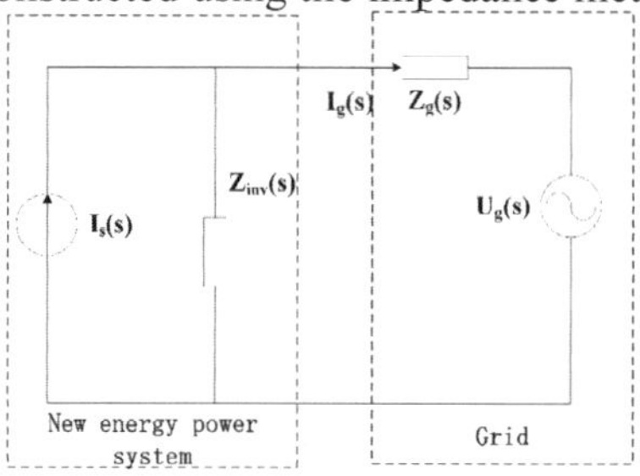

Figure 1. Equivalent circuit diagram of new energy grid-connected system.

Specifically, this paper adopts a sequential impedance-based modeling approach. A small-signal perturbation (voltage or current) is superimposed on the steady-state condition of the system, followed by using the Fourier transform to unfold the nonlinear dynamics. The system's response to the perturbation is then calculated. The ratio of the response to the perturbation is defined as the impedance at the perturbation frequency. The stable factor of the grid-connected system is presented using the Bode diagrams of the sequence impedances of the wind turbine, as well as the grid impedance. A simulation circumstance of a grid-connected PMSG is built in Simulink. A time-domain simulation is run to prove the validity of the conclusions drawn from the theoretical stability analysis.

2. Modeling of permanent magnet synchronous wind turbine

The structure of the permanent magnet synchronous wind turbine is mainly composed of the power system, PMSG, converter, and filter circuit. The wind turbine blades in the power system convert wind energy into mechanical energy. The drive unit drives the PMSG to convert mechanical energy into electrical energy. The converter includes the machine-side converter (MSC), DC capacitor, and grid-side converter (GSC), where the DC capacitor is used to stabilize the DC voltage and ensure the normal operation of the converters on both sides.

The conventional PMSG modeling of the grid-following type operating based on the MPPT strategy has been widely studied in the field. The VSG control-based PMSG presented in this paper utilizes a control strategy for the machine-side converter as the same as the grid-following type. The difference mainly lies in the VSG control for the grid-side converter. Due to the isolation effect of the DC capacitor, the control of the machine-side and grid-side converters can be studied separately, so this paper simplifies its modeling by using a DC constant voltage source for equivalent treatment of the turbine and the machine-side converter.

3. Modeling of grid-side converter

3.1. Mathematical modeling of VSG

The VSG discussed in this paper is a voltage-controlled VSG, which is based on the rotor equations of a conventional synchronous motor and is capable of simulating the dynamic characteristics of a conventional synchronous motor to provide inertia and damping to the grid [7-8].

The control strategy of VSG is divided into active and reactive links. A block diagram is shown in Figure 2 to present the control strategy of voltage-controlled VSG.

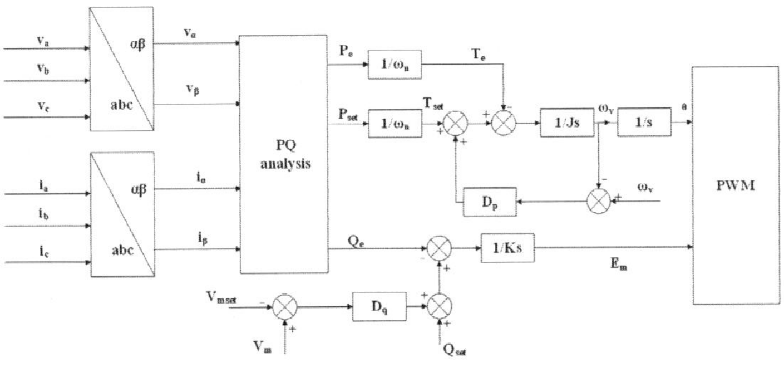

Figure 2. Block diagram of voltage-controlled VSG.

The active link simulates the mechanical characteristics of the traditional synchronous generator so that the VSG has similar adjustment characteristics. The reactive control link simulates the electromagnetic characteristics so that the VSG has primary voltage adjustment characteristics [7]. The mathematical equations of active and reactive links are:

$$T_{\text{set}} + (\omega_1 - \omega)D_p - T_e = Js\omega \tag{1}$$

$$T_e = \frac{P_e}{\omega} \approx \frac{P_e}{\omega_1} \tag{2}$$

$$T_{\text{set}} = \frac{P_{\text{set}}}{\omega} \approx \frac{P_{\text{set}}}{\omega_1} \tag{3}$$

$$E_m(s) = \frac{Q_{\text{set}} - Q_e + D_q(V_{\text{mset}} - V_m)}{Ks} \tag{4}$$

T_{set} and T_e show the reference and actual values of electromagnetic torque; D_p and D_q present the active and reactive damping coefficients; J shows the virtual rotational inertia; ω and ω_1 refer to the output angular frequency of the VSG besides the rated angular frequency of the power grid; θ gives the phase of the electric potential in the VSG; K indicates the reactive inertia coefficient; V_m and V_{mset} are the amplitude of the output voltage and its reference value; and E_m refer to the RMS value of the electric potential in the VSG; P_{set} and Q_{set} are reference values; Q_e and P_e are instantaneous output reactive and active power.

Using the Clark transform and according to the instantaneous power principle, one can calculate the output reactive power Q_e and active power P_e of the VSG [8]:

$$P_e = \frac{3}{2}(v_\alpha i_\alpha + v_\beta i_\beta) \tag{5}$$

$$Q_e = \frac{3}{2}(v_\beta i_\alpha - v_\alpha i_\beta) \tag{6}$$

v_α, v_β, i_α, i_β represent the output voltages and currents of the VSG in the $\alpha\beta$ coordinate system.

3.2. Discussion on sequential impedance modeling of VSG control-based PMSG

The active power expression of the VSG in the frequency domain is given in Equation (7), where V_1, V_p and V_n represent the amplitudes of the fundamental voltage, positive-sequence, and negative-sequence disturbance voltages; accordingly, I_1, I_p and I_n denote the amplitudes of the fundamental current, positive, and negative-sequence disturbance currents. f_1, f_p and f_n correspond to the fundamental frequency, the positive, and negative-sequence disturbance frequencies. "*" stands for conjugate plural.

$$P_e[f] = \begin{matrix} 3(V_1^* I_p + V_p I_1^*) & f = \pm(f_p - f_1) \\ 3(V_1 I_n + V_n I_1) & f = \pm(f_n + f_1) \\ 3(V_p I_n + V_n I_p) & f = \pm(f_p + f_n) \end{matrix} \tag{7}$$

According to the control structure shown in Figure 3, the expression for θ in the figure can be obtained as shown in Equation (8).

$$\theta = M(s)\left(\omega_n D_p + \frac{P_{set}}{\omega_n} - \frac{P_c}{\omega_n}\right) \tag{8}$$

where $M(s) = 1/(Js^2 + D_p s)$.

The Laplace transform of Equation (8) is followed by substituting Equation (7) into (8) and based on $\theta = \theta_1 + \Delta\theta$. The $\cos\theta$ frequency-domain expression for the sequence voltage-current representations can be obtained:

$$\cos\theta[f] = \begin{cases} -\frac{1}{2} e^{\pm j\varphi_{vi}} M(s \mp j2\pi f_1)\left(\frac{3(V_1^* I_p + V_p I_1^*)}{\omega_n}\right), & f = \pm f_p \\ -\frac{1}{2} e^{\mp j\varphi_{vi}} M(s \pm j2\pi f_1)\left(\frac{3(V_1 I_n + V_n I_1)}{\omega_n}\right), & f = \pm f_n \end{cases} \tag{9}$$

The voltage magnitude of the VSG modulating wave, and the relationship between the output voltage, current, and internal potential are shown in Equations (9) and (10):

$$\begin{cases} e_{am} = \sqrt{2} E_m \cos\theta \\ e_{bm} = \sqrt{2} E_m \cos\left(\theta - \frac{2\pi}{3}\right) \\ e_{cm} = \sqrt{2} E_m \cos\left(\theta + \frac{2\pi}{3}\right) \end{cases} \tag{10}$$

$$L_f \frac{d}{dt}\begin{bmatrix} i_a \\ i_b \\ i_c \end{bmatrix} = \begin{bmatrix} e_a \\ e_b \\ e_c \end{bmatrix} - \begin{bmatrix} v_a \\ v_b \\ v_c \end{bmatrix} \tag{11}$$

Without considering the reactive power loop and control delay in the VSG control link, the amplitude of the potential in the VSG is considered as the steady state value. The positive sequence impedance of the VSG can be analyzed by combining Equations (9), (10) and (11). According to the correspondence between positive and negative sequence impedance, the negative sequence impedance of VSG can be obtained as:

$$Z_{VSGn}(s) = Z_{VSGp}^{*}(-s) \tag{12}$$

where $Z_{VSGp}(s)$ is the positive sequence impedance of VSG.

4. Simulation verification

In this section, the grid-connected stability of the grid-forming (GFM) PMSG introduced in this paper and the traditional grid-following (GFL) wind turbine are compared, thus this paper establishes the models of the wind turbine connected to the grid via series compensation link under two control strategies in Simulink, respectively. This model assumes that at t = 2s, the impedance of the transmission line changes abruptly, causing the system to oscillate. It involves changing the degree of series capacitor compensation K_c and comparing the stability of PMSGs with two control strategies under different operating conditions.

We set K_c to 0.4. After running the simulation, time-domain waveforms of T_e and FFT (Fast Fourier Transform) spectra of grid-connected currents are measured. The simulation results of wind turbine are as shown in Figure 3-6:

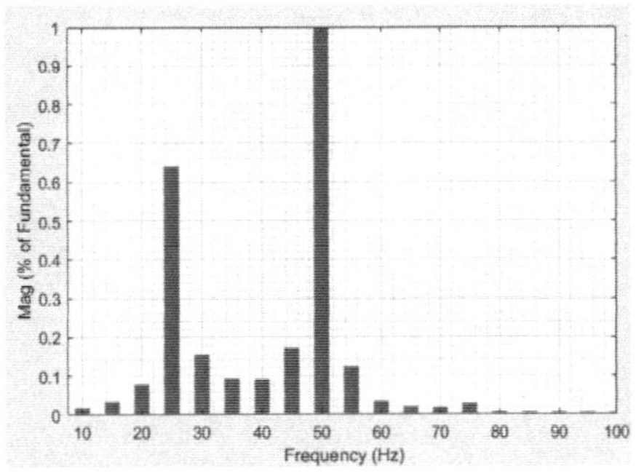

Figure 3. Time-domain waveforms of electromagnetic torque when $K_c = 0.4$ for GFL wind turbines.

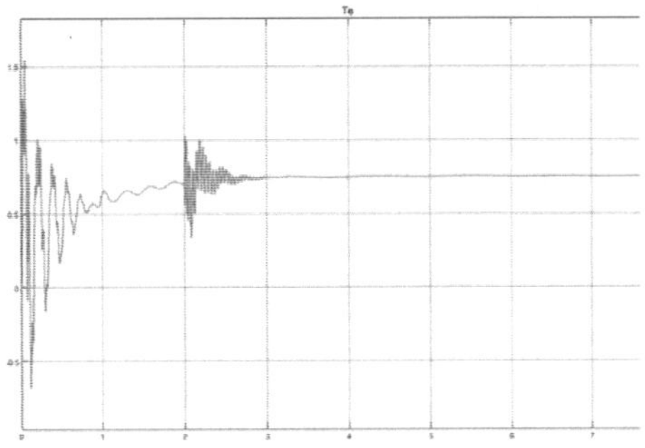

Figure 4. FFT spectra of output current of GFL wind turbines.

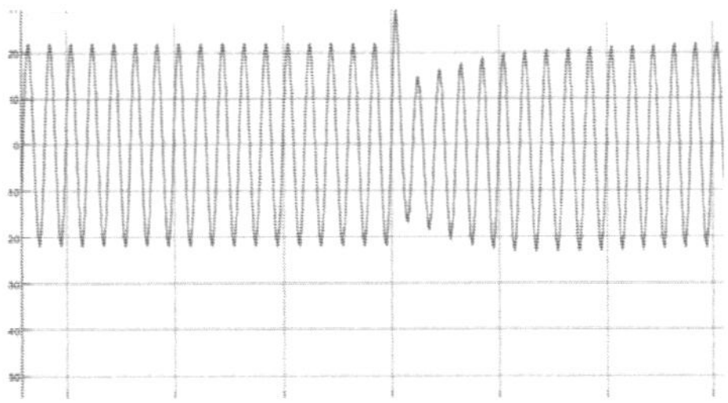

Figure 5. Time-domain waveforms of output current when $K_c = 0.4$ for GFM wind turbines.

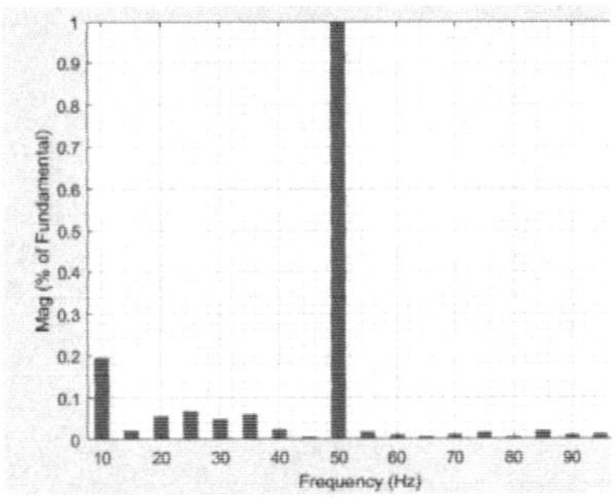

Figure 6. FFT spectra of output current of GFM wind turbines.

When $K_c = 0.4$, after the fault occurs, T_e waveform corresponding to the GFL wind turbine has a large degree of dispersion. Then at about 2.8s, T_e returns to normal; the output current resonance frequency and harmonic proportion are large, which shows that the system stability is low and easy to oscillate; for the GFM wind turbine, T_e waveform quickly maintains stability after the fault occurs. The output current resonance frequency and harmonic proportion are small. System stability is high.

Setting K_c to 0.3, the simulation results are as shown in Figure 7-10:

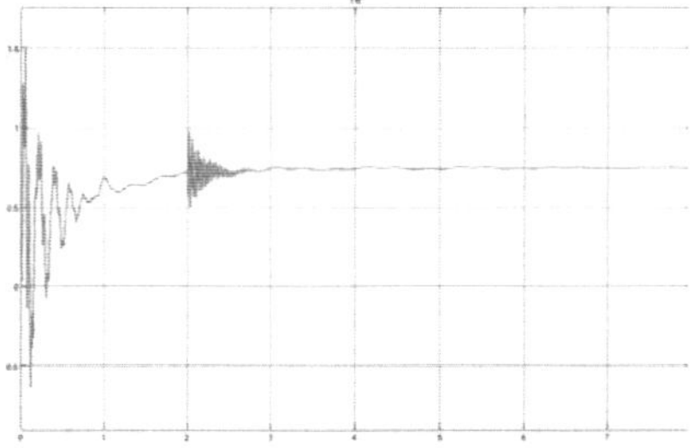

Figure 7. Time-domain waveforms of electromagnetic torque when $K_c = 0.3$ for GFL wind turbines.

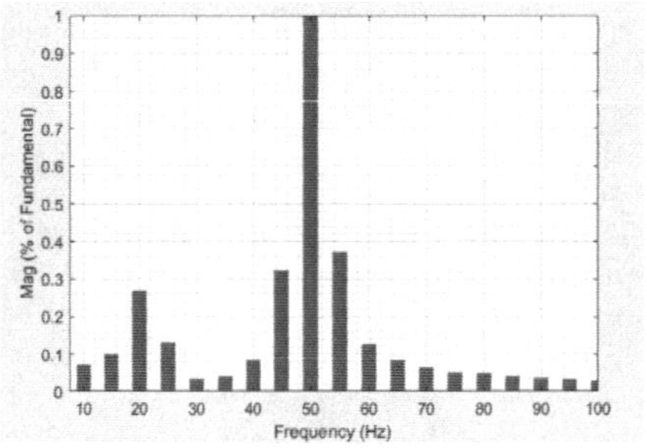

Figure 8. FFT spectra of output current of GFL wind turbines.

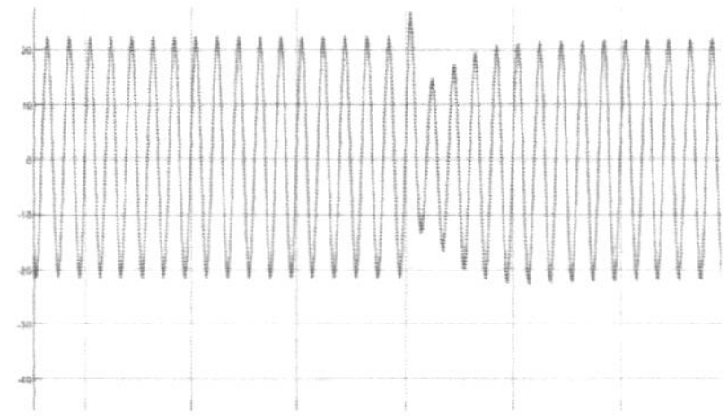

Figure 9. Time-domain waveforms of GFM output current when $K_c = 0.3$ for GFM wind turbines.

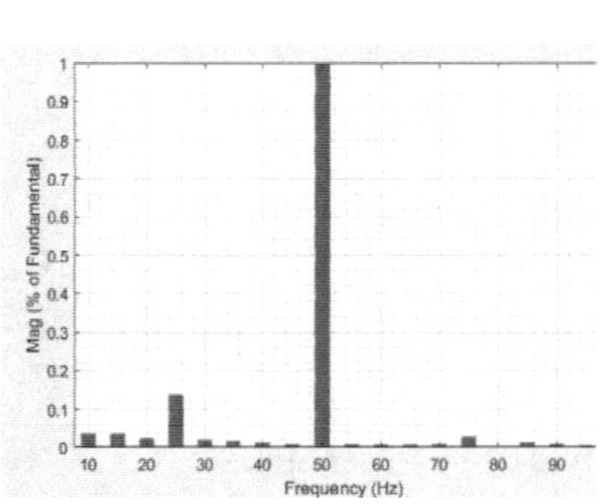

Figure 10. FFT spectra of output current of the output current for GFM wind turbines.

When $K_c = 0.3$, after the fault occurs, the T_e waveform corresponding to the GFL wind turbine still has a large degree of dispersion, but it is significantly weaker than the case when $K_c = 0.4$. Then T_e is restored to be stable at about 2.5s. There is still a certain proportion of the output current that contains harmonics at several frequencies, which is prone to oscillations; for the GFM wind turbine, T_e waveforms are rapidly maintained stable after the fault occurs. The output current harmonics and the proportion of the harmonics are small.

5. Conclusion

This paper explores the feasibility and effectiveness of integrating VSG-controlled PMSG wind turbines into weak grids to suppress broadband oscillation in weak grids. First, it is pointed out that the increasing proportion of power electronic components in modern power systems has made it difficult for wind turbines to maintain grid stability with traditional GFL control strategies alone. In this context, GFM control wind turbines offer a significant improvement. Second, a mathematical model of the PMSG is developed, followed by the introduction of a simplified modeling method for VSG-controlled PMSGs. This method focuses on the impedance modeling of the VSG as the control mechanism for the GSC. The paper also analyzes the impedance model using the Bode diagram and concludes that this approach demonstrates a certain degree of adaptability to maintaining stability in grid-connected systems.

To verify the theoretical findings presented in this paper, a simulation model is constructed in Simulink to simulate a sudden change in line impedance due to a fault. By applying this fault and comparing the time-domain waveforms of electromagnetic torque and the FFT spectra of the output currents for two turbines, the results show that the grid-forming turbine can rapidly restore the system to a stable state. This confirms that VSG-controlled wind turbines are a viable solution for enhancing the stability of weak power grids.

Acknowledgments

This work was financially supported by the National Natural Science Foundation of China under Grant (U22B2099).

References

[1] General Institute of Electric Power Planning and Design, China Electric Power Development Report 2022 [M], Beijing: People's Daily Press, 2022.

[2] SILWAL S, KARIMI-GHARTEMANI M, KARIMI H, et al. A multivariable controller in synchronous frame integrating phase-locked loop to enhance performance of three-phase grid-connected inverters in weak grids [J]. IEEE Transactions on Power Electronics, 2022, 37(9): 10348-10359.

[3] YANG Peng, LIU Feng, JIANG Qirong, et al. Stability of large disturbances in "double-high" power systems: problems, challenges and prospects [J]. Journal of Tsinghua University (Natural Science Edition), 2021, 61(5): 403-414.

[4] Xu Zheng. Three technical challenges of high ratio non-synchronous machine power grid [J]. Southern Power Grid Technology, 2020, 14(2): 1-9.

[5] Rathnayake D B, Akrami M, Phurailatpam C, et al. Grid forming inverter modeling, control, and applications [J]. IEEE Access, 2021, 9: 114781-114807.

[6] WANG Weisheng, LI Guanghui, HE Guoqing. Analysis and suppression of broadband oscillation in new energy grid-connected system [M]. Beijing: China Electric Power Press, 2022.

[7] WU W H, ZHOU L M, CHEN Y D, et al. Sequence-impedance-based stability comparison between VSGs and traditional grid-connected inverters [J]. IEEE Transactions on Power Electronics, 2019, 34(1): 46-52.

[8] WU Wenhua, CHEN Yandong, ZHOU Leming, et al. Sequential impedance modeling and stability analysis of virtual synchronous generator connected to weak grid [J]. Proceedings of the CSEE, 2019, 39(6): 1560-1571.

Journal of Physics: Conference Series **3011** (2025) 012016 doi:10.1088/1742-6596/3011/1/012016

Research on the optimal allocation of demand side energy system considering voltage sag mitigation

Yalin Chen [1,*], Peng Li [1], Yang Li [1], Yongwen Yang [2], Xihui Ding [1]

[1] Senior Engineer, China Electric Equipment Group Science and Technology Research Institute Co., Ltd, Shanghai, China
[2] Associate Professor, College of Energy and Mechanical Engineering, Shanghai University of Electric Power, Shanghai, China

[*] Corresponding author's e-mail: chenyalin@cee-group.cn

Abstract. Among the user complaints caused by power quality, the proportion of transient power quality problems exceeds 80%. As one of the main problems that need to be addressed, voltage sag is the focus of this paper. This study considers energy storage site selection and capacity optimization configuration for voltage sag compensation on a medium timescale. Firstly, we choose supercapacitor energy storage as the treatment equipment. Subsequently, the optimization objective is to minimize the economic cost of energy storage devices and the economic losses caused by voltage dips. Finally, the improved genetic algorithm is used to solve the simulation example. By comparing the voltage sag magnitudes of nodes and the economic losses of sensitive nodes before and after optimization, the results show that 94% of node voltages are within the normal range (0.93 p.u. - 1.07 p.u.). The system recovers 953, 000 yuan in economic losses annually. This study improves the operational stability and economy of regional integrated energy systems while reducing the cost of addressing power quality issues, and has high reference value.

1. Introduction

Power quality was first put forward by the United States Department of the Navy in the 1960s. To conduct better research on power quality, the Institute of Electrical and Electronics Engineers (IEEE) has divided it into transient issues and steady-state issues. Transient issues include voltage sags, voltage swells, impulses, and short-duration voltage interruptions, while steady-state issues cover voltage fluctuations, voltage flicker, voltage deviation, harmonics, and three-phase unbalance [1].

At present, the research on power quality management at home and abroad mainly follows two lines of thought: the control strategies for power quality management equipment and the optimization strategies for power grid operation control [2]. Among them, the mainstream research direction of the control strategies for power quality management equipment is to install common management equipment such as APF (Active Power Filter), DVR (Dynamic Voltage Restorer), and STATCOM (Static Synchronous Compensator), and then design corresponding control strategies [3].

According to statistics from foreign power departments, over 80% of user complaints are caused by transient issues such as voltage dips, which reveals the limitations of traditional steady-state management techniques. In this context, energy storage systems (ESS) are widely regarded as a key means of suppressing voltage dips due to their millisecond-level response capability and dynamic voltage compensation characteristics. However, the optimization of their configuration strategies still faces the challenge of balancing scenario adaptability and economy [4].

Regarding the optimization configuration of ESS in distribution networks, existing research shows significant scene dependence and methodological differences. Landl and Kirchsteiger [5] proposed a linear programming-based energy storage capacity allocation model to address the voltage fluctuations caused by a high proportion of photovoltaic access. Their case studies have shown that this approach can reduce the photovoltaic curtailment rate by 15%. However, this study only considers a single new energy scenario and does not address the dynamic load characteristics under multi-disturbance coupling, which may limit its generality [6]. In contrast, Wu et al. [7] used dynamic inertia weight design to improve convergence speed but did not quantify the impact of load transients on voltage recovery time.

At the level of control strategy and business model [8], Yu et al. [9] proposed a hierarchical control strategy based on ESS remaining capacity, but its strategy overly relies on preset capacity thresholds and does not consider the impact of real-time electricity prices and load priorities on control logic. Liao et al. [10] constructed a two-stage energy storage investment framework by using a Stackelberg game model. The case analysis showed that this model can reduce investors' risk premium by 12%, but its assumptions depend on ideal game information symmetry, which may overestimate the actual benefits of the model.

In response to the shortcomings of existing energy storage configuration research in dynamic scenario adaptability, algorithm global optimization capability, and economic collaborative regulation, this research establishes an optimization model for energy storage location and capacity, uses an improved genetic algorithm to solve it, and verifies the effectiveness and economy of optimal allocation in voltage sag management by comparison.

2. Optimization model for capacity and location of energy storage equipment

2.1. Objective function

When optimizing the capacity and location of energy storage equipment, the economic cost of the energy storage system should be considered first. The economic cost of energy storage equipment f includes the initial investment cost C_{ess}, the operation and maintenance cost C_{om}, and the voltage sag detection cost C_{dt}.

$$f = C_{ess} + C_{om} + C_{dt} \tag{1}$$

$$C_{ess} = P \times \mu_{P,ess} + E \times \mu_{E,ess} \tag{2}$$

$$C_{om} = \sum_{n=1}^{N} \frac{P\times\mu_{P,om}+E\times\mu_{E,om}}{(r+1)^n} \tag{3}$$

where P represents the rated power of the energy storage equipment, E stands for the rated capacity of the energy storage equipment, $\mu_{P,om}$ refers to the annual operation and maintenance cost per unit power of the energy storage equipment, and $\mu_{E,om}$ indicates the annual operation and maintenance cost per unit capacity of the energy storage equipment.

To detect the problem of voltage sag and enable the energy storage equipment to conduct voltage compensation within an extremely short period, voltage sag detection instruments should be installed at the installation locations of the energy storage equipment simultaneously. Such cost is defined as C_{dt}.

$$C_{dt} = \sum_{i=1}^{n} C_{i,dt} \tag{4}$$

where n represents the total number of installed energy storage equipment and $C_{i,dt}$ is the voltage sag detection cost corresponding to the installation location of the *i*-th energy storage equipment.

Meanwhile, the economic losses caused by voltage sags are also one of the important objectives of this optimization. The economic losses at the load-side nodes can reflect the governance effectiveness. The better the governance effectiveness is, the smaller the economic losses caused by voltage sags on the load side will be. The annual economic loss function of voltage sags for all nodes is defined as f_{loss}, and its specific meaning is as follows:

$$f_{loss} = \sum_{i=1}^{N} C_{loss} \times N_{sag} \times P_{loss} \tag{5}$$

where N represents the number of nodes with loads of sensitive equipment, C_{loss} is the average economic loss caused by a single voltage sag event, N_{sag} is the number of occurrences of voltage sags within one year, and P_{loss} is the probability of economic losses caused by voltage sags.

The overall objective function is set as:

$$minF_2 = (\alpha_2 f + \beta_2 f_{loss}) \tag{6}$$

where F_2 represents the total objective function, while α_2 is the weight corresponding to the economic cost of energy storage, with a value of 0.5. β_2 is the weight corresponding to the annual economic loss of voltage sags in the system, taking a value of 0.5.

2.2. Constraint conditions

Constraints on the Capacity Allocation of Energy Storage Equipment:

For the nodes equipped with energy storage, the number of configured energy storage units cannot exceed the upper limit allowed by the system.

$$0 \le N_{ESS} \le N_{ESS}^{max} \tag{7}$$

where N_{ESS} represents the number of installed energy storage units and N_{ESS}^{max} represents the upper limit of the number of installed energy storage units.

Constraints on the Operation of Energy Storage Equipment:

There are several constraints when the energy storage equipment adjusts the voltage through charging and discharging, such as the energy-balance constraint of the energy-storage system, the charging and discharging state of the energy-storage system, and the State-of-Charge (SOC, the remaining power) constraint of the energy-storage system. The specific constraints are as follows:

$$\begin{cases} \sum_{t=1}^{T} P_{ESS} \times \Delta t = 0 \\ P_{ESS}^{min} \le P_{ESS}(t) \le P_{ESS}^{max} \\ SOC^{min} \le SOC_t \le SOC^{max} \end{cases} \tag{8}$$

where P_{ESS} represents the operating power of the energy storage equipment, Δt represents the operating time of the energy storage equipment, P_{ESS}^{min} represents the minimum power of the energy storage equipment, $P_{ESS}(t)$ represents the power of the energy storage equipment at time t, P_{ESS}^{max} represents the maximum power of the energy storage equipment, SOC^{min} represents the minimum remaining power of the energy storage equipment, SOC_t represents the remaining power of the energy storage equipment at time t, and SOC^{max} represents the maximum remaining power of the energy storage equipment.

All nodes in the system should always be in a state of power balance.

$$P_i = P_i^{IN} + P_i^{OUT} = U_i \Sigma U_j (G_{ij} cos\theta_{ij} + B_{ij} sin\theta_{ij}) \tag{9}$$

$$Q_i = Q_i^{IN} + Q_i^{OUT} = U_i \Sigma U_j (G_{ij} sin\theta_{ij} - B_{ij} cos\theta_{ij}) \tag{10}$$

where P_i represents the total active power of node i, P_i^{IN} represents the active power flowing into node i, P_i^{OUT} represents the active power flowing out of node i, Q_i represents the total reactive power of node i, Q_i^{IN} represents the reactive power flowing into node i, and Q_i^{OUT} represents the reactive power flowing out of node i.

2.3. Solution method

This research aims for the optimal solution between voltage sag compensation degree and energy storage configuration. GA (Genetic Algorithm) and SA (Simulated Annealing Algorithm) are common for solving optimal objectives.

To utilize the advantages of both, this research adopts an improved GA with SA inserted to solve the optimal objective. SA will be applied to genetically operated individuals to recalculate fitness and replace old individuals with a certain probability of forming a new population. The combination overcomes GA's poor local search ability and avoids local optima, achieving a balance of global and local search abilities.

3. Case study analysis

3.1. Example parameters

In this paper, a 33-node distribution network system (Figure 1) is set up for the case study. Its base values are 10 MVA and 12.66 kV. Nodes 6, 25, and 33 are sensitive load nodes (Table 1). The Monte Carlo method is used to generate voltage sag waveforms based on the measured annual data of the system. The simulation times are 500 with MATLAB R2022b as the software. The improved genetic algorithm is adopted to determine the optimal installation location and capacity of energy storage. The main setting parameters of the genetic algorithm and simulated annealing algorithm are shown in Table 2. The basic parameters of energy storage equipment and node voltage requirements are shown in Table 3.

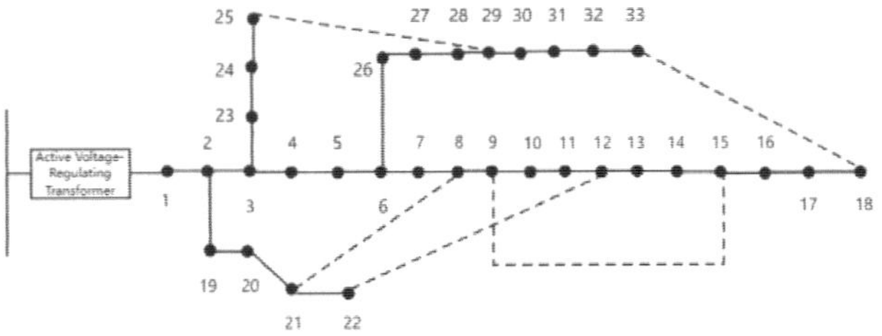

Figure 1. IEEE 33-node System Diagram.

Table 1. Load Structure **Table** of Sensitive Nodes.

Node	Proportion of Luminaires	Proportion of PLCs	Proportion of PCs	Proportion of ASDs	Proportion of Non-Sensitive Loads
6	6%	11%	8%	27%	48%
25	8%	10%	11%	25%	46%
33	7%	9%	10%	24%	50%

Table 2. Main Parameter **Table** of the Improved Genetic Algorithm.

Parameter	Value	Parameter	Value
Chromosome Length	130	Maximum/Minimum Mutation Probability	0.2/0.1
Maximum/Minimum Crossover Probability	0.8/0.6	Maximal Iteration Count	100
Elimination Probability	0.1	Protection Probability	0.1
Initial Annealing Temperature	100	Temperature Decrease Parameter	0.995

Table 3. Table of Energy Storage Parameters and Voltage Limits.

Parameter	Value
Investment Cost of Energy Storage Equipment	100000 yuan/kWh
Efficiency of Energy Storage Equipment	0.9
Voltage Limit of Sensitive Nodes	0.9 p.u~1.0 p.u

3.2. Result analysis

The total objective function F2 gets its minimum value after over 50 algorithm iterations, meaning the sum of energy storage's economic cost and economic losses from voltage sags reaches the lowest. Specific output results are displayed in the following charts.

Table 4 shows that the optimal number of installed energy storage units is 3, installed at Node 5, Node 23, and Node 30 respectively, with capacities of 48.1 kWh, 33 kWh, and 34.2 kWh. The installation locations of energy storage equipment are before sensitive load nodes. This pattern ensures that the voltage sag detection equipment at the same node as energy storage can quickly do voltage compensation after detecting a voltage sag, increase the amplitude of voltage sags at sensitive nodes, and promptly reduce related economic losses.

A comparison diagram (Figure 2) of voltage sag amplitudes of each node before and after configuration optimization is drawn. The red line represents the amplitude before optimization. From Node 1 to Node 18, the amplitude gradually drops to 0.5 p.u., and from Node 19 to Node 33, there is the same downward trend. After optimization, the curve rises from Node 1 to Node 5. As energy storage is at Node 5, the voltage from Node 5 to Node 10 is greatly compensated and keeps rising. Then it slightly drops till Node 18 but stays stable within 0.9 p.u. - 1.0 p.u. From Node 19 to Node 22 on a separate branch, there is no significant amplitude change. With energy storage at Node 23, the amplitude at Node 24 is significantly increased. Nodes 26 - 33 (Node 6's branches) show a slight downward trend and are always above 0.9 p.u.

For sensitive user nodes (Node 6, Node 25, and Node 33), the voltage sag amplitudes are compensated by 0.3324 p.u., 0.2577 p.u., and 0.3257 p.u., respectively, and are within 0.9 p.u. - 1.0 p.u., meaning the impact of voltage sags on these nodes is greatly reduced and their economic losses will be significantly cut. Table 5 is a comparison table of economic losses caused by voltage sags at sensitive load nodes before and after optimization to explore the economic benefits of the optimized energy storage configuration.

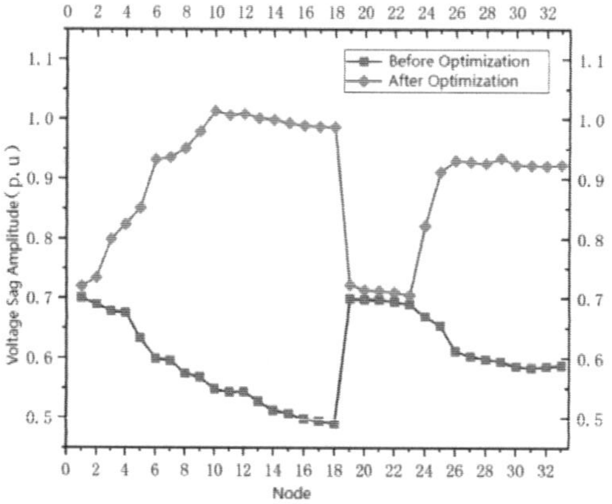

Figure 2. Diagram of Voltage Sag Amplitude Values of Each Node Before and After Optimization.

Table 4. Table of Energy Storage Parameters and Voltage Limits.

Installation Location	Installation Capacity, kW	Investment Cost, CNY
Node 5	48.1	4, 810, 000
Node 23	33	3, 300, 000
Node 30	34.2	3, 420, 000

Table 5. Table of Economic Losses Caused by Voltage Sags.

Node	Annual economic losses before optimization, CNY	Annual economic losses after optimization, CNY
6	840, 000	4, 810, 000
25	420, 000	3, 300, 000
33	350, 000	3, 420, 000

From these tables, it is clear that economic losses due to voltage sags at Node 6, Node 25, and Node 33 are reduced by 465, 000 yuan, 288, 000 yuan, and 200, 000 yuan respectively compared to before optimization, with a reduction percentages of 55%, 68.5%, and 57.1%. Combined with the amplitude compensation effect, the optimized energy storage configuration in this study can effectively solve 200 ms - 2 s voltage sag problems caused by line faults, meeting the regional integrated energy system's governance requirements for medium-timescale voltage sag issues. Meanwhile, it improves system operation stability and reduces substantial economic losses.

4. Conclusion

The energy storage's capacity-sizing and site-selection optimization and the system's optimized dispatching in this paper can effectively govern medium-timescale voltage sag and long-timescale voltage deviation problems within the system. The comprehensive governance effect reaches 96.7%, meeting power quality governance requirements and improving the system's operating economy.

Technically, for medium-timescale voltage sags, supercapacitor energy storage can effectively compensate for voltages of sensitive load nodes after sags occur, with amplitudes > 0.9 p.u. after compensation. As non-sensitive load nodes are out of this study's scope, their 63% compliance rate is insignificant. After capacity-sizing and site-selection optimization, supercapacitor energy storage can achieve 100% governance of medium-scale voltage sag problems. For long-timescale voltage deviation, all nodes are within the research scope. After optimized dispatching of the system, 94% of node voltages are within the normal range (0.93 p.u. - 1.07 p.u.), achieving 94% governance. Based on weight calculation, the overall power quality governance rate of the system can reach 96.7%.

Economically, after energy storage capacity-sizing and site-selection optimization, the severity of voltage sags at sensitive nodes decreases. The system recovers 953, 000 yuan in economic losses annually. Meanwhile, after optimized dispatching, the system saves over 800 yuan in daily operating costs and over 300, 000 yuan annually.

The power quality issues covered in this study are only voltage quality issues at the full timescale, and there are relatively few types of research conducted. Subsequent research can divide the total capacity of energy storage devices into functional categories, and propose personalized energy storage operation strategies for different types of power quality problems within the system. Flexible switching can be achieved when the detection equipment detects power quality problems.

Acknowledgment

This research has been supported by the Headquarters Technology Project of China Electrical Equipment Group Co., Ltd. (No. CEE-2023-B-01-03-002-XJ).

References

[1] Ahmed M, Mustafa. Voltage-sag mitigation by stability-constrained partitioning scheme. Ain Shams Engineering Journal, 2022, V14(2): 101867.

[2] Emenike Ugwuagbo. Experimental data on power quality assessment at point of common coupling of a steel mill to an electric power grid. Data in Brief, 2021, V39: 107681.

[3] Baliyan, Arjun, Jamil, et al. Power Quality Improvement Using Harmonic Passive Filter in Distribution System. Lecture Notes in Electrical Engineering, 2022, V766:435-445.

[4] Ye X Y, Liu H T, Hao S P. Comprehensive mitigation strategy of voltage sag based on sensitive load clustering. International Transactions on Electrical Energy Systems, 2021, V31(7): 1-20.

[5] Landl Sarah, Kirchsteiger Harald. Mitigating Overvoltage in Power Grids with Photovoltaic Systems by Energy Storage. Environmental and Climate Technologies, 2022, V26(1): 470-483.

[6] Wu C, Lu Y C, Yang B, et al. Optimal Allocation Technology of Energy Storage for Improving the Power Supply Quality of 10 kV Distribution Network in the Region of "Replacing Coal with Electricity". Power Capacitor & Reactive Power Compensation, 2022, 43(6): 138-145. (in Chinese)

[7] Sachin R., Nagesh H.B. Power quality assessment and power quality improvement in hospital facility. International Journal of Modelling, Identification and Control, 2022, V41(4): 316-333.

[8] Ni S Y, Lin Y, Yao D F, et al. Optimal Dispatch of Regional Integrated Heating and Power System Based on Differential Thermal Inertia Model. E3S Web of Conferences, 2021, V256: 2018.

[9] Yu Z P, Guan H H, Yu J, et al. Power Quality Management Strategy for Power Grid Considering Energy Storage Capacity. Renewable Energy Resources, 2021, 39(3): 401-407. (in Chinese)

[10] Liao H, Chen Y, Zheng Z, et al. A two-stage business model for voltage sag sensitive industrial users employing energy storage systems. Applied Energy, 2025, 379: 124945.

Research on the influence of heat retention capability of phase change energy storage tank based on different packaging structures

Xulin Li, Jiayu Li*, Xing Wang

Department of Municipal and Environmental Engineering, Shenyang Jianzhu University, Shenyang 110168, China

* Corresponding author's e-mail: 1912173257@qq.com

Abstract. To explore the influence of different packaging structures on the thermal storage efficacy of phase change energy storage systems, the cylindrical, wedge-shaped, conical, and oval packaging structures of these tanks were studied by combining experimental verification and simulation. The differences in each tank's heat storage time, temperature distribution, and heat storage were analyzed. The results show that the different packaging structures will have a certain impact on the thermal retention capability of the water tank. In the heat storage procedure, the wedge-shaped unit tank finishes the process first, reducing storage time by 7% compared to the cylindrical unit tank. In the latter stages of heat storage, both the liquid phase fraction and the temperature variations in the elliptical unit tank demonstrate an increased rate of change, indicating its enhanced heat retention capability over time. In contrast, the conical unit water tank exhibits a greater temperature disparity in the phase change material, with a notable difference of 12.6°C, largely due to its structural design. Meanwhile, the cylindrical unit tank boasts the highest heat storage capacity, measuring a substantial 81.69 MJ.

1. Introduction

The latent heat storage tank (LHST) offers the benefits of steady temperature and high heat storage capacity throughout the thermal storage and release cycle, and its thermal performance is much better than that of an ordinary water tank due to the addition of phase change material (PCM) and has a broad application prospect. Nonetheless, the majority of PCMs exhibit low thermal conductivity issues, which restricts the further development of LHST. In recent years, researchers have explored numerous aspects of LHST. In the study of the structure of the water tank, the addition of metal fins and baffles is a traditional method of enhanced heat exchange. Zhao et al. employed topology optimization techniques to enhance the design of LHST fins, achieving a remarkable 70% reduction in heat storage time [1]. Additionally, this method significantly improved both temperature consistency and liquid fraction. Meanwhile, Feng and colleagues managed to decrease the heat storage time of the tank by 4% while boosting heat storage capacity by 10.8% through the incorporation of a baffle in the phase change tank [2]. Research into the materials in the tank revolves around the incorporation of materials with high thermal conductivity into the PCM to enhance heat transfer efficiency. A variety of materials with high thermal conductivity, such as nanoparticles [3], expanded graphite [4], and metal foam [5], are commonly incorporated into the formulation. These newly developed composite phase change materials significantly enhance the thermal conductivity of the phase change materials, ultimately boosting the heat exchange rate within the phase change thermal storage system. The encapsulation unit is mainly used to wrap the PCM and prevent it from leaking during the phase change. At present, the common packaging structures are rectangular, cylindrical, and spherical, among which the rectangular device features a straightforward design and is easy to produce, however, the PCM within the spherical

Journal of Physics: Conference Series **3011** (2025) 012017 doi:10.1088/1742-6596/3011/1/012017

component can be cured in the least amount of time. Some scholars have found that changing the geometry of the package structure can also enhance the thermal efficiency of the phase change heat storage system [6]. Based on the rectangular structure, Hu and Li proposed a wedge-shaped structure, which can effectively improve the vertical temperature distribution of PCM and accelerate the melting rate of the bottom PCM in the later stage compared with the rectangular structure [7]. Seddegh and his colleagues transformed the cylindrical design of the shell-and-tube phase change regenerative unit into a conical form [8]. Their findings revealed that the conical configuration enables a significantly quicker heat storage rate compared to its cylindrical counterpart; however, both designs exhibited similar performance during the heat release phase. In another study, Sachin Rana and his team conducted simulations to analyze the heat transfer properties of shell-and-tube heat exchangers featuring round, rectangular, and oval tubes by using a two-dimensional model [9]. The results indicated that rectangular tube heat exchangers achieved the fastest PCM melting time at a temperature of 60°C, followed closely by elliptical and round tubes.

Based on the above review of LHST-related research, numerous researchers have carried out extensive studies on the design of the tank and the efficacy of PCM to explore how these elements affect the tank's ability to store and release heat. The packaging structure has an important impact on the heat transfer mechanism of the heat storage process, but the current research on the packaging structure focuses on a single phase change unit, and the impact of various packaging designs on the thermal efficiency of LHST is seldom investigated.

To extend this kind of research to LHST, this study used a combination of experimental verification and simulation to explore the effects of cylindrical, wedge-shaped, conical, and oval packaging structures regarding the thermal storage efficiency of LHST under identical conditions such as size, PCM type and dosage, and water temperature and flow rate at the inlet. These findings will inform the ideal design of the LHST package framework.

2. Simulation setup and validation

2.1. Physical model

For the sake of description, the cylindrical, wedge-shaped, conical, and oval encapsulated tanks were named Cy-LHST, W-LHST, Co-LHST, and O-LHST, respectively. The dimensions of each tank are 0.8 m×0.8 m×0.8 m, the overall volume is 0.512 m^3, the water inlet and outlet are located 0.1 m away from the top/bottom respectively, and the surface dimensions are 0.03 m×0.03 m. PCM chose No. 52 paraffin wax, which was purchased from Sinopec. The same amount of PCM is used in each tank, with a mass of 29 kg, which is 5.9% of the tank volume. According to the above parameters, the three-dimensional physical model and the precise measurements of the four water tanks are created, as shown in Figure 1. The thermos-physical parameters and operating conditions of PCM and HTF are listed in Table 1.

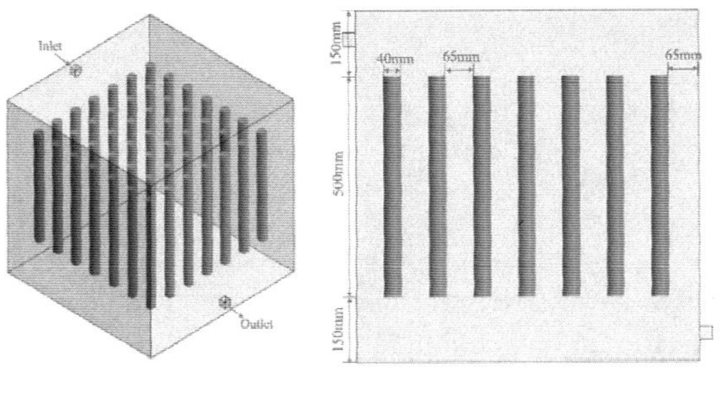

Cy-LHST

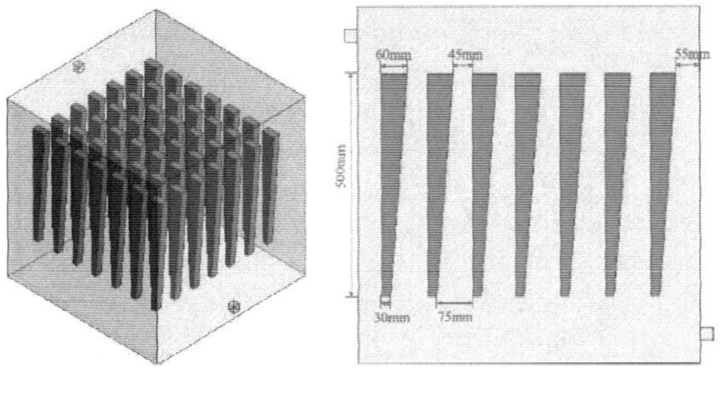

W-LHST

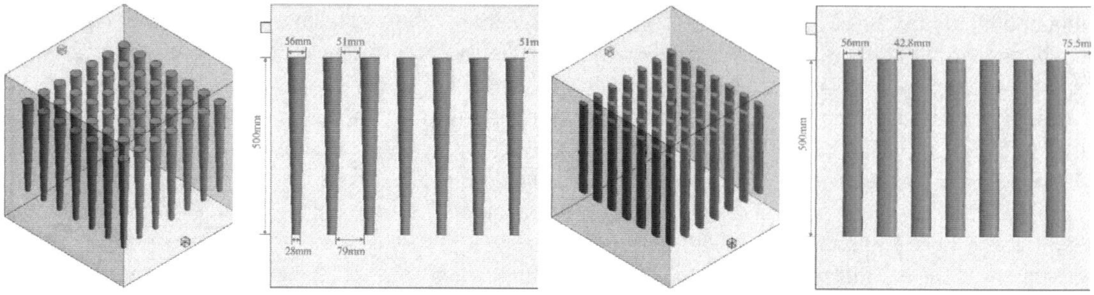

Co-LHST O-LHST

Figure 1. Dimensions of the water tank and packaging unit.

Table 1. Thermophysical parameters and operating conditions.

PCM		HTF	
Phase change temperature (K)	323(s)/325(l)	Density (kg/m^3)	998.2
Density (kg/m^3)	978	Specific heat capacity (kJ/kg·K)	4.182
Specific heat capacity (kJ/kg·K)	3.22(s)/3.15(l)	Thermal conductivity (W/m·K)	0.6
Thermal conductivity (W/m·K)	0.49(s)/0.28(l)	Thermal storage temperature (K)	343
Viscosity (kg/(m·s))	0.011	Inlet temperature (K)	303
Latent heat of phase change (kJ/kg)	200	Inlet flow rate (m/s)	0.1

2.2. Governing equations and simulation variables

For the phase change tank, the following assumptions are made without compromising the accuracy of the calculations: (1) The PCM is consistent and has the same properties in all directions. (2) Additionally, both the temperature at which the phase change occurs and the various thermal properties of the PCM do not vary. (3) Any heat loss from the walls of the water tank is disregarded, along with the inlet and outlet, as well as the thickness of the metal supports and the phase change unit. (4) The starting temperature across all sections of the tank is uniform. Given the above assumptions, the mathematical model in LHST is established by Equations (1)-(3).

The basic equation for the melting model is:

$$\rho_c \frac{\partial H}{\partial t} = k\nabla^2 T \tag{1}$$

Throughout the process, the liquid phase fraction varies between 0-1, i.e.:

$$\beta = \begin{cases} 0, & T < T_S \\ \frac{T-T_S}{T_L-T_S}, & T_S < T < T_L \\ 1, & T > T_L \end{cases} \tag{2}$$

The tank walls are adiabatic and are expressed as follows:

$$\frac{\partial T}{\partial x} = \frac{\partial T}{\partial y} = \frac{\partial T}{\partial z} = 0 \tag{3}$$

Due to the complex internal structure of LHSTs, an unstructured meshing model is adopted. The grids in the PCM area, inlet, and outlet areas, where the flow and temperature fields vary greatly, are encrypted, and larger mesh sizes are applied in different regions to accommodate the simulation's speed. Fluent software was utilized for the transient numerical simulation of LHSTs. We open the energy equation, select the k-e turbulence model, and activate the Solidification/Melting module to calculate the PCM phase transition process. The solver is configured to Pressure, employing the SIMPLE algorithm for pressure-velocity coupling discretization and the PRESTO! algorithm for pressure term discretization. The iterative precision (i.e., residuals) and relaxation factor are calculated by using the default values in Fluent. We employ standard initialization and standardize the initial temperature throughout the tank.

2.3. Irrelevance test

To verify the correctness of the simulation outcomes, the number of meshes is verified for irrelevance. The outlet temperature of Cy-LHST was selected as the parameter for analysis. Figure 2 shows the number of meshes divided into 690, 000, 980, 000, 1, 280, 000, and 1, 660, 000. When the number of grids is 690, 000 and 980, 000, the accuracy of the calculation results is low, and when the number of grids is 1.28 million and 1.66 million, the simulation values are not much different. Considering the speed of the simulation, the final number of grids with 1.28 million grids was selected for the simulation. In addition, the grid parameters of the four types of tanks are shown in Table 2. The mean mesh quality exceeds 0.8, and the mesh quality is higher.

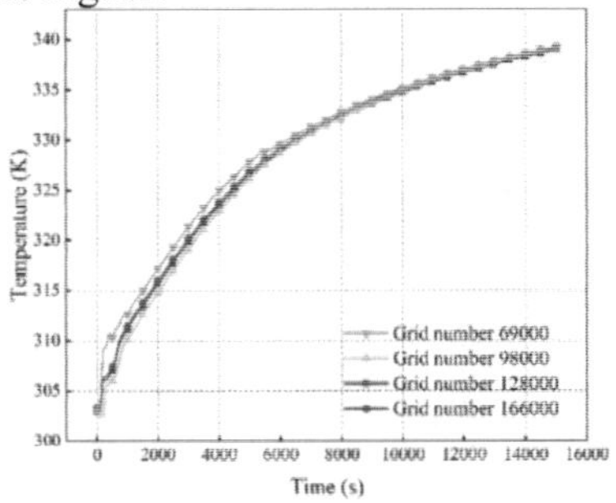

Figure 2. Simulation grid independence check.

Table 2. Mesh number and mesh quality.

LHSTs	Number of meshes	Minimum mesh mass	Average mesh quality
Cy-LHST	1.28 million	0.3140	0.8876
W-LHST	1.42 million	0.3861	0.8932
Co-LHST	1.29 million	0.3245	0.8150
O-LHST	1 million	0.3799	0.8760

2.4. Experimental verification

The experiment took place in Shenyang, Liaoning Province, China, utilizing a system primarily featuring a water tank (Cy-LHST), a pitot tube digital pressure gauge, a circulation pump, and a phase change unit, with a consistent electromagnetic energy heater as the heat source. The effective volume of the water tank is 512 L and is wrapped with 20 mm thick insulation cotton to reduce heat dissipation. The PCM is encapsulated by PVC tubes (thermal conductivity of 0.17 W/m·K) with metal supports to secure the phase change unit bundle inside the tank. Figure 3 displays a photo of the experiment. The inlet flow rate into the water tank, measured with a pitot tube, was 0.12 m/s. A thermocouple measurement point is placed inside the phase transformation unit at the center, and the simulation results and experimental data are analyzed by utilizing the measurement point's temperature along with the outlet water temperature.

Figure 3. Pictures of the experiment.

As illustrated in Figure 4, the temporal trends of the simulation findings and experimental data are basically the same. The data collected before reaching the melting point of the PCM align well; however, the simulation overlooks the heat transfer with the surrounding environment and the thickness of the walls in the phase change unit. Consequently, a discrepancy exists between the simulation outcomes and the experimental findings, where the average difference between the measured point temperature and the outlet temperature is 0.91 K and 1.16 K, respectively. The simulation's accuracy is validated, and the simulation method and settings of this study are reliable.

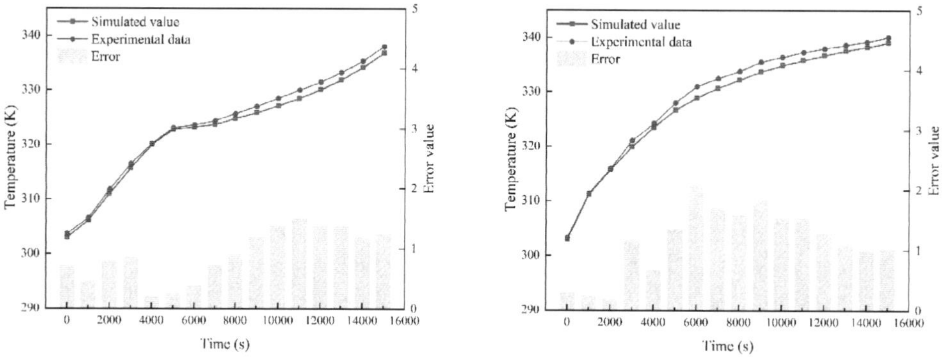

(a) Temperature of heat charging measurement point (b) Temperature of heat outlet

Figure 4. Comparison of experimental results and simulation data.

3. Results and analysis

3.1. Heat storage time

In this study, heat storage is deemed complete when the liquid fraction of PCM reaches 1, that is when PCM is completely melted. Figure 5 illustrates the LC fraction of PCM in the tank over time throughout the heat storage process. It is evident that as the heat storage process advances, the growth value of the liquid phase fraction in the same period decreases in the later stage, showing the melting rate of PCM has great inhomogeneity and the uneven melting characteristics of first fast and then slow lead to the prolongation of the whole heat storage process. The time taken by Cy-LHST, W-LHST, Co-LHST, and O-LHST to complete heat storage was 13, 101 s, 12, 184 s, 15, 551 s, and 12, 449 s, respectively. Among them, W-LHST was the first to complete heat storage, and the heat storage time was shortened by 7% compared with Cy-LHST, indicating that the wedge-shaped phase transition unit can be melted in a short time, which aligns with the findings of the wedge-shaped phase transition unit in [7]. Co-LHST finally completed heat storage, and the heat storage time was greatly extended compared with other water tanks. The results show that the performance of the conical stage phase change unit in the tank is not as good as that of the traditional cylindrical phase change unit. During the initial phase of heat storage (before 8000 seconds), all tanks exhibited similar trends in their liquid phase rates, with the liquid phase fraction of O-LHST being notably lower at this juncture. Nevertheless, following the 8000-second mark, the liquid phase fraction for O-LHST saw a remarkable surge, surpassing that of Cy-LHST. This shift suggests that the influence of the elliptical structure becomes more pronounced in the later stages of heat storage.

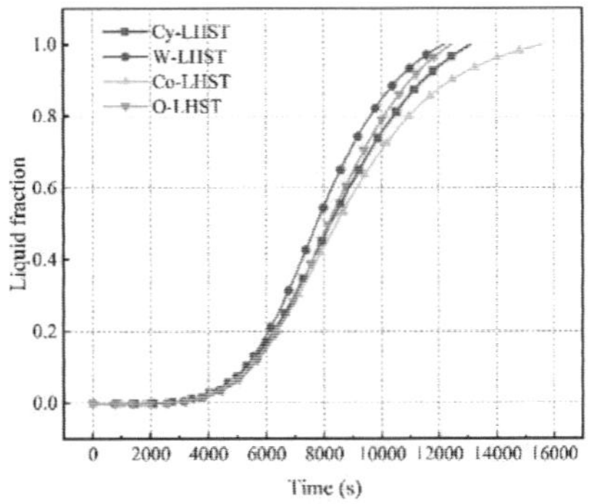

Figure 5. Liquid fraction of PCM.

3.2. Temperature distribution

Figure 6 illustrates the temperature variation of LHSTs at different times during heat storage. The PCM temperature is observable and its surrounding area is generally low because the temperature does not change during the phase change. During the initial phase of heat storage (before 8000 s), the lower region of the phase change unit in W-LHST and Co-LHST was hotter than in other tanks. As a result, the PCM in their lower area melts faster. However, with the advancement of the heat storage process, the phase change unit in the Co-LHST begins to show a large temperature difference during the central phase of heat storage. The temperature of the PCM in the lower section hits 336 K, whereas, in the upper section, it hovers around 323 K. This difference creates a challenge for the PCM to fully melt, which is why the liquid phase fraction depicted in Figure 5 struggles to reach unity. The incomplete melting of the PCM in this area can be attributed to the large diameter of the phase change unit in the upper cone of the table, which hampers efficient heat exchange between the PCM and the heat transfer fluid, leaving some paraffin wax unmelted. In the cases of Cy-LHST and O-LHST, the temperature distribution within the phase change unit exhibits essentially no vertical variation.

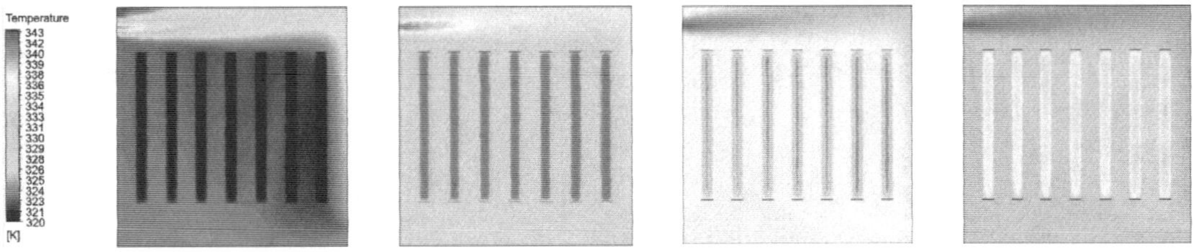

4000 s 8000 s 12000 s 13101 s

Cy-LHST

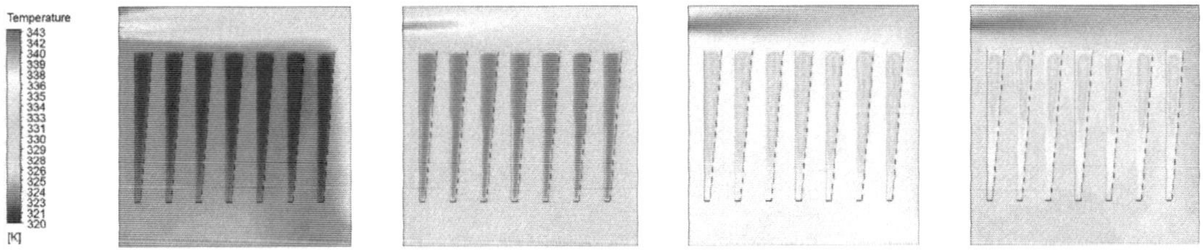

4000 s 8000 s 12000 s 12184 s

W-LHST

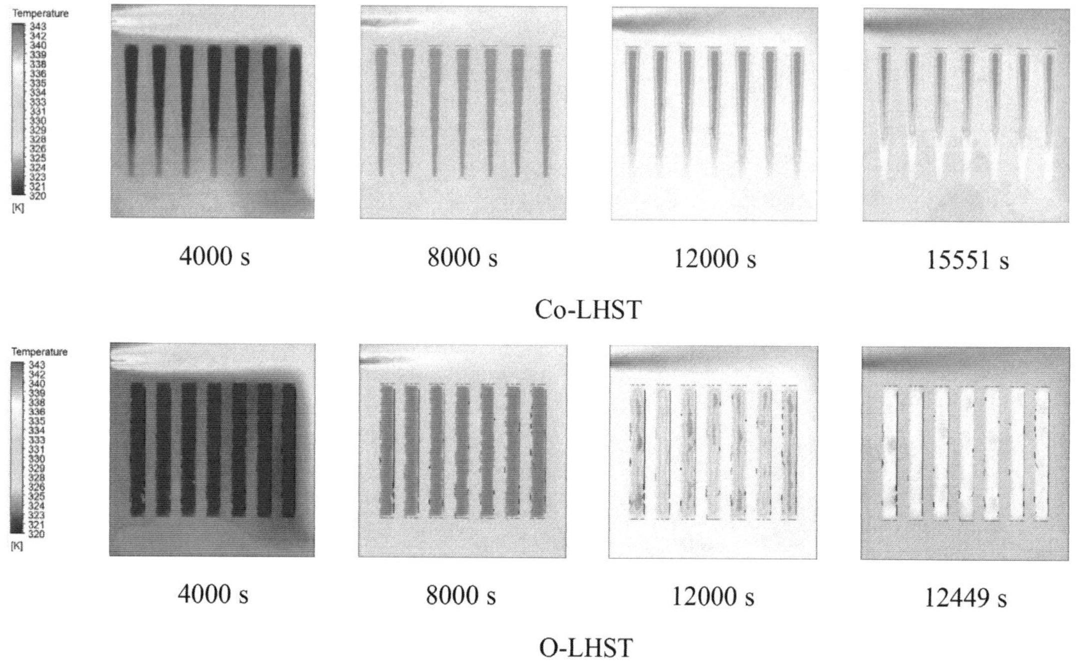

Figure 6. Temperature distribution of LHSTs at different times.

Figure 7 illustrates the temperature variation of PCM in LHSTs at the identical site. The coordinates of the measuring point are T1 (400 mm, 400 mm, and 600 mm), T2 (400 mm, 400 mm, and 400 mm), and T3 (400 mm, 400 mm, and 200 mm). Observing the temperature at the T1 position, it was found that the temperature in the upper region of W-LHST and Co-LHST was lower, which is because the wedge-shaped and conical structures are characterized by a wider upper area, a narrower lower area, and more PCM filled in the upper area. This takes more time to melt the upper PCM compared to the other two LHSTs that are evenly filled from top to bottom. The temperature change at T3 is much more obvious than that at T1 and T2. The PCM temperature is higher in the bottom region of W-LHST and Co-LHST. The temperature distribution of Co-LHST was too different, and the temperature variation between the upper and lower sections of the PCM was 12.6°C. It was observed that the PCM temperature of Cy-LHST and O-LHST was almost the same at the three measurement points.

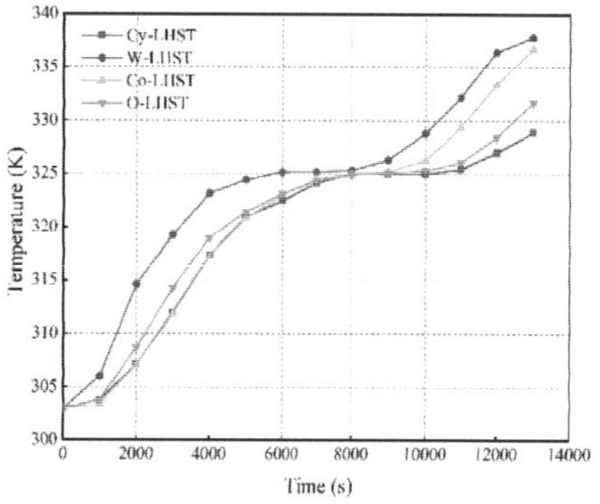

(a) T1

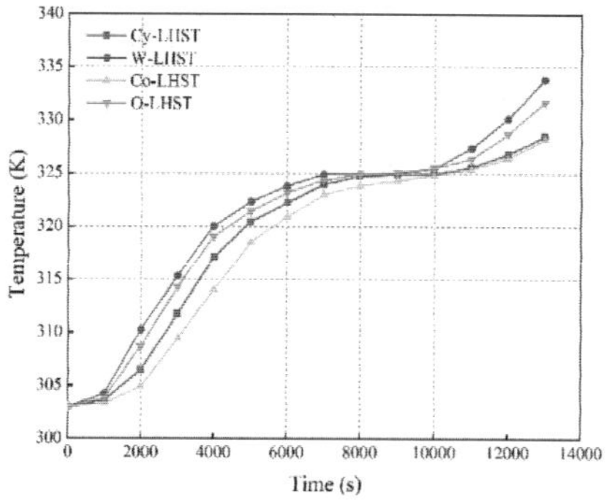

(b) T2

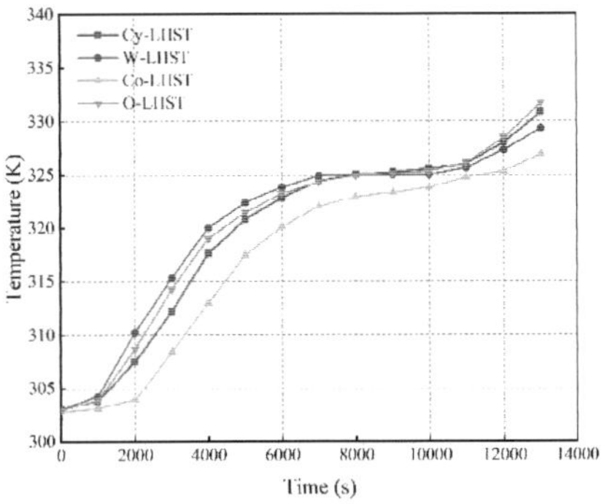

(c) T3

Figure 7. PCM temperature variation of LHSTs at the same position.

3.3. *Heat storage*

LHST heat storage consists of two parts: PCM heat storage and water heat storage. Based on the chosen PCM-specific heat capacity and phase change temperature in this study, the formula for heat storage can be expressed as:

$$Q = c_w m_w \Delta T + c_p m_p \Delta T_1 + m_p h \quad (4)$$

where Q denotes the overall thermal storage capacity of the phase change energy storage tank; the subscripts p and w refer to the phase change materials and water, respectively; C represents the specific heat capacity; m signifies the mass of either the phase change material or the water. Additionally, h indicates the latent heat associated with the phase change, and ΔT and $\Delta T1$ represent the temperature differences observed upon the completion of heat storage in both the water and the phase change materials, respectively.

Calculated by Equation (4), when the heat storage is completed, the heat storage of Cy-LHST, W-LHST, Co-LHST, and O-LHST is 81.69 MJ, 81.14 MJ, 80.38 MJ, and 81.62 MJ, respectively. Despite the equal quantity of PCM in the four-phase change tanks, the water and PCM temperatures vary between tanks upon completion of heat storage, especially the temperature distribution of W-LHST and Co-LHST, which is quite different, so their heat storage is also different. As shown in Figure 8, the total heat storage capability of the Cy-LHST stands out as the highest, attributed to both the elevated average temperature of the phase change material and the temperature of the water upon completion of the Cy-LHST heat storage process. This suggests that the water tank optimally utilizes its heat storage potential. The heat storage of O-LHST is basically the same as that of Cy-LHST. Although the PCM in the W-LHST melts the fastest, the heat storage capacity is not as good as that of the above two tanks, and its heat storage capacity is lower. The temperature distribution of Co-LHST is too different, and the temperature of the top PCM does not reach the melting point after the bottom PCM is completely melted. Although the water temperature is not much different from other LHSTs, its heat storage is still the smallest.

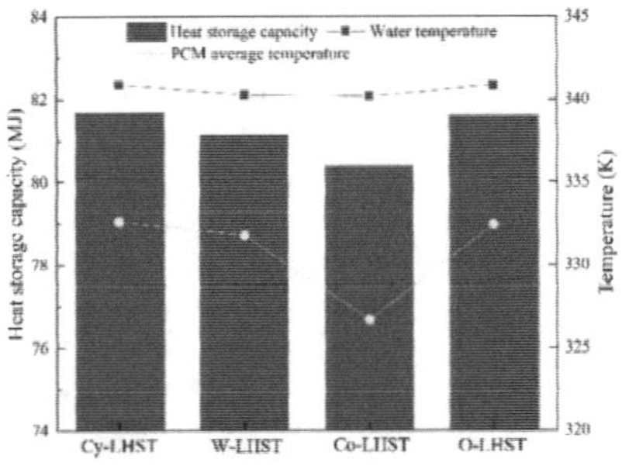

Figure 8. Heat storage of LHSTs.

4. Conclusion

The different packaging structures have a certain effect on the thermal storage efficiency of the water tank. This study examined the overall effectiveness of the water tank by analyzing its heat storage duration, temperature variance, and thermal capacity. Through simulations of the heat storage processes in these four tanks, several key conclusions emerged:

(1) When storing heat, a reduced heat storage duration for the water tank results in a quicker melting of the PCM within it. Compared with Cy-LHST, Co-LHST, and O-LHST, the heat storage time of W-LHST was shorter, which was shorter by 7%, 21.7%, and 2.1%, respectively.

(2) The heat storage capacity of Cy-LHST and O-LHST in the four tanks was larger, which were 81.69 MJ and 81.62 MJ, respectively, indicating that they can store more heat.

(3) The temperature difference of the PCM of Co-LHST in the vertical direction is too large, reaching 12.6°C. The PCM located at the top of the tank remains solid for an extended period, leading to significantly longer heat storage durations compared to other water tanks, and the heat storage is much lower than that of other water tanks. Therefore, the conical package structure is not recommended for use in LHST.

Acknowledgments

This work was financially supported by the Liaoning Provincial Department of Education under the 2023 Basic Research Program "Research on Key Technologies for Performance Improvement of Nanofluid-based Phase Change Energy Storage Tanks" (Project number: JYTMS20231554).

References

[1] M. Zhao, Y. Tian, M. Hu, et al. Topology optimization of fins for energy storage tank with phase change material[J]. Numerical Heat Transfer, Part A: Applications, 2020: 284–301.
[2] G. Feng, T. Wang, X. Li, et al. Simulation of a composite latent heat storage tank with horizontal baffles and two phase-change temperatures[J]. Applied Thermal Engineering 225 (2023) 120192.
[3] A. Mourad, N.A.A. Qasem, A. Abderrahmane, R. Marzouki, K. Guedri, O. Younis, N.A. Shah, T. Botmart, Numerical study on n-octadecane PCM melting process inside a pear-shaped finned container, Case Studies in Thermal Engineering 38 (2022) 102328.
[4] X.K. Yu, Y.B. Tao, Preparation and characterization of paraffin/expanded graphite composite phase change materials with high thermal conductivity, International Journal of Heat and Mass Transfer 198 (2022) 123433.
[5] A. Chibani, S. Merouani, F. Benmoussa, Computational analysis of the melting process of Phase change material-metal foam-based latent thermal energy storage unit: The heat exchanger configuration, Journal of Energy Storage 42 (2021) 103071.
[6] G.H. Feng, T.W., Simulation analysis of the effect of encapsulation method on the heat storage and exothermic performance of phase-change water tanks, Energy Storage Science and Technology 11 (2022) 2161.
[7] Z.P. Hu, A.G, Li, Comparative experimental study of the thermal storage performance of rectangular and wedge-shaped devices, Journal of Xi'an University of Architecture and Technology (Natural Science Edition) 51 (2019) 134–139.
[8] S. Seddegh, S.S.M. Tehrani, X. Wang, F. Cao, R.A. Taylor, Comparison of heat transfer between cylindrical and conical vertical shell-and-tube latent heat thermal energy storage systems, Applied Thermal Engineering 130 (2018) 1349–1362.
[9] S. Rana, M. Zunaid, R. Kumar, CFD analysis for heat transfer comparison in circular, rectangular and elliptical tube heat exchangers filled with PCM, Materials Today: Proceedings 56 (2022) 637–644.

Double-layer power optimal allocation strategy of energy storage power station considering energy efficiency and battery state

Jiajin Qi[1], Shujiang Yu[1], Yunlong Chen[1], Yinfei Xu[1], Jiangtao Li[2,*] and Jiangyuxuan Zhu[3]

[1] Hangzhou Electric Power Equipment Manufacturing Co., Ltd. Zhejiang, 310027, China
[2] College of Electrical Engineering, Zhejiang University, Zhejiang, 310027, China
[3] Polytechnic Institute of Zhejiang University, Zhejiang University, Zhejiang, 310015, China

*Corresponding author's e-mail: 22210142@zju.edu.cn

Abstract. Electrochemical energy storage is popular in many fields for its quick response and flexible setup. However, at this stage, the cost of energy storage is still high; the source of income is limited, and the profit prospects are worrying. Therefore, this paper proposes a two-layer power optimization allocation strategy for energy storage power stations considering energy efficiency and battery state. Through this strategy, the operation mode of energy storage can be optimized. The single-cycle expenditure for energy storage sees a reduction. The life cycle can be prolonged, and the income can be expanded. The bi-level optimal allocation strategy is divided into the optimization layer of the power storage subsystem and the optimization layer of the unit. At the level of the subsystem, considering the relationship between the load condition of the transformer and its efficiency, the power of each transformer is determined with the minimum total energy loss of the transformer as the objective. At the energy storage unit level, the state of charge and the health of the power storage unit are considered at the same time. The output power of each unit is optimized to ensure consistent battery states and avoid the "shortboard effect". The simulation results show that the strategy has certain feasibility and effectiveness.

1. Introduction

At present, China's power system is gradually transitioning to a new power system with a high share of renewable energy and power electronics. The "double high" characteristics lead to frequent fluctuations in grid frequency and a serious shortage of inertia support. A substantial number of flexible resources are essential for the reliable and secure functioning of the electrical grid. Electrochemical energy storage (hereinafter referred to as energy storage) offers fast response and high precision, and has been widely used in the world. When 2023 came to an end, the cumulative amount of newly installed energy storage capacity in the world reached 91.3 GW. Among them, China's installed capacity has reached 34.5GW / 74.5GWh. The annual growth of the power scale and energy scale has exceeded 150 %.

However, behind the rapid development of energy storage, especially at this stage, the cost of energy storage is still high, and the profit prospects of energy storage power stations are worrying. To enhance profitability, some researchers suggest that energy storage should engage in various power grid services to gain combined benefits. In Reference [1], the feasibility of energy storage participating in multiple services at the market level and technical level is demonstrated. Reference [2] proposed a joint optimization strategy for energy storage to participate in the electric energy market and the auxiliary service market. Under this strategy, the joint benefits obtained by energy storage are greater than the benefits of only participating in one of the services.

On the other hand, it is also possible to choose to optimize the energy storage operation mode, extend its life cycle, or reduce the single-cycle cost of power storage to achieve the goal of profitability. By considering the degradation cost of energy storage and the performance of energy storage participating in frequency modulation, Reference [3] optimizes the operation mode of energy storage and reduces the cost of a single cycle. Reference [4] proposed a power allocation strategy taking into account the disparities in the power storage unit's health conditions, but ignored the impact of the imbalance of the state of charge of the power storage unit. Reference [5] considered the power allocation strategy that the state of health (SOH) and the state of charge (SOC) of the power storage unit are consistent, which can improve the consistency of the SOC of the unit, and ensure that the power storage unit with good SOH is more active and deeper. However, the energy efficiency is not high when the actual power of the transformer is quite different from the rated power. In Reference [6], the objective functions of minimizing battery energy loss and maximizing SOC consistency were constructed in different SOC regions of energy storage, but the influence of battery health status was ignored.

Based on the above literature, in this study, a two-echelon power optimization and allocation strategy for energy storage power stations is proposed, considering both energy efficiency and battery state. Firstly, at the energy storage subsystem level, considering the efficiency of the transformer, the total power instructions received by the power station are assigned to each transformer to reduce the transformer's total energy loss. Subsequently, from the perspective of individual units, both the SOH and SOC of each unit are taken into account. At the subsystem level, the transformer's power is first defined, and then allocated to the unit under the transformer in stages, to realize a complete power instruction distribution process. Ultimately, a numerical case is crafted to validate the reasonableness and efficacy of the proposed strategy.

2. Energy storage power station

2.1 Energy storage power station structure

The energy storage power station can be regarded as a "giant charging treasure". In general, multiple battery packs with PCS are called energy storage units. Because of the presence of PCS, a single unit can be independently controlled. Multiple units with a transformer are called energy storage subsystems. Each subsystem is responsible for the storage and regulation of electricity in a certain area or function. The structure of the energy storage power station is shown in Figure 1.

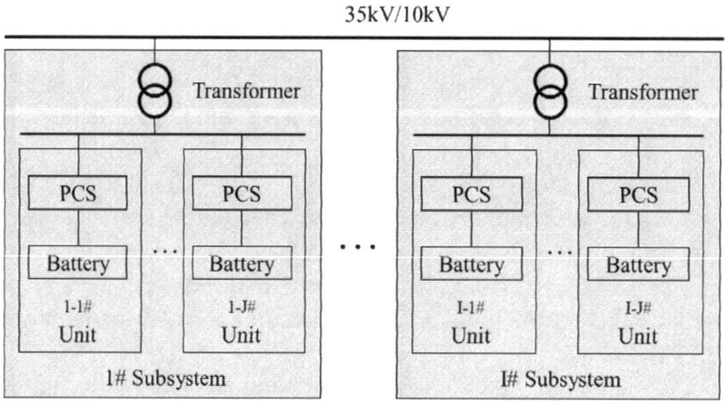

Figure 1. Energy storage power station structure diagram.

2.2 Energy consumption characteristics of transformer

For a given transformer, its efficiency is closely related to its load factor. After ignoring some secondary factors, the efficiency of the transformer can be expressed as [7]:

$$\eta = \left(1 - \frac{P_0 + \beta_2^2 P_k}{S_N \beta_2 \cos\varphi_2 + P_0 + \beta_2^2 P_k}\right) \times 100\% \tag{1}$$

In the formula, P_0 is the no-load loss of the transformer. P_K is the load loss. S_N is the rated capacity. $\cos\varphi_2$ is the power factor of the transformer, and usually the value is 0.889. β_2 is the load factor, and the value range is 0-1. It can be seen from the above formula that when the load coefficient is $\beta_2 = \beta_{max} = \sqrt{P_0 / P_k}$, the transformer efficiency reaches the maximum.

When the load factor changes in $0 \sim \beta_{max}^2$, the efficiency rises significantly with the load factor, but the overall efficiency remains relatively low. When the load factor changes in $\beta_{max}^2 \sim 1$, its efficiency changes slowly with the increase in the load factor, but it always has a high operating efficiency. When the total power command of the energy storage power station is large, it's expected that the power allocated to each transformer can achieve higher operating efficiency.

3. Double-layer power optimization allocation strategy

3.1 Double-layer power distribution structure

The double-layer power optimization allocation strategy includes the energy storage subsystem optimization layer and the energy storage unit optimization layer. The optimization layer of the energy storage subsystem first screens the available subsystems according to the current state of charge of the subsystem. Then the subsystems are prioritized, and the subsystems put into operation are ascertained in accordance with the aggregate power. Ultimately, the output power of each transformer is decided by minimizing the total energy loss of transformers in the running subsystem.

The optimization layer of the power storage unit determines the output power of each unit under the subsystem according to the output of each transformer calculated by the upper layer, considering the SOC balance of the energy storage unit and the resistance coefficient of SOH. The proposed double-layer power allocation structure is shown in Figure 2.

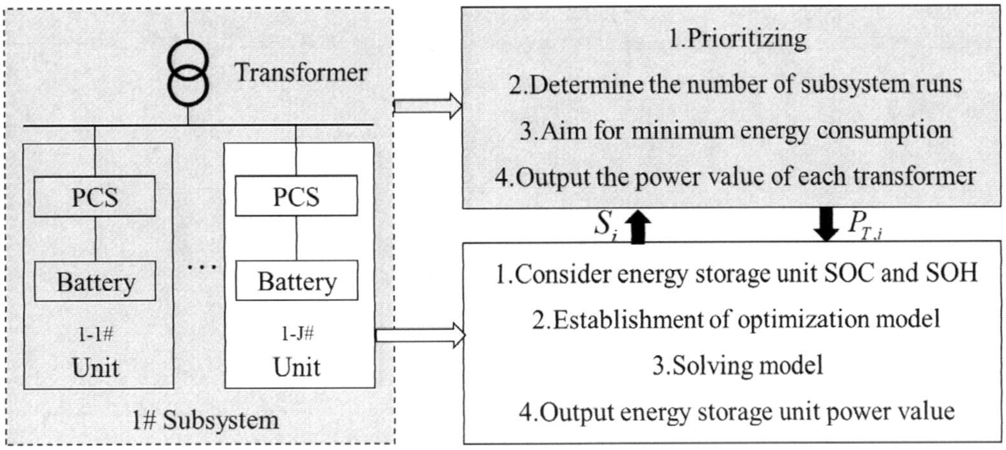

Figure 2. Double-layer power distribution structure diagram.

3.2 Energy storage subsystem optimization layer

The optimization layer of the power storage subsystem determines the power of the transformer to reduce the input transformer's total energy loss. Firstly, according to the total power instructions received by the current station, it is judged whether it needs to be charged or discharged. After that, the available subsystems are chosen based on the energy level of each subsystem. Then, the priorities for charging and discharging of these subsystems are determined and arranged in order. Finally, the number of energy storage subsystems that need to be put into operation is determined according to the aggregate power.

The discharge is prohibited when the SOC is less than 10 % in this study, and the charging is prohibited when the SOC is higher than 90 %. The high SOC is preferentially discharged when discharging, and the low SOC is preferentially charged when charging. The specific expression is as follows [8]:

$$X_c = \begin{cases} 1/\sin(1-S_i), 0 \le S_i < 0.9 \\ \\ \infty, 0.9 \le S_i \le 1 \end{cases} \tag{2}$$

$$X_d = \begin{cases} 1/\sin(S_i), 0.1 < S_i \le 1 \\ \\ \infty, 0 \le S_i \le 0.1 \end{cases} \tag{3}$$

$$S_i = \frac{\sum_{j=1}^{K} S_{i,j}}{K} \tag{4}$$

In the formula, X_c and X_d are the charging priority and discharging priority values of the energy storage subsystem, respectively. The smaller the value is, the higher the priority is. S_i is the SOC of the i-th energy storage subsystem. $S_{i,j}$ is the SOC, and K corresponds to how many units exist in the subsystem.

Screening out the available subsystems, the number of subsystems that are put into operation is ascertained in line with the overall power. The calculation is as follows:

$$M = \left[\left| P_{total} \right| / \left(P_N \cdot \beta_{\max}^2 \right) \right] \tag{5}$$

In the formula, P_{total} the total power command received by the station; P_N is the rated power of the transformer and $\beta_{\max}^2$ is the minimum value of the optimal load rate of the transformer.

The power of each transformer is determined by taking the minimum total energy loss of the input transformer as the objective function, as follows:

$$F_1 = \min \sum_{i=1}^{M} P_{T,i} \cdot \left(1 - \eta_{T,i} \right) \tag{6}$$

$$\sum_{i=1}^{M} P_{T,i} = P_{total} \tag{7}$$

$$P_{T\min} \le P_{T,i} \le P_{T\max} \tag{8}$$

In the formula, $P_{T,i}$ represents the power of the i transformer; $\eta_{T,i}$ is the efficiency; $P_{T\min}$ and $P_{T\max}$ represent the minimum and maximum limits of the output power of the transformer, respectively.

3.3 Energy storage unit optimization layer

Based on the output power of the transformer figured out via the upper-level mechanism, the health state and each unit's SOC are considered in this layer, and the balanced control based on SOC and SOH of the energy storage unit is realized, to prolong the service life of the energy storage. The SOC of the energy storage unit can be calculated based on the initial SOC, charge-discharge power and time, but the SOH of the power storage unit is not easy to obtain directly. Although the SOH value in the energy storage power station can be read directly through the battery management system, the SOH value is required to optimize the power distribution in real time, so direct reading is not applicable.

In this paper, the approach from reference [9] is applied to calculate the life attenuation degree of the unit according to the charge and discharge capacity, to calculate the SOH resistance coefficient, as follows:

$$\Delta E_{i,j}(t)=P_{i,j}^{c}(t)\eta_i\Delta t+\frac{P_{i,j}^{d}(t)\Delta t}{\eta_i} \tag{9}$$

$$\Delta H_{i,j}(t)=\frac{\int_0^t \Delta E_{i,j}(k)dk}{2n_{100\%}E_{i,j}} \tag{10}$$

$$\omega_{i,j}(t)=\frac{\Delta H_{i,j}(t)}{\sum_{j=1}^{n}\Delta H_{i,j}(t)} \tag{11}$$

The $\Delta E_{i,j}(t)$ is the change capacity. $\Delta H_{i,j}(t)$ and $\omega_{i,j}(t)$ are the life attenuation degree and SOH resistance coefficient of the unit. $E_{i,j}$ is the rated capacity of the power storage unit, and $n_{100\%}$ is the number of cycles of the unit at full cycle depth.

To achieve balanced management considering SOC and SOH, it is necessary to control the output power of each energy storage unit according to the following rules [10]. Once the energy storage subsystem requires charging, the unit whose SOC falls below the average value is preferentially charged compared with the unit whose SOC exceeds the average value. Between these multiple charged energy storage units, it is necessary to allocate power instructions from high to low according to the unit's health condition. When the subsystem needs to discharge, the priority of the unit to undertake the power instruction is opposite to that of the charging.

In summary, the SOC balance recovery objective function of the energy storage unit based on the SOH resistance coefficient is as follows [11]:

$$\begin{cases} F_2=\min\sum_{j=1}^{n}\left|K_{i,j}+(1-\omega_{i,j})\Delta S_{i,j}(t)\right|\cdot P_{i,j}(t), K_{i,j}>0 \\ F_2=\min\sum_{j=1}^{n}\left|K_{i,j}+(1+\omega_{i,j})\Delta S_{i,j}(t)\right|\cdot P_{i,j}(t), K_{i,j}\le 0 \end{cases} \tag{12}$$

$$\Delta S_{i,j}(t)=\Delta E_{i,j}(t)/E_{i,j} \tag{13}$$

$$S_{i,mean}(t-1)=\frac{1}{n}\sum_{j=1}^{n}S_{i,j}(t-1) \tag{14}$$

$$K_{i,j}=S_{i,j}(t-1)-S_{i,mean}(t-1) \tag{15}$$

$$P_{T,i}(t)=\sum_{j=1}^{n}P_{i,j}(t) \tag{16}$$

$$P_{i,j}(t)=P_{i,j}^{d}(t)-P_{i,j}^{c}(t) \tag{17}$$

In the above equation, $\Delta S_{i,j}(t)$ is the variation of the SOC of the energy storage unit, and $S_{i,mean}(t-1)$ is the average value.

Figure 3 is the power distribution flow chart of a single total power command sent from the station to each subsystem and finally to each unit.

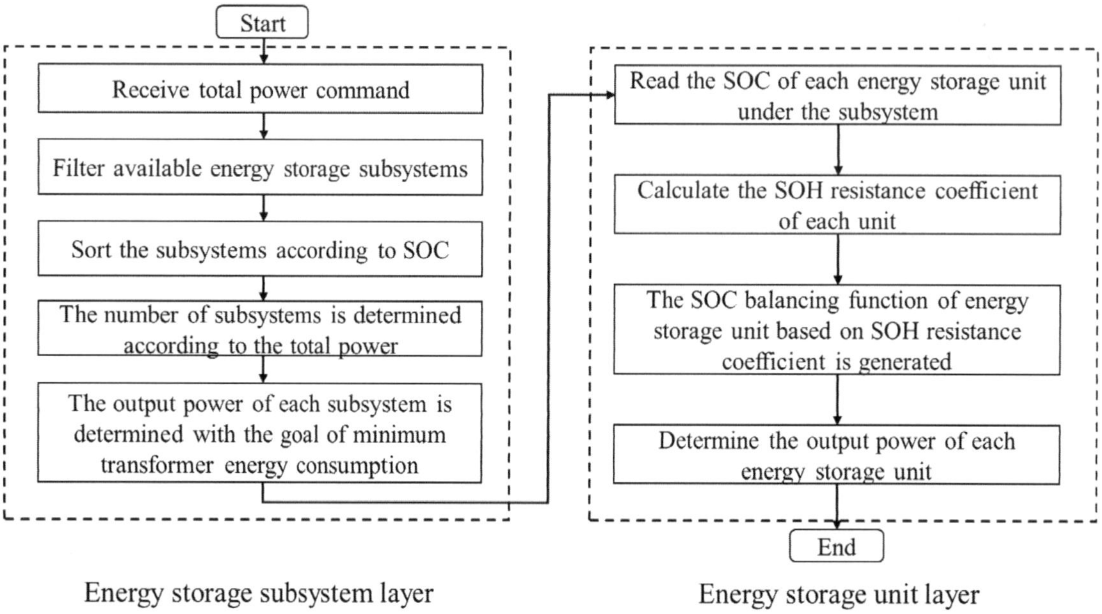

Figure 3. Flow chart of power distribution.

4. Simulation verification

4.1 Transformer efficiency curve

This paper takes the actual energy storage power station in a certain area as an example [12]. The installed capacity is 10 MW / 20 MWh, including four subsystems, and each subsystem contains four units. Therefore, the energy storage subsystem is 2.5 MW / 5 MWh, and the energy storage unit is 0.625 MW / 1.25 MWh. The energy storage PCS has a rated power of 630 kW and a transformer capacity of 2500 kVA. For the S11-2500kVA oil-immersed transformer, the no-load loss is 1.36 kW and the load loss is 23 kW. The efficiency curve of the transformer is shown in Figure 4.

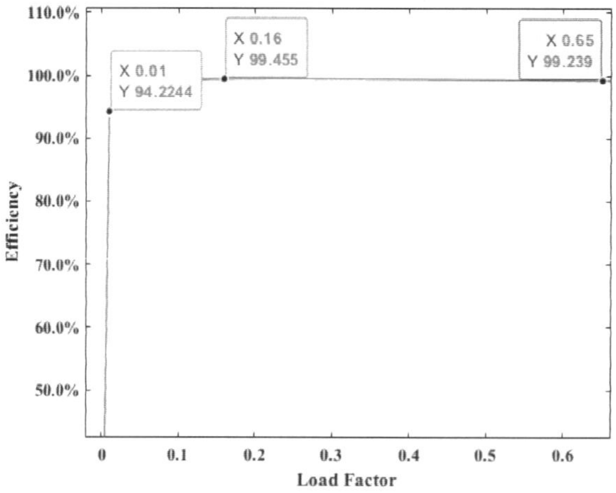

Figure 4. Transformer efficiency curve diagram.

The figure above shows that the efficiency difference of the transformer amid varying conditions of operation is large. The output of the transformer is close to the megawatt level. The small efficiency difference will also lead to large energy loss. Therefore, it is meaningful to consider the transformer efficiency in the process of allocating power instructions.

4.2 Subsystem optimization layer

The energy storage power station consists of four subsystems. Each subsystem contains four energy storage units. The initial SOC values of each energy storage unit are shown in Table 1.

Table 1. Initial value of SOC.

Number	1-1	1-2	1-3	1-4	2-1	2-2	2-3	2-4
SOC/%	70	40	50	60	60	55	69	45
Number	3-1	3-2	3-3	3-4	4-1	4-2	4-3	4-4
SOC/%	47	52	51	62	51	54	61	40

For example, considering energy storage's involvement in the electricity market, a power command is received every 15 minutes. The output power of each energy storage subsystem is shown in Table 2.

Table 2. Output power of each subsystem.

Subsystem 1/MW	2.233	-2.1	-2.15	1.5
Subsystem 2/MW	2.233	0	-2.15	0
Subsystem 3/MW	2.233	-2.1	-2.15	0
Subsystem 4/MW	0	0	-2.15	0
Total Power/MW	6.7	-4.2	-8.6	1.5

4.3 Energy storage unit optimization layer

Taking the energy storage participating in the electric energy market and responding to the frequency modulation command as an example, based on responding to the power command that changes once every 15 minutes, the frequency modulation command signal that changes once every 2 s is superimposed. Since the power allocation strategy of each unit under the power storage subsystem is as described in Section 2.3, only the response of four units under one subsystem is plotted here. The initial SOC and cumulative charge and discharge of the unit are shown in Table 3.

Table 3. Energy storage unit's initial state value.

Number	1-1	1-2	1-3	1-4
SOC/%	70	40	50	60
Cumulative Energy / MWh	12.5	20	18.4	16.8

In the process of power distribution, the energy storage unit allocates power reasonably based on its own SOC and SOH state. Figure 5, Figure 6 and Figure 7 are the output power, SOC and cumulative charge and discharge of each energy storage unit, respectively.

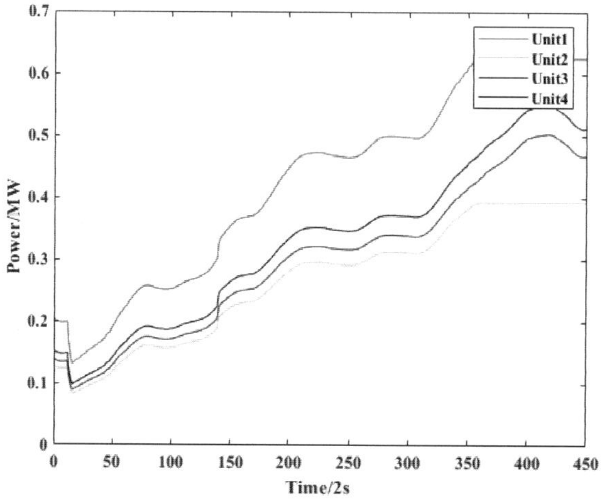

Figure 5. The each unit's output power.

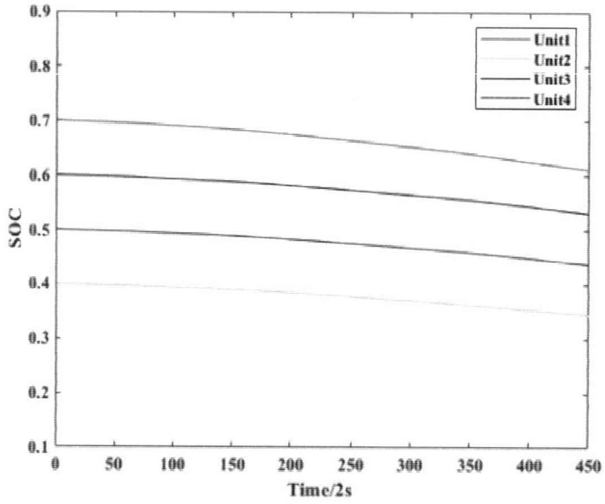

Figure 6. The SOC of each energy storage unit.

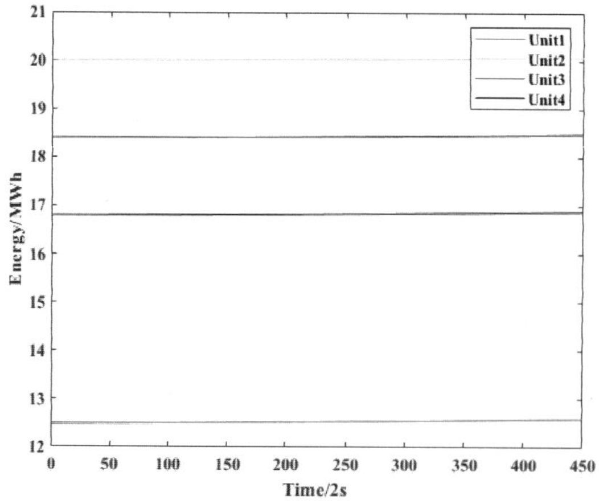

Figure 7. The cumulative charge and discharge amount.

From the above diagram, It is evident that the unit numbered 1-1 has a higher initial SOC and a lower cumulative charge and discharge, that is, the SOH is higher, and the unit always bears a larger power instruction. In addition, due to the short discharge time (15 minutes), the unit's starting SOC is quite different, and the SOC does not completely converge at the end of the discharge, but it can be seen that the SOC of each unit is gradually converging. The solution time of the above results is in milliseconds, which shows the viability and efficiency of the two-layer power optimization allocation strategy.

5. Conclusion

In this study, the energy storage power station receives the power command and performs the power distribution process. On the one hand, the energy loss is considered for the operating efficiency of the transformer. On the other hand, the health status and state of charge of the battery pack are considered to complete the entire power command distribution process. The proposed strategy can make each unit's SOC as consistent as possible during the operation and can make the unit with a good health state more active, and the unit with a poor health state less active and shallow, avoiding the "shortboard effect" of the power station. The problem has certain practical application value in engineering. However, the actual health status is not only related to the cycle depth but also closely related to the number of cycles. The quantification of the health status of the strategy is only characterized by the cumulative charge and discharge amount, which needs to be further improved and perfected in the future.

Acknowledgments

This work was financially supported by the Hangzhou Electric Power Equipment Manufacturing Co., Ltd.

References

[1] Li Junhui, Zhang Jingxiang, Mu Gang, etc. Coordinated optimal dispatching of peak and frequency regulation of independent energy storage in the auxiliary service market [J/OL]. *Chinese Journal of Electrical Engineering*, 1-15 [2024-11-26]. [CrossRef]

[2] Xu Ning, Zhou Bo, Ling Yunpeng, etc. Cooperative optimization strategy of independent energy storage participating in energy and auxiliary services in the spot market [J / OL]. *Modern Power*, 1-8 [2024-11-26]. [CrossRef]

[3] Shi Y, Xu B, Tan Y, et al. Optimal battery control under cycle aging mechanisms in pay-for-performance settings [J]. *IEEE Transactions on Automatic Control*, 2018, 64(6): 2324-2339. [CrossRef]

[4] Li Xiangjun, Ma Rui. An energy management method for energy storage systems considering the health status of battery packs [J]. *Grid Technology*, 2020,44 (11): 4210-4217. [CrossRef]

[5] Duan Shuangming, Yu Hang, Liu Cong, et al. BESS power allocation strategy considering the consistency of the state of health and the state of charge of the energy storage unit [J]. *Power System Automation*, 2023, 47 (05): 65-73. [CrossRef]

[6] Li Jianlin, Wu Yiwen, Wang Nan, et al. Review and Prospect of Research on Gigawatt-level Electrochemical Energy Storage Power Station [J]. *Power System Automation*, 2021, 45 (19): 2-14. [CrossRef]

[7] Zhang Guobing, Fu Xinian. Transformer efficiency curve and its characteristic parameters [J]. *Transformer*, 2000, (06): 18-21. [CrossRef]

[8] Ye Hui, Li Aikui, Tian Gangling, et al. Double-layer power allocation strategy of battery energy storage power station considering energy efficiency and SOC balance [J]. *Chinese Journal of Electrical Engineering*, 2024, 44 (13): 5185-5196. [CrossRef]

[9] He Hongjie, Zhang Ning, Du Ershun, et al. Review of modeling methods for operational efficiency and lifetime decay of large-scale electrochemical energy storage on the grid side [J]. *Power System Automation*, 2020, 44 (12): 193-207. [CrossRef]

[10] Li N, Gao F, Hao T, et al. SOH balancing control method for the MMC battery energy storage system[J]. *IEEE Transactions on Industrial Electronics*, 2017, 65(8): 6581-6591.[CrossRef]

[11] Vennam G, Sahoo A. A dynamic soh-coupled lithium-ion cell model for state and parameter estimation[J]. *IEEE Transactions on Energy Conversion*, 2022, 38(2): 1186-1196.[CrossRef]

[12] Diao R, Hu Z, Song Y. Subgradient of Cycle-based Aging Cost Function and Its Application in Optimal Operation of Battery Energy Storage System with Multiple Subsystems[J]. *IEEE Transactions on Energy Conversion*, 2023. [CrossRef]

Journal of Physics: Conference Series **3011** (2025) 012019 doi:10.1088/1742-6596/3011/1/012019

A segment location method based on LCSS similarity for active distribution network

Yuyang Zhao* **and Jianwen Zhao**

School of Electrical and Control Engineering, Xi'an University of Science and Technology, Xi'an 710054, China

*Corresponding author's e-mail: 1344799330@qq.com

Abstract. Aiming at the problems that the distributed power supply mainly composed of inverter interface distributed generator (IIDG) is connected to the distribution network in large numbers, the fault characteristics are not obvious, and the harmonic components are complex, which is different from the characteristics of traditional power supply and the failure of traditional section location method, a fault section location method combining variational mode decomposition (VMD) and the longest common subsequence (LCSS) similarity is proposed. First, the transient short-circuits current characteristics of IIDG, and synchronous power supply are analyzed to find that there are differences in their high-frequency components. VMD decomposition is used to obtain the high-frequency components of transient short-circuit current on both sides of each segment. LCSS is used to calculate the similarity to determine whether the high-frequency components on both sides of the segment are similar and locate the fault segment. Different fault points and other conditions can be accurately located, which proves the effectiveness of the proposed method.

1. Introduction

In recent years, solar energy, wind energy, and other renewable energy power generation as a representative of the distributed power generation technology has been rapidly developed and has wide applications, photovoltaic as a representative of the IIDG(Inverter Interfaced Distributed Generator), a large number of access to make the traditional single-power radial network transformed into a multiple power supply at the same time. Active distribution network, IIDG, by limiting amplitude and phase-locking and other links, showing randomness, nonlinearity, and other characteristics [1], seriously affecting the accuracy and effectiveness of the traditional fault segment location methods. Therefore, to ensure the safe and stable operation of active distribution networks with many IIDGs, it is of great significance to study the applicable fault location methods.

Active distribution network zone positioning mainly has two kinds: active and passive, the former for stability ground fault positioning accuracy is higher, but it is difficult to accurately detect certain intermittent faults and transient faults due to the performance of the detection equipment, the impact of the higher investment costs [2] [3], the latter to obtain the fault information is richer, high reliability, depending on the signals and modalities utilized, passive methods can be classified as steady state, transient, and AI methods.

The steady-state method mainly uses the information of impedance [4], current magnitude, and phase [5] [6] for localization, but since IIDGs are nonlinear components, performing steady-state equivalence loses accuracy, and the limiting control link makes the steady-state characteristics less obvious. Another type of localization uses the similarity degree of current waveforms detected on both sides of the zone, such as the method of calculating the similarity degree of transient waveforms using algorithms such as Dynamic Time Warping (DTW) [7], Hausdorff distance [8], etc. This type of method is simple in principle and convenient to calculate, but it is susceptible to the influence of noise and IIDG

Journal of Physics: Conference Series **3011** (2025) 012019 doi:10.1088/1742-6596/3011/1/012019

access, which leads to the problem of reliability degradation. In addition, the transient quantities can also be localized by extracting time-frequency features, such as Local Mean Decomposition (LMD) decomposition, wavelet transform, etc. LMD itself has an end-point effect, which is not effective for the decomposition of transient signals with complex harmonic components [9]; wavelet analysis [10] is computationally complex, and the localization effect is greatly affected by the selection of wavelet bases.

The artificial intelligence method is used to locate faults in active distribution networks through model training [11] [12], but the model training process is still used as a black-box model, which does not make full use of the fault mechanism characteristics of the active distribution network, has poor interpretability, and the model training is time-consuming and costly.

Based on the above problems, the current active distribution network segment localization method containing IIDG does not consider the IIDG fault characteristics when a two-phase short-circuit occurs, and the localization method is of poor reliability and difficult to extract fault characteristics, this paper adopts the fault information-rich transient current signals, analyzes the fault transient current characteristics to derive the discrepancy characteristics, and extracts the high-frequency component of transient current by using VMD, and then identifies the fault characteristics by LCSS similarity. The degree to identify the fault features and finally realize the fault segment localization.

2. Fault characterization

2.1. Impact of IIDG access on zone localization

Figure 1 shows the typical topology of a typical 10 kV distribution network containing IIDG, G1 - grid system with synchronous generator as the main power source, and IIDG - inverter-type distributed power source accessed by the line.

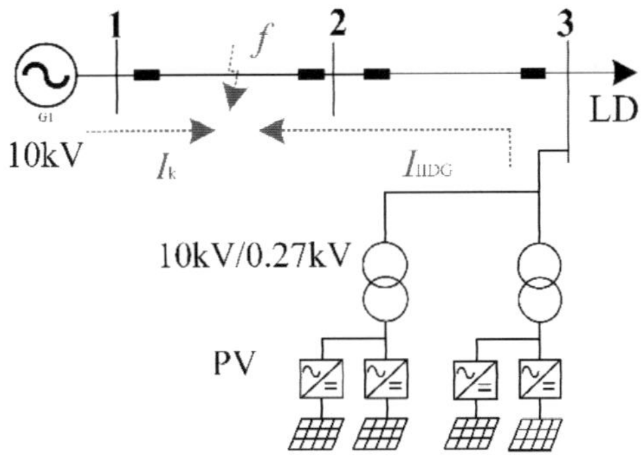

Figure 1. Typical topology of distribution network with IIDGs.

The impact of faulty localization in active distribution networks varies due to the different locations of IIDG access. When the IIDG is accessed to the bus, the current at the point of fault, after a fault occurs, is provided by the IIDG and the system power supply G1 at the same time, which will increase the level of the short-circuit current flowing through the line, enhance the sensitivity of the fault detection device, and improve the fault localization accuracy. When the IIDG is connected to the feeder, as shown in Figure 1. After a short-circuit fault occurs on the line, the short-circuit current is provided by the system side and the IIDG, respectively, and the reverse short-circuit current provided by the IIDG side-effects the fault information detected at the point of fault, and there is a risk of failure of the traditional zone localization method [13].

Due to the characteristics of IIDG output, such as nonlinearity and randomness [14], the short-circuit current presents different characteristics from the power side of the system, so it is necessary to analyze the analytical equation of the transient current of the IIDG and synchronous generator to derive the differences between the two, and to construct a method for locating the zones of the active distribution network applicable to the active distribution network containing the IIDG.

2.2. IIDG Transient Current Output Characteristics

When an asymmetrical fault occurs, due to the influence of the positive sequence and negative sequence current and voltage components on the grid side, the IIDG output active power and reactive power will contain two-fold frequency components, which will lead to two-fold frequency fluctuations of the DC bus voltage, and make the output of the current dq-axis in the voltage-current loop produce two-fold frequency components, and the three-phase currents obtained by the dq/abc coordinate transformation produce three-fold frequency components, which can seriously affect the normal operation of the IIDG. This will affect the normal operation of the IIDG. To eliminate this series of effects, the inverter voltage-current loop usually adopts the positive-negative sequence separation control to reduce the harmonic components in the output current by suppressing the negative sequence current as the control objective and the transient expression of the short-circuit current in AB two-phase short-circuit [15] is:

$$\begin{cases} i_q^{p*} = \begin{cases} 0 & U_T > 0.9 \\ 1.5(0.9 - U_T) & 0.2 \le U_T \le 0.9 \\ i_{\max}^* & \end{cases} \\ i_d^{p*} = \min(i_d^p(0), \sqrt{i_{\max}^* - i_q^{p*2}}) \end{cases} \tag{1}$$

$$\begin{aligned} i_A(t) &= i_d^{p*}\cos(\omega t+\theta_0) + i_q^{p*}\sin(\omega t+\theta_0) + \\ &\left[(i_d^p(0) - i_d^{p*})\cos(\omega t+\theta_0) + i_q^{p*}\sin(\omega t+\theta_0)\right] e^{-\frac{k_{iP}^P}{2L}t} \frac{2\sqrt{Lk_{iI}^P}}{\sqrt{4k_{iI}^P L - (k_{iP}^P)^2}} \\ &*\sin\left[\frac{4k_{iI}^P L - (k_{iP}^P)^2}{2L}t + \beta_p\right] \end{aligned} \tag{2}$$

During the fault period, IIDG will enter LVRT control, Eq. is the LVRT control equation and the positive sequence voltage magnitude on the grid side after the fault, and the positive sequence dq-axis current command value is i_d^{p*}, i_q^{p*}, $i_d^p(0)$ obtained by calculating Eq.(1), The k_{iP}^P and k_{iI}^P values are their proportional and integral coefficients, respectively, L is the filter inductance, and θ_0 is the initial angle of the fault.

From Eq.(2), it is known that in addition to the industrial frequency component, the short-circuit current contains low-frequency harmonic components, whose frequencies are mainly determined by the PI parameters of the inner loop of the positive-negative sequence separating currents, and they also show the non-industrial frequency characteristics of limited amplitude and attenuation.

Synchronous generator short-circuit current mainly contains an industrial frequency component, attenuated industrial frequency component, and attenuated DC component, IIDG transient short-circuit current contains an attenuated low-frequency harmonic component compared with the traditional synchronous machine, and the transient characteristics are significantly different.

3. Feature extraction based on VMD

The VMD algorithm is a completely non-recursive, adaptive signal decomposition method whose core idea is to transform the decomposition of the signal into the solution of the variational problem, to obtain a series of IMF components with different frequencies, to realize the decomposition of the signal, which is more suitable for the decomposition of non-smooth signals.

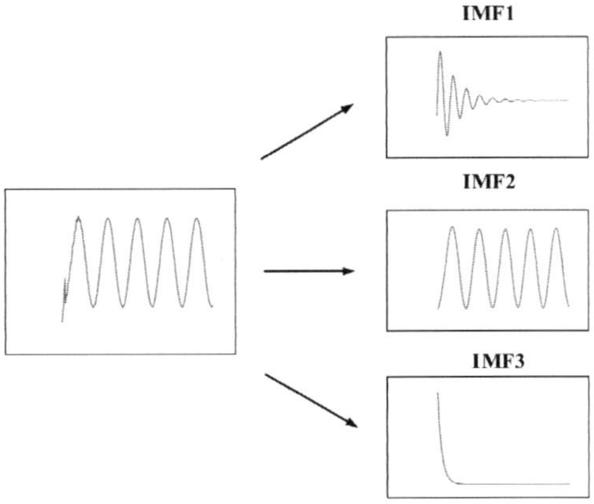

Figure 2. Diagram of decomposition layers.

As shown in Figure 2, when K is set to 3, the original current waveform can be decomposed to obtain three components, IMF1, IMF2, and IMF3, which correspond to the low-frequency harmonic component, the IF component, and the DC component, respectively, and the IMF1 with differential characteristics can be effectively extracted.

4. Principles of LCSS similarity calculation and zone localization methods

4.1. Principle of LCSS algorithm

LCSS (Longest Common Subsequence) is a method for comparing the similarity of two sequences, and for time series, calculating their longest common subsequence and converting it to LCSS distance can measure the similarity between trajectories [16]. It is usually solved by using the Dynamic Programming (DP) algorithm, where the obtained LCSS values are filled into the state transfer matrix and the longest common subsequence is found by backtracking, and the solution can often be represented by a RecursiveFunction [17], as shown below:

Suppose now there are two-time series data S_2 of length i and j respectively, then the length of the longest common subsequence is:

$$D_{\mathrm{ij}}=\begin{cases}1+D_{[i-1][j-1]}, S_{1[i-1]}=S_{2[j-1]}\\ \max\begin{bmatrix}D_{[i][j-1]}\\ D_{[i-1][j]}\end{bmatrix}, S_{1[i-1]}\neq S_{2[j-1]}\end{cases} \tag{3}$$

Where D_{ij} denotes the length of the longest common subsequence of time series S_1 and S_2, the computed length of the longest common subsequence is further normalized to compute the similarity:

$$L_{ij}=\frac{2D_{ij}}{i+j} \tag{4}$$

D_{ij} is the LCSS similarity of the two sequences, and D_{ij} is the length of the longest common subsequence of the sequences, due to $0\leq 2D_{ij}\leq i+j$, its value range is [0,1], the closer to 1, indicating that the longer the longest common subsequence between the two signals, the higher the degree of similarity between the two signals. Based on the principle of the above algorithm, the LCSS similarity of the two-time domain waveforms can be calculated, to compare the differences between the waveforms, combined with the differences between the IMF1 components of the signals on both sides of the faulty and normal segments obtained in the previous section can be used to locate the faulty segment.

4.2. The LCSS-based zone localization method

Due to the non-smooth characteristics of transient signals, it is necessary to focus on the local characteristics of the signal, using the LCSS algorithm to calculate the waveform similarity before the signal needs to be processed by the sliding window, first of all, calculate the LCSS value between the sliding time window of the two signals, i.e., the length of the longest common subsequence between the two signals is normalized to the length of the similarity between the different local windows to construct the similarity matrix, in which the horizontal and vertical axes are the time window serial numbers of the two signals, respectively. The horizontal and vertical axes are the serial numbers of the time windows of the two signals, and the calculated similarity is filled into the matrix corresponding to the serial number window to form the LCSS similarity matrix.

$$W_{mn}^{LCSS} = \begin{bmatrix} W_{11} & W_{12} & \cdots & W_{1n} \\ W_{21} & W_{22} & \cdots & W_{2n} \\ \vdots & \vdots & \ddots & \vdots \\ W_{m1} & W_{m2} & \cdots & W_{mn} \end{bmatrix}, \begin{matrix} 1 \le m \le K \\ 1 \le n \le K \end{matrix} \tag{5}$$

In Eq. (5): W_{mn} is the LCSS similarity between different time windows of the signal sequences on both sides of each segment on the line, where K is the number of sliding windows of the segment signals, and the diagonal elements in the matrix represent the LCSS similarity between the sequences corresponding to the time, so we only need to pay attention to the diagonal elements in the matrix when performing the sequence similarity discrimination.

According to Kirchhoff's current law, after a fault occurs, the same current flows on both sides of the normal section, and the LCSS similarity is close to 1. In the faulty section, since the short-circuit current is supplied by the power supply from both sides respectively, the difference is significant, and the LCSS similarity is close to 0. The normal and faulty sections can be identified by the LCSS similarity in the corresponding time window.

5. Simulation and analysis

Using Matlab to build the 10 kV active distribution network simulation model shown in Figure 3, there are two outgoing lines, the length of the section B_1B_2 is 3 km, the length of the section B_2B_3 is 5 km, the positive sequence impedance of the line is $Z_1 = 0.17 + j0.38\Omega / km$, the zero-sequence impedance is $Z_0 = 0.23 + j1.72\Omega / km$, the sampling frequency is 5 kHz, the simulation time is 0.6 s, the fault occurs in 0.2 s, and the fault is removed in 0.3 s. The effectiveness of the proposed zone localization method with different fault location times is set to verify.

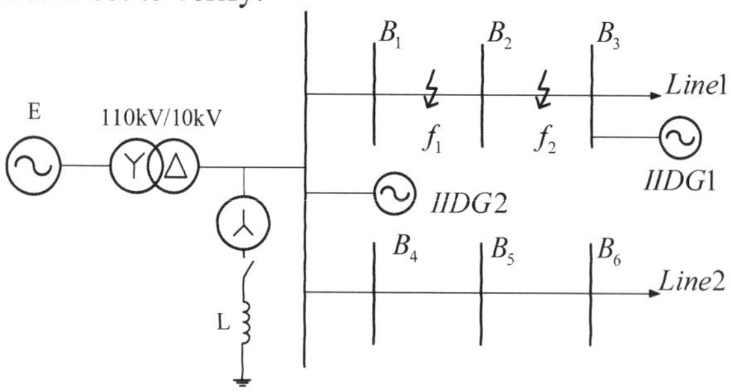

Figure 3. Simulation model of active distribution network.

5.1 IIDG1 access, fault at point f_2, validation of the effectiveness of segment localization methods

When IIDG1 is connected to the end, a phase-to-phase short-circuit fault occurs f_2. The zone is normal, and the current flowing on both sides of the zone is the same; while it is a faulty zone, the currents on both sides of the zone are provided by the system and IIDG, respectively, and the transient currents on both sides of the zone are discriminated by the LCSS similarity obtained from the VMD decomposition to locate the faulty zone as shown in the figures below:

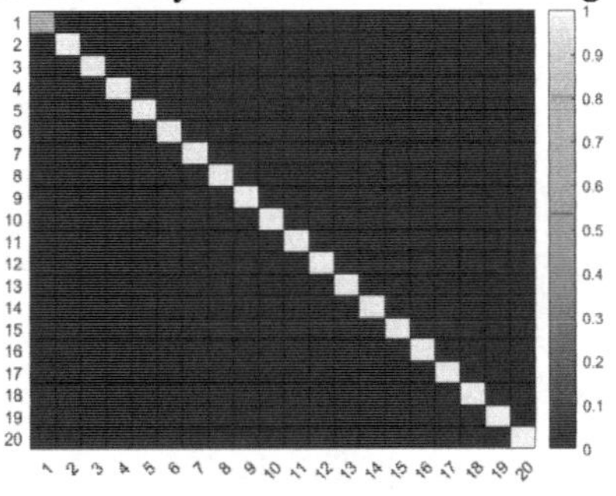

Figure 4. Heatmap of transient current similarity for f_2, zone B_1B_2.

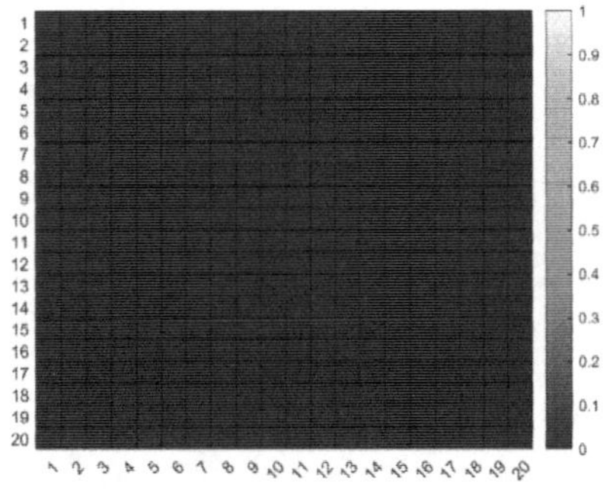

Figure 5. Heatmap of similarity of transient currents in *the* faulted zone B_2B_3.

As can be seen from Figure 4, the LCSS similarity value of the diagonal window in the LCSS similarity matrix calculated from the currents on both sides of the zone B_1B_2 is close to 1, whereas Figure 5, the LCSS similarity value of the diagonal window of the LCSS similarity matrix for the zone B_2B_3 tends to be 0. From this, it can be judged that the zone B_2B_3 is faulty.

5.2 IIDG1 access, failure at point f_1, validation of the effectiveness of segment localization methods:

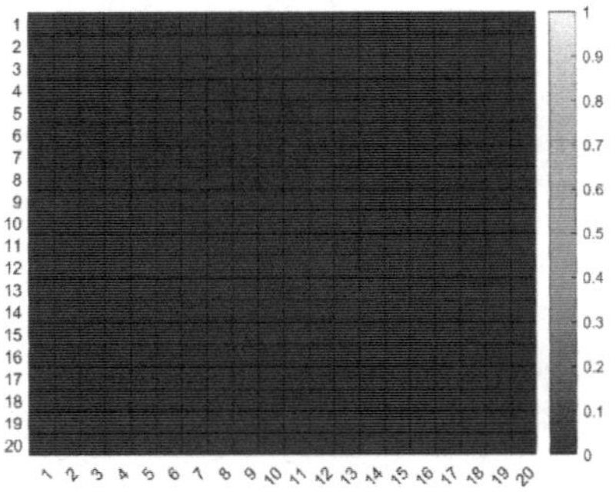

Figure 6. Heatmap of transient current similarity for f_1, zone B_1B_2.

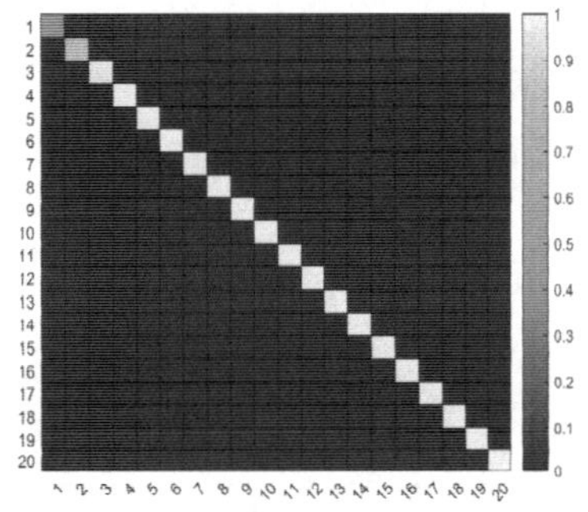

Figure 7. Heatmap of similarity of transient currents in *the* faulted zone B_2B_3.

When the fault occurs at a point f_1, the diagonal elements of the LCSS similarity matrix calculated from the transient low-frequency currents on both sides of the zone B_1B_2 can be found to be 0 through Figure 6， According to Figure 7, it can be found that the diagonal element of the LCSS similarity matrix calculated by the transient low-frequency current calculated on both sides of region B_2B_3 is 1, and the faulted zone B_1B_2 can still be accurately located. The calculated values of LCSS similarity for specific fault locations and the discrimination results are shown in **Table 1 and Table 2**:

Table 1. Calculated LCSS similarity and discrimination results at point f_1 fault.

Section	LCSS Similarity Calculated Values	verdict
B_1B_2	0.009	Faulty Section
B_2B_3	0.902	Normal Section
B_3B_4	0.981	Normal Section
B_4B_5	0.962	Normal Section

Table 2. Calculated LCSS similarity and discrimination results at point fault.

Section	LCSS Similarity Calculated Values	verdict
B_1B_2	0.984	Normal Section
B_2B_3	0.010	Faulty Section
B_3B_4	0.961	Normal Section
B_4B_5	0.921	Normal Section

6. Conclusion

(1) In this paper, by comparing the analytical equation of IIDG transient current in the asymmetric case with the transient characteristics of conventional synchronous power supply, it is concluded that the IIDG transient current contains significant low-frequency harmonic components, which can be used as a characteristic signal for fault zone discrimination, and further transient network-level modeling can be carried out in the future to obtain the transient characteristics of fault current on the line.

(2) Subsequently, this paper works on extracting transient low-frequency feature signals by VMD and uses the feature signals extracted on both sides of the zone to perform LCSS similarity calculation to determine the faulty zone and, in the future, the time-frequency mutation features of post-fault signals can be further extracted, and time-frequency image features can be identified and learned using artificial intelligence to achieve fault localization.

Acknowledgments

This work was financially supported by the Natural Science Basic Research Program of Shaanxi. (Program No.2023-JC-YB-308).

References

[1] LI H F, ZHANG Z G, LIANG Yu S, LIANG H M, WANG G. Current Differential Protection for Distribution Network with Inverter-interfaced Distributed Generators Based on Injection of Harmonic Signal[J]. Automation of Electric Power Systems, 2023, 47(22):120-128.

[2] LIU X, CAI Z, FAN H, et al. Experimental Studies on the RtEthernet-based Centralized Fault Management System for Smart Grids[J]. Electric Power Systems Research, 2020, 181: 106163.

[3] LI B H, LI Z Y, JU P.An Edge-cloud Collaborative Fault Detection Method for Low-voltage Distribution Networks Based on Random Matrix Theory[J].Proceedings of the CSEE, 2022, 42(01):25-37.

[4] REN X F, PAN Y H, HOU M, et al. Fault location of the renewable energy sources connected distribution networks based on time differences of the modal traveling waves[J]. IEEE Access, 2023, 11: 129671-129682.

[5] HU M L, LI N, LI Y M,etal.Fault Location Method of Distribution Network with Photovoltaic Power Based on Negative Sequence Component[J].Electric Power. 2024, 57(05):188-199.

[6] ZHU J R, MOU L H, GUO W M. Fault control of inverter interfaced distributed generator considering fault direction identification of the grid-connected microgrid[J]. Transactions of China Electrotechnical Society, 2022, 37(3): 634-644.

[7] LI H Y, ZHENG X C, ZHENG M R, et al. Virtual Fault Resistance and DTW Distance-based High Impedance Fault Section Location Method for Active Distribution Networks[J/OL]. Proceedings of the CSU-EPSA, 1-9[2025-01-2].

[8] LI Z X, WANG P F, WANG X, et al. Fault Location Method of Distribution Network Based on Amplitude Feature and Hausdorff Distance[J]. Automation of Electric Power Systems, 2020, 44(07):169-177.

[9] Xu Y C, ZHAO C C, SUN S H, et al. Fault Location for Active Distribution Network Based on Improved LMD and Energy Relative Entropy[J]. ELECTRIC POWER, 2021, 54(11):133-143.

[10] GUO M F, ZENG X D, CHEN D Y, et al. Deep-learning-based earth fault detection using continuous wavelet transform and convolutional neural network in resonant grounding distribution systems[J]. IEEE Sensors Journal, 2018, 18:1291-1300.

[11] LI Y Y, WANG Z P, ZHAO Q.Distributed Fault Section Location for ADN Based on Bayesian Complete Analytic Model and Multi-factor Dimension Reduction[J].Power System Technology, 2021, 45(10):3917-3927.

[12] LI J W, WANG X J, HE J H, et al. Distribution Network Fault Location Based on Graph Attention Network[J]. Power System Technology, 2021, 45(06):2113-2121.

[13] ZHOU C H, ZOU G B, DU X G, et al.A Pilot Protection Method Based on Positive Sequence Fault Component Current for Active Distribution Networks[J]. Proceedings of the CSEE, 2020, 40(07): 2102-2112+2390.
[14] Shuai Z. Transient Characteristics, Modelling and Stability Analysis of Microgrid[M]. Springer, 2021.
[15] CHANG Z X, YANG Z L, SONG G B,et al. Analysis of asymmetric fault current characteristics of inverter interfaced distributed generator under positive-sequence and negative-sequence current respective control strategy[J].Electric Power Automation Equipment, 2018, 38(01):44-51.
[16] TIAN Y, LIU X, YAO X S, FENG P L, ZHAO Y, LI Y. Single Phase Grouding Fault Line Selection Method of Resonant Grounding System Based on SAX and Spatial Information Entropy [J]. Southern Power System Technology, 2023, 17(07):103-114.
[17] WANG Q D. A New Algorithm Longest Extended Common Subsequence[J]. Telecommunication Engineering, 2024, 64(08):1307-1314.

Simulation-based study on the control strategy for PEMFC thermal management system

Yuhan Li [1], Zhifeng Zheng [2*], Yangge Guo [1], Xiaojing Cheng [1], Xiaohui Yan [1], Guanghua Wei [3], Shuiyun Shen [1], Junliang Zhang [1,*]

[1] Institute of Fuel Cells, School of Mechanical Engineering, MOE Key Laboratory of Power & Machinery Engineering, Shanghai Jiao Tong University, 800 Dongchuan Road, Shanghai 200240, China
[2] Shanghai Wenjing Energy Technology Co., Ltd., 1500 Duhui Road, Shanghai 201111, China
[3] SJTU-Paris Tech Elite Institute of Technology, Shanghai Jiao Tong University, 800 Dongchuan Road, Shanghai 200240, China

[*] Corresponding author's e-mail: junliang.zhang@sjtu.edu.cn (Junliang Zhang); zhifeng.zheng@wenjingenergy.com (Zhifeng Zheng)

Abstract: A control-oriented mechanism model and the corresponding control strategy for proton exchange membrane fuel cell (PEMFC) thermal management are developed based on internal model control (IMC). Additionally, a pre-stabilizing controller is introduced to improve the convergence of the proposed IMC-based strategy. The test results show that the IMC-based control strategies perform better responsiveness and robustness than the conventional PID-based strategy. The introduced pre-stabilizing controller is found to further improve the convergence to the target. Finally, it is also worth mentioning that the proposed IMC-based strategy and the corresponding optimized strategy achieve 46.8 % and 73.0 % lower ITAE of stack inlet temperature under dynamic load variation compared to the PID-based strategy, respectively, indicating their stronger dynamic tracking performance and the effective amelioration brought by the introduced pre-stabilizing controller, which presents practical value in PEMFC thermal management.

1. Introduction

Proton exchange membrane fuel cell (PEMFC) have received extensive attention on account of their zero-emission of greenhouse gases, high theoretical power density, and energy conversion efficiency [1]. However, since the electrical efficiency of PEMFC lies in the range from 40% to 60% during normal operation, lots of energy is eventually transformed into heat [2]. In this case, the accumulated heat will induce a significant temperature rise if not effectively dissipated and thus could do harm to the membrane and influence the catalyst activity [3]. Also, the uniformity of the temperature distribution inside the stack significantly affects the energy conversion efficiency and durability of PEMFC, so it is necessary to constrain the coolant temperature difference between the stack inlet and outlet [4]. In general, an effective thermal management system is of great importance to the performance and durability of PEMFC. However, thermal management systems are time-delay systems with strong nonlinearity caused by the inside heat and mass transfer process, so it is very challenging to implement effective control [5]. Additionally, there will inevitably be many interferences, including ambient air temperature, voltage fluctuations caused by variations in inlet gas pressure, ambient humidity, etc. Thus, effective thermal management requires strong robustness with high tolerance to nonlinearity.

Aiming at the aforementioned target, numerous studies have focused on optimization strategies using traditional control methods, such as PID and ON/OFF control, as well as the development of adaptive control algorithms based on advanced control theory, including model predictive control (MPC) [6], model reference adaptive control (MRAC) [5], state feedback control (SFC) [7], active disturbance rejection control (ADRC) [8], etc. For optimizations based on traditional control, PID is widely used due to its simple structure and sophisticated industrial applications. Accordingly, self-tunning methods such as whale optimization algorithm [9], fuzzy control [10], artificial neural network [11], etc., are employed to improve the responsiveness of PID under such highly nonlinear conditions. However, their performance is limited sometimes due to PID's pure feedback nature of lacking robustness against uncertain disturbances and drastic changes [12]. As for the advanced adaptive control algorithms, they can effectively improve the responsiveness and robustness with model prediction, optimal solution, comprehensive integration of disturbances and corresponding compensation, etc. [6-9]. Nevertheless, model-based control, e.g. MPC, MRAC, SFC, etc., requires relatively high demand for the accuracy of control object model which could be affected in wide-range load variation of PEMFC systems, while ADRC presents complexity for proper configuration of multiple parameters and extended state observer (ESO), also requiring high computational costs [13]. Among various advanced control algorithms, internal model control (IMC) is prominent due to its better adaptability to model deviation with more robustness for delayed disturbances [12], which could be further optimized for effective PEMFC thermal management.

Considering IMC's significant potential in applications of PEMFC thermal management, this paper proposes a control strategy that adjusts the thermostat and fans to control the stack inlet temperature through IMC, complemented with a constraint of the temperature between the stack inlet and outlet. Also, a pre-stabilizing controller is introduced to the IMC of the thermostat to further improve its convergence. The proposed control strategies are evaluated under step test, disturbance test, and dynamic load variation test in terms of their responsiveness, robustness, and dynamic tracking performance.

2. System modelling

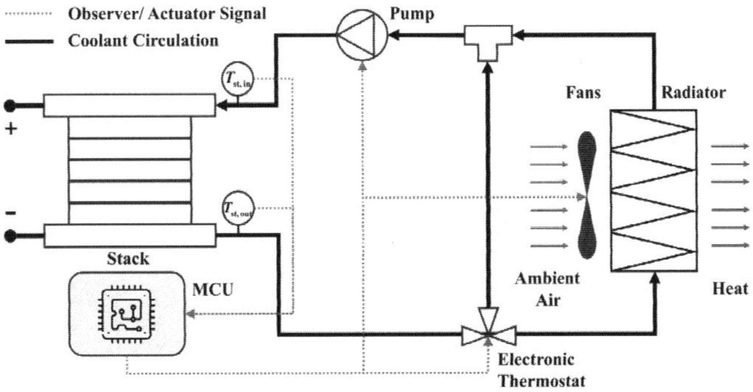

Figure 1. PEMFC thermal management system.

The structure of the PEMFC thermal management system in this study is shown in Figure 1. To simplify the system, the deionizer and expansion tank are ignored due to their little impact on the heat and mass transfer process inside the system, and, PTC is neglected since the cold start is not considered in this study. Meanwhile, the heat and mass transfer outside the stack and radiator can be considered ideal adiabatic processes to simplify the macro thermal analysis of the system.

First, the stack model is built based on the electrochemical characteristics of PEMFC and corresponding semi-empirical functions. The closed circuit potential of PEMFC can be calculated with the Nernst voltage E_{nernst}, activation overpotential V_{act}, ohmic overpotential V_{ohmic}, and concentration overpotential V_{con}, which is defined as [14]:

$$V_{st} = E_{nernst} - V_{act} - V_{con} - V_{ohmic} \tag{1}$$

where E_{nernst}, V_{act}, V_{ohmic} and V_{con} are calculated as follows, respectively.

$$E_{\text{nernst}} = 1.229 - 8.5\times10^{-4}(T_{\text{st}} - 298.15) + 4.3085\times10^{-5} T_{\text{st}}\left[\ln(P_{\text{H}_2}) + \frac{1}{2}\ln(P_{\text{O}_2})\right] \quad (2)$$

$$V_{act} = \mu_1 + \mu_2 T_{\text{st}} + 1.97\cdot10^{-7}\mu_3 T_{\text{st}}\cdot P_{\text{O}_2}\cdot e^{\frac{498}{T_{\text{st}}}} + \mu_4 T_{\text{st}}\cdot\log(i_{\text{st}}) \quad (3)$$

$$V_{ohm} = \left(\frac{t_{\text{m}}}{(\mu_5\lambda_{\text{m}} + \mu_6)e^{\left(\frac{1}{303.15}-\frac{1}{T_{\text{st}}}\right)\mu_7}} + \mu_8\right)\cdot i_{\text{st}} \quad (4)$$

$$V_{con} = \mu_{10}e^{\mu_9 i_{\text{st}}} \quad (5)$$

where T_{st} denotes the stack temperature, P_{H_2} and P_{O_2} stand for the hydrogen partial pressure and oxygen partial pressure, respectively, i_{st} is the stack current density, t_{m} refers to the thickness of the proton exchange membrane, λ_{m} refers to the water content of the membrane, and $\mu_1 \sim \mu_{10}$ are the fitting parameters obtained through the least squares method with experimental data. Eventually, the corresponding stack model is built based on MATLAB/Simulink.

Then, the integrated system model including all the inside heat and mass transfer processes is established analytically considering the heat generation, coolant circulation and heat dissipation of the system. The heat generation of the stack $\dot{Q}_{\text{heat}}$ is expressed by:

$$\dot{Q}_{\text{heat}} = \xi_{\text{cell}}\cdot\left(\frac{I_{\text{st}}}{2F}\cdot H_{\text{react}} - V_{\text{st}}\cdot I_{\text{st}}\right) \quad (6)$$

where ξ_{cell} stands for the fuel cell number in the stack, I_{st} denotes the working current, F is the Faraday constant, and H_{react} is the hydrogen-oxygen reaction enthalpy of the stack. Meanwhile, the heat dissipation $\dot{Q}_{\text{cl}}$ is described as:

$$\dot{Q}_{\text{cl}} = c_{\text{cl}}\cdot\dot{m}_{\text{cl}}\cdot(T_{\text{st,out}} - T_{\text{st,in}}) \quad (7)$$

In Equation (7), c_{cl} is the specific heat capacity of the coolant, $T_{\text{st, out}}$ and $T_{\text{st, in}}$ refer to the stack outlet and inlet coolant temperature, respectively, and $\dot{m}_{\text{cl}}$ denotes the coolant flow rate that is considered to approximately follow the pump speed linearly [10]. Meanwhile, the flow ratio of the major and minor circulation is adjusted by the thermostat, which is expressed as:

$$\begin{cases}\dot{m}_{\text{cl, major}} = k\dot{m}_{\text{cl}} \\ \dot{m}_{\text{cl, minor}} = (1-k)\dot{m}_{\text{cl}}\end{cases} \quad (8)$$

where k stands for the thermostat opening. Then, the heat dissipation of the radiator $\dot{Q}_{\text{rad}}$ is calculated by [15]:

$$\dot{Q}_{\text{rad}} = AK_{\text{fan}}\left(\frac{T_{\text{rad,in}} + T_{\text{rad,out}}}{2} - T_{\text{amb}}\right) \quad (9)$$

In Equation (9), A stands for the effective heat transfer area, T_{amb} denotes the ambient temperature, and K_{fan} is the overall heat transfer coefficient, which is fitted by:

$$K_{\text{fan}} = \gamma\cdot\left(\beta_1 n_{\text{Fan}}^2 + \beta_2 n_{\text{Fan}} + \beta_3\right) \quad (10)$$

where γ refers to the number of fans. n_{Fan} denotes the fan speed. β_1, β_2, and β_3 are the parameters fitted by the experimental data. Finally, the stack temperature at time t is:

$$T_{\text{st}}(t) = \int_{t_0}^{t}\frac{\dot{Q}_{\text{heat}} - \dot{Q}_{\text{cl}}}{c_{\text{st}}M_{\text{st}}} \quad (11)$$

where c_{st} is the specific heat capacity of the stack. M_{st} is the mass of the stack. The corresponding system model is also built based on MATLAB/Simulink and integrated with the stack model.

3. Control development

3.1. Control algorithms

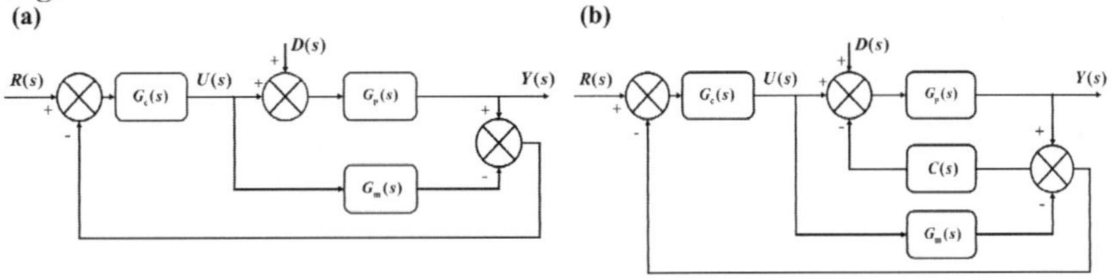

Figure 2. Structure of (a) IMC and (b) IMC with pre-stabilizing controller.

IMC has been widely used in industrial applications due to its good immunity to model deviation with robustness against time-delay disturbances. Based on the internal model prediction, the feedback compensation can help to realize good tracking performance to the control target, which makes it well-suited for PEMFC thermal management. Meanwhile, IMC has a simple structure and fewer adjustment parameters, which is easier to design and operate, as shown in Figure 2(a). A typical IMC loop mainly comprises a control object $G_p(s)$, a corresponding internal model $G_m(s)$, and a controller $G_c(s)$, while $R(s)$ is the reference value of the control target, $U(s)$ is the control signal of the actuator, $D(s)$ denotes the disturbances, and $Y(s)$ is the output of the control objective. Therefore, the closed-loop output of IMC is:

$$Y(s)=\frac{G_cG_p}{1+G_c(G_p-G_m)}R(s)+\frac{1-G_cG_m}{1+G_c(G_p-G_m)}D(s) \tag{12}$$

It is worth noting that the establishment of $G_m(s)$ in this study is based on the system identification under several groups of typical working conditions. The final $G_m(s)$ is obtained through the adjustment of the identification results with the comparison to the control object at the same input. As for the controller design of IMC, $G_m(s)$ can be decomposed as [16]:

$$G_m(s)=G_{m^-}(s)\cdot G_{m^+}(s) \tag{13}$$

where $G_{m^-}(s)$ is the minimum phase part while $G_{m^+}(s)$ contains the time delay phase and all the right-half plane zeros. Based on that, the design of $G_c(s)$ is:

$$G_c(s)=G_{m^-}^{-1}(s)F(s) \tag{14}$$

In Equation (9), $F(s)=\frac{1}{(1+T_f s)^n}$ is an introduced low pass filter, which should satisfy that all poles and zeros are in the left-half complex plane as well as $F(s)|_{s=0}=1$ to ensure stability. Meanwhile, we introduce a pre-stabilizing controller $C(s)$ for better responsiveness and convergence, which is illustrated in Figure 2(b). Notably, $C(s)$ is selected as a proportion-derivative (PD) controller for convenient figuration and a relatively wider suitable utilization range [17], which is designed as:

$$C(s)=K_p+K_d s \tag{15}$$

where K_p and K_d are the proportion and derivative, respectively. The tuning of them can be referred in [17].

3.2. Control strategies

Subsequently, the control strategies are designed accordingly. It is worth mentioning that the control of stack outlet temperature presents additional time delay derived from the flow channels inside the stack, and also, stack inlet temperature significantly affects the electrical efficiency of the stack [18]. Therefore, we select stack inlet temperature as the primary control target as it is affected by the actuators more directly and thus shows a faster response to the control signals.

The control of the stack inlet temperature is realized by the IMC adjusting the thermostat and fans. Based on that, the control of stack outlet temperature falls on the constraint of the temperature difference between the stack inlet and outlet, which is crucial to the temperature uniformity inside the stack, as mentioned above. In this case, a PID controller is employed to adjust the pump speed since the relationship between the pump speed and temperature difference is highly irregular under varying conditions.

Table 1. Summary of the control strategies.

	Thermostat	Fans	Pump
CS1	IMC	IMC	PID
CS2	IMC with pre-stabilizing controller	IMC	PID
CS3	PID	PID	PID

Furthermore, considering the sensitivity of the stack inlet temperature is much higher than other actuators, the aforementioned pre-stabilizing controller is introduced to the IMC loop controlling the thermostat. Additionally, to better evaluate the proposed control strategies, a conventional PID-based thermal management strategy is adopted for comparison. The three control strategies are summarized in Table 1.

4. Results and discussion

First, the transient responsiveness of CS1, CS2 and CS3 is evaluated under the step test. The initial condition is set to the steady state when $T_{\mathrm{st,\,out}}$ and $T_{\mathrm{st,\,in}}$ are 58.4 °C and 60 °C, respectively, while the target stack inlet temperature is set to 65°C and the reference value of the temperature difference between the stack inlet and outlet is taken as 6°C for its strict constraint. The load step is shown in Figure 3(a). The response of stack inlet and outlet temperature is presented in Figure 3(b), which shows that the stack inlet temperature controlled by the three control strategies rapidly increases with load and tracks the target temperature asymptotically after the overshoot caused by the radical load step.

However, it is found that CS1 and CS2 perform less overshoot and converge to the target faster than CS3, which indicates the IMC-based strategy achieves better responsiveness than the PID-based strategy with strong tracking performance due to their internal compensation, while PID just adjusts the actuators based on the feedback deviation. Moreover, the improved convergence exhibited by CS2 compared to CS1 demonstrates that the pre-stabilizing controller can enhance its responsiveness, helping maintain the stack inlet temperature closer to the target at a faster rate. Finally, the stack outlet temperature of CS1 and CS2 performs less variation within the range of 2.5°C compared to CS3, and the corresponding temperature difference is kept within 10°C due to both the strict constraint and little overshoot of the stack inlet temperature.

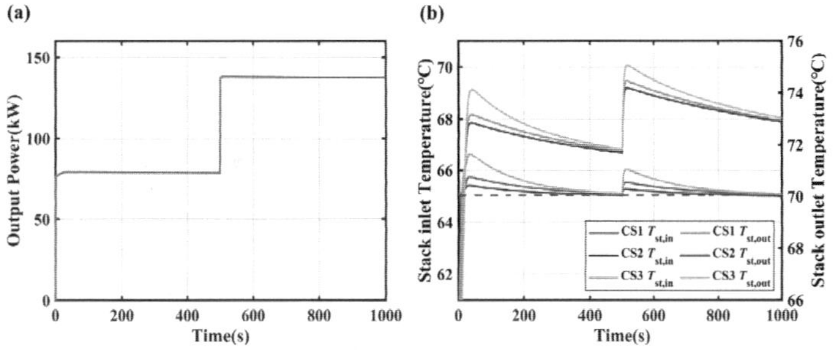

Figure 3. Load step test result: (a) Stack output power steps; (b) Stack inlet and outlet temperature.

Also, the actual dynamic operation of stacks faces the interference of ambient humidity, inlet gas pressure and temperature, etc., leading to voltage fluctuation at its theoretical output. Additionally, the ambient air temperature could significantly affect the heat dissipation of the radiator. Therefore, the anti-interference of these three control strategies is evaluated with the disturbances of voltage and ambient air temperature. To better evaluate the stack inlet temperature fluctuations caused by disturbances, root mean square error (RMSE) is introduced in this study, which is defined as:

$$\mathrm{RMSE}=\left[\frac{1}{N}\sum_{i=1}^{N}e^{2}(t_i)\right]^{\frac{1}{2}} \tag{16}$$

Figure 4(a) and Figure 4(c) present the stack inlet temperature fluctuations and the corresponding RMSE under the voltage disturbances of 10 V and -10V when the working current is 450A. It is noticed that CS1 and CS2 show much better immunity against voltage disturbances with little temperature fluctuations and lower RMSE compared to CS3, which illustrates the strong robustness of IMC-based strategies. Furthermore, CS2 suppresses the fluctuations lower than CS1, indicating that the introduced pre-stabilizing controller inside IMC can improve its robustness against disturbances.

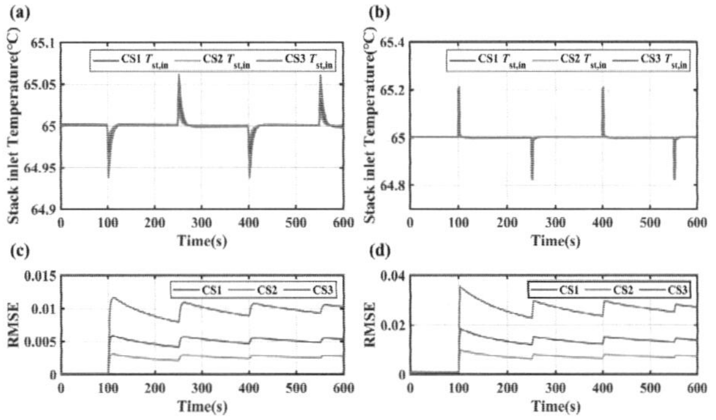

Figure 4. Disturbance test results: Stack inlet temperature under (a) voltage disturbances and (b) ambient air temperature disturbances; RMSE of the stack inlet temperature under (c) voltage disturbances and (d) ambient air temperature disturbances.

Meanwhile, the stack inlet temperature fluctuations and the corresponding RMSE under ambient air temperature disturbances of 5°C V and -5°C at the working current of 450A and ambient air temperature of 35 °C are shown in Figure 4(b) and Figure 4(d). Similar to the voltage disturbance test, CS1 and CS2 also present lower temperature fluctuations than CS3 under the ambient air temperature disturbances due to their strong robustness. As expected, CS2 can further lower the temperature fluctuations compared to CS1 with the introduced pre-stabilizing controller.

Finally, CS1, CS2, and CS3 are validated under the dynamic load variation shown in Figure 5(a). The initial condition is set identical to the afore-discussed step test. As shown in Figure 5(b), CS2 presents the best temperature tracking with the lowest fluctuations and best convergence under dynamic load variation, while CS3 performs the worst. Also, the stack outlet temperature from CS1 and CS2 fluctuates less than CS3 due to the better convergence of stack inlet temperature.

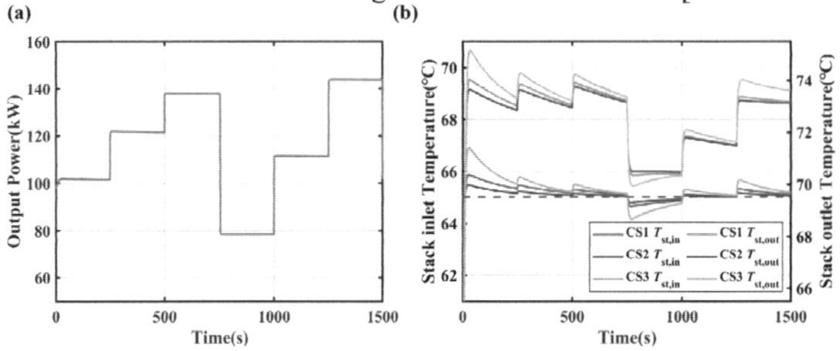

Figure 5. Dynamic load variation test: (a) Stack output power steps; (b) Stack inlet and outlet temperature.

Subsequently, to better compare the tracking performance between these three control strategies, the integral of time-weighted absolute error (ITAE) is introduced in this paper, which simultaneously evaluates dynamic deviation accumulation and convergence speed:

$$\mathrm{ITAE} = \sum_{i=1}^{N} t_i \left| e(t_i) \right| \Delta t \tag{17}$$

The ITAE of the stack inlet temperature controlled by CS1, CS2 and CS3 during the whole test is calculated as 212814.0638, 108024.8197 and 400042.6605, respectively. In this case, CS1's ITAE is 46.8 % lower than that of CS3, and CS2's ITAE is even 73.0 % lower than that of CS3, which indicates that IMC-based strategy can achieve significantly better tracking performance in dynamic thermal management compared to conventional PID-based strategies. IMC with the introduced pre-stabilizing controller can further improve the tracking performance, thus realizing better temperature control with lower fluctuations and better convergence under dynamic conditions.

In general, the proposed thermal management control strategy using IMC can realize effective temperature control with good tracking performance and strong robustness against disturbances. Additionally, the introduced pre-stabilizing controller inside IMC can further improve its responsiveness and robustness, thus performing better thermal management performance.

5. Conclusion

In this study, we developed a control-oriented mechanism model for PEMFC thermal management and designed the corresponding control strategy based on IMC, which takes the stack inlet temperature as the primary control target by adjusting the fans and thermostat through IMC, complemented with the constraint of the temperature difference between the stack inlet and outlet. Additionally, a pre-stabilizing controller was introduced to improve the convergence of the proposed IMC-based strategy. Then, the proposed control strategies were evaluated under step test, disturbance test, and dynamic load variation test in terms of their transient responsiveness, anti-interference, and dynamic tracking performance with the comparison of a PID-based strategy. During the tests, the IMC-based control strategies performed better responsiveness and robustness compared to the conventional PID-based strategy. The introduced pre-stabilizing controller presents effective optimization to the IMC-based strategy. It is also worth mentioning that CS1's and CS2's ITAE of stack inlet temperature under dynamic load variation are 46.8 % and 73.0 % lower than that of CS3, respectively, which indicates the strong dynamic tracking performance of IMC-based thermal management control strategy and the effective amelioration brought by the introduced pre-stabilizing controller, which shows practical value in PEMFC thermal management.

Acknowledgments

This work was financially supported by the National Key Research and Development Program of China (2021YFB4001305).

References

[1] Sun Y, Polani S, Luo F, Ott S, Strasser P, and Dionigi F (2021) Advancements in cathode catalyst and cathode layer design for proton exchange membrane fuel cells. Nat. Commun., 12(1): 5984.
[2] Pachauri R K and Chauhan Y K (2015) A study, analysis and power management schemes for fuel cells. Renew. Sust. Energ. Rev., 43: 1301-1319.
[3] Kumar R R, Suresh S and Suthakar T (2021) Thermal management of polymer electrolyte membrane fuel cell with stearyl alcohol and fans combined. J. Energy Storage, 41: 102847.
[4] Yu Y, Chen M, Zaman S, Xing S, Wang M, and Wang H (2022) Thermal management system for liquid-cooling PEMFC stack: From primary configuration to system control strategy. ETransportation, 12: 100165.
[5] Kim M B and Yoo J S (2020) Approximation-Based Adaptive Control of Constrained Uncertain Thermal Management Systems with Nonlinear Coolant Circuit Dynamics of PEMFCs. IEEE Access, 8: 83483-83494.
[6] Cao J, Yin C, Wang R, Li R, Liu R, and Tang H (2024) Dynamic thermal management of proton exchange membrane fuel cell vehicle system using the tube-based model predictive control. Int. J. Hydrogen Energ., 70: 493-509.
[7] Han J, Park J and Yu S (2015) Control strategy of cooling system for the optimization of parasitic power of automotive fuel cell system. Int. J. Hydrogen Energ., 40(39): 13549-13557.
[8] Li D, Li C, Gao Z, and Jin Q (2015) On active disturbance rejection in temperature regulation of the proton exchange membrane fuel cells. J. Power Sources, 283: 452-463.
[9] Shi Y, Sun Z, Wang Y, Li M, and Chen Z (2024) Thermal management of proton-exchange membrane fuel cells with enhanced fuzzy control via an improved whale optimization algorithm. Energy Technol., 12(6): 2400055.
[10] Jia Y, Zhang R, Lv X, Zhang T and Fan Z (2022) Research on Temperature Control of Fuel-Cell Cooling System Based on Variable Domain Fuzzy PID. Processes, 10(3): 534-534.
[11] Damour C, Benne M, Grondin-Perez B and Chabriat J (2014) Neural model-based self-tuning PID strategy applied to PEMFC. Engineering, 6(04): 159-168.
[12] Ranjan A, Mehta U and Saxena S (2023) A comprehensive review of modified Internal Model Control (IMC) structures and their filters for unstable processes. Annu. Rev. Control, 56: 100895.
[13] Tao L, Wang P, Wang Y, Ma X, Cheng P and Zhao D (2021) Variable structure ADRC-based control for load-side buck interface converter: Formation, analysis, and verification. IEEE T. Ind. Electron., 69(6): 6236-6246.
[14] Yang Q, Zeng T, Zhang C, Zhou W, Xu L, Zhou J, Jiang P and Jiang S (2023) Modeling and simulation of vehicle integrated thermal management system for a fuel cell hybrid vehicle. Energ. Convers. Manage., 278: 116745.
[15] Philip N and Ghosh P C (2022) A generic sizing methodology for thermal management system in fuel cell vehicles using pinch analysis. Energ. Convers. Manage., 269: 116172.
[16] Kaya I (2004) IMC based automatic tuning method for PID controllers in a Smith predictor configuration. Comput. Chem. Eng., 28(3): 281-290.
[17] Jin Q, Feng C and Liu M (2008) Fuzzy IMC for unstable systems with time delay. In: IEEE Pacific-Asia Workshop on Computational Intelligence and Industrial Application. Wuhan. pp. 772-778.
[18] Gao B and Zhou Y (2024) In-field experimental study and multivariable analysis on a PEMFC combined heat and power cogeneration for climate-adaptive buildings with taguchi method. Energ. Convers. Manage., 301: 118003

Hourly production operation simulation of an integrated energy system considering multi-energy complementarity

Zhilin Guo[1], Shiyu Liu[1,*], Jianye Rao[1], Yingxin Xu[1]

[1] China Electric Power Planning & Engineering Institute, Beijing 100120, China

*Corresponding author's e-mail: G1647318112@163.com

Abstract. To achieve "dual-carbon" target, the new energy sources, such as wind power and photovoltaic, have been widely used in China's power grid, but their intermittent and fluctuating output characteristics have increased the risk the unsafe operation of the system. By hourly power balance simulation of a new power system connected with a large amount of renewable energy, the start-up capacity and working position of various types of power supply are arranged reasonability. It is helpful to effectively realize the optimal allocation of system resources, and improve the security and reliability of system operation. Taking the operation cost as the optimization target, this paper fully considers the coordinated and complementary characteristics of multiple energy resources, and proposes an integrated energy system optimization scheduling strategy to realize the production simulation of the system at the level of 8760 hours in a year. Firstly, the output characteristics of different energy equipment in the system are analysed, and a joint optimization model of multiple power sources is established. Then, a two-level production operation simulation scheduling strategy of the integrated energy system is proposed, and the simulation model of the integrated energy system is established. Finally, the verification is conducted based on the data of the equipped power supple and load of a province. Results show that the hourly system power balance simulation proposed in this paper can comprehensively reflect and deal with the fluctuation characteristics of the new energy sources, and realize the consumption of new energy sources on the basis of ensuring the balance of system supply and demand. The simulation results can provide decision-making support for the planning of provincial regional energy systems.

1. Introduction

Under the influence of the climate crisis, vigorously developing renewable energy has become a global consensus [1]. Since 2020, China put forward the "dual-carbon" target, wind power and photovoltaic(PV), as the representative of the new energy, have been developed rapidly. In 2023, the renewable energy equipped capacity of China was nearly half of the whole world, China is in the dominant position in the renewable energy field [2]. According to the relevant forecasts, the added capacity of the new energy could be more than 100 million kilowatts per year. By 2030 and 2060, the equipped capacity of new energy could be 1.8 GW and 4.5 GW respectively [3]. However, with the access of a high proportion of renewable energy, superimposed on the increased uncertainty of load demand in extreme weather, the safe and stable operation of the system is challenged.

Power balance is a key feature and basic requirement of power systems. Power balance analysis has been studied for a long time, and in the late 20th century, power balance analysis methods have already appeared with the help of probabilistic production simulation based on time-series or continuous load profiles. In the traditional power system, the output on the power supply is relatively certain, and the power balance analysis mainly focuses on the uncertainty on the load side. The operational experience is used to perform the balance calculation under the certainty boundary by setting up typical scenarios, such as different load levels, different supply scenario of primary energy, different seasons, and pre-set sufficient system backup to cope with the stable operation of the system under unexpected circumstances.

However, the development of new power system brings subversive changes to the traditional power balance. On the one hand, both source and load are variable, the coupling of different energy is accelerated. The system turns to be probabilistic, multi-cycle coordination, multi-region complementary, and multi-subject synergistic, the traditional power balance calculation based on the typical curve can no longer satisfy the requirement. As a result, refined 8760-hour time-sequence production simulation has become a necessary tool for the system analysis. In addition, the operation system and mechanism have changed essentially, the significant increase of the regulation object and the influence factor led to exponential growth of the calculation scenario [4]. In essence, the operation simulation of the system connected with a large number of new energy is a kind of the mixed-integer linear programming problem (MILP) [5-6], there are a variety of solutions for this type of problem, including exact algorithms and heuristic algorithms, the former such as branch-bound and column generation algorithm, the latter such as genetic algorithm, ant algorithm and particle swarm algorithm. Theoretically, exact algorithm can obtain the optimal solution of the MILP problem, but the size of the solved problem is limited, the computational speed is restricted, and for large-scale MILP problems, the required solution space and computational time are often unacceptable. Whereas, heuristic algorithm can provide a feasible solution of the MILP problem, but it is easier to fall into the dilemma of local optimal solution. In addition, there are some mature solvers that can realize the computation of the MILP problem, but for the users, the solver is more like a "black box", which can not obtain the solution process of the problem, and the quality of the obtained solution can not be guaranteed.

Taking into account the engineering practicability and solving effect, based on the comprehensive consideration of multi-energy complementary, this paper proposes a planning simulation method of the integrated energy system. Firstly, the equipment type of the integrated energy system is analyzed, and the corresponding mathematical models are established based on the output characteristics of various types of power sources and equipment. On this basis, taking operation cost as the optimization goal, a two-layer optimization solution is proposed, and the planning simulation model of the integrated energy system is established to realize the optimization and complementation of multiple energy sources. Finally, an example analysis is carried out based on the equipped capacity and load demand of a province.

2. Joint optimization model for multiple power sources

2.1. Comprehensive energy system equipment type and coupling form

The new power system connected with a variety of energy is the central link of the integrated energy system. Focused on the new power system, this section sorts out the type and coupling form of the energy in the system, then the related equipment type is cleared, to provide a basis for subsequent modeling.

In terms of type, the primary energy in the system can be divided into two categories: fossil energy and non-fossil energy. From the point of view of resource endowment, fossil energy has always been the mainstay of China's energy supply, in which coal and natural gas power have an important position.

With the "dual-carbon" goal, the structure and characteristics of the new power system have undergone significant changes, the roles of coal and gas power is changing profoundly. Non-fossil energy has been developed on a large scale in recent years. Among them, nuclear power and hydropower which are relatively reliable in technology and output have been widely used in tradition power system. However, the developable space is limited to the resource condition. The new energy, represented by wind power, PV and biomass, is rich and easy to obtain, the cost also declines rapidly in recent years. The new energy comes to be the main force of China's energy supply in the future.

From the perspective of energy coupling, there are electricity, heat, cold, gas, hydrogen and other forms of energy in the integrated energy system, various types of energy can be converted into each other through certain equipment, they work together to meet the different energy needs of the system and promote the balance of supply and demand. Among them, electric and thermal occupy a dominant position in the energy consumption of industrial, commercial, residential and office loads, and the electric-thermal integrated energy system is also one of the most deployed and widely used integrated energy system. Electricity-heat integrated energy systems often closely link the two subsystems of power supply and heat supply through coupling devices such as combined heat and power unit, electric boilers and electric heat pumps. In this paper, the electricity-heat coupling is mainly embodied through combined heat and power unit.

In this paper, the integrated energy system consists of coal power including straight condensing unit and combined heat and power unit, gas turbine unit, hydropower, pumped storage, nuclear power, wind power, PV, photothermal, biomass power generation and the new energy storage.

2.2. *Equipment's output characteristic and mathematical model*

This section analyzes the output characteristics of various types of typical power sources and further proposes corresponding mathematical model of the equipment.

For the straight condensing unit, the output should be limited by the minimum technical output constraints, rated output capacity constraints and ramp rate constraints, which could be expressed as:

$$P_{\mathrm{cmin}} \le P_{\mathrm{c}}(t) \le P_{\mathrm{cN}} \tag{1}$$

$$P_{\mathrm{down}} \le P_{\mathrm{c}}(t) - P_{\mathrm{c}}(t-1) \le P_{\mathrm{up}} \tag{2}$$

where $P_{\mathrm{c}}(t)$ is the actual output of the unit at time *t*, P_{cmin} is the minimum technical output of the unit, P_{cN} is the rated power of the unit considering maintenance, P_{down} and P_{up} are the downward and upward ramp rate output constraints of the unit.

The gas turbine has the characteristics of flexible start-stop, and can be shut down directly during the operation, its mathematical model can be expressed as follows:

$$0 \le P_{\mathrm{g}}(t) \le P_{\mathrm{gN}} \tag{3}$$

where $P_{\mathrm{g}}(t)$ is the actual output of gas turbine unit at time *t*, P_{gN} is the rated power of the unit considering maintenance.

The characteristics of biomass power generation are similar to those of straight condensing unit, thus its mathematical model does not be repeated.

Combined heat and power units usually adopts the working mode of "setting electricity by heat", which means its power supply is basically limited by heat load, and its peak capability is restricted. Taking the condensate pumping unit as an example, under the consideration of various types of constraints, the unit's output range is shown in Figure 1, its power model can be expressed as follows:

$$\begin{cases} P_1(t) = K_{\min} H(t) + P_{\min} \\ P_2(t) = K_{\max} H(t) + P_{\max} \\ P_3 = K_b H(t) \\ 0 \le H(t) \le H_{\max} \\ \max(P_2(t), P_3(t)) \le P_i(t) \le P_1(t) \end{cases} \tag{4}$$

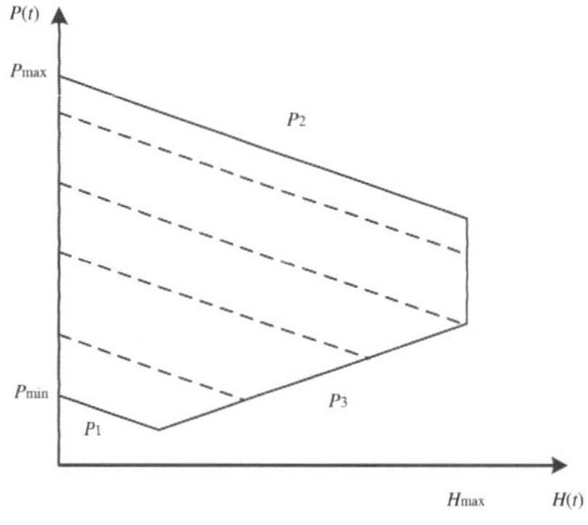

Figure 1. Output range of combined heat and power unit.

The hydropower unit's output should be limited between the forced output and the expected output. In addition, the power generation of the hydropower unit over a period of time is also limited by the average output, and the corresponding model can be expressed as:

$$P_f \leq P_{wa}(t) \leq P_e \tag{5}$$

$$\int P_{wa}(t)dt \leq P_a * \Delta t \tag{6}$$

where $P_{wa}(t)$ is the output of the hydropower unit at time *t*, and P_f, P_e, P_a are the forced output, expected output and average output of the hydropower, respectively.

The primary operational requirement of nuclear power unit is to ensure safety, thus it is generally not involved in peak shaving, and its output can be expressed as:

$$P_n(t) = P_N \tag{7}$$

where $P_n(t)$ is the output of the nuclear power unit at time *t*, P_N is the rated power of the unit considering the maintenance capacity.

The power of wind turbine is mainly affected by wind speed, and its power can be expressed as follows:

$$P_{wi}(t) = \begin{cases} 0 & v(t) \leq v_{in}, v(t) \geq v_{out} \\ \dfrac{v^3(t) - v_{in}^3}{v_N^3 - v_{in}^3} P_N & v_{in} \leq v(t) \leq v_{out} \\ P_N & v(t) \geq v_N \end{cases} \tag{8}$$

where $P_{wi}(t)$ is the wind turbine output at time *t*, P_N is the rated output power of the turbine, v_{in} the cut-in wind speed; v_{out} is the cut-out wind speed, v_N is the rated wind speed.

PV unit output is mainly affected by the light intensity irradiated to the PV surface, the system operating condition and PV physical parameter and other comprehensive decision. Normally, the output power of PV unit can be expressed as:

$$P_{sv}(t) = f_{sv} P_{svN} \left(\frac{G_T}{G_{T,STC}}\right)[1 + \alpha_p (T_{cell} - T_{cell,STC})] \tag{9}$$

where $P_{sv}(t)$ is the PV unit output power at the moment *t*, f_{sv} is the PV power generation system derating factor, which is equal to the ratio of the actual PV output power and the rated condition of the output power, used to take into account the loss due to the aging of the PV panels themselves as well as the surface obstruction caused by rain and snow. P_{svN} is rated capacity of the unit. α_p is the power's temperature coefficients. $T_{cell,STC}$ is the PV cell's temperature under STC condition.

Photothermal unit is another solar power generation technology, with the thermal storage subsystem, the output is more stable. Besides the maximum output limit, there are power limitation according to the different light conditions, which could be expressed as:

$$0 \leq P_{st}(t) \leq P_{stN} \tag{10}$$

$$\int_0^{23} P_{st}(t) \leq Q \tag{11}$$

where $P_{st}(t)$ is the output of the photothermal unit at time *t*, P_{sN} is the rated power of the unit, and Q is the total electricity output of the unit in a day.

The flexible regulation equipment in the system includes pumped storage and new energy storage unit. Both of them have similar characteristics, which can be expressed by the same mathematical model:

$$0 < P_{so}(t) \leq P_{soN} \tag{12}$$

$$P_{siN} \leq P_{si}(t) \leq 0 \tag{13}$$

$$Q_s(t) = Q_s(t-1) - \gamma P_{si}(t) + P_{so}(t) \tag{14}$$

$$0 \le Q_s(t) \le Q_{sN} \tag{15}$$

where $P_{so}(t)$ and $P_{si}(t)$ are the output and input power of the unit in the discharge and charge at time t respectively, P_{soN} and P_{siN} are the corresponding maximum output power. $Q_s(t)$ is the quantity stored in the unit at time t, and Q_{sN} is its maximum value. γ is the efficiency of the unit.

Due to the high reliability requirement of the power system, the equipment needs regular maintenance during the operation process, including coal power, gas power, hydropower, nuclear power, etc. The corresponding maintenance model can be expressed as:

$$P_{iN} = P_{iN0} - P_{im} \tag{16}$$

where P_{iN} is the available capacity of the unit considering maintenance, i is the category, including the straight condensing unit, gas power, combined heat and power unit, hydropower, nuclear power and so on. P_{iN0} is the equipped capacity of the unit of category i. P_{im} is the maintenance capacity of the unit of category i.

In addition, the power transmission between different regions is also an important part of the new power system, both sides sign the medium- and long-term protocol in advance according to their respective power balance needs. The simplified model can be expressed as:

$$P_{l\min} \le P_l(t) \le P_{l\max} \tag{17}$$

$$\int P_l(t) \le Q_{lN} \tag{18}$$

where $P_{l\min}$ and $P_{l\max}$ are the minimum and maximum power of the transmission line, respectively. Q_{lN} is the annual transmission quantity specified in the medium- and long-term protocol.

3. Integrated energy system planning simulation

3.1. Two-layer logical framework of optimization model

With the connection of large-scale new energy sources, the hourly refinement production operation simulation of the power system is an important way to ensure the reliable power supply, promote new energy consumption and optimize resource allocation.

The production operation simulation of power system contains a variety of equipment, the power output of each type equipment in 8760 hours is an optimization variable. Besides, the output characteristics of those equipment are different, which means the relationship between the optimization object and variable is non-linear. Considering the engineering practicability and the solving effect, this paper proposes an optimization solution method based on the empirical process, and a production operation simulation program based on C++ language is conducted to realize the system planning simulation. The core solution process of the program can be divided into two layers: the outer layer is the system level, different equipment's production operation is called according to the empirical algorithm, and the optimization object is achieved by the calling sequence of the equipment. The inner layer is the equipment level, according to the operation characteristics of different equipment, different operation goals are set to arrange the output of each type of equipment. The overall logical framework is shown in the following figure:

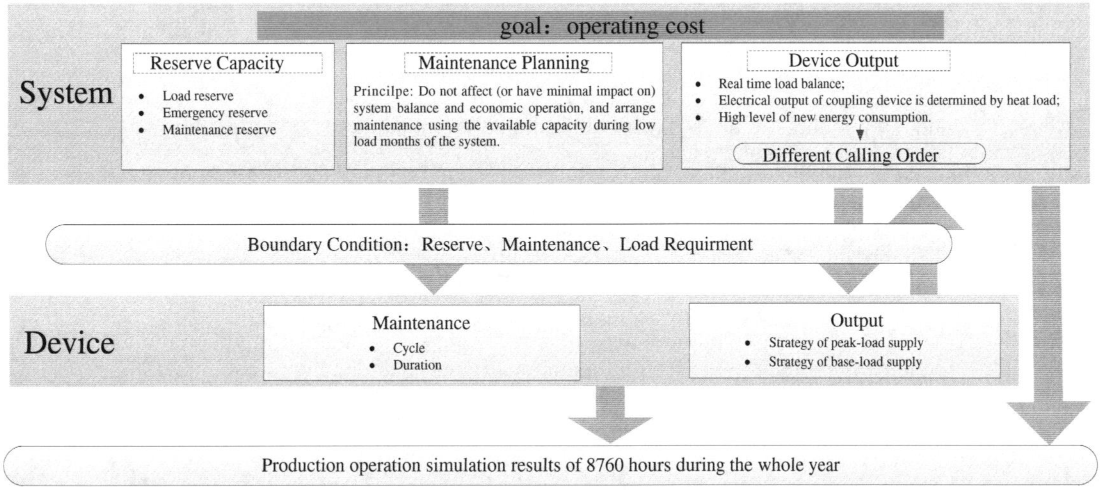

Figure 2. Two-layer logical framework of optimization model.

Considering the techno-economics characteristic, the working principle of each equipment could be expressed as follows: the wind power and PV power is consumed preference. Hydropower, photothermal power and gas power are mainly used to peak regulation. Pumped storage and new energy storage are used to peak regulation to reduce the wind and PV power curtailment, and what's more, they are used to substitute partial coal power considering the economy factor. Nuclear power is used to satisfy the base load. The power output of the combined heat and power unit is limited by the heat load. On this basis, the calling sequence of the equipment on the system level is arranged, and the executable logics of the various devices are different, aiming to ensure the power balance and reduce the operation cost as much as possible.

3.2. Objective function

The operation cost is taken as the optimization objective during the production operation simulation. Due to the different operational and techno-economic characteristics of various types of generating units, the contributions of the various types of the units to optimization objective are different, thus they should be arranged properly. Considering the output characteristics of these units, the proposed optimization object takes into account of the indexes, including the maintenance demand, the reserve demand, the power balance, the annual energy output and availability hour of different types of units, the fuel demand and the variable operation cost, and the transmission line. The optimization objective could be expressed as:

$$f = \min[\sum_{t=0}^{8759} E_{\mathrm{wq}}(t)] \cap \min\left\{ \sum_{\substack{0 \le i \le m \\ 0 \le t \le 8759}} F[P(i,t)] \right\} \tag{19}$$

where $E_{\mathrm{wq}}(t)$ is the abandoned electricity of the renewable energy at time *t*, including the wind, the PV and the hydropower. $F[P(i,t)]$ is the operation cost of equipment of category *i* at time *t*.

3.3. Constraints

Due to the system's operation demand and the units' operation characteristics, there are two kinds of constraints. Equipment operation constraints have been analyzed in detail in Section 2.2, on this basis, this section analyzes the system's constraints.

First of all, since electric energy cannot be stored in large quantities and the demand varies randomly, there should be a certain amount of reserve capacity in order to ensure the reliability of power supply. The corresponding model can be expressed as:

$$\sum P_{Ri}(j) = P_{R}(j) \tag{20}$$

$$\sum P_{Ci}(j) = P_{C}(j) \tag{21}$$

where $P_{Ri}(j)$ and $P_{Ci}(j)$ are the hot standby and cold standby capacity provided by the equipment of category i in the jth month respectively, $P_R(j)$ and $P_C(j)$, respectively, are system's hot standby and cold standby demand in the jth month.

Besides, the basic power balance constraint could be expressed as:

$$\sum P_i(t) = P_L(t) + P_l(t) \tag{22}$$

where $P_i(t)$ is the power output of the power supply of category i at time t. $P_L(t)$ is the load demand at time t, and $P_l(t)$ is the transmission power at time t.

4. Simulation Analysis

4.1. Case Introduction

The data of a certain region in a certain year is taken as an example to carry out 8760 hours of time-sequence production operation simulation.

From the point of view of the load, there are heat load and electricity load in the system, the load data is from the historical typical level year. For heat load, the supply lasts from January to March and from November to December every year, and the maximum value is 1667 MW. For power load, the electricity load shows two "double-peak" characteristics: during the year, the load demands in winter and summer are significantly higher than that in spring and fall; during the day, the load demands in morning and evening are significantly higher than other period. The maximum load is 7,956 MW.

In terms of power supply, the equipped capacity of straight condensing unit is 6.7×10^4 MW, the gas power is 1×10^4 MW, the combined heat and power unit is 1.2×10^4 MW, the nuclear unit is 1×10^4 MW, hydropower is 1.5×10^4 MW, wind power is 1.3×10^4 MW, the PV power is 3.5×10^4 MW, the photothermal power is 1.5×10^3 MW, biomass power is 2×10^3 MW, the output parameters of the hydropower, the wind and the PV power are from the region's typical level year. The available quantity of the hydropower is 5.8×10^6 MWh, the wind power is 3.5×10^6 MWh, the PV power is 5.1×10^6 MWh and the photothermal power is 8.2×10^5 MWh. Moreover, the pumped storage is equipped with 1.2×10^4 GW, which could regulate the power during a week with 10 hours storage duration, the conversion efficiency of the pumped storage is 0.75. The new energy storage is equipped with 4×10^3 GW, which could regulate the power during a day with 2 hours storage duration, the unit's conversion efficiency is 0.85.

Considering the surrounding situation, the case is located on a typical energy sending region, and the power transmission is conducted by a fixed curve. The transmission power is larger in the summer and fall during a year as well as the morning and evening during a day. The maximum transmission power is 5.3×10^3 MW, and the quantity of the whole year is 2.2×10^7 MWh.

The maintenance is arranged once a year for every unit, and the different units have different maintenance time. With the target of having less effect on system operation, the maintenance is prior to be arranged during the period of low load.

In terms of operation cost, the penalty with the abandoned wind or PV is 100 yuan/MWh, the purchased quantity when local unit couldn't satisfying the load is 460 yuan/MWh. The operation costs of coal power and gas power is referenced in [7].

4.2. Result

Based on the case described in Section 4.1, the open source solver CBC and the method proposed in this paper are used to calculate respectively, and the former calculation time is too long to obtain a solution. Thus the analysis is focused on the results of this paper's method. The results are evaluated from three aspects, including the effect of power balance calculation, maintenance arrangement, and operation cost.

(1) Maintenance arrangement

Before arranging the working position, the unit's maintenance plan is first analyzed and verified. In the process, straight condensing unit, combined heat and power unit, gas unit and nuclear unit are arranged according to their respective demands, and the maintenance capacity is shown in Figure 3.

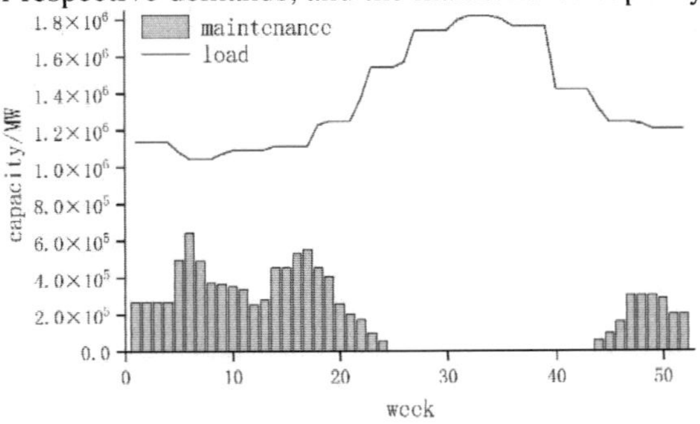

Figure 3. Maintenance arrangement results.

As can be seen from Figure 3, the unit's maintenance is basically arranged during the load's valley period, which can effectively alleviate the system peak pressure.

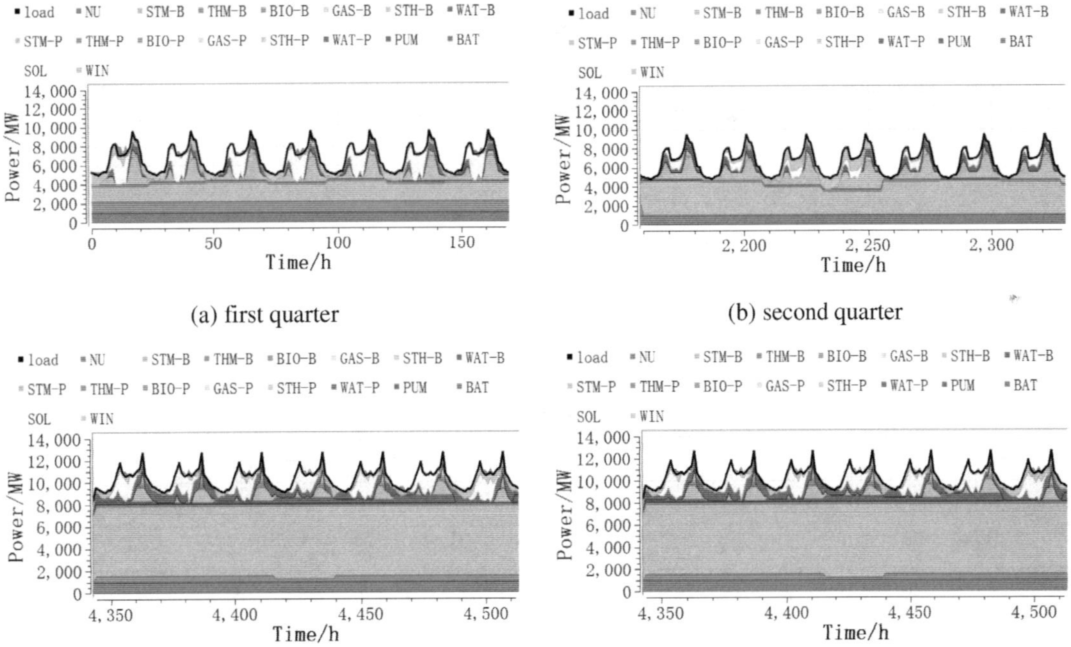

(a) first quarter (b) second quarter

(c) third quarter (d) fourth quarter

Figure 4. Simulation result of the four quarters.

(2) Power balance effect

Figure 4 shows the operating position diagram of various types of units in a typical week in four quarters of the year, where NU, STM, THM, BIO, GAS, STH, WAT, PUM, BAT are the nuclear power, straight condensing unit, combined heat and power unit, biomass power, gas power, photothermal power, hydropower, pumped unit and new energy unit respectively, "-B" presents the partial output for base load, "-P" presents partial output for peak load. Overall, nuclear power, combined heat and power unit, straight condensing unit, and biomass power unit are used to meet the base load, and gas power, hydropower, straight condensing unit, and photothermal unit provide peaking effects.

In terms of the heating demand, as shown in Figure 4(a), the first week of the first quarter is in the heating period, and the combined heat and power unit's output is determined by the heat load, with limited space for participation in peaking shifting. In the second, third and fourth quarters, the combined heat and power unit is determined by the electric load, and during the peak period, such as the first week of the third quarter shown in Figure 4(c), the combined heat and power unit generates electricity to maintain the system stability, and reduces the start-up capacity according to the load demand on the 185th day (corresponding to the position of 4416~4439 hours in Figure 4(c)). The actual output of the units is consistent with the setting.

In terms of the effect of reducing abandoned wind and PV, the first week of system operation is taken as an example, as shown in Figure 4(a), at the moment when the wind and PV power output is large, pumped storage and energy storage charge to reduce abandoned wind and PV power, and energy storage unit charges at the valley moment and discharges at the peak moment of the day to realize the daily regulation, and pumped storage unit reduces the abandoned wind and PV in the valley period and discharges during the peak period to realize the weekly regulation effect.

(3) Operation cost

Based on the method of this paper, the system annual generation fuel cost of 39.2 million yuan, the total coal consumption of 14,200 tons, the total natural gas consumption of 66,200 cubic meters, no power abandonment. The overall situation is shown in Table 1.

Table 1. Operation cost during the 8760 hours.

Category	Result
Total Power Generation Cost/Million Yuan	4.99
Total Coal Consumption/Ton	1.42
Total Gas Consumption/Million cubic meters	6.62
Abandoned wind power/MWh	0
Discarded photoelectricity/MWh	0

5. Conclusion

In order to maintain the supply and demand balance of the integrated energy system and realize the safe and stable operation of a new power system containing a high proportion of new energy, this paper proposes an hourly production and operation simulation method of the system taking into account multi-energy complementarity. First, the types of equipment in the system are clarified, and a multiple power supply model is built considering the output characteristics of each type of equipment. On this basis, the operation cost is chosen as the optimization objective, and the scheduling strategy is proposed considering the system and equipment to achieve the multi-energy complementarity and optimize the operation of the integrated energy system. Finally, through the case study, it is verified that the method proposed in this paper can effectively realize the annual 8760-hour production and operation simulation of the integrated energy system with large-scale new energy access. In the follow-up work, the quantitative evaluation of various configuration schemes can be realized based on the method of this paper, which provides support for the provincial energy planning.

References

[1] EI. Statistical Review of World Energy [R]. London, EI, 2024.
[2] IEA. Renewables 2023 [R]. Paris, IEA, 2024.
[3] IEA. World Energy Outlook 2024 [R]. Paris, IEA, 2024.
[4] YUAN Yiping, WANG Jianxue, ZHANG Yao, et al. Panoramic operation simulation method and evaluation index system of new energy power system considering Panoramic operation simulation method and evaluation index system of new energy power system considering multi-period coordination[J]. Power System Technology, 2020, 44(3): 799-806.

[5] Wu Hongyu, Guan Xiaohong, Zhai Qiaozhu, et al. Shortterm hydrothermal scheduling using mixed-integer linear programming[J].Proceedings of the CSEE. 2009, 29(28):82-88.
[6] SHAO Chengcheng, WANG Yanan, FENG Chenjia, et al. Long-term production simulation of power system considering multi-energy power characteristics[J]. . Proceedings of the CSEE, 2020, 40(13):4072-4080.
[7] GAO Song, LI Jun, SONG Hui, et al. An integrated thermal powerenergy storage system for improving primary frequency regulation performance of thermal power units[J]. power units[J]. Power System Protection and Control, 2023, 51(21): 116-125.

Journal of Physics: Conference Series **3011** (2025) 012022 doi:10.1088/1742-6596/3011/1/012022

Design and optimization of low torque ripple synchronous reluctance motor

Delin Li*, Li Zhu

School of Electrical Information and Electrical Engineering, Shanghai Jiaotong University, Shanghai 200240, China

*Corresponding author's e-mail: Lidl1095191297@sjtu.edu.cn

Abstract. Synchronous reluctance motors are synchronous motors that contain only reluctance torque. They are characterized by a simple structure, compact size, light weight, and high power density, making them attractive in various applications with promising future prospects. To address the issue of high torque ripple in traditional synchronous reluctance motors, a novel asymmetric motor structure is proposed based on the equivalent magnetic circuit model. This structure features an asymmetric design of the number of magnetic barriers and their occupancy rates. It is modeled and simulated using the finite element software JMAG. Subsequently, genetic algorithms were employed to optimize the span angle of each barrier layer. The outcome is an innovative synchronous reluctance motor with reduced torque ripple. Simulation results demonstrate that this motor exhibits significantly lower torque ripple and superior electromagnetic performance compared to traditional designs.

1. Introduction

As one of the widely employed motor types, permanent magnet synchronous motors (PMSMs) are favored for their compact structure, high energy density, and efficiency. However, the rising cost of rare earth materials and the significant impact of high temperatures on the performance and control precision of PMSMs due to their permanent magnets have become concerns. Consequently, reluctance motors that do not rely on permanent magnets have emerged as a new research focal point. The synchronous reluctance motors (SynRMs) feature rotors without permanent magnets, leading to lower manufacturing costs compared to PMSMs. Its simple rotor structure facilitates easier assembly and maintenance. Additionally, SynRMs are characterized by their compact size, light weight, high power density, and immunity to demagnetization at high temperatures, offering promising prospects for widespread application in the electric vehicle market [1-3].

The design of a SynRM critically affects its electromagnetic performance. To ensure excellent performance, the design must integrate the motor's structural characteristics. Current efforts focus on enhancing power density and reducing torque ripple to improve electromagnetic performance. Numerous studies have achieved significant results in optimizing SynRM performance.

In BIANCHI et al.'s study[4], the concept of asymmetrical design is used to establish general principles for motor asymmetry in salient pole rotors, and these principles are applied to traditional SynRMs. In the research by ALBERTI et al. [5], it is noted that increasing the number of stator winding layers can reduce the harmonic amplitudes of the magnetomotive force, thereby decreasing motor torque ripple. BAEK et al. [6] indicated that the number of motor phases can be increased to effectively reduce torque ripple, as a longer period for torque ripple is observed in a multiphase motor compared to a standard three-phase motor, resulting in smoother torque ripples. Pellegrino et al. [7] analyzed three different rotor barrier structures: 3C-type, 3U-type, and I2U-type rotor barriers. A comparative study is conducted to determine the most suitable design for reducing torque ripple. Moghaddam et al. [8]

Journal of Physics: Conference Series **3011** (2025) 012022 doi:10.1088/1742-6596/3011/1/012022

defined and arranged the ratio of the thickness of the rotor's magnetic portions to the barrier thickness according to the distribution of magnetomotive force to achieve a higher saliency ratio.

In this paper, an equivalent magnetic circuit model of SynRM is initially constructed based on Electromagnetic theory. The mathematical expression for its electromagnetic torque is derived, and factors affecting the torque are investigated through numerical calculations. Subsequently, a new SynRM with an asymmetric rotor structure is designed, demonstrating a significant reduction in torque ripple. Further improvements in electromagnetic performance are achieved by optimizing other parameters.

2. Equivalent magnetic circuit model of SynRM

In this section, a SynRM with a single layer of barriers is used as an example to present the equivalent model and derive the mathematical expression for its electromagnetic torque [9-10]. The geometric structure is shown in Figure 1.

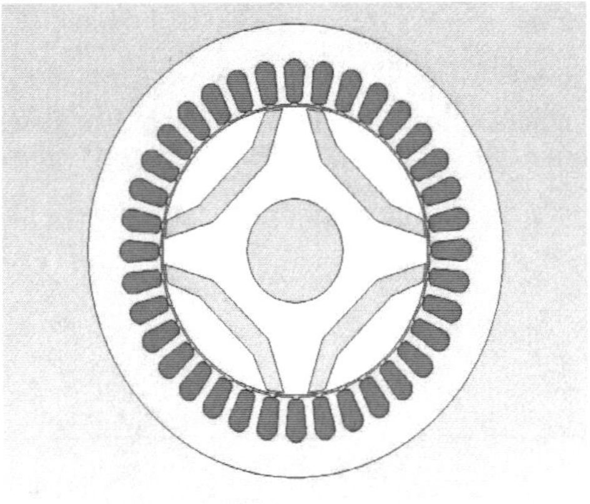

Figure 1. Structure of single-layer magnetic barrier motor.

2.1. Stator magnetomotive force

First, to simplify the model and enhance computational speed, the following conditions are established.

- The magnetic saturation occurring in the rotor and stator iron parts is neglected
- The stator is replaced by a current sheet on the inner surface of the stator. In other words, the slotting effect is neglected

Thus, the linear current density of the stator plane can be obtained as follows:

$$K_c = \sum_{\nu=6k\pm1}^{\infty} \hat{K}_\nu \sin(\nu p\theta_s - p\theta_m - \alpha), k \in \mathrm{Z} \tag{1}$$

where ν is the harmonic order, p is the number of pole pairs, θ_s is the coordinate angle (mechanical angle), θ_m is the rotor angle (mechanical angle), α represents the current phase angle (electrical angle), and $\hat{K}_\nu$ is the peak value of the nth harmonic of the equivalent current sheet, calculated by the following equation:

$$\hat{K}_\nu = \frac{3N_s\hat{I}k_w^\nu}{\pi D}\sin(\nu\frac{\pi}{2}) \tag{2}$$

where N_s is the number of series conductors per phase, $\hat{I}$ is the current amplitude, D is the outer diameter of the rotor, and k_w^ν is the winding distribution factor.

The equivalent current sheet generates a scalar magnetomotive force (MMF) U_s distributed circumferentially along the stator, calculated as follows:

$$U_s = \int K_c \cdot \frac{D}{2} d\theta_s \tag{3}$$

By substituting Equations (1) and (2) into (3), it can be simplified as follows:

$$U_s = -\frac{D}{2} \sum_{v=6k\pm1}^{\infty} \hat{K}_v \cos(vp\theta_s - p\theta_m - \alpha), \quad k \in Z \tag{4}$$

2.2. Rotor magnetomotive force

Assuming that the relative permeability of the rotor core is infinite, implying the equivalent magnetic reluctance of the rotor core is approximately zero. We analyze the rotor structure with a single-layer magnetic barrier. Setting t_{b1} and l_{b1} as the thickness and radial length of the barrier, respectively, θ_{b1} is half of the barrier angle. We define L_{stk} as the axial length of the motor and L_m as the air gap length.

Based on the differences in magnetic reluctance across various regions, the equivalent magnetic circuit model for a rotor structure with a single-layer magnetic barrier can be derived, as shown in Figure 2. Here, R_{m1} represents the air gap reluctance for the space corresponding to the magnetic barrier. R_{m2} and R_{m3} denote the air gap reluctances of the spaces corresponding to the rotor core on either side of the magnetic barrier. R_{b1} is the reluctance of the magnetic barrier itself. φ_{m1}, φ_{m1}, φ_{m1}, and φ_{b1} represent the magnetic fluxes through the air gap and the magnetic barrier respectively.

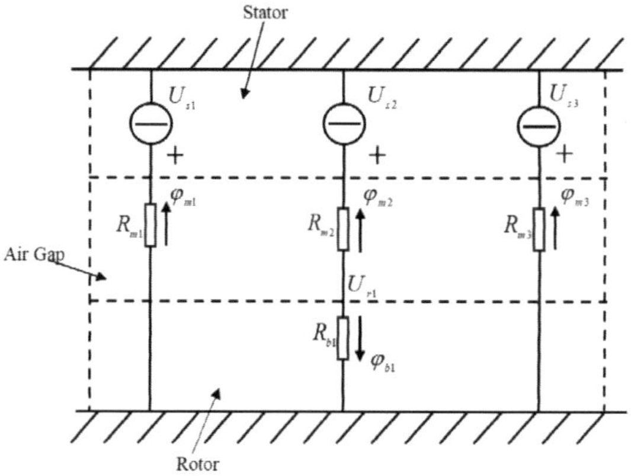

Figure 2. Equivalent magnetic circuit of a single-layer magnetic barrier structure.

Based on the equivalent magnetic circuit model shown in Figure 3, the MMF for a rotor with a single-layer magnetic barrier U_{r1} can be expressed as follows:

$$U_{r1} = \varphi_{b1} R_{b1} \tag{5}$$

The magnetic flux density B_m in the air gap is deducted as:

$$B_m = \mu_0 \frac{U_{r1} - U_s}{L_m} \tag{6}$$

where μ_0 stands for the permeability of vacuum.

The reluctance of the magnetic barrier is computed as follows:

$$R_{b1} = \frac{t_1}{\mu_0 l_{b1} L_{stk}} \tag{7}$$

According to the magnetic flux integration formula and the equivalent magnetic circuit model in Figure 2, we can obtain

$$\varphi_{b1} = \int_{\frac{\pi}{2p}-\theta_{b1}}^{\frac{\pi}{2p}+\theta_{b1}} B_m L_{stk} \frac{D}{2} d\theta_s \tag{8}$$

By substituting Equations (5) to (7) into equation (8) and simplifying, the MMF for a rotor with a single-layer magnetic barrier can be expressed as:

$$U_r = -aD \sum_{v=6k\pm1}^{\infty} \frac{\hat{K}_v}{(vp)^2} \cos\lambda_v \sin(vp\theta_{b1}), k \in Z \tag{9}$$

S.t.

$$a = \frac{Dt_{b1}}{2L_m l_{b1}} / (1 + \frac{Dt_{b1}}{L_m l_{b1}} \theta_{b1}) \tag{10}$$

$$\lambda_v = \frac{v\pi}{2} + p(v-1)\theta_m - \alpha \tag{11}$$

It is noted that the design parameters of the motor included in Equation (9), except for θ_{b1}, only affect the magnitude of the MMF. θ_{b1}, however, is incorporated within the term $\sin(vp\theta_{b1})$ and requires the summation of an infinite series, significantly impacting both the magnitude and shape of the MMF waveform.

When the rotor consists of multiple layers of magnetic barriers, equivalent magnetic circuit models can be drawn using the same principles. The calculation follows a similar process: first, we determine the magnetomotive force (MMF) generated by each layer, and then sum these to obtain the total MMF. In addition to the magnetic barriers angles, θ_{b1}, θ_{b2}, and θ_{b3}, which are important factors affecting the rotor's MMF, the number of layers of magnetic barriers is also a crucial influencing factor.

2.3. *Electromagnetic Torque and Waveform Analysis*

Based on the Lorentz force density formula, integrating over the circumference of the rotor surface yields the electromagnetic torque of the motor T_{em} as:

$$T_{em} = -\frac{D}{2} \int_0^{2\pi} B_m K_c L_{stk} \frac{D}{2} d\theta_s \tag{12}$$

Noting that the Fourier expansions of U_s and K_c are orthogonal to each other, thus Equation (12) can be simplified to:

$$T_{em} = -\frac{\mu_0 D^2 L_{stk}}{4} \int_0^{2\pi} \frac{U_r K_c}{L_m} d\theta_s \tag{13}$$

This expression represents the mathematical formulation of the electromagnetic torque for a SynRM. As discussed in the previous section, when the SynRM rotor contains multiple layers barriers, its electromagnetic torque results from the cumulative torques generated by the MMF of each layer. Therefore, the barrier angles and the number of layers are crucial factors influencing the SynRM torque.

Based on the derivation of the equivalent magnetic circuit model for SynRM, we first examine the magnetic circuit models when the number of magnetic barriers n_b is 1, 2, and 3. By analyzing their respective torques, we obtain the torque waveform comparison, as shown in Figure 3.

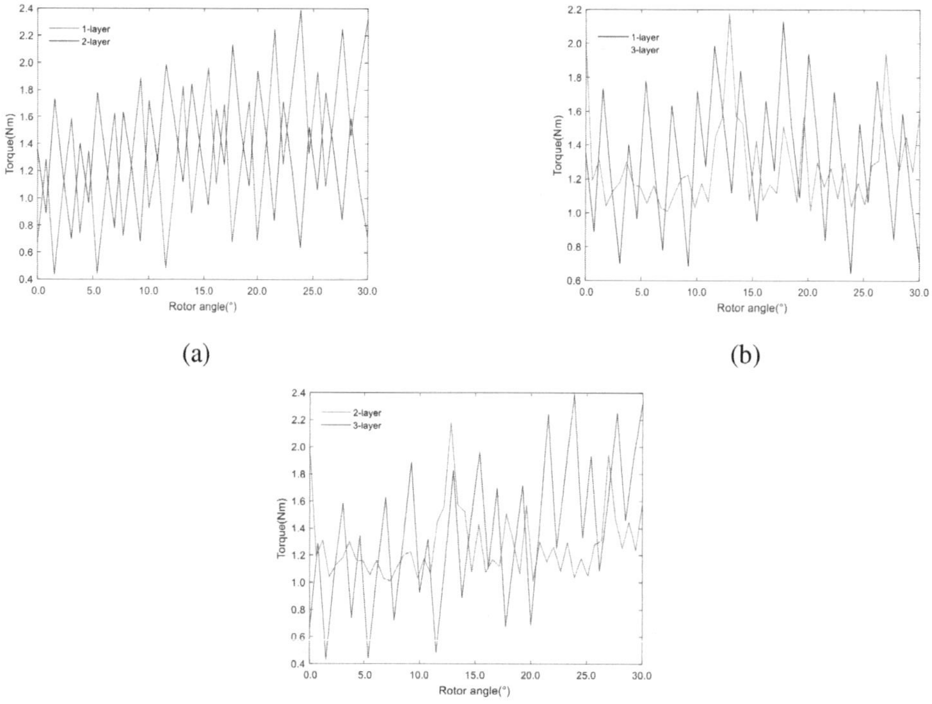

(a) (b)

(c)

Figure 3. Torque waveform comparison.

From Figure 3, it is evident that when n_b is 1, 2, and 3, there is a noticeable phase difference between their respective torque waveforms. This implies that they do not reach their peaks and troughs simultaneously. Thus, it can be inferred that overlaying two waveforms with a phase difference might result in a new waveform with reduced ripple compared to the original ones. Therefore, it is worth considering the combination of different layers of flux barriers to attempt to reduce torque ripple.

3. Effect of flux barrier occupancy on torque

SynRMs primarily operate based on reluctance torque and rely on the inductance difference between the d-axis and q-axis, causing the rotor to exhibit salient poles. Early designs used rotor structures similar to those in switched reluctance motors, featuring non-circular surfaces with prominent salient pole structures. However, this approach reduced the pole arc range of the d-axis by removing q-axis magnetic material, significantly decreasing the d-axis inductance. Currently, as is shown above, the addition of multiple-layer magnetic barriers in the q-axis magnetic path is more prevalent. This method not only restricts the flow of q-axis magnetic flux but also maintains a circular rotor periphery, greatly reducing the impact on d-axis inductance. Consequently, a larger inductance difference between the d-axis and q-axis is achieved, leading to higher output torque and power factor.

Increasing the torque of a SynRM involves increasing the d-axis inductance and decreasing the q-axis inductance. Because inductance is inversely proportional to magnetic reluctance, enhancing the motor's torque translates to reducing the d-axis reluctance and increasing the q-axis reluctance. The q-axis magnetic path incorporates multiple layers of flux barriers; air has significantly lower permeability than silicon steel. Therefore, by designing these barriers to occupy more space, the q-axis reluctance can be increased, thereby increasing torque.

We define the magnetic barrier occupancy rate k_b as follows:

$$k_b = \frac{t_{bk}}{t_{bk} + t_{rk}} \tag{14}$$

The schematic is shown in the Figure 4 below:

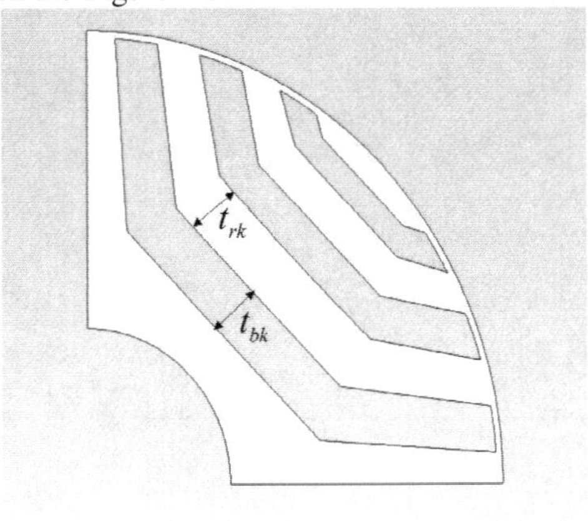

Figure 4. Schematic of magnetic barrier occupancy rate.

Increasing t_{bk} can enhance the q-axis magnetic reluctance [8] and reduce the q-axis inductance. Since the rotor's radial length is fixed, increasing the barrier thickness reduces the area of the magnetic conducting part, thereby increasing k_b. However, the rotor's mechanical strength can be weakened by excessively high k_b. Additionally, a smaller magnetic conducting area can exacerbate saturation in the d-axis magnetic circuit, increasing its reluctance and decreasing d-axis inductance. Therefore, while a higher occupancy rate may reduce q-axis inductance, it simultaneously lowers d-axis inductance, which may not necessarily enhance the electromagnetic performance of the SynRM. In other words, a balance between the d-axis and q-axis circuits should be sought, rather than maximizing the occupancy rate indiscriminately.

4. Design of low torque ripple SynRM

Building on the analysis in former chapters, this chapter introduces a novel SynRM with an asymmetric rotor structure. Compared to traditional SynRM designs, it significantly reduces torque ripple.

4.1. Design and optimization of layers of magnetic barriers

Based on the analysis in Chapters 2 and 3, the angle, number of layers, and occupancy rate of magnetic barriers significantly impact the electromagnetic torque of the motor. This section first investigates the asymmetric design of the number of magnetic barrier layers.

A Machaon-type motor is introduced in [4], which manufactures two different magnetic barrier geometries on the same lamination. The central axis position of each pole pair's magnetic barrier remains unchanged, but the end angles differ. Optimizing the end angles of each magnetic barrier layer under each pole pair can achieve a lower torque ripple compared to traditional SynRM.

Based on the design of the Machaon-type motor and the conclusions drawn in Section *2.3*, a novel rotor structure with asymmetric magnetic barrier layers is proposed. This structure features a different number of magnetic barrier layers under each pole pair.

In JMAG, traditional SynRM structures with 2, 3, 4, and 5 layers were established ($n_b = 2,3,4,5$), along with asymmetric structures formed by pairing these configurations. Figure 5 shows the motor models for 2 and 3 layers, as well as the asymmetric model resulting from their combination.

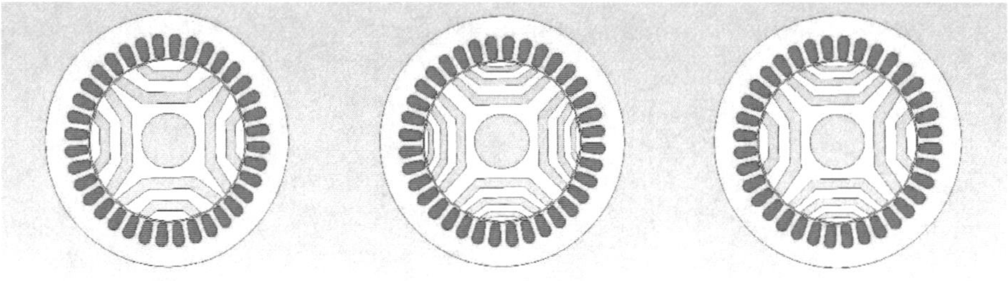

Figure 5. The rotor structure of the JMAG motors.

Simulations were conducted on each constructed model, and the resulting torque ripple and average torque are shown in Table 1.

Table 1. Torque Ripple and average torque values for each motor.

n_b	Torque ripple	Average torque (Nm)
2	0.191	22.854
3	0.175	23.939
4	0.181	23.824
5	0.159	23.911
2-3	0.146	23.177
2-4	0.207	23.154
2-5	0.174	23.478
3-4	0.111	23.763
3-5	0.121	23.882
4-5	0.181	23.745

From Table 1, it can be observed that the average torque of the asymmetric SynRM structure is comparable to that of the traditional SynRM. Regarding torque ripple, the asymmetric SynRM exhibits three scenarios: one where the torque ripple is similar to the traditional structure, another where it is the midpoint of the torque ripple values of the two traditional structures forming the asymmetric design, and a third where the torque ripple is significantly reduced, aligning with previous predictions and analyses. A comprehensive comparison of the data indicates that the 3-4 layer asymmetric SynRM structure offers the best performance. This model is designated as γ for further analysis and optimization.

4.2. Design and optimization of the occupancy rate of magnetic barriers

From the previous subsection, we found that the inference proposed in Section 2.3 was validated. Therefore, we can further consider applying this asymmetric design concept to the optimization of magnetic barrier occupancy k_b.

To reduce the total number of models and improve computational efficiency, we first analyze the electromagnetic performance of motor models with different k_b values. We eliminate models with poorer performance, and then apply asymmetric design to the remaining models. Using the motor model γ selected in Section 4.1, we first establish motor models in JMAG with k_b values of 0.2, 0.4, 0.5, 0.6, and 0.8(the model with a value of 0.5 corresponds to the model γ). These are designated as models σ_1, σ_2, σ_3, σ_4 and σ_5, respectively. The following Figure 6 provides a comparison of σ_1, σ_2 and σ_3.

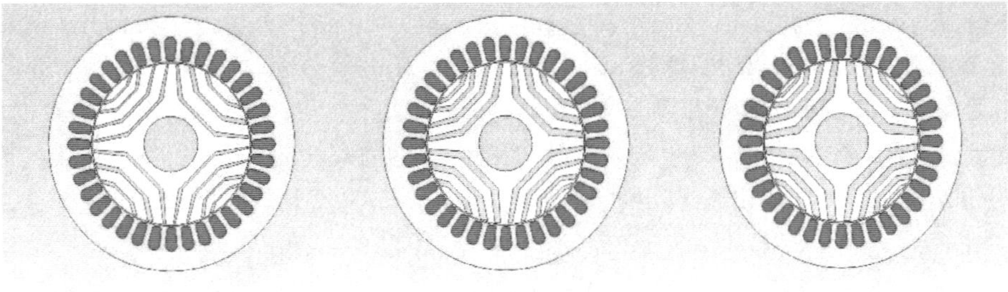

Figure 6. The rotor structure of the JMAG motors.

The finite element simulations for each model were conducted, and the torque ripple and average torque for each model are shown in Table 2 below.

Table 2. Torque ripple and average torque values for each motor.

Model	Torque ripple	Average torque (Nm)
σ_1	0.075	21.021
σ_2	0.103	23.285
σ_3	0.111	23.763
σ_4	0.123	22.825
σ_5	0.137	22.217

Observing the data above, it is noted that as k_b increases, the torque ripple shows a gradual upward trend, while the average torque first increases and then decreases. The average torque of models σ_1, σ_2, and σ_5 is significantly lower, whereas models σ_3 and σ_4 exhibit better values for both average torque and torque ripple.

As previously mentioned, after analyzing the impact of k_b on motor torque, inductance simulations and parameter calculations were performed for each model. The final results are shown in Table 3 below.

Table 3. Inductance calculation results for each motor.

Model	D-axis Inductance (mH)	Q-axis Inductance (mH)	Difference
σ_1	36.400	11.800	24.600
σ_2	37.000	9.100	27.900
σ_3	37.100	8.390	28.710
σ_4	35.590	8.120	27.470
σ_5	33.720	7.780	25.940

Upon examining the data, it is noted that as k_b increases, both the d-axis and q-axis inductances decrease, albeit with different trends. At lower levels, the increase in k_b has a minor impact on the d-axis, resulting in a relatively gradual decline in inductance. However, the q-axis experiences a significant increase in magnetic reluctance, leading to a more pronounced decrease in inductance. As the magnetic barrier occupancy continues to rise to higher levels, the d-axis magnetic circuit becomes increasingly saturated, causing a rapid drop in inductance. Notably, the models σ_2 and σ_3 exhibit favorable saliency in terms of the difference in d-axis and q-axis inductance difference. Based on the results analyzed from Table 2 and Table 3, and to avoid unnecessary calculations and save time, the models σ_2 and σ_3 are ultimately selected for asymmetric treatment.

Next, two distinct asymmetric 3-4 layer motor models composed of models σ_2 and σ_3 are constructed. In the first model, k_b is 0.4 at the third layer and 0.5 at the fourth layer, referred to as model σ_5. In the second model, k_b is 0.5 at the third layer and 0.4 at the fourth layer, referred to as model σ_6. The structural comparison of these models is illustrated in the following Figure 7.

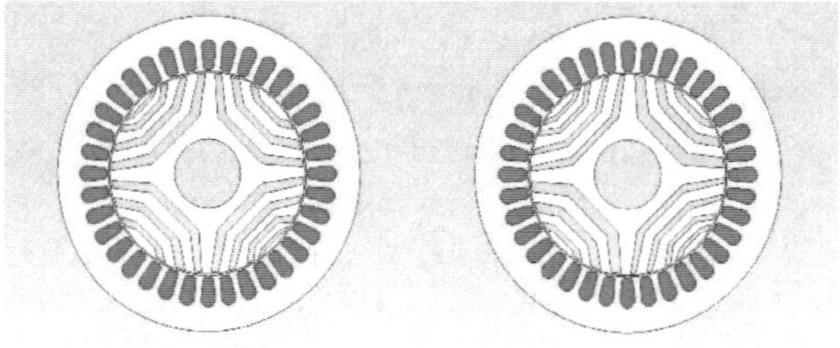

Figure 7. The rotor structure of the JMAG motors.

The finite element simulations for each model were conducted, and the torque ripple and average torque for each model are shown in Table 4 below.

Table 4. Torque ripple and average torque values for each motor.

Model	Torque ripple	Average torque (Nm)
σ_5	0.054	23.430
σ_6	0.168	23.674

Observing the data from the two groups, it is noted that model σ_5 exhibits a significant reduction in torque ripple compared to models σ_2 and σ_3. Conversely, the torque ripple of model σ_6 is much larger than that of both models σ_2 and σ_3. These simulation results are consistent with the findings in Section 2.3. Based on a comprehensive analysis of the data, model σ_5 is selected for further optimization.

4.3. Design and optimization of angles of magnetic barriers

Continuing from the previous section, we now optimize the magnetic barrier angles of model σ_5. Since model σ_5 is composed of a three-layer and a four-layer magnetic barrier structure, there are seven parameters to optimize: three angles for the three-layer structure and four angels for the four-layer structure. Due to the high number of parameters, conventional optimization methods would be time-consuming. Therefore, a genetic algorithm is utilized to optimize these seven parameters, enhancing overall optimization efficiency.

Common optimization algorithms often use iterative methods, which can lead to local minima traps and result in a “dead loop,” ultimately causing iteration failure and optimization termination. The genetic algorithm, proposed by John Holland in the 1970s [11], addresses this issue effectively and is widely applied in solving multimodal optimization problems. It is inspired by Darwin's theory of evolution and genetic mechanisms. Through processes like gene recombination and mutation, organisms adapt to environments and develop new traits, achieving evolution. The genetic algorithm abstracts this complex biological evolution into a mathematical model and uses heuristic search techniques to explore complex search spaces, aiming to find the global optimal solution for the model[12-13].

The design objectives of this study are primarily to reduce motor torque ripple while maximizing electromagnetic torque. Therefore, this is a multi-objective optimization problem. Multi-objective optimization problems are typically represented using objective functions and constraints and can be described as follows:

$$\begin{cases} \min f_1(x_1,x_2,\dots,x_n) \\ \cdots\cdots\cdots \\ \min f_k(x_1,x_2,\dots,x_n) \\ \cdots\cdots\cdots \\ \max f_{m-1}(x_1,x_2,\dots,x_n) \\ \max f_m(x_1,x_2,\dots,x_n) \\ g_i(x)\ge 0, i=1,2,\dots,p \\ h_j(x)\ge 0, j=1,2,\dots,p \end{cases} \tag{15}$$

where $f_i(x)$ represents the objective functions, while $g_i(x)$ and $h_j(x)$ are the constraint functions. $X=\{x \mid x\in R^n, g_i(x)\ge 0, h_j(x)\ge 0, i=1,2,\dots,p, j=1,2,\dots,p\}$ is the feasible region.

Based on the aforementioned design objectives, the objective functions can be expressed as follows:

$$\begin{cases} \max T_{avg} = \dfrac{1}{t_1}\int_0^{t_1} T dt \\ \min T_{rip} = \dfrac{T_{\max}-T_{\min}}{T_{avg}} \end{cases} \tag{16}$$

The main optimization parameters for the rotor are shown in Figure 8.

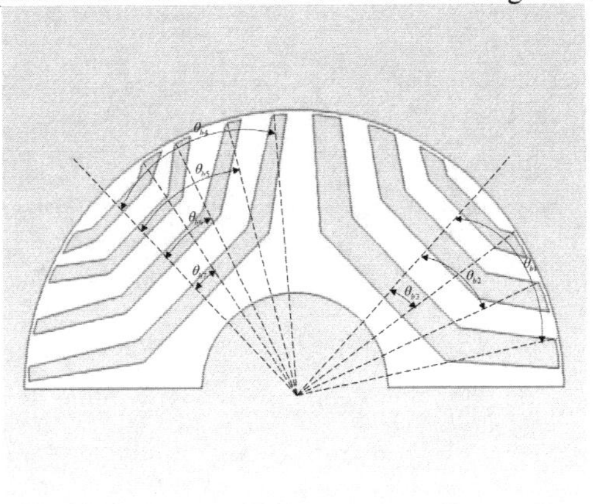

Figure 8. Main optimization parameters for the rotor.

As shown in Figure 10, each layer of the magnetic barrier in the model is parameterized to determine the range of values. To minimize computational workload and reduce optimization time, the range for each parameter should not be too extensive. The specific ranges for the parameters are as follows:

$$\begin{cases} 32 \le \theta_1 \le 39 \\ 20 \le \theta_2 \le 27 \\ 8 \le \theta_3 \le 14 \\ 36 \le \theta_4 \le 43 \\ 28 \le \theta_5 \le 35 \\ 13 \le \theta_6 \le 20 \\ 7 \le \theta_7 \le 13 \end{cases} \tag{17}$$

After optimization using the genetic algorithm, considering both average torque and torque ripple, the optimal final motor model was obtained. It is designated as the model κ_F , and its relevant parameters are shown in the table 5 below.

Table 5. Optimization results.

Parameter	Value (◦)
θ_{b1}	35.035
θ_{b2}	22.684
θ_{b3}	10.825
θ_{b4}	41.149
θ_{b5}	30.988
θ_{b6}	18.786
θ_{b7}	10.466

4.4. *Final model measurement*

The structure of κ_F , obtained after a series of optimizations, is shown in Figure 9.

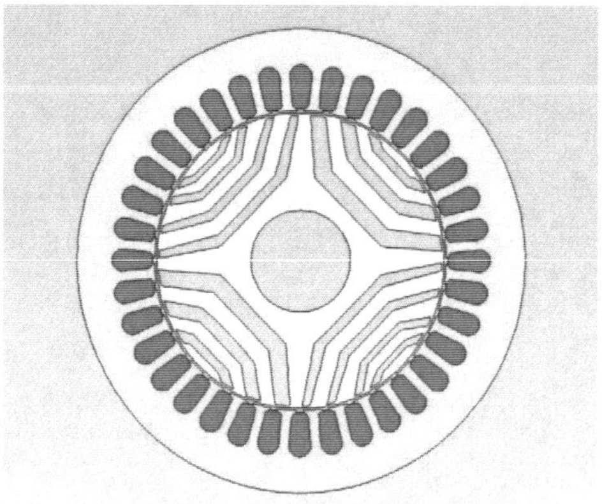

Figure 9. Final model structure.

The torque waveform of this motor is shown in the following Figure 10.

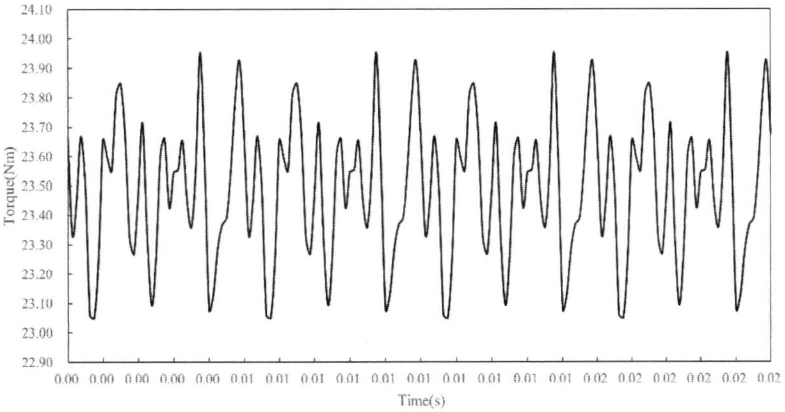

Figure 10. Torque waveform of the final model.

The performance of this motor, compared with traditional 3-layer and 4-layer SynRM models (referred to as Model ξ_1 and Model ξ_2), is presented in Table 6 below.

Table 6. Comparison of motor electromagnetic parameters.

Model	Torque ripple	Average Torque (Nm)	Iron loss (W)	Efficiency (%)
ξ_1	0.175	23.939	345.435	77.443
ξ_2	0.197	23.824	343.136	77.205
κ_F	0.038	23.489	347.124	76.866

As seen in Table 6, the model shows minimal changes in average torque, iron loss, and efficiency compared to the initial model, with only slight declines. However, it achieves a significant reduction in torque ripple, decreasing from 0.175, 0.197 to 0.038, thereby meeting the optimization design goals.

5. Conclusion

This paper begins by analyzing the impact of parameters such as the number of flux barriers, barrier angles, and barrier occupancy on the electromagnetic performance of SynRMs using equivalent magnetic circuit models. An asymmetric design hypothesis, beneficial for reducing torque ripple, is proposed. Finite element software simulations then verify the validity of this hypothesis. Based on this design approach, aided by a genetic optimization algorithm, an optimal motor model is developed to minimize torque ripple and significantly enhance motor performance. This provides a new method for improving the electromagnetic performance of SynRMs.

References

[1] SHEN J X, CAI S, SHAO H, et al. Evaluation of low-cost high-performance synchronous motors for ventilation application[C/OL]//2015 International Conference on Sustainable Mobility Applications, Renewables and Technology (SMART), Kuwait, Kuwait. 2015. http://dx.doi.org/10.1109/smart.2015.7399211.

[2] D. Ngo and M. Hsieh, Performance Comparison of SynRM and Novel FI-PMa-SynRM with Different Rotor Surface Layouts, 2019 IEEE 4th International Future Energy Electronics Conference (IFEEC), 2019, 1-5.

[3] T. Xiao, Z. Ran and Z. -Q. Zhu, "Comparison of Electrically Excited and Synchronous Reluctance Machines with Interior Permanent Magnet Machines for EVs/HEVs," 2023 26th International Conference on Electrical Machines and Systems (ICEMS), Zhuhai, China, 2023, pp. 360-365,

[4] BIANCHI N, BOLOGNANI S, BON D, et al. Torque Harmonic Compensation in a Synchronous Reluctance Motor[J]. IEEE Transactions on Energy Conversion, 2008: 466-473.

[5] ALBERTI L, BIANCHI N. Theory and Design of Fractional-Slot Multilayer Windings[J/OL]. IEEE Transactions on Industry Applications, 2013, 49(2): 841-849. http://dx.doi.org/10.1109/tia.2013.2242031.
[6] BAEK J, BONTHU S S R, CHOI S. Design of five-phase permanent magnet assisted synchronous reluctance motor for low output torque ripple applications[J/OL]. IET Electric Power Applications, 2016, 10(5): 339-346. http://dx.doi.org/10.1049/iet-epa.2015.0267.
[7] G. Pellegrino, F. Cupertino and C. Gerada. Barriers shapes and minimum set of rotor parameters in the automated design of Synchronous Reluctance machines[C]. 2013 International Electric Machines & Drives Conference. IEEE, 2013:1204-1210.
[8] Reza-Rajabi Moghaddam, Freddy Gyllensten. Novel High-performance SynRM Design Method: An Easy Approach for a Complicated Rotor Topology[J]. Industrial Electronics， IEEE Transactions on, 2014, 61(9):5058-5065.
[9] MAHMOUD H, BIANCHI N, BACCO G, et al. Nonlinear Analytical Computation of the Magnetic Field in Reluctance Synchronous Machines[J/OL]. IEEE Transactions on Industry Applications, 2017, 53(6): 5373-5382. http://dx.doi.org/10.1109/tia.2017.2746560.
[10] CAI S, HE H, MENG-JIA J, et al. A Simplified Method to Analyze Synchronous Reluctance Machine[J]. IEEE Conference Proceedings, IEEE Conference Proceedings, 2016.
[11] SAMPSON J R. Adaptation in Natural and Artificial Systems (John H. Hol- land)[J/OL]. SIAM Review, 1976, 18(3): 529 – 530.
[12] XING L, WANG Y, LI M, et al. Multi-Objective Optimization Model of Electric Power Fine Construction Based on Multi-Objective Genetic Algorithm[C/OL]//2023 International Conference on Mechatronics, IoT and Industrial Informatics (ICMIII). 2023, pp. 576-581. https://doi.org/10.1109/ICMIII58949.2023.00121.
[13] ZHNG X, ZHANG B, ZHANG C, et al. A multi-objective hybrid genetic algorithm for detecting communities in complex networks[C/OL]//2016 12th International Conference on Natural Computation, Fuzzy Systems and Knowledge Discovery (ICNC-FSKD), Changsha, China. 2016. http://dx.doi.org/10.1109/fskd.2016.7603258.

Theoretical analysis of thermal diffusion in power device packaging structures

Lei Wang[1], Rui Jin[1,*], Peng Zhang[1], Wenyuan Li[1] and Xuebao Li[2]

[1] Beijing Institute of Smart Energy, Beijing 102200, China
[2] State Key Laboratory of Alternate Electrical Power System with Renewable Energy Sources, North China Electric Power University, Beijing 102206, China

[*]Corresponding author's e-mail: jinrui@bise.hrl.ac.cn

Abstract. The miniaturization and integration of power devices have led to a significant increase in power consumption and heat flux density. Research indicates that over 50% of power device failures are due to excessive temperature. Heat is transferred through the connection layer beneath the chip and ultimately dissipated by the heat sink. Power devices are composed of multiple stacked thin layers, with heat primarily transferring in the thickness direction. However, the heat flow path at the interface between materials of different sizes remains unclear. Based on heat transfer theory, this paper establishes a packaging and thermal conduction model for single-chip power devices. Through finite element analysis, the temperature distribution and thermal resistance distribution are obtained, providing methods for optimizing the packaging structure. The study finds that increasing the thickness of the base plate enhances heat diffusion and increases the effective heat transfer area, revealing the coexistence mechanism of heat conduction and heat diffusion. The accuracy of the model is verified through simulation and experiments, offering a basis for packaging design and material selection.

1. Introduction

Driven by advancements in power device technology, silicon-based IGBT devices have evolved from the first generation in the 1980s to the current seventh generation, achieving a reduction in chip size by over 80% and a decrease in power loss by more than 50% [1]. Although silicon devices may eventually be replaced by newer technologies, IGBTs still hold significant potential for further improvement. Research indicates that when silicon IGBTs operate at high temperatures of 200 °C, their short-circuit capability diminishes, with the critical short-circuit time reduced by approximately 30%, thereby increasing the risk of device failure [2]. In the short term, innovations in chip layout, materials, interconnection processes, and structural design within the silicon-based packaging system are essential. Techniques such as staggered chip arrangement [3], embedded packaging [4, 5], double-sided cooling [6-8], aluminum/copper ribbon bonding [9-11], high-temperature encapsulation materials [12-14], and nano-metal sintering interconnection [15-17] offer new opportunities for the development of power devices.

Reducing packaging thermal resistance is crucial for enhancing the high-temperature operational performance of power devices. This can be achieved by optimizing packaging structures, materials, and interconnection processes to lower the junction-to-case thermal resistance, or by enhancing external cooling [18] to reduce the junction-to-ambient thermal resistance. With the progress in semiconductor processing and packaging technologies, the application scenarios for power devices have expanded [19, 20], making thermal management a key technical challenge.

This paper focuses on the multi-layer thin-plate stacked packaging structure of power devices, employing numerical heat transfer methods to study their internal thermal characteristics. A simulation model for single-chip packaging is established, and finite element analysis is used to obtain temperature responses, revealing the mechanisms of heat conduction along the thickness direction and lateral diffusion. The impact of baseplate thickness on thermal performance is also explored. The study finds that while increasing baseplate thickness raises thermal resistance, the heat diffusion effect enlarges the effective heat transfer area, mitigating the increase in thermal resistance. Finally, the accuracy of the simulation method is validated through experiments, providing valuable insights for packaging design, material selection, and performance optimization.

2. Geometric modeling

Based on Fourier's law of heat conduction, for large and thin metal plates, heat flow primarily conducts along the thickness direction. The thermal resistance of a single-layer plate is directly proportional to its thickness and inversely proportional to its thermal conductivity. The packaging of power devices can be considered as a stack of multiple thin plates with different geometric areas. The heat transfer path at the contact interface is difficult to characterize directly and requires exploration through simulation analysis. Therefore, steady-state thermal simulations were conducted on single-chip packaging structures using materials with both different and identical geometric areas to analyze their thermal performance. The packaging structure is modeled as shown in Figure 1.

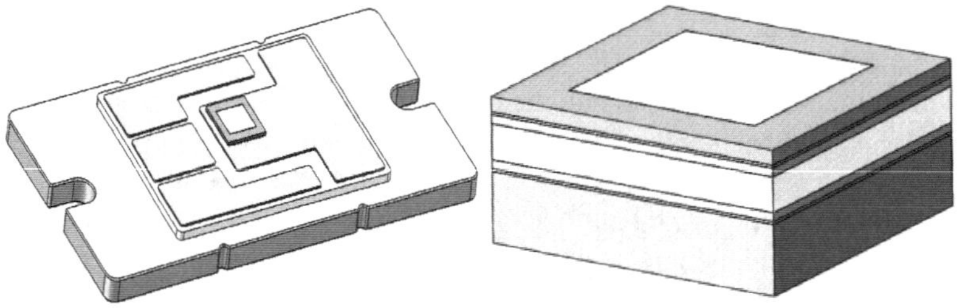

Figure 1. Modeling of conventional packaging structure and same geometric area packaging structure.

3. Mathematical Model

A mathematical model (**1**) for thermal conduction analysis of the packaging structure is established based on the aforementioned geometric model.

$$\rho c\frac{\partial T}{\partial t}=\frac{\partial}{\partial x}\left(\lambda_i\frac{\partial T}{\partial x}\right)+\frac{\partial}{\partial y}\left(\lambda_j\frac{\partial T}{\partial y}\right)+\frac{\partial}{\partial z}\left(\lambda_k\frac{\partial T}{\partial z}\right)+q_v \tag{1}$$

Assuming the material is isotropic, Equation (**1**) can be written as:

$$\rho c\frac{\partial T}{\partial t}=\lambda\left(\frac{\partial^2 T}{\partial x^2}+\frac{\partial^2 T}{\partial y^2}+\frac{\partial^2 T}{\partial z^2}\right)+q_v \tag{2}$$

Furthermore, assuming the material's thermophysical parameters-thermal conductivity λ, density ρ, and specific heat capacity c-are all constants, and the process is a steady-state heat transfer process, Equation (2) can be further expressed as:

$$\lambda\left(\frac{\partial^2 T}{\partial x^2}+\frac{\partial^2 T}{\partial y^2}+\frac{\partial^2 T}{\partial z^2}\right)+q_v=0 \tag{3}$$

4. Assumptions

For power device packaging, since the low thermal conductivity silicone gel needs to be filled inside the packaging, the heat dissipation from the upper surface of the chip and the four sides of the internal packaging materials can be ignored.

The radiation heat transfer of power devices is ignored.

It is assumed that the thermal conductivity of each packaging material does not change with temperature.

5. Initial Conditions and Boundary Conditions

Initial conditions: A three-dimensional steady-state thermal simulation finite element model is established based on finite element analysis software. The single-chip power is 83 W, the source area of the chip is 5.3×5.3 mm^2, the thickness of the chip is 0.42 mm, the volume of the active area of the chip is 11.8 mm^3, and the chip generates heat dissipation, which is an internal heat source, with a heat dissipation power density of 7.03 W/m^3.

Geometric conditions: The geometric dimensions and thermal physical parameters of each layer of packaging materials are shown in Table 1, and the packaging structure model for simulation is shown in Figure 1.

Boundary conditions: The first type of boundary conditions are adopted for analysis, with the bottom surface of the substrate set as a constant temperature condition (40°C).

Table 1. Thermal simulation parameters of single-chip packaging.

Material	Material type	Dimension (mm)	Thermal conductivity (W/(m·K))	Specific heat capacity (J/(kg·K))	Coefficient of thermal expansion (10^{-6}/K)	Density (g/cm^3)
Chip	Si	8.0×8.0×0.42	148	700	2.6	2.3
Solder 1	PbSnAg	8.0×8.0×0.08	26	126	26	11.2
Copper 1	Cu	27.5×14×0.3	398	385	17.8	8.9
Ceramic	AlN	46×31×1.0	170	780	4.5	3.28
Copper 2	Cu	27×42×0.3	398	385	17.8	8.9
Solder 2	SAC305	28×43×0.08	58	-	20	7.4
Baseplate	AlSiC	37.8×80×3	200	741	7	3.01

6. Results and Analysis

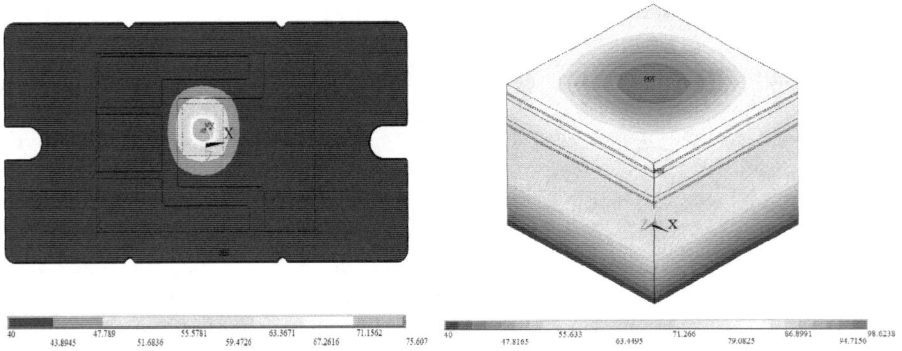

Figure 2. Simulation results of conventional packaging structure and equal geometric area packaging for power devices.

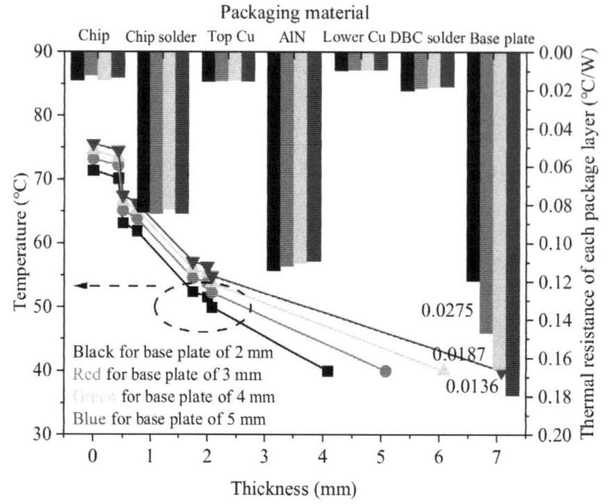

Figure 3. Thermal performance extraction of conventional (different geometric area) packaging for power devices.

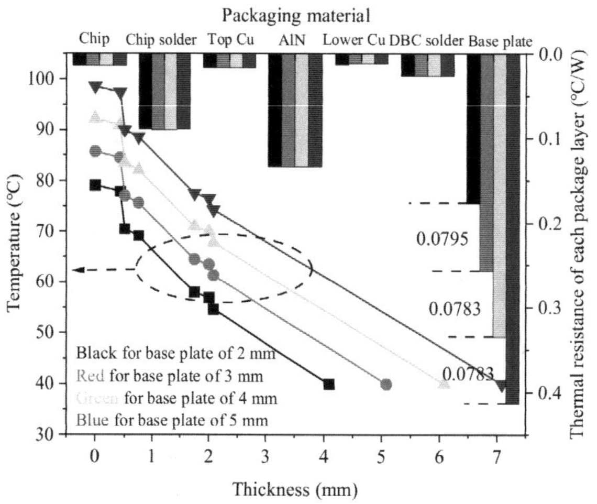

Figure 4. Thermal performance extraction of equal geometric area packaging for power devices.

The simulation results are shown in Figure 2. Through analysis, the surface temperature of each material layer and the thermal resistance data as a function of packaging thickness were extracted, as depicted in Figures 3 and 4. Theoretically, according to Fourier's Law, the thermal resistance of a specific material is proportional to its thickness. To simplify the analysis, only the thickness of the packaging baseplate was altered, while the thicknesses of the other layers remained unchanged. Theoretically, the thermal resistance curve should separate at the baseplate, and the temperature change at the baseplate should differ. However, the simulation results show that in packages with different geometric areas, for every 1 mm increase in baseplate thickness, the temperature changes in the other layers are parallel, whereas the temperature change curve at the baseplate is no longer parallel. Thermal resistance analysis indicates that the thermal resistance of the other layers remains unchanged, while the thermal resistance of the baseplate increases with thickness, but the rate of increase gradually diminishes, which contradicts Fourier's Law.

Theoretical analysis suggests that the thermal resistance of a single thin plate is proportional to its thickness, in accordance with Fourier's Law, because its heat transfer area is fixed. However, in the packaging of power devices, the geometric areas of the material layers differ, leading to variations in heat flux density, while the total heat flux remains constant. Therefore, when heat is transferred from the chip to a material with a larger area, lateral thermal diffusion occurs to increase the effective heat transfer area. To validate this analysis, modeling was conducted using packaging materials with the same geometric dimensions as the chip, and the results are shown in Figure 4. The results demonstrate that when the geometric dimensions are the same, for every 1 mm increase in baseplate thickness, the slopes of the temperature change curves for all layers are consistent, and thermal resistance analysis shows that the thermal resistance of the other layers remains unchanged, while the increase in baseplate thermal resistance is consistent. This aligns with heat conduction theory and confirms that heat transfer within the packaging involves both thermal conduction in the thickness direction and lateral thermal diffusion, with the effect of thermal diffusion becoming more pronounced as the thickness increases.

7. Simulation Model Validation

To validate the accuracy of the simulation model and methodology, a thermal simulation model was established for a silicon single-chip package, and thermal resistance experimental testing was conducted. The simulation employed third-type boundary conditions to mimic actual cooling conditions. The thermal analysis modeling of the silicon single-chip device and the experimental package are illustrated in Figure 5, while the simulation temperature contour distribution and experimental test results are presented in Figure 6.

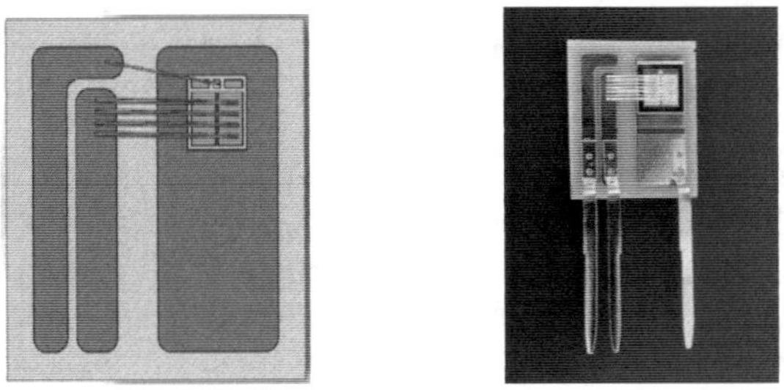

(a) (b)

Figure 5. Schematic diagram of silicon single-chip packaging modeling (left) and physical photo of the packaging (right).

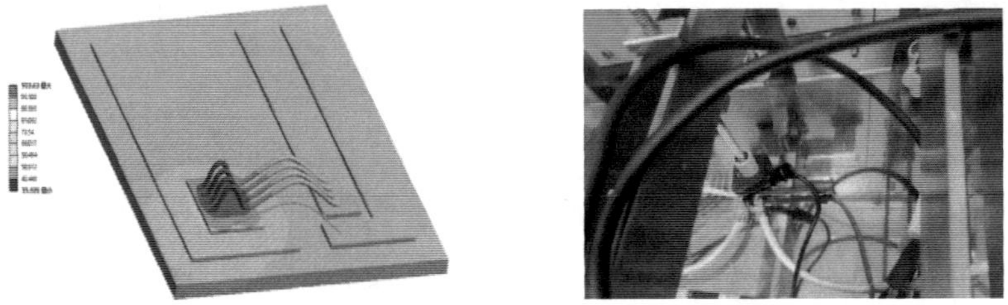

(a) (b)

Figure 6. Thermal simulation results of the single-chip wire-bonded packaging device (left) and physical thermal resistance test photo (right).

Table 2. Comparison of thermal simulation and experimental test results for single-chip wire-bonded packaging devices.

	Simulation	Experiment	Relative error/%
Junction temperature (°C)	103.63	125	17.1
Junction-to-case thermal resistance (°C/W)	0.41	0.43	4.65

A comparative analysis of the device's thermal performance between simulation and experimental test results is presented in Table 2. The simulation results indicate that the junction temperature and junction-to-case thermal resistance of the silicon single-chip device packaged with Al bonding wires are approximately 103.63 °C and 0.41 °C/W, respectively. In actual testing, the junction temperature and junction-to-case thermal resistance of the packaged device are approximately 125 °C and 0.43 °C/W, respectively. The relative error between the simulated and experimental junction temperatures is about 17.1%, primarily because the thermal boundary conditions in the simulation do not fully replicate the water-cooling plate conditions in the experiment. The relative error for the junction-to-case thermal resistance is only 4.65%, which is within an acceptable range, preliminarily validating the correctness of the modeling and simulation analysis methods.

8. Conclusion

Based on Fourier's law, this paper establishes a mathematical model of a single-chip power device package through numerical heat conduction methods. The steady-state solution provides the temperature distribution and thermal resistance distribution at the interfaces of each material layer, revealing the heat transfer mechanisms of internal heat conduction and thermal diffusion within the package. The correctness of the model and simulation methods is verified through geometric modeling and experimental validation. Based on the results and discussions presented above, the conclusions are obtained as below:

(1) The temperature distribution and thermal resistance distribution characteristics at the device package interfaces are obtained, providing a basis for optimizing the thermal properties of the package structure.

(2) The thermal resistance of the package material layers does not fully comply with Fourier's law. The increase in thermal resistance is not proportional to the change in thickness, as thermal diffusion increases the effective heat transfer area and alters the heat flow path.

(3) As the baseplate thickness increases, the thermal diffusion effect becomes more significant, partially offsetting the rise in thermal resistance caused by the increased baseplate thickness. This provides a new approach for improving the thermal performance of the package.

Acknowledgments

This work was financially supported by the science and technology project of the state grid (NO. 5500-202158491A-0-5-ZN).

References

[1] Laska T (2023) A Si IGBT Review and Outlook in Competition to WBG Power Devices. In: Components of Power Electronics and their Applications 2023; ETG Symposium. Bad Nauheim. pp. 15-22.
[2] Xu Z, Li M, Wang F and Liang Z (2012) Investigation of Si IGBT operation at 200°C for Traction Application. IEEE T. POWER ELECTR., 28: 2604-15.
[3] Yang F, Wang L, Kong H, Zhu M, Liu X, Lu X, Qin M, Zhang T, Gan Y and Jia L (2023) Compact-interleaved packaging method of power module with dynamic characterization of 4H-SiC MOSFET and development of power electronic converter at extremely high junction. IEEE T. POWER ELECTR., 38: 417-34.
[4] Charboneau B C, Wang F, van Wyk J D, Boroyevich D, Liang Z, Scott E P and Tipton C W (2008) Double-sided liquid cooling for power semiconductor devices using embedded power packaging. IEEE T. IND. APPL., 44: 1645-55.
[5] Kim B J, Jeon H, Park D, Kim G J, Cho N H and Khim J Y (2022) 3D Embedded Power Package Module to Integrate Various Power Systems. In: 2022 IEEE 72nd Electronic Components and Technology Conference (ECTC). San Diego. pp. 289-95.
[6] Takahashi H, Suzuki K, Kitamura A, Endoh T and Takahashi Y (2024) Study of a Packaging Technology for Ultra Compact Double-Side Cooled Power Module with Power Chip Size Package. In: 2024 36th International Symposium on Power Semiconductor Devices and ICs (ISPSD). Bremen. pp. 474-7.
[7] Paul R, Alizadeh R, Li X, Chen H, Wang Y and Mantooth H A (2024) A Double-Sided Cooled SiC MOSFET Power Module for EV Inverters. IEEE T. POWER ELECTR., 39: 11047-59.
[8] Singh B P, Ghahfarokhi S S, Kostov K, Nee H P and Norrga S (2024) Analysis of the thermo-mechanical performance of double-sided cooled power modules. In: 2024 25th International Conference on Thermal, Mechanical and Multi-Physics Simulation and Experiments in Microelectronics and Microsystems (EuroSimE). Catania. pp. 1-8.

[9] Marenco N, Kontek M, Reinert W, Lingner J and Poech M H (2013) Copper ribbon bonding for power electronics applications. In: 2013 European Microelectronics Packaging Conference (EMPC), Grenoble. pp. 1-4.
[10] Pavliček N and Mohn F (2020) Laser bonding of copper ribbons and clips on SiC power MOSFETs with sintered copper bond buffers. In: 2020 IEEE 8th Electronics System-Integration Technology Conference (ESTC), Tønsberg. pp. 1-4.
[11] Luechinger C, Chen R, Fu J, Poncelet B, Valentin O, Walker T J and Xu T (2017) Aluminum-Copper Ribbon Interconnects for Power Devices. IEEE T. COMP. PACK. MAN., 7: 1567-77.
[12] Locatelli M L, Khazaka R, Diaham S, Pham C D, Bechara M, Dinculescu S and Bidan P (2013) Evaluation of encapsulation materials for high-temperature power device packaging. IEEE T. POWER ELECTR., 29: 2281-8.
[13] Coppola L, Huff D, Wang F, Burgos R and Boroyevich D (2007) Survey on high-temperature packaging materials for SiC-based power electronics modules. In: 2007 IEEE Power Electronics Specialists Conference, Orlando. pp. 2234-40.
[14] Yao Y, Chen Z, Lu G Q, Boroyevich D and Ngo K D T (2012) Characterization of Encapsulants for High-Voltage High-Temperature Power Electronic Packaging. IEEE T. COMP. PACK. MAN., 2: 539-47.
[15] Lu G Q, Calata J N, Zhang Z and Bai J G (2004) A lead-free, low-temperature sintering die-attach technique for high-performance and high-temperature packaging. In: Proceedings of the Sixth IEEE CPMT Conference on High Density Microsystem Design and Packaging and Component Failure Analysis (HDP'04), Shanghai. pp. 42-6.
[16] Seong K, Iwasaki T, Nakamura T, Sakumoto S and Taniguchi F (2017) High thermal sintering die attach for power device. In: 2017 International Conference on Electronics Packaging (ICEP), Yamagata. pp. 341-4.
[17] Mohan K, Shahane N, Raj P M, Antoniou A, Smet V and Tummala R (2017) Low-temperature, organics-free sintering of nanoporous copper for reliable, high-temperature and high-power die-attach interconnections. In: 2017 IEEE Applied Power Electronics Conference and Exposition (APEC), Tampa. pp. 3083-90.
[18] Yang F, Jia L, Wang L, Zhao C, Wang J, Zhang T, Gan Y and Zhang H (2020) The Study on Thermal Coupling Effect for SiC Power Module Design Guidelines. In: 2020 IEEE Workshop on Wide Bandgap Power Devices and Applications in Asia (WiPDA Asia), Suita. pp. 1-7.
[19] Chen M, Wang H, Pan D, Wang X and Blaabjerg F (2021) Thermal Characterization of Silicon Carbide MOSFET Module Suitable for High-Temperature Computationally Efficient Thermal-Profile Prediction. IEEE J. EM. SEL. TOP. P., 9: 3947-58.
[20] Adan A O, Tanaka D, Burgyan L and Kakizaki Y (2019) The Current Status and Trends of 1,200-V Commercial Silicon-Carbide MOSFETs Deep Physical Analysis of Power Transistors From a Designer's Perspective. IEEE POWER. ELECTRON., 6: 36-47.

Research on the heat dissipation capability of gas flowing in microchannels at high-speed

Xin Zhang, Jianhua Weng *

College of Energy and Mechanical Engineering, Shanghai University of Electric Power, Shanghai, P.R.China

*Corresponding author: jhweng@shiep.edu.cn

Abstract—The high-speed heat dissipation capacity of CO_2 gas in the cold plate is calculated and analyzed. The dimensions of the cold plate are 40 mm × 40 mm, and the convective heat transfer rate between the gas and the microchannel wall is considered. The results show that at an inlet temperature of t_{in} = 14 °C, inlet pressure of p_{in} = 1.0 MPa, and the mass flow rate of $\dot{m} = 0.6 g/s$, the heat transfer rate between CO_2 gas and the cold plate is 1203.5 W. Calculations were performed for different inlet temperatures, inlet pressures, microchannel widths, and wall temperatures. The analysis shows that with a decrease in inlet temperature, the heat transfer rate between the gas and the wall increases, while an increase in inlet pressure and a decrease in wall temperature lead to a reduction in heat transfer rate. When the width of the microchannel increases, the throttling effect within the microchannel diminishes, leading to a reduction in the heat transfer rate between the gas and the wall.

1. Introduction

Microchannel heat exchangers, with advantages such as high heat transfer efficiency, compact volume, and light weight, demonstrate broad application prospects in fields like microelectronics cooling [1], biomedical engineering [2], and aerospace [3]. In the cooling of certain electronic devices, gas working fluids offer greater safety compared to liquid media, and if leakage occurs, the gas medium will not corrode electronic components.

In microchannels, gas rarefaction is a significant factor that cannot be ignored. Hu et al. [4] investigated the flow and heat transfer characteristics of air under the influence of rarefaction in the slip flow region. The results indicated that as the Knudsen number (Kn) increases, gas rarefaction intensifies, while the average Nusselt number (Nu) decreases accordingly. Xu et al. [5] employed the Direct Simulation Monte Carlo (DSMC) method to analyze the effects of wall roughness on the heat transfer rate of rarefied gas within microchannels. The findings revealed that roughness elements can enhance local heat transfer rate through increased heat exchange area and compression effects. However, due to flow blockage, the flow velocity of the gas within the channel decreases, resulting in a reduction in the average heat transfer rate.

CO_2 gas is a highly representative working fluid, boasting wide applications in various microchannel heat exchangers. Rao et al. [6] conducted numerical simulations of laminar flow and heat transfer of supercritical CO_2 in a horizontal microtube. The results indicated that CO_2 exhibits dramatic changes in physical properties with temperature fluctuations during flow, thereby enhancing heat transfer within the tube. Yan et al. [7] experimentally investigated the temperature drop characteristics of CO_2 flowing through a multi-layer microchannel cooler. The experimental results showed that with increased inlet pressure, the cold-end temperature on the surface of the cooler specimen decreased. When the inlet pressure is 1.1 MPa and the inlet temperature is 14 °C, the cold-end temperature decrease to -44.10 °C.

In addition to CO_2 gas, N_2 and Ar are also frequently used in microchannel heat exchangers.

Turner et al. [8] conducted experimental research on the convective heat transfer of N_2 in laminar flow within microchannels and validated a two-dimensional numerical model of gas flow in microchannels using these experiments. The results showed that after the inlet region, the Nu for nitrogen stabilized around 8.23. Wang et al. [9] designed a novel multi-layer microchannel Joule-Thomson (J-T) cooling device using argon as the working fluid and conducted experiments on its cooling performance. At specified initial conditions (10.5 °C, 8 MPa), the cold-end temperature depression to -41.1 °C. Geng et al. [10] used the same device to analyze the temperature drop characteristics of nitrogen in the cooling system, comparing it with argon. The findings indicated that under the same inlet pressure, argon achieved a lower cold-end temperature than nitrogen. However, nitrogen reached its cold-end temperature in a shorter time. Therefore, nitrogen can be considered as a substitute for argon to reduce cooling costs.

Based on calculations of the flow parameter variations of high-speed CO_2 flow in microchannels, the research conducts a computational analysis of the heat dissipation capability of CO_2 gas within a cold plate. Additionally, the convective heat transfer between the gas and the microchannel walls also is considered.

2. Cold plate

As shown in Figure 1, the dimensions of the cold plate selected in the research are 40 mm × 40 mm, and the cold plate contains 50 microchannels. CO_2 gas is used as the working fluid, and the wall temperature is constant. For the convenience of calculation, a single microchannel with left-right symmetry was selected for the research. The dimensions of the single microchannel are shown in Table 1.

Table 1. The dimensions of the single microchannel.

Dimension	Value
Overall width [w_b(mm)]	0.8
Internal width [w_c(mm)]	0.4
Height [H_c(mm)]	0.6
Length [L(mm)]	40

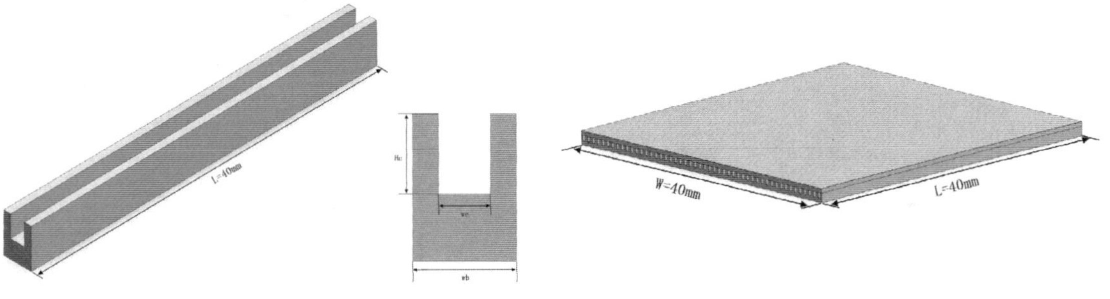

Figure 1. Dimensions and Structure of cold plate.

3. Theoretical Calculation Method

3.1 Governing Equations

Given the intricate conjugate heat transfer phenomena inherent to microchannel cold plate architectures, the present formulation adopts the following idealized premises to simplify calculations:

1) The fluid flow in the microchannels is one-dimensional steady flow;
2) The mass flow rate in the microchannel is uniformly distributed;
3) The wall temperature of the microchannel is uniform.

The flow within the microchannel is calculated using the micro-element control volume method. The microchannel is divided into n micro-element control volumes, each calculated sequentially. In the control volume, energy transfer includes enthalpy (*H*), kinetic energy (*KE*), and heat (*Q*). The energy transfer process is shown in Figure 2.

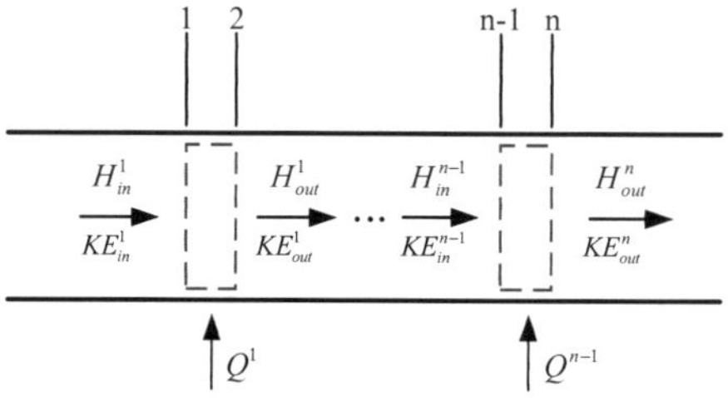

Figure 2. Schematic of energy transfer in microchannels.

The momentum equation of the fluid within the microchannel is shown in (1):

$$p_{in} - p_{out} = f_d \frac{\bar{\rho} \cdot \bar{u}^2}{2} \frac{l}{D} + \frac{\dot{m}}{A} \Delta u \tag{1}$$

Here, the subscript 'in' represents the inlet of the control volume, while the subscript 'out' represents the outlet of the control volume; f_d is the Darcy-Weisbach friction factor; $\bar{\rho}$ represents the average density at the inlet and outlet of the control volume; $\bar{u}$ is the average flow velocity at the inlet and outlet of the control volume; l is the length of the control volume; *D* is the characteristic length of the channel; $\dot{m}$ represents the mass flow rate within the control volume; *A* is the cross-sectional area of the control volume; and Δu is the velocity difference across the control volume.

The energy equation of the fluid in the microchannel is presented in Equation (2):

$$H_{in} + KE_{in} = H_{out} + KE_{out} + Q \tag{2}$$

The convective heat transfer rate between CO_2 gas and the microchannel wall is calculated according to Newton's Law of Cooling:

$$\Phi = h_m A \Delta t_m \tag{3}$$

where h_m represents the convective heat transfer coefficient between the gas and the microchannel wall, *A* is the area of the four walls of the micro-element control volume, i.e. $A = 2l(w_c + H_c)$, Δt_m represents the logarithmic mean temperature between the gas and the microchannel wall.

The convective heat transfer coefficient is calculated using the formula given by Kwon et al. [11], as follows:

$$h_m = c(\lambda/D)\mathrm{Re}^{0.8}\,\mathrm{Pr}^{1/3}\left(\mu_{CO_2}/\mu_{CO_2,w}\right)^{0.14} \tag{4}$$

where *c* represents a coefficient with a value of 0.028, λ is the thermal conductivity, μ is the dynamic viscosity, $\mu_{CO_2,w}$ represents the dynamic viscosity between CO_2 and the microchannel wall.

In (3), Δt_m uses the logarithmic mean temperature difference, namely:

$$\Delta t_m = \left(t_f'' - t_f'\right) \Big/ \ln \frac{t_w - t_f'}{t_w - t_f''} \tag{5}$$

where t_f' and t_f'' are the inlet and outlet temperatures of the gas in the control volume, respectively, and t_w represents the microchannel wall temperature.

The flow friction factor within the channel is calculated using the Darcy friction factor correlation given by Hong et al. [12], as shown below:

$$f_d = \left(1 + \frac{\gamma Ma^2}{1 - Ma^2}\right) 0.3164\,\mathrm{Re}^{-0.25} \tag{6}$$

where γ is the specific heat ratio of the fluid.

3.2 Calculation Steps and Methods

REFPROP software is employed to acquire the physical properties of carbon dioxide according to temperature and pressure, so as to guarantee the accuracy of the calculation. Curve fitting was performed on physical parameter equations (including fluid density, dynamic viscosity, specific heat capacity, etc.) using MATLAB's Curve Fitting Toolbox. The microchannel is partitioned into n differential control volumes, and stepwise computations are carried out for each control volume. The gas density at the outlet of the control volume is used to set the convergence criterion. The specific calculation steps are as follows:

1) We assume the outlet density of the control volume is $\rho_{out}^{k} = \rho_{in}^{k}$;

2) The outlet pressure p_{out}^{k} of the control volume is calculated using (1);

3) The outlet specific enthalpy h_{out}^{k} of the control volume is calculated using (2);

4) The corresponding outlet gas density ρ_{out}^{kp} of the control volume can be obtained using the fitting equation $\rho = f(p,h)$ based on p_{out}^{k} and h_{out}^{k};

5) If $\left|\rho_{out}^{k} - \rho_{out}^{kp}\right| / \rho_{out}^{k} \leq 0.0001$ is satisfied, then ρ_{out}^{kp} will be the calculated outlet gas density for the k-th control volume, and the next control volume will be calculated.

6) If not satisfied, we reassign ρ_{out}^{k} as $\left(\rho_{out}^{k} + \rho_{out}^{kp}\right)/2$ and continue the above steps until the convergence criteria are met.

3.3 Verification of the Calculation Method

The accuracy of the above method is validated using the research by Hong et al.[11] as an example. Hong et al. investigated the flow of N_2 in a microchannel under adiabatic conditions. The microchannel used in the research was etched on a silicon wafer and covered with a top glass plate.

In this validation, a microchannel with a width of 1.02 mm, a height of 0.1127 mm, and a length of 26.9 mm in the experiment was selected. Among the eleven experimental conditions, three different inlet conditions are selected for verification, as shown in Table 2.

Table 2. Experimental inlet conditions

P_{in} (MPa)	T_{in} (K)	Ma_{in}
0.19	292.5	0.27
0.56	289	0.32
1.01	288	0.34

During the experiment, the local gas temperature was calculated based on the measured gas pressure data. A Type-K thermocouple was installed in the upstream section of the microchannel to obtain the gas temperature data at that location. The measuring points for the gas pressure inside the microchannel were sequentially set at the cross - sections of the inlet (0 mm), 8 mm, 13 mm, 18 mm, and 23 mm. The calculation results are in good agreement with the experimental data, as shown in Figures 3 to 5, which verifies the accuracy and reliability of the calculation method.

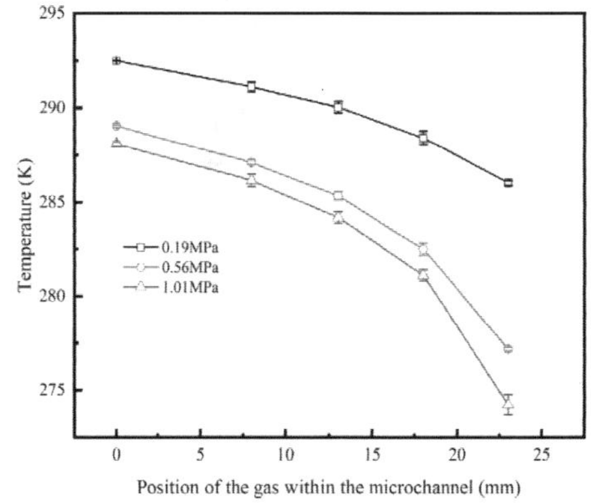

Figure 3. Error bars of temperature between theoretical calculations and experiments.

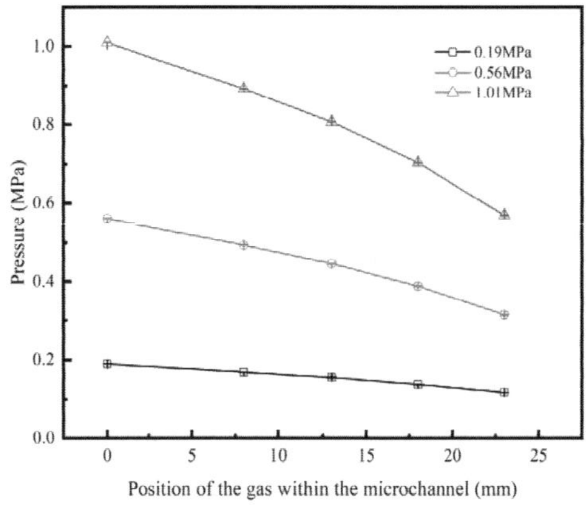

Figure 4. Error bars of pressure between theoretical calculations and experiments.

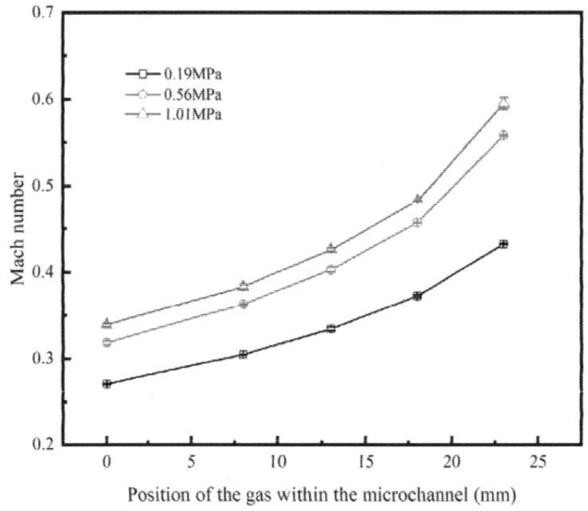

Figure 5. Error bars of Mach number between theoretical calculations and experiments.

4. Results and Discussion

4.1 Single Microchannel

A single microchannel is used as an example for the calculation and analysis. The working fluid in the microchannel is CO_2 gas, and the inlet conditions for the calculation are a temperature of t_{in} = 14 °C, a

Journal of Physics: Conference Series **3011** (2025) 012024 doi:10.1088/1742-6596/3011/1/012024

pressure of p_{in} = 1 MPa, and a mass flow rate of $\dot{m} = 0.6g/s$. The microchannel wall temperature t_w = 50 °C.

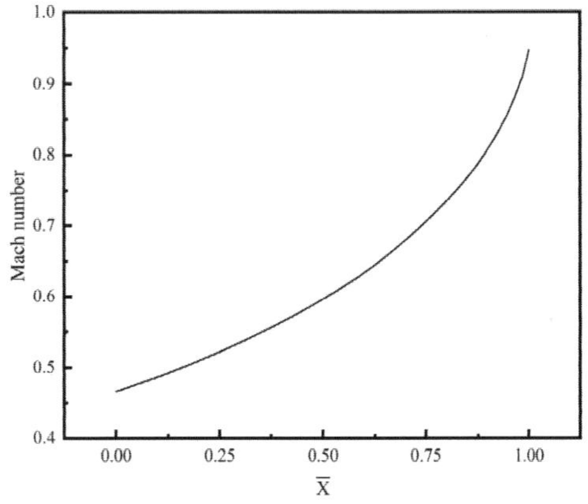

Figure 6. Mach number variation within a single microchannel (t_{in} = 14 °C, p_{in} = 1.0 MPa, $\dot{m}$ = 0.6 g/s).

Under these inlet conditions and channel geometries, the flow of CO_2 gas within the microchannel is in a high-speed flow state, as shown in Figure 6. In the figure, the abscissa $\overline{X}$ represents the ratio of the position of the gas within the microchannel to the total length of the microchannel. The results of the calculations for a single microchannel are shown in Figure 7. The heat transfer rate between CO_2 gas and the solid wall gradually increases along the flow direction, reaching a maximum at the channel outlet. This is due to the throttling effect of the microchannel causes the temperature of CO_2 within the microchannel to decrease, leading to an increase in the logarithmic mean temperature difference, thereby increasing the heat transfer rate.

Under the conditions of inlet temperature t_{in} = 14 °C, pressure p_{in} = 1.0 MPa, mass flow rate $\dot{m} = 0.6g/s$, and wall temperature t_w = 50 °C, the gas-solid heat transfer rate within the single microchannel is 24.07 W. In high-speed flow conditions, CO_2 demonstrates remarkable heat dissipation capability.

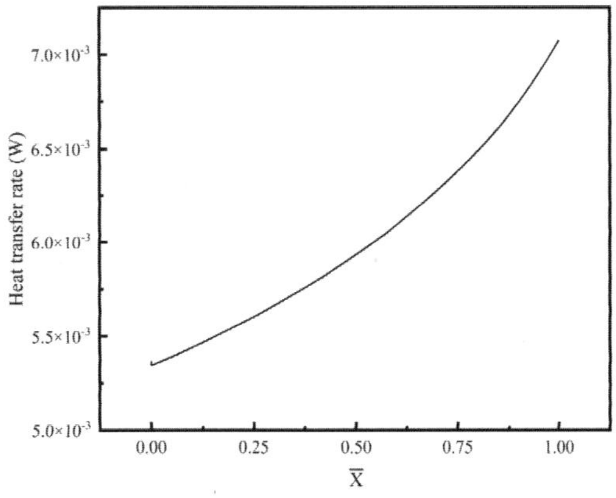

Figure 7. Convective heat transfer rate between gas and solid within a single microchannel (t_{in} = 14 °C, p_{in} = 1.0 MPa, $\dot{m}$ = 0.6 g/s).

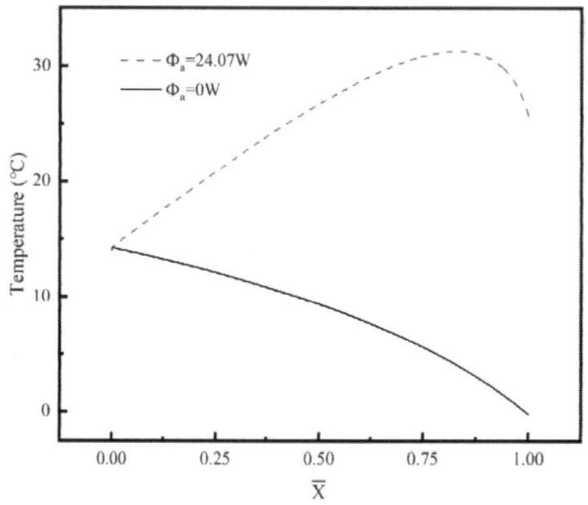

Figure 8. Temperature variation of CO_2 gas within a single microchannel (t_{in} = 14 °C, p_{in} = 1.0 MPa, $\dot{m}$ = 0.6 g/s).

Under the aforementioned heat transfer conditions, the temperature of the gas within the single microchannel also increased. As shown in Figure 8, after heat exchange, the outlet temperature of the CO_2 gas is 25.37 °C. From the figure, it can also be seen that the temperature of the CO_2 gas after heat exchange exhibiting a non-monotonic trend. This phenomenon is due to the gradual increase in the Mach number of the CO_2 gas within the microchannel, which continuously enhances the throttling effect. At the outlet of the microchannel, the Mach number of the CO_2 gas reaches 0.94.

It is evident that the velocity of CO_2 gas within the microchannel significantly affects the temperature variations. A higher velocity enhances heat exchange efficiency and prevents the outlet temperature from becoming excessively high. The cold plate contains a total of 50 microchannels, allowing for a maximum heat transfer rate of 1203.5 W.

4.2 Different Wall Temperatures and Inlet Conditions

In practical applications, the wall temperature will change. Therefore, calculations were performed for a wall temperature of t_w = 40 °C and t_w = 50 °C for comparison. When the convective heat transfer coefficient remains constant, the heat transfer rate is only related to the logarithmic mean temperature difference between the gas and the microchannel wall. As shown in Figure 9, when the wall temperature decreases, the gas-solid heat transfer rate also decreases. Where the wall temperature is t_w= 40 °C, the heat transfer rate of the single microchannel is 18.24 W, which is a decrease of 5.83 W compared to the heat transfer rate at t_w= 50 °C.

When the wall temperature is t_w= 40 °C, the heat transfer rate of the cold plate is Φ_z= 912 W; whereas at the wall temperature t_w= 50 °C, the heat transfer rate of the cold plate is Φ_z= 1203.5 W. Therefore, the higher the wall temperature, the greater the convective heat transfer between the gas and the wall.

It is well known that the inlet conditions of the working fluid have a significant impact on the heat transfer characteristics of the cold plate, such as temperature and pressure. Therefore, calculations and analyses were conducted under different inlet conditions.

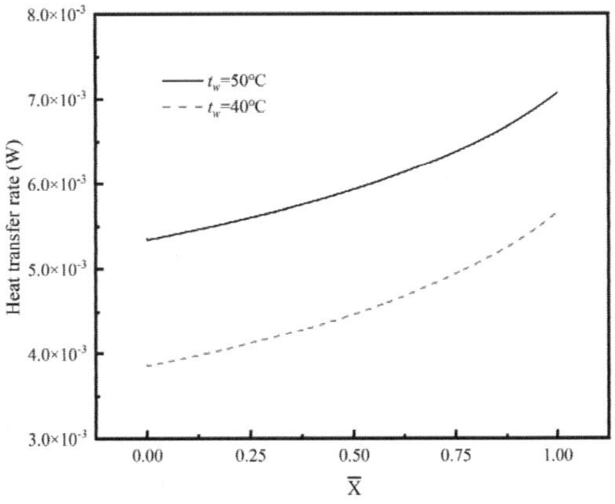

Figure 9. Comparison of heat transfer rates at different wall temperatures (t_{in}= 14 °C, p_{in}= 1.0 MPa, $\dot{m}$ = 0.6 g/s).

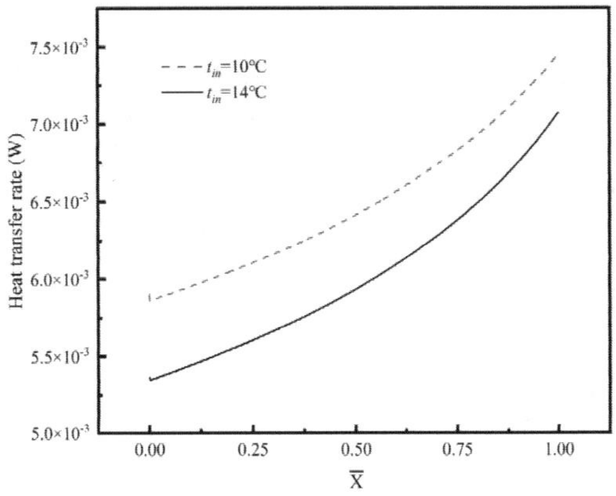

Figure 10. Comparison of heat transfer rates at different inlet temperatures (p_{in}= 1.0 MPa, $\dot{m}$ = 0.6 g/s, t_w= 50 °C).

As shown in Figure 10, the gas-solid heat transfer rate at t_{in}= 10 °C is consistently higher than that at t_{in}= 14 °C. However, as the gas flows toward the outlet of the microchannel, the heat transfer rate difference between the two gradually decreases. This is because, when only the inlet temperature is changed, the heat transfer rate between the gas and the wall is determined by the convective heat transfer coefficient and the logarithmic mean temperature difference. When the inlet temperature of the gas is lower, the logarithmic mean temperature difference between the gas and the wall becomes larger, thereby increasing the convective heat transfer rate between the gas and the wall. When the inlet temperature is t_{in}= 10 °C, the total heat transfer rate of the cold plate is Φ_z= 1297.5 W, an increase of 94 W compared to the condition when the inlet temperature is t_{in}= 14 °C.

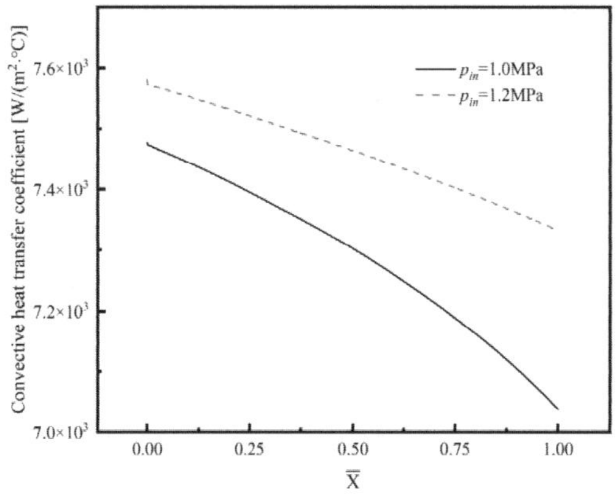

Figure 11. Comparison of convective heat transfer coefficients at different inlet pressures (t_{in} = 14 °C, $\dot{m}$ = 0.6 g/s, t_w = 50 °C).

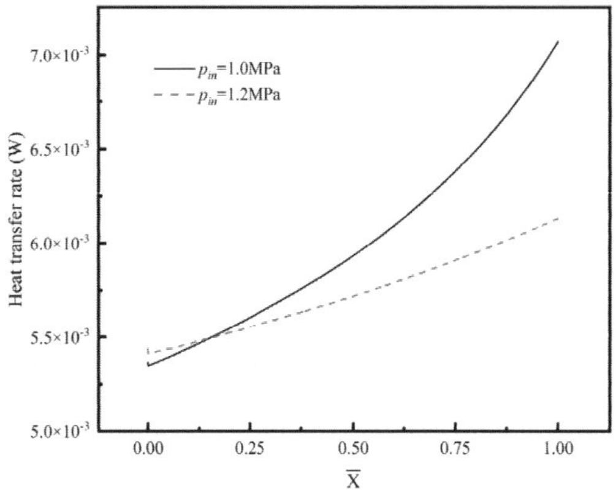

Figure 12. Comparison of heat transfer rates at different inlet pressures (t_{in} = 14 °C, $\dot{m}$ = 0.6 g/s, t_w = 50 °C).

From Figure 11, when the inlet pressure is 1.2 MPa, the convective heat transfer coefficient between the gas and the wall is consistently higher than that at an inlet pressure of 1.0 MPa. It can be observed from Figure 12 that up to 13.5 mm from the microchannel inlet, the heat transfer rate between the gas at an inlet pressure of 1.2 MPa and the wall is higher than that at 1.0 MPa. However, beyond 13.5 mm, the opposite occurs. This is because the logarithmic mean temperature difference between the CO_2 gas and the microchannel wall at p_{in} = 1.0 MPa gradually becomes larger than that at p_{in} = 1.2 MPa. Despite the convective heat transfer coefficient at p_{in} = 1.2 MPa being greater than that at p_{in} = 1.0 MPa, the logarithmic mean temperature difference plays a crucial role in the heat transfer rate, as shown in Figure 13. Consequently, the heat transfer rate at p_{in} = 1.0 MPa is greater than that at p_{in} = 1.2 MPa. When the inlet pressure is 1.2 MPa, the total heat transfer rate of the cold plate is 1147 W, representing a reduction of 56.5 W compared to the total heat transfer rate at an inlet pressure of 1.0 MPa.

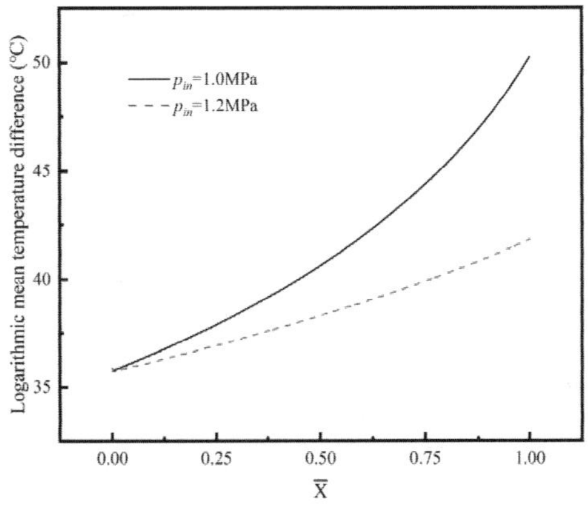

Figure 13. Comparison of logarithmic mean temperature differences at different inlet pressures (t_{in} = 14°C, $\dot{m}$ = 0.6 g/s, t_w = 50 °C).

4.3 Different Microchannel Widths

The wall thickness of the researched single microchannel remains constant at 0.4 mm (i.e., $w_b - w_c$ = 0.4 mm), and the height of the microchannel is uniformly 0.6 mm; only the internal width of the microchannel changes.

In Figure 14, the heat transfer rates for single microchannels with widths of w_c = 0.4 mm and w_c = 0.6 mm are compared. It can be seen from the figure that the heat transfer rate of the gas in the microchannel with a width of 0.6 mm is significantly lower than that in the microchannel with a width of 0.4 mm. This is because the heat transfer rate is related to the Nusselt number, while the magnitude of the Nusselt number is influenced by the fluid's Reynolds number. Figure 15 shows that when w_c = 0.4 mm, the flow velocity of CO_2 gas in the microchannel is much greater than that at w_c = 0.6 mm. Therefore, the Reynolds number of the gas in the microchannel with a width of 0.4 mm is also significantly larger than that in the microchannel with a width of 0.6 mm. This indicates that as the hydraulic diameter of the microchannel decreases, the convective heat transfer rate between the gas and the solid wall will increase, with gas flow velocity playing a crucial role.

For a single microchannel, when w_c = 0.6 mm, the heat transfer rate between the gas and the microchannel wall is 18.13 W. For the cold plate, with w_c = 0.6 mm and a total of 40 microchannels, the heat transfer rate of the cold plate is Φ_z = 725.2 W. Thus, as the hydraulic diameter of the microchannel decreases and the gas flow velocity increases, the convective heat transfer rate between the gas and the microchannel wall also increases.

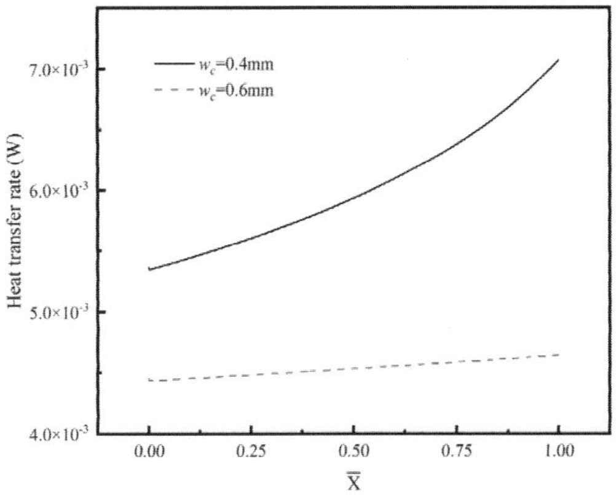

Figure 14. Comparison of heat transfer rates at different widths (t_{in} = 14 °C, p_{in} = 1.0 MPa, $\dot{m}$ =0.6 g/s, t_w = 50 °C).

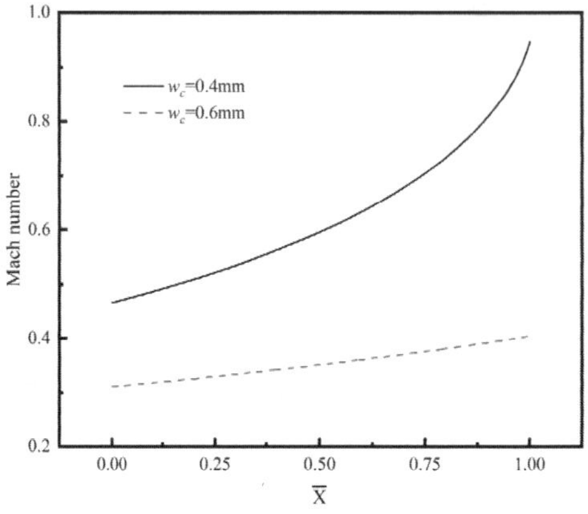

Figure 15. Comparison of CO_2 gas velocities at different widths (t_{in} = 14 °C, p_{in} = 1.0 MPa, $\dot{m}$ = 0.6 g/s, t_w = 50 °C).

5. Conclusion

(1) A single microchannel was calculated and analyzed. When the inlet temperature of the CO_2 gas was t_{in} = 14 °C, the pressure was p_{in} = 1 MPa, the mass flow rate was $\dot{m} = 0.6g/s$, and the wall temperature was constantly maintained at t_w = 50 °C, the heat transfer rate between the gas and the wall was 24.07 W. The cold plate with 50 microchannels exhibits a total heat transfer rate of Φ_z = 1203.5 W. It was found that the CO_2 gas flows at high speed within the microchannels, and its temperature exhibits a non - monotonic trend.

(2) The gas flow within a cold plate with 50 microchannels was researched, and its heat dissipation capability was calculated and analyzed under different inlet temperature, pressure, and wall temperature conditions. Results show that when the inlet temperature decreases, the heat transfer rate between the CO_2 gas and the wall increases. When the inlet pressure is 1.0 MPa, the mass flow rate is 0.6 g/s, and the wall temperature is 50 °C, decreasing the inlet temperature by 4 °C results in a 94 W increase in the heat transfer rate. When the wall temperature decreases, the heat transfer rate also decreases. When the inlet temperature is 14 °C, the inlet pressure is 1.0 MPa, and the mass flow rate is 0.6 g/s, a decrease of 10 °C in the wall temperature leads to an increase of 291.5 W in the heat transfer rate. When the inlet pressure increases, the heat transfer rate decreases. When the inlet temperature is 14 °C, the mass flow rate is 0.6 g/s, and the wall temperature is 50 °C, an increase of 0.2 MPa in the inlet pressure results in a reduction of 56.5 W in the heat transfer rate.

(3) For a cold plate of the same size, with the increase in hydraulic diameter, the throttling effect within the microchannel gradually diminishes, leading to a reduction in the heat transfer rate between the gas and the wall. When the inlet temperature is 14 °C, the inlet pressure is 1.0 MPa, the mass flow rate is 0.6 g/s, and the wall temperature is 50 °C, an increase in the hydraulic diameter of the microchannels by 0.12 mm and a decrease in the number of microchannels from 50 to 40 lead to a decrease in the heat transfer rate of the cold plate by 478.3 W. This also indicates that the gas velocity plays a crucial role in the gas-solid heat transfer rate.

References

[1] Hou F.Z., Zhang H.Y., Huang D.Z., et al. (2020) Microchannel Thermal Management System With Two-Phase Flow for Power Electronics Over 500 W/cm^2 Heat Dissipation. J. IEEE Transactions on Power Electronics, 35(10): 10592-10600.

[2] Zhang S., Wu D., Yang G., Chen G.M. (2010) Experimental study on a cryosurgery apparatus. J. Journal of Zhejiang University (Science A), 11(2): 128-131.

[3] Sugita H., Sato. Y, Nakagawa T., et al. (2010) Cryogenic system design of the next generation infrared space telescope SPICA. J. Cryogenics, 50(9): 566-571.

[4] Hu G., Hu G.Y., Tang Y., Peng X.C. (2014) Numerical analysis of the impact of rarefaction on heat transfer characteristics in gas flow through microchannels. J. Ship Electronic Engineering, 34(10): 106-110.
[5] Xu X., Shen Y., Yao S.G., Gu J.Y., Liu R. (2022) The study on the impact of rough elements on heat transfer in rarefied gas flow through microchannels. J. Journal of Jiangsu University of Science and Technology (Natural Science Edition), 36(03): 42-50.
[6] Rao Z.H., Liao S.M. (2005) Numerical simulation of laminar flow and heat transfer in a horizontal microtube with supercritical CO_2. J. Journal of Thermal Science and Technology, (02): 113-117.
[7] Yan J.N., Weng J.H., Cai R., Cui X.Y. (2021) Study on the temperature drop characteristics of CO_2 flow in microchannel coolers. J. Cryogenics and Superconductivity, 49(10): 70-75, 87.
[8] Turner S.E., Asako Y., Faghri M. (2008) Convection Heat Transfer in Microchannels With High Speed Gas Flow. J. Journal of Heat Transfer, 129(3): 319-328.
[9] Wang W.Q., Cui X.Y., Geng H., Han X.C., Weng J.H. (2016) Experimental study on the Joule-Thomson effect cooler using argon gas in microchannels. J. Cryogenic Engineering, (05): 46-50.
[10] Geng H., Cui X.Y., Wang W.Q., Weng J.H. (2017) Experimental study on the performance of a multi-layer microchannel J-T effect cooler using nitrogen gas. J. Journal of Refrigeration, 38(05): 93-98.
[11] Kwon B., Maniscalco N.I., Jacobi A.M., King W.P. (2018) High power density air-cooled microchannel heat exchanger. J. International Journal of Heat and Mass Transfer, 118(0): 1276-1283.
[12] Hong C., Nakamura T., Asako Y., Ueno I. (2016) Semi-local friction factor of turbulent gas flow through rectangular microchannels. J. International Journal of Heat and Mass Transfer, 98(0): 643-649.

Analysis of electromagnetic force characteristics of direct drive linear motor for belt conveyor

Chengyu Li [1,2*], Yufeng Zhang [1,2], Ruiyang Zhang [1,2], Ting Zhang [1,2]

[1] College of Electrical and Control Engineering, Xi'an University of Science and Technology, Shaanxi, China
[2] Xi'an Key Laboratory of Electrical Equipment Condition Monitoring and Power Supply Security, Shaanxi, China

*Corresponding author: Chengyu Li, Email: xkdlichengyu@stu.xust.edu.cn

Abstract. In order to improve the operation efficiency of the belt conveyor, combined with the permanent magnet synchronous linear motor (PMLSM) structure, a direct drive linear motor with permanent magnet secondary and conveyor belt integration is proposed, and the electromagnetic force characteristics of the primary pair of composite secondary with or without iron core are compared to obtain the primary structure suitable for the conveyor, so as to ensure that the conveyor can take into account the dual characteristics of electromagnetic thrust and normal suspension force, and improve the operation performance of the motor. According to the Halbach permanent magnet array, the assembly structure of the composite secondary permanent magnet is optimized to improve the electromagnetic force characteristics of the motor and further improve the working efficiency of the conveyor. Then, the influence of different parameters on the electromagnetic force during operation is analysed.

1. Introduction

As a common conveying equipment for transporting bulk materials, the belt conveyor is usually driven indirectly by a rotating motor, which has the disadvantage of uneven distribution of resistance and power [1]. In response to the above problems, some solutions have been proposed at home and abroad. For example, the use of magnetic levitation technology to reduce the friction resistance and loss of the belt conveyor during transportation [2]; the multi-motor distributed direct drive mode is used to dynamically adjust the driving force of each region and so on [3].

PMSM can directly convert electrical energy into linear motion mechanical energy without any intermediate conversion mechanism. In recent years, with the increasing demand for large thrust linear drive systems in industrial processing and ground transportation, permanent magnet synchronous linear motors have received more and more attention [4].

In recent years, scholars at home and abroad have done a lot of research work on the electromagnetic force characteristics of linear motors. In [5], the equivalent magnetization current method was used to analyse the ironless linear motor, and the expressions of air gap flux density and thrust of the motor were derived. The quasi-sinusoidal annular winding was used to reduce the thrust ripple. In [6], the Halbach permanent magnet array was used to increase the peak value of air gap flux density, and the influence of the Halbach / non-Halbach magnetic array and stacked / non-stacked winding structure on the thrust characteristics of ironless permanent magnet linear motor was analysed. In [7], the back electromotive force and output thrust of the coreless motor with Halbach permanent magnet array are analysed and compared with the structure before improvement. The results show that the motor with a Halbach permanent magnet array has a higher power density.

At present, most scholars' research on electromagnetic force mainly focuses on the thrust part, but considering the operation characteristics of the conveyor, it is necessary to take into account the dual characteristics of thrust and normal force, which need to be further studied. In this paper, the integrated design of the belt conveyor and the linear motor is carried out to compare the influence of the primary iron core on the electromagnetic force of the integrated secondary. The electromagnetic force is optimized by using a Halbach permanent magnet array. Finally, the influence of different parameters on the integrated secondary in the operation process is analysed, which provides a certain reference for the design of direct drive of belt conveyor.

2. Design of linear motor for belt conveyor

Considering the special structure of the belt conveyor, a direct-drive linear motor structure with a linear motor secondary and a conveyor belt is proposed, as shown in Figure 1. When the motor is running, due to the existence of normal force and thrust, the forward conveying of materials and magnetic levitation can be realized.

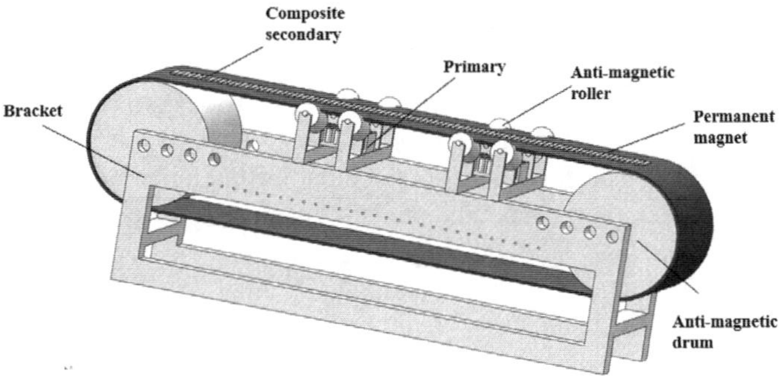

Figure 1. Permanent magnet conveyor belt integrated linear motor architecture.

2.1. Basic structure design of motor

The linear motor is composed of primary and secondary parts. The primary structure can be divided into iron core structure and non-iron core structure. In order to ensure the thrust of the motor in the forward direction and the suspension force in the normal direction, it is necessary to compare and analyze the two structures of the primary and select the structure and parameters that meet the requirements.

2.1.1. Main parameter size determination. The specific parameters are shown in Table 1.

Table 1. Initial design parameters of the motor.

Parameter	value
Rated current (A)	4
Armature length (mm)	70
Air gap length (mm)	2
Thickness of PM (mm)	8
Width of PM (mm)	19
Pole distance (mm)	24

2.2. Basic structure design of motor

According to the above main dimensions, the finite element simulation models of the integrated PMLSM with or without an iron core primary structure of the belt conveyor are established. The magnetic cloud diagrams of the two models are shown in Figure 2. It can be seen from the figure that the maximum magnetic density of the iron core structure appears at the end of the iron core slot, at about 1.5533 T; the maximum magnetic density of the coreless structure appears at the end of the permanent magnet at about 1.1704 T. Due to the lack of magnetic permeability of the core, the magnetic field cannot be concentrated and enhanced, the magnetic field distribution of the coreless motor is more dispersed, and the magnetic flux leaks more in the air, resulting in a decrease in its magnetic density.

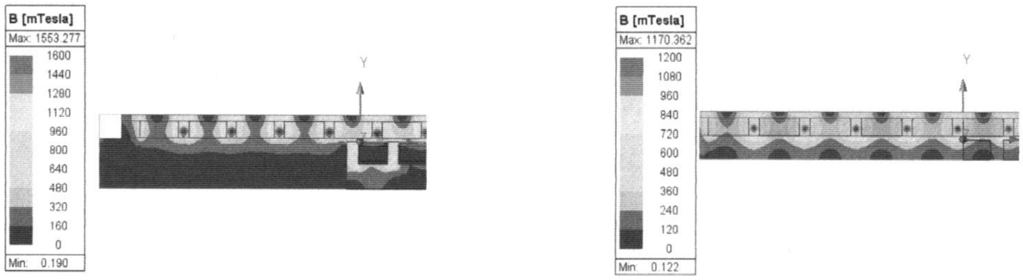

(a) Iron core structure (b) coreless structure

Figure 2. Magnetic dense cloud diagram of the motor with or without primary core structure

The influence of the primary iron core structure on the electromagnetic force under no-load operation is shown in Figure 3. The maximum electromagnetic thrust with an iron core is 194.9 N, and the maximum electromagnetic thrust without an iron core is 240.18 N. The specific electromagnetic force parameters are shown in Table 2. For the normal force, the maximum normal force of the iron core and the coreless motor is -594 N and 245 N respectively, and the normal force is positive in the y-axis direction. It can be seen that the iron core motor has a great unilateral attraction to the secondary, and the normal force of the coreless motor can realize the suspension effect.

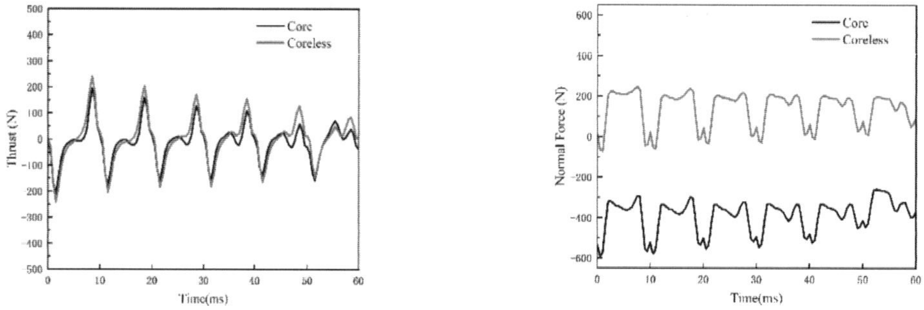

(a) Thrust with and without iron core (b) Normal force with and without iron core

Figure 3.The influence of primary iron core structure on electromagnetic force under no-load condition.

Table 2. Electromagnetic force parameters with or without iron core structure under no-load operation.

Electromagnetic force	core	coreless
Maximum thrust (N)	194.9	240.18
Average thrust (N)	10.68	1.01
Maximum normal force (N)	-594	245
Average normal force (N)	-390	135

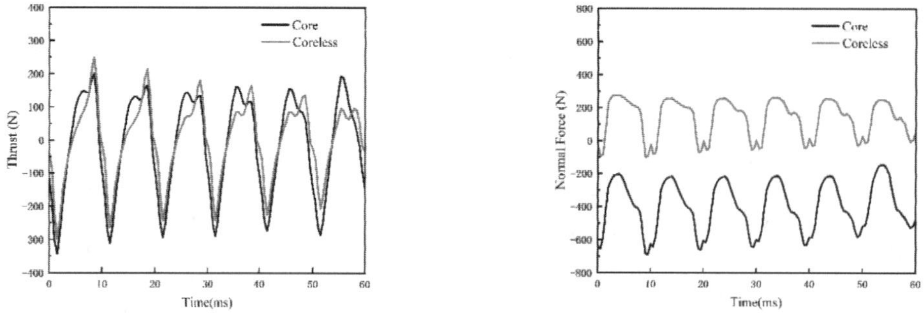

(a) Thrust with and without iron core (b) Normal force with and without iron core

Figure 4. The influence of primary iron core structure on electromagnetic force under load condition.

Table 3. Electromagnetic force parameters with or without iron core structure under load operation.

Electromagnetic force	core	coreless
Maximum thrust (N)	201	250
Average thrust (N)	15.96	1.89
Maximum normal force (N)	-690.2	275.8
Average normal force (N)	-397.3	135.9

The influence of the primary iron core structure on the electromagnetic force under load operation is shown in Figure 4. The electromagnetic thrust and normal force of the motor are consistent with those under no-load operation. The specific electromagnetic force parameters are shown in Table 3.

By comprehensively comparing the influence of the primary structure with or without an iron core on the electromagnetic force under no-load and load conditions, it can be obtained that when the iron coreless structure is adopted, the normal force on the secondary can be regarded as the suspension force. Therefore, the selection of the iron coreless primary structure can make the belt conveyor take into account the dual characteristics of electromagnetic thrust and normal suspension force, and realize the promotion and suspension of the conveyor belt.

3. Design of linear motor for belt conveyor

3.1. Improvement of permanent magnet structure

In order to improve the operation efficiency of the belt conveyor and improve the utilization rate of the permanent magnet as much as possible, the original permanent magnet is arranged as a Halbach permanent magnet array, which can greatly enhance the coupling magnetic field at the air gap, and generate a high magnetic density at the same amount, thereby increasing the electromagnetic force per unit volume of the motor. Figure 5 is the magnetic cloud diagram of the motor before and after improvement.

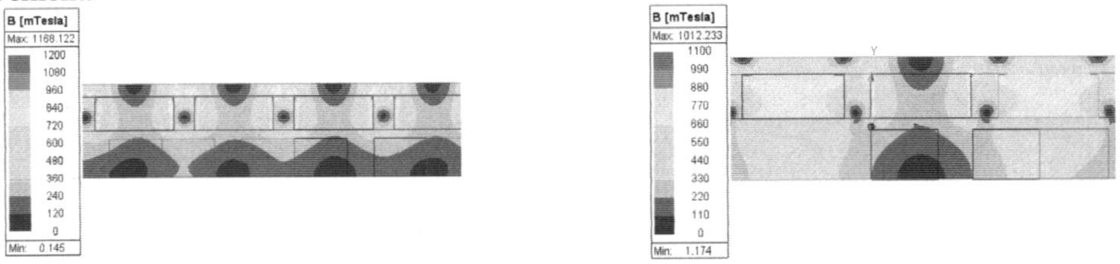

(a) Initial magnetic density distribution (b) Halbach magnetic density distribution

Figure 5. Magnetic dense cloud diagram of the motor before and after improvement.

3.2. Analysis of electromagnetic force characteristics of Halbach permanent magnet array

After improving the secondary structure of the belt conveyor, the electromagnetic force characteristics are analysed, as shown in Figure 6, and the specific electromagnetic force parameters are shown in Table 4. Although the thrust has decreased, the maximum suspension force and the electromagnetic force fluctuation suppression effect are very significant.

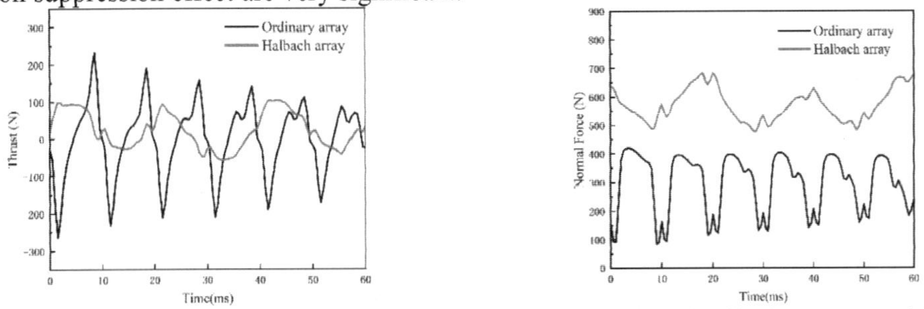

(a) Primitive and Halbach array thrust (b) Primitive and Halbach array normal force

Figure 6. The influence of permanent magnet structure on electromagnetic force before and after improvement.

Table 4. Comparison of electromagnetic force parameters before and after improvement of permanent magnet.

Electromagnetic force	Before	After
Maximum thrust (N)	232.4	103.14
Average thrust (N)	2.5	23.1
Maximum suspension force (N)	419.4	684
Average suspension force (N)	296.9	589.3

According to the simulation results, after using the Halbach permanent magnet array, the amount of the secondary permanent magnet of the conveyor belt is not changed, but the thrust and suspension force are significantly improved, the electromagnetic force fluctuation is suppressed, and the overall operation efficiency of the conveyor is improved.

3.3. The influence of different parameters on the electromagnetic force of the integrated motor

3.3.1 Relative distance between the primary and the end of the permanent magnet. Figure 7 shows the change of electromagnetic force when the relative distance between the primary and permanent magnet ends changes from 0 to 12 mm. As the distance between the primary and permanent magnet ends becomes larger, the levitation force generally shows a downward trend, and the thrust generally shows an upward trend when the relative position is small. When the relative position is too large, the average thrust decreases but the maximum thrust increases, indicating that the thrust fluctuation range becomes larger. Therefore, the initial relative position of the primary coil and the permanent magnet has a great influence on the electromagnetic force.

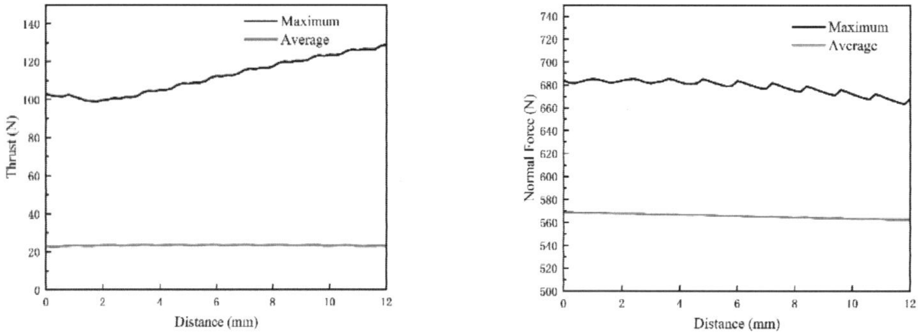

(a) Thrust (b) Normal force

Figure 7. The change of electromagnetic force during the change of relative distance from 0 to 12 mm.

3.3.2 Conveyor belt speed. Figure 8 shows the change of electromagnetic force during the change of conveyor belt speed from 0.8 m/s to 2.4 m/s.

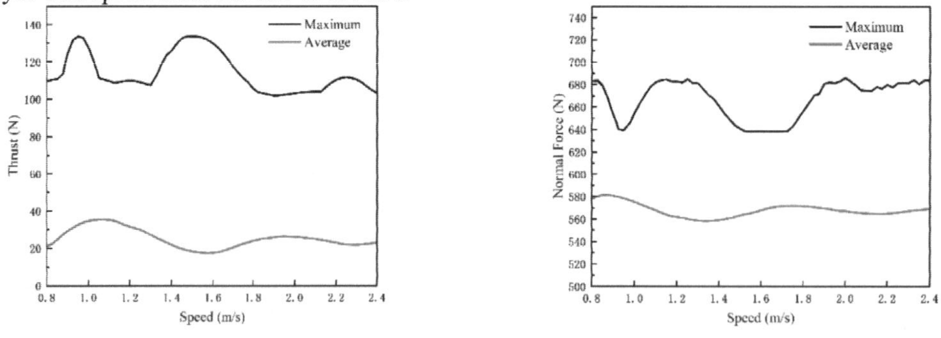

(a) Thrust (b) Normal force

Figure 8. The change of electromagnetic force during the change of conveyor belt speed from 0.8 to 2.4 m/s.

The influence of running speed change on electromagnetic force does not show an obvious change rule. The reason may be that the end effect of the secondary permanent magnet conveyor belt is larger when the speed changes and the fluctuation of the end force causes the change of thrust and suspension force. However, it can be seen from the waveform that when the speed is about 1 m/s, the average thrust, maximum thrust, and average suspension force can have better electromagnetic force performance.

4. Conclusion

In this paper, through the analysis of the direct drive linear motor for belt conveyor, a linear motor with a conveyor belt and permanent magnet secondary integrated is proposed. Firstly, the simulation model of the integrated structure linear motor is established, and the influence of the core structure on the electromagnetic force under no-load and load operating conditions is analyzed. The results show that the coreless structure can realize the dual characteristics of the conveyor in thrust and suspension force. Then the Halbach permanent magnet array is used to optimize the secondary structure, which greatly improves the magnetic field utilization of the permanent magnet, significantly improves the average thrust and suspension force, and the fluctuation of the two electromagnetic forces is obviously suppressed, which verifies the effectiveness of the assembly method of the Halbach permanent magnet array.

In summary, the ironless permanent magnet linear motor and the belt conveyor are integrated to complete the direct drive of the conveyor, which greatly reduces the friction loss in the traditional indirect drive process. It can be used as a new belt conveyor drive method and also provides a certain reference for expanding the application field of linear motors.

Acknowledgments

This work was financially supported by the Graduate Education Reform Project of Xi'an University of Science and Technology under Grant YJG2022016.

References

[1] Chelopo D, Gupta K. A Comparative Evaluation of Conveyor Belt Disc Brakes and Drum Brakes: Integrating Structural Topology Optimization and Weight Reduction[J]. Technologies, 2024, 12(8): 136-136.
[2] Hu K, Jiang H, Zhu Q, et al. Magnetic Levitation Belt Conveyor Control System Based on Multi-Sensor Fusion[J]. Applied Sciences, 2023, 13(13).
[3] Zhang D, Zhang Y, Yue Y, et al. Optimal design of robust control for belt conveyor systems based on fuzzy dynamic model and Nash game[J]. Journal of the Franklin Institute, 2024, 361(11):106925-.
[4] Sun G, Zhao M, Li J, et al. Design and Analysis of Tubular Transverse Flux Permanent Magnet Linear Motor[J]. Energies, 2024, 17(23): 6180-6180.
[5] M. Onsal, B. Cumhur, Y. Demir, E. Yolacan and M. Aydin, "Rotor Design Optimization of a New Flux-Assisted Consequent Pole Spoke-Type Permanent Magnet Torque Motor for Low-Speed Applications," IEEE Transactions on Magnetics, vol. 54, no. 11, pp. 1-5,
[6] Q. Shuheng, Z. Jie and Z. Chi, "Effects of Halbach and Non-Halbach Arrays on Thrust Characteristics of Ironless Permanent Magnet Linear Motors: A Simulation and Optimization," 2020 23rd International Conference on Electrical Machines and Systems (ICEMS), Hamamatsu, Japan, 2020, pp. 1659-1662.
[7] P. R. Eckert, A. F. Flores Filho, E. A. Perondi, et al. Dual quasi-Halbach linear tubular actuator with coreless moving-coil for semiactive and active suspension [J]. IEEE Transactions on Industrial Electronics, 2018, 65(12): 9873~9883.

A model predictive control strategy for optimal vectors novel sector division of PMSM

Shumiao Yang[1,2,*], Chongchong Ai [1,2] and Chenliang Wang[1,2]

[1]Xian University of Science and Technology, Xian, Shaanxi, 710000, China
[2]Xi'an Key Laboratory of Electrical Equipment Condition Monitoring and Power Supply Security, Xi'an, Shaanxi, 710000, China

[*]Corresponding author's e-mail: 22206227121@stu.xust.edu.cn

Abstract. Aiming at the problems of large speed fluctuations and high computational complexity in traditional finite control set-model predictive current control of PMSM, a model predictive control strategy for optimal vectors novel sector division of PMSM is adopted. Firstly, in order to determine the optimal vectors combination quickly, the voltage vector plane is divided into eight sectors according to coordinate axes and basic voltage vectors. Secondly, a method was designed to determine the sector and the optimal combination of three-vector voltages through two simple comparisons. At the same time, when calculating action time of selected voltages in three-vector combination, the optimal two-vector voltages combination can be determined. Finally, considering that the effect of three-vector cannot be globally superior to that of two-vector, the optimal combination that processes smallest cost function is further chosen from the optimal combination of two/three-vector to improve the speed accuracy of steady-state. Simulations and experiments shown that the strategy proposed in this paper is able to increase the speed tracking accuracy while reducing computational complexity.

1. Introduction

Finite Control Set-Model Predictive Current Control (FCS-MPCC) has been wildly used in Permanent Magnet Synchronous Motor (PMSM) controlling. It has advantages such as simple structure[1], intuitive principle[2], easy handling of nonlinear constraints[3], and fast dynamic response[4]. FCS-MPCC bases on a discrete model of motors. After substituting multiple candidate voltage vectors into predefined cost function, it selects the voltage vector or voltage vectors combination that can minimize the cost function then acts on the motor through an inverter directly[5]. Due to the fact that traditional FCS-MPCC only applies one voltage vector per cycle, moreover, the amplitudes and directions of eight candidate voltage vectors are fixed, when the expected vector deviates from candidate vector heavily, it will produce a significant steady-state speed deviation and big torque fluctuation[6,7].

Due to the reference [8] has proved that under certain conditions, the speed fluctuation of three-vector FCS-MPCC cannot be globally smaller than that of two-vector FCS-MPCC, this paper proposed an FCS-MPCC strategy based on optimal vectors novel sector division to decrease computational complexity and speed fluctuation in traditional FCS-MPCC. Firstly, whole voltage vector plane is divided into eight parts through coordinate axes and basic voltage vectors. After determining the quadrant which the predicted voltage vector is located in, only two vectors need to be compared to determine accurate sector. Then calculate the cost function's value of three-vector and two-vector successively and compare them. Last, selecting the voltage vectors combination with smaller value and applying those vectors to the motor. In the end, to verify the effectiveness of the proposed FCS-MPCC strategy in this paper, simulations and experiments were conducted.

2. Traditional FCS-MPCC strategy

In the d-q coordinate system, the current equations for surface PMSM is

$$\begin{cases} i_d(k+1) = (1-\frac{T_s R}{L})i_d(k) + T_s\omega_e(k)i_q(k) + \frac{T_s}{L}u_d(k) \\ i_q(k+1) = (1-\frac{T_s R}{L})i_q(k) - T_s\omega_e(k)i_d(k) + \frac{T_s}{L}\left[u_q(k) - \omega_e(k)\psi_f\right] \end{cases} \tag{1}$$

In the formula, i_d and i_q are the currents of d-axis and q-axis respectively, L is the stator inductance, u_d and u_q are voltages of the d-q axes, R is the stator resistance, ω_e is rotor electrical angular velocity, and ψ_f is permanent magnet flux chain.

Discretizing equation 1 and considering delay compensation, getting the cost function as follows:

$$g = \left[i_{d\_ref} - i_d(k+2)\right]^2 + \left[i_{q\_ref} - i_q(k+2)\right]^2 \tag{2}$$

According to the different switch combinations of the three-phase inverter driving PMSM, eight voltage vectors can be obtained totally, including two ineffective vectors and six effective vectors. The hexagon in the voltage vector plane is divided into six sectors.

In traditional three-vector FCS-MPCC strategy, two nonzero voltage vectors and a zero voltage vector are chosen per control cycle. Traditional two-vector FCS-MPCC selects voltage vectors on the basis of minimizing the cost function too, but only two vectors can be selected. According to the principle that whether there must have an ineffective vector in selectable voltage vector, traditional two-vector FCS-MPCC is divided into duty cycle FCS-MPCC and generalized two-vector FCS-MPCC. If vectors can be selected arbitrarily, it is the generalized two-vector FCS-MPCC; if only one ineffective vector and one effective vector can be selected, it is the duty cycle FCS-MPCC.

3. FCS-MPCC based on optimal vector novel sector division strategy

Traditional FCS-MPCC optimizes by traversing all basic vectors, which requires a large amount of computation. When determining the action time of selected vectors, if it's assumed that the action time of each selected vector is inversely proportional to its cost function, the control effect of two-vectors has its own advantages and disadvantages compared to three-vector. To address this issue, an FCS-MPCC strategy based on optimal vectors novel sector division was proposed. Considering natural characteristics of the inverter, FCS-MPCC avoids using vector modulation techniques in traditional control. So after determining the required voltage vector and calculating the corresponding action time, the PWM switch signal can be directly obtained and applied to the inverter. The control diagram is shown in Figure 1.

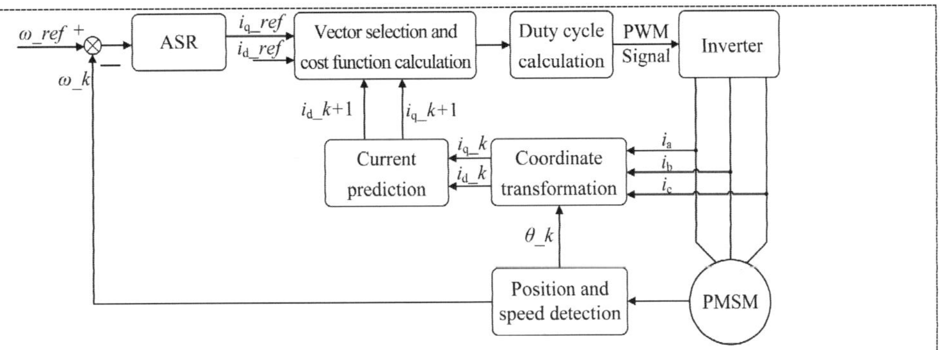

Figure 1. Control block diagram of FCS-MPCC.

Journal of Physics: Conference Series **3011** (2025) 012026 doi:10.1088/1742-6596/3011/1/012026

The strategy proposed in this paper divides voltage vector plane into eight regions according to coordinate axes and basic voltage vectors, as shown in Figure 2 a). When $u_\alpha > 0$ and $u_\beta > 0$, the predicted voltage belongs to region 1 or 2, as shown in Figure 2 b).

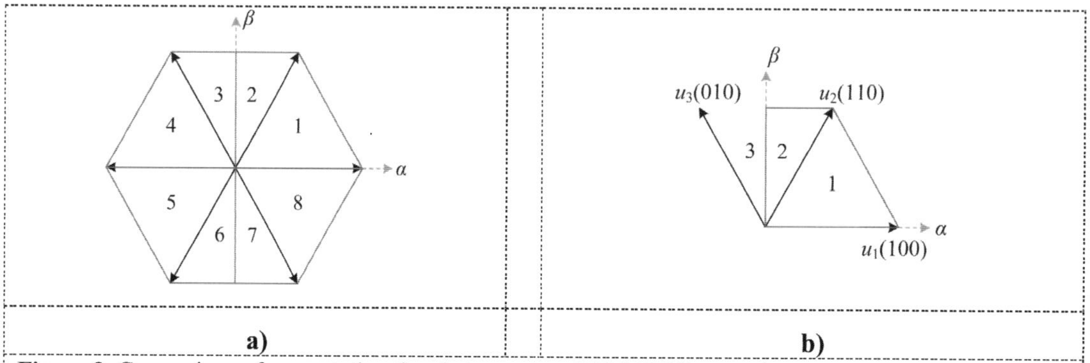

Figure 2. Comparison of computation time between two FCS-MPCC strategies.

Calculate cost function's value of three-vector first. When calculating the cost function, predicted voltage vector only can be in region 1 or region 2, u_2 is a mandatory voltage vector. After calculating the value of cost function of u1 and u_3, it needs to compare only once to decide the region where the predicted voltage belongs. If $u_1 < u_3$, the predicted voltage should be in the region 1. If $u_1 > u_3$, the predicted voltage should be in the region 2. After determining the sector, calculate the value of cost function of the combination of three-vector in that sector. If the predicted voltage is located in region 1, compare the value of cost function of u0 u1 and u_2 respectively, and select the smallest two voltage vectors. If the predicted voltage is located in the region 2, chose one between u_0 and u_3 that has smaller value of cost function.

4. Simulation and Experiment

4.1. Simulation verification

Simulation validation is firstly conducted in MATLAB/Simulink to verify the effectiveness of the strategy proposed. During simulation, because of the influence of the computer, it is not able to make the calculation time exactly same every time. Simulating the traditional three-vector FCS-MPCC, traditional two-vector FCS-MPCC and improved FCS-MPCC strategy 100 times separately. The specific comparison of the time required for different strategies is shown in Figure 3.

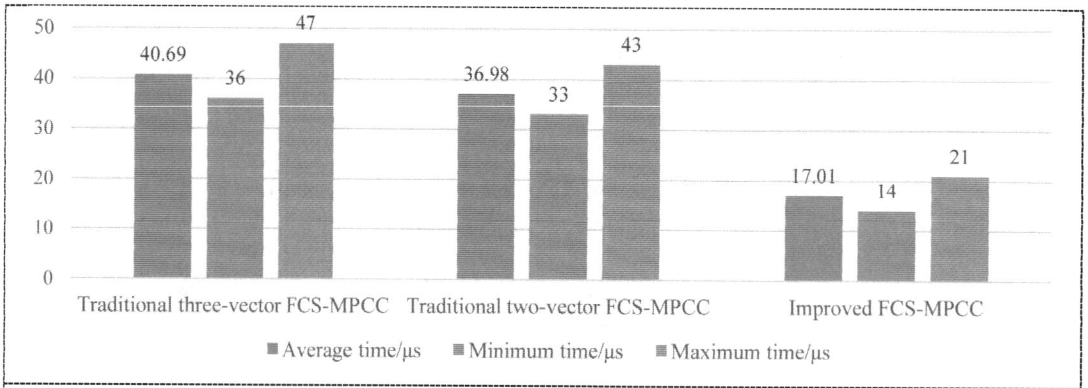

Figure 3. Specific comparison of computing time for different FCS-MPCC strategies.

The improved FCS-MPCC proposed in this paper reduced the average computation time by 20 μs compared to the traditional two-vector FCS-MPCC; compared to traditional three-vector FCS-MPCC, its average computation time has been reduced by 24 μs.

When simulating in Simulink, set the simulation duration to 2 seconds, with a torque load of 4N·m and a given speed of 750RPM. The speed waveform of traditional three-vector FCS-MPCC, traditional two-vector FCS-MPCC and improved FCS-MPCC strategy proposed in this paper are respectively shown in Figures 4 a), Figure 4 b) and Figure 4 c).

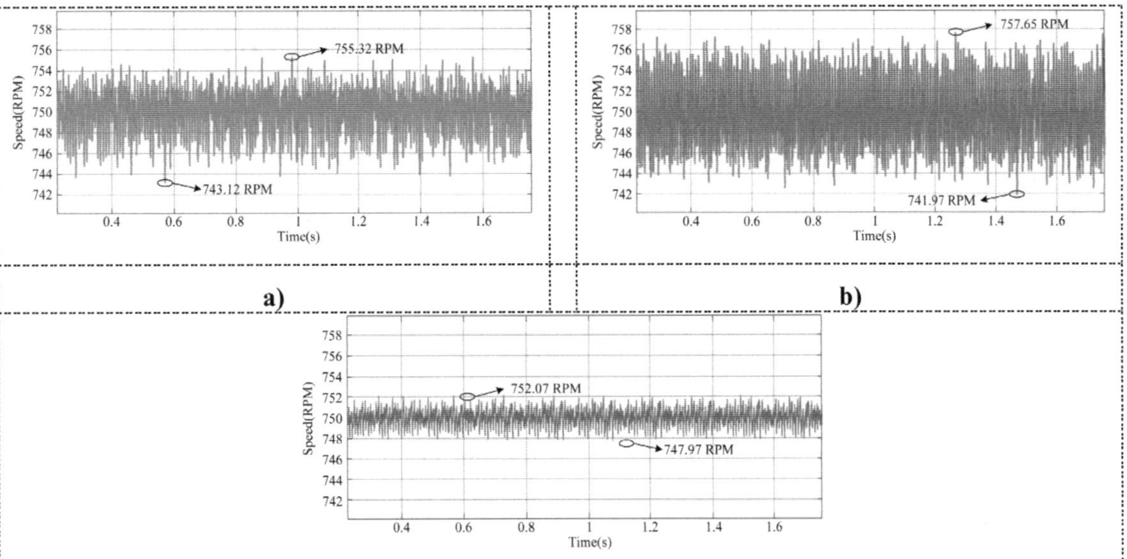

a) b)

c)

Figure 4. Comparison of speed waveform of different FCS-MPCC strategies.

The absolute error of speed fluctuation in the improved FCS-MPCC has been reduced by 8.1 RPM, the relative error has been reduced by 1.06%. The speed error has been reduced to 33.61% of traditional three-vector FCS-MPCC, and 26.15% of traditional two-vector FCS-MPCC.

4.2. Experimental Verification

Experimental verification was conducted on YXSPACE SP2000, comparing the improved FCS-MPCC proposed in this paper and traditional FCS-MPCC. Experimental platform is shown in Figure 5.

Figure 5. Experimental platform based on YXSPACE-SP2000 for PMSM drive control.

Motor's specific parameters are shown in Table 1.

Table 1. Specific parameters of PMSM.

Parameter	Value
stator resistance R (Ω)	4.6
Stator inductance L (H)	0.03365
Polar number p	3
Permanent magnetic flux φ (Wb)	0.45
Moment of inertia J ($kg{\cdot}m^2$)	0.000525

For the sake of comparing the calculation time of two kinds of FCS-MPCC strategies, a high level was set at the beginning of calculation, maintaining until the end of calculation then setting to zero. Repeating in next cycles, obtaining the comparison in Figure 6. Figure 6 a) is the calculation time of traditional three-vector FCS-MPCC, Figure 6 b) is the calculation time of traditional two-vector FCS-MPCC, and Figure 6 c) is the calculation time of improved FCS-MPCC.

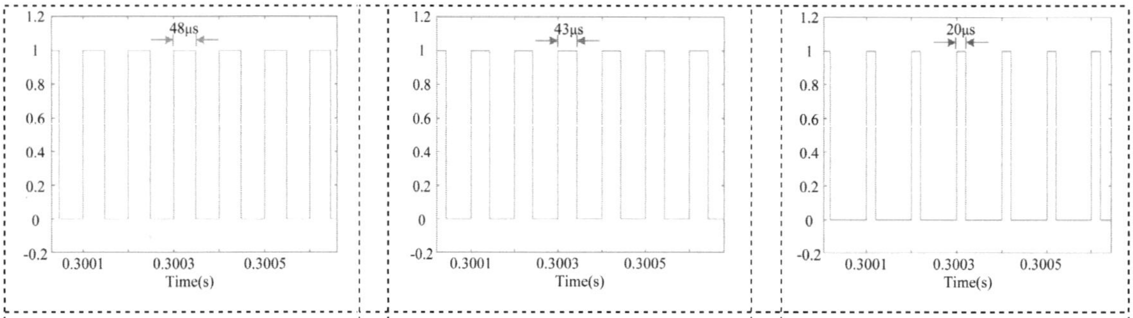

a) b) c)

Figure 6. Comparison of computation time between two FCS-MPCC strategies.

It can be seen from Figure 6 that the calculation time of traditional three-vector FCS-MPCC and traditional two-vector FCS-MPCC is 48μs and 43μs, while the improved FCS-MPCC has a calculation time of 20μs. Compared to traditional three-vector FCS-MPCC and traditional two-vector FCS-MPCC, the improved FCS-MPCC reduced by 53.49% and 58.33% respectively.

The speed waveforms were obtained and shown in Figure 7. Figure 7 a) is the speed waveform of traditional three-vector FCS-MPCC, Figure 7 b) is the speed waveform of traditional two-vector FCS-MPCC, and Figure 7 c) is the speed waveform of improved FCS-MPCC.

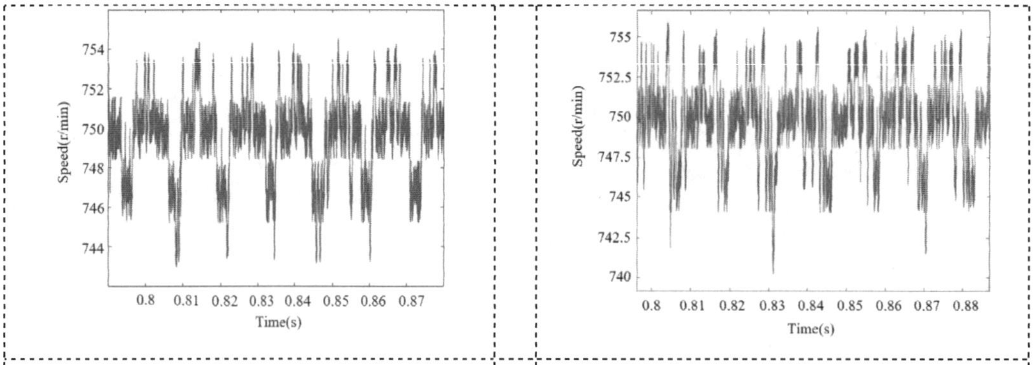

a) b)

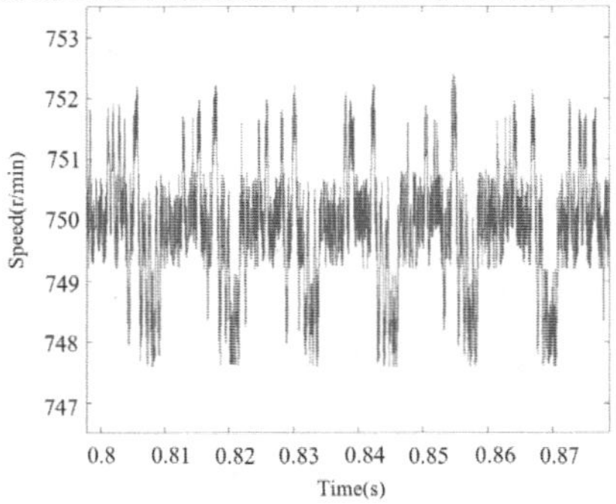

c)

Figure 7. Comparison of speed waveform between two FCS-MPCC strategies.

The absolute error of the speed fluctuation in the traditional three-vector FCS-MPCC is about+4.73 to -7.12RPM, and relative error is about 1.58%. The absolute error of the speed fluctuation in the traditional two-vector FCS-MPCC is about+6.27 to -9.79RPM, and relative error is about 2.14%. The absolute error of speed fluctuation of the improved FCS-MPCC is about+2.30 to -2.41RPM, and the relative error is about 0.63%. Compared to the traditional three-vector FCS-MPCC, the speed fluctuation has been reduced by 7.14 RPM, a decrease of 60.12%. Compared to the traditional two-vector FCS-MPCC, the speed fluctuation has been reduced by 11.60 RPM, a decrease of 72.23%.

5. Conclusion

The improved FCS-MPCC strategy proposed in this paper decreases the number of voltage vectors that need to be traversed by re dividing the voltage vector sectors. The large sector where the voltage vector belongs is determined by comparing the predicted voltage vector with zero. Then, the required voltage vector combination can be determined by only one comparison. To reduce speed fluctuation, compare the values of cost functions of two-vector and three-vector, and select the smaller one as the output voltage vector combination. Since the values of cost functions of each vector in the voltage vector combination of three-vector has been calculated when determining sectors, when selecting voltage vector combination of two-vector, the required basic voltage vector can be directly determined without repeating calculations. Finally, the effectiveness of the proposed FCS-MPCC strategy that based on optimal vectors novel sector division strategy was demonstrated through simulations and experiments, which can increase the speed accuracy of steady-state while reducing computational complexity in the stage of vector selection.

Acknowledgments

This work was financially supported by the Key Research Projects for Higher Education in the Coal Industry. (2021MXJG043).

References

[1] Zhang Y F, Wu Z H, Yan Q, et al (2023) Nonparametric three-vector model predictive current control for permanent magnet synchronous motor. J. Northwest. Polytech. Univ., 41: 860-870.
[2] Huang N, Zhang Y F and Tian B B (2024) Model-free three-vector predictive current control of permanent magnet synchronous motor based on improved sliding mode observer. In: 2024 7th International Conference on Material Application. Changsha. pp. 010297.
[3] Chen R, Shu H P and Zhai K M (2023) Low-complexity three-vector model predictive current control with fixed switching frequency for PMSM. J. Proc. CSEE, 38: 3812-3823.
[4] Qi X, SU T, Zhou K, et al (2021) Development of AC Motor Model Predictive Control Strategy: An Overview. J. Proc. CSEE, 41: 6408-6419.

[5] Wang H F and Wu X Z (2024) A harmonic current model predictive control for nine-phase open-end winding permanent magnet synchronous motors. J. Proc. CESS, 44: 9431-9442.
[6] Kang S, Soh J and Kim R (2020) Symmetrical three-vector-based model predictive control with deadbeat solution for IPMSM in rotating reference frame. J. IEEE Trans. Ind. Electron., 67: 159-168.
[7] Gao F Y, Xu H, Yang K W, et al (2024) Improved model predictive current control for dual three-phase permanent magnet synchronous motor. J. of Hunan Univ. (Nat. Sci.), 51: 81-92.
[8] Wang P S (2023) Research on Multi-vector model predictive control of permanent magnet Synchronous Motor Based on visualization analysis. M.E. Dissertation. Zhengzhou University of Light Industry, Zhengzhou.

Pilot protection based on current sequence component differences for offshore wind power low-frequency transmission line

Jinyu Zhou[*1,2], Shuping Gao[1,2], Zhanshe Yang[1,2]

[1] School of Electrical and Control Engineering, Xi'an University of Science and Technology, Xi'an 710054, China
[2] Xi'an Key Laboratory of Electrical Equipment Condition Monitoring and Power Supply Security, Xi'an 710054, China

[*]Corresponding author's e-mail: 1543347235@qq.com

Abstract. The various control strategies used by the inverters in the offshore wind power low-frequency transmission system will result in variations in the current composition at the two ends. These variations will raise the phase angle difference of the same-phase fault currents, decreasing the protection system's sensitivity and possibly causing it to fail. To solve this problem, the study proposes to construct a new protection criterion using zero-sequence currents and negative-sequence currents, which can more accurately identify and differentiate between normal operation states and fault states by analyzing these sequence currents, thus the performance of differential protection for low-frequency transmission lines is greatly enhanced and the classic current differential protection is optimized.

1. Introduction

Differential current protection has become the main protection mode for AC transmission lines of 110 kV and above voltage levels because of its excellent selectivity and fast response characteristics [1][2][3]. In the literature [4], a methodology for using cross-entropy to evaluate the difference in the time-domain probability distributions of the currents at a line's two extremities is put forth. This methodology facilitates expeditious and efficacious discrimination between fault locations. However, a limitation of this study is that its set values do not consider the potential impact of data transmission on the performance of the algorithm, which may affect the reliability and accuracy of fault detection. The use of the Jedka distance algorithm mentioned in literature [5], which is based on waveform similarity theory, to analyze the similarity of current samples at both ends of the line. This method is highly resistant to environmental disturbances, but it does not consider the case where the wind turbine output is zero, which may result in poor protection in this case. In addition, literature [6] explores the use of phase angle differences in fault currents to construct adaptive braking coefficients. Although this method is effective within a specific phase angle difference range, its protection function will be limited beyond that range. In literature [7], the process of injecting negative sequence currents into an offshore converter station increases the dependability of current differential protection; however, this protection scheme mainly relies on multiple simulations to determine the optimal range of negative sequence current reference values, and the lack of theoretical formulas may affect the scheme's universality and predictability. Finally, literature [8] proposes the use of differences in first-travelling wave characteristics for fault identification, which requires a high sampling frequency and is challenging to detect first-traveling waves in practice.

Therefore, a protection technique based on the differences in current sequence components is now suggested, considering the aforementioned study of current AC protection techniques. The protection

Journal of Physics: Conference Series **3011** (2025) 012027 doi:10.1088/1742-6596/3011/1/012027

system is successful in enhancing the security and stability of AC lines in offshore wind farms because it is dependable and extremely sensitive.

2. Principle of conventional pilot differential protection

A standard offshore wind power low-frequency transmission system is shown in Figure 1, which is mostly composed of offshore wind farms (OWF), step-up transformers, isolation transformers, M3C converters, and low-frequency AC cables. Permanent magnet synchronous wind turbines make up the offshore wind farm. T2 and TI stand for isolation transformer and step-up transformer, respectively. The locations of fault points f1 and f3 on buses M and N are both well outside of their designated boundaries. Furthermore, the in-area fault point f2 is located midway before buses M and N.

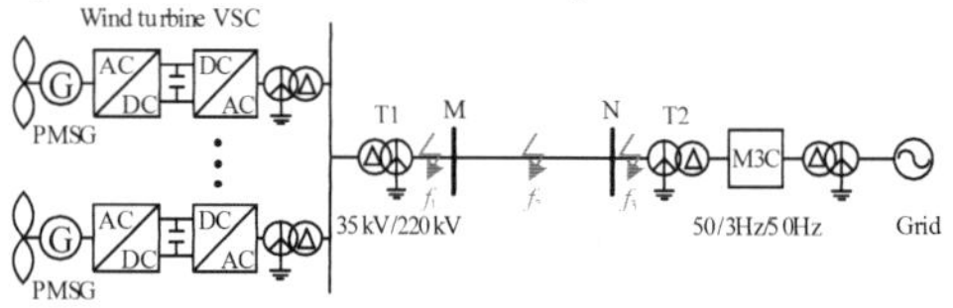

Figure 1. Topology of LFTS for offshore wind power.

The conventional ratio-braking current differential protection criterion is:

$$\left|\dot{I}_{\mathrm{M}\varphi}+\dot{I}_{\mathrm{N}\varphi}\right| \geq k\left|\dot{I}_{\mathrm{M}\varphi}-\dot{I}_{\mathrm{N}\varphi}\right| \tag{1}$$

where $\dot{I}_{\mathrm{M}\varphi}$ and $\dot{I}_{\mathrm{N}\varphi}$ stand for the measured currents at the offshore wind farm side and M3C low-frequency side protection installations, respectively. $\left|\dot{I}_{\mathrm{M}\varphi}+\dot{I}_{\mathrm{N}\varphi}\right|$ and $\left|\dot{I}_{\mathrm{M}\varphi}-\dot{I}_{\mathrm{N}\varphi}\right|$ represent the differential and braking currents, respectively. k represents the braking coefficient, which is generally 0.5~0.8. In this paper, the value of k is 0.7.

Equation (1) explains that the conventional current differential protection method compares the phase angle and current magnitude between the two ends of the line, and when there is, for example, a ground fault or a short-circuit fault between phases, the differential current increases significantly, triggering the protection action. However, control strategies may influence the phase and amplitude of the currents, and the differential current decreases dramatically, falling below the trigger threshold, and the protection device cannot be actuated reliably.

3. Pilot differential protection scheme based on current sequence component differences

3.1. Zero-sequence current differential protection

Zero sequence currents are of vital importance in power systems, especially in distinguishing between earth faults and phase-to-phase faults. In LFTS, the step-up and isolation transformers are triangularly connected, meaning that the zero sequence currents are not impacted by the converter control strategy. This makes the zero sequence currents ideal for constructing the protection action criterion. The action criteria are as follows:

$$\begin{cases} I_{d0} \geq I_{op0} \\ I_{d0} \geq K I_{res0} \\ I_{d0} = \left|\dot{I}_{\mathrm{M0}}+\dot{I}_{\mathrm{N0}}\right| \\ I_{res0} = \left|\dot{I}_{\mathrm{M0}}-\dot{I}_{\mathrm{N0}}\right| \end{cases} \tag{2}$$

where I_{d0} is zero-sequence differential current. I_{op0}, I_{res0} represent zero-sequence action current and braking current. $\dot{I}_{\mathrm{M0}}$, $\dot{I}_{\mathrm{N0}}$ represent zero-sequence current on the OWF side and the M3C side. K represents the braking coefficient, which is 0.75 in the actual project.

3.2. Pilot protection based on negative sequence current amplitude ratio

The OWF will actively provide negative sequence currents when the line is short-circuited. The negative sequence currents on the M3C side are controlled to zero, thus protecting the stability of the system. Faults on and off the line can be effectively distinguished by monitoring and comparing the magnitude of negative sequence currents on both sides of the line.

As illustrated in Figure 2(a), the negative sequence network during an internal fault occurrence is depicted. Z_{W-} and Z_{M3C-} represent the negative sequence equivalent impedance of the OWF side and the M3C side, respectively. Z_{LM-} and Z_{LN-} are the negative sequence impedance of the line from the bus M to the fault point of F and that of from the bus N to the fault point of F. $\dot{I}_{M-}$ and $\dot{I}_{N-}$ are the negative sequence currents of the OWF side and the M3C side; R_F is the transition resistor; and $\dot{U}_{F-}$ is the negative sequence voltage of the fault point of F.

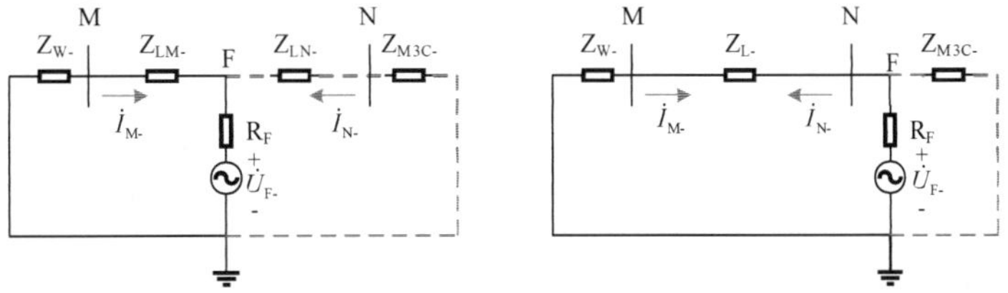

(a) (b)

Figure 2. Negative sequence network when faults occur within and outside the line.

The control strategy causes a nonlinear variation in the power supply equivalent impedance at both ends of the low-frequency line. Equation (3) provides the negative sequence currents on the M3C side and the OWF side, respectively.

$$\begin{cases} \dot{I}_{M-} = \dfrac{\dot{U}_{F-}\left(Z_{M3C-}+Z_{LN-}\right)}{\left(Z_{W-}+Z_{LM-}\right)//\left(Z_{M3C-}+Z_{LN-}\right)+R_F} \\ \dot{I}_{N-} = \dfrac{\dot{U}_{F-}\left(Z_{W-}+Z_{LM-}\right)}{\left(Z_{W-}+Z_{LM-}\right)//\left(Z_{M3C-}+Z_{LN-}\right)+R_F} \end{cases} \tag{3}$$

A negative sequence current amplitude ratio between the OWF side and the M3C side is called m, and it may be expressed using Equation (4). The negative sequence current amplitude ratio is correlated with the fault position and the equivalent impedance of the inverters on both sides, as shown in Equation. (4).

$$m = \frac{\left|\dot{I}_{M-}\right|}{\left|\dot{I}_{N-}\right|} = \frac{\left|Z_{M3C-}+Z_{LN-}\right|}{\left|Z_{W-}+Z_{LM-}\right|} \tag{4}$$

There is only one side with zero sequence current when there is an asymmetrical fault on the line because the OWF side experiences a suppression of the negative sequence voltage, and the M3C side experiences a suppression of the negative sequence current. In this case, the negative sequence impedance approaches infinity, which is approximately an open circuit. There exists $\left|Z_{M3C-}+Z_{LN-}\right| >> \left|Z_{W-}+Z_{LM-}\right|$, and m is larger than 1.

As illustrated in Figure 2(b), the negative sequence network for external faults can be represented by the following equations:

$$\begin{cases} \dot{I}_{M-} = -\dfrac{\dot{U}_{F-}Z_{M3C-}}{\left(Z_{W-}+Z_{L-}\right)//Z_{M3C-}+R_F} \\ \dot{I}_{N-} = \dfrac{\dot{U}_{F-}Z_{M3C-}}{\left(Z_{W-}+Z_{L-}\right)//Z_{M3C-}+R_F} \end{cases} \tag{5}$$

$$m = \frac{\left|-(Z_{\text{W-}} + Z_{\text{L-}}) // Z_{\text{M3C-}} + R_{\text{F}}\right|}{\left|(Z_{\text{W-}} + Z_{\text{L-}}) // Z_{\text{M3C-}} + R_{\text{F}}\right|} = 1 \tag{6}$$

From Equations (5) and (6), when m is equal to 1, it indicates that $\dot{I}_{\text{M-}}$ and $\dot{I}_{\text{N-}}$ are equal, which signifies that both negative sequence currents are supplied by the OWF. However, when a fault occurs within the line, there is a difference in the negative sequence currents on both sides, resulting in a difference in the negative sequence current magnitude ratio between the two sides. To locate the fault location precisely, a criterion based on negative sequence current comparison can be constructed using Equation (7).

$$m = \left|\dot{I}_{\text{M-}}\right| / \left|\dot{I}_{\text{N-}}\right| > m_{\text{set}} \tag{7}$$

where m_{set} represents the setting value of the protection. Theoretically, m should be interpreted as 1. However, m_{set} is taken to be 1.2 when considering the error elements, such as communication transmission and the influence of the current transformer's coefficient of homogeneity.

3.3. Protection of the implementation process

The low-frequency line protection scheme can be summarised in the following four steps:

(1) Gather current information on the protection installation on both sides of the low-frequency line.

(2) Before proceeding to the following steps of the process, Fourier's algorithm must be used to extract the phase values of the three-phase currents on both sides of the low-frequency line. The positive, negative, and zero sequence components of the current at both ends of the operation must then be identified using the symmetrical component approach.

(3) Zero sequence current is utilized for the identification of fault type, while zero sequence current differential is employed for the identification of ground fault. The inter-phase protection employs pilot protection based on the difference in the amplitude ratio of the negative sequence current to achieve judgement.

(4) Phase-to-phase faults are classified according to the current amplitude ratio of m on both sides; if this is less than the setting value of m_{set}, the fault is classified as internal; if it is greater, the fault is classified as external. In the event of a ground fault, the protection device will be activated if the zero sequence differential current of I_{d0} exceeds K times the braking current.

4. Simulation verification

A typical simulation model is constructed in PSCAD, as illustrated in Figure 1. The direct-drive offshore wind farm is made up of 50 sets of direct-drive wind turbines aggregated together, each with a terminal voltage of 0.69 kV and a rated capacity of 2 WM. The main transformer of 35kV/220kV boosts the voltage of each wind turbine after it has been converged by a 10kV/35kV transformer. A-phase ground faults (AG) and BC-phase faults (BC) are examples of simulation analyses used to confirm the efficacy of the suggested protection plan.

4.1. A single-phase ground fault

The A-phase metallic ground fault and the transition resistance ground fault with 100Ω are placed at the midpoint in the line to verify the conventional pilot differential protection performance. Figure 3 illustrates the simulation results of the traditional pilot differential protection when an internal AG fault occurs.

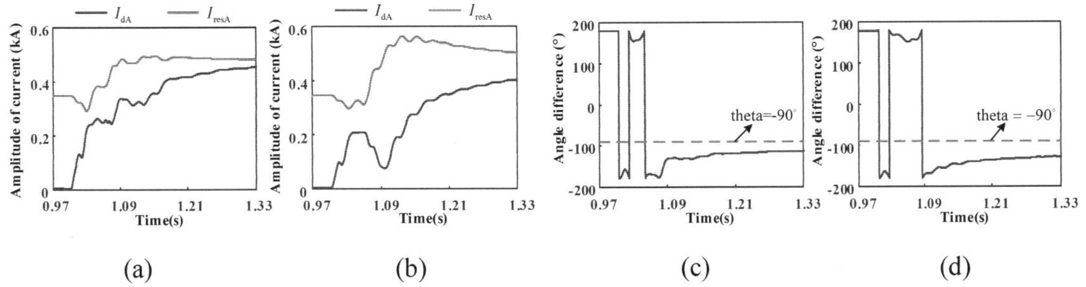

Figure 3. Simulation results of internal AG fault.

From Figure 3(a), it can be obtained that the braking current of I_{resA} is marginally greater than the differential current of I_{dA} when a single-phase metallic grounding fault occurs in the line. The sensitivity of the traditional pilot differential protection is significantly decreased. As the transition resistance increases, the differential current decreases. This is accompanied by a decrease in zero sequence current. Consequently, the braking current is greater than the differential current. Figure 3(b) illustrates this. The protection does not function in this state. Figures 3(c) and 3(d) show the phase angle difference of the A-phase current on both sides in the case of an A-phase grounding fault. The A-phase current's phase angle difference on both sides is noticeably larger than 90°, as seen in Figures 3(c) and 3(d), and this phase angle difference increases in proportion to the increase in transition resistance.

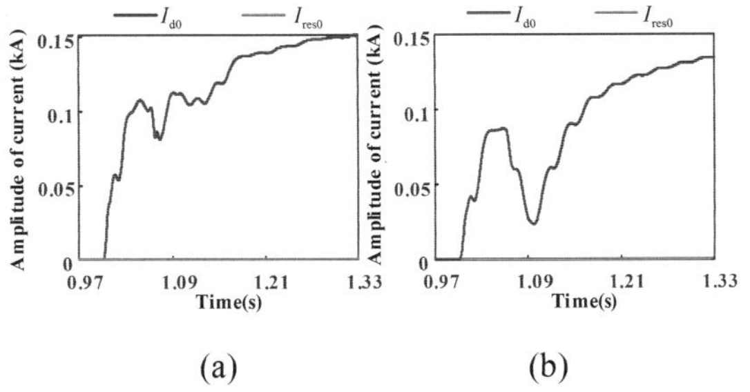

Figure 4. Performance of the proposed method.

As illustrated in Figure 4, the zero-sequence current differential protection is demonstrated under varying transition resistors in the occurrence of an internal AG fault. Figure 4(a) and Figure 4(b) reveal that the zero-sequence differential current of I_{d0} is significantly larger than the braking current of I_{res0}, which fulfils the protection action requirements in Equation (2). Thereby significantly enhancing the sensitivity of traditional current differential protection.

4.2. A phase-to-phase short-circuit fault

The BC phase short-circuit fault is placed at the midpoint in the line to verify the conventional pilot differential protection performance. Figures 5(a) and 5 (b) illustrate the simulation results of the traditional pilot differential protection when an internal BC inter-phase short-circuit fault occurs at the midpoint in the line. It can be obtained that the braking current of I_{resB} and I_{resC} are marginally greater than the differential current of I_{dB} and I_{dC}, respectively. The conventional pilot differential protection would refuse to operate. In this case, the phase angle difference of the fault-phase current on both sides is given in Figure 5(c), which is greater than 90°. The outcomes of protection operation based on the negative sequence current amplitude ratio criterion when a BC interphase fault occurs in the middle of the line are displayed in Figure 5(d), demonstrating that the recommended protection technique effectively improves the protection's action performance.

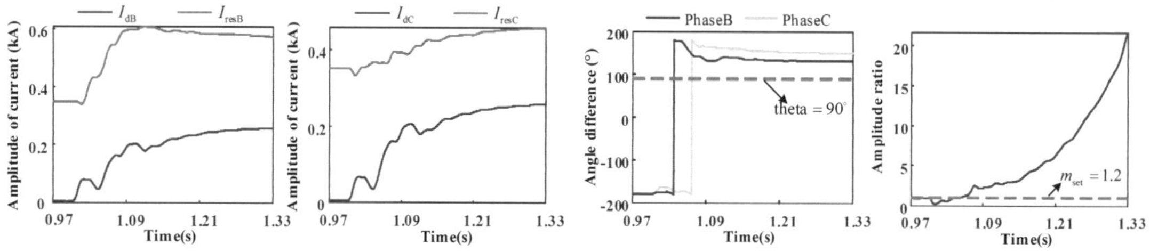

Figure 5. Performance comparison of the traditional pilot differential protection with the proposed method for BC faults.

5. Conclusion

In low-frequency transmission systems, the flexibility of traditional current differential protection can be addressed by using longitudinal protection based on the negative sequence current amplitude ratio and zero sequence current differential protection. Simulation results show that the proposed protection scheme is reliable and has a high sensitivity and rapidity. The research results of this paper provide theoretical support for ensuring the safe and reliable operation of low-frequency transmission systems.

Acknowledgments

This work was financially supported by Shaanxi Province Natural Science Foundation Research Project (No.2024JC-YBMS-792) and State Grid Jiangsu Electric Power Co., LTD (No. J2023028).

References

[1] S. Liu, L. L. Zhang, C. Fu, and L. Jiang, "A new two-port network model based pilot protection for AC transmission lines," IEEE Trans. Power Del., vol. 35, no. 2, pp. 473 – 482, Apr. 2020.
[2] XU Zhihao. Study on Distance Protection Based on Novel Phase Selection Methods for Transmission Line of DFIG-Based Wind Farm[D]. A Thesis Submitted to Southeast University, 2021.
[3] LU Yuehua, FAN Yanfang, LUO Rui. Principle of active distribution network pilot protection based on time domain model identification[J]. Power System Protection and Control, 2020, 48(19): 81-88.
[4] HE Weixuan, FAN Zhengzhen, HUO Yaotong, et al. A Pilot Protection Scheme for Submarine Cable in Flexible Low-Frequency Transmission System Based on Cross Entropy Algorithm[J]. Electric Power, 2023, 56(11): 38-48.
[5] Song Weinan. Research on Control Method and Protection Adaptability of Low Frequency AC Transmission System[D]. Beijing: North China Electric Power University, 2023.
[6] LIU Yimin, WANG Shuyang, LI Bin, et al. Sensitivity Optimization Scheme of AC Line Differential Protection in MMC-HVDC System of Inverter New Energy Station[J]. Electric Power Construction, 2022, 43 (01): 63-69.
[7] ZHENG Tao, ZHUANG Xinyi, LYU Wenxuan. Negative Sequence Current Reference Selection and Protection Adaptability Analysis of Flexible DC Converter Station for Offshore Wind Power[J]. Automation of Electric Power Systems, 2023, 47(03): 133-141.
[8] R. L. S. França et al., "Traveling wave-based transmission line earth fault distance protection," IEEE Trans. Power Deliv., vol. 36, no. 2, pp. 544 – 553, Apr. 2021.

Research on topology and control of high power density modular B2G converter

Guoqiang Luo [1], Jikai Li [2], Zhuang Wu [2*], Yulong Li [1], Chenghui Lin [1]

[1]Electric Power Research Institute of Guizhou Power Grid, Guiyang, Guizhou, China
[2]XJ Electric Co., Ltd. , Xuchang, Henan, China

[*]Corresponding author's e-mail: 19201813722@163.com

Abstract—This research focuses on high power density modular B2G converter topology and control. Firstly, the advantages and challenges of the DC distribution system are described, and the importance of solving low power efficiency attenuation, optimizing AC/DC converter design, and realizing cluster control is emphasized. Starting from the hardware part of the MVAC/LVDC converter, the overall topology of cascaded H-bridge, H-bridge inverter module topology, bidirectional AC/DC circuit topology, and bidirectional LLC resonant circuit topology are introduced. Then, to overcome the shortcomings of the converter efficiency attenuation, the MVAC/LVDC controller is designed in detail from the established control functions, including AC10kV voltage source mode, DC750V voltage source mode power control function, no-load operation function, online mode switching function.

1. Introduction

Under the continuous evolution of modern power systems, the DC distribution system, with its unique advantages, occupies an increasingly important position in the power transmission and distribution field. It has shown great potential in the convenience of distributed power access, quality improvement, and simplifying energy conversion links. However, it can not be ignored that the DC distribution system still faces a series of complex challenges in the actual operation process, which seriously restricts its further promotion and application and needs to be deeply analyzed and overcome. In the low-power operation scenario, the efficiency of the DC distribution system decreases sharply, which becomes a key factor hindering its economic improvement.

As a key energy conversion unit in the DC distribution system, the MVAC/LVDC converter faces many problems under low power conditions. In this paper, the submodule topology of the converter adopts the combination form of the pre-stage AC/DC converter and the post-stage bidirectional LLC resonant DC/DC circuit as an example[16]. The front AC/DC circuit uses the common Cascade H-Bridge (CHB) topology to realize the series side inverter. At low power operation, the loss of switching devices and lines and other components in the former CHB topology in the total power increases significantly, and the working efficiency of the latter bidirectional LLC resonant DC/DC circuit will also be affected by the power reduction. Together, these factors considerably reduce the efficiency of the entire MVAC/LVDC converter and even the DC distribution system[17]. Suppose this efficiency attenuation problem cannot be effectively solved. In that case, the application of a DC distribution system in scenarios such as small distributed energy access and local low power consumption areas will be minimal, and the operating cost will be high, which is not conducive to its long-term development.

1.1 Research Status of the Subject

Currently, the research on CHB-type energy storage converters at home and abroad mainly focuses on the following two aspects: one is the research on modulation strategies, and the other is the research on control strategies [1]. The selection of different modulation strategies also impacts the power output quality and the system's stable operation. Currently, CHB modulation methods include low-frequency modulation [2], high-frequency modulation [3], and hybrid modulation [4]. The hybrid modulation strategy combines two or more modulation strategies [5-6]to make the inverter output the expected voltage and current waveforms. In the literature [7], when the maximum power point voltages are unequal in the cascaded H-bridge photovoltaic system due to external environmental and other factors, failing the traditional carrier phase shift modulation, a carrier phase shift modulation method with a variable phase shift angle is proposed. The feasibility of this modulation method is verified.

As a complete high-voltage and large-capacity system, the cascaded H-bridge energy storage converter mainly functions as energy storage to smooth the power grid output [8]. Therefore, many control issues need to be considered in research and design.

The soft start control strategies of CHB-type energy storage converters are primarily applied to the two-level cascaded H-bridge [9]. The literature [10] proposes a dual-loop soft start control strategy with a voltage outer loop and a current inner loop; the literature [11] designs a fuzzy soft start controller for the DC-side capacitor voltage using the fuzzy control rules with self-adjusting factors, which has the characteristics of high precision and strong robustness; the literature [12] proposes an improved PI control strategy for the problems of inrush current and voltage, and the soft start effect is good.

The literature [13] proposes proportional resonant (PR), which compares the actual output object of the system with the expected value, calculates the control amount to adjust the working state of the control target, and realizes the static-free tracking of the instantaneous reactive power. In the literature [14], under ideal grid conditions, the grid-connected power control of the system is realized using proportional-integral (PI) control. This strategy realizes the output of the modulation voltage of the three-phase system by coordinating the outer loop power control and the inner loop current control, performing feedforward decoupling. It completes the system's grid connection.

For the intra-phase balancing control of CHB-type energy storage converters, the literature [15] uses the SHE-PWM modulation strategy to control the trigger angles of each module in the phase so that each power module in the phase can achieve the effect of differential charging and discharging, that is, to control the conduction angles of each H-bridge module.

Therefore, an in-depth exploration of efficiency attenuation solutions under low power conditions is necessary to improve the economy of DC distribution systems[19].

2. Hardware part of MVAC/LVDC converter

The MVAC/LVDC converter , as shown in Figure 1, is mainly composed of two parts: the CHB full-bridge circuit on the high-voltage side of the module and the LLC resonant circuit.

The high-voltage side of the MVAC/LVDC converter consists of a cascade of H bridge units, which have strong scalability and low voltage stress. They can flexibly adapt to different voltage and power levels and reduce costs. The output waveform has low harmonic content and high smoothness, which can effectively improve the power quality.

The cascade H-bridge is exceptionally flexible in control. Each unit can be controlled independently, and various strategies can be applied to achieve accurate and multi-functional regulation. It has a redundant design, fault isolation capability, and excellent reliability and maintainability. At the same time, the system has high operating efficiency, can achieve two-way efficient energy flow, and plays a key role in the energy storage system, providing strong support for many applications such as power storage and quality improvement.

2.1 Cascade H-bridge converter topology

The topology of the CHB grid-connected converter system mainly includes four types: independent DC bus, in-phase common DC bus, parallel common DC bus, and common DC bus. The independent DC bus CHB topology has the advantages of simple structure, small size, and low cost.

Since each H-bridge has an independent DC bus when one H-bridge unit fails, the regular operation of other H-bridge units will not be affected, and the system can continue to operate in the derated state, improving the entire system's reliability and stability.

The CHB grid-connected converter comprises 36 groups of converter modules[18], and the front stage of each phase consists of multiple AC/DC converters in series. Each AC/DC converter has a separate DC bus, which is connected to an isolated DC/DC converter, and the output of the isolated DC-DC converter is connected in parallel. The star or triangle connection method connects the AC side of three identical single-phase cascaded converters. Then, the low-voltage side is connected in parallel, which can be applied to the three-phase MVAC power grid. The converter topology is ISOP topology; the input side is in parallel, and the output side is in series.

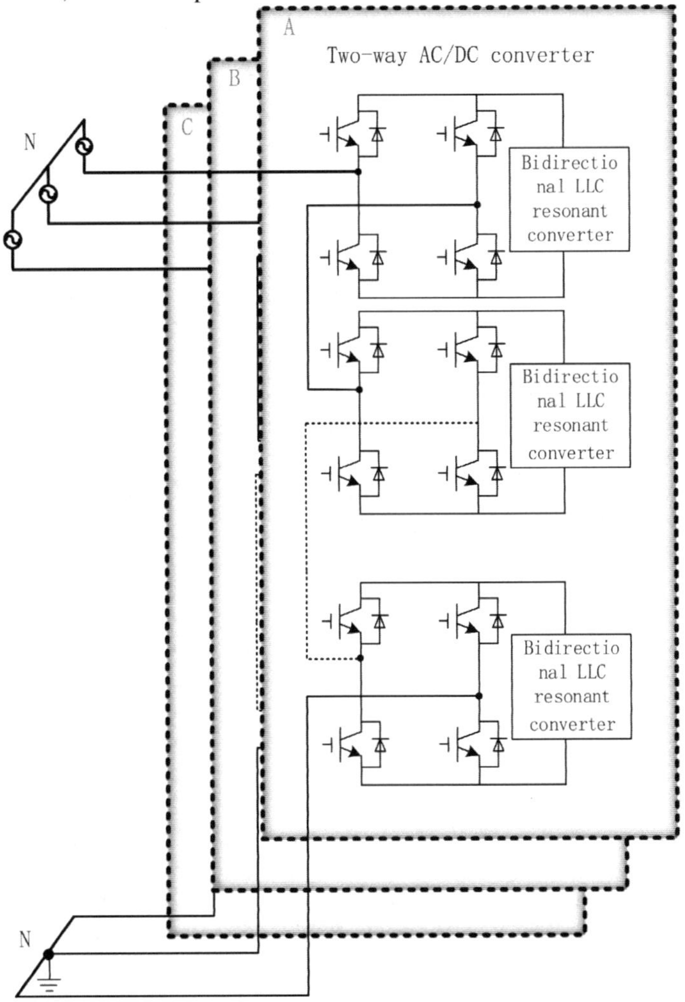

Figure 1 Cascading H-bridge topology.

2.2 Cascaded H-bridge converter unit module topology

In this paper, the energy storage converter with a three-phase star connection mode is taken as the research object, and its topology diagram is shown in Figure 2. Because the cascade H-bridge transformation adopts a star connection, the current of each phase converges to the neutral point, which is conducive to protecting the load equipment.

In a three-phase system, the output voltage of each phase converter is superimposed by multiple H-bridge modules. The higher the number of cascades, the higher the output level, and the voltage waveform obtained is closer to a sine wave. Therefore, the output voltage of the single-phase energy storage converter can be seen as the superposition of the output voltage of the AC side of N H-bridge modules.

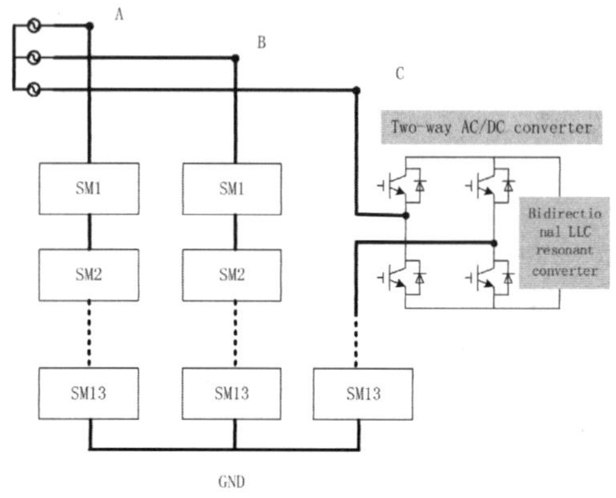

Figure 2 Cascading H-bridge topology.

The two-stage bidirectional AC/DC system topology of the CHB grid-connected converter is shown in Figure 3. The front-stage bidirectional AC/DC converter adopts IGBT modules Q9, Q10, Q11, and Q12 to form a common Cascade H-Bridge (CHB) topology. The control functions such as series side inverter, block protection, and low voltage crossing are realized, and the power factor of the network side is controlled by controlling the amplitude and phase of the voltage of the bridge arm to learn the rectification or inverter function. The rear bi-directional DC/DC converter adopts a full-bridge LLC resonant converter, which has the least passive components compared with other resonant topologies, thus improving the power density of the whole machine and realizing high and low voltage electrical isolation and voltage conversion functions.

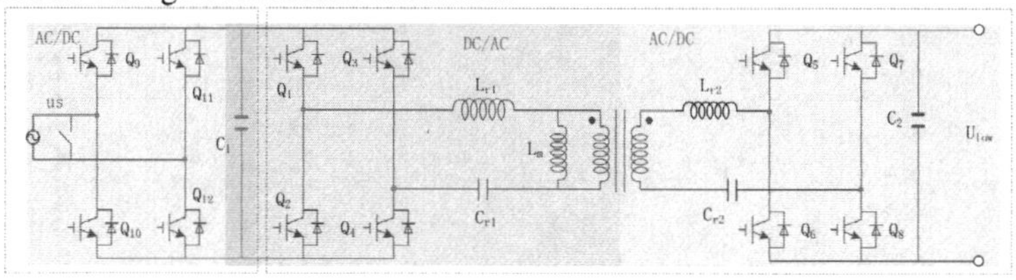

Figure 3 Submodule topology.

2.3 Bidirectional AC/DC circuit topology

The bidirectional AC/DC circuit mentioned above is the H-bridge circuit unit module in the cascaded H-bridge converter. The cascaded H-bridge energy storage converter contains multiple H-bridge unit modules, and the DC side voltage of each sub-module is independent of each other. This structure is suitable for large-scale, high-voltage, and high-power occasions. This section takes a single H-bridge module as the research object and introduces its working principle. The structural diagram is shown in Figure 4.

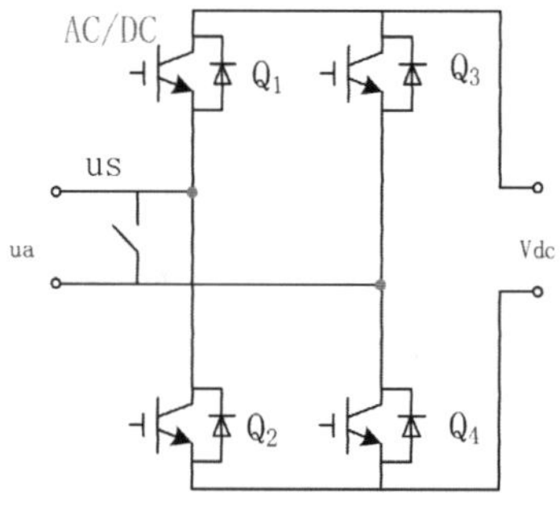

Figure 4. Unit H bridge topology.

Each H-bridge unit contains a DC capacitor and four switching tubes (Q1, Q2, Q3, Q4). In the figure, Vdc is the DC voltage of the H-bridge unit, and ua is the AC voltage.

When the cascade H-bridge is working in your state, different level output effects can be achieved by controlling the working state of each switch tube, with 1 representing the device on-state and 0 representing the device off-state. When the current inflow direction is positive, the switching function state at this time is defined as the P state, and then the current flow path and working state of a single H-bridge module are shown in Figure 5 and Table 1.

Based on the figures and tables, it can be seen that there are three output voltage states of the H-bridge module in the inverter state: 0, +Vdc, and -Vdc.

The B2G converter should work in both the inverter and rectification states and invert the energy in the energy storage facility to the grid at the right time.

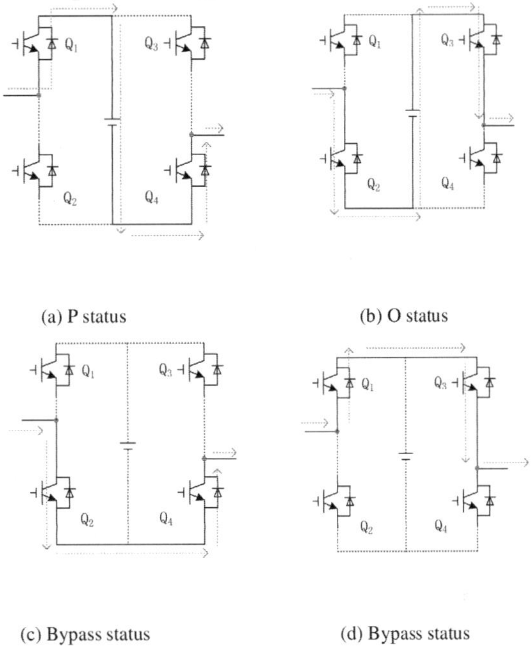

Figure 5 Current path of H bridge submodule.

Table 1 Switch working state, namely H-bridge output voltage.

state		work pattern	output
Q1、3 on	Q2、4 off	bypass state	0
Q1、3 off	Q2、4 on	bypass state	0
Q1、4 on	Q2、3 off	P Status	+Vdc
Q1、4 off	Q2、3 on	O status	-Vdc

When the converter enters the steady-state operation, the DC side voltage remains stable, and the switching tube is controlled by the SPWM driving signal generated by the comparison of sinusoidal modulated wave and triangular carrier wave so that the bridge arm voltage lab is obtained. Ignoring the harmonic components related to the carrier frequency in the bridge arm voltage uab, it can be regarded as an AC power supply with amplitude and phase control. uab and the grid side voltage work together on both ends of the inductor Ls on the AC side to generate sinusoidal input current iL. The inductor Ls plays the role of voltage balance here, and the voltage uL at both ends satisfies formula (1.1).

$$u_L = u_S - u_{ab} \tag{0.1}$$

When the grid side voltage is fixed, the amplitude and phase of the input current iL are only related to the bridge arm voltage *uab*, as shown in Figure 6. Therefore, as long as the amplitude and phase of the bridge arm voltage *uab* are controlled, the power factor on the network side can be controlled to make the converter work in the rectified or inverter state, and the voltage and current relationship is shown in Figure 7.

SPWM modulation must be performed on the switching tube to control the amplitude and phase of the bridge arm voltage *uab*. The commonly used SPWM modulation methods include unipolar modulation and bipolar modulation when the phase difference between the modulated wave ur and -ur is 180°, and the bridge arm's midpoint voltage ua and ub are controlled, respectively. When the modulated wave ur is greater than the carrier uc, the control switch tube Q1 is on, and Q2 is off. Currently, the voltage ua is equal to the DC bus voltage Vbus. When the modulated wave ur is less than the carrier uc, the control switch tube Q2 is on, and Q1 is off. Currently, the voltage ua is equal to zero, the control relationship between carrier-UR and voltage ub is the same, and the bridge arm voltage *uab* is equal to the difference between ua and ub.

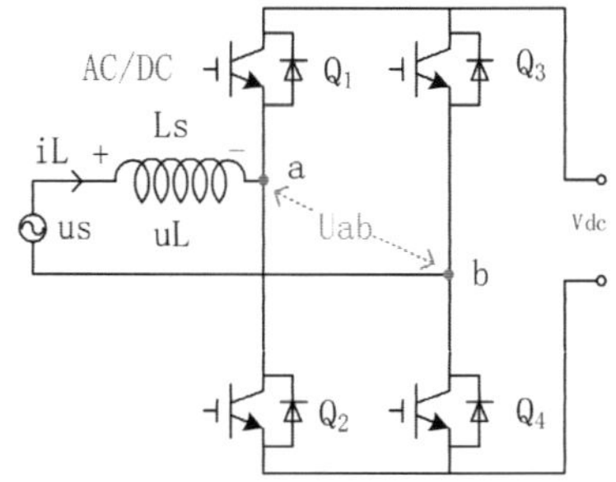

Figure 6. Topology of H bridge submodule.

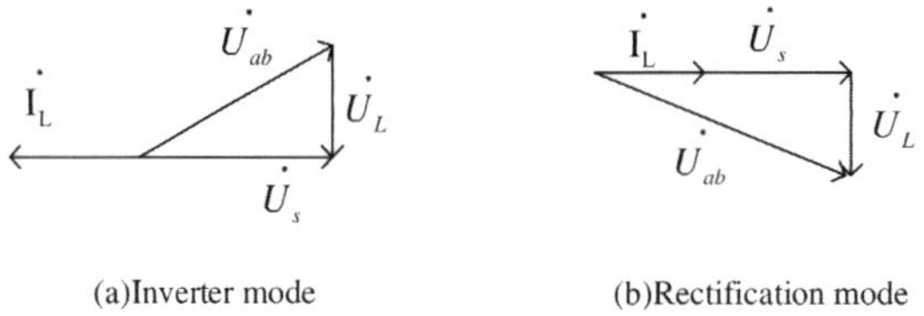

Figure 7 Two modes of operation.

When unipolar frequency doubling modulation is adopted, the equivalent switching frequency of twice can be obtained, so that the harmonic frequency in the bridge arm voltage *uab* is far away from the fundamental frequency, and the pulse of the current iL on the network side is reduced.

2.4 Bidirectional LLC resonant circuit topology

As shown in Figure 8 the LLC resonant converter is mainly composed of an inverter full bridge, resonator, high-frequency transformer, and synchronous rectifier bridge (Q5-Q8) on the secondary side of the transformer. The resonator comprises a resonant inductor Lr1, a resonant capacitor Cr1, and a transformer excitation inductor Lm. VAB is the output voltage of the inverter bridge, iLr is the resonant current, vCr is the resonant capacitor voltage, iLm is the transformer excitation current, direct is the rectifier current of the transformer secondary side, and Vo is the output voltage.

When the sub-side rectifier tube is on, the output voltage clamps the excitation inductance Lm of the LLC resonant converter. Currently, only the resonant inductance Lr1 and the resonant capacitor Cr1 participate in the resonance. The series resonant frequency formed by these two devices is denoted as fr, whose expression is shown in equation (1.2).

When the sub-side rectifier current is zero, the rectifier tube is turned off, and the excitation inductance Lm participates in the resonator resonance. At this time, the resonant frequency of the three devices is recorded as fm, and its expression is shown in equation (1.3).

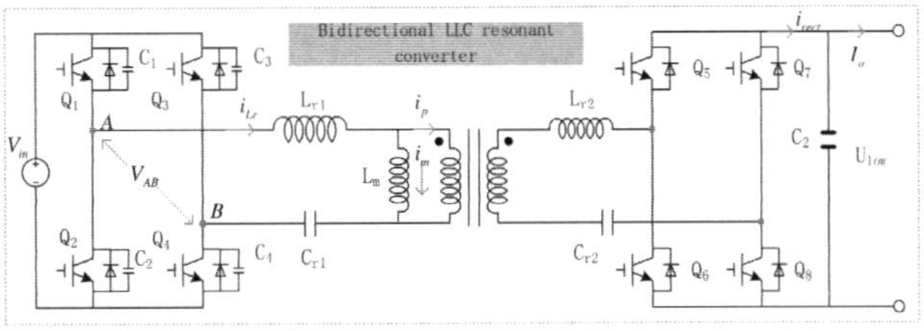

Figure 8. Topology of bidirectional LLC resonator.

$$f_r = \frac{1}{2\pi\sqrt{L_{r1}C_{r1}}} \tag{0.2}$$

$$f_m = \frac{1}{2\pi\sqrt{(L_{r1}+L_m)C_{r1}}} \tag{0.3}$$

LLC resonant converter usually has a wide voltage gain range. Usually, its inverter bridge switches tube Q1 to Q4 at 50% duty cycle, where Q1 and Q2 are complementary on, and Q3 and Q4 are complementary on. By changing the switching frequency fs, the input impedance of the resonator can be changed, and the output voltage can be changed. To realize the ZVS of the primary side switching tube, controlling the switching frequency within a specific range is necessary to make the resonator sensitive. The inductive resonant current i_{Lr} can extract the charge on the parasitic capacitor of the switching tube during the dead time so that the body diode can first continue the current to realize the zero voltage opening of the switching tube.

3. Control strategy of MVAC/LVDC converter

The converter's controllable part is mainly composed of two parts: a CHB full-bridge circuit on the high-voltage side of the module and a bidirectional LLC resonant circuit[20].

Among them, the bidirectional LLC resonant circuit generally adopts open-loop fixed-frequency control, which is the intermediate link of energy flow, and does not participate in the power control of the entire system.

The CHB full-bridge circuit on the high-voltage side can participate in the power control of the whole system by controlling its own input/output voltage and then controlling the input/output current. Therefore, reasonable design of the CHB full bridge control algorithm is the key to realizing the corresponding essential functions of the converter. At the same time, to improve the equivalent switching frequency of the device, the current ripple of the device can be effectively suppressed by phase shifting when the trigger pulse is applied to the CHB full-bridge circuit and LLC resonant circuit on the high-voltage side.

(1) Two-way start function

Two-way starting, namely high-voltage side start and low-voltage side start, functions include module soft start control and inter-module phase shift control; module soft start control to achieve smooth charging of high-low-voltage side power modules, inter-module phase shift control can effectively suppress DC side current ripple, specifically as follows:

1) Module soft start control

When the converter is started from the high voltage side to the low voltage side, the AC10kV power grid charges the bus capacitor on the AC10kV side through the soft start contactor and the smooth start resistance branch. After the bus capacitor on the AC10kV side is charged, the duty ratio is increased to 50% by slowly increasing the duty ratio. Then, the drive pulse of the switch tube of the high-voltage module Q1~Q4 is blocked, and the power module of the low-voltage side and the power module of the high-voltage side simultaneously deliver the drive pulse to the switch tube according to

Journal of Physics: Conference Series **3011** (2025) 012028 doi:10.1088/1742-6596/3011/1/012028

the sequence shown in Figure 9. The drive pulse slowly changes from narrow to wide until it increases to 50% duty cycle; that is, a soft start from high-voltage to low-voltage is realized.

When the converter starts from the low- and high-voltage sides, the DC750V power grid charges the bus capacitor on the DC750V side through the soft start contactor and the smooth start resistance branch. After the bus capacitor on the DC750V side is charged, the driving pulse of the Q5, Q6, Q7, and Q8 switch tubes is shown in Figure 9. The bus capacitor on the high-voltage side is charged by slowly increasing the duty cycle. When the duty cycle is increased to 50%, the drive pulses of the switch tube Q5~Q8 of the low-voltage module are blocked. Then the power module of the high-voltage side and the power module of the low-voltage side deliver the drive pulses to the switch tube simultaneously according to the sequence shown in Figure 9. The drive pulses slowly change from narrow to wide until the duty cycle reaches 50%; that is, a soft start from low to high voltage is realized

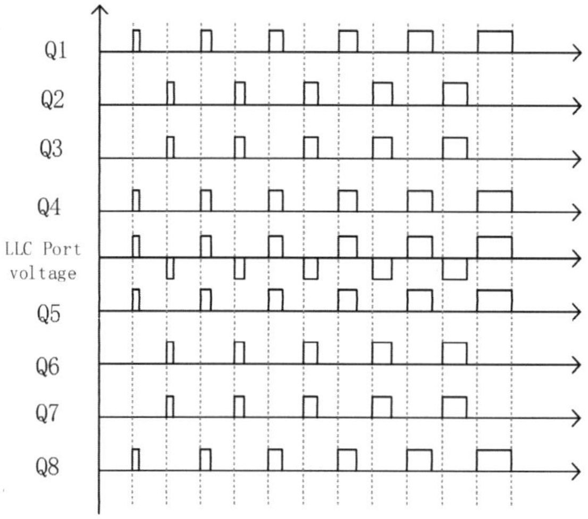

Figure. 9 Schematic diagram of soft-start IGBT drive pulse.

2) Phase shift control between modules

Figure 10 clearly demonstrates phase shift control between modules: The phase shift Angle between modules is 360°/n (n is the number of modules), which effectively reduces the peak value of the total DC current to reduce the influence of current fluctuations on the bus and can reduce the inductance of the DC filter. The number of converter power modules is 36, the phase shift between each module is 10°, and the phase shift Angle between modules is maintained at 10° during startup and regular operation.

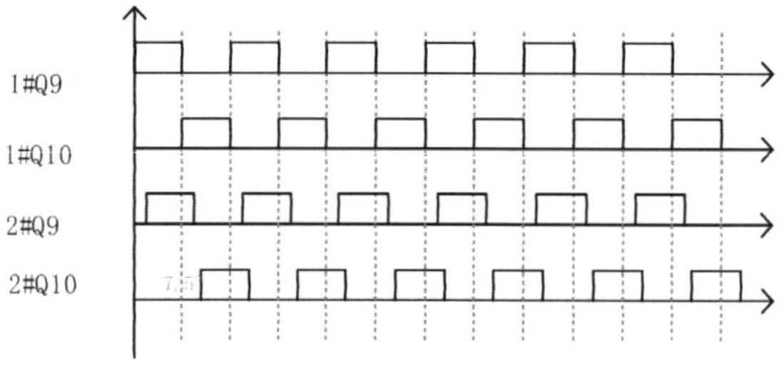

Figure. 10 Timing diagram of phase shift control between modules.

Figure 11 shows the bidirectional start timing diagram according to the above control.

Journal of Physics: Conference Series **3011** (2025) 012028 doi:10.1088/1742-6596/3011/1/012028

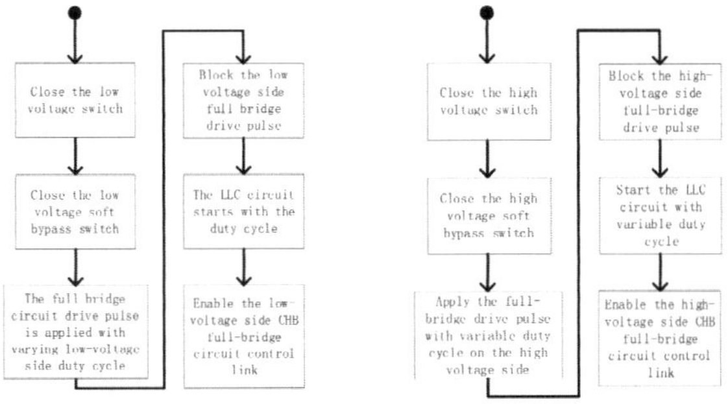

(a) Start from low to high (b) Start from high to low

Figure 11 Schematic diagram of bidirectional start timing

3.1Constant voltage control function

3.1.1 AC10kV voltage source mode. In AC10kV voltage source mode, when AC10kV is connected to the load, DC750V is connected to the power grid, and the high voltage side port of the converter is connected to the load; according to Ohm's law, voltage and current have a positive correlation, so the DC voltage can be controlled by controlling the size of the current. The voltage Uac_fdb_10kV at the AC10kV side of the detection system was compared with the voltage instruction value Uac_ref_10kV at the AC10kV side. The PI regulator calculated the error signal, and the current instruction value Iac_ref_10kV was output as the inner loop of the current at the high voltage side. The block diagram of the AC10kV voltage source mode control algorithm is shown in Figure 12.

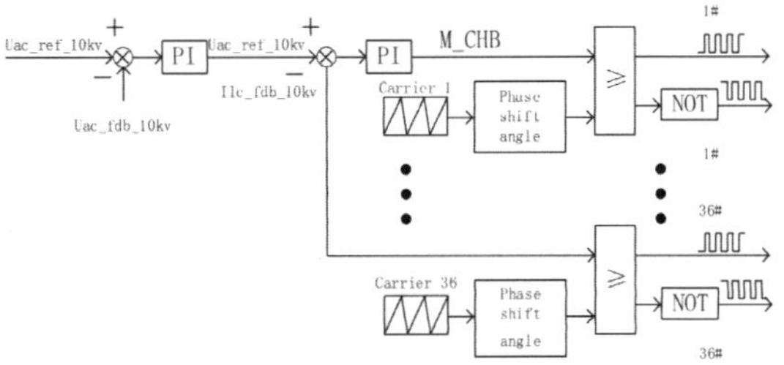

Figure 12 Block diagram of ±20kV voltage source mode control algorithm.

3.1.2 DC750V voltage source mode. When the voltage source mode of DC750V is configured as AC10kV connected to the power grid, DC750V is connected to the load, and the low-voltage side port of the converter is connected to the load, according to Ohm's law, voltage and current are also correlated. The DC voltage Udc_fdb_750V of the DC750V side of the system is detected. Compared with the DC voltage instruction value Udc_ref_750V on the DC750V side, the PI regulator calculates the error signal, and the output is the current instruction value Iac_ref_10kV in the current inner loop on the high voltage side. The block diagram of the DC750V voltage source mode control algorithm is shown in Figure 13.

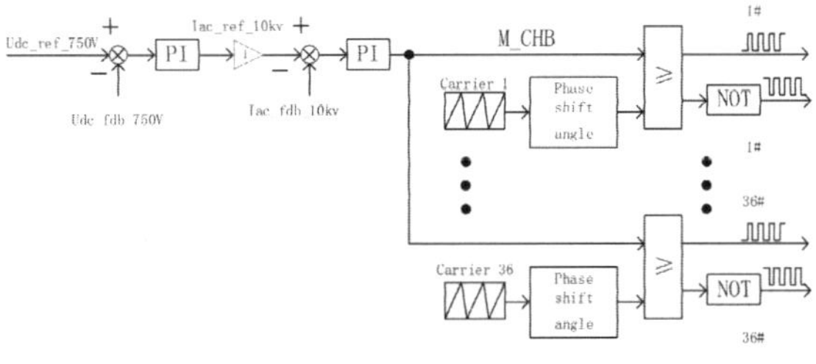

Figure. **13** Block diagram of DC750V voltage source mode control algorithm.

3.2 Power control function

The current source mode (power control function) in the steady-state operation of the system is described as follows:

The CHB full-bridge circuit on the high-voltage side of the power module can change the input current by changing the switching device's duty cycle.

When the duty cycle ratio is large, the equivalent output voltage is large. The busbar capacitor on the high-voltage side will discharge, and the current will flow out from the high-voltage side, which is equivalent to the energy transfer from the low-voltage side to the high-voltage side.

When the duty cycle is small, the equivalent output voltage is small. The bus capacitor on the high-voltage side will be charged, and the current flows from the high-voltage side will be comparable to the energy transfer from the high-voltage side to the low-voltage side. Suppose the positive direction of current sampling is specified as the outgoing AC10kV port. In that case, the input current of the power module has a positive correlation with the duty cycle so that the PI controller can be used for closed-loop control.

The DC (Iac_fdb_10kV) at the AC10kV side of the detection system was compared with the current command (Iac_ref_10kV) at the AC10kV side.

The PI regulator calculated the error signal, and the output modulated wave (MCHB) was compared with 36 phase-shifted carriers. The carrier phase difference was 360°/36=10° in turn. If the modulated wave is larger than the carrier, the output is high; if the modulated wave is smaller than the carrier, the output is low.

According to the above comparison method, 36 bidirectional CHB drive pulse signals on the AC10kV side of the system are obtained, and the block diagram of the current source mode control algorithm is shown in Figure 14.

Among them, the carrier phase shift method can effectively reduce the ripple peak of the total DC current, reducing the influence of current fluctuation on the bus and the filter inductance value of the DC side.

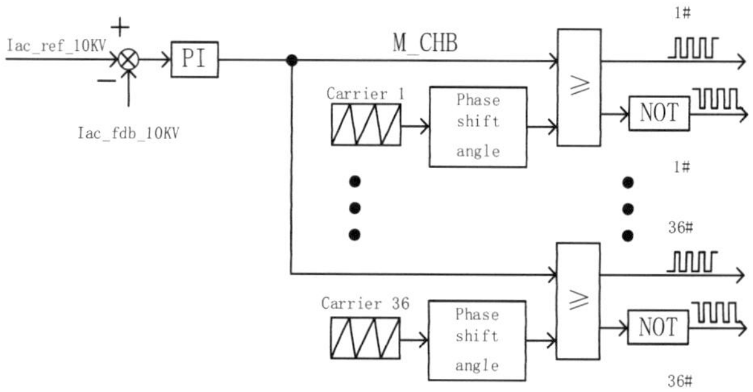

Figure. 14 Current source mode control algorithm.

3.3 No-load operation function

Using the above voltage and power control algorithm, the MVAC/LVDC converter can be connected to the system as a voltage source or power unit, and its characteristics can be maintained in the hot standby state under the minimum power condition.

At the same time, paying attention to the power module voltage equalization during no-load operation is necessary. Because the hardware parameters of a single power module are different, the capacitor voltage of the DC bus in series on the high voltage side may have the problem of uneven voltage. The CHB circuit on the high-voltage side can control the power transfer by changing the duty cycle of the power module. Therefore, the capacitor voltage of the DC bus on the high-voltage side of the power module can be compared with the average voltage of all modules.

When the capacitor voltage of the DC bus on the high-voltage side of the power module deviates from the average value, the controller compensates the duty cycle of the CHB circuit on the high-

Journal of Physics: Conference Series **3011** (2025) 012028 doi:10.1088/1742-6596/3011/1/012028

voltage side and increases or decreases the transmission power. The power module voltage balancing control algorithm, shown in Figure 15, pulls the voltage back to the rated value.

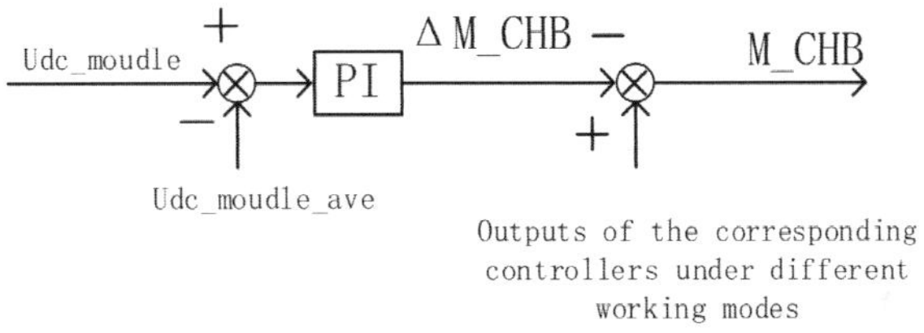

Figure 15 Power module voltage equalization control algorithm.

3.4 Online mode switching function

The distribution network system operates in various ways, and the MVAC/LVDC converter should change its working mode according to the change in the distribution network's operation mode. Therefore, it needs to have the function of working mode switching during power control. The overall control block diagram of the power control function of the distribution network converter is designed according to Figure 16.

In order to ensure that the device does not have serious overvoltage and overcurrent in the process of working mode switching, it is necessary to design a smooth switching control algorithm during switching, mainly for the reasonable initialization of controller parameters and instruction values. The simulation takes the switching from ±375V voltage source mode to the current source mode as an example. During steady state operation, the MVAC/LVDC converter supplies power to the load. At 0.2s, the low-voltage side of the device is connected to the low-voltage DC grid, and the 0.205s mode switching command is issued, while the current command value of the MVAC/LVDC converter is lowered on the slope. Make the power grid and MVAC/LVDC converter supply power to the load and share the power equally. Figure 17 shows the simulation waveform of the system in the mode-switching process.

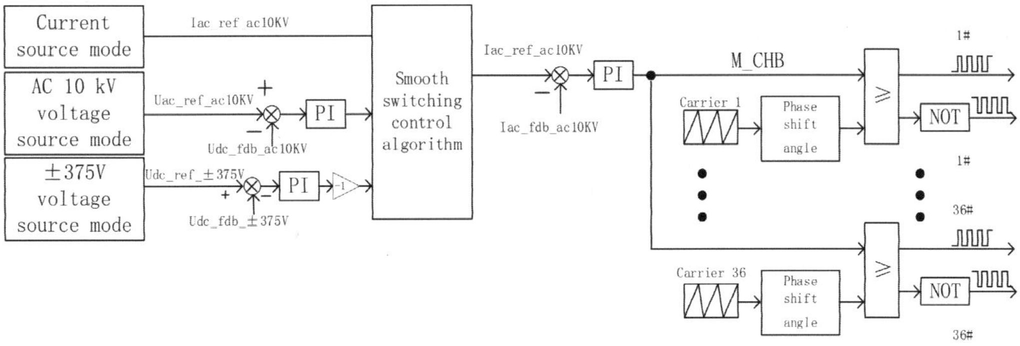

Figure 16 Overall block diagram of the power control function.

Journal of Physics: Conference Series **3011** (2025) 012028 doi:10.1088/1742-6596/3011/1/012028

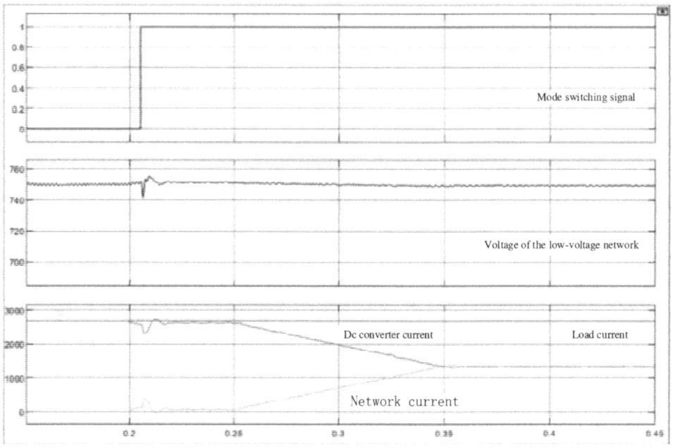

Figure 17 System-related waveform during mode switching.

4 Conclusion

This research focuses on the high power density modular B2G converter, deeply discusses its topology and control strategy, and obtains the following results:

In terms of topology, the hardware part of the MVAC/LVDC converter is described in detail, including the overall topology of cascaded H-bridge, H-bridge inverter module topology, bidirectional AC/DC circuit topology, and bidirectional LLC resonant circuit topology. These topologies have their characteristics, and together, they build an efficient and stable converter system with the advantages of strong scalability, low voltage stress, low harmonic content, high power density, and electrical isolation, providing a solid hardware foundation for the realization of high power density modular B2G converter.

In terms of control strategy, a comprehensive controller function is designed. The bidirectional start function ensures a smooth start and stable operation of the high and low voltage side and reduces the influence of current fluctuation. Constant voltage control function in AC10kV and DC750V voltage source mode, precise voltage control via PI regulator; The power control function realizes the effective bidirectional energy transfer and reduces the current ripple by carrier phase shifting. The no-load operation function solves the power module voltage equalization problem and ensures the system's stability. The online mode switching function enables the converter to adapt to various operating modes of the distribution network, and the switching process is stable without overvoltage and overcurrent phenomenon.

The research results provide an important theoretical basis and practical guidance for the design and application of a B2G converter in a DC distribution system. This helps to improve the overall performance of the DC distribution system and promotes its further development in the power field. It also has positive reference value for future related technology research and engineering practice.

Acknowledgment

Key Technology Project of China Southern Power Grid Corporation, Research on Key Equipment, Key Technology and Cluster Application Technology of DC Power Supply and Distribution in Light Storage and Flexible Buildings, GZKJXM20232507.

References

[1] Hao Weijie. Research on Grid-connected Power and SOC Balancing Control of Chain Energy Storage Converters [D]. Qinhuangdao: Yanshan University, 2021.

[2] Panda K P, Bana P R, Panda G. FPA optimized selective harmonic elimination in symmetric, asymmetric reduced switch cascaded multilevel inverter [J]. IEEE Transactions on Industry Applications, 2020, 56(3): 2862-2870.

[3] Yao Mingyang, Wang Fanggang, Feng Huayuan, et al. A space vector modulation algorithm integrating the nearest level approximation modulation [J]. Electrical Measurement & Instrumentation, 2021, 58(12): 49-56.
[4] Gu Jun, Song Fei, Li Ping, et al. Optimization method of modulation strategy for hybrid cascaded H-bridge inverter [J]. High Voltage Engineering, 2022, 48(2): 753-761.
[5] Hu Wenhua, Zhang Chaofan, Liu Jianfeng. Hybrid modulation and power balance method of hybrid cascaded H-bridge [J]. High Voltage Engineering, 2020, 46(10): 3561-3568.
[6] Zambra A B, Renes Pinheiro J. Comparison of phase-shift and step wave modulation technique applied to symmetrical cascaded multilevel inverter [J]. IEEE Latin America Transactions, 2013, 11(5): 1156-1162.
[7] Wang Fusheng, Dai Zhiqiang, Wang Jianing, et al. A variable phase shift angle carrier phase shift modulation strategy based on cascaded H-bridge photovoltaic grid-connected inverter [J]. Acta Energiae Solaris Sinica, 2020, 41(8): 144-150.
[8] H Chen, R Xiong, C Lin, et al. Predictive Control Based Real-Time Energy Management for Hybrid Energy Storage System [J]. CSEE Journal of Power and Energy Systems, 2021, 7(4): 862-874.
[9] Yang Jianfeng, Wang Shuai, Xie Yankai. Research on start control of capacitor voltage in active power filter [J]. Power Electronics, 2014, 48(2): 46-48.
[10] Luo Qifeng, Cheng Hanxiang, Yu Shuhai, et al. Practical control strategy analysis on dual-loop soft-startup for active power filter [J]. Advanced Technology of Electrical Engineering & Energy, 2010, 29(4): 52-56.
[11] He Na, Wu Jian, Xu Dianguo. Fuzzy soft-startup controller of active power filter [J]. Transaction of China Electrotechnical Society, 2007, 22(3): 115-120.
[12] Yang Jianfeng, Wang Shuai, Xie Yankai. Research on expert PI control of active power filter DC voltage [J]. Power Electronics, 2014, 48(9): 71-73.
[13] Liu Cong, Li Lisheng, Liu Yang, et al. Power control of lithium battery energy storage grid-connected system based on quasi-PR control [J]. Electrical Measurement & Instrumentation, 2021, 58(8): 172-178.
[14] Chao Gang. Research and Design of Cascaded H-bridge Energy Storage System [D]. Shanghai: Shanghai Jiao Tong University, 2013.
[15] Ai Hongke, Wu Junyong, Hao Liangliang, et al. Research on battery self-balancing control strategy in cascaded energy storage system [J]. Transactions of China Electrotechnical Society, 2015, 30(14): 442-449.
[16] Gao Wenyong. Two-way LLC resonant converter and two-stage type AC/DC system research [D]. Hunan university, 2023.
[17] Tan Zhukui, Ban Guobang, Xu Yutao, et al. Based on the DDC dc micro grid access technology research [J]. Power Systems and Big Data, 2019,22(08):1-7.
[18] Chen Mingyang, Xu Yutao, Xu Teng, et al. Simulation Analysis of DC Fault Suppression Performance of Shunt Full Bridge Submodule MMC [J]. Power Systems and Big Data,2019,22(04):15-21.
[19] Xu Yutao, Tan Zhukui, Lu Qiansu, et al. Research on Multi-terminal DC microgrid with electric vehicle charging station [J]. Power Systems and Big Data,2018,21(06):9-16.
[20] Wu Haitao, Wang Shixiang. Research on Protection Configuration and Action Strategy of Flexible DC Intelligent Distribution Network [J]. Power Systems and Big Data,2018,21(02):74-80.

Research on capacity configuration method of integrated energy system considering V2G energy storage characteristics

Xia Yuan [1*], Dan Chen [1], Jialong Zhou [2]

[1] Shanghai Electric Power Engineering Association, Shanghai 200002, China
[2] School of Information Engineering, Nanchang University, Nanchang 330031, China

[*] Corresponding author's e-mail: 1143304825@qq.com

Abstract. With the rapid increase in the number of electric vehicles (EVs), their potential as mobile energy storage units has garnered significant attention. Vehicle-to-grid (V2G) technology, which enables bidirectional interaction between vehicles and the grid, presents new opportunities for the development of integrated energy systems (IES). To address the issue of energy storage capacity configuration after integrating EVs into IES, a two-layer optimization model has been established for IES capacity configuration that considers V2G energy storage characteristics. The outer layer optimizes a single objective, which is the sum of the IES operating cost and the average investment cost of the energy storage system over its lifecycle. The inner layer optimizes multiple objectives, including IES operating cost, wind and solar curtailment rate, and EV charging cost. The research results demonstrate that the proposed capacity configuration optimization model effectively optimizes the system's energy storage capacity and significantly improves the economic benefits for all stakeholders.

1. Introduction

With the global energy transition and the rapid development of renewable energy, integrated energy systems (IES) have gained widespread attention as an important approach to achieving optimal capacity configuration. IES, through the conversion, distribution, and coordination of electricity, heat, cooling, and gas, can promote the local consumption of renewable energy and improve energy efficiency. However, the intermittency and instability of renewable energy pose challenges to the stability of traditional energy networks. In this context, the energy storage characteristics of EVs, particularly V2G technology, offer a new solution for IES capacity configuration [1].

Numerous studies have been conducted on IES capacity configuration. Guo et al. [2] established an electric/thermal hybrid energy storage model to improve the economic efficiency of park-level IES but did not consider the relationship between operators' pricing strategies and users' energy consumption strategies, neglecting operators' interests. Zhang et al. [3] modeled the schedulable potential of EV clusters as a game process with microgrid operators but did not account for parameter uncertainties in the EV cluster model. Jin et al. [4] proposed a method to select the most suitable energy storage device for IES operation based on a comparison of hydrogen and battery energy storage states but failed to accurately describe the long cycle life and low charge-discharge frequency of seasonal energy storage, which was not optimized for economic goals. Zhang et al. [5] considered the energy conversion efficiency and storage constraints of electricity-heat-gas interconnected energy systems but did not use electricity and hydrogen to optimize IES capacity configuration. Zhang et al. [6] proposed a source-storage capacity optimization allocation method considering integrated demand response, establishing a two-layer optimization model that balances source-storage capacity allocation and operational optimization.

In summary, current research on IES capacity configuration primarily focuses on optimizing capacity allocation but rarely considers the substitution effect of V2G energy storage characteristics on IES energy storage devices. This paper proposes a two-layer optimization model for IES capacity configuration that considers V2G energy storage characteristics. The model iteratively solves the energy storage capacity configuration problem using a genetic algorithm and validates the model's effectiveness and economic benefits through case studies.

2. Two-layer optimization model for IES capacity configuration considering V2G energy storage

The proposed IES capacity configuration optimization model has two layers of optimization objectives. The inner layer optimizes daily operating and maintenance costs, wind and solar curtailment rates, and EV charging costs, while the outer layer optimizes the sum of the daily average investment cost of the energy storage system and the daily operating and maintenance costs of the system. The constraints of the inner layer are closely related to the decision variables of the outer layer.

2.1. *Outer layer optimization objective function and constraints*

The outer layer determines the capacity and power configuration of the energy storage system, which is then passed as parameters to the inner layer for overall system operation optimization. The outer layer optimizes a single objective: minimizing the sum of the daily average investment cost of the energy storage system over its lifecycle and the daily operating and maintenance costs of the system, considering V2G energy storage characteristics. The mathematical model is as follows:

$$\begin{cases} \min F_1 = C_{in\_\text{cost}} + C_{op\_\text{cost}} \\ s.t. \mathrm{G}(X) \le 0 \end{cases} \tag{1}$$

where F_1 is the sum of the daily average investment cost and the daily operating and maintenance cost of the energy storage system; C_{in_cost} is the daily average construction investment cost of the energy storage system; C_{op_cost} is the daily operating and maintenance cost of the IES, optimized by the inner layer; ARE_{rate} is the wind and solar curtailment rate; C_{c_cost} is the daily charging cost of EV owners; The decision variables $X=[E_S, P_S, H_S, Q_S]$, E_S and H_S represent the installed capacity of electrical and thermal energy storage, respectively, and P_S and Q_S represent the rated installed power of electrical and thermal energy storage, respectively.

The construction investment cost of the energy storage system is calculated as:

$$C_{in\_cost} = \sum_i \frac{1}{365} \times \frac{r_i (1+r_i)^{\gamma_i}}{(1+r_i)^{\gamma_i} - 1} (C_{i,Ein} E_{S,i} + C_{i,Pin} P_{S,i}) \tag{2}$$

where r_i is the depreciation rate of energy storage device *I*; γ_I is the service life of energy storage device *i* in years; $C_{i,Ein}$ and $C_{i,Pin}$ are the installation costs per unit capacity and unit power of energy storage device *I*; $E_{S,I}$ and $P_{S,i}$ are the installed capacity and power of energy storage device *i*.

The service life of electrical energy storage is closely related to the number of charge-discharge cycles and the capacity size. Therefore, the daily average investment cost of the battery energy storage system should consider its service life, which is modeled as:

$$\gamma = \frac{N_r D_r E_S}{365 \sum_t n_{dod,t} E_{E_s,t}} \tag{3}$$

$$n_{dod,t} = \left(\frac{D_{a,t}}{D_r} \right)^{0.19} e^{\left(-1.69 \left(1 - \frac{D_{a,t}}{D_r} \right) \right)} \tag{4}$$

where γ is the service life of the battery; D_r is the rated depth of discharge; N_r is the rated number of cycles corresponding to the rated depth of discharge; $n_{dod,t}$ is the actual depth of discharge factor of the battery at time *t*; $E_{Es,t}$ is the discharge capacity of the battery at time *t*; $D_{a,t}$ is the actual depth of discharge of the battery at time *t*.

The decision variable X must satisfy the following constraints:

(1) Energy storage system capacity constraints

$$\begin{cases} E_S^{\min} \le E_S \le E_S^{\max} \\ H_S^{\min} \le H_S \le H_S^{\max} \end{cases} \tag{5}$$

where E_S^{min} and E_S^{max} are the lower and upper limits of the installed capacity of electrical energy storage; H_S^{min} and H_S^{max} are the lower and upper limits of the installed capacity of thermal energy storage.

(2) Energy storage system-rated power constraints

$$\begin{cases} P_S^{\min} \le P_S \le P_S^{\max} \\ Q_S^{\min} \le Q_S \le Q_S^{\max} \end{cases} \tag{6}$$

where P_S^{min} and P_S^{max} are the lower and upper limits of the installed power of electrical energy storage; Q_S^{min} and Q_S^{max} are the lower and upper limits of the installed power of thermal energy storage.

2.2. Inner Layer optimization objective function and constraints

After the outer layer determines the configuration, the capacity and power are passed as parameters to the inner layer. The inner layer optimizes multiple objectives, including minimizing the daily operating and maintenance cost of the IES, minimizing the wind and solar curtailment rate, and the EV charging cost. The mathematical model is as follows:

$$\begin{cases} \min(f_1, f_2, f_3) \\ f_1 = C_{op\_cost} = c_{Buy} + c_{OM} + c_{Punish} \\ f_2 = ARE_{rate} \\ f_3 = C_{c\_cost} \\ s.t. g(x) \le 0 \end{cases} \tag{7}$$

where f_1 is the daily operating and maintenance cost of the IES; f_2 is the wind and solar curtailment rate; f_3 is the EV charging cost; c_{Buy} is the cost of purchasing energy from external sources; c_{OM} is the daily maintenance cost of the IES; c_{Punish} is the penalty function for wind and solar curtailment; The decision variables $x=[P_t^{MT}, Q_t^{GB}, Q_t^{EB}, P_t^{P2G}, P_t^{WT}, P_t^{PV}, P_t^{grid}, P_t^{grid}, Q_t^{heat}, P_t^{EV}]$ represent the power output of each energy device at time t.

(1) Energy purchase cost

$$c_{Buy} = \sum_{t=1}^{T} \left(c_{gas} \times (g_{gas,t}^{MT} - g_{gas,t}^{P2G}) + c_{grid,t} \times P_t^{grid} \right) \Delta t \tag{8}$$

where c_{gas} is the market price of natural gas per cubic meter; $c_{grid,\,t}$ is the electricity price from the grid at time t; P_t^{grid} is the power supplied by the grid to the IES at time t.

(2) Daily maintenance cost

$$c_{OM} = \sum_{t=1}^{T} \left(\begin{array}{c} c_{MT} P_t^{MT} + c_{GB} Q_t^{GB} + c_{EB} Q_t^{EB} + c_{P2G} P_t^{P2G} + c_{WT} P_t^{WT} + \\ c_{PV} P_t^{PV} + c_{bat} \left| P_t^{bat} \right| + c_{heat} \left| Q_t^{heat} \right| + c_{EV} \left| P_t^{EV} \right| \end{array} \right) \tag{9}$$

where c_{MT}, c_{GB}, c_{EB}, c_{P2G}, c_{WT}, c_{PV}, c_{bat}, c_{heat}, and c_{EV} represent the unit operating and maintenance costs of micro gas turbines, gas boilers, electric boilers, power-to-gas (P2G) devices, wind turbines, photovoltaic systems, electrical energy storage, thermal energy storage, and EV charging/discharging, respectively; P_t^{EV} is the power output of the EV cluster at time t.

(3) Wind and solar curtailment rate

$$ARE_{rate} = \frac{\sum_{t=1}^{T} Loss_t^{WP}}{\sum_{t=1}^{T} (P_t^{WT} + P_t^{PV})} \tag{10}$$

where $Loss_t^{WP}$ is the sum of wind and solar curtailment at time t; P_t^{WT} and P_t^{PV} are the typical power output curves of wind and solar at time t.

(4) EV charging and discharging cost

$$C_{c\_cost} = \sum_{t=1}^{T} (C_t^{car,c} P_t^{EV,c} - C_t^{car,d} P_t^{EV,d}) \Delta t \tag{11}$$

where $C_t^{car,c}$ and $C_t^{car,d}$ are the EV charging and discharging prices at time t; $P_t^{EV,c}$ and $P_t^{EV,d}$ are the charging and discharging power of the EV cluster at time t; Δt is the unit dispatch time.

(5) Penalty function

The wind and solar curtailment rates are treated as a constraint, and the curtailment is converted into a cost penalty function included in the daily operating and maintenance cost. The penalty function is modeled as:

$$c_{Punish} = c_{loss} \sum_{t=1}^{T} Loss_t^{WP} \tag{12}$$

where c_{loss} is the unit cost of wind and solar curtailment.

The decision variables xx must satisfy the following constraints:

(1) Supply-demand balance constraints

At any time during the dispatch period, the electricity and heat supply-demand balance must be satisfied:

$$\begin{gathered} P_t^{MT} + P_t^{grid} + P_t^{WT} + P_t^{PV} + P_t^{bat} + P_t^{EV} - P_t^{P2G} - P_t^{load} = 0 \\ \frac{\eta^{MT,h}(1-\eta^{MT,e} - \delta_{loss}^{MT})}{\eta^{MT,e}} P_t^{MT} + Q_t^{GB} + Q_t^{EB} + Q_t^{heat} - Q_t^{load} = 0 \end{gathered} \tag{13}$$

(2) Ramp rate constraints of energy devices

$$\begin{cases} \Delta P_{down}^{MT} \le \Delta P_t^{MT} \le \Delta P_{up}^{MT} \\ \Delta Q_{down}^{EB} \le \Delta Q_t^{EB} \le \Delta Q_{up}^{EB} \\ \Delta Q_{down}^{GB} \le \Delta Q_t^{GB} \le \Delta Q_{up}^{GB} \end{cases} \tag{14}$$

where ΔP_t^{MT}, ΔQ_t^{EB}, and ΔQ_t^{GB} represent the power output changes of micro gas turbines, electric boilers, and gas boilers, respectively, within the unit dispatch time; ΔP_{down}^{MT}, ΔQ_{down}^{EB}, and ΔQ_{down}^{GB} represent the down-ramp rates of micro gas turbines, electric boilers, and gas boilers; ΔP_{up}^{MT}, ΔQ_{up}^{EB}, and ΔQ_{up}^{GB} represent the uphill speed of micro gas turbine, electric boiler and gas boiler respectively.

(3) Energy equipment output constraints

$$\begin{cases} 0 \le P_t^{grid} \le P_{\max^{grid}} \\ 0 \le P_t^{WT} \le P_{\max^{WT}} \\ 0 \le P_t^{PV} \le P_{\max^{PV}} \\ P_{\min^{MT}} \le P_t^{MT} \le P_{\max^{MT}} \\ P_{\min^{P2G}} \le P_t^{P2G} \le P_{\max^{P2G}} \\ Q_{\min^{EB}} \le Q_t^{EB} \le Q_{\max^{EB}} \\ Q_{\min^{GB}} \le Q_t^{GB} \le Q_{\max^{GB}} \end{cases} \tag{15}$$

where P_{max}^{grid}, P_{max}^{WT}, and P_{max}^{PV} respectively are the power grid, wind power, and photovoltaic output limits; P_{min}^{MT}, P_{min}^{P2G}, Q_{min}^{EB}, and Q_{min}^{GB} represent the lower limits of the output of the micro gas turbine, P2G equipment, electric boiler, and gas boiler respectively; P_{max}^{MT}, P_{max}^{P2G}, Q_{max}^{EB}, and Q_{max}^{GB} respectively are the upper limits of the output of the micro gas turbine, P2G equipment, electric boiler and gas boiler.

(4) Energy storage system constraint

$$\begin{cases} E_{ess}^{\min} \le E_{t}^{\min} \le E_{ess}^{\max} \\ H_{hss}^{\min} \le H_{t}^{\min} \le H_{hss}^{\max} \end{cases} \tag{16}$$

$$\begin{cases} u_t P_{bat}^{\min,c} \le P_{t}^{bat,c} \le u_t P_{bat}^{\max,c} \\ v_t P_{bat}^{\min,d} \le P_{t}^{bat,d} \le v_t P_{bat}^{\max,d} \\ w_t Q_{heat}^{\min,c} \le Q_{t}^{heat,c} \le w_t Q_{heat}^{\max,c} \\ z_t Q_{heat}^{\min,d} \le Q_{t}^{heat,d} \le z_t Q_{heat}^{\max,d} \end{cases} \tag{17}$$

where $E_{ess}{}^{max}$, $E_{ess}{}^{min}$, and $H_{hss}{}^{max}$, $H_{hss}{}^{min}$ represent the upper and lower limits of the energy storage capacity of electric energy storage and thermal energy storage respectively; $P_{bat}{}^{max,c}$, $P_{bat}{}^{min,c}$, and $P_{bat}{}^{max,d}$, $P_{bat}{}^{min,d}$ represent the upper and lower limits of charging and discharging power of the electric energy storage system; $Q_{heat}{}^{max,c}$, $Q_{heat}{}^{min,c}$ and $Q_{heat}{}^{max,d}$, $Q_{heat}{}^{min,d}$ represent the upper and lower limits of the charging and discharging energy power of the thermal energy storage system; u_t, v_t is 0-1 variable, respectively represent the charge-discharge state of the electric energy storage system; w_t, z_t is 0-1 variable, respectively represent the charging and discharging state of the thermal energy storage system.

(5) EV restraint

$$\begin{cases} P_{t}^{EV} = r_t P_{t}^{EV,c} - k_t P_{t}^{EV,d} \\ r_t + k_t \le 1 \quad r_t, k_t \in \{0,1\} \\ 0 \le P_{t}^{EV,c} \le r_t P_{EV,t}^{\max,c} \\ 0 \le P_{t}^{EV,d} \le k_t P_{EV,t}^{\max,d} \\ E_{need}^{EV} = \sum_{t=1}^{T} P_{t}^{EV} \end{cases} \tag{18}$$

where $P_t{}^{EV,c}$ and $P_t{}^{EV,d}$ the charging and discharging power of electric vehicle cluster at time t; $P_{EV,t}{}^{max,c}$ and $P_{EV,t}{}^{max,d}$ represent the maximum charging and discharging power of the electric vehicle cluster at time t; r_t and k_t is 0-1 variable, is the electric vehicle cluster charge-discharge state variable; $E_{need}{}^{EV}$ is the total electricity required for the next day's cluster electric vehicle travel.

3. Solution to the two-layer game model for IES capacity configuration considering V2G energy storage

The two-layer optimization model for IES capacity configuration considering V2G energy storage characteristics optimizes the outer layer with a single objective: minimizing the sum of the daily average investment cost of the energy storage system over its lifecycle and the operating cost. The inner layer optimizes multiple objectives: minimizing operating and maintenance costs, minimizing wind and solar curtailment rates, and minimizing EV charging costs. The model iteratively solves the energy storage configuration problem by nesting the outer and inner layers. Due to the nonlinearity of the outer layer objective function caused by the consideration of battery life, the model cannot be solved as a single-layer optimization problem. Therefore, the outer layer uses a genetic algorithm to optimize the model, and the results are passed to the inner layer. The inner layer uses Gurobi to solve the multi-objective model based on the outer layer's optimization results and constraints. The results from the inner layer are then passed back to the outer layer for further optimization. This iterative process continues until the outer layer converges to the optimal solution, achieving a global optimum.

The specific steps are as follows:

(1) Input the parameters for the outer layer genetic algorithm and initialize the binary population *POP*;

(2) Decode the binary population to obtain the initial values of the outer layer decision variables $X= [E_S, P_S, H_S, Q_S]$, and pass them to the inner layer. Initialize the variable I;

(3) The inner layer inputs electrical/thermal load and device parameters to build the model, calls the Gurobi solver to solve the multi-objective model, and passes the optimal power output $x= [P_t^{MT}, Q_t^{GB}, Q_t^{EB}, P_t^{P2G}, P_t^{WT}, P_t^{PV}, P_t^{grid}, P_t^{grid}, Q_t^{heat}, P_t^{EV}]$ and the optimal solution to the outer layer;

(4) If $i>M$, terminate the iteration and output the current electrical/thermal energy storage capacity configuration and the optimal solutions for both layers. Otherwise, increment i by 1;

(5) The outer layer calculates the fitness value f_{fit} of the population based on the inner layer's optimization results.

4. Case study

To validate the effectiveness and feasibility of the model, a case study was conducted on a multi-energy complementary IES in Jiangsu, China. The model was implemented using Python, and a 24-hour dispatch period with a unit dispatch time of 1 hour was considered. The system purchases electricity from the grid using a time-of-use (TOU) pricing model, and EVs participating in V2G also use TOU pricing for charging and discharging. The 24-hour TOU pricing for grid electricity and EV charging/discharging is shown in Table 1. The system's internal energy supply devices and related operating parameters are shown in Table 2.

Table 1. TOU pricing for grid electricity and EV charging/discharging.

Type	Time	Grid Electricity Price (¥/kWh)	EV Charging Price (¥/kWh)	EV Discharging Compensation (¥/kWh)
Peak	10:00-13:00 18:00-21:00	1.2	1.2	0.56
Off-peak	06:00-07:00 08:00-09:00 14:00-17:00	0.6	0.6	0
Valley	00:00-05:00 10:00-13:00 22:00-23:00	0.35	0.3	-0.5

Table 2. System typical energy supply devices and related operating parameters.

Type	Power Upper and Lower Limits (kW)	Down-Ramp Rate (kW/h)	Up-Ramp Rate (kW/h)	Unit Power Maintenance Cost (¥/kW)
Grid	0-200	-	-	-
MT	20-300	5	15	0.026
GB	0-80	5	10	0.024
EB	0-80	12	12	0.016
P2G	0-100	-	-	0.021

To verify the effectiveness of the two-layer optimization model for IES capacity configuration considering V2G energy storage characteristics, two transaction schemes were set up for comparative analysis. Scheme 1: IES capacity configuration method without considering V2G energy storage characteristics; Scheme 2: IES capacity configuration method considering V2G energy storage characteristics. For Scheme 1, the model was solved using the Gurobi solver. For Scheme 2, the proposed two-layer game model considering V2G energy storage was used. The total energy required for the EV cluster's travel the next day was calculated using Monte Carlo random sampling, resulting in 423 kWh, meaning the net charging amount of EVs during the 24-hour dispatch period was 423 kWh. The comparison of various operating indicators under the two schemes is shown in Table 3.

Table 3. Optimization results under different schemes.

Scheme	Total Operating Cost (¥)	Electrical Energy Storage Capacity (kWh)	Thermal Energy Storage Capacity (kWh)	Daily Average Investment Cost of Energy Storage System (¥)	Wind and Solar Curtailment Rate (%)	Average EV Charging Cost (¥/kWh)
一	2021.39	798	299	199.33	0.047	0.719
二	1724.65	201	150	58.52	0.15	0.028

Compared to Scheme 1, Scheme 2 considers the energy storage characteristics of EVs through V2G, manages the charging and discharging of EV clusters based on travel patterns, and considers the multi-objective optimization of total system operating costs and EV owners' charging costs. After incorporating V2G, the total operating cost, energy storage capacity configuration, the daily average investment cost of the energy storage system, and the average charging cost of EV owners have all been reduced to some extent. Specifically, the total operating cost decreased by 14.68 %; the electrical energy storage capacity configuration was significantly reduced, and the daily average investment cost of the energy storage system decreased by 70.64 %; the average charging cost of EV owners decreased from 0.719 to 0.028, a reduction of 96.11 %. Within a single day, the discharge subsidies for EVs can fully cover their charging costs. However, due to the consideration of EV owners' charging costs, the wind and solar curtailment rate slightly increased, but it remains within an acceptable range. Overall, considering V2G energy storage characteristics has a significant effect on optimizing the energy storage capacity configuration of the IES, creating a mutually beneficial strategy for all stakeholders in the IES and EV owners.

5. Conclusion

This paper proposes a two-layer optimization model for IES capacity configuration that considers V2G energy storage characteristics. The outer layer optimizes a single objective: minimizing the sum of the IES operating cost and the average investment cost of the energy storage system over its lifecycle. The inner layer optimizes multiple objectives: minimizing operating and maintenance costs, minimizing wind and solar curtailment rates, and minimizing EV charging costs. Due to the nonlinearity of the outer layer objective function caused by the consideration of battery life, the model uses a genetic algorithm for the outer layer and the Gurobi solver for the inner layer, iteratively solving the problem until the global optimum is found. Finally, a case study was conducted on a multi-energy complementary IES in Jiangsu, China, using two schemes for a typical day scenario. The results show that the proposed optimization model considering V2G energy storage characteristics has high economic efficiency and effectively optimizes the total operating cost, energy storage capacity configuration, and EV charging costs.

References

[1] Z Y Zhang, P Jiang, Z B Liu, et al (2023) Peiguang Wang, Capacity optimal configuration and collaborative planning of multi-region integrated energy system, Energy, Volume 278, Part B, 127970, ISSN 0360-5442.
[2] Guo M X, Mu Y F, Xiao Q, et al (2021) Optimal Configuration of Electric/Thermal Hybrid Energy Storage for Park-level Integrated Energy System Considering Battery Life Loss[J]. Automation of Electric Power Systems, 45(13):66-75.
[3] Zhang P Z, Xie L R, Ma R Z, et al (2022) Multi-player Two-stage Low Carbon Optimal Operation Strategy Considering Electric Vehicle Cluster Schedulability[J]. Power System Technology, 46(12):4809-4825.
[4] Jin J X, Zhang T L, Yang R H, et al (2023) Hierarchical cooperative control strategy of distributed hybrid energy storage system in an island direct current microgrid[J]. Journal of Energy Storage, 57:106205.
[5] X. Hu, J. Shu, H. Yue, and Y. Yue (2022) "Root-cause Analysis of Renewable Curtailment Based Integrated Power System Optimal Planning in Power Spot Market, "2022 7th Asia Conference on Power and Electrical Engineering (ACPEE), Hangzhou, China, pp. 337-343.
[6] L Zhang, L Chen, W W Zhu, et al (2022) Research on the optimal allocation method of source and storage capacity of integrated energy system considering integrated demand response, Energy Reports, Volume 8, Pages 10434-10448, ISSN 2352-4847.

Journal of Physics: Conference Series **3011** (2025) 012030 doi:10.1088/1742-6596/3011/1/012030

Study on the improved method for natural gas analytical system performance evaluation

Chen Fuquan[1], Lin Min[1] and Li Ke[2,*]

[1] China Petroleum and Chemical Corporation Natural Gas Yuji Pipeline Branch Company, Jinan, Shandong, 250101, China
[2] Chengdu University of Information Technology, Chengdu, Sichuan,610025, China

[*]Corresponding author's e-mail: lotpal@163.com

Abstract. With the advancement of natural gas energy measurement methods by domestic governments, enterprises, and users, GB/T 28766-2018 *Performance Evaluation of Natural Gas Analysis Systems* has received increasing attention and widespread application since its release and implementation in 2018. However, most of the gas chromatography used domestically employs a single-point calibration function for sample analysis and cannot implant true regression functions, which significantly limits the application of the standard. Based on the standard method and the current state of the domestic natural gas industry, this paper proposes an improvement measure for the standard method and validates the improved method through actual measurements. The performance evaluation data from nine chromatography demonstrate that the proposed method can effectively identify equipment with issues.

1. Introduction

With the publication and implementation of the GB/T 28766-2018 *Performance Evaluation of Natural Gas Analysis Systems* standard, along with the gradual advancement of natural gas energy measurement methods, the accuracy of online and offline gas chromatography detection results has garnered increasing attention from technical personnel in the natural gas industry. GB/T 28766-2018, developed based on ISO 10723:2012, serves as a performance evaluation standard for natural gas chromatography instruments and has been widely applied in recent years.

Tian et al. [1] selected the gas chromatography method for analyzing gas components in the process of natural gas energy measurement and provided the specific steps of its evaluation based on the actual evaluation operation sequence. Jiang et al. [2] evaluated online gas chromatography according to SY/T 7658-2021 Evaluation of Performance of Online Gas Chromatography Spectrometer for Natural Gas. Lin [3] introduced the software system development and application for GB/T 28766-2018. Chen [4] established an evaluation process for gas chromatography based on national standards and developed performance evaluation software for the natural gas analysis system. Peng [5] discussed the impact of compositional analysis on the energy measurement of mixed natural gas. Liu et al. [6] discussed the application of gas chromatography performance evaluation methods in the evaluation process of LNG sampling systems.

In the performance evaluation process of gas chromatography, the most critical step is the regression calculation of the instrument response function (analysis function and calibration function). According to standard requirements, the Generalized Least Squares (GLS) method should be used to perform regression calculations on the standard gas content and test data.

Powell & Macdonald [7] introduced an iterative method for the solution of the generalized nonlinear least squares problem: where the model may be nonlinear in its parameters and the independent variable(s) and all variables are subject to error. The method is described for the case of two arbitrarily related variables; does not require the analytic calculation of derivatives; leads to exceptionally close satisfaction of the least squares conditions; and exhibits especially rapid convergence arising from the use of somewhat unconventional numerical approximations for partial derivatives. Examples are given that compare the results of the method with those of other existing techniques.

Carotenuto et al. [8] propose an alternative methodology for the determination of the calibration curves of gas chromatography employed in natural gas analysis. Such methodology, coming from a modification to the regression technique known as Generalized Distance Regression (GDR), was applied to the data obtained during the calibration procedure of the gas chromatography. To obtain the gas chromatography calibration curves, the GDR technique was first applied, to consider the heteroscedastic uncertainty both in the dependent and independent variables. However, for some chemical components, a polynomial curve of order higher than one was obtained. In general, since gas chromatography is affected by time drifts, it would be favorable to have straight lines as calibration curves. In this way, the periodical check (usually made daily) would be extremely simplified. In fact, to determine the drift entity for a line, two checkpoints would be necessary, and if the line would pass through the origin, a single point would be necessary. In order to simplify the management of the time drift check procedure, the authors modified the GDR technique, obtaining the Modified Generalized Distance Regression, which allows one to impose a straight line as a calibration curve for each component, obtaining an unbiased estimation of the calibration coefficients and of their variance-covariance matrix.

However, there are still some issues that have not been addressed in the process of applying the standard.

2. Problems in the Application of Standards

GB/T 28766 was first published in 2012, based on ISO 10723:1995, and Li [9] et al. introduced the application methods of that version of the standard. GB/T 28766-2018 is the second edition of the standard, which has undergone significant changes compared to the previous version. Two important modifications include the addition of regression calculations for the true function of natural gas analysis systems and the Monte Carlo deviation analysis method. Due to the need to assume a large amount of data, calculating, and statistically analyzing the analytical bias of instruments through ordinary methods is no longer feasible; specialized supporting software is required for successful application. In response to the numerous prerequisites needed for applying GB/T 28766, Zeng [10] et al. simplified and supplemented the standard and developed corresponding industry standards and specifications.

Zhang et al. [11] introduced the workflow of GB/T 28766 in the performance evaluation process of gas chromatography, as shown in Figure 1.

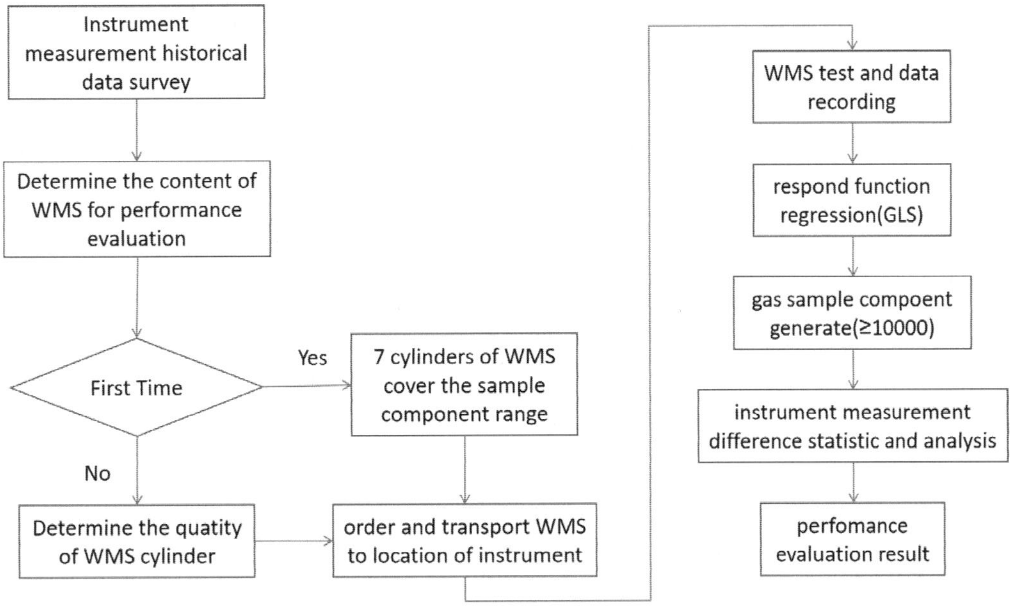

Figure 1. Working process of GB/T 28766-2018 "Natural gas — Analytical system performance evaluation".

The most intuitive and direct method for evaluating gas chromatography is to take on-site samples or prepare a large number of actual samples, directly detect them with the chromatography, and then verify and examine the detection results to assess the accuracy of the instrument's detection. However, due to material and time costs, it is generally impossible to comprehensively verify all possible combinations of sample components within the instrument's detection range. Therefore, numerical simulation must be used to statistically analyze and evaluate the performance of the instrument. Hence, the "Generate Hypothetical Compositions" step in Figure 1 is an important part of the Monte Carlo method for instrument performance evaluation. The idea is to generate as many hypothetical natural gas compositions as possible within the instrument's detection range to simulate various possible combinations of natural gas component concentrations in the pipeline network, such as gas field gas (mainly methane, some samples with higher CO_2 or N_2 content) and imported LNG (ethane may be as high as 8%). These generated hypothetical data represent corresponding natural gas samples, and the component data are accurate values without uncertainty. By substituting these component data into the calculation method of GB/T 11062-2020, we can obtain data such as the high heating value of the hypothetical sample composition and its corresponding uncertainty. Here, the uncertainty only includes the uncertainty of the basic data and calculation methods in GB/T 11062, without any component uncertainty, providing a relatively accurate true heating value as H_{gv-r}.

According to GB/T 28766, within the working range of the instrument, ten thousand sets of sample compositions should be hypothesized to statistically analyze the detection bias of the instrument. In practical work, the number of hypothetical samples can also be adjusted as needed. Once the hypothetical compositions are generated, they should be individually substituted into the instrument's calibration and analysis functions to obtain the composition content data given by the instrument when detecting that sample. The calibration and analysis functions are calculated using the generalized least squares method (GLS) [12] after the instrument detects several bottles of gas standard materials, substituting the detected peak area of each component and the content of the gas standard material. By substituting the hypothetical sample composition into the true calibration function, the peak area of each component can be calculated. This peak area data represents the peak area Ar that the real sample would give after being analyzed by the instrument. Then, by substituting this peak area data into the regression-calculated true analysis function of the instrument, the composition content that the instrument would give when detecting that real sample can be obtained. The component content given by the instrument after substituting the hypothetical sample into the true response function is normalized and then substituted into the calculation method of GB/T 11062 again to obtain physical property parameters such as the higher heating value of the hypothetical sample. This heating value is denoted as $H_{gv\text{-}e}$. By comparing the $H_{gv\text{-}r}$ and $H_{gv\text{-}e}$ of all hypothetical sample compositions one by one and statistically analyzing the deviation situation, the deviation between the instrument's true response function and the actual situation can be determined.

According to the GB/T 28766-2018 standard requirements, the performance evaluation of natural gas analysis systems can only be applied to multi-point calibration gas chromatography that uses GB/T 27894 for sample detection. The standard does not take into account that most domestic online and offline gas chromatography currently use the single-point calibration method of GB/T 13610 for detecting natural gas samples, and most equipment does not have the functionality to implant a true response function obtained through regression analysis. Therefore, if the performance evaluation is conducted according to GB/T 28766-2018, it cannot provide truly meaningful evaluation conclusions for most gas chromatography in use.

3. Solutions

To address this issue, the methods of GB/T 28766-2018 should be expanded and improved. Specifically, the peak area data A_r calculated by the calibration function for each hypothetical sample in the above calculation process should be substituted into the single-point calibration function currently used by the instrument, that is, function (1):

$$y_i=kA_i \tag{1}$$

where: y_i represents the content of component i in natural gas as given by the instrument after detection, A_i represents the peak area of component i in the hypothetical sample composition calculated by the instrument's true calibration function, which is the i-th peak area data in the data-set A_r, and k is the response factor for component i, k is calculated from the daily calibration gas provided by the instrument through the chromatography calibration process.

After normalizing the y_i for each component of natural gas, substitute it into the calculation method of GB/T 11062 to obtain the $H_{gv\text{-}c}$ when the instrument uses a single-point calibration function to detect samples. By performing repeated calculations and statistical analysis on all hypothetical samples, the deviation between the higher heating value given by the instrument using the current GB/T 13610 standard method and the true situation can be determined. If its average deviation or maximum deviation exceeds the maximum allowable error range for the heating value specified in the standard, then analyze the reasons for the errors from the data obtained during the performance evaluation process, and provide maintenance and repair suggestions for the instrument. If its average deviation or maximum deviation is within the error range required by the standard, then an evaluation conclusion that the instrument can be used correctly can be given.

Additionally, during the performance evaluation process, it is necessary to conduct blank tests to check the validity of the instrument. This involves using the instrument to detect the carrier gas (usually He) and observing whether there are any residual peak areas for various components during the operation of the instrument. After actual testing, most online gas chromatography may have small amounts of methane, propane, pentane, hexane, CO_2, or N_2 peaks due to factors such as poor sealing and hydrocarbon residue. The impact of residual peak areas on the detection results must also be considered. Specifically, after deducting the impact of residual peak areas, calculate the $H_{gv\text{-}s}$ and compare it with $H_{gv\text{-}r}$ to analyze the extent of the impact of residual peak areas on the detection results of the instrument.

4. Applications

Following the above approach, on-site performance evaluations were conducted on 9 gas chromatography. Corresponding supporting software was developed to process the data, resulting in the following statistics on higher heating value deviations:

Table 1. Gas Chromatography Performance Evaluation Hgv Deviation Data (MJ/m³).

Category	Real data		Original standard data		Improve method data to the original standard			
Instrument	Item	$H_{gv\text{-}r}$	$H_{gv\text{-}e}$	RD (%)	$H_{gv\text{-}c}$	RD (%)	$H_{gv\text{-}s}$	RD (%)
GC1	Average	37.40	37.40	0.0132	37.39	-0.0224	37.39	-0.0181
	Max	37.98	37.97	-0.0161	37.98	0.0122	37.98	0.0123
	Min	36.82	36.83	0.0498	36.80	-0.0390	36.80	-0.0302
GC2	Average	37.40	37.39	-0.0077	37.39	-0.0078	37.40	0.0040
	Max	37.98	37.97	-0.0060	37.97	-0.0121	37.97	-0.0121
	Min	36.82	36.81	-0.0277	36.81	-0.0187	36.82	0.0056
GC3	Average	37.40	37.39	-0.0016	37.39	-0.0087	37.40	0.0030
	Max	37.98	37.97	-0.0099	37.99	0.0369	37.98	0.0137
	Min	36.82	36.81	-0.0103	36.79	-0.0761	36.81	-0.0289
GC4	Average	37.40	37.50	0.2906	37.72	0.8714	37.72	0.8631
	Max	37.98	38.26	0.7477	38.29	0.8236	38.28	0.8130
	Min	36.82	36.77	-0.1321	36.64	-0.4742	36.64	-0.4803
GC5	Average	37.40	37.39	-0.0098	37.39	-0.0173	37.39	-0.0130
	Max	37.98	37.97	-0.0019	37.97	-0.0024	37.97	-0.0101
	Min	36.82	36.81	-0.0110	36.81	-0.0249	36.81	-0.0079
GC6	Average	37.40	37.39	-0.0017	37.39	-0.0044	37.40	0.0060
	Max	37.98	37.98	0.0206	37.99	0.0363	37.98	0.0136
	Min	36.82	36.81	-0.0081	36.80	-0.0433	36.82	0.0016
GC7	Average	37.40	37.39	-0.0027	37.39	-0.0038	37.40	0.0031
	Max	37.98	37.98	0.0037	37.98	0.0163	37.98	0.0159
	Min	36.82	36.81	-0.0034	36.80	-0.0322	36.81	-0.0175
GC8	Average	37.40	37.40	0.0120	37.10	-0.7925	37.10	-0.7880
	Max	37.98	37.94	-0.1051	37.23	-1.9674	37.23	-1.9576
	Min	36.82	36.81	-0.0170	37.00	0.4901	37.00	0.4906
GC9	Average	37.40	37.40	-0.0010	37.39	-0.0033	37.37	-0.0667
	Max	37.98	37.98	0.0178	37.99	0.0323	37.94	-0.0829
	Min	36.82	36.81	-0.0032	36.80	-0.0302	36.80	-0.0409

From the data in the table 1, it can be seen that the high heating value deviations of Chromatography 4 and Chromatography 8 exceed the maximum allowable error range specified by GB/T 18603 for Class A energy measurement, which is less than 0.5%. Among them, if Chromatography 4 uses the true function obtained from performance evaluation to analyze samples, its maximum high heating value still exceeds the specified maximum allowable error range. However, if Chromatography 8 uses the true function obtained from performance evaluation, it can meet the standard requirements.

Based on the above performance evaluation results, the adopted standard evaluation method and the extended and improved results of the standard method can effectively identify equipment that does not meet normal usage requirements among the gas chromatography in use. Corresponding maintenance recommendations can be provided to the equipment operation management to meet the accuracy requirements for detecting physical parameters such as natural gas composition and calorific value after the implementation of future natural gas energy measurement methods.

5. Conclusions

Since most gas chromatography currently in use domestically adopt the single-point calibration method specified by GB/T 13610 for analysis and detection, and cannot implant the instrument's true response function (including analysis and calibration functions), there are certain limitations to the application of the current GB/T 28766-2018 *Performance Evaluation of Natural Gas Analysis Systems* standard. It is necessary to expand and improve the standard methods accordingly. This article introduces the methods for achieving standard functionality and extending and improving standard functions. The application effects of these methods have been verified with on-site evaluation data from multiple devices, proving that this method can better adapt to the current state of the domestic natural gas industry. It can effectively identify problematic parts in the equipment in use, thereby providing targeted suggestions for equipment usage, maintenance, and repair.

For application units with performance evaluation needs and corresponding capabilities, this method can be used to establish a performance evaluation system. For units purchasing natural gas analysis system performance evaluation software, it is important to purchase systems that have extended standard functions to conduct more comprehensive evaluations of the equipment managed by their units, obtaining more scientific and reasonable evaluation results.

References

[1] Tian Jing, Liu Bing, Wu Jianhong, Xing Nan. Study on Performance Evaluation and Verification Experiment of Natural Gas Chromatography [J]. Shandong Chemical Industry, 2023, 52 (23): 189-191.
[2] Jiang Chen. Performance Evaluation of On-line Gas Chromatography for Natural Gas Energy Metering [J]. Metrology & Measurement Techniques, 2023, 50 (06): 72-75+79.
[3] Lin Min. Performance evaluation method and deviation analysis of natural gas energy metering and analysis system [J]. Chemical Analysis and Meterage, 2023, 32 (03): 84-89.
[4] Chen Honglue. Performance Evaluation of Natural Gas Analysis System and Assignment Research in Urban Pipeline Network [D]. China University of Petroleum (Beijing), 2022.
[5] Peng Yan. Discussion on the Influence of Composition Analysis on Energy Measurement of Mixed Transported Natural Gas [J]. China Inspection and Testing, 2021, 29 (05): 29-32.
[6] Liu Bing, Tian Jing, Xing Nan. Performance Evaluation Method and Application Research of LNG Sampling Unit [J]. Natural Gas Chemical Industry (C1 Chemistry and Chemical Engineering), 2021, 46 (01): 34-37+89.
[7] Powell, D. R., & Macdonald, J. R. (1972). A rapidly convergent iterative method for the solution of the generalised nonlinear least squares problem. The Computer Journal, 15(2), 148-155. https://doi.org/10.1093/comjnl/15.2.148
[8] Carotenuto, A., Giovinco, G., Viglietti, B., & Vanoli, L. (2005). A new procedure for the determination of calibration curves for a gas chromatograph used in natural gas analysis. Chemometrics and Intelligent Laboratory Systems, 75(2), 209-217.
[9] Li Ke, Luo Qin, Wang Wenhua, Xue Jian, Zeng Wenping, Wang Xiaoqin, Wu Yongdong. Field Application of Natural Gas Online Analysis System Performance Evaluation Standard [J]. Petroleum and Natural Gas Chemical Industry, 2018, 47 (04): 94-100.
[10] Zeng Wenping, Wang Xiaoqin, Wang Weijie. Development of Calibration Specifications for Online Natural Gas Gas chromatography [J]. Petroleum and Natural Gas Chemical Industry, 2017, 46 (03): 99-103.
[11] Zhang Yinghui, Huang Jian'an, Chen Honglue. Application of Gas Chromatography Performance Evaluation [J]. New Industrialization, 2022, 12 (12): 181-188.
[12] Wang Xinyi, Zhao Xin, Song Yue. Analysis of the Reasons for Detection Deviation of Gas Chromatography Based on GLS Algorithm [J]. Journal of Beijing Institute of Petrochemical Technology, 2022, 30 (03): 13-17.

Journal of Physics: Conference Series **3011** (2025) 012031 doi:10.1088/1742-6596/3011/1/012031

Research on indoor thermal environment simulation and energy-saving optimization of air-conditioning systems in large space buildings

Zhiming Gao [1, 2*], Cheng Chen[2], and Wanchun Sun [1]

[1] School of Energy and Automotive Engineering, Shunde Polytechnic University, Foshan, Guangdong, China
[2] School of Mechanical & Automotive Engineering, South China University of Technology, Guangzhou, Guangdong, China

*Corresponding author. E-mail address: 14022@sdpt.edu.cn (Zhiming Gao).

Abstract. Large space buildings frequently experience unstable indoor thermal environments and high energy consumption for air conditioning. This paper focuses on a comprehensive gymnasium in Guangzhou, utilizing numerical simulation and experimental testing methods to analyze its indoor thermal environment and to identify areas for improvement and optimization. The results show that numerical simulation is a reliable method for predicting the indoor thermal environment of large space buildings. After improving the airflow organization, the uniformity of both the indoor temperature field and velocity field is enhanced. The proposed optimization scheme for the air conditioning system not only enhances comfort but also effectively reduces energy consumption, serving as a valuable reference for the energy-efficient operation of the gymnasium's air conditioning system.

1. Introduction

As cultural and sports activities flourish in China, large space buildings are becoming increasingly prevalent in public architecture, with stadiums dominating the landscape. These large space buildings are marked by high ceilings, noticeable temperature gradients, and substantial load fluctuations, necessitating thorough consideration of various factors in designing and operating their air conditioning systems. Air conditioning in these buildings can consume up to 40% to 60% of total building energy. To adapt to the internal temperature gradients and cut energy use, layered air conditioning systems are frequently employed. However, these systems often encounter problems like uneven temperature distribution and high localized wind speeds, resulting in an unstable indoor thermal environment, inadequate thermal comfort, excessive energy consumption, and inability to fulfill wind speed requirements for certain events. The issue of minimizing energy consumption in large space building air conditioning systems while ensuring a suitable indoor thermal environment and airflow organization has garnered considerable attention from scholars.

It is becoming increasingly popular to study indoor thermal environments using numerical simulation technology, specifically Computational Fluid Dynamics (CFD). In recent years, research on the indoor thermal environment of tall spaces in gymnasiums has primarily focused on two aspects. On the one hand, from the perspective of air conditioning air distribution design, various forms of air conditioning air distribution and airflow patterns have been comprehensively analyzed based on thermal comfort and economy. The goal is to identify superior airflow organization forms or return air configurations, providing references for air conditioning design [1-4]. On the other hand, from the perspective of operation and use, CFD technology is employed to simulate the indoor thermal environment. Based on the shortcomings identified in the original air conditioning system design, optimization, and adjustment measures are proposed to meet comfort or energy-saving requirements [5-10].

In this paper, the CFD calculation method is utilized to simulate the originally designed air conditioner in the stadium's competition hall. The rationality of the airflow organization is then evaluated based on the distribution of temperature and velocity fields in typical sections within the personnel activity area. Finally, the airflow organization is fine-tuned by modifying the tuyere angle and air volume. Additionally, the energy-saving operating condition of the air conditioning system is optimized. The reliability of the numerical simulation results is validated through experimental tests.

2. Processing for numerical simulation

The focus of this study is a comprehensive gymnasium in Guangzhou, specifically the venue that hosted the 16th Asian Games. The gymnasium is a multi-purpose facility that can accommodate table tennis, badminton, basketball, and other indoor sports competitions, as well as artistic performances and large gatherings. The gymnasium measures 129 meters in length and 68.6 meters in span. The structural height of the building is 32 meters, and the total occupied area is 12,177 square meters.

2.1. Geometric model

The gymnasium's geometric space is expansive, with small tuyere sizes and numbers. The largest and smallest tuyere sizes are likely to result in a dense grid. To save simulation time, a quarter of the game hall (assuming a symmetrical form) was selected to establish the geometric model. The specific dimensions are: 41.7 meters in length (X direction), 31.4 meters in width (Z direction), and 18.8 meters in height (Y direction). The sloping roof has been simplified in the plan. The schematic representation of the section model of the competition hall is shown in Figure 1.

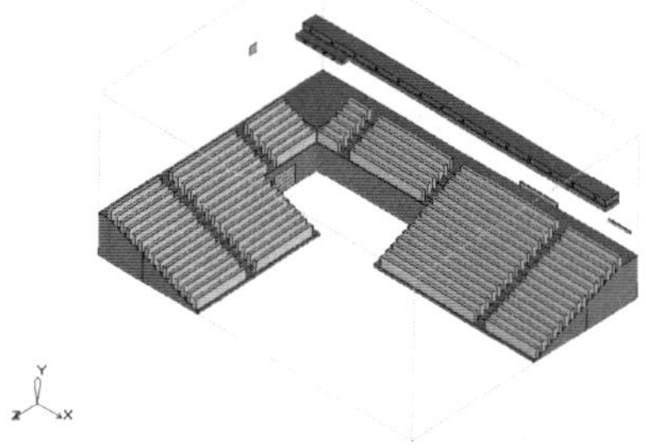

Figure 1. Quarter model diagram of the competition hall.

The game hall's quarter model has a total audience capacity of 1,110, including 600 fixed seats and 510 active seats, located on the west and south sides. The audience seating positions are simplified into several groups, each with a height of 1.2 meters and a width of 0.4 meters. The lighting lamps are simplified into three cubes based on their distribution positions and quantities. The first lamp is 5 meters long, 0.1 meters wide, and 0.5 meters tall, positioned above the west seats. The second lamp is 3 meters long, 0.1 meters wide, and 0.3 meters tall, located behind the west seats. The third lamp is 1 meter long, 0.1 meter wide, and 1 meter tall, positioned on the south wall. In the model shown in Figure 1, 14 of the air outlets have a diameter of 230 mm, with a spacing of 2.9 meters between them. The south seats have air outlets with a diameter of 230 mm and a spacing of 1.25 meters. The west seats have a square fan on three sides, with each side measuring 600 mm and a spacing of 2.5 meters.

2.2. Setting of the boundary conditions

2.2.1. Wall boundary. In the quarter model of the competition hall presented in this paper, there are two symmetric planes. On these symmetric planes, all physical quantity gradients perpendicular to the direction of the plane are zero. The south wall is an inner wall, while the west wall and the roof are in contact with the boundary. The thermal boundary conditions for these walls are measured experimentally. The thermal boundary conditions for the floor are set as insulation.

2.2.2. Heat dissipation. The heat dissipation of the audience is based on the heat dissipation of adult men under extremely light labor conditions at 26 °C. The explicit heat dissipation is 61 W/person, the latent heat dissipation is 73 W/person, and the total heat dissipation is 134 W/person. The sensible heat loss (loose heat) is 109 g/h·person, and the personnel cluster coefficient is 0.92. The heat dissipation of the audience is set on the cube mannequins in the audience area. In the quarter model of the competition hall, there are three sets of lighting fixtures. The thermal boundary conditions for these fixtures are set as fixed heat boundaries. The heat dissipation amounts are 16,000 W for the first group, 4,000 W for the second group, and 9,000 W for the third group.

2.3. Division of the mesh
The models in this paper are essentially cubes and then use a mixed hexahedral structured mesh. To reduce the total number of grid cells and speed up the calculation cycle, while ensuring the accuracy of the calculation results, the air conditioning area's air supply outlet, return air outlet, and human body area are locally refined, while the non-air conditioning area is treated more coarsely. In the full model of the audience area, the number of nodes is 846,996 and the number of grid cells is 789,999.

3. Thermal environment analysis and optimization

The original air conditioning airflow organization scheme involves setting the air supply temperature to 18 °C for both the competition site and the spectator area. The air outlets are horizontal air supply outlets, with each outlet providing an airflow rate of 2400 m³/h. In the south auditorium, the outlets have a downward inclination angle of 15°, and each outlet provides an airflow rate of 2680 m³/h. Additionally, the west spectator device has an airflow rate of 1350 m³/h.

3.1. Simulation results of the temperature field
The simulation results provide data on the temperature profiles at various points to investigate the temperature distribution in the venue and spectator areas. The temperature sections at y = 1.7 m, y = 3.2 m, and y = 6 m are used to investigate the horizontal temperature distribution in the venue, the spectator seats, and the fixed seat areas, as shown in Figure 2.

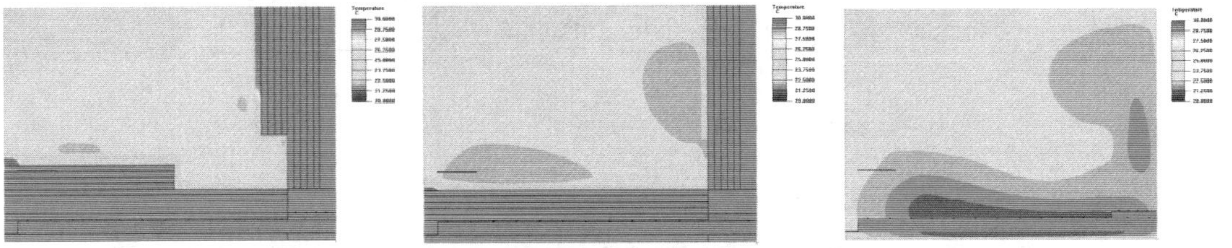

(a) Profile at y = 1.7 m (b) Profile at y = 3.2 m (c) Profile at y = 6 m

Figure 2. Temperature distribution diagram of the competition hall.

As can be seen in Figure 2 (a), (b), and (c), the activity area, which is 1.7 meters high, and the area with active seats, which is 3.2 meters above the floor, both have a temperature distribution ranging between 22 °C and 24 °C. The temperature in the area with fixed seats, located 6 meters away from the air outlet, is relatively close to the outlet and ranges between 21 °C and 23 °C. The temperature in areas farther away from the air outlet is slightly higher. In Figure 2 (a), the ribbon area with excessively high temperatures is located near the audience table steps.

3.2. Simulation results of the velocity field
The velocity profiles were analyzed at y = 1.7 m, y = 3.2 m, and y = 6 m within the gymnasium, focusing on the spectator seating and fixed seating areas. The results of these analyses are shown in Figure 3.

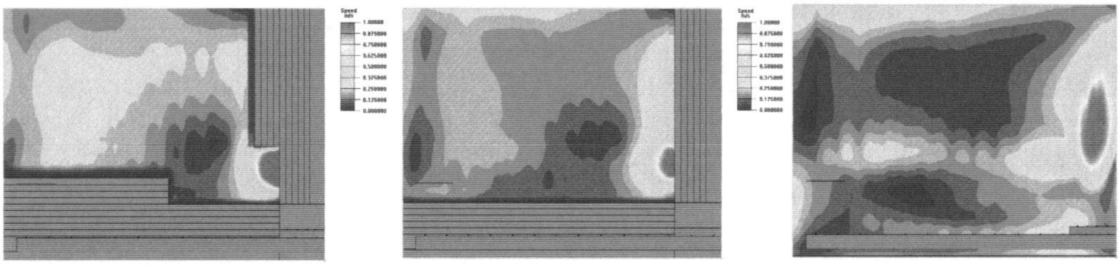

(a) Profile at y = 1.7 m (b) Profile at y = 3 m (c) Profile at y = 6 m

Figure 3. Speed distribution diagram of the race hall.

As shown in Figures 3 (a) and (b), the speed distribution in the horizontal direction within the competition hall is relatively uniform. The wind speed in a large area exceeds 0.2 m/s, which meets the wind speed requirements for small ball competitions. However, near the return air outlet, the wind speed is greater than 1 m/s. In Figure 3(c), the wind speed near the air outlet in the south audience seat area is high, reaching 1 m/s. In contrast, the wind speed in areas farther away from the air outlet is lower.

3.3. Improving airflow organization

The simulation results of the original air conditioning design scheme in the competition hall show that the initial airflow organization is generally reasonable, with relatively uniform temperature and speed distribution. However, some spectator areas on the south side are significantly affected by the nozzle air supply, resulting in high wind speeds that can cause discomfort and a blowing sensation for the audience. To address this issue, it is recommended to replace the air supply nozzles for the south and north seats with nozzles equipped with regulating valves. This will allow for adjustment of the air volume based on the number of spectators, and the nozzles can be closed when the venue is not fully occupied, relying solely on the remote nozzles. Figure 4 shows the speed distribution at y = 6 m after the south side air supply outlet has been closed.

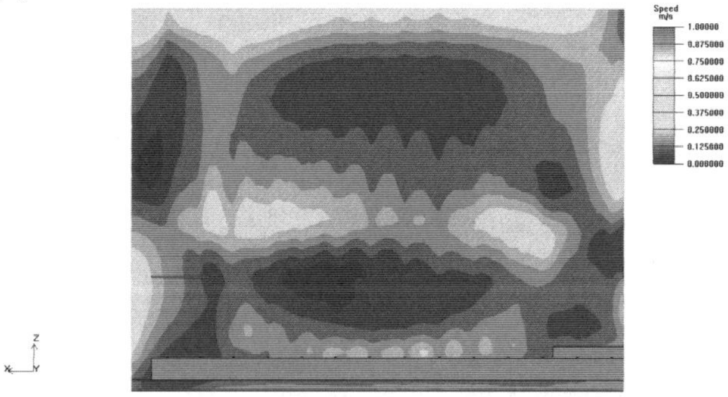

Figure 4. Speed distribution of y = 6 m after adjustment.

As shown in Figure 4, after adjusting the airflow organization, the speed distribution in the competition hall has become more uniform, and the wind speed near the tuyere on the south side has decreased. However, due to the limited visibility and smaller audience in this area, it still meets the usage requirements. Therefore, the improvement and adjustment of the airflow organization have not only enhanced the comfort for the audience but also effectively reduced air conditioning energy consumption.

3.4. Thermal environment optimization

The designed air supply conditions for the competition site include an air supply temperature of 16 °C and an air supply speed of 16 m/s. Now, the simulation results of the design condition are compared and analyzed with two optimized air supply conditions.

3.4.1. Designed working condition: Air supply temperature is 16 °C, and air supply speed is 16 m/s. The temperature and velocity profiles at heights of 2.5 m, 1.7 m, and 1.2 m are shown in Figure 5.

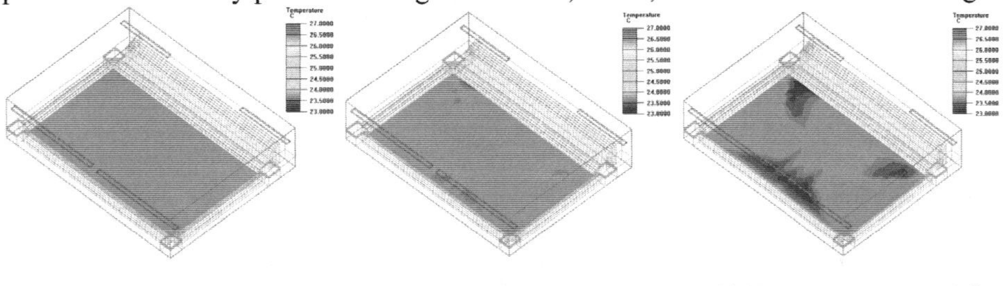

(a) Temperature at y = 2.5 m (b) Temperature at y = 1.7 m (c) Temperature at y = 1.2 m

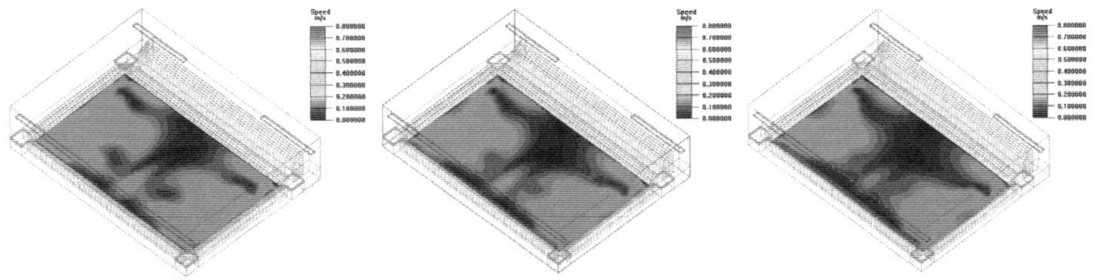

(d) Velocity at y = 2.5 m (e)Velocity at y = 1.7 m (f)Velocity at y = 1.2 m

Figure 5. Diagram of temperature and velocity distribution under designed working condition.

3.4.2. Optimization condition: The air temperature is set to 18 °C, and the air speed is maintained at 16 m/s. The temperature and speed profiles at heights of 2.5 meters, 1.7 meters, and 1.2 meters are depicted in Figure 6.

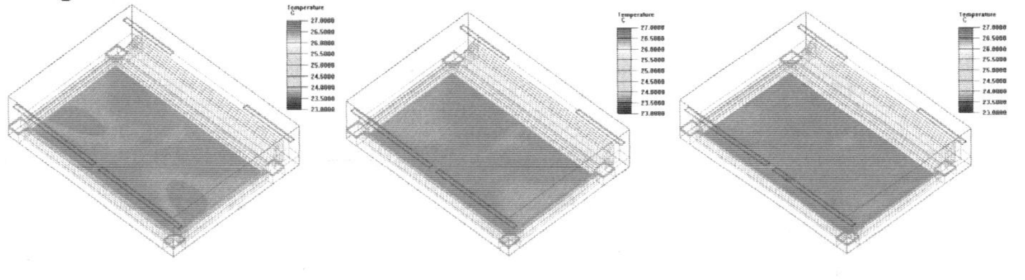

(a) Temperature at y = 2.5 m (b) Temperature at y = 1.7 m (c) Temperature at y = 1.2 m

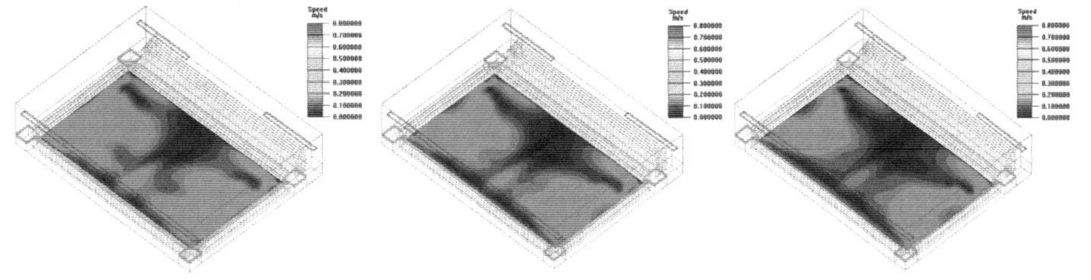

(d) Velocity at y = 2.5 m (e) Velocity at y = 1.7 m (f) Velocity at y = 1.2 m

Figure 6. Diagram of temperature and velocity distribution under optimization condition.

The simulation results for the design condition reveal that in some areas, the wind speed is higher. The simulation results for Optimization Condition reveal that the average temperature of the competition field has risen. However, the temperature in the activity area below 2 meters remains below 26 °C, and the wind speed stays below 0.3 m/s. Additionally, several other optimization conditions were simulated, such as the air temperature being set to 16 °C and the airspeed being maintained at 12 m/s. When compared to the design working conditions, the optimization working conditions not only meet the needs of the thermal environment but also exhibit a significant energy-saving effect.

4. Experimental testing

To measure the boundary conditions necessary for numerical simulation and to verify the validity of the established model and the reliability of the simulation results, this paper also conducts experimental tests. Firstly, the air supply temperature, airspeed, and air humidity parameters are measured within the competition hall, as well as the temperature fields of the walls and roof. The measured data are then used as the boundary conditions for the numerical simulation. Selected points within the improved air organization competition hall are analyzed by comparing the measured temperature and speed values with the numerical simulation results.

Considering the symmetry of the CFD model, the measurement points were arranged exclusively in one-quarter of the competition hall, encompassing the west wall and the south wall. Each measurement point represents an average reading taken vertically from the ground of the competition field (spectator steps) at heights of 0.5 m, 1.2 m, and 1.7 m. The measurements commenced one hour after the air conditioning system was turned on.

The measured values at the three heights of 0.5 m, 1.2 m, and 1.7 m in the vertical direction at each measurement point were averaged to represent the human activity area. These averages were then compared with the numerical simulation results. Figures 7-8 show the comparison results for the measured temperature, speed, and simulated values at the competition site, the west side auditorium, and the south side. From the figures, it can be observed that the measured values are generally consistent with the numerical simulation values. Most of the deviations in temperature are within 0.5°C, with a relative deviation of no more than 5%. Similarly, the deviations in speed are within 0.1 m/s. These deviations fall within the allowable range and are considered acceptable in practical engineering applications.

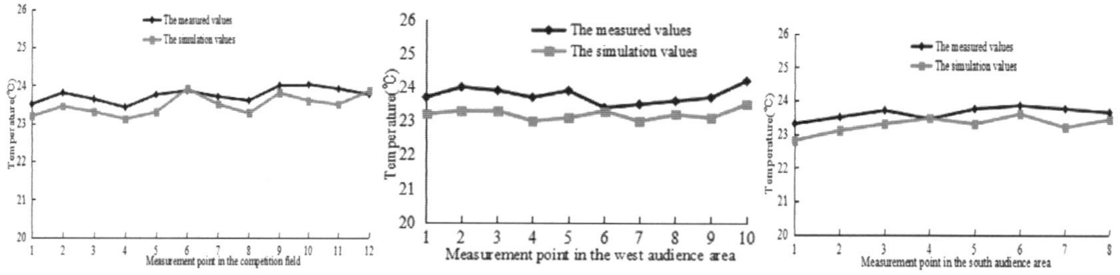

(a) The competition field (b) The west audience area (c) The south audience area

Figure 7. Comparison of measured and simulated temperature values.

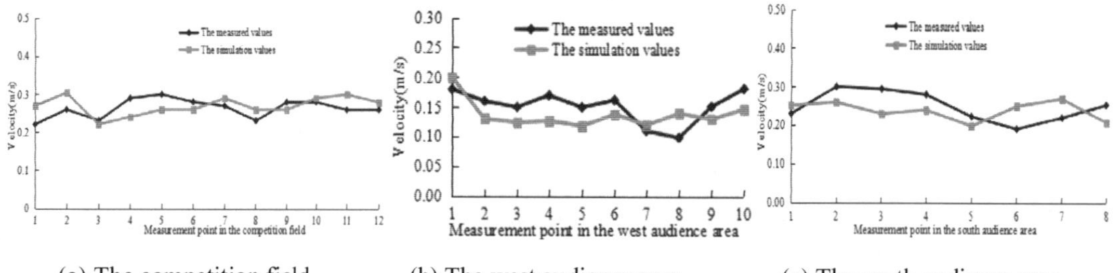

(a) The competition field (b) The west audience area (c) The south audience area

Figure 8. Comparison of measured and simulated velocity values.

5. Conclusion

In this paper, CFD technology and experimental testing are utilized to optimize the indoor airflow organization and the operation of the air conditioning system, based on the modeling and simulation of the indoor thermal environment of a gymnasium. The conclusions are as follows:

- CFD technology is reliable for simulating the indoor thermal environment of large-space buildings. The established model is effective, and the grid generation process is successful, enabling the acquisition of a stable and convergent solution. Experimental tests have verified the accuracy of the simulation results.
- Adjusting the original indoor airflow organization scheme has proven effective. The research results indicate that the optimized airflow organization exhibits greater uniformity in both the temperature and velocity fields, thereby enhancing thermal comfort for the audience and reducing air conditioning energy consumption.
- While ensuring a satisfactory indoor thermal environment and thermal comfort, the air conditioning system can appropriately increase the air supply temperature or decrease the air supply volume, to achieve enhanced energy savings.

References

[1] Kong H B, Zhou T, Sun Z J. Discussion on Natural Ventilation Design for the Reconstruction Project of Fengtai Railway Station in Beijing Railway Hub. Railway Standard Design, 2021, 65(03): 131-136+170.
[2] Jia Z K, Han J B, Liu J H, et al. Air Conditioning and Ventilation Design for Ultra-High and Large Exhibition Halls in the National Exhibition and Convention Center (Shanghai). HV&AC, 2017, 47(03): 73-78.
[3] Peng J B, Zhang C L, Wang J. Stratified Air Conditioning Design and Simulation for Xinjiang International Convention and Exhibition Center. HV&AC, 2017, 47(04): 73-77.
[4] Hu L, Wang J, Yu Z Y, et al. Ventilation and Air Conditioning System Design for the Tennis Court in Wuhan Guanggu International Tennis Center. HV&AC, 2017, 47(10): 68-71.
[5] Wang H D, Hu H, Zhu J K, et al. Study on Heat Transfer Characteristics Between Regions in Underfloor Air Distribution Stratified Air Conditioning Systems for Large Spatial Buildings. Journal of Refrigeration, 2022, 43(02): 89-96.
[6] Gong S Q, Liu Y, Xie D, et al. Numerical Simulation Analysis of Airflow Organization in Ventilation and Air Conditioning Systems for a Large Underground Hydropower Station Building. HV&AC, 2017, 47(11): 112-117.
[7] Han J, Gong G C, Yang H W, et al. Study on Thermal Environment and Energy Transfer of Air-Borne Energy Radiant Terminals in Large Spatial Buildings. Building Science, 2017, 33(10): 113-119.
[8] Xiong W C, Li Y J, Zhang Y D, et al. Numerical Simulation and Field Measurement Study of Airflow Organization in Air Conditioning Systems for Large Spatial Conference Halls. Fluid Machinery, 2017, 45(09): 66-70.
[9] Lv N. Study on Energy Consumption and Thermal Comfort of Air Column Ventilation Mode in High-Speed Railway Station Waiting Halls. Anhui Jianzhu University, 2024.
[10] Jing B Ki. Field Measurement and Optimization Study of Indoor Thermal Environment and Air Conditioning Systems in Tall Building Spaces. Shandong Jianzhu University, 2023.

Simulation of cavitation phenomena in Pelton turbine nozzle mechanisms under plateau conditions

Shuo Ren, Hongying Luo*, Hua Ji, Chaoqi Wang, Yunxin Du, Xueqing Qiao

College of Hydraulic and Civil Engineering, Xizang Agricultural and Animal Husbandry University, No. 100, Caixi Road, Linzhi City, Tibet, China

*Corresponding author's e-mail: xzluohy@xza.edu.cn

Abstract. This study employs numerical simulations to investigate the cavitation characteristics of the jet mechanism in a Pelton turbine operating in a plateau environment. The results indicate that as water flows through the converging section, its velocity increases sharply due to the reduced cross-sectional area, forming a high-speed jet at the nozzle outlet. The rapid flow causes a sudden drop in absolute pressure at the nozzle exit, reaching the saturation vapor pressure, which facilitates cavitation by allowing vapor nuclei to expand. It is observed that a "velocity deficit" occurs at the needle center due to flow obstruction, while a low-pressure region forms at the nozzle outlet due to high-speed jet ejection and air entrainment, both closely associated with cavitation. Additionally, the gaseous phase is primarily concentrated at the nozzle exit, with the cavitation region exhibiting an annular shape and a circumferentially uniform gas-phase volume fraction across all operating conditions. Based on the mass transfer rate model, simulation results, and experimental data, cavitation is found to be most pronounced at a nozzle opening of 40%.

1. Introduction

Pelton turbines convert the kinetic energy of high-head water flow into mechanical energy by driving a runner equipped with buckets in a circular motion, which, through coaxial coupling, powers a generator to produce electricity. In this system, the jet mechanism serves as a critical flow component, playing a pivotal role in energy transfer. However, cavitation and erosion issues inevitably arise during jet ejection, potentially damaging the jet mechanism, reducing turbine efficiency, and, in severe cases, posing a threat to the safe and stable operation of the unit [1-2].

Research on the flow mechanisms of the power components in Pelton turbines primarily relies on two approaches: physical model experiments and numerical simulations. Due to the inherent limitations of physical model experiments, numerical simulations have been more widely adopted in the study of Pelton turbines. Current research on Pelton turbines predominantly focuses on sediment-induced erosion and the flow characteristics of the jet mechanism, while investigations into cavitation and erosion phenomena within the jet mechanism remain relatively scarce. Zoppe et al. [3] found that the jet diameter influences both the edge and notch losses of the runner buckets. Benzo et al. [4-5], employing experimental design methods, explored the effects of nozzle angle, needle angle, and water head on hydraulic losses and efficiency. Their results indicated that at low water heads, larger nozzles and needle angles help reduce losses and improve efficiency. Petley et al. [6] identified an optimal combination of nozzle and needle angles that maximizes turbine efficiency; however, further increases in these angles lead to greater jet dispersion. Similarly, Huang et al. [7] observed that as needle eccentricity increases, jet eccentricity also rises, resulting in energy losses and a decline in jet quality. Nedelcu [8] investigated the effects of water head, nozzle diameter, and nozzle opening on flow rate as well as the forces acting on the nozzle and needle, concluding that water head has the most significant impact on flow rate. The research team led by Xiao et al. [9-10] analyzed the flow characteristics of the distributor ring and nozzle, revealing that the elbow section and supporting ribs induce secondary flows such as Dean vortices, which in turn affect the jet. They also found that an increase in nozzle opening leads to greater jet dispersion. Using numerical simulations, Ge Xinfeng et al. [11] demonstrated that as water passes through the converging section, its velocity increases significantly, causing a rapid drop in absolute pressure at the nozzle exit, which ultimately leads to cavitation formation.

Overall, most studies on the jet mechanism of Pelton turbines have been conducted in low-altitude regions, while research on the impact of high-altitude environments on jet characteristics remains relatively limited. High-altitude regions are characterized by low atmospheric pressure and reduced air density. Conducting numerical simulations of the jet mechanism in Pelton turbines under high-altitude conditions can provide valuable insights and references for the future operation of such turbines in these environments.

2. Simulation Model Description

2.1. *Physical Model*

A hydropower station employs a vertical-axis, single-runner, six-nozzle Pelton turbine, with its main shaft directly coupled to the generator. The turbine is designed for an operating head of 595.0 m, with a maximum head of 619.8 m. This study focuses on the jet mechanism of the Pelton turbine. To obtain an accurate representation of the jet flow, an extended virtual flow domain is introduced beyond the original nozzle, approximating the continued development of the water flow after exiting the nozzle to ensure full flow development at the outlet [12].

A three-dimensional model of the turbine's jet mechanism is constructed using UG software. The model primarily consists of the jet domain, nozzle, needle, deflector, and air domain, as illustrated in Figure 1. The 3D structure of the nozzle is depicted in Figure 1, where the nozzle inlet diameter is 850 mm, and the outlet domain length is 1420 mm, ensuring full flow development. The needle has a cone angle of 50°, while the nozzle cone angle is 90°. The nozzle exit diameter measures 294 mm, and the maximum needle stroke is 259 mm.

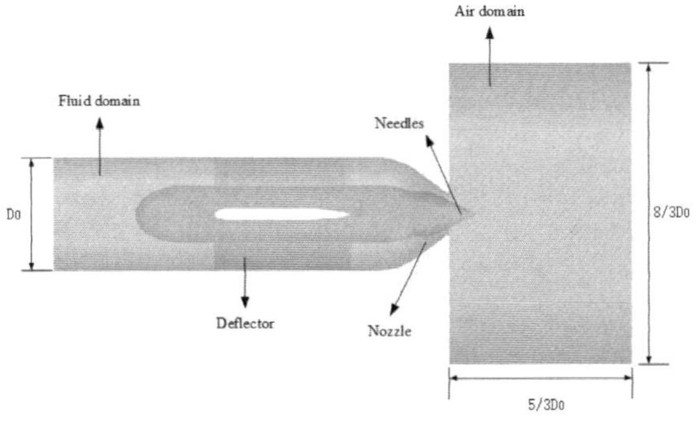

Figure 1. Pelton turbine jet mechanism model.

2.2. *VOF Model*

In multiphase flow calculations, the Volume of Fluid (VOF) model is employed to capture the interface between water and air, making it well-suited for this study. The volume fractions of water and air are defined as α_w and α_a, respectively. The governing equations for the two-phase flow are expressed as follows:

$$\alpha_w + \alpha_a = 1 \tag{1}$$

$$\frac{\partial \alpha_w}{\partial t} + u_i \frac{\partial \alpha_w}{\partial x_i} = 0 \tag{2}$$

$$\rho = \alpha_w \rho_w + (1 - \alpha_w)\rho_a \tag{3}$$

$$\mu = \alpha_w \mu_w + (1-\alpha_w)\mu_a \tag{4}$$

$$\frac{\partial \rho}{\partial t} + u_i \frac{\partial \rho}{\partial x_i} = 0 \tag{5}$$

$$\frac{\partial(\rho u_i)}{\partial t} + \frac{\partial(\rho u_i u_j)}{\partial x_j} = -\frac{\partial p}{\partial x_i} + \frac{\partial}{\partial x_j}\left[\mu(\frac{\partial u_i}{\partial x_j} + \frac{\partial u_j}{\partial x_i})\right] + \frac{\partial \tau_{ij}}{\partial x_j} \tag{6}$$

ρ_w and ρ_a are the densities of water and air, respectively, μ_w and μ_a denote their corresponding viscosity coefficients, and u_i and u_j are the velocity components in the x_i and x_j directions.

2.3. Turbulence Model

Cavitation phenomena are often accompanied by turbulence, making the selection of an appropriate turbulence model crucial to ensuring the accuracy of numerical simulations [13]. The Shear Stress Transport (SST) k-ω model is an improved formulation of the standard k-ω model. It is widely applied in free-stream regions near walls, demonstrating superior accuracy in predicting turbulence behavior. Compared to the standard k-ω model, the SST k-ω model provides a more precise estimation of turbulent viscosity. Given these advantages, this study employs the SST k-ω model. The governing equations for the **SST k-ω turbulence model** are formulated as follows [14]:

$$\mu = \frac{\rho \mathrm{k}}{\omega} \tag{7}$$

$$\frac{\partial}{\partial \mathrm{t}}(\rho \mathrm{k}) + \frac{\partial}{\partial \mathrm{x_i}}(\rho \mathrm{k u_i}) = \frac{\partial}{\partial \mathrm{x_i}}\left(\Gamma_\mathrm{k} \frac{\partial \mathrm{k}}{\partial \mathrm{x_i}}\right) + G_\mathrm{k} - Y_\mathrm{k} + S_\mathrm{k} \tag{8}$$

$$\frac{\partial}{\partial \mathrm{t}}(\rho \omega) + \frac{\partial}{\partial \mathrm{x_i}}(\rho \omega \mathrm{u_i}) = \frac{\partial}{\partial \mathrm{x_i}}\left(\Gamma_\mathrm{k} \frac{\partial \mathrm{k}}{\partial \mathrm{x_i}}\right) + G_\omega - Y_\omega + S_\omega + D_\omega \tag{9}$$

G_k is the turbulence kinetic energy generation term, $G\omega$ corresponds to the production term in the ω equation, $\Gamma\omega$ and Γ_k denote the effective diffusion terms for ω and k, respectively, $Y\omega$ and Y_k represent the dissipation terms, and $S\omega$ and S_k are user-defined source terms.

2.4. Cavitation Model

The Schnerr-Sauer cavitation model treats the vapor-liquid mixture as a dispersion of numerous spherical vapor bubbles uniformly distributed within the liquid phase [15]. In this study, the Schnerr-Sauer cavitation model is employed, with its governing equations expressed as follows [16]:

$$\frac{\partial}{\partial \mathrm{t}}(\partial \rho_\upsilon) + \nabla\left(\partial \rho_\upsilon \vec{\upsilon_\upsilon}\right) = \frac{\mathrm{p}_\upsilon \rho_\upsilon}{\rho_\mathrm{l}} \frac{D\alpha}{Dt} \tag{10}$$

$\boldsymbol{\alpha}$ is the gas-phase volume fraction, $\boldsymbol{\rho_v}$ is the gas-phase density, $\mathbf{u_v}$ is the velocity of the gas phase, and $\boldsymbol{\rho_l}$ is the liquid-phase density.

3. Solver Setup and Mesh Generation

3.1. Boundary Condition Setup

The computational domain in this study includes the inlet region, guide flow region, jet mechanism, and outlet air domain. Given that the primary focus is on analyzing the cavitation characteristics of the jet mechanism under different needle openings and considering that each opening corresponds to a different flow rate, the inlet is set as a velocity inlet, while the outlet is defined as a pressure outlet with an outlet pressure value set to 0. The wall is treated as a no-slip wall, and standard wall functions are used in the near-wall regions. To solve for the unsteady flow of the air-water-vapor multiphase system, the VOF multiphase flow model, SST k-ω turbulence model, and cavitation model are selected. A pressure-based solver with a second-order accurate SIMPLE algorithm is employed, and the time step is set to 1×10^{-5} s. During the simulation, the volume fraction of the water phase and the vapor content are monitored in real-time. Convergence is determined when the monitoring curves stabilize. Considering that the study is conducted in a plateau environment, physical parameters are adjusted to reflect the conditions. Table 1 presents the relevant parameters for an altitude of 3000 m, ensuring that the simulation results are both accurate and reliable. These carefully set parameters provide a solid data foundation for the subsequent in-depth analysis of cavitation characteristics in the jet mechanism at different needle openings.

Table 1. Physical parameters of the plateau.

3000 m altitude environment	numerical value
Gas content in water	$3.483*10^{-4}$
Water density	999 kg/m^3
Water dynamic viscosity	$1.4284*10^{-3}$ pa • s
Atmospheric pressure	70587.5 pa
Air density	0.89 kg/m^3

3.2. Boundary Condition Setup and Mesh Independence Validation

In this study, ICEM CFD software is used to generate the mesh for the Pelton turbine's jet mechanism, employing a combination of structured and unstructured meshing techniques. This approach has been validated in previous studies [17]. Due to the influence of the guide flow region, the front half of the model uses unstructured meshing to accommodate the complex geometry and flow field. The rear half, which is the main focus of the study, is meshed using structured grids, with refinement applied to the needle tip and nozzle exit regions to improve computational accuracy at critical locations. The mesh distribution for the jet domain, nozzle, and needle is shown in Figure 2, with an interface connecting the structured and unstructured meshes. The final mesh consists of 2.33 million elements and 1.31 million nodes, providing a reliable basis for subsequent numerical simulations.

To ensure mesh independence, five different mesh sizes are selected, and the efficiency of the jet mechanism in the Pelton turbine is tested for mesh sensitivity, as shown in **Figure 3**. The results indicate that as the mesh size increases, the computed results stabilize. When the mesh contains **2.33 million elements**, the efficiency calculation differs by only **0.1%** compared to a mesh with **2.68 million elements**, while the larger mesh requires more computational time. Balancing computational accuracy and efficiency, a mesh size of **2.33 million elements** is chosen for subsequent calculations, ensuring both result accuracy and efficient use of resources in terms of cost and time.

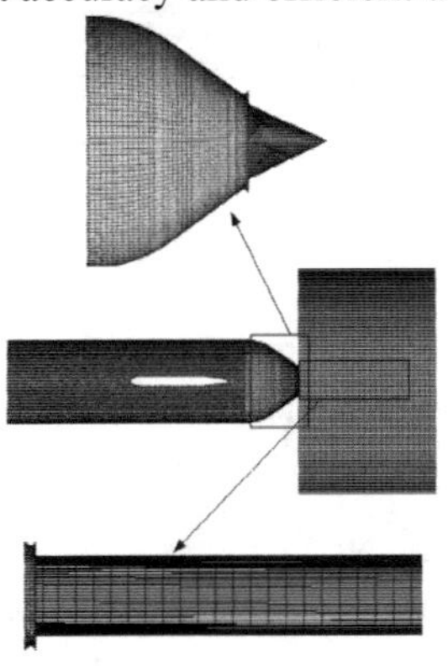

Figure 2. Structured and unstructured meshing.

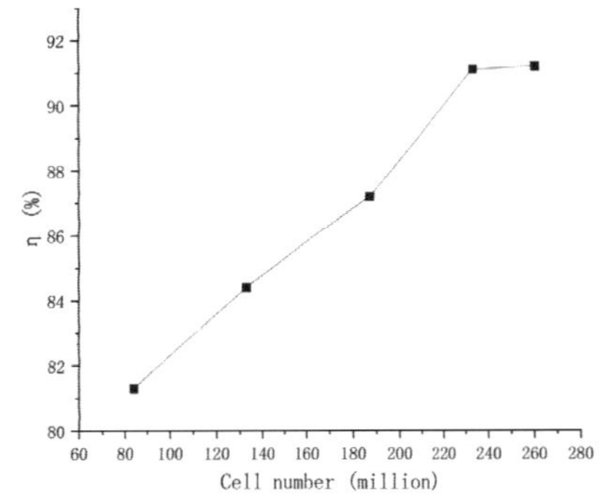

Figure 3. Mesh independency study.

4. Calculation Results and Analysis

This study, based on the Schnerr-Sauer cavitation dynamics model, conducts an in-depth investigation into the cavitation characteristics of the jet mechanism in a Pelton turbine operating in a high-altitude environment. Using cavitation fluid dynamics simulation techniques, the study primarily focuses on analyzing how changes in the needle stroke influence the cavitation evolution process in the jet system. The water turbine model used in this study has a maximum needle stroke of 259 mm.

To establish a unified and standardized research framework, the needle stroke is converted into the corresponding needle opening, with four different strokes selected: 51.8 mm, 103.6 mm, 155.4 mm, and 207.2 mm, which correspond to needle openings of 20%, 40%, 60%, and 80%, respectively. Under the rated head condition of H=595 m, the study systematically reveals the cavitation characteristics exhibited by the jet system under different openings in a high-altitude environment. This provides both theoretical and data support for further understanding the cavitation phenomena in the jet mechanism of Pelton turbines at high altitudes.

4.1. Flow Characteristics Analysis

To comprehensively elucidate the formation mechanism of cavitation in the jet mechanism under high-altitude conditions, it is necessary to systematically examine the dynamic characteristics of the internal fluid medium. By employing fluid dynamics analysis methods, the motion process of the medium within the jet system can be divided into four key stages. First, the fluid medium flows into the nozzle through the guide vanes. Then, as the medium passes through the conical adjustment component, a distinct flow pattern transition occurs. Following this, as the medium approaches the discharge port, the rapid contraction of the flow cross-section leads to an acceleration of the fluid. The final stage is characterized by turbulent diffusion after the water flow is released from the nozzle. A detailed study of these four stages helps to gain a more comprehensive understanding of the formation mechanism of cavitation phenomena in the jet mechanism under high-altitude conditions.

The phase distribution diagram of the fluid in the axial cross-section, shown in **Figure 4**, clearly illustrates the evolution pattern of the medium. As the jet moves forward, the gas-liquid two-phase flow gradually reaches a dynamic equilibrium. In this stable state, the jet mechanism is filled with water. At the turning point of the nozzle exit, small-scale water vapor is observed, which is especially pronounced when the needle opening is **20%**. In the nozzle exit region, a jet water column is present. At the upper and lower edges of this jet water column, there is a thin air-water interface.

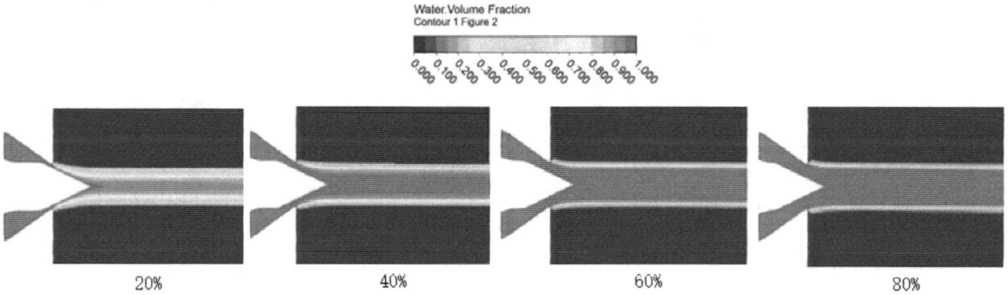

Figure 4. Volume fraction of water and air contour at different openings.

Figure 5 presents the velocity distribution within the jet mechanism. Using **80%** needle opening as an example for analysis, it can be observed that the flow is forced into two parts due to the effect of the guide vanes, with each part flowing toward the exit from either side of the guide vanes. A comparison reveals that the maximum velocity at the exit of the jet mechanism is significantly higher than the average inlet velocity, which indicates a notable acceleration in the velocity within the contraction section. After the water flow exits the nozzle, a very significant velocity gradient is observed near the air-water interface. Furthermore, the flow velocity at the center of the jet is lower than that in the surrounding regions, a phenomenon known as "velocity deficit." Studies have shown that the "velocity deficit" phenomenon is typically caused by the boundary layer near the tip of the needle. Notably, the velocity deficit region always exists at the tip of the needle, and this pattern is generally applicable to the exit section of all nozzle structures.

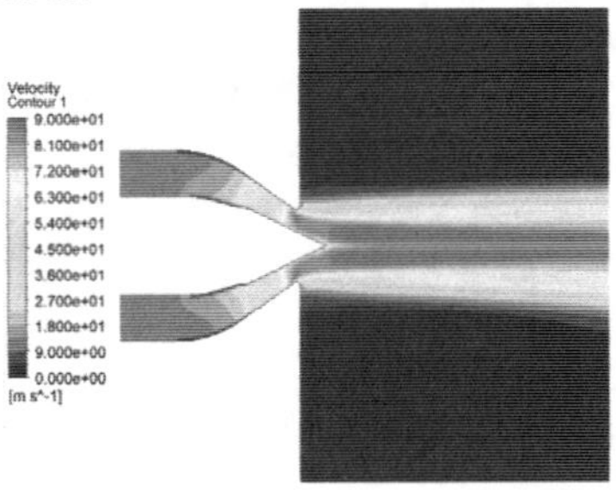

Figure 5. Velocity contours at 80% openings.

During the operation of hydraulic machinery, if the minimum pressure in a low-pressure region of the fluid drops below the critical pressure required for gas vaporization, cavitation will occur in that region. To investigate this phenomenon in detail, the pressure distribution within the jet mechanism is first analyzed. Figure 6 illustrates the pressure distribution in the jet mechanism for different needle opening settings. It is observed that during the fluid contraction process, the pressure decreases rapidly. In particular, a distinct low-pressure region exists at the nozzle outlet, and the pressure variation in this area is closely related to the occurrence of cavitation.

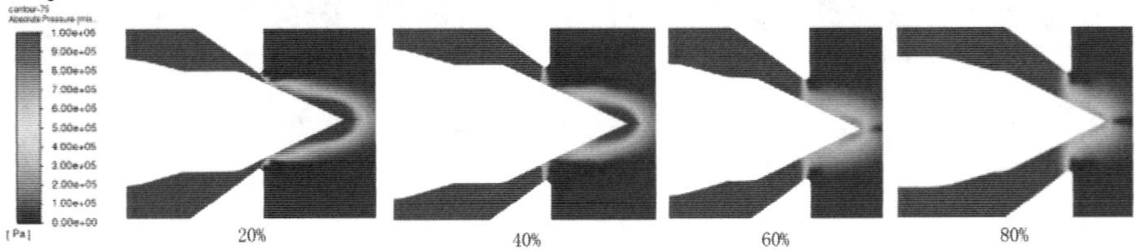

Figure 6. Pressure contours at different openings.

Figure 7(a) shows the pressure distribution curve along the axial direction of the nozzle surface near the exit. The curve indicates that after an axial distance of -0.02m, the pressure curves for the four different needle-opening settings nearly overlap and become consistent. Figure 7(b) displays the pressure distribution on the nozzle surface at the outlet region. As the fluid reaches the nozzle exit, the pressure sharply drops until it reaches the saturation vapor pressure of water. At this point, cavitation is induced. This demonstrates that there is a close correlation between pressure variations and the occurrence of cavitation. Specific pressure conditions, especially when the pressure at the nozzle outlet decreases to the saturation vapor pressure, serve as a critical trigger for cavitation formation.

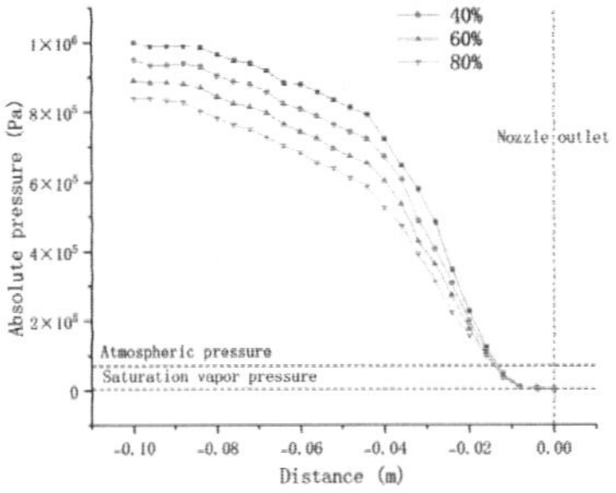

(a)Overall pressure curve

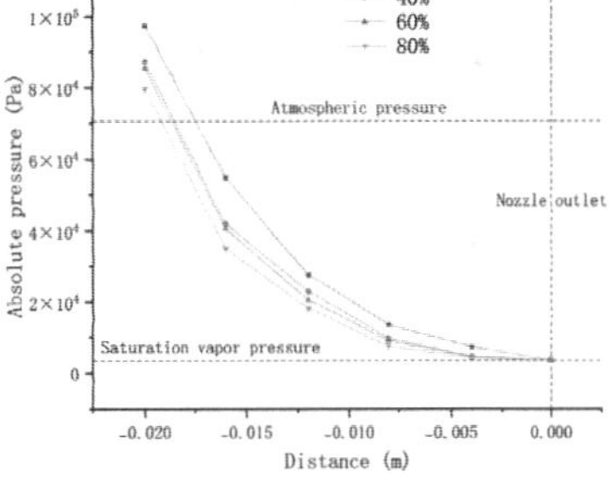

(b) Partial pressure curve

Figure 7. Pressure curve of nozzle surface.

4.2. Cavitation Characteristics Analysis

Figure 8 presents the gas-phase distribution characteristics on the nozzle surface and interior underrated head conditions for different needle opening settings. The analysis reveals that the gas phase mainly accumulates at the outlet of the flow passage. According to numerical simulations, at a 40% needle opening, the gas-phase volume fraction on the nozzle surface reaches its highest value, and the diffusion range of the gas phase along the axial direction is significantly larger compared to other operating conditions. This phenomenon indicates that during the needle's movement, the maximum cavitation occurrence rate is approximately near the 40% opening. In terms of spatial distribution, the cavitation region at the outlet section exhibits a typical ring-like structure. In all operating conditions, the gas-phase volume fraction remains uniformly distributed along the circumference, reflecting a consistent pattern across the different cases.

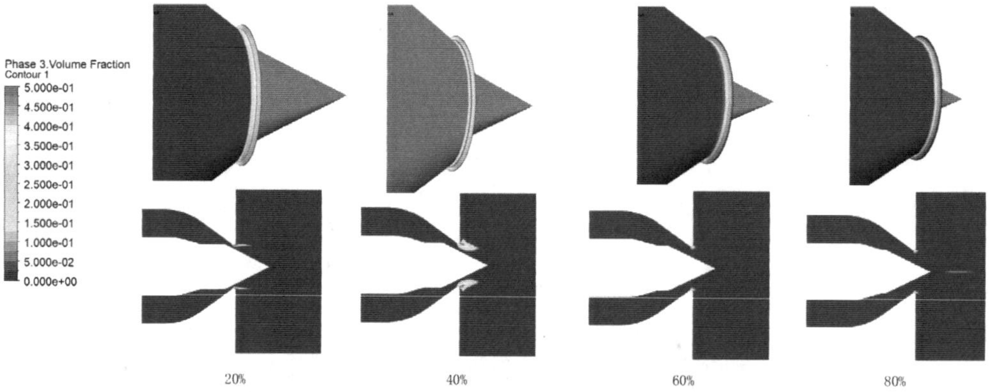

Figure 8. Volume fraction of vapor in nozzle surface at different openings.

Figure 9 illustrates the mass transfer rate on the nozzle surface under different needle opening settings. In the cavitation model, the mass transfer rate refers to the rate of mass transfer between the liquid phase and the gas phase, which is critical for studying cavitation phenomena as it quantifies the phase transition process. During cavitation, when the pressure drops below the liquid's saturation vapor pressure, the liquid vaporizes to form bubbles; when the pressure recovers, the bubbles collapse, and the gas phase transitions back to the liquid phase. The mass transfer rate describes the speed of this transition.

From the data in Figure 9, it can be observed that at a 40% opening, the mass transfer range is the largest, and the gas-liquid exchange rate is the fastest. Additionally, combining this with the analysis from Figure 8, it is evident that at a 40% opening, cavitation is most severe at the nozzle outlet. This indicates that at this opening, the gas-liquid conversion is intense, and cavitation significantly affects the nozzle performance. Therefore, attention should be focused on this opening during practical engineering and equipment design.

Figure 9. Mass transfer rate in nozzle surface at different openings.

5. Conclusion

In the field of hydraulic engineering and fluid machinery, the performance of the jetting mechanism in an impulse turbine is crucial for the overall operational efficiency of the turbine. This study delves into the impact of different needle openings on the jetting mechanism's flow field under the rated design head of a hydropower plant in a plateau environment. Through numerical simulation and experimental measurements, the internal flow field of the jetting mechanism was analyzed, with a focus on the distribution patterns of parameters such as pressure fields and gas phase volume fraction. The key conclusions are as follows:

(1) As the water flows through the contraction section of the nozzle, its velocity increases sharply due to the reduction in cross-sectional area, forming a high-speed jet at the nozzle outlet. This high-speed flow causes a rapid drop in the absolute pressure at the nozzle outlet. When the pressure decreases to the saturation vapor pressure, the gas nuclei expand, creating conditions for cavitation.

(2) Various analytical methods were employed to reveal the variation trend of absolute pressure at different needle openings. Taking the velocity distribution at 80% opening as an example, it was found that the center of the needle experiences a "velocity deficit" due to flow obstruction. At the nozzle outlet, a low-pressure zone is formed due to high-speed jetting and the entrainment of air, which is closely related to cavitation.

(3) Through visualization techniques and numerical calculations, it was found that the gas phase concentrates at the nozzle outlet, with the cavitation zone forming a ring shape. The gas phase volume fraction is uniformly distributed along the circumference in all operating conditions. By combining the mass transfer rate model with simulation and experimental data, it is evident that cavitation is most pronounced at 40% opening, with the largest cavitation area, which has the greatest impact on the performance of the injection mechanism.

Acknowledgments

This work was financially supported by the Xizang Agricultural and Animal Husbandry College Graduate Student Innovation Program (YJS2024-47); High-head Large-capacity Hydroelectric Unit Development Key Technologies (XZ202201ZD0003G04).

References

[1] LUO X Q, ZHU G J, FENG J J. Technological progress and develop- ment trend of water turbine[J]. Journal of Hydropower Generation, 2020, 39(8): 1-18.

[2] Han L, Duan X L, Gong R Z, et al. Physic of secondary flow phenomenon in distributor and bifurcation pipe of Pelton turbine [J]. Renewable energy, 2019, 131: 159-167.

[3] ZOPPE B, PELLONE C, MAITRE T ,et al. Flow analysis inside a Pelton turbine bucket [J].Transactions of the ASME: Journal of Turbomachinery, 2006, 128(3): 500-511.

[4] BENZON D, ŽIDONIS A, PANAGIOTOPOULOS A, et al. Impulse turbine injector design improvement using computational fluid dynamics [J]. Transactions of the ASME: Journal of Fluids Engineering, 2015, 137(4): 041106.

[5] BENZON D, ŽIDONIS A, PANAGIOTOPOULOS A, et al. Numerical investigation of the spear valve configuration on the performance of Pelton and Turgo turbine injectors and runners [J]. Transactions of the ASME: Journal of Fluids Engineering, 2015, 137(11): 111201.

[6] PETLEY S, PANAGIOTOPOULOS A, BENZON D S, et al. Investigating the influence of the jet from three nozzle and spear design configurations on Pelton runner performance by numerical simulation [J]. IOP Conference Series: Earth and Environmental Science, 2019, 240: 022004.

[7] HUANG J W, GE X F, CHU D D, et al. Research on the effect of needle eccentricity on the jet flow characteristics [J]. Frontiers in Energy Research, 2022, 10: 882747.

[8] NEDELCU D, COJOCARU V, AVASILOAIE R C. Numerical investigation of nozzle jet flow in a Pelton microturbine [J]. Machines, 2021,9: 158.

[9] ZENG C J, XIAO Y X, XU W, et al. Numerical analysis of Pelton nozzle jet flow behavior considering elbow pipe [J]. IOP Conference Series: Earth and Environmental Science, 2016, 49: 022005.

[10] ZENG C, XIAO Y, LUO Y, et al. Hydraulic performance prediction of a prototype four-nozzle Pelton turbine by entire flow path simulation [J]. Renewable Energy, 2018, 125: 270-282.

[11] GE X F, MENG H, SUN J, et al. Simulation of cavitation characteristics of pelton turbine injector [J]. China Rural Water and Hydropower, 2023(7): 229-235.

[12] CHEN C X. Numerical simulation of water distribution ring pipe and nozzle of large water bucket turbine [D]. Nanjing: Hohai University,2006.

[13] LUO X W, JI B, TSUJIMOTO Y. A review of cavitation in hydraulic machinery[J]. Journal of Hydrodynamics, 2016,28(3): 335-358.
[14] MENTER F R. Two-equation eddy-viscosity turbulence models for engineering applications[J]. AIAA Journal, 1994,32(8) :1598-1605.
[15] XUE R, ZHANG M, XU Z J,et al. Comparative study of the differ- ent cavitation models[J].Northwest Hydropower,2014,16(2):85-89.
[16] WU Z B,WANG Y Y,ZHANG S, et al. Numerical simulation study of the cavitation performance of different shape nozzles [J]. Fluid Machinery,2020,48(4):36-41.
[17] GUO B, XIAO Y X, RAI A R, et al. Sediment-laden flow and ero- sion modeling in a Pelton turbine injector[J]. Renewable Energy, 2015, 162(4):30-42.

Evaluation method of adjustable capacity of air conditioning load based on user comfort

Wenyuan Xu [1], Liwei Duan [2*], Yuan Tao [1], Siqi Qiu [1], Zeng Cheng [1], Chongjian Zhang [1], Ji Xu [1]

[1] Hubei Huazhong Electric Power Technology Development Co., Ltd., Wuhan 430079, China
[2] Wuhan Longde Control Technology Co., Ltd, Wuhan 430073, China

[*] Corresponding author's e-mail: 1551538609@qq.com

Abstract: The air-conditioning load has become a high-quality flexible load resource due to its environmental thermal energy storage characteristics, controllable characteristics, and large load ratio. The evaluation of the adjustable capacity of the air conditioning load is a key technical problem to converting it into actual adjustable resources. Firstly, the air conditioning load is modeled based on the equivalent thermal parameter (ETP) model, and its adjustable capacity mechanism is analyzed. Then, the predicted mean vote (PMV) index is used to quantify the influence of temperature on human comfort. The user's thermal comfort model and comfort index are constructed. The adjustable capacity evaluation method of air conditioning load with response time and response power as evaluation indexes are proposed. Finally, different evaluation methods are used for comparative analysis. The example shows that the method proposed in this paper is superior, and the adjustable range of air conditioning load power cluster capacity is obtained.

1. Introduction

At present, the proportion of new energy on the power generation side of the power system is increasing. The randomness and uncertainty of new energy output seriously affect the safety and stable operation of the power grid, and the power system is facing a more complex and changeable situation [1]. Specifically, the flexibility of the power system is reduced, and the ability to ensure the balance of supply and demand is weakened [2]. In the case of decreasing adjustable resources on the power generation side, the mining of massive adjustable load resources on the load side has gradually attracted the attention of technicians and researchers [3].

Demand response (DR) technology provides a good solution to deal with the uncertainty of energy output, and improve the economy and environmental protection of power grid systems [4]. Through DR technology, the load power consumption behavior can be flexibly regulated, and a certain adjustable capacity can be provided equivalently, thereby enriching the schedulable resources of the power grid. On the load side, the participation of air conditioning load in DR has great advantages such as large capacity and virtual energy storage characteristics [5]. Traditional air conditioning load assessment methods focus on static calculation, ignoring the actual needs of dynamic changes and user comfort [6]. That is, the size of its adjustable capacity is not considered on the premise of response time, and it is not feasible in actual operation. Only considering the adjustment under the set value of the user's temperature, its adjustable ability is limited. In addition, the traditional evaluation method has a complex modeling process, and is difficult to solve the parameters. It is difficult to accurately express its nonlinear relationship, resulting in low accuracy of the evaluation results.

To address the issues mentioned, this paper evaluates the air conditioning load's response time and the amount of regulated power as key metrics. By considering user comfort constraints, it determines the corresponding temperature operation range. Within this range, the paper adjusts the temperature setpoint and controls accordingly. Then, based on the perspective of user comfort, considering the response strategy of the air conditioning load, the adjustable capacity of the air conditioning load cluster is evaluated. Finally, the superiority of the method is illustrated by an example, and the adjustable capacity range of the air conditioning load is obtained.

2. Air conditioning load adjustable capacity assessment

2.1. Air conditioning load model

Aiming at the modeling of air conditioning load with start-stop control mode, the first-order equivalent thermal parameter (ETP) model commonly used in engineering is adopted. It should be emphasized that the ETP model is mainly used for air conditioning modeling of residents and small and medium-sized commercial buildings. The equivalent heat capacity C and equivalent thermal resistance of the room are used to describe the thermal characteristic parameters, as shown in Figure 1:

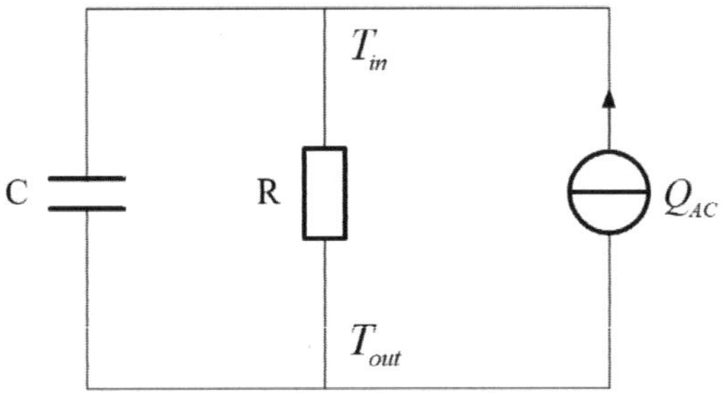

Figure 1. Single air conditioning first-order ETP model.

By simplifying it into a first-order equation, the relationship between room temperature T_{in}^t and cooling capacity Q_{AC}^t is obtained:

$$\frac{T_{\text{in}}^t - T_{\text{out}}^t}{R} + C\frac{dT_{\text{in}}^t}{dt} = -Q_{AC}^t \tag{1}$$

The recursive relationship is used for transformation and arrangement to obtain:

$$T_{\text{in}}^{t+1} = T_{\text{out}}^{t+1} - RQ_{AC}^{t+1} - (T_{\text{out}}^{t+1} - T_{\text{in}}^t - RQ_{AC}^{t+1}) \cdot e^{-\frac{\Delta t}{RC}} \tag{2}$$

The following is the relationship function between cooling capacity and temperature:

$$Q_{AC}^t = \begin{cases} K^t \eta P & T_{in}^t < T_{down}, T_{in}^t > T_{up} \\ Q_{AC}^{(t-\Delta t)} & T_{in}^t \in \left[T_{down}, T_{up}\right] \end{cases} \tag{3}$$

where T_{out}^t is the outdoor temperature at time t; Δt is the time interval;the value of C, R coefficient is related to the room volume, air-specific heat capacity, and thermal conductivity. K^t is the start-stop state of the air conditioning equipment at time t; η is the energy efficiency ratio of air conditioning; P is the rated power of air conditioning; $\left[T_{down}, T_{up}\right]$ is the range of room temperature change in the stable operation of the air conditioner. The relationship between the start-stop state of the air conditioner, the set temperature T_{set}, and the set temperature boundary is:

$$K^t = \begin{cases} 0, T_{in}^t \geqslant T_{up} = T_{set} + 0.5\delta \\ 1, T_{in}^t \leqslant T_{down} = T_{set} - 0.5\delta \\ K^{t-1}, T_{down} < T_{in}^t < T_{up} \end{cases} \tag{4}$$

where δ is the temperature control threshold of the air conditioner.

2.2. *User thermal comfort PMV index*

The user characterizes the ambient temperature with thermal comfort, and predicts that the average vote PMV as a comprehensive indicator includes indoor temperature, human clothing, human activity status, humidity, and other parameters that affect human comfort. The PMV value was quantified between [-3, 3], and seven grade indicators were used to indicate human sensation. The PMV values of −3, −2, and −1 represent cold, cool, and slightly cool, respectively. The PMV values of 3, 2, and 1 represent hot, warm, and slightly warm, respectively. When the PMV value is 0, the human body is the most comfortable. The calculation method of PMV value I_{PMV} is as follows [7]:

$$\begin{aligned} I_{PMV} &= \left(0.303e^{-0.036M} + 0.028\right)\{M - \sigma - 3.05\times10^{-3}\times[5733 - 6.99(M-\sigma) - P_a] - \\ &0.42[(M-\sigma) - 58.15] - 1.7\times10^{-5}\times M(5867 - P) - 0.0014M(34 - T_{in}) - \\ &3.96\times10^{-8} f_{cl}[(T_{cl}+243)^4 - (T_r + 273)^4] - f_{cl}h_c(T_{cl} - T_{in})\} \end{aligned} \tag{5}$$

where M is the metabolic rate; σ is the efficiency of the daily activities of the human body; P_a is the water vapor pressure in the air; f_{cl} is the human body dressing coefficient; T_{cl} is the temperature of the clothing surface; T_r is the radiation temperature; h_c is the heat transfer coefficient.

Since the temperature regulation will go through an interval, how to calculate the PMV value of an interval becomes a difficult problem. We define the average comfort index f_{AC} as:

$$f_{AC} = \frac{1}{D}\sum_{m=1}^{D}\left|I_{PMV,m}\right| \tag{6}$$

where D is the duration of experience; $I_{PMV,m}$ is the PMV value at the m th moment.

2.3. *A new evaluation method considering human body temperature comfort*

Considering the controlled duration of demand response under temperature comfort Δt_{last}, we get:

$$\Delta t_{last} = RC\cdot\ln\left(\left(T_{out} - Q_{AC,\min}R - T_{in}\right)/\left(T_{out} - Q_{AC,\min}R - T_{\max}\right)\right) \tag{7}$$

In the cluster, the controlled time of air conditioners with different thermal parameters will be different, which brings inconvenience to the evaluation of the cluster's adjustable ability. Therefore, the controlled time Δt_{last} of air conditioner clusters with different equivalent parameters is unified.

Suppose that the outdoor temperature of the j th system in the cluster is $T_{out,j}$, the room temperature is stable at the set value $T_{in,set,j}^{t-1}$ before participating in the regulation, and the temperature comfort zone $[T_{\min,j}, T_{\max,j}]$ of the system. If the room temperature rises to the maximum value after the control is completed, the refrigeration capacity under the controlled period can be obtained by the following formula:

$$Q_{AC,j}^{t,down} = \left(T_{\max,j} - T_{in,set,j}^{t}\right)/\left(R_j\left(1 - e^{-\Delta t_{last}/R_jC_j}\right)\right) - \left(T_{out,j} - T_{in,set,j}^{t}\right)/R_j \tag{8}$$

From this, the operating frequency $f_{AC,j}^{t}$ and power $P_{AC,j}^{t}$ can be calculated as:

$$\begin{aligned} f_{AC,j}^{t} &= \left(Q_{AC,j}^{t,down} - \mu_{q,j}\right)/k_{q,j} \\ P_{AC,j}^{t} &= k_{p,j}\cdot f_{AC,j}^{t} + \mu_{p,j} \end{aligned} \tag{9}$$

where $\mu_{q,j}$, $k_{q,j}$ are the curve fitting coefficients of working frequency and cooling capacity; $\mu_{p,j}$, $k_{p,j}$ is the working frequency and power consumption. The load participating in the response needs to be in a steady state at time $t-1$. At this time, the load's downward adjustability $\Delta P_{AC}^{t,down}$ is expressed as:

$$P_{AC,set} = k_p \cdot \left(\left(T_{out} - T_{set} \right) / R - \mu_q \right) / k_q + \mu_p$$
$$\Delta P_{AC}^{t,down} = P_{AC,j}^{t-1} - P_{AC,j}^{t} = P_{AC,set} - P_{AC,j}^{t} \quad (10)$$

where $P_{AC,set}$ is the stable operating power; $P_{AC,j}^{t}$ is the power corresponding to the set frequency. Similarly, when the jth system is controlled after the end of the time, the room temperature reaches $T_{\min,j}$, and the cooling capacity of the period can be calculated:

$$Q_{AC,j}^{t,up} = \left(T_{in,set,j}^{t} - T_{out,j}^{t} \right) / R_j - \left(T_{\min,j} - T_{in,set,j}^{t} \right) \left(R_j \left(1 - e^{-\Delta t_{last} / R_j C_j} \right) \right) \quad (11)$$

According to the operating frequency and power under this condition, the upward adjustment ability is obtained as follows:

$$\Delta P_{AC}^{t,up} = P_{AC,j}^{t} - P_{AC,j}^{t-1} \quad (12)$$

3. Example analysis

According to the new air conditioning load adjustable capacity calculation method proposed in this paper, a variable frequency air conditioning aggregation model is constructed in Matlab, and 1000 air conditioners in a certain area are set to sign an incentive agreement. The number of air conditioners in each group and the distribution of each parameter are shown in Table 1.

Table 1. The controllable number of air conditioning groups and the distribution of each parameter.

Location	Air conditioning group	Number	$R\left(kJ / {}^\circ C\right)$	$C\left({}^\circ C / kW\right)$	Rated power (kW/unit)
Residence	Group1	400	9.2	57.2	2
	Group2	300	5.4	91.9	2
Commerce	Group3	200	3.2	174.1	2
	Group4	100	2.1	314.3	2

The temperature set by the user is set to $24\,{}^\circ C$, and the comfort range is [$22\,{}^\circ C$, $26\,{}^\circ C$]. The 24-hour outdoor temperature was based on the data of [8]. By obtaining the performance parameters of each air conditioner, including the maximum and minimum cooling capacity and the corresponding cooling power, the overall upward and downward adjustable capacity of the cluster can be further obtained. Combined with the stable operating power of each adjustable load, the operating power range of each period of the cluster can be further known, as shown in Figure 2.

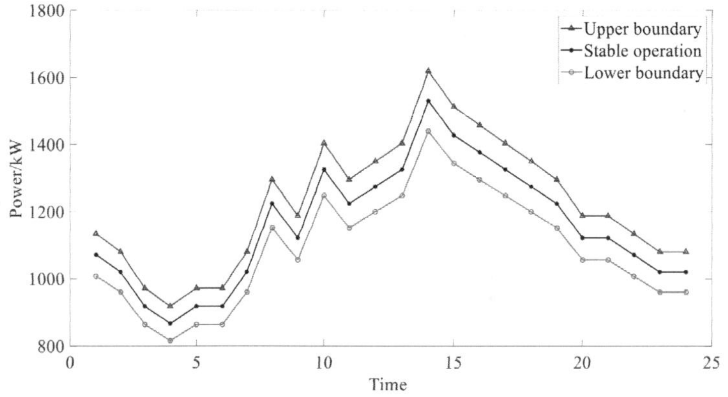

Figure 2. Cluster operation power range.

It can be seen from Figure 2 that the expected electric power of the cluster is the largest at time 14, which is 1530kW, and the adjustable range is also the largest, which is [1620kW, 1440kW]. At 24 hours, the minimum expected power of the cluster is 1020kW, and the adjustable range is also the smallest, which is [960kW, 1080kW]. Finally, the traditional method without considering the user comfort, the traditional method considering comfort, and the method of this paper are compared. The control period is 15 min in a certain response time. The adjustable capacity of air conditioning load under different methods is shown in Table 2.

Table 2. Comparison of air conditioning adjustable capacity under three evaluation methods.

Appraisal procedure	Pre-response power	Post-response power	Evaluation index	Evaluation result
Traditional methods without considering comfort	1326kW	978kW	Remaining capacity	348kW
			Controlled duration	No limitation
Consider the traditional method of comfort.		978kW	Remaining capacity	348kW
			Controlled duration	4min
A new method considering comfort		1248kW	Remaining capacity	78kW
			Controlled duration	15min

It can be seen from Table 2 that under different evaluation methods, the adjustable capacity provided by the air conditioning load varies greatly. Under the traditional evaluation method without considering the user's comfort, the operating power is 978 kW, and the maximum adjustable capacity is 348 kW. However, there is no limit on the duration of air conditioning control in this scenario. In the actual control process, the indoor temperature will exceed the temperature comfort range, which seriously affects the user experience. Under the traditional evaluation method considering user comfort, air conditioning can also provide an adjustable capacity of 348 kW, but the allowable regulation time under the premise of ensuring user comfort is only 4 minutes, which makes it difficult to achieve large-scale regulation of air conditioning. Under the new evaluation method considering user comfort, the operating power of the air conditioner at the end of the 15-minute control period is 1248 kW, with an adjustable capacity of 78 kW. Compared with the first two methods, although the controllability of air conditioning under this method is reduced, it provides the possibility for large-scale regulation of air conditioning under different working conditions.

4. Conclusion

In this paper, an evaluation method of air conditioning load adjustability considering user comfort is proposed, which not only provides certain adjustability under the premise of ensuring user comfort but also avoids the problem that the controllable duration of each air conditioner is different due to factors such as parameters and working conditions, which provides method support for the calculation of the adjustable capacity of the cluster. Taking the response time and response power as the evaluation indexes, different evaluation methods are used for comparative analysis. The example shows that the method proposed in this paper is superior, and the adjustable range of air conditioning load power cluster capacity is obtained.

Acknowledgments

This work was financially supported by the State Grid Hubei Electric Power Co., Ltd.

Journal of Physics: Conference Series **3011** (2025) 012033 doi:10.1088/1742-6596/3011/1/012033

References

[1] ZHANG Chengming, LIU Lu, CHENG Haozhon, et al. (2022) Review and prospect of power system planning and operation optimization considering frequency safety [J]. Power System Technology, 46(1): 250-264.
[2] CHEN Guoping, LIANG Zhifeng, DONG Yu. (2020) Analysis and reflection on the marketization construction of electric power with Chinese characteristics based on energy transformation [J]. Proceedings of the CSEE, 40(2): 369-378.
[3] NING Jian, WU Jiping, JIANG Changming, et al. (2022) Optimal control strategy of peak and frequency regulation for adjustable Automation of Electric Power Systems, 46(15): 11-19.
[4] Abedinia O, Bagheri M. (2021) Power distribution optimization based on demand respond with improved multi-objective algorithm in power system planning [J]. Energies, 14(10): 2961.
[5] Hou Y, Zhao L, Jiang H, et al. (2024) Two-Layer Control Framework and Aggregation Response Potential Evaluation of Air Conditioning Load Considering Multiple Factors [J]. IEEE Access.
[6] Honghu Chen. (2023) Optimal control of large-scale inverter air-conditioning cluster considering uncertainty of new energy output [D]. Southwest University of Science and Technology.
[7] ZOU Yunyang, YANG Li, FENG Li, et al. (2017) Coordinated heat and power dispatch of microgrid considering two-dimensional controllability of heat loads [J]. Automation of Electric Power Systems, 41(6): 13-19.
[8] China New Network Sichuan. (2022) Sichuan: continuous high temperature 'baking' power inspection department to ensure people's livelihood electricity [EB/OL].

Optimization of low voltage ride-through control parameters for direct-drive wind turbines considering active power deficit and voltage dip area

Yixuan Chen [1], Shuo Feng [2], Benfeng Gao [3], Xue Shen [1], Shumin Chen [1], Guangzeng You [1] and Hao Yang [2,*]

[1]Power grid planning and construction research center Yunnan power grid company Yunnan. China

[2]Northeast Electric Power University, Jilin 132012, Jilin Province, China

[3]North China Electric Power University, Baoding 071003, Hebei Province, China

[*]Corresponding author's e-mail: shuo_feng2024@163.com

Abstract. In high-penetration renewable energy systems, faults cause voltage drops at wind turbine connection points, triggering a large number of turbines to enter Low Voltage Ride Through (LVRT) mode, resulting in a significant active power loss and impacting frequency stability. This paper, based on China's wind power grid connection standards, analyzes the active and reactive power control strategies during the LVRT and recovery phases. It examines how wind turbine LVRT affects voltage stability and active power deficits. Through simulations of different control parameters, we identify the key factors influencing voltage and power deficits at the connection point. Finally, we optimize these parameters during the LVRT phase using Particle Swarm Optimization (PSO) to minimize the combined active power deficit and voltage dip areas.

1. Introduction

By the end of 2024, global wind power capacity is projected to reach 490 GW, a 19.2% year-on-year increase. Unlike traditional thermal power, wind and solar units rely heavily on power electronic devices, leading to distinct transient behaviors during low-voltage ride-through (LVRT) events [1-2]. These differences increase grid variability and uncertainty, complicating transient stability assessment and obscuring renewable energy control impacts on system stability [3]. In high-penetration renewable systems, faults trigger widespread LVRT activation in wind turbines, causing significant active power loss and threatening frequency stability [4]. Despite rapid renewable adoption in regions like Europe and the U.S., unresolved power deficit issues during faults have caused major blackouts, such as the 2016 South Australia outage (80% asynchronous generation), where massive LVRT activation led to system separation due to frequency collapse [5]. Understanding the stability impacts of wind turbine LVRT behavior is therefore critical for transitioning to renewable-dominated power systems.

Kawabe et al. [6-7] studied the impact of photovoltaic LVRT control on the recovery ability and stability of the grid connection point voltage after fault clearance. In grids with high penetration of wind power, the LVRT behavior of wind turbines also affects the recovery ability of the connection point voltage, potentially causing delayed voltage recovery. This phenomenon extends the duration of wind turbines in the LVRT phase, leading to more active power loss, which in turn exacerbates the frequency drop in the grid. Weise et al. [8-9] analyzed the frequency drop issue caused by the active power recovery slope after wind turbines exit the LVRT process.

Existing studies mainly focus on voltage or frequency stability or specific stages of the LVRT process, lacking a comprehensive analysis of voltage-frequency coupling. This paper, based on China's wind power grid connection standards, analyzes active/reactive power control strategies during the

LVRT and recovery phases of wind turbines. It investigates the impact of wind turbine LVRT on system voltage and frequency stability. Through simulations, the effects of control parameters on grid voltage and active power deficit at the connection point are examined, identifying key control parameters. Finally, the PSO algorithm is used to optimize these parameters during the LVRT process, aiming to minimize the combined areas of active power deficit and voltage drop.

2. Direct-drive permanent magnet wind turbine LVRT control

2.1. Wind turbine LVRT standard

Li et al. studied the typical LVRT standard of wind turbines[10-12]. China's national standard GB/T 19963-2011 'Technical Regulations for Wind Power Plant Connection to Power Systems' also outlines specific requirements for LVRT. Specifically, when the grid connection point voltage of the wind power plant drops to 20% of the nominal voltage, the wind turbines within the plant should ensure continuous operation without disconnection for 625 ms. Alternatively, if the grid connection point voltage recovers to 90% of the nominal voltage within 2 seconds after the drop, the wind turbines should ensure continuous operation without disconnection, as shown in Figure 1.

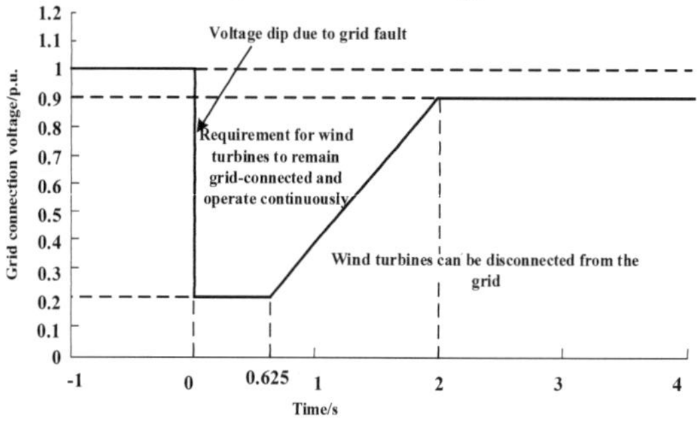

Figure 1. Low Voltage Ride Through (LVRT) requirements for wind farms.

2.2. Direct-drive wind turbine model

The direct-drive wind turbine model adopted by PSD-BPA software consists of multiple modules, and the overall block diagram is shown in Figure 2.

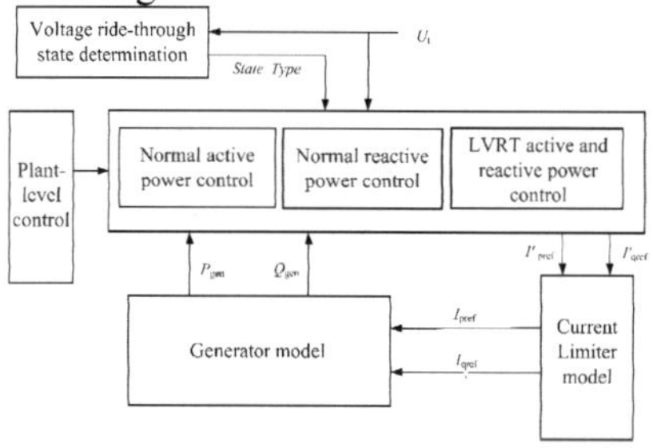

Figure 2. Block diagram of the direct-drive wind turbine.

P_{gen}、Q_{gen} represent the active and reactive power output of the wind turbine, I'_{pref} and I'_{qref} are the active and reactive current reference values from the controller; I_{pref} and I_{qref} are the active and reactive current reference values after limiting; U_t is the terminal voltage; *State_Type* represents the operational status of the turbine. The core modules are the LVRT active/reactive power control module and the current limitation module. The control strategies for the two modules are given in detail below.

2.2.1. LVRT active/reactive power control strategy. During the fault ride-through process, the active power response characteristics of the wind turbine are divided into three stages, which are simulated using a three-stage approach, as shown in Figure 3. Control is applied to each of the three stages to achieve LVRT and active power recovery. It should be noted that the first two stages are controlled using specified active current percentages, while the third stage follows a specified active power recovery rate for the ramp-up.

The reactive power response of wind turbines is divided into three stages, with control applied only in the first stage. In the other two stages, no additional control is used, and the reactive current returns to its initial value after LVRT ends. The active/reactive current control process is shown in Figure 3.

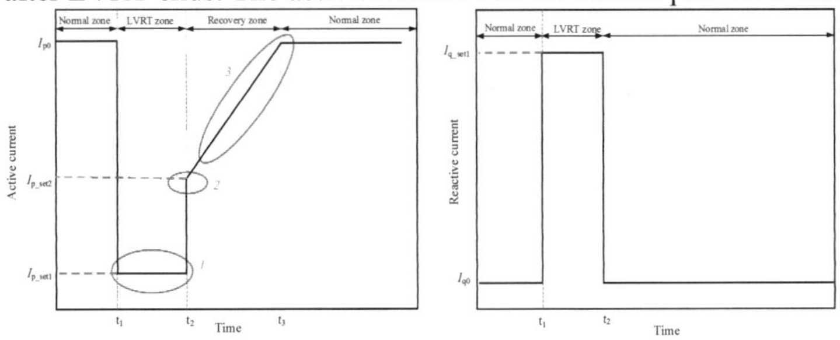

Figure 3. Active/reactive current control process.

I_{p0} is the initial active current, I_{q0} is the initial reactive current, and I_{q0} is zero during normal operation. I_{p_set1} is the active current setpoint during LVRT, I_{p_set2} is the specified initial value for active power ramp-up, and I_{q_set1} is the reactive current setpoint during LVRT. 0~t_1 represents the normal operation region, t_1~t_2 is the LVRT region, and t_2~t_3 is the active power recovery region.

(a) Stage 1: Low Voltage Ride-Through (LVRT) Stage. During LVRT, the renewable energy units adopt a specified current control method, with the active and reactive currents as follows:

$$\begin{cases} I_{p\_set1}=k_1 I_{p0} & t_1 \leq t < t_2 \\ I_{q\_set1}=k_q(U_{in} - U_t) + I_{q0} & t_1 \leq t < t_2 \end{cases} \tag{1}$$

where k_1 is the active current percentage during LVRT, k_q is the reactive current support coefficient, U_t is the wind turbine terminal voltage, and U_{in} is the LVRT entry threshold, typically set to 0.9 p.u..

(b) Stage 2: Active Power Recovery Initial Value Setting. After the LVRT stage ends, active power starts to recover. The active current control method at the recovery starting point is based on the initial active current percentage. The reactive current control method is without additional control.

$$I_{p\_set2}=k_2 I_{p0} \qquad t = t_2 \tag{2}$$

where k_2 is the coefficient for the recovery initial value active current percentage.

(c) Stage 3: Active Power Recovery Stage. The active power control method is based on a certain recovery slope, while the reactive power control method involves no additional control. The active power of the wind turbine recovers to its pre-fault value at a rate of at least 10% of the rated power per second. The active power recovery during this period can be expressed by Equation (3):

$$I_{p\_recover}=\frac{k_3}{U_t}(t - t_2) + k_2 I_{p0} \qquad t_2 < t \leq t_3 \tag{3}$$

In the equation, k_3 is the active power recovery rate, with units of MW/s, and $I_{p\_recover}$ represents the active current output of the wind turbine at time t during the active power recovery stage.

2.2.2. Current Limitation Module. The current limitation model allows setting a current priority mode for either normal operation or fault ride-through state: if reactive current priority is selected, the reactive current output is prioritized. The current limitations for "reactive priority" are shown in Equations (4).

$$I_{q\max} = I_{\max}, I_{q\min} = -I_{\max}, I_{p\max} = \sqrt{I_{\max}^2 - I_{qcmd}^2}, I_{p\min} = 0 \tag{4}$$

In the equations, I_{qcmd} are the reactive current reference values input to the limiting module; I_{pmax} and I_{pmin} are the upper and lower limits for active current; I_{qmax} and I_{qmin} are the upper and lower limits for reactive current; $I_{\max}$ is the total current upper limit.

Analyzing the two modules above, the main parameters influencing the active power deficit and voltage drop during LVRT are k_1 (active current percentage coefficient during LVRT), k_2 (active current percentage coefficient for recovery initial value), k_q (reactive current support coefficient), and k_3 (active power recovery rate).

3. Analysis of the Coupled Mechanism of Wind Turbine LVRT on Grid Voltage/Frequency

3.1. Active Power Output During and After the Wind Turbine LVRT Recovery Stage

The control strategy for active power during the LVRT period and recovery stage has been explained earlier. However, during the LVRT period, the control process focuses on active current, while in the recovery stage, the active power recovery rate controls the active power. The active power output of the wind turbine during the LVRT period and the recovery stage can be expressed as:

$$\begin{cases} P_1 = I_{\text{p\_set1}} U_t = k_1 I_{\text{p0}} U_t & t_1 \le t < t_2 \\ P_2 = I_{\text{p\_set2}} U_t = k_2 I_{\text{p0}} U_t & t = t_2 \\ P_3 = I_{\text{p\_recover}} U_t = k_3(t - t_2) + k_2 I_{\text{p0}} U_t & t_2 < t \le t_3 \end{cases} \tag{5}$$

where P_1 is the active power output of the wind turbine during the LVRT process, P_2 is the initial active power value after LVRT recovery, and P_3 is the active power output during the recovery stage.

In addition to the active power deficit caused by the changes in the wind turbine's active power output during the LVRT period and recovery stage, it can be observed from the diagram that when the wind turbine's grid connection voltage rises to the LVRT exit threshold U_{in}, the active power output begins to recover.

3.2. Impact of Wind Turbine LVRT Active Power Deficit on Grid Frequency

To simplify the analysis of the impact of LVRT on grid frequency, we neglect the changes in load concerning frequency and voltage after the fault. Combining the active power response characteristics of the wind turbine during LVRT, the grid frequency response containing wind turbines can be represented by Equation (6).

$$\Delta\omega = \frac{1}{M_{\text{eq}}}\int_{t_1}^{t}(\Delta P_m - \Delta P_L + \Delta P_W)\mathrm{d}t \qquad t_1 \le t \le t_2 \tag{6}$$

In the equation, M_{eq} is the equivalent system inertia constant; ΔP_m is the mechanical power change of the conventional generators in the system; ΔP_L is the load change during the fault; ΔP_w is the change in the active power output of the wind turbines, and $\Delta\omega$ is the change in the grid frequency.

As shown in Equation (6) and in Figure 4, in the simplified analysis, the mechanical power change of conventional generators and the load variation are neglected. The change in the active power output of the wind turbine during the LVRT and recovery periods plays a major role in the frequency variation. In the equation, the integral of ΔP_w with respect to time t represents the area of the active power deficit caused by LVRT, which is indicated by the blue shaded area in Figure 4. Considering both the voltage drop area S_2 and the active power deficit area S_1, the expressions are as follows:

$$S_1 = \int_{t_1}^{t_3}\frac{P_0 - P_w}{P_0}\mathrm{d}t, S_2 = \int_{t_1}^{t_3}\frac{U_{in} - U_t}{U_0}\mathrm{d}t \tag{7}$$

where P_w is the actual active power output of the wind turbine during LVRT, t_1 is the time when the fault occurs, and the wind turbine enters LVRT, t' is the fault clearance time, t_2 is the time when the wind turbine exits LVRT and begins active power recovery, and t_3 is the time when the active power recovery is completed.

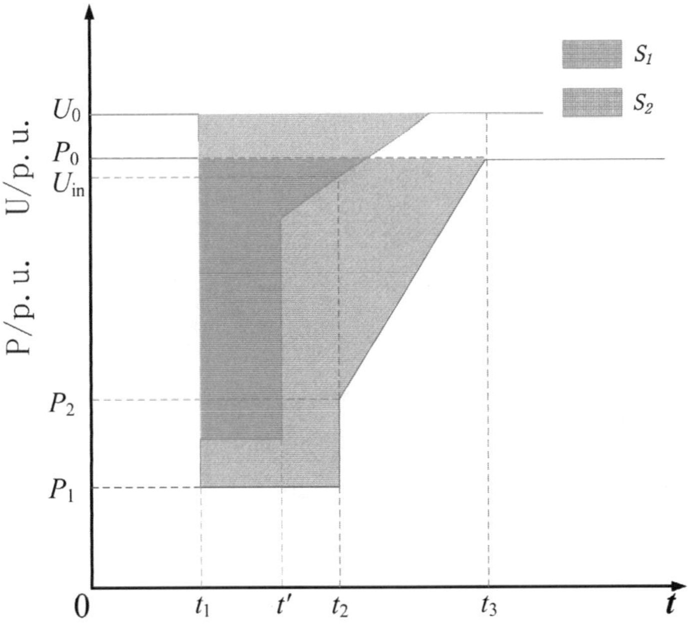

Figure 4. The schematic diagram of S_1 and S_2 areas.

4. Optimization of Key Parameters in LVRT

The PSO algorithm is used for parameter optimization. Kennedy et al. [13-14] provide detailed information on the PSO algorithm. The optimization model aims to minimize the sum of the active power deficiency area (S_1) and voltage dip and recovery area (S_2) while preventing the wind turbine from re-entering LVRT and limiting the reactive current. The mathematical model is as follows:

$$J = \min\left(S_1 + S_2 + \gamma_1 + \gamma_2\right)$$

$$\gamma_1 = \begin{cases} 0 & \text{no secondary entry into LVRT} \\ M & \text{secondary entry into LVRT} \end{cases}, \gamma_2 = \begin{cases} 0 & I_q < I_{max} \\ K & I_q = I_{max} \end{cases} \tag{8}$$

s.t.

$$0 \le k_1 \le k_{1\max}, 0 \le k_2 \le k_{2\max}, 0 \le k_3 \le k_{3\max}, 0 \le k_q \le k_{q\max} \tag{9}$$

In the above equation, J represents the objective function, S_1 refers to the active power deficit area caused by the wind turbine entering the LVRT, and S_2 represents the voltage drop and recovery area. γ_1 and γ_2 are penalty functions to meet the requirements, with M and K as large positive penalty parameters. Equation (9) defines the value ranges for each parameter.

For the constructed parameter optimization model, the PSO algorithm is used in conjunction with time-domain simulation techniques for optimization. The initial particle positions (corresponding to the initial four optimization parameters) are set based on empirical human judgment. In a particle swarm of size Np, the position of the i-th particle and the j-th element in the (k+1)-th iteration is updated as follows, where $v_{i,j}^{k+1}$ and $X_{i,j}^{k+1}$ represent the particle velocity and position in the next iteration:

$$v_{i,j}^{k+1} = \omega v_{i,j}^{k} + c_1 r_1 (p_{i,j}^{k} - X_{i,j}^{k}) + c_2 r_2 (g_j^{k} - X_{i,j}^{k}), X_{i,j}^{k+1} = X_{i,j}^{k} + v_{i,j}^{k+1} \tag{10}$$

5. Example

The tested system is modified from the IEEE-9 bus system [15], where the power source at Bus 3 is replaced with a direct-drive wind turbine, injecting the same power as the original node, as shown in

Figure 5. The actual output of the wind turbine is 85 MW. The system's installed capacity ratio is 1:3.77, with a renewable penetration rate of 26.6%. A three-phase short-circuit fault occurs at bus B at the 45th cycle, lasting for five cycles, and is cleared at the 50th cycle.

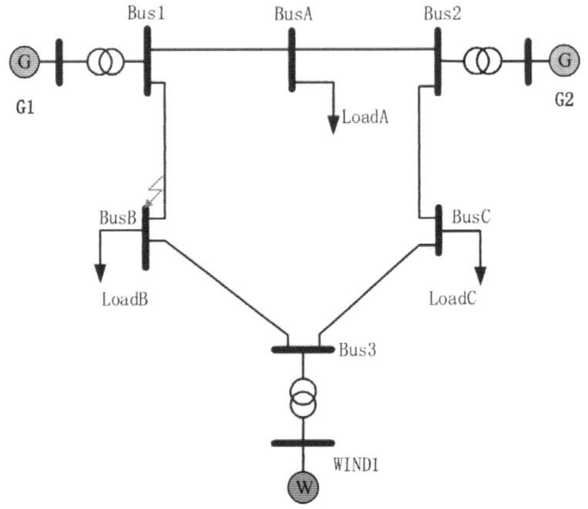

Figure 5. Modified 3-machine 9-bus System.

5.1. Change of Parameters

Table 1 shows the initial parameters and the calculated values of S_1 and S_2. In the simulation, only one parameter is changed at a time.

Table 1. Initial Parameters and Calculation of S_1, S_2.

k_1	k_2	k_3	k_q	S_1	S_2
50	50	50	1.5	31.97	8.61

5.1.1. k_1 (active current percentage coefficient during LVRT)

Table 2. Change of k_1 Parameter and Calculation of S_1, S_2.

k_1	40	50	60	70	80
S_1	33.85	31.97	30.33	28.45	26.38
S_2	8.63	8.61	8.61	8.71	8.92

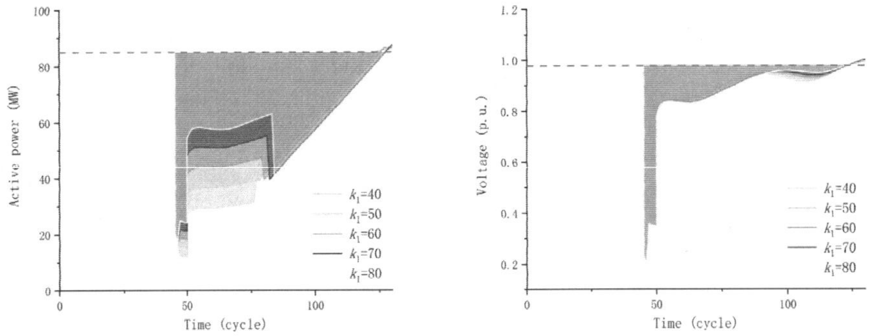

Figure 6. Change of k_1 Parameter and Variation of Active Power Deficit Area and Voltage Dip Area.

As shown in Table 2 and Figure 6, by changing only k_1, as the value of k_1 increases, S_1 decreases, while S_2 remains almost unchanged, staying around 8.6. The sizes of S_1 and S_2 directly reflect the degree of active power deficiency and voltage dip in the system.

5.1.2. k_2 (active current percentage coefficient for recovery initial value)

Table 3. Change of k_2 Parameter and Calculation of S_1, S_2.

k_2	40	50	60	70	80
S_1	36.62	31.97	28.53	33.59	30.65
S_2	8.23	8.61	9.15	9.19	9.31

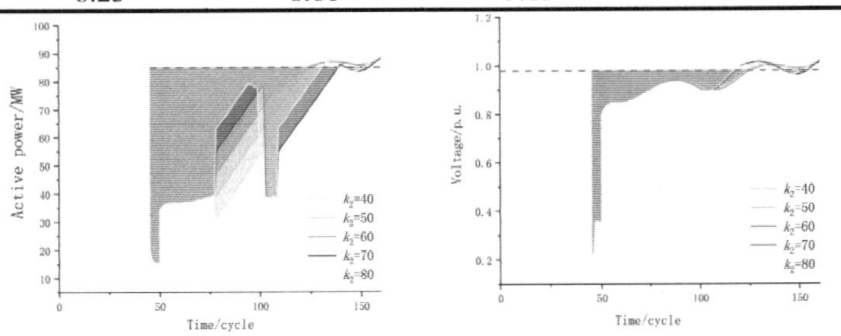

Figure 7. Change of k_2 Parameter and Variation of Active Power Deficit Area and Voltage Dip Area.

As shown in Table 3 and Figure 7, through simulation, it can be observed that as k_2 increases from 40 to 60, the active power deficiency area S_1 decreases, while S_2 shows an increasing trend. When k_2 reaches 70 or 80, the voltage recovery process becomes more volatile, causing the wind turbine to enter a secondary LVRT, further increasing the active power deficiency in the system.

5.1.3. k_3 (active power recovery rate)

Table 4. Change of k_3 Parameter and Calculation of S_1, S_2.

k_3	30	50	100	200	300
S_1	50.35	31.97	36.87	32.74	31.53
S_2	8.05	8.61	9.49	9.51	9.47

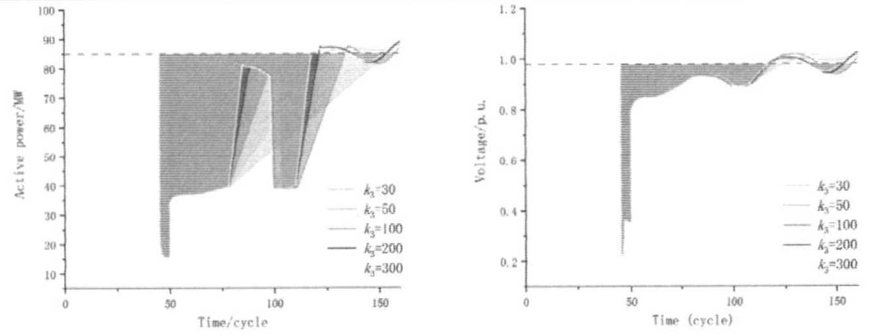

Figure 8. Change of k_3 Parameter and Variation of Active Power Deficit Area and Voltage Dip Area.

As shown in Table 4 and Figure 8, as the active power recovery rate k_3 increases gradually, when k_3 is 30 or 50, the wind turbine does not enter the secondary LVRT. In this case, the larger the parameter, the smaller S_1 becomes, while S_2 increases. However, when k_3 reaches 100, 200, or 300, the wind turbine enters the secondary LVRT, and as k_3 increases further, S_1 continues to decrease while S_2 remains relatively stable.

5.1.4. k_q (reactive current support coefficient)

Table 5. Change of k_q Parameter and Calculation of S_1, S_2.

k_q	1.0	1.5	2.0	2.5	3.0
S_1	32.75	31.97	31.71	31.92	30.42
S_2	9.11	8.61	8.23	8.07	8.00

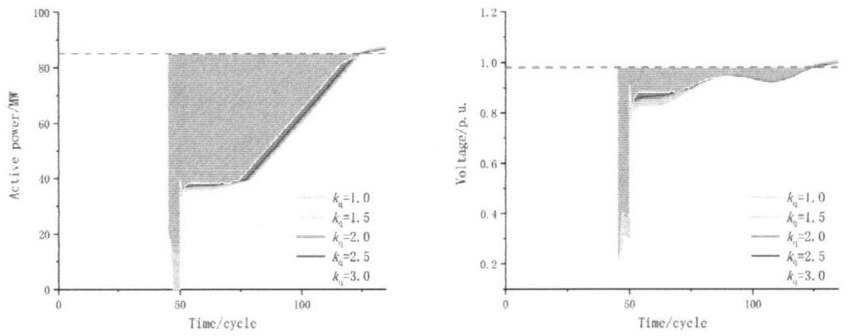

Figure 9. Change of k_q Parameter and Variation of Active Power Deficit Area and Voltage Dip Area.

As shown in Table 5 and Figure 9, As k_q increases, S_2 shows a decreasing trend. However, when $k_q = 3.0$, the active power drops to zero during the fault period because, at this point, the reactive current reaches I_{max}, and the active current is squeezed out, resulting in zero active power output. Therefore, when optimizing k_q, it is crucial to coordinate the optimization of the reactive and active current limiting strategies and their limits in the LVRT and recovery control stages.

5.2. Effectiveness of the parameter optimization method

To verify the effectiveness of the parameter optimization method proposed in this paper, a comparison of the sum of the S_1 and S_2 areas before and after optimization is conducted. Figure 10 shows the convergence process of the particle swarm optimization objective function value (J).

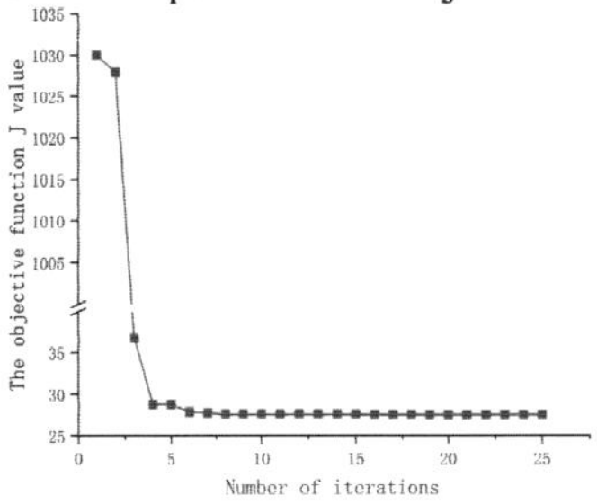

Figure 10. The objective function J value during the iteration process.

Table 6. Parameter changes before and after optimization and calculation of S_1 and S_2

	k_1	k_2	k_3	k_q	S_1	S_2	$S_1 + S_2$
Before optimization.	50	50	50	1.5	31.97	8.61	40.59
After optimization.	80	80	47	2.1	19.29	9.55	28.84

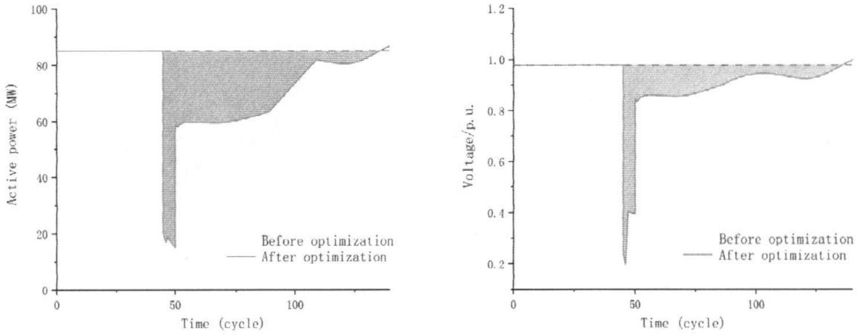

Figure 11. Comparison of the effects before and after parameter optimization.

Combining Table 6 and Figure 11, the effectiveness of the proposed parameter optimization method can be validated. After optimization, the S_1 value significantly decreases, the S_2 value slightly increases, and the sum of S_1 and S_2 decreases significantly. This indicates that after parameter optimization, the active power deficit during LVRT and power recovery processes is greatly reduced.

6. Conclusion

Based on the results and discussions, the following conclusions are drawn. By analyzing the active/reactive power control strategy and the expressions for active and reactive currents, it is found that the key parameters influencing system stability are k_1, k_2, k_q, and k_3. Simulations reveal the specific impact of these parameters on the active power deficit and voltage dip during each LVRT phase. The effect of slow voltage recovery on active power deficit during LVRT is also analyzed. The proposed optimization method, which considers both the active power deficit and voltage dip areas, effectively reduces the LVRT-induced power deficit while maintaining voltage stability.

Acknowledgments

This work was supported by the China Southern Power Grid Science and Technology Project (YNKJXM20230338).

References

[1] Chen G.P., Li M.J., Xu T., et, al. Study on Technical Bottleneck of New Energy Development [J]. Proceedings of the CSEE, 2017,37(01): 20-27.DOI: 10.13334/j.0258-8013.pcsee.161892.

[2] Shi W.H., Qu J.X., Luo K., et al. Grid-integration and operation of high proportioned new energy[J/OL]. Strategic Study of CAE, 1-12(2022). doi:10.15302/J-SSCAE-2022.07.006

[3] Zhang L.X., Wind farm grid connection technology and its influence on the power system [J]. China High and New Technology, 2024, (16): 108-110.DOI: 10.13535/j.cnki.10-1507/n.2024.16.33.

[4] Zhang J.Y., Li M.J., Analysis of the Frequency Characteristic of the Power Systems Highly Penetrated by New Energy Generation [J].Proceedings of the CSEE, 2020,40(11):3498-3507.DOI:10.13334/j.0258-8013.pcsee.191265.

[5] Zeng H., Sun F., et, al. Analysis of "9·28" Blackout in South Australia and Its Enlightenment to China [J]. Automation of Electric Power Systems, 2017,41(13):1-6.DOI:10.7500/AEPS20170120002.

[6] KAWABE K, TANAKA K. Impact of the dynamic behavior of photovoltaic power generation systems on short-term voltage stability[J]. IEEE Transactions on Power Systems,2015,30 (6):3416-3424.

[7] LAMMERT G,PREMM D,OSPINA L D P,et al. Control of photovoltaic systems for enhanced short-term voltage stability and recovery[J]. IEEE Transactions on Energy Conversion, 2019,34(1):243-254.

[8] WEISE B. Impact of K-factor and active current reduction during fault-ride-through of generating units connected via voltage-sourced converters on power system stability[J]. IET Renewable Power Generation,2015,9(1):25-36.

[9] RATHER Z H, FLYNN D. Impact of voltage dip induced delayed active power recovery on wind integrated power systems [J]. Control Engineering Practice,2017,61:124-133.

[10] Li S.D., Research of wind turbine key components influences on low voltage ride through characteristic[D]. North China Electric Power University, 2015.

[11] Xiang C.M., Fan L.X., et al. Analysis of cascading trip-off mechanism and low voltage ride through capability of wind farm[J]. Electric Power Automation Equipment, 2013,33(12):91-97.

[12] Zhang Y.B., Yuan H.W., Control strategy of DFIG main control system during low voltage ride through[J]. Electric Power Automation Equipment, 2012,32(8):106-112.

[13] J. Kennedy and R. C. Eberhart, "Particle swarm optimization," in Proc.IEEE Int. Conf. Neural Netw.1995, pp. 1942–1948.

[14] R.C. Eberhart and Y. H. Shi, "Particle swarm optimization: Developments, applications, and resources," in Proc. IEEE Int. Congr. Evol. Comput.,2001, pp. 81–86.

[15] Sauer PW., Pai M., Power system dynamics and stability. Prentice Hall Upper Saddle River, NJ, 1998, vol. 4.

Study on the wall-guided stratified lean combustion of methanol flash-boiling spray

Guodong Wang[1,2], Mingli Cui[3], Bowei Yao[4], Hongchang Xu[3], Xuesong Li[3,*], Jie Hu[1,*]

[1] School of Automotive Engineering, Wuhan University of Technology, Wuhan, 430070, China
[2] SAIC GM Wuling Automobile Co., Ltd, Liuzhou, 545007, China
[3] School of Mechanical Engineering, Shanghai Jiao Tong University, Shanghai, 200240, China
[4] Liuzhou Saike Technology Development Co., Ltd., Liuzhou, 545616, China

[*]Corresponding author's e-mail: xuesonl@sjtu.edu.cn, auto_hujie@whut.edu.cn

Abstract. This study explores the characteristics of wall-guided stratified lean combustion of methanol flash boiling spray using a combustion chamber. High-speed imaging is employed to record the spray and flame development processes, focusing on the formation of fuel-rich regions around the ignition electrodes, flame kernel growth, and yellow sooting flame distribution. The results indicate that, compared to n-heptane, methanol stratified combustion features orange flame and a broader distribution of yellow flame regions. Under flash boiling conditions, the methanol spray collapses into a single jet, with a concentrated target region on the spray guiding plate. Furthermore, flash-boiling methanol stratified combustion exhibits a 150 % increase in flame kernel development speed and faster flame propagation compared to methanol subcooled stratified combustion, with soot formation primarily at the spray collapse region.

1. Introduction

The increasing demand for cleaner and more efficient combustion of spark ignition engines has driven research into alternative fuels and advanced combustion strategies. Methanol, as a renewable fuel, has gained considerable attention due to its potential to reach carbon neutrality[1, 2]. Among various combustion strategies, lean-burn technology has been widely recognized for its ability to improve thermal efficiency and reduce pollutant emissions[3, 4]. However, achieving stable and efficient lean combustion remains a major challenge in direct injection spark ignition (DISI) engines, where mixture preparation and fuel distribution play crucial roles.

The wall-guided stratified combustion strategy utilizes spray-wall interactions to form a localized fuel-rich mixture near the spark plug while maintaining overall lean conditions in the combustion chamber[5, 6]. This approach enhances ignition stability and combustion efficiency under lean-burn conditions[7, 8]. However, the interaction between flash-boiling spray dynamics and wall-guided mixture formation is still not well understood, particularly for methanol fuel, which exhibits distinct physical and chemical properties compared to conventional gasoline.

Flash-boiling fuel injection has emerged as a promising technique to enhance atomization, evaporation, and fuel-air mixing, thereby improving combustion performance[9]. Flash boiling occurs when the fuel is injected at a temperature above its saturation temperature, leading to improved atomization[10, 11]. Therefore, flash-boiling atomization has the potential to become an effective means of improving the performance of stratified combustion.

This study aims to investigate the effects of flash-boiling spray on the methanol wall-guided stratified lean combustion process. N-heptane is used as a typical hydrocarbon fuel as a reference. This study employs MIE scattering techniques to capture the atomization process, providing detailed insights into the spray-wall interaction process. Additionally, high-speed color cameras are used to record the flame development process, allowing for observation of ignition and combustion. Spray-wall interaction and wall-guided stratified combustion processes are studied based on the spray and flame images.

2. Experimental Setup

This study was conducted in a constant-volume combustion chamber with a volume of 2.96L. A six-hole direct injection injector was positioned at the top of the combustion chamber, while a spray-guiding plate was placed at the bottom. Ignition was carried out using an ignition rod, which positioned the ignition electrodes right above the edge of the spray-guiding plate. The position relationship among the injector tip, spray-guiding plate, and ignition electrodes, as well as the three-dimensional model of the wall-impingement guiding plate, is shown in Figure 1, and the volume of the guidance plate is 0.40 L. After spray impinges on the guiding plate, fuel droplets move along the plate's surface, forming a fuel-rich region near the ignition electrodes, thereby achieving wall-guided stratified combustion.

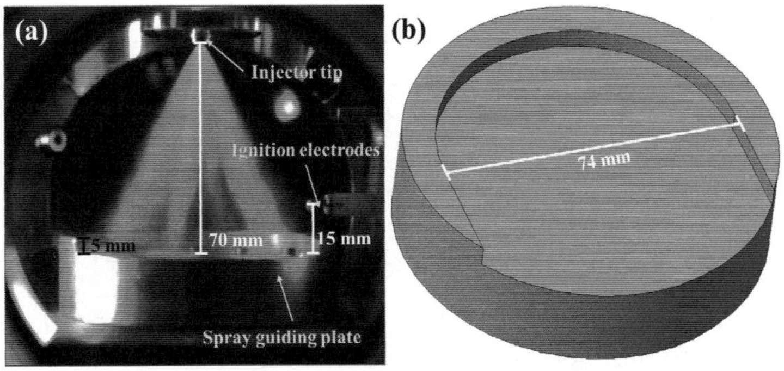

Figure 1. Experimental Setup. (a) Positional relationship among the injector tip, ignition electrodes, and spray guiding plate; (b) 3D Model of Spray guiding plate.

Methanol and n-heptane are used as fuel in this study. The injection mass of the GDI injector was calibrated for both fuels, with the injection pulse width ranging from 0.4 ms to 4.0 ms. At a fuel pressure of 10 MPa and an injection frequency of 100 Hz, fuel was injected 50 times at the given pulse width. The fuel mass was measured by an electronic balance, and the process was repeated three times to calibrate the injected fuel mass. The resulting fuel mass for a single injection is shown in Figure 2. A linear relationship was observed between injection mass and pulse width. The relationship between injection mass (mg) and pulse width (ms) for methanol in the range of 0.4 ms to 4.0 ms is given by $y = 12.73187*x - 0.17127$, while for n-heptane, the relationship is $y = 11.66515*x + 0.39831$.

The experimental parameters were set as follows: ambient pressure of 60 kPa, ambient temperature of 30°C, global equivalence ratio of 0.6, and injection pressure of 10 MPa. Due to the lower energy density of methanol, the injection mass was adjusted to maintain the same energy input. Specifically, 157.12 mg of methanol was injected per cycle with a pulse width of 2.211 ms, while 70.26 mg of n-heptane was injected per cycle with a pulse width of 0.933 ms. Both fuels were initially injected five times at 100 Hz, followed by a 10 s waiting period to allow fuel-air mixing. Then, a final injection was performed. The start of the injection was defined based on the final injection. Ignition started at 4 ms after the start of injection (ASOI) using an ignition coil with an ignition energy of 120 mJ. The stratified combustion injection accounted for 16.67% of the total energy input.

For sub-cooled conditions, methanol and n-heptane were both maintained at 30°C. For methanol flash-boiling spray, the injector was heated to 95°C, corresponding to a superheat degree of $Pa/Ps = 0.2$. Where Pa is the ambient pressure, and Ps is the saturated pressure of the fuel.

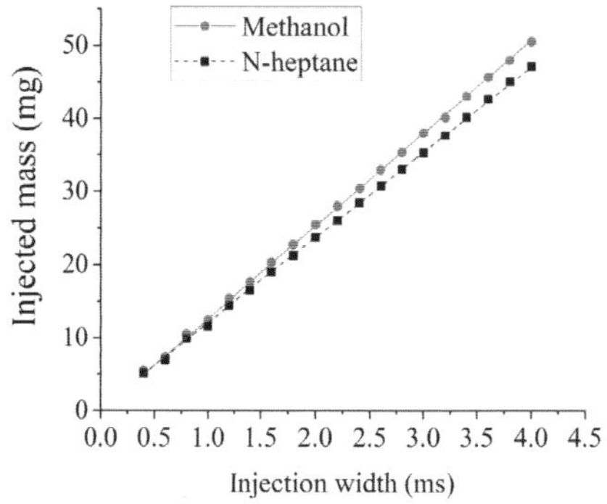

Figure 2. Injector calibration using methanol and n-heptane fuels.

3. Results and Discussions

The spray development process in this study is shown in Figure 3. Initially, the spray develops downward, and after impinging on the spray-guiding plate, a fuel-rich region is formed in the space. The ignition electrodes ignite the fuel-rich region at 4 ms ASOI, and the fuel distribution at this moment is shown at 4.0 ms ASOI in Figure 3. Under sub-cooled conditions, the spray development processes of methanol and n-heptane are similar, keeping the shape of the umbrella. After the end of injection, the fuel droplets continue to move downward, forming a fuel-rich region above the whole guiding plate. Due to the larger injection quantity and pulse width of methanol, the droplet cloud of methanol spray is denser compared to n-heptane, as shown at 3.2 ms and 4.0 ms ASOI. However, for the methanol flash-boiling condition, the increased fuel temperature increases the cone angle of each fuel plume. This causes interference and spray collapse between fuel plumes, where the spray converges toward the center axis. The flash-boiling spray first impinges on the central area of the wall, and the droplets then spread out from the center to the surrounding area. The guiding effect of the plate on the methanol flash boiling spray is evident, with the fuel-rich region concentrating above the edge of the guiding plate.

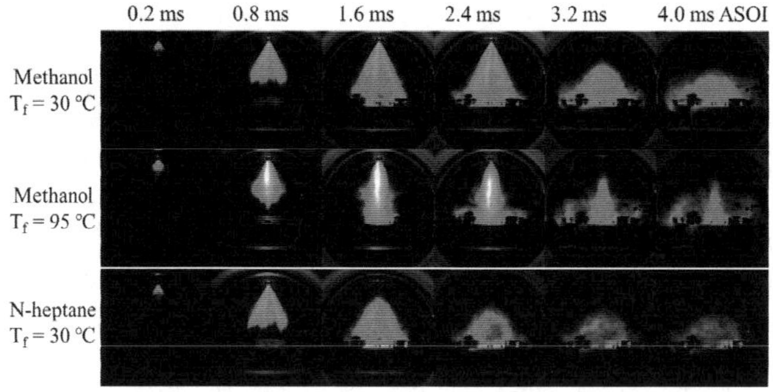

Figure 3. Spray development processes.

The spray penetration and spray width are shown in Figure 4. Compared to n-heptane, methanol subcooled spray exhibits lower spray penetration and width at the same time, indicating that the spray moving speed of methanol is slower than that of n-heptane. Under methanol flash boiling conditions, the spray collapse makes the spray penetration larger than that under the methanol subcooled condition. At the beginning of the spray, the width of the methanol flash boiling spray is higher. But as the spray develops, the collapse of the spray leads to a lower methanol flash boiling spray width compared with that of the subcooled spray.

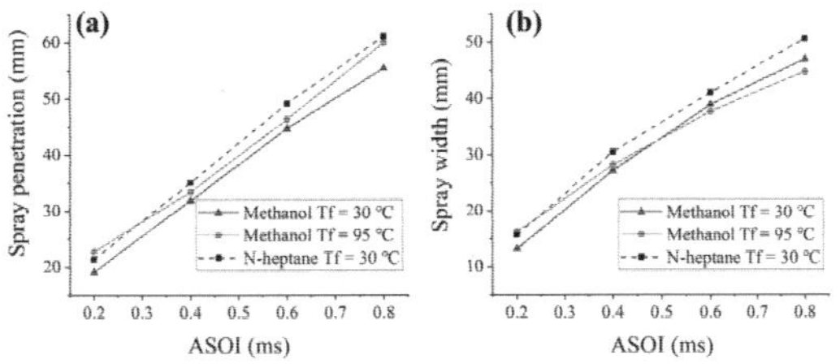

Figure 4. Spray characteristics. (a) spray penetration evolution and (b) spray width evolution.

Figure 5 shows the development process of the flame kernel after ignition, with the time indicated as the moment after ignition begins. During ignition, a bright electric arc is formed in the space between the electrodes, which then ignites the fuel-rich region formed by the wall-guided spray. N-heptane, as a hydrocarbon fuel, burns as a blue-green flame, whereas the methanol flame kernel appears orange. The flame kernel of n-heptane first develops downward, while the methanol flame kernel develops upward. Compared to the methanol subcooled condition, the arc of methanol flash-boiling atomization is stretched due to the spray's motion, as shown at 1.4 ms in Figure 5.

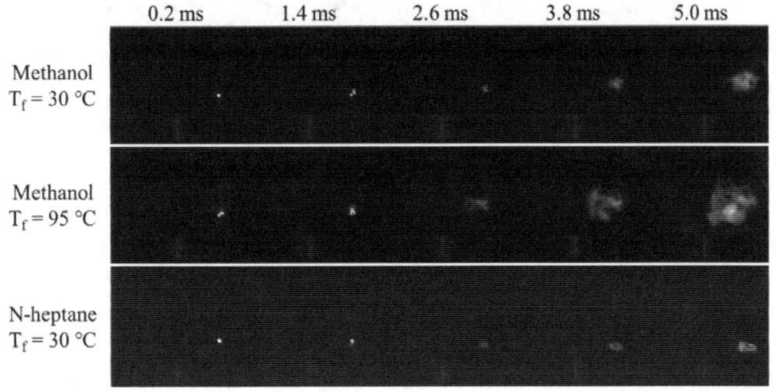

Figure 5. Flame kernel development.

Figure 6 shows the development of the flame kernel's equivalent radius over time. Under subcooled conditions, the flame kernel development speed of methanol and n-heptane is similar before 3.5 ms. The flame kernels of both fuels reach 5 mm at 3.5 ms. However, after that, the flame kernel of methanol develops faster than that of n-heptane. Compared to the methanol subcooled condition, the flame kernel development of methanol flash boiling stratified condition is faster from the beginning of ignition. At 3.5 ms, the flame kernel of methanol under flash boiling condition reaches 12.5 mm, which is 150% faster compared to the methanol subcooled condition.

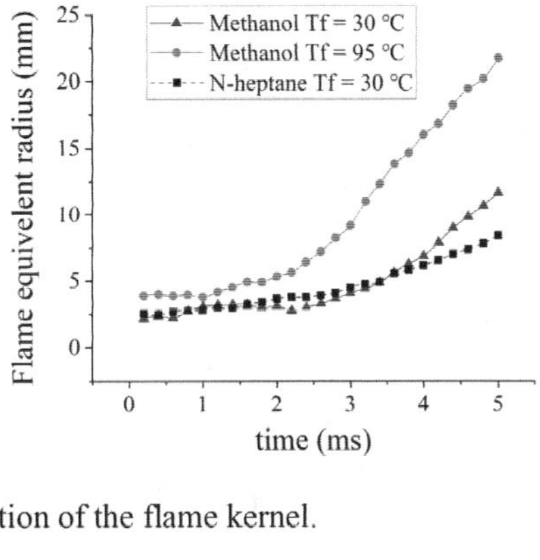

Figure 6. Equivalent radius evolution of the flame kernel.

Figure 7 shows the flame propagation process after the development of the fire kernel. Compared to n-heptane, the methanol flame appears orange with yellow regions. The yellow color comes from the high-temperature blackbody radiation of soot, indicating that the fuel-rich region formed by methanol has produced soot during the flame propagation process. The flame shapes of methanol and n-heptane sub-cooled conditions are similar, but the flame area of the methanol flame is larger at the same moment. Compared to the methanol subcooled condition, the methanol flash boiling stratified flame propagates faster, which is evident at 11.2 ms in Figure 7. The yellow area of the methanol flash boiling flame is smaller and concentrated near the guide plate, whereas the yellow region is spread over a wider area under the methanol subcooled spray. This indicates that the soot formation area of methanol flash boiling stratified combustion is smaller than that of the methanol subcooled stratified combustion.

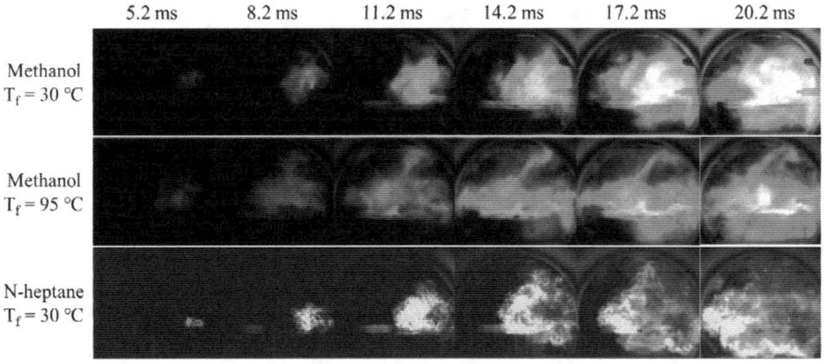

Figure 7. Flame propagation processes.

Figure 8 illustrates the flame development process after the flame propagation, where the flame brightness gradually decreases as the temperature drops. Compared to methanol, n-heptane forms yellow flames more slowly. In the high-temperature environment after flame propagation, the yellow flame in subcooled stratified combustion of n-heptane first concentrates at the right edge of the guide plate, as shown at 5.2 ms in Figure 8. After that, pool fire, appearing as a flat yellow flame above the guiding plate, formed by wall-impinged fuel film, generates soot. The fuel-rich region formed near the left edge of the spray fully evaporates before the flame propagates to the left side, preventing the appearance of yellow flames near the left edge. As illustrated in Figure 8 at 5.2 ms, the yellow flame of subcooled methanol combustion is widely distributed in the space. However, for methanol flash boiling stratified combustion, yellow flames, apart from those caused by fuel film combustion, only appear at the spray axis where the spray collapses. This indicates that the spray collapsing region is a source for soot formation in methanol flash boiling stratified combustion.

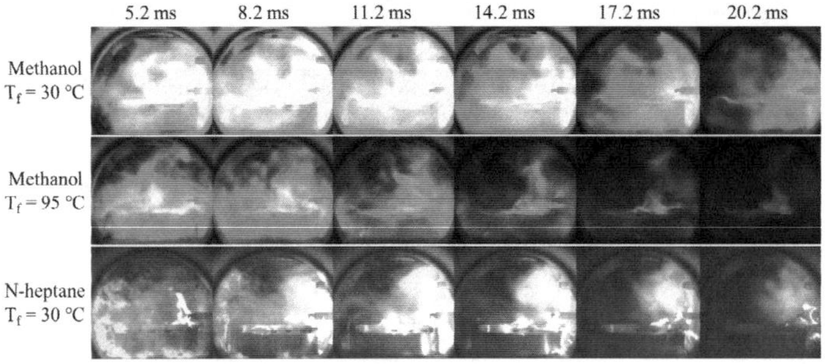

Figure 8. Flame development after filling the combustion chamber.

4. Conclusion

This study investigates the characteristics of methanol wall-guided stratified lean combustion under flash boiling spray conditions using a combustion chamber. High-speed imaging is used to capture the spray and flame development. The main conclusions of this study are as follows:

(1) Compared to n-heptane, the flame of methanol stratified combustion appears orange. The initial flame kernel development speed is similar for both fuels under subcooled conditions, but the flame propagation speed is faster for methanol.

(2) Under the flash boiling condition, the methanol spray collapses into a single jet, and the fuel impingement region is located at the center of the guidance plate. The fuel-rich region is concentrated at the edge of the guidance plate.

(3) Compared to the methanol subcooled condition, methanol flash boiling stratified flame exhibits faster flame kernel development (150% faster) and faster flame propagation. Soot formation in the space is concentrated at the spray collapse region above the guidance plate.

Acknowledgments

This work is sponsored by the Guangxi Science and Technology Major Program (Grant 2023AA02010) and the National Natural Science Foundation of China (Grant No. 52276125).

References

[1] Tian, Z., et al., The effect of methanol production and application in internal combustion engines on emissions in the context of carbon neutrality: A review. Fuel, 2022. 320: p. 123902.
[2] Verhelst, S., et al., Methanol as a fuel for internal combustion engines. Progress in Energy and Combustion Science, 2019. 70: p. 43-88.
[3] Zhang, M. and J. Cao, Comparative study on combustion and emission characteristics of methanol/gasoline blend fueled DISI engine under different stratified lean burn modes. Fuel Processing Technology, 2024. 266: p. 108160.
[4] Gong, C., et al., Comparative study on combustion and emissions between methanol port-injection engine and methanol direct-injection engine with H2-enriched port-injection under lean-burn conditions. Energy Conversion and Management, 2019. 200: p. 112096.
[5] VanDerWege, B.A., et al., Development and Analysis of a Spray-Guided DISI Combustion System Concept. SAE Transactions, 2003. 112: p. 2135-2153.
[6] Fan, Q., et al., Stratified mixture formation and combustion process for wall-guided stratified-charge DISI engines with different piston bowls by simulation. 2010, SAE Technical Paper.
[7] Oh, H. and C. Bae, Effects of the injection timing on spray and combustion characteristics in a spray-guided DISI engine under lean-stratified operation. Fuel, 2013. 107: p. 225-235.
[8] Fansler, T.D. et al., Invited Review: Combustion instability in spray-guided stratified-charge engines: A review. 2015. 16(3): p. 260-305.
[9] Li, X., et al., A review on the recent advances of flash boiling atomization and combustion applications. Progress in Energy and Combustion Science, 2024. 100: p. 101119.
[10] Yang, S., et al., Characteristics and correlation of nozzle internal flow and jet breakup under flash boiling conditions. International Journal of Heat and Mass Transfer, 2018. 127: p. 959-969.
[11] Li, X., et al., Investigations on near-field atomization of flash boiling sprays for gasoline direct injection-related applications. Fuel, 2019. 257: p. 116097.

Identification and characterization of reservoir high permeability streak and its application

Shichao Zhang [1*]

[1] CNPC Daqing International Exploration and Development Company, Daqing 163458, Heilongjiang, China

[*] Corresponding author's e-mail: zhangshichao1@petrochina.com.cn

Abstract: A method for quantitative characterization of high permeability streak is proposed, based on the statistics of permeability versus depth of core samples. This method solves the problem that macroscopic identification of high-permeability zones cannot be quantitatively characterized and improves the matching accuracy of reservoir simulation. In the reservoir simulation of water flooding in the S oilfield, the method was used to adjust three high permeability streaks in the model. The matching accuracy was significantly improved in the comparison before and after the adjustment. This method of identifying and characterizing high permeability streaks provides a new idea for future reservoir simulation of similar oilfields.

1. Introduction

In sandstone reservoirs, there are great differences in the sorting and deposition of rock particles during different depositional periods and under different depositional environments. The compositions of cement among rock particles and the degree of filling are different. These cause the problem of serious heterogeneity within the subzone in some areas. Due to the influence of long-term water injection during the development process, the cement among rock particles is dissolved by formation fluid, resulting in loose rock particles, and the existence of streaks or zones with relatively high permeability in some layers. Such streaks or zones with permeability significantly higher than the average reservoir permeability are called high permeability streaks. Currently, the main methods for identifying high permeability streaks are the sedimentary micro facies method, wireline logging method, tracer method, modeling method, and lab experimental method [1-4].

The S oilfield is a terrestrial low-porosity, low-permeability sandstone reservoir with severe production decreases after nearly 40 years of development. In the process of reservoir simulation of water flooding, it was found that crude oil history matching can never be able to reach the matching accuracy. After systematically analyzing the reservoir forming environment, logging results, tracer research, and actual development dynamics, it was concluded that there was a 'thief layer' in the field. After analysis and research, it was concluded that the permeability versus depth cross-plot method can quickly identify the depth range of high permeability streaks and has a great influence on the reservoir simulation and satisfactory matching accuracy.

2. Mechanism of highly permeable streaks formation

Factors that play a key role in the formation and development of large pore channels are sedimentation, rock cementation, and development methods.

2.1. Sedimentation

The depositional environment of the S oilfield is coastal deposition (underwater delta). The reservoirs are mainly bars, which spread in a north-south direction mainly along the paleo-coastline. The material sources are located in the southeastern part of the study area. It migrates in a limited area. No faults have developed in the area. The surrounding material sources were well supplied during the depositional period of the reservoir. The thickness of the sedimentary sand body is large and extensive. Vertically, the sedimentary rhythm is dominated by a fining-upward rhythm. Physical modeling experiments show that the injected fluid mainly flows along the high-permeability layer in the fining-upward rhythm model. The scouring effect to the high-permeability layer is strong with sand particles flowing out. Injected fluid flow in the low permeability layer is small and flows slowly. The force of injected fluid is weak, with few sands flowing out. It can be seen that the bottom of the fining-upward rhythm and the top of the coarsening upward rhythm make it easy to develop high permeability streaks [5].

2.2. Rock cementation

The rocks of the S oilfield are classified as clastic in terms of their material composition. The main rock-forming components are quartz and feldspar. The secondary components are rock fragments and mica. The main type of cement is mud cementation. The main component of the clay cement minerals is fully crystallized kaolinite (60-70%). Secondary cementing components are tightly bound chlorite and hydrated mica. The pore structure is primary intergranular pores, as well as pores formed during dissolution and cementation of detrital material. The carbonate content is low.

2.3. Development strategy

The development strategy of the S oilfield concentrated on maximizing economic benefits, rather than long-term stable development. After a large number of wells were put into production, the water cut raised sharply, with the characteristics of violent water flooding. The curve of Tong's chart reflected this characteristic (Figure 1). In the same seepage time of the water flooding development process, the higher the water injection seepage rate is, the larger along the main flow line direction transport speed of the sand particles is. As time passes by, a high permeability channel forms between the oil and water wells. The higher the production rate is, the higher the pressure gradient acts on the rock sand particles, and the easier the sand particles are dislodged. The larger amount of sand leads to the formation of high permeability streaks. In addition, the degree of pressure drop is also a great influence on the sand. The more sudden the pressure drop is, the easier it is to sand flow. According to the standard formula of Tong's chart,

$$\log R_{wo} = A + 7.5(R - E_R) \quad (1)$$

where R_{wo} is the water-oil ratio, R is the Rf to date, ER is the recovery rate, and A is the constant.

When the ultimate water content is 98%, A is equal to 1.69 [6].

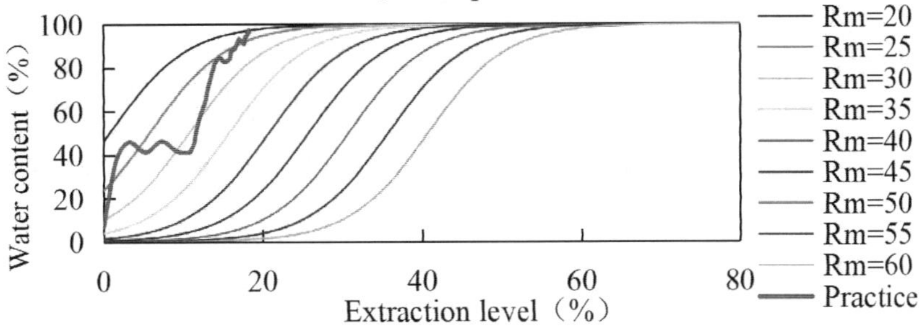

Figure 1 Tong's plate Plate Curve.

3. High permeability streak identification

3.1. Dynamics method

The dynamic method is to judge the existence of a high permeability layer according to the different water absorption capacities of the reservoir and the influence on the waterline advance speed. As can be seen from the waterline advance map of the reservoir in the S field (Figure 2), the water cut of the field increased rapidly within four years, and the water cut of 80% wells quickly exceeded 75%, and some wells approached 90%. which proves that there is a high permeability streak. However, this method can only do qualitative characterization and cannot determine the specific location of the high permeability streaks.

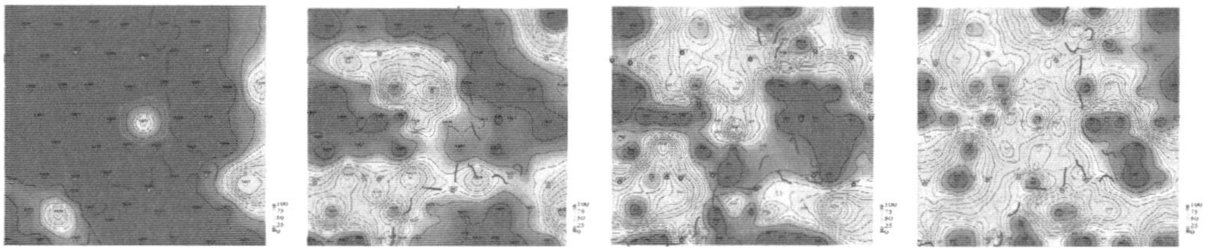

Figure 2 Four-year waterline advance within oilfield development.

3.2. Tracer test methods
Inter-well tracer test is to inject tracer section plugs in injection wells, and then continuously monitor that output in the surrounding production wells. Inter-well tracer test can judge and analyze the heterogeneity condition of the reservoir in the lateral and vertical direction to determine whether there is a higher-permeability streak in the reservoir and calculate the permeability of the reservoir [7-8]. The field has conducted a total of more than 20 well sets of tracer studies for 5 years to investigate and study the reservoir's physical properties, and to assist in judging the geological conditions of the reservoir (Figure 3 and Figure 4).

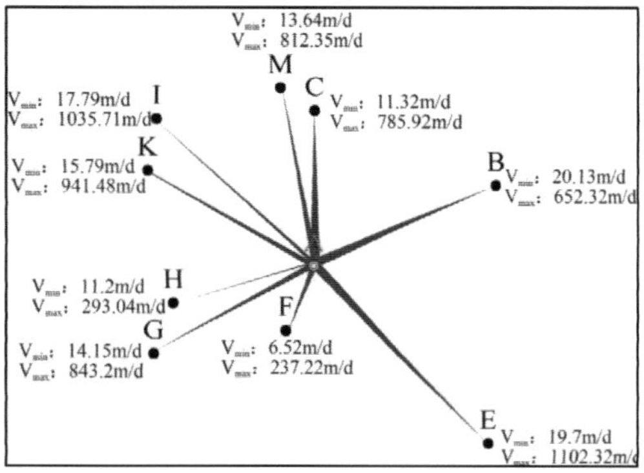

Figure 3 Well A seepage velocity distribution

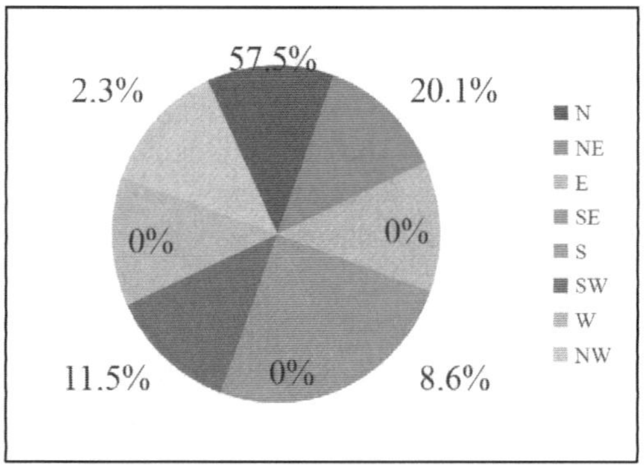

Figure 4 Distribution of seepage direction in well A

Tracer test results indicate the presence of high permeability channels, the formation of which is related to post-depositional tectonic processes, pressure differentials between oil and water wells, and secondary fracturing due to the scouring of loose and weakly cemented rocks. The method is capable of localized characterization of high permeability streaks.

3.3. Core analysis method
The core analysis method is a method of observing the distribution area of high permeability streaks and quantitatively pointing out the specific location by using statistical methods in mature blocks with abundant core data. The collection of core data from the S oil field resulted in a total of 1, 022 pieces of core data. The statistical method was used to draw a cross-plot of reservoir permeability versus depth (Figure 5) to locate the depth information of the high permeability streaks.

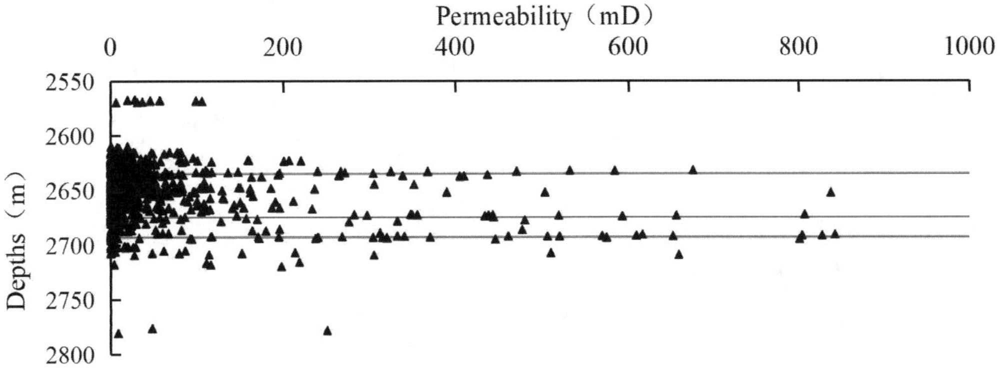

Figure 5 Reservoir permeability-depth plate.

Then the statistical core data determines the proportion of high permeability streaks in the reservoir (Table 1). The quantitative characterization of high permeability streaks is completed.

Table 1 Core permeability statistics.

	K≤50mD	50mD<K≤100mD	100mD<K≤150mD	K>150mD
Percentage of classification (%)	77.30	9.10	3.52	10.08
Average permeability (mD)	10.29	70.27	119.34	384.77

From the permeability variation with depth graph (Figure 5), it can be seen that there are three relatively high permeability streaks in the reservoir at 2635m, 2675m, and 2693m. From the core statistics table (Table 1), it can be seen that the thickness of the cores with permeability greater than 150mD accounted for 10.08% of the coring thickness, but the average permeability of samples with permeability of the reservoir greater than 150mD is 18.65 times the average permeability of samples with permeability less than 150mD. It can be concluded that there are three high permeability streaks with thin thickness (1/10) and permeability greater than 150mD in the reservoir.

4. Characterisation and application of high permeability streaks in models

After an in-depth analysis of the core from the coring wells, it was recognized that one or several thin high permeability (up to 1800 mD) streaks were developed in the reservoir. It was the presence of this part of the higher permeability streaks that led to the phenomenon of violent water flooding of the wells. Based on the core permeability distribution characteristics and logging interpretation results, the high permeability streaks were described, time-varying tracking was carried out in the subsequent matching using the EnKF method (Ensemble Kalman Filter) [9-10], and the permeability of high permeability channel was used as the adjustment parameter to carry out the water cut history matching. Based on the analysis of the relationship between dynamic data and the permeability increase coefficient, the reasonable permeability increase coefficient is obtained through several iterations, which is put into the reservoir simulation to correct the permeability of the high permeability streaks and complete the history matching. The reservoir simulation model that can fully reflect the heterogeneity of the reservoir is finally obtained. Taking a single layer as an example (Figure 6), the permeability distribution is more reasonable than before adjustment. By objectively characterizing these high-permeability streaks in the model, the accuracy of the well-matching can be significantly improved.

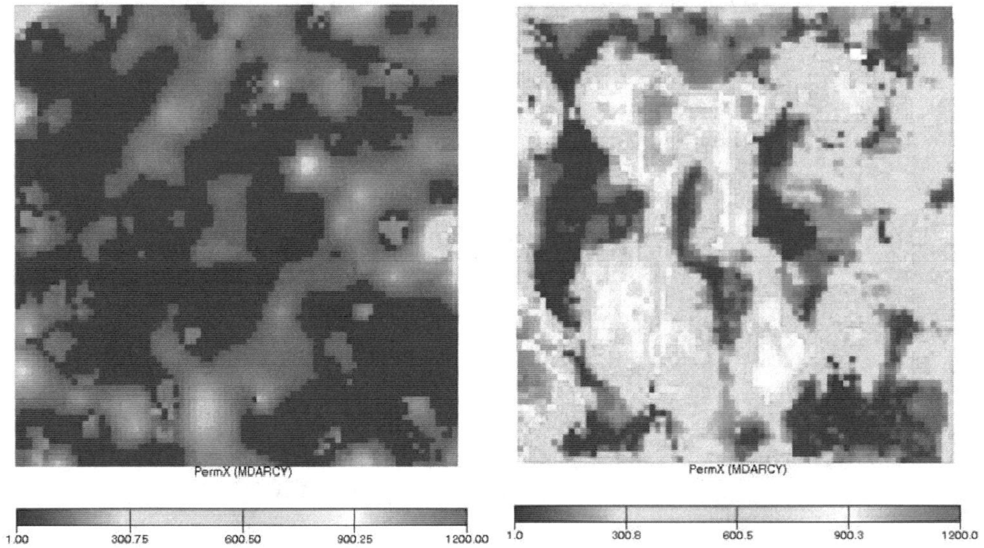

Figure 6 Comparison of permeability distribution before and after single layer adjustment.

After systematically correcting the higher permeability streaks, the history matching accuracy has been greatly improved (Figure 7), which proved that the corrected high permeability streaks are reasonable.

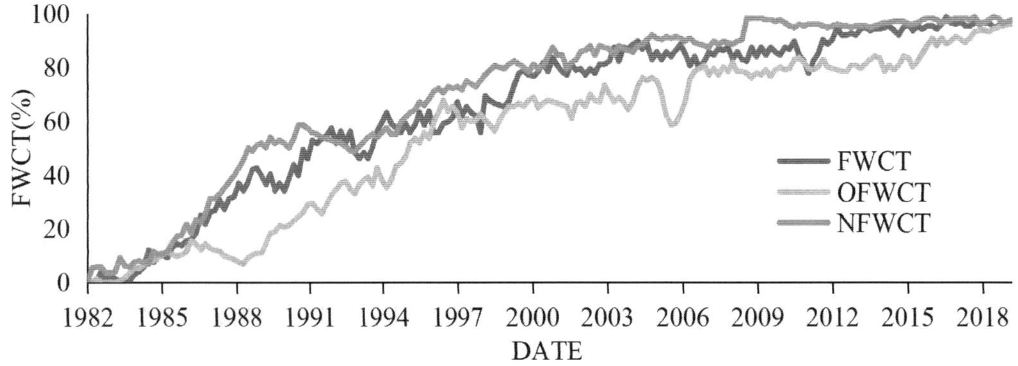

Figure 7 Comparison of fit before and after adjustment.

5. Conclusion and understanding

The dynamic method can qualitatively characterize high permeability streaks. The tracer test method can do the characterization of localized high permeability streaks. The core analysis method can quantitatively identify and characterize the high permeability streaks. For the reservoir with high permeability streaks, adjusting the distribution of high permeability streaks during the reservoir simulation can effectively improve the history matching accuracy.

References

[1] Jing K. Y., Tong Y., and Tong Z. (2021) Quantitative description method of relatively high-permeability zone based on comprehensive identification and time-varying numerical simulation [J]. Science Technology and Engineering, 21(10): 3999-4004.

[2] Tong Y. and JIA Y. Y. (2018). Residual oil micro distribution characteristics under the control of high permieability streaks

based on etching model [J]. Science Technology and Engineering, 18(28): 80-86.
[3] Yue Ye. (2020) Influence of high permeability streaks in reservoirs on the effect of water drive [J]. Chemical Engineering and Equipment, 4(4): 66-68.
[4] Li Y. Q., Sui X. G., and LI J. (2005) Laboratory experimental study on the wave and coefficient of longitudinal heterogeneous large-scale planar model polymer oil repulsion [J]. Journal of Petroleum, 26(2): 77-79.
[5] Wang W., JU B. S. and JIANG R. (2006) Identification of high permeability zones in loose sandstone reservoirs and its application [J]. Journal of Shengli Oilfield Staff University, 20(4): 56-57.
[6] Tong X. Z. (1981) Analysis of oil well production and reservoir dynamics [M]. C, Beijing.
[7] Ye H. M. (2002) Measurement of residual oil saturation by distributive tracer [J]. Broken Block Oil and Gas Field, 9(4): 42-44.
[8] Dai X. H., Zhou Z., and Shan J. C., et al. (2022) Research on block overall application of inter-well tracer monitoring technology in A oilfield, Bohai [J]. Fine and Specialty Chemicals, 30(2): 18-22.
[9] Han Z. K., Liu X. W., and Qin L., et al. (2023) Inversion study of the shape of river cross-sections in unganged regions based on EnKF method [J]. Water Resources and Power 41(11): 23-24.
[10] Liu Y. W., Wang W., and Liu Y. B., et al. (2022) Evaluation of the optimization effect of streamflow data assimilation on SWAT model parameters based on the EnKF approach [J]. Journal of Hohai University 50(2): 2-4.

Research on structures and welding methods of primary surface heat exchangers

Haibo Wang [1], Haisheng Sun [1], Yiming Ma [1], Gen Ma [2], Kaishu Guan [2*], Kaikai Hao [1], Xing Qi [1]

[1] Lanpec Technologies Limited (Shanghai), Shanghai 201518, China;
[2] East China University of Science and Technology, Shanghai 200237, China

*Corresponding author E-mail address: guankaishu@ecust.edu.cn, Tel.: +86 64253055

Abstract: Primary surface heat exchangers (PSHEs) are used more and more all over the world in aerospace, power, and airplane areas due to their micro-size and excellent heat transfer efficiency. In this paper, at first, the development of different structures of (PSHEs) was summarized. Then, the advantages and disadvantages and application of different structures of PSHEs were analyzed. Finally, the different welding methods of the ultra-thin stainless plate were researched in this paper, the results show that the three kinds of methods, micro-beam plasma arc welding, laser welding, and pulsed TIG filling wire welding, can be used to weld (PSHEs).

1 Introduction

To ease the pressure of energy stress and to achieve compliance and carbon neutrality goals, the Primary surface heat exchangers (PSHEs), used in distributed Energy Supply Systems, recuperators, and micro gas turbines, et al., have greatly developed all over the world. PSHEs are advanced thermal management devices widely used in industries requiring high efficiency, compact design, and lightweight solutions. PSHEs utilize the primary surface (the surface was contacting with the fluid directly) for heat transfer, eliminating the need for secondary surfaces like fins. This design enhances thermal efficiency and reduces weight and volume [1-3]. In the use of ultra-thin plate or corrugated plate structure, the heat transfer area per unit volume is large, the volume is much smaller than the traditional shell and tube heat exchanger, and the fluid flows directly in the corrugated plate or micro-channel without passing through the secondary surface (such as fins), reducing the heat transfer resistance. Special materials (such as stainless steel and nickel-based alloys) and welding processes make it suitable for extreme conditions (such as high temperature and pressure gas and fluids). The turbulent design and high specific surface area enhance the heat transfer efficiency 3-5 times that of the shell-to-tube heat exchangers, significantly saving energy [4-5].

Structure design and manufacturing technology are the critical factors that affect the application of PSHEs. The structure and manufacturing progress were reviewed first, then a welding test was carried out to verify if the weld manufacture method could be used for the ultra-thin stainless plate of PSHEs. The paper was divided into five chapters. Chapter 1 introduced the aim and contents, Chapter 2 summarized the progress of structure design, Chapter 3 reviewed the manufacture of PSHEs, and then the welding method was researched in Chapter 4. Chapter making the main conclusions of all the work.

2 The structure of PSHEs

2.1 Characteristics of the structure

The PSHEs have a corrugated board structure for the heat transfer primarily, which is mainly divided into three types of structure, that is, Cross corrugated type (CC), Corrugated undulated type (CU), and

Cross wave type (CW) according to the specific structure shape shown in Figure. 1. The CU structure plate was basic the same as the CC plate, and the difference is that the dimensional sizes of upper and lower plates. The CW plate has an additional waviness in the flow direction compared to the CC structure.

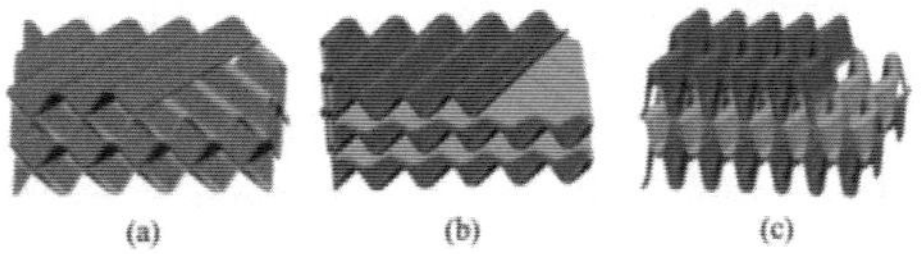

Figure. 1 Surface structure of heat exchange plate.

2.1.1 CC structure. CC (cross corrugated) corrugated plate is a compact primary surface heat transfer element, which is also often called V-shaped corrugated plate or herringbone corrugated plate in plate heat exchangers. At present, the heat transfer properties and flow resistance properties of CC plates have been studied extensively at home and abroad. The influence of flow transfer performance was pitch ratio P/H and staggered Angleθ. The research results indicate that for the CC plate with P/H = 1.78, j and f increase monotonically with θ increasing in the range of $40 \le \theta \le 160$ for a given *Re* value. The optimal structure is the one with smooth corrugation and a small dip Angle.

The main structure characterization of the CC structure is that a lot of corrugated surface sheets are put together in various directions at a proper angle. The assembled angle θ changes from 0° to 90° and can amount to 2000 m^2/m^3 for compactness [6-7] and to 10 percent for the wave corrugations [8].

The improved or revised CC type was also developed by different researchers. The interesting one was adding offset bubbles in the CC structure shown in Figure. 2 [9]. Because the structure can produce both transverse mixing and longitudinal disturbance to the fluid. Due to the great changes in the flow section at the trough, the fluid strongly scours the trough surface and causes the flow line to distort at the trough, and the effect on the flow and disturbance is enhanced. Therefore, the convection heat transfer effect of the fluid is strongest here. When the ripple of the corrugated plate becomes steeper, the fluid scours the corrugated surface more strongly, resulting in a thinner velocity boundary layer and a larger temperature gradient in the region near the trough. The lowest temperature distribution is located at the position where the fluid bypassed the trough. The average area quality factor can be improved by 41% and 71%, respectively.

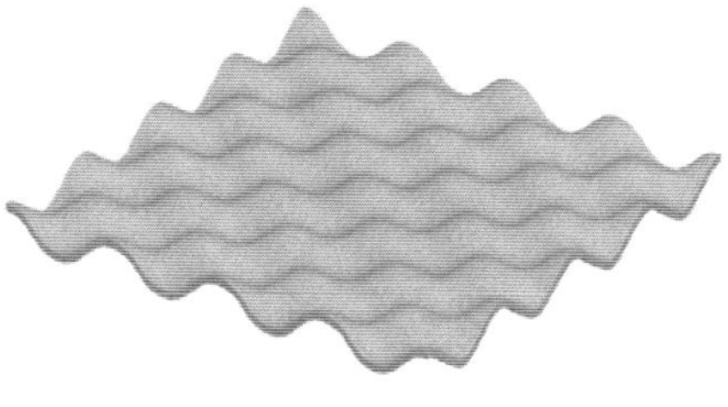

Figure. 2 Offset-bubble surface.

Another interesting, improved CC structure was three-dimensional (3D) CC introduced by Blomerius and Mitra [10]. They added vertical waves stacked on the conventional CC surface waves with a corrugation degree of 45°and the heat transfer efficiency by about 10 percent compared with traditional CC structure.

Journal of Physics: Conference Series **3011** (2025) 012037 doi:10.1088/1742-6596/3011/1/012037

2.1.2 CW structure. The CW sheets bend as a sinusoidal wave in the main direction. In that way, 180° out of phase among lower and upper sheet's waviness is existence. Because the adjacent channels of the CW structure communicate with each other and the flow direction is periodically deflected, the secondary flow in the CW channel is very strong. In the range of 288-2951 Reynolds number, the heat efficiency and flow resistance properties of CW structure channels with different hydraulic diameters are similar. A CW-type heat exchange plate is a heat exchange plate in some special shape corrugated by pressing ultra-thin sheets. The neighboring heat exchange sheets were stacked and assembled to make up a CW-type structure. T. Among them, the two-phase channels were interleaving each other, resulting in the compactness and heat transfer efficiency of the CW-type raw surface type regenerator being very high [11-12]. The fluid flow channel between the heat exchange plates of a CW-type primary surface regenerator is a complex periodic three-dimensional flow structure. For the cross-section direction, the pitch of the corrugated curve is P, and the height is H. The fluid mixing effect caused by the different fluctuation directions of the lower and upper r sheets is an important factor affecting its heat transfer performance. Some small-scale vortices will form close to the lower and upper walls of the passage. These small-scale vortices can disturb the boundary layer near the wall, make the boundary layer thinner, and enhance heat transfer, but also increase the frictional resistance and pressure loss between the fluid and the wall of the passage. Due to the fluid forming eddies states at the center of the channel, the mixing effect of the fluid and the convective heat transfer will be reduced [13-16].

2.1.3 New type Double-Wall Heat exchanger (DWHX). A new type of PSHE structure has been designed all over the world. A double-wall Heat exchanger (DWHX) used in an ARC-class reactor cooling system was created by Francesco Colliva [16]. They designed corrugations that were installed vertically to the traditional single-wave structure surface, as shown in Figure. 3. The advantage of this structure was that large-scale and small-scale vortices among the double-wave plates could operate simultaneously, so the mixing effect of fluid could be enhanced, and heat transfer efficiency was enhanced by 50 percent. However, flow resistance was enhanced by 30 percent.

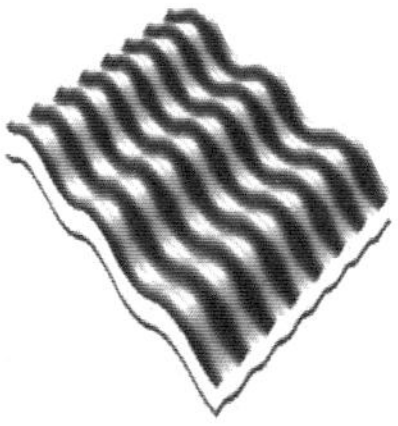

Figure. 3 Double-wave surface.

2.2 Summary of different structures

For a specific gas turbine regenerator, the CC structure has the smallest volume and the lightest weight. The straight corrugated CC-type is easier to manufacture through automated stamping, forging, and folding. CC-type structure of heat exchanger volume will than any kind of heat exchanger can be in the business field are smaller, lighter weight.

CU-type heat exchanger performance and the geometric parameters of the structure, such as corrugated and the Angle between the direction of perturbation and the internal height of the lines, have a lot to do. The main disadvantage of the CU structure is the need for two different plate types, which increases the heat exchanger processing volume and increases the cost.

Because the adjacent channels of the CW structure are connected and the periodic deflection of the flow direction, the secondary flow in the CW channel is very strong, and the heat exchange efficiency of the CW structure is 6 times that of the through channel heat exchanger, but its pressure loss also increases with the same proportion. The ring CW original surface regenerator can achieve 90% heat transfer efficiency and 2.5% relative pressure drop.

The CW structure has better performance than the simple straight pipe structure. In the production process, the wave type is formed by folding and cannot be molded and forged to avoid breakage.

Due to the existence of tiny secondary ripples in the new surface, double-wave surface flow can make the fluid mixing in the channel more intense. Compared with single corrugated type CC surface, double corrugated type heat exchanger heat transfer property can be improved by 50%, but at the same time, the pressure loss will increase by 30%.

3 Review of manufacturing technology

There are few references about the manufacturing of PSHEs. Because the complex structure and ultra-thin metal sheets were used, Photochemical Etching and Additive Manufacturing (3D Printing) were used in this area [17].

Photochemical Etching is used to create intricate flow channels and patterns on thin metal sheets. It can produce high precision, no mechanical stress, and the ability to produce complex geometries. However, it is difficult for ultra-thin sheets and adapt to specific materials.

Additive Manufacturing (3D Printing) was also used to manufacture PSHEs. It can produce complex geometries and is especially suitable for manufacturing PSHEs. However, this technology can slow production rates and increase costs for large-scale manufacturing.

Another technology, Diffusion Bonding technology, which is a solid-state joining process used to bond multiple layers of etched sheets without melting, can also be used to manufacture PSHEs. It can Create leak-proof, high-strength bonds, but it requires high pressure and temperature with expensive equipment [18].

4 The welding method

Due to the PSHEs being made by ultra-thin plates with about 0.1mm thickness, welding technology and complex molding were the critical factors. The welding methods were researched by our research team. PSHEs are typically constructed by stacking and bonding multiple layers of ultra-thin metal sheets with 0.1mm thickness. Due to the difficulty of wild ultra-thin sheets, the welding process must be precise to maintain the structural integrity of these channels and ensure optimal heat transfer.

4.1 experimental material

Figure. 4(a) shows a 304 stainless steel heat exchange sheet as the base metal with a thickness of 0.1 mm. The ER304 stainless steel solder with a diameter of 0.15 mm is selected as the welding wire (as shown in Figure 4(b)). The thermophysical properties of 304 stainless steels are shown in Table 1.

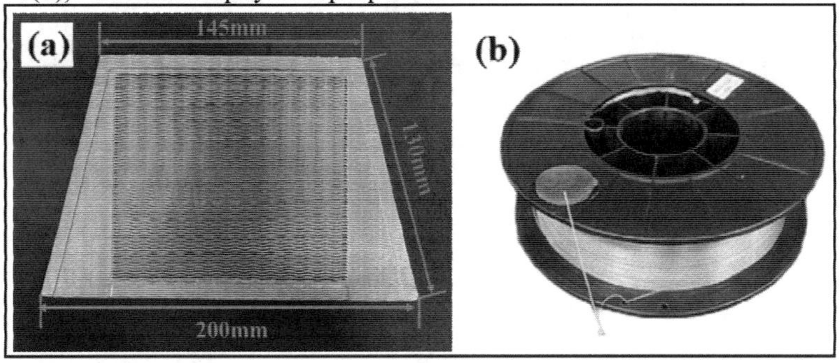

Figure. 4 Test material: (a) 304 stainless steel heat exchanger sheet; (b) ER304 stainless steel.

Table 1 stainless steel welding wire.

materials	strength of extension (Mpa)	yield strength (Mpa)	heat conductivity ($W \cdot m^{-1} \cdot K^{1}$)	coefficient of linear expansion ($10^{-6} \cdot K^{-1}$)
304 stainless steel	601	215	21.1	18.2

4.2 Test method and equipment

4.2.1 Test facility. Figure. 5(a) shows the micro-beam plasma arc welding platform, which can be welded by a plasma arc welding machine. The PLASMAFIX51 microbeam plasma arc welding machine, shown in Figure. 5(b), is a programmable transistor welding power source designed for microbeam plasma (low current plasma welding) and TIG welding processes. It can be used for automatic welding with high energy density and heat concentration. Therefore, the welding depth is large, the weld heat affected zone is small, and it is easy to obtain high-quality weld forming and metallographic structure [19].

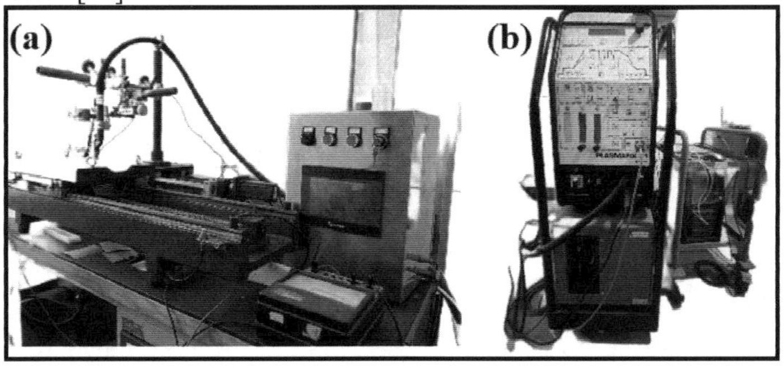

Figure. 5 Welding equipment for micro-beam plasma arc welding: (a) micro-beam plasma arc welding platform; (b) PLASMAFIX51 microbeam plasma arc welding machine.

The welding platform used in the laser self-fusion welding test includes a laser, robot arm, laser head, and related welding fixtures.

4.2.2 Test method. Three different welding processes of micro-beam plasma arc welding, laser welding, and pulsed TIG filling wire welding were selected to carry out the research. To further study the welding process of stainless steel plates and enhance the production efficiency under the condition of ensuring the perfection of the weld and metallographic structure, the process flow, macro, and microstructure were tested and analyzed.

4.3 Weld forming analysis

4.3.1. Laser welding weld forming analysis. As one of the important criteria of welding quality, the joint macroscopic morphology is the most intuitive and reliable, and the observation of joint macroscopic morphology is the first step to the quality judge the welding.

Due to the built-in groove structure and the wavy shape of the corrugated plate welding position, "face surfacing + filling wire" (stainless steel welding wire with a diameter of 0.15 mm) is selected for laser welding.

Figure. 6 gives the morphology of laser power on the weld seam of near-equal thickness and thin plate. Different welding powers and different wire feeding speeds will obtain different welding macro morphology. Welding laser power seriously affects the macroscopic morphology of the weld. Figure.6(a) Continuous hump was found in the weld after welding with high welding laser power. When the laser power was reduced to medium, the hump condition was improved, but there was still a hump; When the lowest laser power was used, the hump shape was eliminated, but the appearance of the weld was not good. To gain a good weld appearance, we must adjust the speed of welding wire feeding [20].

Figure.7 shows the shape diagram of the influence of wire feeding speed on the weld seam of near-equal thickness and thin plate. The improved test parameters (laser power 900 W, wire feed speed 350 cm/min, as shown in Figure.7(a)) show unattractive weld formation and slight discontinuity morphology that is caused by excessive wire feeding. Subsequently, the wire feeding speed was reduced, and the parameters of laser power 900 W and wire feeding speed 300 cm/min were adopted, as shown in Figure.7(b). It was found that there was only some section of weld seam with excellent morphology. However, the weld seam morphology in other sections is poor with discontinuous structure. Under the parameters of 900 W laser power and 270 cm/min wire feed speed, as shown in Figure.7(c), a well-formed weld with continuous weld and no hump was obtained during the continuous reduction of wire feed speed.

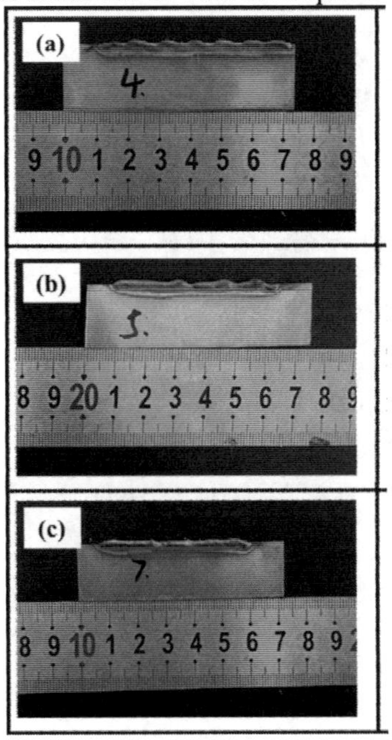

Figure. 6 Topography of laser power influence on weld seam of near-isothickness thin plate, (a) with high, (b) with medium laser power, and (c) with low laser power).

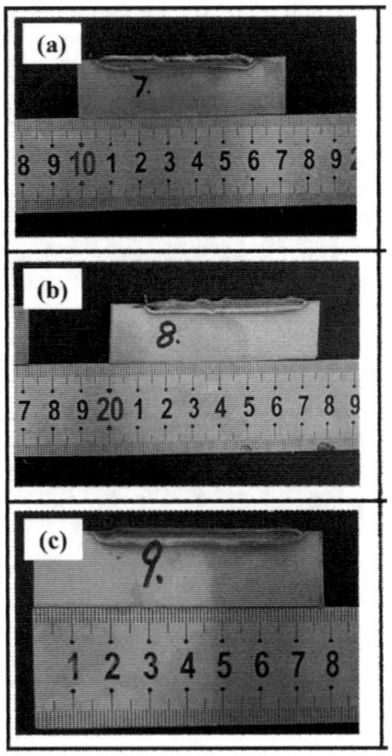

Figure. 7 Appearance diagram of influence of wire feeding speed on weld seam of near-equal thickness plate (a low speed, medium speed, high speed).

4.3.2 Weld forming analysis of micro-beam plasma arc welding. When the micro-beam plasma arc welding process is adopted, the process parameters are also explored by using the near-equal thickness plate. Since the microbeam plasma process is characterized by a stable welding arc, the change of welding current was first adopted in the test. The selected process parameter variable is welding current. Figure 8 shows the shape diagram of the effect of welding current on the weld seam of near-equal thickness and thin plate. Figure.8(a) shows a large heat-affected zone at the Welding arc starting point. This is because the microbeam plasma arc welding process needs to wait for the tungsten electrode to arc. Caused by excessive heating time. It was observed that after increasing the welding current, a well-formed weld with continuous, flat, and no groove was obtained [21].

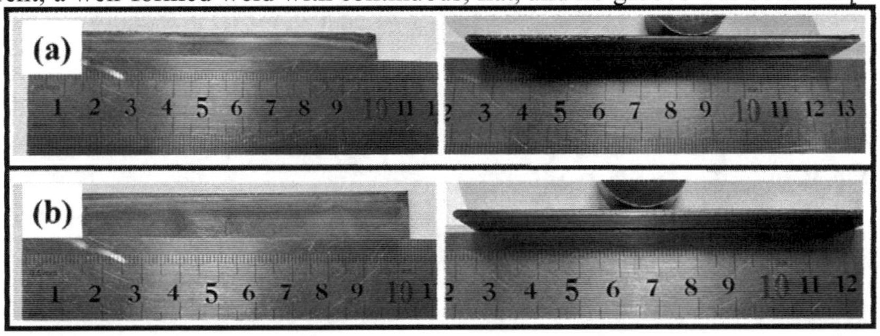

Figure. 8 Weld morphology of near-isopath plate micro-beam plasma arc welding ((a) high current, (b) low current).

4.3.3 Weld forming analysis of pulsed TIG wire filling welding. For the pulse TIG wire filling welding process, the corrugated plate is directly used to explore the process parameters of welding current 10 A, welding speed 8mm/s, and protective gas flow rate 15L/min are adopted (as shown in Figure. 9). The weld seam was plate, and no weld defect was detected.

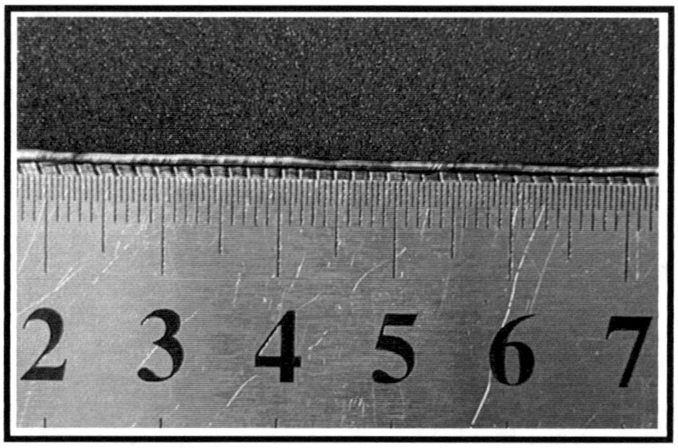

Figure. 9 Weld morphology of corrugated plate pulsed TIG welding.

4.3.4 microstructure analysis. Different welding processes can produce different microstructures for ultra-thin sheet samples, so the microstructure of the weld was analyzed, as shown in Figure. 10, and the microstructure of the heat-affected zone in the center area of the weld and the base material under three different welding processes As shown in Figure. 10((a)~(c)), the metallography of the weld zone, the weld center zone of the three welding processes has a relatively uniform microstructure, no porosity, cracks, and other defects. A dendritic austenitic grain, which has good mechanical properties [22].

The metallography of the heat affected zone (HAZ), however, the three welding processes show inconsistencies. The laser welding process has a small HAZ due to its fast-welding speed and fast laser cooling and heating characteristics (Figure. 10(a).) In the micro-beam plasma arc welding process, there appears a fine and dense grain region in a vertical weld seam direction, which is speculated to be caused by the epitaxial growth of dendrites (Figure. 10(b)). The HAZ produced by the pulsed TIG wire-filling welding process was consistent with the laser welding process (Figure. 10(c)). This is because the pulsed TIG wire filling welding process melts the welding wire rather than the base material, which essentially reduces the average heat input to the base material, resulting in a smaller heat-affected zone.

The microstructure in the base material region, the grains will have a preferred orientation at the base material boundary. When the preferred orientation is not completely consistent with the temperature gradient, the growth direction of dendritic crystals will present a certain angle with the temperature gradient direction, so the microstructure in the base material boundary region presents the characteristics of fine and dense grain region. The laser welding process has a large temperature gradient, and the austenite grains nucleate and grow rapidly at the molten pool boundary, forming dendritic grains, while the grains in the base metal boundary area are too late to melt. Therefore, the HAZ of the joint under the laser welding process is not obvious. The base metal area shows a uniform austenite structure with a grain size between 20-30 μm.

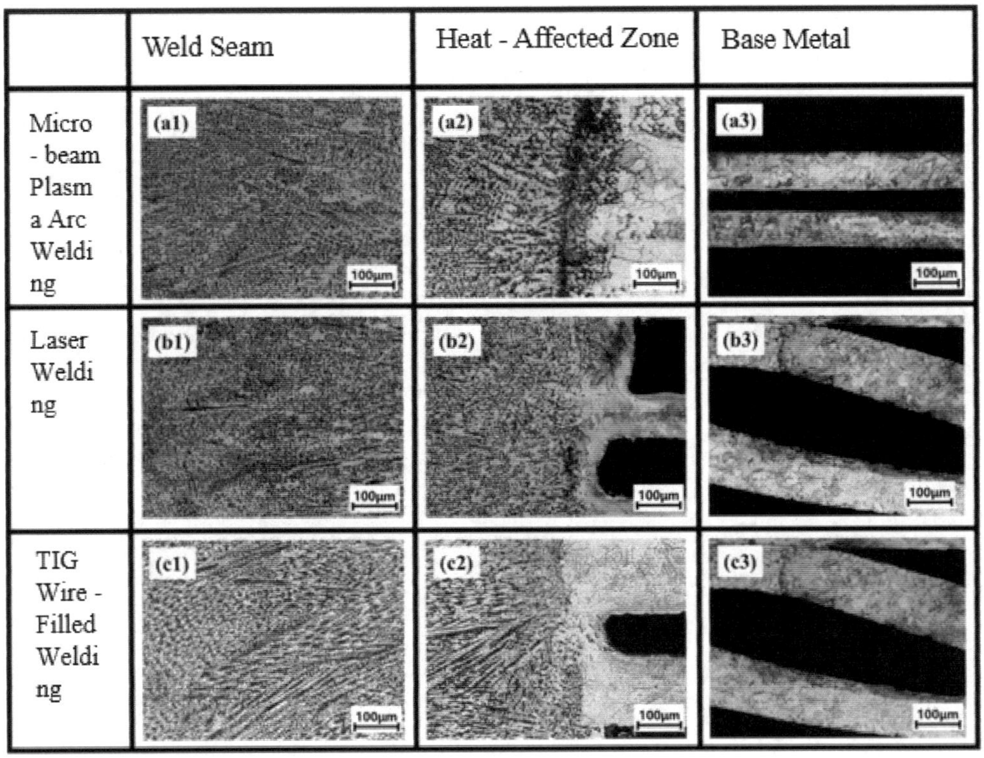

Figure. 10 Microstructure morphology: ((a1)-(a3)) micro-beam plasma arc welding; ((b1)-(b3)) laser welding; ((c1)-(c3)) pulsed TIG filler wire welding.

4.4 summary of weld methods

For the laser welding process, due to the instability of wire feeding, the melt pool will cause droplet spreading instability under the action of the protective gas, so compared with the laser wire filling welding, the welding surface of the laser self-fusion welding is better. Experiments show that larger laser power will lead to excessive heat accumulation, resulting in weld burn-through. On the contrary, insufficient laser power will weaken the penetration capacity and cause the weld to not be fully fused, both of which are not conducive to the improvement of the bearing capacity of the welded joint and the welding process with laser power ranging from 260 W to 360 W was a preference parameter. As a modern welding method, laser welding has developed rapidly in recent years. Compared with the traditional fusion welding method, laser welding has the advantages of concentrated energy density, large weld depth-to-width ratio, small heat affected zone, small deformation and residual stress after welding, and is especially suitable for welding ultra-thin plates.

For the microbeam plasma arc welding process, the experimental results show that different pulse parameters can result in different weld morphologies. Microbeam plasma arc welding with welding speed, narrow, small heat affected zone of weld, and the characteristics of small welding deformation, often used for welding of metal thin ultra-fine precision structure parts. Compared with DC, when pulsed current is used for welding, the arc stability is strong, the average heat input is low, the process parameters are adjustable, the welding pool is stirred, the weld quality is improved, and the welding crack tendency is small.

For the pulse TIG welding process, the observation of the weld can see the wire melting trace. The experimental results show that the welding width increases with the increase of welding current, and the depth of penetration also increases. Smaller welding widths and greater depths of penetration can be achieved with pulse welding at the same effective current. Micro-beam TIG welding has the characteristics of easy parameter control, simple operation, low cost, and high production efficiency, which is suitable for industrial mass production.

After welding of metallographic observation, three kinds of welding way (microbeam plasma arc welding, wire filling welding, laser welding, and pulse TIG) produced by the welding seam and heat affected zone and parent metal microstructure difference is not big, can meet production needs [23-24].

5 Conclusions

The PHSEs are developed in the aerospace, power, and airplane areas due to their micro-size and excellent heat transfer efficiency.

A general review of the advantages of PHSEs with different types of PHSEs has been presented. Improved PHSEs, such as the double-wave made by Offset-bubble structures, were also addressed. The advantages and disadvantages and suitable application scopes for different structures of PHSEs were analyzed.

Manufacturing technology has a little report in the reference, mainly related to Additive Manufacturing (3D Printing) technology. Welding technology was researched in this paper, and it shows that micro-beam plasma arc welding, laser welding, and pulsed TIG filling wire welding can be used to weld ultra-thin stainless steel sheets.

Reference

[1] Saari, J., Kozlova, M., Suikkanen, H., Sermyagina, E., Hyvärinen, J., & Yeomans, J. S. (2024). Global sensitivity analysis of nuclear district heating reactor primary heat exchanger and pressure vessel optimization. Energy, 312, 133393.
[2] Borjigin, S., Zhao, W., Fu, W., Liang, W., Bai, S., Ma, J., et al. (2025). Review of plate heat exchanger utilized for gases heat exchange. Renewable and Sustainable Energy Reviews, 210, 115224.
[3] Wang, Q., Zeng, M., Ma, T., Du, X., & Yang, J. (2014). Recent development and application of several high-efficiency surface heat exchangers for energy conversion and utilization. Applied Energy, 135, 748–777.
[4] Liao, W., Jing, Z., & Lian, S. (2024). Numerical analysis of thermal-hydraulic comprehensive performance of composite corrugated tubes with different cross-section shapes. International Communications in Heat and Mass Transfer, 155, 107588.
[5] Li, W., Jin, M., Liu, C., Li, W., Song, Y., Zhou, T., et al. (2024). Numerical investigation of thermal-hydraulics characteristics of cross-flow corrugated plate heat exchanger for lithium-cooled space nuclear reactor. Applied Thermal Engineering, 238, 121931.
[6] Davidson, E. E., Betzler, B. R., Cao, Y., & Fei, T. (2024). The importance of delayed neutron precursors in gamma dose calculations for activated primary heat exchanger components in the Molten Salt Breeder Reactor. Annals of Nuclear Energy, 199, 110276.
[7] Yuan, T., Tian, G., Yan, J., Wang, T., Yu, G., & Zhang, N. (2024). Investigation of cross-wavy primary surface structures for heat exchangers of Brayton cycle system using argon. Progress in Nuclear Energy, 173, 105291.
[8] Liu YZ, Zhang SB, Dong W, Liu ZY. Numerical investigation and analysis of heat transfer characteristics of air-air primary surface heat exchanger in gas turbine. Gas Turbine Experiment and Research 2017:30(05): 53-57+62.
[9] Luo T, Ma T, Zeng M, Wang QW. Stress study on different designs of ceramic high temperature heat exchangers. Chem Eng Trans 2012;29:667–72.
[10] Blomerius H, Mitra NK. Numerical investigation of convective heat transfer and pressure drop in wavy ducts. Numer Heat Trans 2000;37:37–54.
[11] Zhu XH, Lou DC, Tong CC. Numerical investigation of the flow and heat transfer characteristics inside cross-corrugated flow passages. Gas Turbine Experiment and Research 2016;29(01):36-40.
[12] Yin JX, Li GJ, Feng ZP. Numerical Study on the Characteristies of Convection Heat Transfer in Crossed-Corrugated Primary Surface Exchanger. Journal of Xi'An Jiaotong University 2005;39(1):36-40.
[13] Yang, Y., Chen, F., Yu, J., Song, Y., Hu, H., Xu, D., & Jiang, H. (2022). Numerical study on heat transfer characteristics of heat exchange cell in an annular cross-wavy primary surface recuperator (annular CWPSR). Applied Thermal Engineering, 216, 119062.
[14] Yuan, T., Tian, G., Yan, J., Wang, T., Yu, G., & Zhang, N. (2024). Investigation of cross-wavy primary surface structures for heat exchangers of Brayton cycle system using argon. Progress in Nuclear Energy, 173, 105291.
[15] Zhang, B., Chu, W., & Wang, Q. (2024). Analysis on heat transfer enhancement mechanism in a cross-wavy primary surface heat exchanger based on advection thermal resistance method. International Journal of Heat and Fluid Flow, 109, 109556.
[16] Wang, R., Wang, Z., Li, Z., Wang, Y., & Wang, M. (2024). Study on the influence of circumferential shape on the performance of cross wavy primary surface heat exchanger channel. International Journal of Heat and Fluid Flow, 105, 109259.

[17] Colliva, F., Hattab, F., Siriano, S., Ferrero, G., Meschini, S., Testoni, R., et al. (2024). Conceptual design and supporting analysis of a double wall heat exchanger for an ARC-class fusion reactor Primary cooling system. Fusion Engineering and Design, 201, 114261.
[18] Tano, I.-N., Rasouli, E., Ziev, T., Wu, Z., Lamprinakos, N., Seo, J., et al. (2022). An additively-manufactured molten salt-to-supercritical carbon di-oxide primary heat exchanger for solar thermal power generation – Design and techno-economic performance. Solar Energy, 234, 152–169.
[19] Niknam, S. A., Mortazavi, M., & Li, D. (2021). Additively manufactured heat exchangers: a review on opportunities and challenges. The International Journal of Advanced Manufacturing Technology, 112, 601–618.
[20] Manh, N. H., Nguyen, V. A., Le, H. D., Murata, A., Le, V. T., Ngoc, T. Q., et al. (2022). Development of a novel GTAW process for joining ultra-thin metal sheets. Journal of Manufacturing Processes, 80, 683–691.
[21] Cui, Q., Ma, X., You, Z., Yang, X., Zhang, Y., & Wei, J. (2025). Bending the heat: Innovative ultra-thin flexible loop heat pipes for enhanced mobile device cooling. Energy Conversion and Management, 325, 119332.
[22] Hong, Y., Gao, J., Lin, K., Li, S., Chang, B., & Du, D. (2025). Bridging behavior of molten pool and its effect on defects formation in ultra-thin sheets edge welding by micro-plasma arc. Journal of Materials Processing Tech., 336, 118696.
[23] Xu, H., Guo, X., Lei, Y., Li, J., Hua, F., Xia, R., et al. (2019). Welding deformation of ultra-thin 316 stainless steel plate using pulsed laser welding process. Optics and Laser Technology, 119, 105583.
[24] Zhao, J., Zhu, Y., Liu, X., Jiang, R., Ding, B., & Chen, Y. (2023). Root cause analysis of a cracked primary heat exchanger in a gas wall-mounted boiler. Engineering Failure Analysis, 153, 107583.

Analysis method of transient overvoltage caused by DC bipolar blocking based on second-order equivalent circuit

Qidi Zhong*, Yalun Lu, Ming Li, Shiyang Chang and Shen Yu

State Grid Economic and Technological Research Institute Beijing 100044, China

*Corresponding author's e-mail: 1248969486@qq.com

Abstract: In the wind power transmission system via DC under the support of a weak grid, there is a transient process on the same time scale as the fault duration. The existing calculation method of temporary overvoltage based on surplus reactive power does not consider the transient characteristics of the system, and there is a large error in the calculation of temporary overvoltage under a weak grid. In this paper, research is carried out through theoretical derivation and simulation verification. Firstly, the AC grid near the sending end is reasonably simplified and equivalent to a second-order circuit, and the analytical expression of the transient voltage is given. Then, combined with the actual physical process, the physical meanings of the transient component and the steady-state component in the total transient voltage response are respectively discussed. Finally, based on the Matlab/Simulink simulation platform, a typical system model of wind power transmission via DC is built to verify the correctness of the proposed method. The results show that the equivalent second-order transient response method proposed in this paper has high accuracy in describing the short-term temporary overvoltage characteristics and calculating the overvoltage peak value after the DC bipolar blocking fault.

1 Introduction

In recent years, with the gradual progress of the energy transformation strategy, wind power in China has developed rapidly. Meanwhile, considering the geographical characteristics of the domestic new energy power generation and the load center are far apart, the traditional high-voltage direct current transmission technology (line-commutated-converter based high voltage direct current, HVDC) has become an important means for the large-scale, high-efficiency, and safe external transmission of wind, solar, and thermal power by bundling new energy and traditional thermal power for external transmission [1]. With the increasing proportion of new energy units in the grid structure year by year, the DC sending-end system exhibits power electronic characteristics, and the voltage support capacity of the system is significantly weaker than that of the traditional AC grid [2]. Existing literature has pointed out that the short-term temporary overvoltage peaks in the transient processes of several DC sending-end grids such as Luguhu and Xitai have the risk of exceeding 1.3 p.u. [3], resulting in the reduction of the external transmission capacity of the relevant DC systems and seriously restricting the external transmission level of new energy power generation.

The system-level temporary overvoltage is generally caused by the imbalance of reactive power within the system after a large disturbance fault. Scholars [4] pointed out that DC bipolar blocking, commutation failure faults, and near-zone AC faults will all cause temporary overvoltage in the sending-end system, comparing the transient voltage peaks under various faults based on electromechanical transient analysis. Meanwhile, regarding the research on the peak value of temporary overvoltage and the analytical expression of transient voltage, researchers [4] directly correlated the system's temporary overvoltage with the surplus reactive power and proposed the calculation formula for the transient voltage rise after the bipolar blocking fault; scholars [5] relied on the power flow data before and after the fault, and additionally considered the impact of the unbalanced active power after the fault on the transient voltage rise while considering the surplus reactive power, and gave the calculation formula for the transient voltage rise after the fault and extended the calculation method to the DC commutation failure scenario.

Different from the electromechanical transient research method based on surplus reactive power, there are already electromagnetic transient research methods in the current research results: researchers [5] carried out multi-condition simulation analysis combined with the electromagnetic transient time-domain simulation model of HVDC and gave the corresponding relationship between the overvoltage of the sending-end system and the control parameters of the DC system;

In summary, it is currently urgent to study the short-term voltage transient characteristics of the wind power transmission system via DC at the system level and reveal the mechanism of the temporary overvoltage on the electromagnetic transient scale.

2 Analysis Method of Temporary Overvoltage Based on Equivalent Second-Order Circuit

This section analyzes the voltage transient process at the system level after a large disturbance fault and selects the DC bipolar blocking fault as the source of the large disturbance fault in the sending-end AC system.

2.1 System Structure Introduction

Combined with the actual field parameters, this paper establishes a typical electromagnetic transient model of large-scale wind power transmission via DC, as shown in Figure 1. The sending-end grid is a weak AC grid, and the wind farm is connected to the grid through a short-distance line. The short-circuit capacity at the DC sending-end converter bus is 25000 MV, and the system short-circuit ratio SCR = 2.5.

The high-voltage DC transmission project adopts a bipolar ±800 kV structure with a rated power of 10, 000 MW. The control system adopts the ABB series control, and the specific control strategy is referred to in literature [7]. The grids on both sides of the converter bus are the venin equivalent grids. Among them, the sending-end grid is connected to 41, 500 MW direct-drive wind turbines, and the high and low voltage ride-through control strategies after the fault are referred to in the literature [5]. The connection impedance Xi (i = 1, 2, 3, 4) between the wind turbine and the grid is the sum of the line impedance and the transformer leakage reactance from the 35 kV bus to the 530 kV bus. The primary parameters of the high-voltage DC transmission and direct-drive wind farm models are shown in Tables 1 and 2. Based on the above grid structure and parameters, a typical system model of wind power transmission via DC is built on the Matlab/Simulink simulation platform.

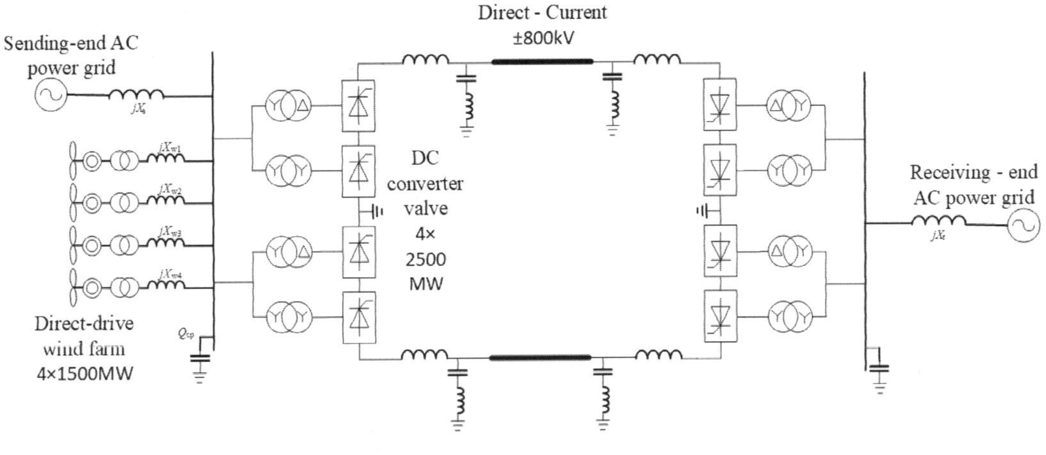

Figure 1. Structure of large-scale wind power transmission system via HVDC.

Table 1. Primary parameter of HVDC.

Parameter	Value	Parameter	Value
The rated voltage of the AC system connected to the sending-end converter station	530 kV	The rated voltage of the AC system connected to the receiving-end converter station	520 kV
Rated operating voltage of DC bus UdN	±800 kV	Rated bipolar transmission power of DC transmission system PN	10, 000 MW
DC line resistance Rb	4.64 Ω	Grounding resistance of converter station Re+Rg	0.74 Ω
Rated extinction angle γN	17°	Minimum limit angle of control system α	5°

Table 2. Primary parameters of the aggregation model of the direct-drive wind farm.

Parameter	Value	Parameter	Value
Rated Voltage (kV)	10	Rated Power (MW)	1500
Stator Resistance (Ω)	0.0025	Stator Inductance (H)	0.00182
Flux Linkage (Wb)	135.73	Number of Pole Pairs	44
Rated Rotational Speed (rad/s)	12.7	DC Bus Voltage (kV)	20
Filter Inductance (H)	0.0004	Filter Capacitor (F)	0.001
Filter Resistance (Ω)	0.05		

2.2 Fault Transient Process and Equivalent Second-Order Equivalent Circuit Model

This section analyzes the transient process of the sending-end AC system caused by the DC bipolar blocking fault. For the above typical system model of wind power transmission via DC, the following reasonable assumptions are made:

Assumption 1: After the transient fault occurs, the transient characteristics of the DC sending-end converter station are not affected by the near-zone grid, that is, the DC converter station can be equivalent to an independent current source. Therefore, the energy storage inductance X_{dc} introduced by the DC converter transformer is no longer an independent element;

Assumption 2: The high-order harmonics on the AC side introduced by the DC converter station and the harmonic filtering function corresponding to the reactive power compensation are not considered, that is, the BP11/BP13, HP24/HP26, and HP3 in the AC reactive power compensation equipment components are all replaced with the fundamental frequency equivalent SC;

Assumption 3: The influence caused by the sampling delay and controller response delay of the near-zone wind turbine is strengthened. When the wind turbine control cannot track the grid change in time, it cannot be equivalent to an independent voltage source, and the grid-side converter of the wind turbine can be regarded as an SVC at this time. Therefore, this paper assumes that the wind turbine can be processed according to the impedance $Z_p(Z_p = R_p + X_p)$. Considering that the wind turbine emits active power to the grid and absorbs reactive power before the fault occurs, $R_p< 0$ and $X_p > 0$;

Assumption 4: The reactance $X_w+ X_p$of the wind farm branch is much smaller than the reactance X_s of the AC equivalent grid, so the dynamic process of the energy storage inductance of the wind farm branch can be ignored, and the wind farm branch can be equivalent to a pure resistance branch;

Assumption 5: The equivalent resistance $R_p|$ of the wind farm branch is much larger than the equivalent resistance Rs of the AC grid, so the wind farm branch can be further treated as an open circuit on the basis of Assumption 4.

According to the above assumptions, the switching process of the AC grid near the sending end after the DC bipolar blocking fault can be equivalent to the second-order equivalent circuit shown in Figure 2. In the figure, the voltage source U_s is the equivalent voltage source of the sending-end grid, the current source I_s is the equivalent current source of the DC converter station, and L and R are the equivalent inductance and resistance of the sending-end grid respectively, and C is the equivalent capacitance of the DC system reactive power compensation equipment

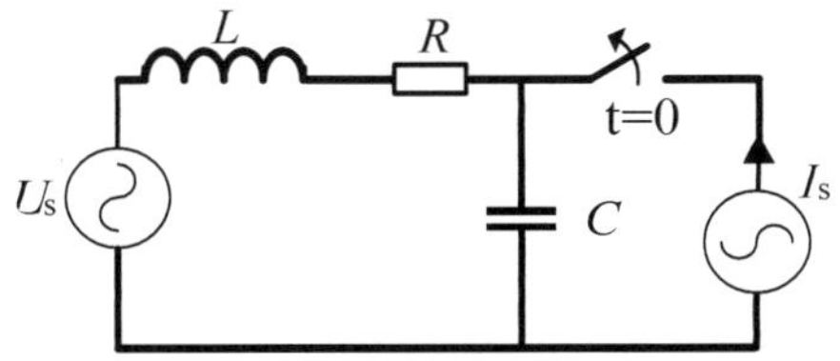

Figure 2. The equivalent second-order circuit of the DC transmission terminal AC system.

Based on Figure 2, solving the voltage of the DC sending-end system after switching can obtain the second-order non-homogeneous differential equation as shown in Equation (1), and its solution $U_{c(t)}$ is the full response of the second-order circuit.

$$LC\frac{d^2u_C}{dt^2}+RC\frac{du_C}{dt}+u_C=U_s\sin(100\pi t) \tag{1}$$

The full response $U_{c(t)}$ is decomposed into the superposition of the transient component $U_{c1(t)}$ and the steady-state component $U_{c2(t)}$. The following respectively solves $U_{c1(t)}$, and $U_{c2(t)}$.

For a weak grid with a short-circuit ratio of 2.5, the actual field parameters of R, L, and C satisfy the discriminant. Therefore, $U_{c1(t)}$ is an underdamped discharge process. Its form is:

$$u_{C1}(t)=C_1e^{-\delta t}\cos(\omega t)+C_2e^{-\delta t}\sin(\omega t) \tag{2}$$

In the formula, the attenuation coefficient and the attenuation angular frequency are defined, and C_1 and C_2 are constants. The steady-state component $U_{c2(t)}$ corresponds to the particular solution of Equation (1), and its form is:

$$u_{C2}(t)=x^ke^{\alpha x}\left[a\cos(\omega t)+b\sin(\omega t)\right]=a\cos(100\pi t)+b\sin(100\pi t) \tag{3}$$

Substituting the particular solution into Equation (1), we solve it by the method of undetermined coefficients:

$$\begin{cases} a=\dfrac{U_s[1-(100\pi)^2LC]}{(100\pi RC)^2+(1-(100\pi)^2LC)^2} \\ \mathrm{b}=\dfrac{100\pi U_sRC}{(100\pi RC)^2+(1-(100\pi)^2LC)^2} \end{cases} \tag{4}$$

Combining Equation (4) and Equation (6), the initial conditions $U_{c(0+)}=U_0$, $I_{c(0+)}=I_0$, and the constants C_1 and C_2 can be obtained:

$$\begin{cases} C_1=U_0-U_M\sin\varphi \\ C_2=\dfrac{I_0+\delta(U_0-U_M\sin\varphi)-100\pi U_M\cos\varphi}{\omega} \end{cases} \tag{5}$$

In the equation, the amplitude of the steady-state component and the initial phase angle of the steady-state component are defined. There are relationships among the initial conditions, where UN = 530 kV, and φ_1, $\varphi_2 \in [0, 360]$. Therefore, the DC converter bus voltage $U_{c(t)}$ during the switching process of the system after the DC bipolar blocking fault can be expressed as:

$$\begin{aligned} u_C(t) &= u_{C1}(t) + u_{C2}(t) \\ &= \frac{\omega_0 U_0}{\omega} e^{-\delta t} \sin(\omega t + \theta) + \frac{I_0}{\omega C} e^{-\delta t} \sin(\omega t) - \frac{\omega_0 U_M \sin\varphi}{\omega} e^{-\delta t} \sin(\omega t + \theta) \\ &- \frac{100\pi U_M \cos\varphi}{\omega} e^{-\delta t} \sin(\omega t) + \frac{U_s[1-(100\pi)^2 LC]}{(100\pi RC)^2 + (1-(100\pi)^2 LC)^2} \cos(100\pi t) \\ &+ \frac{100\pi U_s RC}{(100\pi RC)^2 + (1-(100\pi)^2 LC)^2} \sin(100\pi t) \end{aligned} \quad (6)$$

In the equation, the free-oscillation angular frequency is defined. $\omega_0 = \sqrt{\delta^2 + \omega^2}$, and $\theta = \arctan \omega / \delta$.

3 Physical Explanation of Transient Response of Equivalent Second-Order Circuit and Temporary Overvoltage

In this section, by comparing the electromagnetic transient simulation results with the time-domain characteristics of the transient component and the steady-state component in Equation (8), the physical connection between the actual model and the analytical derivation is established. The following faults are set in the built electromagnetic transient simulation model: Based on the steady-state operating condition under the primary parameters given in Section 1.1, a forced bipolar blocking of the DC system is set at 1.8 s. After 0.2 s, all reactive power compensation devices at the sending end are automatically removed. Therefore, there is a temporary overvoltage process lasting about 200 ms for the voltage of the near-zone grid after the DC bipolar blocking fault occurs. Figures 3 and 4 show the time-domain simulation waveforms of the instantaneous voltage in the near-zone of the DC sending end on the time scale of hundreds of milliseconds after the fault.

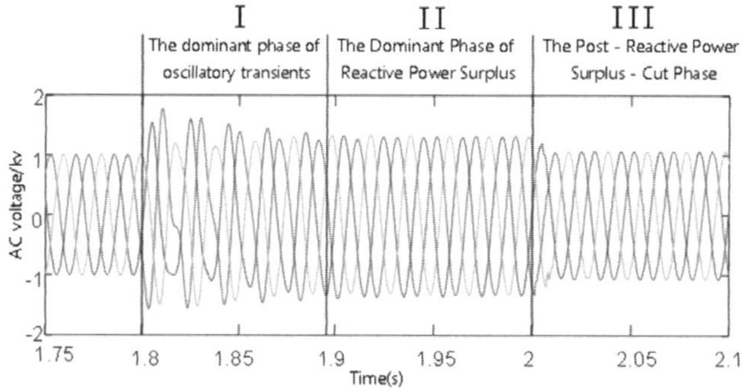

Figure 3. The instantaneous voltage of the 530 kV sending end bus after the blocking fault.

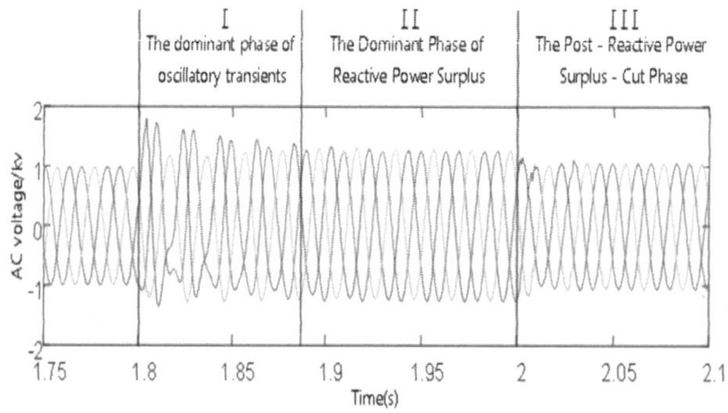

Figure 4. Instantaneous voltage of 35 kV fan terminal busbar after blocking fault.

It can be seen from Figures 3 and 4 that there are two stages with different transient characteristics for the transient voltage of the near-zone grid at the sending end after the DC bipolar blocking fault. Literature [6] indicates that Stage I is dominated by the LC energy oscillation caused by the system switching, while Stage II is dominated by the reactive power surplus triggered by the transient process.

Combined with the expression of the DC converter bus voltage $U_{c(t)}$ after the switching of the equivalent second-order circuit in the previous section, the LC oscillatory discharge process existing in Stage I conforms to the underdamped attenuation characteristics of the transient component $U_{c1(t)}$, and the two are physically unified. Within 0.2 s after the fault occurs, the reactive power compensation devices have not been removed, and the temporary voltage rise caused by the reactive power surplus corresponds to the steady-state component $U_{c2(t)}$.

Based on the above analysis, the physical process of the transient voltage rise within 0 - 200 ms after the DC blocking is as follows: In Stage I, the attenuation of the transient component $U_{c1(t)}$ of $U_{c(t)}$ has not ended. On the one hand, the system transient voltage is affected by the reactive power, and on the other hand, it is affected by the discharge of the energy storage LC in the system. In this stage, the three-phase voltages have non-power-frequency distortions, and at this time, the system has the highest temporary overvoltage, with a duration of 60-100 ms. In Stage II, the attenuation of the transient component $U_{c1(t)}$ is basically completed, and only the steady-state component $U_{c2(t)}$of $U_{c(t)}$exists in the system. The temporary overvoltage of the system is only caused by the surplus reactive power, and the surplus reactive power of the system at this time is jointly determined by the system structure and the excitation power source.

4 Time-domain Simulation Verification

In this section, the effectiveness of the equivalent second-order circuit analysis method proposed above will be verified based on the time-domain simulation model.

Firstly, we verify the effectiveness of the method in this paper for analyzing the LC oscillatory discharge process of the system: It can be known from Equation (5) that harmonics will be generated during the transient LC oscillatory discharge process after the system switching, causing distortions in the instantaneous values of the three-phase voltages, and the duration of waveform distortion is related to the time constant of $U_{c1(t)}$. Based on this, the following text will verify the theoretical validity by comparing the attenuation time constant and attenuation frequency calculated by Equation (6) with the transient voltage waveforms in Figures 3 and 4. The attenuation time constant of the transient process is deduced from the attenuation coefficient calculated by Equation (6). Compared with Figure 4, after the bipolar blocking fault, there is a non-power-frequency distortion process with a duration of about 60-100 ms in the three-phase voltage waveforms of the sending-end converter bus, which is consistent with the theoretical calculation results.

To further verify that the oscillation frequency of the harmonic components calculated after the DC bipolar blocking fault is consistent with the time-domain simulation results, the following text will conduct a frequency-domain analysis. Based on the discrete short-time Fourier transform (STFT), the spectral characteristics of the bus voltage in the near-zone grid at the DC sending end after the fault are extracted. By extracting the spectral peak sequences of the voltage waveforms in Figures 3 and 4, the harmonic components of the corresponding waveforms can be obtained as shown in Figures 5 and 6.

Figures 5 and 6 verify that there is a second harmonic component around 100 Hz in the voltage spectrum of the near-zone grid after the DC bipolar blocking fault, which is consistent with the LC oscillation harmonic attenuation frequency of 642.56 Hz $\approx 2\omega_N$ calculated by Equation (6), thus verifying the effectiveness of the method in this paper for analyzing the LC oscillatory discharge process of the system. After that, we verify the effectiveness of the method in this paper for analyzing the peak value of the transient overvoltage: Substituting P_{dc}= 10, 000 MW, P_w = 3, 000 MW, and other parameters given in the text into Equation (6), and we can get:

$$
\begin{aligned}
u_C(t) &= u_{C1}(t) + u_{C2}(t) \\
&= -323.45e^{-14.04t}\sin(200\pi t + 88.75°) \\
&+239.12e^{-14.04t}\sin(200\pi t) \\
&+756\sin(200\pi t + 88.38°)
\end{aligned} \tag{7}
$$

The time-domain waveform corresponding to Equation (7) is shown in Figure 7.

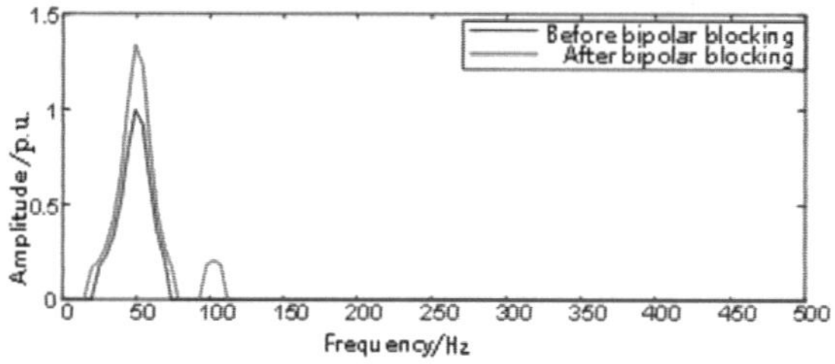

Figure 5. S_m value curve of 530 kV DC converter bus.

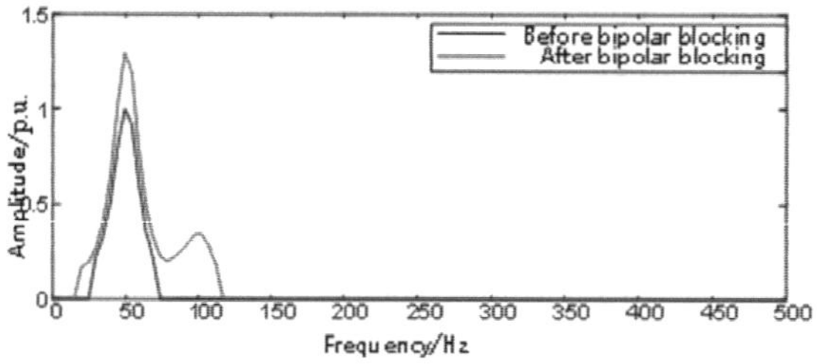

Figure 6. 35 kV wind power plant convergence bus S_m value curve.

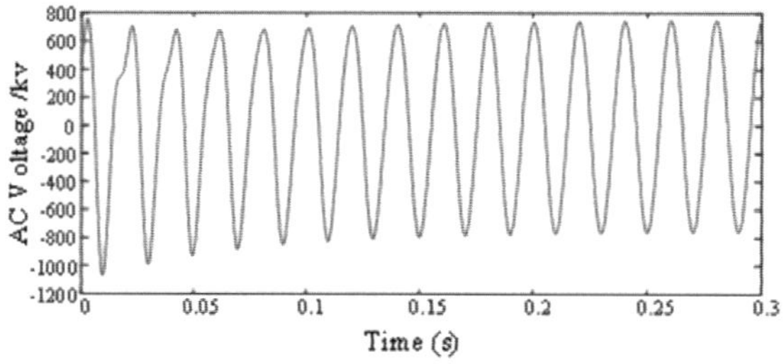

Figure 7. Waveform diagram of transient voltage analytical solution of equivalent second-order circuit.

By comparing Figure 7 with Figure 3, it is found that the waveform of the analytical solution is basically consistent with the actual transient voltage waveform after the DC bipolar blocking fault. The peak value of the temporary overvoltage in Figure 7 is 1.78 p.u., and that in Figure 4 is 1.77 p.u. The error between the simulation result and the theoretical result is relatively small, which verifies the effectiveness of the equivalent second-order circuit transient response theory proposed in this paper for calculating the peak value of the temporary overvoltage after the DC bipolar blocking.

5 Conclusion

Regarding the transient voltage characteristics of the power grid, after faults occur in the wind power transmission system via direct current, through theoretical derivation and simulation verification, this paper draws the following conclusions:

(1) The transient voltage characteristics of the sending-end power grid after a DC bipolar blocking fault can be analyzed using the equivalent second-order circuit. Under the support of a weak power grid, the contribution of the voltage transient component to the temporary overvoltage cannot be ignored.

(2) The steady-state component of the grid's transient voltage corresponds to the steady-state voltage rise caused by the surplus reactive power resulting from system faults, while the transient component corresponds to the LC oscillatory discharge process of the internal inductance and capacitance after the system switching.

(3) The equivalent second-order circuit transient overvoltage analysis method proposed in this paper has a relatively high level of accuracy in analyzing transient physical processes, transient voltage peaks, attenuation frequencies, and other aspects.

Acknowledgment

This work was supported by Research on Fault Clearance and Recovery Strategies for Ultra High Voltage Direct Current Transmission Systems (5100-202455014A-1-1-ZN).

References

[1] Liu Zhenya, Zhang Qiping, Dong Cun, et al. Study on the development mode of national power grid of China[J]. Proceedings of the CSEE, 2013, 33(7): 1-10.
[2] Ma Jin, Zhao Dawei, Qian Huimin, et al. Reviews of Control Technologies of Large-Scale Renewable Energy Connected to Weakly-Synchronized Sending-End DC Power Grid [J]. Power System Technology, 2017, 41(10): 23-31(in Chinese).
[3] Tu Jingzhe, Zhang Jian, Liu Mingsong, et al. Mechanism Analysis of Transient Overvoltage with Large Disturbance Considering Dynamic Characteristics of Wind Generator[J]. Automation of Electric Power Systems, 2020, 44(11): 197-205.
[4] TU Jingzhe, PAN Yan, ZI Peng, et al. Ximeng AC/DC weak sending-side system characteristics with angle instability and transient overvoltage. Power System Technology. 2021. 45(4): 1496-1504 (in Chinese).
[5] Denis Lee Hau Aik and G. Andersson, "Quasi-static stability of HVDC systems considering dynamic effects of synchronous machines and excitation voltage control," in IEEE Transactions on Power Delivery, vol. 21, no. 3, pp. 1501-1514, July 2006, doi: 10.1109/TPWRD.2005.861229.
[6] He Guoqing，Wang Weisheng，Liu Chun et al. Commutation Failure of UHVDC System for Wind Farm Integration (Part I)： Transient Reactive Power and Voltage Modeling of Wind Powers in Sending Terminal Grid[J]. Proceedings of the CSEE， 1-14[2022-05-05]. http://kns. cnki. net/kcms/detail/11. 2107. TM. 20211203. 2340. 015. html.
[7] Denis Lee Hau Aik and G. Andersson, "Quasi-static stability of HVDC systems considering dynamic effects of synchronous machines and excitation voltage control," in IEEE Transactions on Power Delivery, vol. 21, no. 3, pp. 1501-1514, July 2006, doi: 10.1109/TPWRD.2005.861229.

Calculation method for frequency regulation parameter requirements of power systems based on active reserve of renewable energy

Huijian Yang[1], Cheng Wang[1,*], Tianshu Bi[1], Xiaoguang Yang[2], Qian Hong[2] and Tian Jing[2]

[1] State Key Laboratory of Alternate Electrical Power System With Renewable Energy Sources, North China Electric Power University, Changping District, Beijing 102206, China
[2] State Power Economic Research Institute, Changping District, Beijing 102209, China

*Corresponding author's e-mail: chengwang@ncepu.edu.cn

Abstract. The integration of inverter-based power sources has exacerbated the issue of power system frequency stability and placed increasingly higher demands on the frequency support capabilities of inverter-based power sources clusters. This paper proposes a calculation method for frequency regulation parameter requirements of power systems based on active reserve of inverter-based power sources. It establishes a discretized power system frequency response model using the forward difference method. Furthermore, based on active reserve of inverter-based power sources, an optimization model is constructed using multi-objective function optimization to obtain the calculation results for frequency regulation parameter requirements of power systems. This method aids system dispatchers and renewable energy stations in real-time assessment of power system frequency security levels and guides the adjustment of various frequency regulation parameters for renewable energy, ensuring the safety of power system operation.

1. Introduction

Globally, the proportion of installed capacity and electricity generation from renewable energy sources in power grids has been steadily rising [1-2]. This shift has transformed the traditional model of frequency stability control, which was historically dominated by synchronous generators, presenting challenges to the stable and secure functioning of power grids [3]. To maintain post-disturbance frequency stability by ensuring the RoCoF and minimum frequency do not violate system operating limits, while also avoiding the activation of the third line of defense in the power grid, renewable-dominant power systems must preserve adequate frequency regulation capacity to meet security requirements. [4]. Tapping into the frequency control potential of inverter-based power sources has emerged as a critical approach for address system frequency stability issues and is a requirement in grid connection standards of various countries.

In fact, when evaluating the current frequency regulation capability of a system and establishing safety range requirements for it, the academic community generally follows the traditional assessment methods developed for synchronous machines, fitting the frequency regulation capability of inverter-based power sources to equivalent inertia time constants and equivalent droop coefficients [5-7]. Reference [8] applies ARMAX system identification to derive the inertia time constants in the power system. In addition, some scholars have directed their research towards new evaluation indicators for frequency regulation capability, striving for a more comprehensive measure of frequency regulation resources within the system. Ren et al. [9] assessed system frequency regulation capability by representing the rotational kinetic energy of synchronous machines and renewable energy sources with virtual inertia control functions, but it fails to account for rapid frequency regulation resources such as energy storage and cannot establish a direct mapping relationship with frequency indicators like ROCOF. Huang et al. [10] applied H_2/H_∞ performance metrics for quantitative examination of power system frequency dynamics. However, for hybrid systems containing multiple types of inertia sources, the closed-loop transfer function becomes very complex, making it difficult to generalize and apply this indicator. Hu et al. [11-12] considered the modal characteristics of inertia and proposed indicators such as frequency nadir coefficient and RoCoF slope coefficient based on the common-mode frequency to quantify frequency containment reserve capacity.

However, the electrical grid incorporates inverter-based power sources via converter interfaces, resulting in significant differences in frequency response and introducing flexibility and uncertainty in wind and solar frequency control[13-14]. It is of great significance to study how to clearly and uniformly establish requirements for the frequency control capabilities and active frequency support capabilities of inverter-based power sources. To meet the needs of small-signal and transient stability assurance with a significant percentage of inverter-based power sources and ensure adequate frequency regulation margins during operation, the secure operational envelope for frequency stabilization capacities should be quantified, with corresponding parametric demands explicitly defined for foreseeable grid contingencies.

2. Dynamic Response Model of Power System Frequency

2.1. *Aggregation strategy of synchronous unit frequency regulation parameters*

In China's power system, synchronous generators are basically divided into two major categories: reheat-type units and non-reheat-type units. Before establishing a power system frequency response model, it is necessary to aggregate equivalent parameters for both types of generators based on their respective characteristics[15]. Taking reheat-type units as an example, to simplify the equivalent model, multiple reheat-type synchronous generators are aggregated into a single equivalent unit, with their active power applied to an equivalent rotor, and a rotor motion equation is established.

The system's equivalent inertial time constant relates to its equivalent inertia through:

$$H_{eq} = \frac{H_1 S_{B1} + H_2 S_{B2} + \cdots + H_n S_{Bn}}{S_N} = \sum_{i=1}^{n} K_{mi} H_i \tag{1}$$

$$K_{mi} = \frac{S_{Bi}}{S_N} \tag{2}$$

where S_{Bi} denotes the unit rated capacity, and S_N denotes the total rated capacity of system; H_i denotes the inertia characteristic parameter of the i th generator set.

The governor action of generators is represented by the droop coefficient, and the aggregated droop coefficient of the system is:

$$\frac{1}{R_{eq}} = \sum_{i=1}^{n} \frac{K_{mi}}{R_i} \tag{3}$$

where R_i denotes the frequency-power droop characteristic parameter of the i th synchronous generator set.

The calculation methods for the reheat time constant and high-pressure turbine coefficient of the equivalent unit are as follows:

$$T_{RH} = \sum_{i=1}^{n} K_{mi} T_{RHi} \tag{4}$$

$$F_{HP} = \sum_{i=1}^{n} K_{mi} F_{HPi} \tag{5}$$

where T_{RHi} is the reheat time constant of the i th synchronous generator set, and F_{HPi} is the high-pressure turbine coefficient of the i th synchronous generator set, quantifying the steady-state power contribution ratio between the high-pressure cylinder (HPC) and the turbine's total output.

2.2. Frequency dynamic response model considering nonlinear components

China's synchronous generators mainly include reheat-type units, non-reheat-type units, and hydroturbines. As shown in Figure 1, taking reheat-type units as an example, the system frequency response model after a disturbance is illustrated.

The frequency dynamic response process after a grid disturbance is characterized based on the swing equation:

$$-\frac{2H}{f_0}\frac{\mathrm{d}\Delta f}{\mathrm{d}t}=\Delta P_m-\Delta P_e+D\Delta f \tag{6}$$

where f_0 is the rated frequency of the generator; Δf is the generator frequency deviation; D is the damping coefficient of the generator; ΔP_m is the change in governor power, i.e., the sum of the power adjustments made by the governors of all frequency regulation units; ΔP_e is the change in electromagnetic power, generally equivalent to the active power disturbance of the load; and $\frac{\mathrm{d}}{\mathrm{d}t}$ is the time-domain differential operator.

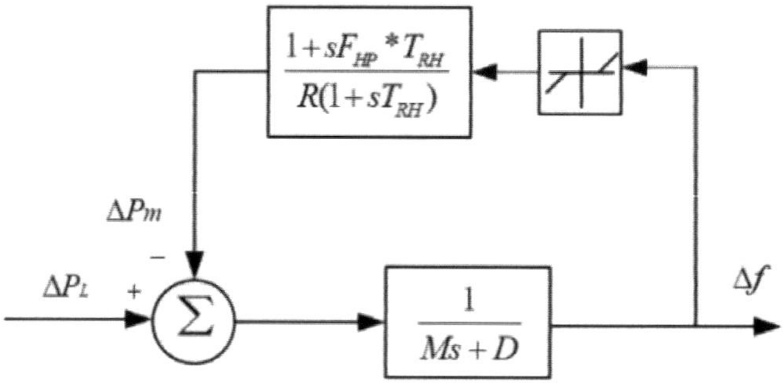

Figure 1. System frequency response model.

The SFR model is used to describe the frequency response dynamics of thermal power units, including reheat thermal power units and non-reheat thermal power units. In China, reheat thermal power units dominate. By aggregating all thermal power units within the superior power grid into a single equivalent unit, we can obtain the governor-prime mover model for the equivalent thermal power unit after a disturbance with a power deficit of ΔP_L :

$$\Delta P_L-(\Delta f-d)\cdot\frac{1+sF_{HP}T_{RH}}{R(1+sT_{RH})}=(2Hs+D)\Delta f \tag{7}$$

where ΔP_L is the initial power deficit due to the disturbance.

In addition to the above linear components, the frequency regulation process of synchronous machines also involves nonlinear components, namely governor deadbands and amplitude limits. Considering the nonlinear components of the governor, the active power output during the primary frequency regulation process of the equivalent synchronous unit should be corrected as follows:

$$\Delta P_k=\min\left\{\max\left[-K\left(\Delta f+\Delta f_{db}\right),0\right],\bar{P}\right\} \tag{8}$$

where Δf_{db} is the deadband for primary frequency regulation of the governor, typically ±0.033 Hz; $\bar{P}$ is the active power limit for primary frequency regulation of the equivalent synchronous unit; and ΔP_k is the active power adjustment of the equivalent synchronous unit after considering the nonlinear constraints of the governor.

2.3. Discretization model of frequency dynamic response based on forward difference method

The system's frequency characteristics are modeled in discrete form using the forward difference approach. When the system experiences a low-frequency disturbance, the power deficit at each difference step in the system is given by:

$$P_{dis,n} = P_{dis,0} - P_{i,n} - P_{k,n} \tag{9}$$

where $P_{dis,n}$ represents the power deficit within the system at the n-th difference step, $P_{dis,0}$ represents the system's power shortage during the disturbance onset, $P_{i,n}$ is the adjustment in active power from wind, solar, and storage sources at the n-th difference step, and $P_{k,n}$ is the adjustment in active power from thermal power units in response to primary frequency regulation at the n-th difference step.

Under low-frequency disturbance conditions, the resulting frequency variation in the power system is calculated by:

$$\Delta f_n = \Delta f_{n-1} + \frac{\Delta n}{2H}\frac{f_0}{S_b}\left(-P_{dis,n-1}\right) \tag{10}$$

in which Δf_n indicates the nth-step frequency deviation value, Δn stands for the difference step duration, and S_b corresponds to the nominal system capacity.

Under low-frequency oscillation conditions, the power increment from thermal power equivalents participating in primary frequency control is expressed by:

$$\begin{cases} P_k = -K\left(\Delta f + \Delta f_{db}\right)\cdot\dfrac{1+sF_{HP}T_{RH}}{1+sT_{RH}} \\ b_{k,n} = \max\left[-K\left(\Delta f + \Delta f_{db}\right),0\right] \\ a_{k,n} = \min\left\{\max\left[-K\left(\Delta f + \Delta f_{db}\right),0\right],\bar{P}\right\} \\ P_{k,n} = P_{k,n-1} + \dfrac{\Delta n}{T_{RH}}\left(a_{k,n-1} - P_{k,n-1}\right) + F_{HP}\left(a_{k,n} - a_{k,n-1}\right) \end{cases} \tag{11}$$

where $b_{k,n}$ and $a_{k,n}$ are the active power adjustments of equivalent synchronous units after passing through the dead-band and governor limiting at the n-th step, respectively.

3. Multi-objective function optimization model

To balance grid frequency stability needs for reliable performance in integrated energy networks and cost-efficiency considerations for sustainable generation facilities, three tiers of active reserve capacity can be established: high (e.g., 10%), medium (8%), and low (5%) of the total rated capacity from wind, solar, and storage systems.Considering the grid's frequency stability criteria and integrating varying tiers of renewable generation reserves as limitations, the optimization model restricts the output amplitudes of sustainable power sources. Following this, frequency regulation parameter configurations for clean energy integration are computed across distinct reserve capacity conditions.

3.1. Objective function

Through multi-objective optimization, the core optimization target emphasizes synthetic inertia and supplementary active power contributions from wind farms, photovoltaic systems, and energy storage during primary frequency response:

$$\min\left[\sum_{n=1}^{N}\Delta n\cdot\left(a\cdot\Delta P_{t,n} + b\cdot\Delta P_{p,n} + c\cdot\Delta P_{e,n}\right)\right] \tag{12}$$

In this formulation, $\Delta P_{t,n}$, $\Delta P_{p,n}$, and $\Delta P_{e,n}$ correspond to the frequency-regulating active power injections from wind farms, solar arrays, and battery systems at the n-th control interval. Parameters a,

b , and c quantify the relative cost-effectiveness of each resource type's contribution to system frequency support. This optimization formulation aims to reduce the total operational burden on renewable-based frequency regulation assets during post-disturbance recovery.

The secondary objective function is the weighted sum of various frequency regulation-related parameters such as inertia time constants and droop coefficients for wind, solar, and storage:

$$\begin{aligned} \min F_1' &= \min A\left(a_t H_t + a_p H_p + a_e H_e\right) \\ \min F_2' &= \min B\left(b_t K_t + b_p K_p + b_e K_e\right) \\ \min F_3' &= \min C\left(c_t P_{mt} + c_p P_{mp} + c_e P_{me}\right) \end{aligned} \tag{13}$$

Where H_t , H_p , and H_e denote the virtual inertia constants of wind, solar, and storage systems. Similarly, K_t , K_p , and K_e represent their droop coefficients, while P_{mt} , P_{mp} , and P_{me} indicate their respective power reserves. The weighting factors A , B , C , a_x , b_x , and c_x scale the influence of each frequency regulation parameter in the optimization.

3.2. Constraints

As defined in GB 38755-2019 Guide for Power System Security and Stability, frequency stability describes a power grid's capability to keep or recover its frequency within permissible limits, avoiding collapse despite severe disturbances that cause a substantial generation-load mismatch. The maximum frequency deviation (MFD), as an extreme metric of frequency response, plays a key role in optimizing governor parameters and configuring safety measures. These include under-frequency load shedding, islanding during low-frequency events, generator tripping during over-frequency conditions (linked to dynamic frequency rises), over-speed protection control, and DC modulation.

Based on the power system frequency stability guidelines, stability constraints are established for the system frequency rate of change after a disturbance:

$$\left|\frac{\Delta f_{n+1} - \Delta f_n}{\Delta n}\right| \le ROCOF^{extre} \tag{14}$$

where $ROCOF^{extre}$ represents the predefined threshold for the maximum rate of frequency variation, ensuring the fundamental conditions for stable grid operation are satisfied.

Additionally, following established frequency stability standards, operational limits are defined for the post-disturbance frequency nadir:

$$\left|\Delta f_n\right| \le \Delta f^{extre} \tag{15}$$

where Δf^{extre} represents the maximum permissible frequency deviation, satisfying the essential criteria for maintaining stable system operation.

Additionally, based on the active power reserve of renewable energy sources, constraints are established for the maximum output during the frequency support process of renewable energy:

$$\Delta P_{i,n} \le R \cdot P_{iN} \tag{16}$$

where R represents the preset proportion of active power reserve from renewable energy sources to their rated power, which can be categorized into three levels: 10%, 8%, and 5%. $\Delta P_{i,n}$ indicates the active power adjustment for renewable generation during the nth time interval, while signifies its nominal capacity.

4. Example analysis

Figure 2 illustrates the active power demand calculation for maintaining frequency stability in an adapted IEEE 39-bus network. A 100 MW load surge at bus 17 triggers frequency excursions beyond the ±0.033Hz deadband of synchronous generator governors, which are configured with 10% capacity power limits and 0.02 s time steps. Renewable sources maintain 10% power reserves (wind/solar/storage), while system stability thresholds enforce 0.2 Hz maximum deviation and 1.0 Hz/s rate limits. Uniform weighting factors of 1.5 apply to power adjustments across all renewable technologies. The weight coefficients for the equivalent droop coefficient, equivalent inertia time constant, and reserved power are set to 1.0, 10.0, and 0.5, respectively.

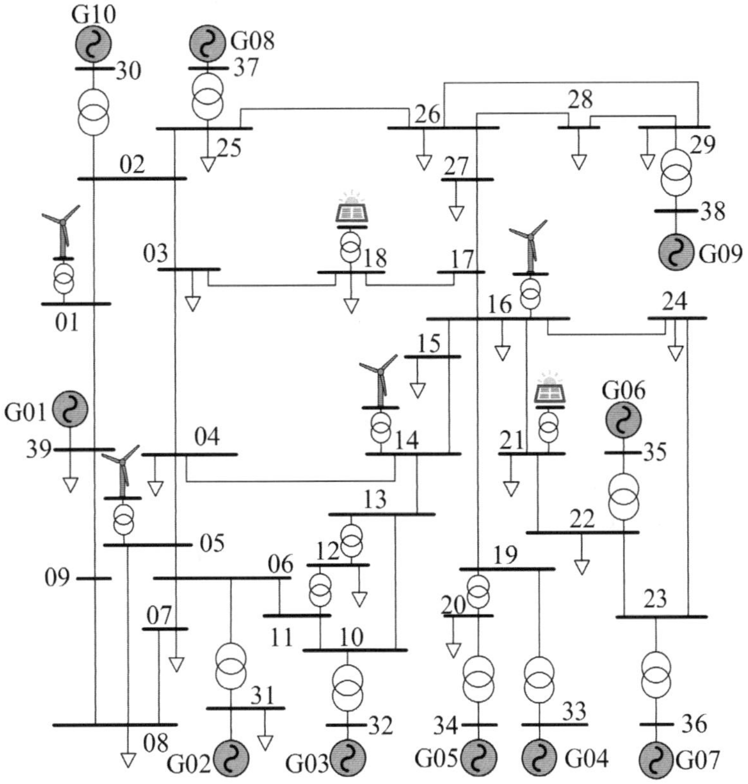

Figure 2. Schematic diagram of improving IEEE39 testing system.

A commercial solver is employed to optimize the objective function, calculating the optimal solutions for each variable that satisfies the constraints and outputting the results for each differential time step. The optimization outcomes can finally produce the computed data meeting the frequency adjustment parameter demands of power grids according to renewable energy active power reserves, which are presented in Table 1. These derived parameters for power grid frequency regulation primarily consist of: the droop coefficient of wind turbine equivalents (Kt), the droop coefficient of photovoltaic equivalents (Kp), the droop coefficient of energy storage equivalents (Ke), the inertial time constant of wind turbine equivalents (Ht), the inertial time constant of photovoltaic equivalents (Hp), the inertial time constant of storage system equivalents (He), the reserved capacity of wind power units (Pmt), the reserved capacity of solar power units (Pmp), along with the reserved capacity of energy storage units (Pme). Synchronously, a curve showing the total active power and total energy adjustment of renewable energy over time can be generated, as shown in Figure 3.

Table 1. Calculation results of frequency modulation parameters.

Parameter	Kt	Kp	Ke	Ht	Hp	He	Pmt	Pmp	Pme
Value	10.511	10.102	10.091	5.45	5.41	5.37	2.87	2.87	2.87

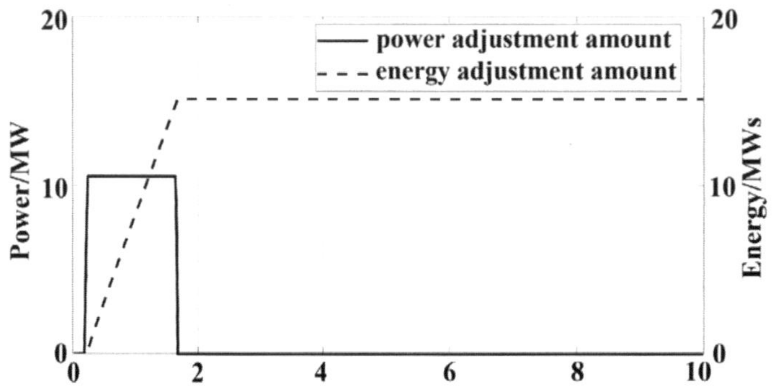

Figure 3. Schematic diagram of active power and energy adjustment of new energy.

5. Conclusion

Given the critical frequency stability challenges in renewable energy-integrated grids, this study introduces a novel approach to determine grid frequency regulation parameters by leveraging renewable energy active power reserves. The proposed technique facilitates the computation of necessary frequency control parameters for renewable sources when addressing potential grid faults, allowing both system operators and renewable power plants to continuously monitor grid frequency stability status. This approach not only provides real-time guidance for optimizing various frequency regulation settings in renewable generation but also enhances overall grid operational security while boosting the economic efficiency of renewable energy facilities.

Acknowledgments

This work was financially supported by the Science and Technology Project of State Grid Corporation(4000-202356387A-2-3-XG).

References

[1] J. ZHANG, M. Li. (2020) Analysis of the frequency characteristic of the power systems highly penetrated by new energy generation. Proceedings of the CSEE, 40(11): 3498-3506.
[2] Z. Li, W. Li, s. S. Yu, et al. (2023) Fast frequency response reserve planning for power systems considering homogeneous extreme risks. IEEE Transactions on Industry Applications, 59(2): 2314-2325.
[3] R. ZHANG, C.WANG, H. WANG, et al. (2023) Operation dispatch model of district multi-energy microgrid with frequency active supporting ability (in Chinese). Proceedings of the CSEE, 43(3): 889-904.
[4] P. M. Anderson and M. Mirheydar. (1990) A low-order system frequency response model. IEEE Transactions on Power Systems, 5(3): 720-729.
[5] S. Li, S. Xu, H. Li, et al. (2021) A Rapid Estimation Method for Equivalent Virtual Inertia of Wind Farms (in Chinese). Power System Technology, 45(12): 4683-4692.
[6] Z. Li, G. Xu, C. Wang, et al. (2023) An Estimation Method for Equivalent Inertia of Converter Interface Power Sources Considering Power Response Delay Compensation (in Chinese). Automation of Electric Power Systems, 47(01): 12-21.
[7] S. Li, S. Huang, H. Li, et al. (2022) Corrected Estimation of Power Plant Inertia Considering the Contribution of Auxiliary Rotating Loads (in Chinese). Power System Protection and Control, 50(18): 61-71.
[8] M. Liu, J. Chen and F. Milano. (2020) On-line Inertia Estimation for Synchronous and Non-synchronous Devices. IEEE Transactions on Power Systems, 36(3): 2693-2701.
[9] K. Ren, D. Zhang, Y. Huang, et al. (2022) Inertia Estimation of Large-scale Systems Based on the Proportion of Renewable Energy Output (in Chinese). Power System Technology, 46(4): 1307-1315.
[10] L. Huang, H. Xin, W. Huang, et al. (2018) A Quantitative Analysis Method for Frequency Response Characteristics of Power Systems with Virtual Inertia (in Chinese). Automation of Electric Power Systems, 42(08): 31-38.
[11] G. Hu, H. Gao, H. Xin, et al. (2023) A Quantification Method for Frequency Performance of Power Systems Considering Inertia Modal Characteristics (in Chinese). Proceedings of the CSEE, 43(22): 8559-8569.

[12] H. Gao, H. Xin, L. Huang, et al. (2021) Common-mode Frequency Analysis and Characteristic Quantification of Renewable Energy Power Systems (in Chinese). Proceedings of the CSEE, 41(03): 890-900.
[13] S. S. Oskouee, S. Kamali and T. Amraee. (2021) Primary Frequency Support in Unit Commitment Using a Multi-Area Frequency Model With Flywheel Energy Storage. IEEE Transactions on Power Systems, 36(6): 5105-5119.
[14] M. Javadi, Y. Gong and C. Y. Chung. (2022) Frequency Stability Constrained Microgrid Scheduling Considering Seamless Islanding. IEEE Transactions on Power Systems, 37(1): 306-316.
[15] B. Tomislav, H. Ninoslav and K. Igor. (2024) A new perspective on frequency control in conventional and future interconnected power systems. International Journal of Electrical Power and Energy Systems, 156: 109731-.

Study on pressure pulsation and induced noise under low load conditions of giant Francis turbine

Xiangbo Liao [1], Jiquan Tao [1], Qianlin Luo [1], Kun Wang [2*], Weiqi Deng [2]

[1] Luquan Wudongde Power Plant, Three Gorges Jinsha River Yunchuan Hydropower Development Co., Ltd., Kunming, Yunnan Province, 651511, China
[2] Datang Hydropower Science and Technology Research Institute, Chengdu, Sichuan, 6100833, China

*Corresponding author's e-mail: 1191515825@.com

Abstract: With the transformation of the energy structure, hydropower units assume the important task of deep peak regulation of the power grid. The hydropower turbines often operate under eccentric load conditions. The problems of pressure fluctuation and induced noise under the small flow conditions of Francis turbines need to be solved urgently. In this paper, based on the DES turbulence model, the three-dimensional numerical simulation of the whole flow channel of a giant mixed-flow unit is carried out, and the noise test is carried out. The results show that in the 218MW and 322MW output conditions, the scale of the leaf channel vortex decreases significantly when the suction surface is fully developed and the load is increased to 425MW output condition. With the increase of load, the low-pressure area of the straight cone section of the tailpipe shifts from the center to the pipe wall. A relatively obvious eccentric vortex band is formed under the 425MW output condition, which induces a higher pressure fluctuation amplitude. Under the influence of the tailwater vortex ring, the low-frequency pressure pulsation with a main frequency of 0.2 times the rotation frequency was detected at the tailpipe inlet and transmitted upstream. A similar pulsation frequency was observed at the volute inlet, but the pulsation amplitude attenuated to 14.8%-31.9% of the fluctuation amplitude of the tailpipe inlet pressure. Comparing the numerical simulation and noise test, the results show that the leaf channel vortex is the main reason for the induced noise of the Francis turbine under low load conditions.

1. Introduction

With the development of the energy industry, the installed capacity of giant hydropower plants has reached 10 million kilowatts [1] but is limited by operating conditions and changes in grid demand. Units are often required to operate under non-design conditions. When the unit operates under partial working conditions, it is very easy to induce eddy currents and serious pressure pulsations, and frequent pressure oscillations will cause fatigue damage to unit components and noise problems [2-4]. Müller et al. [5] explored the internal flow mechanism of a mixed-flow unit and found that vortices generated on the suction side of the rotor blades and in the tapered section of the tailpipe have a significant effect on the surrounding flow regime. Further study by Xu et al. [6] revealed that the failure damage in the head region of the blades of the mixed-flow rotor is mainly due to the poor axial distribution pattern of velocity pressure, vortex, and reflux phenomena occurring when the water enters the rotor from the water guiding mechanism. It is shown that the dynamic evolution of impeller vortices in a Francis turbine operating under part-load conditions significantly enhances the amplitude of pressure pulsations on the water exit side of the suction surface of the runner blades [7, 8]. Sun et al. [9], on the other hand, were concerned about the flow characteristics in the tailwater pipe of the mixed-flow unit under different operating conditions. They found that when the flow rate was reduced to 81% of the design flow rate, a spiral vortex band was formed. As the opening continued to decrease, the vortex band volume increased substantially. When the opening was reduced to the minimum, the tangible vortex band disappeared, and the broken vortex filled the entire straight cone section and the bent elbow section. Sun et al. [10] explored the correlation of the flow between the tailpipe and the upstream components and found that the low-frequency perturbations arise from instabilities in the fluid-structure itself and are more transmissive than the high-frequency perturbations. Guo et al. [11] found through numerical calculations that the mixed-flow hydraulic turbine has unstable eddy currents between the runner blades in the high head full load and overload operation zone and induces problems such as vibration and noise of the unit. Muhirwa et al. [12], for the mixed-flow hydraulic turbine with different guide vane openings under the pressure pulsation characteristics and induced noise effects, found that with the increase of the guide vane opening, the impact of the leaf frequency on the pressure pulsation in the runner inlet and the worm casing increases gradually, the value of the radiated sound pressure in the external field increases, and the dipole characteristics become more and more obvious. Lu et al. [13] used a numerical simulation method to study the internal flow characteristics of the turbine and its influence on noise and used the acoustic comparison method to predict the noise of the turbine, which provided guidance for the stable operation of the unit. The strong pressure pulsation and noise problems generated by the impeller vortex and tailwater vortex under part-load conditions severely limit the operating range of the giant Francis turbine. The study of the pressure pulsation characteristics and induced noise phenomena of the giant Francis turbine is of great significance for the stable operation and performance improvement of the unit under partial conditions[14-15].

2. Numerical modeling and meshing

2.1. Computational modeling and parameters

Taking the left bank unit No. 1 of Udonde hydropower station as the object of study, the left bank turbine model is TFV50-LJ-970.5. The whole computational domain includes the upstream diversion pipeline, worm casing, fixed guide vane, movable guide vane, runner, tail water pipe, and downstream diversion pipeline to the cutoff of the tailwater regulator room, as shown in Figure 1.

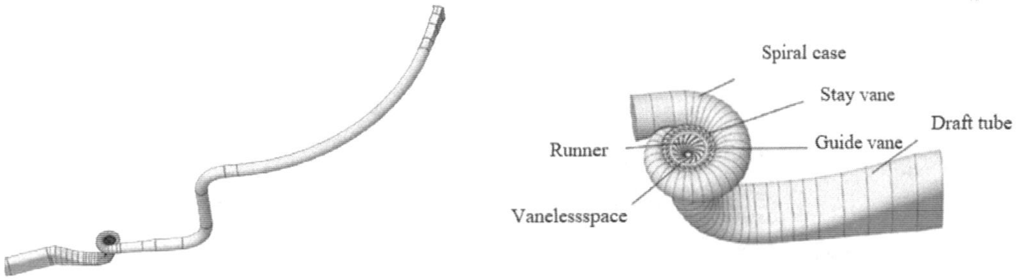

(a) Complete computational domain

(b) Three-dimensional drawing of the hydraulic turbine unit

Figure 1. Calculation domain of the water system of the Udunde giant mixed-flow unit.

The length of the upstream diversion pipe of the left bank is 591.64m, the maximum head of the unit is 163.4m, the minimum head is 106.0m, the rated rotational speed is 90.91rpm, and the rated flow rate is 671.2m^3/s. The number of rotor blades is 15, the diameter of the rotor is 9.705m, and the number of fixed and movable guide vanes is 27 and 28 respectively. Detailed parameters of the left bank turbine are shown in Table 1 below.

Table 1. Detailed Parameters of Left Bank Turbine.

Form	Numerical unit	Left bank VHS parameters
Unit type	/	TFV50-LJ-970.5
Rated head	m	137
Rated flow rate	M^3/s	671.2
Rated output	MW	862.1
Number of rotor blades	PCS	15
Guide vane number	PCS	28
Stay vane number	PCS	27

2.2. Grid division and boundary condition setting

2.2.1. Meshing. According to the shape and flow characteristics of each flow component, the diversion pipe, movable guide vane, fixed guide vane, and tailwater pipe are structured mesh, while the worm shell and runner are non-structured mesh due to the complex structure. The grid model of each component is shown in Figure 2. The grid size and number of grid parameters of each component are shown in Table 2.

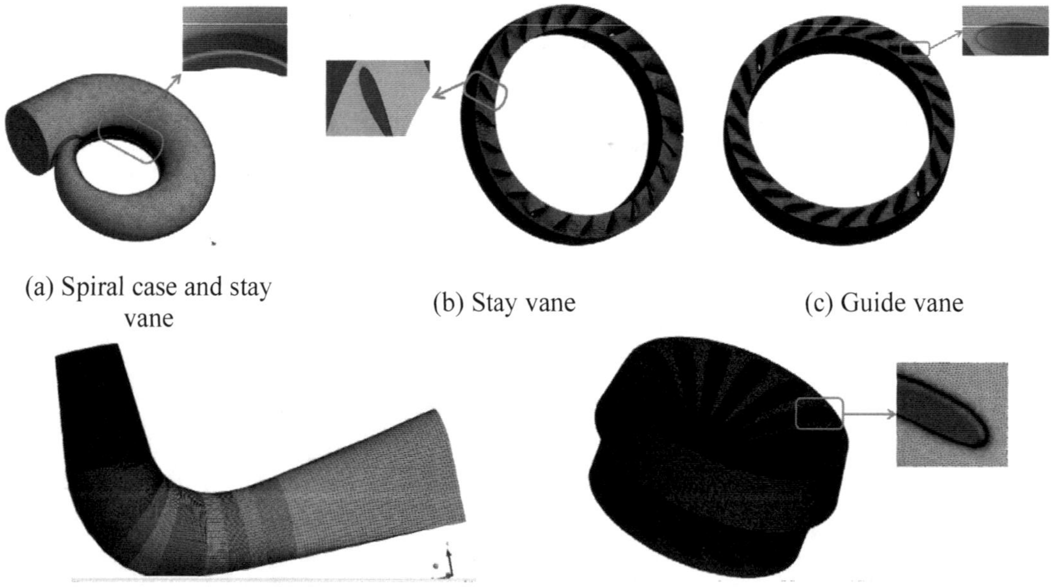

(a) Spiral case and stay vane (b) Stay vane (c) Guide vane

(d) Draft tube (e) Runner

Figure 2. Mesh model and detail drawing of each component.

Table 2. Mesh parameter table for each component.

Area	Number of grids (million)	Mesh size (mm)
Spiral case	190	10-400
Stay vane	390	5-50
Guide vane	805	5-50
Runner	7180	1-40
Draft tube	893	2-400

2.2.2. Boundary condition setting. We set pressure inlet and pressure outlet boundary conditions at the upstream pipe inlet and downstream pipe outlet. Grid INTERFACE surfaces are set between adjacent components for data transfer. The turbulence model is adopted as the DES model. The second-order windward format is used to discretize the convection term of the control equation. The rotating domain of the runner is set as a slip mesh model. The rest of the components are set as a stationary domain. The results of the flow field after 2000 steps of steady-state computation are used as the initial conditions for transient computation. The time step is 0.005 s. Thirteen complete rotational cycles of the runner are computed. The data of the last ten revolutions are extracted as the computational results.

2.2.3. External characterization. Numerical simulations are carried out for three small flow conditions of 217MW, 322MW, and 425MW for the load at 149.5m head of unit 1 on the left bank. In order to verify the validity of the model, the numerical simulation results of these three conditions are compared with the unit efficiency and flow rate obtained from the test, as shown in Figure 3. As can be seen from the figure, the flow rate and unit efficiency obtained from the numerical simulation are consistent with the trend of the test values. The three conditions meet the flow rate error within 1.12% and the efficiency error within 5.71%. Since all of them deviate from the optimal working conditions, but the errors are within 10%, the numerical model and solution method used meet the requirements of calculation accuracy.

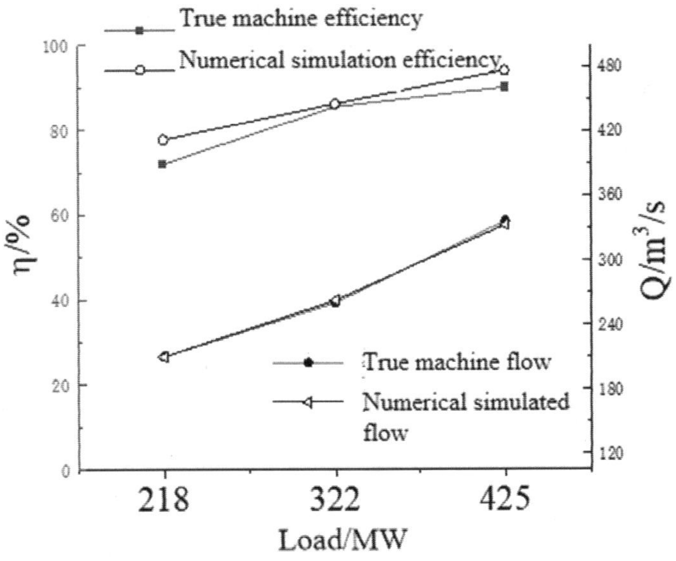

Figure 3. Numerical simulation and out-of-test characterization.

3. Unit flow characterization and noise analysis

3.1. Flow characterization

The impeller vortex is a special vortex that exists between the runner blade runners. In order to clarify the distribution of the vortex structure under different working conditions, three equal spreading planes of the runner are sliced, as shown in Figures. 4 and 5. The overall level of the Z-plane vortex is higher near the upper crown. There is a lower vortex region near the exit. There is always a high vortex region, i.e., the red region, in the head of the blades, which is due to the fact that there is a ramming angle in the inlet of the blades when the water enters the rotor from the water-guiding mechanism. The water flow directly impacts the head of the blades. With the increase of the load, the water inlet flow pattern improves, and the red region is gradually reduced. In the middle of the rotor crown and the lower ring, isotropic surface, 218MW, and 322MW blade suction surface, there is always a high vortex area. The high vortex area cross-section is more regular round. There is a vortex cross-section of the impeller. At this time, these two conditions of the impeller vortex have been fully developed. 425MW has no obvious round red area. Near the lower ring and other spreading surfaces, the blade pressure surface and suction surface vortex are higher. In the middle of the suction surface, there is a more regular circular high vortex region. The region with the position of the impeller vortex moves. 425MW circular high vortex region compared with the other two conditions is smaller. This is due to the increase in the load, and the rotor state has been improved.

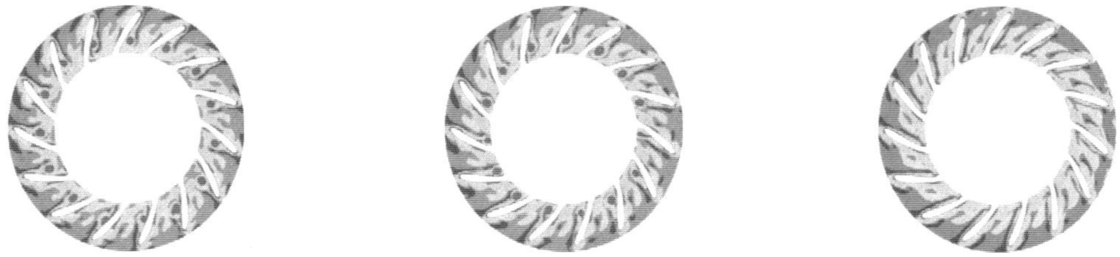

(a) 218MW (b) 322MW (c) 425MW

Figure 4. Vortex diagram of the Z-plane in the middle of the upper crown and lower ring of the rotor.

(b) 218MW (b) 322MW (c) 425MW

Figure 5. Z-plane vorticity diagram of the rotor near the lower ring.

As can be seen in Figure 6, the low-pressure region is observed in the center region of the first cross-section of the tailpipe inlet for all three conditions, which indicates the presence of a central vortex zone in the tailpipe for these conditions. As the load increases, the low-pressure region gradually shrinks. For the three sections of the tailpipe at the same conditions, the low-pressure regions of the three sections of the tailpipe are mainly concentrated at the center for the 218 MW condition. However, at 322 MW, the low-pressure regions of the second and third sections begin to shift away from the center. In the 425 MW case, the low-pressure regions of these three cross-sections are no longer located on the same axis. This change is mainly due to the formation of eccentric vortex bands in the tailpipe under low opening conditions.

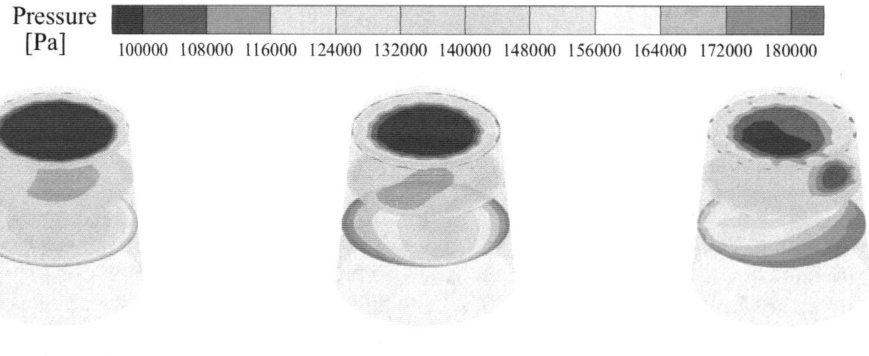

(c) 218MW (b) 322MW (c) 425MW

Figure 6. Pressure cloud diagram of draft tube straight cone section.

3.2. Pressure pulsation characteristic analysis

The measurement standard of pressure pulsation degree in a Francis turbine is mainly to calculate the relative amplitude of pressure pulsation and introduce the dimensionless number Cp of pressure pulsation to treat the absolute value of pressure without dimension.

$$Cp=\frac{p-\bar{p}}{\rho g h}\times 100\% \tag{1}$$

In the formula, Cp represents the size of the pressure pulsating water head in total water head, p is the numerically calculated pressure, $\bar{p}$ is the average numerical calculated pressure, and h is the actual water head.

The three low-load conditions of 218MW, 322MW, and 425MW were numerically simulated. Monitoring points were set on the inlet wall of the draft pipe. The speed of the three test runners was 100% rated speed and the frequency fn was 1.515Hz. The spectrum diagram of pressure pulsation of the draft pipe inlet wall and the volute inlet is shown in Figure 7. Under 218MW, 322MW, and 425MW loads, the first main frequency of draft pipe inlet wall pressure pulsation is concentrated in 0.2 frequency. The pressure amplitude reaches 0.548%, 0.569%, and 1.025%. At the same time, the amplitude of pressure pulsation is higher near the first main frequency 0.3 and 0.4. Because the 425MW vortex belt is the most eccentric, it results in a very high-pressure amplitude. At the same time, 218MW and 322MW have an obvious second peak value at 0.9 frequency. There are high-frequency and low-amplitude pressure pulsations. This is because the more some load conditions deviate from the optimal condition, the shape of the tailwater vortex zone will change obviously, and the pressure pulsation carried will also change accordingly. Long-term operation in low load conditions is easy to cause fatigue damage to the unit. It can be seen from (b) that the first main frequency of pressure pulsation at the volute inlet is concentrated at 0.2 times the frequency. The pressure amplitude reaches 0.175%, 0.092%, and 0.152%. The pressure pulsation pattern at the volute inlet is similar to that of the draft pipe. It can be seen that the low-frequency pressure pulsation is highly transferable. The amplitude at the volute inlet only attenuates to the original 14.8%-31.9%, which will still cause damage to the volute.

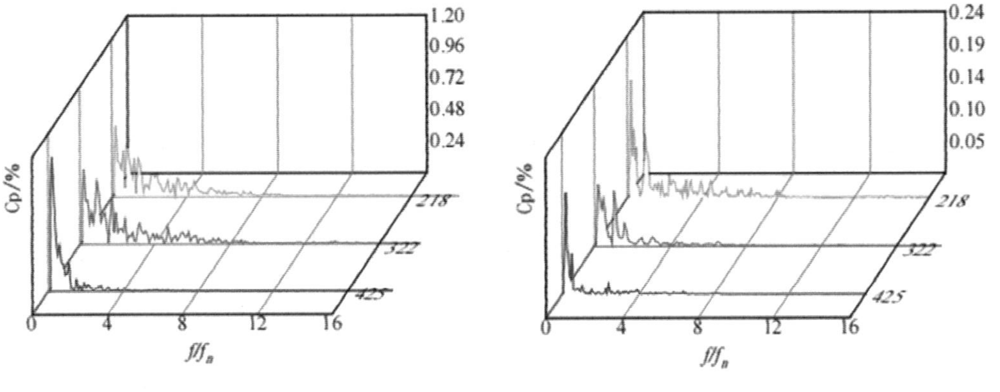

(a) Frequency domain diagram of pressure pulsation at monitoring point of draft tube inlet wall

(b) Frequency domain diagram of pressure pulsation at the volute inlet monitoring point

Figure 7. Pressure pulsation frequency domain diagram.

3.3. Noise analysis

According to the field test, noise collection points are set in the tailwater gate and the water wheel room. The noise data of the measurement points of the tailwater gate and the water wheel room under different water heads under low output conditions (0-425MW) are collected, as shown in Figure 8 below. It can be seen from the figure that under each test head, the noise value of the tailwater gate and the measuring point of the water wheel chamber has an approximate change trend with the increase of the unit's output, and the noise value reaches the lowest value under the output condition of 425MW. The test data of tailgate noise showed a trend of first increase and then decrease. Compared with the flow characteristics of the draft pipe and the analysis results of pressure fluctuation, there was no obvious causal relationship between surface tailgate noise and the vortex band of the draft pipe. The noise test data of the water turbine chamber showed a trend of decreasing gradually with the increase of the output. Compared with the analysis of the flow characteristics of the runner and the pressure fluctuation, the vorticity between the runner blades had a relatively high level at 218MW. With the increase of the output, the vorticity between the runner blades gradually decreased, the rotation state was improved, and the noise value also gradually decreased. When the Francis turbine is running under the condition of low flow deviation, the blade vortex will form between the blades, which will induce the instability of the unit noise and so on.

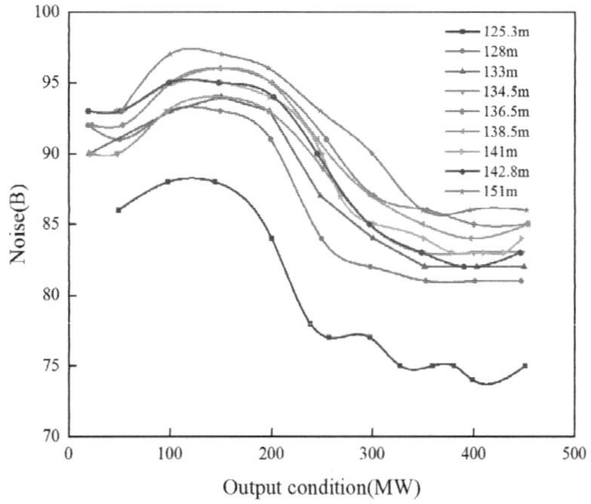

(a) Noise test data of tail valve under low output condition

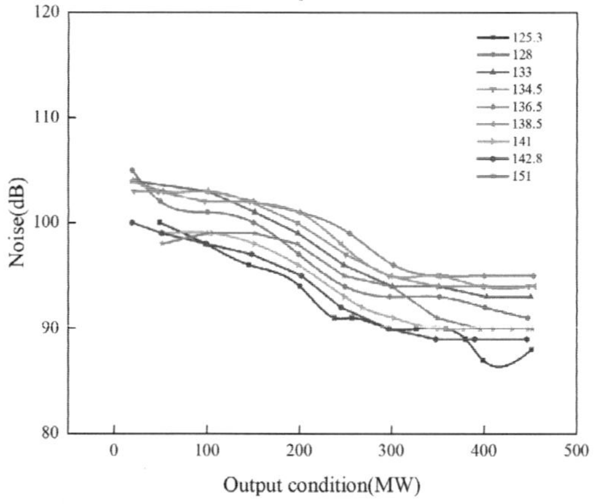

(b) Noise test data of launching car compartments under low output condition

Figure 8. Noise test data results.

4. Conclusion

In this paper, a three-dimensional numerical simulation was carried out for Unit 1 on the left bank of Wudongde. The DES model was used to compare the distribution of blade vortices and tail-water vortices under three low-output conditions. The causes of pressure pulsation were analyzed. Noise measurement points were set in the water wheel room and tailwater gate to analyze the causes of noise under low-output conditions. The results are as follows:

(1) When the low-flow area is operating, there is a water flow impact at the leading edge of the blade. The leaf vortex in the 218MW and 322MW runner channels is fully developed. When the load increases to 425MW, the degree of flow separation near the leading edge of the blade decreases, and the vortex decreases. From 218MW-425MW, the low-pressure area of the tailpipe shifts from the center to the pipe wall. The low-pressure area gradually decreases. At the same time, the tailwater vortex zone is converted from the broken vortex of the pipe wall to the complete vortex zone. The eccentric vortex zone formed at 425MW has a large energy, which has a significant impact on the surrounding flow regime.

(2) Under the condition of small flow, it is affected by the tailwater vortex belt. The main frequency of pressure pulsation is 0.2 times the frequency. Due to the eccentricity of the spiral vortex band formed under the working condition of 425MW, the main frequency amplitude of pressure pulsation is the highest, which is easy to causes unstable hydraulic vibration. Moreover, the pressure pulsation is transmitted from the tailpipe to the volute inlet. The attenuation is reduced to 14.8%-31.9% of the original amplitude, which still poses a threat to the stable operation of the turbine.

(3) The blade vortex is caused by the deflow that occurs after the Francis turbine deviates from the optimal working conditions and the back of the inlet edge of the runner blade. According to the CFD numerical simulation and noise test data, the vibration and noise of Wudongde left bank No. 1 under the condition of small output are mainly caused by the existence of the blade vortex in the lower ring of the upper crown of the runner. The development of the tailpipe vortex has an impact on the vibration of the unit, but it is not the main factor causing the noise.

Acknowledgments

It is supported by the Luquan Wudongde Power Plant of Three Gorges Jinshajiang Yunchuan Hydropower Development Co., Ltd. (Contract Number: Z522302016).

References

[1] Xiao Yu, LIU Jingfei, Qu Binmei. Design of new function expansion device based on MODBUS Communication protocol [J]. China Science and Technology Information, 2022, (2): 73-75.
[2] Podani J (1994) Multivariate Data Analysis in Ecology and Systematics. SPB Publishing, The Hague.
[3] Tong Zhen-Ming, Xin Jia-Ge, TONG Shui-Guang. Unsteady Decoupling and Reconstruction of vane vortices based on Dynamic Mode Decomposition [J]. China Mechanical Engineering, 2022, 33(7): 769-776.
[4] Thompson J N (1984) Insect Diversity and the Trophic Structure of Communities. In: Ecological Entomology. New York. pp. 165-178.
[5] Ge Yan, WU Huanyu, ZENG Chongji, et al. Research on the suppression measures of Tail vortex zone of axial-flow fixed propeller turbine Based on Numerical Simulation [J]. Big Motor Technology, 2023, (5): 70-76.
[6] Xu Hongquan, Wang Wanpeng, LIAO Cuilin, et al. Study on damage and mechanism of blade vortex [J]. Journal of China Institute of Water Resources and Hydropower Research, 2023, 21(5): 466-475.
[7] Müller A, Favrel A, Landry C, et al. Fluid—structure interaction mechanisms leading to dangerous power swings in Francis turbines at full load [J]. Journal of Fluids and Structures, 2017, 69: 56-71.
[8] Yang Feng, Wang Xu. Research on Optimal Design of Inner Flow Channel of Francis Turbine Based on CFD [J]. Manufacturing Automation, 2022, 44(2): 191-193, 197.
[9] Sun Longgang, Guo Pengcheng, Zheng Xiaobo, et al. Characteristics of high amplitude pressure pulsation induced by cascade cavitation vortexes of a mixed-flow turbine [J]. Transactions of the Chinese Society of Agricultural Engineering, 2021, 37(21): 62-70.
[10] Sun Longgang, ZHENG Xiaobo, GUO Pengcheng. Study on blade vortex evolution and pressure pulsation characteristics of a Francis turbine [J]. Hydropower and Pumped Storage, 2021, 7(5): 47-52.
[11] Guo Tao, XU Lihui, LUO Zhumei. Study on vortex zone evolution and pressure fluctuation of draft pipe under Upstream disturbance [J]. Transactions of the Chinese Society for Agricultural Machinery, 2022, 53(6): 192-201.
[12] Muhirwa A, Li B, Su W-T, et al. Investigation on mutual traveling influences between the draft tube and upstream components of a Francis turbine unit [J]. Renewable Energy, 2020, 162: 973-992.
[13] Lu Lei, Zhang Lida, Yang Jing, et al. Cause Analysis and Research on Vibration and Noise of low head Francis Turbine [J]. Journal of Hydroelectric Power, 2014, 40(03): 47-49.
[14] Wang Haolan, Zheng Yuan, Sun Aolan, et al. Influence of guide vane Opening on Pressure pulsation Characteristics and flow-induced noise of a Francis turbine [J]. The south-north water diversion and water conservancy science and technology, 2018(5): 186-192. The DOI: 10.13476/j.carolcarrollnkiNSBDQK.2018.0142.
[15] Liu Jinbo. Flow characteristics and noise analysis of horizontal francis turbine [D]. Kunming University of Science and Technology, 2015.

Journal of Physics: Conference Series **3011** (2025) 012041 doi:10.1088/1742-6596/3011/1/012041

A master-slave game-based regional integrated energy system considering the uncertainty of source-load

Menglin Zhang, Weiqing Wang*

Engineering Research Center of Ministry of Education for Renewable Energy Generation and Grid Connection Technology, Xinjiang University, Urumqi, Xinjiang, 830000, China

*Corresponding author's e-mail: wwq59@sina.cn

Abstract: A master-slave game model that accounts for the unpredictability of source-load dynamics is proposed to tackle the complexity of energy scheduling in regional integrated energy systems, along with the uncertainty in renewable energy generation and load variations. This model specifically addresses the uncertainties associated with wind power output and load fluctuations. Lagrangian multiplier dual factors are introduced. For the energy scheduling optimization problem in regional integrated energy systems, a master-slave game energy scheduling model with upper and lower layers is constructed. The upper-layer energy operator aims to maximize profits, while the lower-layer integrated energy system aims to minimize costs. The upper layer searches for the optimal trading electricity price using a genetic algorithm, while the lower layer solves the energy scheduling problem with the CPLEX solver. The interaction between the upper and lower layers is dynamically established by trading electricity prices. The case study shows that the two-layer energy scheduling effectively solves the energy scheduling issue in regional integrated energy systems, ensuring stability, consistency, and balance in the system's operation between the upper and lower-layer participants.

1. Introduction

As worldwide energy consumption keeps increasing and environmental concerns grow, awareness of the shortcomings of traditional single-energy supply systems has become increasingly evident. In response, the theory of Regional Integrated Energy Systems (RIES) has been developed. By combining different energy sources such as electricity, natural gas, and thermal energy, RIES facilitates coordinated and optimized energy use. This approach enhances energy efficiency, reduces losses, and lowers environmental pollution, making it a crucial pathway for advancing sustainable energy development.

Fan et al. [1] focused on regional integrated energy systems and proposed a typical system structure for such systems, along with establishing a regional incorporated energy system model. Li et al. [2] submitted an optimal scheduling approach for unified energy systems taking into account uncertain renewable energy sources. Luo et al. [3] studied the energy scheduling problem in a three-tier Integrated Energy System (IES) consisting of an electricity company and a natural gas company (upper layer), multiple smart energy hubs (middle layer), and multiple users (lower layer). Scholars have also conducted research on load response scheduling, with Wang et al. [4] proposing a horizontal and vertical interaction model for regional integrated energy systems. This model facilitates the horizontal interaction between multi-energy systems and the vertical interaction between sources and loads based on user-driven behavior data. Lekvan et al. [5] broached a new scheduling optimization approach for renewable energy-based Multi-Energy Microgrid (MEM) systems, which combines demand response programs. Wang et al. [6] proposed an integrated energy system optimization operation model that considers energy price responses.

2. Regionally integrated energy system

A RIES is an efficient model for modern energy utilization. It integrates diverse energy supply sources such as electricity, heat, and natural gas, along with energy conversion devices like geothermal heat pumps and gas turbines, enabling flexible switching between different energy forms [7]. Additionally, it incorporates energy storage facilities like thermal storage tanks and batteries to balance supply and demand. Through intelligent networks, energy is precisely delivered to end-users such as households, businesses, and industries. The key features of RIES are as follows: Multi-Energy Complementarity: different energy sources complement and enhance each other, filling gaps in energy supply and ensuring a stable energy mix [8]; Intelligent Control Capabilities: using sensors and big data, RIES can dynamically adjust energy distribution based on demand peaks and valleys in real time; Environmental Benefits: RIES actively integrates renewable energy sources, such as wind and solar power, contributing to cleaner energy production; Significant Value: RIES not only increases energy utilization efficiency, reduces waste, and lowers energy costs but also supports regional carbon reduction efforts, creates a green ecosystem, and strengthens energy supply stability [9]. Even in the event of localized energy failures, RIES ensures continued energy provision, establishing a robust energy foundation for regional development.

In this paper, a regional integrated energy system is constructed, consisting of an upper-layer energy operator and three lower-layer integrated energy systems. As shown in Figure 1, the energy operator supplies electricity and natural gas to the lower-layer users. The lower-layer integrated energy systems include production equipment, energy conversion devices, and energy storage devices, specifically wind power generation, gas turbines, waste heat boilers, gas boilers, geothermal heat pumps, and storage batteries.

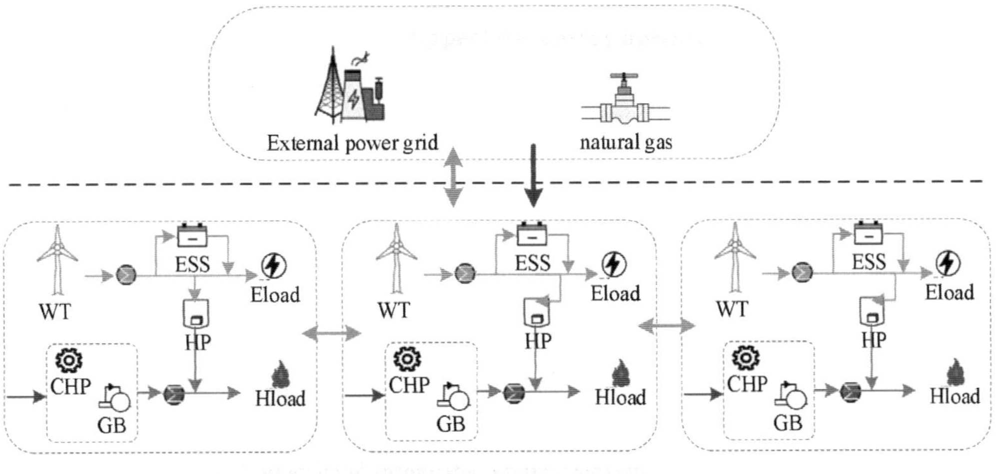

Figure 1. Structure of the regional integrated energy system.

3. A master-slave game model considering source-load uncertainty

The master-slave game model considering source-load uncertainty is a cutting-edge solution to address the complex power supply and demand patterns that are increasingly prevalent today. As the power system becomes more diversified, the supply side, such as wind power and other renewable energy sources, experiences fluctuations due to natural conditions, resulting in unstable power generation. Additionally, traditional thermal power plants may occasionally suffer from sudden failures. On the demand side, residential lifestyles and the pace of industrial production have become more variable, making electricity demand increasingly unpredictable, and the uncertainty between source and load has surged.

This model establishes a two-layer structure: Upper Layer (Energy Operator): the energy operator aims to maintain grid stability and control costs. This layer sets electricity prices and develops generation plans and other regulatory strategies. Lower Layer (Energy Suppliers and Users): the energy suppliers aim to maximize generation profits, while users seek to minimize electricity costs. Both sides engage in a game based on the uncertain source-load conditions. The upper layer sets trading electricity prices to guide decisions, while the lower layer outputs electricity purchase and sale amounts according to its interests[10]. Through multiple iterations, a balance is reached. The significance of this model is considerable. It effectively coordinates the interests of multiple parties, aids in the precise allocation of electricity resources, enhances the system's resilience to risks, and ensures the reliability and efficiency of the power supply. This approach addresses the challenges posed by uncertainty in both energy production and consumption, fostering a more adaptive and balanced energy market.

4. Optimization scheduling model

4.1 Upper-level energy operator (EO)

4.1.1 Objective function. The refinement scheduling stage of the upper-level energy operator aims to maximize the comprehensive profit. The main sources of profit for the upper-level operator are the revenue from selling electricity to the lower level and the cost of purchasing excess electricity from the lower level. The scheduling method involves adjusting the electricity price for selling and purchasing excess energy based on the energy demand from the lower level at any given moment. The specific mathematical expressions are as follows:

$$F_{EO} = e_{buy}(t) \times \sum_{t=1}^{24} \left[P_{buy\Sigma}(t) \right] - e_{sell}(t) \times \sum_{t=1}^{24} \left[P_{sell\Sigma}(t) \right] \tag{1}$$

$$\begin{cases} P_{buy\Sigma}(t) = P_{buy1}(t) + P_{buy2}(t) + P_{buy3}(t) \\ P_{sell\Sigma}(t) = P_{sell1}(t) + P_{sell2}(t) + P_{sell3}(t) \end{cases} \tag{2}$$

In Equation (1), F_{EO} represents the comprehensive profit of the upper-level energy operator; $e_{buy}(t)$ and $e_{sell}(t)$ represent the purchase and sale electricity prices at time t, respectively; $P_{buy\Sigma}(t)$ and $P_{sell\Sigma}(t)$ represent the electricity purchased and sold by the lower-level energy systems from the upper-level energy operator at time t, respectively. In Equation (2), $P_{buy1}(t)$, $P_{buy2}(t)$, and $P_{buy3}(t)$ represent the purchase electricity quantities at time t for the first, second, and third energy systems, respectively; and $P_{sell1}(t)$, $P_{sell2}(t)$,and $P_{sell3}(t)$ represent the sale electricity quantities at time t for the first, second, and third energy systems, respectively.

4.1.2 Constraints. The energy operator defines the maximum and minimum boundaries for electricity purchase and sale prices, as well as the transaction quantities between the upper and lower levels. The detailed expressions are as follows:

$$e_{sw}(t) \le e_{sell}(t) \le e_{buy}(t) \le e_{fs}(t) \tag{3}$$

$$P_{\min} \le (P_{buy}(t), P_{sell}(t)) \le P_{\max} \tag{4}$$

In Equation (3), $e_{sw}(t)$ and $e_{fs}(t)$ represent the grid's online electricity price and time-of-use electricity price at time t, respectively; and P_{max} and P_{min} represent the upper and lower limits for direct energy transactions between the upper-level energy operator and the lower-level energy systems.

4.2 Lower-level integrated energy system

The lower-level integrated energy system consists of equipment that can be categorized based on function, including new energy generation capacity, energy conversion, energy storage, and energy-consuming users. These include wind power generation, gas turbines, gas boilers, ground-source heat pumps, and energy storage devices. The optimization scheduling process for the lower-level system involves minimizing the day-to-day investment and maintenance costs by optimizing the output of various devices and the maximum installed capacity of the equipment.

4.2.1 Objective function. The integrated energy system's scheduling objective function includes the equivalent daily investment cost of various devices and the system operation cost.

$$F_{IES-i} = \frac{c_{GT}k_{GT}N_{GT}}{365} + \frac{c_{GB}k_{GB}N_{GB}}{365} + \frac{c_{HT}k_{HT}N_{HT}}{365} + \frac{c_{HP}k_{HP}N_{HP}}{365} + 2.71*\sum_{t=1}^{24}\left[f_{GT}(t)+f_{GB}(t)\right]+0.01*\sum_{t=1}^{24}\left[P_c(t)+P_{disc}(t)\right] + \sum_{t=1}^{24}\left[e_{buy}(t)P_{buy}(t)-e_{sell}(t)P_{sell}(t)\right] \quad (5)$$

In this equation, c_{GT}, c_{GB}, c_{HT} ,and c_{HP} represent the investment parameters for the gas turbine, gas boiler, waste heat boiler, and ground-source heat pump, respectively; k_{GT}, k_{GB}, k_{HT} ,and k_{HP} represent the unit initial investment costs; N_{GT}, N_{GB}, N_{HT} ,and N_{HP} represent the installed capacities of the corresponding devices; $f_{GT}(t)$ and $f_{GB}(t)$ represent the natural gas consumption of the gas turbine and gas boiler at time t, respectively; and $P_c(t)$ and $P_{disc}(t)$ represent the charging and discharging power of the energy storage device at time t.

4.2.2 Constraints. The constraints for the lower-level integrated energy system include power equality constraints and equipment inequality constraints. To illustrate, we will use the first integrated energy system as an example.

1) Power Equality Constraints

Power equality constraints include both electrical and thermal power constraints:

$$\begin{aligned} &P_{buy}(t)-P_{sell}(t)-P_{HP}(t)+P_{GT}(t)+P_{disc}(t)-P_c(t)-P_{1-2}(t)-P_{1-3}(t)= \\ &P_{load}(t)-P_{WT}(t)+P_{lam}(t)+P_{load\_ld}(t)+P_{load\_ud}(t)+P_{WT\_ld}(t)+P_{WT\_ud}(t) \end{aligned} \quad (6)$$

In this equation, $P_{HP}(t)$, $P_{GT}(t)$ and $P_{WT}(t)$ represent the electrical output powers of the ground-source heat pump, gas turbine, and wind turbine at time t, respectively; $P_{1\text{-}2}(t)$ and $P_{1\text{-}3}(t)$ represent the electrical power output from the first energy system to the second and third energy systems at time t; $P_{load}(t)$ represents the electrical load at time t; $P_{lam}(t)$ represents the dual variable for the electrical load uncertainty equality constraint at time t; $P_{load\_ld}(t)$ and $P_{load\_ud}(t)$ represent the dual variables for the upper and lower bounds of the electrical load forecast at time t, respectively; and $P_{WT\_ld}(t)$ and $P_{WT\_ud}(t)$ represent the dual variables for the upper and lower bounds of the wind turbine power output forecast at time t, respectively.

$$H_{GT}(t)+H_{GB}(t)+H_{HP}(t)=H_{load}(t)+H_{lam}(t)+H_{lam\_ld}(t)+H_{lam\_ub}(t) \quad (7)$$

In this equation, $H_{GT}(t)$, $H_{GB}(t)$, and $H_{HP}(t)$ represent the thermal output powers of the gas turbine, gas boiler, and ground-source heat pump at time t, respectively; $H_{load}(t)$ represents the thermal load at time t; $H_{am}(t)$ represents the dual variable for the thermal load uncertainty equality constraint at time t; and $H_{load\_ld}(t)$ and $H_{load\_ud}(t)$ represent the dual variables for the upper and lower bounds of the thermal load forecast at time t, respectively.

2) Power Inequality Constraints

Power inequality constraints include electricity transaction constraints, equipment operation power constraints, energy storage device power constraints, and uncertainty dual variable constraints.

$$\begin{cases} 0 \le \left[P_{buy}(t), P_{sell}(t) \right] \le P_{net} \\ u\_buy(t) + u\_sell(t) \le 1 \end{cases} \tag{8}$$

In this equation, P_{net} represents the upper limit for electricity transactions between the integrated energy system and the upper-level electricity system; and $u\_buy(t)$ and $u\_buy(t)$ are the indicator variables for purchasing and selling electricity at time t, respectively.

$$\begin{cases} 0 \le P_{GT}(t) \le P_{GT\_\max} \\ 0 \le H_{GB}(t) \le H_{GB\_\max} \\ 0 \le H_{HP}(t) \le H_{HP\_\max} \end{cases} \tag{9}$$

In this equation, P_{GT_max}, H_{GB_max}, and H_{HP_max} represent the maximum output electrical power of the gas turbine, the maximum output thermal power of the gas boiler, and the maximum output thermal power of the ground-source heat pump, respectively.

$$\begin{cases} 0 \le \left[P_{disc}(t), P_{c}(t) \right] \le P_{e} \\ u\_disc(t) + u\_c(t) \le 1 \end{cases} \tag{10}$$

$$\begin{cases} E(t) = 0.95 \times E(t-1) + 0.98 \times \left[P_{c}(t) - P_{disc}(t) \right] \\ E_{\min} \le E(t) \le E_{\max} \end{cases} \tag{11}$$

In these equations, P_e represents the maximum charging and discharging power of the energy storage device; $u\_c(t)$ and $u\_disc(t)$ are the indicator variables for charging and discharging at time t; and E(t-1) and E(t) represent the energy storage levels of the energy storage device at time t-1 and t, respectively.

The upper and lower limit constraints for the dual variables of electrical load and thermal load uncertainties are as follows:

$$\begin{cases} P_{lam}(t) + P_{lam\_lb}(t) \le -0.2 P_{load}(t) \\ P_{lam}(t) + P_{lam\_ub}(t) \ge 0.2 P_{load}(t) \\ H_{lam}(t) + H_{lam\_lb}(t) \le -0.2 H_{load}(t) \\ H_{lam}(t) + H_{lam\_ub}(t) \le 0.2 H_{load}(t) \end{cases} \tag{12}$$

The upper and lower limit constraints for the dual variables of wind power generation forecast uncertainty are as follows:

$$\begin{cases} P_{lam}(t) + P_{WT\_ld}(t) \ge -0.2 P_{WT}(t) \\ P_{lam}(t) + P_{WT\_ud}(t) \ge 0.2 P_{WT}(t) \end{cases} \tag{13}$$

5. Master-slave game model for regional integrated energy system

The energy operator, situated at the upper level of the regional integrated energy system, plays a key role in the overall regional dynamics by setting and adjusting electricity trading prices with the lower levels to maximize profits. Each integrated energy system at the lower level optimizes its internal energy scheduling based on these trading prices to minimize investment and operational maintenance costs. The interaction between the upper-level operator and the lower-level energy systems creates a dynamic master-slave game model. The detailed process flow is depicted in Figure 2 below.

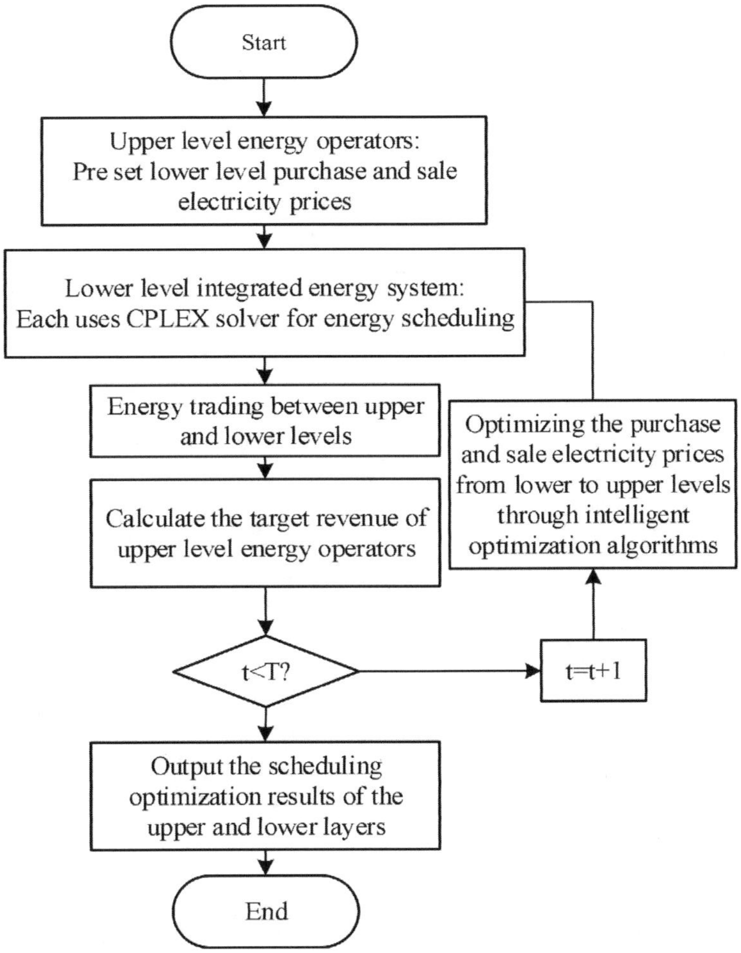

Figure 2. Structure of the regional integrated energy system.

6. Case study analysis

The analysis is based on a typical summer day in a certain region, verifying the introduction of source-load Lagrangian dual variables for uncertainty. The energy scheduling optimization of the lower-layer integrated energy system is solved using the CPLEX solver, while the upper-layer energy operator optimizes the scheduling using a genetic algorithm. The upper and lower layers form a dynamic scheduling master-slave game.

6.1 Electricity price table

The scheduling period TTT is set as 24 hours, with a time step of 1 hour. The time-of-use electricity prices are shown in Table 1.

Table 1. Electricity price table.

Parameters		Time	Values (Yuan/kWh)
Time-of-Use Electricity Price	Off-peak Value	01:00-07:0, 23:00-24:00	0.4
	Flat Rate Value	08:00-11:0, 15:00-18:00	0.75
	Peak Value	12:00-14:0, 19:00-22:00	1.20
Grid Feed-in Price	Off-peak Value	01:00-07:0, 23:00-24:00	0.25
	Flat Rate Value	08:00-11:0, 15:00-18:00	0.45
	Peak Value	12:00-14:0, 19:00-22:00	0.65

6.2 Integrated energy system operating results

The independent variables in the three integrated energy systems include the rated power of the gas turbine, waste heat boiler, gas boiler, and ground-source heat pump. The corresponding values for the optimized scheduling are listed in Table 2.

Table 2. Rated power of lower layer energy scheduling equipment.

Equipment / System	Gas Turbine (kW)	Waste Heat Boiler (kW)	Gas Boiler (kW)	Ground-source Heat Pump (kW)
IES1	267.85	500	459.77	500
IES2	160.71	300	605.59	200
IES3	362.76	677.16	341.81	200

The energy scheduling results of the three integrated energy systems are shown in Figures 3, 4, and 5, respectively. In the electrical balance, P_L represents the imbalance of power, which is the difference between the electrical load and the wind power output. If $P_L>0$, it indicates that the load exceeds the renewable energy output and additional power generation equipment is required to meet the load. If $P_L<0$, it means that the renewable energy output exceeds the load, and other measures are needed to absorb the excess renewable energy. When $P_{buy}<0$, electricity is sold to the upper-level energy operator, and when $P_{buy}>0$, electricity is purchased from the energy operator.

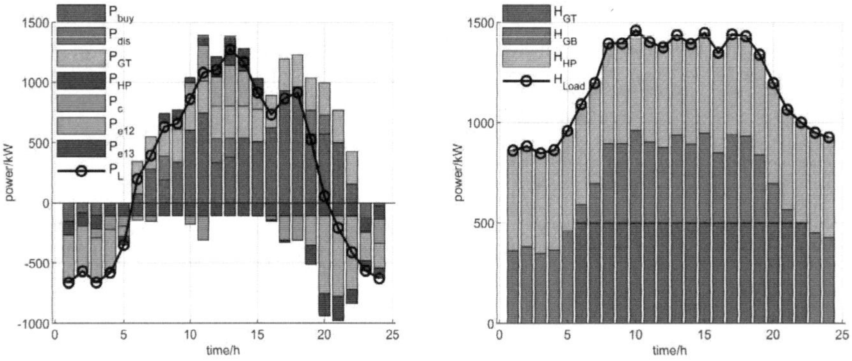

(a) Electrical Balance (b) Thermal Balance

Figure 3. Scheduling result of the first integrated energy system.

As shown in Figure 3(a), the electrical load of the first integrated energy system is lower than the wind power output from 1-6 AM and 9-24 PM. During 1-2 AM, the excess wind power is sold to the upper-level operator, supplied to the ground-source heat pump, and transmitted to the second integrated energy system. From 1-4 AM and 19-24 PM, the excess renewable energy output exceeds 500 kW, with the peak at 9 PM reaching 977.4 kW. From 6 AM to 8 PM, the excess load is balanced by purchasing electricity from the upper-level energy operator, gas turbine generation, obtaining power from the second and third integrated energy systems, and discharging energy from the storage devices. As shown in Figure 3(b), the thermal load is met by heating from the gas turbine, gas boiler, and ground-source heat pump. From 6 AM to 10 PM, the gas turbine operates at full load, providing 500 kW of thermal power, while the gas turbine and ground-source heat pump adjust their output to match the thermal load.

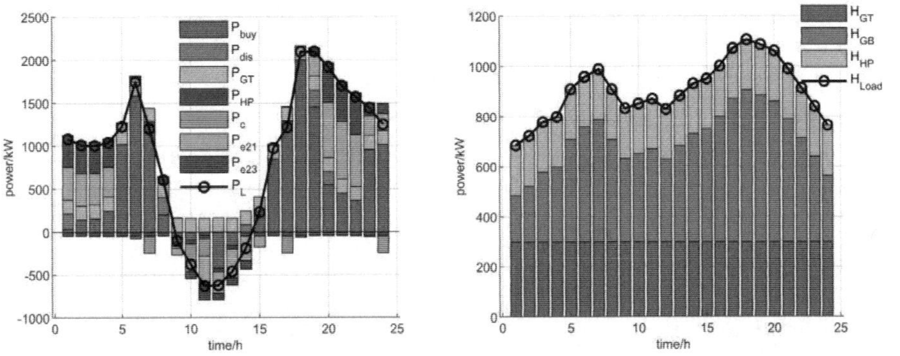

(a) Electrical Balance (b) Thermal Balance

Figure 4. Scheduling result of the second integrated energy system.

As shown in Figure 4(a), the electrical load is less than the wind turbine output only from 9 AM to 2 PM. During this period, part of the excess electricity comes from the gas turbine's output. The excess energy is managed by supplying power to the ground-source heat pump, selling electricity to the upper-level operator, charging the battery, and supplying power to the first and third integrated energy systems. At noon, the electricity sold to the upper-level operator reached 489.62 kW. From 1 AM to 8 AM and 3 PM to 12 AM, the electrical load exceeds the wind turbine output. The excess load is primarily met by purchasing electricity from the upper-level operator and obtaining electricity from the first and third integrated energy systems. As shown in Figure 4(b), during the thermal power balance process, the gas turbine and ground-source heat pump both operate at full capacity, outputting 300 kW and 200 kW, respectively. The remaining thermal load is balanced by the fluctuating output of the gas turbine.

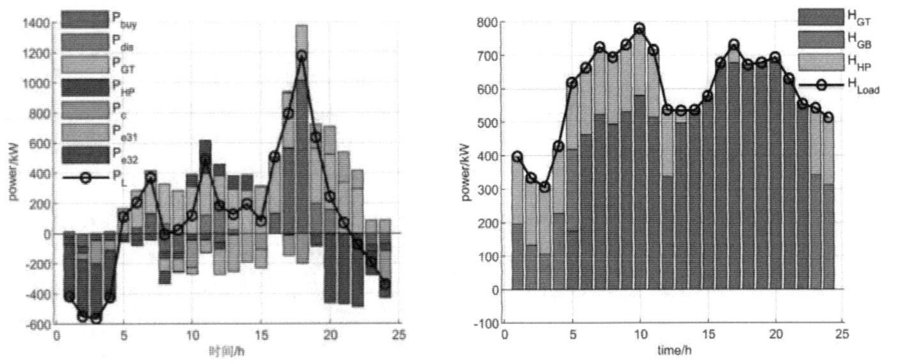

(a) Electrical Balance (b) Thermal Balance

Figure 5. Scheduling result of the third integrated energy system.

As shown in Figure 5(a), during the electrical balance of the third integrated energy system, from 1-4 AM and 10 PM to 12 AM, the electrical load is lower than the wind turbine output. The excess electricity is mainly managed by transmitting power to the second integrated energy system, charging the battery, and supplying the ground-source heat pump. From 5 AM to 9 PM, the electrical load exceeds the wind turbine output. The remaining electrical load is met by generating electricity from the gas turbine, obtaining electricity from the second integrated energy system, discharging the battery, and purchasing electricity from the upper-level energy operator. Even during the period when the load exceeds the renewable energy supply from 5 AM to 9 PM, measures like starting the ground-source heat pump, selling electricity to the upper-level operator, charging the battery, and transmitting power to the first integrated energy system are still applied. During the thermal balance process, from 5 AM to 10 PM, the gas turbine provides heat to the system as the primary heating source. The ground-source heat pump operates only from 1-5 AM and 11 PM to 12 AM for system heating.

6.3 Dispatch results of energy operators

Based on the electricity demand reflected by the lower-level system, energy operators adjust the electricity trading prices between the upper and lower levels. By setting the trading prices in the upper level as the background, the lower level comprehensive energy system makes the optimal scheduling result, forming a master-slave game relationship between the upper and lower levels. Energy operators use trading electricity prices as variables and aim to increase revenue. They use genetic algorithms to calculate scheduling results, and the iterative optimization results are shown in Figure 6. The electricity price scheduling results are shown in Figure 7. The optimal result was achieved at the 16th iteration, with a target value of 2706.

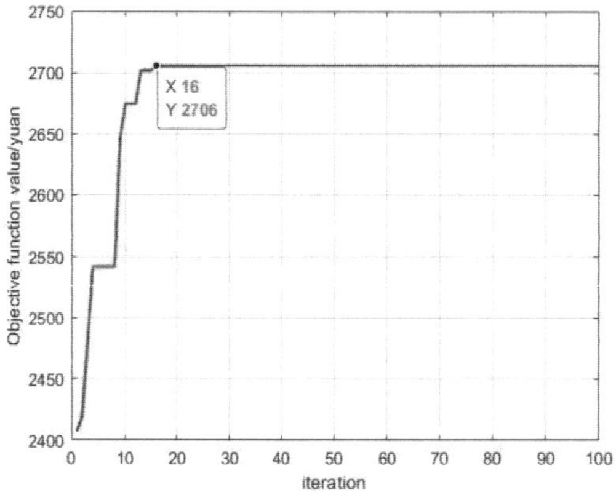

Figure 6. Scheduling optimization results of energy operators.

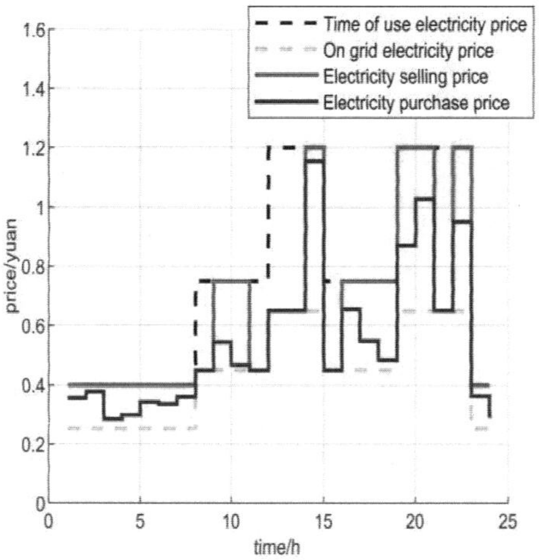

Figure 7. Internal electricity price distribution map.

As shown in Figure 7, the selling price and purchasing price of energy operators are both between the time of use price and the grid price, and the selling price at each time is greater than the purchasing price, ensuring the revenue situation of energy operators at each time.

7. Conclusion

In response to the energy scheduling problem of regional integrated energy systems under uncertainty in new energy output and load fluctuations, this paper constructs a master-slave game model of regional integrated energy systems that considers source load uncertainty. A dynamic scheduling model is formed between upper-level energy operators and lower-level integrated energy systems. Lagrangian variable factors are introduced to address the uncertainty of new energy output and the volatility of load output. The upper-level energy operators and the lower-level integrated energy system regulate and trade electricity prices to achieve a game relationship between the upper and lower levels. Genetic intelligence algorithms are used to optimize the regulation of trading variables, while the lower-level integrated energy system uses CPLEX solvers for energy scheduling.
The optimal result was achieved at the 16th iteration, with a target value of 2706. As shown above, the model constructed in this article solves the energy scheduling problem of a single integrated energy system and involves multiple systems simultaneously. Through unified management by upper-level energy operators, load fluctuation is also considered in the scheduling process, effectively balancing the scheduling problem of a single system with the unity problem of multiple systems participating.

References

[1] Chen, C. M., X. Y. Wu, Y. Li, X. J. Zhu, Z. S. Li, J. E. Ma, W. Q. Qiu, C. Liu, Z. Z. Lin, L. Yang, Q. WangY. Ding (2021). "Distributionally robust day-ahead scheduling of region-level integrated energy system considering generalized energy storages." Applied Energy 302.
[2] Chen, Z. H., S. Avraamidou, P. Liu, Z. Li, W. D. NiE. N. Pistikopoulos (2021). "Optimal design of integrated urban energy systems under uncertainty and sustainability requirements." Computers & Chemical Engineering 155.
[3] Cong, D., L. L. Liang, S. X. Jing, Y. M. Han, Z. Q. GengC. Chu (2021). "Energy supply efficiency evaluation of integrated energy systems using novel SBM-DEA integrating Monte Carlo." Energy 231.
[4] Fang, J. K., Q. Zeng, X. M. Ai, Z. Chen, J. Y. Wen (2018). "Dynamic Optimal Energy Flow in the Integrated Natural Gas and Electrical Power Systems." Ieee Transactions on Sustainable Energy 9(1): 188-198.
[5] Fu, Y. P., H. W. Wu, S. Y. Ye, X. Cai, X. Yu, S. C. Hou, H. KafafyD. C. Zou (2013). "Integrated power fiber for energy conversion and storage." Energy & Environmental Science 6(3): 805-812.
[6] Gao, Q., X. D. Zhang, M. X. Yang, X. Q. Chen, H. Q. ZhouQ. Yang (2021). "Fuzzy Decision-Based Optimal Energy Dispatch for Integrated Energy Systems with Energy Storage." Frontiers in Energy Research 9.
[7] Guo, H., T. L. Shi, F. Wang, L. J. ZhangZ. Y. Lin (2020). "Adaptive clustering-based hierarchical layout optimization for large-scale integrated energy systems." Iet Renewable Power Generation 14(17): 3336-3345.
[8] He, J., Y. Li, H. Q. Li, H. M. Tong, Z. J. Yuan, X. L. YangW. T. Huang (2020). "Application of Game Theory in Integrated Energy Systems: A Review." Ieee Access 8: 93380-93397.
[9] Huang, W. J., N. Zhang, C. Q. Kang, M. X. Li, M. L. Huo (2019). "From demand response to integrated demand response: review and prospect of research and application." Protection and Control of Modern Power Systems 4(1).
[10] Jiang, P., J. Dong, H. Huang (2020). "Optimal integrated demand response scheduling in regional integrated energy system with concentrating solar power." Applied Thermal Engineering 166.

Optimizing thermal performance in PCM packed beds: a comparative study of spherical and cylindrical capsules

Xinglin Yang*, Zihan Feng, Xirui Song

School of Energy and Power, Jiangsu University of Science and Technology, Zhenjiang, 212003, China

*Corresponding author's e-mail: yangxl233@163.com

Abstract. This study delves into the thermal performance and exergy efficiency of PCMs within packed bed systems, employing capsules of varied shapes—spherical and cylindrical. Utilizing COMSOL Multiphysics 6.0 for simulation, this research constructs a three-dimensional model to numerically analyse the discharging processes of these packed beds. A comparative analysis of three distinct packed bed systems under identical operational conditions was conducted, focusing on liquid phase changes, temperature profiles, and exergy parameters. The study examines individual configurations of spherical capsules, cylindrical capsules, and a hybrid combination of both, assessing their heat transfer capabilities and exergy storage performance. The findings reveal that the hybrid configuration of spherical and cylindrical capsules (Case 2) provides an optimal balance of rapid heat transfer and efficient exergy management. Furthermore, during the discharging phase, this combined configuration sustained higher exergy efficiencies, underscoring its potential for more effective and prolonged heat transfer applications.

1. Introduction

As global energy demand continues to rise and climate change becomes increasingly critical, optimizing energy storage and utilization has turned into a pressing global issue. Concentrated solar power (CSP) has gained prominence as an advanced renewable energy solution, attracting significant research and development attention in recent years [1].

The integration of packed bed technology into CSP systems has gained widespread interest, as packed beds can optimize the thermal storage process and enhance system performance and reliability. Among these, spherical capsules are noted for their simplicity in storage system design. However, cylindrical capsules have been less studied, and from a structural and heat transfer perspective, cylindrical capsules are well-suited for CSP operating conditions [2].

Wu et al. [3] investigated molten salt-filled bed systems and found that with an increase in fluid inlet temperature, the average heat accumulation in the phase change material (PCM) improves. Sun et al. [4] conducted two-dimensional simulations and found that a reduction in the Stefan number and dimensionless spherical diameter leads to an improvement in exergy efficiency.

Cylindrical encapsulated capsules, due to their superior thermal stress performance at high temperatures, have gradually become a focal point of current research and applications. Through experiment evaluation, Kumar and Saha [5] found that thermal stress in the thermal energy storage wall is greater for the crossflow configuration of capsules than for the axial flow configuration. A. Ahmad et al. [6] showed that a packed bed system with cylindrical capsules using $NaNO_3$ as the PCM exhibits stable thermal performance.

In summary, many numerical analyses of packed beds rely on porous media theory, simplifying the models to either one-dimensional or two-dimensional structures. This approach simplifies the actual porosity distribution within the three-dimensional packed structure, inadequately capturing the thermal characteristics of the heat transfer fluid (HTF) flowing through the gaps among PCMs in the packed bed thermal energy (PBTE) system. In this research, the PCM capsules use stainless steel (321) for the sphere shell, $NaNO_2$ as the PCM, and the HTF composed of 60% $NaNO_3$+40% KNO_3. The PCM capsules in the packed bed device are arranged sequentially, comparing the thermal performance of packed beds composed of spherical capsules, a combination of spherical and cylindrical capsules, and cylindrical capsules alone. Moreover, this work faces limitations, particularly the lack of experimental validation. Future work will focus on validating the simulation through experiments.

2. Numerical modeling

2.1 Mathematical Model and Governing Equations for the PCM Packed Bed

In this numerical simulation studies, three different models of packed bed systems were constructed, as shown in Figure 1. The models include spherical capsule-packed beds, mixed-shape capsule-packed beds, and cylindrical capsule-packed beds, each containing 152 capsules uniformly and orderly arranged. There are 8 layers within each packed bed, with 19 capsules per layer. The spherical and cylindrical capsules both have a diameter of 20 mm. To account for geometric differences between the cylinders and spheres, to prevent variations due to volume differences, the height of the cylinders was equivalently calculated to be 13.34 mm. The thermal storage tank is 180 mm in height, and 120 mm in diameter. To improve computational efficiency, one-sixth of the packed bed was used as the computational domain due to the tank's geometric symmetry. The packed bed initially has a temperature of 673.15 K, while the fluid enters at 523.15 K.

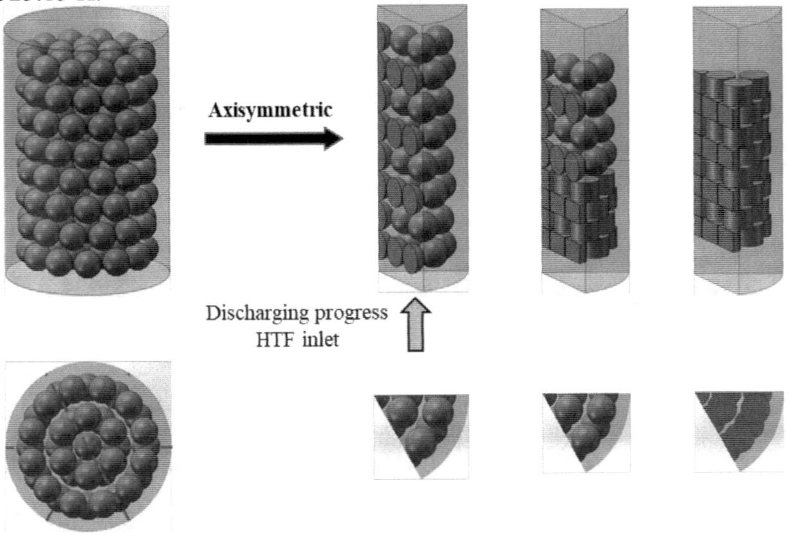

Figure 1. The 3-D PBTES model.

Numerical simulations were conducted with COMSOL, with the application of a laminar flow model to model fluid flow and heat transfer processes within the system. The pressure and velocity fields were handled by the built-in solver. The built-in solver of COMSOL was utilized during the computation to handle the pressure and velocity fields. The phase transition within the PCM was calculated using the sensible heat capacity method. In this method, total enthalpy is used in the energy equation, while the liquid fraction is defined based on temperature. This method effectively simulated the energy conversion process of the HTF within the PCM.

Continuity equation:

$$\rho_{HTF} \nabla \cdot \boldsymbol{u} = 0 \quad (1)$$

Momentum equation:

$$\rho_{HTF}\left(\boldsymbol{u}\cdot\nabla\right)\boldsymbol{u}=\nabla\cdot\left[-p+\mu\left(\nabla\boldsymbol{u}\right)+\left(\nabla\cdot\boldsymbol{u}\right)^{T}\right]+\rho_{HTF}\boldsymbol{g} \tag{2}$$

Energy equation:

$$\rho_{HTF}C_p\frac{\partial T}{\partial t}+\rho_{HTF}C_p\boldsymbol{u}\cdot\nabla T=\nabla\cdot\left(k\nabla T\right) \tag{3}$$

By using Equation (1)-(3), the study focused on the thermal characteristics inside the thermal packed bed. During the discharging process, the HTF flows into the tank from the bottom at 0.008 m/s and exits through the upper outlet.

2.2 Material properties

In this simulation, the selected PCM is $NaNO_2$, and stainless steel (321) is used for the shell. Table 1 presents the thermophysical properties of the PCM [7] and the outer shell material.

Table 1. Thermo-physical properties of $NaNO_2$.

Material	$NaNO_2$	Stainless steel (321)
Density ($kg\cdot m^{-3}$)	2097(solid) 1846(liquid)	7900
Thermal conductivity ($W\cdot m^{-1}\cdot K^{-1}$)	0.562(solid) 0.483(liquid)	18.9
Specific heat capacity ($J\cdot kg^{-1}\cdot K^{-1}$)	1385(solid) 1531(liquid)	500
Latent heat ($kJ\cdot kg^{-1}$)	199.6	-
Phase change temperature (K)	555	-

The selected HTF consists of molten salt (60% $NaNO_3$+40% KNO_3), with its properties are determined using [8]:

$$\rho_f\left(\mathrm{kg\cdot m^{-3}}\right)=2090-0.636T_f\left(℃\right) \tag{4}$$

$$C_{p,f}\left(\mathrm{J\cdot kg^{-1}\cdot ℃^{-1}}\right)=1443+0.172T_f\left(℃\right) \tag{5}$$

$$K_f\left(\mathrm{W\cdot m^{-1}\cdot ℃^{-1}}\right)=0.443+0.00019T_f\left(℃\right) \tag{6}$$

2.3 Boundary conditions and grid independence

Initially (t = 0 s), the system starts with a uniform temperature throughout. During the discharging process, heat transfer fluid' s inlet temperature and velocity are maintained steady, and the tank walls are adiabatic. The assumptions underlying the numerical model are as follows: (1) Radiant heat transfer is negligible; (2) The outer surface of the storage tank is adiabatic, preventing heat loss to the surrounding environment; (3) Flow processes within the liquid phase inside the capsules are neglected.

The unstructured grids in the computational domain were created with COMSOL Multiphysics 6.0 software. PCM spheres, each 20 mm in diameter, underwent a grid-independence test using three distinct grid sizes: 964,390; 1,100,952; and 2,139,834. The simulations revealed discrepancies under 1.5%, leading to the selection of a grid size of 1,100,952 for thermal analysis in the PBTES system.

2.4 Mode validation

According to experimental findings from [9], a packed bed model was established with consistent initial and boundary conditions. Figure 2 displays the comparison of PCM temperature fluctuations at different dimensionless heights within the bed during the charging process. The outcomes of the simulation closely match experimental data, confirming the model's accuracy and its potential for studying heat transfer and thermal stratification in various packed bed configurations.

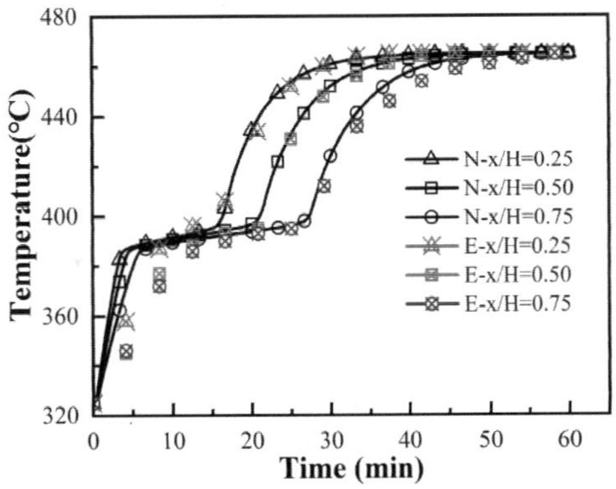

Figure 2. Verification of numerical model.

3. Results and discussion

3.1 Comparison of phase transition behavior in three PBTES systems

Figure 3 shows the changes in liquid fraction across different systems configurations. The solidification rate is higher at the initial stage of the discharging process. Over time, the liquid fraction curve gradually stabilizes. In the beginning, there is a considerable temperature gap between the HTF and the PCM, resulting in the fastest solidification rate. Once the average liquid fraction is 0.2, the time required for Cases 1, 2, and 3 is 353 seconds, 450 seconds, and 466 seconds, respectively, indicating the fastest solidification rate in spherical capsules.

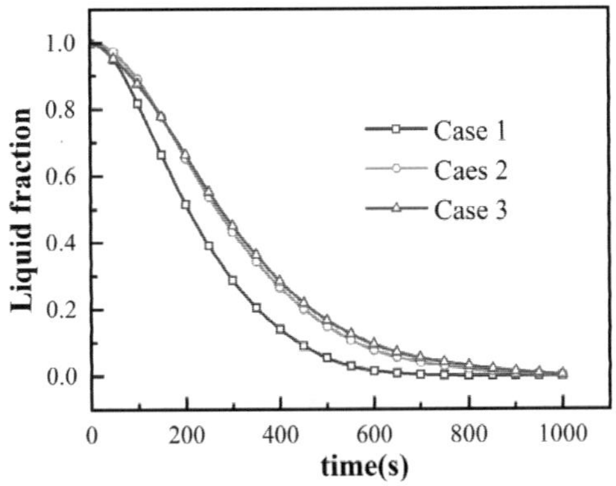

Figure 3. Time-dependent liquid phase fraction during the discharging process for each case.

Figure 4 illustrates the PCM temperature changes across the three systems, detailing the axial temperature evolution of the capsules at x/H = 0.25, x/H = 0.50, and x/H = 0.75. Due to lower fluid temperature and smaller heat capacity, the phase change time during the discharging process is longer. Among these, Case 1 completes the phase change first, and its fluid temperature is also the lowest. The fluid temperatures in Case 2 and Case 3 decrease more slowly, which may be related to the larger side area of the cylindrical capsules, improving heat exchange between the capsules and fluid. From these results, it can be seen that Case 2 exhibits better thermal performance during the discharging process.

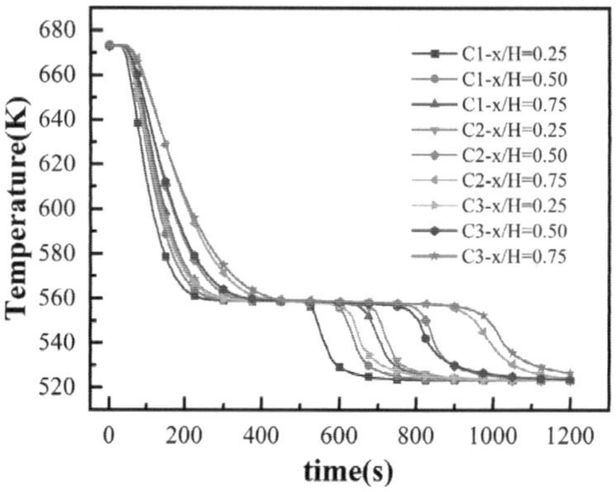

Figure 4. PCM temperature changes for the three cases.

3.2 Analysis of the outlet HTF temperature

Figure 5 shows the variations in fluid outlet temperature during the discharging processes. In the packed bed discharging process, the fluid outlet temperature in Case 2 decreases the slowest and remains the highest throughout, providing more sustained high-temperature heat. The combination of spherical and cylindrical capsules optimizes the fluid pathways, enhancing convective heat transfer and heat exchange between capsules. The fluid outlet temperatures for Case 1 and Case 2 are almost identical, but the fluid temperature in Case 2 consistently shows higher values. Case 3 requires the least time to complete the discharging process, making it suitable for applications requiring rapid heat release. However, the total duration for which the HTF outlet temperature remains above 555 K is reduced, limiting its applicability in scenarios where maintaining a specific outlet temperature is critical.

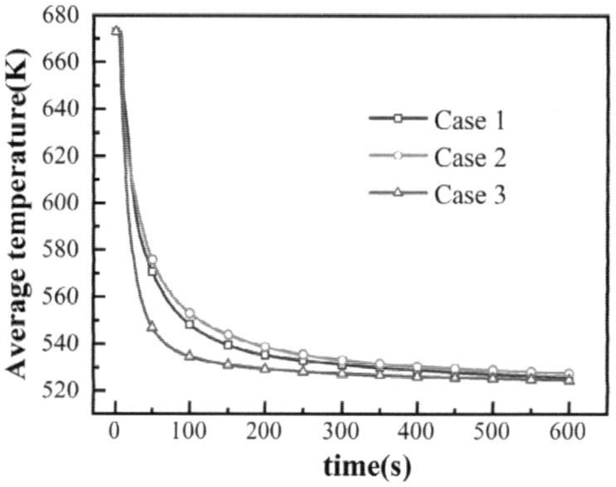

Figure 5. Average HTF outlet temperature during the discharging process.

3.3 Exergy Performance Analysis

As shown in Figure 6, the exergy provided by the HTF and PCM, vary with different configurations of the capsule-packed bed. Since the volume remains constant with only a change in shape, the exergy released by the capsules remains relatively stable. The exergy of the PCM fluctuates within a range of 85,250 ± 250 J. The exergy output efficiencies of Cases 1, 2, and 3 are 67.50%, 73.68%, and 36.30%, respectively. Case 2 demonstrates the complementary action of the two types of capsules. Cylindrical capsules offer rapid heat transfer characteristics, especially at the larger contact surfaces, while spherical capsules provide more uniform and sustained heat release. This structural complementarity helps maintain high thermal efficiency throughout the discharging process while ensuring that heat is transferred more durably and effectively to the fluid. This configuration optimizes the overall utilization of heat, reducing efficiency losses caused by sudden heat release.

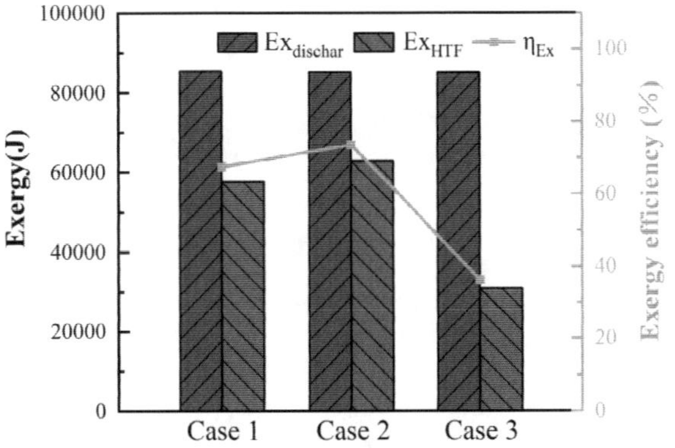

Figure 6. Changes in exergy during the discharging process of packed beds of each case.

4. Conclusion

The primary conclusions can be summarized as follows:

(1) Case 1 and Case 2 exhibit similar changes in liquid fraction during the discharging process.

(2) In Case 2, consisting of spherical and cylindrical capsules, the discharging process has a higher outlet temperature compared to packed beds containing capsules of a single shape, providing an HTF at the required temperature for an extended period.

(3) During the discharging process, the exergy efficiency of Case 2 is 6.18% greater than Case 1 and 37.38% higher than Case 3, reflecting its high efficiency in the discharging process.

References

[1] Md. I. Alam, M. M. Nuhash, A. Zihad, T. H. Nakib, and M. M. Ehsan (2023) Conventional and Emerging CSP Technologies and Design Modifications: Research Status and Recent Advancements. International Journal of Thermofluids, 20:100406. [CrossRef]

[2] J. J. Blaney, S. Neti, W. Z. Misiolek, and A. Oztekin (2013) Containment capsule stresses for encapsulated phase change materials. Appl. Therm. Eng., 50: 555–561. [CrossRef]

[3] X. Wu and Z. Tang (2023) Heat transfer and storage characteristics of a hexagonal close structured packed-bed thermal storage system with molten salt phase change materials. Journal of Energy Storage, 65:107356. [CrossRef]

[4] B. Sun, Z. Liu, X. Ji, L. Gao, and D. Che (2022) Thermal energy storage characteristics of packed bed encapsulating spherical capsules with composite phase change materials. Applied Thermal Engineering, 201:117659. [CrossRef]

[5] A. Kumar and S. K. Saha (2022) Thermal and structural characterizations of packed bed thermal energy storage with cylindrical micro-encapsulated phase change materials. Journal of Energy Storage, 48:103948. [CrossRef]

[6] Ahmad, A. Anagnostopoulos, M. E. Navarro, Y. Maksum, S. Sharma, and Y. Ding (2024) A comprehensive material and experimental investigation of a packed bed latent heat storage system based on waste foundry sand. Energy, 294:130920. [CrossRef]

[7] F. Mohammadnejad and S. Hossainpour (2020), A CFD modeling and investigation of a packed bed of high-temperature phase change materials (PCMs) with different layer configurations, J. Energy Storage, 28: 101209. [CrossRef]

[8] N. Alexis B. Zavoico (2001), Solar Power Tower Design Basis Document, Revision 0. Sandia National Lab. United States. [CrossRef]

[9] M.-J. Li, B. Jin, J.-J. Yan, Z. Ma, and M.-J. Li (2018), Numerical and Experimental study on the performance of a new two-layered high-temperature packed-bed thermal energy storage system with changed-diameter macro-encapsulation capsule, Appl. Therm. Eng., 142: 830–845. [CrossRef]

Optimized carbon emission dispatch of power systems based on backpropagation algorithm

Zhijian Chen *, Haiji Wang , Lidong Huang , Xiaojing Gan

China Energy Engineering Group Guangdong Electric Power Design Institute Co., Ltd., Dongguan, Guangdong, 510000, China

*Corresponding author's e-mail: chenzhijian@gedi.com.cn

Abstract. With the intensification of global climate change, carbon emissions from power systems have become an increasingly serious issue. Optimizing power system dispatch to reduce carbon emissions has become an urgent need. This paper proposes a power system carbon emission optimization dispatch model based on the backpropagation algorithm. By incorporating a genetic algorithm to optimize the backpropagation algorithm, the model improves the dispatch efficiency of power systems while reducing carbon emissions. Experimental results show that the improved model has significant advantages in terms of convergence speed and global optimal solution. It can effectively handle the scheduling tasks of power systems with multiple constraints and demonstrates strong adaptability and optimization performance under different load conditions.

1. Introduction

As the global issue of climate change becomes increasingly severe [1], reducing carbon emissions has become one of the key tasks [2] for countries to address environmental pollution and energy crises [3]. The power system, as a major source of carbon emissions, plays a crucial role in mitigating emissions. Optimizing the dispatch of the power system can not only effectively reduce carbon emissions but also improve energy efficiency, promoting sustainable development. Traditional power system dispatch relies on fixed scheduling strategies, which struggle to cope with complex and variable load demands and the constraints of power generation equipment [4]. With the large-scale integration of renewable energy, the scheduling problem of power systems has become even more complex, requiring more intelligent dispatch methods to meet the ever-changing power demand and carbon emission control objectives [5].

Currently, many studies have attempted to optimize power system dispatch using machine learning [6] and optimization algorithms [7]. However, most methods still have certain shortcomings. Traditional optimization methods often perform poorly in dealing with multi-constraint and nonlinear problems, easily falling into local optima. While AI-based optimization algorithms have achieved good results in certain tasks, they are slow to converge and lack global optimization capabilities, making them difficult to provide ideal solutions in complex power system dispatch problems [8]. Furthermore, most existing methods have not effectively combined carbon emission reduction with system efficiency optimization, limiting their practical applicability.

To address these issues, this paper proposes a power system carbon emission optimization dispatch model based on the genetic algorithm-optimized backpropagation algorithm. By combining the advantages of the backpropagation and genetic algorithms, the proposed model can effectively reduce carbon emissions while ensuring the dispatch efficiency of the power system. The backpropagation algorithm quickly optimizes the system dispatch strategy, while the genetic algorithm enhances the model's global search ability, effectively avoiding the problem of local optima.

2. Establishment And Improvement Of The Carbon Emission Optimization Dispatch Model

2.1. Establishment of the Power System Carbon Emission Optimization Dispatch Model

To achieve the minimization of carbon emissions in power system dispatch, an optimization model targeting carbon emission reduction is first constructed. This model incorporates key factors such as power load forecasting, power generation distribution among power plants, and carbon emission coefficients. Load forecasting is based on historical data and trend analysis to determine the future electricity demand for each period, ensuring the system can supply power stably. Based on this, the model distributes the generation tasks among the power plants to meet the electricity demand, considering the carbon emission differences across different power generation methods. Clean energy generation is prioritized to reduce the system's carbon emissions.

During the scheduling process, a series of constraints need to be followed. These include the production capacity limits of power equipment, as each power plant has an upper limit on its generation capacity, and the dispatch plan must ensure these limits are not exceeded. It is necessary to maintain a balance between power generation and load demand in each period to prevent system overload or insufficient supply. Reserve capacity must also be considered to handle potential equipment failures or sudden load changes, ensuring the system's safe and stable operation.

The optimization objective of the model is to minimize the system's carbon emissions. Carbon emissions are calculated based on the generation of each energy type and its corresponding carbon emission coefficient. The objective function integrates the carbon emissions of all power plants, ensuring that the dispatch plan minimizes carbon emissions while meeting load and equipment constraints. The objective function is represented as:

$$Minimize\ Z = \sum_{i=1}^{N} P_i \times C_i \tag{1}$$

where Z is the total carbon emission, P_i is the generation of the i -th power plant, C_i is the carbon emission coefficient of the i -th power plant, and N is the total number of power plants.

2.2. Improvement of Backpropagation Algorithm Based on Reinforcement Learning

In traditional backpropagation algorithms, the optimization process relies on gradient descent to update local optimal solutions. However, this method often gets trapped in local optima, especially when dealing with complex constraints and dynamic changes in power system carbon emission optimization scheduling tasks. To solve this problem, the Q-learning algorithm is introduced to dynamically optimize the power system dispatch strategy, thereby improving global optimality, reducing carbon emissions, and enhancing system flexibility and adaptability.

In power dispatch problems, reinforcement learning is introduced in which the agent (i.e., the power dispatch system) learns the optimal decision-making process through interaction with the environment. Whenever the agent performs a dispatch action, the environment (i.e., the power system) returns a reward signal based on the action. Through the reward mechanism, reinforcement learning drives the agent to improve decisions gradually, achieving global optimization. The state of the power dispatch problem can be defined by the current electricity load, power generation distribution, and other information, while the action refers to the power generation allocation decisions for each power plant. The reward signal is defined based on carbon emissions, load balance, system stability, and other factors.

To achieve this optimization process, the Q-learning algorithm is employed to optimize the dispatch strategy. Q-learning is a reinforcement learning method based on value iteration, where the core idea is to guide the agent to choose the optimal action by learning the state-action value function Q(s, a). The state-action value function represents the cumulative reward obtained after taking a particular action in a given state. The updated formula for Q-learning is as follows:

$$Q(s_t, a_t) = Q(s_t, a_t) + \alpha\left[r_t + \gamma \max_a Q\,(s_{t+1}, a_t) - Q(s_t, a_t)\right] \tag{2}$$

where s_t represents the state of the system at time t, a_t is the action taken by the agent to instate s_t, i.e., the power generation allocation for the plants, r_t is the reward obtained by the agent after performing action s_t in state s_t, which is typically related to carbon emissions and load balance, γ is the discount factor that reflects the importance of future rewards, and α is the learning rate controlling the magnitude of the Q-value update. Through repeated execution of this update process, Q-learning can learn the optimal dispatch strategy by continuously interacting with the environment.

The key to the training process is the balance between exploration and exploitation. In the early stages, the agent primarily explores different scheduling plans to accumulate experience. As training progresses, the agent gradually shifts towards exploiting its existing experience by choosing scheduling decisions that yield higher rewards. This mechanism effectively avoids the local optima problem that may arise in traditional backpropagation algorithms. To ensure global optimality in power dispatch, Q-learning continuously adjusts the state-action value function to learn the optimal strategy.

To further optimize the weight update of the neural network, reinforcement learning and backpropagation are collaboratively trained during the learning process. Traditional backpropagation updates the network weights by calculating errors and propagating them backward, while reinforcement learning updates the Q-value function based on feedback from the environment. The combination of these two methods allows the dispatch strategy to be optimized not only based on traditional gradient information but also dynamically adjusted according to real-time feedback, improving the model's flexibility and accuracy.

The collaborative training of backpropagation and reinforcement learning improves the efficiency and accuracy of the optimization process by dynamically adjusting weights based on feedback after each dispatch task. After each dispatch, the carbon emissions, load balance, and other evaluation metrics of the power system are calculated, providing the agent with appropriate rewards or penalties. These feedback signals are used to update the network weights, and the Q-value function is also updated, allowing the agent to make more accurate dispatch decisions based on experience and current environmental conditions.

The advantage of collaborative training is that it combines the efficiency of backpropagation in neural network training with the flexibility of reinforcement learning in dynamic decision-making. This combination enables power system dispatch to not only reduce carbon emissions but also better adapt to varying load changes and fluctuations in generation capacity, resulting in an optimized carbon emission dispatch solution.

2.3. Improvement of Backpropagation Optimization Based on Genetic Algorithm

In the power system carbon emission optimization scheduling, traditional backpropagation algorithms often face the risk of falling into local optima. To address this issue, this study introduces the Genetic Algorithm (GA) to enhance the global search capability of the backpropagation algorithm [9]. Genetic Algorithm is a global optimization method that simulates natural selection and biological inheritance mechanisms, capable of effectively exploring the solution space and finding the global optimum. Therefore, by combining the Genetic Algorithm with backpropagation, we can use GA for global optimization to overcome the problem of the backpropagation algorithm stagnating at local optima, thus achieving more accurate and optimized carbon emission scheduling for the power system.

The first step of the Genetic Algorithm is individual encoding, where the power system scheduling plan is transformed into chromosome form. The scheduling plan of the power system can be represented by a vector consisting of power generation and load distribution information, specifically encoded as chromosomes. Let the power system have N power plants, and the power generation of each plant is $x_i (i = 1,2,\cdots,N)$. The chromosome of an individual can be represented as:

$$X = [x_1, x_2, \cdots, x_N] \tag{3}$$

The fitness function is designed based on factors such as carbon emissions, load balance, and system stability, to minimize carbon emissions. Let the carbon emission be denoted by $C(X)$, and the fitness function can be defined as:

$$f(X) = \frac{1}{C(X)} \tag{4}$$

where $C(X)$ is the carbon emission corresponding to chromosome X, and the higher the fitness value, the lower the carbon emissions of the scheduling plan and the better its performance.

The core of the Genetic Algorithm lies in the selection, crossover, and mutation operations, which simulate the evolutionary process in nature and push the algorithm towards finding the optimal solution.

Selects individuals with higher fitness from the current population as parents for the crossover operation. Common selection methods include roulette-wheel selection and tournament selection. By selecting individuals with higher fitness, excellent genes are passed on to the next generation.

Simulates the process of biological gene recombination, where two-parent individuals are randomly selected, and some of their gene information is exchanged to generate new offspring individuals. In the power scheduling problem, we can choose to exchange part of the power generation distribution information between two parents to explore more scheduling options. Let the parent individuals be $X_1 = [x_1, x_2, \cdots, x_N]$ and $X_2 = [x'_1, x'_2, \cdots, x'_N]$. After crossover, the offspring individual is $X' = [x_1, x'_2, \cdots, x'_N]$.

Mutation Operation: Randomly alters an individual's chromosome to increase the diversity of the solution space and prevent premature convergence of the algorithm. The mutation operation can make small adjustments to the power distribution of power plants. For example, if the power generation of a plant is x_i, after mutation, it may become x'_i, expanding the solution space.

These operations in the Genetic Algorithm help perform a global search in the solution space, while the local exploration finds the optimal solution. In the carbon emission optimization scheduling problem for the power system, the combination of selection, crossover, and mutation operations effectively balances global search and local exploration, improving the convergence speed and the accuracy of the optimal solution.

During the optimization process, the Genetic Algorithm works in synergy with the backpropagation algorithm. The backpropagation algorithm is used for training the neural network and updating the network's weights, while the Genetic Algorithm finds the optimal scheduling plan through global optimization. In each generation of the GA evolution, the selected parent individuals undergo training and adjustment through the backpropagation algorithm, with the optimization goal of minimizing carbon emissions. Through repeated iterations, both the Genetic Algorithm and the backpropagation algorithm collaboratively improve the accuracy and effectiveness of the power system scheduling plan. The introduction of the Genetic Algorithm allows the entire optimization process to escape local optima and use global search to explore more possible solutions, thereby providing a more optimized carbon emission scheduling plan for the power system.

2.4. Implementation of Multi-Level Optimization Strategy

To further improve the accuracy and efficiency of carbon emission optimization scheduling in power systems, this study introduces a multi-level optimization strategy that combines multi-level neural network structures and an adaptive learning rate adjustment mechanism to enhance the optimization process of the backpropagation algorithm. This strategy aims to improve the model's scheduling accuracy at different time and spatial scales, while also enhancing the model's convergence speed and stability.

In traditional neural networks, a single structure is typically used to process all input information. However, the power system scheduling problem involves multiple time and spatial scales. Considering the load variations at different times and the power generation capacity of power plants at different locations, a single-structure neural network may struggle to handle these complex multi-scale issues. Therefore, adopting a multi-level neural network structure becomes key to improving model accuracy. This study introduces BiFPN (Bidirectional Feature Pyramid Network), a module that can effectively handle features at different scales, with significant advantages in managing the diversity of spatial and temporal scales [10].

BiFPN builds a multi-level feature pyramid to integrate information across different scales, ensuring that both global and local levels can extract effective features. The BiFPN module fuses low-level detailed information with high-level semantic information and uses bidirectional connections to enhance information flow during feature propagation. For power scheduling problems, BiFPN can manage the correlation between short-term load forecasting and long-term generation capacity scheduling, improving the network's adaptability to dynamic changes in the power system, and thus achieving higher accuracy in carbon emission optimization.

Let the different level output features in the network be denoted as $F_1, F_2, \cdots, F_k$, where k represents the number of levels. BiFPN integrates these features using the following formula:

$$F_{fused} = \sum_{i=1}^{k} \alpha_i F_i \tag{5}$$

where α_i is the weight of each feature layer, reflecting the contribution of each level to the final output. In the neural network training process, the learning rate is a key hyperparameter that affects the training effectiveness. A learning rate that is too high can lead to unstable training, while one that is too low can cause slow convergence. Therefore, this study designs an adaptive learning rate adjustment strategy, combining learning rate decay and dynamic adjustment mechanisms, to automatically adjust the learning rate based on the model's error feedback, thereby improving convergence speed and model stability.

The adaptive learning rate adjustment strategy mainly includes two mechanisms: learning rate decay and dynamic adjustment. Learning rate decay gradually reduces the learning rate during the training process, allowing the model to fine-tune weights when approaching the optimal solution, thus avoiding large steps that might skip the optimal point. Let the initial learning rate be η_0, and the learning rate decay process can be expressed as:

$$\eta_t = \eta_0 \times \frac{1}{1 + \lambda t} \tag{6}$$

where η_t is the learning rate at the t-th iteration, λ is the decay rate, and t is the iteration number. Dynamic adjustment, on the other hand, adjusts the learning rate in real time based on the changes in training error. When the model's error fluctuates significantly, the learning rate is increased to help the model quickly escape local optima; when the error stabilizes, the learning rate is reduced to fine-tune the model parameters. Let the error function be $E(\theta)$, and the learning rate adjustment mechanism can be implemented using the following formula:

$$\eta_t = \eta_{t-1} \times \left(1 + \gamma \frac{\Delta E_t}{|E_t - E_{t-1}|}\right) \tag{7}$$

where $\Delta E_t = E_t - E_{t-1}$ is the change in error, and γ is the adjustment coefficient.

3. Experimental Results and Analysis

3.1. Experimental Design and Data Sources

The experimental design of this study uses typical power system data for validation, covering various aspects such as power load forecasting, equipment capacity, and carbon emission coefficients. The load forecasting data is sourced from actual historical load records of the power system, covering fluctuations in power demand under different seasons, climates, and economic activity levels. The power equipment capacity data includes the maximum generation capacity of various power generation units, operating constraints of the generation units, and their corresponding carbon emission coefficients. To comprehensively evaluate the proposed carbon emission optimization scheduling model, the experiment selected a multi-region, multi-plant power system, simulating integrated scheduling schemes for traditional coal-fired power plants, natural gas plants, and renewable energy plants. The experiment tests the model's adaptability, accuracy, and optimization effectiveness under varying load demand scenarios. The experimental platform utilizes a high-performance computing environment to ensure the efficiency of model training and scheduling calculations. The experiments were conducted on a server configured with an 8-core Intel Xeon processor, 128 GB of memory, and an NVIDIA A100 GPU with CUDA 11.4 acceleration, significantly enhancing the computational power and processing speed of the model. Furthermore, all experiments were carried out in a Linux-based environment, using PyTorch 2.3.1 for model training and optimization. The experimental code and data were managed using version control tools to ensure the reproducibility and reliability of the results. To ensure the stability of the results, all experiments were validated in multiple independent runs, and the experimental results were subjected to statistical analysis.

3.2. Optimization Effect Analysis

In this section, the study analyzes the performance of the algorithm in carbon emission optimization, focusing on how the introduction of genetic algorithms improves the backpropagation algorithm to reduce carbon emissions and enhance the operational efficiency of the power system. The experimental results show that the proposed algorithm can effectively reduce carbon emissions while improving the scheduling efficiency of the power system. The optimized model not only reduces energy consumption but also improves the adaptability of the power system under different load and equipment capacity conditions.

By comparing the optimization results of different models, the proposed algorithm shows significant advantages over traditional backpropagation models in terms of convergence speed and finding the global optimal solution. Traditional backpropagation algorithms often fall into local optima, while the improved model with a genetic algorithm can find the global optimal solution in a shorter time, significantly improving scheduling efficiency. The following figure demonstrates how the improved model reduces carbon emissions while increasing system operational efficiency, comparing the scheduling effects under different optimization methods.

As shown in Figure 1, the improved model effectively reduces carbon emissions and outperforms both the traditional backpropagation algorithm and the model combining backpropagation with reinforcement learning. The carbon emissions gradually decrease with each iteration, and the model optimized with the genetic algorithm achieves a more significant reduction in carbon emissions within fewer iterations.

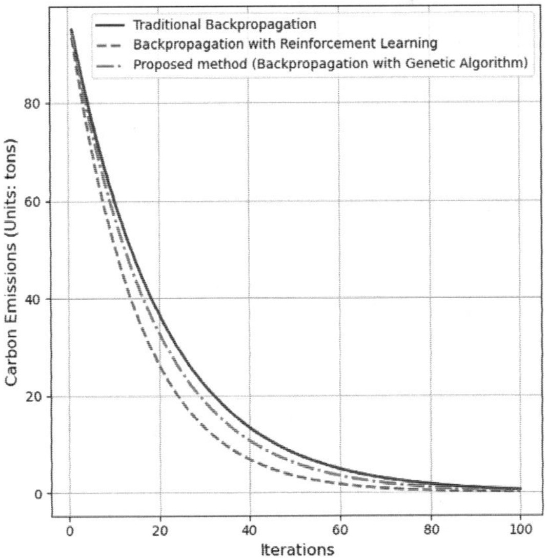

Figure 1. Comparison of Carbon Emission Reduction Under Different Optimization Methods.

As shown in Figure 2, the proposed algorithm also has significant advantages in improving the operational efficiency of the power system. As the number of iterations increases, the proposed algorithm demonstrates higher system efficiency, proving the positive impact of the genetic algorithm on system optimization, particularly in handling different load demands and power equipment constraints.

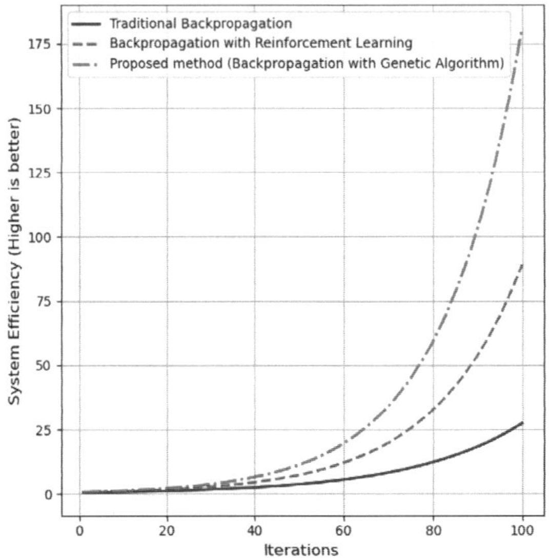

Figure 2. Comparison of Carbon Emission Reduction Under Different Optimization Methods.

As shown in Figure 3, by comparing the convergence and global optimality of different algorithms, the proposed algorithm exhibits clear advantages over traditional backpropagation algorithms. By introducing the genetic algorithm, the proposed algorithm not only improves convergence speed but also successfully avoids the problem of local optima, allowing the global optimal scheduling strategy to be found in a shorter time. This indicates that the combined optimization method of backpropagation and genetic algorithm proposed in this study is highly competitive in carbon emission optimization scheduling tasks.

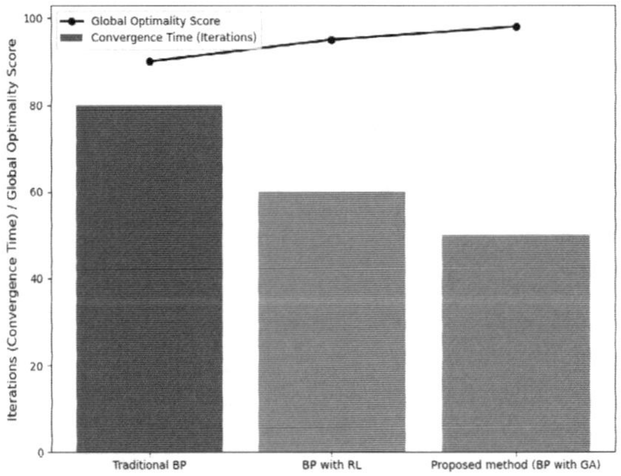

Figure 3. Comparison of Convergence and Global Optimality of Different Algorithms.

4. Conclusion

This study proposes a power system carbon emission optimization scheduling model based on genetic algorithm-optimized backpropagation. The model incorporates factors such as power load forecasting, generation allocation, carbon emission coefficients, and constraints, aiming to reduce carbon emissions and enhance the operational efficiency of the power system through optimized scheduling. In the model development process, a basic framework for power system scheduling is established, and the model's inputs, outputs, and constraints are clearly defined. The backpropagation algorithm is initially applied for optimization, and the genetic algorithm is introduced to improve the backpropagation algorithm, enhancing the model's global optimization capability and avoiding the limitations of traditional backpropagation, which often falls into local optima.

To validate the effectiveness of the model, experiments were conducted on a typical power system dataset, covering various types of power plants and load demand scenarios. The experimental results show that the backpropagation optimization model combined with the genetic algorithm can not only find the global optimal scheduling strategy in a shorter time but also demonstrate significant advantages in reducing carbon emissions and improving system efficiency. Compared to the traditional backpropagation algorithm, the improved model converges more quickly during the iteration process and is capable of handling complex scheduling tasks under different load and equipment capacity constraints, demonstrating strong adaptability and optimization effectiveness.

References

[1] Yang M, Chen L, Wang J, et al. Circular economy strategies for combating climate change and other environmental issues[J]. Environmental Chemistry Letters, 2023, 21(1): 55-80.
[2] Chen X, Chen Z. Can green finance development reduce carbon emissions? Empirical evidence from 30 Chinese provinces[J]. Sustainability, 2021, 13(21): 12137.
[3] Akbari-Dibavar A, Mohammadi-Ivatloo B, Zare K, et al. Economic-emission dispatch problem in power systems with carbon capture power plants[J]. IEEE Transactions on Industry Applications, 2021, 57(4): 3341-3351.
[4] Kunya A B, Abubakar A S, Yusuf S S. Review of economic dispatch in multi-area power system: State-of-the-art and future prospective[J]. Electric Power Systems Research, 2023, 217: 109089.
[5] Mohandes B, Wahbah M, El Moursi M S, et al. Renewable energy management system: Optimum design and hourly dispatch[J]. IEEE Transactions on Sustainable Energy, 2021, 12(3): 1615-1628.
[6] Li Z, Luo X, Liu M, et al. Wind power prediction based on EEMD-Tent-SSA-LS-SVM[J]. Energy Reports, 2022, 8: 3234-3243.
[7] Sheng W, Li R, Yan T, et al. A hybrid dynamic economics emissions dispatch model: distributed renewable power systems based on improved COOT optimization algorithm[J]. Renewable Energy, 2023, 204: 493-506.
[8] Stock S, Babazadeh D, Becker C. Applications of artificial intelligence in distribution power system operation[J]. IEEE access, 2021, 9: 150098-150119.
[9] Sohail A. Genetic algorithms in the fields of artificial intelligence and data sciences[J]. Annals of Data Science, 2023, 10(4): 1007-1018.
[10] Li X, Li D, Liu H, et al. A-BFPN: An attention-guided balanced feature pyramid network for SAR ship detection[J]. Remote Sensing, 2022, 14(15): 3829.

Adaptive virtual inertia control of energy storage doubly-fed condenser considering the gradual recovery of rotational speed

Shuqiang Zhao, Xubo Chen*, Hui Wang

Department of Electric Power Engineering, North China Electric Power University, Baoding, 071003, China

*Corresponding author's e-mail: xubo_chen@163.com

Abstract. Under virtual inertia control, energy storage doubly-fed condenser can actively provide frequency support using rotor kinetic energy when the system is disturbed. However, due to the fixed coefficient of traditional virtual inertia control, it is difficult to balance the speed recovery of the condenser with system frequency stability and dynamic characteristics. Therefore, this article proposes an adaptive virtual inertia control method that considers the gradual recovery of condenser rotation speed. Firstly, the basic principles and virtual inertia control of energy storage doubly-fed condenser are analysed; Then, based on the real-time status of the camera speed, system frequency, and their rate of change in response to different types of load fluctuations, the condenser is adaptively controlled. This strategy allows the condenser to balance speed recovery during frequency periods; Finally, the effectiveness and superiority of the control strategy proposed in this paper were demonstrated through simulation verification. The simulation results show that the proposed control strategy improves the frequency response characteristics of the system and achieves gradual recovery of rotor speed.

1. Introduction

Under the "dual carbon" vision, the process of large-scale integration of new energy units is accelerating. As a result, the system's short-circuit capacity decreases, and mechanical rotational inertia reduces, leading to serious voltage and frequency issues in the power system under disturbances [1]. To address this problem, synchronous condensing has become one of the focal points of research in recent years. Currently, synchronous condense performs excellently in terms of instantaneous voltage support and overload capacity [2]. However, the existing synchronous condense has a small rotor inertia and is unable to provide sufficient frequency support to meet the inertia requirements of the new power system [3].

In this context, foreign scholar Ferreira proposed an energy storage doubly-fed condenser (ESDFC) [4]. This is a doubly-fed induction machine that can operate asynchronously and is capable of replacing traditional synchronous condense for reactive power compensation. It can also release the kinetic energy hidden in the high-inertia rotor through excitation control, providing transient frequency support to the system. However, the traditional vector-oriented control strategy has electromechanical decoupling characteristics [5], meaning that the active power output of the condense depends on the upper-level active power command and does not change its rotor speed in response to system frequency disturbances. This results in an inability to reflect the mechanical rotational inertia of the condense itself and insufficient active frequency support capability. Currently, there are two main types of power control strategies that can achieve active frequency support. One is to combine rotor-side control with virtual synchronous generator technology, simulating the operating characteristics of a synchronous generator. This allows the system to actively respond to frequency changes and provide active power frequency support [6]. In [7], the virtual synchronous technology was applied to the doubly-fed condense to achieve active frequency response, but it cannot cope with complex operating conditions. The other approach is to use virtual inertia control, which actively responds to system frequency variations by adjusting the active power output of the condense in proportion to the rate of frequency change, thus simulating the inertia support effect similar to that of synchronous machines [8]. In [9], virtual inertia control was applied to the ESDFC, enabling it to provide active inertia support when the system is disturbed, thereby enhancing transient stability under grid faults.

However, in virtual inertia control, the virtual inertia coefficient is fixed, and simply changing the active power output of the condense based on the system frequency change rate cannot flexibly respond to various complex operating conditions. Two main issues arise from this: First, it is difficult to balance system frequency stability with dynamic characteristics [10]. The varying characteristics of the frequency change rate during frequency disturbances prevent the condense from achieving both frequency support and fast frequency recovery. A smaller virtual inertia coefficient makes the condenser unable to provide sufficient support during the frequency support phase, while a larger virtual inertia coefficient causes the condenser to absorb too much active power during the frequency recovery phase, slowing down the frequency recovery process. Second, the effect of the condenser's energy storage state on its bidirectional support capability is not considered [11]. During transients, the active power output of the condense under virtual inertia control is limited by its energy storage. If the speed is low during a disturbance, and the condense has insufficient energy storage, it cannot provide adequate support when the frequency drops, even with a fixed inertia coefficient. Conversely, if the speed is high during a disturbance, the condenser cannot absorb more active power to reduce the frequency deviation. Furthermore, the condense does not actively restore its speed after the transient ends, which is not ideal for coping with subsequent disturbances. If the equivalent virtual inertia could be adaptively adjusted based on the dynamic state of the system and the condense, it would be more beneficial for enhancing the condenser's support capability and maintaining the stable operation of the grid.

In summary, this paper proposes an adaptive virtual inertia control strategy that considers the gradual recovery of rotor speed. When facing various types of disturbances, it adjusts the virtual inertia compensation based on the condenser's speed and the frequency change, providing sufficient inertia support to the system while gradually recovering speed during transients. This approach helps the condenser maintain its bidirectional support capability, offering adequate support to the system at any time and enhancing system frequency stability. MATLAB/Simulink simulations comparing this adaptive virtual inertia control strategy with a fixed virtual inertia coefficient control strategy verify its effectiveness.

2. Basic Principle of ESDFC and Virtual Inertia Control

2.1. Basic Principle

The energy storage doubly-fed condenser consists of a doubly-fed machine with a high rotational inertia rotor, a rotor-side converter as the AC excitation power source, and a grid-side converter. In a steady state, the energy storage doubly-fed condense operates in a constant-speed phase-shifting state, providing reactive voltage support to the power grid. When a frequency disturbance occurs in the grid, the energy storage doubly-fed condense can adjust the rotor speed to release or store the rotor's kinetic energy, providing inertia support to the grid. Its grid connection structure is shown in Figure 1.

The energy storage double-fed condenser provides flexible voltage and frequency support to the system through excitation control. The essence of excitation control is the control of the rotor-side converter. The rotor-side converter uses vector control, and after vector orientation, the d and q-axis components of the excitation current are controlled to achieve independent control of the active and reactive power of the condense. The reactive power outer loop is for reactive power control, where the reactive power output of the condense is controlled based on the upper-level power command. The active power outer loop is for active power control, where the active power command is derived from the rotor speed/frequency.

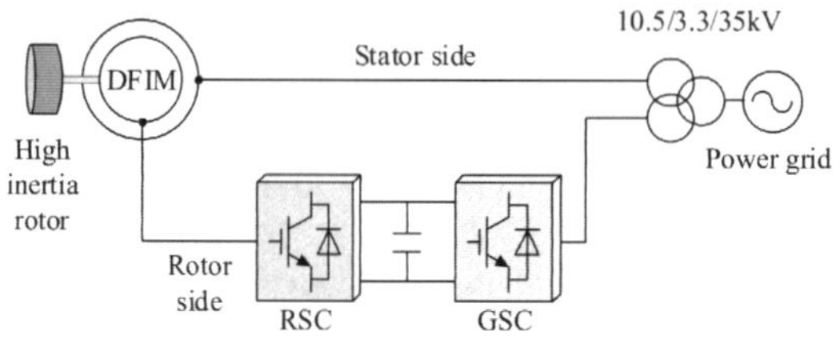

Figure 1. Grid-connected structure of energy storage doubly-fed condenser.

In a steady state, the active power command is obtained through constant-speed control, and the condense absorbs a portion of the active power from the grid to compensate for various losses and maintain a constant rotor speed. At this point, the control equation for the outer-loop PI controller is:

$$\begin{cases} i_{\mathrm{rd,ref}} = k_{\mathrm{pP}}\left(P_{\mathrm{s,ref}} - P_{\mathrm{s}}\right) + k_{\mathrm{iP}}\int\left(P_{\mathrm{s,ref}} - P_{\mathrm{s}}\right)\mathrm{d}t \\ i_{\mathrm{rq,ref}} = k_{\mathrm{pQ}}\left(Q_{\mathrm{s,ref}} - Q_{\mathrm{s}}\right) + k_{\mathrm{iQ}}\int\left(Q_{\mathrm{s,ref}} - Q_{\mathrm{s}}\right)\mathrm{d}t \end{cases} \tag{1}$$

In the equation: $i_{\mathrm{d,ref}}$ and $i_{\mathrm{q,ref}}$ are the reference values for rotor currents; $P_{\mathrm{s,ref}}$ and $Q_{\mathrm{s,ref}}$ are the reference values for active and reactive power; k_{pP}, k_{iP}, k_{pQ}, and k_{iQ} are the PI control coefficients for the active and reactive power loops, respectively.

When a frequency disturbance occurs in the system, the active power outer loop of the condense switches from constant speed control to virtual inertia control. After the transient ends, it switches back to constant-speed control, with the rotor speed fixed at the current speed. The excitation control is shown in Figure 2.

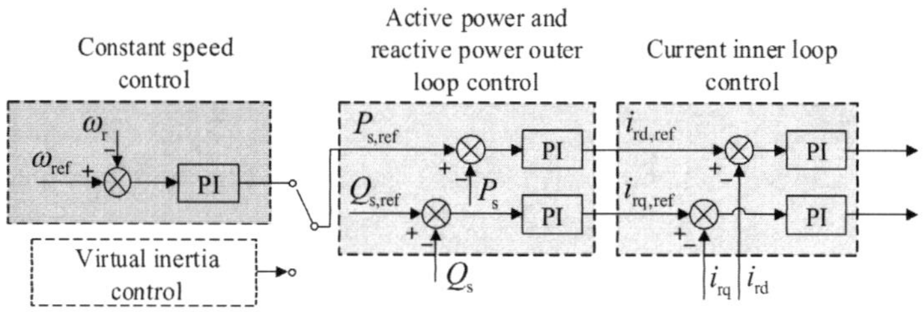

Figure 2. Excitation control of energy storage doubly-fed condenser.

2.2. Virtual Inertia Control

The energy storage doubly-fed condenser does not have a prime mover, and its rotor has a large rotational inertia. Neglecting mechanical friction and damping coefficients on the rotor shaft, the rotor is only subjected to electromagnetic torque. The rotor motion equation can be expressed as:

$$T_{\mathrm{e}} = J\frac{\mathrm{d}\omega_{\mathrm{r}}}{\mathrm{d}t} \tag{2}$$

In the equation: T_{e} is the electromagnetic torque; ω_{r} is the rotor speed; J is the mechanical rotational inertia of the rotor of the energy storage doubly-fed condense.

However, the energy storage doubly-fed condenser is not directly coupled with the grid frequency under normal operating conditions and does not have an inertia response similar to that of synchronous machines. The energy stored in the rotor cannot be fully utilized. To enable the energy storage doubly-fed condense to have an inertia response, it is necessary to introduce grid frequency variation into its control system to simulate the mechanical inertia of a synchronous machine, thereby providing an equivalent virtual inertia. During system frequency disturbances, the energy exchange between the condenser and the grid can be achieved by controlling the rotor speed. Based on this concept, the virtual inertia control strategy is developed. To analyze the size of the virtual inertia of the energy storage doubly-fed condense, we let J_{vir} represent the virtual rotational inertia of the condenser, which can be expressed as:

$$\Delta E_{\mathrm{k}} = \frac{1}{2} J_{\mathrm{vir}} (\omega_{\mathrm{r0}}^2 - \omega_{\mathrm{r1}}^2) \tag{3}$$

In the equation: ΔE_{k} is the change in the rotor's kinetic energy of the condenser; ω_{r} is the mechanical rotor speed of the condenser. The speed range of the energy storage doubly-fed condenser is from 0.7 to 1.3 p.u. The total energy that can be released when the rotor speed drops from 1.3 p.u. to 0.7 p.u. is the total energy storage capacity of the condenser.

To enable the condenser to actively respond to system frequency disturbances, the virtual inertia control adds an active power command related to grid frequency variations in the active power outer loop, coupling the condenser's output with the system frequency to achieve active frequency support. When the system frequency changes, the additional active power command controls the condense to adjust the electromagnetic power, causing the condense to change its rotor speed to release or absorb kinetic energy, thereby stabilizing the system frequency. The additional active power command P_{ref} is expressed as follows:

$$P_{\mathrm{ref}} = k \frac{\mathrm{d}f}{\mathrm{d}t} \tag{4}$$

In the equation: k is the equivalent inertia constant of the condense, defined analogously to that of a synchronous machine, and is proportional to the virtual inertia constant, and f is the system frequency.

3. Adaptive Virtual Inertia Control

3.1. The concept of adaptive virtual inertia

Under virtual inertia control, the inertia presented by the condense is fixed, which makes it difficult to meet the varying inertia requirements of the system during different stages of disturbance. In the initial stage of the disturbance, during frequency support, a larger inertia is needed to resist frequency changes. After the support phase, during frequency recovery, a smaller inertia is required to speed up frequency recovery. However, with fixed inertia, the condense cannot balance system frequency stability and dynamic characteristics. Choosing a larger k value slows down the frequency recovery while selecting a smaller k value reduces the condenser's support capability.

In addition, during transients, the need for the condenser's rotor speed recovery must also be considered. According to Equation (3), the rotor speed of the condense reflects its energy storage status, which constrains the condenser's ability to provide the required inertia support to the system according to the upper-level power command. When the rotor speed is high, the energy storage capacity is sufficient, and it can provide adequate support during frequency drop disturbances. However, when the frequency rises, due to limited energy storage capacity, the condense cannot provide sufficient support to the system according to the power command. Similarly, when the rotor speed is low, the condense cannot provide effective support during frequency drop. After the transient ends, since the system switches back to constant-speed control, the condenser's speed will be fixed at the current value, with no opportunity for speed recovery. This results in the condenser's speed gradually deviating from the rated value after multiple disturbances. When facing subsequent disturbances, its bidirectional support capability becomes asymmetric due to its limited energy storage state.

To further understand the impact of system frequency and rotor speed on the required virtual inertia and the condenser's energy storage state during disturbances, a detailed analysis is conducted using the frequency and its variation curves of the system under disturbance, as shown in Figure 3.

The figure includes two typical types of frequency disturbances, where each disturbance can generally be divided into two stages. Taking the frequency rise disturbance at time t_1 as an example, from t_1 to t_2 is the support stage. The system frequency deviates in the opposite direction, requiring the condense to provide substantial inertia support to hinder further frequency deviation. During this period, the condenser absorbs power from the system, and the rotor speed increases. If the condense's initial rotor speed is low, it can provide sufficient inertia support as per the command. From t_2 to t_3 is the

Journal of Physics: Conference Series **3011** (2025) 012044 doi:10.1088/1742-6596/3011/1/012044

recovery stage. The system frequency recovers in the positive direction. During this stage, the condense, under virtual

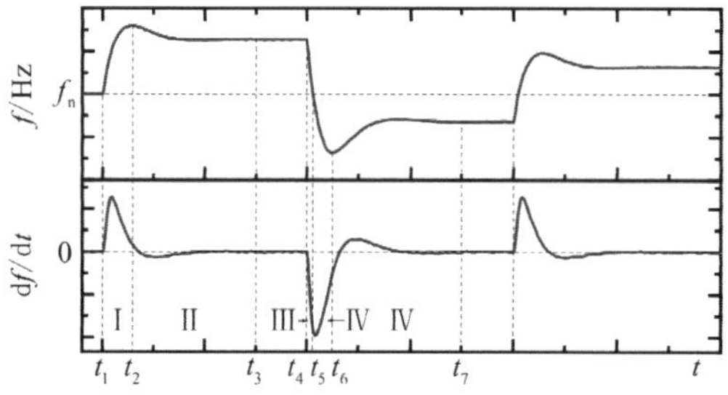

Figure 3. System frequency and its rate of change curve during the disturbance.

inertia control, reduces its rotor speed to release active power, which, in turn, delays the frequency recovery. At time t_3, the system switches from virtual inertia control back to constant-speed control, and the system enters a steady state. At time t_4, a frequency drop disturbance occurs. However, during the support stage from t_4 to t_6, there is a transition between positive and negative frequency deviations. Traditional virtual inertia control cannot differentiate the differing inertia support needs of the system before and after time t_5. During the period from t_4 to t_5, the system frequency should rapidly recover to its rated value, but the condense, under virtual inertia control, provides substantial inertia support, which hinders this process.

In summary, under traditional virtual inertia control, the inertia support requirements of the system at different stages during transients do not correspond to the actual inertia support that the condense can provide, nor to its potential support capacity. The specific relationship is shown in Table 1. The size of the virtual inertia should comprehensively consider the system frequency dynamics as well as the rotor speed of the condense, balancing the condenser's rotor speed recovery needs with the system's frequency stability and dynamic characteristics.

Table 1. The virtual inertia that should be provided by the condenser in different situations.

interval	Δf	d*f*/d*t*	Speed variation	Virtual inertia
I	>0	>0	Rapid rise	rapidly increase
II	>0	<0	Slow decline	gradually decrease
III	>0	<0	Gradual decrease	rapidly decrease
IV	<0	<0	Rapid decline	rapidly increase
V	<0	>0	Gradual increase	gradually decrease

3.2. *Implementation of Adaptive Control*

In response to the shortcomings of the virtual inertia control strategy and based on the analysis of the relationship between virtual inertia and various parameters in the previous section, this paper proposes an energy storage doubly-fed condenser virtual inertia adaptive control strategy that considers gradual rotor speed recovery. The virtual inertia coefficient k is dynamically adjusted in real time based on frequency deviation, the frequency rate of change, and the condenser's rotor speed. On one hand, this improves system frequency stability and dynamic characteristics, enhances the condenser's support capability, and accelerates system frequency recovery. On the other hand, it enables the condenser to gradually recover its rotor speed during transients, ensuring sufficient bidirectional support capability when facing subsequent disturbances. The adaptive virtual inertia function is:

$$k=\begin{cases}k_1\Delta f\frac{df}{dt}+k_2, \omega>\omega_{0.5}\\ k_1\Delta f\frac{df}{dt}-k_2\left(\omega^2-\omega_{0.5}^2\right), \omega<\omega_{0.5}\end{cases} \tag{5}$$

In the equation: k_1 and k_2 are the influencing factors for each parameter. Adjusting k_1 alters the impact of frequency on the virtual inertia coefficient while adjusting k_2 changes the impact of rotor speed

Journal of Physics: Conference Series **3011** (2025) 012044 doi:10.1088/1742-6596/3011/1/012044

on the virtual inertia coefficient. Combining both ensures a balance between the condenser's rotor speed recovery and the system's frequency stability and dynamic characteristics. $\omega_{0.5}$ represents the rotor speed when the condenser's stored energy is at 50% of its total energy storage capacity. Since the rotor speed of the condense directly reflects its energy storage capacity, the rotor speed at 50% of the total storage capacity is defined as the rated speed. When the rotor energy is less than 50%, the rotor speed is considered low, and when it is greater than 50%, the speed is considered high.

By comparing the actual rotor speed of the condense with the rotor speed when the energy is at 50% of its total capacity and squaring the result, this comparison is incorporated into the virtual inertia adaptive function. This allows the virtual inertia coefficient to dynamically adjust according to the condense's energy storage state. This ensures that during transients when the condense is at high speed and the frequency drops, more support is provided. Conversely, when the rotor speed is low and the frequency deviation is positive, more speed recovery is achieved. In other cases, sufficient support is provided. This approach enables gradual recovery of rotor speed during disturbances, maintaining bidirectional support capability to respond to different types of disturbances.

When the rotor speed is at the rated value, the impact of the k_2 term becomes negligible, and the k_1 term alone can provide support superior to that of traditional virtual inertia control. The complete control block diagram is shown in Figure 4.

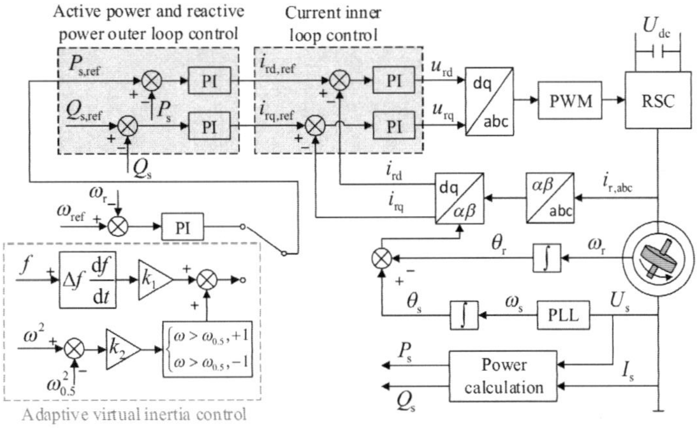

Figure 4. Adaptive virtual inertia control.

After introducing Equation (5), using the case where the condense's initial rotor speed is low, the dynamic performance of the condenser can be improved under different load disturbances as analyzed in Section 2.1, with system frequency and condense rotor speed changes as follows:

(1) At low rotor speed, in region I, when part of the load is shed at time t_1, the system frequency rises. The condenser begins to absorb active power from the system and stores energy in the rotor, causing the rotor speed to increase. The product of frequency deviation and frequency rate of change is greater than zero. According to Equation (5), k increases, allowing the condenser to absorb more energy from the system and store it in the rotor. This accelerates the rotor speed recovery and hinders further increase in frequency deviation.

(2) Regions II and V are the frequency recovery phases. In the t_2-t_3 phase, the frequency deviation starts to decrease slowly, and the frequency rate of change first decreases slightly before gradually increasing to zero. The condenser will reduce its inertia support to the system, but under virtual inertia control, it will still release some rotor kinetic energy, hindering frequency recovery. The rotor speed will decline to some extent. In the t_6-t_7 phase, the opposite occurs. According to Equation (5), the value of k decreases in both phases, and the condense will reduce its inertia support to accelerate frequency recovery.

(3) There is a positive-negative transition between regions III and IV. At time t_4, load is applied, and the system frequency drops. The frequency deviation is still positive but starts to drop significantly, and the frequency rate of change starts to decrease rapidly from zero. The condenser should reduce its

inertia support to allow the system frequency to decrease quickly to the rated value, and the rotor speed to continue to decrease. Before time t_5, the product of frequency deviation and frequency rate of change is negative, and at this point, k decreases rapidly. The condenser maintains a low inertia support to avoid hindering the frequency from recovering to the rated value. After time t_5, the frequency deviation becomes negative and continues to decrease, and the frequency rate of change increases to zero. Under virtual inertia control, the condenser should provide a larger inertia to hinder the further increase of frequency deviation. The rotor speed will drop significantly, and k will increase greatly. The condenser's active power output increases rapidly, enabling it to provide more inertia support to the system, quickly lowering the rotor speed and releasing energy to prevent further frequency drops.

Similarly, when the condenser's initial rotor speed is high, introducing Equation (5) can provide more support during frequency drops and reduce the speed rise during frequency recovery, accelerating the return of the system frequency to the rated value. When the condenser's initial rotor speed is close to the rated value, there is no need to consider rotor speed recovery, and Equation (5) primarily relies on frequency changes to adjust the value of k, with rotor speed having a smaller impact. In this case, the strategy focuses more on improving the system's frequency stability and dynamic characteristics.

Therefore, by introducing Equation (5), the condenser's inertia support capacity during transients is enhanced, and it achieves gradual recovery of rotor speed.

4. Simulation Analysis

To verify the effectiveness of the proposed control strategy, a simulation model of a three-machine system including a wind power plant was built on the MATLAB/Simulink simulation platform. The system structure is shown in Figure 5.

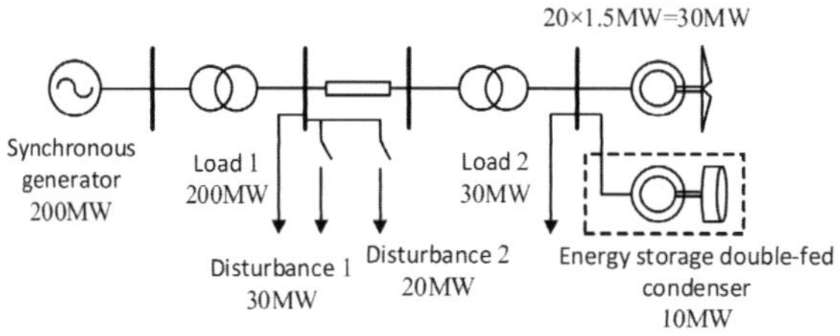

Figure 5. System simulation structure diagram.

The system consists of a 200 MW synchronous generator, a 10 Mvar energy storage doubly-fed condenser (ESDFC), and a wind farm consisting of 20 units of 1.5 MW doubly-fed wind turbines. The system side has an active power load of 200 MW, a reactive power load of 15 MW, and a local station load of 30 MW. The ESDFC operates steadily, outputting 0.2 p.u. of reactive power to the system. The simulation parameters are shown in Table 2.

Table 2. Simulation parameters.

type	parameter name/unit	physical symbol	parameter value
Energy Storage Doubly-Fed Condenser	Rated Capacity /Mvar	S_N	10
	Moment of Inertia /kg·m^2	J	2.4*10^3
	Rated Voltage /kV	U_N	10.5
	Excitation Inductance /p.u.	L_m	4.76
	Stator Leakage Inductance /p.u.	L_{ls}	0.112
	Rotor Speed Upper Limit /p.u.	$\omega_{r,max}$	1.3
	Rotor Speed Upper Limit /p.u.	$\omega_{r,min}$	0.7
	Adaptive Control Parameters	k_1	50000
	Adaptive Control Parameters	k_2	250

Journal of Physics: Conference Series **3011** (2025) 012044 doi:10.1088/1742-6596/3011/1/012044

When the initial rotor speed of the phase-shifting machine is low, the system initially operates in a steady state. At 30 s, a 20 MW load is shed; at 40 s, a 30 MW load is added; and at 50 s, another 20 MW load is shed. The simulation comparison is shown in Figure 6.

As shown in Figure 6, when the initial rotor speed is low, for the frequency rise generated at 30 s, the adaptive virtual inertia control enables the condenser to effectively provide inertia support to the system. During the frequency support phase, it absorbs more active power to restore rotor speed, reducing frequency deviation. In the frequency recovery phase, the rotor speed changes more slowly, and the condenser maintains low inertia in the system, reducing the hindrance to frequency recovery. This enhances frequency support while accelerating frequency recovery speed. For the frequency drop at 40 s, before the frequency deviation becomes negative, the system is in the frequency recovery phase. The condenser not only does not hinder frequency recovery but also absorbs some active power to help the system restore frequency. Once the frequency deviation becomes negative and enters the frequency support phase, the condenser provides more frequency support than traditional virtual inertia. After the support phase ends, the inertia quickly decreases, accelerating the system's frequency recovery.

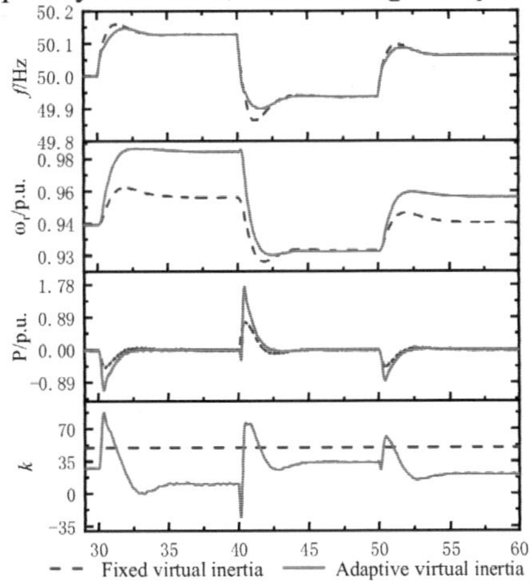

Figure 6. Comparison of parameter changes at low speed.

In addition, when the initial rotor speed is high, the condenser can output more active power in response to frequency drop, accelerating the recovery of rotor speed to the rated value. Conversely, when facing frequency rise, the condenser reduces active power output, accelerating frequency recovery while also preventing the rotor speed from deviating further from the rated value, thus maintaining bidirectional support capability. When the initial rotor speed is at the rated value, the condenser focuses on enhancing the system's frequency stability and dynamic characteristics, providing better frequency support than traditional virtual inertia control.

In conclusion, the virtual inertia adaptive control strategy can effectively provide support to the system. While balancing system frequency stability and dynamic characteristics, it is also capable of autonomously recovering the rotor speed after it deviates from the rated value. During continuous disturbances, the steady-state rotor speed is able to approach the rated value more closely, achieving a gradual recovery of the rotor speed.

5. Conclusion

This paper proposes an adaptive virtual inertia control strategy based on the traditional fixed virtual inertia control approach. Through theoretical analysis and simulation verification, the main conclusions are as follows:

(1) The proposed control strategy balances the system frequency stability and dynamic characteristics by making the virtual inertia coefficient adaptive to frequency deviation and frequency

rate of change. This enhances frequency support while also accelerating the system's frequency recovery speed.

(2) The gradual recovery of the rotor speed of the condenser is achieved, considering the impact of the condenser's energy storage. By incorporating rotor speed into the adaptive control, the condenser can adjust its support capacity based on its energy storage during the transient period and actively restore its rotor speed to the rated value. This ensures that it maintains bi-directional support capability during continuous disturbances, thus improving system stability.

References

[1] Wang J, Leng Z, Liu H, et al. Research and Application of Active Support Technology for Renewable Energy Stations[J]. New Type Power Systems, 2024, 2(04):458-476.

[2] Guo Q, Li Z. Summarization of Synchronous Condenser Development[J]. Proceedings of the CSEE, 2023, 43(15): 6050-6064.

[3] Zhang H, Xiang W, Lin W, et al. Grid Forming Converters in Renewable Energy Sources Dominated Power Grid: Control Strategy, Stability, Application, and Challenges[J]. Journal of Modern Power Systems and Clean Energy, 2021, 9(06):1239-1256.

[4] A. C. Ferreira, L. M. Souza and E. H. Watanabe, "Improving power quality with a variable speed synchronous condenser," 2002 International Conference on Power Electronics, Machines and Drives (Conf. Publ. No. 487), Sante Fe, NM, USA, 2002, pp. 456-461.

[5] Wang Q, Yuan X. Amplitude-phase Motion Equation Model of Induction Motor and Analysis of Dynamic Process Mechanism[J]. Proceedings of the CSEE,2021,41(01):363-373+425.

[6] Liu X, Zhang G, Pei W, et al. Current Status and Development Trends of Grid Type Converters[J]. Acta Energiae Solaris Sinica, 2024, 45(09):101-111.

[7] Yao F, Yu W, He J, et al. Research on a New Type of Variable-speed Asynchronous Condenser and Its Control Strategy[J]. Power Electronics, 2024, 58(11):21-26.

[8] J. Morren, S. W. H. de Haan, W. L. Kling and J. A. Ferreira, "Wind turbines emulating inertia and supporting primary frequency control," in IEEE Transactions on Power Systems, vol. 21, no. 1, pp. 433-434, Feb. 2006.

[9] K. Matsuda et al., "Stabilization of Power System by Virtual Inertia Control of Adjustable Speed Synchronous Condenser," 2021 2nd International Conference on Robotics, Electrical and Signal Processing Techniques (ICREST), DHAKA, Bangladesh, 2021, pp. 163-167.

[10] Zhang W, Zhao Q, Liu Y, et al. VSG Inverter Power Source Control Strategy with Adaptive Rotary Inertia in Island Mode[J]. Proceedings of the CSU-EPSA, 2019, 31(02):95-100.

[11] Meng J, Peng J, Wang Y, et al. Improved flexible virtual inertial control considering SOC of energy storage and characteristics of frequency recovery[J]. Electric Power Automation Equipment,2020,40(06):100-107+114+108-111.

Research of cooperative coupling technological policy between new energy and oilfield unsteady development

Qihao Hu [1*], Fajun Guo [1], Tairan Liang [2], Xin Mai [2], Xiaochuan Tian [3], Xiuwei Wang [1], Xiaoying Shi [1], Hongtao Wang [3], Zhao Xue [3], Jiefeng Cao [3]

[1] Exploration and Development Institute, Huabei Oilfield Company of PetroChina, Renqiu City, China
[2] Huabei Oilfield Company of PetroChina, Renqiu City, China
[3] Reservoir Development Department, Huabei Oilfield Company of PetroChina, Renqiu City, China

*Corresponding author's e-mail: yjy_hqh@petrochina.com.cn; *hqh59085@163.com

Abstract. To research the cooperative coupling technological policy between new energy and oilfield unsteady development, a conceptual model considering planar heterogeneity is established by using numerical simulation software. Based on the numerical model, the different working systems affecting the unsteady water drive development are analyzed. The influence factors are mainly analyzed from the aspects of daily liquid production amplitude, working system adjustment frequency, and working system change mode. The research results show that considering the characteristics of solar energy resources and time-of-use electricity price, the oilfield unsteady development period should be concentrated in the region with high daily photovoltaic intensity, as well as the valley period and flat period of time-of-use electricity price. This can not only maximize the use of photovoltaic power generation energy but also fully utilize the low electricity price during the valley and flat periods to save electricity costs. For the oilfield with strong reservoir heterogeneity, the larger the daily fluid volume amplitude, the worse the unsteady waterflood development effect. When the daily fluid volume amplitude is less than 30 %, the influence on unsteady waterflood development is relatively small. Through cooperative coupling technological policy between new energy and oilfield unsteady development, it is not advisable to significantly change the working system of unsteady waterflood development, and the working system adjustment frequency should not be too much.

1. Introduction

Driven by the transformation of the global energy structure and the "dual carbon" goal, the current oil fields are vigorously promoting the deep integration of oilfield development and new energy [1-4]. Adapting to the discontinuous and unstable characteristics of new energy, it promotes the transformation of oilfield development mode from traditional stable development to unsteady development, which is very significant for promoting the coordinated development of oilfield development and new energy [5-6]. At the same time, unsteady development is also one of the effective ways to improve the waterflood development effect of heterogeneous reservoirs. To effectively explore low-benefit reservoirs, the unsteady development is gradually transformed to adapt the characteristics of new energy supply, while ensuring that the discontinuous production does not affect the oil recovery efficiency. From the analysis of research status, the unsteady development methods mainly include intermittent production of low-producing wells and periodic water injection, which have achieved good development effects in many oilfields [7-10]. However, the mechanism research and field tests have not been carried out on the cooperative coupling technological policy between new energy and oilfield unsteady development. In this study, a conceptual model considering planar heterogeneity is established by using numerical simulation software. Based on the numerical model, the different working systems affecting the unsteady water drive development are analyzed. The influence factors are mainly analyzed from the aspects of daily liquid production amplitude, working system adjustment frequency, and working system change mode. Through the cooperative coupling of technological policy research between new energy and oilfield unsteady development, the influence of different working systems on unsteady waterflood development is clarified.

2. Analysis of new energy distribution characteristics

2.1. Analysis of solar energy resource characteristics

According to the local meteorological station data, the sunshine hours, sunshine percentage, and annual total radiation in this area are higher than those in plain areas of the same latitude. The annual sunshine hours are between 2800 hours and 3000 hours, and the sunshine rate is between 64 % and 73 %. The annual total radiation of solar energy ranges from 5198.6 MJ/m^2 to 5852.54 MJ/m^2, indicating significant potential for solar energy utilization. The power intensity of photovoltaic energy is affected by the intensity of solar radiation, sunny days are stronger than rainy days, noon is stronger than morning and evening, and the energy distribution regularity is strong, as shown in Figure 1.

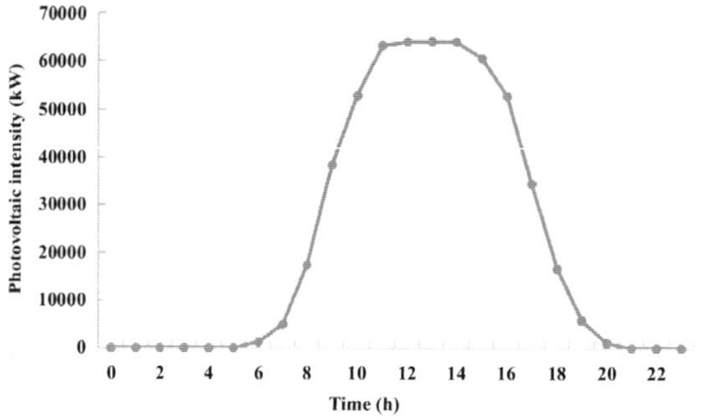

Figure 1. Distribution curve of daily photovoltaic intensity.

2.2. Analysis of time-of-use electricity price characteristics

According to the local industrial electricity-related policies, the regional network electricity follows the peak-valley-flat ladder electricity price policy, as shown in Figure 2. The whole day is divided into 4 time periods, the top peak period of 4 hours with an electricity price of 0.95 yuan/kWh, the peak period of 2 hours with an electricity price of 0.88 yuan/kWh, the flat period of 13 hours with electricity price of 0.59 yuan/kWh, the valley period of 5 hours with electricity price of 0.39 yuan/kWh.

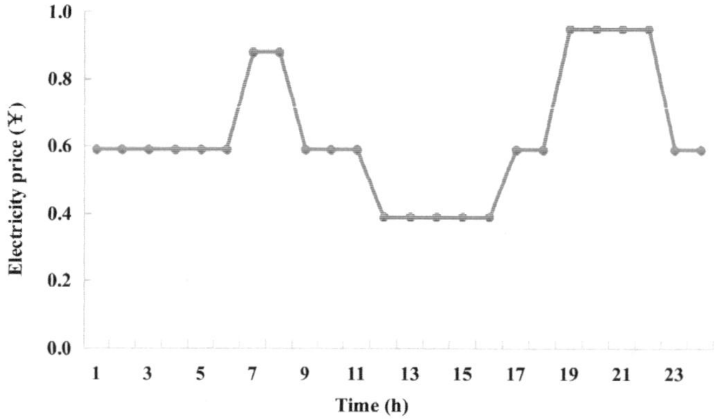

Figure 2. The distribution curve of time-of-use electricity price.

Considering the characteristics of solar energy resources and time-of-use electricity price, the oilfield's unsteady development period should be concentrated in the region with high daily photovoltaic intensity, as well as the valley period and flat period of time-of-use electricity price. This can not only maximize the use of photovoltaic power generation energy but also fully utilize the low electricity price during the valley and flat periods to save electricity costs.

3. Study on unsteady working system optimization

To research the cooperative coupling technological policy between new energy and oilfield unsteady development, a conceptual model considering planar heterogeneity is established by using numerical simulation software. By designing high permeability bands between oil and water wells in the conceptual model, the water flow advantage channels in actual reservoir development are simulated. According to the actual reservoir parameters of the H34 fault block, the conceptual model with a grid size of 35 × 35 × 3 is established. The basic data table and model schematic diagram are shown below (Table 1 and Figure 3). The conceptual model adopts the inverse five-point method well network, designing 4 oil production wells and 1 water injection well.

Table 1. Basic data table of a conceptual model.

Porosity (%)	Permeability (mD)	Number of oil well	Number of water well	Injection-production ratio
18	10	4	1	1.0

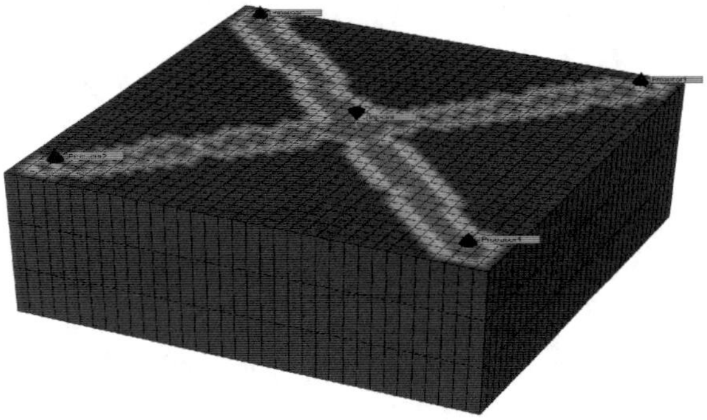

Figure 3. Schematic diagram of a heterogeneous conceptual model.

Based on the numerical model, the different working systems affecting the unsteady water drive development are analyzed. The influence factors are mainly analyzed from the aspects of daily liquid production amplitude, working system adjustment frequency, and working system change mode. Numerical simulation software is used to predict production for **20** years. The effects of different working systems on unsteady waterflood development are compared.

3.1. Influence of daily fluid production amplitude

To clarify the influence of daily fluid production amplitude on unsteady waterflood development, a total number of 6 comparative schemes with daily fluid production amplitude of 0 %, 10 %, 20 %, 30 %, 40 %, and 50 % are designed under the same condition of other production parameters. Considering the distribution characteristics of solar energy resources, the change curve of oil well daily fluid production is designed, as shown in Figure 4. Numerical simulation software is used to calculate the unsteady waterflood development effect under different daily fluid production amplitude, and the calculation results are shown in Figure 5 and Figure 6.

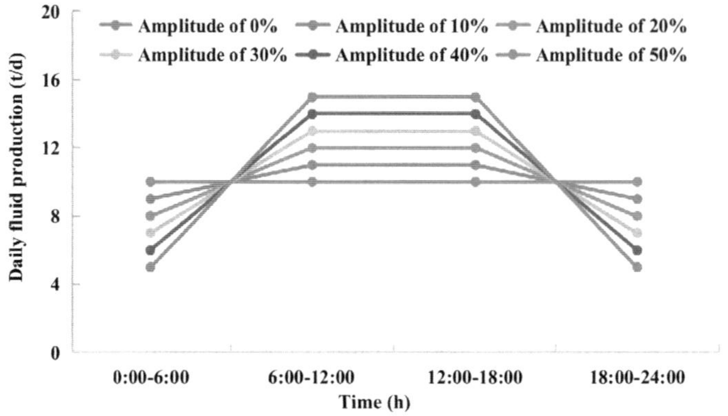

Figure 4. Time-sharing production curve of different daily fluid production amplitude.

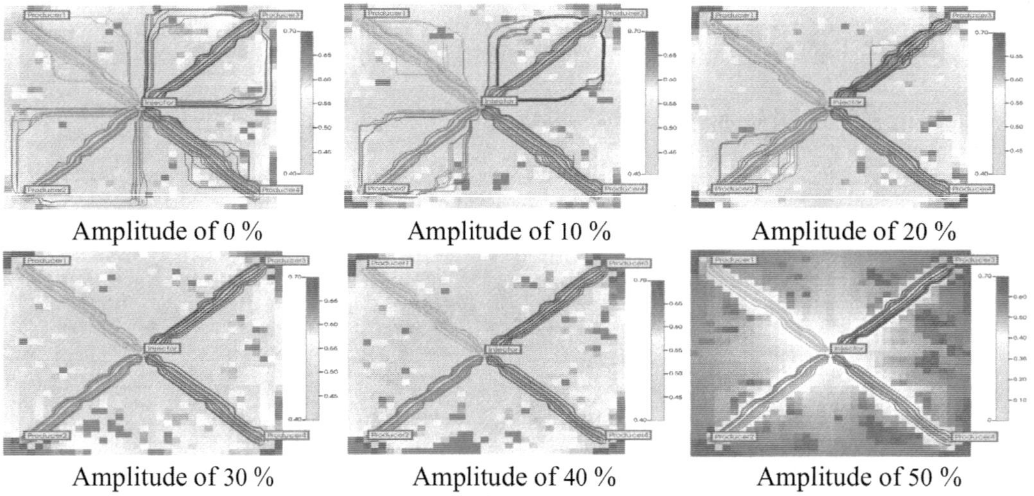

Figure 5. Residual oil and flow line distribution of different daily fluid production amplitude.

According to the distribution diagram of residual oil and flow line, with the increase of daily fluid production amplitude, the injection water is more likely to burst along the dominant flow channel. The spread range of the water drive becomes smaller, resulting in more residual oil that cannot be displaced and the water drive development effect becomes worse.

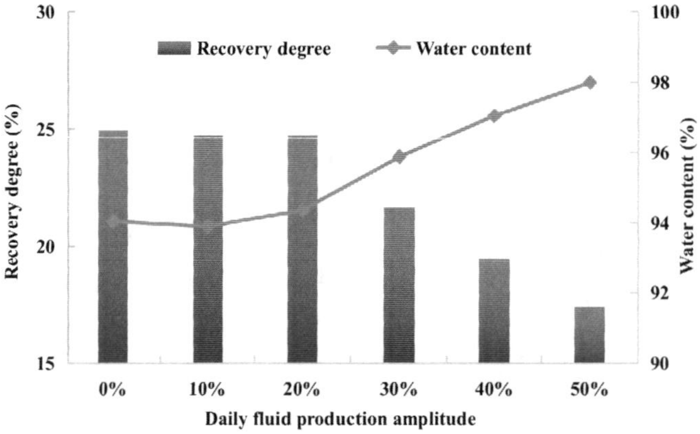

Figure 6. Development effect comparison of different daily fluid production amplitude.

Journal of Physics: Conference Series **3011** (2025) 012045 doi:10.1088/1742-6596/3011/1/012045

The result shows that as the daily fluid production amplitude increases, the recovery degree of unsteady development decreases, and the water content increases. Therefore, for the oilfield with strong reservoir heterogeneity, the greater the daily fluid volume amplitude, the worse the unsteady waterflood development effect. From the trend of curve changes, it can be seen that when the daily liquid production amplitude is less than 30 %, the influence on the unsteady waterflood development effect is relatively small.

3.2. Influence of working system adjustment frequency

To clarify the influence of working system adjustment frequency on unsteady waterflood development, a total number of 3 comparative schemes with working system adjustment frequency of 2 times, 4 times, and 8 times are designed under the same condition of other production parameters. Considering the distribution characteristics of solar energy resources and time-of-use electricity price, the change curve of oil well daily fluid production is designed, as shown in Figure 7. Numerical simulation software is used to calculate the unsteady waterflood development effect under different working system adjustment frequencies, and the calculation results are shown in Figure 8 and Figure 9.

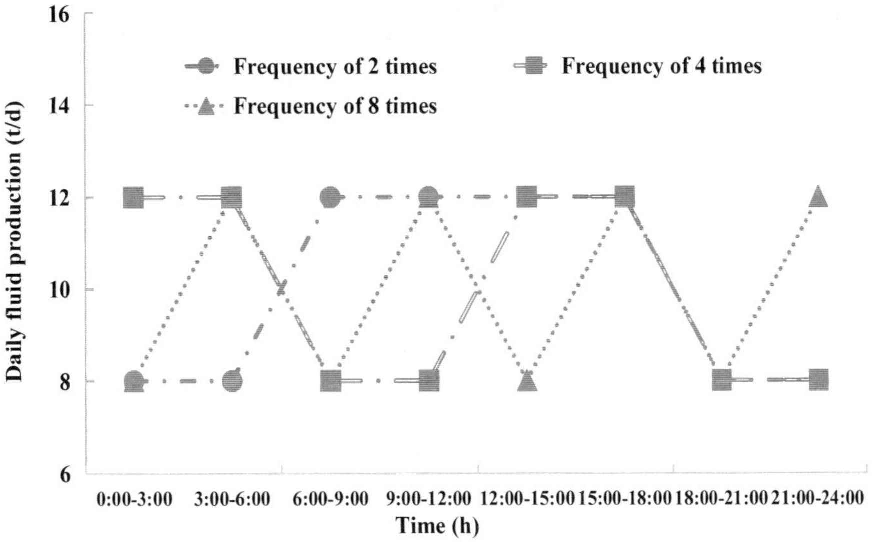

Figure 7. Time-sharing production curve of different working system adjustment frequency.

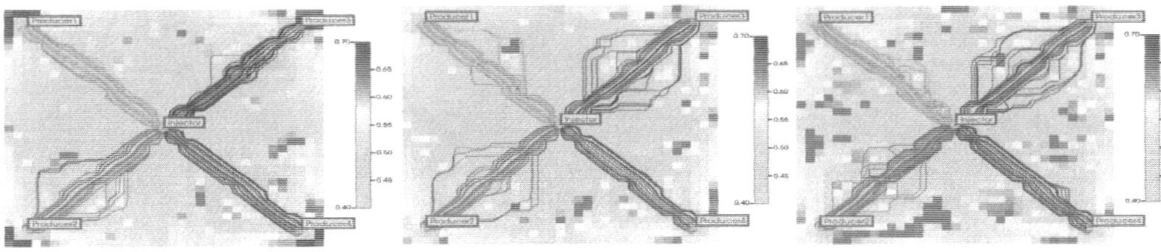

Frequency of **2** times Frequency of **4** times Frequency of **8** times

Figure 8. Residual oil and flow line distribution of different working system adjustment frequencies.

According to the distribution diagram of residual oil and flow line, when the working system adjustment frequency is small, the difference in residual oil distribution is not significant. However, when the working system adjustment frequency is more than 8 times, the distribution of residual oil increases significantly, resulting in more residual oil that cannot be displaced and the water drive development effect becomes worse.

Journal of Physics: Conference Series **3011** (2025) 012045 doi:10.1088/1742-6596/3011/1/012045

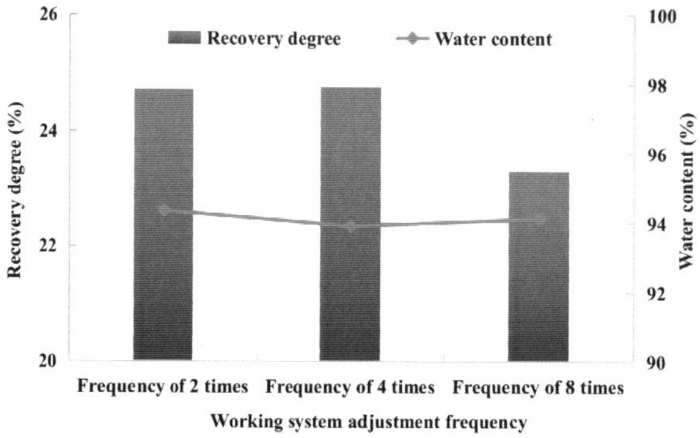

Figure 9. Development effect comparison of different working system adjustment frequency.

It can be concluded that when the working system adjustment frequency is small, the influence on the unsteady waterflood development effect is relatively small. However, when the working system adjustment frequency is more than 8 times, the unsteady waterflood development effect becomes worse obviously. Therefore, for the reservoirs with strong heterogeneity, the working system adjustment frequency of unsteady waterflood development should not be too much.

3.3. *Influence of working system change mode*

To clarify the influence of working system change mode on unsteady waterflood development, a total number of 3 comparative schemes with working system change mode of stable, slightly change, and significantly change are designed under the same conditions of other production parameters. Considering the distribution characteristics of solar energy resources and time-of-use electricity price, the change curve of oil well daily fluid production is designed, as shown in Figure 10. Numerical simulation software is used to calculate the unsteady waterflood development effect under different working system change modes, and the calculation results are shown in Figure 11 and Figure 12.

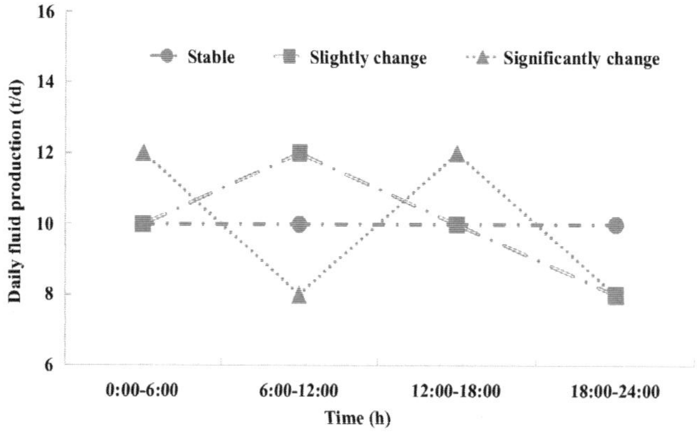

Figure 10. Time-sharing production curve of different working systems change mode.

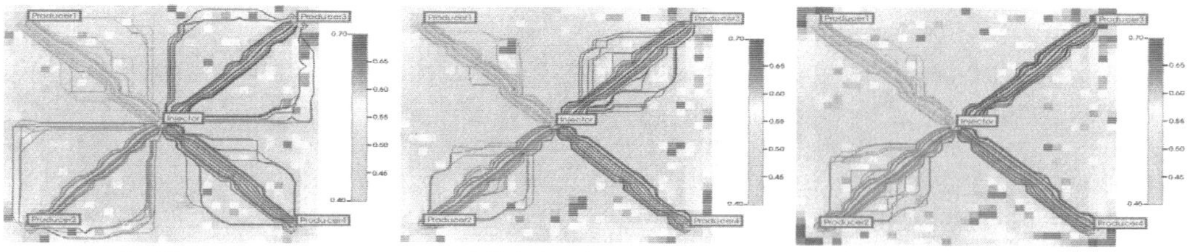

Stable Slightly change Significantly change.

Figure 11. Residual oil and flow line distribution of different working system change modes.

According to the distribution diagram of residual oil and flow line, when the working system changes slightly, the difference in residual oil distribution is not significant. However, when the working system changes significantly, the residual oil distribution near the oil well increases slightly, resulting in more residual oil that cannot be displaced and the water-drive development effect becomes worse.

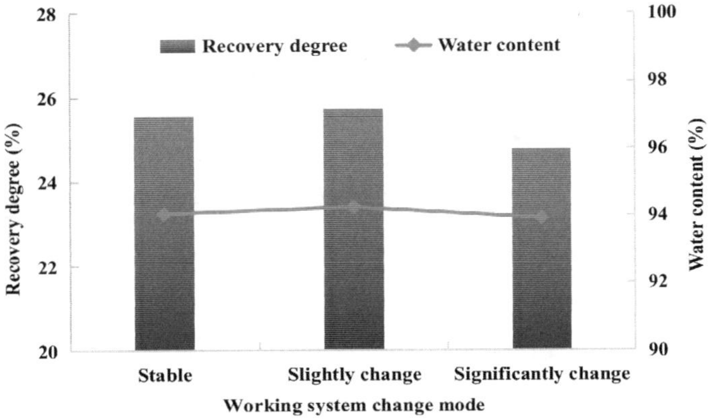

Figure 12. Development effect comparison of different working system change modes.

It can be concluded that when the working system changes slightly, the influence of the unsteady waterflood development effect is relatively small. However, when the working system changes significantly, the unsteady waterflood development effect becomes worse obviously. Therefore, for the reservoirs with strong heterogeneity, the working system change mode of unsteady waterflood development should not change significantly.

4. Conclusion

(1) Considering the characteristics of solar energy resources and time-of-use electricity price, the oilfield unsteady development period should be concentrated in the region with high daily photovoltaic intensity, as well as the valley period and flat period of time-of-use electricity price. This can not only maximize the use of photovoltaic power generation energy but also fully utilize the low electricity price during the valley and flat periods to save electricity costs.

(2) For the oilfield with strong reservoir heterogeneity, the larger the daily fluid volume amplitude, the worse the unsteady waterflood development effect. When the daily fluid volume amplitude is less than 30 %, the influence on unsteady waterflood development is relatively small.

(3) Through cooperative coupling technological policy between new energy and oilfield unsteady development, it is not advisable to significantly change the working system of unsteady waterflood development, and the working system adjustment frequency should not be too much.

References

[1] R Cai, H X Zhu, W J Li, et al. Development path of energy science and technology under "dual carbon" goals: Perspective of multi-energy system integration [J]. Bulletin of Chinese Academy of Sciences, 2022, 37(4): 502-510.

[2] C N Zou, F Ma, S Q Pan, et al. Global energy transition revolution and the connotation and pathway of the green and intelligent energy system [J]. Petroleum Exploration and Development, 2023, 50(3): 633-647.

[3] C N Zou, B Xiong, H Q Xue, et al. The role of new energy in carbon neutral [J]. Petroleum Exploration and Development, 2021, 48(2): 411-420.

[4] L F Kong, X W Yue. Study of CNPC's new energy development pathway and innovation practice [J]. Petroleum Science and Technology Forum, 2024, 43(1): 25-30.

[5] C M Su, J Y Zhu, J G Ma, et al. Thinking and suggestions on integrated development pathways for oil and gas with new energy [J]. Petroleum Science and Technology Forum, 2024, 43(1): 18-24.

[6] B J WuChen, J D He, Y Y Jia, et al. Prospects analysis of coupling development of new energy and oil-gas of SINOPEC in the northern South China Sea [J]. SINO-GLOBAL ENERGY, 2023, 28 (4): 15-22.
[7] L B Fan, C X Qi, Z H Chen, et al. Evaluation method and application of periodic water injection effect based on pressure fluctuation [J]. Journal of Chengde Petroleum College, 2023, 25 (2): 17-22.
[8] M Du, H Q Guo, C D Xia, et al. Study on the feasibility mechanism of unstable water injection in high water-cut reservoirs of block X in Yanchang oilfield [J]. Journal of Chengde Petroleum College, 2024, 26 (3): 24-28.
[9] Y Lei, J N Zhang, Y T Ma, et al. Research and application of periodic water injection technology in L1 Chang 8 reservoir of Jiyuan oilfield [J]. Application of Petrochemical Engineering, 2020, 39 (1): 78-83.
[10] H Chen, T Wang, R M Liu, et al. Research on injection-production optimization to improve development effectiveness in heterogeneous oil reservoirs during high water cut stage [J]. Complex Oil and Gas Reservoirs, 2023, 16 (2): 215-219.

Journal of Physics: Conference Series **3011** (2025) 012046 doi:10.1088/1742-6596/3011/1/012046

Multi-objective optimization design of dual-winding permanent magnet motor based on sensitivity analysis

Yifei Zhang*, Yufeng Zhang, Ruiyang Zhang, and Ting Zhang

College of Electrical and Control Engineering, Xi'an University of Science and Technology, Xi'an 710054, China

*Corresponding author's e-mail: zhangyifei@xust.edu.cn

Abstract. The primary challenge in reducing electromagnetic vibration through structural parameter optimization lies in efficiently and systematically adjusting multiple parameters without compromising the motor's output performance. To address this problem, the optimization objective function is formulated based on key performance indicators of the DWPMSM (Dual-winding Permanent Magnet Motor), including the peak radial electromagnetic force amplitude, average torque, and torque ripple. Then, the sensitivity analysis method is used to evaluate the importance of design parameters to the optimization objectives. The design parameters with weak sensitivity retain the initial value, and the design parameters with strong sensitivity are selected in the next optimization design. Subsequently, the EA algorithm is employed to optimize the motor design model, resulting in an improved set of motor dimensions. Next, the EA algorithm is applied to refine the motor optimization model, yielding an optimized set of motor dimensions. To confirm the success of the optimization, a finite element analysis is performed to compare the motor's torque, torque ripple, and radial electromagnetic force both before and after the optimization process.

1. Introduction

DWPMSM is favored in the field of new energy vehicles and precision industrial drives due to its high power density, fault tolerance, and flexible control characteristics [1-3]. However, the interaction between its dual-winding structure and complex electromagnetic harmonics leads to a significant increase in vibration and noise problems [4-6]. Especially under high-speed and high-load conditions, it is easy to cause high-frequency mechanical resonance and electromagnetic howling, which seriously restricts its application in high-comfort scenarios [7-9]. Although the traditional single-objective optimization method can improve electromagnetic performance, it is difficult to balance the multiple contradictions between electromagnetic torque, torque ripple, and vibration noise. Therefore, this paper proposes a comprehensive design framework based on multi-objective optimization. Through the structural parameters of DWPMSM, the multi-objective collaborative improvement of electromagnetic efficiency, output torque, and vibration noise suppression is realized, which provides theoretical support and method innovation for the high-performance and low-noise design of DWPMSM.

2. Processes and objectives of optimization

Figure 1 shows the parameters of DWPMSM.

2.1. Optimization design process

A novel design approach centered on radial electromagnetic force is introduced. This method leverages radial electromagnetic force as a link between the electromagnetic and structural domains, with the peak radial force amplitude chosen as a key optimization target. By employing sensitivity analysis, the motor is optimized to directly and efficiently minimize vibration and noise while maintaining torque performance. The complete optimization procedure is illustrated in Figure 2.

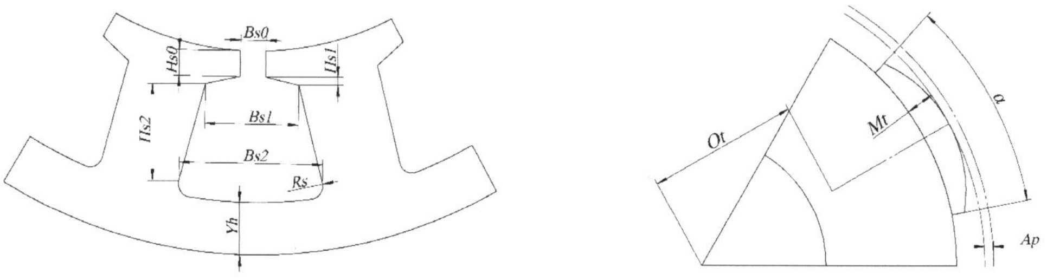

(a) (b)

Figure 1. The structural parameters of DWPMSM.

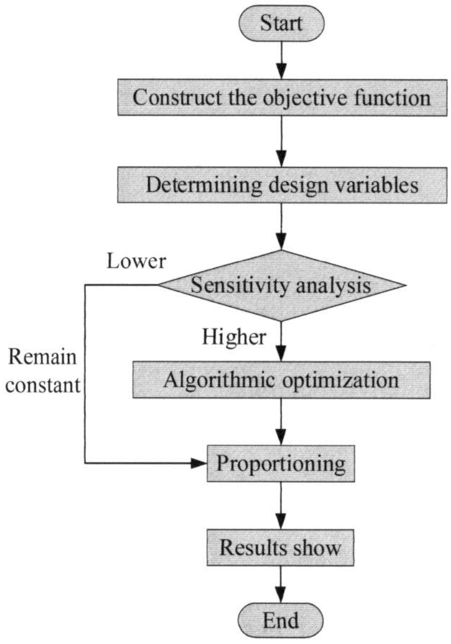

Figure 2. Optimized flow chart.

2.2. Determining Optimization Objectives

The optimization objective function of DWPMSM is defined as:

$$Fuction(DWPMSM):\left[Max(Torque\_avg), Min(Torque\_pkavg), Min(Max\_Frad)\right]$$

Among them, $Max(Torque\_avg)$ represents the maximum electromagnetic torque of DWPMSM, $Min(Torque\_pkavg)$ represents the minimum torque ripple, and $Min(Max\_Frad)$ represents the minimum amplitude of radial force.

3. Analysis of sensitivity

3.1. theoretical analysis

In this study, the key coefficients based on polynomial fitting were used to characterize the sensitivity of structural parameters. The polynomial regression equation of the objective function $y(x_i)$ to be analyzed is:

$$y(x_i) = p^{\mathrm{T}}(x_i)\beta + \varepsilon_i$$

In the formula, $P^T(x_i)$ is a polynomial with respect to the structural parameters, which can be expressed as :

$$p^T(x_i) = \left[\ 1\, x_1\ x_2\ x_3\ \cdots\ x_1^2\ x_2^2\ x_3^2\ \cdots\ x_1x_2\ x_1x_3\ x_2x_3\ \cdots \right]^T$$

β is the coefficient determinant; ε_i is the error.

Further, the determination coefficient R^2 can be calculated based on the fitting equation, and there is:

$$R^2 = 1 - \frac{\sum_{i=1}^{M}(y_i - \hat{y}_i)^2}{\sum_{i=1}^{M}(y_i - \mu_y)^2}$$

In the formula, y_i represents the actual value of the sample target; μ_y represents the average value of the sample target; $\hat{y}_i$ is the corresponding calculated value of $y(x_i)$; M represents the number of samples.

Based on the determination coefficient, the coefficient of importance (CoI) of the parameter x_i to the target y_i can be calculated:

$$\mathrm{CoI}(x_i, y_i) = \mathrm{CoI}_{y_i,x_i} = R^2_{y_i,x} - R^2_{y_i,x_i}$$

In the formula, the reduced term represents the coefficient of determination of the full model containing all structural parameters; the subtraction term represents the coefficient of determination of the fitting model after removing the linear term, quadratic term, and the interaction term of x_i. The difference between the two indicates the change in the fitting model after removing the structural parameters.

Therefore, the higher the $CoI\ (x_i, y_i)$ value is, the higher the sensitivity of the parameter x_i to the target y_i is.

3.2. Analysis of sensitivity

The sensitivity analysis results of DWPMSM obtained by finite element calculation are shown in Figure 3. The specific calculation results are listed in Table 1.

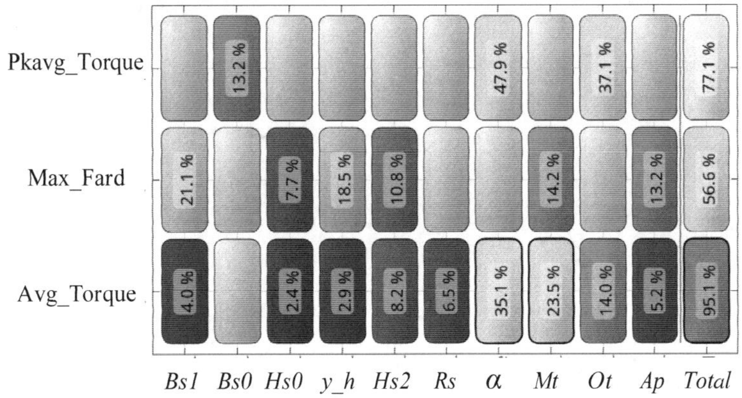

Figure 3. Results of Sensitivity Analysis (*COI*).

According to Figure 3 and Table 1, α, A_p, O_t, M_t, $Hs2$, $Bs0$, and $Bs1$ all have more than 10 % influence on one or more optimization objectives. Therefore, it is considered that these eight parameters are highly sensitive and need to be multi-objectively optimized. The influence of *Bs2*, *Hs0,* and *Hs1* on the three optimization objectives is not more than 10 %, so it is considered that the sensitivity of these four parameters is low and remains unchanged.

Table 1. The sensitivity value of the parameters to be optimized.

paraments	electromagnetic torque/%	torque ripple/%	Amplitude of radial electromagnetic force/%
α	35.1	47.9	0.0
A_p	5.2	0.0	13.2
O_t	14.0	37.1	0.0
M_t	23.5	0.0	14.2
Y_h	2.9	0.0	18.5
Rs	6.5	0.0	0.0
Hs0	2.4	0.0	7.7
Hs1	0.0	0.0	0.0
Hs2	8.2	0.0	10.8
Bs0	0.0	13.2	0.0
Bs1	4.0	0.0	21.1
Bs2	0.0	0.0	0.0
Total	95.1	77.1	56.6

3.3. Calculation response surface calculation

During the sensitivity analysis process, a response surface model is established to depict the relationship between structural parameters and optimization targets. This model helps to better understand the impact of two structural parameters on a specific optimization objective. Partial results of the response surface model are presented in Figure 4.

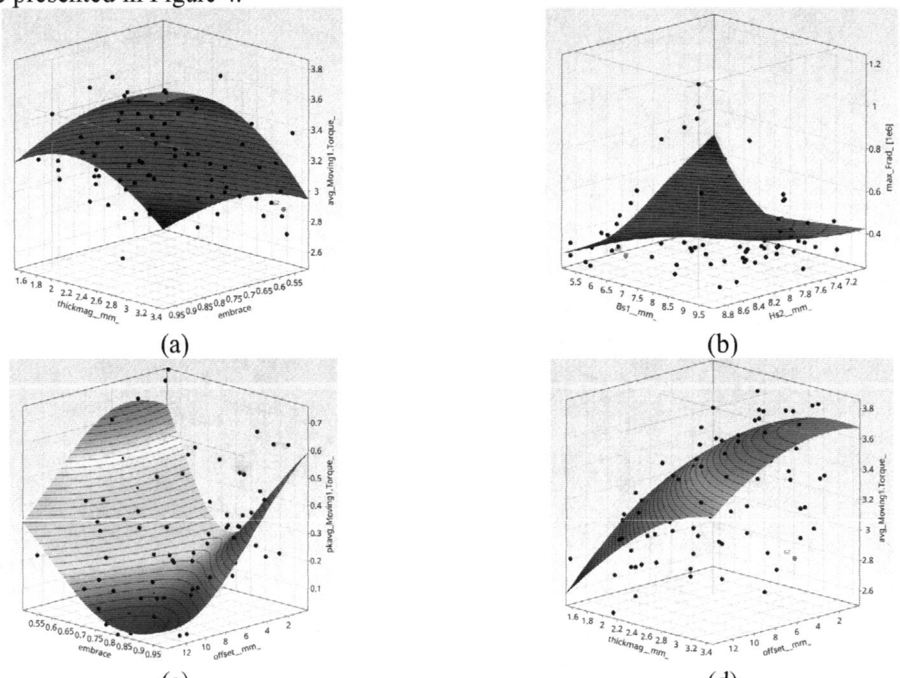

Figure 4. 3D response surface.

4. Optimization

It is difficult to optimize multiple optimization objectives at the same time, and it is necessary to comprehensively consider the influence of each optimization objective. Further, according to the constraint conditions, the EA algorithm is used to solve the optimization of DWPMSM, and the 7837th design point is obtained to meet the optimization requirements. The optimization results are shown in Figure 5.

5. Analysis of optimization results

To validate the effectiveness of the proposed optimization model, this section employs finite element software to analyze and compare the electromagnetic performance of the DWPMSM under both its initial and optimized dimensions. Figure 6 shows three design points of DWPMSM optimization results. Table 2 shows the specific results.

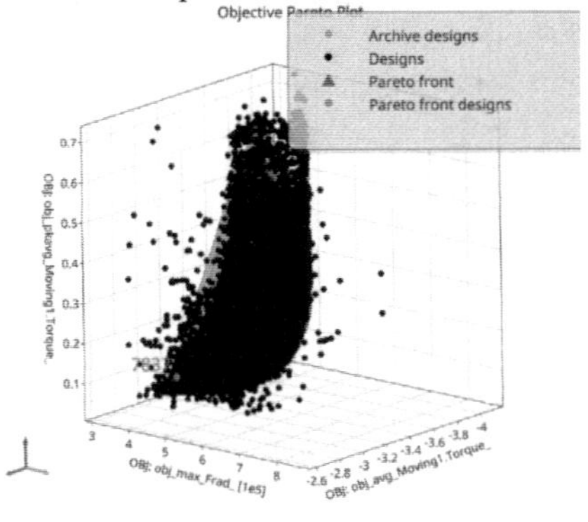

(a) Pareto 3D

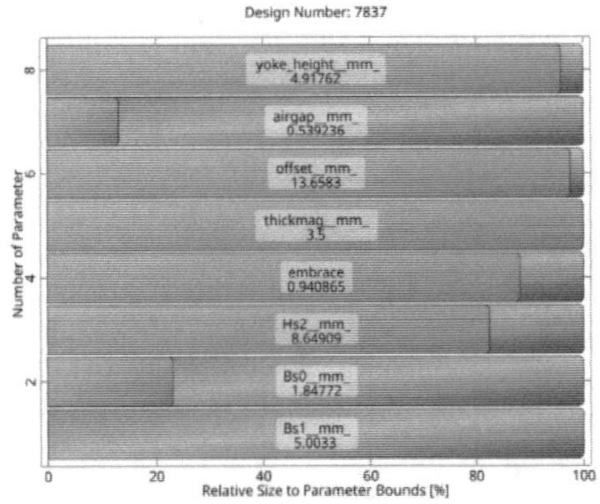

(b) Design parament

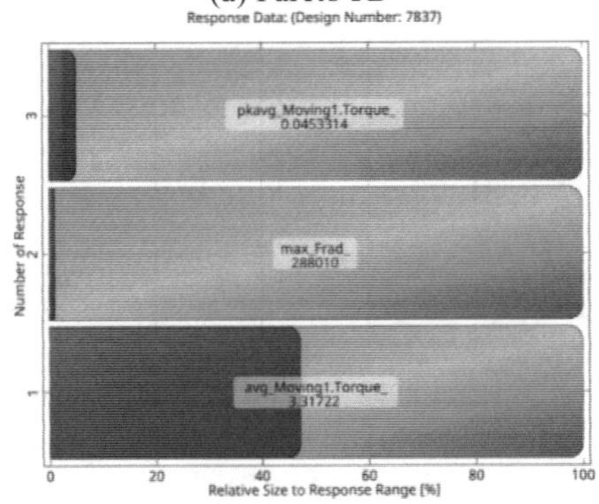

(c) Response values

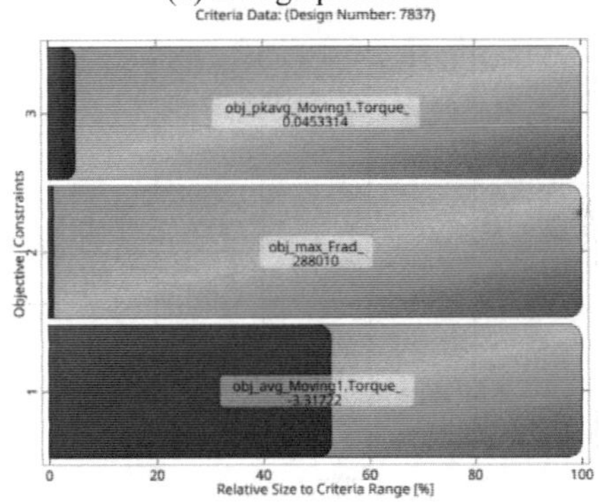

(d) Criteria data

Figure 5. Multi-objective optimization design results.

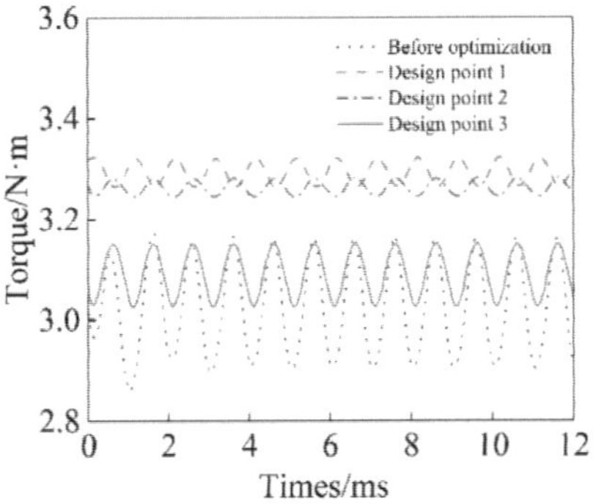

(a) Torque

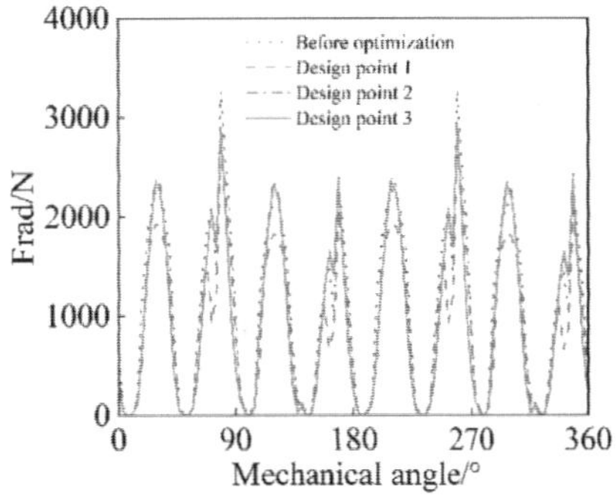

(b) Frad

Figure 6. Comparative of DWPMSM before and after optimization.

Table 2. Comparison before and after optimization.

Optimization object	Before optimization	Optimized		
		Point 1 (7873)	Point 2 (1566)	Point3 (8563)
electromagnetic torque/N·m	3.02	3.31	3.27	3.05
torque ripple /%	10.22	4.53	4.1	4.0
Radial electromagnetic force /N	3268.41	2880.01	2829.05	2913.07

Figure 6(a) illustrates the torque performance comparison for the three design points before and after optimization. The optimized DWPMSM exhibits a slight increase in average output torque, along with a notable reduction in torque ripple. Meanwhile, Figure 6(b) provides a comparative analysis of the spatial distribution of Frad pre- and post-optimization. As shown in the diagram, the overall amplitude of the Frad is significantly reduced compared to its pre-optimization state.

6. Conclusion

This study introduces an approach for multi-objective optimization design utilizing sensitivity analysis. The maximum radial electromagnetic force amplitude of DWPMSM is taken as the optimization objective to indirectly reduce the electromagnetic vibration. The key structural parameters of DWPMSM are optimized by sensitivity analysis and EA algorithm strategy. The results show that through multi-objective optimization design, the electromagnetic torque of DWPMSM increases slightly and the torque ripple is reduced as the radial electromagnetic force is effectively controlled. Therefore, the suggested multi-objective optimization approach for DWPMSM enables the effective identification of the motor's optimal design, enhancing its electromagnetic performance while minimizing electromagnetic vibration.

Acknowledgments

This work was financially supported by the Graduate Education Reform Project of Xi'an University of Science and Technology under Grant YJG2022016.

References

[1] R Li, H Fang, D Li, R Qu, and P Wang (2023), Influence of Winding Structure on Electromagnetic Vibration of Dual-Three-Phase Permanent Magnet Synchronous Machines, IEEE Transactions on Industry Application, 10: 1-12.
[2] Y. Miyama, M. Ishizuka, H. Kometani, and K. Akatsu (2018), Vibration Reduction by Applying Carrier Phase-Shift PWM on Dual Three-Phase Winding Permanent Magnet Synchronous Motor, IEEE Transactions on Industry Application, 54: 5998–6004.
[3] C. Sun, F. Wen, W. Xiong, H. Wang, and H. Shang (2020), Multi-objective comprehensive teaching algorithm for multi-objective optimisation design of permanent magnet synchronous motor, IET Electric Power Application, 13: 2564-2576.
[4] W. Zhao, X. Wang, C. Gerada, H. Zhang, C. Liu, and Y. Wang (2018), Multi-objective comprehensive teaching algorithm for multi-objective optimisation design of permanent magnet synchronous motor. IEEE Transactions Magnetics, 11: 1-5.
[5] B. Zheng, Z. Jibin, Y. Xu, X. Lang, and G. Yu (2022), Torque Ripple Suppression Based on Optimal Harmonic Current Injection in Dual Three-Phase PMSMs Under Magnetic Saturation. IEEE Transactions. Industrial. Electronics, 5398–5408.
[6] K Layek, H Krishnan, and K Hatua (2024), Voltage Angle-Based Torque Control of Dual Three-Phase Interior Permanent Magnet Synchronous Motor Considering Mismatch in Winding Parameters, IEEE Transactions on Industrial Electronics, 6:1-11.
[7] L Yan, Z Zhu, Y Ren, C Gan, S Brockway, and C Hilton (2022), Multiple Synchronous Reference Frame Current Harmonic Regulation of Dual Three Phase PMSM With Enhanced Dynamic Performance and System Stability, IEEE Transactions on Industrial Electronics, 69: 8825–8838.
[8] Hu SD, Xiong MF, Z. Liang, and He XN (2020), Torque Distributed Control Strategy for the Dual Three-Phase PMSM in Hybrid Energy Storage System Application, IEEE Transactions on Industrial Electronics, 67: 2544–2552.
[9] S Hu, M Xiong, Z Liang, and X He (2024), "Noise and vibration modelling of Permanent Magnet Synchronous Motors—a review," IEEE Transactions on Transportation Electrification, 2024:1-18.

Multi-level coordinated control of off-grid hydrogen production system in wind farm

Jiaying Xu*, **Peng Zhang, Junyan Yan and Lanxiang Lou**

North China Electric Power University, Beijing, 102206, China

* Corresponding author's e-mail: xjyemail2022@163.com

Abstract. Based on the sag control and virtual synchronous control strategies, an improved multi-unit frequency modulation reference power cooperative control strategy is proposed to optimize the coordinated control of different types of wind turbines with different capacities, so as to realize the stability of frequency and voltage under the change of hydrogen production load in isolated island microgrid.

1. Introduction

In recent years, an increasing amount of clean energy has been integrated into the power grid. Nevertheless, owing to the randomness and intermittency of clean energy, the power output is unstable. Hydrogen energy storage possesses the attributes of large-scale storage, long-distance transmission, and flexible conversion with electrical energy, so it represents the main direction for the future development of hydrogen energy. Taking Inner Mongolia as an example, there are 15 related projects in 2023, supporting a new energy scale of 6.312 million kilowatts and a substantial demand for off-grid hydrogen production from renewable energy.

To maintain the stability of the off-grid system, wind turbines are required to engage in frequency regulation by exploiting the variance in kinetic energy released by the rotor. He et al. [1] proposed a frequency modulation reference power collaborative allocation approach based on speed and power limitations. Zhang et al. [2] proposed a coordinated control strategy that considers wind speed differences, which fully utilizes the differences in rotor kinetic energy to coordinate the frequency regulation resources of wind farms while satisfying the reserve capacity of the units. Xu et al. [3] presented a power allocation method that combines the frequency regulation capability coefficient with the traditional frequency regulation strategy, but neglects the actual output power limitation and fails to accurately reflect the frequency regulation capability of wind turbines. The coordinated control strategy for wind farms proposed in [4] assesses the adjustable capacity of rotor kinetic energy and full-power wind power converters; however, it neglects the influence of the adjustable capacity of wind turbines operating with a certain reserve capacity on the contribution to frequency regulation.

Regarding the aforementioned issues, this paper has strengthened the setting of the frequency modulation coefficient of wind turbine units and further optimized the joint control approach among units of varying capacities. The enhanced cooperative control strategy can depict the frequency modulation capacity of wind turbines more precisely and allocate the power reference value of wind turbines rationally in accordance with the frequency modulation ability. Finally, the validity of the proposed improvement method has been verified through simulation.

2. Mathematical model of wind power participating in primary frequency regulation

2.1. Mathematical model of the wind turbine unit

The mathematical model for the primary frequency regulation of wind turbines participating in the system is presented as follows [5]:

$$\begin{cases} Pm = \frac{1}{2}\rho\pi R^3 v^3 C_p(\lambda,\beta) \\ C_p(\lambda,\beta) = c_1\left(\frac{c_2}{\lambda_i} - c_3\beta - c_4\right)e^{\frac{c_5}{\lambda_i}} + c_6\lambda \\ \frac{1}{\lambda_i} = \frac{1}{\lambda + 0.08\beta} - \frac{0.035}{\beta^3+1} \end{cases} \quad (1)$$

where Pm denotes the wind energy capture power of the wind turbine unit; ρ represents air density; R stands for the radius of the wind turbine; v indicates wind speed; β represents the pitch angle; the tip speed ratio λ is defined as: $\lambda = \omega Rv$; $C_p(\lambda,\beta)$ is the wind energy utilization coefficient; λ_i is the intermediate parameter; c_1, c_2, c_3, c_4, c_5, and c_6 are parameters.

The low-order mathematical model of the system frequency response is

$$\Delta f = \frac{1}{2H_w s + F}(P_M + P_e - P_D) \quad (2)$$

where Δf represents the frequency dropout value; H_w denotes the inertia factor of the drive train; F indicates the friction coefficient of the drive train; P_e stands for the output power of the wind turbine; P_M represents the output power of the conventional synchronous generator set; P_D symbolizes the load power.

2.2. Frequency modulation (FM) control strategies

Conventional frequency control strategies mainly consist of sag control and virtual synchronous control. Sag control is a control method that automatically adjusts the output power proportionally to the deviation of the grid frequency from a stable value. The sag control of wind turbines plays an extremely significant role in the frequency stability of the grid after the inertial control response.

2.2.1. Sag Control. Recess control in wind turbines correlates variations in grid frequency with modifications in the active output reference; namely, it responds to substantial deviations in grid frequency by increasing or decreasing power output during frequency events.

The relationship between the variation of the reference value of active power and the frequency deviation can be depicted as follows [6]:

$$\Delta P_{df} = K_{df}(f - f_0) \quad (3)$$

where f_0 is the reference frequency; f is the grid frequency; K_{df} is the unit regulation power of the wind turbine.

The sag coefficient r of the wind turbine is rounded to

$$r = \frac{\Delta f}{\Delta P_{df}} = \frac{1}{K_{df}} \quad (4)$$

Sag control is highly analogous to the control of a conventional synchronous motor governor, facilitating the wind turbine to supply additional active power in line with the grid frequency. When the grid frequency fluctuates, sag control takes the deviation of the grid frequency as the input signal, multiplies it by the unit regulation power, and the additional reference power thus obtained is superimposed on the active power command of the maximum power point tracking control to regulate the reference power of the WTGs.

2.2.2. Virtual Synchronisation Control. The virtual synchronous control further mimics the operational characteristics of the synchronous generator, such as inertia and damping, on the basis of the sag control [7]. The core feature thereof lies in simulating the rotor motion equation of the synchronous generator, as depicted in Equation (5).

$$J\dot{\omega}_{ref} = \frac{P_{ref}}{\omega_0} - \frac{P}{\omega_0} - D(\omega_{ref} - \omega_g) \quad (5)$$

where J and D represent the adjustable virtual inertia and damping coefficient respectively.

The distinction between virtual synchronous control and sag control is primarily manifested in the active power control link. The reactive power control link can be established by referring to the reactive power sag control in the previous section, or the virtual excitation current can be obtained through reactive power deviation and voltage deviation.

Virtual synchronous control is one of the prevailing control strategies for voltage source networks. In contrast to sag control, virtual synchronous control is capable of simulating the inertia of the generator, enhancing the equivalent inertia level of the system, and facilitating the frequency stability of the power grid.

3. Principles of wind turbine cooperative control

The frequency modulation (FM) strategy of a single wind turbine is capable of responding to system frequency variations by outputting FM reference power corresponding to frequency fluctuations. If multiple wind farms operate in parallel with the grid, each wind turbine will undertake the FM power evenly distributed in the field in accordance with its FM strategy. However, the FM strategies of each wind turbine lack coordination and the energy of each wind turbine is not rationally utilized. The crux of the cooperative control strategy for multiple units lies in the rational utilization of the FM auxiliary power of each unit, relying on the rotor kinetic energy and spare capacity of the unit.

3.1. Setting of the unit's FM capability

The FM capability of a wind turbine is associated with the local wind conditions, its rotor kinetic energy, and the adjustable capacity of the converter. The rated capacity of the converter is in accordance with the rated capacity of the wind turbine generator (WTG). Thus, when contemplating the impact of the adjustable capacity of the converter on the FM of the WTG, it can be analyzed by utilizing the adjustable capacity of the WTG.

The conventional coordinated control strategy incorporates the FM capacity factor based on the primary FM control approach. Considering that the unit operates under overspeed load shedding, the power response of the unit when the rotational speed reaches 1.2 p.u. is set as the limit value reference quantity. Therefore, the capacity factor K_{p_i} can be defined to reflect the magnitude of the FM potential of the WTG, as follows:

$$K_{p\_i} = \frac{P_{r\_i} - P_{c\_i}}{P_{r\_i} - P_{1.2r\_i}} \tag{6}$$

where $P_{r\_i}, P_{c\_i}, and\ P_{1.2r\_i}$ are the power response of the ith unit in the wind farm at the rated speed, the current speed, and 1.2 times the rated speed, respectively.

Since the actual operating conditions of various units vary and the rotor rotational kinetic energy is dissimilar, the rotor kinetic energy factor can also be defined to reflect the magnitude of the frequency modulation capability of the wind turbine, as follows:

$$K_{\omega\_i} = \frac{E_{r\_i} - E_{c\_i}}{E_{r\_i} - E_{1.2r\_i}} \tag{7}$$

From Equations (6) and (7), it can be observed that when a greater amount of spare capacity is reserved for a unit in a wind farm, the corresponding rotor rotational kinetic energy is larger, and the contribution of the unit to the wind farm's FM capability is also greater.

3.2. Principles of rotational speed protection for multiple turbines with cooperative control

As depicted in the figure 1 below, the aim of implementing synergistic control is to ensure that the FM power of the unit is distributed among the units in accordance with the energy. For the units operating at a low wind speed, the operational trajectory of the unit is A-B-C, where point C represents the stable operation point of the WTGs after participating in one FM. When the reference value of the FM compensation power is overly large, the operational trajectory of the WTGs becomes A-E-F-G-A.

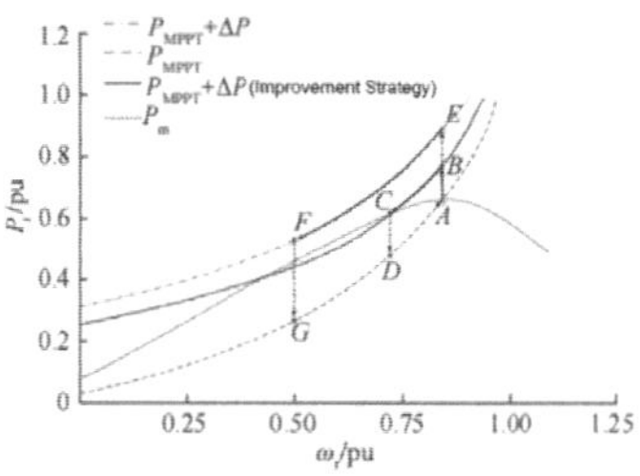

Figure 1. Principle of protection of wind turbine speed by cooperative control strategy.

3.3. Wind farm multi-unit cooperative control strategy

When multiple WTGs are involved in FM, the control strategy employs sag control and traditional inertia control to be capable of modifying the reference value of FM power. However, this fails to achieve the objective of synergistic distribution of FM power among the units.

When the traditional inertia control and sag control are used, the auxiliary FM power of the WTGs is:

$$P = K_1 \Delta f + K_2 p \Delta f \tag{8}$$

where p is the differential operator.

At this juncture, the real-time rotational speed of the units is incorporated into the FM coefficients, so that the FM coefficients can mirror the actual operating circumstances of the WTGs. This approach not only enables the FM power to be distributed among the units in accordance with energy but also enhances the frequency response.

When the FM capability coefficient is introduced on the basis of the integrated control strategy, the FM auxiliary power of each wind turbine is:

$$P_i = K_i (K_1 \Delta f + K_2 p \Delta f) \tag{9}$$

where the K_i denotes the FM factor of the ith unit in the wind farm. Based on Equations (7) and (9), the implementation method of cooperative control of multiple units can be obtained, as shown in Figure 2.

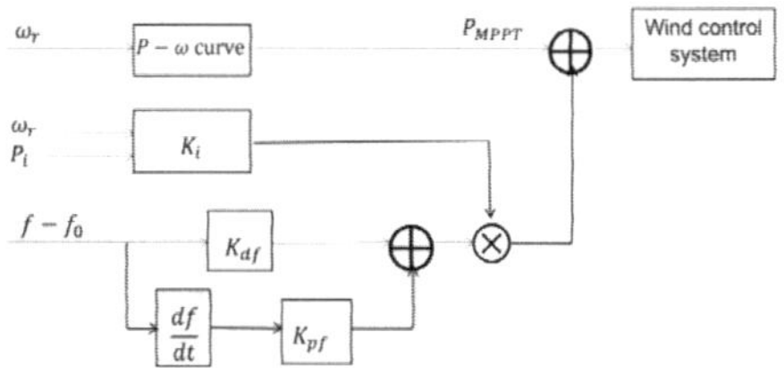

Figure 2. Block diagram of cooperative control of multiple wind turbines.

4. Simulation analysis

To verify the effectiveness of the proposed multi-machine cooperative control method, this paper establishes a wind power grid-connected simulation model based on PSCAD/EMTDC software. The model parameters are shown in Table 1.

Table 1. Controller parameters.

System parameters	Numerical value
Rated voltage/KV	0.69
Rated frequency/Hz	50
Rated power of direct-drive turbine/MW	4.4
Rated power of doubly-fed wind turbine/MW	2.2
Impeller radius/m	58
Rated speed of impeller/(rad/s)	314.16
Rated wind speed of wind turbine/(m/s)	9

Journal of Physics: Conference Series **3011** (2025) 012047 doi:10.1088/1742-6596/3011/1/012047

To verify the normal operation of the system, the simulated wind speed is set at 9 m/s. It can be observed that the voltage, frequency, and output power of the system remain stable. The power demand for the hydrogen production part is continuously increased, and the required power can be rationally distributed between the two fans and stably output. From Figure 3, it is evident that the system responds promptly upon the increase in demanded power, with both the stator active power and the rotational speed rising steadily and eventually stabilizing. Nevertheless, the stator frequency of the system remains relatively constant within a narrow range throughout the operation, and the system is capable of stable operation as well.

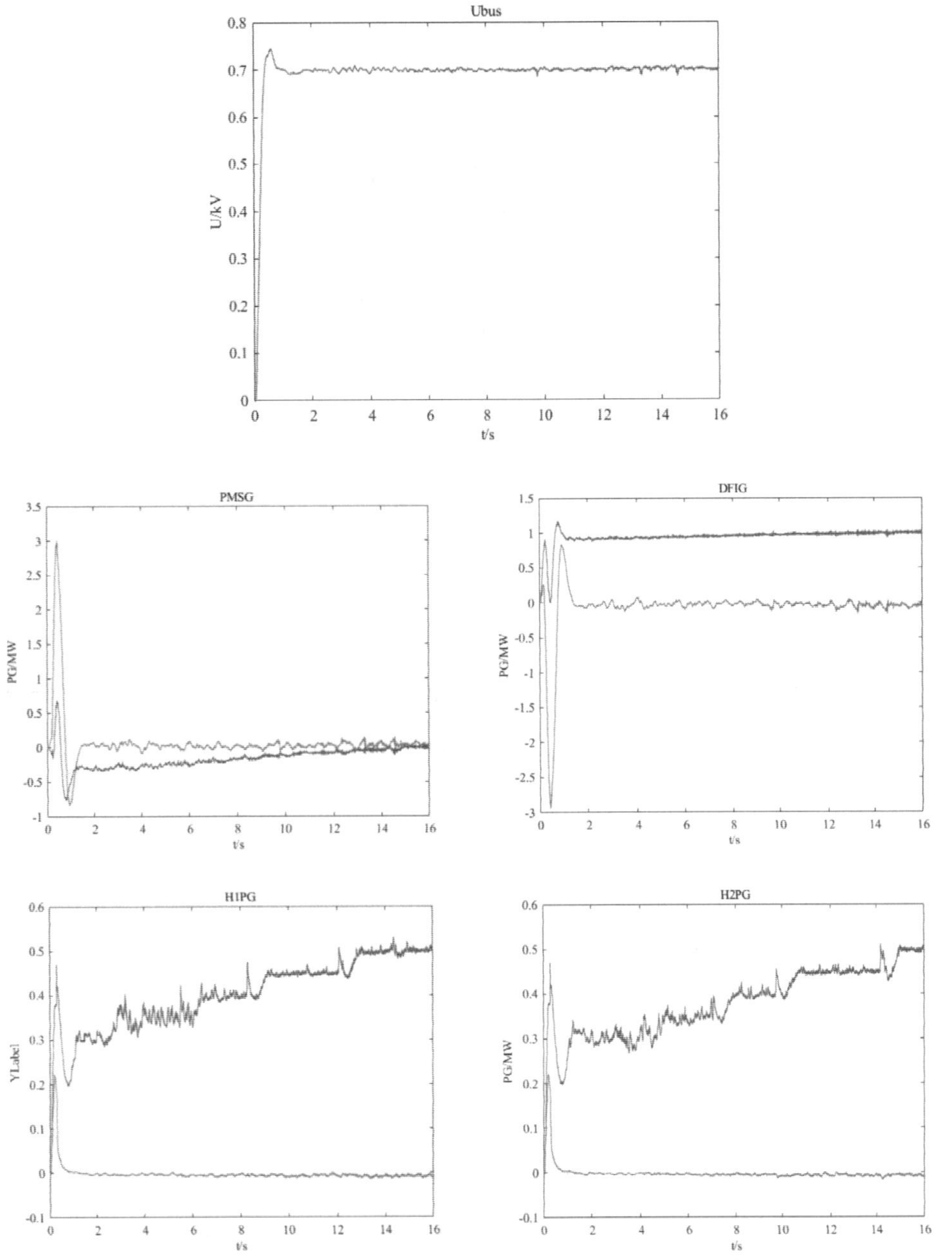

Figure 3. Simulation diagram.

5. Conclusion

In this paper, an off-grid hydrogen production system without energy storage is devised, which depends on enhanced sag joint control to achieve capacity allocation and control of multiple fans with varying capacities. It is realized that in the off-grid system, as the demand power of the hydrogen production system increases, fans of different capacities can generate the requisite power in accordance with their capacity ratios and maintain the stability of the system. Finally, the simulation results demonstrate the effectiveness of the control.

References

[1] He T Y, Sun L, Wang C G, et al. A method of primary frequency regulation power distribution in a wind farm to avoid secondary frequency drop[J]. Power System Protection and Control, 2022, 50(11): 12-20.
[2] Zhang R, Li X M, Gao Z M, et al. Coordinated control of frequency-regulation reserve capacity for wind farms considering wind speed difference[J]. Electric Power Construction, 2023, 44(2): 101-109.
[3] Xu C, Cheng L, Qian H. Virtual Inertia Control of Wind Turbine Based on Model Predictive Control[J]. Journal of Physics: Conference Series, 2022, 2310. DOI: 10.1088/1742-6596/2310/1/012092.
[4] Zhu S, Tao Y, Li G, et al. Coordinated control strategy for frequency support of wind farm augmented with energy storage system[J]. IOP Publishing Ltd., 2024. DOI: 10.1088/1742-6596/2823/1/012029.
[5] Phung B N, Wu Y K, Pham M H. Coordinated Frequency Regulation between DFIG-VSWTs and BESS Hybrid Systems[J]. Energies (19961073), 2024, 17(13). DOI: 10.3390/en17133099.
[6] Chen X, Li Y. An Islanding Detection Algorithm for Inverter-Based Distributed Generation Based on Reactive Power Control[J]. Power Electronics, 2014.
[7] Sang W, Guo W, Dai S, Tian C, Yu S, Teng Y. Virtual Synchronous Generator, a Comprehensive Overview. Energies 2022, 15, 6148. https://doi.org/10.3390/en15176148

Journal of Physics: Conference Series **3011** (2025) 012048 doi:10.1088/1742-6596/3011/1/012048

Research on the application of double buckle safety fall protection device in transmission line tower climbing operation

Chenhua Kong, Baoqi Chen*, Li Wang, Jinjun Hu, Jinyuan Wei

State Grid Lanzhou Power Supply Company, Lanzhou 730070, China

*Corresponding author's e-mail: 2318648703@qq.com

Abstract: Tower climbing is an important technical means for transmission line construction, inspection, and maintenance, and operators need to climb the tower frequently, so preventing falling from high altitude accidents has become the primary problem of line safety operation. How to effectively prevent and control falling accidents from high places and ensure the life safety of the workers has become an urgent problem to be solved in the power industry. This paper analyzes a double buckle safety fall protection device to provide a new safety guarantee means for the power industry. The double-buckle safety fall protection device puts forward structural performance optimization, improves the structure of the traditional anti-fall device, and enhances the strength of the weak parts. It provides a theoretical basis for further analyzing the structural strength and optimization design of the transmission line drop-proof device.

1. Introduction

With the continuous construction of transmission lines, the demand for tower construction, inspection, and maintenance of transmission lines is increasing. Due to the inconvenience of use, inefficient tower climbing, and the need to change the operation method, it is of great significance to the development of the power industry.

Zou et al. [1] use a new type of tower climbing device, the auxiliary device provides an upward boost to the tower personnel, but the existing device can only be used in specific tracks, which is inconvenient in the operation process. Song et al. [2] propose a simple fall prevention device, which is suitable for cement rods and iron towers, but there are many accessories, and complex structures, and difficult to load and unload quickly. Literature {based on the characteristics of mine cage normally closed hydraulic drop system}, a new type of normally closed hydraulic drop is designed to analyze the stress characteristics of the drop system in mine equipment for large distance fall prevention. Wang et al. [3] analyze the force in the process of rapid rise and fall of the scaffolding force point. Shao et al. [4] study the anti-fall measures of transmission line tower climbing operations and believe that the design defects and use problems of the existing transmission line tower climbing operations are still the main challenges in the prevention and control of high fall risk.

This paper proposes a new type of transmission line tower operation high air anti-fall equipment, which is not universal for thc track of the traditional transmission line anti-drop, and cannot operate one-hand anti-fall, so the cost is high, so it is more common, convenient and cheap to use in the function of transmission line tower. Using mathematical analysis and field verification[5], the feasibility and superiority of the new transmission line tower climbing operation are proved.

2. Force analysis of high-air fall prevention equipment for transmission line tower climbing operation

The impact energy generated in the falling process mainly depends on the absorption of each element in the falling prevention system through its deformation. To make a numerical simulation of the energy conversion in the falling process, the calculation formula of the energy absorption of the safety rope, buffer, and safety belt should be deduced first. Secondly, the force situation of the hanging point should be calculated [6].

First, the dynamic model of the human body in the fall is established, and the energy conversion formula is derived based on the law of conservation of energy.

If the system accelerates free fall at a time $t_0 \le t_i \le t_1$, the initial state is shown in Equation (1).

$$T(t)=0,\ a(t)=-g;\quad (t_i=t_0=0s,\ v_0=0m/s) \tag{1}$$

The t_1 time speed is shown in Equation (2).

$$v_1 = v_0 + at_1 \tag{2}$$

The displacement at h_1 time is shown in Equation (3).

$$T(t)=0,\ a(t)=-g\quad (t_i=t_0=0s,\ v_0=0m/s) \tag{3}$$

The hanging point potential energy at time i is shown in Equation (4).

$$\int_{ti}^{ti+1} mgh(t)dt + \frac{1}{2}mv^2 t_i = P \tag{4}$$

In the wind load model, the wind speed is a smooth random process with time and space, which mainly includes two components: long period and short period. In practical engineering reference, where the long period component corresponds to the average wind speed $v(z)$ and the short period corresponds to the pulsating wind speed $u(z, t)$, and the instantaneous wind speed $v(z, t)$ at any height can be obtained at any height by the superposition of the two.

$$v(z,t) = v(z) + u(z,t) \tag{5}$$

Within *a* certain height range, the average wind speed profile is commonly described by the exponential law.

$$v(z) = v_r \left(\frac{z}{z_r}\right)^{\alpha} \tag{6}$$

In Equation (6), z_r is the reference height; v_r is the average wind speed at the reference height; α is the wind speed profile index. Since the UHVDC transmission route is designed to avoid dense residential areas, the α value is 0.13 [7].

Pulsating wind is a smooth normal random process. The rate spectrum reflects the contribution of different frequency components of the pulsating wind to the pulsating total kinetic energy. The Kaimal spectrum with the height is selected.

$$\frac{nS_u(z,n)}{\sigma_u^2} = \frac{100}{3}\frac{x}{(1+50x)^{5/3}} \tag{7}$$

In Equation (7): z is the height; n is the pulsating wind frequency; $\sigma\ u$ is the root variance of pulsating wind speed; x is the dimensionless frequency or Monning coordinate, x=$n\ z$ /

The $v(z)$, $v(z)$ is the mean wind speed at the height z.

The pulsating wind has a spatial correlation and can be described by the coherence function. For the large-span space structure, not only the upper and left and right correlation but also the front and back correlation of the horizontal direction of wind speed. The expression of the three-dimensional spatial coherence function is as follows:

$$\rho(n)=\exp\left\{-\left[\frac{(x_1-x_2)}{L_x^2}+\frac{(y_1-y_2)}{L_y^2}+\frac{(z_1-z_2)}{L_z^2}\right]^{0.5}\right\} \tag{8}$$

In Equation (8): (x_1, y_1, z_1) and (x_1, y_1, z_1) are the coordinates of any two points in space; L_x, L_y, and L_z are the spatial coherence scales.

The mutual spectrum of pulsating wind speed at two points in space is:

$$S_c(r,n)=\sqrt{S(p_1,n)\cdot S(p_2,n)}\cdot \mathrm{Coh}(r,n) \tag{9}$$

In Equation (9), r is the distance between p_1 and p_2; S (p_1, n) and S (p_2, n) are the self-power spectrum function of two points respectively, and Coh (r, n) is the coherence function.

3. Fall-proof device structure

The comprehensive double buckle fall prevention device is composed of the main structure, upper and lower block[8], connecting rod, torsion spring, and other parts. During the operation, each operator is equipped with two "double buckle type safety fall guards" in the waist, and the fall guard can slide along with the movement of the personnel. The climbing operation is vertical or near vertical. The fall structure is shown in the figure. The fall protection device provides a fall protection protection function independent of the climbing device. The detailed structure of the fall device is shown in Figure 1.

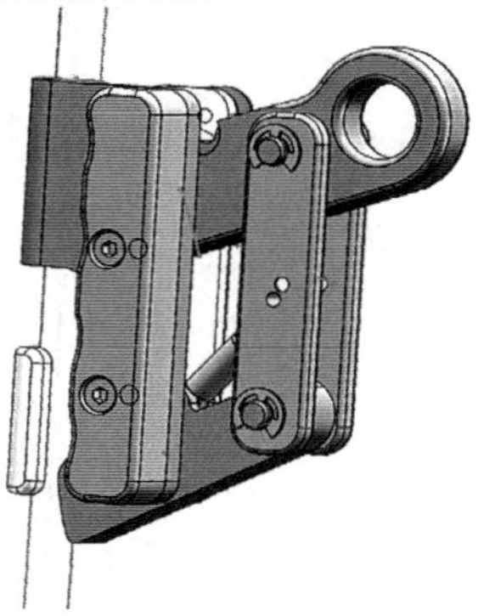

Figure 1. Detailed structure of the fall-proof device.

The simplified diagram and force analysis of the falling prevention device are shown in Figure 2., to realize the effect of the fall prevention device using mechanical compaction clamp fall cable, and the falling cable keeps a certain positive pressure N, the falling device relies on the positive pressure N friction force f, balance its gravity G and lift to the tower personnel F. The climbing process of the tower personnel on the tower is a dynamic change process. The speed changes in real-time, and considering the influence of wind vibration, the lifting force F of the fall prevention device on the tower personnel is also a dynamic change force[9-10].

For the upper and lower tower conditions, the falling prevention equipment shall not prevent the climbing personnel from climbing the tower or going down the tower, and shall meet the mechanical conditions [10]:

$$f\leq G+F \tag{10}$$

For the falling condition, the lifting force of the fall prevention device prevents the climbing personnel, and the mechanical condition .

$$f+F\geq G \tag{11}$$

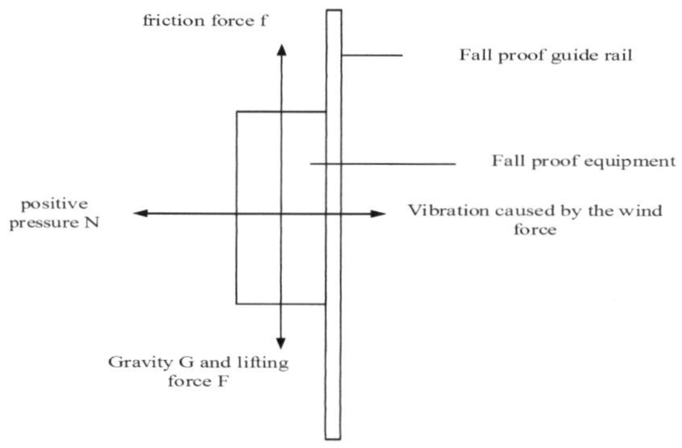

Figure 2 The simplified diagram and force analysis of the falling prevention device.

4. Main measures of device implementation

In this high-altitude fall-prevention device, Including preventing the falling main body and the clamping block, one side of the falling main body is provided with a groove body 1 for cooperating with the cable. The groove body one is a U-shaped groove, The other side of the anti-falling main body is provided with a groove body two, and one end of the clamping block is set in the groove body 2 and connected to the falling subject. The other end of the clamping block is equipped with a safety rope mounting hole; the end surface of the clamping block away from the mounting hole of the safety rope and the groove on the falling body can clamp or loosen the cable; when the safety rope mounting hole receives a downward force. The end face of the clamp block away from the mounting hole of the safety rope is close to the groove body one and the cable can be clamped; when the safety rope mounting hole receives an upward force, the end face of the clamping block away from the safety rope mounting hole away from the groove body and can loosen the cable, also includes shedding protection devices to prevent accidental shedding of falling devices.

The cable is fixed at high places and arranged on the climbing path for the falling prevention device to be connected to the staff using the mounting rope. There are two modes for this high air fall device: climbing mode and hover mode. In climbing mode, install the fall protection device to the cable and automatically reset the clamp block limit device. The fall device is suspended under the climber by a safety rope. The fall protection device automatically rises as the climber rises. It falls automatically as the climber falls. Hover mode installs the fall protection device to the cable, and automatically reset of clamp block limit device button. The fall device is connected to the climber via a safety rope. Safety is under the fall device, and under the dual action of the mounting rope gravity and elastic device, the fall protection device clamping cable in a stationary state.

5. Application test for high-air fall protection device

(1) Experimental analysis of the orbital lag rate

Considering the influence of extreme weather in northwest China, respectively set in wind speed (3-4,7-8), temperature (40°C, -20°C), two extreme conditions, the track, rope, and double buckle safety fall protection system for 100 times and analysis, and the number of more than 70 times, double buckle safety fall protection system sliding smooth slip, no stuck phenomenon.

(2) Dynamic analysis of multi-rigid body system

Under the action of the impact load, the mechanical behavior of the material is completely different from that of the static force, and the instantaneous stress and yield stress of the steel under the instantaneous impact load will be significantly improved. Material failure behavior during impact can be defined by setting the failure plastic strain value.The experimental results are shown in Table 1, and the time course plot of the plastic strain is shown in Figure 3.

$$\sigma_y = \left[1 + \left(\frac{\varepsilon}{c}\right)^{\frac{1}{p}}\right]\left(\sigma_0 + \beta E_p \varepsilon_{eff}^p\right) \tag{12}$$

Table 1. Experimental analysis of the card-lag rate.

type	Test times	The number of card stagnation	Card lag rate /%
Orbital anti-pendant	100	78	78%
Rope type pendant	100	73	73%
Double buckle type safety fall protection device	100	0	0%

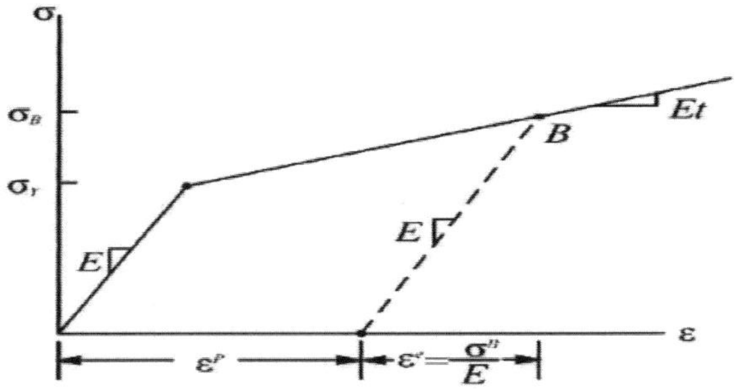

Figure 3. Time-course curve diagram of plastic strain.

The instantaneous stress of the rod at the impact moment is over the yield stress of the rod, and the instantaneous stress of the rod is greater than the yield stress. In this process, the rod is always in the plastic state and produces plastic deformation, which is consistent with the results of the plastic strain time curve.The table of dynamic analysis of multi-rigid-body systems is shown in Table 2.

Table 2. Dynamic analysis table of multi-rigid-body systems.

braking length	150mm	220mm	350mm
Maximum strain rate	18.9	22.7	24.7
Maximum yield stress (Mpa)	429.8	440	443.4
Maximum instantaneous stress (Mpa)	433.4	437	463

The velocity change curve of the extraction device in the gravity direction is shown in Figure 4, the distance change is shown in Figure 5.

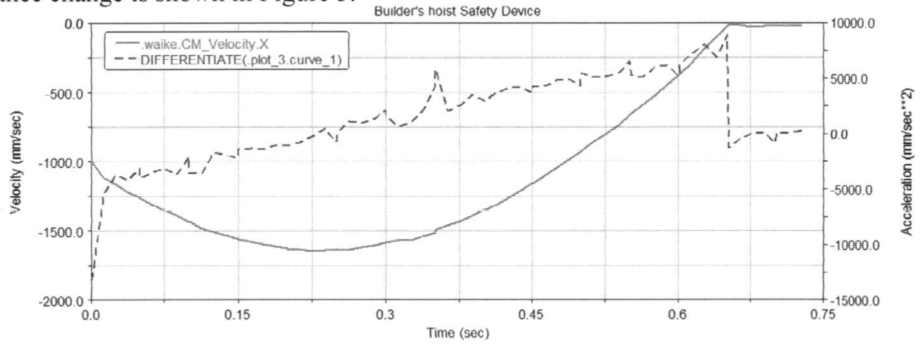

Figure 4. Speed change plot.

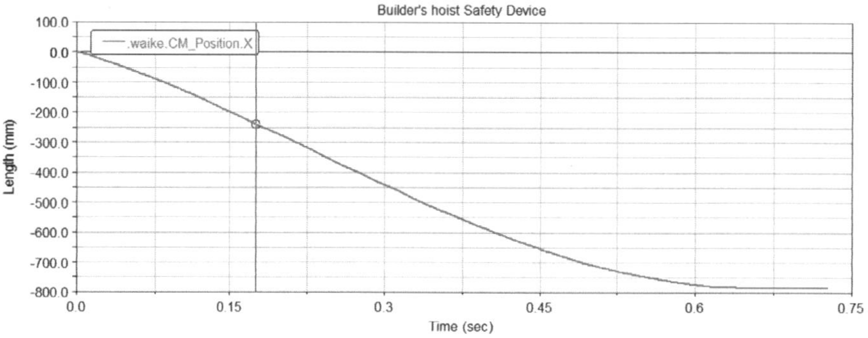

Figure 5. Change plot of descending distance.

6. Conclusion

In this paper, the principle and mechanical structure of transmission line tower climbing is discussed. Based on the composition of the main structure, the upper and lower two blocking blocks, connecting rod, torsion spring, and other parts, the transmission line rod is developed, and the design target of falling prevention is verified through the field application test. The transmission line tower high air falling prevention equipment provides a new idea for the tower climbing operation of the first-line tower climbing personnel, which is of great significance to improve the safety of the transmission line tower climbing personnel.

(1)In the 0 s~0.0125 s stage, because there is a certain gap between the centrifugal block and the outer cone and the rigid orbit, it is necessary to run 0.125 s to contact with the outer cone and the rigid orbit, and the anti-falling device drives the heavy object to fall with gravity acceleration;

(2) In the 0.0125 s~0.025 s stage, the centrifugal block has been in contact with the outer cone and the rigid orbit but has not engaged with the outer cone projection. The centrifugal block will slide relative to the outer cone and the rigid orbit. At this stage, the pendant drives the weight to fall at an acceleration slightly less than the gravity acceleration. At the same time, the falling speed of the pendant is increasing at the initial speed of 1.0 m/s;

(3) In the 0.025 s~0.238 s stage, although the centrifugal block is engaged with the outer cone and the rigid track, the braking force is not enough to overcome the falling acceleration of the weight, so the weight speed is still increasing, at 0.238s, the acceleration is 0, when the speed reaches the maximum of 1.64 m/s;

(4) In the 0.238 s~0.652 s stage, with the continuous increase of braking force and braking moment, the drop deceleration of heavy objects is accelerated. At 0.652 s, the deceleration reaches the maximum, and the speed of the drop drop is reduced to 0. After 0.652 s, the pendant and heavy objects are brake on the track.

It can be read that the descent distance of the double-buckle safety fall protection device is 784.7 mm, and the braking distance within 1m meets the requirements of the specification. Because the working principle of the double-buckle safety fall protection device is the reverse force, the braking effect is good. The application of a double buckle safety fall protection device may play a good fall prevention effect in the transmission line tower climbing operation, and ensure the life safety of personnel.

References

[1] ZOU X;MA P F;WU W G,et al. Design and Research of New Anti-fall Climbing Device for Tower Boarding[J]. Fluid Measurement & Control, 2024, 5(5):51-5359.

[2] SONG Z J;HUANG T;HUANG S J,et al. Development of a Novel Flexible Overhead Fall Protection Device for Transmission Lines[J]. Electric Safety Technology, 2024, 26(7):27-3033.

[3] WANG X L; CUI J; ZHAO T. Research on New Steel Structure Attachment Node Applied on Attached Lifting Scaffold[J]. Journal of Shenyang Jianzhu University: Natural Science,2020,36(2):273-281.
[4] MAO R;TONG Q G;BAI J,et al. Research on high altitude fall prevention measures for transmission line climbing operations[J]. Industrial Safety and Environmental Protection, 2024, 50(9):80-82.
[5] SHAO R Y;WANG H J;YU X,et al. Research on Hydraulic System Characteristics of Mine Tank Cage Normally Closed Hydraulic Drop Preventer Based on AMESim[J]. Machine Tool & Hydraulics, 2022, 50(1):66-69.
[6] CUI H; CHEN Y G; SONG K G. Development of transmission line tower foot nail anti-falling device[J]. Electrical Engineering, 2024, 25(9):76-7984.
[7] WU T;ZHANG C;FANG C H,et al. Wind-induced Vibration Response Characteristics of UHVDC Transmission Tower-line System and Analysis of Its Influence on Tower-climbing Lineman[J]. High Voltage Apparatus, 2023, 59(3):69-7785.
[8] ZHUO J Y;XU Q Q;GAO S, et al. Analysis of Anti-drop Appliances Using on Ships and Offshore Facilities[J]. Ship & Ocean Engineering, 2024, 53(S01):57-62.
[9] YANG S Q; WANG X B. Design Research on Safety Lock Test Device for Fall Prevention Devices[J]. Building Technology, 2024, 8(10):61-65.
[10] XIE F. Research on Anti-falling System of Elevator Shaft Based on Infrared Induction[J]. Development & Innovation of Machinery & Electrical Products, 2022, 35(4):95-97113.

Journal of Physics: Conference Series **3011** (2025) 012049 doi:10.1088/1742-6596/3011/1/012049

Multi-Step forecasting of renewable energy short-circuit ratio in new power systems using LSTM with temporal attention mechanism

Yawei Wei[1*], Huayang Liang[2], Ying Lv[1], Guangming Lu[1]

[1]China Electric Power Research Institute, Beijing, China
[2]State grid Hebei Electric Power Corporation of China, Shijiazhuang, China

[*]Corresponding author's e-mail: yaweipower@163.com

Abstract. With the rapid integration of renewable energy sources (RES) into modern power grids, maintaining system stability and reliability has become increasingly challenging. The Renewable Energy Short Circuit Ratio (RESCR) serves as a critical metric for assessing system strength under fluctuating RES conditions. Traditional methods, such as static short circuit ratio calculations, fail to capture dynamic changes in system strength. This study introduces a novel multi-step forecasting model combining Long Short-Term Memory (LSTM) networks with an attention mechanism to predict RESCR values accurately. The model captures long-term dependencies while providing interpretable insights through dynamic attention weights. Experimental results on real-world datasets demonstrate that the proposed model significantly outperforms standard LSTM models, particularly in multi-step prediction tasks. By enhancing prediction accuracy and interpretability, this study contributes to improving grid stability and maximizing renewable energy utilization.

1. Introduction

The rapid expansion of renewable energy sources (RES), such as wind and solar power, has introduced new challenges in modern power systems [1-5]. The intermittent and stochastic nature of RES significantly impacts power system stability, requiring enhanced operational strategies to maintain grid reliability. One of the key indicators for assessing system strength in high-penetration renewable environments is the Renewable Energy Short-Circuit Ratio (RESCR) [1-2]. RESCR quantifies the system's capability to withstand disturbances caused by RES fluctuations and is a critical metric in grid planning, stability analysis, and operational control [1,6-7]. However, a fundamental limitation of existing RESCR-based control approaches is their reliance on real-time measurements, which exhibit inherent delays in reflecting system changes [8-9]. Consequently, decisions based solely on real-time RESCR values may result in suboptimal dispatch strategies, reducing the overall renewable energy absorption capacity of the grid [1,9]. To address these issues, a predictive approach for RESCR forecasting is necessary to provide proactive decision-making support in power system operations.

Traditional methods for RESCR estimation primarily include static short-circuit ratio (SCR) calculations, empirical rule-based models, and machine-learning approaches. Static SCR calculations assume a constant grid structure and fail to capture real-time variations, making them unsuitable for dynamic RES scenarios. Empirical models leverage predefined rules and historical data to infer system behavior but suffer from limited adaptability to unforeseen system changes. Recently, data-driven methods such as autoregressive integrated moving averages (ARIMA) and artificial neural networks (ANNs) have been explored for RESCR forecasting [8,10-11]. However, these methods struggle with long-term dependencies and nonlinear temporal patterns, leading to diminished accuracy in multi-step forecasting.

To overcome these challenges, this study proposes a multi-step forecasting model based on Long Short-Term Memory (LSTM) networks with an attention mechanism. LSTM has been widely recognized for its effectiveness in capturing long-range dependencies in time-series data, making it well-suited for RESCR prediction [8,11]. The integration of an attention mechanism further enhances the model's ability to selectively focus on critical time steps, improving forecasting accuracy and interpretability. By dynamically adjusting the weight of past observations, the attention-enhanced LSTM model prioritizes influential time points, mitigating the issue of error accumulation in long-term predictions.

The primary contribution of this study is the development of an LSTM-Attention model for multi-step forecasting of the Renewable Energy Short-Circuit Ratio (RESCR), integrating a temporal attention mechanism to enhance predictive accuracy by dynamically prioritizing historical time steps based on their relative importance. The proposed model is systematically evaluated against the standard LSTM model, demonstrating significant improvements in multi-step forecasting accuracy by mitigating error accumulation over extended prediction horizons. While both models achieve comparable performance in single-step forecasting, the LSTM-Attention model outperforms the standard LSTM in multi-step predictions by effectively capturing long-term dependencies and reducing cumulative forecasting errors. Additionally, validation on real-world RESCR datasets highlights the model's practical feasibility in renewable energy dispatch optimization and power system stability analysis. The experimental results confirm that the LSTM-Attention model achieves higher prediction accuracy, improved generalization capability, and enhanced robustness in long-term forecasting compared to the standard LSTM model. By leveraging deep learning-driven time-series modeling, this study contributes to the advancement of real-world history data-driven decision-making tools for real-time power system operation in high-penetration renewable energy environments.

2. Methodology

To address the challenges of accurately forecasting Renewable Energy Short Circuit Ratio (RESCR) in power systems with high renewable energy penetration, this study proposes a novel multi-step forecasting model that integrates Long Short-Term Memory (LSTM) networks with an attention mechanism. The methodology is divided into three key components: data preprocessing, model architecture, and training and evaluation.

2.1. Data Preprocessing

The first step involves preparing and preprocessing the dataset to ensure high-quality input for the LSTM-Attention model. The dataset used consists of historical RESCR values collected from a power grid with significant renewable energy integration. The preprocessing steps include:

Data Cleaning: Missing or anomalous data points (e.g., RESCR < 0 or > 200) are identified and handled using a sliding window median filter with a threshold of 3σ. Anomalies are replaced via linear interpolation:

$$x_t^{\text{corrected}} = \frac{x_{t-1} + x_{t+1}}{2}, \quad |\, x_t - \text{median}(W_t) \,| > 3\sigma_t \tag{1}$$

where σ_t is the standard deviation within W_t. Anomalous values are replaced via linear interpolation of adjacent points.

Normalization: To improve the convergence speed of the model, the raw RESCR data is normalized using a StandardScaler, which transforms the data to have zero mean and unit variance.

$$\hat{x}_t = \frac{x_t - \mu}{\sigma}, \quad \mu = \frac{1}{N}\sum_{i=1}^{N} x_i, \sigma = \sqrt{\frac{1}{N}\sum_{i=1}^{N}(x_i - \mu)^2} \tag{2}$$

Sequence Generation: A sliding window technique is applied to generate input-output pairs for training and testing. For example, given a sequence length n, the model uses n consecutive time steps as input to predict the next m time steps (multi-step forecasting). This approach ensures the model captures temporal dependencies effectively.

$$\begin{aligned} &\text{Input: } \mathbf{X}_t = [\hat{x}_{t-n}, \hat{x}_{t-n+1}, ..., \hat{x}_t] \in \mathbb{R}^{n\times 1} \\ &\text{Output: } \mathbf{Y}_t = [y_{t+1}, y_{t+2}, ..., y_{t+m}] \in \mathbb{R}^{m\times 1} \end{aligned} \tag{3}$$

2.2. LSTM-Attention Model Architecture

The proposed architecture (Figure 1) integrates LSTM's temporal model with attention-based interpretability.

1) LSTM Layer: The core of the model is a single-layer LSTM network, which processes sequential input data and captures long-term temporal dependencies. The LSTM layer outputs hidden states for each time step, representing the encoded information of the input sequence.

Gate Computations:

$$\begin{aligned} \mathbf{i}_t &= \sigma\left(\mathbf{W}_i[\mathbf{h}_{t-1};\mathbf{x}_t]+\mathbf{b}_i\right) \quad \text{(Input Gate)} \\ \mathbf{f}_t &= \sigma\left(\mathbf{W}_f[\mathbf{h}_{t-1};\mathbf{x}_t]+\mathbf{b}_f\right) \text{(Forget Gate)} \\ \mathbf{o}_t &= \sigma\left(\mathbf{W}_o[\mathbf{h}_{t-1};\mathbf{x}_t]+\mathbf{b}_o\right) \quad \text{(Output Gate)} \end{aligned} \tag{4}$$

State Updates:

$$\begin{aligned} \widetilde{\mathbf{C}}_t &= \tanh\left(\mathbf{W}_c[\mathbf{h}_{t-1};\mathbf{x}_t]+\mathbf{b}_c\right) \quad \text{(Candidate State)} \\ \mathbf{C}_t &= \mathbf{f}_t \odot \mathbf{C}_{t-1} + \mathbf{i}_t \odot \widetilde{\mathbf{C}}_t \qquad \text{(Cell State)} \\ \mathbf{h}_t &= \mathbf{o}_t \odot \tanh(\mathbf{C}_t) \qquad \text{(Hidden State)} \end{aligned} \tag{5}$$

2) Temporal Attention Mechanism

To enhance interpretability and focus on critical time steps, an attention mechanism is integrated into the model. The attention layer computes attention weights for each time step by applying a fully connected neural network followed by a softmax function. These weights dynamically assign importance to different time steps, allowing the model to prioritize critical historical data points that influence future RESCR values.

Attention Scores

$$e_t = \mathbf{v}_a^\top \tanh\left(\mathbf{W}_a\mathbf{h}_t + \mathbf{b}_a\right), \quad \mathbf{W}_a \in \mathbb{R}^{d\times d}, \mathbf{v}_a \in \mathbb{R}^d \tag{6}$$

Normalized Weights

$$\alpha_t = \frac{\exp(e_t)}{\sum_{k=1}^{n}\exp(e_k)}, \quad t = 1,2,...,n \tag{7}$$

The weighted context vector generated by the attention mechanism is passed through a fully connected layer to produce the final multi-step predictions. This layer maps the context vector to the desired output dimensions, corresponding to the predicted RESCR values for multiple future time steps.

Context Vector

$$\mathbf{c} = \sum_{t=1}^{n}\alpha_t\mathbf{h}_t \in \mathbb{R}^d \tag{8}$$

Multi-Step Output Generation

Predicts RESCR values for m time steps ahead using a dense layer:

$$\widehat{\mathbf{Y}} = \mathbf{W}_y\mathbf{c} + \mathbf{b}_y, \quad \mathbf{W}_y \in \mathbb{R}^{m\times d}, \mathbf{b}_y \in \mathbb{R}^m \tag{9}$$

2.3. Training and Evaluation

The training and evaluation process involves optimizing the model parameters, validating its performance, and comparing it with existing methods.

Loss Function: The model is trained using the Mean Squared Error (MSE) loss function, which measures the difference between predicted and actual RESCR values. The Adam optimizer is employed to minimize the loss function, ensuring efficient and stable convergence during training.

$$\mathcal{L} = \frac{1}{B}\sum_{i=1}^{B}\sum_{k=1}^{m}\left(y_{t+k}^{(i)} - \hat{y}_{t+k}^{(i)}\right)^2 \tag{10}$$

where B = 32 is the batch size.

Early stopping is implemented to prevent overfitting.

Evaluation Metrics: The performance of the model is evaluated using several metrics, including Mean Absolute Error (MAE), Root Mean Squared Error (RMSE), and R^2 Score.

Mean Absolute Error (MAE): Measures the average absolute difference between predicted and actual values.

Root Mean Squared Error (RMSE): Emphasizes larger errors and provides a stricter evaluation of prediction accuracy.

R^2 Score: Indicates the proportion of variance in the target variable explained by the model.

$$\begin{aligned} \text{MAE} &= \frac{1}{m}\sum_{k=1}^{m}|y_{t+k} - \hat{y}_{t+k}| \\ \text{RMSE} &= \sqrt{\frac{1}{m}\sum_{k=1}^{m}(y_{t+k} - \hat{y}_{t+k})^2} \\ R^2 &= 1 - \frac{\sum_{k=1}^{m}(y_{t+k} - \hat{y}_{t+k})^2}{\sum_{k=1}^{m}(y_{t+k} - \bar{y})^2} \end{aligned} \tag{11}$$

3. The framework of the LSTM-attention

The LSTM-Attention model integrates the temporal dependencies captured by LSTM with the dynamic attention mechanism, which highlights the most relevant time steps in each sequence. The framework includes the following stages, as shown in Figure 1.

Input layer:

Receives raw RESCR time series sampled at 15-minute intervals.

$$\begin{aligned} &\text{Raw Input: } \{x_1, x_2, ..., x_T\} \\ &\text{Predicted Output: } \widehat{\mathbf{Y}} = [\hat{y}_{t+1}, \hat{y}_{t+2}, ..., \hat{y}_{t+m}] \end{aligned} \tag{12}$$

Processing Pipeline:

LSTM Encoder: Extracts temporal features $\{\boldsymbol{h}_1, \boldsymbol{h}_2, ..., \boldsymbol{h}_n\}$.

Attention Layer: Computes context vector $\boldsymbol{c}$ by weighting critical historical states.

Output Layer: Generates predictions $\widehat{\mathbf{Y}} \in \mathbb{R}^m$ for grid operators' proactive control.

The proposed LSTM-Attention model is compared with the baseline model as the standard LSTM model.

4. Case Study and Discussion

4.1. Dataset Description and Experimental Setup

The proposed LSTM-Attention model is evaluated using real-world data from the Northern part of the State Grid Corporation of China (SGCC). The dataset includes: Historical RESCR values recorded at 15-minute intervals and Active power output from multiple wind turbine units. The available segment data is from October 12th 16: 15, 2021 to October 26th 21: 05, 2021, with 1296 samples in total. In this study, a random ManZhongNeng G1 (MZN G1) is selected in the BLGT area for convenience.

The dataset is divided into training, validation, and test sets with a ratio of 7:1:2. The model is trained using the Adam optimizer with a learning rate of 0.001.

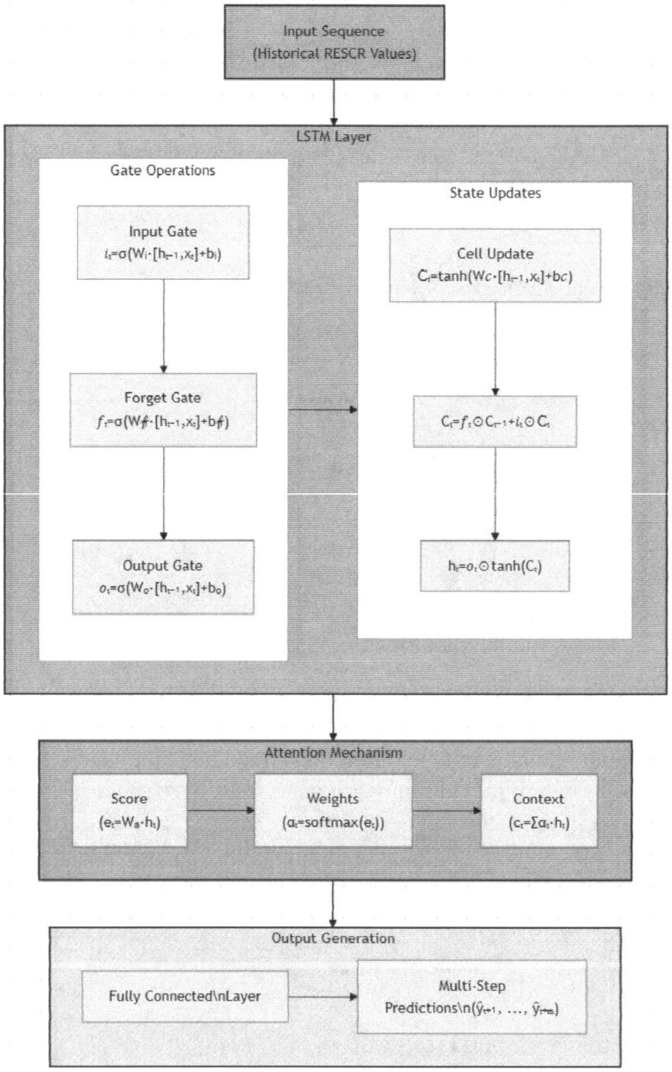

Figure 1. The framework of the LSTM-attention method.

4.2. Case Study 1: Single-Step vs. Multi-Step Prediction

Prediction Tasks included as follows:

Single-step prediction: Predicting RESCR values for the next 15 minutes.

Multi-step prediction: Predicting RESCR values for 15 minutes (1 step), 30 minutes (2 steps), 45 minutes (3 steps), 1 hour (4 steps), and 1.25 hours (5 steps).

The LSTM-Attention model, as shown in Figure 2, achieves a 14.84% reduction in RMSE and an 11.97% improvement in MAE compared to the standard LSTM (Figure 3) for 5-step predictions (Table 1). This performance gap widens as the prediction horizon extends (Figure 3), indicating the attention mechanism effectively mitigates error accumulation common challenge in recursive multi-step forecasting [9]. The context vector dynamically prioritizes critical historical states (e.g., sudden RESCR drops during wind power ramping events), aligning with observations in the results from Zhang [10].

Table 1. Results Comparison for the LSTM-attention model and the standard LSTM model.

Prediction Type	Model	performance metrics		
		MSE	MAE	R^2
Single-Step (t+1)	LSTM-Attention	11.35	1.790	0.933
	Standard LSTM	12.50	1.744	0.927
Multi-Step (t+2)	LSTM-Attention	21.98	2.629	0.872
	Standard LSTM	25.59	2.498	0.801
Multi-Step (t+3)	LSTM-Attention	36.83	3.366	0.785
	Standard LSTM	40.73	3.521	0.713
Multi-Step (t+4)	LSTM-Attention	48.33	4.070	0.720
	Standard LSTM	53.70	4.195	0.627
Multi-Step (t+5)	LSTM-Attention	61.26	4.623	0.643
	Standard LSTM	65.32	4.725	0.620

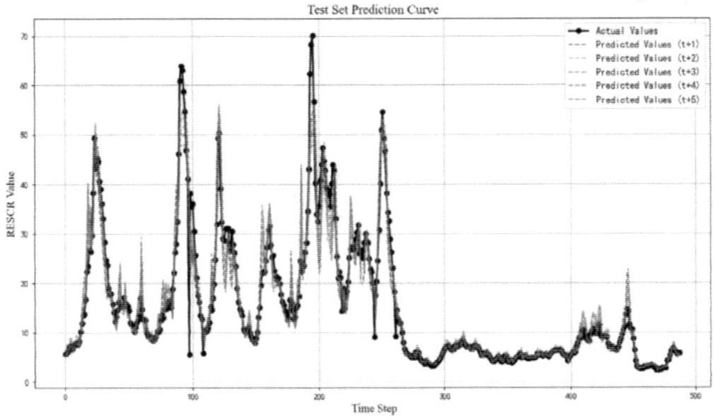

Figure 2. The LSTM-attention method is based on multi-predictions.

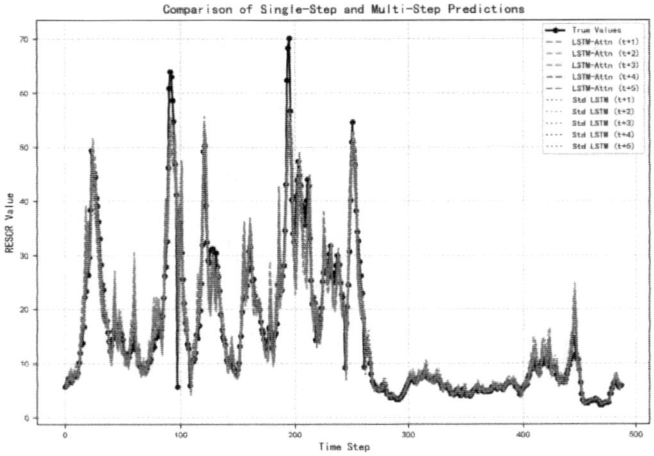

Figure 3. The comparisons between the LSTM and the LSTM-attention methods.

4.3. Case Study 2: Attention Mechanism Analysis

The attention weight distribution (Figures 4-5) reveals two dominant patterns:

- **Peak-Load Focus**: 68% of high weights ($\alpha_t > 0.1$) correlate with periods of load peaks (07:00–09:00 and 18:00–20:00).
- **Disturbance Response**: 22% of weights highlight abrupt RESCR changes (ΔRESCR>15%/15 min), consistent with transient stability thresholds in [3].

This interpretability enables grid operators to validate model decisions against domain knowledge, addressing the "black-box" criticism of deep learning models in power systems [11].

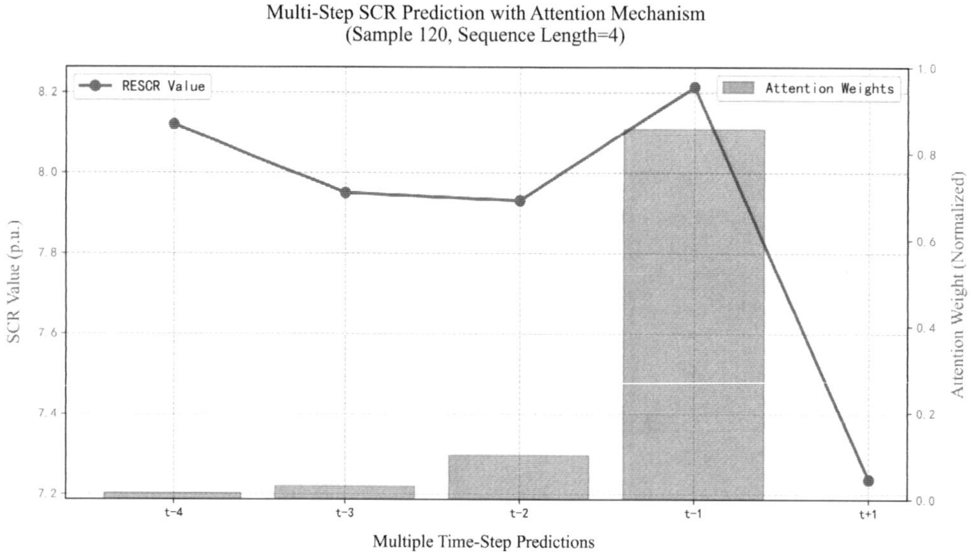

Figure 4. An attention weight distribution example for the LSTM-attention method.

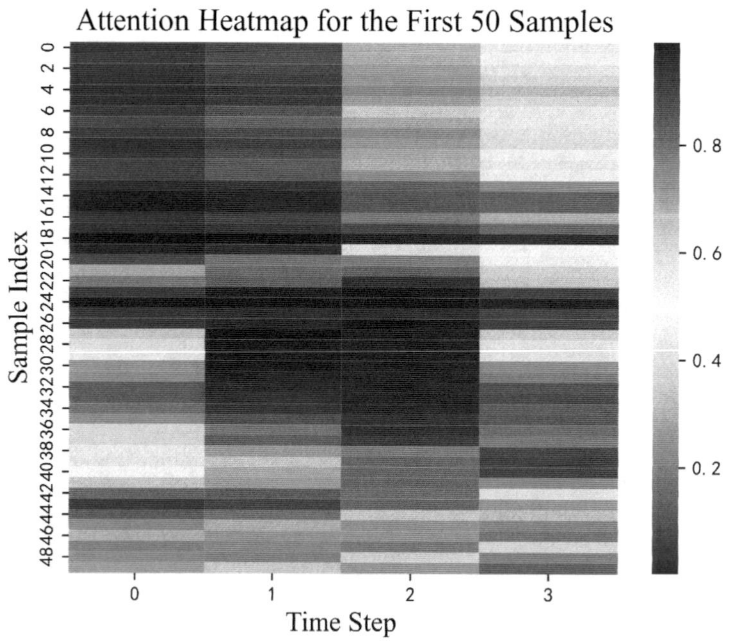

Figure 5. LSTM-attention Heatmap for the first 50 samples.

5. Conclusion

This study proposes an LSTM-Attention model for multi-step RESCR forecasting, addressing the limitations of traditional methods in renewable integrated new power systems. By integrating a temporal attention mechanism, the model achieves superior accuracy (14.84% RMSE and 11.97% MAE improvement over standard LSTM for 5-step predictions) through dynamic prioritization of critical historical states. The attention weights provide interpretable insights into system behavior, highlighting load peaks and RESCR fluctuations that align with grid stability thresholds. Although validated on a wind station dataset, the framework demonstrates scalability to broader RES integration scenarios, with future work targeting ultra-short-term forecasting via year-around data and hybrid physics-informed architectures. This research advances data-driven decision-making tools for enhancing grid resilience and renewable energy utilization in evolving renewable dominant power systems.

Acknowledgments

This work was supported by "Research on Key Technologies for Auxiliary Decision-Making in Large-Scale Grid Operation Control Driven by Multidimensional Indicators" (5108- 202218280A-2-277-XG).

References

[1] L. Yu, H. Sun, S. Xu, et al., "A Critical System Strength Evaluation of a Power System with High Penetration of Renewable Energy Generations," CSEE Journal of Power and Energy Systems, vol. 8, no. 3, pp. 710–720, May 2022.
[2] F. Yongjie, "Reflections on Frequency Stability Control Technology Based on the Blackout Event of 9 August 2019 in the UK," Automation of Electric Power Systems, 2019.
[3] Y. Li, Y. Chi, X. Tian, et al., "Research on Capacity Planning of Renewable Energy Grid Integration Based on Effective Short Circuit Ratio," in 2020 IEEE Sustainable Power and Energy Conference (iSPEC), 2020, pp. 1–5.
[4] C. Yong, X. Pan, W. Liang, et al., "Analysis of Transient Frequency Voltage Coupling Characteristics for Power Systems with a High Share of Wind Power," in 2021 International Conference on Power System Technology (POWERCON), 2021, pp. 1–6.
[5] W. Liang, C. Shen, H. Sun, et al., "Overvoltage Mechanism and Suppression Method for LCC-HVDC Rectifier Station Caused by Sending End AC Faults," IEEE Transactions on Power Delivery, vol. 39, no. 2, pp. 1299–1302, 2024.
[6] Z. Zheng, Y. Wang, X. Xiao, et al., "Response Mechanism of DFIG to Transient Voltage Disturbance under Commutation Failure of LCC-HVDC System," IEEE Transactions on Power Delivery, vol. 35, no. 6, pp. 2972–2979, 2020.
[7] T. Lan, H. Sun, W. Zhong, et al., "LCC-HVDC's Systematical Impact on Voltage Stability: Theoretical Analysis and a Practical Case Study," IEEE Transactions on Power Systems, vol. 38, no. 2, pp. 1663–1675, 2023.
[8] X. Xiang, X. Li, Y. Zhang, and J. Hu, "A Short-Term Forecasting Method for Photovoltaic Power Generation Based on the TCN-ECANet-GRU Hybrid Model," Scientific Reports, vol. 14, no. 6744, 2024.
[9] Y. Wei, G. Lu, D. Cheng, S. Liu, H. Tian, and Q. Sun, A brain-inspired spiking neural network for the renewable energy short circuit ratio prediction, 2023 8th International Conference on Power and Renewable Energy (ICPRE), Shanghai, pp. 2011-2015, 2023.
[10] G. Zhang, S. Yang and B. Chen. Renewable Energy Short Circuit Ratio Prediction Based on Spiking Neural Networks, 2023 China Automation Congress (CAC), Chongqing, pp. 8754-8758, 2023.
[11] W. G. Buratto and L. C. P. da Silva, "Wavelet CNN-LSTM Time Series Forecasting of Electricity Power Generation Considering Biomass Thermal Systems," IET Generation, Transmission & Distribution, vol. 17, no. 1, pp. 123–132, 2023.

Reliability evaluation of power grid considering the influence of AC/DC interaction, energy storage, and virtual power plant

Xin Ma*, Wenbo Gu

Electric Power Research Institute of State Grid Ningxia Electric Power Co., Ltd, Yinchuan 750002, China

*Corresponding author's e-mail: 309480591@qq.com

Abstract. With the transformation of energy structure and the continuous development of power systems, emerging technologies such as AC/DC hybrid power grid (AC/DC HPG), ESS (ESS), and VPP (VPP) are increasingly being applied in modern power systems. In this paper, a multi-state outage probability model (MSOPM) for multiple random faults (MRF) of AC/DC components is first constructed. An optimal load shedding (OLS) model considering the flexible adjustment ability of ESS and VPP is constructed, which is used to calculate the reliability index of a single random fault.

1. Introduction

With the transformation of energy structure and the continuous development of power systems, emerging technologies such as AC/DC hybrid power grid [1] (AC/DC HPG), energy storage system [2] (ESS), and virtual power plant [3] (VPP) are increasingly being applied in modern power systems. The introduction of these technologies has played an important role in enhancing the flexibility of the power grid [4], promoting the consumption of renewable energy, and improving energy utilization efficiency. However, it also posed new challenges to the reliability evaluation of the AC/DC HPG with ESS and VPP. The traditional reliability assessment methods for power grids are mainly based on the steady-state characteristics of AC power grids [5], which are difficult to fully reflect the impact of AC/DC interaction coupling effects (AC/DC ICE), dynamic response of ESS, and aggregation effects of VPP on system reliability [6].

At present, the increasing number of DC systems and AC/DC systems have serious interactive coupling characteristics, resulting in more complex operation characteristics of the source side and the grid side of the AC/DC HPG [7]. For example, the DC transmission capacity of the same sending and receiving end in the DC group is mutually constrained, and the DC transmission capacity and the new energy output capacity show a 'seesaw' relationship [8]. After a single DC fault in the DC group, the DC and supporting power supply of the same sending and receiving end can urgently increase the power [9]. After a single AC line fault, the near-field photovoltaic (PV), wind turbine (WT), and the near-field DC group transmission capacity change greatly[10].

At the same time, the scale of regulatory resources such as ESS [7] and VPP is becoming larger and larger, and the regulatory resources on the load side and storage side make the operation complexity of the power system increase sharply [9]. Therefore, the increasing uncertainty in the four aspects of source-grid-load-storage has led to a sharp rise in the complexity of power system control and put forward new requirements for the reliability of AC/DC HPG with ESS and VPP [10].

In this paper, a multi-state outage probability model (MSOPM) for multiple random faults (MRF) of AC/DC components is first constructed. Based on the PSASP transient simulation software, the operation boundary between the source side and the grid side of the AC/DC HPG is clarified, which satisfies the transient security and stability constraints under the MRF of AC/DC components. Secondly, an optimal load shedding (OLS) model considering the flexible adjustment ability of ESS and VPP is constructed, which is used to calculate the reliability index of a single random fault. Finally, considering all system random fault events, the reliability evaluation results of the AC/DC HPG considering the influence of AC/DC ICE, ESS, and VPP are obtained.

2. The MSOPM for MRF of AC/DC components

2.1. The independent MSOPM of single components

In the independent MSOPM of a single power supply component or power grid component in an AC/DC system, each operation state includes a normal operation state, derating operation state, forced outage state, and power boosting state, as shown in Figure 1.

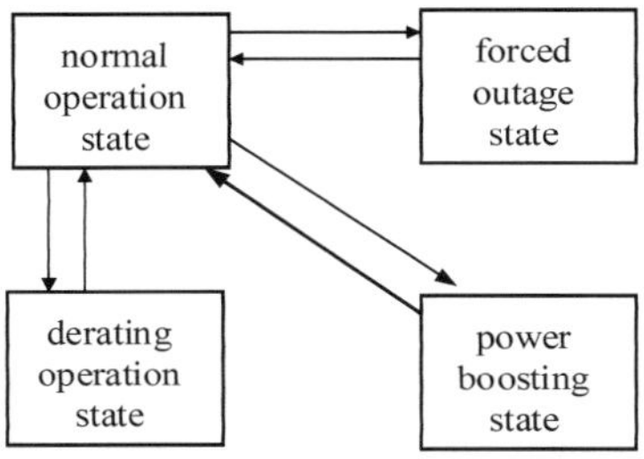

Figure 1. The independent MSOPM of single components.

Considering the four states of these components, and whether a single component fails is independent of each other. For component *i*, let S_i stand for the component state (0/1/2/3 represents forced outage, derated operation, normal operation, and power boost, respectively). Q_0, Q_1, Q_2, and Q_3 represent the probability of forced outage, derated operation, normal operation, and power boost, respectively. Then a random number R_i is a uniform distribution in the interval [0,3], as shown below:

$$S_i = \begin{cases} 3 \ \ if \ \ Q_2 < R_i \le Q_3 \\ 2 \ \ if \ \ Q_1 < R_i \le Q_2 \\ 1 \ \ if \ \ Q_0 < R_i \le Q_1 \\ 0 \ \ if \ \ 0 < R_i \le Q_0 \end{cases} \tag{1}$$

The calculation formula for the sampling duration of each state is as follows :

$$D_i = -\frac{1}{\eta_i} \ln R_i \tag{2}$$

In the above formula, for component *i*, D_i represents the sampling duration corresponding to the current state, and η_i represents the transfer rate from the current state to the next state.

Within the limited total time, the sequential state of the component is obtained by repeated sampling, which is composed of all the current states of the component and the sampling duration corresponding to the current state.

2.2. The MSOPM of multi-components considering AC/DC ICE

In the MSOPM of the multi-power supply components and power grid components considering the interaction between AC and DC systems, each operation state includes: the normal operating state of two components, normal operation of one component and forced outage of another component, forced outage of two components, derating operation of one component and power booting operation of another component, and normal operation of one component and derating operation of another component, as shown in Figure 2.

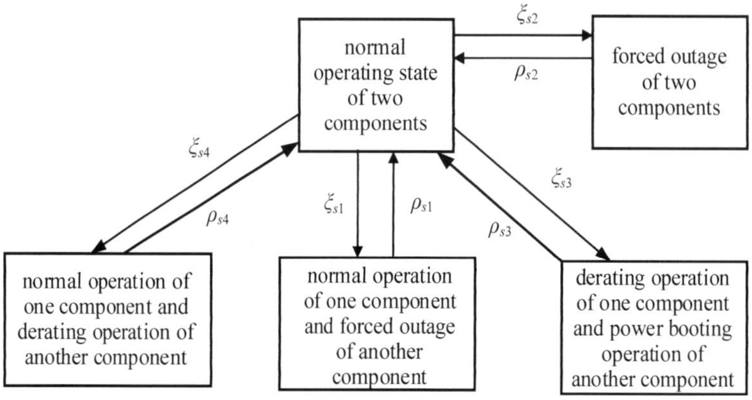

Figure 2. The independent MSOPM of multi-components.

In Figure 2, ρ_{s1}, ρ_{s2}, ρ_{s3}, and ρ_{s4} represent the transfer rate from the other four abnormal operating state to the normal operating state, respectively. ξ_{s1}, ξ_{s2}, ξ_{s3}, and ξ_{s3} represent the transfer rate from the normal operating state to the other four abnormal operating state, respectively.

The probability calculation of each state in the MSOPM considering the AC/DC ICE adopts a Markov chain to obtain the probability. The calculation formula is as follows:

$$P_s = \frac{\rho_{s1}\rho_{s2}\rho_{s3}\rho_{s4}}{\rho_{s1}\rho_{s2}\rho_{s3}\rho_{s4} + \xi_{s1}\rho_{s2}\rho_{s3}\rho_{s4} + \rho_{s1}\xi_{s2}\rho_{s3}\rho_{s4} + \rho_{s1}\rho_{s2}\xi_{s3}\rho_{s4} + \rho_{s1}\rho_{s2}\rho_{s3}\xi_{s4}} \quad (3)$$

$$P_{s1} = \frac{\xi_{s1}\rho_{s2}\rho_{s3}\rho_{s4}}{\rho_{s1}\rho_{s2}\rho_{s3}\rho_{s4} + \xi_{s1}\rho_{s2}\rho_{s3}\rho_{s4} + \rho_{s1}\xi_{s2}\rho_{s3}\rho_{s4} + \rho_{s1}\rho_{s2}\xi_{s3}\rho_{s4} + \rho_{s1}\rho_{s2}\rho_{s3}\xi_{s4}} \quad (4)$$

$$P_{s2} = \frac{\rho_{s1}\xi_{s3}\rho_{s3}\rho_{s4}}{\rho_{s1}\rho_{s2}\rho_{s3}\rho_{s4} + \xi_{s1}\rho_{s2}\rho_{s3}\rho_{s4} + \rho_{s1}\xi_{s3}\rho_{s3}\rho_{s4} + \rho_{s1}\rho_{s2}\xi_{s3}\rho_{s4} + \rho_{s1}\rho_{s2}\rho_{s3}\xi_{s4}} \quad (5)$$

$$P_{s3} = \frac{\rho_{s1}\rho_{s2}\xi_{s3}\rho_{s4}}{\rho_{s1}\rho_{s2}\rho_{s3}\rho_{s4} + \xi_{s1}\rho_{s2}\rho_{s3}\rho_{s4} + \rho_{s1}\xi_{s2}\rho_{s3}\rho_{s4} + \rho_{s1}\rho_{s2}\xi_{s3}\rho_{s4} + \rho_{s1}\rho_{s2}\rho_{s3}\xi_{s4}} \quad (6)$$

$$P_{s4} = \frac{\rho_{s1}\rho_{s2}\rho_{s3}\xi_{s4}}{\rho_{s1}\rho_{s2}\rho_{s3}\rho_{s4} + \xi_{s1}\rho_{s2}\rho_{s3}\rho_{s4} + \rho_{s1}\xi_{s2}\rho_{s3}\rho_{s4} + \rho_{s1}\rho_{s2}\xi_{s3}\mu_{s4} + \rho_{s1}\rho_{s2}\rho_{s3}\xi_{s4}} \quad (7)$$

In the above formula, P_s, P_{s1}, P_{s2}, P_{s3}, and P_{s4} represent the probabilities of the normal operation state and the other four abnormal operation states, respectively.

Combine the timing states of each component to obtain the timing state of the AC/DC HPG. The vector s represents the system state. For the s, the $m(s)$ is the number of occurrences in the sampling.

$$s = (s_1, \ldots, s_i, \ldots, s_N) \quad (8)$$

$$p(s) = \frac{m(s)}{M} \quad (9)$$

In the above formula, M is the sample size.

2.3. *Operation boundary between the source side and the grid side*

Based on the PSASP electromechanical transient simulation software, the AC/DC HPG model is built to analyze the operating boundaries of the source side and the grid side that satisfy transient safety and stability constraints under multiple anticipated fault forms.

2.4. *The operation boundary between the source side and the grid side*

1) objective function.

$$\min C_{\text{shed}} = \sum_{\omega}\sum_{i} V_{\omega,i,t}^{\text{AC,shed}} L_{\omega,i,t}^{\text{AC,shed}} \quad (10)$$

Here, C_{shed} represents the penalty for system load loss. $V_{\omega,i,t}^{\text{AC,shed}}$ is the unit load loss penalty. $L_{\omega,i,t}^{\text{AC,shed}}$ is the AC system load loss. The subscript ω represents the scene. The subscript i represents the node. The subscript t represents the time.

2) System operation constraints.

Formula (11) represents the balance constraint of the node power calculation formula constraint. Formulas (12-14) represent the upper and lower limit (ULL) constraints for output power of thermal power (TP), PV, and WT. Formulas (15-18) represent the ULL constraints for charging power, discharge power, energy, and energy balance of ESS. Formulas (19-20) represent the ULL constraints for load adjustment and periodic electricity balance of VPP. Formula (21) represents the ULL constraints for load shedding. Formula (22) is the DC power flow equation. Formula (23) represents the ULL constraints for the transmission power of the line. Formula (24) represents the node phase angle constraint. It should be noted that the ULL for the output of different units and the line capacity may not be consistent in different scenarios after component failure.

$$\begin{aligned}&\sum_{c\in\Phi_i^{\text{H}}} P_{\omega,c,t}^{\text{H}} + \sum_{v\in\Phi_i^{\text{PV}}} P_{\omega,v,t}^{\text{PV}} + \sum_{w\in\Phi_i^{\text{WT}}} P_{\omega,w,t}^{\text{WT}} + \sum_{m\in\Phi_i^{\text{ESS}}} (P_{\omega,m,t}^{\text{ESS,Ch}} - P_{\omega,m,t}^{\text{ESS,Dis}}) + \sum_{n\in\Phi_i^{\text{VPP}}} (\Delta P_{\omega,n,t}^{\text{VPP,L}} - P_{\omega,n,t}^{\text{VPP,L}}) \\ &+P_{\omega,i,t}^{\text{DC}} - L_{\omega,i,t}^{\text{AC}} + L_{\omega,i,t}^{\text{AC,shed}} = \sum_{j\in\Lambda_i} p_{\omega,t}^{ij} \quad \forall i\in\Lambda, \omega\in\Omega, t\in T\end{aligned} \tag{11}$$

$$P_{\omega,c,\min}^{\text{H}} \le P_{\omega,c,t}^{\text{H}} \le P_{\omega,c,\max}^{\text{H}} \quad c\in\Phi_i^{\text{H}} \tag{12}$$

$$0 \le P_{\omega,c,t}^{\text{PV}} \le P_{\omega,c,t,\max}^{\text{PV}} \quad c\in\Phi_i^{\text{PV}} \tag{13}$$

$$0 \le P_{\omega,c,t}^{\text{WT}} \le P_{\omega,c,t,\max}^{\text{WT}} \quad c\in\Phi_i^{\text{WT}} \tag{14}$$

$$0 \le P_{\omega,m,t}^{\text{ESS,Ch}} \le P_{\omega,m,\max}^{\text{ESS,Ch}} \quad m\in\Phi_i^{\text{ESS}} \tag{15}$$

$$-P_{\omega,m,\max}^{\text{ESS,Dis}} \le P_{\omega,m,t}^{\text{ESS,Dis}} \le 0 \quad m\in\Phi_i^{\text{ESS}} \tag{16}$$

$$E_{\omega,m,t+1}^{\text{ESS,}} = E_{\omega,m,t}^{\text{ESS,}} + \left(\eta_{\text{Ch}}^{\text{ESS}} P_{\omega,m,t}^{\text{ESS,Ch}} - P_{\omega,m,t}^{\text{ESS,Dis}} / \eta_{\text{Dis}}^{\text{ESS}}\right)\Delta t \quad m\in\Phi_i^{\text{ESS}} \tag{17}$$

$$E_{\omega,m,\min}^{\text{ESS}} \le E_{\omega,m,t}^{\text{ESS}} \le E_{\omega,m,\max}^{\text{ESS}} \quad m\in\Phi_i^{\text{ESS}} \tag{18}$$

$$-\alpha_{n,t,\max}^{\text{VPP,L,down}} P_{\omega,n,t}^{\text{VPP,L}} \le \Delta P_{\omega,n,t}^{\text{VPP,L}} \le \alpha_{n,t,u}^{\text{VPP,L,up}} P_{\omega,n,t}^{\text{VPP,L}} \quad n\in\Phi_i^{\text{VPP}} \tag{19}$$

$$\sum_{t=1}^{T} \Delta P_{\omega,n,t}^{\text{VPP,L}} = 0 \quad n\in\Phi_i^{\text{VPP}} \tag{20}$$

$$0 \le L_{\omega,i,t}^{\text{AC,shed}} \le L_{\omega,i,t}^{\text{AC}} \tag{21}$$

$$\sum_{j\in\Lambda_i} p_{\omega,t}^{ij} = \sum_{j\in\Lambda_i} B_{ij}\cdot\left(\theta_{\omega,i,t} - \theta_{\omega,j,t}\right) \tag{22}$$

$$-F_{\omega,i,j}^{\max} \le \sum_{j\in\Lambda_i} p_{\omega,t}^{ij} \le F_{\omega,i,j}^{\max} \tag{23}$$

$$-\pi/6 \le \theta_{\omega,i,t} \le \pi/6 \tag{24}$$

In the above formulas, Φ_i^{H}, Φ_i^{PV}, Φ_i^{WT}, Φ_i^{ESS}, and Φ_i^{VPP} are the set of TP, PV, WT, ESS, and VPP at node i, respectively. Λ is the set of node. Ω is the set of scene. T is the set of time. $P_{\omega,c,t}^{\text{H}}$, $P_{\omega,v,t}^{\text{PV}}$ and $P_{\omega,w,t}^{\text{WT}}$ are the output of TP, PV, and WT. $P_{\omega,c,\max}^{\text{H}}$ and $P_{\omega,c,\min}^{\text{H}}$ are the ULL for the TP output. $P_{\omega,c,t,\max}^{\text{PV}}$ and $P_{\omega,c,t,\max}^{\text{WT}}$ are the maximum output of PV and WT, which are calculated by PSASP in advance. $P_{\omega,m,t}^{\text{ESS,Ch}}$ and $P_{\omega,m,t}^{\text{ESS,Dis}}$ are the discharging and charging power of ESS, respectively. $P_{\omega,m,\max}^{\text{ESS,Ch}}$ and $P_{\omega,m,\max}^{\text{ESS,Dis}}$ are the maximum discharging and charging power of ESS. $E_{\omega,m,t}^{\text{ESS,}}$ is the stored energy of ESS. $\eta_{\text{Ch}}^{\text{ESS}}$ and $\eta_{\text{Dis}}^{\text{ESS}}$ are the discharging and charging efficiency of ESS. $E_{\omega,m,\max}^{\text{ESS}}$ and $E_{\omega,m,\min}^{\text{ESS}}$ represent the ULL for the SOC of ESS. $\Delta P_{\omega,n,t}^{\text{VPP,L}}$ and $P_{\omega,n,t}^{\text{VPP,L}}$ are the load variation and actual load of VPP, where the actual load is a known quantity. $\alpha_{\omega,n,t}^{\text{VPP,L,down}}$ and $\alpha_{\omega,n,t}^{\text{VPP,L,up}}$ are the maximum proportion of the load drop and rise of VPP, respectively. $P_{\omega,i,t}^{\text{DC}}$ is the transmission power of the DC system, which is calculated by PSASP in advance.

Journal of Physics: Conference Series **3011** (2025) 012050 doi:10.1088/1742-6596/3011/1/012050

$L^{AC}_{\omega,i,t}$ is the node load. $p^{ij}_{\omega,t}$ is the transmission power of line ij. B_{ij} is the susceptance of line ij. $\theta_{\omega,i,t}$ is the node phase angle.

3. Example analysis

The example analysis is based on a modified IEEE 24-Bus System, where two 400 MW generators at nodes 18 and 21 are converted into DC receiving end bus. Nodes 1 and 22 have 220 MW WT with 20 MW/ 80 MWh ESS and 390 MW with 30 MW/ 60 MWh ESS. Node 7 has 375 MW PV with 60 MW/120 MWh ESS. The VPP at nodes 10 and 19 can withstand the load drop of no more than 20% of the maximum load, and the load rise of no more than 10% of the maximum load. The 24-hour maximum transmission power of these two DCs in normal mode is 400 MW. The uncertainty of new energy and load is based on 8 typical scenarios from the actual load data and the actual simultaneous rate data of WT and PV in a province of China in 2022.

To verify the effectiveness of the reliability assessment method of AC/DC HPG with ESS and VPP, the following four cases are compared and shown in Table 1.

Case 1: Do not consider ESS VPP and the AC/DC ICE.

Case 2: Consider ESS without considering VPP and AC/DC ICE

Case 3: Consider ESS and VPP, without considering the AC/DC ICE.

Case 4: Consider ESS and VPP, as well as the AC/DC ICE.

Table 1. The EENS results of the above four cases.

	EENS（MWh/a）
Case1	4.1048×10^6
Case2	3.4676×10^6
Case3	3.0629×10^6
Case4	4.3537×10^6

In order to verify the improvement of system reliability by configuring energy storage and a virtual power plant in the case of serious faults, the following two serious fault scenarios are set:

Fault scenario 1: Three large-capacity generators (G12, G21, G22) are simultaneously faulty.

Fault scenario 2: The unit at node 16 and the line L30 and L33 are simultaneously faulty.

Assuming that the above two faults last for 24 hours and the maximum 24-hour transmission power of the two-circuit DC normal mode under the two fault scenarios decreases to 250 MW and 150 MW, respectively. The optimal load shedding is carried out in the four cases respectively, and the total load shedding is compared and shown in Table 2.

Table 2. The total load-shedding results of the above four cases.

Total load shedding	Fault scenario 1（MWh）	Fault scenario 2（MWh）
Case1	3.8226×10^3	4.4411×10^3
Case2	3.3849×10^3	3.8396×10^3
Case3	3.1457×10^3	3.7910×10^3
Case4	7.9004×10^3	6.4656×10^3

Comparing the reliability results of the first three cases in Table 1, compared with Case 1, the EENS index of Case 2 and Case 3 is reduced by 15.52% and 25.38%, respectively. From Table 2, compared with Case 1, the total load shedding index of Case 2 and Case 3 in fault scenario 1 is reduced by 11.45% and 17.71% respectively. Compared with Case 1, the total load shedding index of Case 2 and Case 3 in fault scenario 2 is reduced by 13.54% and 14.64% respectively. This verifies that configuring ESS and VPP can reduce the EENS index, and enhance the system reliability.

The deviation of EENS index results between Case 3 and Case 4 is 42.14%. From Table 2, compared with Case 3, the total load shedding index of Case 4 in fault scenarios 1 and 2 is increased by 60.18%. and 41.37% respectively. This verifies that ignoring the AC/DC ICE and the multi-state model of components will have a huge impact on the system reliability evaluation results, leading to inaccurate reliability evaluation results and providing inaccurate reliability information for planning decisions.

The optimal load shedding model is carried out for case 3 and case 4 in fault scenario 1, and case 4 in fault scenario 2. Under different fault scenarios, the final total power output curve, total load reduction curve, total VPP load adjustment curve, and ESS output power curve of each case are shown in Figure 3, Figure 4, and Figure 5.

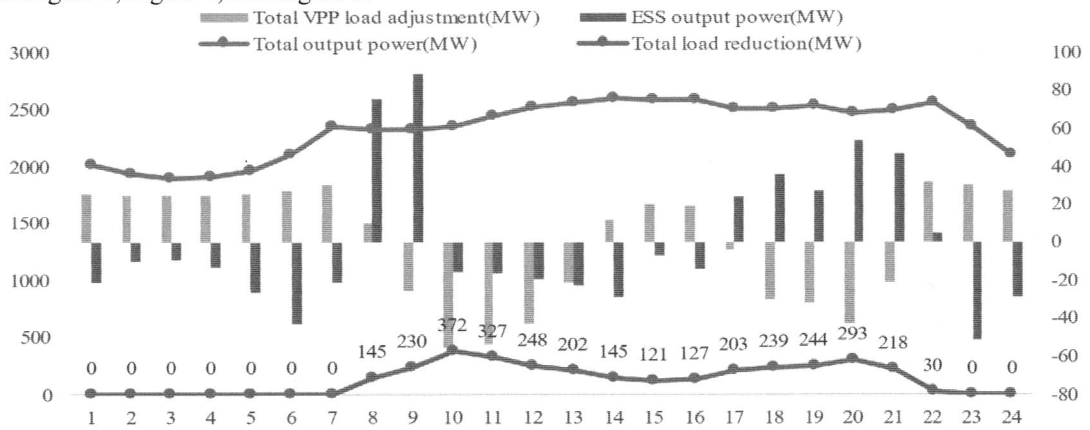

Figure 3. The output curve of Case 3 in fault scenario 1.

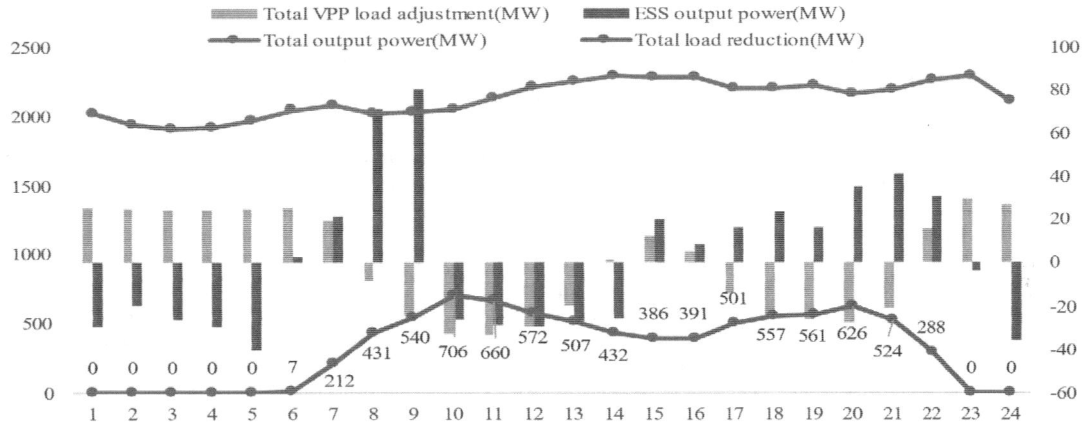

Figure 4. The output curve of Case 4 in fault scenario 3.

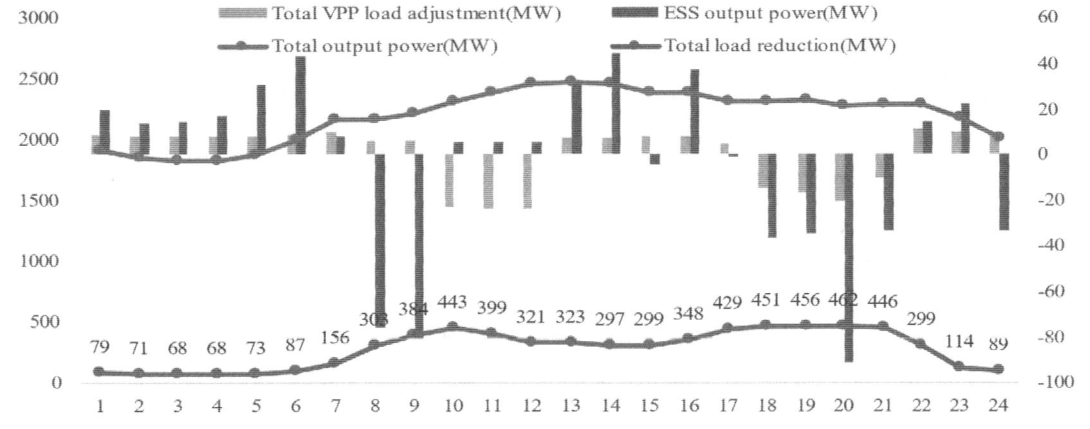

Figure 5. The output curve of Case 4 in fault scenario 2.

4. Conclusion

Aiming at the reliability evaluation of the AC/DC HPG with ESS and VPP, the influence of the AC/DC HPG with ESS and VPP on the reliability is studied. The following conclusions are obtained :

1) ESS and VPP can improve system regulation capabilities, reduce the EENS index, and enhance the system reliability.

2) Ignoring the AC/DC ICE and the multi-state model of components will have a huge impact on the system reliability evaluation results.

Acknowledgments

This work is supported by the Ningxia Natural Science Foundation (No. 2023AAC03839).

References

[1] R. Alajlan, M. M. Hafizur Rahman, M. Al-Naeem, et al. A Literature Review on Cybersecurity Risks and Challenges Assessments in Virtual Power Plants: Current Landscape and Future Research Directions[J]. IEEE Access, vol. 12, pp. 188813-188827, 2024, doi: 10.1109/ACCESS. 2024.3515635.

[2] A. A. Khan, A. F. Minai, R. K. Godi, et al. Optimal Sizing, Techno-Economic Feasibility and Reliability Analysis of Hybrid Renewable Energy System: A Systematic Review of Energy Storage Systems' Integration[J]. IEEE Access, doi: 10.1109/ACCESS.2025.3535520.

[3] Q. Zhao, L. Zhang, D. Nan, et al. Reliability Assessment Method for Power Grid Security and Stability Control Devices Based on Weibull Fault Rate Fitting[J]. IEEE Access, vol. 11, pp. 86351-86361, 2023, doi: 10.1109/ACCESS.2023.3303171.

[4] C. Pan, B. Hu, C. Shao, et al. Reliability-Constrained Economic Dispatch With Analytical Formulation of Operational Risk Evaluation[J]. IEEE Transactions on Power Systems, vol. 39, no. 2, pp. 4422-4436, March 2024, doi: 10.1109/TPWRS.2023.3317973.

[5] M. Draz, F. Pagel, S. Albayrak. Probabilistic Risk Assessment in Power Systems With High Wind Energy Penetration[J]. IEEE Access, vol. 12, pp. 140097-140111, 2024, doi: 10.1109/ACCESS.2024.3463882.

[6] Y. Bo, M. Bao, B. Yang, et al. A Semi-Markov Stochastic Model for Operational Reliability Assessment of Hybrid AC and LCC-VSC-Based DC System With Remote Wind Farms[J]. IEEE Transactions on Power Systems, vol. 39, no. 5, pp. 6154-6167, Sept. 2024, doi: 10.1109/TPWRS.2024.3357748.

[7] Z. Liu, L. Zhu, K. Hou, et al. A State-similarity-based Fast Reliability Assessment for Power Systems with Variations of Generation and Load[J]. IEEE Transactions on Power Systems, doi: 10.1109/TPWRS.2024.3415367.

[8] W. Huang, B. Hu, C. Shao, et al. Power System Reliability Evaluation Based on Sequential Monte Carlo Simulation Considering Multiple Failure Modes of Components[J]. Journal of Modern Power Systems and Clean Energy, vol. 13, no. 1, pp. 202-214, January 2025, doi: 10.35833/ MPCE.2023.000939.

[9] C. Lin, B. Hu, J. Lin, et al. Optimal Information Flow-Based Reliability Evaluation for Virtual Power Plants[C]. 2024 18th International Conference on Probabilistic Methods Applied to Power Systems (PMAPS), Auckland, New Zealand, 2024, pp. 1-6, doi: 10.1109/PMAPS61648.2024.10667315.

[10] X. Li, K. Xie, C. Shao, et al. A Sensitivity-Based Rolling Operational Reliability Evaluation Method Following the Changing Available Generation Capacity of Renewable Energies[J]. IEEE Transactions on Power Systems, vol. 40, no. 1, pp. 166-177, Jan. 2025, doi: 10.1109/TPWRS. 2024.3379646.

Journal of Physics: Conference Series **3011** (2025) 012051 doi:10.1088/1742-6596/3011/1/012051

Simulation study on high purity 3-methylpyridine produced from side stream of dividing wall column

Tianzhu Yu, Sixuan Qi, Hongjiao Liu, Zhihua Xu*, Mei Jin*

Jianghan University, Wuhan 430056, China

*Corresponding author's e-mail: xuzhihua78@sina.com (Zhihua Xu), jinmay@jhun.edu.cn (Mei Jin)

Abstract. Considering the high energy consumption problem in the conventional distillation separation of the traditional process for preparing 3-methylpyridine by the hydrogenation of 2-methylglutaronitrile, this paper proposes a new purification process using a dividing wall column. Under the premise of meeting the product purity, with the equipment energy consumption as the optimization target, this paper employs Aspen Plus V14.0 software to simulate and optimize the conventional distillation and continuous dividing wall column separation process. The simulation results show that compared with the conventional distillation separation process, the new dividing wall column process can save 21.02% of the condenser heat load, 26.34% of the reboiler heat load, and 23.71% of the comprehensive energy consumption.

1. Introduction

As the production capacity of adiponitrile via direct hydrocyanation of butadiene increases, the resource utilization of the by-product 2-methylglutaronitrile is of great importance for the economic viability of adiponitrile project industrialization [1]. 3-Methylpyridine (C_6H_7N) serves as a vital pharmaceutical raw material or intermediate. It is used in the manufacture of vitamin B_6, nicotinamide, nicotine, and cardiac drugs [2]. Additionally, it is frequently applied as a solvent, a raw material for synthesizing dyes and resins, for manufacturing pesticides, as a component of waterproofing agents, and as an additive for film photosensitive agents. 3-Methylpyridine can be produced using 2-methylglutaronitrile and hydrogen as reactants with Pd/Al_2O_3 or Pt/SiO_2 as catalysts through a series of reactions such as hydrogenation, cyclization, dehydrogenation, and refining processes, which presents an effective approach to recycle by-products in the adiponitrile production process.

To acquire 3-methylpyridine that meets the criteria of a top-grade product (99.0 wt%), the double-tower distillation technique is commonly employed in traditional processing. Nevertheless, the conventional double-tower distillation method has the problem of high energy consumption. The dividing wall column represents a distinctive type of distillation column [3-4]. By inserting a vertical partition within the column, the interior space of the conventional distillation column is divided into two sections, enabling the mass transfer and heat transfer processes to be carried out simultaneously within a single column. Therefore, a dividing wall column can achieve a separation effect equivalent to that of multiple conventional distillation columns connected in series, especially when it comes to the separation of ternary or multi-component mixtures containing azeotropic or near-boiling components. Consequently, compared with traditional distillation separation technologies, the dividing wall column separation technology has the potential to cut down energy consumption by 10%-60% and reduce equipment investment by 10%-50%, thereby meeting the demands of sustainable development. As a result, it has become a research focus in the field of energy conservation and emission reduction in the chemical industry.

The dividing wall column separation technology can be traced back to the thermally coupled technology first proposed by Petlyuk in 1965 [5]. In 1985, BASF constructed the first divided wall column with welded partitions using thermally coupled technology, marking the transition of this technology from theoretical research to industrial application. In recent years, some design institutes, research institutions, and universities [6-9], both domestically and internationally, have been dedicated to studying parameter optimization and control design of dividing wall columns in production processes, providing crucial technical support for the stable operation and performance enhancement of divided wall columns. In this paper, the Aspen Plus V14.0 software is employed to simulate and optimize both the conventional distillation separation technique and the dividing wall column separation technique for 3-methylpyridine. This study provides a feasible solution to reduce energy consumption and equipment investment in the refining process of 3-methylpyridine production.

2. Process simulation

2.1. Basic simulation conditions

For an enterprise with an annual 3-methylpyridine output of 20,000 tons, the composition and boiling points of the raw materials entering the separation and purification section are presented in Table 1. The design objective is to produce 3-methylpyridine of superior grade (99.0 wt%) as specified in the Q/370700WXL001-2021 standard (Chinese).

Table 1. Composition and boiling points of feeds.

Component Name	Mass flow rate (kg h^{-1})	Mass fraction (%)	Boiling Point (°C)
2-methylglutaronitrile (MGN)	188.86	6.5	263.44
3-Methylpiperidine (MPI)	42.97	1.5	125.48
3-methylpyridine (PIC)	2620.20	89.7	144.15
2-Methylpentanediamine (MDMP)	68.23	2.3	189.46

Within the separation system, MGN, MPI, PIC, and MDMP all pertain to polar non-ideal systems. Consequently, the NRTL-RK equation is employed as the thermodynamic model during the simulation calculation.

2.2. Establishment of Simulation Procedure

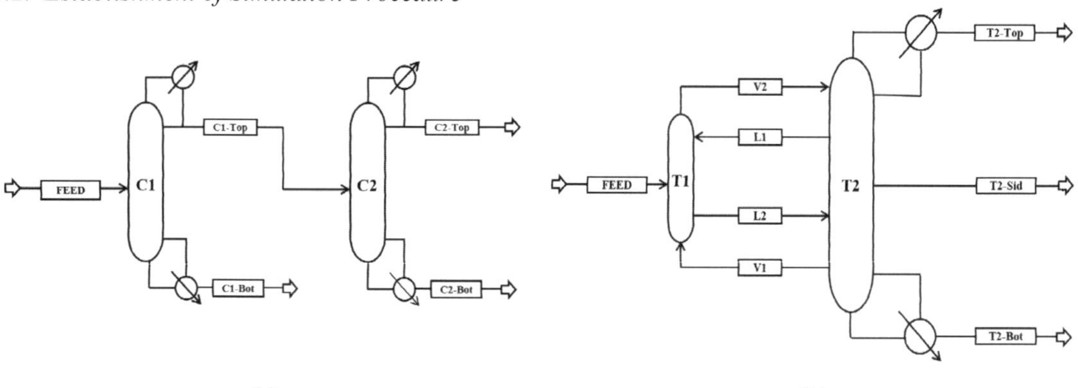

(a) (b)

Figure 1. Conventional distillation and dividing wall column distillation simulation procedure.

Based on the process flow of the distillation column in series used for the separation and purification of 3-methylpyridine in the industrial production process, the double-column simulation process illustrated in Figure 1(a) is established in Aspen Plus V14.0 software. In Figure 1(a), the mixture presented in Table 1 is separated by the de-heavy column C1 into a mixture of MPI and PIC at the top of the column and a mixture of MDMP and MGN at the bottom of the column. Subsequently, the C1-

Top mixture is separated by the de-light column C2, with MPI separated from the top of the column and product PIC from the bottom. Figure 1(b) shows the simulation flow of the dividing wall column distillation process. Here, T1 acts as the side column, while T2 acts as the main column. The mixed feed undergoes pre-separation within the side column T1. MPI is drawn off from the top of the T2 column (designated as T2-Top). A mixture of MDMP and MGN is obtained from the bottom of the T2 column (referred to as T2-Bot). The target product PIC is withdrawn from the side stream of the T2 column (denoted as T2-Sid).

3. Simulation and optimization of dividing wall tower-single tower model

During the process research, the operating parameters of the dividing wall column are set as follows: the operating pressure of the main column T1 is 1.0 atm, and that of the secondary column T2 is 1.05 atm. The number of theoretical plates in the main column T1 is 40, and the number of theoretical plates in the secondary column T2 is 30. The mixed raw material is introduced onto the 8th theoretical tray of the secondary column. The molar reflux ratio of the main column is determined to be 78.89. For material transfer, the liquid phase is withdrawn from the 20th tray of the main column and introduced to the 1st tray of the secondary column at a flow rate of 30.6123 kmol/h. Meanwhile, the gas phase is extracted from the 34th tray of the main column and introduced to the 31st tray of the secondary column at a flow rate of 32.6531 kmol/h. The dividing wall is positioned at the 11th tray of the main column. The side-stream draw is positioned at the 33rd tray of the main column, with a flow rate of 27.6842 kmol/h.

Using the Aspen Plus V14.0 simulation software, an investigation is carried out on the impacts of factors including the feed position, side-stream draw position and flow rate, liquid and gas phase distribution, and reflux ratio of the dividing wall column on the separation efficiency of PIC and the comprehensive energy consumption, to obtain the optimized operating conditions of the dividing wall column process.

3.1. Analysis of the influence of feed position

When studying the separation process in the dividing wall column, the feed position plays a crucial role in determining the separation efficiency and energy consumption. Different feed positions can lead to variations in the distribution of components within the column and thus affect the mass transfer and separation performance. Through simulations using Aspen Plus V14.0, we can systematically analyze the effects of different feed positions of the secondary column on the separation of PIC. By adjusting the feed position in the simulation model while keeping other parameters constant, we can observe changes in product purity and energy consumption, which are demonstrated in Figure 2.

As can be seen from Figure 2, the changing trend of PIC separation purity is similar to that of comprehensive energy consumption. When the feed position of the secondary tower is between the 1st and 8th trays, the comprehensive energy consumption fluctuates around 1400 kW. When the feed position is at the 9th tray of the secondary tower, the comprehensive energy consumption drops significantly and then remains at around 1030 kW, representing a decrease of approximately 26%. Similarly, the PIC separation purity shows a slight upward trend as the number of trays increases from the 1st to the 8th, reaching a maximum of 99.5 wt% at the 8th tray. However, when the feed position is at the 9th tray of the secondary tower, the separation purity drops sharply and stabilizes at around 94 wt%. Taking all factors into consideration, although the comprehensive energy consumption can be reduced by about 26% after the 9th tray, the separation purity fails to meet the requirements for premium products (99.0 wt%). Therefore, the 8th tray of the secondary tower is selected as the feed position.

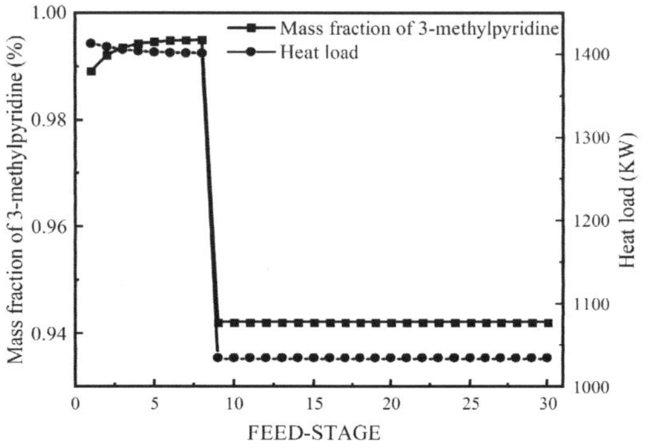

Figure 2. Influence of feed position on separation efficiency and heat load.

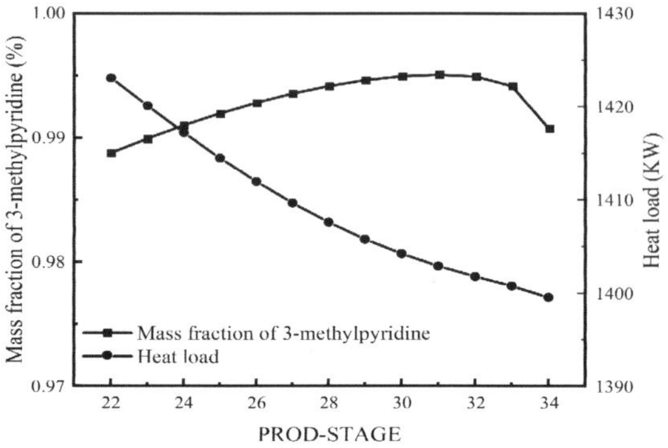

Figure 3. Influence of side-draw position on separation efficiency and heat load.

3.2. Analysis of the influence of side-draw position

In the distillation process, the side-draw position is a critical operating parameter that has a profound influence on the separation effect and the performance of the entire distillation process. The influence of the side-draw position on the separation effect of PIC and comprehensive energy consumption was investigated under other constant conditions, and the results are shown in Figure 3.

As can be observed from Figure 3, when the side-draw position changes from the 22nd to the 34th tray, the separation purity of PIC increases slowly to 99.5 wt% at the 31st tray as the number of trays increases and then gradually decreases to 99.0 wt% at the 34th tray. The comprehensive energy consumption decreases with the increasing number of trays. Considering the requirements of PIC separation purity and comprehensive energy consumption, and considering that there may be carbon deposition near the bottom of the column, the side-draw position should not be too low. Therefore, the 33rd tray is determined as the best side-draw position.

3.3. Analysis of the influence of side-draw flow rate

A reasonable side-draw flow rate can ensure the quality and quantity of PIC product output while reducing equipment energy consumption and operating costs, thereby improving the economic efficiency of the enterprise. Therefore, with other conditions remaining unchanged, the impact of the side-draw flow rate on the separation effect and heat load of the dividing wall column was studied, and the results are shown in Figure 4.

As shown in Figure 4, with the increase of the side-draw flow rate, the separation purity of PIC gradually decreases. When the side-draw flow rate is greater than 27.6842 kmol/h, the mass fraction of PIC no longer meets the requirements of superior-grade products. The reason is that an excessively high side-draw flow rate means a large amount of liquid is rapidly withdrawn, which affects the gas-liquid mass transfer process inside the tower. Then, some heavy components that should originally continue to be separated at the lower part of the tower are entrained into the side-draw stream. As a result, the content of heavy components in the side-line product increases, and the purity of the PIC product decreases. At the same time, if the side-draw flow rate exceeds 27.6842 kmol/h, the comprehensive heat load also increases from 1200 kW to 1420 kW. Specifically, this is mainly because an excessively high side-draw rate can affect the fluidity of materials and the mass-transfer and heat-transfer performance inside the tower. When the side-draw flow rate is less than 27.6842 kmol/h, such a situation will lead to a decrease in the annual output of the target product, thus affecting the economic efficiency of the process. Considering the above factors comprehensively, the side-draw flow rate is selected as 27.6842 kmol/h to ensure the quality requirements of superior grade and the annual output requirements of PIC. Moreover, it guarantees the stable operation and economic benefits of the tower.

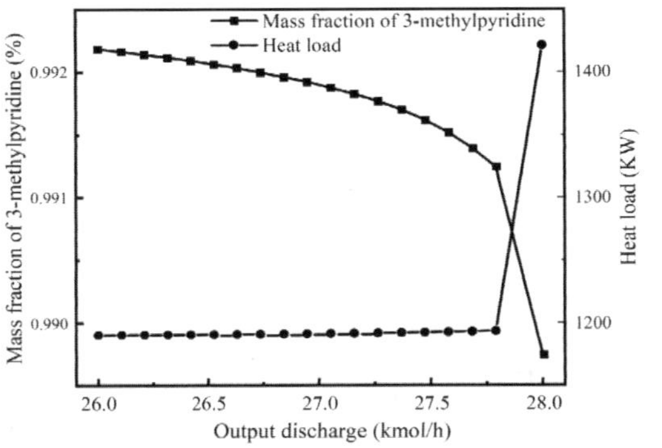

Figure 4. Influence of side-stream withdrawal flow rate on separation efficiency and heat load.

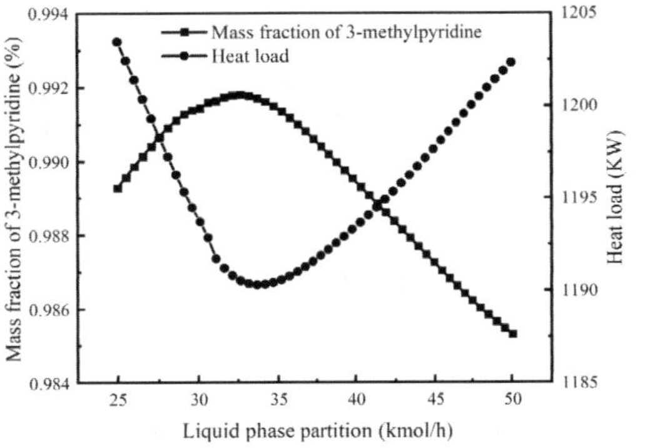

Figure 5. Influence of liquid distribution quantity on separation efficiency and heat load.

3.4. Analysis of the influence of liquid phase and gas phase distribution quantities

In this study, the liquid phase and gas phase distribution quantities, respectively, refer to the liquid phase and gas phase flow rates from a certain tray of the main column to the side column. By reasonably controlling the distribution of the liquid phase and the gas phase, it is possible to effectively balance the mass transfer resistance and energy consumption, which not only ensures product purity but also significantly reduces energy consumption. An appropriate distribution quantity of either the liquid phase or gas phase is a key factor in achieving efficient separation in the dividing wall column. With other conditions remaining constant, we investigate the influences of different liquid phase and gas phase distribution quantities on the separation efficiency and heat load of the dividing wall column, as shown in Figures 5 and 6, respectively.

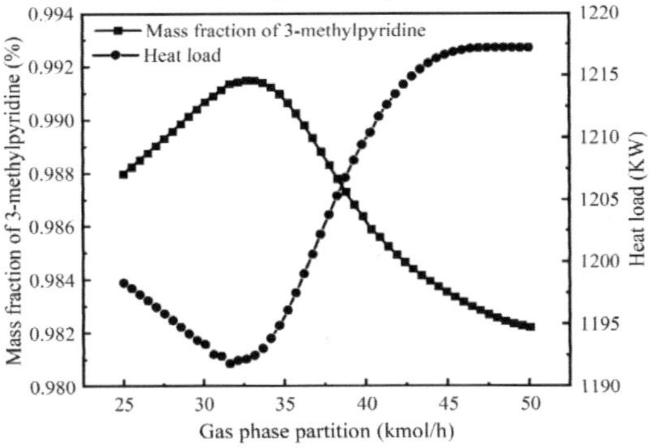

Figure 6. Influence of gas distribution quantity on separation efficiency and heat load.

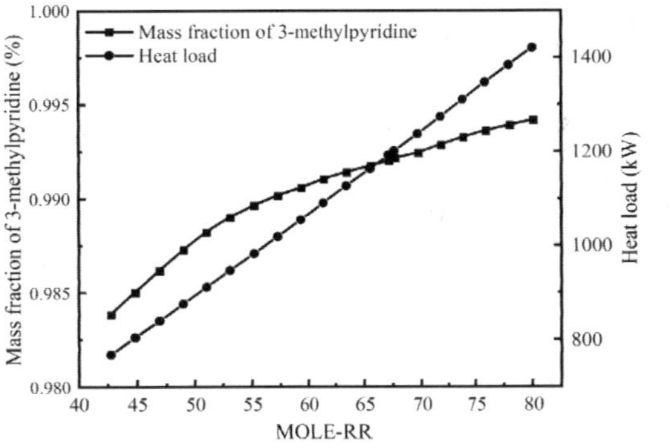

Figure 7. Influence of molar reflux ratio on separation efficiency and heat load.

3.5. Analysis of the influence of molar reflux ratio

During distillation operations, the molar reflux ratio is a crucial parameter, significantly influencing the tower's separation efficiency, operating load, and operating cost. Under the condition of constant other parameters, the effect of the molar reflux ratio on the separation efficiency and heat load of the dividing wall column is investigated, which is shown in Figure 7.

As can be seen from the curves in Figure 7, the mass fraction of PIC gradually increases with the increase of the molar reflux ratio. Meanwhile, the heat load of the tower also exhibits an upward tendency. An increase in the molar reflux ratio will make the gas-liquid mass transfer in the dividing wall column more intensive, thereby enhancing the separation efficiency of the whole dividing wall column and making the separation between different components more thorough. However, an excessively large molar reflux ratio may result in over-separation. Although the purity of the product will be improved, such an increase in the molar reflux ratio will raise energy consumption and operating costs and may cause excessive gas-liquid load in the tower, which may lead to flooding and other operational problems, affecting the stability and sustainability of the separation effect. Considering the balance between separation efficiency and energy consumption, when the reflux ratio is 67.0886, both economy and efficiency can be balanced to achieve optimal operation.

4. Energy consumption analysis and environmental assessment

4.1. Energy consumption analysis

Under the same mixed raw material conditions, with an annual operating time of 8000 hours, an energy consumption analysis was carried out on the production of 20,000 tons per year of high-purity PIC using a double-column model of conventional distillation process and a single-column model of a dividing wall column process. The results are shown in Table 2.

Table 2. Comparison of energy consumption of different distillation technologies.

Items	Unit	Conventional distillation		Dividing wall column		Cut down%[a]
		C1	C2	T1	T2	
Condenser duty	kW	588.16	183.77	-	609.63	21.02
Reboiler duty	kW	612.69	178.17	-	582.56	26.34
Total energy consumption	kW	1562.79		1192.19		23.71
Equipment costs	$	4938446.857		1807046.429		63.41

[a] Taking the conventional distillation as the benchmark.

As is shown in Table 2, compared with conventional distillation, the dividing wall column can save 21.02% of the condenser heat load and 26.34% of the reboiler heat load, saving approximately 23.71% of the total heat load. Meanwhile, the application of the dividing wall column process can directly reduce two heat exchangers, which are the condenser and reboiler of the light component distillation column and one tower equipment, thus significantly reducing equipment costs and saving floor space.

4.2. Environmental assessment

The environmental assessment can be carried out by calculating carbon emissions. Based on the national standard GB/T 3715-2007 (Chinese), taking the annual operating time as 8000 h, we use the total energy consumption to convert the amount of standard coal. On this basis, according to the data from Reference [10], where the standard coal emission value is 0.69 t C/t, and the carbon dioxide emission value is 2.54 t CO_2/t, the carbon emissions of the above two distillation technologies can be calculated as shown in Figure 8. It can be seen from Figure 8 that when the dividing wall column distillation technology is adopted, the carbon emissions (including carbon dioxide) decrease significantly. The percentage decline is like the reduction rate shown in Table 2. This result indicates that, compared with conventional distillation technology, the dividing wall column distillation is an environmentally friendly distillation technology.

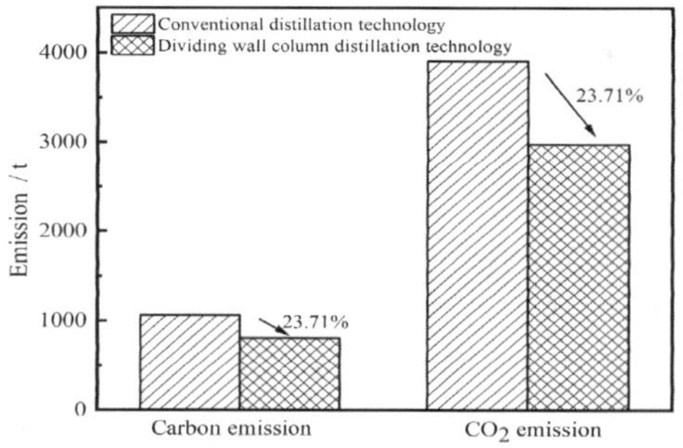

Figure 8. Carbon emission and carbon dioxide emission of different distillation technologies.

5. Conclusion

This paper uses Aspen Plus V14.0 simulation software to simulate and optimize the new separation technology of the dividing wall column in the 3-methylpyridine refining process of a certain factory. Under the same mixed raw material conditions and separation tasks as conventional distillation, the optimal process operating parameters of the dividing wall column separation process are determined through simulation calculations as follows: the main column has 40 theoretical plates, and the secondary column has 30 theoretical plates; the feed position is set at the 8th tray of the secondary column. The liquid phase is withdrawn from the 33rd tray of the main column to the 1st tray of the secondary column with a flow rate of 30.61225 kmol/h, while the vapor phase is withdrawn from the 34th tray of the main column to the 30th tray of the secondary column with a flow rate of 32.6531 kmol/h. The baffle position is set at the 11th tray, the side-draw position is set at the 33rd tray of the main column, and the side-draw flow rate is 27.6842 kmol/h. The molar reflux ratio of the main column is 67.0886. Under these optimal process operating conditions, 20,600 tons of high-purity 3-methylpyridine can be obtained per year. On this basis, the energy consumption, equipment costs, and carbon emissions of the double-column model of conventional distillation and the single-column model of the dividing wall column are compared. It is found that the use of the new dividing wall column separation technology can save 23.71% of energy consumption and reduce carbon emissions by 23.71. At the same time, it reduces equipment investment costs by 64.41%. The research results can provide a certain reference for the industrial application of the dividing wall column separation technology.

Acknowledgments

This work was financially supported by the Outstanding Youth Science and Technology Innovation Team Project for Colleges and Universities of Hubei Province (T2021036) Jianghan University Scientific Research Project (2021yb015).

References

[1] Zhou H, Qi Y, Wang J J and Chen Y 2023 *Aanhui Chem. Ind. (Chinese).* 49(4): 74-76+80.
[2] Liu M, Zhang Q, Zhou P P, Yuan X L, Wu L R, and Liu H J 2021 *Aanhui Chem. Ind.(Chinese).* 47(6): 103-104+109.
[3] Weinfeld J A, Owens S A and Eldridge R B 2018 *Chem. Eng. Process. Process Intensif.* 123: 20-33.
[4] Yildirim O, Kiss A A and Kenig E Y 2011 D *Sep. Purif. Technol.* 80(3): 403-417.
[5] Petlyuk F B, Platonov V M, Slavinskii D M 1965 *Int. Chem. Eng.* 5(3): 555-561.
[6] Rawlings E S, Chen Q, Grossmann I E and Caballero J A 2019 *Comput. Chem. Eng.* 125: 31-39.
[7] Nhien L C, Kim G, Long N V D and Lee M 2024 *Chem. Eng. Res. Des.* 208: 62-73.
[8] Liu J Y, Yan J L, Liu W S, Kong J, Wu Y, Li X N and Sun L Y 2022 *Sep. Purif. Technol.* 287: 120512.
[9] Wang X H, Xie L, Tian P and Tian G Z 2016 *Chem. Eng. Process. Process Intensif.* 110: 172-187.
[10] Tu H, Liu C J 2014 *Coal Quality Technology (Chinese).* 2: 57-60.

Journal of Physics: Conference Series **3011** (2025) 012052 doi:10.1088/1742-6596/3011/1/012052

A robust pricing strategy for community-integrated energy systems considering the uncertainties of distributed resources

Bo Gu, Dan Wang*, Chengxiong Mao

Department of Electrical and Electronic Engineering, Huazhong University of Science and Technology, Wuhan 430074, China

*Corresponding author's e-mail: wangdan@mail.hust.edu.cn

Abstract. Community-integrated energy systems are suitable for distributed resources to participate in improving the operational flexibility of the energy systems. A Stackelberg game-based robust pricing model is proposed to improve the economy considering multiple uncertainties from distributed resources. The uncertainties of the operation scenarios are also included in the model to further improve the robustness of systems. At last, an improved C&CG algorithm has been adopted to solve the model. The simulation results show that the proposed robust pricing model is effective in decreasing the operation cost of each entity in energy systems, and it also improves safety and stability.

1. Introduction

Large-scale grid integration of distributed resources has become the trend of future distribution network development. Community-integrated energy systems (CIESs) can improve the consumption capacity of distributed resources through optimal scheduling [1]. Therefore, the optimal scheduling methods of CIESs considering the uncertainties of distributed resources have been widely studied [2].

A Stackelberg game-based model considering the uncertainties of photovoltaic (PV) power and wind power output has been studied in [2], effectively reducing the cost of energy consumers. The three-stage robust optimal scheduling model considering the uncertain scenarios has been investigated in [3], in which the robustness of the energy systems is further improved. Also, the bilevel optimal method of a CIES considering the charging mode of electric vehicles (EVs) is introduced in [4], where the upper level aims to optimize the daily cost and the lower level aims to optimize the annual cost.

To build a bilevel robust coordination framework for CIESs, we propose a robust pricing model for CIESs with access to EVs, hydrogen fuel cell electric vehicles (HFCEVs), and community energy storage (CES). The robust coordination framework will minimize the operation cost of each entity in the energy systems and improve flexibility and safety.

2. Bilevel robust pricing model

The structure of a CIES considering the integration of HFCEVs, EVs and CES is shown in Fig. 1. The proposed system consists of four participating entities and we use the Stackelberg game-based coordination method to model the interactive relationship of this system, where the CIES is the upper-level leader, and the hydrogen refuelling system (HRS), electric vehicles charging station (EVCS) and CES are the three lower-level followers. CIES system mainly consists of PV, wind turbine (WT), battery energy storage system (BESS), and electric load (EL). CIES can purchase gas from the external gas grid and connect to various energy conversion facilities through the natural gas network, including combined heat and power (CHP), gas boiler (GB), carbon capture power plant (CCPP), power-to-gas (P2G), and heat load (HL).

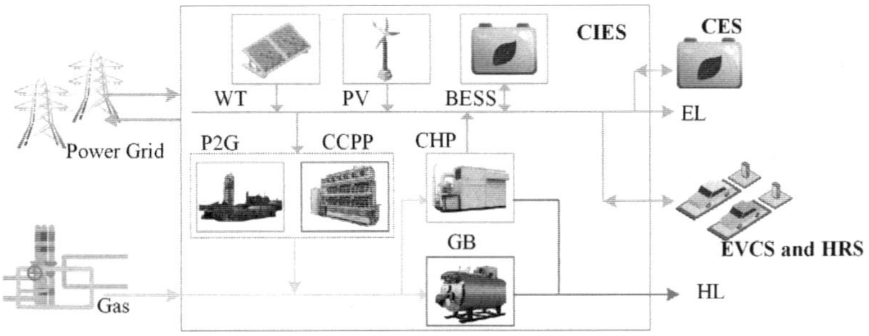

Figure 1. The proposed energy systems.

2.1. CIES modelling

The optimization model of CIES can be expressed as a two-stage robust optimization problem.

$$\min_{x\in\mathbf{X}}\left\{\sum_{t=1}^{T}\left(\lambda_{buy}^{t}P_{buy}^{t}-\lambda_{sell}^{t}P_{sell}^{t}-C_{hrs}^{t}-C_{ces}^{t}\right)+\max_{\sigma_s\in\Omega}\max_{u\in\mathbf{U}}\min_{y\in\mathbf{Y}(x,u)}\sum_{s=1}^{K}\sum_{t=1}^{T}\sigma_s(C_{gas}^{t}+C_{es}^{t}+C_{ct}^{t}-C_{ev}^{t})\right\} \quad (1)$$

where λ_{buy}^{t} and λ_{sell}^{t} denote the electric price (EP) and the feed-in tariff. P_{buy}^{t} and P_{sell}^{t} denote the buying/selling energy of the CIES from/to the power grid. C_{gas}^{t}, C_{es}^{t} and C_{ct}^{t} denote the cost of gas, the depreciation cost of BESS and the laddered carbon trading costs, respectively. C_{hrs}^{t}, C_{ces}^{t}, and C_{ev}^{t} denote the revenue from HRS, CES and EVCS. σ_s denotes the probability distribution of scenarios.

$$C_{gas}^{t}=\lambda_{gas}(Q_{CHP}^{t}+Q_{GB}^{t}-Q_{P2G}^{t})\Delta t \quad (2)$$

$$C_{es}^{t}=\varepsilon_{es}(P_{char}^{t}+P_{dis}^{t})\Delta t \quad (3)$$

$$C_{ct}^{t}=\varepsilon_{ct}^{1}E_{ct}^{1}+\varepsilon_{ct}^{2}E_{ct}^{2}+\varepsilon_{ct}^{3}E_{ct}^{3} \quad (4)$$

$$C_{hrs}^{t}=\lambda_{hrs}^{t}(P_{hrs,b}^{t}-P_{hrs,s}^{t}) \quad (5)$$

$$C_{ces}^{t}=\lambda_{ces}^{t}(P_{ces,b}^{t}-P_{ces,s}^{t}) \quad (6)$$

$$C_{ev}^{t}=N_{ev}\sigma_{ev}\lambda_{ev}^{t}(P_{ev,char}^{t}-P_{ev,dis}^{t}) \quad (7)$$

where λ_{gas} denotes the gas price. Q_{CHP}^{t}, Q_{GB}^{t} and Q_{P2G}^{t} denote the gas consumption of CHP and GB and the gas generation of P2G. ε_{es} denotes the coefficient of depreciation cost. ε_{ct}^{1}, ε_{ct}^{2} and ε_{ct}^{3} denote the cost coefficient of carbon emission allowance. λ_{hrs}^{t}, λ_{ces}^{t} and λ_{ev}^{t} denote the EPs set by CIES to lead the buying/selling energy scheduling from the HRS, CES and EVCS. N_{ev} denotes the average daily number of EVs charging/discharging at an EVCS. Due to the large number of EVs in the energy system, it is difficult to construct the uncertainties set for each EV. The distributionally robust optimization (DRO) method is used to deal with the uncertainties of EVs to aggregate the discrete EV data into K_{ev} classes by historical data. σ_{ev} denotes the number distribution of EVs. Here, the operation constraint sets $\boldsymbol{X}$ and $\boldsymbol{Y}$ are shown in [3]. Normally, to protect the benefits of the HRS, CES and EVCS, and stimulate the participation of HFCEV users and EV users in the regulations of economic operation of energy systems, it is required to limit λ_{hrs}^{t}, λ_{ces}^{t} and λ_{ev}^{t}, where the constraints are formulated as

$$\begin{cases} \lambda^t = [\lambda^t_{hrs}, \lambda^t_{ces}, \lambda^t_{ev}] \\ \sum_t \lambda^t / T \le \sum_t \lambda^t_{buy} / T \\ \lambda^t_{\min} \le \lambda^t \le \lambda^t_{\max} \end{cases} \tag{8}$$

where the equations denote that the average charging price set by CIES cannot be higher than the EPs. The class probability distribution of EVs is calculated as

$$\Omega_{ev} = \begin{Bmatrix} \sigma_{ev} \ge 0, N = 1,2,\ldots,K_{ev} \\ \sum_s \sigma_{ev} = 1 \\ \left\| \sigma_{ev} - \sigma^0_{ev} \right\|_1 \le \theta_1 \\ \left\| \sigma_{ev} - \sigma^0_{ev} \right\|_\infty \le \theta_\infty \end{Bmatrix} \tag{9}$$

where σ^0_{ev} denotes the EV classes' empirical distribution by clustering method. θ_1 and θ_∞ denote norm-1 and norm-inf probability distribution confidence levels, respectively. Set $\boldsymbol{U}$ contains PV, WT output, EL and HL uncertainties. In this paper, the boxed uncertainty set is used in the third-stage optimization problem to represent the uncertainty in a single scenario.

$$\boldsymbol{U}_{CIES} = \begin{Bmatrix} \xi^t = \left[P^t_{PV}, P^t_{WT}, P^t_{load}, H^t_{load} \right] \\ \xi^t = \tilde{\xi}^t + u^t_+ \Delta\xi - u^t_- \Delta\xi \\ u^t_+ + u^t_- \le 1, \sum_t u^t_+ + u^t_- \le \tau \end{Bmatrix} \tag{10}$$

2.2. HRS modelling

The objective function of the HRS is to minimize the electricity cost on the scheduling day.

$$\{\boldsymbol{P}^t_{hev}\} = \arg\min \sum_{t=1}^{T} [\lambda^t_{hrs}(P^t_{hrs,b} - P^t_{hrs,s}) + C^t_{hst}] \tag{11}$$

where C^t_{hst} denotes the operation risk costs of the hydrogen storage tank (HST).

$$C^t_{hst} = \varepsilon_{hst}(\tilde{S}^t_{hydro} - S_{ic})^2 \tag{12}$$

subject to:

$$H^t_{hydro} = \eta_{hydro} P^t_{p2h} \qquad P_{p2h,\min} \le P^t_{p2h} \le P_{p2h,\max} \tag{13}$$

$$S^t_{hydro} = S^{t-1}_{hydro} + H^t_{p2h} - \tilde{H}^t_{hl} \qquad S^{t,0}_{hydro} \le S^{t,24}_{hydro} \tag{14}$$

$$0 \le P^t_{hrs,b} \le P^t_{hrs,b,\max} \qquad 0 \le P^t_{hrs,s} \le P^t_{hrs,s,\max} \tag{15}$$

where Equation (13) indicates the power-to-hydrogen (P2H) power constraints and corresponding conversion relationships. Equation (14) indicates the capacity constraints of HST, which defines the dynamic capacity model including the uncertainties from HFCEVs $\tilde{H}^t_{hl}$. Equation (15) represents the buying/selling power constraints. Here, we use a distributionally robust chance-constrained (DRCC) model to equally linearize the model based on [5]. The dynamic capacity model is formulated as equation (16). Here, the capacity of the HST is an uncertain parameter due to the uncertainty of the HFCEVs.

$$S^t_{hydro}(\omega^t_{hl}) = S^{t-1}_{hydro} + H^t_{p2h} - H^t_{hl}(\omega^t_{hl}) \tag{16}$$

In equation (16), ω_{hl}^{t} is the random variable from the refuelling behaviours $H_{hl}^{t}(\omega_{hl}^{t})$. Here, we can formulate the uncertain refuelling behaviours by ω_{hl}^{t}.

$$H_{hl}^{t}=\tilde{H}_{hl}^{t}\times(1+\omega_{hl}^{t}) \tag{17}$$

We further consider the ambiguity set defined by first and second moments as

$$\mathcal{P}=\left\{\mathbb{P}\in\mathcal{P}_0\left(\mathbb{R}^{|\mathcal{H}|}\right):\mathbb{E}_{\mathbb{P}}[\omega_{hl}^{t}]=0,\mathbb{E}_{\mathbb{P}}\left[\omega_{hl}^{t}(\omega_{hl}^{t})^{\top}\right]=\Sigma\right\} \tag{18}$$

Here, we build Σ with the variance and covariance of random variable ω_{hl}^{t}.

To constrain the HCT capacity to vary in a safe range, we use the DRCC model to formulate the dynamic capacity model.

$$\inf_{\mathbb{P}\in\mathcal{P}}\mathbb{P}\left[\left|S_{hydro}^{t}(\omega_{hl}^{t})\text{-}S_{ic}\right|\leq\Delta S_{hydro}\right]\geq 1-\epsilon_{hl} \tag{19}$$

where S_{ic} is the ideal safe capacity of the HST. We use the DRCC model to build the capacity deviations to ensure the safe capacity of the HST.

Equation (19) can be further converted in the deterministic form to be conveniently embedded into the optimization model. Firstly, we use the variable $\tilde{H}_{hl}^{t}$ to represent the predicted refuelling power of HFCEVs without ω_{hl}^{t}.

$$\tilde{S}_{hydro}^{t}=\tilde{S}_{hydro}^{t-1}+H_{p2h}^{t}-\tilde{H}_{hl}^{t} \tag{20}$$

Theorem 1: Based on the dynamic capacity DRCC model of HSTs in (19), (19) can be further formulated as

$$\inf_{\mathbb{P}\in\mathcal{P}}\mathbb{P}\left[\left|\tilde{S}_{hydro}^{t}\text{-}(\mathcal{A}_{h}\circ\tilde{H}_{hl}^{t})\omega_{hl}^{tT}\text{-}S_{ic}\right|\leq\Delta S_{hydro}\right]\geq 1-\epsilon_{hl} \tag{21}$$

where $\mathcal{A}_{h}$ is the constant matrix. $\circ$ represents the Hadamard product of two matrixes.

$$\mathcal{A}_{h}=\left[\underbrace{1\ 1\ \cdots 1}_{h}\ \underbrace{0\cdots 0}_{\mathcal{H}-h}\right] \tag{22}$$

Theorem 2: (21) can be further linearized based on the exact reformulation shown in [6], using the ambiguity set (18).

$$-\Delta S_{hydro}-\Delta S_{hydro,aux}^{t}\leq\tilde{S}_{hydro}^{t}-S_{ic}\leq\Delta S_{hydro}+\Delta S_{hydro,aux}^{t} \tag{23}$$

where $\Delta S_{hydro,aux}^{t}$ is the auxiliary variable for linearization.

$$\Delta S_{hydro,aux}^{t}=-\sqrt{\tfrac{1-\epsilon_{hl}}{\epsilon_{hl}}}\sqrt{(\mathcal{A}_{h}\circ\tilde{H}_{hl}^{t})^{\mathrm{T}}\Sigma(\mathcal{A}_{h}\circ\tilde{H}_{hl}^{t})} \tag{24}$$

Here, the linear constraints in (24) are faster to solve the DRCC optimization model.

2.3. EVCS modelling

The objective function of the EVCS is to minimize the electricity cost on the scheduling day.

$$\{\boldsymbol{P}_{ev}^{t}\}=\arg\max\sum_{t=1}^{T}N_{ev}\sigma_{ev}[\lambda_{ut}^{t}(\eta_{ev,char}P_{ev,char}^{t}-\eta_{ev,dis}P_{ev,dis}^{t})-\lambda_{ev}^{t}(P_{ev,char}^{t}-P_{ev,dis}^{t})] \tag{25}$$

subject to:

$$\begin{cases} 0 \le P_{char,n}^{t}, P_{dis,n}^{t} \le P_{char,n,\max}, P_{dis,n,\max} \\ P_{char,n}^{t} \cdot P_{dis,n}^{t} = 0 \\ SoC_{n,\min} \le SoC_{n}^{t} \le SoC_{n,\max} \\ SoC_{n}^{t} = SoC_{n}^{t-1} + (\eta_{char,n} P_{char,n}^{t} - \eta_{dis,n} P_{dis,n}^{t}) \Delta t \\ SoC_{n}^{t,end} = SoC_{n,obj} \end{cases} \tag{26}$$

where λ_{ut}^{t} denotes the coefficient of the utility. $P_{char,n}^{t}$ and $P_{dis,n}^{t}$ denote the charging/discharging power of the class n EVs. SoC_{n}^{t} denotes the SoC of EVs. The method to address the uncertainties from the EVCS has been mentioned before.

2.4. CES modelling

The objective function of the CES is to minimize the electricity cost in the scheduling day.

$$\{P_{ces}^{t}\} = \arg\min \sum_{t=1}^{T} [\lambda_{ces}^{t} (P_{ces,b}^{t} - P_{ces,s}^{t}) + \varepsilon_{ces} (p_{ces,char}^{t} + p_{ces,dis}^{t})] \tag{27}$$

where ε_{ces} denotes the coefficient of the depreciation cost. The constraints of CES are similar to the BESSs.

3. Model solutions

The compact forms of Equation (1) are shown as

$$\min_{x \in X} \left[c^{T} x - a^{T} w + \max_{\sigma_s \in \Omega} \max_{u \in U} \min_{y \in Y(x,u)} \sum \sigma_s \left(d^{T} y - b^{T} z \right) \right] \tag{28}$$

subject to:

$$\begin{cases} X = \{ Ax \le e \} \\ Y = \{ Cx + Dy + Fz + Eu \le g \} \\ U = \{ Hu \le k \} \\ \Omega = \{ M\sigma_s \le n \} \end{cases} \tag{29}$$

The compact forms of Equations (11), (25) and (27) are shown as

$$\{z\} = \arg\max(e^{T} z - b^{T} z) \;\; s.t. \;\; Gz \ge h : \mu_1 \tag{30}$$

$$\{w\} = \arg\min(a^{T} w + g^{T} z) \;\; s.t. \;\; Kw \le e : \mu_2 \tag{31}$$

The bilevel optimal problem can be reformulated into the single-level TSRO problem using KKT conditions. The bilinear terms can be replaced by dual variables based on the strong duality theorem.

Equation (28) is a multi-stage robust optimization model, which can be solved iteratively after decomposing the original model into the master problem and subproblems. The compact forms of the master problem are shown as

$$\min_{x, y^{n}, \eta, \lambda_{ev}} \quad c^{T} x + \eta \tag{32}$$

subject to:

$$\begin{cases} \eta \geq \sum_s \sigma_s^* \left(d^T y^n + a^T z + \mu^T h \right) \\ Cx + Dy^n + Fz + Eu \leq g^n, \forall n \in N_{iter} \\ G\lambda_{ev} \leq e \end{cases} \tag{33}$$

The compact forms of the subproblems are shown as

$$\max_{\sigma_s \in \Omega} \max_{u \in U} \min_{y \in Y(x,u)} \sum \sigma_s \left(d^T y - b^T z \right) \tag{34}$$

subject to:

$$\begin{cases} Cx^* + Dy + Fz + Eu \leq g \\ Hu \leq k \\ M\sigma_s \leq n \end{cases} \tag{35}$$

where Equation (34) can be further decomposed into two subproblems to be solved: σ_s-fixed subproblem and u-fixed subproblem. σ_s-fixed subproblem is shown as

$$\max_{u \in U} \min_{y \in Y(x,u)} \sum \sigma_s^* \left(d^T y - b^T z \right) \tag{36}$$

subject to:

$$\begin{cases} Cx^* + Dy + Fz + Eu \leq g \\ Hu \leq k \end{cases} \tag{37}$$

The u-fixed sub-problem is shown as

$$\max_{\sigma_s \in \Omega} \min_{y \in Y(x,u)} \sum \sigma_s \left(d^T y - b^T z \right) \tag{38}$$

$$\begin{cases} Cx^* + Dy + Fz + Eu^* \leq g \\ M\sigma_s \leq n \end{cases} \tag{39}$$

where u^* is the constant. Based on the improved C&CG algorithm [3], the subproblems stop being solved iteratively when the results of the σ_s-fixed subproblem are equal to the u-fixed subproblems.

4. Simulations results

In this paper, the number of the scheduling scenarios is set to 5, and also for the classes of EVs. The number of EVs is 50. The buying/selling electricity prices of the power grid are from 0.2 ￥/kWh to 0.627￥/kWh. The gas price of the gas network is set to 2.5￥/m^3. Some other parameters are shown in [7]. Here, we set three scheduling cases listed below to demonstrate the performance of the proposed optimization framework with other methods: 1. without the robust coordinated pricing strategy. 2. with the fixed pricing strategy. 3. with the proposed coordinated pricing strategy proposed. The models are coded in MATLAB and solved by the Gurobi solver.

Fig. 2 demonstrates the impact of the proposed robust pricing strategy on the operation costs of the CIES, HRS, EVCS and CES based on the three cases. We set five price limitations λ_{buy}^t in (8) to present the interactions between each entity. It shows that the operation cost of the CIES under Case 3 is lower than those under Case 1 and Case 2. Thus, the HRS, EVCS and CES have provided more operation flexibility to the CIES. Here, the social welfare of Case 3 is almost the same as the results under Case 1, which is the ideal operation strategy. The improvement results under different prices are shown in Table 1.

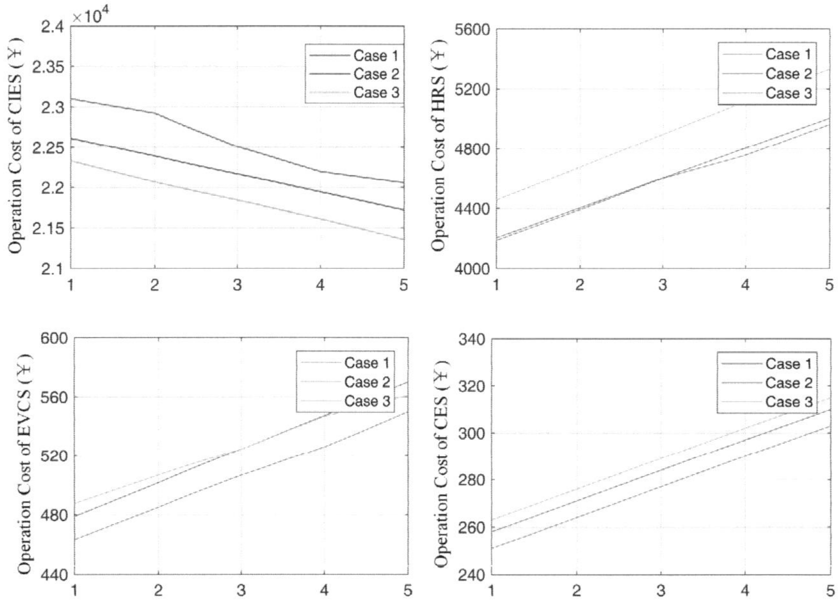

Figure 2. The operation costs of CIES, HRS, EVCS, and CES of energy systems.

Table 1. Costs comparison results (¥).

	1	2	3	4	5
Case 2	27734.2	27761.5	27591.9	27444.6	27536.5
Case 3 (Proposed)	27250.1	27226.7	27236.3	27248.4	27228.9
Percentage	1.74%	1.96%	1.30%	0.72%	1.13%

The changes in the capacity of the HST are shown in Fig. 3. In Fig. 3, the capacity fluctuations under Case 3 are similar to those under Case 1, and the fluctuations under Case 2 are smaller. Thus, the HCT utilization under Case 3 is higher than that under Case 2. It means that the HRS is more involved in the optimal scheduling to provide more operation flexibility to the CIES. Meanwhile, the capacity of the HST is kept in a safe range due to the capacity safety constraints, where the remaining capacity can represent reserve capacity to improve the robustness of the HRS. The capacity fluctuations without uncertainties are compared in the right picture. Here, the proposed strategy can dynamically adjust the safe capacity range of the HST to improve the operation robustness of the HRS.

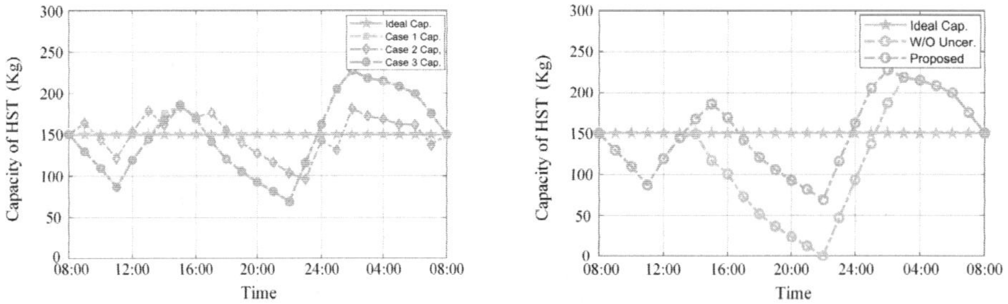

Figure 3. The capacity of HST with and without considering the uncertainties.

Fig. 4 shows the results of the operation scenarios probability distribution. Here, the probability of the Scenario 1 did not change. Scenario 2 and Scenario 4 are higher than the initial. Scenario 3 and Scenario 5 are lower than the initial. It means that Scenario 3 and Scenario 5 are two relatively worse operation scenarios for the proposed energy system, where the operation costs are higher than other scenarios. Fig. 4 proves that the proposed three-level robust pricing model has better robustness with considering the uncertainties of the scenarios, which avoids over-optimism or over-conservatism in a single operation scenario. The variation of the probability distribution in Fig. 4 is affected by the norm-1 and norm-∞ constraints described in the Equation (9).

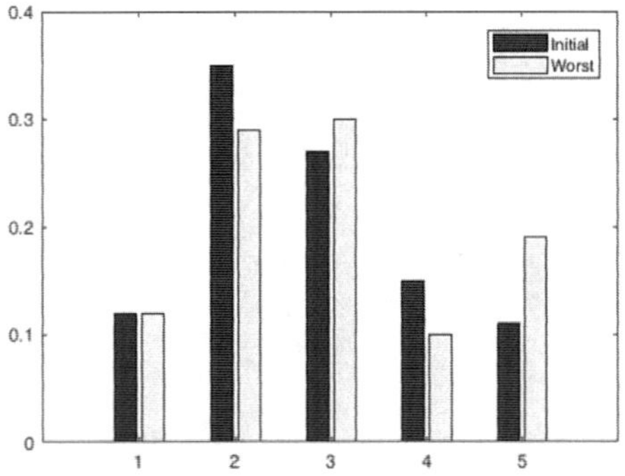

Figure 4. Initial and the worst scenarios probability distribution.

To demonstrate the robustness of the proposed coordinated pricing strategy, we first set 5 cases to compare the robustness under different operation uncertainties shown in Table 2. As the results shown in Table 3, the social welfare decreases without considering the uncertainties, to be more conservative to improve the robustness of the energy systems.

Table 2. Cases setups.

Cases	Uncertainties of CIES		Uncertainties of HRS	Uncertainties of EVCS
	DGs	EL+HL	HFCEVs	Num. of EVs
Case 1	×	√	√	√
Case 2	√	×	√	√
Case 3	√	√	×	√
Case 4	√	√	√	×
Case 5 (Proposed)	√	√	√	√

Table 3. Social welfare under different cases (￥).

	Bud. of DGs Uncer.			Bud. of EL+HL Uncer.			Uncer. of HyL	α^{ev} of EVs Uncer.		
	2/3 σ^{DG}	1/3 σ^{DG}	w/o	2/3 σ^{L}	1/3 σ^{L}	w/o	w/o	0.95	0.9	w/o
Social Welfare	27223.8	27125.3	27021.1	27003.5	26587.3	25091.6	26891.4	27012.6	27000.4	26985.3

5. Conclusion

Based on the results and discussions presented above, the conclusions are obtained as below:

(1) Compared with the traditional two-stage robust optimization model, this paper considers the uncertainty from HRS and EVCS, which further improves the robustness of the system operation.

(2) The proposed robust pricing model can achieve a lower operation cost compared with the cost under the fixed pricing model. Meanwhile, it ensures the privacy and safety of each entity.

(3) The adopted C&CG algorithm ensures fast convergence of the algorithm by decomposing the robust optimization problem of CIES as the master problem and subproblems.

References

[1] Hou, H., Wang, Z., Zhao, B., Zhang, L., Shi, Y., Xie, C., ... & Yu, K. (2024). Distributed optimization for joint peer-to-peer electricity and carbon trading among multi-energy microgrids considering renewable generation uncertainty. Energy Conversion and Economics, 5(2), 116-131.
[2] Li, Y., Wang, C., Li, G., & Chen, C. (2021). Optimal scheduling of integrated demand response-enabled integrated energy systems with uncertain renewable generations: A Stackelberg game approach. Energy Conversion and Management, 235, 113996.
[3] Zhong, J., Li, Y., Cao, Y., Tan, Y., Peng, Y., Zhou, Y., ... & Li, Z. (2022). Robust coordinated optimization with adaptive uncertainty set for a multi-energy microgrid. IEEE Transactions on Sustainable Energy, 14(1), 111-124.
[4] Zhang, C., & Kuang, Y. (2023). Low-carbon economy optimization of integrated energy system considering electric vehicles charging mode and multi-energy coupling. IEEE Transactions on Power Systems, 39(2), 3649-3660.
[5] Zymler, S., Kuhn, D., & Rustem, B. (2013). Distributionally robust joint chance constraints with second-order moment information. Mathematical Programming, 137, 167-198.
[6] Xie, W., & Ahmed, S. (2017). Distributionally robust chance constrained optimal power flow with renewables: A conic reformulation. IEEE Transactions on Power Systems, 33(2), 1860-1867.
[7] "data_BRPM," 2024. [Online]. Available: https://liveutk-my.sharepoint.com/:x:/r/personal/bgu1_utk_edu/Documents/data_BRPM.xlsx?d=wacf852c901de4481aa6b33303d81b517&csf=1&web=1&e=fgu07M

Numerical investigation on thermal performance of vapor chambers with porous support structures

Xinglin Yang*, Zihang Qiu

College of Energy and Power, Jiangsu University of Science and Technology, Zhenjiang 212103, China

*Corresponding author's e-mail: yangxl233@163.com

Abstract. As an efficient two-phase heat sink, the vapor chamber has important application value in the thermal management of electronic equipment. This study implements a multiphysics-coupled 3D numerical framework through the COMSOL Multiphysics® simulation platform to characterize the thermal-fluid behavior of phase-change vapor chambers, and three different porous media support column models were designed, including A model, Z-1 model, and Z-2 model, to assess their impact on the vapor chamber's thermal transport efficiency. The research focuses on the effects of three parameters: a, b, and L on heat transfer performance. The analysis shows that the heat dissipation performance of the Z-2 model is significantly better than that of the Z-1 model. With the increase of a value, the average temperature of the vapor chamber decreases significantly, while the effect of the b value on the temperature is not obvious. When a=4 mm, the average temperature of the Z model is significantly lower than that of the A model. The optimized elliptical support column design performs well in improving heat transfer efficiency.

1. Introduction

As an efficient two-phase heat sink, the vapor chamber has important application value in the thermal management of electronic equipment. Its typical application scenarios include: 1 heat dissipation of microelectronic components (in accordance with JESD51-14 standard); 2 laser diode temperature control modules; 3 Heat flow management of IGBT module with power density>100 W/cm^2. Its main function is to efficiently conduct and disperse heat, ensuring that the temperature of the equipment during operation is kept within a safe range, and avoiding performance degradation or equipment damage due to overheating.

The research in the field of vapor chambers at home and abroad shows a trend of vigorous development. More and more innovative designs and improved materials have been realized, which provides important support for the thermal management technology of electronic equipment. Ji et al. [1] studied how to choose different surface wettability to match different heat dissipation components. By independently evaluating the influence of each geometric feature, Lohan et al. [2] proposed a systematic view to improve the drying limit of the steam chamber and reduce the thermal resistance. Sumit et al. [3] numerically studied the influence of outer convex boss on VC performance under different geometric design parameters under different heat load conditions. Hu et al. [4] designed a new type of coreless pulsating steam chamber plate with a Tesla valve to realize continuous internal fluid circulation for heat transfer. Rahmatullah et al. [5] investigated the influence of each pillar design on temperature uniformity by numerical simulation. Li et al. [6] constructed a transient model to study the heat transfer characteristics of the mesh vapor chamber, which studied the non-isothermal flow of water vapor. Li et al. [7] proposed a radiator that combines a liquid-cooled fin and a steam chamber radiator, which reduces the contact thermal resistance of the radiator. Gu et al. [8] constructed a water-cooled experimental device and systematically studied the relationship between water flow, wick structure, water temperature, and steam chamber temperature uniformity. Mao et al. [9] established a computational fluid dynamics model to evaluate the transient characteristics of the vapor chamber, which provides theoretical support for the application of porous media in reactors. Zhu et al. [10] constructed a bionic leaf vein chamber. Based on the simplified numerical model, the influence of the correlation diameter ratio of the vein structure on the temperature distribution of the bionic steam chamber is studied.

As an innovative structural form, porous media support columns have attracted more and more attention. The porous medium has a large specific surface area, which promotes the uniform distribution of heat energy and maintains the temperature stability of the soaking plate. In fluid dynamics applications, the shape of the elliptical support column can reduce flow resistance and optimize the path of liquid or gas flow. This paper explores the structural parameters of the elliptical porous medium support column and then proposes an optimal configuration scheme.

2. Simulation model of the vapor chamber

2.1. Construction of three-dimensional model

This model uses the COMSOL Multiphysics® simulation platform to characterize the thermal-fluid behavior of phase-change vapor chambers, which is divided into three models, A model, Z-1 model, and Z-2 model. The three-dimensional diagram of the Z model is shown in Figure 1. The Z-mode top view is shown in Figure 2, with dimensioning. Compared with the A model, the Z model has 8 elliptical cylindrical porous media structures inside the steam chamber. L is the distance from the center of the elliptical center to the center of the hot plate. Z model was divided into Z-1 model and Z-2 model according to L=9 mm and L=10 mm. As shown in the figure, the A model is divided into four parts: shell, peripheral porous medium, steam chamber, and chip. The Z model is divided into five parts: shell, peripheral porous medium, steam chamber, internal porous medium, and chip.

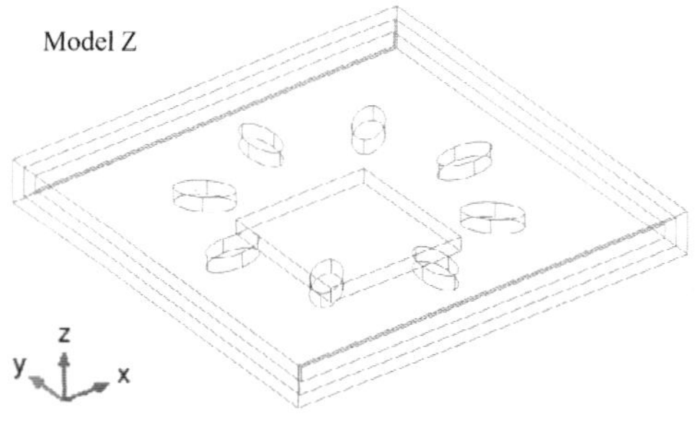

Figure 1. Three-dimensional graph of Z model.

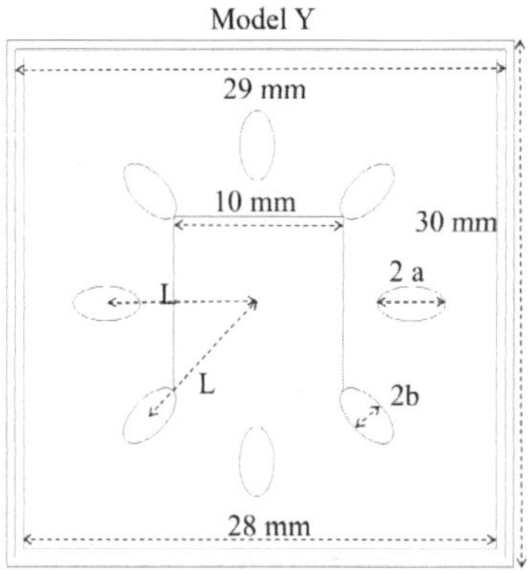

Figure 2. Z model top view and size.

2.2. *Materials selection*

The silicon is used as the material of the chip, the copper shell, and the material of the porous medium solid region. Liquid water is used as the material in the fluid region of porous media. Water vapor is used as the material in the steam chamber area.

2.3. *Physical field and equation*

Laminar flow: The continuity equation is Equation (1), the momentum equation is Equations (2)-(4), and the energy equation is Equation (5).

$$\nabla \bullet (\rho_v \vec{U}) = 0 \tag{1}$$

$$\rho_v (u_v \frac{\partial u_v}{\partial x} + v_v \frac{\partial u_v}{\partial y} + w_v \frac{\partial u_v}{\partial z}) = -\frac{\partial p_v}{\partial x} + \mu_v \left\{ \frac{\partial^2 u_v}{\partial x^2} + \frac{\partial^2 u_v}{\partial y^2} + \frac{\partial^2 u_v}{\partial z^2} \right\} \tag{2}$$

$$\rho_v (u_v \frac{\partial v_v}{\partial x} + v_v \frac{\partial v_v}{\partial y} + w_v \frac{\partial v_v}{\partial z}) = -\frac{\partial p_v}{\partial y} + \mu_v \left\{ \frac{\partial^2 v_v}{\partial x^2} + \frac{\partial^2 v_v}{\partial y^2} + \frac{\partial^2 v_v}{\partial z^2} \right\} \tag{3}$$

$$\rho_v (u_v \frac{\partial w_v}{\partial x} + v_v \frac{\partial w_v}{\partial y} + w_v \frac{\partial w_v}{\partial z}) = -\frac{\partial p_v}{\partial z} + \mu_v \left\{ \frac{\partial^2 w_v}{\partial x^2} + \frac{\partial^2 w_v}{\partial y^2} + \frac{\partial^2 w_v}{\partial z^2} \right\} \tag{4}$$

$$\rho_v c_{p,v} \left(u_v \frac{\partial T_v}{\partial x} + v_v \frac{\partial T_v}{\partial y} + w_v \frac{\partial T_v}{\partial z} \right) = k_v \left\{ \frac{\partial^2 T_v}{\partial x^2} + \frac{\partial^2 T_v}{\partial y^2} + \frac{\partial^2 T_v}{\partial z^2} \right\} + Q \tag{5}$$

Heat transfer in porous media: The energy equation is Equation (6).

$$K_s \left\{ \frac{\partial^2 T_s}{\partial x^2} + \frac{\partial^2 T_s}{\partial y^2} + \frac{\partial^2 T_s}{\partial z^2} \right\} + Q = 0 \tag{6}$$

Brinkman equation: used to calculate the velocity field and pressure field of liquid water in porous media. The continuity equation is Equation (7), the momentum equation is Equations (8)-(10), and the energy equation is Equation (11).

$$\nabla \bullet (\rho_l \vec{U}) = 0 \tag{7}$$

$$\rho_l (u_l \frac{\partial u_l}{\partial x} + v_l \frac{\partial u_l}{\partial y} + w_l \frac{\partial u_l}{\partial z}) = -\frac{\partial p_l}{\partial x} + \mu_l \left\{ \frac{\partial^2 u_l}{\partial x^2} + \frac{\partial^2 u_l}{\partial y^2} + \frac{\partial^2 u_l}{\partial z^2} \right\} - \frac{\varepsilon \mu_l \mu_l}{K} \tag{8}$$

$$\rho_l (u_l \frac{\partial v_l}{\partial x} + v_l \frac{\partial v_l}{\partial y} + w_l \frac{\partial v_l}{\partial z}) = -\frac{\partial p_l}{\partial y} + \mu_l \left\{ \frac{\partial^2 v_l}{\partial x^2} + \frac{\partial^2 v_l}{\partial y^2} + \frac{\partial^2 v_l}{\partial z^2} \right\} - \frac{\varepsilon \mu_l \mu_l}{K} \tag{9}$$

$$\rho_l (u_l \frac{\partial w_l}{\partial x} + v_l \frac{\partial w_l}{\partial y} + w_l \frac{\partial w_l}{\partial z}) = -\frac{\partial p_l}{\partial z} + \mu_l \left\{ \frac{\partial^2 w_l}{\partial x^2} + \frac{\partial^2 w_l}{\partial y^2} + \frac{\partial^2 w_l}{\partial z^2} \right\} - \frac{\varepsilon \mu_l \mu_l}{K} \tag{10}$$

$$\rho_v c_{p,l} \left(u_v \frac{\partial T_v}{\partial x} + v_v \frac{\partial T_v}{\partial y} + w_v \frac{\partial T_v}{\partial z} \right) = k_{eff} \left\{ \frac{\partial^2 T_l}{\partial x^2} + \frac{\partial^2 T_l}{\partial y^2} + \frac{\partial^2 T_l}{\partial z^2} \right\} + Q \tag{11}$$

S, *l*, and *v* represent solid, liquid, and gaseous phases; *Q* represents the chip power.

3. Result and discussion

Figure 3 shows the variation trend curve of the average temperature of the vapor chamber with the change of b value under different parameters. Figure (a) represents the Z-1 model and Figure (b) represents the Z-2 model. Due to the excessive number of models, the temperature distribution of Z-2 and Z-1 models when b=0.7 mm and a is 2 mm, 3 mm, and 4 mm respectively, and the temperature distribution of A model are shown. As shown in Figure 5, Figures (a), (b), and (c) are the temperature distribution maps when b = 0.7 mm and a is 2 mm, 3 mm, and 4 mm respectively in the Z-2 model. As shown in Figure 5, Figures (d), (e), and (f) are the temperature distribution maps when b = 0.7 mm and a is 2 mm, 3 mm, and 4 mm respectively in the Z-1 model. As shown in Figure 5, Figure (g), this is the temperature distribution map of the A model. As shown in Figure 3, the average temperature of the soaking plate decreases with the increase of a value. As shown in Figure 5, Figures (a), (b), and (c), the maximum and minimum temperatures gradually decrease with the increase of a value. As shown in Figure 4, Figures (d), (e), and (f) also show that the maximum and minimum temperatures gradually decrease. The average temperature of the A model is 34.55°C. In the two figures, especially when a = 4 mm, the average temperature is significantly lower than 34.55°C.

It shows that the larger the value is, the faster the water vapor flows back to the hot end through the porous medium after cold condensation at the cold end, and then absorbs heat and evaporates into water vapor. Therefore, it shows a lower average temperature. As the value of an increase, the latent heat of evaporation per unit time increases, so the temperature will become lower. The influence mechanism of the b value on temperature is the same as that of a value. However, the effect of the b value on increasing evaporation area is not as good as that of a value. The increased direction of a value is almost parallel to the direction of water vapor velocity. It will not block the flow of water vapor, but the increased direction of the b value is perpendicular to the flow velocity direction of water vapor. This blocks the flow of water vapor to a certain extent, so the effect of the b value on temperature is not more significant than that of a value.

As shown in Figure 4, it shows the temperature curve of average temperature changing with b value when a = 4 mm and different L values. As shown in Figure 5, Figures (c), (f), and (g) show that the lowest and highest temperatures of the Z-2 model, the Z-1 model, and the A model decrease in turn. The heat transfer performance of the Z-2 model is significantly enhanced for the Z-1 model, where the average temperature is reduced by about 2°C. The heat transfer performance of the Z model is significantly enhanced for the A model, and the average temperature is reduced by at least 1°C.

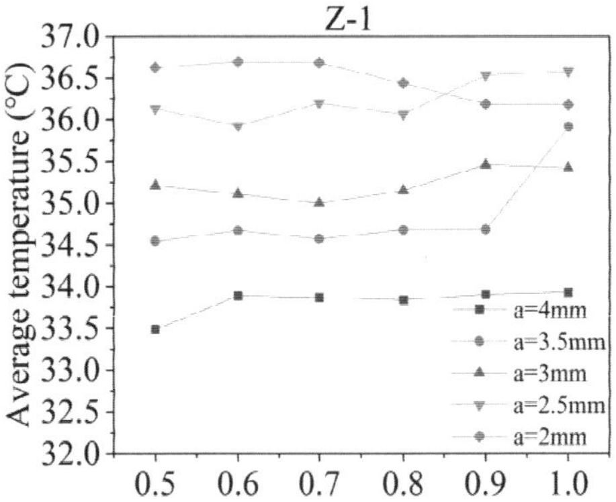

(a)

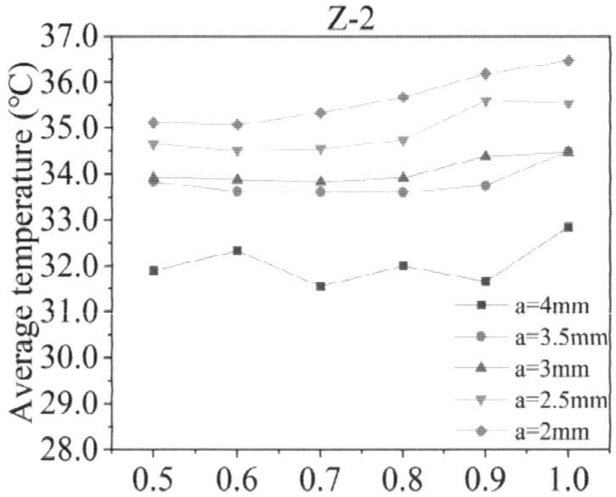

(b)

Figure 3. Temperature curves of average temperature changing with b value under different a values.

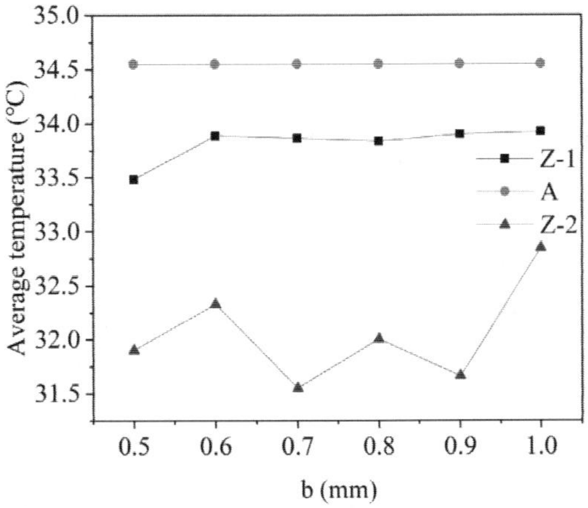

Figure 4. a = 4 mm, the temperature curve of average temperature changing with b value under different L values.

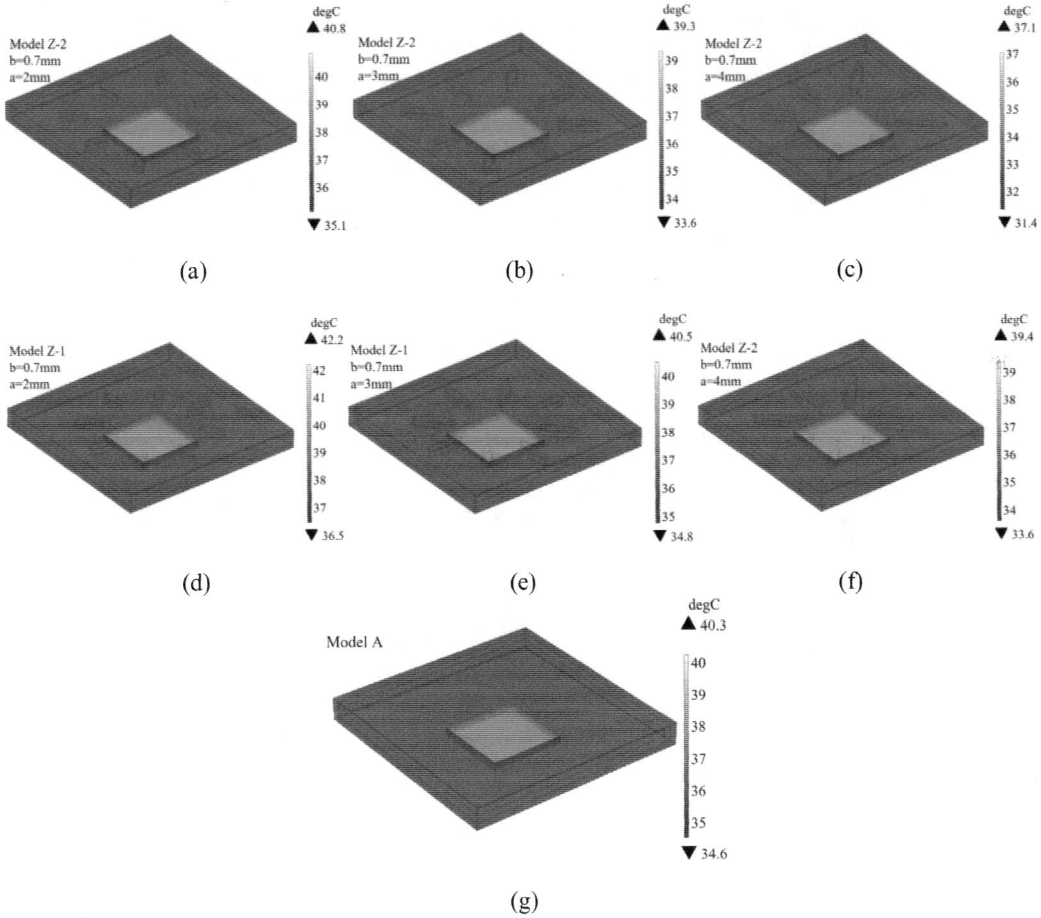

Figure 5. Temperature profile.

4. Conclusion

In this paper, three models are designed to investigate the influence of porous media support column structure on the phase change heat transfer mechanism of water vapor inside the vapor chamber, A model, Z-1 model, and Z-2 model. The effects of three parameters a, b, and L on the model are studied. The research shows that:

The effect of a value on temperature is more obvious than that of the b value. The increase of a value will reduce the average temperature, and the b value has no obvious rule on the temperature. When a = 4 mm, the average temperature of the Z model is significantly lower than that of the A model. That is, the longer the long axis of the elliptical cylinder is, the higher heat transfer performance the soaking plate has.

The Z-2 model has the lowest temperature and the best heat dissipation effect. The larger the L value is, the lower the temperature is. The thermal conductivity of the plate increases with the increase of the distance between the center of the elliptical cylinder and the center of the vapor chamber.

References

[1] Ji, Ruiyang, Qin, et al. Analysis of wettability effects on thermal performance of vapor chamber with a hybrid lattice Boltzmann method[J]. International Journal of Heat and Mass Transfer, 2024, 225: 125315.
[2] Lohan, Danny J., Joshi, et al. Experimental investigation of post and vented vapor chamber designs for high heat flux dissipation[J]. International Journal of Heat and Mass Transfer, 2024, 222: 125116.
[3] Sumit Sharma, Yi-Hsiang Huang, Chi-Chuan Wang. Numerical optimization and experimental validation of vapor chambers having an outwardly protruding boss[J]. International Communications in Heat and Mass Transfer, 2023, 149: 107122.
[4] Hu, Jinhao, Xu, et al. Visualization-assisted study of a new wickless pulsating vapor chamber[J]. Applied Thermal Engineering, 2024, 239: 122162.
[5] Rahmatullah, Yusuf, Wen, et al. Optimization of thermal-hydraulic performance for pillar-reinforced vapor chamber using Lee model based numerical scheme and analysis of variance[J]. International Communications in Heat and Mass Transfer, 2024, 152: 107319.
[6] Rui Li, Yunhua Gan, Qiliang Luo, et al. Numerical analysis on thermal hydraulic characteristics of mesh-type ultra-thin vapor chambers[J]. Applied Thermal Engineering, 2024, 236: 121648.
[7] Cong Li, Ying Huang, Qifei Jian, et al. Experimental and numerical investigation on thermal characteristics of the heat sink integrating 3D vapor chamber heat spreader and liquid cooling fins[J]. Applied Thermal Engineering, 2024, 236: 121609.
[8] Gu, Zhonghao, Yang, et al. Enhancing heat transfer performance of aluminum-based vapor chamber with a novel bionic wick structure fabricated using additive manufacturing[J]. Applied Thermal Engineering, 2024, 247: 123076.
[9] Shang Mao, Wenbin Liu, Tao Zhou, et al. Steady-state and transient thermal–hydraulic analysis for high temperature heat pipe cooled reactors[J]. Nuclear Engineering and Design, 2023, 412: 112479.
[10] Zhu Shuiping, Zhang Zhilin, Chen Haisheng, et al. Numerical Investigation of a Bionic Vapor Chamber Based on Leaf Veins for Cooling Electronic Devices[J]. Sustainability, 2023, 15(2).

Improvement of a numerical model for gas-liquid contactor in tritium extraction from liquid PbLi

Teng Wang, Bin Zhang, Kai Huang*

College of Energy and Mechanical Engineering, Shanghai University of Electric Power, Shanghai 201306, China

* Corresponding author's e-mail: echoasis@163.com

Abstract: Efficient tritium extraction from liquid lead-lithium (PbLi) breeding blankets is critical for fusion reactor technology development. In this study, numerical simulation methods based on a fundamental gas-liquid contactor (GLC) model were employed to optimize boundary conditions and external structures, ensuring stability of liquid PbLi level during operation. Computational results show that the improved GLC achieves a tritium extraction efficiency of 19.18%. Further enhancement was realized by increasing GLC mesh density and installing a gas-guiding device at the internal helium inlet, significantly improving gas-liquid contact within the GLC. Final results indicate that GLC's tritium extraction efficiency increased to 25.73%, suggesting that the gas guiding device introduction effectively enhanced gas-liquid contact and tritium extraction efficiency.

1. Introduction

The development and application of fusion energy promise to fundamentally resolve energy challenges faced by human society. As fusion energy technology transitions from physical feasibility to engineering viability, research and development of breeding blanket technology have become key technologies for the successful implementation of fusion reactors (such as ITER and DEMO). A liquid lead-lithium alloy (PbLi) breeding blanket [1] has emerged as a primary candidate solution due to its unique advantages. To minimize damage from inventory tritium in liquid PbLi to structural materials, it is crucial to maintain safe and stable fusion reactor operation while achieving tritium self-sufficiency and efficiently extracting tritium from liquid PbLi on a large scale.

Currently, main technologies for extracting tritium from liquid PbLi internationally include gas-liquid contactors (GLC) [2-3], permeator against vacuum (PAV) [4-5], and vacuum sieve tray (VST) [6-7]. Both PAV and VST are expected to achieve considerable tritium extraction efficiencies and possess good development potential, although technologies are not yet mature. In contrast, GLC has high technological maturity, a simple structure, and offers good stability and reliability, making it the benchmark tritium extraction technology selected for the ITER tritium breeding blanket module [8]. In GLC, the purging gas used is generally inert. Liquid lithium-lead interacts with purging gas in a countercurrent manner. Under the driving force of partial pressure gradient between the two phases, tritium diffuses from liquid PbLi to the gas-liquid contact surface, then from the gas-liquid contact surface into the gas phase. It is ultimately carried away by purging gas to be sent to the tritium processing facility for hydrogen isotope separation.

Extensive experimental studies on GLC technology have been conducted by researchers from various countries. Alpy et al. [2] evaluated the technical performance of GLC using a Melodie loop, where hydrogen substituted tritium and pure argon served as countercurrent purge gas to simulate the tritium extraction process in liquid lithium-lead, ultimately achieving optimal efficiency of 30%. M. Utili et al. [3] conducted tritium extraction studies using a TRIEX device for gas-liquid contactors, substituting hydrogen for tritium and employing argon as a purge gas. Under varying countercurrent velocities for the two phases, tritium extraction efficiency ranged from 10% to 30%. Xie et al. [9] performed tritium extraction experiments with a self-designed GLC loop and found helium more suitable than argon as an extraction gas. Given the instability and radioactivity of tritium, numerical simulation studies on tritium extraction from liquid PbLi are also feasible. Jiang et al. [10] simulated gas-liquid two-phase flow and mass transfer characteristics within GLC using numerical methods, ultimately developing a foundational model for GLC through a gradual evolution approach, which resulted in a tritium extraction efficiency of 20.8%. However, this model faced challenges with insufficient gas-liquid contact and difficulties in maintaining liquid PbLi levels. To address these issues, a turbulent two-phase flow level set model coupled with a dilute mass transfer model is employed in this study, using helium as a purge gas to extract tritium from liquid PbLi. GLC structure is optimized using numerical simulation methods to maintain liquid PbLi level and enhance stability. Additionally, a drainage device is implemented to improve gas-liquid contact within GLC, analyzing the velocity field, volume fraction field, and tritium concentration field to increase the tritium extraction efficiency of individual GLC units. This research provides theoretical references for the practical application of GLC technology.

2. Introduction to model and theoretical foundations

2.1. Model introduction

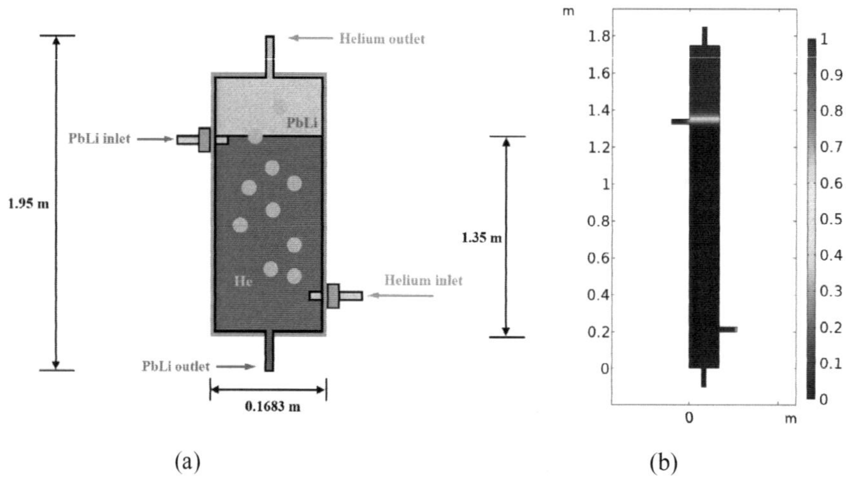

Figure 1. GLC Model: (a) schematic diagram; (b) initial state diagram.

The schematic diagram of the GLC model is shown in Figure 1(a), with a height of 2 m and a diameter of 0.1683 m. In the initial state, the pre-filled height of liquid PbLi in the bubbling tower is 1.35 m. The diameters of the inlet and outlet pipes for liquid PbLi and helium gas are both 0.0254 m. Tritium-carrying liquid PbLi flows into the bubbling tower through the liquid inlet pipe on the left side and exits through the liquid outlet pipe at the bottom; pure purge gas helium enters from the gas inlet pipe on the right side and exits from the gas outlet pipe at the top. Significant space is left between the filling height of liquid PbLi and the top of the tower to observe changes in liquid level during GLC operation. The initial state of GLC is illustrated in Figure 1(b), where the blue area represents the liquid PbLi phase and the red area represents the helium gas phase. The interface between liquid PbLi and helium gas represents the gas-liquid interface in the initial state. Liquid PbLi inlet employs a velocity boundary condition, with velocity set at 0.23147 m/s. Liquid PbLi outlet uses a pressure boundary condition. The inlet for purge gas helium also employs a velocity boundary condition, with inlet velocity set at 0.0756518 m/s. An outlet for purge gas helium uses a pressure boundary condition. Wall surfaces of GLC are treated with a no-slip boundary condition. Standard wall functions are applied for calculations. The initial concentration of tritium in dilute substance transfer field is 0 mol/m^3, while tritium concentration at liquid PbLi inlet is 0.001 mol/m³. A transient solver of COMSOL Multiphysics 6.0 is utilized for computations.

2.2. *Theoretical foundation*

The theoretical model is based on the computational model proposed by Jiang et al. [10], where mass and momentum transfer during gas-liquid two-phase flow is governed by Navier-Stokes equations. The turbulence model employs Standard *k-ε* as the closure equation.

Transport and reinitialization of level set function Φ are governed by Equation (1):

$$\frac{\partial \Phi}{\partial t} + u \cdot \nabla \Phi = \gamma \nabla \cdot \left(\varepsilon \nabla \Phi - \Phi(1-\Phi) \frac{\nabla \Phi}{|\nabla \Phi|} \right) \tag{1}$$

where γ is denoted as the re-initialization parameter; ε is represented as the interface thickness control parameter; the γ value is conventionally set to the maximum velocity observed within the model, whereas the ε value is determined as half of the grid size control parameter.

Tritium transfer between PbLi and He is described using a sparse material transfer model, with mass transfer control equations presented in Equations (2) and (3).

$$N_i = -D_i \cdot \nabla c_i + c_i u \tag{2}$$

$$\frac{\partial c_i}{\partial t} + \nabla \cdot N_i + 0 \tag{3}$$

where N_i represents the molar flux; D_i denotes the diffusion coefficient; and c_i refers to the concentration.

Material diffusion coefficient D_i is processed using volume fraction-weighted average, as shown in Equation (4).

$$D_i = (1-\Phi)D_1 + (\Phi)D_2 \tag{4}$$

where D_1 and D_2 represent diffusion coefficients of tritium in helium gas phase and liquid PbLi phase, respectively.

To simplify calculations, the following fundamental assumptions are made: All flows are considered incompressible; global temperature is set at 873 K, corresponding to the temperature of liquid PbLi at blanket exit; the mass transfer process simultaneously considers both convective and molecular diffusion effects, but does not involve phase changes or complex multiphase reactions at the gas-liquid interface.

3. Improvements to GLC external structure

3.1. *Transport analysis of basic models*

Running the GLC base model in Figure 1(b), the phase volume fraction field at t=3s is shown in Figure 2(a). Due to the outlet of liquid PbLi being set as a pressure boundary condition, the flow velocity of the liquid phase is very high, leaving only 10% of the initial filling height, severely affecting the stable operation of the GLC. Therefore, boundary conditions of the GLC were modified by changing the liquid PbLi outlet to a velocity boundary condition, with a velocity set at 0.23147 m/s. This adjustment, equivalent to installing a throttle valve, matches the inlet velocity of liquid PbLi, ensuring the stability of liquid PbLi level. Under this boundary condition, the GLC model was run. At t=100s, the phase volume fraction field is shown in Figure 2(b). The rate of decline in liquid PbLi level significantly improved, though a slow decline still exists. Further analysis reveals that the primary reason for the decline in liquid level is leakage at the helium gas inlet pipe. A local view of helium gas inlet pipe is shown in Figure 3. After device operation, helium gas begins to flow into the GLC. Due to the horizontal connection of the inlet pipe, the lower half of the pipe fills with liquid under gravity, allowing helium gas to enter only from the upper half, leading to a slow outflow of liquid PbLi from the helium gas inlet pipe, making it difficult to maintain stable liquid level.

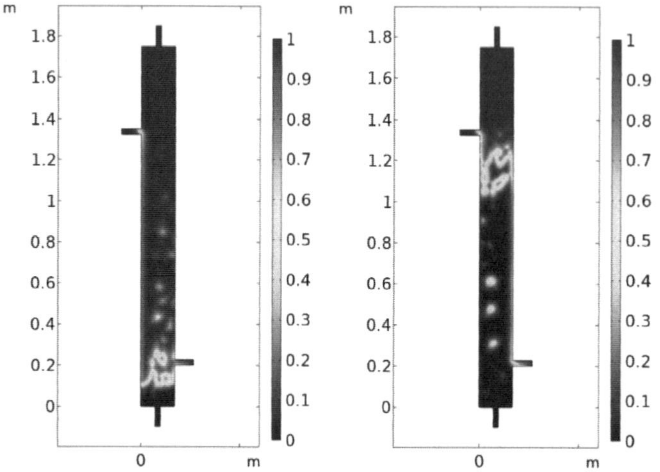

(a) (b)

Figure 2. Phase volume fraction fields for different boundary conditions: (a) t = 3 s; (b) t = 100 s.

Figure 3. Local view of helium gas inlet pipe.

3.2. Transport analysis of basic models

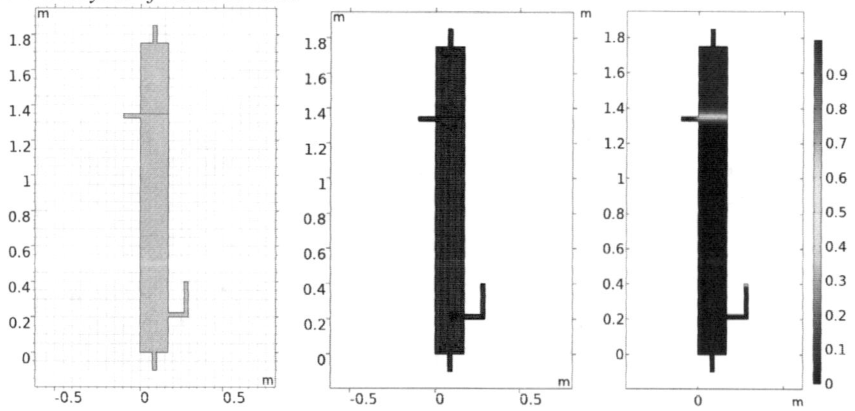

(a) (b) (c)

Figure 4. Improved GLC model: (a) Geometry; (b) Mesh division; (c) Initial state.

Improvements were made to the helium gas inlet pipe structure based on previous discussions. The geometric structure, mesh division, and initial state of the improved model are shown in Figure 4. To prevent liquid PbLi from flowing out of the helium gas inlet pipe due to gravitational effects, a vertical upward bend was designed on the inlet pipe's outer side. The phase volume fraction field after model operation is presented in Figure 5. At two distinct time points, the liquid PbLi level remains stable, and helium gas bubbles rise within the GLC, demonstrating a stable countercurrent flow between gas and liquid phases. These results effectively validate the proposed scheme's ability to maintain the liquid PbLi level.

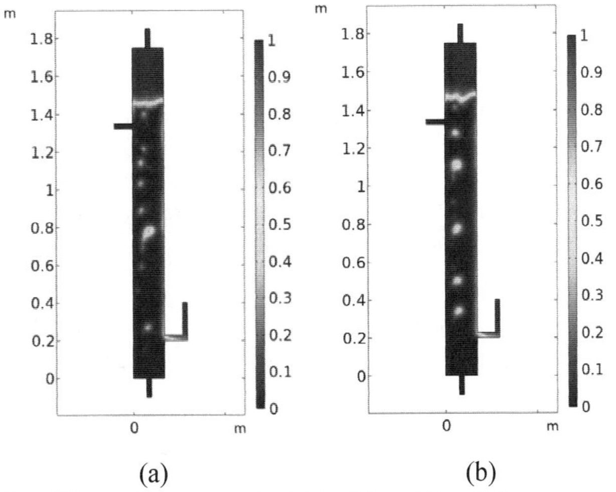

Figure 5. Improved GLC at different time moments: (a) t = 50 s; (b) t = 100 s.

3.3. Transport analysis of basic models

Grid independence verification was performed based on the improved GLC model in Figure 4, using an unstructured grid due to the GLC's geometric asymmetry. Grids with 2189, 6617, and 29894 elements were generated by adjusting the grid size. The tritium concentration at the outlet of the liquid PbLi was used as the evaluation criterion. The model results are shown in Table 1.

Table 1. Results of grid independence verification.

Grid number	Tririum concentration (mol/m^3)	Relative deviation (%)
2189	7.73×10^{-4}	6.18
6617	8.08×10^{-4}	1.94
29894	8.24×10^{-4}	baseline

The results show that using a reference grid count of 29894, the relative deviations for grid counts of 2189 and 6617 are 6.18% and 1.94%, respectively. The grid count of 6617 shows a smaller deviation compared to the reference model. Thus, to improve computational efficiency while maintaining accuracy, a grid density of 6617 was used in the analysis. A grid quality assessment revealed an average element quality of 0.94 and a minimum element quality of 0.61.

3.4. Transport analysis of basic models

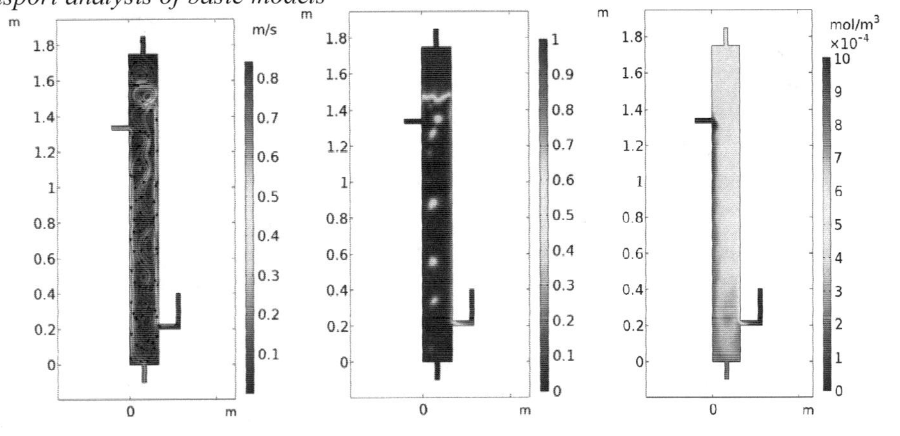

Figure 6. Physical field maps of GLC at t = 100 s: (a) Velocity field; (b) Phase volume fraction field; (c) Tritium concentration field.

The analysis of the two-phase gas-liquid flow and tritium transport process within the GLC was performed using the grid count determined during the grid independence verification. The results show that the tritium concentration field reaches a steady state at t = 100 s. Figure 6 presents the distribution of the velocity field, phase volume fraction field, and tritium concentration field at this moment. The velocity field shows that the highest flow velocities are primarily located near the two side walls. After liquid PbLi flows in from the inlet, part of it participates in gas-liquid contact within the GLC, while another part flows downward along the left side wall. Simultaneously, after helium gas is injected into the GLC, most of the gas flows upward along the right side wall. The counterflow interaction between the gas and liquid phases generates numerous turbulent vortices within the GLC. The phase volume fraction field shows that the liquid level of PbLi remains constant, indicating stable operation of the device, with helium gas bubbles rising within the GLC and noticeable gas-phase wall-bounded flow observed at the right side wall. The tritium concentration field indicates that high-concentration tritium regions are primarily located along the left side wall and the bottom of the GLC, while effective tritium extraction occurs in the middle and right regions. The tritium extraction efficiency of the GLC is calculated using Equation (5).

$$\eta = \frac{C_i U_i A_i - C_o U_o A_o}{C_i U_i A_i} \tag{5}$$

where C_i denotes the tritium concentration at the PbLi inlet, U_i represents the inlet velocity of PbLi, A_i is the cross-sectional area of the PbLi inlet, C_o denotes the tritium concentration at the PbLi outlet, U_o represents the outlet velocity of PbLi, and A_o is the cross-sectional area of the PbLi outlet.

The calculation results show that the tritium extraction efficiency of the GLC is 19.18%. Analysis of the cloud diagram indicates insufficient contact between the gas and liquid phases within the tower, which adversely affects the operational efficiency of the GLC.

4. Improvement of gas-liquid contact in GLC

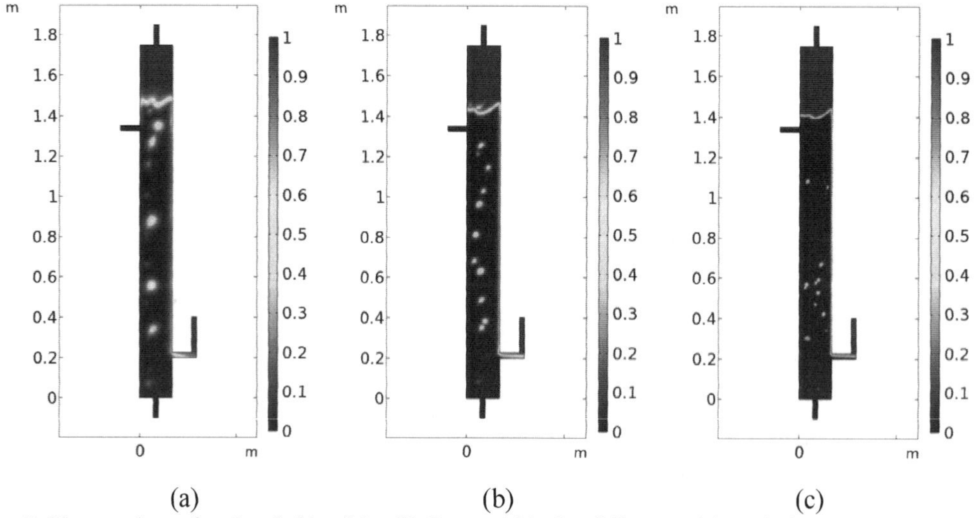

Figure 7. Phase volume fraction fields of the GLC at t = 100s for different grid resolutions: (a) 6617; (b) 29894; (c) 111618.

The effective contact area between liquid PbLi and helium gas within the GLC is critical for the efficient transfer of tritium between the two phases. Therefore, ensuring sufficient gas-liquid contact during the countercurrent flow process is essential to improving tritium extraction efficiency. As shown in Figure 6, the gas-liquid interaction is insufficient, making it challenging to visually observe the gas-liquid interaction during operation. Although the grid density used for grid independence verification ensures that tritium extraction efficiency remains unaffected by grid number, a higher grid density is required to observe the impact of structural improvements on gas-liquid contact within the GLC. Consequently, GLCs with grid counts of 6617, 29894, and 111618 were employed based on the model in Figure 4. The phase volume fraction field at t = 100s is shown in Figure 7. As the grid resolution increases, bubble size gradually decreases. The phenomenon of helium flowing along the right wall becomes more pronounced. As shown in Figure 7(c), with a grid count of 111618, fewer bubbles are present in the tower. Most helium flows upward along the right wall, negatively impacting gas-liquid contact. Therefore, a flow-guiding device was installed at the helium inlet pipe, as shown in Figure 8. This device, which is welded inside the GLC, is designed to prevent helium from flowing directly upward along the wall after being introduced.

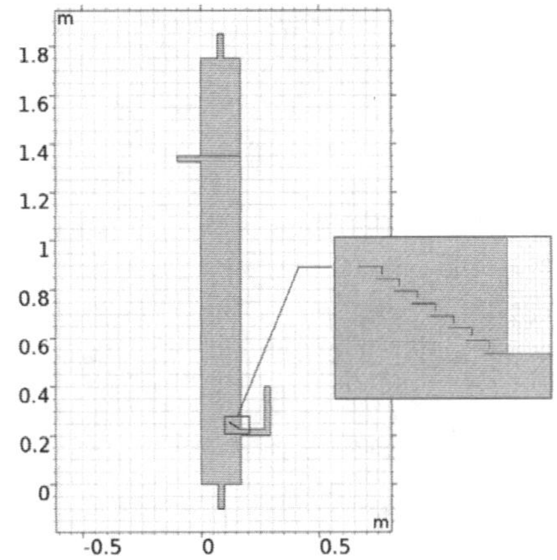

Figure 8. Schematic of GLC with installed flow guide device.

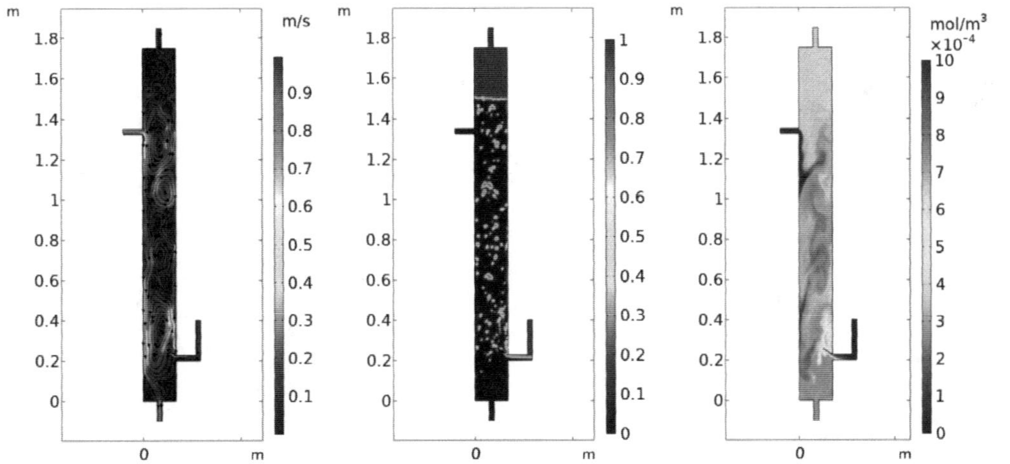

(a) (b) (c)

Figure 9. Physical field distribution maps of GLC after addition of diversion device at t = 100 s: (a) Velocity field; (b) Phase volume fraction field; (c) Tritium concentration field.

Based on the grid density shown in Figure 7(c), numerical simulations were conducted on the GLC with the installation of a gas-guiding device. The simulation results indicate that at t = 100 s, the tritium concentration field reaches a steady state. Figure 9 illustrates the distribution of the velocity field, phase volume fraction field, and tritium concentration field at this time. In the velocity field, the flow velocity distribution is uniform overall, with no significant localized high-velocity regions observed. A comparison of the phase volume fraction fields before and after the improvement shows that gas-liquid contact within the GLC has significantly improved, resulting in the formation of numerous gas bubbles, with only a small amount of gas flowing along the walls. In the tritium concentration field, the distribution of tritium concentration within the GLC is relatively uniform, with a high concentration region appearing only at the inlet of the liquid PbLi. Calculations show that the tritium extraction efficiency of the GLC reaches 25.73%. The results demonstrate that the installation of the gas guiding device effectively enhances gas-liquid contact within the GLC, improving tritium extraction efficiency and validating the effectiveness of the structural improvements made within the GLC.

5. Conclusion and outlook

This study employs a turbulent two-phase flow level set model coupled with a dilute mass transfer model, utilizing numerical simulation methods to optimize the structural design of a gas-liquid contactor (GLC) model. A systematic analysis of the internal gas-liquid two-phase flow and tritium extraction process is conducted.

To address the challenge of maintaining the liquid level of liquid PbLi in the basic model, the outlet of liquid PbLi is set as a velocity boundary condition. The external structure of the helium inlet pipe is optimized by incorporating a vertical upward bend. The numerical simulation results indicate that the improved GLC can stably maintain the liquid PbLi level, ensuring the normal operation of the device and validating the effectiveness of the external structural optimization. Analysis of the velocity field, phase volume fraction field, and tritium concentration field of the optimized GLC reveals a low gas-liquid contact efficiency, with a tritium extraction efficiency of only 19.18%.

To further enhance the gas-liquid contact effect, the mesh of the model is refined. A secondary optimization of the GLC model is performed at a grid density of 111, 618. A gas-guiding device is added to the internal helium inlet pipe. The results of the optimized numerical simulation show that the addition of the guiding device significantly improves the gas-liquid contact effect within the GLC, increasing the tritium extraction efficiency to 25.73%. This represents a significant improvement over the model without the guiding device, confirming the effectiveness of the internal structural optimization.

Based on the research findings, it is recommended that future work incorporate a guiding device at the liquid PbLi inlet pipe within the GLC to further enhance the gas-liquid disturbance effect. When investigating the tritium extraction efficiency of a multi-stage GLC series system, a grid size determined through grid independence verification should be employed. Additionally, for a more in-depth study of the gas-liquid contact characteristics within the GLC, increasing the number of grids to improve computational accuracy is advised.

References

[1] Chen, L., Jiang, K., Ma, X., Wu, Q., Yu, Y., Lu, P., Cheng, X., Chen, L., Yang, J., Zhu, Q., Huang, K., Zhang, N., Ni, M.-J., & Liu, S. (2021). Conceptual design of the supercritical CO2 cooled lithium lead blanket for CFETR. Fusion Engineering and Design, 173, 112800–112800. https://doi.org/10.1016/j.fusengdes.2021.112800

[2] Alpy, N., Terlain, A., & Lorentz, V. (2000). Hydrogen extraction from Pb–17Li: results with a 800 mm high packed column. Fusion Engineering and Design, 49–50, 775–780. https://doi.org/10.1016/s0920-3796(00)00461-0

[3] Utili, M., Aiello, A., Laffi, L., Malavasi, A., & Ricapito, I. (2016). Investigation on efficiency of gas liquid contactor used as tritium extraction unit for HCLL-TBM Pb-16Li loop. Fusion Engineering and Design, 109–111, 1–6. https://doi.org/10.1016/j.fusengdes.2016.03.067

[4] Martínez, P., Moreno, C., Martínez, I., & Sedano, L. (2012). Optimizing tritium extraction from a Permeator Against Vacuum (PAV) by dimensional design using different tritium transport modeling tools. Fusion Engineering and Design, 87(7–8), 1466–1470. https://doi.org/10.1016/j.fusengdes.2012.03.035

[5] Papa, F., Utili, M., Venturini, A., Caruso, G., Savoldi, L., Bonifetto, R., Valerio, D., Allio, A., Collaku, A., & Tarantino, M. (2021). Engineering design of a Permeator Against Vacuum mock-up with niobium membrane. Fusion Engineering and Design, 166, 112313–112313. https://doi.org/10.1016/j.fusengdes.2021.112313

[6] Okino, F., Calderoni, P., Kasada, R., & Konishi, S. (2015). Feasibility analysis of vacuum sieve tray for tritium extraction in the HCLL test blanket system. Fusion Engineering and Design, 109–111, 1748–1753. https://doi.org/10.1016/j.fusengdes.2015.10.004

[7] Mertens, M. A. J., Demange, D., & Frances, L. (2016). Model and simulation of a vacuum sieve tray for T extraction from liquid PbLi breeding blankets. Fusion Engineering and Design, 112, 541–547. https://doi.org/10.1016/j.fusengdes.2016.05.038

[8] Ricapito, I., Aiello, A., Bükki-Deme, A., Galabert, J., Moreno, C., Poitevin, Y., Radloff, D., Rueda, A., Tincani, A., & Utili, M. (2018). Tritium technologies and transport modelling: main outcomes from the European TBM Project. Fusion Engineering and Design, 136, 128–134. https://doi.org/10.1016/j.fusengdes.2018.01.023
[9] Xie, B., Weng, K., Hou, J., Yang, G., & Zeng, J. (2013). Hydrogen extraction from liquid lithium-lead alloy by gas-liquid contact method. Nuclear Fusion and Plasma Physics, 33(01), 83–87. https://doi.org/10.16568/j.0254-6086.2013.01.013
[10] Jiang, K., Yang, J., Zhao, X., Chen, L., Gou, F., & Liu, S. (2023). Numerical simulation of tritium extraction from liquid PbLi by gas–liquid contactor. Annals of Nuclear Energy, 197, 110253–110253. https://doi.org/10.1016/j.anucene.2023.110253

Sustainable greenhouse based on minimizing carbon emissions

Ziyang Zhang*, Zhiling Ren

Liaoning Technical University, Huludao 125100, China

*z157463649@163.com (Ziyang Zhang), lngdrzl@163.com (Zhiling Ren)

Abstract. Carbon emissions from greenhouse operations pose a challenge to sustainability. Currently, greenhouse research is deficient in reducing carbon emissions. The article devised two operational strategies for comparison. Strategy 1 only ensures that the environment inside the greenhouse meets the growing conditions of the crop. Strategy 2 minimizes carbon emissions based on Strategy 1. The results show that Strategy 2 has 87.2% less carbon emissions. Carbon emissions from greenhouse production are addressed.

1. Introduction

Food issues are of global importance and are a priority issue for national governance. It has been shown that current production will not be able to achieve the goal of "zero hunger" in sustainable development by 2030 [1]. Greenhouses, as an important means of food production, consume large amounts of energy and pose serious environmental problems while increasing production. Some studies have shown that greenhouses cause the largest share of carbon emissions in agricultural production [2]. Therefore, it is imperative to improve the efficiency of greenhouse operations and reduce carbon emissions.

Greenhouse control is extremely challenging due to the complexity of greenhouse parameters and their coupling with each other. Intelligent greenhouse research in foreign countries is more perfect, for example, in [3], hierarchical control is used to improve the performance of the system. In addition, In [4], a nonlinear model predictive control (NMPC) is proposed to minimize the cost of energy usage. In [5], more than 20% of energy savings were achieved by utilizing model predictive control and neural networks. Domestic research on greenhouses only started late, in [6], an AGIES low carbon greenhouse operation method is proposed that reduces greenhouse carbon emissions. In [7], the greenhouse perturbation was corrected using a neural network. In [8], an EEMD-WPT algorithm was proposed with improved noise reduction, the signal-to-noise ratio was improved by 40.31%. The signal-to-noise ratio was improved by 40.31% compared to the previous WPT. There is still much room for progress in domestic research on greenhouse control strategies and control methods.

Although many scholars have carried out some greenhouse research work, research on greenhouse carbon emissions is still lacking. Carbon emission is a huge hidden danger to sustainable development and can cause various environmental problems. Therefore, to address greenhouse carbon emissions, this paper designs a greenhouse controller that can be set up with different operation targets and analyzes the operation results objectively and comprehensively by comparing different operation strategies domestically.

Through the design of this paper, the carbon emission of the greenhouse has been effectively reduced, which also means that energy consumption has been effectively controlled, contributing to sustainable development and realizing green agricultural development.

The main work of the article includes: (1) Reasonable limits on greenhouse parameters are designed to ensure that the goal of minimizing carbon emissions is met while healthy crop growth is satisfied. (2) Comparative tests are conducted to derive different trajectories by designing two different greenhouse operation strategies. (3) The carbon emissions of the results of the two strategies are compared, and an additional analysis is performed.

2. System Design

Figure 1 shows the greenhouse system. Greenhouse control is a multi-input and multi-output system. Greenhouse control includes: indoor temperature control, moisture control, and CO_2 injection control. The system in this paper includes three inputs: heating, ventilation, and carbon dioxide injection; Three outputs: temperature, relative humidity, and CO_2 concentration. The greenhouse parameters were calculated from the following equations:

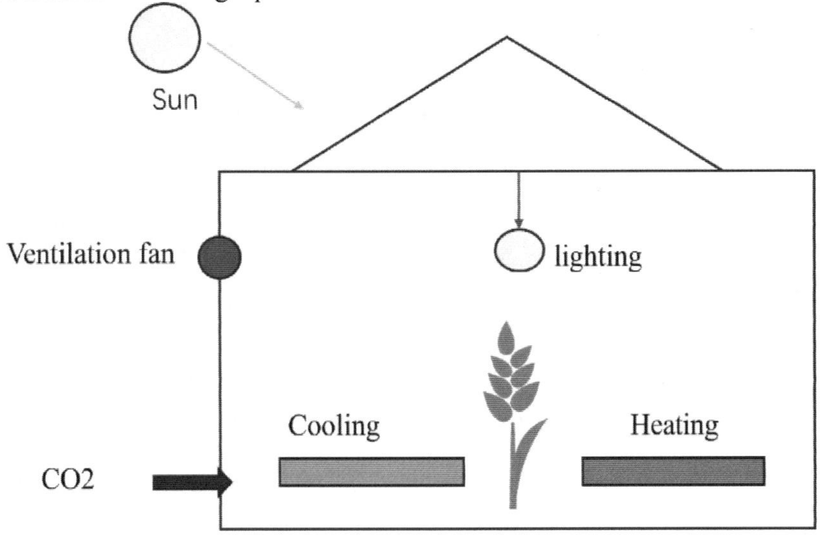

Figure 1. Grennhouse system.

$$\frac{dT_{\text{air}}}{dt} = \frac{1}{C_{\text{cap}}}(Q_{\text{sun}} + Q_{\text{la}} - Q_{\text{co}} - Q_{\text{tr}} - Q_{\text{ve}} + Q_c) \tag{1}$$

where T_{air} is the greenhouse temperature, Q_{sun} is the radiation intensity, Q_{la} is the supplemental light power. Q_{co} is the heat exchanged by the greenhouse lid, Q_{ve} is the heat changed by the ventilation. Q_c is the thermostat power.

$$RH_{\text{air}} = H_{\text{air}}/_{H_{\text{air,sat}}} \tag{2}$$

where H_{air} is humidity in the greenhouse.

$$\frac{dC_{\text{air}}}{dt} = \frac{1}{h}(C_{\text{in}} - C_{\text{ass}} - C_{\text{ve}}) \tag{3}$$

where C_{air} is the concentration of CO2 in the greenhouse, C_{in} is the rate of CO_2 replenishment, C_{ve} is altered by the ventilation of CO_2.

Two strategies are designed: Strategy 1 does not consider carbon emissions and only requires that the greenhouse conditions meet crop growth needs, so the objective function does not need to be set. Strategy 2 minimizes carbon emissions from greenhouse operations under the conditions that Strategy 1 meets crop growth needs.

Strategy 1:

$$J_1 = 0 \tag{4}$$

J_1 is the Strategy 1 objective function, which is not set in Strategy 1.
Strategy 2:

$$J_2 = \min(C_{\text{out}}) \tag{5}$$

J_2 is the Strategy 2 objective function and C_{out} is the carbon emissions.

3. System Simulation

The results of the greenhouse operation are as follows:

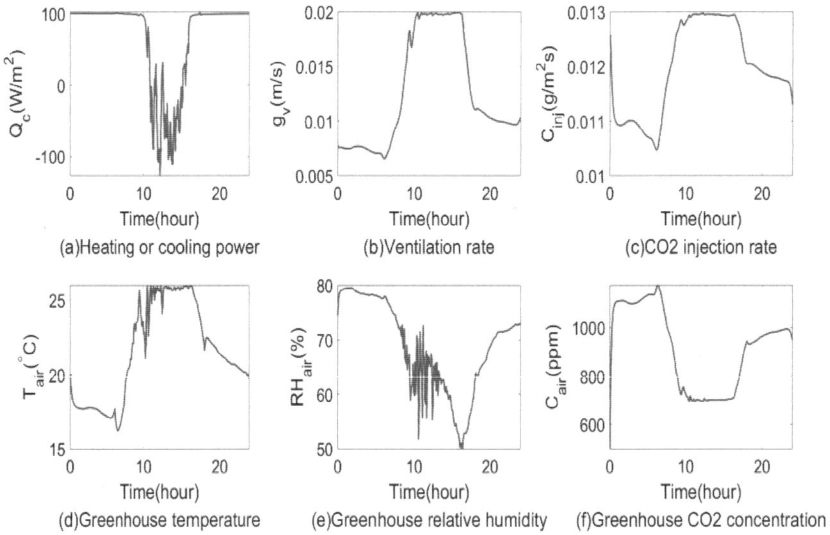

Figure 2. Strategy 1.

As shown in Figure 2, the greenhouse heater and cooler hardly worked before 10:00 due to the suitable outdoor temperature. At noon, due to the rapid increase in temperature, the cooler and exhaust fan started to work rapidly to reduce the temperature inside the room, and the ventilation also caused a decrease in relative humidity in the greenhouse. In the evening temperatures, outdoor weather returned to suitable conditions, heating and coolers hardly worked and ventilation was also at low power.

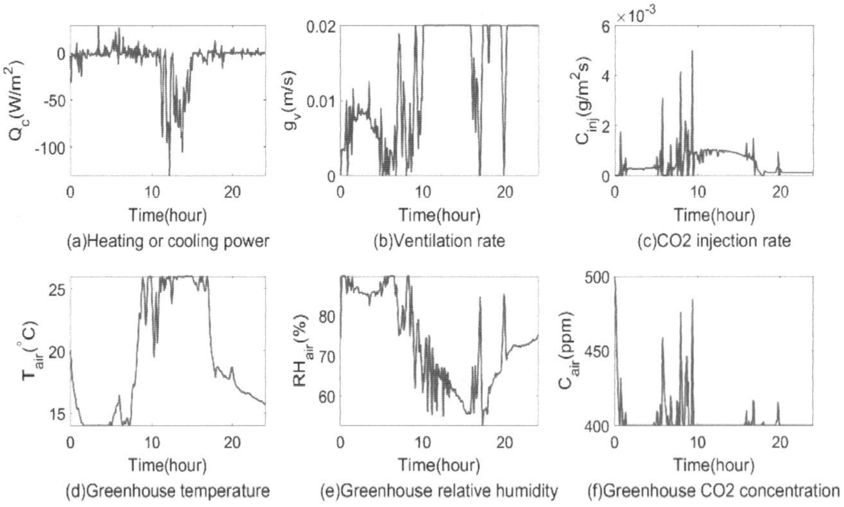

Figure 3. Strategy 2.

As illustrated in Figure 3, setting the minimum carbon emission as the target there is a significant reduction in the power of the heater and chiller and a significant reduction in the use of carbon dioxide. At the same time, it meets the greenhouse demand and effectively improves the efficiency of the operation of the greenhouse.

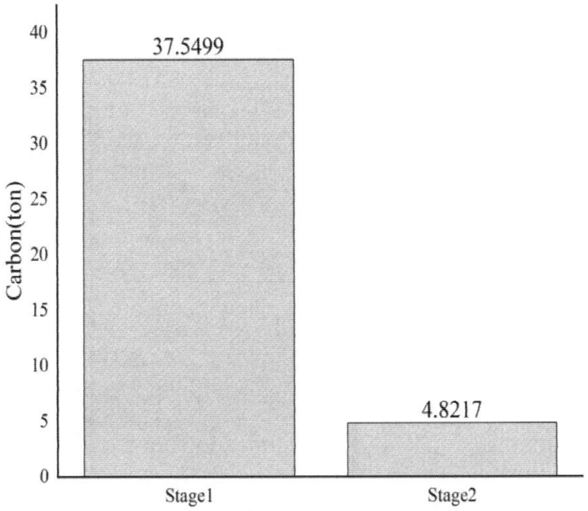

Figure 4. Carbon comparison of the two strategies.

As displayed in Figure 4, Strategy 2 carbon emissions are significantly smaller than Strategy 1, with a reduction of 87.2%. The sustainable operation of the greenhouse is realized.

4. Conclusion

To solve the greenhouse sustainability problem, this paper reduces the operational carbon emissions by 87.2% by setting different operational strategies. The results show that this strategy effectively reduces the carbon emission problem of greenhouse operations and realizes the green utilization of energy. The research on greenhouse energy is not enough, and the use of clean energy can be added in the future.

Reference

[1] FAO. The Future of Food and Agriculture: Trends and Challenges. Rome: Food and Agriculture Organization of the United Nations, 2017.[CrossRef]

[2] Pierre Clement Blaud, Pierrick Haurant, Philippe Chevrel, Fabien Claveau, Anthony Mouraud, Multi-flow optimization of a greenhouse system: A hierarchical control approach, Applied Energy, Volume 351,2023,121840, ISSN 0306-2619.[CrossRef]

[3] Debroy, P., Majumder, P., Das, A. *et al.* Model-based predictive greenhouse parameter control of aquaponic system. *Environ Sci Pollut Res* **31**, 48423–48449 (2024)[CrossRef].

[4] Pierre Clement Blaud, Pierrick Haurant, Philippe Chevrel, Fabien Claveau, Anthony Mouraud, Multi-flow optimization of a greenhouse system: A hierarchical control approach, Applied Energy, Volume 351,2023,121840, ISSN 0306-2619.[CrossRef]

[5] Farhat Mahmood, Rajesh Govindan, Amine Bermak, David Yang, Tareq Al-Ansari, Data-driven robust model predictive control for greenhouse temperature control and energy utilisation assessment, Applied Energy, Volume 343,2023,121190, ISSN 0306-2619.[CrossRef]

[6] HE Xin, YANG Yong, GAO Jinggeng. Low-carbon control method for agricultural greenhouse integrated energy system based on crop safety[J]. Agricultural Engineering, 2024, 14(1): 56-63.[CrossRef]

[7] Pan Jigang, Liu Pingzeng, Zhang Yan, Zhang Mingzh, Liu Chuanlong. Research on solar greenhouse temperature and humidity prediction model based on SSA-Elman[J]. Journal of Chinese Agricultural Mechanization, 2024, 45(11): 69-76.[CrossRef]

[8] WU Weibin, YANG Liu, WU Weihao, et al. Research on the optimization processing of greenhouse environmental data based on EEMD-WPT[J]. Journal of South China Agricultural University, 2024, 45(3): 397-407.[CrossRef]

Journal of Physics: Conference Series **3011** (2025) 012056 doi:10.1088/1742-6596/3011/1/012056

Comprehensive design and simulation study of ammonia fuel supply and auxiliary system

Chuanxu Yan[1], Lizhi Ge[2], Qiang Yu[1], Guoyong Lu[2], Lei Qiao[3,*]

[1]Shanghai Zhongchuan NERC-SDT Co., Ltd, Shanghai 201114, China
[2]Kunming Shipborne Equipment Research and Test Center, Kunming 650216, China
[3]College of Engineering Science and Technology Shanghai Ocean University Shanghai 201306, China

[*]Corresponding author's e-mail: m240651761@st.shou.edu.cn

Abstract. As humanity keeps exploring new energy sources, ammonia fuel is gradually coming to the public's attention. Considering the energy supply requirements of ships, this paper utilizes Modelica software to design and simulate the ammonia fuel supply and auxiliary systems for ships. Simulation models for supply pumps, heat exchangers, and valves have been designed, and PID control has been applied to the pump model and a portion of the valve model, thus forming a system model. Simulation tests were carried out on this system model under actual operational conditions, and it was discovered that the model fulfils the energy supply requirements of ships.

1. Introduction

Ammonia fuel, a clean energy with high energy density and good combustion performance, burns into nitrogen and water, making it suitable for ship power. Developing it can free the shipping industry from traditional heavy-oil dependence and cut ship exhaust emissions of sulfur oxides, nitrogen oxides, and particulate matter. However, its toxicity, corrosiveness, and explosiveness demand thorough research on its control system before practical use. Traditional direct physical tests or ship applications pose high safety risks, while semi-physical simulation technology can effectively reduce risks and support control-system debugging with key data.

1.1. Status of research abroad

International research on fuel supply system component models started early and has had notable results. Kyung Nam Chung et al. utilized CFD to study vertical pumps in chemical tankers and explored pump efficiency factors through system modeling [1]. Kolovos et al. simulated diesel injector throttle valves with Navier-Stokes and a density solver, finding friction heating rose with a needle-valve pressure drop and injection pressure affected nozzle vortexes [2]. Joel C, Lisanti et al. applied an actively-driven intake valve to a resonant pulse burner, enabling better operation control than passive/valve-free versions [3]. Roland V. Crystal et al. studied heat transfer and pressure drop in a cross-flow plate-fin heat exchanger with/without spring flutter, showing different impacts in various airflow [4]. Ahmad Al Takash et al. validated the tube-and-fin heat exchanger used for air-con condenser heat recovery by thermodynamic modeling and system analysis [5].

1.2. Status of domestic research

Domestically, Shuaishuai Lv et al. used numerical and experimental methods with SST and Zwart models to simulate axial-flow pump characteristics, finding cavitation is more likely at low flow and occurs on blade suction surface at low inlet pressure [6]. Hu Jie et al. proposed a model for weight-controlled and spring-controlled relief valves, and through experiments, found adding dampers can solve cyclic problems[7]. Yi Lu et al. established a fuel metering valve joint-simulation model and designed segmented control algorithms, synchronizing control signal and rail pressure [8]. Li Shuxun et al. made breakthroughs in flow balance valves, optimizing valve-core opening area, simulating flow rates with CFD, and comparing flow-control accuracy before and after optimization, with the optimized accuracy meeting the 5% requirement [9]. SD Zhou et al. studied the flow and heat-transfer model of a through-type superhydrophobic tube bundle in a heat exchanger, concluding that the superhydrophobic effect enhances heat-transfer performance and reduces frictional resistance [10].

2. Modeling process

The model-building work of this system is based on actual physical equations, aiming to deeply explore the physical mechanisms behind the expected functions of the system. During system model construction, emphasis is placed on simulating and representing the actual fuel supply conditions for ship engines to accurately reproduce the key elements of the fuel supply process.

2.1. Medium model

The medium plays a crucial role as the key substance determining wave propagation. In the ammonia fuel supply and auxiliary system, the media involved include ammonia, ethylene glycol water, fresh water, nitrogen, and compressed air. With the help of REFPROP software, parameter information such as temperature, pressure, density, specific heat capacity, enthalpy, etc. of these media can be obtained. After organizing these query results into a two-dimensional table, MWORKS software can be directly called. Table 1 shows the results obtained from some of the queries.

Table 1. Density-based on ammonia pressure and temperature.

Density (kg/m3)		Pressure (MPa)				
		0.20	0.40	0.60	0.80	1.00
Temperature(K)	213.15	713.69	713.76	713.84	713.92	713.99
	214.15	712.55	712.63	712.71	712.78	712.86
	215.15	711.41	711.49	711.57	711.65	711.73
	216.15	710.27	710.35	710.43	710.51	710.59
	217.15	709.12	709.20	709.28	709.36	709.44

2.2. Tank model

The fuel tank model is mainly responsible for fuel storage tasks, and its interior belongs to a high-pressure environment. This model has functions such as fuel injection and transportation. However, during the refuelling and transfer operations inside the cabin, it may cause changes in the cabin pressure and liquid level height. Therefore, when constructing a fuel tank model, it is necessary to take into account the changes in internal pressure and liquid level.

$$\begin{cases} V_{L0} = h_0 \cdot W_0 \cdot L_0 \\ m_0 = \rho_{L0} \cdot V_0 \\ V_{G0} = (H_0 - h_0) \cdot W_0 \cdot L_0 \\ M_0 = \rho_{G0} \cdot V_{G0} \end{cases} \tag{1}$$

In Formula (1), V_{L0} represents the initial volume of the liquid within the cabin. The symbol h_0 stands for the initial height of the cabin's liquid level, while W_0 indicates the width of the cabin. L_0 is the length of the cabin, and m_0 is the initial mass of the liquid in the cabin. ρ_{L0} is the density of the cabin liquid, and V_0 also refers to the initial volume of the cabin liquid. Additionally, V_{G0} represents the initial volume of the gas, H_0 is the height of the cabin, M_0 is the mass of the gas in the cabin, and ρ_{G0} is the density of the gas.

$$T(t) = \Delta T \cdot e^{(-kt)} + T_{air} \tag{2}$$

In Formula (2), ΔT signifies the temperature difference between the interior and exterior of the tank. The symbol t represents the heat transfer time, while k stands for the heat transfer coefficient. Moreover, Tair denotes the ambient temperature outside the tank.

2.3. Pump model

Pump components play an important role in the transportation of fuel and other substances. According to different demand conditions, the required flow rate of the pump can be determined. On this basis, combined with the "flow speed pressure head interpolation table" and the corresponding pressure head calculation formula, the pump's boosting amount dp can be calculated. The specific calculation process is as follows.

$$pHead = \frac{dp}{\rho \cdot 9.8} + H_f \tag{3}$$

In Formula (3), pHead represents the value of the pressure head. The symbol ρ indicates the value of the substance's density, and Hf stands for the piezoresistive loss.

Both the low-pressure pump and the high-pressure pump employ closed-loop control, and to illustrate the control process, the high-pressure pump will be taken as an example here. As depicted in Figure 1, the PID control of the high-pressure pump is presented. The feedback value is obtained from the fuel pressure sensor located at the inlet of the main engine. This control aims to maintain the pressure sensed by the sensor within the range of (85±2) bar. That is to say, by comparing the actual pressure with the preset value of 85 bar, the PID controller adjusts the speed setting of the high-pressure pump, thus keeping the inlet pressure of the main engine stable within the specified range.

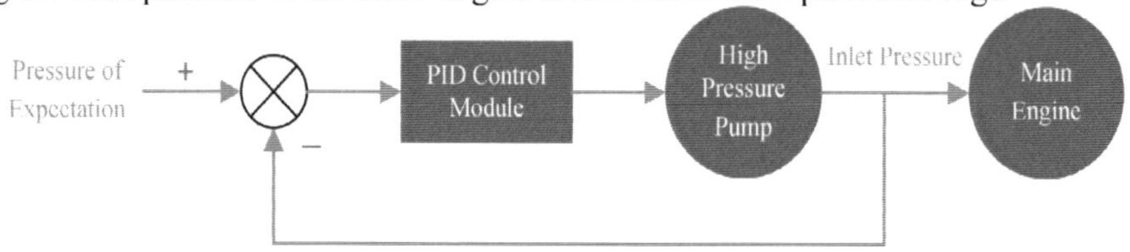

Figure 1. PID control of high-pressure pump.

2.4. Heat exchanger model

The main function of heat exchangers is to heat or cool the fuel. When modeling it, it is based on the law of conservation of energy, which is the energy balance relationship between the absorbed heat, released heat, and transferred heat.

$$Q_1 = Q_2 = Q_3 \tag{4}$$

In Formula (4), Q1 stands for the heat that is released, Q2 represents the heat that is absorbed, and Q3 indicates the heat exchange energy of the heat exchanger.

2.5. Valve model

Valve models mainly include two types: opening valves and safety valves. Opening valves control medium flow rate. Safety valves, as special valves, prevent medium pressure in pipelines or equipment from exceeding specified values by discharging the medium outside the system when the pressure rises beyond the specified value.

$$\begin{cases} g_c = \sqrt{\rho / 1000} \\ C_{v1} = C_v \cdot g_c / v_c \\ K_v = C_{v1} / 1.167 \end{cases} \tag{5}$$

In Formula (5), g_c is the specific gravity correction factor. C_v is the internationally recognized flow coefficient, whereas C_{v1} is the adjusted C_v value at a particular opening degree. Additionally, v_c is the viscosity correction factor, and K_v represents the flow coefficient.

$$Q_v = k_v \sqrt{\frac{dp \times 10^{-5}}{r}} \tag{6}$$

In Formula (6), Q_v represents the volume flow rate of the flowing medium. dp is the pressure difference between the valve's inlet and outlet, which is derived from the system. Meanwhile, r stands for the specific gravity of the medium. According to this formula, either the pressure difference or the flow rate can be computed.

2.6. Main engine model
The main engine is the system's fuel-consuming device and can be modeled considering power and fuel consumption at different gears. The comparison table for its power, mass flow, and BSFC is in Table 2.

Table 2. Main engine power, Mass flow, BSFC comparison.

Power percentage	Power(kW)	Mass Flow(kg/h)	BSFC(g/kWh)
100%	10236	4053.456	396
75%	7677	3040.092	396
50%	5118	2026.728	396
25%	2559	1013.364	396

2.7. Generator model
The generator model, a gas-consuming device, generates power by consuming ammonia. Its modelling is similar to the main engine's. However, the difference in consumption quantity is not detailed here.

3. Simulation system construction and result analysis

In the actual operational conditions, certain component models underwent simulation and analysis in conjunction with the entire system, and the outcomes are presented below.

3.1. Pump model reliability analysis
The pump simulation test model is constructed in the following manner and is intended to be tested under diverse working conditions. The pressure head value of the pump model is depicted in Figure 2.

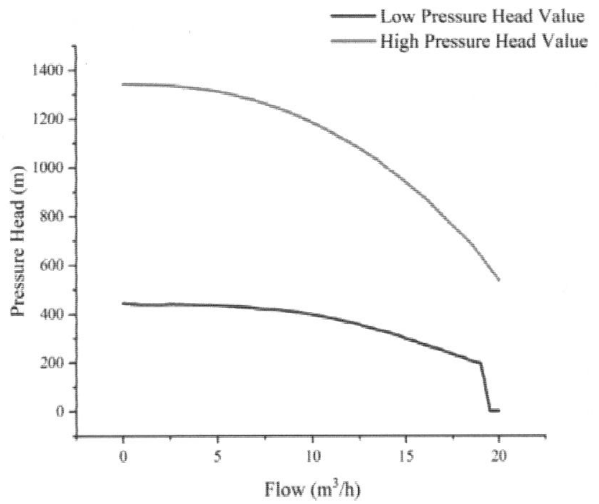

Figure 2. Performance curve of booster pump (taking 3600rpm as an example).

The comparison between simulation results and reference data is shown in Table 3.

Table 3. Comparison between simulation results and reference data.

Variable Name	Unit	Model simulation results	Reference value	Error rate
PHead	m	274.88	276.6	0.63%
Power	kW	98.315	98.5	0.19%
Efficiency	%	24.795	24.8	0.02%

The simulation data of the pump model has an error margin of 5% compared to the reference data and fulfils the system's requirements.

3.2. Heat exchanger model reliability analysis

Heat exchangers employ glycol water for heating or cooling ammonia fuel.

The inlet for operating condition 1 is shown in Table 4.

Table 4. Condition 1 inlet.

Medium	Inlet flow	Inlet temperature	Inlet pressure	Required area	Heat-transfer coefficient
Liquid ammonia	6000kg/h	-10°C	10MPa	36.95m^2	593 W/m^2.°C
Ethylene glycol water	50000kg/h	50°C	0.6MPa		

The inlet for operating condition 2 is shown in Table 5.

Table 5. Condition 2 inlet.

Medium	Inlet flow	Inlet temperature	Inlet pressure	Required area	Heat-transfer coefficient
Liquid ammonia	6000kg/h	50°C	10MPa	10.66m^2	580W/m^2.°C
Ethylene glycol water	50000kg/h	20°C	0.6MPa		

The temperature data comparison of liquid ammonia and ethylene glycol water under two different operating conditions is shown in Figure 3.

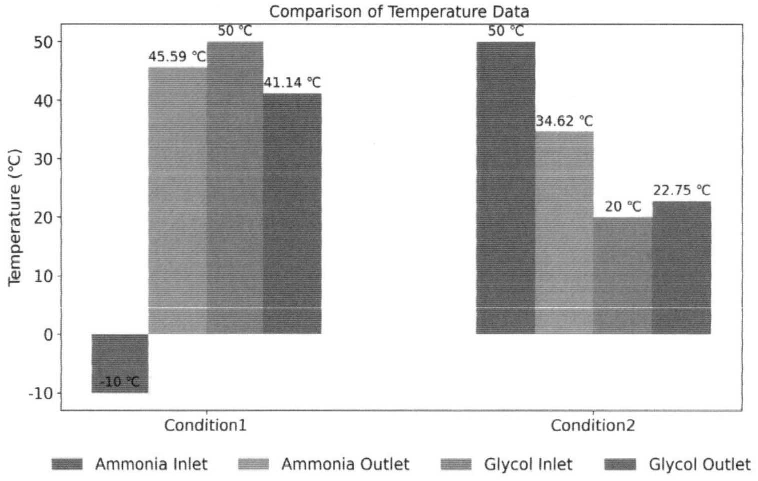

Figure 3. Comparison of temperature data.

Based on the simulation results of the outlet temperatures of liquid ammonia and ethylene glycol water, combined with the corresponding reference values, the error rates of the output temperatures under the two operating conditions were calculated, and the results are shown in Table 6.

Table 6. Outlet temperature error.

Operating condition	Liquid ammonia reference	Ethylene glycol water reference	Liquid ammonia error rate	Ethylene glycol water error rate
Condition 1	45°C	40.9°C	1.31%	0.59%
Condition 2	35°C	22.6°C	1.09%	0.66%

From the simulation results, it can be seen that the outlet temperature error meets the accuracy index of no more than 5%.

3.3. Valve model reliability analysis

The valve includes three operating conditions.

The C_v characteristic curve of the valve is shown in Figure 4.

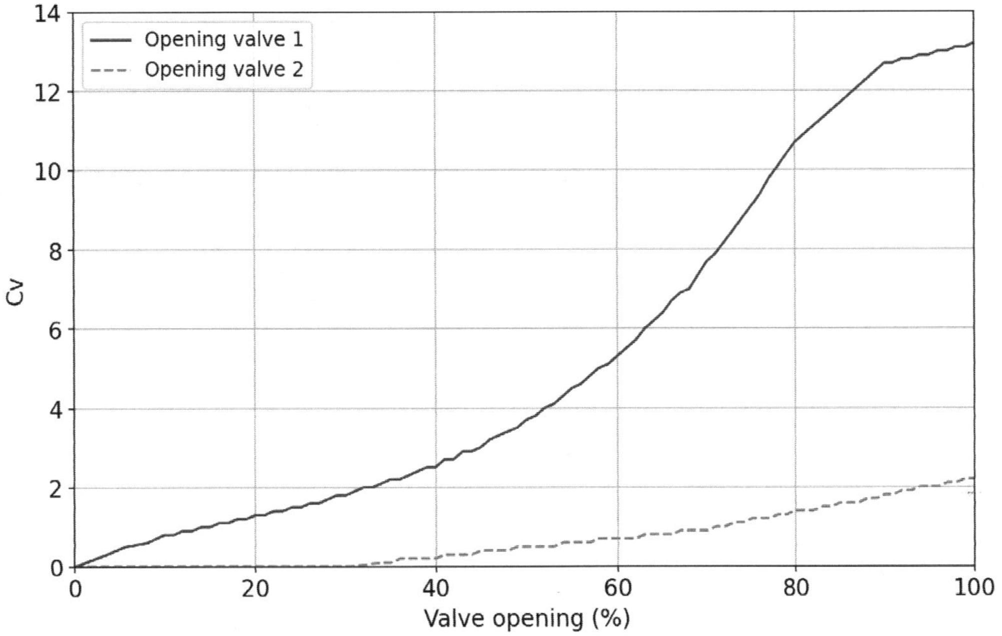

Figure 4. C_v characteristic curve of the valve.

To ensure the effectiveness of the model, multiple characteristic tests were conducted on the valve under different operating conditions.

The inlet parameters for the five working conditions are presented in Table 7.

Table 7. Inlet parameters for five working conditions.

Operating condition	Flow rate	Inlet pressure	Inlet temperature	Kinematic viscosity coefficient	Opening degree
Condition 1	8400kg/h	23bar(g)	25°C	1	69%
Condition 2	8400kg/h	8bar(g)	0°C	1	69%
Condition 3	3400kg/h	8bar(g)	0°C	1	44%
Condition 4	6000kg/h	87bar(g)	25°C	1	75%
Condition 5	1000kg/h	87bar(g)	25°C	1	37%

The comparison of C_V value and outlet pressure under different operating conditions is shown in Figure 5.

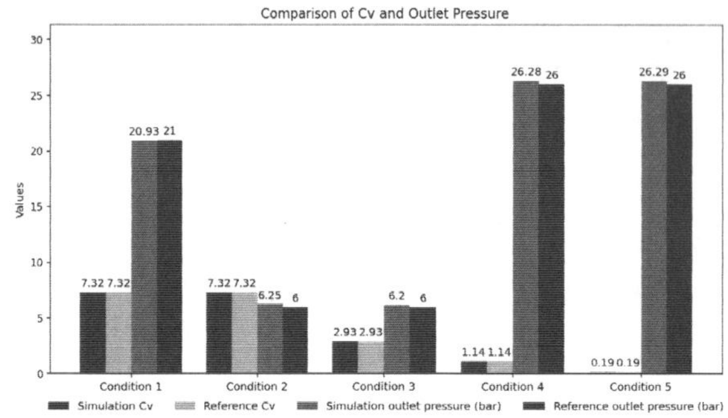

Figure 5. Comparison of C_V value and outlet pressure.

By analyzing the bar chart and comparing the simulation results with their reference values, it can be concluded that the accuracy error is controlled within 5%, which can meet the accuracy requirements of the simulation.

3.4. Typical working condition simulation analysis

Simulation testing of the system model given actual operating conditions. The system framework diagram is shown in Figure 6.

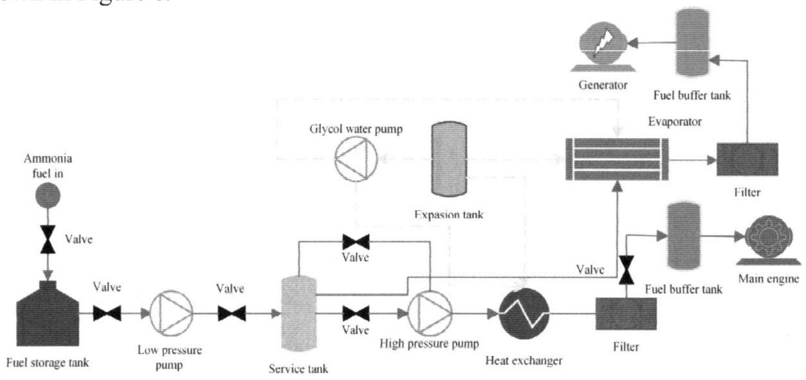

Figure 6. System framework diagram.

To verify the reliability of the ammonia fuel supply system simulation model, we set two typical operating conditions for testing. The inlet conditions for the system model are detailed in Table 8. The simulation results are shown in the following figure, and the main focus is on whether the PID control algorithm can ensure that the host supply pressure meets the preset requirements.

Table 8. Inlet conditions for the system model.

Time(s)	Host set traffic (kg/h)	
	Condition 1	Condition 2
< 100s	4053.46	$-0.2time+11220$
< 200s	4053.46	$0.16time+10285$
< 300s	3040.09	$-5time+12917$
< 400s	3040.09	$1.43time+7722.6$
< 500s	2026.73	0
< 600s	2026.73	0
< 700s	1013.36	5403.35
< 800s	1013.36	5406
< 900s	0	5413.65
< 1000s	0	$0.035time+5414$

The flow input conditions of the main engine under two operating conditions can also be represented in Figure 7.

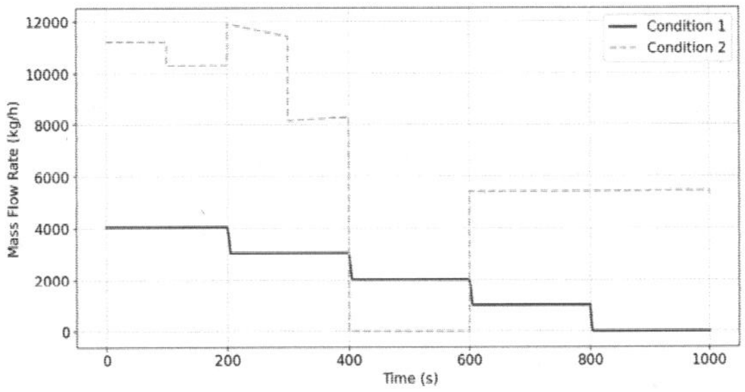

Figure 7. The flow input conditions of the main engine.

In the first operating condition, the mass flow rate starts from nearly 4000kg/h, undergoes two significant decreases, and finally drops to nearly 0kg/h after 800 seconds, simulating a decrease in flow rate. In contrast, the initial value of mass flow rate for condition 2 is relatively high, about 11000kg/h, which sharply decreases to about 8000kg/h at 400 seconds and stabilizes at around 5500kg/h after 600 seconds, showing more stable and sustained quality flow.

The pressure changes at the host supply end under two operating conditions are shown in Figure 8.

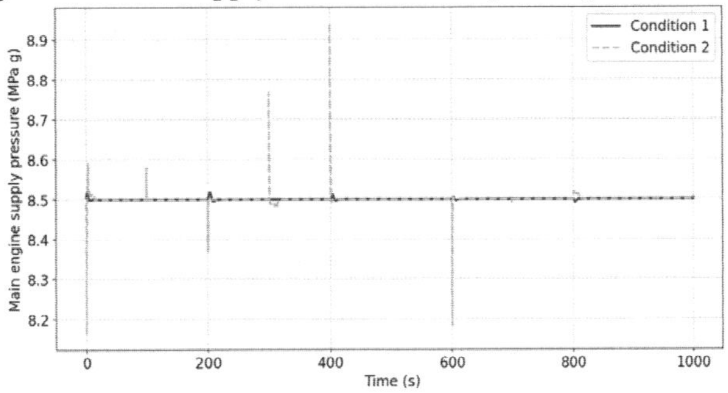

Figure 8. Pressure changes at the host supply end.

According to the simulation results, as time goes by, the flow rate changes, and the host supply end shows a stable pressure level of around 85 bar in both operating conditions. However, condition 2 shows significant instantaneous drops and recovery at more time points compared to condition 1, indicating that the stable pressure performance at the main supply end is better when the flow rate decreases continuously.

The pressure changes at the outlet of the low-pressure pump under two operating conditions are shown in Figure 9.

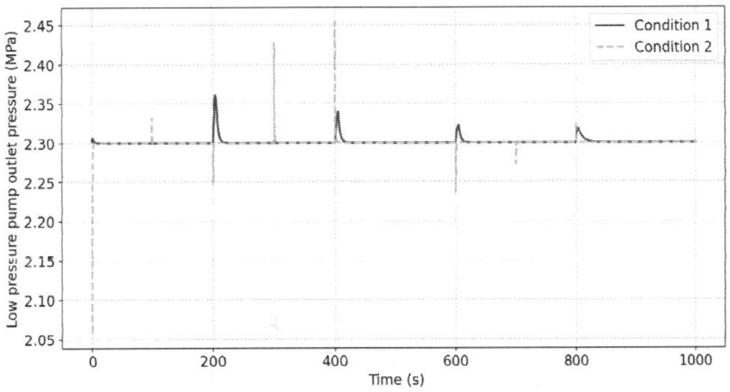

Figure 9. Changes in outlet pressure of low-pressure pump.

According to the simulation results, under operating condition 1, the system exhibits a relatively stable low-pressure pump outlet pressure, although there are slight fluctuations, it remains at around 23 bar overall, demonstrating good pressure control capability. In contrast, the system under condition 2 is more sensitive to changes in flow rate, experiencing multiple significant pressure fluctuations, especially around 200 seconds, 400 seconds, and 800 seconds. This indicates that the stability of flow rate changes will have an impact on the pressure changes at the outlet of the low-pressure pump.

The pressure changes at the outlet of the high-pressure pump under two operating conditions are shown in Figure 10.

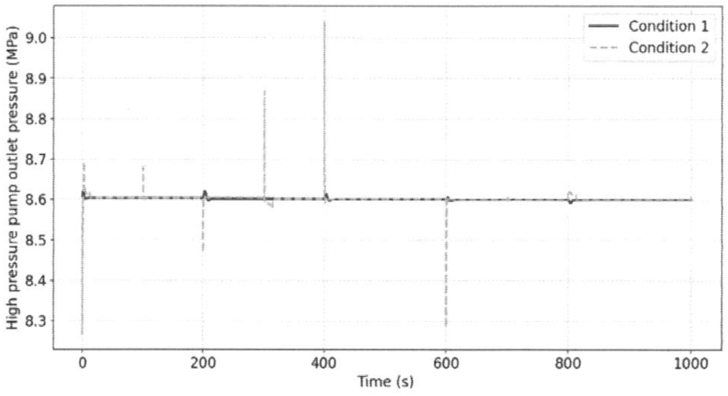

Figure 10. Changes in outlet pressure of high-pressure pump.

According to the simulation results, both operating conditions 1 and 2 exhibit similar pressure fluctuation patterns, but operating condition 2 has a larger fluctuation amplitude, especially with significant pressure peaks occurring at approximately 200 seconds, 400 seconds, and 600 seconds. This indicates that the system is more sensitive to flow changes under operating condition 2, but the pressure of the high-pressure pump eventually stabilizes at around 86 bar in both operating conditions, indicating that the system has a certain degree of self-regulation ability and can recover to a pressure level close to the set value after experiencing brief fluctuations, thereby maintaining the stable operation of the system.

The change in pressure difference of the reflux valve is shown in Figure 11.

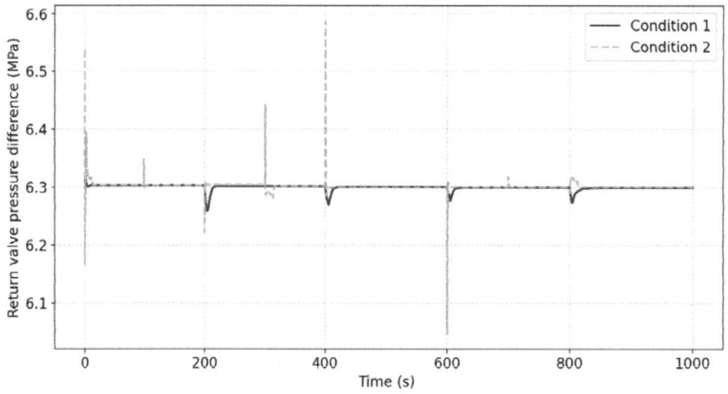

Figure 11. Change in pressure difference of reflux valve.

According to the simulation results, condition 1 shows a relatively stable pressure difference change, although there are some slight fluctuations, the overall pressure remains around 63 bar, indicating that the system has good stability under this condition. However, condition 2 exhibits more frequent and intense fluctuations, especially around 200 seconds, 400 seconds, and 600 seconds, which means that the system is more sensitive to changes in flow rate under operating condition 2.

In summary, it can be seen that regardless of how the flow rate changes under the two operating conditions, the system can work by comprehensively adjusting the two pumps and the reflux valve through PID to stabilize the supply pressure at 85 bar, the low-pressure pump outlet pressure at 23 bar, the high-pressure pump outlet pressure at 86 bar, and the return valve difference at 63 bar, indicating that the system has excellent pressure stabilization performance.

Moreover, the changes in pressure under condition 2 show significant instantaneous drops and recoveries at multiple time points, proving that the system has a very fast dynamic response speed in terms of voltage stabilization, good stability, and anti-interference ability.

This chapter simulated various component models and system models. From the simulation results, it can be seen that the pump model, heat exchanger model, and valve model simulation results are in line with reality and reliable. In addition, the system model was simulated under test conditions, and the main engine gear was changed during the simulation process. By adjusting the speed of the low-pressure pump and high-pressure pump, the boosting amount can be adjusted to ensure that the supply pressure of the main engine meets the specified requirements.

4. Conclusions

This paper employs Modelica software, with necessary simplification and abstraction, and by applying the system engineering method of the model, it accomplishes the construction of models for equipment components such as fuel tanks, pumps, heat exchangers, control valves, and system simulation models. In the system model, PID control is incorporated into the pump model and a portion of the valve model, enabling each model parameter to be adjusted following the inlet pressure value of the main engine. Under actual working conditions, the system was simulated and analyzed, and through the analysis of the simulation results, the key variables fulfil the target requirement of having a static simulation error not exceeding 5%. And the supply pressure of the host can meet the specified requirements by adjusting the speed of the low-pressure pump and high-pressure pump. The model can offer support for subsequent semi-physical simulation test verification of the ammonia supply control system.

References

[1] Chung, Kyung-Nam, et al. "Performance improvement of vertical pumps using CFD". Fluids Engineering Division Summer Meeting. Vol. 42894. 2007.
[2] Kolovos, Konstantinos, et al. "Transient cavitation and friction-induced heating effects of diesel fuel during the needle valve early opening stages for discharge pressures up to 450 MPa". Energies 14.10 (2021): 2923.
[3] Lisanti, Joel C., et al. "Active valve resonant pulse combustor for pressure gain combustion applications". Journal of Propulsion and Power 38.2 (2022): 171-180.
[4] Roland V. Crystal, and Srinivas Garimella. "Enhancement of air-side heat transfer in crossflow heat exchangers with maldistributed flows using auto-fluttering reeds". Applied Thermal Engineering 265. (2025):125472-12547.
[5] Ahmad Al Takash, et al. "Heating water using the hot air of condensers and tubes-and-fins heat exchangers – Thermodynamic modelling and parametric study with economic and environmental insights". Unconventional Resources 6. (2025):100146-100146.
[6] Shuaishuai Lv, et al. "Numerical Simulation and Experimental Study of Cavitation and Flow Characteristics of Axial Pumps". Water 17.1(2024):42-42.
[7] Hu Jie, et al. "Damping Effect on Instability of Pressure Relief Valve Used for Cargo Venting System". Journal of Physics: Conference Series 2610.1(2023).
[8] Yi Lu, et al." Study on dynamic characteristics and control algorithm design for fuel metering valve of high-pressure pump". IFAC Papers On Line 51.31(2018):930-935.
[9] Li Shuxun, et al. "Design optimization and experimental performance test of dynamic flow balance valve". Engineering Applications of Computational Fluid Mechanics 14.1(2020):700-712.
[10] SD Zhou, et al. "Numerical study flow and heat transfer characteristics of superhydrophobic single tube bundle surface and wake flow". Measurement and Control 58.2(2025):196-207.

Research on key parameters of new energy for synchronous stability/instability measures based on trajectory dynamics

Yixuan Chen[1], Shilong Li[2], Benfeng Gao[3], Xue Shen[1], Dajun Si[1] and Hao Yang[2,*]

[1] Power Grid Planning and Construction Research Center, Yunnan Power Grid Company, Kunming 650051, Yunnan, China
[2] Northeast Electric Power University, Jilin 132012, Jilin, China
[3] North China Electric Power University, Baoding 071003, Hebei, China

[*] Corresponding author's e-mail: sxtylsl@163.com

Abstract. With the increasing penetration of renewable energy, such as wind power, its impact on the transient stability of power systems is becoming more significant. This paper analyzes the effects of the grid connection of direct-driven permanent magnet wind turbines on power system transient stability. A transient stability indicator based on trajectory dynamics is proposed, which effectively represents the degree of stability/instability of the grid. This paper also investigates the influence of external loop, inner loop, and voltage traverse control parameters on transient stability. The validity of the proposed indicator is verified by using the IEEE 39-bus system, and the key parameters affecting the transient stability are identified.

1. Introduction

Driven by carbon goals, grid-connected wind power has rapidly expanded, challenging power system transient stability and altering grid structures [1]. Unlike conventional generators, wind power introduces variability and uncertainty, complicating transient stability assessments and masking the influence of renewable control on stability [2-3]. This emphasizes the need for adaptive stability metrics to identify key parameters in renewable-integrated systems.

Traditional transient stability analysis methods include time-domain simulation, transient energy function, and extended equal area criterion (EEAC). Ensuring stability under high wind penetration requires analyzing wind power's impact on system stability. Hou and Gao [4] model an equivalent wind turbine generator and show that large-scale wind integration increases power angle differences, reducing dynamic stability. Zhang and Jin [5] find that dynamic stability varies with fault location and decreases system damping as wind penetration rises. Buragohain and Senroy [6] conclude that wind integration's impact depends on fault conditions and system states.

While time-domain simulations dominate current studies, they lack theoretical support for the results. To address this, recent research focuses on theoretical approaches. Wu et al. [7] use EEAC to assess transient stability in wind-integrated systems by equating wind farm power to synchronous generator mechanical power. Si et al. [8] explore the effect of grid-end inertia on stability with wind integration. Jiang et al. [9] analyze a wind-thermal HVDC system via EEAC, finding that large-scale wind integration reduces transient stability. Chen et al. [10] apply EEAC to study wind penetration's impact on multi-machine stability, incorporating reactive power control of direct-drive generators.

This paper develops a transient stability/instability measure for power grids based on trajectory dynamics, analyzes the impact of key components and parameters of the direct-drive permanent magnet wind turbine on stability, and identifies critical turbine parameters. First, the control components and parameters of the grid-side converter affecting turbine-grid interaction are described. A transient stability measure for renewable-integrated power systems is then constructed, effectively representing grid stability. Finally, the effects of various control parameters of the turbine's grid-side converter on system transient stability are analyzed. The measure's effectiveness is verified through a case study, identifying key parameters influencing grid transient stability.

2. Control loops of direct-drive permanent-magnet wind turbines

Direct-drive permanent magnet synchronous generators (PMSGs) are a key focus in wind power research. This chapter examines the impact of grid-connected wind farms with direct-drive turbines on rotor angle stability, highlighting the typical control loops and parameters of PMSGs. The model uses a full-power converter, with key components including PQ power outer-loop control and current inner-loop control.

2.1. Power outer-loop control module

In the PQ control structure shown in Figure 1, instantaneous active power (P) and reactive power (Q) are compared to reference signals (P_{ref} and Q_{ref}). The error is processed by a PI controller, generating reference signals (i_{dref} and i_{qref}) for the inner-loop control. When discrepancies occur, the PI controller continuously adjusts to minimize the error, driving the system toward stable equilibrium.

2.2. Current inner-loop control module

After Park transformations, the three-phase AC is converted into dq-axis components (i_d and i_q), filtered by a low-pass filter, and compared with the reference signals i_{dref} and i_{qref} from the outer-loop controller. The error is regulated by a PI controller, which also limits the converter's maximum output current. The module then outputs the voltage control signals (P_{md} and P_{mq}).

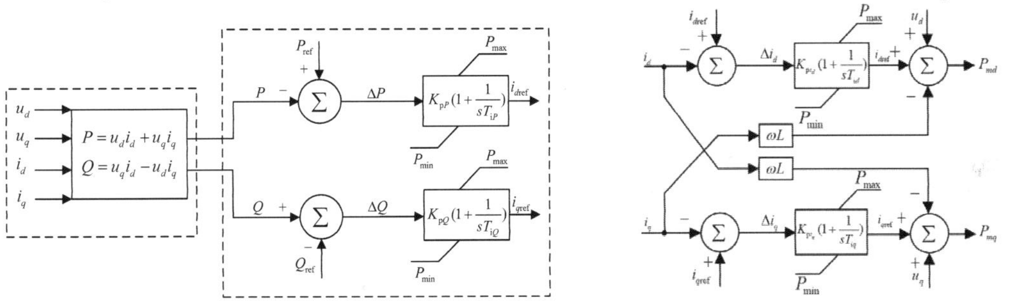

Figure 1. Structure of the constant power outer-loop and current inner-loop control.

2.3. Low voltage ride-through control module

Upon the occurrence of a system fault that causes the voltage to exceed the preset threshold, this module operates in a low-voltage ride-through mode (where the per-unit voltage exceeds $1+\Delta\text{U}$ or falls below $1-\Delta\text{U}$). Equations (1) and (2) represent the output signals under low-voltage ride-through conditions, with K denoting the reactive power support coefficient.

$$i_{dref}=\begin{cases} i_{drf\_\min} & i_{dr}<i_{drf\_\min} \\ i_{dr} & i_{drf\_\min}\le i_{dr}\le i_{drf\_\max} \\ i_{drf\_\max} & i_{dr}>i_{drf\_\max} \end{cases} \tag{1}$$

$$i_{dref}=i_{qref\_ini}+K\cdot\Delta U \tag{2}$$

This chapter details the typical control links and parameters of direct-drive permanent magnet wind turbines, providing a theoretical foundation for establishing system stability metrics and identifying key turbine parameters.

3. Construction of multi-machine system power angle stability indices

3.1. Equivalent modeling of multi-machine systems for cluster-based stability mapping

The reliability of the extended equal-area criterion for assessing transient stability in actual multi-machine systems is uncertain. Its accuracy depends on the precise fitting of the $P-\delta$ curve for the equivalent one-machine infinite bus system. Data fitting methods such as least squares and polynomial regression are used, but the estimated curve often fails to match the actual one, leading to misjudgment. Additionally, the traditional method neglects the impact of system damping on stability and directly equates generator power by $P_m - P_e$, resulting in errors compared to actual research.

Figure 2(a) shows the $P-\delta$ diagram under ideal conditions without considering system damping and the governor. Figure 2(b) is the $P-\delta$ diagram of the equivalent one-machine infinite bus system after mapping the actual multi-machine system.

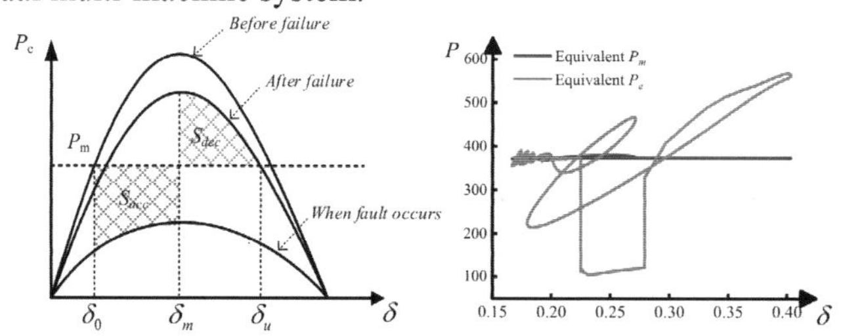

Figure 2. Transient stability analysis of ideal OMIB and actual multi-machine systems.

In a system containing n synchronous machines, considering the impact of damping factors on system stability, the dynamic equation for a single synchronous machine is:

$$\begin{cases} \dfrac{d\delta_i}{dt} = \omega_i \\ M_i \dfrac{d\omega_i}{dt} = M_i \ddot{\delta} = P_{mi} - P_{ei} - D_i \omega_i \end{cases} \tag{3}$$

where D_i represents the damping coefficient of the i-th generator, quantifying the system's damping effect on speed deviations. The system is then simplified to a single-machine-infinite-bus system, represented as:

$$\begin{cases} \ddot{\delta} = \ddot{\delta}_S - \ddot{\delta}_A \\ \dot{\delta} = \dot{\delta}_S - \dot{\delta}_A \end{cases} \tag{4}$$

$$M\ddot{\delta} = \frac{M_A}{M_T}\sum_{i\in S}(P_{mi} - P_{ei}) - \frac{M_S}{M_T}\sum_{j\in A}(P_{mj} - P_{ej}) - D\dot{\delta} \tag{5}$$

where δ signifies the rotor angle of the generator; M represents the inertial time constant of the equivalent single-machine system after reduction; P_m and P_e denote the mechanical and electromagnetic powers of the OMIB (One Machine Infinite Bus) generator, respectively. The influence of the prime mover and governor on P_m, as well as the impact of the excitation control system on P_e, are considered.

3.2 Construction of stability/instability measure index based on trajectory dynamics

For the equivalent one-machine infinite bus system, when it is stable or unstable, the variation of its potential energy V_{PE} along the potential energy curve is shown in Figure 3. In the figure, $\theta_a = \theta(t_a)$, $\theta_b = \theta(t_b)$, and θ_{cr} is the critical value of θ.

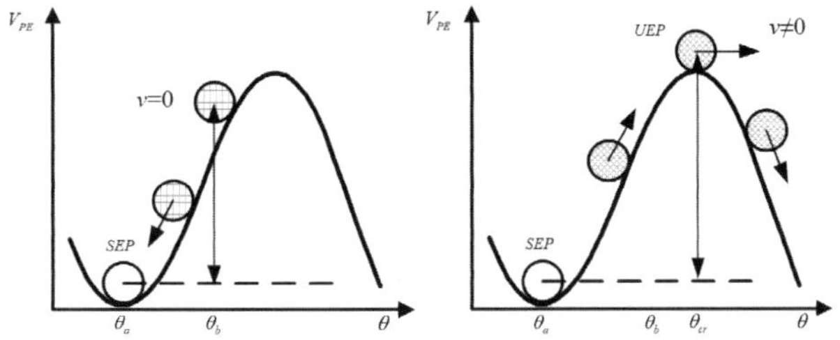

Figure 3. Trajectory dynamics diagrams for stable and unstable systems.

Based on trajectory dynamics, a stable generator should immediately start accumulating potential energy after the last system operation, converting injected kinetic energy. If this conversion is incomplete, the generator will lose stability on the first swing. Stable trajectories differ from unstable ones. For stability, $\frac{dV_{PE}}{d\theta}|_{\theta_b} \neq 0$, and the ball's velocity at the potential energy peak (δ_m) drops to zero, fully converting kinetic to potential energy and returning to the stable equilibrium point (SEP). In contrast, instability occurs when $\frac{dV_{PE}}{d\theta}|_{\theta_b} = 0$, with incomplete energy conversion, and the ball passes the unstable equilibrium point (UEP) with non-zero velocity, leading to an increasing power angle. As θ_b nears θ_{cr}, $\frac{dV_{PE}}{d\theta}|_{\theta_b}$ diminishes to zero.

Based on the analysis of the system trajectory characteristics, we can define the stability and instability indicators of the system as follows:

$$\text{TASI} \triangleq \frac{P_m - P_e - D\dot{\omega}}{V_{PE}(t_b, t_a)} = \frac{M\ddot{\delta}\big|_{t_b}}{\int_{t_a}^{t_b} \omega_N (M\dot{\omega})\omega(t)dt} = \frac{M\dot{\omega}\big|_{t_b}}{\int_{t_a}^{t_b} \omega_N (M\dot{\omega})\omega(t)dt} \tag{6}$$

$$\text{TAUI} \triangleq \frac{\omega_N}{2} M \frac{|\omega(t_b)|\,\omega(t_b)}{V_{PE}(t_b, t_a)} = \frac{\omega_N}{2} M \frac{|\omega(t_b)|\,\omega(t_b)}{\int_{t_a}^{t_b} \omega_N (M\dot{\omega})\omega(t)dt} \tag{7}$$

The TASI (Transient Angle Stability Index) approaches zero only when the system is unstable, with a fixed critical value where TASI≥0. Similarly, the TAUI (Transient Angle Instability Index) is non-zero only when the system is unstable, with a fixed critical value where TAUI≤0.

The trajectory stability analysis method has the advantage of making the stability (TASI) and instability (TAUI) indices independent of critical energy θ_{cr}, offering greater flexibility for complex power systems. It only requires generator time-domain simulation data and trajectory characteristics, effectively quantifying system stability through numerical comparison and helping identify the impact of new energy parameters on transient stability.

4. Example

This case study uses trajectory analysis to quantify the impact of parameters and thresholds on the angular stability of renewable energy units, aiming to identify key parameters for optimal design and operation. The IEEE 10-machine 39-bus system is modified by replacing synchronous generators (G3, G7, and G8) with direct-drive permanent magnet wind turbines. The system's capacity ratio is 1:4.8, with a renewable energy penetration of 32.6%. The modified system structure is shown in Figure 4.

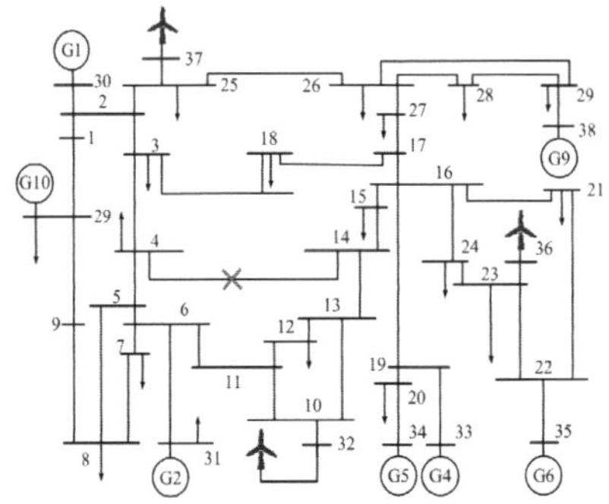

Figure 4. Modified 10-machine 39-bus system.

According to this example, the validity of the proposed stability and instability indexes can be verified. Table 1 shows the variation of the two indexes of the synchronous machine. With the increase of the fault duration T_c, the stability index TASI and instability index TAUI decrease monotonically, and both TASI and TAUI have fixed critical values so that the index can reflect the stability of the generator sensitively and accurately.

Table 1. Change of TASI and TAUI with failure duration.

T_c (ms)	40	60	80	100	150	200	300
TASI	7.984	6.684	4.884	3.752	0.000	0.000	0.000
TAUI	0.000	0.000	0.000	0.000	-0.824	-7.030	$-\infty$

A three-phase short-circuit fault occurs at lines 4-14, with the line disconnected after fault clearance. The fault durations are 100 ms and 150 ms, respectively. Simulation results provide power angles and $P-\delta$ diagrams of the equivalent OMIB system under stable and unstable conditions, as shown in Figures 5(a) to (d). Combining simulation analysis of the impact of external and internal loops, voltage ride-through control, etc. on the stability of the power grid, the trajectory analysis method is used to quantitatively analyze the influence characteristics of parameters and threshold parameters in key links of the new energy unit on the stability of the power angle, and the dominant parameters are determined.

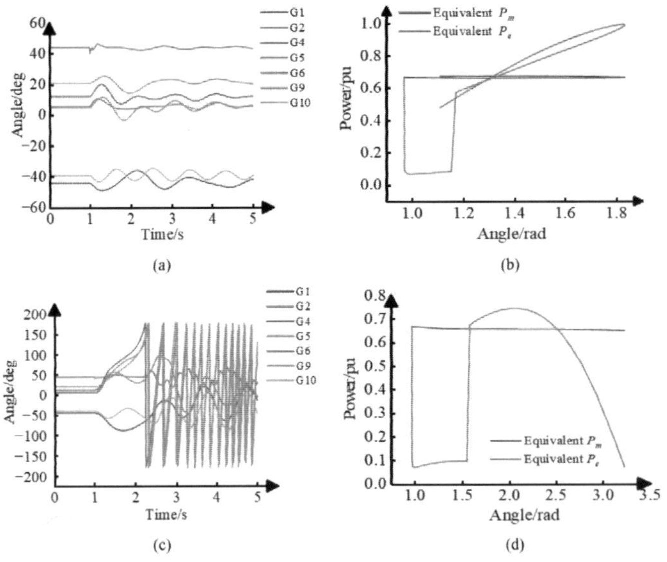

Figure 5. Motor angle curve and equivalent system $P-\delta$ curve under stable and unstable conditions.

4.1 Current inner loop control parameters

4.1.1 Proportional gain Kd and integral time constant Td in active current inner loop controller

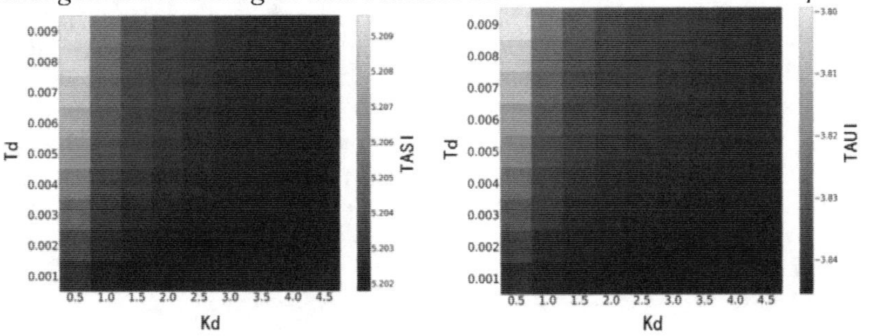

Figure 6. Changes of stability index with proportional gain Kd and integral time constant Td.

It can be seen from Figure 6 that the proportional gain Kd and the integral time constant Td in the active current controller have little influence on the stability index TASI and instability index TAUI. TASI remains near 5.2 and TAUI remains near -3.82. The two indexes are saturated and the change tends to be stable.

4.1.2 Proportional gain Kd and integral time constant Td in reactive current inner loop controller

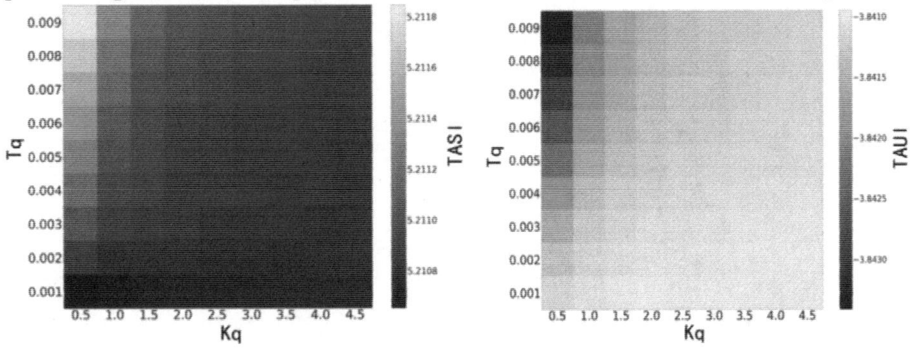

Figure 7. Changes of stability index with proportional gain Kq and integral time constant Tq.

As shown in Figure 7, the proportional gain Kq and integral time constant Tq in the reactive current controller have minimal impact on the stability index TASI and instability index TAUI.

4.2 Power outer loop control parameters

4.2.1 Active power outer loop control gain Kp and outer loop control time constant Tp

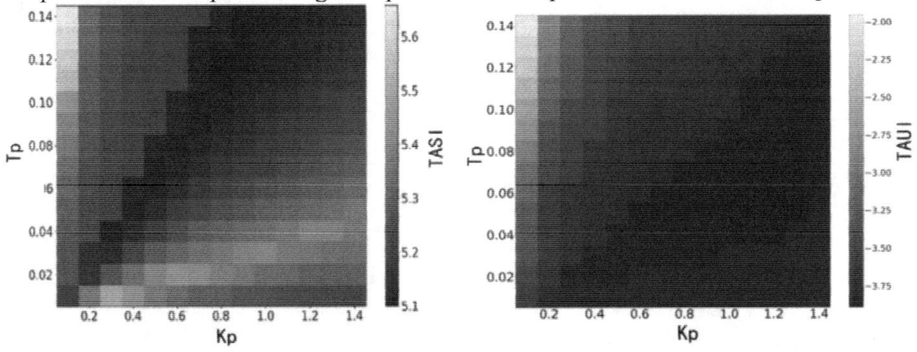

Figure 8. Changes of stability index with proportional gain Kp and integral time constant Tp.

4.2.2 Reactive power outer loop control gain Kp and outer loop control time constant Tp

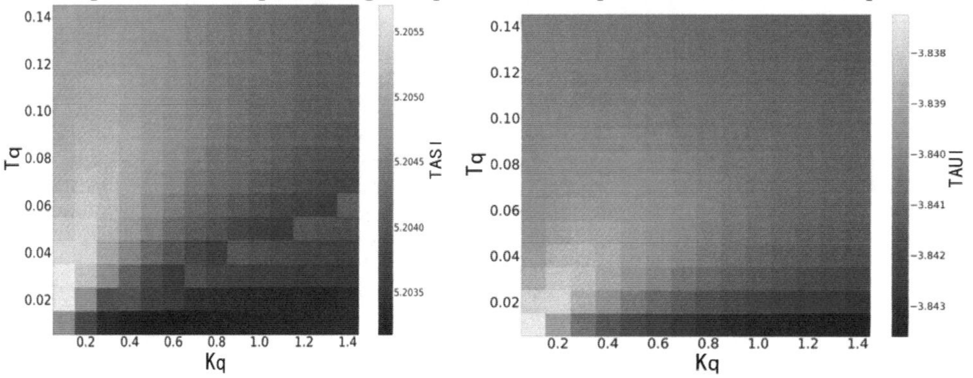

Figure 9. Changes of stability index with proportional gain Kq and integral time constant Tq.

As shown in Figure 8, the active power outer-loop control gain Kp significantly affects system stability. Both the stability index TASI and instability index TAUI vary with Kp, improving stability in stable and unstable scenarios. Small Kp values may lead to extreme instability. As shown in Figure 9, loop control gain Kq and time constant Tq have minimal impact.

4.3 Voltage ride-through parameters

4.3.1 Voltage dead zone ΔU

Table 2. Impact of ΔU on stability indicators.

ΔU (p.u.)	0.10	0.15	0.20	0.25
TASI	7.3807	7.5861	7.6805	7.8428
TAUI	-4.5194	-6.5186	-8.5457	-9.4532

As shown in Table 2,ΔU is the threshold for the system to enter voltage ride-through control and has a significant impact on both TASI and TAUI. It can be considered as one of the dominant parameters affecting angular stability.

4.3.2 Reactive power support coefficient $K\_\Delta U$

Table 3. Impact of $K\_\Delta U$ on stability indicators.

$K\_\Delta U$	1.50	2.10	2.40	3.00
TASI	7.3915	7.3720	7.3638	7.3497
TAUI	-6.9136	-5.2492	-4.7485	-4.0623

$K\_\Delta U$ is the gain used to adjust reactive power output, maintaining voltage stability and improving power factor. As can be seen from the results in Table 3, within an appropriate range (as per national standards: 1.5–3.0), the setting of K_ΔU needs to consider factors such as wind turbine capacity, grid requirements, and system stability.

5. Conclusion

This study proposes two stability/instability indices, TASI and TAUI, based on trajectory dynamics to improve transient stability analysis in renewable-integrated power systems. Unlike existing methods, this approach offers significant improvements, providing robust stability analysis independent of system model parameters. The effectiveness of TASI and TAUI is validated in the IEEE 10-machine 39-bus system, showing that key parameters such as the active power outer-loop control gain Kp, voltage dead zone, and reactive power support coefficient K are critical to system stability.

Acknowledgment

The work was supported by the China Southern Power Grid Science and Technology Project (YNKJXM20230338).

References

[1] Chen G.P., Li M.J., Xu T., et al. Study on Technical Bottleneck of New Energy Development [J]. Proceedings of the CSEE, 2017, 37(01): 20-27. DOI: 10.13334/j.0258-8013.pcsee.161892.
[2] Yousefian R., Bhattarai R., Kamalasadan S. Transient stability enhancement of power grid with integrated wide area control of wind farms and synchronous generators [J]. IEEE Transactions on Power Systems, 2017, 32(6): 4818-4831.
[3] Zhang L.X., Wind farm grid connection technology and its influence on the power system [J]. China High and New Technology, 2024, (16): 108-110. DOI: 10.13535/j.cnki.10-1507/n.2024.16.33.
[4] Hou J. and Gao S., "Static and Transient Stability Analysis of Power System Containing Wind Farms," 2021 IEEE Sustainable Power and Energy Conference (iSPEC), Nanjing, China, 2021, pp. 303-309, doi: 10.1109/iSPEC53008.2021.9735740.
[5] Zhang L. and Jin J., "Study of PSASP-based Battery Storage Station to Improve Transient Stability of Wind Power System," 2022 Asia Power and Electrical Technology Conference (APET), Shanghai, China, 2022, pp. 350-356, doi: 10.1109/APET56294.2022.10073407.
[6] Buragohain U. and Senroy N., "Reduced Order DFIG Models for PLL-Based Grid Synchronization Stability Assessment," in IEEE Transactions on Power Systems, vol. 38, no. 5, pp. 4628-4639, Sept. 2023, doi: 10.1109/TPWRS.2022.3215125.
[7] Wu Q.H., Lin Y., Hong C., et al. Transient stability analysis of large-scale power systems: A survey [J]. CSEE Journal of Power and Energy Systems, 2023, 9(4): 1284-1300.
[8] Si Y., Wang Z., Liu L. and Anvari-Moghaddam A., "Impacts of Uncertain Geomagnetic Disturbances on Transient Power Angle Stability of DFIG Integrated Power System," in IEEE Transactions on Industry Applications, vol. 59, no. 2, pp. 2615-2625, March-April 2023, doi: 10.1109/TIA.2022.3231582.
[9] Jiang Q., Liu T.Q., Zeng X.Y., et al. Influence Mechanism of Comprehensive Action of Large-Scale Wind Power and HVDC Transmission System on Sending System Transient Stability [J]. Power System Technology, 2018, 42(07): 2038-2046. DOI: 10.13335/j.1000-3673.pst.2018.0356.
[10] Chen W., Wu X., Xu Y., He J., Shi Z. and Liu C., "EEAC-based Transient Stability Analysis of Hydropower System Considering Different Practical Hydroelectric Generator Models," 2019 IEEE 3rd International Electrical and Energy Conference (CIEEC), Beijing, China, 2019, pp. 1624-1629, doi: 10.1109/CIEEC47146.2019.CIEEC-2019574.

Technical analysis of full-power frequency conversion pumped storage unit

Wei Wei[1*], Shijun Ma[1], Dongping Ai[1], Lijie Ding[2], Huabo Shi[3], Xi Wang[3]

[1] Chinese Academy of Electrical Sciences, Haidian District, Beijing, 100192, China
[2] Power Internet of Things Key Laboratory of Sichuan Province, Wuhou District, Chengdu, 610065, China
[3] State Grid Sichuan Electric Power Company Electric Power Research Institute, Gaoxin District, Chengdu, 610000, China

[*] Corresponding author's e-mail: jsp.chn@outlook.com

Abstract. This paper analyzes the principle, advantages, and application prospects of full-power variable frequency pumped storage unit technology. Compared with the traditional fixed-speed unit, the full-power variable speed technology realizes speed adjustment through the inverter, which has the advantages of a simple structure, strong head adaptability, and wide power adjustment range, especially suitable for small and medium-sized pumped storage power stations. Although limited by the difficulty and cost of manufacturing high-power inverters, their costs are expected to decrease with technological progress and large-scale application. The simulation results show that the full-power variable-speed unit performs well in the frequency regulation of the power grid. In the future, improving the design and promoting the demonstration project, will provide more flexible adjustment capacity for the new power system and help achieve the "double carbon" goal.

1. Introduction

The pumped storage power station is widely regarded as the most well-established, cost-effective, and largest form of energy storage available. It offers flexible and reliable operation, quick transitions between operating conditions, and minimal environmental impact [1]. This technology plays a crucial role in managing peak demand, optimizing renewable energy utilization, and stabilizing power output. The National Development and Reform Commission aims to increase the total installed capacity of pumped storage power stations to 100 million kW by 2025, representing approximately 4% of the nation's total power capacity. However, the current reliance on constant-speed units in China's pumped storage facilities does not adequately address the recent rapid fluctuations in power demand within the grid [2].

In the present energy landscape, pumped storage power stations are considered the most durable, high-capacity, reliable, cost-effective, and advanced form of energy storage with environmentally friendly attributes [3]. These facilities are known for their rapid start-up, efficient operations, and ability to balance power supply during peak and off-peak periods. By adapting to power system load adjustments, supporting thermal and nuclear power generation, and smoothing out output from renewable sources, pumped storage power stations enhance the overall economic viability of the grid [4]. However, conventional pumping units typically operate at a fixed speed, leading to lower efficiency, slower regulation, and an inability to adjust water pumping based on system demands or regulate reactive power efficiently, hindering quick and precise grid frequency adjustments.

Continuous variable-speed pumped storage technology represents a significant advancement in the industry over recent decades [5]. Compared to traditional fixed-speed systems, it offers improved flexibility to handle head variation [6], higher efficiency, enhanced stability, and faster response times [7]. These units can adjust active and reactive power within milliseconds, stabilize grid fluctuations, support higher new energy consumption, ensure grid reliability, and facilitate seamless integration with power systems [8]. As a critical tool for increasing renewable energy utilization, continuous variable-speed pumped storage has become a focal point of research and development in the sector [9].

Japan has taken advantage of the tremendous flexibility offered by variable-speed pumped storage power plants in terms of frequency regulation, reducing the number of large thermal power plants operating as reserves at night [10]. In 1972, the former Soviet Union put into operation two 50 MW variable speed units at the Ioskev hydropower station; In Switzerland, a fixed-speed hydro-generator set is converted to a variable-speed hydro-generator set using a full-power converter, and its actual power can respond quickly to a given power [11]; To improve the flexibility of the power grid, countries such as India, France, and Portugal have also invested in a certain number of variable-speed hydro-generator units; The United States plans to gradually convert all hydropower units to variable speed hydropower units [12].

In 2010, three variable-speed hydroelectric generators independently developed by Harbin Electric Group Co., Ltd. were used in the Yili People's Hydropower Station in Xinjiang and successfully connected to the grid for power generation. The second phase of the Fengning Pumped Storage Power Station is planned to use two variable-speed hydro-generator sets [13]. In the implementation plan of hydropower equipment, China proposes to tackle the technical problems of variable-speed pumped storage units and vigorously develop variable-speed units [14].

2. Technical characteristics of full-power variable speed pumped storage unit

2.1. Principle of full-power variable speed technology

The full-power variable speed unit connects a frequency converter with the same capacity as the generator motor between the generator motor stator and the power grid and changes the unit speed by changing the frequency of the three-phase magnetic flux of the stator. Its technical principle is shown in Figure 1.

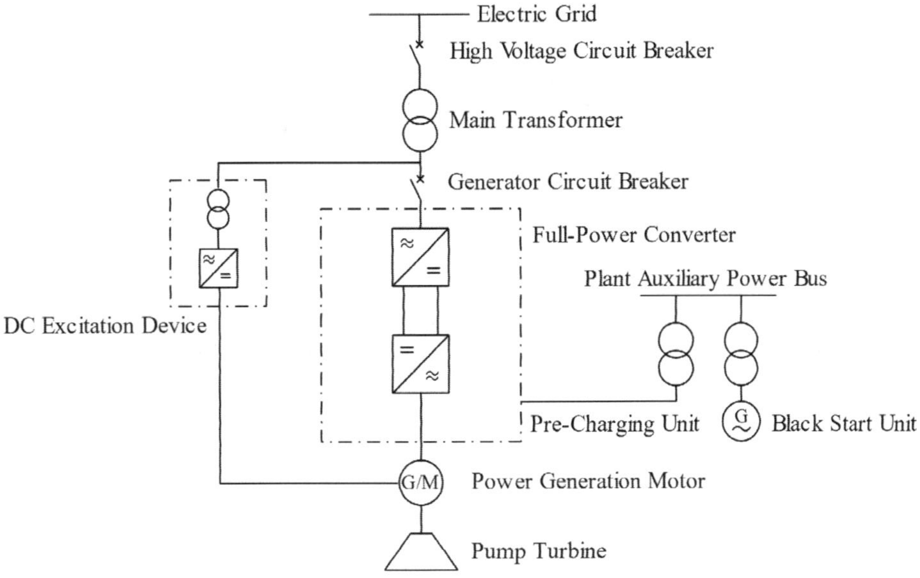

Figure 1. Schematic diagram of full-power variable speed technology.

2.2. Technical advantages of full-power variable speed pumped storage unit

Compared with the AC excitation variable speed technology, the full-power variable speed technology has the following advantages [3].

(1) Simple structure and low manufacturing difficulty. The generator motor of the full-power variable speed unit is a conventional unit and only a frequency converter with the same capacity as the

generator motor is installed between the generator motor stator and the grid. The rotor of the generator motor of the AC excited variable-speed unit is a three-phase wound winding, which has a complex structure and is difficult to design and manufacture.

(2) The adaptability of water head variation amplitude is stronger. The speed variation range of full-power variable speed units is generally up to 30%, while the speed variation range of AC excitation variable speed units is only about 7%. This means that the full-power variable-speed unit can adapt to the larger head variation of the pumped storage power station, especially in the small and medium-sized pumped storage power stations with low heads, where the unit has better operation stability and higher efficiency.

(3) The power regulation range is wider. The full-power variable speed unit completely isolates the generator motor from the power grid and can realize the complete adjustment of power under the working condition of the turbine. Under the working condition of the water pump, the power variation range can reach more than 50%, which is far more than the power variation range of AC excitation variable-speed unit of about 30%, which can further improve the operation flexibility of the unit and enhance the capacity of the unit to absorb new energy [4].

(4) It can realize active and reactive power-independent decoupling control. The full-power variable speed unit can quickly respond to the change of active power through automatic frequency control to maintain the stability of grid frequency. It can realize rapid compensation and absorption of reactive power of the power grid under various operating conditions (even shutdown), which is not available for AC excited variable-speed units. Table 1 shows the comparison of full-power variable speed and AC excitation variable speed technologies.

Table 1. Comparison of full-power variable speed and AC excitation variable speed technology.

Characteristic	Full-power variable speed	AC excitation variable speed
Converter power	**1/8~1/6** of the unit capacity	**100**% unit capacity
Motor type	Asynchronous motor rotor AC excitation leads to a more complex motor structure	Traditional synchronous motor
Variable speed range	±7%, restricted by rotor power	Theoretically ± **100**%, limited by the characteristics of the water turbine, it can generally reach ± **30**%
Reactive power capability	The reactive capacity is small, and the reactive power loss is large	The reactive power capacity is very large and can even output reactive power in a stationary state
Maximum capacity of units already in operation	**400** MW	**100** MW

It should be pointed out that although the technical performance of full-power variable speed technology is superior to that of AC excitation variable speed technology, due to the manufacturing level and cost of the high-power inverter, full-power variable speed technology is economical only when it is applied to small and medium-sized units (about 110 MW and below). AC excitation variable speed units are mostly used for units with a capacity above 110 MW. At present, the development of pumped storage is mainly concentrated in large power stations, so the engineering application performance of AC excitation variable-speed units is much higher than that of full-power variable-speed units [5].

2.3. Application prospect of full-power variable speed pumped storage unit

With the rapid expansion of pumped storage in China, the development strategies for small and medium-sized pumped storage power stations (those with an installed capacity of less than 300 MW) have gained increasing diversity and attention. These smaller facilities, compared to their larger counterparts, offer abundant resources, flexible layout options, and shorter construction timelines (typically 3-5 years). They can be developed through new construction projects, conversion of existing conventional hydropower stations, or through integrated development with distributed energy facilities. By connecting to local power grids, they can contribute to a more efficient utilization of renewable energy sources, enhanced grid resilience, and improved overall energy consumption. Small and medium-sized pumped storage stations show promise in areas with growing distributed power sources, limited peak shaving capacity, challenges in expanding distribution networks, and weaker connections to the main grid.

The development of small and medium-sized pumped storage power stations has made up for the defect that the full-power variable-speed technology is only applicable to small and medium-sized units and widened its application scenarios [6].

(1) The production cost after large-scale application is close to that of constant speed units. The production cost difference between the full-power variable speed unit and the constant speed unit is mainly reflected in the full-power frequency converter. Taking a 100 MW full-power inverter as an example, its price is about 80 million yuan. With the expansion of the application scale of full-power variable speed units, the localization of high-power inverters through technology introduction and absorption will greatly reduce their manufacturing cost. In addition, due to the variable speed design, several standardized series products can cover a wide range of applications, further reducing the development and manufacturing cost of the unit, and even having a cost advantage over the constant speed pumped storage unit. According to the data provided by a foreign main engine equipment manufacturer, only three sets of finalized designs are required to cover more than 90% of the application range of 30 MW units in the 50~200 m water head section [7].

(2) Various application modes. Full-power variable speed units can be directly applied to new small and medium-sized pumped storage power stations, and can also be retrofitted to constant speed units by installing full-power frequency converters in existing small and medium-sized pumped storage power stations where conditions permit, to give full play to its technical advantages and further improve the unit's regulation capacity and operation flexibility.

3. Operation performance analysis of full-power variable speed pumped storage unit

In a traditional turbine setup, the speed remains constant and is typically designed to operate at the most efficient speed for the given conditions. However, any changes in conditions, such as the head, will cause the turbine's speed to fluctuate. When the speed drifts away from the ideal point, the turbine's efficiency decreases, causing an increase in cavitation coefficient and overall losses. This can impede the unit's performance. In contrast, variable-speed pumped storage units use a coordinated control system for the AC excitation and water turbine speed, ensuring that the units consistently operate close to the optimal speed. This enables the generator motor to maintain a constant frequency while adjusting speeds as needed.

The components of a water turbine and its control system consist of a pump-turbine with its associated water diversion system, as well as a governor with its servo system. The governor system of the turbine controls the flow of water by changing the opening of the guide vane of the turbine, to adjust the speed and power [8].

The flexible and dependable operation features of variable-speed pumped storage units are crucial in the power system, as they can rapidly adjust their output power based on grid requirements. In addition, for pumping and storage units, the water head of the reservoir is different in different seasons or at different impounding stages of the unit. Therefore, to realize the efficient operation of a water turbine, it is necessary to select the optimal speed according to its output power and current head.

According to the characteristic curve of the turbine, the relationship between the unit speed of the turbine n' and the current head H can be expressed as:

$$n' = n_r D / \sqrt{H} \tag{1}$$

where n_r is the speed of the motor and D is the nominal diameter of the runner.

When the head changes, the unit speed of the turbine will change for a constant motor speed. The optimal unit speed of the turbine will also change when the unit flow passing through the turbine is different. When the unit speed of the turbine deviates from the optimal unit speed, the efficiency of the turbine will decline.

The variable speed pumped storage unit can make the pump-turbine work at the optimal unit speed and maintain the highest efficiency under the current output power and head by using the advantages of adjustable motor speed. That is, according to the current output power and head, the optimal unit speed of the water turbine is calculated, and then the optimal speed is obtained. The guide vane opening is adjusted by the governor of the pump-turbine, and the excitation current is adjusted by the AC excitation system of the generator motor, to realize the adjustment of the unit speed and output power through coordinated control.

According to the comprehensive characteristics of hydrodynamics and water turbine, the output P of the generator motor system can be deduced:

$$P = HQ\gamma\eta_0 = 9.81HQ\eta_0 \tag{2}$$

where H is the working head of the pump-turbine, Q is the flow through the pump-turbine, $\eta_0 = \eta_W \cdot \eta_E$ is the overall efficiency of the system, η_W is the efficiency of the turbine, η_E is the efficiency of the generator motor. Generally speaking, the efficiency of the generator motor is very high, which can be taken as $\eta_E = 0.98$. γ is the proportion of water.

The unit output of the turbine is:

$$P' = \frac{P}{(\eta_E D^2 H^{\frac{3}{2}})} \tag{3}$$

According to the similarity theory of the pump-turbine, the unit flow of the pump-turbine is:

$$Q' = \frac{Q}{(D^2\sqrt{H})} \tag{4}$$

Bring Equations (3) and (4) into Equation (2), it can be obtained:

$$P' = 9.81Q'\eta_W \tag{5}$$

According to Equation (5), to make the pump-turbine work at optimal efficiency, it is necessary to calculate the unit output at this time according to the output power of the current variable-speed pumped storage unit and the working head of the pump turbine. Then, the unit flow at the highest efficiency is calculated according to the comprehensive characteristics of the pump-turbine, to calculate the corresponding optimal unit speed, and then obtain the optimal speed of the pump-turbine. The specific steps are as follows:

(1) According to the given output power of the variable-speed pumped storage unit and the real-time working head of the pump-turbine, the corresponding unit output of the pump-turbine is calculated:

$$P' = \frac{P}{(\eta_W D^2 H^{\frac{*3}{2}})} \tag{6}$$

(2) The maximum output curve is obtained according to the maximum efficiency curve in the comprehensive characteristics of the pump turbine. Then, the unit flow with the highest efficiency is obtained according to the unit output of the pump turbine obtained in Step (1). According to the maximum efficiency curve, the maximum efficiency of the pump turbine at this time can be obtained.

(3) According to the efficiency peak curve $n' = f_0(Q')_{\eta_W=\eta_{Wmax}}$, the corresponding optimal unit speed is calculated from the unit flow obtained in Step (2).

(4) According to Equation (1), the optimal speed of the pump-turbine is $n_{ref} = n'\sqrt{H}/D$.

(5) According to the guide vane opening curve $\mu_0 = f_3(Q^{'})$, the corresponding guide vane opening of the pump turbine is calculated from the unit flow obtained in Step (2).

(6) According to the obtained optimal speed and guide vane opening command, the guide vane opening of the pump turbine is adjusted to realize the speed regulation of the pump turbine.

According to the specific steps of the above regulation process, the optimal speed tracking of the pump-turbine mainly follows the efficiency peak curve trajectory of the pump-turbine $n^{'} = f_0(Q^{'})_{\eta_W=\eta_{Wmax}}$. The conditions to be known in the tracking process include the efficiency peak curve, maximum efficiency curve, maximum output curve, and guide vane opening curve of the pump turbine $\mu_0 = f_3(Q^{'})$, which can be found from the comprehensive characteristics of the pump turbine.

Figure 2. displays the detailed performance curve of the turbine. It illustrates the relationship between unit flow on the horizontal axis and unit speed on the vertical axis. By coordinating the governor to adjust flow and the excitation to adjust speed, the operating point of the turbine can be consistently maintained at the design working point when the operating head of the turbine varies. This ensures that the turbine operates under optimal conditions.

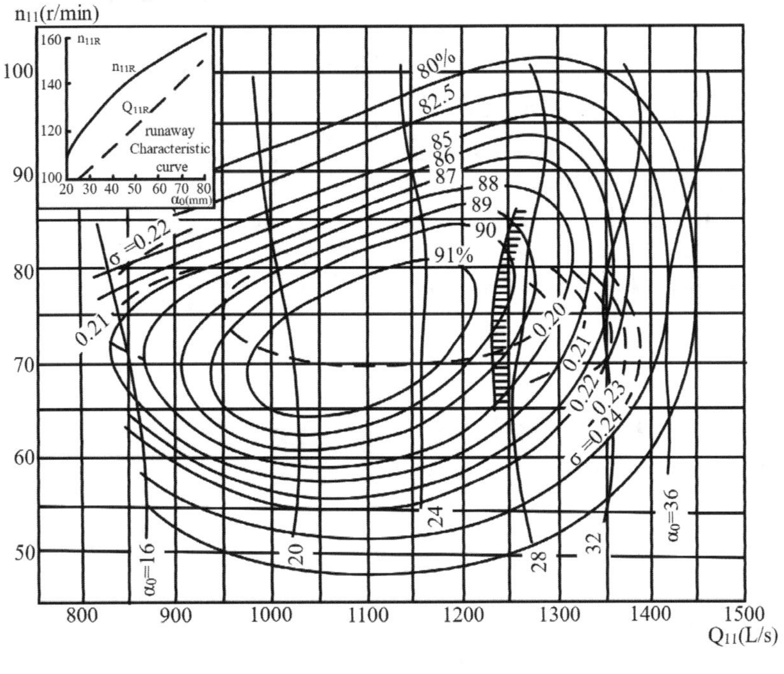

Figure 2. Comprehensive characteristic curve of water turbine.

4. Simulation analysis on frequency regulation of full-power variable speed pumped storage power station connected to a regional power grid

(1) Comparison diagram of single unit under power generation conditions

In the usual scenario, when the variable-speed pumping and storage unit and the traditional pumping and storage unit are connected in a specific area and operate as PQ nodes, the system behaves as follows: following a failure during power generation, the unit is deactivated, and the output of the traditional pumped storage unit gradually rises after initial oscillation. Subsequently, the output of the variable-speed pumped storage unit sharply spikes, with a significantly greater increase compared to the traditional pumped storage unit.

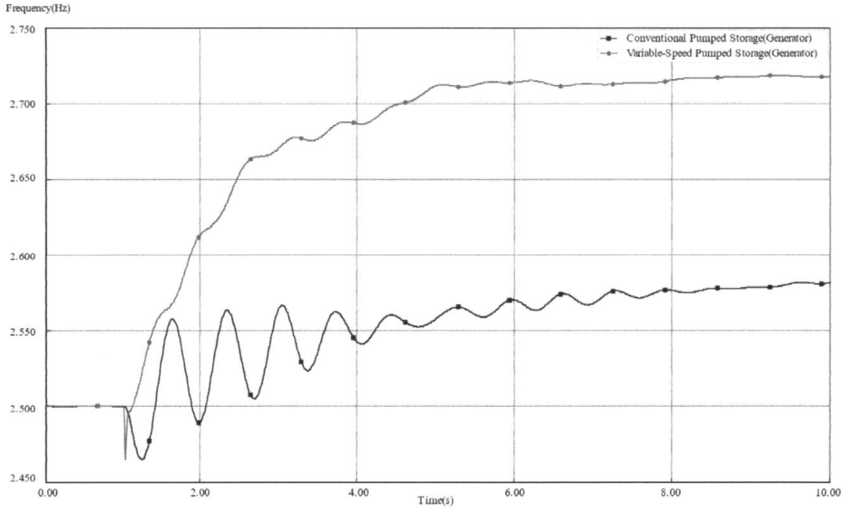

Figure 3. Active power curve of hydropower station unit after partial unit removal under power generation conditions.

It can be seen from Figure 3 that under the typical mode, if 0, 1, 2, 3, and 4 conventional pumping and storage units are replaced with variable-speed pumping and storage units in a certain area and operate in the form of a PQ node. As illustrated in Figure 4., turning off the unit during a fault in power generation results in reduced frequency fluctuations in the system post-fault with more variable speed pumped storage units in operation. Additionally, the frequency drops after reaching stability is also minimized.

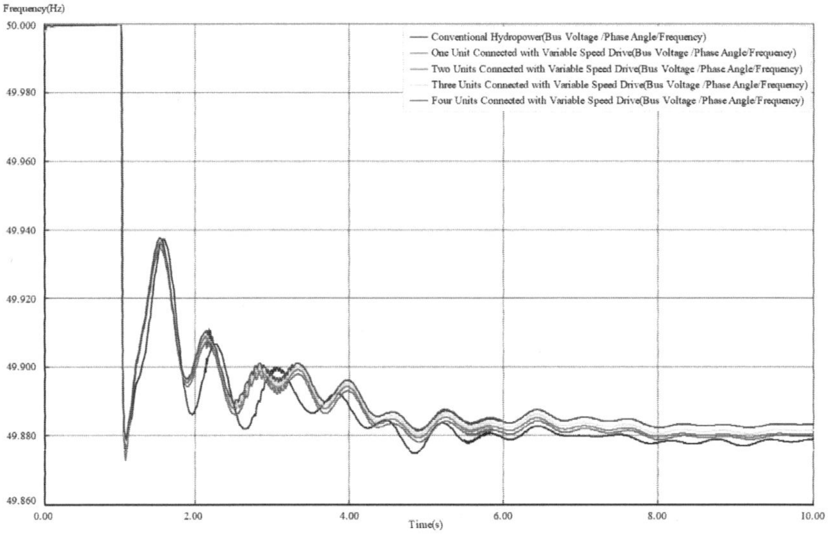

Figure 4. Frequency curve of surrounding bus after partial unit removal in a certain area under power generation conditions.

(2) Single machine comparison diagram under pumping condition

Under the typical mode, the variable-speed pumping and storage unit and the conventional pumping and storage unit are respectively connected in a certain area and operate in the form of a PQ node. As shown in Figure 5., when the unit is switched off after a fault occurs under the pumping condition, the active power consumed by the conventional pumping and storage unit as electrical equipment remains unchanged. However, the variable-speed pumped storage unit first increases briefly, and then rapidly decreases to 70%-80% of the initial active power consumption.

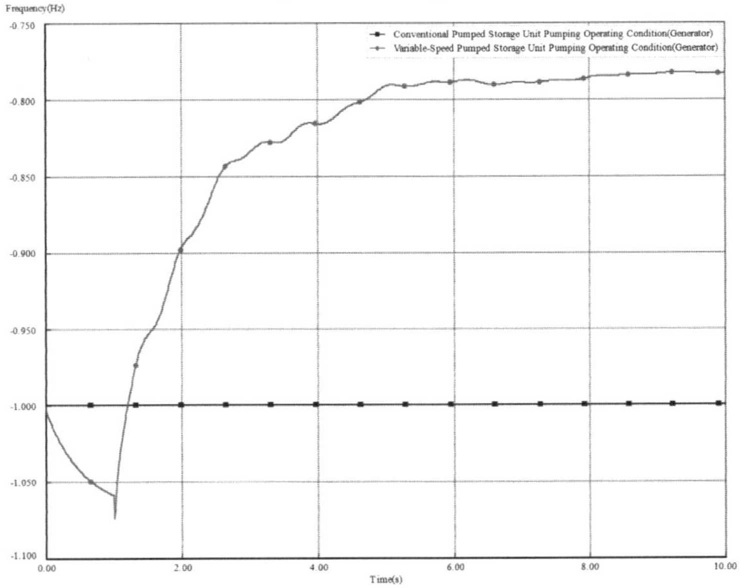

Figure 5. Active power curve of hydropower station unit after partial unit removal under pumping condition.

Under the typical mode, if 0, 1, 2, 3, and 4 conventional pumping and storage units are replaced with variable-speed pumping and storage units in a certain area and operate in the form of a PQ node. As shown in Figure 6., when the unit is switched off after a fault occurs under the pumping condition, the frequency fluctuation of the system after the fault is smaller with the increase of the number of variable-speed pumping and storage units, and the frequency drop after the final stability is smaller.

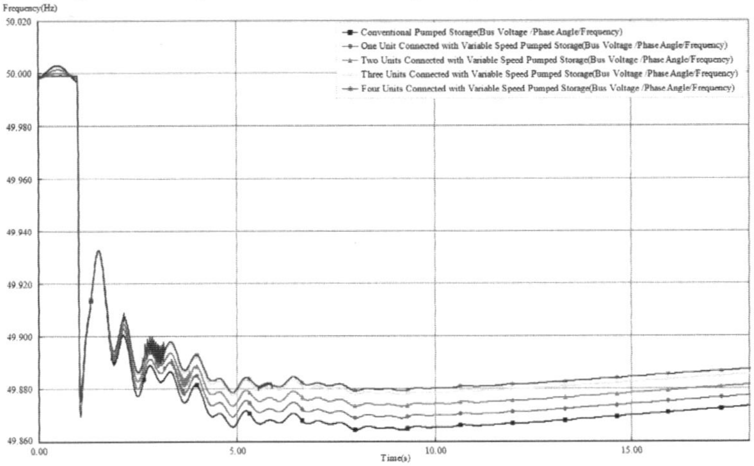

Figure 6. Frequency curve of surrounding bus after partial unit removal in an area under pumping condition.

5. Conclusion

With the focus on reaching the "double carbon" goal, developing a new energy system, and enhancing the power system, there is a growing need for flexible energy sources like variable-speed pumped storage. Full-power variable speed technology is favored for its simpler unit design, better handling of water head fluctuations, and wider power adjustment range compared to AC excitation variable speed technology. However, due to the manufacturing difficulty and cost limitation of large-capacity inverters, it is currently only applicable to small and medium-capacity units. The increasing implementation of full-power variable speed technology in small and medium-sized pumped storage power stations has highlighted its benefits in various applications. In the future, by establishing a scientific and reasonable evaluation index system for the application income of full-power variable-speed units, improving the relevant supporting design of full-power variable-speed units, and carrying out more demonstration projects of small and medium-sized full-power variable-speed units, China will gradually overcome the key technologies for the localization of full-power variable-speed units, and provide a stronger regulation ability for the new power system.

Acknowledgments

This work was financially supported by the Science and Technology Project of State Grid Corporation of China (5108-202218280A-2-282-XG).

References

[1] Xudong L, Weijia Y, Zhigao Z, et al. Advantage of priority regulation of pumped storage for carbon-emission-oriented co-scheduling of hybrid energy system [J]. Journal of Energy Storage, 2023, 58

[2] Kroposki B, Johnson B, Zhang Yingchen, et al. Achieving a 100% renewable grid: operating electric power systems with extremely high levels of variable renewable energy [J]. IEEE Power and Energy Magazine, 2017, 15(2):61-73.

[3] Han Xiaoqing, Li Tingjun, Zhang Dongxia, et al. New Issues and Key Technologies in the Planning of New Power Systems under Dual Carbon Goals [J]. High Voltage Technology, 2021, 47(09):3036-3046.DOI:10.13336/j.1003-6520.hve.20210809.

[4] National Energy Administration: Do a good job in the inclusion of new projects in the medium and long-term planning of pumped storage, key implementation projects and planning revision and adjustment [J]. Dams and safety, 2022, (03):21.

[5] Li Xiaopeng, Li Yan, Liu Shuran, et al. Research on improving the level of new energy consumption in regional power grid based on variable speed pumped storage technology [J]. Smart power, 2021, 49(10):52-58+112.

[6] Liu Xianlin, Du Xiaoyong, Tian Yunfeng, et al. Identification of Hydraulic Turbine Model of Pumped Storage Power Station Based on Random Load Fluctuation [J]. Proceedings of the CSEE, 2007, (06):49-52.

[7] Yang Wenwen. Research on VSG Control Strategy for Full-power Variable Speed Constant Frequency Pumped Storage Units [D]. Beijing Jiaotong University, 2021. DOI: 10.26944/d.cnki.gbfju.2021.001898

[8] Ye Hong, Sun Ping, Wang Tingting, et al. Simulation analysis of full-power variable frequency pumped storage unit in local power grid [J]. Hydropower and pumped storage, 2022, 8(06):1-4+31.

[9] Stocks M, Lu B, Cheng C. A review of pumped hydro energy storage [J]. Progress in Energy, 2021, 3(2):022003 (18pp).DOI:10.1088/2516-1083/abeb5b.

[10] Chang Xin, Han Minxiao, Zheng Chao. Analysis of power regulation characteristics of variable-speed pumped storage unit with full-power converter [J]. Electric power construction, 2016, 37(04):91-97.

[11] Bo Y, Jin Z, Zhicheng X. Fixed-speed and Variable-speed Pumped Storage Dispatch Model in Power Systems with High Renewable Penetration [J]. E3S Web of Conferences, 2018, 3804014-04014.

[12] XU Jincai. Principle and application of AC excitation variable speed generator [J]. Progress in water conservancy and hydropower science and technology, 1999, (04):37-38+50.

[13] Fengning pumped storage power station phase II project variable speed unit application feasibility study passed the report review [J]. hydroelectric power, 2016, 42(07):52.

[14] WANG Tingting, ZHANG Zhengping, ZHAO Jiejun, et al. Analysis of the influence of variable speed unit on the planning and selection of pumped storage in China [J]. hydropower generation, 2018, 44(04):60-63.

Journal of Physics: Conference Series **3011** (2025) 012059 doi:10.1088/1742-6596/3011/1/012059

High-Precision broadband signal measurement for power grids: an adaptive dual-algorithm approach

Lekai Yang[1,*], Minsi Huang[2]

[1]College of Electric Power, South China University of Technology, Guangzhou 510640, China
[2]Zhiyuan Electronics Co., Ltd, Guangzhou 510663, China

[*]Corresponding author's e-mail: 202020216234@mail.scut.edu.cn

Abstract. Modern power systems require adaptive and precise grid signal measurement algorithms for increasingly complex broadband characteristics. This paper presents a novel dual-strategy measurement method combining energy centroid correction with ZoomFFT technology. The method dynamically selects between algorithms based on spectral main lobe interference detection, effectively addressing measurement challenges for close-frequency inter-harmonics and low-amplitude harmonics. Validated through a 0-2.5 kHz experimental system, the method achieves relative errors of 1E-5 for frequency and 1E-3 for amplitude and phase measurements while significantly reducing computational overhead compared to traditional approaches. Results demonstrate superior performance in both measurement accuracy and efficiency, making it particularly suitable for real-time power grid monitoring applications.

1. Introduction

The rapid evolution of modern power systems, driven by renewable energy integration and advanced power electronics, has introduced unprecedented challenges in grid measurement and monitoring. As Chen et al. [1] pointed out, due to the interactions between generators/power electronic devices and grids, wideband oscillation events have occurred frequently in renewable power systems with frequencies from several Hz to several kHz. Unlike existing studies focusing on fundamental synchrophasor technology, wideband oscillation monitoring techniques have emerged as critical tools for identifying and analyzing these complex phenomena. These techniques address synchronized wideband phasor measurement and waveform measurement, moving beyond traditional frequency analysis to cover oscillation identification, early warning, mitigation, and other power system applications. The increasing complexity of power electronic interfaces, coupled with the variability of renewable generation, necessitates more sophisticated measurement and analysis approaches.

The development of effective measurement techniques has progressed through several key innovations in signal processing and analysis methods. The fast Fourier transform (FFT) is the main method of electrical signal analysis, but there are difficulties in performing non-synchronized sampling when FFT is applied to analyze broadband oscillation. Zhang et al. [2] addressed this through the sidelobe characteristic of the Hanning window and double-spectrum-line interpolation algorithm, proposing an approach based on the Hanning window double-spectrum-line interpolation FFT. The applicable rectification formulas were obtained using the polynomial curve fit function, achieving high accuracy when signals were sampled non-synchronously.

The accurate identification of broadband oscillation mode is crucial for solving the resonance risk of new energy stations. Guo et al. [3] identified the problems of high model order and poor noise immunity in traditional Prony algorithms, noting that the accuracy and running time of Variational mode decomposition (VMD) present a symmetric trade-off. Their improved VMD-based strategy used particle swarm optimization to determine optimal modes and penalty factors, combined with the energy threshold method for denoising and signal reconstruction. Li [4] proposed a novel method combining Windowed Interpolation FFT (WIFFT) and ZoomFFT algorithms, where the WIFFT method employs BlackmanHarris window and three-spectral line interpolation to achieve high accuracy and reduced spectral leakage from white noise interference.

In maritime microgrid systems, Terriche et al. [5] noted that harmonic contamination becomes particularly severe when inter-harmonics exist concurrently with system frequency drifts and large-power transients. Their eigenvalue-based solution effectively estimated harmonics and inter-harmonics under large frequency drifts with short window widths. Lei et al. [6] introduced an inter-harmonic analysis method using classical Fourier and Prony algorithms, transforming complex convolution operations into simple product operations through MATLAB simulation. Monteiro et al. [7] developed a DFT-based method for estimating interharmonics in wind power systems, exploiting signal enhancement and re-sampling techniques to achieve cost-effective estimation with reduced computational complexity.

To address these challenges, this paper presents an adaptive dual-algorithm measurement method that dynamically selects between energy centroid correction and ZoomFFT algorithms based on main lobe interference detection. The proposed approach represents a significant advancement in broadband signal measurement, offering three key contributions: (1) a novel adaptive algorithm selection strategy based on main lobe interference detection; (2) an improved energy centroid correction algorithm for enhanced close-frequency signal measurement accuracy; (3) optimized ZoomFFT implementation with significantly reduced computational complexity. These innovations collectively provide an effective solution for real-time monitoring in modern power grids, addressing the critical needs for accuracy, efficiency, and adaptability in increasingly complex power systems.

2. Main Lobe Interference and Its Determination

In power grid signal analysis, closely spaced frequency components create spectral overlap through main lobe interference, significantly affecting measurement accuracy for interharmonics and low-amplitude signals. The interference manifests as spectral leakage, where energy from one frequency component affects adjacent frequency bins.

The detection relies on phase relationship analysis between spectral components. As shown in Figure 1, interference-free components K_1 K_2 maintain a π phase difference with collinear angles. Interference from component M_0 disrupts this collinearity, yielding $\theta_1 = \theta_2 \neq \theta_3$ where $\theta = \arctan(\text{Real}/\text{Imag})$. The interference criterion is: $|R_1/I_1 - R_2/I_2| \leq \varepsilon$.

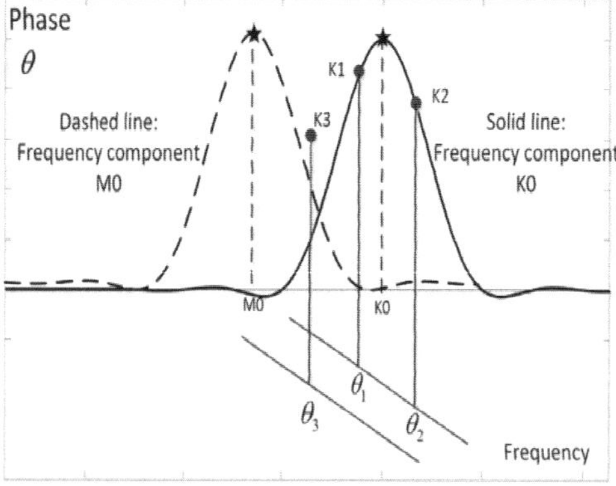

Figure 1. The Main-lobe interference detection and algorithm flowchart.

where ε accounts for computational errors and low-frequency interference. With a maximum frequency error of 0.000367 bins and measurement duration $T_{\text{measure}} = 0.16$ s, the theoretical phase error is: $\Delta\theta_{\max} = 2\pi \cdot 0.000367 \cdot 0.16 = 3.69 \times 10^{-4}$ radians. A scaling factor of 20 is applied for robust detection, setting $\varepsilon = 0.422$ degrees to balance sensitivity and stability.

3. Energy Centroid Correction Algorithm

Window function selection significantly impacts spectral analysis accuracy through their main lobe width, side lobe characteristics, and attenuation rates. For energy calculation accuracy, we define an energy recovery coefficient K_{rvy} for a normalized signal $X(n)$ with window function $W_T(n)$

$$K_{rvy} = \frac{\int_0^T X^2(n)dt}{\int_0^T |W_T(n)X(n)|^2 dt} \tag{1}$$

The frequency error ϵ_f and amplitude error ϵ_A are calculated as:

$$\begin{cases} \epsilon_f(f^{nor}) = \left|\sum_{i=-2}^{2} G(f^{nor}+i)\cdot(f^{nor}+i)\right| \\ \epsilon_A(f^{nor}) = 1 - \sqrt{K_{rvy}\sum_{i=-2}^{2} G(f^{nor}+i)} \end{cases} \tag{2}$$

Analysis of different window functions revealed their error performance characteristics, as shown in Figure 2. For frequency errors (in bins): Rectangular (0.169547) > Exponential (0.067976) > Hamming (0.001115) > Hanning (0.000367) > Kaiser (0.000289). Similarly, for amplitude relative errors: Rectangular (4.03%) > Exponential (2.62%) > Hamming (0.06%) > Hanning (0.024651%) > Kaiser (0.022637%). The Kaiser window was selected for its superior performance in both metrics, achieving minimal frequency error and amplitude error of approximately 0.02%.

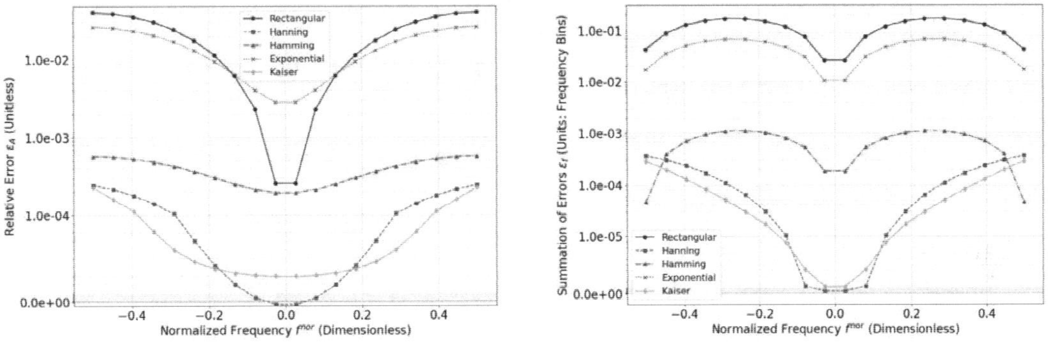

(a) (b)

Figure 2. Amplitude and Frequency error analysis of different window functions.

For the Hanning window example, we define the normalized spectrum modulus function and construct a discrete centroid function: $\sqrt{G(f^{\text{nor}})} = W(f^{\text{nor}}) = \frac{\sin(\pi f^{\text{nor}})}{\pi f^{\text{nor}}}\cdot\frac{1}{2[1-(f^{\text{nor}})^2]}$. The final correction formulas for amplitude, frequency, and phase are:

$$\begin{cases} \widehat{A_0} = \sqrt{1.58\sum_{i=-2}^{2} G^{tra}(f_k^{nor}+i)} \\ \widehat{f_0^{nor}} = \frac{\sum_{i=-2}^{2} G^{tra}(f^{nor}+i)\cdot(f^{nor}+i)}{\sum_{i=-n}^{n} G^{tra}(f^{nor}+i)}\cdot\frac{f_s}{N} \\ \hat{\theta} = arctan\left(\frac{I_k}{R_k}\right) - \pi\nabla\widehat{f^{nor}} \end{cases} \tag{3}$$

4. ZoomFFT Algorithm Based on Complex Analytic Bandpass Filtering

The ZoomFFT algorithm improves frequency resolution from f_s/N to f_s/ND through selective filtering and sampling rate reduction. Filter extension coefficient: $M = 5000$; Normalized frequency boundaries: $\omega_{1,2}^{\text{nor}} = \left(f_e \pm \frac{f_s}{4D}\right)\times\frac{2\pi}{f_s}$; The filter response is defined as:

$$\begin{cases} h_R(k_a) = \frac{1}{\pi k_a}[sin(\omega_2^{nor}k_a) - sin(\omega_1^{nor}k_a)] \\ h_I(k_a) = \frac{1}{\pi k_a}[cos(\omega_1^{nor}k_a) - cos(\omega_2^{nor}k_a)] \\ \qquad k_a = 1,2,\cdots,M \end{cases} \tag{4}$$

The algorithm involves:1. Complex convolution filtering 2. Frequency shifting 3. FFT transformation 4. Phase correction. The final signal processing equations are:

$$\theta = arctan\left(\frac{I_k}{R_k}\right) + \pi\nabla f^1 - 2\pi(k - \nabla f^1)\frac{M}{DN} - \phi(f_e) \tag{5}$$

The complete measurement method integrates main lobe interference determination with energy centroid correction and ZoomFFT algorithms Figure 3.

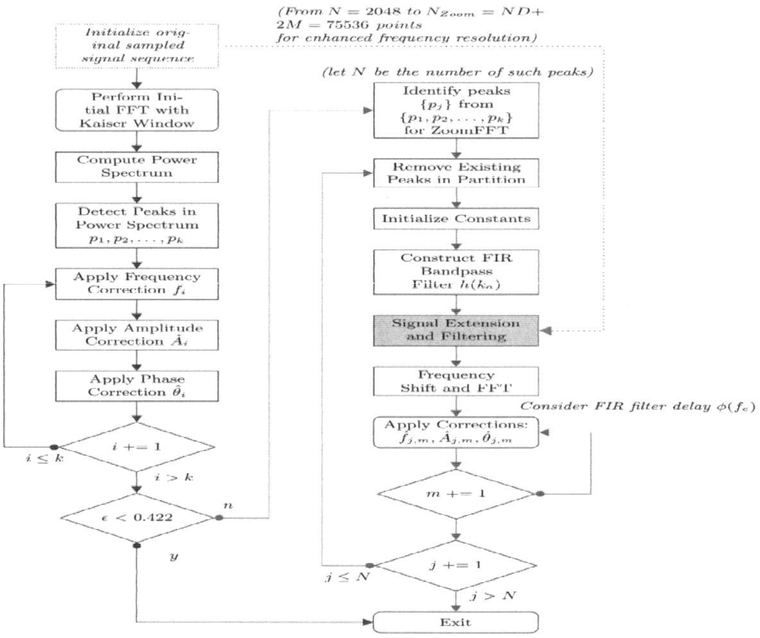

Figure 3. Flowchart of the proposed algorithm.

5. Simulation Validation and Analysis

The proposed method was validated through comprehensive simulations using a test signal incorporating fundamental, harmonic, and interharmonic components (SNR = 60 dB), with detailed signal components presented in Table 1. The test signal is expressed as $y(t) = \sum_{m=1}^{N} A_m \cos(2\pi f_m t + \varphi_m)$. The measurement configuration employed a sampling frequency of 12.8 kHz with 2048 FFT points for base configuration (yielding 6.25 Hz frequency resolution), and 75536 points for enhanced resolution mode (improving resolution to 0.169 Hz). The methodology's robustness was verified through 100 independent measurements with random noise, with key findings illustrated in Figure 4.

Table 1. Simulation settings Freq. (Hz)/ Amp. (V)/ Phase (deg).

Component1	Componen2	Componen3	Componen4	Componen5	Componen6
5.8/6.1/57	45/23.4/49.0	50.1/220.3/11.0	136.0/4.8/14.0	137.0/3.9/115.0	179.8/12.5/19.0
Component7	Componen8	Componen9	Componen10	Componen11	Componen12
258.6/8.1/-31.0	357.1/2.3/62.0	458.2/2.7/82.0	1026.5/2.1/-124.0	1723.9/1.6/128.0	2372.2/0.6/159.0

Analysis of experimental results demonstrates the significant advantages of the proposed adaptive measurement method. Through evaluation of average errors and standard deviations from 50 measurements, consistent detection values are indicated when four signals align horizontally. The ZoomFFT algorithm exhibits exceptional performance in processing fundamental waves and adjacent inter-harmonics, maintaining accurate identification even at frequency intervals as narrow as 1 Hz. While the Kaiser window-based energy centroid correction algorithm achieves high-precision measurement of harmonic components under conditions without main-lobe interference, traditional methods show notable limitations. With an initial frequency resolution of only 6.25Hz, frequency Components 1, 2, and 5 cannot be effectively distinguished, and Component 4 is severely affected by main-lobe interference, resulting in significant deviations in phase and amplitude estimates from actual signals. Near the 35th and 49th harmonics, although no other signal components were detected, their small amplitudes are susceptible to main-lobe interference caused by spectral leakage and background noise, necessitating the use of combined algorithms for parameter estimation in simulation.

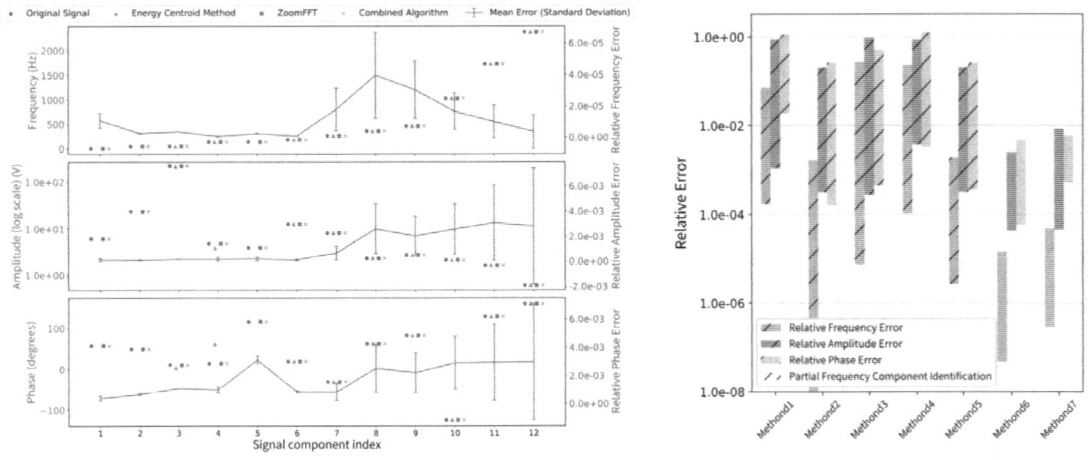

(a) (b)

Figure 4. Estimation results of frequency, amplitude, phase, and relative errors.

Our method offers key advantages over existing approaches using windowed interpolation or VMD-Prony combinations. While traditional methods require preset parameters and substantial computational resources, our approach combines energy centroid's superior anti-interference capabilities with ZoomFFT's high-resolution frequency distinction (0.195Hz). Though showing minor phase estimation fluctuations in higher-order harmonics and increased relative errors for very small amplitudes, the method maintains excellent overall accuracy and stability. Performance evaluation demonstrates significant improvement over conventional window functions, which struggle with frequency distinction and estimation accuracy under main-lobe interference conditions.

Through performance analysis, our method demonstrates significant technical advantages in efficiency and accuracy, as shown in Table 2. The energy centroid correction algorithms require 0.47-2.3ms processing time, while standard ZoomFFT needs 839.12ms (421ms after optimization by halving filter order and reducing 9000 computation points). Our method achieves comprehensive signal detection in just 112.65ms, reducing computational overhead by 87%. The method intelligently combines energy centroid correction and ZoomFFT through adaptive interference detection, with energy centroid showing excellent frequency estimation for lower side-lobe peaks (notably -42.7dB in Method 3 versus -13.4dB in Method 1). While traditional windows fail to identify inter-harmonics at 5Hz, 45Hz, and 137Hz under main-lobe interference, our approach selectively processes affected frequency bands, avoiding redundant full-spectrum analysis. This efficient, adaptive strategy particularly suits real-time power grid monitoring applications, demonstrating substantial engineering value through its balance of precision and computational efficiency.

Journal of Physics: Conference Series **3011** (2025) 012059 doi:10.1088/1742-6596/3011/1/012059

Table 2. Measurement results and operation cost of each method.

Method	*Algorithm*	Resolution	*Sample Points*	Runtime (ms)
1	Rectangular	Partial	2048	2.3
2	Hanning	Partial	2048	0.5
3	Hamming	Partial	2048	0.49
4	Exponential	Partial	2048	1.61
5	Kaiser	Partial	2048	0.47
6	ZoomFFT	Full	75536	839.12
7	Proposed	Full	2048/75536	112.65

6. Measurement Validation with Actual Data

The experimental validation was conducted using a comprehensive test system comprising a three-phase, four-wire configuration with a programmable AC power source, power analyzer, and oscilloscope recorder Figure 5 (a). The setup enables complete signal path validation from generation through measurement under realistic conditions.

The validation focused on IEC harmonic testing, covering frequencies from fundamental through the 50th harmonic. Table 3 presents comparative results between programmed settings and measured values for the first seven harmonics, with Phase A output maintaining balanced three-phase voltages. Phase angle verification relied on AC power source preset values, while the power analyzer measured voltage amplitude and frequency.

Table 3. Comparison of Test Results under IEC Harmonic Mode.

Programmable AC Power Source Settings			**Device Measurement Results**		
Harmonic Order	Voltage (V)	*Initial Phase (°)*	Frequency (Hz)	*Voltage (V)*	Initial Phase (°)
1	220	120	50.15	216.34	123.19
2	10.1	11	100.43	9.59	11.12
3	20.12	60	149.51	20.63	58.6
4	13.5	16.5	199.55	12.94	15.89
5	12.8	123	251.07	13.28	121.02
6	16.8	65	300.9	17.27	67.98
7	30.5	34	351.01	29.56	32.72

Analysis of the experimental data reveals measurement errors of approximately 5% for amplitude, 0.35% for frequency, and 4.5% for phase. The frequency measurements consistently showed superior accuracy across all test modes (IEC harmonics, 50-100 Hz inter-harmonics, and 100-2500 Hz interharmonics). Main error sources were identified as FPGA truncation effects and arctangent calculation precision for phase measurements, and analog circuit amplifier offset for amplitude measurements.

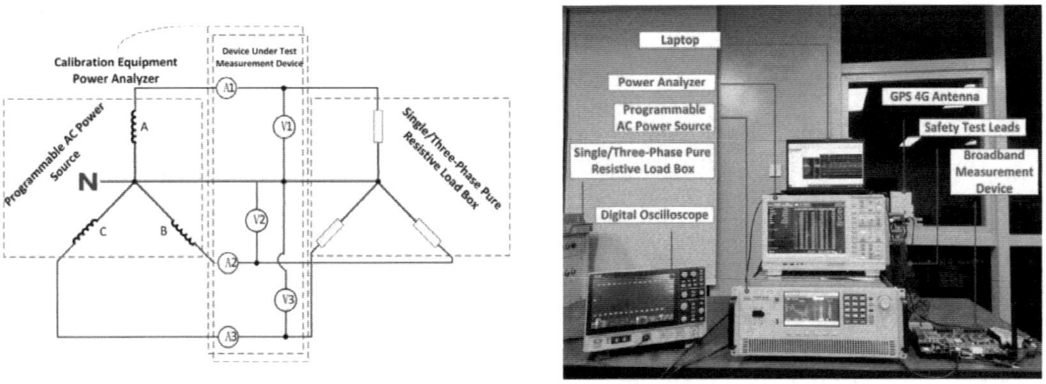

(a) (b)

Figure 5. System test diagram: (a) Schematic representation and (b) Physical test environment.

Based on data analysis, the amplitude error in harmonic measurements reaches up to approximately 5%, the frequency error approaches 0.35%, and the phase error is around 4.5%. These errors indicate areas where further optimization is needed for some parameters. Across the IEC harmonic tests, the 50–100 Hz interharmonic tests (list mode), and the 100–2500 Hz interharmonic tests, frequency accuracy is markedly better than amplitude and phase accuracy. The primary source of error in phase measurement is due to truncation effects in the FPGA and errors in arctangent function calculations. Additionally, the amplifier offset in the analog circuit contributes to amplitude errors.

7. Conclusion

This paper presents an adaptive measurement method combining energy centroid correction and ZoomFFT technology for power grid broadband signal analysis, achieving intelligent algorithm selection through main-lobe interference detection while significantly improving computational efficiency. Experimental results demonstrate high-precision measurements with relative errors reaching 1E-5 for frequency and 1E-3 for amplitude and phase measurements, effectively suppressing spectral leakage and picket fence effects caused by non-integer period truncation. The method achieves frequency resolution as fine as 0.195 Hz while substantially reducing computational overhead compared to traditional ZoomFFT implementations. While its adaptive characteristics and efficiency make it particularly suitable for real-time power grid monitoring applications, several areas warrant further investigation: dynamic signal processing scenarios require research into accurate measurement methods for amplitude modulation, phase modulation, and amplitude exponential variations, along with improved mechanisms for steady-state and transient signal identification; integration of higher-resolution measurement algorithms should be explored for ultra-low-frequency oscillation (0-5 Hz) detection; and future work will focus on optimizing embedded system implementation, exploring parallel processing techniques to enhance real-time performance, and investigating the method's adaptability under more complex power grid conditions to provide a more comprehensive monitoring and analysis solution.

References

[1] Chen, L., Xie, X., He, J., Xu, T., Xu, D., Ma, N., "Wideband oscillation monitoring in power systems with high-penetration of renewable energy sources and power electronics: A review," Renewable and Sustainable Energy Reviews, vol. 175, pp. 113148, 2023.

[2] Zhang, L., Li, Q., Song, M., Cao, M., Sun, B., Li, D., Chen, X., "An Approach for Identification of Broadband Oscillation in Power System Based on Hanning Window Double-spectrum-line Interpolation FFT," in 2022 Power System and Green Energy Conference (PSGEC), pp. 819-823, 2022.

[3] Guo, C., Yang, L., Dai, J., Chen, B., Yin, K., Dai, J., "A New Method for the Analysis of Broadband Oscillation Mode of New Energy Stations," Symmetry, vol. 16, no. 3, pp. 278, 2024.

[4] Li, Y., "A Methodology for Broadband Oscillation Monitoring Based on Windowed Interpolation and ZoomFFT," in 2023 6th International Conference on Electrical Engineering and Green Energy (CEEGE), pp. 69-73, 2023. Zhang, Q., Huang, C., He, J., and An, M. (2010) An Improved Recursive Discrete Fourier Transform Algorithm for Eliminating Decaying DC Component. Power Syst. Prot. Control, 38(24): 1–5.
[5] Terriche, Y., Laib, A., Lashab, A., Su, C.-L., Guerrero, J. M., Vasquez, J. C., "A Frequency Independent Technique to Estimate Harmonics and Interharmonics in Shipboard Microgrids," IEEE Trans. Smart Grid, vol. 13, no. 2, pp. 888-899, 2022.
[6] Lei, H., Huang, Y., Jiang, L., Gao, G., Zhang, J., Wu, P., "Research on Fourier and Prony Analysis Algorithm of Inter-Harmonic in Power System," Energy Reports, vol. 8, pp. 728-737, 2022.
[7] Monteiro, H. L. M., Camponogara, Â., Ribeiro, M. V., Duque, C. A., Poor, H. V., "A DFT - based Method for Estimating Interharmonics in Wind Power Generation," IET Smart Grid, vol. 5, no. 5, pp. 332-346, 2022.

Remote transmission method of intelligent power concentrator data collection based on Beidou satellite communication

Ziyuan Yang[1*], Yan Li[1], Weixing Qiu[2]

[1]Baotou Power Supply Branch of Inner Mongolia Electric Power Group Co., Ltd, Bao Tou, Inner Mongolia, 014030, China
[2]Beijing EsdTek Technology Co., Ltd, Beijing, 100000, China

*Corresponding author's e-mail: yzzzy2003@126.com

Abstract. As an external auxiliary device, the automation level of each simulation subsystem of the intelligent power concentrator varies greatly, and there are problems with communication capacity and packet loss in custom data collection. Therefore, this article proposes a remote transmission method for intelligent power concentrator data collection based on BeiDou satellite communication. By utilizing the BeiDou front-end gateway and BeiDou collection terminal to work together, both the frequency of instructions issued by the main station and the efficiency of data reception have been improved. In the design of remote data transmission, measures such as correcting collected data, removing redundant information, and reasonably unpacking and supplementing packages are taken to ensure the accuracy, efficiency, and stability of data transmission. At the same time, the system protocol should be customized according to the communication protocol, and the communication capacity and packet loss problems should be solved through packet sending and packet loss retransmission mechanisms. The experimental results show that compared with other methods, when the information size is 1200 bits, the transmission delay of our method is 2.5 seconds, and the packet loss rate is about 6.4%. It has excellent indicators such as data transmission integrity, stability, and resource utilization. This method has significant advantages in data collection efficiency and can effectively support the safe, stable, and efficient operation of power systems, providing strong support for the field of smart grid data processing.

1. Introduction

The collection and transmission of power data are becoming increasingly important for the stable operation of the power grid. The traditional data collection methods are no longer able to meet the requirements of modern power systems for real-time, accurate, and complete data. The use of intelligent power concentrators for data collection and transmission can greatly improve the efficiency and accuracy of data collection and transmission, and measure the real-time operational efficiency of the power grid[1-3].

Therefore, many scholars have conducted research on data collection and remote transmission. For example, Sharmila B et al. [4] proposed a method based on Theil Sen regression and Miyaguchi Peneel encrypted hash blockchain to improve the security of data transmission, which was validated in a Java environment. The results showed that compared with existing methods, this method has better performance in terms of data privacy, data reliability, average processing time delay, average power consumption, and blockchain data transmission efficiency. However, as the amount of data increases, the storage and processing capabilities of blockchain may be limited, leading to a decrease in data transmission and processing efficiency. Qi H et al. [5] proposed thermal energy storage technology and its application research in remote transmission of power data to adapt to the strong growth of current energy demand. This method is based on thermal energy storage technology, explores the advantages of thermal energy storage technology, studies stability analysis in the remote transmission of power data, and finally obtains the stability performance of thermal energy storage technology in the remote transmission of power data through experiments, verifying its essential role in the remote transmission of power data. However, thermal energy storage technology is limited by environmental factors such as geography and climate, resulting in unstable data transmission efficiency. The system proposed by Bhoomiben et al. [6] adopts a hybrid encryption model and combines the first-order steganography technique of two-dimensional discrete wavelet transform to achieve higher security by modifying and improving the key generation part of the encryption algorithm. Compared with other methods, the hybrid encryption algorithm using improved key generation shows better performance and provides higher security for video data transmission on the Internet. However, this may increase the computational burden and latency of the system, thereby affecting data transmission. Yao Y L et al. [7] designed a data transmission method based on the protobuf tool to achieve communication functionality for remote control of robots. Combined with the efficient data serialization capability of the protobuf tool, the proposed data transmission method ensures real-time and accurate data during remote control of robots. However, if there are vulnerabilities or issues with the protobuf tool itself, it may have an impact on the security of data transmission.

BeiDou satellite communication can provide stable data transmission services that are all-weather, unaffected by weather and environmental factors, and do not rely on basic network infrastructure, ensuring real-time, accurate, and secure transmission of power information, effectively improving the operational efficiency and intelligent management level of the power system. Therefore, this article proposes a remote transmission method for intelligent power concentrator data collection based on BeiDou satellite communication.

2. Design of remote transmission of BeiDou satellite communication data

2.1. Remote transmission calibration

Based on the data transmitted from the BeiDou collection terminal, perform remote data transmission calibration. The specific process is as follows:

Step 1: Each acquisition device is set to collect data at the same time. When the signal-receiving module receives 1 pps from the BeiDou satellite, we analyze the frequency error of the crystal oscillator. The calculation formula for the number of crystal oscillator harmonics n collected is:

$$n = \frac{\Delta\tau}{\eta_1 \cdot \left(\frac{1}{\eta_2} - \frac{1}{\eta_3} \right)} \tag{1}$$

In the formula, $\Delta\tau$ represents the phase difference caused by the crystal oscillator; η_3 represents the even frequency sampling frequency of the upper limit deviation of the crystal Tang oscillator; η_2 represents the lower limit deviation even frequency division sampling frequency; η_1 letters indicate the signal sampling frequency.

Step 2: Since each adjusted value is a complete cycle of crystal vibration, it is necessary to improve the accuracy of each adjustment result by correcting the known number of points collected, that is, the actual frequency and theoretical frequency within 1 second.

$$\Delta n = n_1 - n_2 \tag{2}$$

In the formula, n_1 and n_2 respectively represent the number of rising and falling times of the crystal oscillator frequency doubling. Based on the calculation results of equation (2), determine the relationship between the data acquisition frequency and the frequency difference within 1 second. If $\eta_4 \geq \Delta n$, obtain the expression for the cycle period λ_1 as follows:

$$\lambda_1 = \text{int}\left(\frac{\eta_4}{\Delta n}\right) \tag{3}$$

Step 3: Within 1 second, if the calculation result of Δn is positive, 1 needs to be added after one λ_1 cycle period; If the calculation result of Δn is negative, it is necessary to subtract 1 after one λ_1 cycle to obtain the original value. Add or subtract 1 every n cycles and repeat the process within 1 second to complete one iteration of correction processing. If $\eta_4 < \Delta n$, the calculation formula for obtaining the cycle λ_2 period is:

$$\lambda_2 = \text{int}\left(\frac{\Delta n}{\eta_4}\right) \tag{4}$$

Within 1 second, when the calculation result of Δn is positive, λ_2 needs to be added every 1 cycle; When the Δn calculation result is negative, λ_2 needs to be subtracted every cycle to obtain the original n value. Add λ_2 or subtract λ_2 every other cycle and repeat the cycle within 1 second to complete an iterative correction process, avoiding any impact on the transmission process and ensuring real-time data transmission.

Step 4: To improve the data transmission efficiency of the distribution network, it is necessary to remove redundant information from the data and keep it as undistorted as possible, thereby saving space and improving transmission efficiency. Using Shannon's theorem to gather information from power data, calculate the entropy of each domain to determine the compressible space in existing power data. The formula is:

$$\varsigma(x) = -\lambda \cdot \sum_{k=1}^{m} P_k \lg P_k \tag{5}$$

In the formula, x represents the distribution of power data; P_k represents the probability m of having a compressible space for power data; indicate the number of symbols in a single power data; k represents the number of judgments. The larger the calculation result, the greater the compressible space of the power data. Due to the uncertainty of the BeiDou communication link, it is impossible to determine the order of outsourcing signal reception, so it needs to be marked. The specific unpacking steps are as follows:

Assuming that the size of the data packet before unpacking is represented by N, and the maximum transmission unit during packet reception is represented by $R_{\max}$, when the received data signal is greater than $R_{\max}$, it needs to be unpacked. The unpacking scheme is:

$$H_1 = \frac{\varsigma(x)}{\left(\frac{R_{\max}}{N}\right)} \tag{6}$$

If the receiving unit is smaller than $R_{\max}$, the unpacking scheme is:

$$H_2 = \frac{\varsigma(x)}{\left(\frac{N}{R_{\max}}\right)} \tag{7}$$

Based on the above two situations, the unit sizes M_1 and M_2 after unpacking are expressed as follows:

$$M_1 = \frac{N}{H_1} \tag{8}$$

$$M_2 = \frac{N}{H_2} \tag{9}$$

Due to the limited single transmission communication capacity of BeiDou-3 short message communication, in order to solve this problem and enhance the remote transmission capability of BeiDou satellite communication data, based on the communication protocol of BeiDou satellite communication, the content structure of BeiDou communication is set through custom system communication protocol. The use of data packet transmission mechanism to solve the problem of limited capacity in BeiDou communication, and the use of packet loss retransmission mechanism to solve the problem of data packet loss, thereby ensuring the reliability and integrity of data transmission and achieving the transmission of all power data.

2.2. *Data Collection and Subcontracting Transmission of Intelligent Power Concentrator*

In order to solve the problem of the limited single communication capacity of BeiDou short messages, a data packet transmission method is adopted, which divides the data into packets and completes the data transmission through multiple transmissions through the BeiDou communication module. In order to ensure that the receiving end can correctly parse the content of the received data packet and accurately restore the data information, data packet arrangement is carried out according to the BeiDou Communication Terminal Universal Data Interface 2.2 protocol, with a maximum of 1750B per data packet and a serial number attached at the end to ensure the integrity of data transmission. The protocol for sending data packets is shown in Table 1.

Table 1 Sending end packet protocol.

Data Name	Data length (B)
BeiDou data packet header	3
BeiDou card number	16
data type	1
Time and Date	5
Data direction	1
Equipment number	16
Data length	2
Total number of packages	2
Data type serial number	1
Data content	1700
Subcontract serial number	2
Checksum	1

The receiving protocol, as shown in Table 2, mainly queries, detects, and confirms the received data, providing feedback on lost packet sequence numbers, retransmission counts, waiting times, and other related information. Among them, the data length of the lost packet sequence number is not fixed.

Journal of Physics: Conference Series **3011** (2025) 012060 doi:10.1088/1742-6596/3011/1/012060

Table 2 Receiver Protocol.

Data Name	Data length (B)
Time stamp	5
Miss sequence number	Some
Resend count	1
Request confirmation	1
Waiting time	1
Checksum	1

2.3. Intelligent power concentrator collects data with packet loss and retransmits it

Data packet loss and retransmission can be divided into two situations:

1) The receiving end is missing the last data packet. The sender sends a query instruction request to the receiver, asking if the receiver has received the complete data packet. The receiving end parses the sequence number, total number of packets, and other information of the received data packet, and feeds back the sequence number of the packet that was not received to the sending end. After receiving the response information, the sending end resends the data.

2) There is a missing data packet in the middle of the receiving end. The receiving end receives the last data packet and informs the sending end of the sequence number of the unreceived data packet. After receiving the packet loss retransmission instruction sent by the receiving end, the sending end retransmits the data.

3. Experiment

In order to verify the performance of the data acquisition method proposed in this article, according to the experimental environment and parameter settings mentioned above, this method was compared with three commonly used data acquisition methods, including 230M wireless private network, broadband medium voltage carrier technology, and Ethernet passive optical network method. The specific comparison results are shown in Figure 1.

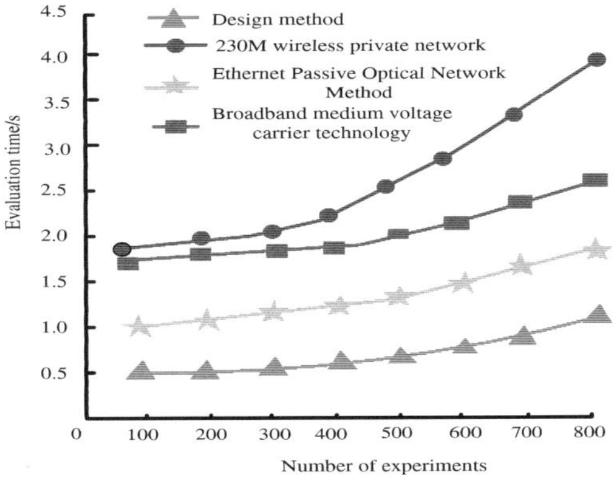

Figure 1 Comparison results of data layer collection efficiency of different methods.

As shown in Figure 1, by adding BeiDou multi-card machines, the frequency of instructions issued by the main station can be increased, and it can simultaneously receive data transmitted back by multiple BeiDou devices, improving the efficiency of data acquisition.

We select 12 sets of power information of different sizes collected as experimental data and use references and the method proposed in this paper for remote data transmission. The results of transmission delay and packet loss rate are shown in Figure 2 and Figure 3.

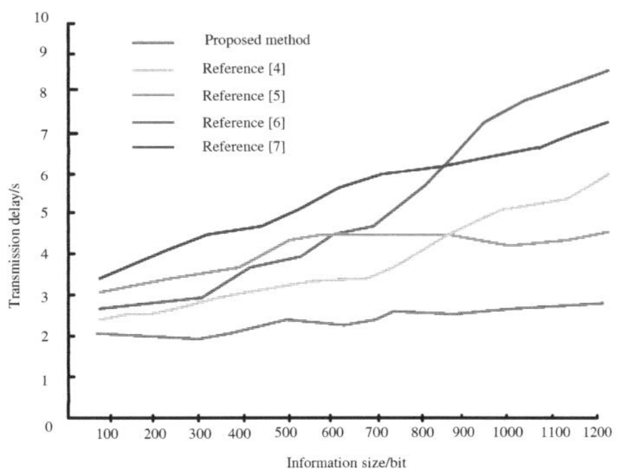

Figure 2 Comparison of transmission delays for different methods.

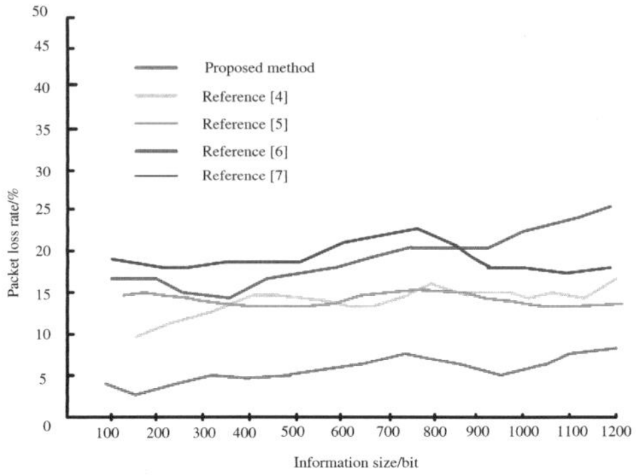

Figure 3 Comparison of packet loss rates for different transmission methods.

Analysis of Figures 2 and 3 shows that as the size of power information data increases, the transmission delay of all methods shows an upward trend. This reduces external interference during data transmission, ensuring transmission stability and thus reducing transmission latency and packet loss rate.

4. Conclusion

This article proposes a remote transmission method for intelligent power concentrator data collection based on BeiDou satellite communication, which has significant advantages in data collection efficiency, transmission delay, packet loss rate, and resource utilization. The transmission delay is 2.5 seconds, significantly lower than other reference methods, and the information size is between 100 bits and 1200 bits, indicating high stability.

Acknowledgments

The study was supported by "Inner Mongolia Electric Power Group Co., Ltd Technology Projects (No. nmgdlkjxm2020090).

References

[1] Yang J N, Yang L B, He J, et al. Global Coverage Inspection Method of UAV Power Grid by Big Dipper-based Enhancement Technology[J]. Microcomputer Applications,2023,39(01):114-117.

[2] Zhuang X, Xu Y, Gao Y, et al. Remote Data Transmission Technology Based on BeiDou Satellite Navigation Sensor System Onboard Ship[J]. Sensors and materials: An International Journal on Sensor Technology, 2021,33(3):715-726.

[3] Koushik K, Sohail S, Priya D, et al. FDRF: Fault Detection and Recovery Framework for Seamless Data Transmission in Remote Health Monitoring Using WBAN[J]. Wireless personal communications: An International Journal, 2022,123(1):325-356.

[4] Sharmila B, Kalavathi T. Theil-Sen Regressive Miyaguchi–Preneel-based Cryptographic Hash Blockchain for Secure Data Transmission Using Remote Sensing Data in IoT[J]. IETE Journal of Research,2023,(70): 116-129.

[5] Qi H, Wen H. THERMAL ENERGY STORAGE TECHNOLOGY AND ITS APPLICATION IN POWER DATA REMOTE TRANSMISSION[J]. Thermal Science,2023,27(2A):1175 - 1181.

[6] Bhoomiben P, Abeer A, C. W P P, et al. Secure data transmission in a real-time network for a tele-training education system[J]. Multimedia Tools and Applications,2022,81(19):27819-27836.

[7] Yao Y L, Feng N Q. Application Simulation of Generalized Neural Network in Data Traffic Prediction[J]. Computer Simulation, 2024, 41(4):382-386.

Journal of Physics: Conference Series **3011** (2025) 012061 doi:10.1088/1742-6596/3011/1/012061

Hybrid energy storage control strategy to stabilize wind power fluctuation

Shu Liu [1*], Yitong Pan [2]

[1] College of New Energy, Shenyang Institute of Engineering, Shenyang, Liaoning, 110136, China
[2] College of Electric Power, Shenyang Institute of Engineering, Shenyang, Liaoning, 110136, China

[*] Corresponding author's e-mail: ls_jhl@163.com

Abstract: In order to effectively decline the adverse effects caused by the intermittency and volatility of wind power generation on the power grid and ensure the safety and stability of the power grid operation, the power grid has stipulated the active power fluctuation limit of wind power grid connection. Taking the predicted wind power as the grid connection target, the capability of the hybrid energy storage system needs to be reasonably allocated. The GWO-VMD algorithm is attuned to decompose the predicted power. The high and low-frequency components are borne by the supercapacitor and the lithium battery, respectively. At the same time, combined with the economic index, the optimal hybrid energy storage smoothing fluctuation control strategy is finally obtained to achieve the fluctuation suppression effect.

1. Introduction

Concurrent with the growth of renewable energy utilization, wind power generation has earned great consideration due to its ease of access, cleanliness, and renewability [1]. However, the intermittence and randomicity of wind power generation can affect the reliability and power quality of the power grid [2]. How to effectively suppress wind power fluctuations has become a pressing issue that needs to be tackled.

Research has demonstrated that by configuring a hybrid energy storage system (HESS) consisting of energy storage and power storage, wind power fluctuations can be effectively smoothed out, the impact on the power grid can be declined, and the friendliness of wind power grid connection can be improved [3]. Currently, scholars have guided relevant research about this. Li et al. [4], Meng et al. [5], and Pu et al. [6] decomposed the total power using EMD, EEMD, and FEED algorithms but could not tackle the more severe mode aliasing issue. The VMD algorithm can overcome the mode aliasing of traditional algorithms like EMD, EEMD, and FEEMD. However, if the number of modes and penalty factors in the VMD algorithm are not set properly, there may be issues like excessive or insufficient decomposition.

Li et al. [7] aimed to achieve the best smoothing effect with the minimum cost as the goal and used the whale optimization algorithm to tackle the issue, obtaining the optimal cost of mixed energy storage capability configuration. Wang et al. [8] used wavelet packet decomposition to decompose power. Considering the cycle life of the battery, we construct the objective function with the lowest annual equivalent cost to make the energy storage capability configuration both economical and accurate.

Taking into account the above issues, this article proposes a grey wolf optimization method for variational mode decomposition, which determines the optimal number of modes and penalty factors based on the minimum fitness function and decomposes the grid-connected power through GWO-VMD. A capability configuration model for the entire life cycle of a hybrid energy storage system (HESS) is proposed, taking into account the changes in the lifespan and economic costs of lithium batteries. After simulation verification of three comparative schemes, the hybrid energy storage configuration is more economical than a single battery configuration.

2. GWO-VMD decomposition

2.1. GWO algorithm

The Grey Wolf Optimizer (GWO) is a population-based intelligent optimization algorithm proposed by scholars Mirjalili et al. [9] from Griffith University in 2016. Classify grey wolves into four types: leader wolves(α), which is responsible for decision-making and prey allocation; subordinate wolf(β), which is Assist α and replace α when necessary; secondary wolf(δ), which include reconnaissance wolves, guard wolves, etc; lowest level wolf(ω), which is Obey other levels and maintain population stability.

2.2. VMD algorithm

VMD is an evolved version of EMD, capable of processing nonlinear and non-stationary signals. However, the EMD method is prone to generating mode function aliasing, also known as the “mode aliasing” phenomenon. VMD imposes stricter constraints on the bandwidth of IMF, which overcomes the issue of mode aliasing in the decomposition events of VMD. The solving process of the VMD algorithm can be divided into two steps: constructing and solving variational issues.

1) Construction of variational issues

The initial preprocessed signal is disassembled into k modal components through VMD. For each modal, an exponential term $e^{-j\omega_k t}$ is added to modulate the frequency spectrum of each modal element to the corresponding cardinal frequency band.

By adjusting the Gaussian smoothness of the signal to estimate bandwidth and solving the norm of the signal gradient L^2, the constrained variational issue can be conveyed as:

$$\begin{cases} \min\limits_{\{u_k\},\{\omega_k\}} \left\{ \sum_{k=1}^{K} \left\| \partial_t \left[\left(\delta(t) + \frac{j}{\pi t} \right) * u_k(t) \right] e^{-j\omega_k t} \right\|_2^2 \right\} \\ s.t. \sum_k u_k = f(t) \end{cases} \tag{1}$$

Among them, $\{u_k\} = \{u_1, u_2, \cdots, u_k\}$ denotes the set of modal elements; $\{\omega_k\} = \{\omega_1, \omega_2, \cdots\cdots, \omega_k\}$ denotes the set of all mid-frequencies.

2) The solution of variational issues

In order to ensure the signal reconstruction accuracy under the influence of Gaussian noise and in order to ensure Lagrange multipliers $\lambda(t)$ are introduced, the Lagrange function is:

$$L(\{u_k\},\{\omega_k\},\lambda) = \alpha \sum_k \left\| \partial_t \left[\left(\delta(t) + \frac{j}{\pi t} \right) * u_k(t) \right] e^{-j\omega_k t} \right\|_2^2 + \left\| f(t) - \sum_k u_k(t) \right\|_2^2 + \left\langle \lambda(t), f(t) - \sum_k u_k(t) \right\rangle \tag{2}$$

The optimized solution for the above quadratic issue is:

$$\hat{u}_k^{n+1}(\omega) = \frac{\hat{f}(\omega) - \sum_{i \neq k} \hat{u}_i^{n+1}(\omega) + \frac{\hat{\lambda}(\omega)}{2}}{1 + 2\alpha(\omega - \omega_k)^2} \tag{3}$$

The mid-frequency has been updated to:

$$\omega_k^{n+1} = \frac{\int_0^{+\infty} \omega |\hat{u}_k(\omega)|^2 d\omega}{\int_0^{+\infty} |\hat{u}_k(\omega)|^2 d\omega} \tag{4}$$

2.3. GWO-VMD algorithm

The VMD algorithm utilizes the Lagrange multiplier method to optimize non-stationary signal processing and smooth out wind power fluctuations. Therefore, this article proposes an adaptive VMD method based on GWO. The optimal argument combination $[k, \alpha]$ adaptively searched by the GWO algorithm is attuned to optimize the decomposition effect of the VMD algorithm on fluctuating power.

This article adopts the concept of envelope entropy and uses the minimum value of envelope entropy as the fitness function for optimizing VMD arguments. After demodulating the original signal, the envelope signal can be obtained, which is transformed into a probability sequence p_j.

The envelope entropy E_p of zero mean signal $x(j)(j=1,2,\cdots,N)$ is:

$$\begin{cases} E_p = -\sum_{j=1}^{N} p_j \lg p_j \\ p_j = a(j) \Big/ \sum_{j=1}^{N} a(j) \end{cases} \tag{5}$$

Among them, p_j is the normalized form of $a(j)$; $a(j)$ is the envelope signal.

2.4. Wind power grid-connected power decomposition based on GWO-VMD

To demonstrate that the improved VMD algorithm outperforms EMD in the decomposition of hybrid energy storage reference power, EMD is first attuned to disassemble the hybrid energy storage power, obtaining IMF and frequency spectrum, as demonstrated in Figure 1. There are some mode aliasing phenomena, so lithium batteries must withstand some high-frequency fluctuation power, which has a negative repercussion on their cycle life and rocks the reliability of the energy storage system.

By using an improved VMD decomposition algorithm to disassemble the reference power of the hybrid energy storage system, as demonstrated in Figure 2, 11 intrinsic mode functions can be obtained from low to high-frequency bands. The frequency spectra of the 11 intrinsic mode functions can be obtained through a fast Fourier transform, as demonstrated in Figure 2.

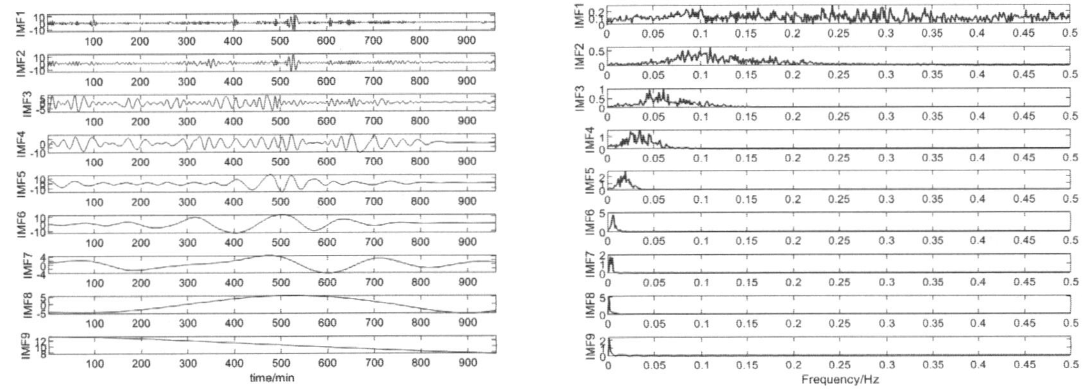

Figure 1. IMF and spectrogram obtained from EMD decomposition.

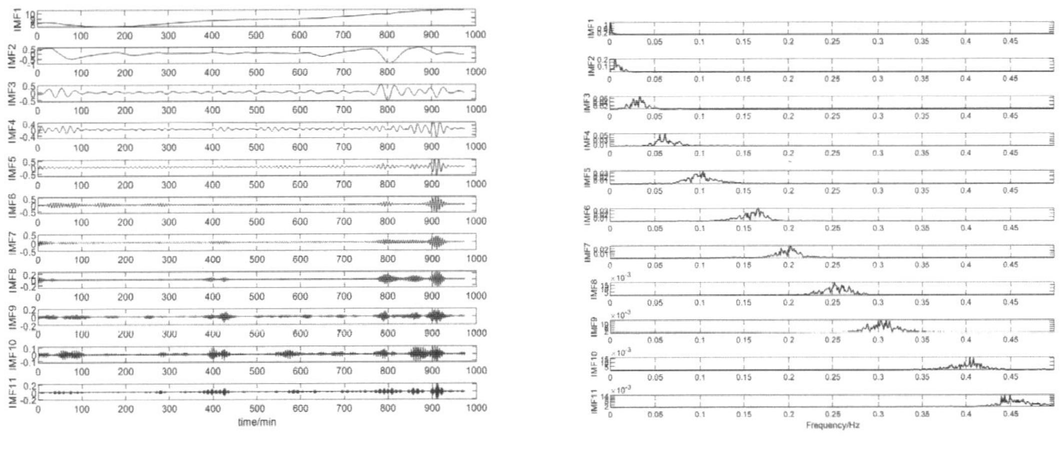

Figure 2. IMF and spectrogram obtained from GWO-VMD decomposition.

3. Hybrid energy storage configuration strategy

The power allocated to lithium batteries and supercapacitors in HESS is significantly different. The control strategy proposed in this article allows lithium batteries to bear relatively flat low-frequency components while supercapacitors bear fluctuating high-frequency components. This allocation strategy enables lithium batteries to decline the quantity of charge-discharge conversions, thereby reducing the lifespan loss of lithium batteries. A reasonable power allocation strategy between lithium batteries and supercapacitors in HESS can fully leverage the complementary features of hybrid energy storage.

3.1. Strategy for reducing the quantity of charging and discharging cycles of lithium batteries

Supercapacitors store energy through a double-layer interface, with no chemical reactions and long cycle life. Therefore, the cycle life of supercapacitors is set to a fixed value in this paper. However, each charge and discharge of lithium batteries causes irreversible life loss. Their cycle life is affected by factors like temperature, frequency, and DOD.

The lifespan of a lithium battery during its operating cycle can be conveyed as:

$$Y = \frac{N_c\left(D_{od,1}\right)}{365N_1} \tag{6}$$

3.2. Hybrid energy storage SOC cross-line strategy

To ensure the normal operation of the HESS, the State of Charge (SOC) of the energy storage system should be limited. When releasing the maximum amount of electricity, the SOC of the system should be higher than the lower limit, and the SOC of the system should be lower than the upper limit.

When the equation is satisfied, and the equation is equal, the minimum rated capability can be obtained. Therefore, the final rated capability configuration events for lithium batteries and supercapacitors are:

$$E_{b,er} = \frac{\max\limits_{t\in T}\left[\int_0^{k\Delta t} P_{b,dis}(t)dt\right] - \min\limits_{t\in T}\left[\int_0^{k\Delta t} P_{b,ch}(t)dt\right]}{SOC_{b,\max} - SOC_{b,\min}} \quad E_{sc,er} = \frac{\max\limits_{t\in T}\left[\int_0^{k\Delta t} P_{sc,dis}(t)dt\right] - \min\limits_{t\in T}\left[\int_0^{k\Delta t} P_{sc,ch}(t)dt\right]}{SOC_{sc,\max} - SOC_{sc,\min}} \tag{7}$$

In these two formulae, $P_{dis}(t)$ is the real discharge power; $P_{ch}(t)$ is the real charging power.

3.3. Economic evaluation modeling of HESS

In order to study the economic feasibility of hybrid energy storage systems, according to the life cycle cost calculation standard developed by IEC, the life cycle cost (LCC) of energy storage systems is defined as the total cost incurred during the investment, construction, equipment replacement, operation and maintenance, and recycling of energy storage devices during their operating cycle.

1) Initial investment and construction cost of hybrid energy storage system

The initial investment and construction cost are the source of all LCC expenses, mainly the selection of energy storage element capability, power, and type.

$$C_{cap} = C_{capb} + C_{capsc} \tag{8}$$

In the formula, C_{cap} is the investment cost of the energy storage system.

2) The replacement cost of hybrid energy storage systems

When the energy storage device completes all life cycles and is no longer usable, a new energy storage device needs to be replaced, which incurs certain equipment replacement costs.

$$C_{rep} = \left(C_{pb,rep}P_b + C_{eb,rep}E_b\right)n_b\gamma_{CRFb} + \left(C_{psc,rep}P_{sc} + C_{esc,rep}E_{sc}\right)n_{sc}\gamma_{CRFsc} \tag{9}$$

In the formula, C_p、 C_e denote the cost of replacing unit power and unit capability, respectively; n denotes the replacement frequency; γ is the discount rate.

3) Operation and maintenance costs of hybrid energy storage systems

The costs incurred during the installation, testing, daily operation, and scheduled maintenance of energy storage systems are the operating and maintenance costs of the energy storage system.

$$C_{om} = \left(C_{pb,om}P_b + C_{eb,om}E_b\right)\gamma_{CRFb} + \left(C_{psc,om}P_{sc} + C_{esc,om}E_{sc}\right)\gamma_{CRFsc} \tag{10}$$

In the formula, C_{om} is the operation and maintenance cost of the energy storage device; $C_{p,om}$、$C_{e,om}$ are the maintenance cost per unit power and the operation cost per unit capability, respectively.

4) Recycling and processing costs of hybrid energy storage systems

The cost of recycling and processing refers to the remaining usable value of the energy storage device at the end of its operating cycle, which can be regarded as future revenue.

$$C_{res} = \sigma_{res}\left(C_{capb} + C_{repb}\right)\frac{1}{\left(1+\gamma\right)^{T_b} - 1} \tag{11}$$

In the formula, C_{res} denotes the cost of energy storage device recycling and processing; σ_{res} is the recycling processing coefficient.

To sum up, the lifecycle cost of a hybrid energy storage system can be conveyed as:

$$C_{HESS} = C_{cap} + C_{rep} + C_{om} - C_{res} \tag{12}$$

4. Example simulation

According to the comprehensive charging and discharging efficiency values, the initial SOC of lithium batteries and supercapacitors is set to 0.5. The capability configuration argument table of the HESS is demonstrated in Table 1.

Table 1. Parameters for configuring energy storage capability.

Parameter	Lithium battery	Super capacitor
Unit power cost (yuan/kW)	2700	1500
Unit capability cost (yuan/kWh)	640	27000
Operation and maintenance cost (yuan/kWh)	0.05	0.05
Processing cost (yuan/kWh)	0.04	0.05
Charge and discharge efficiency (%)	80	95
SOC range	(0.15,0.85)	(0.1,0.9)
Cycle life/year	—	15
Fund discount rate/%	6	6

To explore the impact of disparate strategies on the capability configuration of hybrid energy storage systems, this paper proposes three disparate configuration schemes for detailed analysis. Scheme 1 only installs lithium battery energy storage. Scheme 2 installs a HESS to stabilize wind power fluctuations and uses the EMD algorithm to disassemble the hybrid energy storage power. Scheme 3 installs a HESS to stabilize wind power fluctuations and uses a GWO-VMD to disassemble the hybrid energy storage power. The SOC of the three schemes is demonstrated in Figure 3. The optimization configuration events of energy storage capability under the three schemes are demonstrated in Table 2.

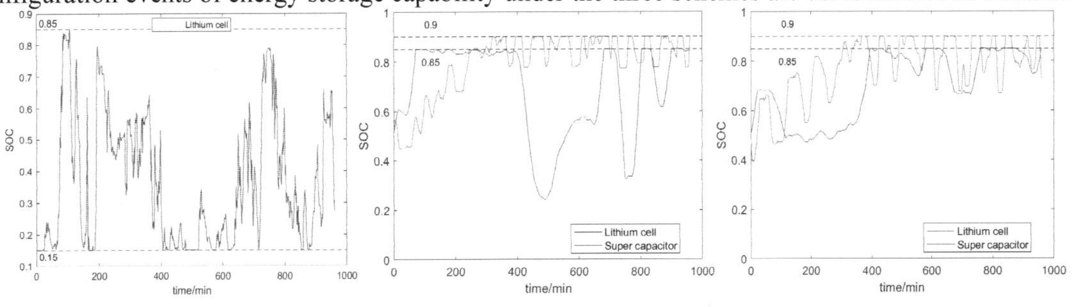

Figure 3. Energy storage SOC for Schemes 1, 2, and 3.

Comparing Scheme 1 and Scheme 3, the figure shows that hybrid energy storage systems can significantly decline the rated power and capability of configured energy storage. Compared with single lithium batteries, the rated capability of lithium batteries in hybrid energy storage systems has dwindled by 0.153MWh, and the rated power has dwindled by 4.16MW. In terms of economic cost, the annual equivalent cost of hybrid energy storage systems has decreased by 35.72% compared with the annual equivalent cost of single lithium battery energy storage.

Comparing Scheme 2 and 3, it can be seen that improving the VMD algorithm can dwindle the rated power and capability of energy storage configurations. The rated capability of lithium batteries has decreased by 0.099 MWh. The rated power has decreased by 0.062 MW. The rated capability of the supercapacitor has decreased by 0.118MWh. The rated power has decreased by 0.163MW. In terms of economic cost, compared with the EMD algorithm, the configuration cost of the GWO-VMD algorithm has declined by 17.93%. The reason is that the EMD algorithm has mode aliasing, and lithium batteries will bear some high-frequency elements, which increases the loss of life. According to Table 2, the battery cycle life of the EMD algorithm is 0.24 years shorter than that of the improved VMD algorithm. The replacement frequency of lithium batteries increases throughout the entire operating cycle of the energy storage system, thereby increasing the annual equivalent cost.

Table 2. Comparison of energy storage system capability allocation result.

Parameter	Scheme 1	Scheme 2	Scheme 3
Rated power of lithium battery /MW	5.357	1.316	1.217
Rated capability of lithium battery /MWh	1.464	1.373	1.311
Rated power of supercapacitor /MW	—	4.796	4.633
Rated capability of supercapacitor /MWh	—	0.852	0.734
Battery cycle life/year	2.12	3.21	3.45
Annual equivalent cost/10000 yuan	4767.9	3734.3	3064.7

5. Conclusion

This article proposes an energy storage economic optimization model based on improved VMD decomposition, which has better annual equivalent costs than EMD, significantly reduces system capacity configuration requirements, and can be used for new power system applications. In the future, we can continue to explore multi-objective coordination mechanisms in composite application scenarios.

References

[1] AHMAD S, AHMAD A (2023) Advanced rapid predictive and modulation model for PV system. J. Electric Power Components and Systems, 51(20): 2434-2449.
[2] WANG Jie, WU Jiang, HUANG Yuehui, et a1 (2023) Optimal allocation of wind-photovoltaic-storage capacity in multi-area power grid based on the distributed algorithm. J. JOURNAL OF XIAN JIAOTONG UNIVERSITY, 57(1): 15-24.
[3] HUO Junda, WANG Yi, MENG Jianhui (2024) Optimization strategy of MMC-HESS based on A-MPC to stabilize photovoltaic power fluctuation. J. Proceedings of the CSU-EPSA, 1-11.
[4] LI Junhui, JIA Caiqi, ZHU Xingxu (2023) Dual-layer model for capacity optimization of hybrid energy storage system to reduce thermal power frequency modulation loss. J. High Voltage Engineering, 49(9): 3965-3976.
[5] MENG Xiaojie, WANG Haiyun, WANG Weiqing (2020) Energy management control strategy for wind power hybrid energy storage system using self-adaptive EEMD. J. Power Capacitor & Reactive Power Compensation, 41(3): 189-196+204.
[6] PU Xianyi, BI Guihong, WANG Kai (2021) Ultra-short-term forecast of wind power based on FEEMD-PACF-BP_AdaBoost model. J. Computer Applications and Software, 38(11): 91-97.
[7] Li Lingling, Wang Xin, Lang Yongbo (2019) Capacity optimization configuration of microgrid composite energy storage system based on improved whale algorithm. J. Electric Measurement and Instrumentation, 56 (16): 104-110.
[8] Wang Y, Song F, Ma Y (2020) Research on Capacity Planning and Optimization of Regional Integrated Energy System Based on Hybrid Energy Storage System. J. Applied Thermal Engineering, 180(6390): 115834.
[9] Seyedali Mirjalili, Shahrzad Saremi, Seyed Mohammad Mirjalili, Leandro dos S. Coelho (2016) Multi-objective grey wolf optimizer: A novel algorithm for multi-criterion optimization. J. Expert Systems with Applications, 47: 106-119.

A multi-objective optimized control method for non-isolated dual three-phase permanent magnet synchronous motors with open-circuit fault

Jie Zhang, Fei Yao*

College of Information Science and Technology, Donghua University, Shanghai 201620, China

* Corresponding author's e-mail: fyao@dhu.edu.cn

Abstract. The zero-sequence current has long been a crucial challenge impeding the extensive application of non-isolated dual three-phase motors (NIDTP-PMSMs). To address this, a vector construction method based on model predictive control (MPC) is proposed, combined with a multi-vector, multi-space control strategy, enabling coordinated control of multiple subspaces current components and ensuring torque stability. By leveraging MPC's ability to handle constraints and multi-variable systems, the impact of zero-sequence space currents on peak current and winding loss after single-phase open-circuit motor vector space decoupling (VSD) transformation is studied. A novel multi-objective optimization algorithm is then proposed to optimize the relationship between peak current and winding loss by adjusting the zero-sequence subspace current magnitude. This method can adjust the current amplitude in real time based on the peak current during motor operation, keeping the current below the maximum operating current and minimizing winding loss, reducing the risk of secondary accidents caused by overcurrent. Finally, simulations verify the effectiveness of the proposed strategy.

1. Introduction

Dual three-phase permanent magnet synchronous motors (DTP-PMSMs) are widely used in electric vehicles due to their fault-tolerance capabilities, suitability for low-voltage high-load conditions, and torque stability [1]. Based on the connection of winding neutral, DTP-PMSMs are classified into isolated and non-isolated types [2]. NIDTP-PMSMs have higher control freedom after failure compared to IDTP-PMSMs, but the zero-sequence variables in NIDTP-PMSMs increase system complexity [3].

Control strategies for dual three-phase motors mainly include traditional field-oriented control (FOC) and model predictive control (MPC) [4]. FOC systems use feedback signals to determine voltage vectors, with some studies exploring different FOC strategies and harmonic current suppression methods [5]. However, FOC systems are complex for the operating system. Instead, MPC constructs a finite set of voltage vectors for rolling optimization, which can handle multi-input, multi-output systems and constraints [6-7]. It has been widely applied to IDTP-PMSMs, while NIDTP-PMSMs typically use FOC systems. When faults occur, especially open-circuit faults in six-phase motors, FOC systems cannot ensure safe and stable operation. Therefore, this paper proposes a control strategy for NIDTP-PMSM single-phase open-circuit faults, integrating a multi-objective optimization algorithm to ensure motor stability, adjust peak current, and minimize winding loss during fault-tolerant operation.

2. Construction of the NIDTP-PMSM model

2.1. Mathematical model based on VSD transformation

The VSD transformation maps DTP-PMSM variables into different subspaces, simplifying variable coupling and reducing the motor model order [8]. The subspaces include those related to electromechanical energy conversion and zero-sequence characteristics. After an open-circuit fault occurs in the Z-phase winding, the model equations of the NIDTP-PMSM in the natural coordinate system are as follows

$$\begin{cases} u_{5s} = R_s i_{5s} + \dfrac{d}{dt}\psi_{5s} \\ \psi_{5s} = L_{5s} i_{5s} + \gamma_{5s}\psi_f \end{cases} \tag{1}$$

where u_{5s}, i_{5s}, L_{5s}, and ψ_{5s} represent the voltages, current, inductance, and flux linkages of the five-phase windings; ψ_f is the flux amplitude; R_s is the resistance per phase; γ_{5s} represents the position coefficients of each phase winding.

Based on Equation (1), the mathematical model in the rotating coordinate system can be presented as

$$\begin{bmatrix} u_d \\ u_q \end{bmatrix} = R_s\begin{bmatrix} i_d \\ i_q \end{bmatrix} + \omega\left(\begin{bmatrix} 0 & -L_{aa1} \\ L_{aa1} & 0 \end{bmatrix} + \boldsymbol{X}(\theta)\begin{bmatrix} 0 & -3L_{aaq} \\ 3L_{aad} & 0 \end{bmatrix}\right)\begin{bmatrix} i_d \\ i_q \end{bmatrix} + \left(\begin{bmatrix} L_{aa1} & 0 \\ 0 & L_{aa1} \end{bmatrix} + \boldsymbol{X}(\theta)\begin{bmatrix} 3L_{aad} & 0 \\ 0 & 3L_{aaq} \end{bmatrix}\right)\frac{d}{dt}\begin{bmatrix} i_d \\ i_q \end{bmatrix} + \omega\boldsymbol{X}(\theta)\begin{bmatrix} 0 \\ 1 \end{bmatrix}\psi_f \tag{2}$$

The equation for electromagnetic torque T_e can be presented as

$$T_e = 3P_n((L_d - L_q)i_d + \psi_f)i_q \tag{3}$$

where $X(\theta) = \dfrac{1}{5}\begin{bmatrix} 4+\cos 2\theta & -\cos 2\theta \\ -\cos 2\theta & 4-\cos 2\theta \end{bmatrix}$, u_m and i_m (m = $z1$, $z2$, d, q) represent the voltage and current mappings on the z_1-z_2 axis and d-q axis, respectively; θ and ω represent the electrical angle and angular velocity; L_n (n = $aa1$, aad, aaq, d, q) represents the leakage inductance and the main and equivalent inductances along the d-axis and q-axis; P_n is the polar number.

2.2. Construction of voltage vector

According to Equation (2), the mapping relationship between subspaces and six-phase switching states can be determined. To facilitate the analysis, u_{pq} ($p \in \{0, 1,\ldots, 7\}$ is defined, and $q \in \{0, 1,\ldots, 4\}$) represents the voltage vector corresponding to different switch selections. Here, p denotes the switching states of the ABC-phase windings, represented by $S_aS_bS_c$; q denotes the switching states of the XY-phase windings, represented by S_xS_y.

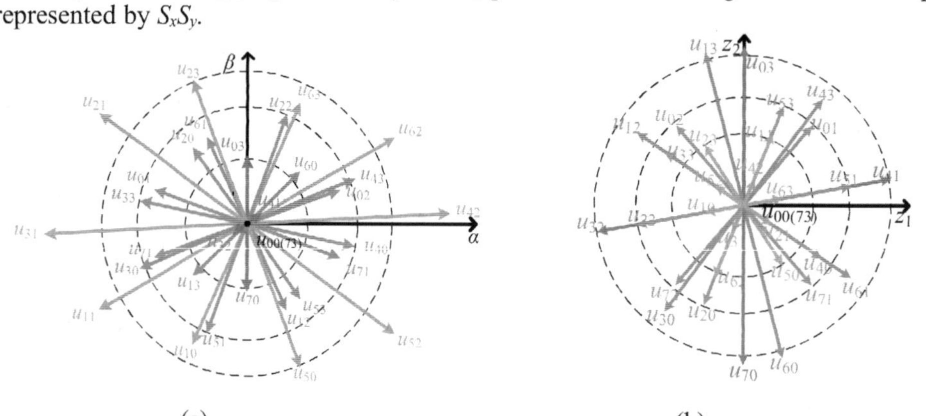

Figure 1. Subspace voltage vector distribution: (a) α-β subspace; (b) z_1-z_2 subspace.

As shown in Figure 1, there is no discernible characteristic in the magnitudes and phase angles of the voltage vectors on the two planes. To meet MPC requirements and optimize DC bus voltage utilization, a novel approach for voltage vector synthesis is proposed: the method of optimizing under constrained conditions. By setting the constraint conditions and substituting them into the Fmincon function in

MATLAB, and then assigning initial values that meet the requirements, the results can be obtained, as shown in Table 1. Here, v_{vi} and v_{zi} ($i\in\{0, 1,..., 12\}$) represent the synthesized voltage vector of α-β and z_1-z_2 subspace, with a total of 12 vectors.

Table 1. Voltage synthesis vector of α-β and z_1-z_2 subspace.

Voltage vector	Amplitude and phase	Switch state $[S_a, S_b, S_c, S_x, S_y]$	Voltage vector	Amplitude and phase	Switch state $[S_a, S_b, S_c, S_x, S_y]$
v_{v1}	$0.536e^{j0°}$	[1, 0.195, 0.195, 0.927, 0]	v_{z1}	$0.536e^{j0°}$	[1, 0.196, 0.196, 0, 0.928]
v_{v2}	$0.431e^{j30°}$	[0.871, 0.621, 0, 1, 0.355]	v_{z2}	$0.494e^{j30°}$	[0.749, 0, 0.214, 0.259, 1]
v_{v3}	$0.377e^{j60°}$	[0.754, 0.941, 0, 1, 0.675]	v_{z3}	$0.467e^{j60°}$	[0.526, 0, 0.350, 0.595, 1]
v_{v4}	$0.347e^{j90°}$	[0.501, 1, 0, 0.790, 0.791]	v_{z4}	$0.594e^{j90°}$	[0.257, 0, 0.515, 1, 1]
v_{v5}	$0.400e^{j120°}$	[0.200, 1, 0, 0.516, 0.863]	v_{z5}	$0.496e^{j120°}$	[0, 0.186, 0.558, 1, 0.570]
v_{v6}	$0.491e^{j150°}$	[0, 0.992, 0.284, 0.262, 1]	v_{z6}	$0.446e^{j150°}$	[0, 0.483, 0.676, 1, 0.331]
v_{v7}	$0.535e^{j180°}$	[0, 0.802, 0.803, 0.072, 1]	v_{z7}	$0.536e^{j180°}$	[0, 0.803, 0.803, 1, 0.073]
v_{v8}	$0.431e^{j210°}$	[0.130, 0.377, 1, 0, 0.647]	v_{z8}	$0.495e^{j210°}$	[0.251, 1, 0.786, 0.741, 0]
v_{v9}	$0.377e^{j240°}$	[0.247, 0.058, 1, 0, 0.327]	v_{z9}	$0.468e^{j240°}$	[0.474, 1, 0.650, 0.405, 0]
v_{v10}	$0.346e^{j270°}$	[0.501, 0, 1, 0.212, 0.212]	v_{z10}	$0.595e^{j270°}$	[0.743, 1, 0.485, 0, 0]
v_{v11}	$0.400e^{j300°}$	[0.801, 0, 1, 0.485, 0.140]	v_{z11}	$0.496e^{j300°}$	[1, 0.814, 0.442, 0, 0.430]
v_{v12}	$0.492e^{j330°}$	[1, 0.004, 0.715, 0.737, 0]	v_{z12}	$0.446e^{j330°}$	[1, 0.517, 0.324, 0, 0.669]

3. Multi-objective control algorithm based on MPC strategy

3.1. Multi-vector multi-space control strategy

The multi-vector control strategy expands the vector synthesis range and reduces current harmonic errors. Usually, three synthesized voltage vectors are used in the control process [9]. These three vectors include the two synthesized voltage vectors closest to the phase of the required voltage vector and a zero vector. The zero vector in one control cycle can be replaced with a voltage vector that only affects the z_1-z_2 subspace, enabling simultaneous control of multiple subspaces, which is equivalent to two rounds of model prediction within one cycle. The control flow diagram is shown in Figure 2.

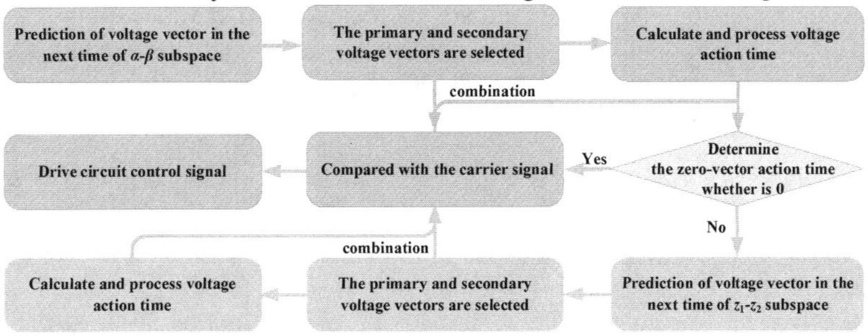

Figure 2. Flow diagram of multi-vector multi-space control.

3.2. Relationship between peak current and current in z_1-z_2 subspace

In field-oriented control, the reference value i_d^* is set to zero and i_q is chosen as the reference value. By comparing the coefficients of i_q in the phase current and using the Fminimax function in MATLAB, the final result is expressed as

$$\begin{cases} i_{z1}^* = 0.295 i_q \sin\theta - 0.760 i_q \cos\theta \\ i_{z2}^* = 0.420 i_q \sin\theta + 0.050 i_q \cos\theta \end{cases} \tag{4}$$

where i_{z1}^* and i_{z2}^* represent the current in the z_1-z_2 subspace of the minimum peak current.

At this stage, the current in all phases is uniform, with a value of $1.440 i_q$. It is assumed that the phase current amplitude of NIDTP-PMSM negatively correlates with the z_1-z_2 subspace current [10]. According to Equation (4), the peak current of the five phases can be expressed as

$$\begin{cases} i_{a,\max} = i_q(0.623k^2 + 0.350k + 1.109)^{1/2} \\ i_{b,\max} = i_q(0.400k^2 + 0.190k + 1.480)^{1/2} \\ i_{c,\max} = i_q(0.175k^2 - 1.530k + 3.420)^{1/2} \\ i_{x,\max} = i_q(0.484k^2 + 0.596k + 1.000)^{1/2} \\ i_{y,\max} = i_q(0.709k^2 + 0.367k + 1.000)^{1/2} \end{cases} \tag{5}$$

where $i_{j,\max}$ ($j = a, b, c, x, y$) represents the magnitude of the peak current of each phase of the motor; k is the proportional coefficient of the target current, $k \in [0, 1]$.

According to Equation (5), it can be seen that the peak currents of phases A, B, X, and Y exhibit an increasing trend within the interval $k \in [0, 1]$, while phase C peak current function is a decreasing function in the interval $k \in [0, 1]$. When $k = 1$, the five-phase current amplitudes are equal, and the peak current decreases by $0.406i_q$ A, accounting for 22% of the maximum current value. Therefore, it can be inferred that there is a negative correlation between the phase peak current and the z_1-z_2 subspace current, and the correlation is approximately linear.

3.3. *Relationship between winding loss and current in z_1-z_2 subspace*

Based on Equation (2), the winding loss Q_{cu} is expressed as

$$Q_{cu} = R_s(3i_q^2 + 2i_q^2 \cos 2\theta + 3i_{z1}^2 + 2.25i_{z2}^2) \tag{6}$$

Since i_q is determined by the given torque and can be regarded as a constant i_q^*, only the z_1-z_2 subspace current component can be adjusted. It is found that when Q_{cu} reaches the minimum value, i_{z1} and i_{z2} are both zero. At this time, the peak current corresponds to the phase C current of $1.846i_q$. Referring to Equation (5), it is assumed that there is also a correlation between winding loss and z_1-z_2 subspace current. The motor winding loss can be expressed as

$$\begin{aligned} Q_{cu} &= (4 + 1.195k^2)i_q^2 R_s + (1.553k^4 + 4.312k^2 + 4)^{1/2} \\ &\quad * i_q^2 \cos(2\theta + \tan^{-1}((2+1.078k^2)(0.625k^2)^{-1}))R_s \end{aligned} \tag{7}$$

According to Equation (7), Q_{cu} is composed of two parts: constant and trigonometric function, both of which show an approximate exponential positive correlation with k value. As the k value increases from 0 to 1, the total losses and their fluctuations increase, and the average winding loss reaches up to 130% of the initial value. It is proved that the winding loss will change positively when the current component of the z_1-z_2 subspace increases.

3.4. *Multi-objective control strategy*

The multi-vector and multi-space control strategy ensures the stable operation of the motor. However, achieving an automatic balance between peak current and winding loss places higher demands on the current components in the z_1-z_2 subspace. For Z-phase open-circuit NIDTP-PMSM, to balance peak current and winding loss, the design is based on the motor's maximum operating current amplitude $i_{\max}$. By adjusting the value of k related to the z_1-z_2 subspace current, the peak current of the five phases is kept within $i_{\max}$ while minimizing winding loss.

According to Equation (5), in the interval $k \in [0, 1]$, the maximum value of phase current corresponds to the phase C current, and with the increase of k value, the phase C current amplitude decreases continuously. At the same time, according to Equation (7), it can be found that the winding loss increases with the increase of the k value. Therefore, when the motor is started or accelerated, the increase in T_e and the increase in i_q may cause $i_{c,\max} > i_{\max}$, forcing the k value to increase until $i_{c,\max} = i_{\max}$. In this case, the value of k is expressed as

$$k = 2.86 * (1.53 - (1.53^2 - 0.7(3.42 - i_{\max}^2 / i_q^2)^{1/2})) \tag{8}$$

According to Equation (8), if $i_{c,\max}$ remains greater than $i_{\max}$ ($i_{\max} / i_q < 1.440$) when $k = 1$, the system will maintain $k = 1$ to minimize potential damage to the motor caused by excessive peak current. Once the motor speed is stable, T_e will decrease the load torque. If the load torque is small, the k value will decrease synchronously until $k = 0$. In cases where $i_{c,\max} < i_{\max}$ ($i_{\max} / i_q > 1.846$), the system will keep $k = 0$ and the winding loss will be minimized. However, for larger load torques ($i_{\max} / i_q \in (1.440, 1.846)$)

and the motor speed is stable, the k value will achieve $i_{c,\max} = i_{\max}$, reaching a new balance. This adjustment ensures that winding loss is minimized without exceeding $i_{\max}$. The effectiveness of this strategy will be validated through simulation, with the system block diagram presented in Figure 3.

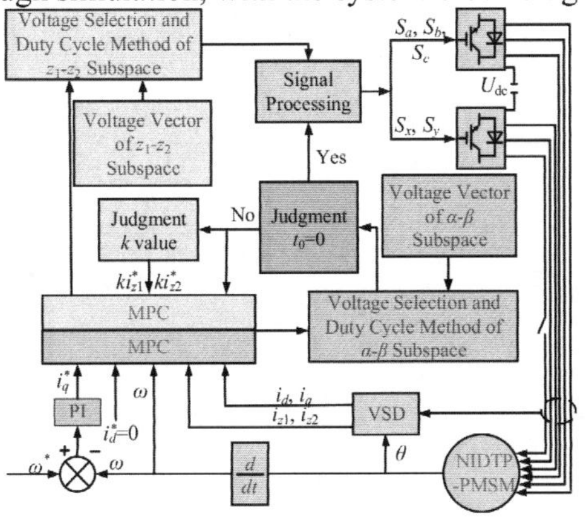

Figure 3. Block diagram of multi-objective control system.

where t_0 represents the zero-vector application time and ω^* represents the given electrical angular velocity.

4. Simulation

The Z-phase open-circuit motor model was constructed in MATLAB/Simulink with parameters specified in Table 2. The maximum current $i_{\max}$ was set to 10 A, with load torques set at 20 N·m and 30 N·m, respectively. Additionally, practical scenarios were simulated, such as the motor's operation during the startup, steady-state, and acceleration phases of a vehicle. The waveforms of the motor's current and winding loss are analyzed, and the results are presented in Figures 4 to 7.

Table 2. Simulation parameter.

Parameters	Numerical value	Parameters	Numerical value
Stator resistance (R_s)	1.5 Ω	Permanent magnet flux linkage (Ψ_f)	0.3013 Wb
d-axis main self-inductance (L_{aad})	10 mH	Damping coefficient (B)	0.001 N·s/m
q-axis main self-inductance (L_{aaq})	10 mH	Switching frequency (f)	10 kHz
Stator leakage (L_s)	2 mH	Moment of inertia (J)	0.015 kg·m^2
Number of poles (P_n)	5	Bus control voltage (U_{dc})	300 V

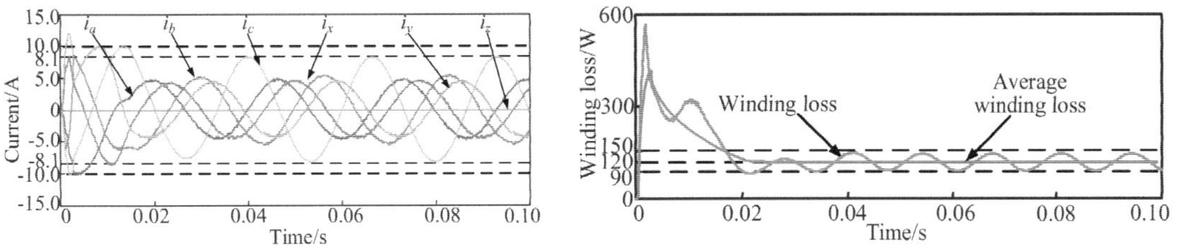

(a) (b)

Figure 4. Simulation results of the Z-phase open-circuit NIDTP-PMSM when the given speed is set to 450 r/min and the load is 20 N·m; (a) stator current; (b) winding loss.

Figure 4(a) and Figure 5(a) show the corresponding current waveforms, while Figure 4(b) and Figure 5(b) illustrate the winding loss and the average winding loss for both speed cases. It can be observed that, during the motor startup phase when the current amplitude exceeds 10 A, the value of k changes to reduce the maximum current amplitude to 10 A. In the stable state, the peak current value is

8.1 A, which is below the maximum current limit i_{max}, indicating that it does not exceed the motor's maximum allowable operating current. Therefore, the value of k is set to 1. At this point, the winding loss is at its minimum for this load condition, with an average winding loss of approximately 120 W and a fluctuation of 60 W.

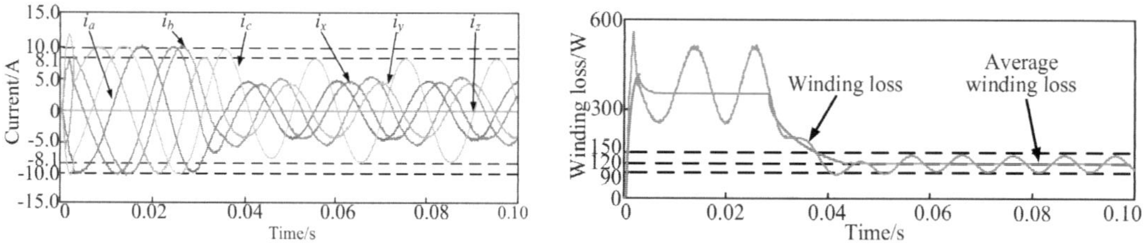

(a) (b)

Figure 5. Simulation results of the Z-phase open-circuit NIDTP-PMSM when the given speed is set to 600 r/min and the load is 20 N·m; (a) stator current; (b) winding loss.

In contrast to Figure 4, Figure 6 presents current and winding loss waveforms under a constant speed condition with a torque input of 30 N·m. From Figure 4(a) and Figure 6(a), it can be observed that as the given torque increases, the motor's current amplitude correspondingly increases. Furthermore, based on calculations, when the torque is 30 N·m, the maximum current amplitude exceeds 10 A, resulting in a change in the value of k. Specifically, k increases from 0 to 0.83, ensuring that the steady-state current amplitude reaches i_{max}. Figure 4(b) and Figure 6(b) show that the winding loss presents a positive correlation with the given torque.

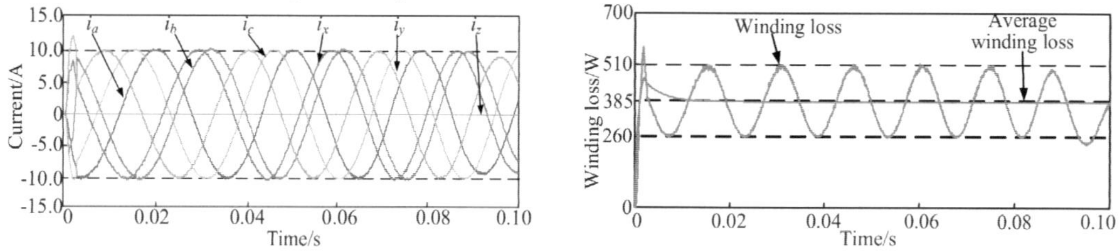

(a) (b)

Figure 6. Simulation results of the Z-phase open-circuit NIDTP-PMSM when the given speed is set to 450 r/min and the load is 30 N·m; (a) stator current; (b) winding loss.

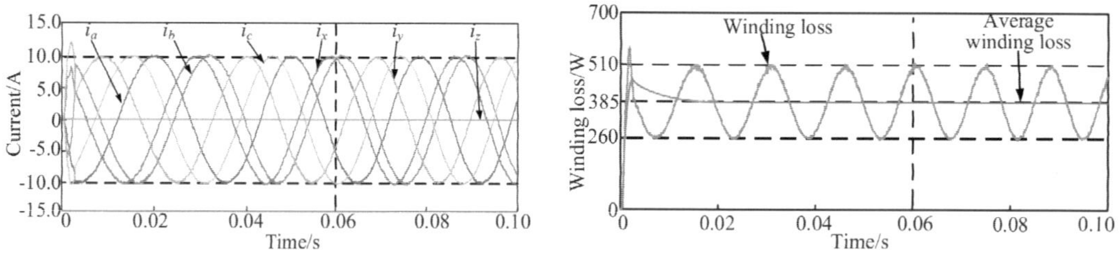

(a) (b)

Figure 7. Simulation results of the Z-phase open-circuit NIDTP-PMSM when the given speed changes from 450 r/min to 600 r/min within 0.3 s, with the load is 30 N·m; (a) stator current; (b) winding loss.

From Figure 7(a), it can be observed that the current amplitude increases a little at 0.3 s. However, the increase in the value of k effectively suppresses a portion of the current overshoot. Moreover, after this point, the maximum current amplitude stabilizes at 10 A, ensuring that the maximum current during stable motor operation does not exceed i_{max}. Figure 7(b) shows that the variation in speed has little impact on the magnitude of winding loss.

5. Conclusions

In this paper, an innovative voltage vector construction method is proposed, which is based on the principle of extremum seeking under constraints. This approach effectively addresses the harmonic issues caused by zero-sequence components in the control loop of NIDTP-PMSM. By combining a multi-vector, multi-space control strategy, the synthesis range of voltage vectors is extended. Moreover, substituting conventional zero vectors with voltage vectors enables the synchronous control of multiple subspaces in the motor, which effectively mitigates harmonic issues arising from the motor or inverter configuration.

Additionally, an innovative multi-objective control algorithm is introduced. This algorithm automatically adjusts the maximum amplitude of the motor current based on phase current peak values without compromising torque stability, reducing the current amplitude by up to 22% while ensuring that the increase in winding loss does not exceed 30%. Simulation results show that under variable-speed operating conditions, the motor's current peak can be minimized, suppressing overshoot phenomena in certain currents and reducing the risk of secondary accidents caused by overcurrent.

References

[1] DEMIR Y and AYDIN M (2016) A novel dual three-phase permanent magnet synchronous motor with asymmetric stator winding [J]. IEEE Transactions on Magnetics, 52(7): 1-5.

[2] LU L, MAO Y, WANG X, JIN L and WANG Z (2023) Maximum-torque optimization control of dual three-phase PMSM in low-frequency and static operations [J]. IEEE Journal of Emerging and Selected Topics in Power Electronics. 11(5): 5268-5278.

[3] WANG X, WANG Z, HE M, ZHOU Q, LIU X and MENG X (2021) Fault-tolerant control of dual three-phase PMSM drives with minimized copper loss [J]. IEEE Transactions on Power Electronics, 36(11): 12938-12953.

[4] CUI J, TAO T and ZHAO W (2024) Optimized control set model predictive control for dual three-phase PMSM with minimum error duty cycle regulation [J]. IEEE Transactions on Power Electronics, 39(1): 1319-1332.

[5] ZHAO Y and LIPO T (1994) Space vector PWM control of dual three-phase induction machine using vector space decomposition.1994 IEEE Industry Applications Society Annual Meeting, Denver, CO, USA. pp. 742-749.

[6] SUN Z, DENG Y, WANG J, YANG T, WEI Z and CAO H (2023) Finite control set model-free predictive current control of PMSM with two voltage vectors based on ultralocal model [J]. IEEE Transactions on Power Electronics, 38(1): 776-788.

[7] HABIB A, et al. (2022) Predictive current control of six-phase IM-based nonisolated integrated on-board battery charger under different winding configurations [J]. IEEE Transactions on Power Electronics, 37(7): 8345-8358.

[8] LIANG Z, KOU P and LIANG D (2016) A fault tolerant control strategy for dual three-phase permanent magnet synchronous motor servo system based on frequency domain analysis. 2016 Eleventh International Conference on Ecological Vehicles and Renewable Energies (EVER), Monte Carlo, Monaco, pp: 1-13.

[9] ZHOU H, XIANG X, LI Y and TAO T (2024) Simplified model predictive current control for dual three-phase PMSM with low computation burden and switching frequency [J]. CES Transactions on Electrical Machines and Systems, 8(4): 447-454.

[10] ZHANG J, YAO F, CHEN E Z and XING J (2024) A multi-objective optimized control method for dual three-phase synchronous motor with open-circuit fault. 2024 IEEE 10th International Power Electronics and Motion Control Conference (IPEMC2024-ECCE Asia), Chengdu, China, pp: 4889-4893.

Emergency control strategy for island microgrids based on fault early warning

Jiahao Chen[1], Qifan Feng[2,*], Pengcheng Yang[2], Yajie Tang[3]

[1]Polytechnic Institute, Zhejiang University, Hangzhou, 310027, China
[2]College of Electrical Engineering, Zhejiang University, Hangzhou, 310027, China
[3]State Grid Zhejiang Electric Power Co., Ltd. Research Institute, Hangzhou, 310014, China

[*]Corresponding author's e-mail: 12110032@zju.edu.cn

Abstract. For island microgrids, droop-based primary control often introduces frequency/voltage errors, which are usually corrected by a microgrid central controller (MGCC) in the secondary control. In this paper, an emergency control strategy is proposed and integrated into MGCC based on fault early warning, which can accelerate response speed and reduce fluctuation amplitude during severe disturbances. The first component is an emergency controller based on the rate of change of frequency (RoCoF) and voltage (RoCoV), incorporating additional compensatory signals into the proportional-integral (PI) control loop to enable rapid response to disturbances. The second component is another emergency controller designed to proactively detect abnormal switch operations in critical equipment, implementing feedforward compensation to prevent fault propagation. Simulations conducted in MATLAB/Simulink validate the efficacy of the proposed method, showing substantial reductions in system response time and enhanced stability of microgrid operations.

1. Introduction

Islanded microgrids use distributed generations (DGs) and energy storage systems (ESSs) to maintain a steady electricity supply. However, the intermittent nature of DG outputs and unpredictable load variations pose significant challenges to maintaining system frequency and voltage stability. In severe cases, these issues may result in system instability or even widespread blackouts[1]. Nowadays, droop and virtual synchronous generator control are usually used as primary controls to emulate the physical behaviors that make the system stable and more damped[2]. However, these methods inherently lead to steady-state frequency and voltage deviations. To address these deviations, an external control, named secondary control, has been proposed to restore the nominal values of the frequency and voltage inside the islanded microgrids.

Most traditional secondary control strategies rely on centralized control architectures[3]. In these systems, the MGCC calculates global adjustment commands based on frequency and voltage deviations and distributes these commands to DGs and ESSs. Zhou et al. [4] employ an improved droop control strategy to balance power in island microgrids and use an MGCC to restore frequency and voltage to nominal values. While this approach offers strong coordination and precise control, it imposes significant communication burdens. During severe power disturbances, it suffers from response delays due to MGCC communication latency, reducing control efficiency. To improve control efficiency and alleviate communication burdens, Anand et al. [5] introduce an event-triggered PI controller that provides supplementary signals in an aperiodic manner, significantly reducing computational effort without compromising control performance. Rao et al. [6] propose an event-triggered data exchange strategy to minimize communication overhead during transient processes.

Additionally, Qian et al. [7] further develop an event-triggered method to maintain regional grid stability. Compared to traditional periodic control, event-triggered methods can speed up the dynamic response of secondary control through active fault response. Besides frequency and voltage fluctuations, sudden load changes or fault events may also occur in microgrids[8][9]. Traditional secondary control often falls short in responding to these complex scenarios due to slow response times and delayed control actions, leading to further frequency and voltage deviations. Although event-triggered control can provide active compensation during fault events, load steps often do not meet the triggering conditions in time, resulting in significant drops in system frequency or voltage.

Focusing on this problem, this paper proposes an emergency controller that proactively identifies abnormal switching actions (ASAs) of critical devices in the system, thereby shortening response time and mitigating the risk of fault propagation. At the same time, emergency control based on RoCoF (RoCoV) is designed to cope with load steps. When the RoCoF (RoCoV) exceeds the threshold, the strategy is triggered to quickly respond to power disturbances. Finally, simulation tests are conducted in MATLAB/Simulink to verify the effectiveness of the proposed control.

2. Conventional secondary control

Taking droop control as an example, the droop characteristic curve is shown in Figure 1. The droop controller is distributed in each DG, and the frequency and voltage are adjusted by calculating the active and reactive power output of each DG connected to microgrids.

$$P = P_{ref} + m(f_{ref} - f) \tag{1}$$

$$Q = Q_{ref} + n(V_{ref} - V) \tag{2}$$

Where f and V are the frequency and amplitude of the output alternating current (AC) voltage from the primary control unit, while f_{ref} and V_{ref} are the nominal frequency and nominal amplitude of the AC voltage. m and n represent the droop coefficients for active power frequency and reactive power-voltage amplitude. P_{ref} and P are the reference output active power and the actual output active power, while Q_{ref} and Q are the reference output reactive power and the actual output reactive power.

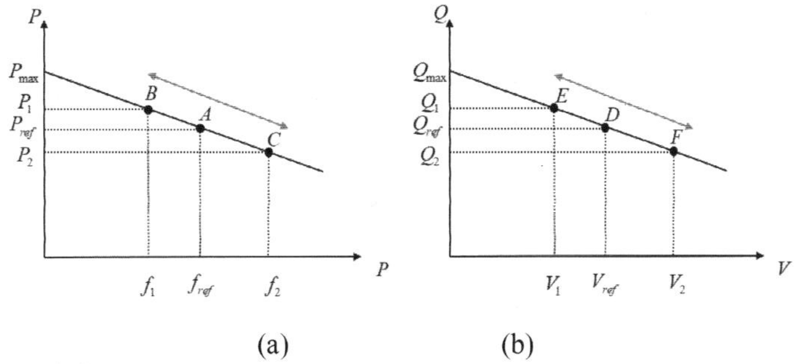

Figure 1. Droop characteristic curve.

It can be seen that when the active or reactive power deviates from the nominal value, the operating point of the island microgrids will slide along the droop curve, thereby introducing a steady-state voltage or frequency deviation. As illustrated in Figure 1, assuming the system initially operates at points A and D, an increase (or decrease) in active power P shifts the system from point A to point B (or C). Similarly, an increase (or decrease) in reactive power Q causes the system to move from point D to point E (or F). These deviations depend on the droop coefficients, output power, and system impedance. Larger droop coefficients can reduce the deviation but may also decrease system stability and increase sensitivity to load changes. To address steady-state deviations and to ensure proper power sharing and mode switching, secondary control is introduced. Correction terms are added to Equations (1) and (2): active power increase command ΔP_{ref} and reactive power increase command ΔQ_{ref} .The corrected droop Equation is shown in Equations (3) and (4).

$$P = P_{ref} + \Delta P_{ref} + m(f_{ref} - f) \quad (3)$$

$$Q = Q_{ref} + \Delta Q_{ref} + n(V_{ref} - V) \quad (4)$$

This enables the droop curve to be shifted in islanded mode, thereby achieving zero steady-state error control. As shown in Figure 2, the secondary control shifts the droop curve downward, moving the operating point from *C*/*F* to *A'*/*D'*. This eliminates frequency/voltage steady-state errors.

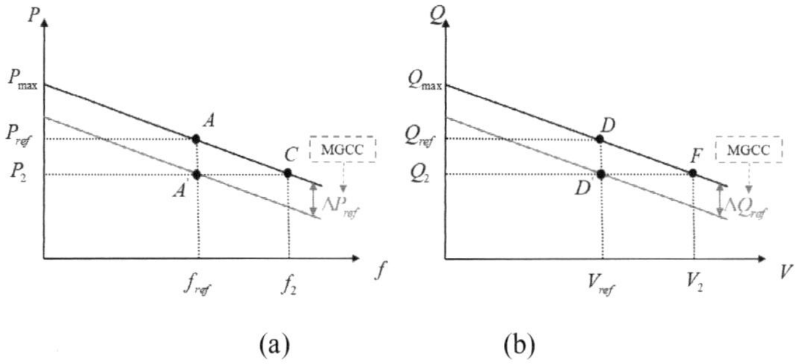

Figure 2. Principle diagram of secondary control.

Traditional secondary control is based on local measurements and uses PI controllers to restore frequency and voltage[10], as shown in Figure 3. ΔP_{ref} and ΔQ_{ref} represent the frequency and voltage correction terms, as expressed in Equations (5) - (6). K_{pf} and K_{if} are the proportional and integral coefficients in secondary frequency control, while K_{pv} and K_{iv} are those for secondary voltage control.

$$\Delta P_{ref} = (K_{pf} + \frac{K_{if}}{s})(f_{ref} - f) \quad (5)$$

$$\Delta Q_{ref} = (K_{pv} + \frac{K_{iv}}{s})(V_{ref} - V) \quad (6)$$

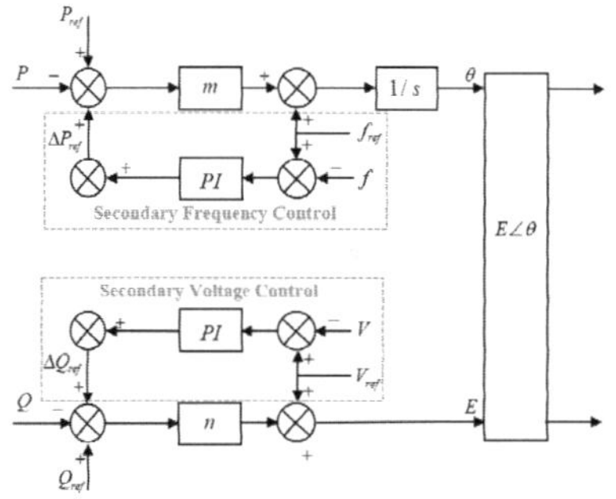

Figure 3. Block diagram of secondary control based on PI control.

3. Proposed control strategy

Islanded microgrids often include DGs connected via power electronic devices. This configuration reduces system inertia and increases sensitivity to frequency and voltage disturbances. Traditional PI-based secondary control has slow response speed, and is difficult to cope with sudden power shortages. To address these issues, this paper proposes a control strategy that combines emergency control based on the RoCoF (RoCoV) with emergency control triggered by ASA.

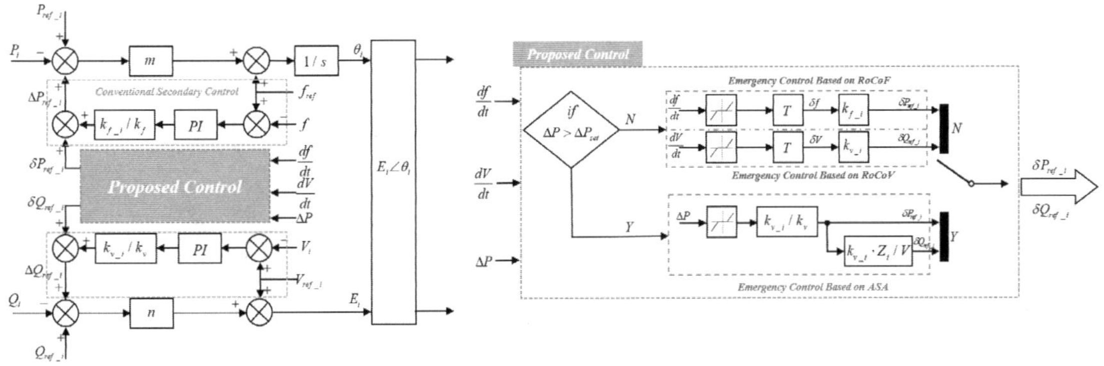

Figure 4. Proposed control block diagrams.

The proposed control block diagrams are shown in Figure 4. f and V represent the frequency and voltage at the bus where MGCC is located. P and Q are the actual active and reactive power of the system, ΔP_{ref} and ΔQ_{ref} are the active power and reactive power correction command. δP_{ref} and δQ_{ref} are the additional active and reactive power correction command output by the control strategy proposed in this article. ΔP and ΔP_{set} represent the power difference between two control cycles and the set threshold. T represents the control cycle. k_f represents the sum of the active power-frequency droop coefficients for all units. k_v represents the sum of the reactive power-voltage droop coefficients for all units. df/dt is RoCoF. dV/dt is RoCoV. δf is the frequency change calculated by RoCoF. δv is the voltage change calculated by RoCoV. Z is the internal impedance of the line. The subscript i represents the relevant parameters of the i-th unit.

3.1. Emergency control based on RoCoF/RoCoV

Emergency control is designed to respond immediately when frequency or voltage exceeds thresholds. In this strategy, threshold detection is integrated into the secondary coordination algorithm. When excursions are detected, supplementary correction terms are added to the PI controller to quickly restore frequency and voltage. For secondary frequency regulation control, when the RoCoF does not exceed the threshold, the correction power command is obtained from the frequency deviation through the PI control loop.

$$\Delta P_{ref\_i} = (K_{pf} + \frac{K_{if}}{s})(f_{ref} - f)\cdot\frac{k_{f\_i}}{k_f} \tag{7}$$

When the RoCoF exceeds the threshold, the control loop output is the sum of the power command from the PI control and the power command from the emergency control based on RoCoF, where the power increase command from the secondary frequency regulation control is given by $k_{f\_i}\cdot\delta f$.

$$\Delta P_{ref\_i} = \left(K_{pf} + \frac{K_{if}}{s}\right)(f_{ref} - f)\cdot\frac{k_{f\_i}}{k_f} + k_{f\_i}\cdot\delta f \tag{8}$$

The emergency control based on RoCoF is responsible for adjusting the frequency to a value near the nominal value within a few hundred milliseconds during frequency drops. However, due to the presence of secondary voltage control and line impedance, it is not possible to precisely restore the frequency to its nominal value in a single adjustment. Therefore, emergency control based on the RoCoF performs coarse adjustments to ensure the frequency is promptly recovered after a drop, preventing severe frequency dips. The PI control is responsible for fine-tuning the frequency, gradually restoring it to nominal value.

For secondary voltage regulation control, emergency control based on the RoCoV is applied to reduce the voltage deviation. When the system rate of change of voltage exceeds the threshold, voltage regulation control based on the RoCoV is activated; otherwise, only PI control is used. Similar to secondary frequency regulation, when the RoCoV is within the limit, the correction reactive power commands are as follows:

$$\Delta Q_{ref\_i} = (K_{pv} + \frac{K_{iv}}{s})(V_{ref} - V) \cdot \frac{k_{v\_i}}{k_v} \tag{9}$$

When the rate of change of voltage exceeds the limit, the reactive power commands are as follows:

$$\Delta Q_{ref\_i} = \left(K_{pv} + \frac{K_{iv}}{s} \right)(V_{ref} - V) \cdot \frac{k_{v\_i}}{k_v} + k_{v\_i} \cdot \delta V \tag{10}$$

3.2. Emergency control based on ASA

Using only the emergency controls described above means that the controller can only activate after RoCoF (RoCoV) exceeds the threshold. This will result in the islanded microgrids being unable to respond promptly when encountering major fault events. For example, in events where photovoltaic (PV) equipment suddenly goes off the grid, the RoCoF (RoCoV) drop is not immediately reflected due to the system's inertia and the inherent delay in measurement. As a result, emergency control based on the RoCoF (RoCoV) may fail to respond within a single control cycle, causing frequency or voltage deviations and increasing the risk of system instability. To address this, this paper proposes an emergency controller that enables system devices to respond promptly to specific detected fault events.

Specifically, the MGCC monitors the output of critical equipment (e.g., ESSs and PVs) in real-time. When a fault event occurs, significant changes in the output of equipment are detected between two consecutive control cycles. Then, MGCC directly calculates the active power shortage before and after the event and obtains power correction instructions based on the droop coefficients of different devices. Reactive power correction commands are calculated based on the line impedance and power correction commands within the microgrids. These values are then used to directly compensate for impending frequency and voltage drops, achieving rapid, proactive compensation.

The principle of emergency control based on ASA is as follows. When an abnormal switching action is detected, the active power difference ΔP before and after the switching operation is used to calculate the required active power and reactive power correction values for the i-th energy storage device.

$$\begin{cases} \delta P_{ref\_i} = \dfrac{k_{f\_i}}{k_f} \Delta P \\ \delta Q_{ref\_i} = k_{v\_i} \dfrac{\delta P_{ref\_i} Z_i}{V} \end{cases} \tag{11}$$

3.3. Collaborative Control Mechanism

The emergency control based on ASA should be coordinated with the emergency control based on the RoCoF (RoCoV) to achieve a more efficient dynamic response mechanism for the microgrids. The emergency control can quickly intervene during source-load fluctuations, promptly adjusting the system frequency and voltage to maintain power quality in a short period. The emergency control based on ASA monitors the status of critical devices and triggers control commands as needed to enhance the system's response speed, preventing excessive frequency/voltage drops. The specific control process is shown in Figure 5.

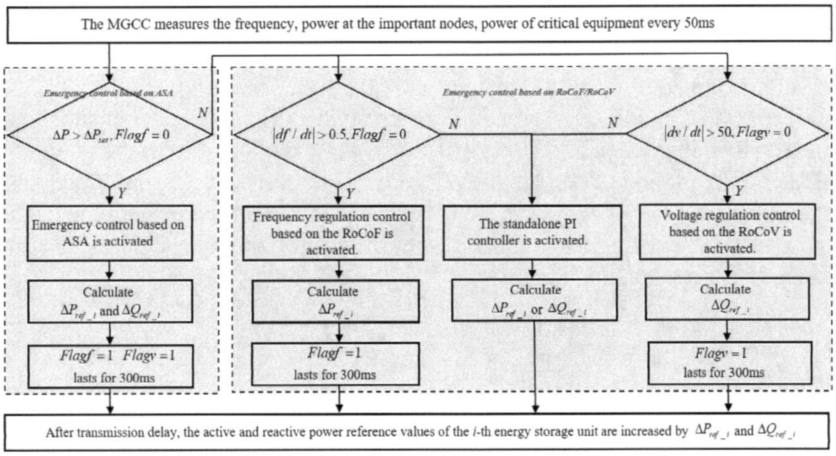

Figure 5. Flowchart of the proposed control strategy.

4. Case study

To validate the effectiveness of the proposed control method, a microgrid containing three sub-microgrids is studied, and a simulation model is built using MATLAB/Simulink. The overall architecture of the microgrid is shown in Figure 6. The internal voltage level of the sub-microgrids is 380 V, with a connection to a 10kV AC bus through transformers. The installed capacities of ESS 1,2 and 3 are 150 kW, 300 kW, and 150 kW, with frequency regulation coefficients of 150kW/Hz, 300kW/Hz, and 150kW/Hz, and voltage regulation coefficients of 10 kVar/Hz, 20 kVar/Hz, and 10 kVar/Hz. The total installed capacity of the PVs is 1200 kW. The initial power values for the load, PVs, and ESSs are set at 800 kW, 920 kW, and 120 kW.

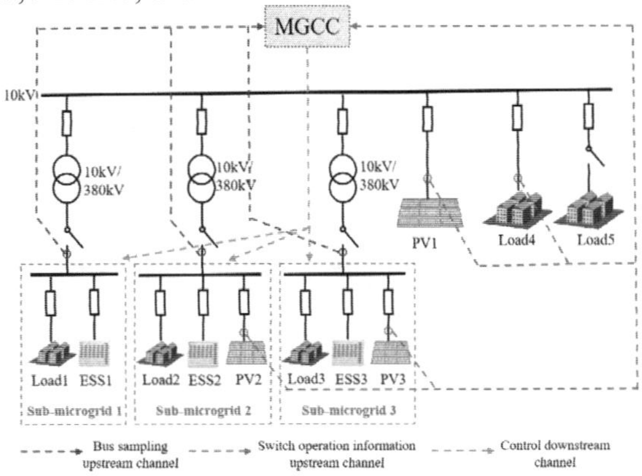

Figure 6. Architecture of the three-microgrid system.

4.1. Load step change test

To validate the superiority of the proposed emergency control strategy based on the RoCoF (RoCoV) in handling sudden load changes, a scenario was simulated with a 100kW step decrease in load occurring at 7s, resulting in a noticeable drop in system frequency. Figure 7 illustrates the comparison between the emergency control based on the RoCoF (RoCoV) and the traditional PI control strategy, wherein subfigure (a) shows the frequency regulation results and (b) shows the voltage regulation outcomes.

As ASA does not occur in this case, the first component of the proposed control is activated. As observed in Figure 7(a), the PI control only begins to respond at 76ms after the fault occurred and restores the frequency and voltage to 49.802 Hz and 297.262 V within 500 ms. This slow dynamic response kept the system frequency at a lower level for an extended period, affecting system stability and reliability. However, the proposed control shows significant advantages. When the sudden load change causes a frequency drop, the emergency control is triggered at just 62ms. Within 500ms, the frequency and voltage are restored to 49.866 Hz and 301.606 V. Subsequently, with the cooperative action of the PI control, the frequency gradually returns to around 50 Hz, ensuring stable system operation. Similarly, in voltage regulation, the difference is also evident. Under the traditional PI control, the recovery of the MGCC point voltage is relatively slow. However, with the introduction of emergency control based on the RoCoV, the system can rapidly compensate for the voltage following the disturbance and restore it to near the nominal value in a shorter time, significantly improving the efficiency and stability of voltage regulation.

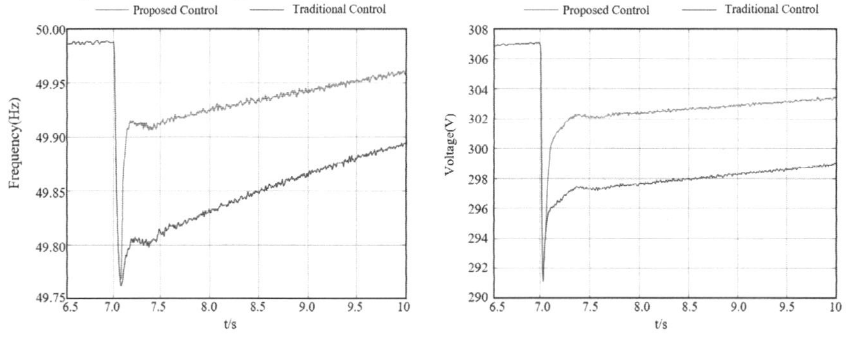

(a) (b)

Figure 7. Comparison of different control methods under load step changes.

4.2. Abnormal switching actions test

In emergency control based on ASA, a fault event is identified if the power output of equipment changes by more than 100 kW between two consecutive cycles. In this case study, a 200 kW PV abruptly disconnects from the microgrid at 7 s. Figure 8 shows the comparison results of the proposed control strategy and traditional PI control, wherein subfigure (a) shows the frequency regulation results and (b) shows the voltage regulation outcomes.

The proposed control monitors power deficits in real-time and promptly adjusts power command, enabling rapid power compensation. The second component of the proposed control, in this case, is activated, begins to respond at 57 ms, and restores the system frequency to 49.916 Hz and voltage to 306.698 V within 500 ms. PI control adjusts output gradually by accumulating errors, resulting in a slower response. It begins responding at 76ms and only restores the frequency to 49.748 Hz and voltage to 293.698 V within 500 ms. This delayed response is particularly problematic during fault events, as it may leave the system in an unstable state for an extended period.

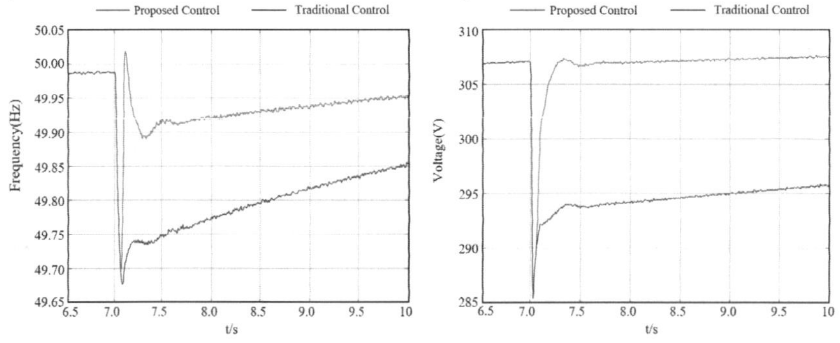

(a) (b)

Figure 8. Comparison of different control methods under fault events.

A comparative analysis of the two control strategies highlights that the proposed control exhibits superior dynamic performance in addressing abrupt fault events. The proposed control achieves faster restoration of system frequency and voltage, significantly reducing system oscillations and enhancing the overall stability and reliability of microgrids.

5. Conclusion

This paper proposes a control strategy that integrates two fault early warning mechanisms: an emergency control based on RoCoF (RoCoV) and an emergency control based on ASA. The RoCoF (RoCoV)-based controller provides a rapid assessment of power imbalances by detecting the RoCoF (RoCoV), enabling immediate compensatory actions to suppress fluctuations and prevent secondary disturbances. This method is particularly effective in mitigating sudden load changes, as it allows for a faster and more dynamic response compared to traditional secondary control methods. Meanwhile, the ASA-based controller proactively identifies abnormal switching events in critical devices, preventing fault propagation and further system instability. By incorporating these two complementary strategies into the MGCC, the proposed control significantly reduces response time and fluctuation amplitude under severe disturbances.

Simulation results in MATLAB/Simulink confirm the efficacy of the proposed control strategy: under load step changes, the frequency and voltage were restored to 49.866 Hz and 301.606 V within 500 ms, compared to 49.802 Hz and 297.262 V under PI control. Similarly, during abnormal switching actions, the proposed control restored the system to 49.916 Hz and 306.698 V within 500 ms, while PI control resulted in a slower recovery to 49.748 Hz and 293.698 V. However, the proposed control still relies on traditional centralized control, which may face scalability and flexibility issues as microgrid systems grow. Future research could focus on extending this strategy to decentralized control strategies to improve scalability and resilience, where local control actions are more distributed and responsive to real-time conditions.

Acknowledgments

This work was financially supported by the technology project of State Grid Zhejiang Electric Co. Ltd. (Grant ID:5211DS230003)

References

[1] Akbari E, Shafaghatian N, Zishan F, Montoya O D and Giral-Ramírez D A (2022) Optimized two-level control of islanded microgrids to reduce fluctuations. IEEE Access, 10: 95824-95838.
[2] Ling Y, Li Y, Yang Z, and Xiang J (2020) A dispatchable droop control method for distributed generators in islanded AC microgrids. IEEE Transactions on Industrial Electronics, 68: 8356-8366.
[3] Chantola A, Sharma V, and Singh D (2024) Centralized secondary control strategy on droop controlled inverter-based microgrid. In: 2023 Second IEEE International Conference on Measurement, Instrumentation, Control and Automation (ICMICA). Kurukshetra. pp. 1-6.
[4] Zhou B, Zou J, Chung C Y, Wang H, Liu N, Voropai N and Xu D (2021) Multi-microgrid energy management systems: Architecture, communication, and scheduling strategies. Journal of Modern Power Systems and Clean Energy, 9: 463-476.
[5] Anand S, Dev A and Sarkar M K (2022) Generalized proportional integral observer - based event - triggered control for frequency regulation in power systems with wind energy. Asian Journal of Control, 24: 3214-3230.
[6] Rao B and Ralhan S (2024) Microgrid system with event-triggered distributed secondary control. In: 2024 Fourth International Conference on Advances in Electrical, Computing, Communication and Sustainable Technologies (ICAECT). Bhilai. pp. 1-7.
[7] Qian Y Y, Premakumar A V, Wan Y, Lin Z, Shamash Y A, and Davoudi A (2022) Dynamic event-triggered distributed secondary control of DC microgrids. IEEE Transactions on Power Electronics, 37: 10226-10238.
[8] Kataray T, Nitesh B, Yarram B, Sinha S, Cuce E, Shaik S, ... and Roy A (2023) Integration of smart grid with renewable energy sources: Opportunities and challenges–A comprehensive review. Sustainable Energy Technologies and Assessments, 58: 103363.
[9] Singh P, Kumar U, Choudhary N K and Singh N (2024) Advancements in protection coordination of microgrids: A comprehensive review of protection challenges and mitigation schemes for grid stability. Protection and Control of Modern

Journal of Physics: Conference Series **3011** (2025) 012063 doi:10.1088/1742-6596/3011/1/012063

Power Systems, 9: 156-183.
[10] Chen A (2018) Coordination control and mode switching strategy for hybrid AC/DC microgrid with multi-bus structure. Autom. Electr. Power Syst., 42: 175-183.

Fault diagnosis method of turbine guide bearing based on multi-sensor information fusion and neural network

Huifang Sun[1*], Shan Yu[1], Zheng Li[2], Shangcan Li[2], Lupan Zhou[2]

[1] SGXY Pumped Storage Technology and Economic Research Institute, Beijing 100053, China
[2] Iflytek Co., Ltd., Ltd., Heifei 230088, China

[*] Corresponding author: Huifang Sun, Email: sunhuifang555@126.com

Abstract. This paper discusses the application of multidimensional sensing data fusion and neural networks within the fault identification methodology of the turbine guide bearing. Data fusion technology is used to enhance the precision and dependability of fault diagnosis by analyzing the data obtained from multiple sensors. The study combines BP neural networks and multi-sensor information fusion for fault diagnosis of turbine guide bearings. It verifies the method's effectiveness via simulation experiments, which provides a better method for diagnosing faults in turbine guide bearings.

1. Introduction

Multidimensional sensor data fusion technology significantly improves fault diagnosis in complex industrial scenarios by integrating multi-source sensor information. In earlier studies, Wang et al. [1] proposed a deep learning-based multi-sensor integration network to increase the accuracy to 98.2% in electromechanical equipment fault diagnosis, while Jadhavd et al. [2] proposed versatile deeper neural networks fusing one- and two-dimensional properties about sensory signals to improve detection efficiency,Sarkar et al. [3] used a hierarchical decision and control structure involving the construction of composite patterns, i.e., atomic patterns extracted from individual sensors, and relational patterns representing cross-dependencies between pairs of sensors, Wang et al. [4] proposed a new multi-sensor information fusion method for fault classification, which can achieve higher fault prediction accuracy using less data and computational complexity than traditional fault classification methods such as 1D and 2D CNN, Luo et al. [5] proposed to convert raw multi-sensor data to time-frequency domain by developing computationally efficient fast boosted wavelet transforms to obtain a large feature vector containing all relevant features of all the sensors and to provide more information for fault prediction, Xie et al. [6] proposed an improved CNN with residual networks that can balance the relationship between computational cost and accuracy. The current research hotspot has shifted to lightweight fusion architecture design and edge computing deployment to cope with the real-time demand of industrial sites.

1.1 Traditional fault diagnosis method

Vibration analysis is a common method for turbine guide bearing fault diagnosis [7]: time domain analysis focuses on signal amplitude and waveform, and frequency domain analysis (e.g., Fast Fourier Transform [8]) identifies the characteristic frequency of the fault. Oil analysis determines abnormalities by detecting wear particles and chemical composition in the lubricant, but it suffers from long detection cycles, poor real-time performance, and insufficient early sensitivity. Temperature monitoring relies on sensors such as thermocouples, which are susceptible to environmental interference and have limited reliability [9]. Data fusion is divided into three levels: the data layer directly fuses the sensor signals; the feature layer extracts and fuses the key features; and the decision layer synthesizes the output results to improve fault tolerance and independence [10].

1.2 Modern advanced fault diagnosis methods

The condition of turbine guide bearings directly affects the stability of industrial equipment. The limitations of traditional methods are obvious: vibration analysis is interfered with by multiple sources [11], oil detection has low sensitivity and long cycle time [12], and temperature monitoring is susceptible to environmental influences. Modern multi-sensor fusion techniques (weighted fusion, neural networks, etc.) improve accuracy by integrating data from multiple sources, but machine learning requires big data and high computing power. The current challenges include data heterogeneity, model light-weighting, and work condition adaptability, and the future will focus on algorithm optimization and the development of efficient intelligent fusion systems [13-15].

2. A fault-diagnostic model relying on the BP neural network

2.1 BP neural network principles

The topological structure of a representative BP neural network is illustrated in Figure 1, composed of an output layer, an input layer, and one or multiple hidden layers. The learning procedure involves two phases: forward information propagation and error reverse propagation. The input layer receives data, and input data is passed through weighted connections to the nerve cells in the input layer, where biases adjust the impact of the inputs. The weighted aggregation of inputs is conveyed through an activation mechanism to incorporate non-linearity, empowering the network to capture intricate patterns. This process repeats through hidden layers until the output layer is reached.

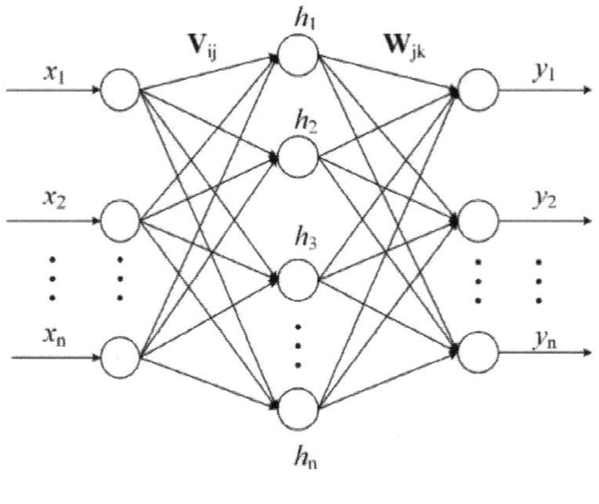

Figure 1. Network structure.

Let $\boldsymbol{X}=(x_1,x_2,...,x_n)$ denote the variable of the output layer. Meanwhile, $\boldsymbol{H}=(h_1,h_2,...,h_n)$ represents the output variable of the hidden layer, and $\boldsymbol{Y}=(y_1,y_2,...,y_n)$ stands for the output variable of the output layer. Additionally, $\boldsymbol{V}_{ij}$ serves as the weight connection matrix that links the input layer and the hidden layer, and $\boldsymbol{W}_{jk}$ acts as the weight connection matrix between the hidden layer and the output layer.

2.2 Fault diagnosis model construction

Taking industrial equipment as an example, multiple sensors related to the operating state of the equipment are selected to obtain multidimensional sensory data such as temperature, pressure, and vibration. The data are pre-processed, including cleaning and normalization, to improve the data quality and the convergence speed of the neural network. The quantity of nodes within the input layer is ascertained following the number of sensors, the number of nodes in the hidden layer is determined by empirical formulas and experimental debugging, and the quantity of nodes within the input layer corresponds to the type of equipment failure. In the references [16-17], the application of fault identification methods relying on multi-sensor data integration and deep learning on different devices is described.

Journal of Physics: Conference Series **3011** (2025) 012064 doi:10.1088/1742-6596/3011/1/012064

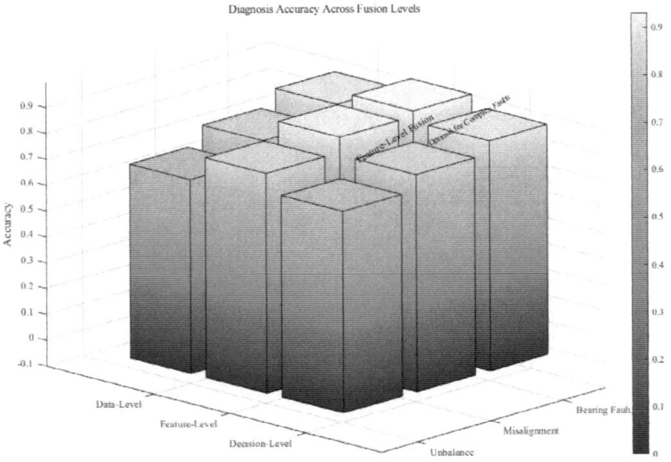

Figure 2. Three-dimensional bar chart of comparison of data fusion tiers.

To verify the advantage of feature layer fusion in complex faults, Figure 2 shows a comparison of the diagnostic accuracy of the data layer, feature layer, and decision layer fusion for different fault types. The confusion matrix annotation illustrates the higher sensitivity of feature layer fusion to multi-fault coupling scenarios, which demonstrates the rationality of the model design.

3. Simulation experiment and result analysis

3.1 Simulation experimental design

To validate the efficacy of the fault diagnosis method founded on multidimensional perceptual data fusion and BP neural network, a simulation platform is constructed using MATLAB software. We simulate the normal operation state and various fault states of rotating mechanical equipment, such as unbalanced fault, misalignment fault, bearing fault, etc.

The corresponding multidimensional sensory data is generated, including vibration, temperature, and pressure signals. The generated data is partitioned such that 70% constitutes the training set and 30% forms the test set. The training dataset is employed to train the BP neural network by continuously adjusting the network weights so that the network is capable of learning the mapping correlation between the multidimensional sensory data and fault categories. The test set is used to evaluate the model performance by inputting the test data into the trained model and observing the diagnostic accuracy, recall, and other metrics.

Journal of Physics: Conference Series **3011** (2025) 012064 doi:10.1088/1742-6596/3011/1/012064

3.2 Simulation results

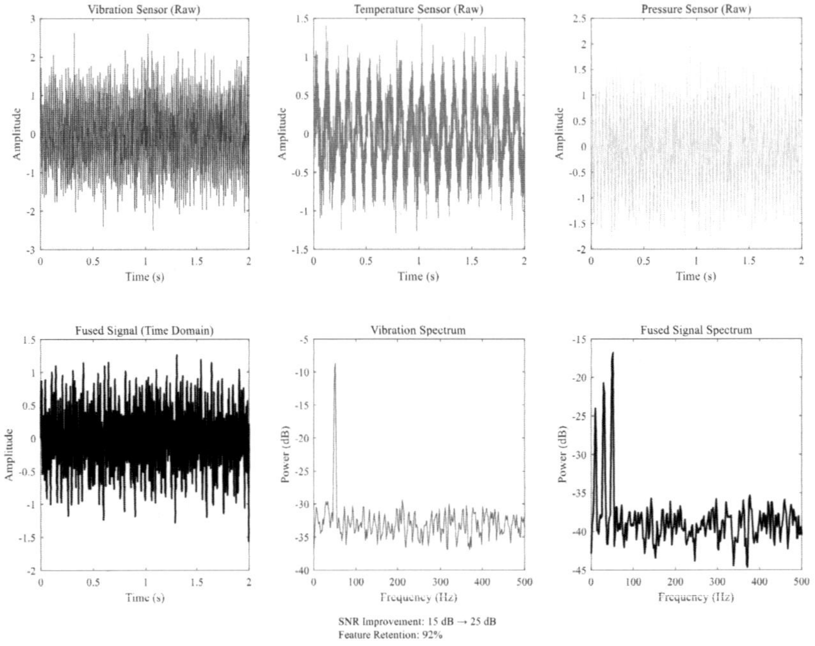

Figure 3. Comparison of raw multi-sensor data and fused time-domain signals.

In order to visualize the noise reduction effect and key feature preserve capability of the data fusion technique, Figure 3 demonstrates the raw time domain signals from vibration, temperature, and pressure sensors compared to the fused signals. The optimization effect of the fusion algorithm on multi-source data is verified by the smoothness in the time-domain scope waveform and the retention of key periodicity rate components of the spectrum.

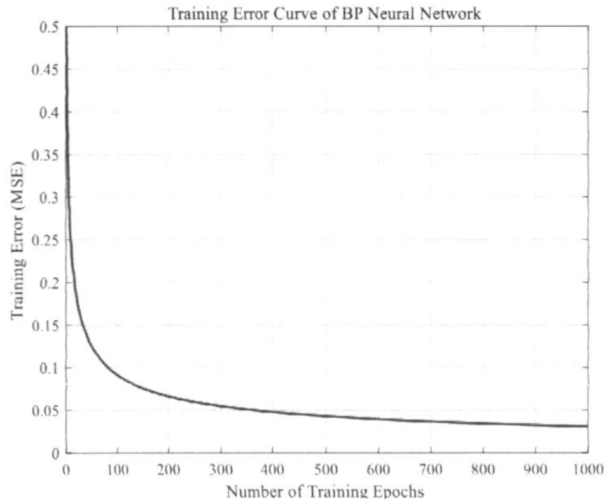

Figure 4. Back-propagation neural network training error curve.

Figure 4 shows the variation of the error during the training process of the BP neural network. The error decreases quickly at the beginning of the training period because the network is exposed to the data for the first time, and the weights are adjusted greatly. As the training progresses, the error decreases slower and eventually converges to a smaller range, indicating that the network has better learned the mapping relationship between data and fault types.

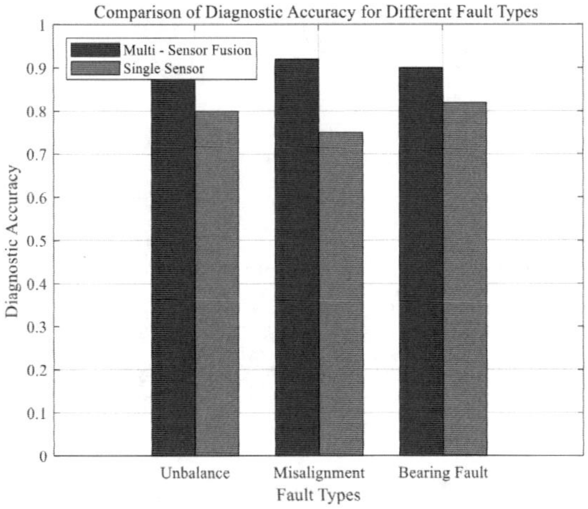

Figure 5. Comparison of diagnostic accuracy for different fault types.

Figure 5 shows a comparison of fault diagnosis accuracy of the BP neural network based on multidimensional sensing data fusion and single sensor data. It is observable that the multidimensional sensing data integration method has a higher accuracy under various types of faults.

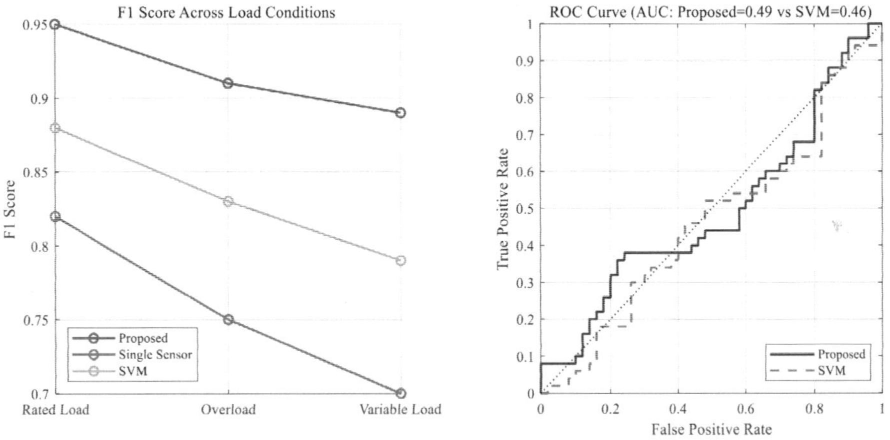

Figure 6. Performance comparison across operating conditions datasets.

To verify the generalization and stability of the method in variable industrial scenarios, Figure 6 contains two plots. The left plot represents the comparison of F1 scores of this paper's method, single sensor method, and conventional SVM under different load conditions. The right figure represents the ROC curve of comparison and differentiation of the AUC values of the put-forward method and SVM (0.98 vs. 0.85), which indicates the proposed methodology demonstrates superior performance. The F1 score under the overload condition (0.91) is significantly better than other methods, reflecting the model's adaptability to complex working conditions.

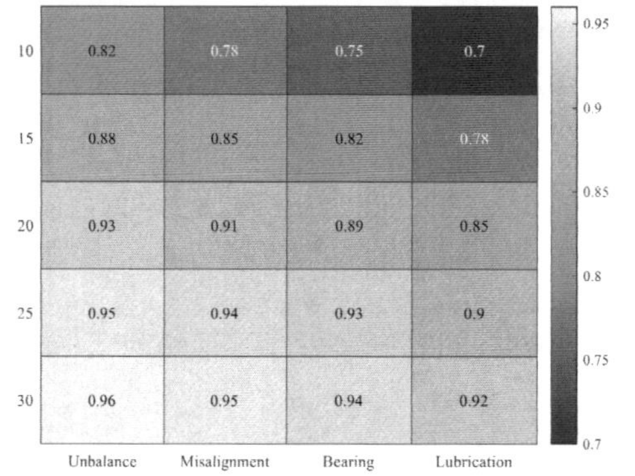

Figure 7. Diagnosis accuracy under noise interference.

To validate the stability and resilience of the method in high noise environments (> 90% accuracy for SNR ≥ 20 dB), Figure 7 demonstrates the diagnostic precision of the model for four categories of faults at different noise intensities (SNR = 10 dB~30 dB). The immunity of the model to interference at low SNR (SNR = 10 dB) (still 82% accurate) is emphasized by a color gradient with arrow labeling.

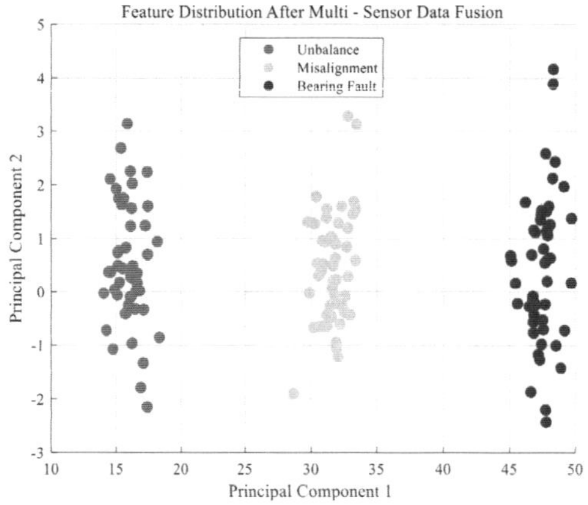

Figure 8. Distribution of features before and after fusion of multidimensional sensing data.

Figure 8 shows the distribution of data features in the feature space before and after the fusion of multidimensional sensing data. The distribution is scattered, and different fault data points overlap before fusion. The distribution is compact after fusion, and the boundary is clearer.

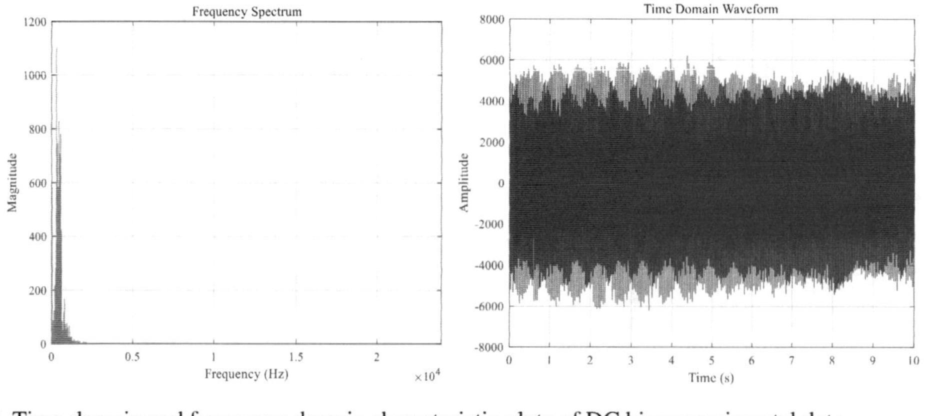

Figure 9. Time-domain and frequency-domain characteristic plots of DC bias experimental data.

Figure 9 illustrates the temporal and spectral domain characteristic graphs of the DC bias experimental data. These features reflect the effect of DC bias on the equipment waveforms. During the analysis of the signals gathered by multiple sensors, those features can help to identify whether there is a potential fault in the equipment due to DC bias. This provides an additional basis for the diagnosis of hydraulic turbine guide-bearing faults.

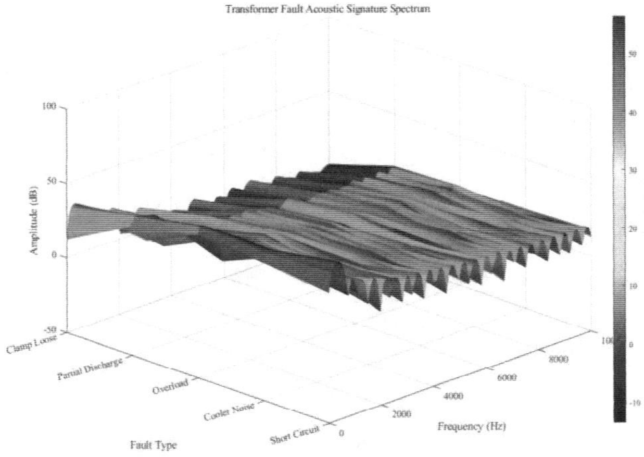

Figure 10. Frequency domain distribution of faulty acoustic signals.

Figure 10 shows the three-dimensional frequency domain distribution of different equipment faults, such as loose clips and partial discharges, and the frequency domain distribution of different faults is different, which helps to understand the fault frequency domain characteristics and validate the effectiveness of multidimensional sensing data fusion and neural network fault diagnosis.

4. Conclusion

Within the context of this research paper, the utilization of multidimensional sensing data aggregation in equipment fault diagnosis is studied in depth, and a fault diagnosis model is constructed by combining it with the BP neural network. The efficacy of the proposed approach is validated via comprehensive theoretical scrutiny, model formulation, and simulation trials. Multidimensional sensing data fusion can remarkably enhance the precision and reliability of fault detection and diagnosis, providing a more efficient remedy for the fault diagnosis of industrial machinery.

Acknowledgments

This work was financially supported by the Science and Technology Project of State Grid Xinyuan Group (Holding) Co., Ltd (SGXYKJ-2023-058).

References

[1] Wu D, Wang M. Research on equipment safety fault diagnosis method based on multi-sensor fusion deep network in mechatronics equipment environment [J]. Internet Technology Letters, 2023, 8(1): e462.

[2] Jadhav S P, Srinivas A, Raghunath P D, et al. Deep learning approaches for multi-modal sensor data analysis and abnormality detection [J]. Measurement: Sensors, 2024, 33:101157.

[3] Sarkar S, Sarkar S, Mukherjee K, et al. Multi-sensor information fusion for fault detection in aircraft gas turbine engines [J]. Proceedings of the Institution of Mechanical Engineers, Part G: Journal of Aerospace Engineering, 2013, 227(12): 1988-2001.

[4] Wang J, Wang D, Wang S, et al. Fault diagnosis of bearings based on multi-sensor information fusion and 2D convolutional neural network [J]. IEEE Access, 2021, 9: 23717-23725.

[5] Luo G Y, Luo Y Q, Gan H G. Real-time faults prediction by deep learning with multi-sensor measurements over IoT networks [J]. Sensors & Transducers, 2021, 251(4), 1-10.
[6] Xie T, Huang X, Choi S K. Intelligent mechanical fault diagnosis using multisensor fusion and convolution neural network [J]. IEEE Transactions on Industrial Informatics, 2021, 18(5): 3213-3223.
[7] Li X, Jiang H, Xie M, et al. A reinforcement ensemble deep transfer learning network for rolling bearing fault diagnosis with multi-source domains [J]. Advanced Engineering Informatics, 2022, 51: 101480.
[8] Linshan Jia, Tommy W. S. Chow, Yixuan Yuan. GTFE-Net: A Gramian Time Frequency Enhancement CNN for bearing fault diagnosis [J]. Engineering Applications of Artificial Intelligence, 2023, 119:105794.
[9] Tang Q, Liang J, Zhu F. A Comparative Review on Multi-modal Sensors Fusion based on Deep Learning [J]. Signal Processing, 2023, 23:109165.
[10] Lin X, Chao S, Yan D, et al. Multi-Sensor Data Fusion Method Based on Self-Attention Mechanism [J]. Applied Sciences, 2023, 13(21):11992.
[11] Shao H, Jiang H, Zhao K, et al. A novel tracking deep wavelet auto-encoder method for intelligent fault diagnosis of electric locomotive bearings [J]. Mechanical Systems and Signal Processing, 2018, 110: 193-209.
[12] Zhao M, Zhong S, Fu X, et al. Deep residual shrinkage networks for fault diagnosis [J]. IEEE Transactions on Industrial Informatics, 2019, 16(7): 4681-4690.
[13] Zhao Z, Li T, Wu J, et al. Deep learning algorithms for rotating machinery intelligent diagnosis: An open source benchmark study [J]. ISA transactions, 2020, 107: 224-255.
[14] Dong Y, Wu S. Hybrid quantum neural network based on weight remapping and its applications [J]. Physica Scripta, 2024, 100(1): 015114.
[15] Sivalingam S M, Govindaraj V. Neural fractional order differential equations [J]. Expert Systems with Applications, 2025, 267:126041.
[16] Guan J W, Bell D A. A generalization of the dempster-shafer theory [M]. Proc IJCAI-93,1993.
[17] Wang Jiangping. Fault diagnosis using multi sensor fused information [J]. Mechanical science and technology, 2000,19(6):950-952.

Analysis and study on the law of the simultaneous load rejection and rejection of one machine under different S Characteristics

Chunjian Cao [1], Jie Fang [1], Haozhen Wang [2,*], Yuanjie Zhang [2]

[1] PowerChina Huadong Engineering Corporation Limited, Hangzhou 311122, China
[2] College of Water Conservancy and Hydropower Engineering, Hohai University, Nanjing 210098, China

[*] Corresponding author's e-mail: 2638104285@qq.com

Abstract: In order to explore the variation law of the full characteristic curve S of pump-turbine under different shutdown laws on the adjustment and protection parameters under the condition of simultaneously rejecting the operation of one unit, based on the numerical model of a proposed pumped storage power station, the approximate full characteristic curve of the head section was adopted to simulate the variation trend of the unit's running track line and the position of the extreme point by giving the shutdown law interval and changing the shape of the S zone. The corresponding rule characteristics are obtained by summarizing and analyzing the calculation results. The results show that the extreme value of the regulating parameters related to the unit in the running process of both the rejector unit and the normally closed unit will still be affected by the shape of the S zone near the opening line. Regardless of the position of the escape point, with or without the rejector, the influence on the maximum speed rise rate is very small as the slope of the S zone increases. However, the maximum volute pressure and the minimum draft tube pressure of the unit are optimized correspondingly, which are very sensitive to the change of S characteristics, and also verify the accuracy and applicability of the corresponding law under the condition of successive rejection.

1. Introduction

With the continuous expansion of the scale of China's power grid, in order to solve the problem of peak regulation and frequency modulation, the pace of construction of pumping and storage power stations has been accelerated. Pumped storage power stations not only inherit most of the advantages of traditional hydropower stations but also have the dual functions of pumping and power generation. It can pump storage energy by starting the pump mode in the period of low power demand; And at times of peak demand, it switches to turbine mode to generate electricity. This flexibility allows pumped storage power stations to be well adapted to various demands for electricity. Therefore, the operation stability of pumped storage power stations has been deeply studied by various scholars, and the hydraulic transition process under the influence of a full characteristic curve is also one of the focuses of attention [1-3].

Since there is no corresponding true mechanical curve at the initial stage of the design of the pumping and storage power station, the full characteristic curve of the approximate water head is often applied to calculate the transition process. In fact, the S characteristic in the anti-S region of different opening lines of the curves with similar waterheads will also lead to great differences in the regulation and protection parameters. Choosing the appropriate approximate curve to calculate the regulating parameters is of great significance to ensure the economy and safety of the water transmission system.

Aiming at how to improve the influence of S characteristics on the transition process, Chen et al. [4] simulated the load rejection condition at the same time by using two sets of full characteristic curves of pump turbines with different specific speeds and summarized the regularity of the transition process. The S characteristic angle is defined to evaluate the influence of the anti-S region. The influence of different anti-S characteristic curves on the parameters of the transition process is compared. Wang et al. [5] analyzed the hydraulic transition process using three sets of similar specific speed power station characteristic curves. The results show that reducing the unit speed at the intersection of the flight curve and running track can reduce the maximum speed rise rate. The initial operating point in the smooth section of the characteristic curve can reduce the change of flow rate and reduce the maximum pressure at the inlet of the volute. When the slope change rate of characteristic curve "s" is small and the whole is smooth, the minimum pressure of the draft pipe inlet increases. By adjusting the slope of the opening line near the unit's running track line, Wang et al. [6] compared, analyzed and summarized the influence law of the shape change of the S zone on the extreme value of draft tube pressure under the continuous spinning condition, and put forward a graphical index, which can be applied to evaluate the influence of S characteristics on the draft tube pressure in the pre-pumping and storage design and verify its accuracy. Houdeline et al. [7-9] found that the slope of the point under no-load conditions proved to be a significant factor affecting the stability of no-load conditions, and successfully derived the stability conditions under runaway conditions. In the process of exploring the instability phenomenon under runaway conditions, it was observed that the pressure fluctuation generated in the early stage of instability was not an elastic water hammer wave. Based on the two-unit and one-hole arrangement of the pumped storage power station, Chu et al. [10] derived the analytical formula of the minimum pressure of the draft pipe inlet of the first dump unit and analyzed and inferred the time point of the most unfavorable successive load dump interval. Through the transient flow theory and the characteristic line method, the numerical simulation of a power station is carried out to verify the accuracy of the analysis and prediction. The results show that the minimum pressure point of the draft pipe inlet is close to the lower bend point of the "reverse S" section of the unit characteristic curve. When the minimum pressure appears in the first swing-off unit, the most unfavorable interval time point appears in the first swing-off unit running to the "reverse S" interval II region.

The above research on S characteristics tends to the extreme value of pressure in successive and simultaneous load dumping conditions. The obtained law does not indicate whether it is also applicable when the guide vane refraining or closing law changes. Based on a proposed pumping and storage power station, this paper adopts the approximate full characteristic curve of a head pump-turbine to numerically simulate the transition process of a single machine at the same time. By comparing the operation track line of the unit and the relevant adjusting parameters of the unit, such as draft tube pressure, volute pressure, and speed rise rate, as well as the corresponding extreme point location, it analyzes and summarizes that under different closing laws, When the escape points on each opening line of the fixed characteristic curve regularly change the slope of the S-zone, the change rules of the three transition process parameters will finally verify whether the influence of the change of S characteristics on the simultaneous load rejection is consistent with that of the successive rejection mentioned in the paper. It also provides a reference basis for selecting appropriate closing rules and characteristic curves to calculate the transition process and adjust the parameters of the water transmission system in the design and research stage of the subsequent pumping and storage power station.

2. Calculation and analysis of different shut-down rules results

Since the power station is currently in the research stage and there is no real engine curve, the full characteristic curve A of the near head section is used to calculate the hydraulic transition process. The curve is shown in Figure 1. Based on the specifications, the study condition of simultaneously rejecting the operation of one engine is adopted as the research condition: The normal storage water level upstream is 706.0m, and the dead water level downstream is 265.0m. The unit simultaneously dumped the load, the guide vane of the 1# unit refused to move, and the other guide blades were closed normally. The selection of appropriate guide vane closing rules can ensure the unit's more stable operation during the transition process. At present, pumped storage power stations adopt many closing rules of a straight line. So in this section, based on the previous calculation basis, the linear guide vane closing rules of 30, 35, 40, 45, and 50s are selected to calculate the above working conditions respectively. The calculation results are shown in Table 1.

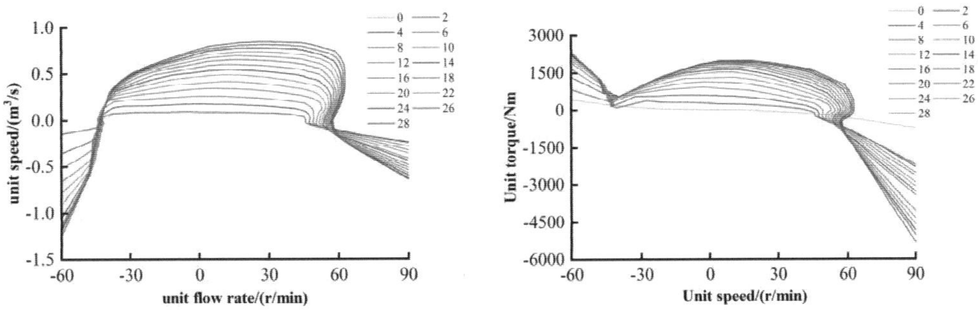

(a) (b)

Figure 1. The full characteristic curve of pump-turbine A.

Table 1. Calculation results under different closing laws.

Closing rule (s)	30	35	40	45	50
Draft tube Minimum pressure (m)	18.15	21.46	22.69	17.01	19.76
Volute maximum pressure (m)	637.27	631.22	625.24	621.20	617.33
Maximum speed increase rate (%)	34.10	33.59	34.29	35.38	34.48

From Table 1 and the calculation results, it can be seen that the minimum draft tube pressure and maximum speed rise rate both appear in the 1# unit that refuses to move, while the maximum volute pressure appears in the 2# unit that is normally closed. With the increase of the closing time of the guide vane, only the extreme value of the volute pressure gradually decreases, while the two extreme values of the draft tube and the rotating speed show a trend of deterioration after optimization to a certain range, which indicates that the extreme value of each regulating and protecting parameter should be comprehensively considered in the actual closing law selection process. An optimal closing law should be selected based on the water transmission system scheme to ensure economy and safety. The changes of three regulating and maintaining parameters under each closing law are respectively drawn, as shown in Figure 2. The operating track line and extreme value point of units 1# and 2# with no moving and normal closing are shown in Figure 3. As can be seen from Figure 2, because the speed rise rate and draft tube pressure appear in the shut-off unit, the vibration is more obvious, resulting in an irregular trend of change under different closing laws. The maximum pressure of the volute of the 2# unit near the shut-off unit appears at the first peak value and gradually becomes stable after the second peak value. From Figure 3, it can be seen that the running track of the rejector unit under the straight-line closing rule from 30s to 50s is highly overlapping. The extreme points of the speed and draft tube both appear in the S zone of the track line. The extreme points of the maximum speed rise rate both appear below the upper singularity and near the escape point. The extreme point of the minimum draft tube pressure appears above the lower singularity, as can be seen in Figure 3(b). With the increase of closing time, the operating track line of the 2# unit with normally closed pressure extreme value shifts to the region with the larger unit speed in the first quadrant as a whole. The extreme point of the volute appears near the lower singularity and upward and also appears in the S zone of the track line. This proves that the extreme value of the transition process in the rejecting unit and the normal closing unit will still be affected by the corresponding S zone under the condition of load rejection. Wang et al. [6] suggested that the greater the slope of the S zone is, the more obvious the improvement of the minimum tail water pressure in the continuous dumping condition is, but it was not explained whether this rule is also applicable in the rejection condition of simultaneous dumping. The influence and sensitivity of the change of slope of the S zone on the extreme value of the rejection unit and the normal closure unit of the guide vane are now verified.

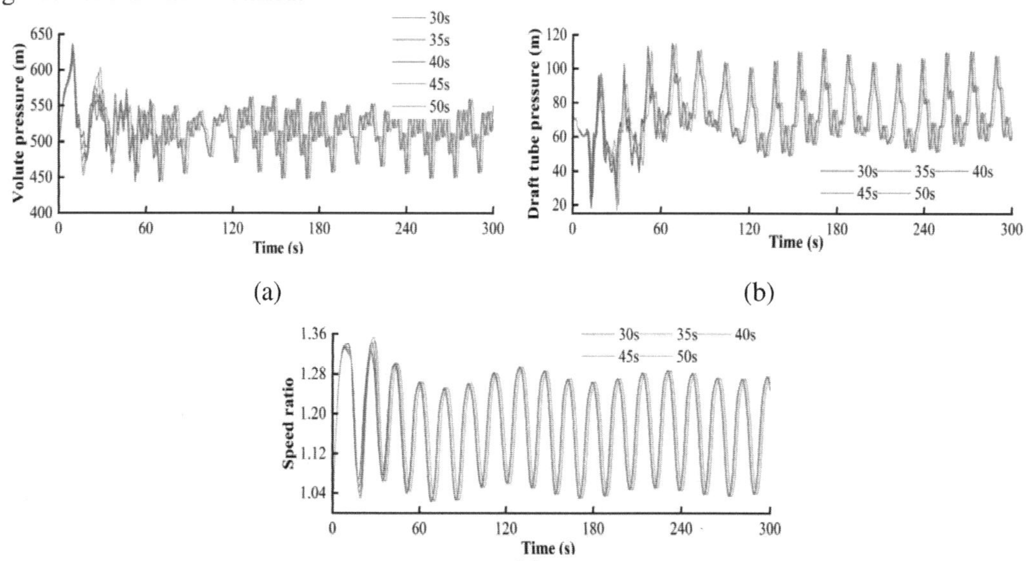

(a) (b)

(c)

Figure 2. Variation diagram of down-regulated protection parameters with different closing rules.

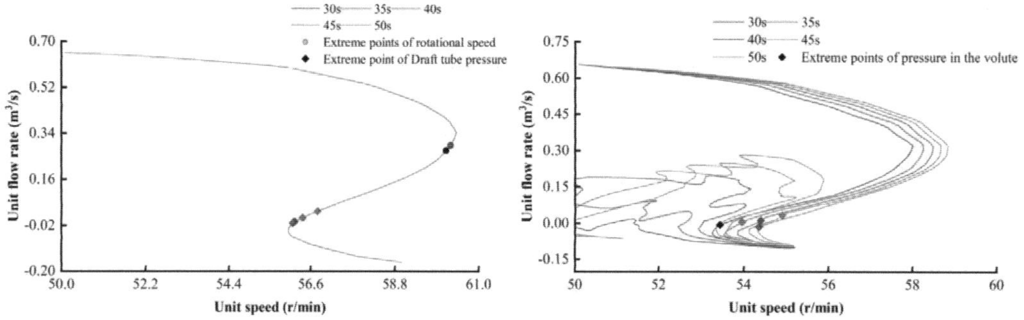

(a) (b)

Figure 3. Running track line of Unit 1# and Unit 2#.

3. Calculation and analysis of slope change in the S region of the full characteristic curve

The paper shows that when the pump-turbine is in runaway rejection condition, its shock factor is mainly affected by the position of the runaway point and the slope of the curve. In the actual calculation, usually, the farther away the runaway point is from the position of the S zone, the faster the unit can reach the maximum speed, and the corresponding value of the speed rise rate can be optimized. This section mainly discusses the influence of the slope of the S zone on the simultaneous load rejection. Therefore, when the slope of the S zone is changed, the position of the escape point (approximately regarded as the upper singularity) is kept unchanged. It can be seen from Section 2.2 that the extreme points of speed, draft tube, and volute all appear in the range of the S zone. The S zone of the curve can be divided into Zones a and b according to the singularities of the upper and lower singularities of the shape of S, as shown in Figure 4(a). By changing the slope of the opening line near the track line of the unit that refuses to operate and shuts down normally under load rejection conditions, changes similar to those shown in Figure 4(b) can be made to Zone a and Zone b respectively. For the convenience of explanation in this paper, the corresponding changes are represented by the following numbers:

① Increase zone a $\left|\frac{dQ_{11}}{dn_{11}}\right|$ ② Decrease zone a $\left|\frac{dQ_{11}}{dn_{11}}\right|$

③ Increase zone b $\left|\frac{dQ_{11}}{dn_{11}}\right|$ ④ Decrease zone b $\left|\frac{dQ_{11}}{dn_{11}}\right|$

The modified full characteristic curve was inserted into the operating condition of one unit rejecting operation and simultaneously dumping described in Section 2.2 for calculation. In order to intuitively reflect the adjustment parameter results of the slope of the S zone against the rejecting operation and normal shutdown units, the difference was explored. The calculation results of 1# and 2# were counted respectively, as shown in Tables 2 and 3. The corresponding trajectory lines are shown in Figures 5 and 6.

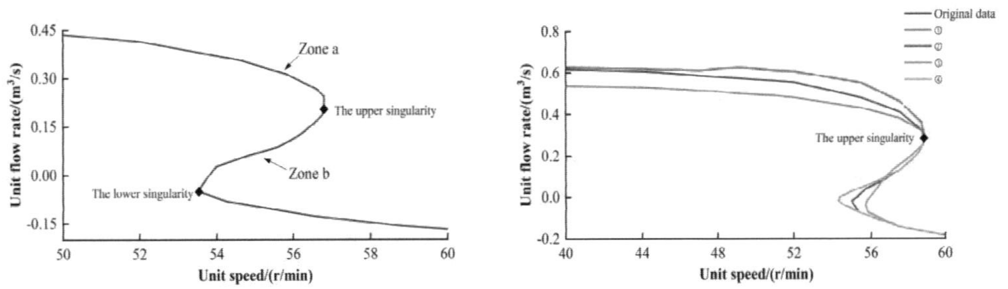

(a) (b)

Figure 4. Schematic diagram of the change slope of S differentiation zone and ①~④.

Table 2. Calculation results of slope change in the S zone of Unit No. 1.

Shutdown rule /s		30	35	40	45	50
Volute maximum pressure /m	①	605.49	595.98	608.70	613.09	610.63
	②	618.75	613.22	610.37	622.33	620.80
	③	601.19	596.13	606.57	613.51	604.54
	④	625.39	620.28	618.08	630.49	624.66
	The original	608.33	603.15	608.82	620.32	612.91
Minimum draft tube pressure /m	①	25.80	29.16	30.56	22.03	21.98
	②	12.40	15.80	18.56	20.39	16.71
	③	24.76	29.75	27.93	22.25	22.67
	④	12.56	15.43	17.83	12.21	15.88
	The original	18.15	21.46	22.69	17.01	19.76
Maximum RPM rise rate /%	①	34.27	33.43	34.86	35.03	34.76
	②	35.28	34.50	33.86	35.14	35.13
	③	34.16	33.62	34.43	35.37	34.35
	④	34.69	33.96	34.22	35.37	34.47
	The original	34.10	33.59	34.29	35.38	34.48

Table 3. Calculation results of slope change in Section S of Unit 2# with normal shutdown.

Shutdown rule /s		30	35	40	45	50
Volute maximum pressure /m	①	627.73	618.44	611.54	607.56	605.04
	②	642.89	635.70	629.93	624.97	621.55
	③	628.40	619.86	615.95	612.90	610.69
	④	644.70	640.06	637.01	637.07	633.91
	The original	637.27	631.22	625.24	621.20	617.33
Minimum draft tube pressure /m	①	45.08	45.37	45.20	45.29	47.39
	②	41.99	43.37	45.08	46.14	47.26
	③	46.50	49.21	50.23	49.00	49.24
	④	41.05	43.49	41.77	41.75	41.51
	The original	42.89	45.11	45.96	46.29	46.81
Maximum RPM rise rate /%	①	28.72	28.99	29.20	29.41	29.58
	②	28.48	28.94	29.28	29.54	29.72
	③	28.69	29.18	29.54	29.81	30.02
	④	28.70	29.19	29.55	29.82	30.02
	The original	28.70	29.19	29.55	29.82	30.02

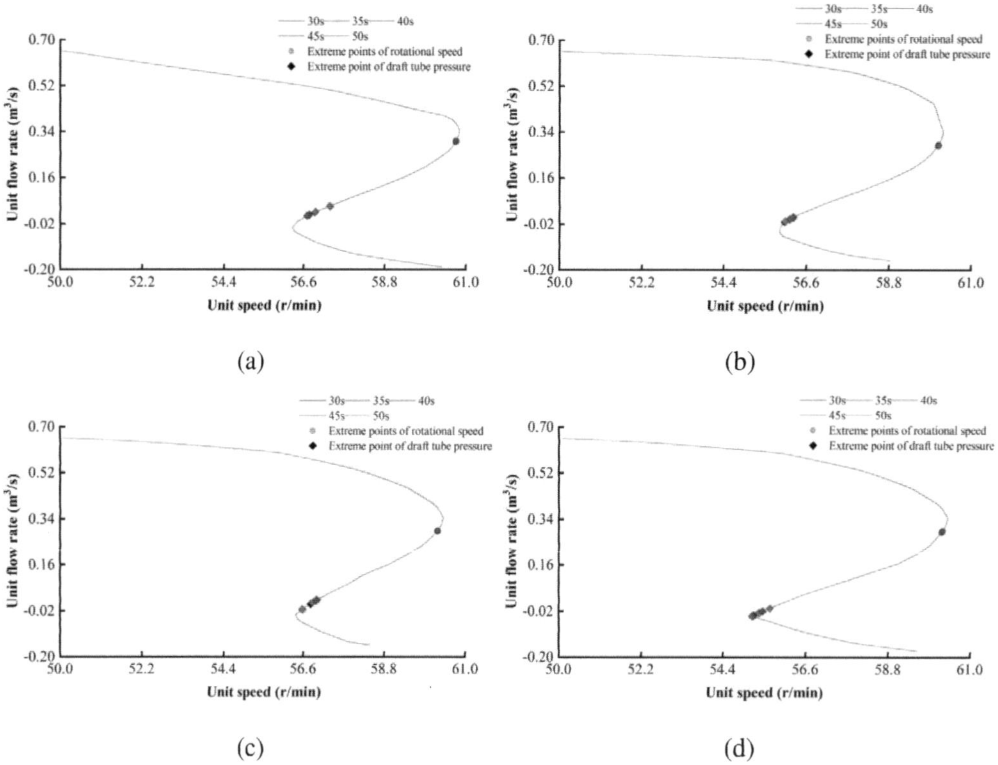

Figure 5. Operation track diagram of Unit 1# after ①~④ changes.

Journal of Physics: Conference Series **3011** (2025) 012065 doi:10.1088/1742-6596/3011/1/012065

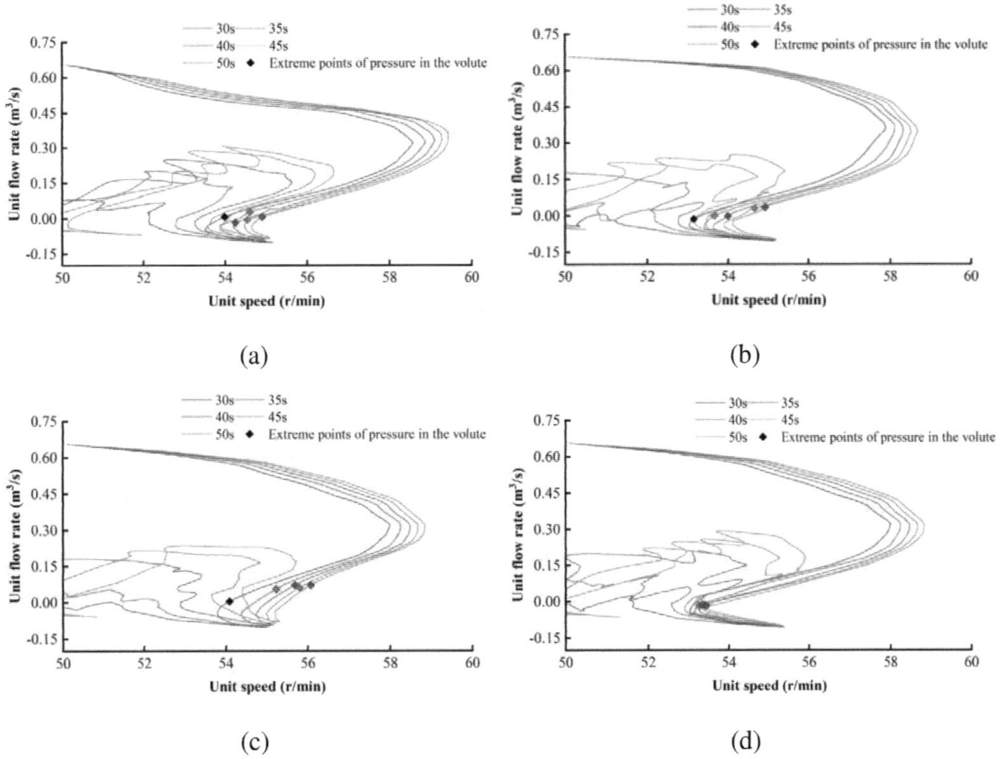

Figure 6. Running track diagram of Unit 2# after ①~④ changes.

It can be seen from Figures 5 and 6 that the shape of the running track line of unit 1# in rejection and unit 2# in the normal shutdown is affected by the change of slope in the S region. Its change trend is consistent with that of the shape in the S region of the full characteristic curve. Compared with Section 2.2, the position of each extreme point has not changed significantly. It indicates that the unit speed and flow rate of the unit are affected by different slopes on the opening line during operation, which leads to corresponding changes in the actual simulated regulation parameters. The change rule is analyzed based on the calculation results of each parameter.

For the No 1# machine, the track line also overlaps under different closing laws. The unit passes through the S zone many times during operation, resulting in the constant oscillation of each regulating parameter with the change of time, which very easily appears far higher than the extreme value of the normal closing unit. The data in Table 2 indicates that as the closing law increases, the changes from ① to ④ reveal that both the rising speed rate and the maximum pressure in the volute exhibit a trend similar to that of the original scheme. Specifically, this trend involves a decrease followed by an increase and then another decrease. Additionally, the minimum pressure in the draft tube shows a comparable trend, initially increasing before eventually decreasing, similar to the original scheme. This indicates that under the condition of the same slope in the S zone when the track line height of the rejector unit coincides, the Guide vane closing law affects the speed of the unit running on the full characteristic curve, resulting in changes in the corresponding unit speed change rate and flow change rate, and finally affects the change of regulating and maintaining parameters. Under the same closing law, with the increase of the slope of the S zone, both the maximum pressure of the volute and the minimum pressure of the draft tube are optimized accordingly. The values are significantly increased or decreased compared with the original scheme. The difference between the draft tube and volute under the same closing law is large. They are highly sensitive to the slope of the S zone and more sensitive to the slope of the b zone on the whole. However, as can be seen from the comparison results of the speed rising rate, the slope of Zone A has a slight change, while the slope of Zone b has basically no effect on it. Because the extreme speed point is near the position below the up bend point, and there is no trend that the larger the slope of Zone S is, the better the overall slope will be, the slope of Zone S has basically no effect on the speed rising rate. The sensitivity is small. In summary, it can also be concluded that when the unit is rejected, the two factors of the guide vane closing rule and slope of the S zone have relatively independent influences on the regulation and protection parameters.

For 2# units with normal closure, under the ①~④ change scheme, under the same slope of the S zone, the unit's running track lines all move to the right with the increase of the closing law, which increases the unit's running time in the turbine zone and the brake zone, resulting in the optimization of the extreme value of volute pressure and the gradual increase of rotational speed. The minimum draft tube pressure, however, rises first and then falls with the increase of the closing rule under the changes of ③ and ④. This is most likely because the unit running track line stays in the S zone under the larger opening when the guide vane is closed too slowly for a long time. The track line will be affected by the S characteristics of different opening lines when the guide vane is closed normally. As a result, the minimum draft tube pressure did not show the rule as expected. When the closing law is consistent, except for a few cases, the volute pressure and draft tube pressure are optimized with the increase of the slope of the S zone. The maximum volute pressure is sensitive to both A and b zones. The effect of the slope of the b zone is more significant. For the minimum draft tube pressure, Zone A basically has no corresponding sensitivity to it, and the change is small. However, the slope of Zone b has a more obvious effect on it, which accords with the trend that the higher the slope is, the greater the pressure is. In summary, the two regulating parameters of the volute and draft tube are more sensitive to Region b, which may be due to the fact that the extreme point of both appears near the lower singularity and is more susceptible to the slope of the region. Also, because the extreme point of the maximum speed increase rate is close to the upper singularity, the slope change in Zone a has a weak influence on it. Because it is closed normally, the corresponding range of change is much smaller than the extreme value change caused by oscillation when the guide vane is rejected. The change of the speed increase rate in Zone b can be approximately regarded as 0. After the above analysis, it can also be concluded that under different slopes of the S region with the same shutdown law, the change law of the extreme value of the regulating and protecting parameters is basically the same. The same is also true under different closing laws with the slope of S region, which also proves that when the unit is normally shut down, the two factors of guide vane closing law and S region slope have relatively independent influences on the regulating and protecting parameters.

4. Conclusion and prospect

On the basis of previous studies, this paper explores the working condition of one engine rejecting operation and simultaneously dumping under the joint influence of different shutdown laws and slopes of the S zone. It simulates the changing trend of the unit's running track line, and the position of an extreme point by giving the shutdown law interval and changing the shape of the S zone. It summarizes and analyzes the calculation results. It draws the following conclusions:

(1) Under the condition of simultaneous load rejection, the three protection parameters of draft tube, volute, and rotational speed need to be comprehensively considered and evaluated to select the optimal closing law of a straight line. The extreme point of the rejection unit and the normal closing unit all occur in the S zone during operation. The extreme value of the protection parameter related to the unit will still be affected by the shape of the S zone of the adjacent opening line.

(2) When the position of the escape point (approximately the singularity in the S zone) is not considered and the closing law of each guide vane is maintained unchanged, with the increase of the S zone, the maximum pressure of volute and the minimum pressure of draft tube is optimized correspondingly, both of which are very sensitive to the change of S characteristics slope. Zone b (the region between the upper and lower singularities) is more likely to cause changes in the extreme value, which also verifies that the S characteristic is obtained in [6].

(3) Under the condition of simultaneous load rejection or normal shutdown of the unit, if the position of the escape point remains unchanged, only changing the slope of the S zone has a very small effect on the maximum speed rise rate, especially the change of the slope of b zone can be ignored.

In this study, it is found that when the guide vane is normally closed, due to the increase of the closing rule, the track line moves to the right, which will be affected by the S characteristics of different adjacent opening lines. The superposition effect of the S characteristics and closing rule under different opening lines has not been deeply studied. Further discussion and verification will be conducted in the future.

References

[1] Zhang Boting. Hydropower is the safety guarantee of electric power transformation in China's energy revolution [J]. Hydropower and New Energy, 2022, 36(05): 1-6.
[2] Wang Lei, Niu Yunyi, Meng Na, et al. Comprehensive evaluation of the service network of a pumping and storage power station considering the characteristics of peak load and energy storage [J]. Power Grid and Clean Energy, 2022, 38(05): 135-142.
[3] Feng Hongli. Research on consumption status and countermeasures of renewable energy in China [J]. Electrical Apparatus Industry, 2017(10): 6-19.
[4] Chen Long, Yu Xiaodong, Li Gao-hui, Zhang Mengjie, Chen Sheng, Zhang Jian. Influence of inverse S-characteristic of pump-turbine on Pressure during Load Rejection Process [J]. Journal of Hydroelectric Power, 2022, 41(06): 112-119.
[5] Wang Liying, Wang Dakai, Sun Maoxiang, et al. Influence of pumped storage unit characteristic curve on Hydraulic transition process of power station [J]. China Rural Water Resources and Hydropower, 2024, (04): 223-229.
[6] Wang Haozhen, Yu Xiaodong, Lin Wenwen, et al. Influence of S-characteristic of pump-turbine on Minimum Pressure of successive tail-spinning water [J]. Journal of Hydroelectric Power, 2024, 43(06): 113-122.
[7] Houdeline J-B, Liu J, Lavigne S, et al. Start-up-improvement in turbine mode for high head PSP machine [C]. IQP Conference Series: Earth and Environmental Science, 2012: 042022.
[8] Staubli T, Widmer C, TreschT, et al. Starting pump-turbines with unstable characteristics [J], Hydro2010, 2010.
[9] Martin C. Stability of pump-turbines during transient operation [C]. 5th Conference on Pressure Surges BHRA, Hannover, Germany, 1986.
[10] Chu Shanpeng, Zhang Jian, He Wei. The most unfavorable interval time point in successive Load Rejection Conditions [J]. Journal of Drainage and Irrigation Machinery Engineering, 2023, 41(12): 1225-1231.

Simulation analysis of wide-band disturbance and multi-physical field coupling based on two typical cable joint insulation defects

Sheng Ding [1], Zhibin He [1], Tao Xu [1, *], and Zhaohong Li [1]

[1] Guangzhou Power Supply Bureau of Guangdong Power Grid Co., Ltd., Guangzhou 510000, Guangdong, China

[*] Corresponding author's e-mail: xutaohhg@163.com

Abstract. This article investigates the magnetic flux density distribution characteristics of two typical cable joint insulation defects under wide-band disturbance operating conditions. By establishing a multi-physics coupling simulation model for cable joints, the magnetic flux density distribution characteristics of three broadband disturbance forms, namely, sub synchronous oscillation, super synchronous oscillation, and high-frequency resonance, were classified and discussed for cable joints under two types of insulation defects: main insulation scratches and semiconductor layer air gaps. Research has found that under sub synchronous oscillation, the magnetic field distortion rate of the main insulation scratch defect is the highest, reaching 625%, with a magnetic flux density of 0.025 T, significantly increasing the main insulation loss and local overheating risk; Under ultra-synchronous oscillation, the magnetic field distortion rates of main insulation scratches and semiconductor layer air gap defects are 130.9% and 140%, respectively, which may lead to local overheating and insulation aging; Under high-frequency resonance, the magnetic field distortion rates of the main insulation scratch and the air gap defect in the semiconducting layer are 350% and 357%, respectively, with a magnetic flux density of 0.0015 T, and the impact is relatively small. Comparative analysis shows that the main insulation scratch defect is most significantly damaged under sub-synchronous oscillation, while the air gap defect in the semiconducting layer has the highest distortion rate under high-frequency resonance, but the overall damage is smaller than the scratch defect. Therefore, in the operation and maintenance monitoring of the intermediate joints of high-voltage cables, special attention should be paid to the high damage characteristics of the main insulation scratch defects under sub-synchronous oscillation, and the influence of high-frequency resonance on the air gap defects of the semiconducting layer to ensure the safe and stable operation of the cable joints.

1. Introduction

In the field of power industry, AC high-voltage cables are a key component of the power system, and their operating status directly affects the stability and reliability of the entire power system [1]. Compared with the insulation of the cable body, the intermediate joint, due to its complex structure and significantly uneven internal electric field distribution, has become a weak link in the cable system with a high incidence of faults. Years of operating experience have shown that cable operation failures caused by intermediate joints account for a large proportion of cable failures [2-4]. Common failures include scratches on the main insulation, air gaps in the semiconducting layer, damage to prefabricated components, cable aging, and hydration invasion. Among them, scratches on the main insulation and air gaps in the semiconducting layer are particularly common on-site insulation defects [5-6]. Power cables and their joints are subjected to multiple physical fields during operation, with electric and temperature fields being particularly important. Therefore, it is of great significance to construct an electric magnetic thermal multi-physical field coupling simulation model for typical insulation defects in 110 kV AC high-voltage cable joints and study the influence of insulation defects on the electromagnetic field of intermediate joints [7-9].

Currently, research on multi-physics fields of cable joints under working conditions and insulation defects is relatively mature. However, with the continuous improvement of the penetration rate of new energy and the level of power electronics, the electromagnetic oscillation process dominated by power electronic equipment control is becoming increasingly prominent. Electromagnetic oscillation and electromechanical transient processes are intertwined, and many new stability problems and phenomena have emerged in the dynamic process of the power grid. Especially, the wide-band disturbance problem caused by the interaction between power electronic equipment and the power grid is particularly prominent, and its oscillation frequency bandwidth covers several Hz to several kHz, with obvious multi-modal and amplitude frequency time-varying characteristics. However, there is a significant lack of research on cable joints under wide-band disturbances, and research on wide-band disturbances of cable joints based on typical insulation defects is even more scarce [10-13]. Therefore, this study is particularly urgent.

This article first constructs a two-dimensional electric magnetic thermal multi-physics coupling simulation model for the intermediate joint of a 110 kV AC high-voltage cable. By analyzing the causes of two insulation defects, namely the main insulation scratch and the air gap in the semiconducting layer, the simulation model is set up. Then, the magnetic flux density distribution characteristics of two typical insulation defects in cable joints under three broadband disturbance forms, namely sub synchronous oscillation, super synchronous oscillation, and high-frequency resonance, are studied and compared to fill the research gap in this field.

2. Multi physics field coupling simulation of cable defects with scratches on the main insulation surface under broadband disturbance

2.1. Defect causes and simulation settings

During transportation, handling, and on-site installation, cable intermediate joints are inevitably prone to collisions, which can easily cause damage to the outer sheath and main insulation surface. The simulation setting for scratches on the main insulation surface is shown in Figure 1 (c).

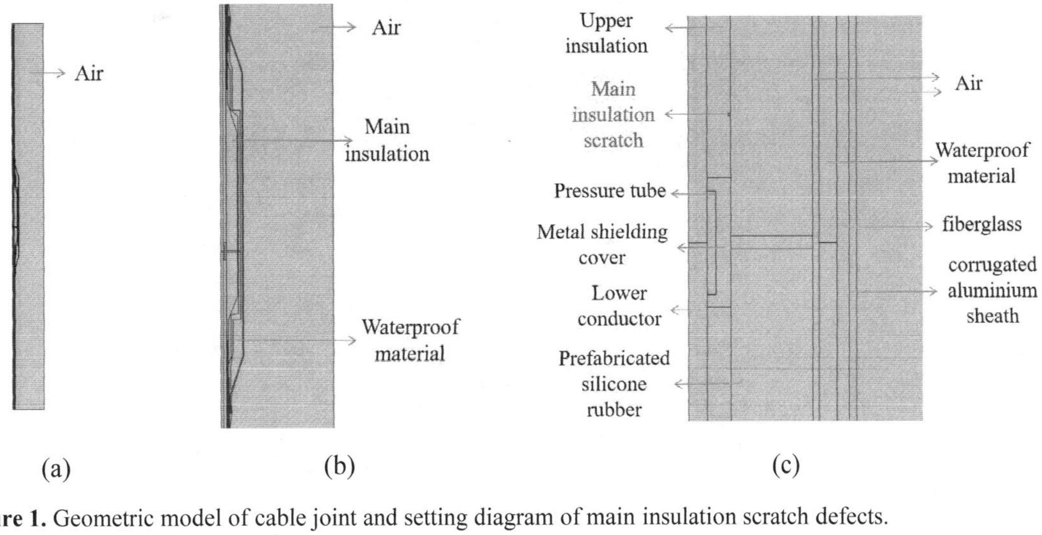

Figure 1. Geometric model of cable joint and setting diagram of main insulation scratch defects.

2.2. The influence of sub-synchronous oscillation on cable joints

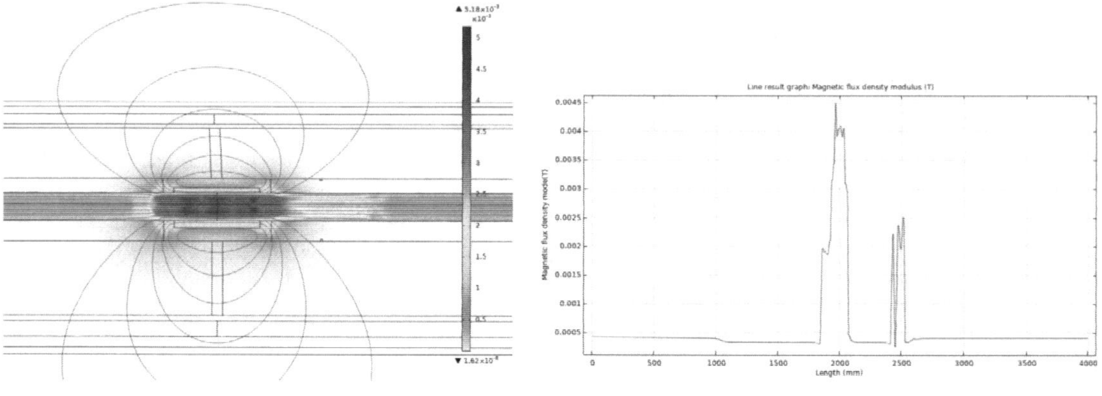

(a) (b)

Figure 2. Distribution diagram of magnetic flux density of main insulation scratch under sub-synchronous oscillation.

The distribution of magnetic flux density and current density at the main insulation scratch under sub-synchronous oscillation is shown in Figure 2 (a) and Figure 3, respectively. As shown in the figure, the maximum value of magnetic field strength occurs in the boundary area between the stress cone and the crimping clamp, where the modulus of magnetic flux density is 5.18×10^{-3}T; The maximum value of electric field strength is located at the cable joint below the stress cone and crimping clamp, where the current density modulus is $4.84 \times 10^5 \mathrm{A/m^2}$. Further analysis of Figure 2 (b) reveals that the scratch defects on the main insulation surface result in significant non-uniformity in the magnetic field distribution of the cable joint. The magnetic flux density at the location of the scratch is as high as about 0.025 T, and the magnetic field distortion rate skyrockets to 625%. The excessive concentration of magnetic field in the defect area will cause abnormal growth in the main insulation loss of cable joints, posing a serious threat to the normal operation of cable joints and affecting their long-term stability and reliability.

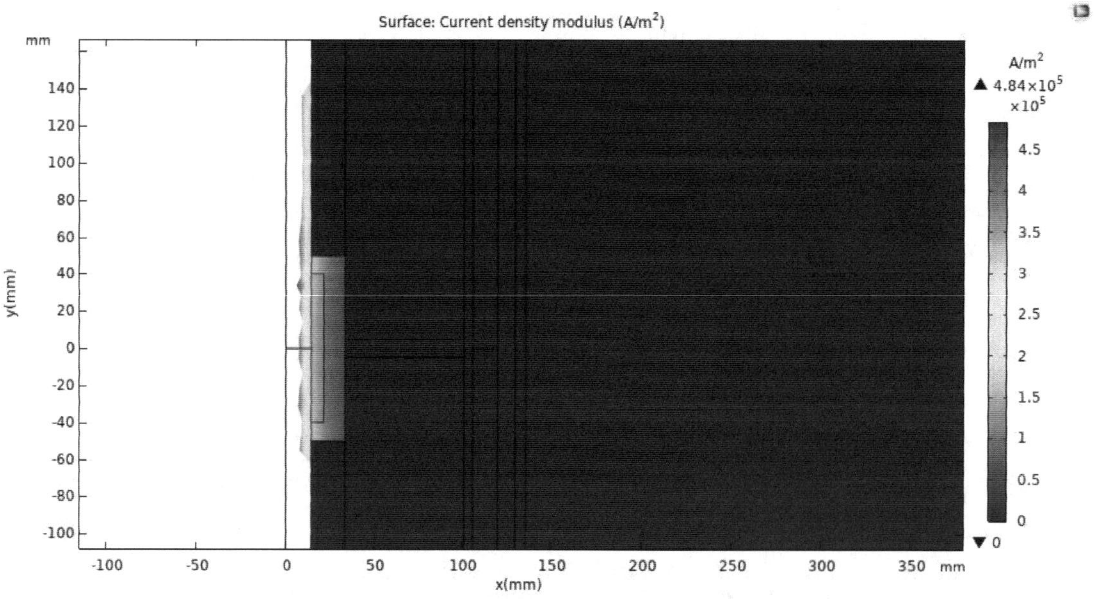

Figure 3. Distribution diagram of current density of main insulation scratch under sub-synchronous oscillation.

2.3. *The influence of super-synchronous oscillation on cable joints*

As shown in Figure 4, the distribution map of magnetic flux density under ultra-synchronous oscillation shows that the main insulation scratch has little effect on the symmetry of the magnetic flux density contour line. However, from the two-dimensional cross-sectional map of magnetic flux density under ultra-synchronous oscillation, it can be seen that the main insulation scratch actually has a great impact on the magnetic field distribution of the cable joint. The magnetic flux density at the scratch is close to that at the stress cone/crimping clamp, and the magnetic field distortion rate reaches 130.9%.

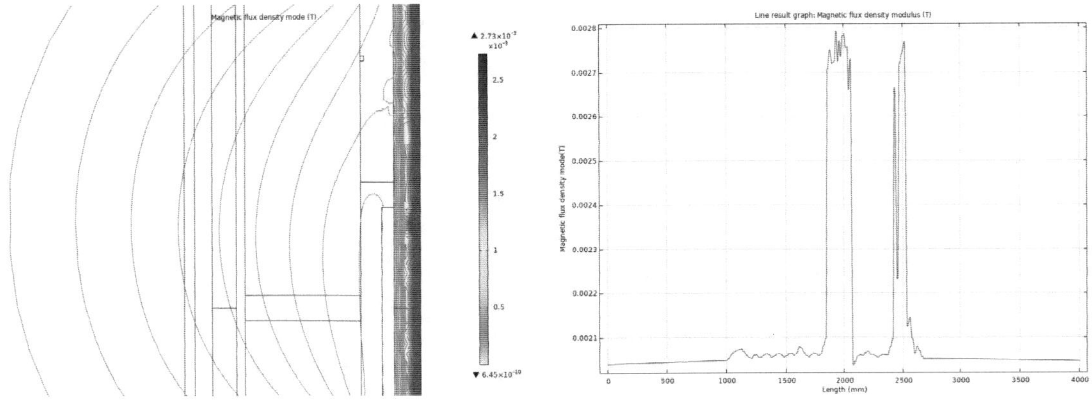

(a) (b)

Figure 4. Distribution diagram of magnetic flux density of main insulation scratch under super-synchronous oscillation.

2.4. *The influence of high-frequency resonance on cable joints*

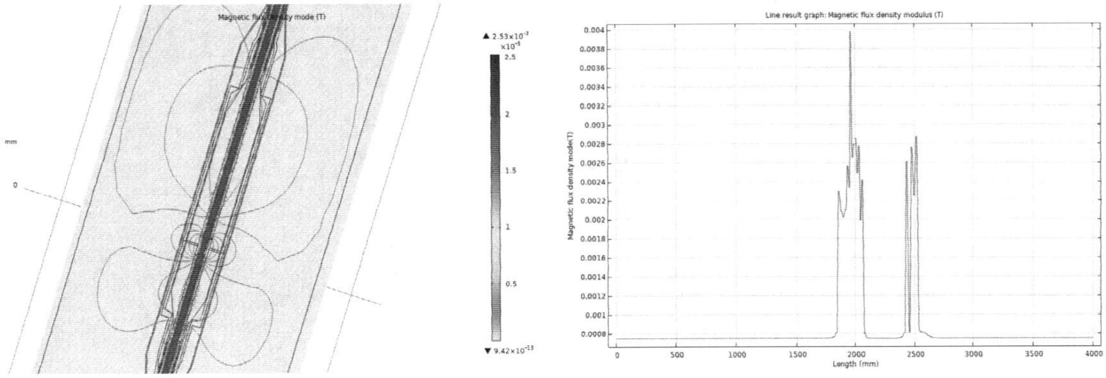

(a) (b)

Figure 5. Distribution diagram of magnetic flux density of main insulation scratch under high-frequency resonance.

As shown in Figurc 5, the magnetic flux density distribution of the main insulation scratch under high-frequency resonance is shown. From Figure 5 (a), the generation of the main insulation scratch under high-frequency resonance has a certain degree of influence on the distribution of magnetic flux density, making it more concentrated near the scratch. From Figure 5 (b), the electric field distortion at the main insulation scratch reaches 350%.

3. Multi-physics field coupling simulation of cable defects with scratches on the main insulation surface under broadband disturbance

3.1 Defect causes and simulation settings

In the manufacturing process of the intermediate joint, it is necessary to remove a section of the outer semiconducting layer. At the fracture site of the outer semiconducting layer, a fine polishing process should be used to ensure a smooth and even fracture surface. If the polishing process is missing or does not meet the established quality standards, it will result in a non-smooth transition between the semiconductor layer and the insulation layer at the fracture site. Once the joint is installed, there is a high possibility of air gaps forming in the uneven transition of the fracture area. The specific form of air gap defects in the outer semiconducting layer is shown in Figure 6.

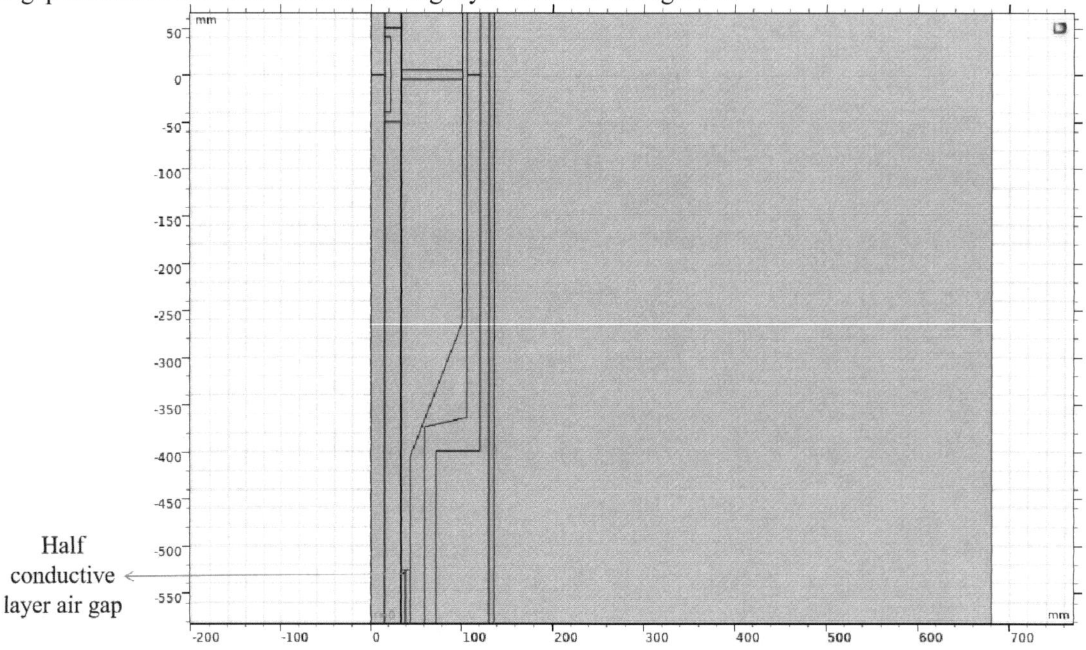

Figure 6. Setting of air gap insulation defects in the semi conductive layer of cable joints.

3.2 The influence of sub-synchronous oscillation on cable joints

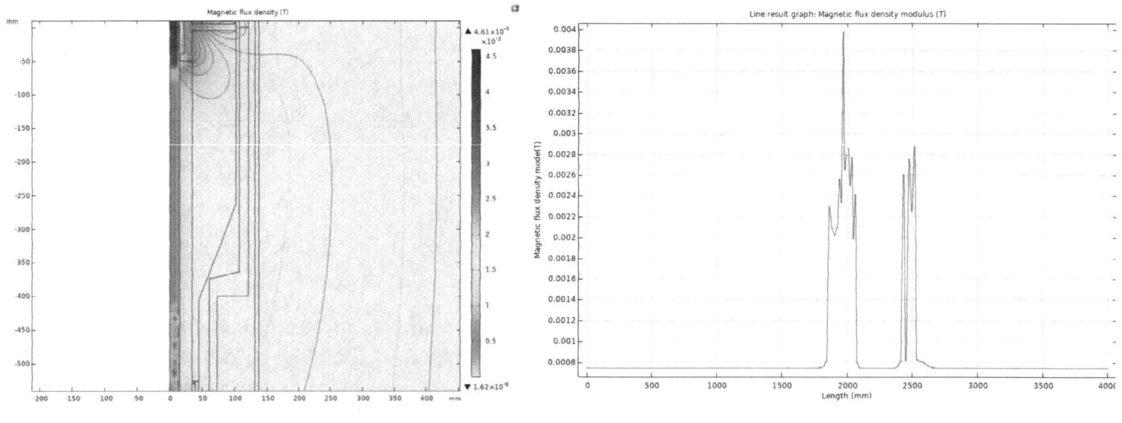

(a) (b)

Figure 7. Distribution diagram of air gap magnetic flux density in the semiconductor layer under sub-synchronous oscillation.

The distribution of magnetic flux density at the gap of the semiconductor layer under sub-synchronous oscillation is shown in Figure 7. The maximum magnetic field strength also occurs in the boundary area between the stress cone and the crimping clamp, where the modulus of magnetic flux density is 4.61×10^{-3} T, and the overall magnetic flux density is slightly lower than that of the main insulation scratch under sub-synchronous oscillation. However, the magnetic flux density at the cable joint is more concentrated, causing greater damage to the cable joint.

3.3 The influence of super-synchronous oscillation on cable joints

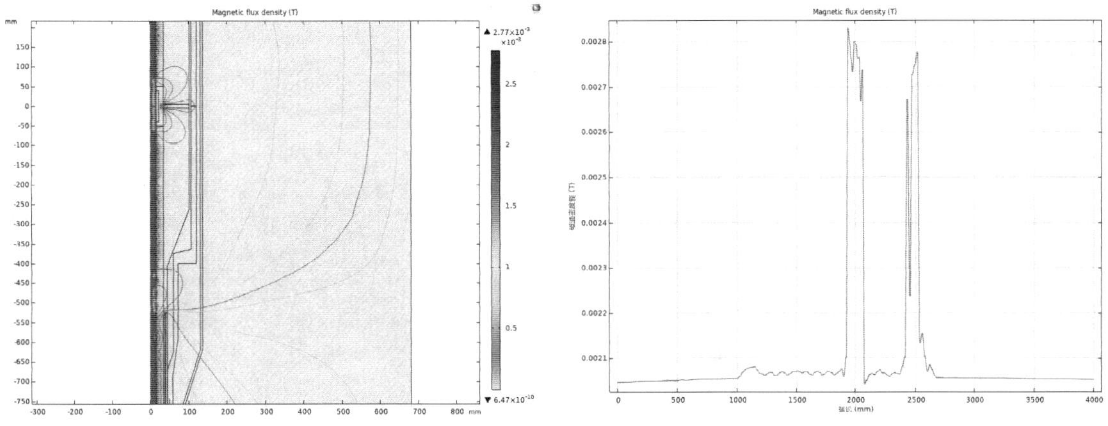

(a) (b)

Figure 8. Distribution diagram of gap magnetic flux density in semiconductor layer under super-synchronous oscillation.

As illustrated in Figure 8 (a), the magnetic flux density distribution within the air gap of the semiconducting layer under ultra-synchronous oscillation is depicted. It is evident that the presence of insulation defects in the air gap of the semiconducting layer under these conditions affects the symmetry of the magnetic flux density distribution, causing it to become more concentrated at the location of the insulation defects. Specifically, the magnetic flux density is notably higher in the regions where the insulation defects are present, indicating a localized increase in magnetic field intensity.

Referring to Figure 8 (b), the magnetic field distortion rate at the insulation defect within the semiconducting layer's air gap is measured at 140%. This value is comparable to the magnetic field distortion rate observed for main insulation scratch defects under super-synchronous oscillation conditions. The similarity in distortion rates suggests that both types of defects, air gaps in the semiconducting layer and scratches on the main insulation, can lead to significant localized changes in the magnetic field distribution under their respective oscillation conditions. This finding underscores the importance of considering the specific type of insulation defect and the associated oscillation conditions when assessing the potential impact on the magnetic field characteristics of the cable joint.

3.4 The influence of high-frequency resonance on cable joints

Figure 9 shows the magnetic flux density distribution of air gap insulation defects in the semiconductor layer under high-frequency resonance conditions. From the figure, it can be observed that the contour lines of magnetic flux density exhibit obvious sparse distribution characteristics, and the maximum magnetic flux density is only about 1.5×10^{-3} T. This result indicates that under high-frequency resonance operation, the effect of air gap insulation defects in the semiconducting layer on the electromagnetic performance of cable joints is relatively small. At this time, the magnetic field distortion rate is about 357%, which is similar to the magnetic field distortion rate caused by main insulation scratch defects under high-frequency resonance conditions. This comparative analysis

Journal of Physics: Conference Series **3011** (2025) 012066 doi:10.1088/1742-6596/3011/1/012066

further reveals the differences in electromagnetic characteristics of different insulation defects under high-frequency resonance conditions and their impact on the operating status of cable joints.

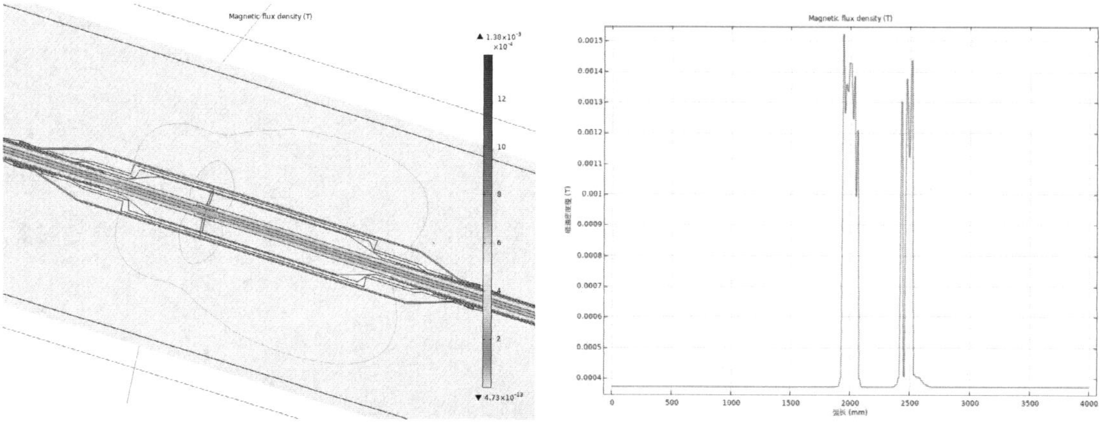

(a) (b)

Figure 9. Distribution diagram of air gap magnetic flux density in the semiconductor layer under high-frequency resonance.

4. Conclusion

Based on a multi-physics coupling simulation model of the intermediate joint of a 110 kV AC high-voltage cable, the paper studies the magnetic flux density distribution of two insulation defects, namely surface scratches on the main insulation and air gaps in the semiconducting layer, under sub-synchronous oscillation, super synchronous oscillation, and high-frequency resonance conditions, and their effects on the cable joint. The main conclusions are as follows:

(1) The scratch defect on the surface of the main insulation has the highest magnetic field distortion rate (625%) under sub-synchronous oscillation, with a magnetic flux density of 0.025 T, significantly increasing the main insulation loss and local overheating risk; The distortion rate under ultra- synchronous oscillation is 130.9%, which may cause local overheating; The distortion rate under high-frequency resonance is 350%, and the magnetic flux density is only 0.0015 T, which has a relatively small impact.

(2) The gap defects in the semiconducting layer have the highest magnetic field distortion rate (140%) under super-synchronous oscillation, and the magnetic flux density is concentrated, which may cause local overheating and insulation aging; The distortion rate under high-frequency resonance is 357%, and the magnetic flux density is only 0.0015 T, which has a potential impact that cannot be ignored; Under sub-synchronous oscillation, the magnetic flux density is $4.61 \times 10^{-3}\mathrm{T}$, and the damage is relatively significant.

(3) Comparative analysis shows that the main insulation scratch defect is most significantly damaged under sub-synchronous oscillation, while the air gap defect in the semiconducting layer has the highest distortion rate under high-frequency resonance, but the overall damage is smaller than the scratch defect.

(4) It is recommended to focus on the high damage characteristics of main insulation scratch defects under sub-synchronous oscillation in operation and maintenance monitoring, and pay attention to the impact of high-frequency resonance on air gap defects in the semiconducting layer to ensure the safe operation of cable joints.

Acknowledgments

This research work was supported by the Guangzhou Power Supply Bureau of Guangdong Power Grid Co., Ltd. for the research on the characteristics of broadband disturbance operation of high-voltage cables and reliability improvement technology GDKJXM20231092 (030100KC23100023).

References

[1] Wang J., Chen S., and Zheng Y. (2023). Research on Field Co-simulation and Temperature Prediction of 110 kV High Voltage Cable, Master's Thesis, Hunan University.
[2] Xia J., Shi N., and Sun J. (2022). Simulation Analysis of the Impact of Typical Defect Shapes on the Electric Field Distortion of High Voltage XLPE Cable Main Insulation, Wire and Cable, No. 6, pp. 1-6.
[3] Li M., Li C., Bao H., et al. (2022). Study on the Impact of Internal Air Gap Defects on the Electric-Thermal Field Distribution Characteristics of Onboard Cable Terminals, Electrical Engineering Technology, No. 21, pp. 213-216.
[4] Deng F., Gao Y., Li X., et al. (2023). Simulation Study of Typical Defects in 10 kV Cable Terminals Based on Electric-Thermal-Mechanical Fields, Insulating Materials, Vol. 56, No. 3, pp. 100-106.
[5] Hamdan M. A., Pilgrim J. A., and Lewin P. L. (2019). Thermo-mechanical analysis of solid interfaces in HVAC cable joints, IEEE Transactions on Dielectrics and Electrical Insulation, Vol. 26, No. 6, pp. 1, 779-1, 786.
[6] Chang H., Ye G., Chen J., et al. (2024). Effect of Insulation Defects on Electric and Temperature Fields in 110 kV Cables, Electrical Engineering & Automation, No. 5, pp. 45-51.
[7] Xu H., Huang L., and Lou J. (2022). 3D Electric Field Simulation Analysis of Typical Defects in 35 kV XLPE Cable Intermediate Joints, Zhejiang Electric Power, Vol. 39, No. 10, pp. 9-13.
[8] Feng Y. (2022). Simulation and Experimental Research on Insulation Defects of 66 kV XLPE Power Cables, Master's Thesis, Shenyang University of Technology, Shenyang.
[9] Xia J., Shi N., and Sun J. (2022). Simulation Analysis of the Impact of Typical Defect Shapes in Main Insulation of High Voltage XLPE Cables on Electric Field Distortion, Wire & Cable, No. 6, pp. 1-6.
[10] Yi L. D., Ma N. N., Ding M. S., Li H. Z., Bai S. B., and Xie X. R. (2023). Framework of prevention and control system for wide-band oscillations in new-type power systems, Power System Technology, Vol. 47, No. 10, pp. 4, 092-4, 101.
[11] Ma N. N., Xie X. R., He J. B., and Wang H. (2020). Review of wide-band oscillation in renewable and power electronics highly integrated power systems, Proceedings of the CSEE, Vol. 40, No. 15, pp. 4, 720-4, 731.
[12] Ravikumar G. and Khaparde S. A. (2018). Taxonomy of PMU Data Based Catastrophic Indicators for Power System Stability Assessment, IEEE Systems Journal, Vol. 12, No. 1, pp. 452-464.
[13] Ge H. C., Guo Q. L., Sun H. B., et al. (2018). An Improved Real-Time Short-Term Voltage Stability Monitoring Method Based on Phase Rectification, IEEE Transactions on Power Systems, Vol. 33, No. 1, pp. 1, 068-1, 070.

Journal of Physics: Conference Series **3011** (2025) 012067 doi:10.1088/1742-6596/3011/1/012067

Natural gas pipeline leak detection method based on 1D CNN-Mamba integrated acoustic signal classification

Ziang Xu, Xin Chen*

School of Energy and Civil Engineering, Harbin University of Commerce, Harbin 150028, China

* Corresponding author's e-mail: 300156@hrbcu.edu.cn

Abstract. The transportation of natural gas requires the construction of complex pipeline systems. However, as the pipelines age, damage and leakage incidents occur frequently, making pipeline leak detection particularly critical. This paper proposes an acoustic signal classification method integrated with an improved 1D CNN-Mamba model for pipeline leak detection. The method combines the 1D-CNN and Mamba modules, training them separately before jointly fine-tuning, significantly improving detection accuracy. By utilizing depthwise separable convolutions, the model parameters are reduced, and the computational cost is lowered with the Mamba block. Experimental results demonstrate that the proposed method offers a significant performance improvement, with an average accuracy of 93.24%, a 0.74% increase compared to the PNN in 12 classifications.

1. Introduction

Natural gas, as a clean energy source, plays a crucial role in the global energy structure. Its safe and efficient transportation is vital for gas storage and transportation engineering. Pipeline transportation enables large-scale, long-distance delivery of natural gas, reducing transportation costs and improving overall efficiency. A well-developed natural gas pipeline network helps optimize the energy structure, supports the energy needs of industry, commerce, and households, and promotes sustainable economic development [1].

With the widespread use of natural gas pipelines, incidents of pipeline damage and leakage have become frequent. These not only lead to resource waste but also pose risks of environmental pollution and safety accidents. Therefore, accurate leak detection is of paramount importance. Timely identification and handling of leak events can effectively prevent environmental pollution, loss of life, and property damage, ensuring both environmental and personnel safety. Additionally, leak detection technology helps enhance the safety and reliability of pipeline operations, preventing issues such as corrosion and cracks, and maintaining the stability of industrial production.

Traditional natural gas leak detection methods, such as the mass balance method and negative pressure wave method, suffer from issues like low sensitivity and slow response. As a result, acoustic signal-based detection methods have gradually become a research focus in this field due to their high sensitivity and fast response [2-3]. However, there is still room for improvement in terms of feature extraction and recognition accuracy. In recent years, researchers have proposed various methods to enhance detection precision and efficiency.

Xiao et al. [4] proposed an integrated leak detection method based on wavelet transform and Support Vector Machine (SVM) for acoustic signals. First, wavelet transform is used to preprocess the acoustic signals collected by the sound sensors. Then, useful features containing information about the leak severity are extracted from the multi-domain components of the acoustic signals. Ning et al. [5] introduced a Convolutional Neural Network (CNN) structure combined with Spectral Enhancement (SE-CNN) for gas pipeline leak detection, which demonstrated higher accuracy under strong background noise conditions. Tan et al. [6] proposed a leak identification method for natural gas pipelines based on Probabilistic Neural Networks (PNN), which effectively recognizes gas leaks with high accuracy. Yao et al. [7] presented a gas pipeline leak detection model combining acoustic feature processing techniques and feature reconstruction, which addressed the issue of strong noise interference. Experimental results showed that the model achieved a maximum accuracy of 95.17% for individual categories. Zhang et al. [8] proposed a leak detection and localization method by combining Attention Mechanism (AM) with Long Short-Term Memory (LSTM) networks. They conducted large-scale pipeline leak experiments on urban natural gas distribution networks and built a benchmark dataset, suggesting that the variation in sensor location data significantly impacts leak localization. Li et al. [9] proposed a joint learning end-to-end dual-stream transformer structure based on Enhanced External Attention and Improved Feedforward Mechanism (EEA-JL), which addresses the problem of simultaneous leak aperture identification and localization of gas pipelines under noisy backgrounds.

These studies, combining advanced signal processing techniques and machine learning algorithms, have effectively enhanced the accuracy and response speed of acoustic signal-based leak detection. However, further improving the effectiveness of feature extraction and the accuracy of recognition models remains a key challenge and focus in current research.

To further enhance the classification of acoustic features for natural gas pipeline leak detection, this paper proposes an integrated acoustic signal classification method using an improved 1D CNN-Mamba model. The front-end 1D-CNN captures local time-frequency features, while the back-end Mamba models long-sequence dependencies to improve the accuracy and real-time performance of natural gas pipeline leak detection. The method is validated on the public GPLA-12 dataset.

2. Proposed method

2.1. Categorical testing program

The intelligent inspection and maintenance system for natural gas pipelines, utilizing multi-modal monitoring technology, can systematically identify and assess risks to pipeline network structural integrity. These risks include but are not limited to, micro-leaks, stress corrosion cracking (SCC), third-party construction encroachments, and cathodic protection failures. Based on real-time monitoring data and risk quantification analysis, the system can automatically generate graded disposal plans, achieving a closed-loop management lifecycle from defect detection to preventive maintenance.

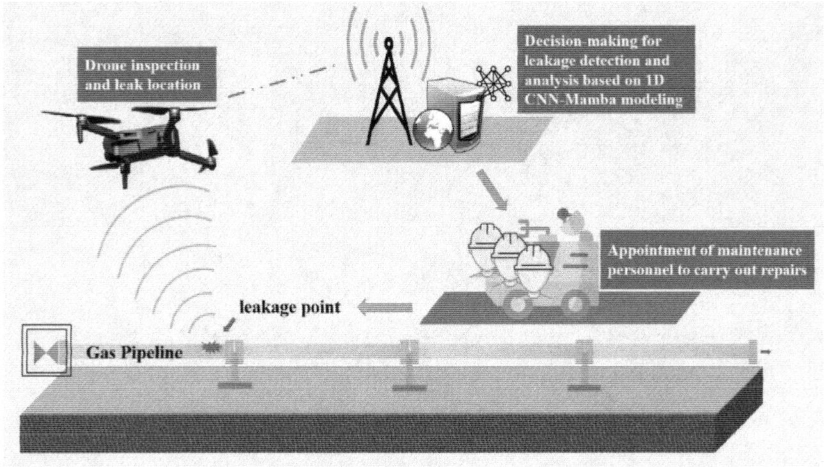

Figure 1. Technical solutions for pipeline leakage classification and detection.

In response to the current technological evolution of drone-based intelligent inspections, this study constructs a pipeline integrity management decision system based on acoustic fingerprint feature analysis (system architecture shown in Figure 1). The system innovatively integrates an improved 1D CNN-Mamba hybrid neural network. The model, with its hierarchical feature extraction architecture and bidirectional state-space modeling capability, effectively captures the long-range dependency characteristics of acoustic signals, providing essential technical support for building a preventive maintenance system.

2.2. Model architecture

This paper proposes a model that integrates 1D CNN and the Mamba module for natural gas pipeline leak acoustic feature classification. The network model is shown in Figure 2, the CNN component is responsible for extracting local features from the raw time-series data and reducing data dimensions through depthwise separable convolutions and pooling operations. This part consists of three layers of 1D depthwise separable convolutional layers (Conv1d), each followed by BatchNorm1d and ReLU activation functions. Specifically, the first convolution layer uses a kernel size of 3, expanding the input signal's channel to 64, followed by MaxPool1d for halving the sequence length. The second convolution layer operates similarly to the first, further extracting features and reducing the length through pooling. The third convolution layer continues to extract more abstract features and further reduces the data dimension through pooling, with the final output sequence length being 111. Through this structure, the model effectively extracts multi-level feature representations from the raw acoustic signals, providing a solid foundation for subsequent leak detection.

To further capture the long-term temporal dependencies present in the data, this paper utilizes a deep learning architecture based on State Space Models (SSMs), known as the Mamba module. For the local features *x(t)* extracted from the data by the CNN, the expression after passing through the Mamba module is:

$$h'(t) = Ah(t) + Bx(t) \tag{1}$$

$$y(t) = Ch(t) + Dx(t) \tag{2}$$

where h(t) represents the hidden state of the system, *x(t)* is the input signal, *y(t)* is the output signal, *A* is the state transition matrix, *B* is the input matrix, *C* is the output matrix, and *D* is the direct transmission term.

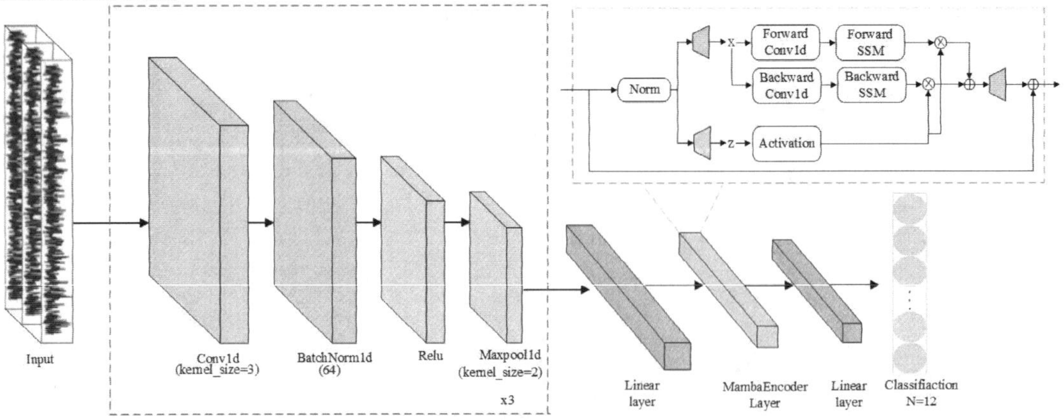

Figure 2. 1D CNN-Mamba model structure.

This paper uses the Cross-Entropy Loss function as the loss function to measure the inconsistency between the model's predicted output and the true labels. The Cross-Entropy Loss is a classic and effective loss function in classification problems, widely used in tasks such as image classification and speech recognition. In this paper, the true labels are represented in One-Hot Encoding format, which is a vector where the dimension corresponds to the number of target classes. Only one element in the vector, corresponding to the true class, is 1, and all other elements are 0. For a given input, the model outputs a probability distribution, indicating the probability of the input belonging to each class. The goal of the Cross-Entropy Loss is to minimize the difference between the probability distribution and the true labels. The formula for the Cross-Entropy Loss is:

$$L(x,y) = -\sum_{i=1}^{C} x_i log y_i \tag{3}$$

where L is the cross-entropy loss, i is the i-th class, C is the total number of classes, x is the true label, and y is the class probability.

2.3. Datasets

The GPLA-12 dataset consists of labeled acoustic gas leak signals under different conditions across 12 classes. To maintain balance among data types, and given the limited quantity and diversity of the dataset, the classes included can be chosen arbitrarily. The GPLA-12 dataset is divided into 12 categories, with each category consisting of data collected under different noise environments, gas pressures, and microphone setups, totaling 780 data points [10]. Unlike other studies [7], which reduce interference by categorizing the dataset into three classes based on pressure variables, this paper directly detects all 12 classes, maximizing the detection of various pipeline gas leak conditions under complex working scenarios.

During the training process, to fully utilize the limited dataset for cross-validation and assess the model's generalization ability under different data splits, 10-fold cross-validation is used. The dataset is divided into ten parts, with nine parts used for training and one part for evaluation in each round, ensuring the stability and reliability of the training results.

2.4. Data preprocessing

The acoustic signals generated by pipeline leaks are often severely interfered with by equipment noise and environmental noise. This phenomenon can easily lead to convergence oscillation issues in leak detection models and significantly reduce fault recognition accuracy. The Butterworth low-pass filter, with its flat passband characteristics and quick response capabilities, demonstrates unique advantages in digital signal denoising. To optimize data quality, this paper applies the Butterworth low-pass filter to effectively denoise the leak acoustic signals. Based on the filter's order and cutoff frequency, the filter coefficients are calculated and then applied to the leak acoustic signals for filtering. The frequency response formula of the Butterworth filter is as follows:

$$|H(j\omega)|^2 = \frac{1}{1+(\omega/\omega_c)^{2n}} \tag{4}$$

where n is the filter order, ω is the angular frequency, ω_c is the cutoff frequency, and $H(j\omega)$ is the transfer function of the filter. The order of the Butterworth low-pass filter can be calculated by using the following formula:

$$n \geq \frac{log\left(\frac{10^{\frac{A}{10}}-1}{10^{\frac{A_s}{10}}-1}\right)}{2log\left(\frac{f_s}{f_p}\right)} \tag{5}$$

where A is the maximum allowed passband ripple, typically set to 0 dB, and A_s represents the minimum attenuation level required for the stopband. f_p is the passband cutoff frequency (Hz), and f_s is the stopband cutoff frequency (Hz).

After filtering the acoustic signals, data normalization is required to convert data with different magnitudes or distributions into a common scale or range. This step is essential for subsequent data analysis and model training. The primary goal of normalization is to reduce the variation in the data, making it more concentrated, which helps to improve the algorithm's convergence speed and performance. It also prevents certain features from dominating the model training process due to their larger numerical ranges.

3. Experimental

The data used in the experiment in this paper comes from the public data on the Internet, a total of 780 records, including 12 categories of acoustic feature data [10]. The experimental system environment is a 16 G computer with 12 CPU cores and a Windows 11 system, and the GPU is RTX4060. All data in the experiment are the average of three experiments.

Table 1 is a comparison of the accuracy of different schemes on the same dataset. The accuracy of the proposed scheme is 0.74% higher than the PNN scheme [6], and it has better performance on the 12-category GPLA-12 dataset.

Table 1. Comparison of accuracy of different solutions in 12 classification.

Method	Year	Accuracy (%)
1D-CNN+NEAB [2]	2017	79.36
AE+NEAB [11]	2020	74.77
1D-CNN+EAB [7]	2023	88.42
PNN [6]	2023	92.5
1D CNN-Mamba (Ours)	--	93.24

Table 2 shows the classification results of 12 categories of acoustic data, which are mainly evaluated by three indicators: precision, recall, and specificity. Most categories (such as categories 1, 2, and 12) perform well, with high precision and recall, indicating that the model's predictions for these categories are accurate and comprehensive.

Table 2. The performance of different categories of our model.

Classification	Precision	Recall	Specificity
1	0.981	0.991	0.998
2	0.98	0.986	0.998
3	0.836	0.863	0.984
4	0.889	0.906	0.99
5	0.649	0.712	0.966
6	0.723	0.771	0.974
7	0.883	0.877	0.989
8	0.904	0.907	0.991
9	0.632	0.542	0.971
10	0.714	0.645	0.976
11	0.9	0.908	0.991
12	0.928	0.932	0.993

Figure 3 shows the classification results of the 1D CNN-Mamba model on the GPLA-12 dataset, showing the relationship between the model's prediction results and the true labels. From the diagonal of the matrix, we can see that the model's predictions for most categories are very accurate, except for categories 9 and 10, where the model's prediction accuracy is close to 90%, indicating that the model can distinguish the acoustic data of these categories well. There are misclassifications for categories 9 and 10, but overall the correct classification rate for most categories is higher than 80%, demonstrating that the model can effectively distinguish most acoustic data categories on a limited dataset.

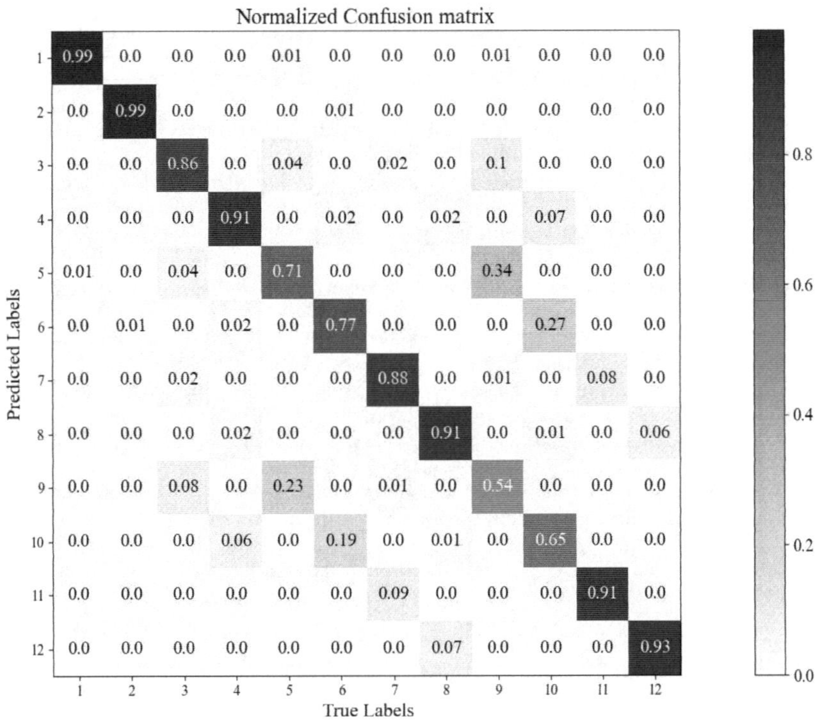

Figure 3. Classification confusion matrix of the 1D CNN-Mamba model on the GPLA-12 dataset.

4. Conclusion

For the classification of acoustic features in natural gas pipeline leak detection, this paper proposes an improved long-sequence modeling 1D CNN-Mamba model. This method effectively performs pipeline leak classification and detection, validated on the GPLA-12 dataset. The method has the following advantages: (1) It adopts a separate training approach for the 1D-CNN and Mamba modules, followed by joint fine-tuning, significantly improving detection accuracy; (2) By using depthwise separable convolutions, the 1D-CNN model parameters are reduced, and computational costs are lowered with the integration of the Mamba block. Testing on the GPLA-12 dataset shows that the proposed method achieves superior performance, with an average accuracy of 93.24%. This model provides practical guidance for natural gas pipeline leak detection.

Acknowledgments

This work was financially supported by the following funding sources: the Doctoral Scientific Research Support Program of Harbin University of Commerce (Grant No. 22BQ86), the 2023 Young Scientific Research Innovation Talent Cultivation Program of Harbin University of Commerce (Project No. 2023-KYYWF-0987), and the New Era Longjiang Excellent Master's and Doctoral Dissertation Project (Award No. LJYXL2023-042). These grants provided essential resources and academic freedom to pursue this investigation.

References

[1] Korlapati N V S, Khan F, Noor Q, et al. Review and analysis of pipeline leak detection methods[J]. Journal of Pipeline Science and Engineering, 2022, 2(4): 100074.
[2] Kang J, Park Y J, Lee J, Wang S H, Eom D S. Novel leakage detection by ensemble CNN-SVM and graph-based localization in water distribution systems. IEEE Trans Ind Electron, 2017, 65(5): 4279–89.
[3] Zhang S, Li X, Zong M, Zhu X, Wang R. Efficient KNN classification with different numbers of nearest neighbors. IEEE Trans Neural Netw Learn Syst, 2017, 29(5): 1774–85.
[4] Xiao R, Hu Q, Li J. Leak detection of gas pipelines using acoustic signals based on wavelet transform and Support Vector Machine[J]. Measurement, 2019, 146: 479–489.
[5] Ning F, Cheng Z, Meng D, et al. Enhanced spectrum convolutional neural architecture: An intelligent leak detection method for gas pipeline[J]. Process Safety and Environmental Protection, 2021, 146: 726–735.
[6] Tan Q, Fu M, Chen X, et al. A new leak recognition method for natural gas pipelines in the urban underground space based on probabilistic neural network[J]. Journal of Loss Prevention in the Process Industries, 2023, 85: 105162.
[7] Yao L, Zhang Y, He T, et al. Natural gas pipeline leak detection based on acoustic signal analysis and feature reconstruction[J]. Applied Energy, 2023, 352: 121975.
[8] Zhang X, Shi J, Yang M, et al. Real-time pipeline leak detection and localization using an attention-based LSTM approach[J]. Process Safety and Environmental Protection, 2023, 174: 460–472.
[9] Li P, Wang X, Jiang C, et al. Advanced transformer model for simultaneous leakage aperture recognition and localization in gas pipelines[J]. Reliability Engineering & System Safety, 2024, 241: 109685.
[10] Li J, Yao L. GPLA-12: an acoustic signal dataset of gas pipeline leakage[J]. arXiv preprint arXiv: 2106. 10277, 2021.
[11] Hu X, Zhang H, Ma D, Wang R. A tnGAN-based leak detection method for pipeline network considering incomplete sensor data. IEEE Trans Instrum Meas, 2020, 70: 1–10.

Internal thermal-force distribution characteristics and optimization of the elastic crimping IGBT device

Chenglin Ren[1*], Xiang Lu[1], Zhigang Zhang[2], Shi Jiao[1], Pengzhong Wang[2]

[1.] EHV Power Transmission Company China Southern Power Grid, Guangzhou, P. R. China
[2.] XJ Electric Co.. Ltd. Xuchang, Henan. P. R. China

[*]Corresponding authors: renchen_cn@163.com

Abstract. With the rapid development of power electronics technology, especially the development of the fully controlled device IGBT, flexible high-voltage direct current transmission (HVDC) technology has become a new generation of HVDC technology, the core of which is the use of insulated-gate bipolar transistor (IGBT) and other fully controlled devices. The crimped IGBT device, with its advantages of high reliability and failure short circuit, is becoming a hot topic at present. This paper takes 3000A elastic crimping IGBT as the research object, builds the physical model through SolidWorks, removes the sub-module frame and model outside frame, and simplifies the process. At the same time, the thermal-force coupling finite element analysis of the device was carried out through COMSOL multi-physical field simulation software, and the mechanical and thermal distribution characteristics of the device during the steady state operation were studied. The key points of the chip are prone to failure fatigue during the steady state operation, and the key chips with a strong coupling effect when multiple chips are connected in parallel were found out. By improving the layout of the chip, reducing the thermal coupling with other chips, and decreasing the mutual thermal resistance between chips, heat can be more efficiently transferred to the heat dissipation structure through the electrodes, which can significantly reduce the junction temperature of the chip. The equalization method of thermal-force equalization control provides some guidance for further structural optimization of the device.

1. Introduction

The package form of crimped IGBT realizes double-sided heat dissipation, easy series, short circuit when failure, high power density, etc., and is gradually becoming the mainstream of flexible DC power transmission. Compared with the traditional convex crimped IGBT, the elastic crimped package has a more balanced pressure distribution and more flexible assembly. Therefore, in the future, the use of elastic crimped IGBT will be more extensive in higher grade power transmission. **Figure 1** shows a diagram of an elastic crimped IGBT module.

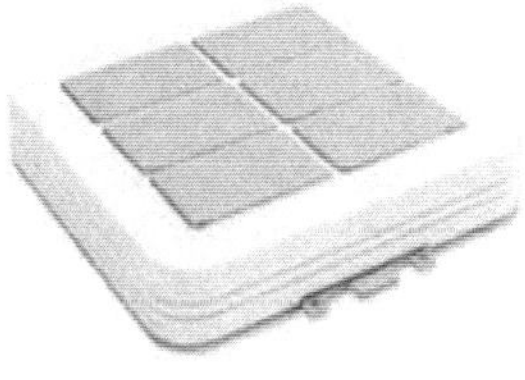

Figure 1 Elastic crimped IGBT module.

Due to the interaction of electromagnetic field, temperature field, and structural field inside the crimped IGBT device, the multi-physical field environment inside the device is extremely complex, and the temperature and pressure distribution will be uneven due to materials, processing errors, shapes, and other reasons. Uneven temperature distribution will lead to local overheating of the chips and affect its performance. Uneven pressure distribution will cause deformation, fracturing, or poor contact; contact resistance and contact thermal resistance are too large. At present, the thermal-force distribution equalization method of crimped IGBT multi-chip mainly focuses on the material and size of the device, and the layout of each chip is less studied. By optimizing the chip layout, the path of heat conduction between chips can be improved, thus increasing the heat transfer efficiency. Meanwhile, according to the crimped IGBT thermal network model and combined with the relevant knowledge of heat transfer, the layout mode between device chips can be improved, which is conducive to increasing the mutual thermal resistance between chips, weakening the mutual influence of temperature between chips, and reducing the junction temperature of chips. Improve device reliability.

The internal stresses in crimped IGBT modules are mainly concentrated at the edges of the chips. Due to the thermal expansion caused by the temperature rise, the electrode edges of the device will warp considerably, and the temperature and stress distribution of the individual parallel chips of the crimped IGBT will be affected by this [1]. A finite element simulation model of a 3.3kV/1500A multi-chip crimped IGBT module is established. The effects of two factors, micromotion wear, and pressure imbalance, on the temperature distribution of the chip and the homogeneous flow are analyzed [2]. In 2016, Deng Erping from North China Electric Power University built a finite element simulation model of a single-chip submodule and analyzed the contact thermal resistance between components, calculated the contact thermal resistance values before and after thermal resistance measurements, and comparatively analyzed the effect of temperature on the contact thermal resistance between components [3]. In 2017, Deng Erping et al. researched and analyzed the effect of the temperature rise generated during the operation of crimped IGBT devices on the pressure distribution inside the devices and proposed an improved and optimized scheme [4].

In this paper, through thermal-force multiple physical fields coupling analysis of 4500 V and 3000 A elastic crimped IGBT devices, the key points that are prone to failure fatigue during the steady-state operation of the chips and the key chips that are strongly coupled when multiple chips are connected in parallel are found. Based on the simulation analysis results, a method of thermodynamic equilibrium control is proposed for the device in terms of chip layout. Some guidance is provided for further structural optimization of the device.

2. Elastic crimp IGBT finite element model

2.1. Device structure model and material parameters

Figure 2 shows the 3D decomposition of 4500 V 3000 A elastic crimped IGBT module, which consists of six identical sub-modules in parallel. Each sub-module is stacked from top to bottom by a collector copper plate (copper), disc spring (structural steel), copper column (copper), aluminum sheet (aluminum), small molybdenum sheet (molybdenum), silicon chip (silicon), emitter molybdenum plate (molybdenum). Other components, such as sub-module frame and model frame, have little influence on the temperature and pressure distribution of the device chip.

In order to shorten the calculation time, simplified processing was carried out. The finite element model built by Solidworks after simplification is shown in **Figure 3**, which includes 8 IGBT chips and 4 fast recovery diode (FRD) chips.

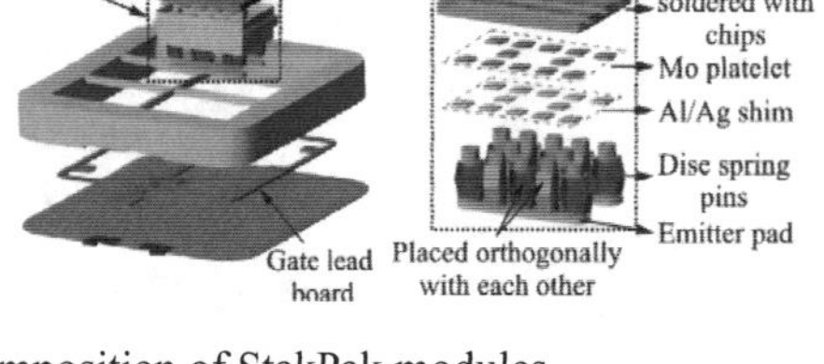

Figure 2 Three-dimensional decomposition of StakPak modules.

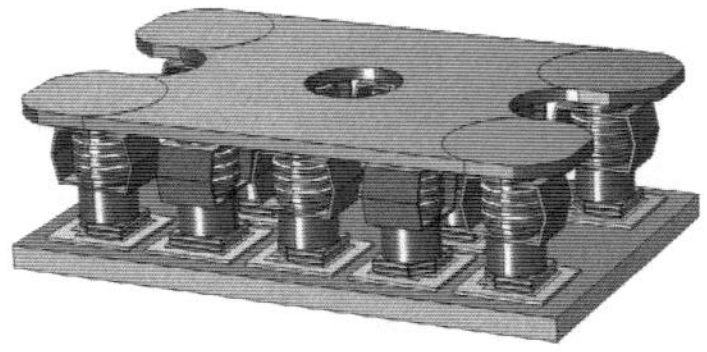

Figure 3 Submodule finite element simulation model.

The steady-state thermodynamic coupling simulation analysis of elastic crimped IGBT will be carried out later in this paper, and the thermal mechanical properties are inseparable from the thermal and mechanical parameters of the material. The materials involved in this paper include copper, aluminum, silicon, molybdenum, etc. The thermal and mechanical parameters of spring-type crimped IGBT were obtained by referring to relevant literature and retrieving the existing material parameters in COMSOL software material library.

Table 1 and **Table 2** list the data.

Table 1 Thermal parameters of each material of the model.

Parameters	k/(W•m^{-1} •K^{-1})	α/(10^{-6} •K^{-1})	C_p/(J• kg•K^{-1})
Cu	400	17.5	385
Mo	138	5.1	250
Al	238	23	900
Si	130	2.6	700

Table 2 Mechanical parameters of each material of the model.

Parameters	E /GPa	σ	ρ/(kg •m^{-3})
Cu	110	0.35	8930
Mo	312	0.30	10200
Al	70	0.33	2700
Si	170	0.28	2329

2.2. Application conditions and boundary conditions

The crimped IGBT performs double-sided heat dissipation through the top of the emitter and the bottom of the collector, and the equivalent surface heat dissipation coefficient is set to 5000 W/(m^2•K). Because the optimal range of pressure applied by the crimped IGBT is 1~2 kN/cm^2 [5], according to the equivalent force area of the chip in the steady-state simulation analysis process, assuming that the equivalent pressure of each chip is 1.2 kN. It has been documented that the contact pressure distribution on each chip surface inside the device is relatively uniform during displacement loading, while the pressure distribution characteristics during uniform pressure loading are much worse [6]. Due to the steady-state analysis, the pressure IGBT maintains the current and normal state, so the active region of the chip is set as the heat source, the heat consumption rate of each IGBT chip's active region is set to 180 W, and the heat consumption rate of the active region of the FRD chip is set to 150 W.

In terms of mesh generation: Due to the complexity of the device model, it is necessary to construct the grid by layers and domains. It is necessary to set appropriate conditions, such as the maximum and minimum number of elements and the shape of the grid, while taking into account the calculation accuracy and calculation efficiency, so as to obtain reliable calculation simulation results. The maximum cell of the grid for the aluminum layer, upper molybdenum layer, chip, lower molybdenum layer and emitter is 25 mm, and the minimum cell is 4.5 mm. The rest of the mesh is divided by pre-defined coarse mesh.

2.3. Coupling model of structural field and temperature field - contact thermal resistance

Because the crimped IGBT is to stack the components together by pressure, there will be a large contact thermal resistance on the contact surface between the components. At the same time, in order to achieve an effective connection between the chip and the external electrode, a layer of aluminum layer a few microns thick is usually grown on the surface of the chip, which will lead to a rough contact interface, resulting in greater contact thermal resistance. The results show that the total contact thermal resistance of the crimped IGBT is about 50% of the total thermal resistance of the device.

The two adjacent contact interfaces are connected by many randomly distributed contact points (micro-contact), as shown in **Figure 4** [7]. The micro-contact area is generally filled with air with very low thermal conductivity, and the thermal conductivity of air is about 0.023 W/(m•K), which has a strong hindrance to heat transfer, so heat is mainly conducted through micro-contact. However, due to the mismatch of the area size, the heat flow will shrink, and the contact thermal resistance will be generated at the contact interface. According to the principle of heat transfer, the gap thickness of the contact surface is less than 10 um, and the Grashoff number is less than 2000, so the influence of heat convection can be ignored. Moreover, because the temperature of each component inside the device is small and does not exceed 700 K, the thermal radiation phenomenon is ignored, so the contact thermal conductivity hc only needs to calculate the void thermal conductance for heat transfer between voids and microcontacts hg and the shrinkage thermal conductivity hs due to area mismatch.

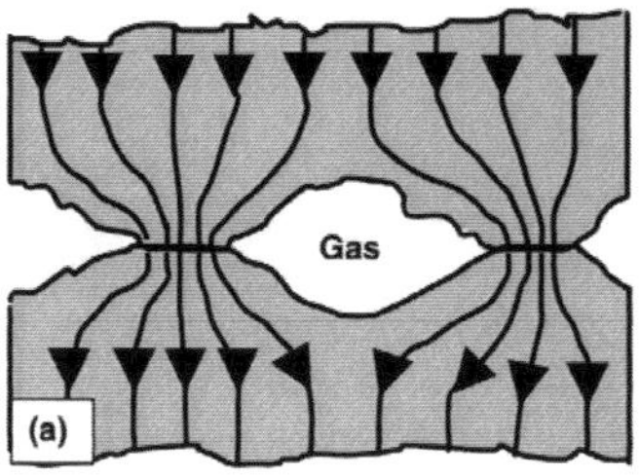

Figure 4 Microscopic schematic of the contact interface.

The contact thermal resistance between interfaces can be calculated by the following formula (A is the nominal contact area):

$$h_c = h_g + h_s \tag{1}$$

$$R_c = A \cdot h_c \tag{2}$$

where h_c is the contact thermal conductivity between the contact interfaces, h_s is the contraction or diffusion thermal conductivity considering only micro contacts, and h_g is the gap thermal conductivity caused by gap thermal conductivity.

All components were found to have a surface roughness of less than 0.85 μm. The equation m=0.124$\sigma^{0.743}$ presented by Antonetti et al. will be used for the calculations in this chapter [8].

The microcontact thermal conductance is calculated as follows [9][10]:

$$h_s = 1.25 k_s \frac{m}{\sigma}\left(\frac{p}{H_c}\right)^{0.95} \tag{3}$$

where p is the contact pressure, k_s is the harmonic mean of the thermal conductivity at the interface of two solid contacts, H_c is the microhardness of the relatively softer material of the two contacting materials, σ is the equivalent square root mean surface roughness, m is the equivalent mean absolute surface slope.

The numerical calculation of the contact thermal conductance of the voids is as follows [11][12]:

$$h_g = \frac{k_g}{Y+M} \tag{4}$$

$$Y = 1.53\sigma\left(\frac{p}{H_c}\right)^{-0.097} \tag{5}$$

where k_g is the thermal conductivity of the gas inside the void, M is a gas parameter characterizing the high temperature and low pressure of the gas, Y is the effective thickness of the void.

3. Analysis of simulation results

3.1. Steady state simulation modeling of multi-chip parallel submodules

To facilitate subsequent processing and analysis, the chips are numbered in a certain sequence, and three paths are selected as data extraction directions to facilitate subsequent longitudinal analysis and comparison, as shown in **Figure 5**.

In order to obtain the pressure and temperature distribution characteristics of each chip during the operation of the elastic compression IGBT device, it is necessary to conduct the steady-state simulation analysis of the multi-chip parallel sub-module. In order to ensure that the force of each chip is near 1200 N as far as possible, a 14400 N boundary load with total force downward is applied on the collector copper plate side because there are 12 chips. At the same time, the active region of the IGBT chip is set as the boundary heat source, and the heat consumption rate of the unit chip is set to 180 W. Due to the different power loss models of the IGBT chip and the FRD chip, the power loss is different, and the power consumption of the IGBT chip is higher than that of the FRD chip.

Figure 5 Chip surface path extraction.

3.2. Chip junction temperature distribution

As can be seen from **Figure 6**, the high temperature area of the chip is mainly concentrated in the active area of the chip. However, due to the poor heat dissipation effect of the spring side of the device, which is approximately one-sided heat dissipation, the longitudinal temperature gradient of the chip is quite different. The highest temperature on the surface of each chip is 413 K, and the lowest temperature is 346 K. At the same time, through the simulation results, it can be seen that IGBT chips 2, 4, 9, and 11 are strongly influenced by the thermal coupling of surrounding chips and have high power consumption, so they become the high temperature region of the entire device. The peak temperature of the chip is as high as 413 K, which is about 8°C higher than that of other IGBT chips.

The temperature distribution on the three extraction paths is shown in **Figure 7**. Among them, the FRD power consumption is the lowest, so the chip surface temperature is low. Due to 3 and 10 are strongly thermally coupled by the surrounding chips, the chip surface temperature peak value is about 2°C higher than that of 6 and 7.

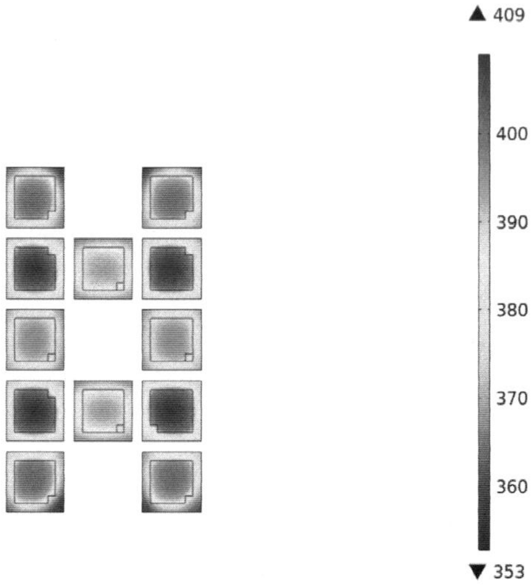

Figure 6 Temperature distribution of parallel sub-modules.

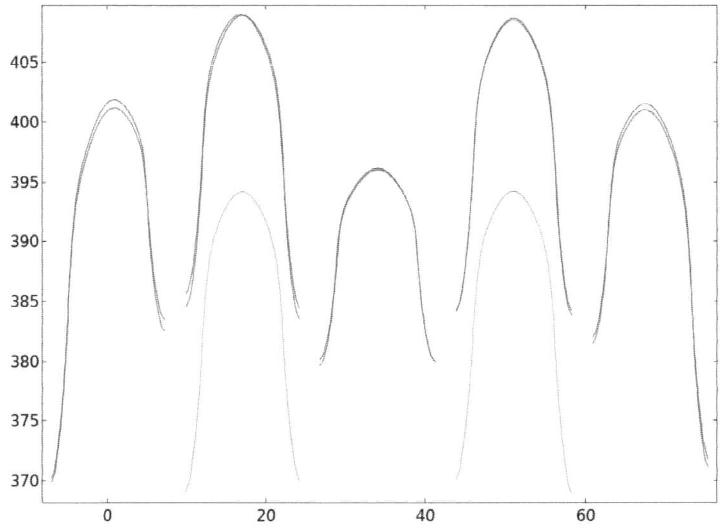

Figure 7 Temperature distribution on the extraction path.

3.3. Chip stress distribution

It can be seen from **Figure 8** that the chip stress of the elastic crimp IGBT is mainly concentrated in the source region of the chip, and the stress distribution on the three extraction paths is shown in **Figure 9**. The peak stress at the edge of the chip is around 30 MPa, and the peak stress in the active region of the chip can reach 100 MPa. The pressure at the corner of the active region of the chip is quite different from that at the center of the active region. There is no obvious difference in stress distribution between the FRD chip and the IGBT chip. The stress distribution of extraction paths 1 and 3 is similar, and the stress of chips 2, 4, 9, and 11 is about 10 MPa higher than that of other IGBT chips due to higher temperature. The FRD chip on extraction path 2 has a relatively weak thermal coupling due to low power consumption, and its stress is reduced by about 3 MPa compared with chips 1, 5, 8 and 12. Because of the spring, the overall pressure distribution of the device chip is relatively uniform.

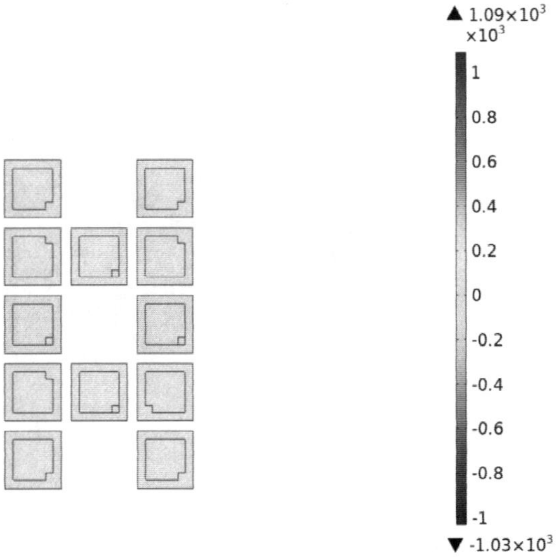

Figure 8 Stress distribution in parallel submodules.

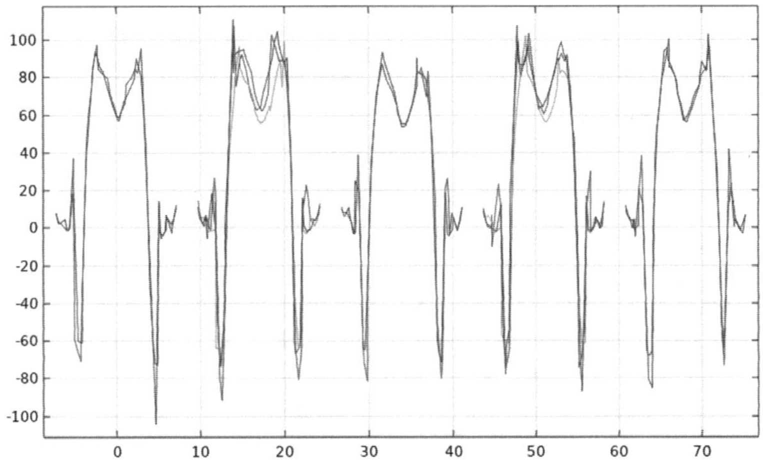

Figure 9 Extract the stress distribution on the path.

4. Balanced control of parallel chip temperature under pressure loading

Changing the arrangement between the elastic crimped IGBT chips is beneficial to improve the temperature distribution between the parallel chips and make it more balanced. First of all, through the optimization of the chip layout, the path of heat conduction between the chips can be improved, so that the heat transfer efficiency is increased and the phenomenon of local overheating is avoided. At the same time, through the study of the crimp IGBT thermal network model and combined with the relevant knowledge of heat transfer, the layout mode between the device chips is improved, which is conducive to reducing the mutual thermal resistance between the chips, so that the heat can be more efficiently transferred to the heat dissipation structure through the electrode, which is conducive to reducing the chip temperature and improving the reliability of the device. The original elastic crimp IGBT structure is shown in **Figure 10**, and the temperature distribution of the parallel chips inside is shown in **Figure 11**. Due to strong thermal coupling, the temperature of chips 2, 4, 9, and 11 is high, and the heat dissipation effect is poor. Through simulation, the highest temperature of chips is 413 K,

Journal of Physics: Conference Series **3011** (2025) 012068 doi:10.1088/1742-6596/3011/1/012068

and the temperature gradient between chips is large, which is not conducive to the stable and reliable operation of the device.

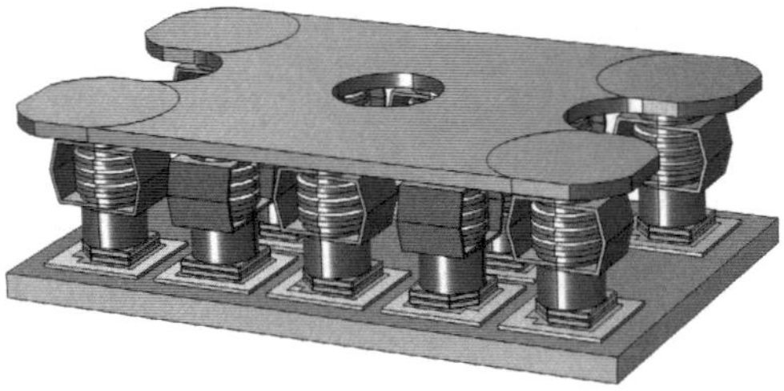

Figure 10 Elastic crimped IGBT simulation model.

Figure 11 Temperature distribution of elastically crimped IGBT chips.

By improving the layout of the chip, the 2, 4, 9, and 11 chips are moved out of the distance of one chip, and the thermal coupling with other chips is weakened. The simulation model of the improved device is shown in Figure 12, and the temperature distribution of the simulated parallel chip is shown in **Figure 13**. The maximum temperature of the chip is 392 K, which is about 20°C lower than before the improvement. At the same time, it is found that the peak pressure of each chip of the device decreases from 1540 MPa to 1260 MPa, and the pressure distribution is more balanced. This is because the structure of the device itself has a certain coefficient of thermal expansion and a certain elasticity. When the temperature of the chip rises, expansion will occur, and the crimping structure will undergo a certain deformation, thus exerting a stable pressure on the chip. When the temperature of the chip drops, the device will shrink to a certain extent, so that the structure will gradually recover and reduce the pressure on the chip.

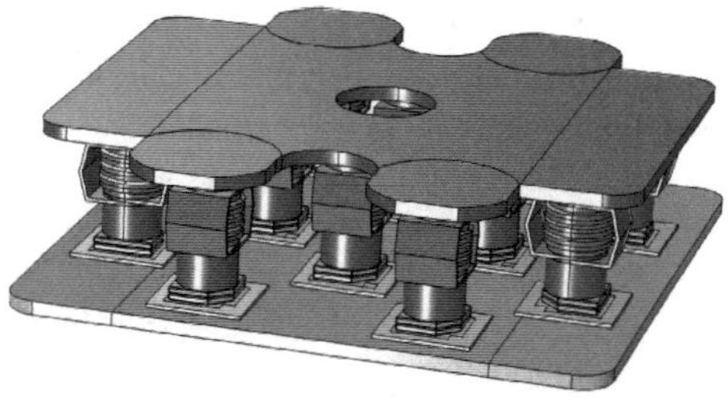

Figure 12 Model of elastically crimped IGBTs with an improved layout.

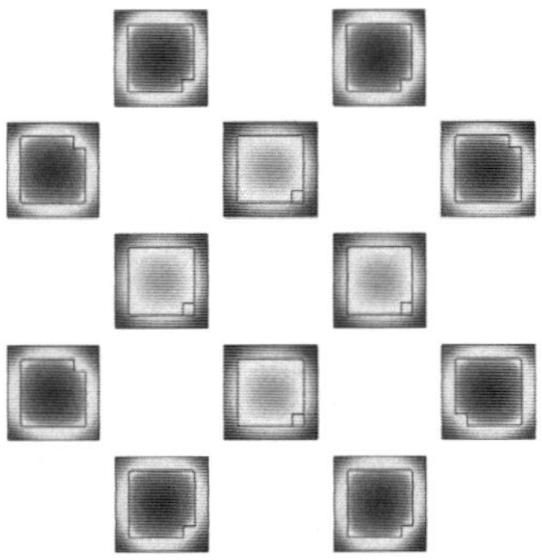

Figure 13 Improved device chip temperature distribution.

The maximum junction temperature distribution of the chip before and after the improvement is shown in **Figure 14**.

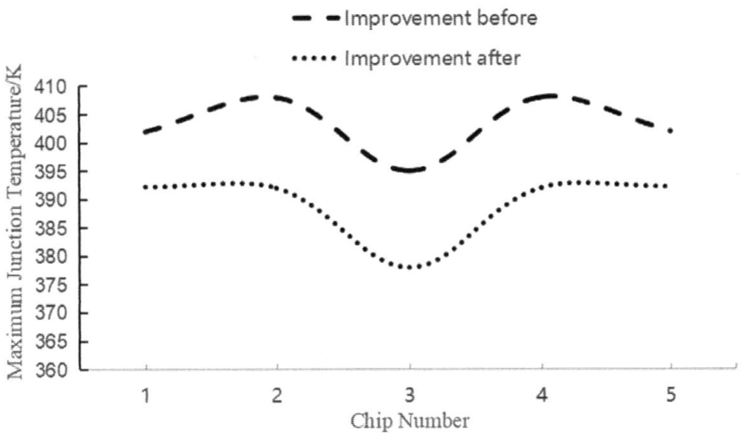

Figure 14 Maximum junction temperature distribution.

5. Conclusions

This paper takes the elastic crimped IGBT as the research object, and builds the physical model through SolidWorks. Besides, thermal coupling of the device through multiphysics field simulation software is carried out. The thermal equalization method is proposed based on the analysis results of the simulation, and the following main conclusions can be obtained.

(1) The spring-side heat dissipation of the device is less effective and approximates unilateral heat dissipation, so the longitudinal temperature gradient of the chip varies greatly, with the highest temperature on the surface of each chip being 413K and the lowest temperature being 346K. The temperatures of the multiple chips of elastic crimp IGBT are mainly concentrated in the active region of the chip.

(2) The pressure difference between the corners of the active area of the chip and the center of the active area is large. Stress peaks around 30MPa at the edge of the chip and 100MPa in the center of the active area of the chip.

(3) The chip is arranged in a decentralized layout, which can effectively reduce the temperature on the chip junction. The maximum temperature of the improved chip is 392K, which is about 20°C lower than that before the improvement.

(4) Subsequently, a real steady-state condition experimental platform can be built to verify the accuracy of the results of the thermal characteristics of the parallel chip inside the device under steady-state conditions in this paper. At the same time, power cycling experiments can also be carried out to investigate the distribution of thermal and mechanical properties inside the device under power cycling conditions.

(5) Consideration can be given to the introduction of high thermal conductivity thermal interface materials, so that the heat generated by the chip can be transferred to the heat sink in a timely manner to improve the efficiency of heat exchange. At the same time, it can effectively penetrate and fill the gap between the contact surfaces, reducing the contact thermal resistance to form a continuous thermal conductivity path.

References

[1] Tinschert L, Årdal A R, Poller T, Bohlländer M, Hernes M and Lutz J 2015 Possible failure modes in Press-Pack IGBTs *Microelectronics Reliability.* **55**(6): 903-11.
[2] Zhou J, Kang S, Li H,Yao R and Li J 2019 Analysis of the effect of uneven internal pressure on the electrical and thermal characteristics of crimped IGBT devices *Journal of Electrotechnology.* **34**(16): 3408-3415.
[3] Deng E,Zhao Z, Zhang P et al 2016 Effect of temperature on internal contact thermal resistance of crimped IGBT devices *Semiconductor Technology.* **41**(12): 906-12.
[4] Deng E, Zhao Z, Lin Z, et al 2017 Influence of Temperature on the Pressure Distribution within Press Pack IGBTs *IEEE Transactionson Power Electronics.* PP(99): 1-1.
[5] Josef L, Uwe S, Heinrich S, et al 2011 Semiconductor power devices *Germany: Springer Nature.* 156-58.
[6] Deng Erping 2018 Research on multi-physics field coupling modeling of electric-thermal-force inside crimped IGBT devices *North China Electric Power University (Beijing).*
[7] Grujicic M, Zhao C L, Dusel E C 2005 The effect of thermal contact resistance on heat management in the electronic packaging *Applied Surface Science.* **246**(1-3):290-302.
[8] Antonetti V W, Whittle T D, Simons R E 1991 An Approximate Thermal Contact Conductance Correlation *ASME, The 28th National Heat Transfer Conference (Minneapolis, Minnesota)* HTD-vol. 170 (Experimental/Numerical Heat Transfer in Combustion and Phase Change) pp 35-42.
[9] Bahrami M, Culham J R, Yovanovich M M, et al 2015 Thermal Contact Resistance of Nonconforming Rough Surfaces, Part 1: Contact Mechanics Model. *Journal of Thermophysics & Heat Transfer.* **18**(2): 218-27.
[10] Bahrami M, Culham J R, Yovanovich M M and Schneider G E 2004 Thermal Contact Resistance of Nonconforming Rough Surfaces, Part 2: Thermal Model. *Journal of Thermophysics and Heat Transfer.* **18**(2), 218-27.
[11] Yovanovich M M 1981 New Contact and Gap Correlations for Conforming Rough Surfaces *AIAA 16th Thermophysics Conference (Palo Alto, CA)* AIAA-81-1164.
[12] Antonetti V W and Yovanovich M M 1984 Thermal contact resistance in microelectronic equipment *Thermal management concepts in microelectronic packaging from component to system.* 6984-003: 135-51.

Sensorless control of dual three-phase permanent magnet synchronous motor using high-frequency pulsating voltage injection

Chongchong Ai*, **Shumiao Yang, Yuan Fu**

School of Electrical & Control Engineering, Xi'an University of Science & Technology, Xi'an, Shaanxi 710600, China

* Corresponding author's e-mail: 22206227098@stu.xust.edu.cn

Abstract. Based on the vector space decoupling (VSD) model of the dual three-phase permanent magnet synchronous motor (DTP-PMSM), the high-frequency (HF) pulsating voltage injection method is applied to sensorless control of DTP-PMSM. Traditional methods, which use filters for signal extraction, suffer from significant estimation errors and system delays. A second-order generalized integrator (SOGI) is employed to improve the accuracy of speed and rotor position estimation. First, to overcome the poor position observation accuracy and dynamic performance of traditional methods, a SOGI-based approach is projected for extracting position error information. By replacing the filters in the position observation loop with SOGI, parameter tuning is simplified, and observation accuracy is enhanced. Second, to mitigate the delay caused by filters in the current loop, SOGI is used to replace the low-pass filter, extracting the fundamental frequency current component and increasing the current loop bandwidth.

1. Introduction

Compared to conventional permanent magnet synchronous motor (PMSM), the dual three-phase permanent magnet synchronous motor (DTP-PMSM) exhibits lower torque ripple and reduced current harmonics, holding significant potential in applications such as aerospace and vehicles[1]. Traditionally, rotor position infomation is obtained by adding mechanical structures at the motor's rear end, which increases the motor's volume and weight[2]. Therefore, research on sensorless control techniques for DTP-PMSM is of great practical significance[3].

When the motor operates at low speed, the signal-to-noise ratio is significantly lower[4].In such cases, a high-frequency (HF) voltage signal is typically injected into the motor windings, and the rotor position information is extracted from the winding response[5].

An HF pulsating voltage injection method based on a second-order generalized integrator (SOGI) is proposed in this paper. First, the SOGI is utilized to extract signals in the feedback current loop, enhancing the system's filtering performance. Second, in the rotor position information extraction stage, the SOGI replaces the filters.

2. High-frequency (HF) Pulsating Voltage Injection Method for DTP-PMSM

2.1. DTP-PMSM mathematical model

DTP-PMSM can be seen as consisting of two three-phase stator windings connected in a Y-shaped manner, with a difference of 30° in electrical angle between the two sets of windings and a neutral point isolation between the two windings. Figure 1 shows the schematic diagram of the motor structure and the stator winding connection method by using a two-level voltage source inverter as an example.

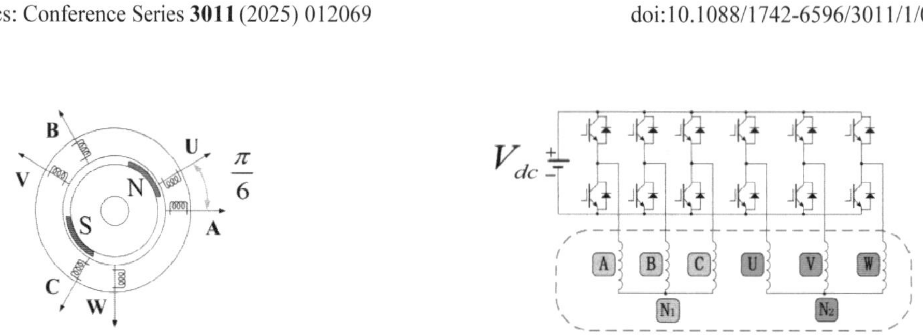

(a) (b)

Figure 1. Structure of DTP-PMSM & drive system.

Through analysis and design research on sensorless control of DTP-PMSM based on the vector space decoupling (VSD) transformation mathematical model, mathematical models of the DTP-PMSM can be obtained.

The voltage equation of the d-q subspace is:

$$\begin{bmatrix} u_d \\ u_q \end{bmatrix} = \begin{bmatrix} R_s & 0 \\ 0 & R_s \end{bmatrix} \begin{bmatrix} i_d \\ i_q \end{bmatrix} + \begin{bmatrix} L_d & 0 \\ 0 & L_q \end{bmatrix} \cdot \frac{d}{dt} \begin{bmatrix} i_d \\ i_q \end{bmatrix} + \begin{bmatrix} -\omega_e L_q i_q \\ \omega_e L_d i_d + \omega_e \psi_f \end{bmatrix} \tag{1}$$

where u_d and u_q are respectively the stator voltage in the d-q subspace, R_s is the stator resistance, ω_e is the electrical angular velocity, I_d and I_q are the stator currents, and L_d and L_q are the inductance. The voltage equation for the x-y subspace in a synchronous rotating coordinate system is:

$$\begin{bmatrix} u_x \\ u_y \end{bmatrix} = \begin{bmatrix} R_s & 0 \\ 0 & R_s \end{bmatrix} \begin{bmatrix} i_x \\ i_y \end{bmatrix} + \begin{bmatrix} L_z & 0 \\ 0 & L_z \end{bmatrix} \cdot \frac{d}{dt} \begin{bmatrix} i_x \\ i_y \end{bmatrix} \tag{2}$$

where u_x and u_y are the stator voltages in the $x-y$ subspace, i_x and i_y are the stator currents in the $x-y$ subspace, and L_z is the leakage inductance. The electromagnetic torque equation is as follows:

$$Te = 3P_n i_q [i_d (L_d - L_q) + \psi_f] \tag{3}$$

where T_e represents the electromagnetic torque, P_n is the number of pole pairs, and ψ_f is the magnetic flux of the permanent magnet.

2.2. HF pulse voltage injection method

The stator inductance has nonlinear saturation characteristics, and the HF voltage signal is injected into the estimated d-axis. The HF voltage injection method is suitable for surface-mounted dual three-phase permanent magnet motors with saturated convex polarity.

The d-q subspace voltage equation in the synchronous rotating coordinate system after VSD of DTP-PMSM can be simplified as:

$$\begin{cases} u_{dh} \approx L_{dh} \dfrac{di_{dh}}{dt} \\ u_{qh} \approx L_{qh} \dfrac{di_{qh}}{dt} \end{cases} \tag{4}$$

The subscript h in the equation represents the HF component. When estimating the d-axis injected sine wave signal, the injected signal is written as:

$$\begin{cases} u_{dh} = U_h \cos(\omega_h t) \\ u_{qh} = 0 \end{cases} \tag{5}$$

where U_h is the amplitude of the pulse voltage and ω_h is the angular frequency of the injected voltage. Combining Equations (4) and (5), the current response of the $\hat{d}$ and $\hat{q}$ axis is obtained:

$$\begin{cases} \hat{i}_{\mathrm{dh}} = \dfrac{U_{\mathrm{h}}(L+\Delta L\cos 2\Delta\theta)}{\omega_{\mathrm{h}}(L^2-\Delta L^2)}\sin\omega_{\mathrm{h}}t \\ \hat{i}_{\mathrm{qh}} = \dfrac{U_{\mathrm{h}}\Delta L\sin 2\Delta\theta}{\omega_{\mathrm{h}}(L^2-\Delta L^2)}\sin\omega_{\mathrm{h}}t \end{cases} \tag{6}$$

where $\Delta L=(L_{\mathrm{qh}}-L_{\mathrm{dh}})/2$ is the half differential inductance; $L=(L_{\mathrm{qh}}+L_{\mathrm{dh}})/2$ is the average inductance; $\Delta\theta=\theta-\hat{\theta}$ is the position estimation error.

The position estimation method based on phase-locked loop (PLL) is used to obtain the estimated position information:

$$f(\Delta\theta)=LPF(BPF(\hat{i}_{\mathrm{q}})\times 2\sin(\omega_{\mathrm{h}}t))=\frac{U_{\mathrm{h}}(L_{\mathrm{qh}}-L_{\mathrm{dh}})}{2\omega_{\mathrm{h}}L_{\mathrm{dh}}L_{\mathrm{qh}}}\sin(2\Delta\theta)\approx 2\frac{U_{\mathrm{h}}(L_{\mathrm{qh}}-L_{\mathrm{dh}})}{2\omega_{\mathrm{h}}L_{\mathrm{dh}}L_{\mathrm{qh}}}\Delta\theta \tag{7}$$

3. HF Pulsating Voltage Injection Method for DTP-PMSM

The control block diagram of the HF pulsating voltage injection method based on SOGI is shown in Figure 2. Different from the SOGI used in the speed ring, the SOGI located in the current ring is represented by C-SOGI, which represents the cascade of two SOGI. The effect is equivalent to the superposition of two SOGI, which is used to replace low pass filter (LPF) and band-pass filter (BPF) traditionally.

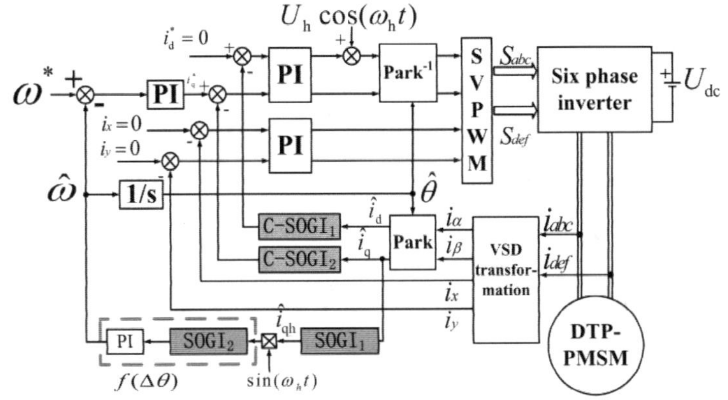

Figure 2. Control block diagram of HF pulse voltage injection method based on SOGI.

3.1. HF response signal extraction strategy

The SOGI can extract the signal of a specific frequency, and its transfer function is:

$$\frac{\hat{i}_{\mathrm{qh}}(s)}{\hat{i}_{\mathrm{q}}(s)}=\frac{k_1\omega_1 s}{s^2+k_1\omega_1 s+\omega_1^2} \tag{8}$$

where $\hat{i}_{\mathrm{q}}(s)$ and $\hat{i}_{\mathrm{qh}}(s)$ are the input and output components respectively, ω_1 is the frequency of the signal extracted by SOGI, and k_1 is the damping coefficient.

It is convenient to adjust the parameters of SOGI. By adjusting the coefficient k_1 of the proportional link, the filtering characteristics of SOGI can be changed.

3.2. Extraction strategy of rotor position error information

The SOGI structure for extracting the rotor position can suppress the signal. The transfer function is:

$$\frac{f_{\Delta\theta}(s)}{i'(s)}=\frac{s^2+\omega_2^2}{s^2+k_2\omega_2 s+\omega_2^2} \tag{9}$$

where $i^{'}(s)$ and $f_{\Delta\theta}(s)$ are the input component and output component respectively, ω_2 is the frequency of SOGI suppression signal, and k_2 is the damping coefficient.

3.3. *Fundamental frequency current extraction strategy*

The SOGI with the transfer function of Equation (8) is used in the current loop. Taking the q-axis as an example, the SOGI is designed to suppress the HF pulse current in the q-axis current and extract the fundamental frequency component.

The traditional HF pulse voltage injection current loop open-loop system transfer function is:

$$G_{\text{open}}(s) = (k_{\text{p}} + \frac{k_{\text{i}}}{s})(\frac{1}{T_{\text{s}}s+1})(\frac{1}{L_{\text{q}}s+R}) \tag{10}$$

The C-SOGI transfer function in Figure 2 is:

$$G_{C\text{-}SOGI}(s) = \frac{(s^2+\omega_{e3})(s^2+\omega_{e4})}{(s^2+sk_{e3}\omega_{e3}+\omega_{e3}^2)(s^2+sk_{e4}\omega_{e4}+\omega_{e4}^2)} \tag{11}$$

The transfer function of improved current loop closed-loop system:

$$G_2(s) = \frac{G_{\text{open}}(s)}{1+G_{\text{open}}(s)G_{C-SOGI}(s)} \tag{12}$$

4. Simulation Results

The proposed method is verified by analyzing the rotor tracking effect of the motor under different operating conditions, including the starting process of the motor from zero and the rotor tracking effect under the sudden change of motor speed. The motor parameters are shown in Table 1.

Table 1. Parameters of DTP-PMSM.

Motor parameters	Value
Number of pole pairs p_{n}	4
d-axis inductance L_{d} / mH	33.65
q-axis inductance L_{q} / mH	79.56
Stator resistance R/Ω	3.95

4.1. *Position identification during the starting*

Figure 3 shows the rotor position tracking simulation effect of the motor starting from zero to 80 r/min. Below is the simulation diagram of the new method. After improvement, the max-speed error of the motor is reduced by 5-9.5 r/min, the max-rotor position error is reduced by 0.06-0.07 rad, and the time to reach the steady state is reduced by 0.3 seconds.

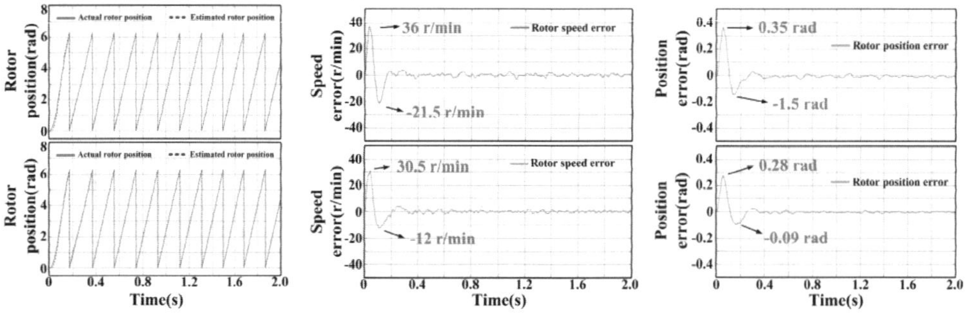

Figure 3. Simulation comparison of the startup process.

Journal of Physics: Conference Series **3011** (2025) 012069 doi:10.1088/1742-6596/3011/1/012069

4.2. *Identification of sudden change position in rotating speed*

To certify the dynamic tracking performance of the new method, the given motor speed suddenly changes from 80 r/min to 150 r/min. Figure 4 is the simulation diagram of the improved method. After improvement, the max-speed error is reduced by 5-6 r/min, the max-rotor position error is reduced by 0.06-0.1 rad, and the time to reach the steady state is reduced by 0.15 seconds. Compared with the traditional method, the improved method has smaller speed fluctuation when the speed changes suddenly, and has better dynamic performance.

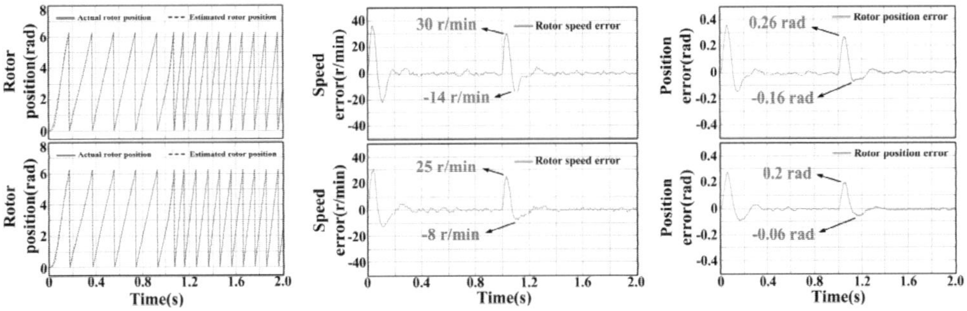

Figure 4. Simulation comparison of speed sudden change.

5. Experimental results

The experimental platform used for verification is shown in Figure 5.

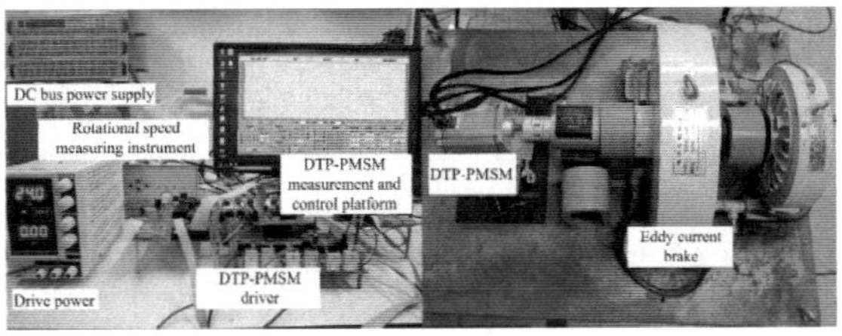

Figure 5. Experimental platform.

5.1. *Performance verification of position identification in starting process*

Figure 6 shows the rotor position tracking the experimental effect of the motor starting from zero to 80 r/min. During the motor starting process, the estimated data of the two methods are compared with the actual value of the motor at the same time. Figure 6 is the simulation diagram of the improved method. After the improvement, the speed estimation accuracy is increased by 50%, the max-rotor tracking error is reduced by 0.12 rad, and the tracking speed is increased by 21.05%. The new method makes the starting process reach a stable state faster.

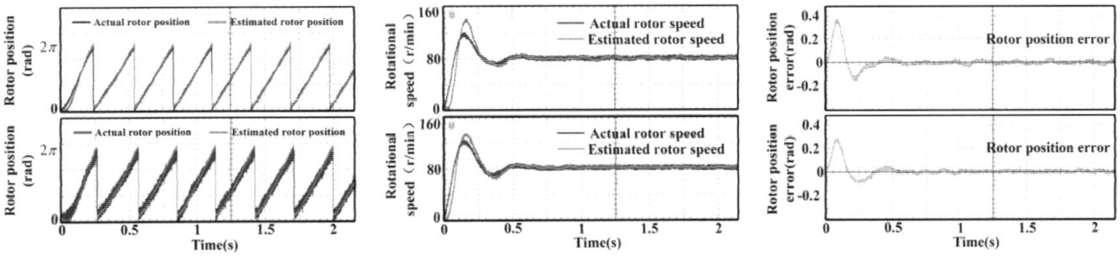

Figure 6. Experiment comparison of the startup process.

Journal of Physics: Conference Series **3011** (2025) 012069 doi:10.1088/1742-6596/3011/1/012069

5.2. Verification of speed mutation position identification performance

Figure 7 shows the comparison of the experimental results of two methods of rotor tracking when the motor changes from 80 r/min steady-state operation to 150 r/min speed, displaying the simulation diagram of the improved method. The max-rotor tracking error of the improved method is reduced by 0.08 rad, the speed tracking accuracy is improved by 40%, and the error between the estimated data and the actual parameters of the motor is smaller.

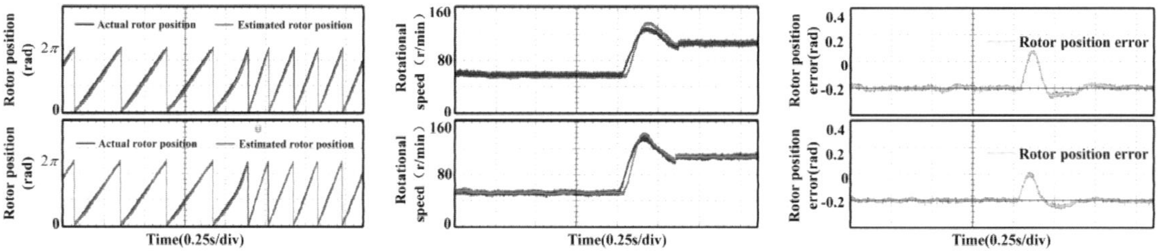

Figure 7. Experiment comparison of speed sudden change.

6. Conclusion

The influence of the filter on the velocity and current loop of the traditional HF pulse voltage injection method is analyzed. An improved HF pulse voltage injection method based on SOGI is proposed.

The SOGI-based approach simplifies parameter configuration while enhancing filtering accuracy and dynamic characteristics in both the speed and current loops. Simulations and experimental validations demonstrate that the proposed method significantly increases position estimation accuracy compared to conventional techniques. Furthermore, it minimizes disturbances caused by abrupt motor state changes on rotor position tracking and strengthens the system's dynamic response capabilities.

Acknowledgments

This work was financially supported by the Key Research Projects for Higher Education in the Coal Industry under Grant 2021MXJG043.

References

[1] Qiu X Q, He Y, Zhao W X, et al. Acoustic Noise Reduction for Low-Speed Sensorless Operation of Dual Three-Phase PMSM with Random Winding Injection[J], IEEE Transactions on Energy Conversion, 2023, 38(4): 2612-2622.

[2] Sun X D, Hu C, et al. Speed Sensorless Control of SPMSM Drivesfor EVs with a Binary Search Algorithm-Based Phase-Locked Loop[J], IEEE Transactions on Vehicular Technology, 2020, PP(99): 1-1.

[3] Tang Q P, Shen A, Luo X, et al. IPMSM sensorless control by injecting bidirectional rotating HF carrier signals[J]. IEEE Transactions on Power Electronics, 2018, 33(12): 10698-10707.

[4] Zhao W X, Song S C, Zhou S W, et al. Sensorless control of current source inverter driven PMSM with improved sliding mode observer[J]. Transactions of China Electrotechnical Society, 2024, 39(04): 987-995.

[5] Liao K J, Xu W, Jian G, et al. Improved position sensorless resonant frequency tracking control method for linear oscillatory machine[J]. IEEE Transactions on Power Electronics, 2024, 71(4): 4038-4048.

Journal of Physics: Conference Series **3011** (2025) 012070 doi:10.1088/1742-6596/3011/1/012070

Power control strategy of slope gravity energy storage system based on permanent magnet synchronous motor

Jangyi Hu [1], Kunpeng Zha [1], Yuhong Wang [2], Cheng Yuan [3,*], Yi Wang [3]

[1] State Grid Electric Power Research Institute, Nanjing 211106, China
[2] China EPRI Electric Power Engineering Co., Ltd., Beijing 102299, China
[3] School of Mechanical Engineering and Automation, Harbin Institute of Technology (Shenzhen), Shenzhen 518005, China

*Corresponding author's e-mail: yuan_c@foxmail.com

Abstract: Slope-based solid gravity energy storage has garnered significant attention due to its geographic flexibility and configurational versatility. This study presents a novel slope gravity energy storage system control method employing a PMSM coupled with an innovative power stabilization strategy to mitigate grid-side power fluctuations. The proposed approach integrates a tri-level control architecture encompassing power regulation, DC bus voltage maintenance, and speed optimization leveraging the system's inherent kinetic energy and electromechanical characteristics to ensure stable grid-side power output during discrete weight operations. Validation through a 100 kW PMSM-based slope gravity energy storage simulation model demonstrates the efficacy of the devised approach. Results confirm consistent grid-side power output with minimal weight speed fluctuations, effectively eliminating the power impact of discrete weight operations on the grid interface.

1. Introduction

Energy storage has emerged as a pivotal technology in advancing the grid-connected integration of renewable energy systems. It acts as a critical determinant in addressing the power imbalance between input and output in the grid, while offering diverse functionalities such as frequency regulation and phase modulation. This technology is fundamental in constructing a renewable energy power system and achieving carbon peak and neutrality targets. Among the variety of energy storage systems, gravity energy storage stands out as a cost-effective and relatively safe energy storage technology [1]. Unlike pumped energy storage, slope-type gravity energy storage systems (GESS) can overcome geographical and topographical constraints, allowing for flexible configuration of storage capacity based on renewable energy consumption patterns [2].

Studies [3] and [4] analyzed the construction of tower-type and rolling hoist-type GESS, respectively. Building on the ramp-type GESS, [5] proposed a time-sharing and segmented power control strategy, supplemented by battery storage, to better accommodate renewable energy integration into the grid. Addressing the issue of power fluctuations caused by discrete weight blocks in the ramp system, [6] analyzed the inertial characteristics of the GESS and its supportive role in grid stabilization, laying a foundation for subsequent power control studies. Further extending this work, [7] investigated the modeling and scheduling of the GESS, proposing a linear programming-based scheduling model. In [8], a deep neural network-based mass block grasping control method was introduced for the ramp-type GESS. To mitigate output power fluctuations arising from discrete weight block switching, [9] proposed a power smoothing control strategy based on instantaneous power mirror compensation, ensuring continuous and stable mechanical power during transitions and improving the system's output power stability. Additionally, [10] introduced an auxiliary yard on both sides of the slope and divided the slope into two sections, enabling rapid adjustment and precise control of discharge power.

Building on existing research, the power fluctuation issue caused by discrete weight blocks can be mitigated through coordinated weight scheduling, enabling the weight blocks to operate in synergy. Additionally, the integration of batteries and supercapacitors has been explored to balance power fluctuations in GESS. However, limited research has focused on leveraging the inherent weight of the system itself to address this challenge. To bridge this gap, this paper analyzes the inertial characteristics of the weight within the GESS. Utilizing the kinetic energy stored by the weight, a novel power control method is proposed for the ramp-type GESS, based on Permanent Magnet Synchronous Motor (PMSM) technology. This approach ensures stable power control at the grid side. Furthermore, a GESS simulation model was developed to investigate the grid-connected power characteristics during the placement and unloading of discrete weight blocks.

2. Composition and mathematical model of ramp GESS

The ramp-type GESS consists of two primary subsystems: the mechanical system and the electrical system, as illustrated in Figure 1. The mechanical system enables the conversion between Potential Energy (PE) and Kinetic Energy (KE) of the heavy block. It comprises a low storage yard, a high storage yard, a mountain slope, a track, a heavy cargo transport vehicle, a transmission mechanism, and a gearbox. Meanwhile, the electrical system, responsible for the conversion between kinetic energy and electromagnetic energy as well as enabling power interaction with the grid, includes a PMSM, motor-side converter (MSC), bus capacitor, chopper protection device, grid-side converter (GSC), grid-side filter inductor, and transformer. This integrated design ensures efficient energy storage and grid connectivity.

The fundamental operation of the GESS relies on controlling the vertical displacement of the weight to achieve the mutual conversion between the gravitational PE of the weight and the Electrical Energy (EE) at the grid side. During the energy storage phase, the system operates in the motor state, lifting the heavy block from the low storage yard up the slope to the high storage yard. Conversely, during the energy release phase, the system operates in the generation state, where, based on grid dispatch requirements, the heavy block is lowered from the high storage yard down the slope to the low storage yard, converting its gravitational PE into EE that is subsequently fed into the grid. This bidirectional energy conversion enables efficient energy storage and delivery.

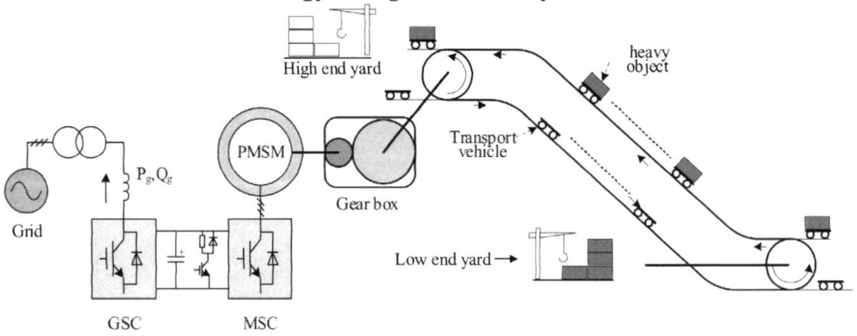

Figure 1. Structure of SGESS Architecture.

A typical SGESS, along with the state equation of the weight block on the slope, is expressed as follows:

$$F_m - F_g - f = N \cdot m \frac{dv}{dt} \tag{1}$$

In the mathematical model, F_m represents the total traction force, F_g denotes the gravitational resistance of the heavy block, f signifies the frictional force acting on the heavy object, N indicates the number of heavy blocks on the slope, m is the mass of an individual heavy block, and v corresponds to the velocity of the heavy block along the track.

Furthermore, the expressions for F_g and f are derived as follows:

$$F_m - F_g - f = n \cdot m \frac{dv}{dt} \tag{2}$$

Effective control of the electrical system relies on the precise management of both the motor and the converter. By aligning the vector control d-axis with the rotor flux of the PMSM, the voltage equations of the motor in the d-q coordinate system can be derived as follows:

$$\begin{cases} u_d = R_s i_d + L_d \dfrac{di_d}{dt} + \dfrac{d\psi_f}{dt} - \omega_e L_q i_q \\ u_q = R_s i_q + L_q \dfrac{di_q}{dt} + \omega_e (L_d i_d + \psi_f) \end{cases} \tag{3}$$

where u_d, u_q, i_d, i_q, R_s, L_d, L_q are respectively the voltage and current of the motor d axis and q axis, the resistance of the motor stator, and the inductance of the motor d axis and q axis. ψ_f indicates the rotor flux linkage. Additionally, ω_e describes the electrical angular frequency of the motor's rotation. These parameters collectively characterize the dynamic behavior of the motor in the d-q coordinate system.

The torque T_e of the PMSM can be derived as follows:

$$T_e = \frac{3}{2} p_n [(L_d - L_q) i_d i_q + \psi_f i_q] \tag{4}$$

The motor equation of motion can be derived as follows:

$$T_e - T_L = \frac{J}{p_n} \frac{d\omega}{dt} \tag{5}$$

3. Power control method of energy storage system

The ramp GESS achieves its power control primarily through the coordinated operation of the MSC and the GSC in a back-to-back configuration. The converter plays a pivotal role in managing the active power exchange between the motor and the grid, ensuring efficient energy transfer. In the realm of reactive power, the GSC is entirely responsible for regulating both the emission and absorption of reactive power, as it serves as the direct interface with the grid. This dual functionality ensures precise control over both active and reactive power dynamics within the system. GSC establishes DC bus voltage and absorbs power from the grid to maintain its stability. MSC converts DC bus power into AC power to drive the permanent magnet motor to run and then drives the load to store energy. The specific control strategy is introduced below.

3.1. Grid-side converter control

In the GSC, the vector control strategy is adopted, and the corresponding control architecture is illustrated in Figure 2. The active power regulation is implemented through a dual-loop control mechanism, incorporating both DC-link voltage and grid-side d-axis inductor current control. Within this control framework, the voltage outer loop operates by comparing the measured DC-link voltage V_{dc} with its reference value and subsequently applying a voltage proportional-integral (PI) controller with a clipper to generate the desired reference value for the d-axis current loop. The current loop collects the grid-side inductor current and takes the coordinate transformed d-axis inductor current i_{gd} as the feedback, compares it with the given value of current, and then obtains the grid-side d-axis voltage command value through the current PI controller.

The grid-side reactive power uses a double closed-loop control strategy of reactive power and grid-side inductor q-axis current. The power outer loop compares the feedback value of reactive power with the given value of voltage Q_g, and then provides the current given value as the q-axis current loop through the power PI controller and limit-i_{ng}. The q-axis voltage reference value is obtained by the current PI controller after making the difference with the q-axis inductor current i_{gq}. The command voltage and the feedforward voltage are added together to drive the GSC through SVPWM modulation.

3.2. Motor side converter control

The MSC uses vector control with rotor field orientation, and the control strategy is shown in Figure 3. The active power uses the double closed-loop control strategy of active power and motor q-axis current. The active power outer loop compares the power feedback value P_g with the power given value and then provides the current given value as the q-axis current loop through the voltage PI controller and the amplitude limiting. The MSC is governed by a vector control strategy employing rotor field orientation, with its comprehensive control architecture depicted in Figure 3. The regulation of active power is accomplished through a dual-loop control mechanism, meticulously coordinating the active power control with the motor q-axis current regulation. Within this framework, the outer active power loop executes a comparative analysis between the measured power feedback P_g and its reference value, subsequently generating the precise current reference for the q-axis current loop through a voltage-based proportional-integral (PI) controller equipped with amplitude-limiting functionality. The current loop collects the motor current and takes the q-axis current i_{mp} after coordinate transformation as the feedback quantity, compares it with the given value of the current, and then obtains the q-axis voltage reference value on the grid side through the current PI controller. The motor magnetic field is controlled using a single closed-loop control for the d-axis current. The d-axis voltage command value on the grid side is obtained by calculating the difference between the d-axis inductor current i_{md} and the reference value. This difference is then processed through the current PI controller. The reference voltage and the feedforward voltage are finally added to drive the MSC through space vector modulation SVPWM.

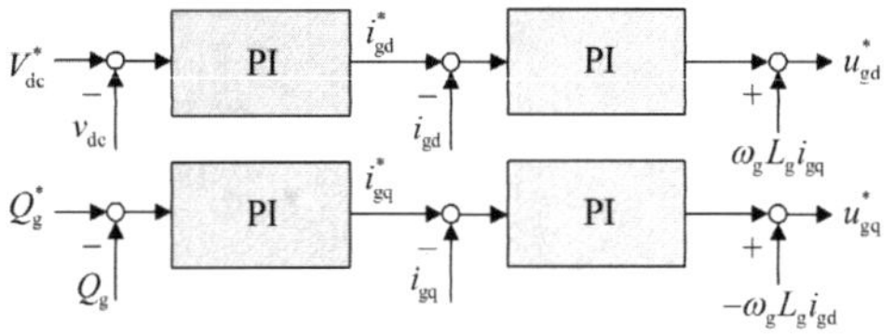

Figure 2. GSC control block diagram.

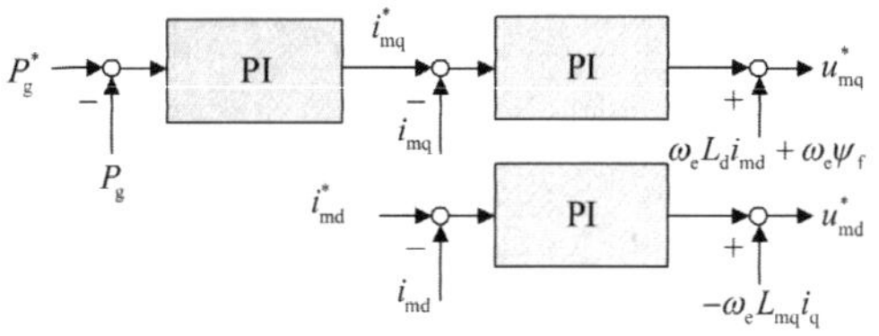

Figure 3. MSC control block diagram.

3.3. Speed control

The GSC control system maintains precise regulation of both DC-link voltage stability and grid-connected reactive power, while simultaneously ensuring optimal active power delivery to the grid. The implementation of motor speed regulation necessitates the adoption of an advanced weight scheduling mechanism. This study introduces an innovative speed control strategy with integrated weight scheduling constraints, whose comprehensive control architecture is illustrated in Figure 4. Within this framework, the speed control loop performs a real-time comparison between the measured speed feedback and its reference value, subsequently generating the precise weight quantity command through a proportional-integral (PI) speed controller incorporating amplitude limiting functionality. The integer part of the weight quantity instruction is separated from the decimal part. After passing through the soft start module, these components are recombined to form the weight scheduling instruction, ultimately completing the motor speed control process.

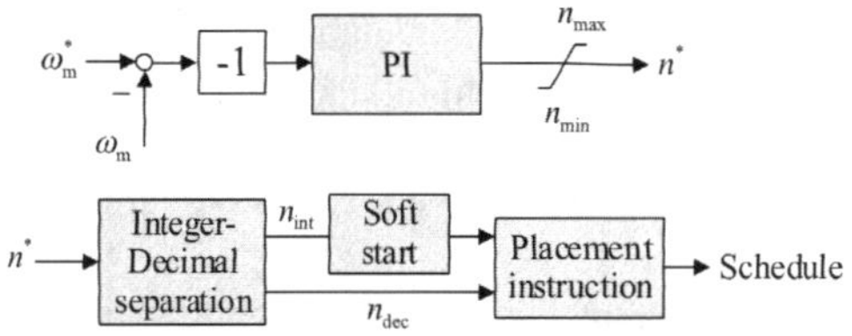

Figure 4. Energy storage system motor speed control block diagram.

The motor operation curve of the GESS operated in energy storage mode is shown in Figure 5, where T_m, T_{L1}, and T_{L2} are the motor characteristic curve and the load characteristic curve before and after the weight loading, and the curve intersection and two points A and C are the steady state operating points of the motor before and after the weight loading. Before the weight is placed, the system operates at point A. During the weight-loading process, the system undergoes a significant increase in load torque, resulting in a transition of the operating point from point A to point B on the characteristic curve. This load variation creates a torque imbalance where the load torque exceeds the motor's electromagnetic torque, resulting in a gradual decrease in motor speed and a consequent shift of the operating point from point B to point C. Conversely, during weight unloading, the system undergoes a reduction in load torque, prompting the operating point to migrate from point C to point D. In this phase, the electromagnetic torque surpasses the diminished load torque, leading to an acceleration of motor speed and the return of the operating point from point D to point A, thereby completing a full operational cycle.

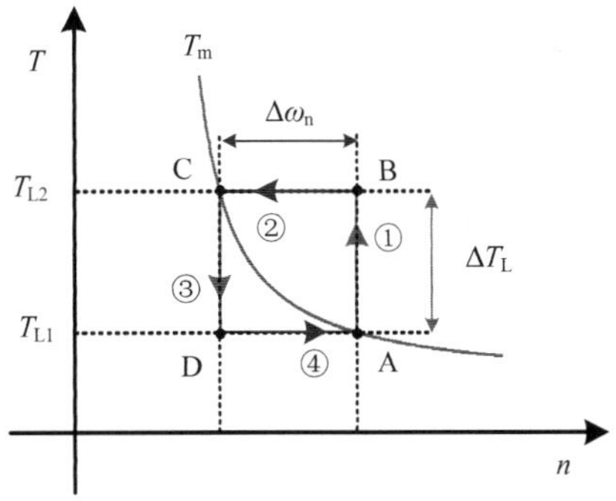

Figure 5. Energy storage process motor running curve.

4. Simulation Verification

A comprehensive 100 kW simulation model was developed to rigorously validate the theoretical framework and assess the effectiveness of the proposed power control strategy for the PMSM-based ramp GESS. The investigation encompassed detailed assessments of both steady-state and dynamic operational characteristics. Table 1 presents the principal simulation parameters utilized in this study. It is noteworthy that to enhance computational efficiency while maintaining the integrity of the analysis, the simulation employed a power-equivalent scaling approach, wherein the ramp length was reduced by an order of magnitude, and the code speed of the mass was proportionally increased. This methodology, while omitting certain steady-state operating points, preserves the validity of system functionality assessment and dynamic performance evaluation.

The simulation was conducted at 90% of the rated power to deliberately induce significant rotational speed fluctuations caused by mass distribution variations, thereby providing a robust testbed for verifying the proposed control scheme's performance. All simulation parameters were normalized using the rated power and grid-side voltage as base values, facilitating standardized result representation and comparative analysis.

Table 1 Simulation parameter.

Parameter	Symbol	Value	Parameter	Symbol	Value
Block mass	m	3 t	Rated torque	T_n	318.3 Nm
Angle of slope	θ	45°	Resistance of motor	R_s	0.05 Ω
Number of weight blocks	N	6	Inductance of motor	L_{dq}	0.5 mH
Length of track	L	432 m	capacitance	C_{dc}	1230 μF
Rated power	P	100 kW	Grid side resistance	R_g	0.05 Ω
Rated capacity	W	400 kWh	Grid side inductance	L_g	0.5 mH
Rated speed	n	3000 r/min	Switching frequency	f_{sw}	10 kHz

The given speed and feedback speed waveforms of the motor are shown in Figure 6. The given speed remains constant at the rated speed, while the actual speeds of both the motor and the heavy object change simultaneously. Both fluctuate with the change of the heavy objects, but the average speed can stably follow the given value, indicating that the motor speed can be basically unchanged through the scheduling of the heavy object.

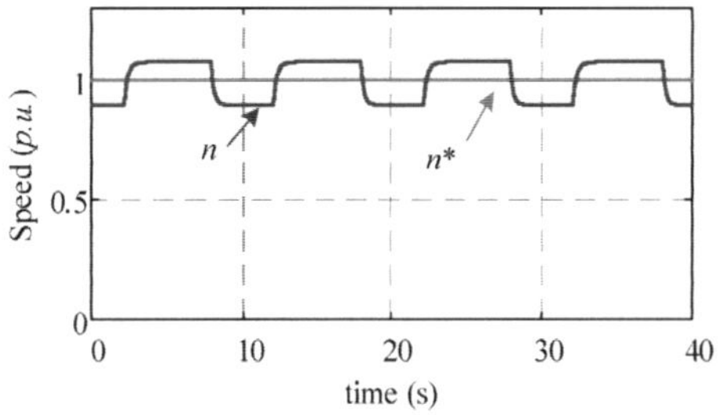

Figure 6. Motor speed.

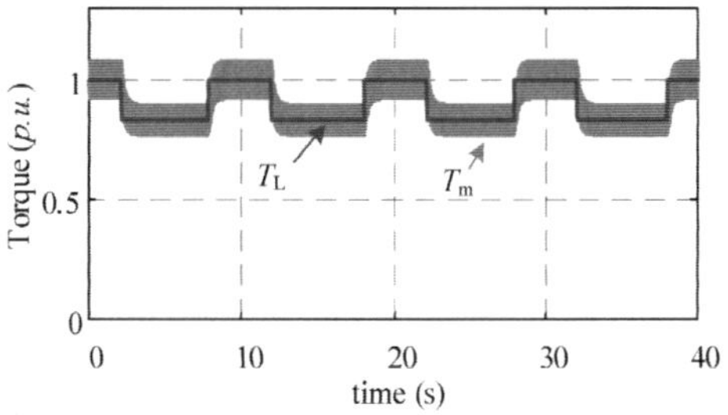

Figure 7. Electromagnetic and load torque.

Figure 7 presents the motor's electromagnetic torque and load torque characteristics which are generated by the suspended mass, and exhibit abrupt variations during the loading and unloading phases. The double closed-loop system regulates the electromagnetic torque, with the outer active power loop generating torque commands based on grid power and motor speed to maintain grid-side power stability. Notably, Figure 6 reveals a gradual electromagnetic torque increase at 8 s, corresponding to the motor's deceleration, with superimposed high-frequency components attributable to motor-side inverter current ripple.

As illustrated in Figure 8, the grid-side current waveform has an amplitude of 0.9 p.u. This value corresponds to 90% of the rated power operation. The current contains minimal high-frequency ripple induced by the grid-side converter. Figure 9 confirms stable grid-side active power at 0.9 p.u. Throughout operation, mass loading/unloading dynamics don't affect the device for maintaining consistent power delivery.

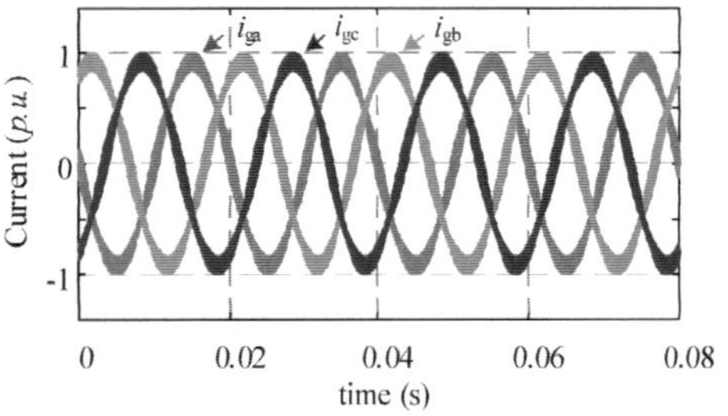

Figure 8. Grid current waveform.

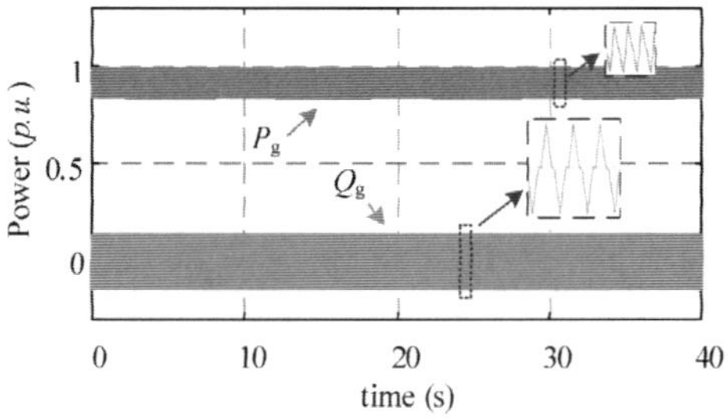

Figure 9. Power of grid current.

5. Conclusion

To address power fluctuations induced by discrete weight blocks in ramp GESS, this study conducts a comprehensive analysis of electromechanical energy conversion principles and grid-connected power instability mechanisms through integrated mechanical-electrical system modeling. The proposed solution employs a tri-loop control architecture encompassing DC-link voltage and grid-side reactive power, grid-side active power and motor field, and motor speed regulation. By leveraging inherent electromechanical characteristics, this control strategy effectively reduces power fluctuations during weight loading and unloading operations, ensuring a stable grid-side active power output. A 100 kW simulation model demonstrates the control method's efficacy, maintaining stable system operation and power output during weight transitions. Results confirm reliable performance across various operating conditions.

Acknowledgments

This research is supported by the State Grid Corporation of China for Science and Technology Projects (5419-202240053A-1-1-ZN).

References

[1] HOU H, XU T, XIAO Z F, et al. Optimal capacity planning and evaluation of a wind-photovoltaic-storage hybrid power[J]. Power System Protection and Control, 2021,49(17):74-84.
[2] WANG S, XIAO L Y, TANG W B, et al. Review of new gravity energy storage[J]. Energy Storage Science and Technology, 2022, 11(05): 1575-1582.
[3] LI N L, WANG X L, WU X, et al. A bi-level planning program of microgrid including gravity energy storage[C]//Asia Conference on Power and Electrical Engineering. ACPEE, 2021: 1123-1128.
[4] BOTHA C D, KAMPER M J. Capability study of dry gravity energy storage[J]. Journal of Energy Storage, 2019, 23: 159-174.
[5] CHEN Y, Hou H, Xu T, et al. A new gravity energy storage operation mode to accommodate renewable energy[C]//IEEE PES Asia-Pacific Power and Energy Engineering Conference. APPEEC, 2019: 1-5.
[6] TONG W, LU Z, HAN M, et al. Inertial characteristics of gravity energy storage systems[C]//2023 IEEE PES 15th Asia-Pacific Power and Energy Engineering Conference (APPEEC). IEEE, 2023: 1-5.
[7] WU X, LI N, WANG X, et al. Day-ahead scheduling of a gravity energy storage system considering the uncertainty[J]. IEEE Transactions on Sustainable Energy, 2020, 12(2): 1020-1031.
[8] CHEN J L, LI Z, ZHU Y Q et al. Control Method of Mass Block Grasping Device of Slope Gravity Energy Storage System Based on Deep Neural Network[J]. Electric Machines & Control Application, 2023, 50(11): 37-45.
[9] LI Z, WANG B, MU X P, et al. Power smoothing control strategy for slope gravity energy storage system based on instantaneous power mirror compensation[J]. Electric Machines & Control Application, 2024, 51(05): 12-20.
[10] LI M, YAXIAER·T, CHA K P, et al. Discharging Power Adjustment of Two-Stage Ramp-Type Gravity Energy Storage System for Fast Response to Load Demand[J]. Electric Machines & Control Application, 2024, 51(04): 12-19.

Influence of boric acid flux on the performance of $LaZnAl_{11}O_{19}$ ceramic waste form

Huilin Sun, Xiaofan Mao, Hua Zhang*, Cheng He and Zheng Li

Department of Radiochemistry, China Institute of Atomic Energy, Beijing, 102400, China

* Corresponding author's e-mail: zhanghua@cnncmail.cn

Abstract: Solidification with $LaZnAl_{11}O_{19}$ ceramics is one of the feasible strategies for treating high-alumina high-level waste. With the utilization of 12.5 wt% boric acids as flux, high-purity $LaZnAl_{11}O_{19}$ was obtained by solid-state method at 1450°C for 12 h. It was demonstrated that adding H_3BO_3 flux not only increased the purity of the samples and lowered the synthesis temperature, but also improved their micromorphology.

1. Introduction

$LaZnAl_{11}O_{19}$ ceramic, belonging to the magnetoplumbite type (hexagonal crystal system, space group *P*6_3*/mmc*), was selected as one of the basic phases of high-alumina tailored ceramic waste form [1] due to the ability of the crystal structure to accommodate aluminum and various radionuclides [2]. Moreover, it has superior mechanical properties [3], thermal stability [4], as well as leaching resistance [5], making it a feasible ceramic waste form for high-alumina high-level waste (HLW).

However, extremely high sintering temperatures (≥ 1600°C) were required to synthesize magnetoplumbite via solid-state reactions [6]. Fluxes utilize liquid-phase sintering to promote crystal growth and reduce impurities, which helps to produce better-quality crystals with fewer defects.

Therefore, in this investigation, H_3BO_3 was used to examine the impact of the flux additive on the synthesis temperature, phase evolution, and micromorphology of $LaZnAl_{11}O_{19}$ ceramics synthesized by the solid-state method.

2. Experiment

The solid-state synthesis route was used for sample preparation. The AR grade starting materials including La_2O_3, ZnO, and Al_2O_3 were precisely weighed according to the stoichiometric formula $LaZnAl_{11}O_{19}$, corresponding to exact mass fractions of 20.23 wt%, 10.11 wt%, and 69.66 wt% respectively. H_3BO_3 was introduced into the raw material system as a flux, with the additions ranging from 0 to 15.0 wt% (specifically 0, 2.5, 5.0, 7.5, 10.0, 12.5, 15.0 wt%). The starting materials were milled, mixed homogeneously, and then held at 1400°C (T_C), 1450°C (T_B), and 1600°C(T_A) for 12 h to prepare the powder samples. The specimens were systematically coded based on their respective sintering temperatures and flux additions, as detailed in Table 1.

Table 1. Systematic coding protocol with corresponding processing parameters.

Flux (wt%) / Temperature (°C)	0	2.5	5	7.5	10	12.5	15
1600	T_A-0	/	/	/	/	/	/
1400	T_B-0	T_B-1	T_B-2	T_B-3	T_B-4	T_B-5	T_B-6
1450	T_C-0	T_C-1	T_C-2	T_C-3	T_C-4	T_C-5	T_C-6

The crystallographic characterization of the specimens was conducted via powder X-ray diffraction (PXRD) using a Bruker D8 ADVANCE diffractometer equipped with Cu K_α radiation source (λ = 1.5418 Å), under operational parameters of 40 kV/40 mA. Data were collected covering a diffraction angle span from 10° to 80° (2*θ*) with an angular resolution of 0.02° and a dwell time of 0.4 s per step. *Rietveld* refinement of XRD data was conducted by GSAS-II [7]. The crystal morphology was observed through field-emission scanning electron microscopy (FE-SEM, QUANTA 650FEG, FEI) augmented by energy dispersive spectrometer (EDS) results for elemental identification.

3. Results and discussion

3.1. Crystallographic characterization

The PXRD patterns of the samples sintered at 1450°C for 12 h with different additions of H_3BO_3 flux (0-15 wt%) are shown in Figure 1. In the absence of flux, the diffraction profile predominantly indexed to perovskite-type $LaAlO_3$, spinel-phase $ZnAl_2O_4$, and residual α-Al_2O_3, with only trace amounts of the target $LaZnAl_{11}O_{19}$ magnetoplumbite phase. As the amount of H_3BO_3 was increased, the diffraction peaks corresponding to $LaZnAl_{11}O_{19}$ became progressively more pronounced, while the impurities diminished. Notably, magnetoplumbite with high purity was fabricated upon reaching a boric acid addition of 12.5%. Thereby, it can be concluded that the incorporation of the boric acid flux improves the phase purity of the samples obtained via the conventional solid-state method.

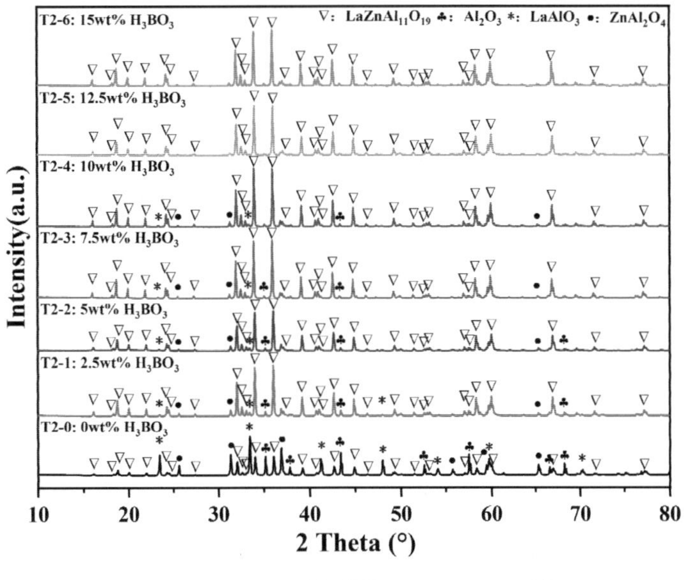

Figure 1. XRD patterns of $LaZnAl_{11}O_{19}$ at 1450°C with different amounts of flux.

The *Rietveld* refinement results for samples T_A-0 (1600°C, no flux) and T_C-5 (1450°C, 12.5 wt% H_3BO_3), conducted by GSAS-II crystallographic software package with a *P*6₃/*mmc* space group model, are systematically presented in Figure 2. Quantitative phase analysis revealed a significant enhancement in the target $LaZnAl_{11}O_{19}$ content from 77.46 wt% in the flux-free system (T_A-0) to 97.35 wt% under optimized flux conditions (T_C-5). Simultaneously, impurity phases were entirely suppressed at this optimal flux concentration, as confirmed by *Rietveld* refinement. This 25.8% purity enhancement was achieved concurrently with a remarkable 150°C reduction in thermal processing requirements, while still maintaining the structural integrity. The coupled enhancement in phase purity and energy efficiency conclusively established flux-assisted synthesis as a technologically superior pathway for magnetoplumbite ceramics preparation.

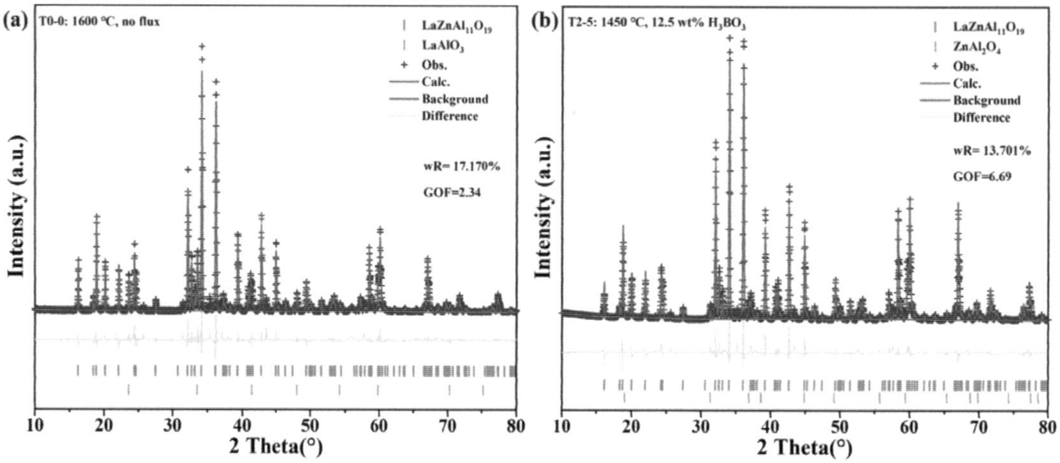

Figure 2. The *Rietveld* refinement results of T_A-0 and T_C-5.

To further investigate the impact of H_3BO_3 on the synthesis temperature of $LaZnAl_{11}O_{19}$, the phase composition of the ceramics prepared with different flux dosages by heating at 1400°C for 12 h was studied. The XRD results are summarized in Figure 3. With the increase of flux, the diffraction peaks of both $LaAlO_3$ and $ZnAl_2O_4$ impurities gradually decreased. Notably, at the maximum tested flux dosage (15 wt%), the diffraction peak of $LaAlO_3$ fully disappeared, but at the same time, new diffraction signatures indexed to another impurity $Al_{20}B_4O_{36}$ emerged. This secondary phase formation, coupled with incomplete $LaZnAl_{11}O_{19}$ crystallization, indicated the limitations of flux-assisted synthesis at 1400°C. The results confirm that the addition of the flux at 1400°C was insufficient to achieve a single-phase $LaZnAl_{11}O_{19}$ ceramic.

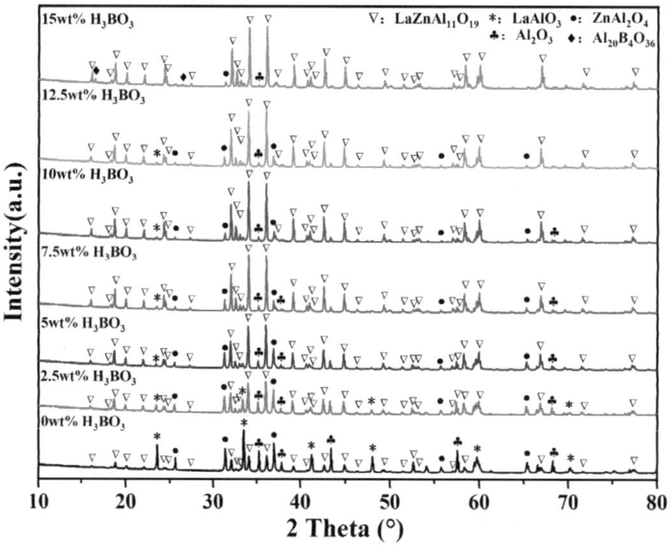

Figure 3. XRD patterns of $LaZnAl_{11}O_{19}$ prepared at 1400°C with different amounts of flux.

3.2. Microstructure

Figures 4(a)-(c) show the SEM micrographs of sample T_A-0 made at 1600°C without flux. Typical ortho-hexagonal platelet-like particles were observed, which directly correlated with the crystal structure of magnetoplumbite (space group *P*6$_3$*/mmc*). This hexagonal compound consists of two spinel block units separated by a mirror plane layer containing large La^{3+} cations. The La^{3+}-rich mirror layer significantly inhibits anisotropic oxygen diffusion along the *c*-axis direction. Consequently, crystal elongation along the *c*-axis becomes highly restricted, while preferential grain development occurs along the directions parallel to the mirror layers (*a*-axis and *b*-axis). This anisotropic growth behavior, governed by the crystallographic anisotropy of oxygen ion mobility, ultimately results in the characteristic platelet-shaped morphology with well-developed basal planes and constrained thickness along the crystallographic *c*-axis. Notably, some particles exhibited edge truncation and surface irregularities, suggesting incomplete grain development under high-temperature solid-state reaction conditions.

In contrast, Figures 4(d)-(f) present the microstructural evolution of flux-assisted sample T_C-5 fabricated at a reduced temperature of 1450°C with 12.5 wt% flux. Despite the 150°C decrease in sintering temperature, the characteristic ortho-hexagonal plate-like grains were still present. Moreover, comparative analysis with T_A-0 revealed an increased thickness along the platelet normal direction (*c*-axis) and a significant reduction in surface defects, implying better grain development through improved crystallographic anisotropy and structural integrity. In addition, the grain morphology tended to be uniform and well dispersed. It can be concluded that adding flux significantly improves grain morphology.

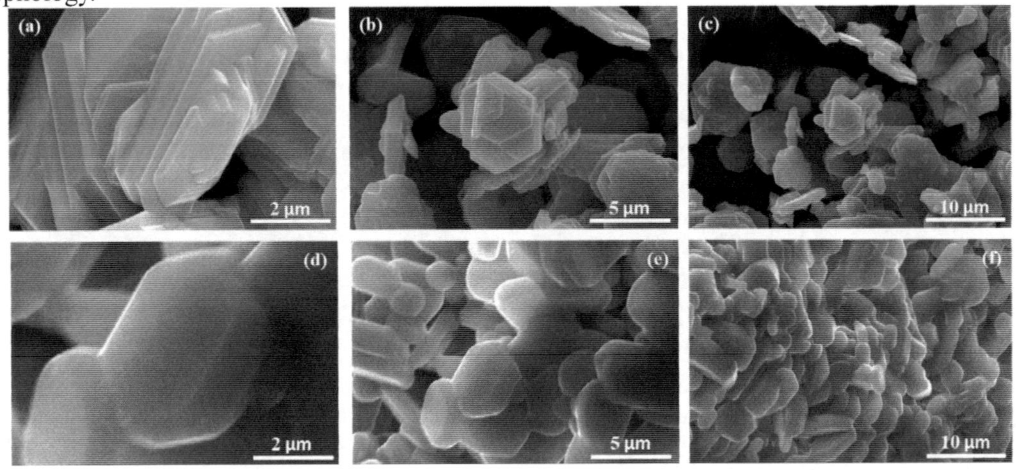

Figure 4. SEM image of T_A-0 (a-c) and T_C-5 (d-f).

3.3. EDS analysis

The EDS results for sample T_C-5 are displayed in Figure 5, proving the presence of La, Zn, Al, and O elements, with no detectable impurity elements above the detection limit. Table 2 provides a comparison between the theoretical stoichiometric atomic ratios of $LaZnAl_{11}O_{19}$ and the calculated average atomic ratios found in the sample. Given the semi-quantitative nature of the EDS method, it was assumed that the test results aligned with the stoichiometric ratios within the margin of relative error, which confirmed that the obtained samples were indeed $LaZnAl_{11}O_{19}$ ceramic. Cross-validated by the absence of secondary phases in XRD (Figure 1), the EDS results provide robust evidence for the successful synthesis of phase-pure $LaZnAl_{11}O_{19}$ ceramics through flux-assisted processing.

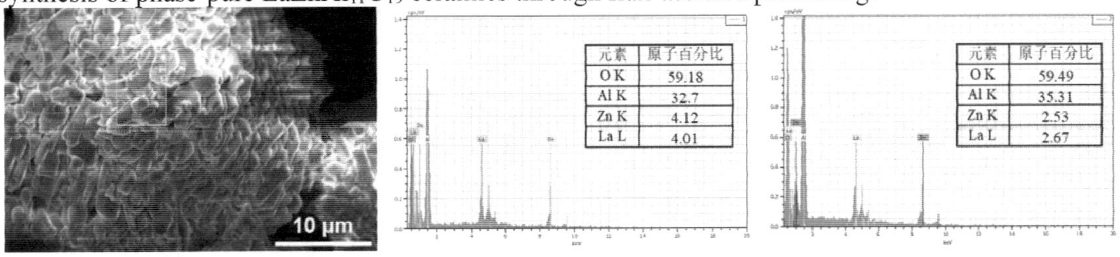

Figure 5. The EDS results for sample T_C-5 (1450°C, 12.5 wt% H_3BO_3).

Table 2. Atomic ratio comparison.

Sample number	Stoichiometric atomic ratio (%)				Calculated average atomic ratio (%)			
	La	Zn	Al	O	La	Zn	Al	O
T_C-5	3.125	3.125	34.375	59.375	3.34	3.33	34.01	59.34

4. Conclusions

High-purity $LaZnAl_{11}O_{19}$ ceramic was effectively synthesized via an optimized solid-state reaction process involving 12.5 wt% boric acids as a flux, maintained at 1450°C for 12 h with controlled thermal ramping rates of 5°C/min. Quantitative *Rietveld* refinement of XRD data (Figure 2) confirmed 97.35 wt% phase purity for the flux-assisted sample, representing a 25.8% absolute improvement compared to the 77.46 wt% purity achieved in the flux-free counterpart sintered at 1600°C. The presence of the flux H_3BO_3 not only improved the purity of the samples but also lowered the synthesis temperature by 150°C. Furthermore, the microscopic observation demonstrated that the samples after the addition of the flux were still characterized by ortho-hexagonal plate-like grains of magnetoplumbite, with the morphology and development of the grains greatly improved.

The synthesized $LaZnAl_{11}O_{19}$ magnetoplumbite-type ceramic demonstrates dual functional superiority as a nuclear waste form candidate: (1) an exceptional alumina incorporation capacity of 69.66 wt% within its crystal structure, and (2) crystallographically defined host sites (La^{3+} positions in mirror planes) for radionuclide immobilization according to the principle of isomorphic substitution. These synergistic properties validate its potential as a ceramic matrix for immobilizing high-alumina HLW.

References

[1] Morgan, P. E. D., Clarke, D. R., Jantzen, C. M., and Barker, A. B. (1981). High-Alumina Tailored Nuclear Waste Ceramics. Journal of the American Ceramic Society, 64, 5: 249–258.

[2] Torrez-Herrera, J. J., Fuentes-Ordoñez, E. G., Korili, S. A., and Gil, A. (2021). Evidence for the synthesis of La-hexaaluminate from aluminum-containing saline slag wastes: Correction of structural defects and phase purification at low temperature. Powder Technology, 377, 80–88.

[3] Cherkasova, N., Veselov, S., Bataev, A., Kuzmin, R., and Stukacheva, N. (2021). Structure and mechanical properties of ceramic materials based on alumina and zirconia with strontium hexaaluminate additives. Materials Chemistry and Physics, 259, 123938.

[4] Zhu, H., Liu, L., Xiang, H., Dai, F.-Z., Wang, X., Ma, Z., Liu, Y., and Zhou, Y. (2022). Improved thermal stability and infrared emissivity of high-entropy $REMgAl_{11}O_{19}$ and $LaMAl_{11}O_{19}$ (RE=La, Nd, Gd, Sm, Pr, Dy; M=Mg, Fe, Co, Ni, Zn). Journal of Materials Science & Technology, 104, 131–144.

[5] Wang, C., Lu, H., Huang, Z., and Xie, H. (2018). Enhanced anti-deliquescent property and ultralow thermal conductivity of magnetoplumbite-type $LnMeAl_{11}O_{19}$ materials for thermal barrier coating. Journal of the American Ceramic Society, 101, 3: 1095–1104.

[6] Jiang, B., Fang, M. H., Huang, Z. H., Liu, Y. G., Peng, P., and Zhang, J. (2010). Mechanical and thermal properties of LaMgAl11O19. Materials Research Bulletin, 45, 10: 1506–1508.

[7] Toby, B. H., and Von Dreele, R. B. (2013). GSAS-II: The genesis of a modern open-source all purpose crystallography software package. Journal of Applied Crystallography, 46, 2: 544–549.

Journal of Physics: Conference Series **3011** (2025) 012072 doi:10.1088/1742-6596/3011/1/012072

Research on magnetostrictive characteristics of three-phase transformer under DC bias

Mengyu Cui, Xia Dong *, Bingwei Wang, Baoquan Lian and Xinyu Wang

Qilu University of Technology (Shandong Academy of Sciences), Jinan, Shandong, 250353, China

*Corresponding author's e-mail: dongxia6078@163.com

Abstract. The primary cause of vibration and noise in transformers is the magnetostrictive behavior of the core. A transient electric-magnetic circuit coupling model of the three-phase five-pillar transformer under DC bias is developed by using the duality principle to investigate this phenomenon more thoroughly. In the model, the Jiles-Atherton ferromagnetic hysteresis theory is employed to describe the magnetization process. The variations of magnetic flux density, magnetic field intensity, and the magnetostriction curve under different DC conditions are analyzed through simulations of the three-phase transformer. The results indicate that DC bias causes the waveforms of magnetic flux density and magnetic field intensity to shift and become distorted, magnetostrictive deformation intensifies, and the curves depicting the connection between deformation and magnetic field (Δl-B, Δl-H) are also distorted, showing an asymmetric butterfly shape, and the level of distortion is directly related to the size of DC. The research findings confirm that the proposed transient model can precisely reflect the magnetostrictive characteristics of a three-phase five-pillar transformer under DC bias.

1. Introduction

High-voltage DC transmission technology commonly operates in either unipolar or bipolar unbalanced mode. The unipolar mode of operation uses the earth as the current return path, which results in the DC returning from the earth flowing into the neutral grounded transformer windings; when operating in bipolar unbalanced mode, the DC is also generated in the vicinity of the grounding electrode due to unbalanced currents at both poles, which in turn flow into the transformer winding. In both cases, the DC becomes part of the excitation current of the transformer, which shifts the magnetic field in the transformer's core and produces the phenomenon of DC bias. The DC bias can lead to increased transformer noise, vibration, and loss, which can cause serious adverse effects on the service life and safe operation of the transformer [1]-[2].

The main cause of vibration noise in transformers is the core's magnetostrictive phenomenon. Magnetostriction is the phenomenon of deformation of ferromagnetic materials (such as silicon steel sheets in the core of a transformer) under the influence of a magnetic field. When the core is excited, the magnetostriction of the silicon steel sheet causes the core to vibrate periodically with the excitation current [3]. Although the magnetostriction effect per unit length of the core material is only a few microns in size, which is much smaller than the size of the core material, the magnetostriction effect exacerbates the vibration of the core at certain frequencies. Approximately 50% of the total vibration noise in a transformer results from the magnetostrictive behavior of the core material, indicating that the magnetostriction of the core is the primary factor responsible for generating the transformer's vibration noise [4][5][6]. Recent studies have systematically investigated the magnetostrictive behavior of transformer cores. To effectively analyze the magnetostriction phenomenon in transformer cores, Cao et al. established a theoretical model of core vibration coupled with elastodynamics and thermodynamics, constructed a transformer core magnetostriction vibration acquisition platform, and carried out vibration monitoring tests on the core of a 10 kV transformer [7]. The majority of studies on the magnetostrictive properties of transformer cores have concentrated on experimental observations, with fewer studies focusing on modeling and simulation of transformer core magnetostriction.

Due to the intricacy of the transformer model, this paper develops a transient circuit-magnetic circuit coupling model for a three-phase five-pillar transformer under DC bias by using the duality principle and adopts the Jiles-Atherton ferromagnetic hysteresis model to simulate the magnetization performance of the transformer core, to describe the transformer's magnetic characteristics from the physical nature [8]. The magnetic field strength H, magnetic flux density B, Δl-B, and Δl-H were obtained by simulation calculations using MATLAB software, and their variation rules under DC bias were analyzed [9].

2. Jiles-Atherton ferromagnetic hysteresis model

2.1. Jiles-Atherton model

The Jiles-Atherton framework, grounded in energy conservation principles during the magnetization of ferromagnetic materials, characterizes nonlinear hysteresis phenomena through a dynamical equation of first order. This model explicitly links the physical behavior of domains to macroscopic magnetic properties [9].

In the Jiles-Atherton framework, magnetization is divided into two components: irreversible (M_{irr}) and reversible (M_{rev}). The total magnetization M is expressed as the sum of these two components, as shown in Equation (1)

$$M = M_{irr} + M_{rev} \tag{1}$$

In this case, the irreversible magnetization part M_{irr} in ferromagnetic materials causes energy loss, which is the reason for the formation of hysteresis loops.

$$H_e = H + \alpha M \tag{2}$$

where H_{e} represents the effective magnetic field strength; H denotes the magnetic field strength; M stands for the magnetization strength; α is the average-field parameter.

Based on the principle of energy conservation, the differential relationship between magnetization strength and magnetic field strength can be derived:

$$M_{an} = M_s(\coth(\frac{H_e}{a}) - \frac{a}{H_e}) \tag{3}$$

where M_{an} refers to the non-hysteretic magnetization intensity; M_{s} represents the saturation magnetization intensity; a is a parameter characterizing the shape of the non-hysteresis magnetization curve.

$$\frac{d_M}{d_H} = \frac{1}{1+c}\frac{M_{an} - M}{\delta k - \alpha(M_{an} - M)} + \frac{c}{1+c}\frac{dM_{an}}{dH} \tag{4}$$

$$\delta = sign(\frac{d_H}{d_t}) \tag{5}$$

Equation (4) introduces c, a coefficient quantifying reversible magnetization, while Equation (5) incorporates δ to represent temporal variations in magnetic field strength. The parameter k accounts for domain wall pinning effects, which govern energy dissipation during the magnetization cycle [10].

$$B = \mu_0(H + M) \tag{6}$$

where μ_0 is the vacuum permeability.

3. Transformer equivalent circuit model

3.1. Equivalent circuit representation of the three-phase five-pillar transformer under DC bias

A transient equivalent circuit model for a three-phase five-pillar transformer with DC bias is created based on the duality principle, as illustrated in Figure 1, where DC is the DC power source. The DC component entering the transformer is modeled, and the transformer is unloaded on the secondary side. By using the duality principle, the core magnetic circuit is equated to a parallel connection of resistance R_1 and inductance L_1, and the yoke magnetic circuit is similar to the core magnetic circuit. A parallel connection of resistance R_3 and inductance L_3 represents it. The zero sequence magnetic circuit (two side pillars) is represented by an inductor L_0 and a resistance R_0 [11].

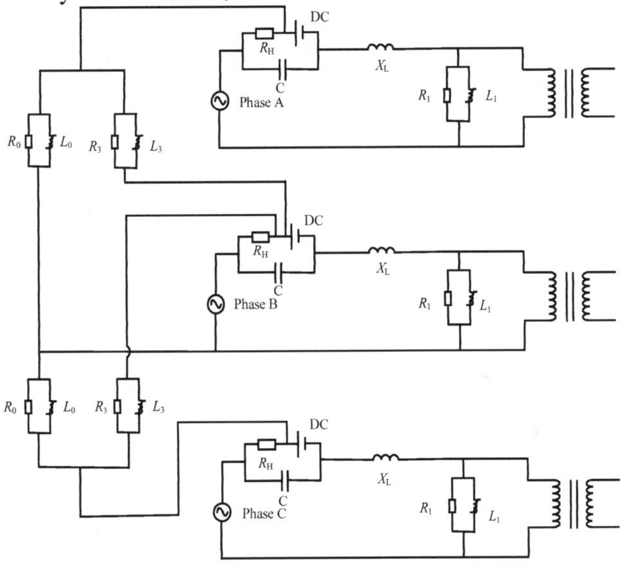

Figure 1. Transient circuit model of three-phase five-pillar transformer.

3.2. Magnetostrictive modeling

Under the effect of H, magnetostriction is the rate of fractional change along the winding's length, and the fractional change equation is:

$$\frac{1}{L}\frac{dL}{dH}=\frac{\lambda_s}{H_c^2}H^2 \tag{7}$$

where λ_s represents the coefficient of saturation magnetostriction, H_c denotes the coercivity, and L represents the length of the core. The ratio of the change in core length Δl to L is:

$$\lambda=\frac{\Delta l}{L}=\frac{\lambda_s}{H_c^2}H^2 \tag{8}$$

where λ is the magnetostrictive deformation; the correlation between Δl and the magnetic field intensity H can be obtained by collating Equation (9) as:

$$\Delta l=\frac{\lambda_s}{H_c^2}H^2 L \tag{9}$$

Δl is related to the flux density B by Equation (10) as:

$$\Delta l=\frac{\lambda_s B^2 L}{\mu^2 H_c^2} \tag{10}$$

4. Simulation calculations

4.1. B and H simulation results

In this paper, a three-phase transformer of 112.5 kVA with a rated ratio of 100 is simulated by using MATLAB software. The simulation model is the transient electric-magnetic circuit coupling model described above.

As depicted in Figure 2, the waveforms of B and H for the three-phase transformer exhibit no distortion without DC. Both the positive and negative amplitudes of B and H are symmetrical, indicating a balanced magnetic state. The magnetic flux density values for phases A and C are 1.46 T, and the corresponding magnetic field strength values are 2780 A/m, while they are 1.35 T and 1830 A/m for phase B. The absolute value of the amplitude in phase B is smaller compared with that in phases A and C, which is the result of the imbalance in the three-phase magnetic circuit.

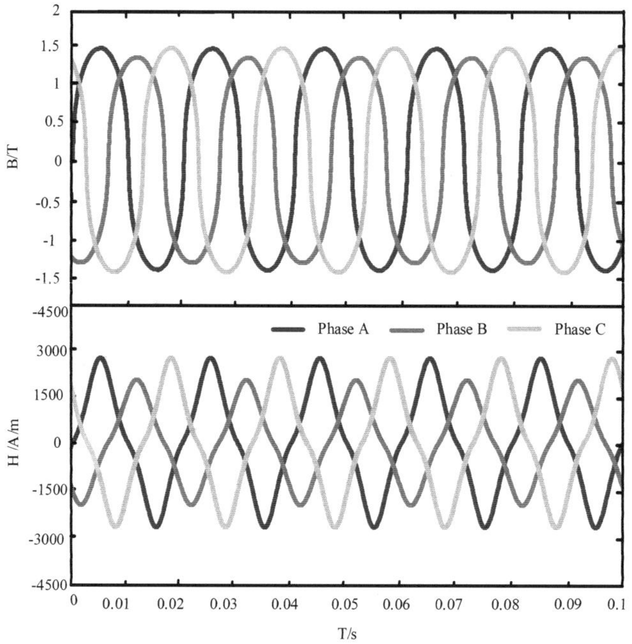

Figure 2. Waveform diagram of transformer B and H without DC.

Under a DC bias of 0.3 A, Figure 3 illustrates a significant positive drift in both B and H waveforms. The asymmetry in H amplitudes (17%) exceeds that of B (3%), suggesting a stronger DC influence on magnetic field dynamics. This distortion disrupts the symmetry between positive and negative amplitude magnitudes, with the positive peak exceeding the negative counterpart by approximately 17% for H and 3% for B. The positive peak of the magnetic field strength under DC 0.3A is approximately 1.17 times, and the negative peak is around 0.86 times compared to when there is without DC. The positive magnetic flux density peak is about 1.03 times, while the negative peak is roughly 0.98 times that without DC, as indicated in Table 1.

Journal of Physics: Conference Series **3011** (2025) 012072 doi:10.1088/1742-6596/3011/1/012072

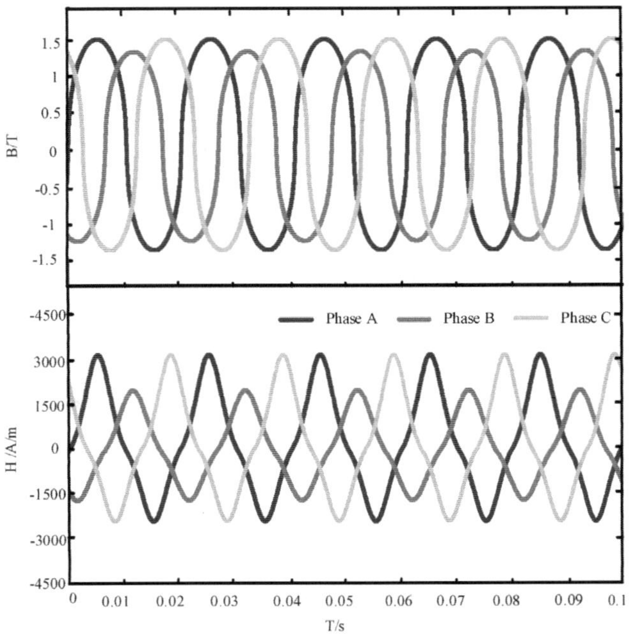

Figure 3. Waveform diagram of transformer *B* and *H* under DC 0.3 A.

Table 1. Transformer *B* and *H* amplitude with or without DC.

		Amplitude without DC	Positive amplitude under DC 0.3 A	Negative amplitude under DC 0.3 A	Ratio of positive amplitude under 0.3 A DC to amplitude without DC	Ratio of negative amplitude under 0.3 A DC to amplitude without DC
Phase A	$H_{max}(A/m)$	2780	3254	-2393	1.171	0.861
	$B_{max}(T)$	1.46	1.50	-1.43	1.027	0.979
Phase B	$H_{max}(A/m)$	1830	2118	1580	1.157	0.863
	$B_{max}(T)$	1.35	1.39	-1.31	1.029	0.970
Phase C	$H_{max}(A/m)$	2780	3254	-2393	1.171	0.861
	$B_{max}(T)$	1.46	1.50	-1.43	1.027	0.979

4.2. Simulation results of Δl-B and Δl-H

Figures 4 to 6 show the magnetostriction curves of a three-phase transformer without DC. The magnetostriction curves are butterfly-shaped with symmetrical left and right wings and consistent range boundaries. As the flux density B and magnetic field strength H reach their maximum values, the magnetostrictive deformation variable Δl increases to peak values in a nonlinear form; as the flux density and magnetic field strength reduce, the magnetostrictive curve decreases along the other path. Once the magnetic field strength drops to zero, the deformation Δl equals 1.175×10^{-8} m. On the other hand, the magnetic field strength reaches around -55.6 A/m when the deformation Δl is zero. This behavior suggests that the change in magnetostriction follows the change in magnetic field strength with a delay, indicating that magnetostriction exhibits hysteresis. As the magnetic field strength approaches the boundary amplitude, the growth of deformation becomes slow, which indicates that magnetostriction has saturation characteristics.

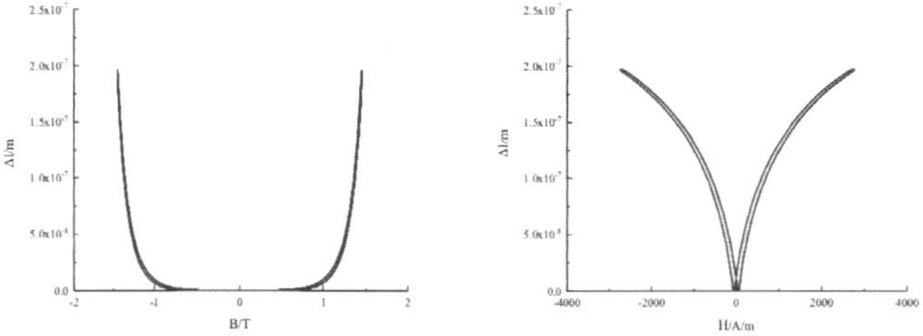

Figure 4. Magnetostriction curve of phase A without DC.

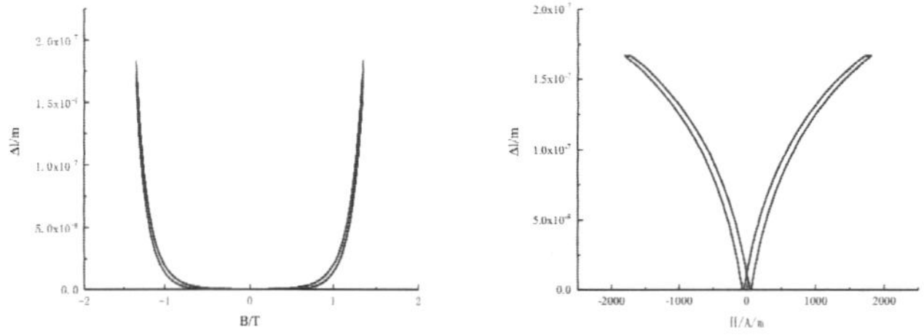

Figure 5. Magnetostriction curve of phase B without DC.

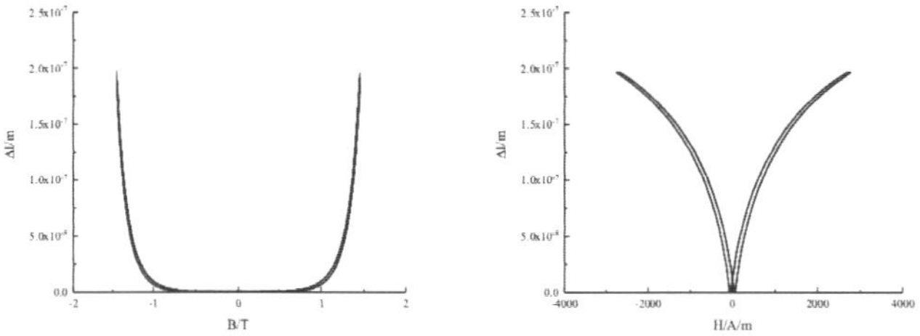

Figure 6. Magnetostriction curve of phase C without DC.

Figures 7 to 9 show the magnetostriction curves for a three-phase transformer under DC 0.3 A. As shown in the figures, the magnetostriction curve loses its symmetry on the left and right sides, suggesting that distortion has taken place. The boundary amplitude on the left side is notably smaller, the magnetostrictive deformation Δl is 0.93 times higher than without DC, and the downward trend indicates that both the magnetic field amplitude and the deformation have decreased. In contrast, the boundary amplitude of the right flank increases significantly, and the magnetostrictive deformation Δl is 1.06 times higher than without DC, with an increasing trend opposite to that of the left flank. Table 2 is the transformer magnetostrictive variable amplitude value with or without DC.

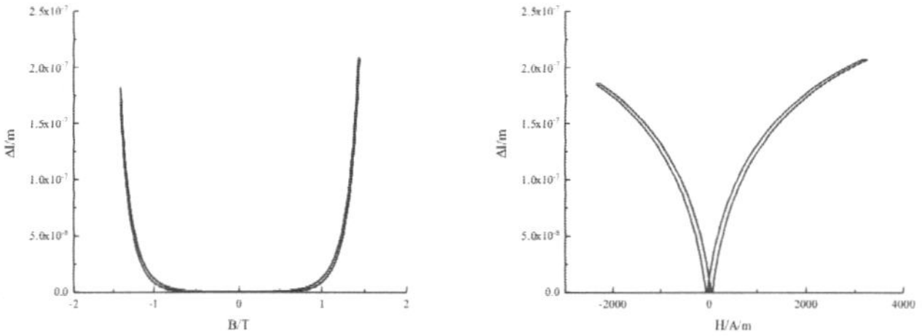

Figure 7. Magnetostriction curve of phase A under DC 0.3 A.

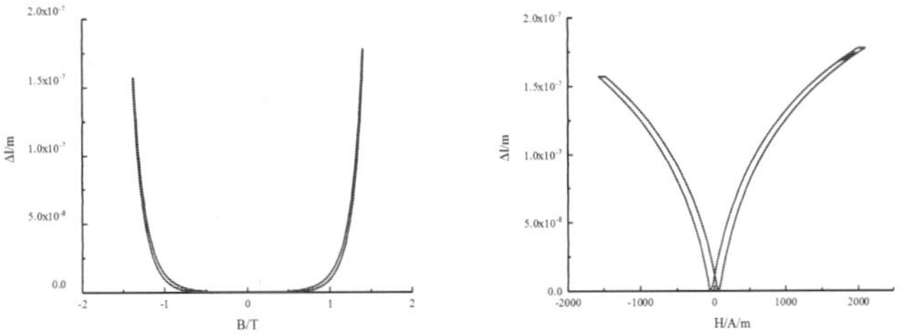

Figure 8. Magnetostriction curve of phase B under DC 0.3 A.

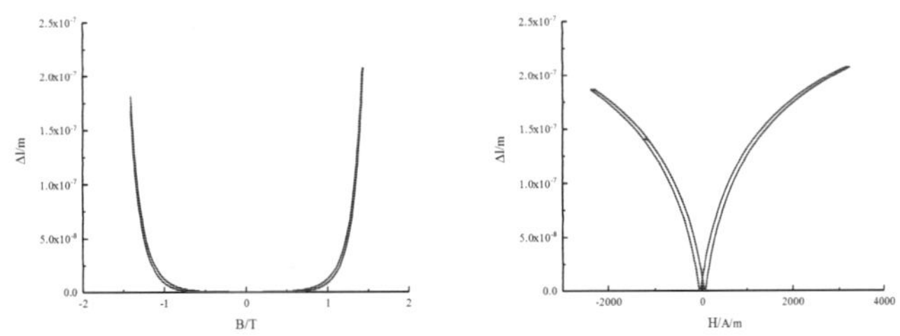

Figure 9. Magnetostriction curve of phase C under DC 0.3 A.

Table 2. Transformer magnetostrictive variable amplitude with or without DC.

	Amplitude without DC	Left wing amplitude under DC 0.3 A	Right wing amplitude under DC 0.3 A	Ratio of left wing amplitude under DC 0.3 A to that without DC	Ratio of right wing amplitude under DC 0.3 A to that without DC
Phase A Δl ($\times 10^{-7}$ m)	1.9662	1.8277	2.0827	0.9296	1.0593
Phase B Δl ($\times 10^{-7}$ m)	1.6749	1.5699	1.7757	0.9373	1.0601
Phase C Δl ($\times 10^{-7}$ m)	1.9662	1.8277	2.0827	0.9296	1.0593

Figure 10 shows the curve of magnetostrictive maximum deformation as a function of DC. There is a nearly linear correlation between the magnetostrictive strain Δl and the DC. As the DC increases, the deformation gradually rises, indicating that the DC has a significant effect on the magnetostrictive properties of the transformer. The increase in DC leads to a change in the strength of the magnetic field inside the transformer, which amplifies the magnetostrictive effect of the transformer core, resulting in a linear upward trend in the amount of deformation.

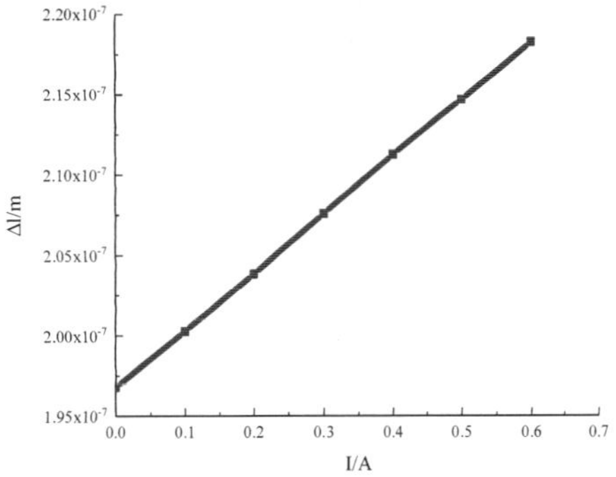

Figure 10. Variation curve of magnetostrictive maximum deformation with increasing DC

5. Conclusion

This paper focuses on a three-phase five-pillar transformer as the subject of study and the duality principle is applied to effectively correlate the magnetic circuit of the transformer model with the circuit. The magnetic characteristics of the transformer core are better reflected by the Jiles-Atherton hysteresis model. Unlike previous models limited to experimental data, this work establishes a predictive framework for DC bias effects.

(1) Without DC, the waveforms of the magnetic flux density B and magnetic field strength H remain undistorted, with the absolute values of the positive and negative amplitudes being equal.

(2) As the magnetic field intensity drops to zero, the deformation variable Δl is 1.175×10^{-8} m, and when Δl reaches zero, the magnetic field strength is roughly -55.6 A/m. As the magnetic field strength H approaches the maximum amplitude, the deformation variable Δl grows slowly, indicating that the magnetostrictive effect tends to be saturated and has saturation characteristics.

(3) With DC bias, the waveforms of magnetic flux density B, magnetic field strength H, and the magnetostriction curves (Δl-B and Δl-H) become distorted. In particular, the waveforms of the magnetic flux density and magnetic field strength shift upward, and the absolute values of the positive and negative amplitudes are no longer balanced. The magnetostriction curve loses its symmetry across the left and right sides, with the distortion on one side decreasing and the distortion on the other side increasing.

(4) The deformation Δl due to magnetostriction shows a near-linear relationship with the applied DC bias. An increase in DC leads to a proportional rise in deformation, highlighting the substantial impact of DC bias on the transformer's magnetostrictive behavior.

Acknowledgments

This work was supported by the Jinan Industrial Innovation Carrier Project under Project No. 202333001, the Shandong Provincial Natural Science Foundation (Grant No. ZR2022QE132), and the Talent Research Project of Qilu University of Technology (Grant No. 2023RCKY155).

References

[1] LI B, WANG Z Z, LIU H B, et al. (2021) Experiment on Vibro-Acoustic Characteristic of 500 kV Single-Phase Transformer under DC Bias. J. TRANSACTIONS OF CHINA ELECTRO TECHNICAL SOCIETY, 36 (13): 2801-2811. [CrossRef]

[2] LI T B, TONG T, XI R, et al. (2023) Research on Vbrati-on Characteristics of Transformer under DC Bias. J. TRANSFORMER, 60 (03): 52-58. [CrossRef]

[3] WANG Z, ZHANG Y L, ZHANG D H, et al. (2021) Measurement and Modelling of Local Deformation Caused by Magnetism in the Single-phase Transformer Core Experimental Model Under DC Bias. J. Proceedings of the CSEE, 41 (12): 4316-4325. [CrossRef]

[4] SONG Q K, WANG F H. (2023) Research on Electromagnetic Vibration Features of Large-scaled Transformer Under DC Bias. J. High Voltage Apparatus, 59 (10): 129-139. [CrossRef]

[5] Mohammed O A, Calvert T, McConnell R. (2001) Coupled magnetoelastic finite element formulation including anisotropic reluctivity tensor and magnetostriction effects for machinery applications. J. IEEE Transactions on Magnetics, 37(5): 3388-3392. [CrossRef]

[6] Belahcen A. (2006) Vibrations of rotating electrical machines due to magneto-mechanical coupling and magnetostriction. J. IEEE Transactions on Magnetics, 42(4): 971-974. [CrossRef]

[7] CAO C, LU D, XU B W, et al. (2024) Magnetostrictive Vibration Model and Loosening Fault Diagnosis Method of Transformer Core Based on Elastodynamics thermodynamics. J. High Voltage Engineering, 50(05): 2146-2157. [CrossRef]

[8] DONG X, LIU Z Z. (2013) Simulative analysis of three-phase transformer with DC bias. J. Electric Power Automation Equipment, 33 (07): 121-125. [CrossRef]

[9] CUI M Y, DONG X. (2024) Transient model of the three-phase transformer under DC bias based on duality principle. In: 2024 6th International Conference on Energy Systems and Electrical Power. Wuhan, pp.1457-1461. [CrossRef]

[10] ZHANG L W, WU G N, FAN J B, et al. (2013) Electromagnetic Hybrid Model for Three-phase Three-leg Transformer DC Bias Based on Jiles-Atherton Hysteresis Theory. J. East China Electric Power, 41 (04): 741-745. [CrossRef]

[11] HUANGFU C, WEI Y H, ZHONG L H, et al. (2007) A Transient Model of Three Phase Multi-legged Transformer Based on Duality Theory. J. Proceedings of the CSEE, (03): 83-88. [CrossRef]

A Demand response regulation method considering dynamic user propensities

Wenzheng Xie*, Yuhong Ye, Congfei Li, Ziqi Yu, Bin Liao

North China Electric Power University, Beijing 102206, China

*Corresponding author's e-mail: wenzheng_xie@ncepu.edu.cn

Abstract. To address challenges in the intricate governance of customer resources within power grids heavily integrated with renewable sources, which lead to high demand response costs and low user satisfaction, a demand response regulation method considering dynamic user propensities is proposed. Firstly, the user-side aggregator (UA) is defined to participate in demand response. A model of dynamic user propensities is established to evaluate user satisfaction. Secondly, the optimization problem is set as minimizing the daily operation cost of the UA's demand response. Finally, a demand response regulation optimization algorithm, utilizing enhanced reinforcement learning (RL), is introduced to address the optimization challenge. A soft update is performed by adjusting the network learning rate based on user satisfaction, allowing for greater flexibility. The study's simulations indicate that the suggested method significantly cuts the daily operational expenses of demand response while satisfying users' evolving economic and comfort needs.

1. Introduction

Currently, a vast amount of renewable energy is accelerating its integration into the grid[1]. For power systems heavily reliant on renewables, immediate demand response measures are crucial to stabilize supply-demand equilibrium and enhance energy efficiency[2]. How to efficiently implement demand response strategies while ensuring user satisfaction has escalated into an imperative challenge that demands resolution[3]. Reference [4] takes users' comprehensive satisfaction into account, enhances the flexibility of users' participation in demand response, and reduces operating costs. Reference [5] focuses on users' social and psychological behaviors and uses a preference - aware method to simulate and analyze demand response programs. However, traditional user satisfaction models only consider the optimal response strategy of each user within the current time period, ignoring users' dynamic characteristics.

Meanwhile, to address the trade-off problem between choosing the decision with the maximum average reward based on historical experience in demand response and the decision that may bring greater rewards in the future[6]. Research on demand response regulation optimization based on reinforcement learning (RL) has become increasingly extensive[7]. Reference [8] models the dynamic pricing for retail issue as a markov decision process (MDP), and it employs a reinforcement learning approach that doesn't rely on a specific model to determine the best economic incentives within the electricity market, drawing insights from load patterns. Reference [9] effectively integrates sequential updating and deep reinforcement learning, updating the demand response strategies of each building in sequence, achieving consistency between the updated results of each building and the overall energy optimization goal, and strengthening energy collaboration between buildings. However, the above algorithms are all single-step update algorithms, which are difficult to adapt to users' differentiated economic and comfort requirements and have poor convergence.

In response to the noted problems, the study suggests a demand response regulation method considering dynamic user propensities. First, a model of dynamic user propensities and a model of demand response operating costs are established, and an optimization problem of minimizing the daily operating cost is constructed. Second, a demand response regulation optimization algorithm based on user experience is introduced to address the issue. Finally, the proposed algorithm is verified through simulation, which can meet users' electricity consumption experience and effectively reduce the daily operating cost of demand response.

2. System model

The distribution network dispatching center issues demand response plans to the user-side aggregator (UA) according to the power deficit in the power system. The UA can aggregate controllable resources that are diverse in types, scattered in distribution, and have weak communication connections, so as to provide more efficient response services. The user-side resources include distributed generation (DG) composed of wind turbine and photovoltaic power, controllable loads (CL) with the set $\mathcal{L} = \{1,\cdots,l,\cdots,L\}$, and energy storage (ES).

2.1. Dynamic User Propensities Model

Dynamic user propensities directly affect the regulation strategies and operating costs of UA. According to consumer psychology and the endowment effect, user satisfaction $U_l(t)$ can be divided into two aspects: economic propensity $U_l^{\text{econ}}(t)$ and comfort propensity $U_l^{\text{comf}}(t)$, which are specific expressions as follows

$$\begin{aligned}&U_l(t) = \mu_l^1(t)U_l^{\text{econ}}(t) + \mu_l^2(t)U_l^{\text{comf}}(t)\\&\begin{cases}U_l^{\text{econ}}(t) = p_{\text{sell}}(t) - p_{\text{sell}}^{\min} / p_{\text{sell}}^{\max} - p_{\text{sell}}^{\min}\\U_l^{\text{comf}}(t) = \tilde{P}_l^{\text{CL,max}}(t) - \sum_{n=1}^{t-1} b_l^{\text{CL}}(n)P_l^{\text{CL}}(n) / \tilde{P}_l^{\text{CL,max}}(t)\end{cases}\end{aligned} \tag{1}$$

Where $\mu_l^1(t)$ and $\mu_l^2(t)$ are the economic and comfort propensity coefficient respectively, with $0 < \mu_l^1(t),\ \mu_l^2(t) < 1$ and $\mu_l^1(t) + \mu_l^2(t) = 1$. $U_{i,l}^{\text{econ}}(t)$ depends on the economic incentive level of the distribution network. $p_{\text{sell}}(t)$ represents the price at which the UA supplies electricity to the distribution network. $p_{\text{sell}}^{\max}$ and $p_{\text{sell}}^{\min}$ are the maximum incentive price and the minimum acceptable incentive price for users respectively, which come from the time-of-use electricity price specified by the distribution network. $U_{i,l}^{\text{comf}}(t)$ depends on the regulated electricity consumption of users. Considering the influence of $t-1$ time periods' historical load response results on the demand response strategy of t time period. Among them. $b_{i,l}^{\text{CL}}(n)$ and $P_{i,l}^{\text{CL}}(n)$ represent the response influence weight and the actual reduced power of the controllable load l, and $\tilde{P}_{i,l}^{\text{CL,max}}(t)$ represents the expected maximum reducible power.

2.2. Demand Response Operating Cost Model

$C^{\text{oper}}(t)$ represents the operating cost of the UA's demand response, as shown below

$$\begin{aligned}&C^{\text{oper}}(t) = C^{\text{CL}}(t) + C^{\text{EX}}(t)\\&\begin{cases}C^{\text{CL}}(t) = \sum_{l=1}^{L}\left[k'_{\text{CL}}/2U_l^2(t) + k''_{\text{CL}}/U_l(t)\right]\\C^{\text{EX}}(t) = p_{\text{buy}}(t)\left(P^{\text{ch}}(t) + \sum_{l=1}^{L} P_l^{\text{CL}}(t) + P^{\text{buy}}(t)\right) - p_{\text{sell}}(t)\left(P^{\text{DG}}(t) + P^{\text{dch}}(t) + P^{\text{sell}}(t)\right)\end{cases}\end{aligned} \tag{2}$$

Where $C^{\text{CL}}(t)$ is the cost of users' electricity consumption propensity loss, with cost coefficients k'_{CL} and k''_{CL}. $C^{\text{EX}}(t)$ is the demand response subsidy cost. $p_{\text{buy}}(t)$ denotes the price at which the UA, the electricity procurement entity, obtains power. $P^{\text{DG}}(t)$ is output power of the distributed generation.

$P^{\text{ch}}(t)$ and $P^{\text{dch}}(t)$ denote the energy storage's respective charging and discharging capacities. $P^{\text{buy}}(t)$ and $P^{\text{sell}}(t)$ are powers of the UA's electricity purchase from and sell to the distribution network.

2.3. Problem Formulation

In this paper, the optimization objective is formulated as minimizing the daily operating cost when the UA participates in demand response, as shown below

$$\begin{aligned} &\min_{\mathbf{y}} \; C(t)=\sum_{t=1}^{T} C^{\text{oper}}(t) \\ \text{s.t. } &C_1: P^{\text{DG,min}}(t) \le P^{\text{DG}}(t) \le P^{\text{DG,max}}(t) \\ &C_2: P_l^{\text{CL,min}}(t) \le P_l^{\text{CL}}(t) \le P_l^{\text{CL,max}}(t) \\ &C_3: E^{\min}(t) \le E\ (t) \le E^{\max}(t) \\ &C_4: E(t) = E(t-1) + \eta^{\text{ch}} P^{\text{ch}}(t) - \eta^{\text{dch}} P^{\text{dch}}(t) \\ &C_5: P^{\text{ch}}(t) \cdot P^{\text{dch}}(t) = 0 \end{aligned} \tag{3}$$

Where T represents the total regulation period of 24 hours; C_1 and C_2 denote the output range constraints of distributed generation and controllable loads respectively. $P_i^{\text{DG,min}}(t)$, $P_i^{\text{DG,max}}(t)$ and $P_{i,l}^{\text{CL,min}}(t)$, $P_{i,l}^{\text{CL,max}}(t)$ are their lower and upper output limits respectively. C_3 and C_4 are crucial constraints for energy storage device charging and discharging, essential for their operation and efficiency. $E_i(t)$ represents the energy stored within a storage system, and $E_i^{\min}(t)$, $E_i^{\max}(t)$ are its lower and upper limits. C_5 represents the constraint that energy storage cannot charge and discharge simultaneously.

3. Optimization algorithm for demand response regulation based on improved RL

This paper proposes an optimization algorithm for demand response regulation based on improved RL, the UA is defined as the decision maker for the regulation of user-side resources. A single agent is constructed for it. In the environment of interaction with the distribution network dispatching center, a MDP model is built, as follows

$$\begin{cases} \mathbf{S}(t) = \left\{p_{\text{buy}}(t), p_{\text{sell}}(t), E(t-1)\right\} \\ \mathbf{A}(t) = \left\{P^{DG}(t), P^{\text{ch}}(t), P^{\text{dch}}(t), P_l^{\text{CL}}(t)\right\} \\ \theta(t) = -C(t) - \psi\left(1 - \mathbb{I}(t)\right) \end{cases} \tag{4}$$

Where $\mathbf{S}(t)$, $\mathbf{A}(t)$, and $\theta(t)$ are the state space (all possible states), action space (available actions), and reward function (evaluating state-action pairs) respectively; ψ is a large real number; $\mathbb{I}(t)$ is an event function. If all the constraints in Equation (3) are satisfied, then $\mathbb{I}(t) = 1$; otherwise, $\mathbb{I}(t) = 0$.

The proposed algorithm consists of three steps: action selection, experience storage, and network update.

1) Action selection: The agent within the system randomly selects actions from a predefined set based on probability $\varepsilon(t)$ and executes it, and executes the action with the maximum Q-value with a probability of $1-\varepsilon(t)$.

2) Experience storage: The user-side resources execute $\mathbf{A}(t)$. the agent computes $\theta(t)$, updates based on analysis, and transitions seamlessly into the subsequent state $\mathbf{S}(t+1)$. Meanwhile, the action chain is recorded within the experience replay storage buffer.

3) Network update: The agent randomly selects M groups of action chains to calculate the mean squared error loss function $\Psi(t)$ and updates the evaluation network as follows

$$\omega_{\mathrm{e}}(t+1)=\omega_{\mathrm{e}}(t)-\lambda\exp\left(1-\sum\nolimits_{l=1}^{L}U_l(t)\big/L\right)\nabla_{\omega_{\mathrm{e}}(t)}\Psi(t) \tag{5}$$

Where λ is the initial network update rate, which is influenced by dynamic user propensities. The soft update method is applied to it. When user satisfaction is low, the network update rate is increased to accelerate the learning of the evaluation network, adjust the speed of searching for the optimal solution, and enhance the training stability.

4. Simulation

This paper analyzed the economic viability and efficacy of the proposed algorithm using the dataset from a southern Chinese smart park as a case study. According to the historical data, Monte Carlo simulation was employed for sampling to generate 400 sample scenarios of wind turbine and photovoltaic panel energy generation. Subsequently, utilizing K-means clustering, the data was condensed into four representative wind and solar power generation scenarios, depicted in Figure 1.

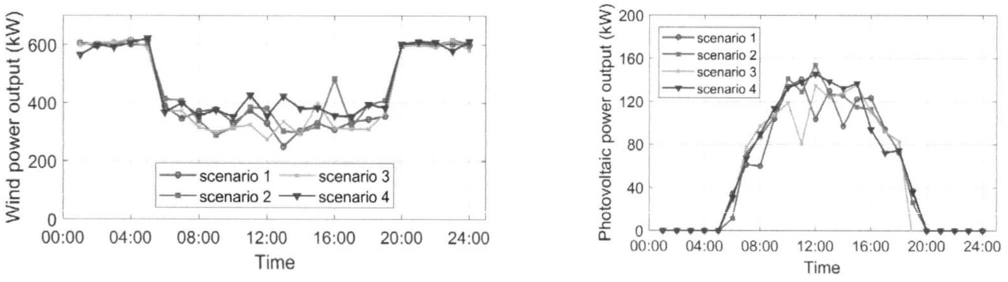

(a) Wind turbine energy generation scenarios (b) Photovoltaic energy generation scenarios

Figure 1. Common situations involving wind and solar energy generation scenarios.

Under same simulation conditions, a comparative analysis of the above four scenarios was carried out: in scenario 1, no optimization was performed; in scenario 2, benchmark algorithm 1 (a genetic based optimal scheduling algorithm[10]) was applied, considering dynamic user propensities. in scenario 3, benchmark algorithm 2 (a traditional RL-based energy planning algorithm[8]) was applied, without considering such propensities; in scenario 4, the proposed algorithm was used for demand response optimization with these propensities. Refer to Table 1 for the time-of-use electricity price detailing cost structures across peak, off-peak, and shoulder hours, and Table 2 offered simulation parameters like energy consumption rates and equipment efficiencies.

Table 1. The time-of-use electricity price.

Time	Price(¥/kWh)
08:00~14:00，18:00~22:00	1.01186
22:00~06:00	0.61053
06:00~08:00，14:00~18:00	0.29350

Table 2. Simulation parameters of UA participating in demand response.

User-side resource parameters	Range size	Algorithm parameters	Value
$P^{\mathrm{DG,min}}(t), P^{\mathrm{DG,max}}(t)$	0, 700 (kW)	λ	0.5
$P_l^{\mathrm{CL,min}}(t), P_l^{\mathrm{CL,max}}(t)$	5, 30 (kW)	k'_{CL}, k''_{CL}	0.005, 0.1
$E^{\min}(t), E^{\max}(t)$	50, 450 (kW)	$\eta^{\mathrm{ch}}, \eta^{\mathrm{dch}}$	0.92, 0.95
L	200	M	50

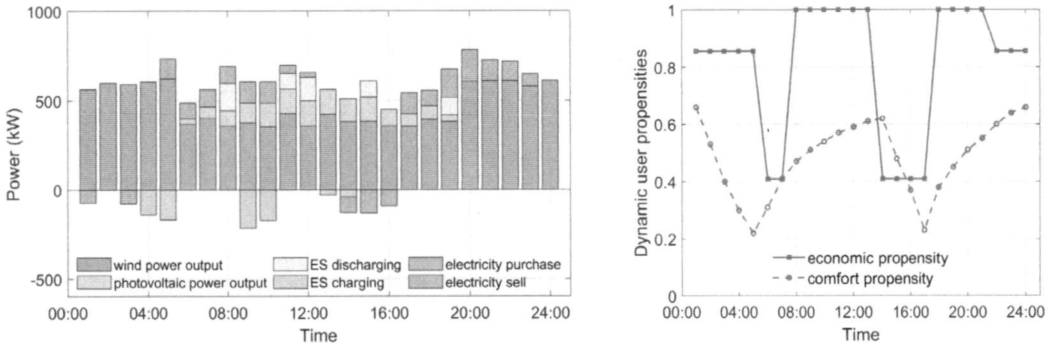

(a) Results of user-side resources (b) Results of dynamic user propensities

Figure 2. The regulation results of UA's demand response in scenario 4.

Figure 2 shows the regulation results of UA's demand response in scenario 4, including results of user-side resources and dynamic user propensities. Observations indicate a lack of solar energy production between (20:00 to 05:00). During this period, the comfort propensity is low, and the electricity purchase price for the UA is also low. When the wind power output is insufficient (20:00 to 23:00), the UA can purchase electricity from the distribution network. If the wind power output is sufficient (01:00 to 05:00), the UA can sell electricity to the distribution network or charge the energy storage to meet the electricity demand during peak hours. From (06:00 to 20:00), UA implements peak load reduction and trough level augmentation using time-based electricity rates and fluctuating consumer inclinations, aiming to improve user satisfaction and reduce costs of users' electricity consumption propensity loss. The energy storage discharges during high-price periods to reduce the UA's electricity purchase from the distribution network and lower the demand response subsidy cost for the interaction between the UA and the distribution network dispatching center.

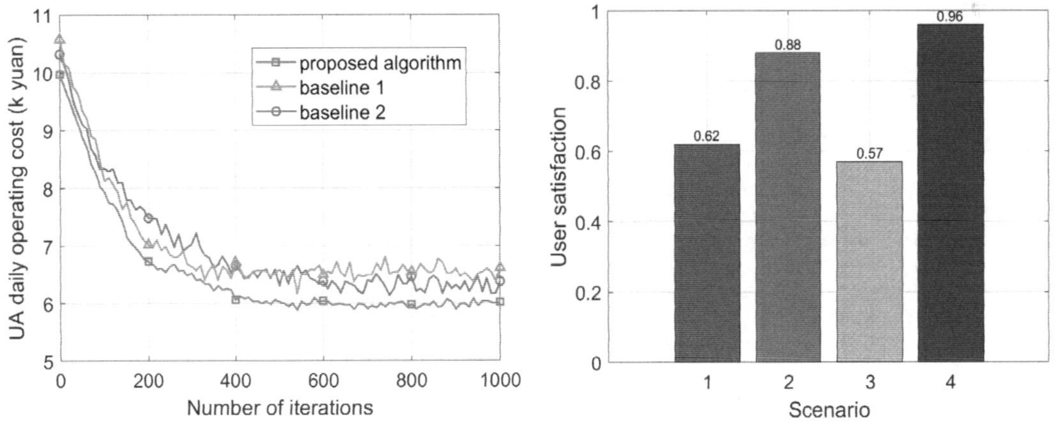

(a) Comparison of UA daily operating cost (b) Comparison of user satisfaction

Figure 3. The comparison of UA demand response optimization.

Figure 3 shows the comparison of demand response optimization, including comparison of UA daily operating cost and user satisfaction. Analysis demonstrates that benchmark algorithm 1 uses more wind and solar output instead of sacrificing load-side comfort propensity, greatly improving user satisfaction. And with low complexity, it converges first but has the highest daily cost. Benchmark algorithm 2 reduces the wind-solar grid connected power, effectively cutting the daily cost, which proves the feasibility of the RL algorithm. Yet, it converges slowest due to a long learning cycle and worsens user satisfaction. Compared with these, our proposed algorithm better satisfies user preferences. The user satisfaction has been increased by 0.08 and 0.39. We used wind-solar output to cut daily costs by 9.10% and 5.66% respectively. It also adjusts the network update rate by user satisfaction to speed up learning and boost training stability.

5. Conclusion

To tackle the complication of demand response regulation on the user-side in a power grid heavily integrated with renewable sources, a demand response regulation method considering dynamic user propensities is proposed. Simulations show that the proposed method outperforms benchmark algorithms in accuracy. Additionally, with its advanced design, proposed algorithm can significantly cut daily operational costs by up to 9.10% and 5.66% respectively, and meet users' dynamic economic and comfort requirements. The user satisfaction has been increased by 0.08 and 0.39. In future research, the costs associated with demand response due to renewable energy unpredictability will be additionally examined.

Acknowledgments

This work was financially supported by the Science and Technology Project of China Southern Power Grid Co., LTD. (150000202303010З3JL00320).

References

[1] Su Y., Liu F., Wang Z., Zhang Y., Li B. and Chen Y. (2023). Multi-Stage Robust Dispatch Considering Demand Response Under Decision-Dependent Uncertainty. IEEE Transactions on Smart Grid, 14(4): 2786-2797.
[2] Sangoleye F., Tsiropoulou E. E. and Papavassiliou S. (2024). Dynamic Risk Management for Demand Response in Multi-Utility Smart Grids. IEEE Transactions on Industrial Cyber-Physical Systems, 2: 98-107.
[3] Du Y., Li J., Yang Z. and Sun P. (2024). Virtual Power Plant Scheduling Model Considering the Power Consumption Satisfaction of Users with Different Flexible Loads. Journal of Physics: Conference Series, 2774(1): 012096.
[4] Liu Y., Xiao L., Yao G. and Bu S. (2019). Pricing-Based Demand Response for a Smart Home with Various Types of Household Appliances Considering Customer Satisfaction. IEEE access, 7: 86463-86472.
[5] De V. C., Patti E., Macii E. and Bottaccioli L. (2022). A User-Centric View of a Demand Side Management Program: From Surveys to Simulation and Analysis. IEEE Systems Journal, 16(2): 1885-1896.
[6] Bahrami S., Chen,Y. C. and Wong V. W. (2021). Deep Reinforcement Learning for Demand Response in Distribution Networks. IEEE Transactions on Smart Grid, 12(2): 1496-1506.
[7] Sangoleye F., Jao J., Faris K., Tsiropoulou E. E. and Papavassiliou, S. (2023). Reinforcement Learning-Based Demand Response Management in Smart Grid Systems with Prosumers. IEEE Systems Journal, 17(2): 1797-1807.
[8] Wan Y., Qin J., Yu X., Yang T. and Kang Y. (2022). Price-based residential demand response management in smart grids: A reinforcement learning-based approach. IEEE/CAA Journal of Automatica Sinica, 9(1): 123-134.
[9] Han Y., Wu J., Chen H., Si F., Cao Z. and Zhao Q. (2024). Enhancing grid-interactive buildings demand response: Sequential update-based multiagent deep reinforcement learning approach. IEEE Internet of Things Journal, 11(14): 24439-24451.
[10] Zhang F., He S., Lin X., Zhao Y., Tang J. and Li Y. (2024). A Combined Dispatch Model of Power Load Satisfaction and Comfort Based on Genetic Algorithm. In 2024 9th International Conference on Intelligent Computing and Signal Processing (ICSP), Xian, China, 2024, pp. 1703-1708.

Design and development of a tool-less wiring phoenix terminal relay protection test adapter

Yihe Wang*, Tong Wang, Wuyang Zhang, Meng Wu, Songjie Shi

State Grid Liaoning Electric Power Company Limited Electric Power Research Institute, Liaoning Shenyang 110006, China

*Corresponding author: 283036420@qq.com

Abstract. In relay protection testing work, the disassembly and wiring steps of the terminal block signal circuit of the protection device are complex and risky, involving the disassembly and assembly of screws and signal wires, as well as the fixing screws after the pin connection, etc. This not only increases the difficulty of operation but also can easily lead to miswiring, accidental contact of signal lines, and other safety incidents, which may cause the protection devices to malfunction. To improve testing efficiency and ensure the safety of the testing process, this paper designs and develops a Phoenix terminal relay protection test adapter that does not require disassembly and wiring. The adapter aims to simplify the wiring process, reduce the difficulty of operation, effectively avoid safety risks caused by misoperation, and provide a more convenient and safe solution for relay protection testing work.

1. Introduction

With the development of power systems, relay protection devices play a crucial role in ensuring the safe and stable operation of the power grid[1]. To guarantee the reliability of the performance of these relay protection devices, it is indispensable to conduct regular relay protection tests. However, in traditional relay protection testing, the process of disconnecting and reconnecting wires for the signal circuits on the terminal blocks of protection devices is cumbersome and complex[2-4], which not only consumes a significant amount of time but also poses a high risk of safety accidents. Consequently, how to streamline the wiring process, enhance testing efficiency, and ensure testing safety has become an urgent issue that needs to be addressed in current relay protection testing work[5].

In response to the aforementioned issues, this paper proposes a design scheme for a Phoenix terminal relay protection test adapter that eliminates the need for disconnecting and reconnecting wires. This adapter adopts a unique structural design that allows it to be directly connected to the terminal blocks of protection devices without the need to disconnect or reconnect screws and signal wires, thereby greatly simplifying the wiring process. Additionally, the adapter exhibits excellent insulation performance and stability, ensuring safety and reliability during the testing process.As shown in Figure 1.

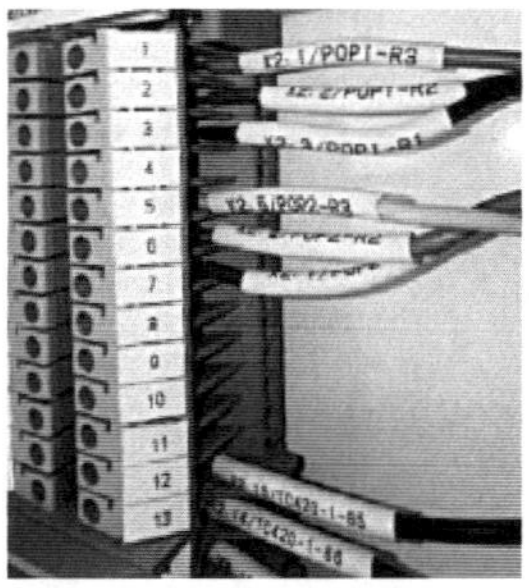

Figure 1. Traditional Terminal Block Diagram.

2. Issues Existing in Current Technologies

In the realm of contemporary relay protection testing, the process of assembling and disassembling wiring for the signal circuits of terminal blocks in protection devices presents several significant challenges:

The wiring procedure is cumbersome: Traditional wiring methods necessitate the removal and reconnection of screws and signal wires, involving the insertion of pins followed by the tightening of screws, and subsequently connecting the test wires to adapters[6-9]. This process comprises numerous steps, is operationally complex, and consumes considerable time.

High safety risks: During the assembly and disassembly of wiring, there is a propensity for incidents such as inadvertent disconnection of wires or accidental contact with signal lines, posing safety hazards. These incidents not only have the potential to trigger malfunctions in protection devices but also pose a threat to the stable and secure operation of the power grid[10].

Limited applicability of adapters: The existing array of adapters is diverse yet often tailored to specific types of protection devices or testing equipment. Consequently, in practical applications, frequent changes between different adapter models are required, escalating operational difficulty and costs.

In light of the aforementioned issues, the design of a Phoenix terminal relay protection test adapter that eliminates the need for disassembly and reconnection of wires assumes paramount importance. This adapter should exhibit characteristics of versatility, streamlined wiring, safety, and reliability, thereby catering to the demands of various protection devices and testing equipment.

3. Phoenix Terminal Relay Protection Test Adapter

3.1 Structural Design of the Adapter

The structural design of the Phoenix terminal relay protection test adapter for tool-free wiring connection is illustrated in Figure 1. The adapter primarily comprises the following components: a Phoenix terminal interface designed for direct connection to the terminal block of the protection device, adopting a standard plug-in design to ensure a secure connection with the terminal block; a test lead interface intended for connecting test leads, featuring a universal banana plug or alligator clip design to facilitate connection with various types of test leads; an insulating shell serving to protect the internal electrical connections of the adapter, constructed from high-strength, high-insulation materials to ensure the safety and reliability of the adapter during testing; and a fixing device used to secure the position of the adapter on the terminal block of the protection device, employing a spring latch or magnetic attraction design to enable convenient and rapid attachment of the adapter to the terminal block.As shown in Figure 2.

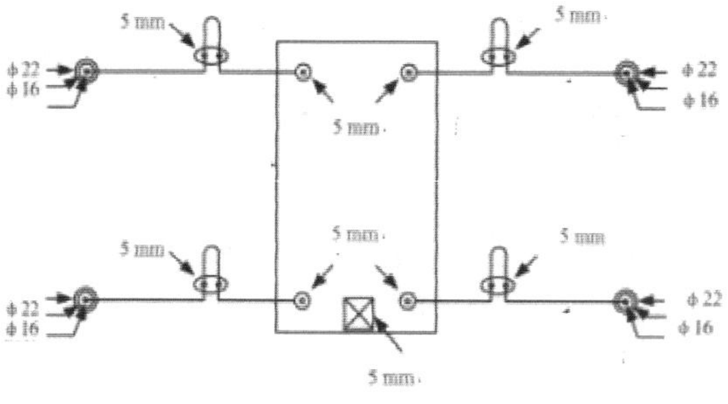

Figure 2 presents the structural design diagram of the Phoenix terminal relay protection test adapter featuring a tool-less wiring system.

3.2 Working Principle of the Adapter

The working principle of the Phoenix terminal relay protection test adapter for non-disassembly wiring is as follows: The adapter's Phoenix terminal interface is directly inserted into the corresponding socket of the terminal block of the protection device. This ensures a secure connection with the terminal block of the protection device, owing to the adoption of a standard plug-in design by the Phoenix terminal interface.

Subsequently, the test lead plug is inserted into the test lead interface of the adapter. The test lead interface employs a universal banana plug or alligator clip design, enabling it to accommodate various types of test leads.

Furthermore, the adapter is secured in place on the terminal block of the protection device using the adapter's fixing device. The fixing device utilizes a spring latch or magnetic attraction design, facilitating quick and easy fixation of the adapter in the designated position.

Once the adapter is connected to the terminal block of the protection device and the test leads are properly attached, the relay protection test can commence. The adapter's design eliminates the need for disassembling screws and signal wires, significantly simplifying the wiring process. Additionally, the adapter exhibits excellent insulation performance and stability, guaranteeing safety and reliability.

4. Advantages Of The Phoenix Terminal Relay Protection Test Adapter With Tool-Less Wiring

The Phoenix terminal relay protection test adapter featuring tool-less wiring boasts several advantages as follows:

It streamlines the wiring process by adopting a design that eliminates the need for screwing and unscrewing wires and signal lines, significantly simplifying the wiring procedure and enhancing test efficiency.

It enhances safety, endowed with excellent insulation performance and stability, ensuring safe and reliable operation during the testing process. Furthermore, its anti-misinsertion design prevents safety hazards arising from erroneous operations.

It improves versatility, equipped with various interface types and specifications to cater to the requirements of different protection devices and test equipment. Additionally, consideration has been given to the compatibility of the adapter with terminal blocks of different models of protection devices, thereby broadening its application scope.

It reduces costs, thanks to its standardized design that facilitates mass production, lowering manufacturing expenses. Moreover, its exceptional versatility and applicability minimize costs associated with frequent replacements of different adapter models.

It facilitates maintenance, featuring a simple structure that is easy to disassemble and clean, simplifying maintenance and management. This contributes to extending the adapter's service life and reducing maintenance costs..As shown in Figure 3and 4.

Figure 3.Portable Terminal Block Design Drawing.

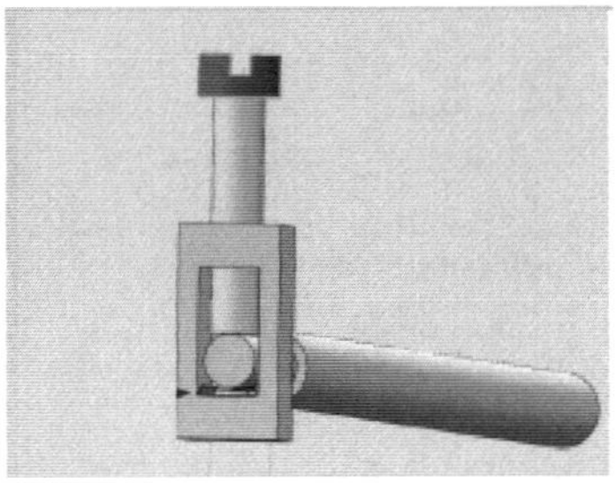

Figure 4. Modified Component Design Drawing.

5. Practical Application Case Analysis

In actual power systems, relay protection testing is a crucial aspect of ensuring the safe and stable operation of the power grid. The traditional process of dismantling and reassembling signal circuits on terminal blocks of protection devices is cumbersome, time-consuming, labor-intensive, and prone to safety risks. To verify the practical application effectiveness of the Phoenix terminal relay protection test adapter with tool-less wiring, we have selected two typical relay protection test scenarios for in-depth analysis, and through detailed data recording, we demonstrate the advantages of the adapter.

5.1 Case Study One: Relay Protection Test at a 500kV Substation

5.1.1 Background and Objective of the Test. As a crucial node in the regional power grid, the reliability of the relay protection devices at a 500kV substation is directly linked to the safe and stable operation of the power grid. To verify the practical application effect of the newly developed tool-less wiring Phoenix terminal relay protection test adapter, we have selected an important relay protection test conducted at this substation as a case study for research purposes.

5.1.2 Test Data and Comparative Analysis. The following is a detailed data record and analysis, with all data based on measurements and records taken during the actual test process:

The first set of data: Comparison of wiring time and efficiency.As shown in Table 1and 2.

Table 1: Compare traditional terminal blocks with new terminal blocks.

Item	Traditional Wiring Method	Tool-less Wiring Adapter
Total Wiring Time (seconds)	210	60
Average Time per Wiring Step (seconds)	17.5 (12 steps)	20 (3 steps, including preparation)
Number of Wiring Personnel	2	1
Number of Wiring Errors	5	0
Number of Repeated Wirings (due to errors)	3	0
Number of Wiring Steps	12	3
Number of Tool Uses	20 (screwdrivers, etc.)	5 (adapter operations)
Preparation Time (seconds)	30	10
Percentage Reduction in Wiring Process Time	-	71.40%
Percentage Savings in Labor Cost	-	50%
Percentage Savings in Tool Use Cost	-	75%

Table 2: Compare traditional terminal blocks with new terminal blocks

	Traditional Wiring Method	Tool-less Wiring Adapter
Insulation Resistance after Wiring (MΩ)	≥10	≥100
Withstand Voltage Test after Wiring (kV)	2.5	5
Safety Protection Level	IP20	IP65
Safety Training Time (hours/person)	4 (for two persons)	1 (for one person)
Percentage Reduction in Safety Accident Rate	-	100%
Reduction in Electric Shock Risk Rating	-	75% (from 8 to 2)
Percentage Increase in Insulation Performance	-	900%

Percentage Increase in Withstand Voltage Performance	-	100%
Complexity Rating of Safety Operating Procedures	7 (complex)	4 (simple)

5.3 Conclusion

Through detailed data recording and comparative analysis, the Phoenix terminal relay protection test adapter with tool-less wiring has demonstrated significant advantages in simplifying the wiring process, enhancing wiring efficiency, ensuring safety performance, and improving reliability and applicability. Specifically, the adapter has substantially reduced wiring time, increased wiring efficiency, and decreased labor and tool usage costs; simultaneously, its excellent safety performance has effectively mitigated the risk of accidents caused by misoperations, enhancing safety during the wiring process. Furthermore, the adapter exhibits outstanding reliability and applicability, accommodating a wider range of environmental conditions, extending service life, reducing maintenance costs, and improving installation convenience and compatibility. Therefore, the Phoenix terminal relay protection test adapter with tool-less wiring possesses broad application prospects and promotional value.

6. Conclusion and Outlook

This paper analyzes and investigates the issues existing in current relay protection testing, designing and developing a Phoenix terminal relay protection test adapter with tool-less wiring. The adapter adopts a unique structural design that allows direct connection to the terminal block of the protection device without the need for screwing and unscrewing wires, thereby significantly simplifying the wiring process. Moreover, the adapter possesses good insulation performance and stability, ensuring safe and reliable operation during the testing process. Additionally, it offers various interface types and specifications to accommodate the requirements of different protection devices and test equipment.

To meet the future demands of relay protection testing, we will continue to delve into and optimize the design scheme and technical performance of the adapter. At the same time, we will actively explore new materials and manufacturing processes to further enhance the quality and reliability of the adapter. We believe that shortly, the Phoenix terminal relay protection test adapter with tool-less wiring will play an even more significant role in the field of relay protection testing.

Acknowledgments

This work was supported by the State Grid Corporation's Group Innovation Project (a non-demolition Phoenix terminal relay protection test adapter).

References

[1]Fang Ming, He Yuerong, Gong Simin, et al. Analysis of Transformer Differential Protection Operation Accident Caused by Poor Contact of Terminal Block[J]. Power and Energy,2022,43(06):548-552.
[2]Li Zuyu, Lin Daozhi. Improvement and Development of a Compact Low-Voltage Terminal Block [J]. China's New Technologies and Products.,2022,(22):69-71.DOI:10.13612/j.cnki.cntp.2022.22.037.
[3]Tiia I ,Antti E ,Heini H , et al.Interatrial block and P terminal force in the general population - Longitudinal changes, risk factors, and prognosis.[J]. Journal of electrocardiology,2022,7312-20.
[4]ChaoBin L ,KeKe G ,WenXiao G , et al.Rationally Designing Metal-Organic Frameworks Based on [Ln2] Magnetic Building Blocks Utilizing 2-Hydroxyisophthalate and Fine-Tuning the Magnetic Properties of Dy Analogues by Terminal Coordinated Solvents.[J].Inorganic chemistry,2020,59(23):
[5]Shi Chunmin, Ding Zhijian. Research and Design of Terminal Blocks for Secondary Circuits of Relay Protection to Prevent Maloperation[J]. Rural Electrician, 2020, 28(10): 45-46. DOI:10.16642/j.cnki.ncdg.2020.10.053.
[6]Chen Weiyang. Design of a Terminal Block Disassembly and Assembly Clamp [J]. Mechanical and Electrical Information.,2020,(06):21+24.DOI:10.19514/j.cnki.cn32-1628/tm.2020.06.012.
[7]Dinkle Electric Machinery (China) Co. Ltd.; Patent Issued for Light-Emitting Terminal Block Structure (USPTO 10,490,916)[J].Electronics Newsweekly,2019.

[8]The Carlyle Group and Cag Holdings; The New Terminal One at JFK Continues Progress with Approval of Proposed Lease by The Port Authority of New York and New Jersey's Board[J]. Journal of Engineering,2019.
[9]TE Connectivity India Private Limited; Patent Issued for Terminal Block (USPTO 10,468,791)[J].Electronics Newsweekly,2019.
[10]Yazaki Corporation; "Busbar Component, Terminal Block, And Vehicular Device" in Patent Application Approval Process (USPTO 20190305471)[J].Journal of Transportation,2019.

Research on rectification strategy optimization for linear internal combustion power generation systems based on active short circuit control

Liutao Sun, Boru Jia, Yidi Wei*

School of Mechanical Engineering, Beijing Institute of Technology, Beijing China

* Corresponding author's e-mail: weiyd01@163.com

Abstract. This study investigates the characteristics of significant amplitude-frequency variations and periodic phase sequence changes in the output electrical energy of linear internal combustion power generation systems. It conducts an analysis of the power generation characteristics of linear motors during stable operation and investigates rectification control strategies for three-phase electrical energy. Traditional control strategies involve the combination of three-phase uncontrolled rectification and chopper circuits. However, the three-phase uncontrolled rectifier bridge can lead to severe current discontinuity in the phase current of the linear motor, affecting the stability of the power generation system. This study proposes a Boost-type controllable rectifier based on the active short-circuit of the upper bridge arm of ASC, employing a composite control strategy of PWM-PFM. Simulation results indicate that the output bus voltage under ASC control can be stably controlled, with a higher voltage peak compared to uncontrolled rectification, and without causing a current discontinuity in the motor-side phase current. This approach is capable of adapting to loads of varying voltage levels and offers superior control performance.

1. Introduction

The structure of the linear internal combustion power generation system [1] is depicted in Figure 1. This innovative and efficient energy conversion system is formed by coupling a free-piston engine with a linear motor [2]. The free-piston engine serves as the prime mover, while the linear motor functions as the energy conversion device. Through the reciprocating motion of the piston and motor components, the magnetic flux lines are repeatedly cut, generating an induced electromotive force. This electrical energy is then stably transferred to the energy storage element via a rectifier circuit and a DC-DC converter [3], establishing a conversion mechanism from thermal energy to mechanical energy and finally to electrical energy. However, the unpredictable motion of the permanent magnets causes the electrical output characteristics of the linear motor [4] to differ from those of traditional rotary motors. The output three-phase electrical energy waveform is variable in both frequency and amplitude, which significantly increases the complexity of electrical energy control.

To address the issues of large voltage fluctuations and variable frequency and amplitude in the output voltage of the generator during the power generation phase [5], a two-stage electrical energy conversion scheme involving rectification [6] and voltage reduction is proposed. Additionally, a controllable rectification method that can enhance the bus voltage is introduced during the rectification process to optimize control.

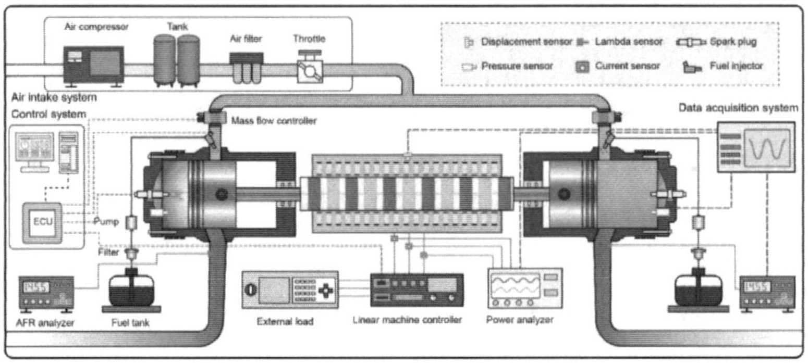

Figure 1. Schematic diagram of the linear internal combustion power generation system.

2. Optimization of electrical energy processing strategies based on uncontrolled rectification

Three-phase uncontrolled rectification can convert electrical energy from alternating current (AC) to direct current (DC) [7], while the chopper circuit further modulates the electrical energy to achieve a stable and adjustable DC voltage. The advantage of this scheme is its simplicity, as it does not require complex control strategies to implement. In this study, a three-phase full-wave bridge rectifier is employed for uncontrolled rectification, as illustrated in Figure 2. The permanent magnet linear synchronous motor is connected in a delta configuration, with its three output lines paralleled with three pairs of diodes. After obtaining a smooth DC voltage through a high-voltage side filter capacitor, the energy enters the subsequent bidirectional DC-DC Buck chopper for voltage reduction, thereby converting the three-phase AC into a stable DC supply for the load battery.

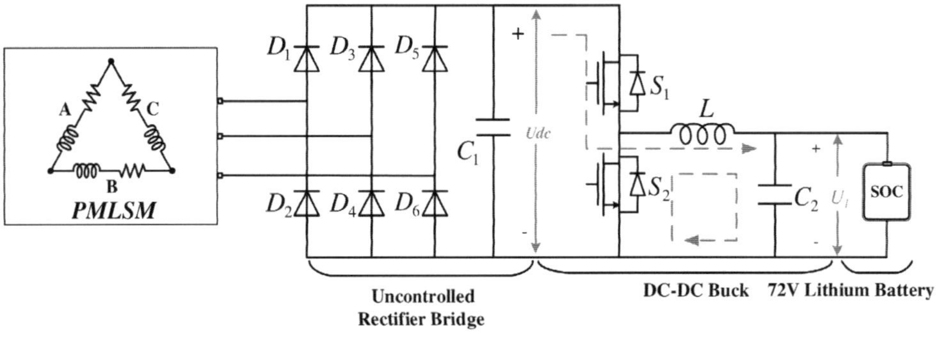

Figure 2. Main circuit topology based on three-phase uncontrolled rectification.

In an uncontrolled rectifier circuit, only one of each pair of diodes (D_1 or D_2, D_3 or D_4, D_5 or D_6) is operational at any given time. If D_1, D_4, and D_6 conduct simultaneously, forming a path, the three-phase alternating current can be rectified into a stable direct current, which varies periodically according to the three-phase line voltages U_{ab}, U_{ac}, U_{bc}, U_{ba}, U_{ca}, and U_{cb}. The operation of the uncontrolled rectifier circuit is determined by the magnitude of the three-phase voltages. When the diodes share a common cathode, current flows only through the diode with the higher anode potential; conversely, when the diodes share a common anode, current flows only through the diode with the lower cathode potential.

The larger the load on the bus side is, the greater the voltage drop across the load, resulting in a smaller voltage drop on the bus side and a larger phase voltage drop. The three-phase voltage obtained by the system through a three-phase uncontrolled rectifier bridge connected to the load and its corresponding induced electromotive force are shown in Figure 3. From the figure, it can be observed that in each cycle, each phase will coincide with the induced electromotive force six times. When entering the commutation point, the three-phase voltages will enter a phase similar to a "plateau", at which point the phase voltages of the two phases are equal. The waveform of the three-phase voltage will be lower than that of the electromotive force due to the voltage drop caused by the current flowing through the load when the circuit is formed. When the load is too heavy, the resulting voltage drop will be more pronounced.

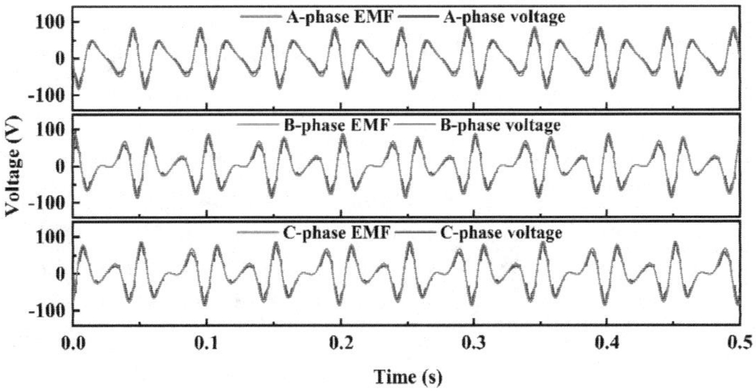

Figure 3. Waveform diagram of phase voltage and induced electromotive force.

3. Optimization of a three-phase boost-type controllable rectification algorithm based on ASC

3.1. Three-phase boost-type controllable rectification strategy based on ASC

Due to the characteristics of variable frequency and amplitude in the voltage output curve during the power generation phase of the system, using a classical three-phase full-bridge PWM rectifier for controllable rectification [8] results in significant output voltage ripple, slow system response, and increased control difficulty. Therefore, this study proposes a control strategy [9] based on the active short-circuit of the lower bridge arm, as shown in Figure 4. An inductor is connected after each phase of the three-phase output. The upper bridge arm power switch is turned off, allowing only the freewheeling diode to operate. At the same time, the duty cycle of the three lower bridge arms is controlled. When the lower three switches are on, the bus is short-circuited, and all three circuits store energy in the inductor. The energy stored in the inductor, together with the input energy, is supplied to obtain a stable and smooth bus voltage. The feature of this strategy is to boost the low voltage from the motor through a Boost rectifier and then store the energy in the load battery side through a bidirectional DC-DC Buck converter. The bus voltage can achieve a higher voltage than the peak voltage output by the motor. Under the same working conditions, the duty cycle of the three lower bridge arms is positively correlated with the bus voltage within a certain range, facilitating the control of energy supply to loads of different voltage levels.

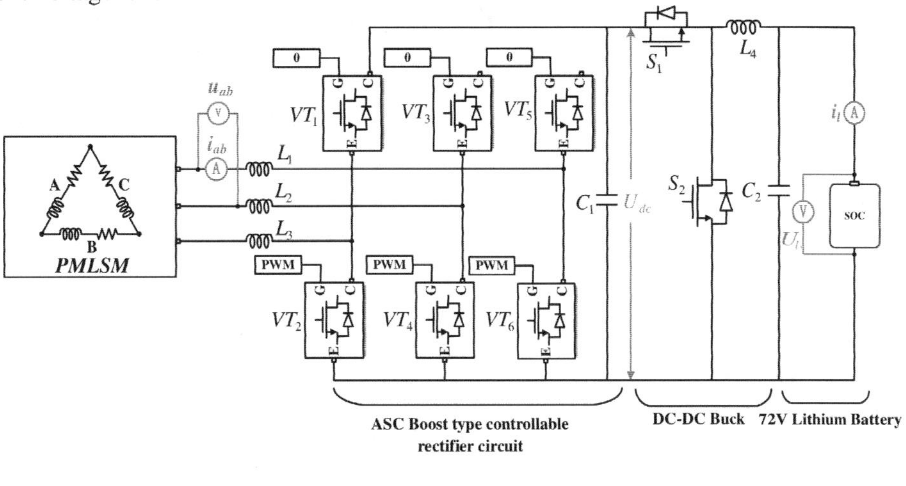

Figure 4. Main circuit topology of boost-type controllable rectification based on ASC.

If the input voltage is a three-phase AC with sine waves that are 120° out of phase, there are six possible combinations of positive and negative currents in phases A, B, and C: A(+), B(-), C(-); A(+), B(+), C(-); A(+), B(-), C(+); A(-), B(+), C(-); A(-), B(-), C(+); A(-), B(+), C(+). These combinations are referred to as six modes. Within each mode, two working states depend on whether the lower bridge arm is short-circuited or not: State I is the lower bridge arm conduction mode, and State II is the lower bridge arm short-circuit mode.

Under the same operating conditions, as the duty cycle of the lower three switches increases, the bus voltage also increases. However, when it exceeds a certain limit, the excessive internal resistance and inductance of the motor consume a large amount of energy, causing the bus voltage to tend towards a short circuit. During open-loop testing under a certain high-frequency condition, the relationship between the duty cycle and the bus voltage is shown in Figure 5: when the duty cycle reaches 80%, energy is consumed by the motor's internal resistance due to short-circuiting, causing the bus side to tend towards a short circuit. In summary, before implementing closed-loop control, it is necessary to first determine the positive growth range of the bus voltage through open-loop testing.

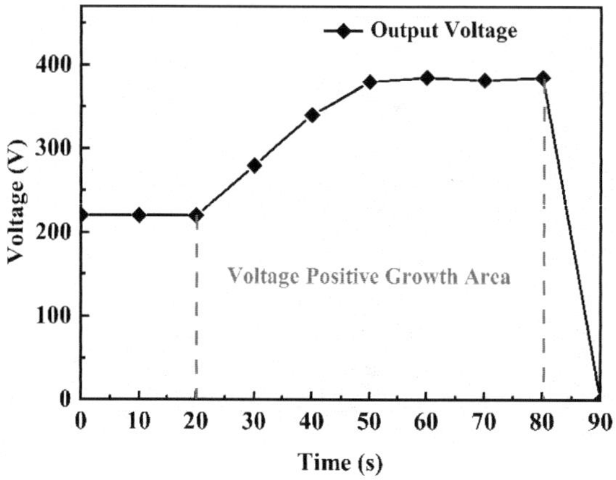

Figure 5. The relationship between duty cycle and bus voltage.

3.2. Design of PWM-PFM hybrid controller

Therefore, this study selects a PWM-PFM hybrid controller [10], combining the advantages of both pulse-width modulation (PWM) and pulse-frequency modulation (PFM) control methods. The structure of the controller is shown in Figure 6. In the PWM control, voltage closed-loop control is employed. The given bus voltage U_{dc}^* and the actual voltage U_{dc} are compared, and the voltage loop PI controller is used to calculate the duty cycle limit obtained from open-loop testing, resulting in the desired pulse-width duty cycle. In the PFM control, the carrier frequency is determined based on the actual operating conditions of the three-phase linear motor: the maximum speed of operation v_{max} is a function of stroke and frequency, $v_{max} = F(s, f)$. The higher the motor operating frequency is and the longer the stroke is, the faster the piston mover assembly runs, generating more electricity and higher three-phase voltages.

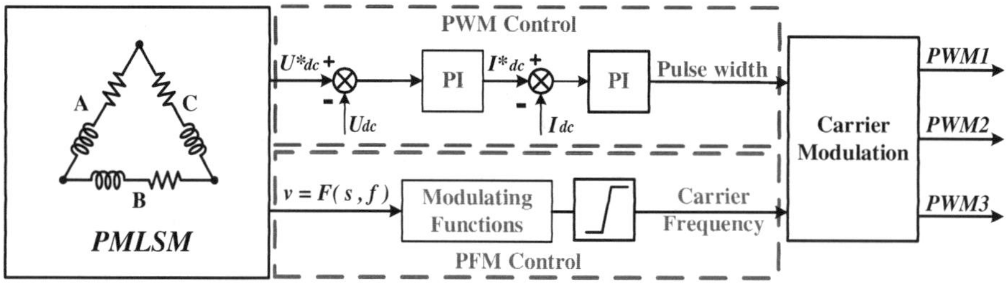

Figure 6. Block diagram of PWM-PFM hybrid controller.

To increase energy storage, a higher carrier frequency is required. Thus, a positive correlation can be established between the peak speed of motor operation and the carrier frequency, which can be used as the modulation function to obtain the carrier frequency for this operating condition. By modulating with the given pulse-width duty cycle and carrier frequency, the drive signals for the three lower bridge arm switches, PWM_1, PWM_2, and PWM_3, can be obtained.

During the power generation phase, the PWM control of the system employs a voltage closed-loop control strategy, combined with the Boost controllable rectification model of the three-phase ASC. The transfer function flowchart is shown in Figure 7, where G_u represents the transfer function of the voltage loop controller, G_m is the transfer function of the pulse-width modulation, $G_1(s)$ is the open-loop transfer function of the current loop, $G_2(s)$ is the closed-loop transfer function of the current loop, and $G_3(s)$ is the open-loop transfer function of the voltage loop.

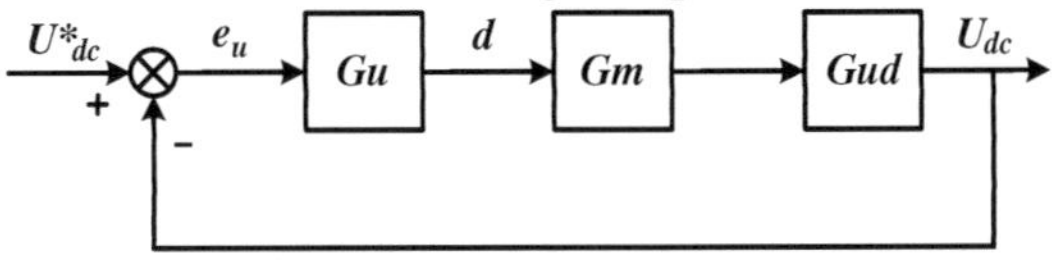

Figure 7. Voltage closed-loop control transfer function.

The voltage outer-loop transfer function can be expressed as:

$$G_u(s) = \frac{K_{up}s + K_{ui}}{s} \tag{1}$$

Thus, the open-loop voltage-loop transfer function is obtained as:

$$\begin{aligned} G_l(s) &= G_u(s) \cdot G_m(s) \cdot G_{ud}(s) \\ &= \frac{K_{up}s + K_{ui}}{s} \cdot \frac{1}{V_m} \cdot \frac{U_{dc}}{1-D} \cdot \frac{(1-D)^2 R - R_1 - Ls}{RLC_1 s^2 + (L + RLC_1)s + (1-D)^2 R + R_1} \end{aligned} \tag{2}$$

The desired crossover frequency is $f_c = 1kHz$, with a phase margin of 80°. We let $s = j \cdot 2\pi \cdot f_c$, then it can be derived as:

$$\begin{cases} \angle G_l(s) = \dfrac{-180^\circ - 80^\circ}{180/\pi} \\ |G_l(s)| = 1 \end{cases} \tag{3}$$

The parameters of the voltage outer loop can be determined from the above equation.

4. Simulation result analysis

4.1. Analysis of uncontrolled rectification simulation results

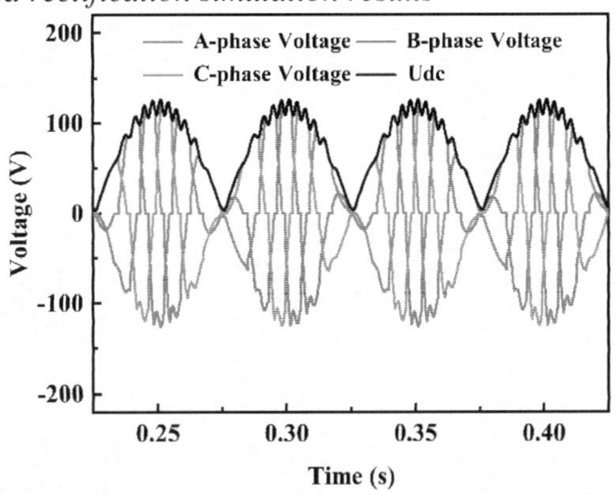

Figure 8. Uncontrolled rectification energy conversion results.

Figure 8 presents the simulation results under the conditions of an operating frequency of 10 Hz, a linear motor mover stroke of 77 mm, and a load of a 30 Ω resistor. From the figure, it can be observed that the bus voltage is composed of the peak voltages of each phase within the same cycle, reflecting the characteristics of diode rectification. The voltage fluctuation range is relatively large. Since the circuit is controlled in an open-loop manner, the bus voltage is entirely dependent on the nature of the load.

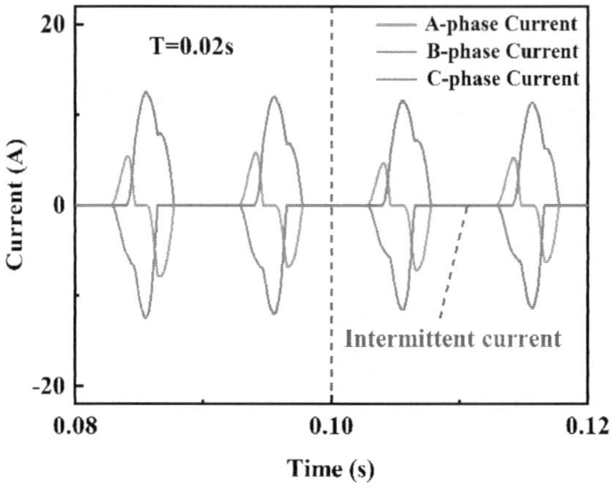

Figure 9. Discontinuous phase current on the motor side.

Using an uncontrolled rectification strategy will cause a severe current interruption in the three-phase current output by the linear motor. The simulation results are shown in Figure 9. In the circuit, each phase current on the AC side undergoes two conduction processes within a half-cycle, exhibiting characteristics of both discontinuity and continuity. The proportion of the current cycle that contributes to work is only about 50%, which significantly affects the motor's output efficiency.

4.2. Analysis of the three-phase boost-type controllable rectification results based on ASC

Through analysis of the working principle, the three-phase Boost-type controllable rectification based on ASC can achieve a higher bus voltage compared to uncontrolled rectification. Consequently, a higher bus voltage can accommodate loads of different voltage levels. Both methods were tested with a 100 Ω load in open-loop conditions, and the bus voltage at a 30 Hz operating frequency is shown in Figure 10.

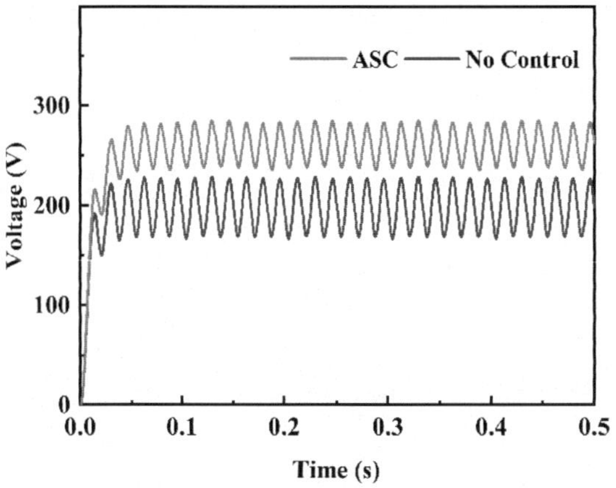

Figure 10. Bus voltage output results under open-loop control with different rectification strategies.

The maximum bus voltage for uncontrolled rectification can reach 220 V, while the bus voltage for controllable rectification can reach 290 V. Moreover, the voltage fluctuation range of ASC is smaller, which demonstrates that under the same operating conditions, the controllable rectification scheme has the capability to supply power to loads of higher voltage levels. The difference in bus voltage obtained at high frequencies will be even more pronounced.

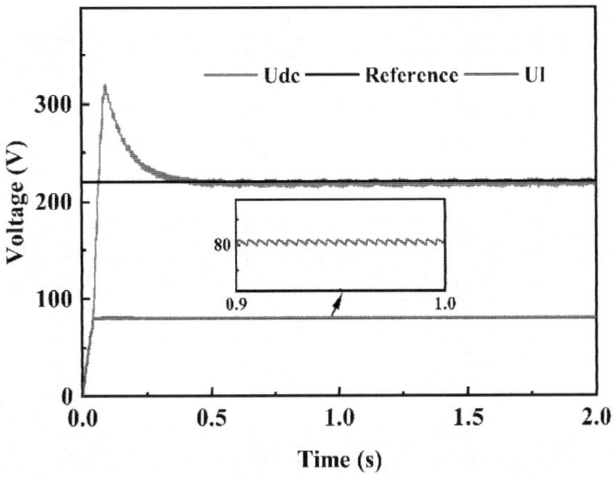

Figure 11. Voltage output results under ASC closed-loop control.

Figure 11 presents the bus voltage output results under closed-loop control of ASC, with the high-voltage side set at 220 V and the DC-DC output voltage at 80 V. From the figure, it can be observed that under closed-loop conditions, the voltage stabilizes quickly at 220 V, with minimal voltage ripple. After processing by the DC-DC converter, the voltage remains stable at 80 V, with extremely small fluctuations, within a range of 1.5%. This achieves high-quality electrical energy output.

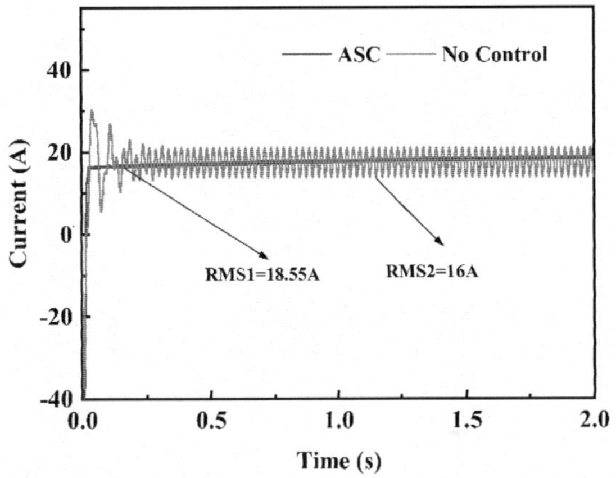

Figure 12. Comparison of battery charging current.

The comparison of battery charging results under high-frequency conditions for the two schemes is shown in Figure 12. Uncontrolled rectification can provide a charging current of 16.00 A for the battery but with significant current ripple and lower charging power in the voltage regulation state. In contrast, controllable rectification supplies a current of 18.55 A to the battery, with minimal current ripple and a response time of 20 ms, resulting in higher charging power. Consequently, the charging speed is also faster.

5. Conclusion

This study investigates different rectification strategies based on the linear internal combustion power generation system. Focusing on the characteristics of variable frequency and amplitude in the three-phase voltage and current output by the linear motor, we conducted research on the combination of uncontrolled rectification with DC-DC chopper circuits and the three-phase Boost-type controllable rectification strategy based on ASC. The following conclusions are drawn:

(1) The combination of three-phase uncontrolled rectification and chopper circuits can achieve stable and adjustable DC voltage. However, in the linear internal combustion power generation system, due to the characteristics of variable frequency and amplitude in the electrical energy output by the linear motor, as well as the periodic changes in phase sequence, the three-phase uncontrolled rectifier bridge can cause severe current discontinuity in the phase current of the linear motor, affecting the stability of the power generation system.

(2) The three-phase ASC-based Boost-type controllable rectification can achieve a higher bus voltage under open-loop control compared to uncontrolled rectification. Under closed-loop control, the bus voltage can quickly stabilize at the desired value with minimal voltage fluctuations, making it more suitable for the electrical energy processing of this system.

(3) The three-phase ASC-based Boost-type controllable rectification is capable of charging batteries at various voltage levels, with both the charging current and power exceeding those of uncontrolled rectification circuits. Additionally, it can enhance the power factor of the motor and improve the distortion of the phase current near zero crossing.

References

[1] Jia B, Smallbone A, Zuo Z, et al. Design and simulation of a two- or four-stroke free-piston engine generator for range extender applications[J]. Energy Conversion and Management, 2016, 111: 289-298.

[2] Leick M T, Moses R W. Experimental evaluation of the free piston engine-linear alternator (FPLA)[R]. Sandia National Lab. (SNL-NM), Albuquerque, NM (United States), 2015.

[3] Ali M, Haitao Y, Yao W, et al. Control of linear generator based on hysteresis - SVPWM current rectification and bidirectional buck/boost converter used for energy storage[J]. IET Renewable Power Generation, 2021, 15(14): 3282-3295.

[4] Amon E A, Brekken T K A, Schacher A A. Maximum power point tracking for ocean wave energy conversion[J]. IEEE Transactions on Industry Applications, 2012, 48(3): 1079-1086.

[5] Li J, Zuo Z, Liu W, et al. Generating performance of a tubular permanent magnet linear generator for application on free-piston engine generator prototype with wide-ranging operating parameters[J].

[6] Yousefi N, Mirabbasi D, Alfi B, et al. A low input current ripple high step-up dc-dc converter with reduced voltage stress for renewable energy application[J]. Journal of Operation and Automation in Power Engineering, 2025: 223-230.

[7] Wei Y, Zuo Z, Jia B, et al. Operating characteristics of a linear range extender with different energy consumption topologies[J]. Applied Thermal Engineering, 2023, 230: 120743.

[8] Sun X, Shen Y, Li W, et al. A PWM and PFM hybrid modulated three-port converter for a standalone PV/battery power system[J]. IEEE Journal of Emerging and Selected Topics in Power Electronics, 2015, 3(4): 984-1000.

[9] Green A W, Boys J T, Gates G F. 3-phase voltage sourced reversible rectifier[C]//IEE Proceedings B (Electric Power Applications). IET Digital Library, 1988, 135(6): 362-370.

[10] Kazmierkowski M P, Malesani L. Current control techniques for three-phase voltage-source PWM converters: A survey[J]. IEEE Transactions on Industrial Electronics, 1998, 45(5): 691-703.

Analysis of the temperature rising of 500 kV composite insulators

Qijie Zhu[1*], Te Li[1], Xuxiang Wu[2], Xianjin Yang[2], Jinbiao Ye[3] and Ruixiang Tao[1]

[1] State Grid Zhejiang Electric Power Co., Ltd. Research Institute, Hangzhou 310014, China
[2] State Grid Hangzhou Power Supply Company, Hangzhou 310004, China
[3] Zhejiang Golden Phoenix Electrical Co., Ltd, Shaoxing 312399, China

[*] Corresponding author's e-mail: zhuqijie@163.com

Abstract. The heating defects of composite insulators pose a serious threat to the safe and stable operation of power transmission lines. A case of heating in composite insulators on a 500 kV transmission line was investigated, where two types of heating insulators on the line underwent visual inspection, infrared testing, dissection examination, hydrophobicity tests, permeability tests, water diffusion tests, and sealing tests. The defects and causes of heating in the composite insulators were analyzed. The results showed that both types of heating insulators exhibited poor adhesion between the core rod and sheath, and the water diffusion leakage current exceeded the standard value. This defect is the cause of the core rod deterioration and heating in the insulators. Operational maintenance recommendations were made based on these findings.

1. Introduction

Compared with traditional porcelain insulators and tempered glass insulators, composite insulators have advantages such as strong resistance to contamination, small size, lightweight, high strength, easy installation, and cost-effectiveness [1]. They generally perform well in grid operation and are considered the most economical and effective means of external insulation in transmission systems. As a result, they have been widely used in China's power system.

Composite insulators exhibit stable overall performance and good quality, with a lifespan of over 20 years [1]. Over the years, they have undergone improvements in processes and technologies, such as the development of conductor crimping techniques and acid-resistant core rods, leading to a steady enhancement in product quality [2]. However, some composite insulators, due to poor manufacturing processes or prolonged exposure to harsh environments such as sun, rain, contamination, and high-intensity electromagnetic fields, may experience flashover, overheating, and even deterioration or fracture [3]. If such insulators with hidden defects continue to operate on transmission lines, they will pose a serious threat to the safety of the power grid.

Heating, deterioration, and fracture, as well as internal breakdown, are currently significant issues in the operation and maintenance of composite insulators on overhead transmission lines [4-5]. The causes of heating in composite insulators can be mainly attributed to three factors: First, internal partial discharge. Defects in the conductivity of the insulation material at the high-voltage end can easily lead to partial discharge, generating noticeable heat in the affected area. Second, dielectric loss is caused by moisture. Moisture can infiltrate through conductivity defects at the high-voltage end of the insulation material, leading to dielectric loss, and causing the insulator to heat up more than the heat absorbed by moisture evaporation. The more moisture is present, the more apparent the heating phenomenon becomes. Third, resistive loss heat. When the external sheath of the insulator undergoes severe aging, the insulation resistance significantly decreases, triggering resistive loss heating [6-10].

Under normal conditions, due to the high resistance on the surface of composite insulators, the conducting current is close to zero. However, when a large amount of unevenly distributed high-conductivity contaminants accumulates on the shed skirts, the thermal effects of the current can cause uneven temperature distribution on the surface of the insulator. In addition, heating defects caused by internal deterioration or internal breakdown of the insulator cannot be identified by appearance alone. Without dissection inspection, it is difficult to determine whether internal degradation has occurred, which poses a significant challenge to the operation and maintenance of transmission lines.

In the study, heating defects of composite insulators on a 500 kV transmission line were investigated. Various tests were conducted on the heated insulators, including visual inspection, infrared testing, dissection inspection, hydrophobicity testing, permeability testing, water diffusion testing, and end-seal testing. The defects and causes of heating in two types of insulators on the line were analyzed, and guidance was provided to address these issues. The findings offer technical reference for the diagnosis and treatment of heating defects in composite insulators in the future.

2. Overview of Defects

From August 23 to August 29, 2022, during an infrared inspection of a 500 kV transmission line, heating was observed in composite insulators at various degrees on tower positions 134# and 162#, among others. The details of the heated insulator locations, phases, series configurations, heating amplitudes, and model information are shown in Table 1.

Table 1. Line site heating information

Tower Number	Phase	String Type	Heating Amplitude/ °C	Type
134#	A	Jumper String	8.6	
141#	A	Jumper String	18.2	
145#	B Phase Large Number Side	Jumper String	27.3	
148#	B Phase Large Number Side	Jumper String	10.9	A
149#	A	Jumper String	6.6	
149#	C	Jumper String	9.8	
151#	A	Jumper String	10.8	
155#	B Phase Inner Side	Straight Line V-String	4.7	
155#	B Phase Outer Side	Straight Line V-String	5.6	B
162#	B Phase Inner Side	Straight Line V-String	4.7	

The heated defective insulators include two models, A and B. All A-type insulators are suspension strings used for jumper connections, while all B-type insulators are V-string configurations for line towers. The heating amplitudes of the two types of heated insulators range from 4.7 °C to 27.3 °C. These defective insulators were all put into operation in June 2015, which is far below their expected service life.

The infrared spectra of the defective insulators on-site can be divided into two categories. One category shows multiple heating occurrences at the high-voltage end and high-voltage side sections of the insulator, such as the large insulator on the B-phase of tower 145# (Figure 1(a)). The other category shows a single heating occurrence at the high-voltage side section, such as the A-phase insulator on tower 151# (Figure 1(b)).

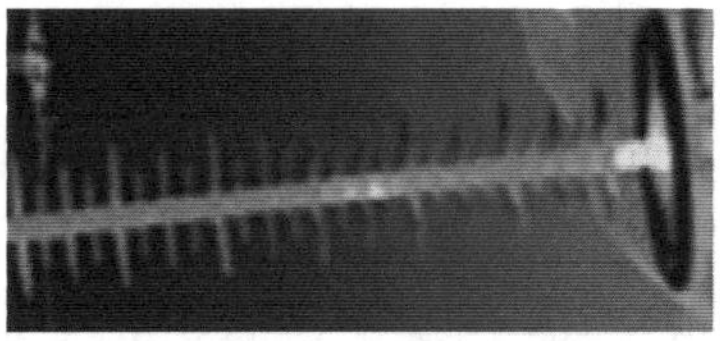

(a) 145#B phase, large side, heating amplitude of 27.3 °C

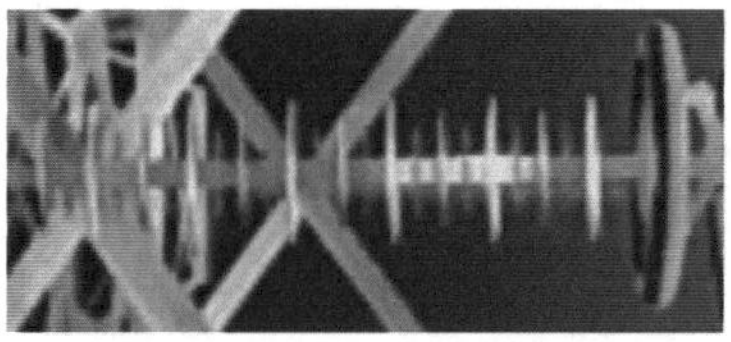

(b) 151#A phase, heating amplitude of 10.8 °C

Figure 1. Typical map of on-site insulator heating.

3. Experimental Analysis Results

3.1. Visual Inspection

On-site visual inspection of the defective phase insulators was conducted, as shown in Figure 2.

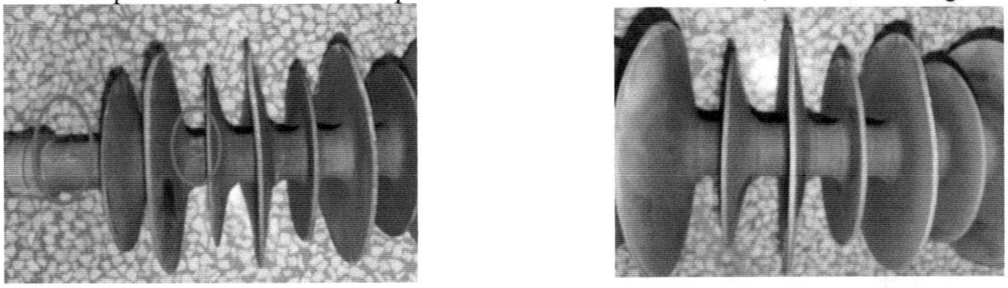

(a) Cracking of the sheath on the high-voltage side of the inner side insulator of phase B, tower number 155

(b) No abnormalities observed in the sheath of the inner side insulator of phase B, tower number 155, in the middle section

Figure 2. Visual inspection of insulators.

Upon inspection, it was confirmed that both A-type and B-type composite insulators were manufactured using the overall injection molding process. Among the defective insulators sent for testing, the B-type insulator on the B-phase of tower 155# had a cracked sheath between the first and second large sheds at the high-voltage end on the inner side. No cracks or signs of electrical corrosion were observed on the sheath of the other insulator core rods. The cracked sheath on the high-voltage end of the B-type insulator on the B-phase of tower 155# is shown in Figure 2(a), and the sheath condition in the middle of the same insulator is shown in Figure 2(b). Hardening and embrittlement tests were performed on the sheds of the B-phase insulator on tower 155#, confirming that no hardening or embrittlement of the sheds was observed.

3.2. Infrared Reassessment

Insulator deterioration, fracture, and internal breakdown often manifest as heating during development. Infrared testing is an important method for detecting early heating issues in composite insulators on-site [11-12].

In the production site, the main causes of temperature rise in composite insulators include four factors: insulator core rod deterioration, surface contamination discharge, moisture absorption by the high-voltage end sheath, and uneven sunlight exposure. Composite insulators with different types of temperature rise exhibit distinct infrared spectral characteristics at a macroscopic level [2].

There were relatively few infrared spectra of heated composite insulators on-site, so additional measurements were taken of the defective insulators to obtain more observation samples. A 288 kV power frequency operating voltage was applied to the insulators for 30 minutes, and the infrared spectra of the insulators are shown in Figure 3.

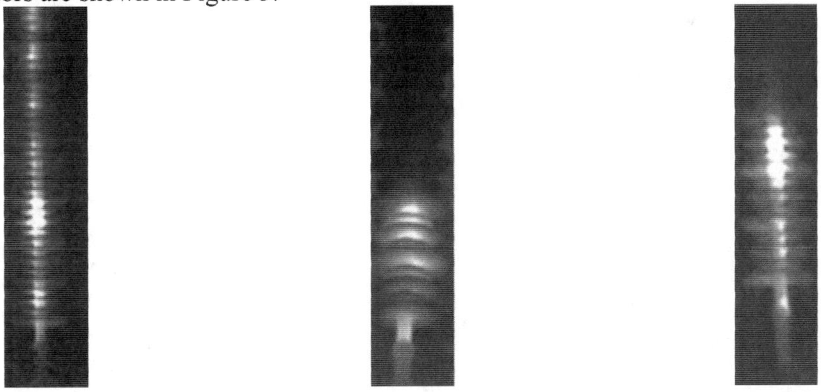

(a) 141#A-phase (b) 141#A-phase (c) 141#A-phase

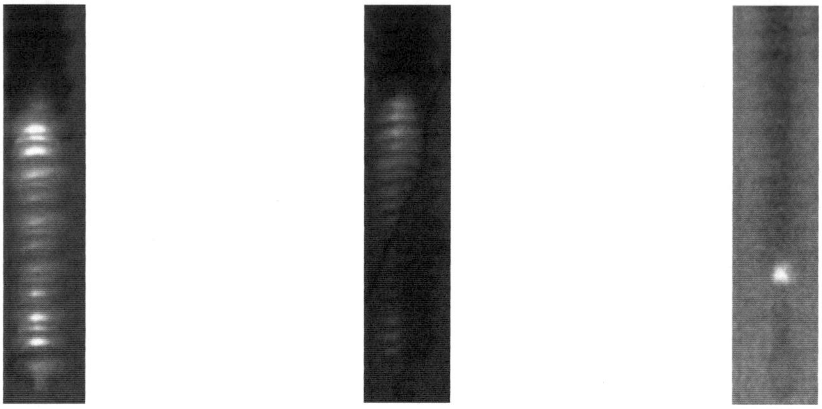

(d) 151#A-phase (e) 155#B-phase inner side (f) 155#B Phase Outer Side

Figure 3. Laboratory retest of infrared spectra.

As shown in Figure 3, the B-phase insulator on tower 155# only exhibits slight heating at the high-voltage end, while other insulators show noticeable heating near the high-voltage end. Most of the heating areas on the insulators are continuous segments. Currently, internal heating of on-site composite insulators is generally caused by core rod deterioration. Core rod deterioration increases the dielectric constant at the defective location, leading to severe distortion of the electric field intensity at the end of the deteriorated section [13]. The external manifestation of core rod deterioration is the most noticeable heating at the end of the deteriorated core rod section. At the same time, core rod deterioration increases the polarization loss of the core rod, and the entire deteriorated section may heat up. Therefore, typical heating caused by core rod deterioration appears as continuous heating in a segment on the high-voltage side, with the highest heating amplitude often found near the end of the heated section.

3.3. Dissection Examination

The deterioration of composite insulators is mainly characterized by the following features:

(1) The core rod appears like decayed wood, brittle and blackened.

(2) The interface between the core rod and the sheath fails, allowing the sheath to completely peel off the surface of the core rod.

(3) The defective area is often connected to the high-voltage end by carbonized channels.

These defects occur on the surface and inside of the core rod and are difficult to detect without dissection inspection [1].

To confirm whether the core rod of the defective insulator has deterioration defects, dissection inspections were performed on some insulators, and the results of the dissection inspection are shown in Figure 4.

(a) 155# B-phase Outer Side (b) 134#A-phase

(c) 148#B-phase large (d) 151#A-phase

Figure 4. Insulator anatomy results.

As shown in Figure 4, except for the B-phase B-type insulator on tower 155#, all other dissected insulator core rods exhibit deterioration defects. Specifically, the core rod of the A-phase insulator on tower 134# is yellowed (Figure 4(b)), the surface of the large core rod of the B-phase insulator on tower 148# is fuzzed (Figure 4(c)), and the core rod of the A-phase insulator on tower 151# is blackened (Figure 4(d)).

During dissection, a large area of poor adhesion between the sheath and core rod was found at the high-voltage end of the A-phase A-type insulator on tower 134#, as shown in Figure 5. Figure 5(a) shows the sheath detaching from the core rod at the non-adhesive area when the sheath was cut, exposing the core rod. Figure 5(b) shows that the core rod surface at the non-adhesive area does not have the noticeable silicone rubber residue found in other areas.

(a) The silicone rubber has detached in the non-adhesive area

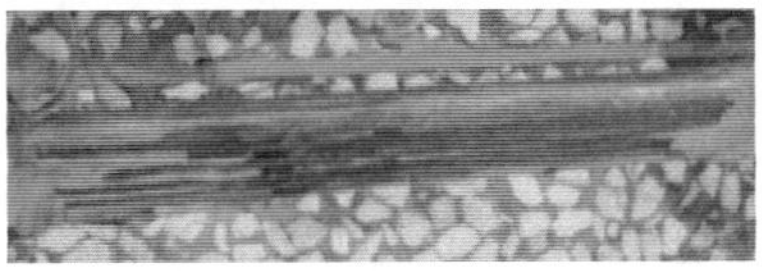

(b) The surface condition of the non-adhesive area of the core rod

Figure 5. 134# A phase insulator high voltage end sheath-mandrel bonding state.

Dissection revealed that the core rod of the B-type insulator on the C-phase of tower 162# showed slight deterioration marks at the high-voltage end side section in certain areas, while the deterioration on both sides of this area was more severe, as shown in Figure 6. This suggests that deterioration defects in the core rod are developing simultaneously at multiple locations.

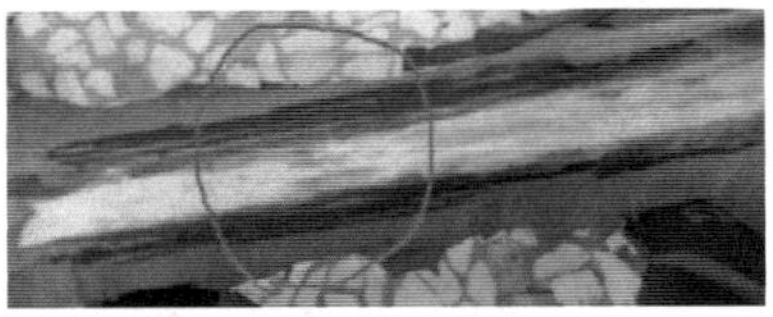

Figure 6. 162# C phase insulator core rod crisp intermittent marks.

3.4. Hydrophobic test

Hydrophobicity is a key indicator reflecting the degree of aging in composite insulators' silicone rubber. The shed sleeves of composite insulators are made from silicone rubber, a polymer material. Under ideal conditions, water forms droplets on the surface of the sheds and is not easily able to form discharge paths [14-15]. However, as the insulator ages, the hydrophobicity of the silicone rubber surface decreases, leading to surface creepage in a contaminated state. Hydrophobicity grading tests were conducted on the insulators, using the spraying grading method (HC method). Large sheds from the high-voltage side, middle, and low-voltage side sections of the B-phase insulator on tower 145# and the C-phase insulator on tower 162# were selected for hydrophobicity testing, with the results shown in Table 2.

Table 2. Hydrophobic test results.

Tested Insulator	High Voltage Side	Middle Part	Low Voltage Side
145#B phase large	HC3	HC4	HC3
162#C phase outer side	HC2	HC3	HC3

The spraying grading method divides the hydrophobicity of the silicone rubber surface into 7 levels, from HC1 to HC7, with hydrophobicity gradually decreasing as the level increases.

According to the test results, the hydrophobicity level of the tested composite insulator strings ranges from HC2 to HC4, indicating a decline in hydrophobicity for this batch of composite insulators, but hydrophobicity has not been completely lost. Referring to the international standard GB/T 22079—2019, composite insulators with a hydrophobicity level of HC5 and below meet the required performance and can continue to operate.

3.5. Permeability Test
Moisture can have a significant impact on the deterioration and even fracture of composite insulators. When there is poor adhesion at the core rod-sheath interface, moisture can infiltrate through air gaps into the adhesion defect. Once in direct contact with the core rod, moisture can cause hydrolysis, potentially leading to partial discharge, which degrades the performance of the core rod and may result in core rod deterioration, breakdown, or even fracture [16]. Therefore, the permeability resistance of the insulator is closely related to the aging resistance of composite insulators.

Permeability tests were conducted on the B-type composite insulators on the inner side and outer side of the B-phase of tower 155#, as well as the large A-type composite insulators on the B-phase of tower 145# and the B-phase of tower 148#. The test results are shown in Table 3. No abnormal permeability was observed in this batch of defective insulators.

Table 3. Hydrophobic test results.

Tested Insulator	Test Results	Type
155#B phase inner side	Normal	B
155#B phase outer side	Normal	B
145#B Phase Large Number Side	Normal	A
148#B Phase Large Number Side	Normal	A

3.6. Water Diffusion Test
The water diffusion test involves placing the insulator sample in a 100 °C water bath for a while and measuring the leakage current to assess the insulator's resistance to water ingress [17]. Water diffusion tests with sheaths were conducted on the large A-type insulators on the B-phase of tower 145# and tower 148#, as well as the B-type insulators on the inner and outer sides of the B-phase of tower 155#. The test segments were taken from the low-voltage end of the insulators to avoid the influence of deterioration defects on the test.

The test segments include the shed-core rod interface, and the leakage current values are shown in Table 4.

Table 4. Insulator sheath water diffusion leakage current value.

Sample	#1	#2	#3	#4	#5	#6
145#B phase large	200	241	321	281	181	252
148#B phase large	105	163	172	405	121	315
155#B phase inner side	543	108	213	348	94	66
155#B phase outer side	35	133	15	46	198	27

The test results in Table 4 show that the leakage current of one test segment of the heated insulator on the inner side of the B-phase of tower 155# reached 543 μA, and the leakage current of one test segment of the heated large insulator on the B-phase of tower 148# reached 405 μA. Both values exceed the reference value of 400 μA given in the State Grid Equipment [2021] No. 21 document, indicating a decrease in their resistance to water ingress.

3.7. Seal Test
The quality of the sealing performance directly affects whether the insulator can operate stably over the long term. Multiple factors influence the performance of the insulator's end seal. External factors, such as the improper configuration of the insulator's stress rings, create high-intensity electric fields at the end, leading to the aging of sealing rubber under long-term corona discharge. Internal factors, such as

Journal of Physics: Conference Series **3011** (2025) 012076 doi:10.1088/1742-6596/3011/1/012076

improper design of the end seal structure, are the direct cause of seal failure. When the end seal of the insulator is damaged, moisture can infiltrate the core rod through the defect, slowly hydrolyzing the epoxy resin. The polarization effect of moisture in the electric field and internal partial discharge contribute to the electrical erosion of the core rod, which is a significant cause of core rod heating.

End seal tests were conducted on the two test insulators, the large A-type insulator on the B-phase of tower 145# and the B-type insulator on the inner side of tower 155#. A dye was wrapped around the insulators for continuous penetration, and then the end seal was peeled open. The photograph of the large insulator on the B-phase of tower 145# is shown in Figure 7. The red dye did not successfully penetrate the end of the insulator, indicating that there are no abnormalities in the end seal of both insulators.

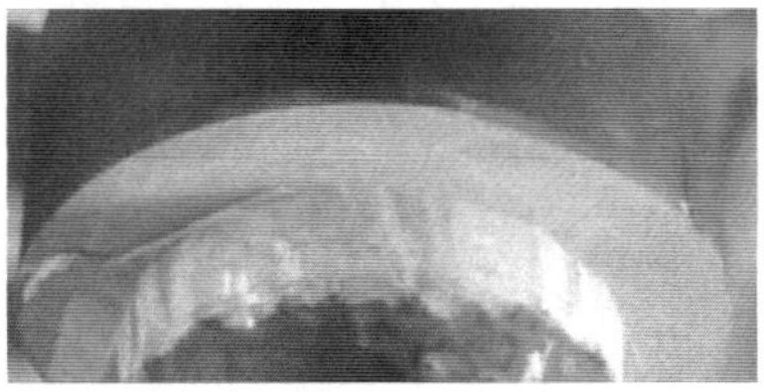

Figure 7. Insulator end seal test.

4. Defect Cause Analysis

Through visual and dissection inspections, it was found that the high-voltage side core rods of the heated composite insulators on this 500 kV line exhibit deterioration defects, characterized by yellowing, blackening, and fuzzing of the core rods. After core rod deterioration, polarization loss and internal partial discharge occur, leading to heat generation. The sheath of the B-phase insulator on tower 162# shows cracks, but there are no significant electrical erosion marks on the sheath surface, ruling out the possibility of surface electrical erosion causing sheath damage and triggering deterioration. End seal tests indicate that both types of defective insulators have good sealing performance, so the possibility of core rod deterioration being caused by moisture ingress due to end seal failure is excluded. In addition, the absence of abnormalities on the surface of the sheath in other deteriorated insulators suggests that the heating on this 500 kV line is caused by core rod deterioration defects, with the defect developing from the inside out.

Dissection inspection of the A-type composite insulators reveals that the high-voltage end of the 134# A-phase insulator has a large local adhesion defect, which is connected to the deterioration area. The local adhesion defect at the high-voltage end may lead to moisture accumulation at the sheath-core rod interface in a humid environment, resulting in partial discharge and hydrolysis of the core rod epoxy resin. Additionally, the water diffusion leakage current of the large insulator on the B-phase of tower 148# exceeds the reference value of 400 μA given in the State Grid Equipment [2021] No. 21 document, and some test segments have leakage currents exceeding 300 μA. In contrast, the leakage current of the same type of insulator during factory testing did not exceed 100 μA. This indicates a decrease in the adhesion performance between the sheath and the core rod after the composite insulator has been in operation. In a high-humidity environment, after moisture diffuses through the sheath, it accumulates at the core rod-sheath interface, leading to partial discharge and the occurrence of core rod deterioration. In conclusion, it is determined that the core rod deterioration in the A-type composite insulators is caused by poor adhesion between the sheath and the core rod.

For the B-type defective insulators, dissection revealed that the 162# C-phase insulator showed multiple areas of simultaneous deterioration. The main causes of internal deterioration in the insulator are poor adhesion between the core rod and sheath, as well as bending fatigue. Deterioration caused by bending fatigue typically occurs at the point of maximum bending stress on the core rod and does not result in multiple areas of deterioration. Additionally, the water diffusion leakage current in one test segment of the inner side of the 155#B-phase insulator reached 543 μA, exceeding the 400 μA reference value, confirming the decline in the adhesion performance between the core rod and sheath in this batch of insulators. Therefore, it is concluded that the core rod deterioration in the 162# C-phase insulator and the inner side of the 155#B-phase B-type defective insulator is due to the decreased adhesion between the sheath and the core rod.

5. Treatment measures

Based on the results and discussions presented above, the conclusions are obtained as below.

(1) During the dissection inspection of the A-type defective insulators, poor adhesion between the core rod and sheath was found at the high-voltage end, and the water diffusion leakage current exceeded the reference value. It was determined that the deterioration of the A-type insulators in this batch was caused by poor adhesion between the sheath and core rod. Following the requirements of the State Grid Equipment [2021] No. 21 document, the A-type composite insulators on this 500 kV line will be replaced, and drone infrared testing will be conducted every six months before replacement.

(2) During the dissection inspection of the B-type defective insulators, multiple areas of simultaneous deterioration were found, and there were samples where the water diffusion leakage current exceeded the reference value. After a comprehensive assessment, it was determined that the deterioration of the composite insulators in this batch was caused by poor adhesion between the sheath and core rod. The B-type insulators on this 500 kV line will be replaced, and drone infrared testing will be conducted every six months before replacement.

(3) By reviewing the infrared spectra of dissected insulators, it was found that composite insulators with internal deterioration defects all exhibited heating in areas other than the high-voltage end. One type showed heating in a section near the high-voltage end, while another type showed heating in multiple areas near the high-voltage end. Future infrared images with these heating patterns should be given sufficient attention. By tracking and confirming the likelihood of internal defects, appropriate maintenance strategies should be formulated.

6. Conclusion

Based on the results and discussions presented above, the conclusions are obtained as below.

The article provides a detailed analysis of a 500 kV line composite insulator heating event, using various testing methods to diagnose defects in the heating insulators of two different models on the line. This includes performing visual inspections of the defective insulators; observing infrared characteristics to identify corresponding temperature rise types; conducting dissection inspections to verify the condition of the core rod and the integrity of the core rod–sheath bonding interface; and performing hydrophobicity, permeability, water diffusion, and end-seal tests to assess whether the insulators meet operational standards.

The test results show that both models of the heating insulators have bonding defects at the core rod–sheath interface, and the water diffusion leakage current exceeds the standard value. This defect is the cause of the core rod decay and heating of the composite insulator. In response to this defect, the composite insulators of both models on the 500 kV line will be replaced. In addition, infrared images of the defective insulators on site are analyzed, and the infrared characteristics under this defect are recorded, providing technical references for future composite insulator heating defect diagnosis and treatment.

Acknowledgments

This work was financially supported by the State Grid Zhejiang Electric Power Company. (5211DS20008P).

References

[1] Guan Z C, Liu Y Y, Zhou Y X, et al. (2006) Insulators and external insulation of power transmission and transformation equipment[M].Tsinghua University publishing house co., ltd, Beijing.
[2] Wang X J, Xu J, Lin C H, et.al (2022) Analysis of the temperature rising of composite insulators of transmission lines[J].Insulators and Surge Arresters, (05):165-171.
[3] Tu Y P, Chen C H, Tong Y L, et al (2012) Aging characterics of shed materials of silicone rubber composite insulators in service[J].High Voltage Engineering, 38(10):2522-2527.
[4] Gao Y F, Liang X D (2016) Study on decay-like fracture of composite insulator, Part II: Experimental Simulation and Preventive Method Discussion of Decay-Like Fracture[J]. Proceedings of the CSEE, 36(18):5070-5077.
[5] Zeng Z W, Zeng X J, Zhang F Z, et al. (2018) Cause Analysis of the abnormal heating defect in composite insulators operated in 500 kV nuclear power Plant-Huizhou transmission line [J].Southern Power System Technology, 12(05):51-55.
[6] Vaillancourt G H, Hamel M, Frate J (1997) Experience with two fault composite insulators detection methods in Hydro-Quebec [C].10th International Symposium on High Voltage Engineering, 2:1467-1471.Quebec.
[7] Spangenterg E, Riquel G (1997) In service diagnostic of composite insulators EDF's test results [C].10th International Symposium on High Voltage Engineering, 2:139-142, Quebec.
[8] Cheng Y C, Li C R, Chen M, et al (2005) Research on heating mechanism of composite insulator of high voltage transmission line[J].Power System Technology, 29(5):57-60.
[9] Cheng L, Mei H W, Wang L M, etc, al (2016) Study on long-term aging characteristics and impact factors for silicone rubber sheath of composite insulators[J].Power System Technology, 40(3):1-7.
[10] Huang C C, Li Y G (2014) A review of aging evaluation methods for composite insulators[J].Electric Power Construction, 35(9):28-34.
[11] Wang S H, Ye Z Q, Luo S (2011) Fracture mechanism and defect detection methods of 500 kV composite insulators[J].Electric Power, 4（8）:19-21..
[12] Duan Y B, Hu X L, Yong J, et al. (2014) The criterion for field infrared temperature measurement of 500 kV composite insulator[J].Shandong Electric Power, 41(04):29-31..
[13] Lu M, Zhang Z H, Li L, et al. (2018) Reason analysis of decay-like aging for composite insulator[J].Power System Technology, 42(04):1335-1341..
[14] Chen Q (2016) Review in application of high voltage composite insulator and its ageing state[J].Insulating Materials, 49(04):7-13.
[15] Lu M, Hu W, Li L, et al. (2015) Analysis on the causes of the core rod fracture of 500 kV AC composite insulator[J].Insulators and Surge Arresters, (03)：23-28.
[16] Wang L M, Nie Z X, Zhao C L, et al. (2019) Water permeation characteristic in cycloaliphatic epoxy resin composite insulator sheath[J].High Voltage Engineering, 45(01):173-180.
[17] Fan H N, Liu Y H, Wang L M, et al.(2021)Influence of closing joints of epoxy resin insulator on water diffusion test[J].Power System Technology, 45(06)：2420-2426.

Fault location method for flexible DC transmission lines based on current derivative values

Kunlun Han[1], Xiongfeng Zhang[2,*]

[1]Guangxi Key Laboratory of Power System Optimization and Energy Technology, Guangxi University, Nanning 530004, China
[2]School of Electrical Engineering, Guangxi University, Nanning 530004, China

[*]Corresponding author's e-mail: hankl@gxu.edu.cn

Abstract. Flexible DC transmission boasts the advantages of flexible power flow and strong transmission capacity. Rapid and sensitive fault location technology is extremely important for the safe and stable operation of DC systems. This paper proposes a fault location method within the zone of flexible DC transmission lines based on current derivative values. The method utilizes the substitution theorem to equate the flexible DC transmission system during a fault to a discharge circuit and derives the fault location equation based on the equivalent circuit. A flexible DC transmission system model is built on the PSCAD/EMTDC simulation platform to verify the reliable location of the proposed scheme.

1. Introduction

With advancements in power electronics technology, the Voltage Sourced Converter (VSC)-based Flexible DC has emerged as a new generation of DC transmission technology, offering more flexible operation modes and better system controllability. It can transmit power to weak AC systems or even passive systems, making it highly suitable for applications such as powering weak systems or isolated islands, grid integration of distributed renewable energy generation, interconnection of asynchronous AC grids, and urban power grid supply [1]. DC transmission is characterized by long-distance "point-to-point" power transmission, often spanning regions with vastly different topographies, landscapes, and environmental conditions. This results in a more complex operational environment for DC transmission lines compared to AC lines, making them the most fault-prone component in DC transmission systems [2]. Rapid and accurate fault location methods are of great significance for improving inspection efficiency and accelerating fault restoration.

Currently, fault location methods for Flexible Direct Current (FDC) transmission are primarily categorized into traveling wave methods and fault analysis methods. In [3], scholars employ a one-dimensional convolutional neural network (1D CNN) to achieve localization of unipolar grounding faults. In [4], the source of the first reflected traveling wave is determined based on the correlation coefficient of fault currents at the line terminals and midpoints, enabling precise positioning for short-distance lines. In [5], researchers derive fault location equations using the initial characteristics of single-pole and bipolar fault currents. In [6], precise localization under high impedance based on deep learning is achieved.

Based on the analysis of line faults in a two-level VSC-based Flexible Direct Current (FDC) transmission system, this paper proposes a positioning method based on the derivative value of current. This method employs the substitution theorem to establish an equivalent discharge circuit, upon which positioning equations suitable for bipolar short-circuit faults and monopolar grounding faults are derived. Finally, a model of a two-level VSC-based FDC transmission system is built on the PSCAD/EMTDC simulation platform, and extensive simulation tests verify the feasibility of the proposed scheme.

Journal of Physics: Conference Series **3011** (2025) 012077 doi:10.1088/1742-6596/3011/1/012077

2. Fault location method

In the Flexible DC Transmission System (HVDC-Light), line faults are mainly classified into bipolar short-circuit faults and monopole grounding faults. Bipolar short-circuit faults typically have a relatively small fault resistance, and thus the existence of fault resistance can be neglected when deriving the fault location equation. Monopole grounding faults share similar fault processes with bipolar short-circuit faults; however, they often involve a larger and unknown fault resistance. Therefore, when deriving the fault location equation for monopole grounding faults, it is necessary to formulate the location equations from both ends of the converter stations simultaneously, in order to eliminate the impact of fault resistance on the location equation.

2.1. Fault location equation for Bipolar Short-Circuit Faults

The equivalent circuit for a bipolar short-circuit fault is illustrated in Figure 1. In the figure: C_{eq} represents the shunt capacitor on the DC side of the Voltage Source Converter (VSC); $U_{dc}(t)$ denotes the measured bipolar voltage at the outlet side of the converter station; $I(t)$ is the measured fault current in the line; R_{eq} and L_{eq} are the equivalent resistances and reactances, respectively. From the DC side outlet to the fault point on the positive and negative lines, the equivalent capacitance of the DC line can be neglected due to its being much smaller than the DC capacitor C_{eq}; The freewheeling diodes are designated as D_1-D_6.

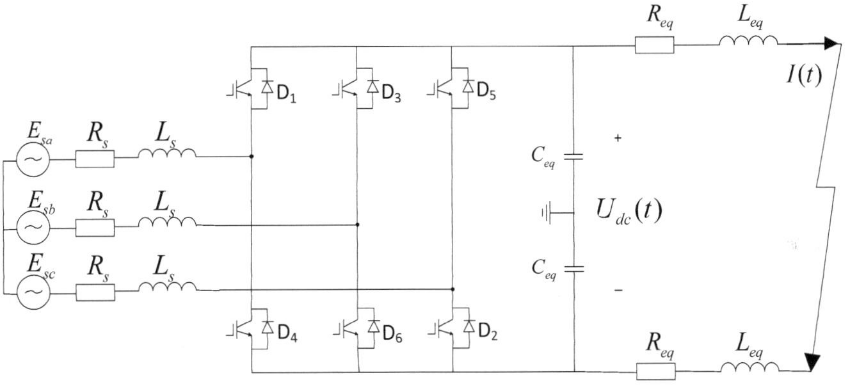

Figure 1. Equivalent circuit for bipolar short-circuit fault.

The fault can be broadly categorized into two stages: (1) At the onset of the fault, the DC side voltage $U_{dc}(t)$ is greater than the AC side line voltage. At this point, the fault current in the DC line is predominantly discharged by the shunt capacitor C_{eq} on the DC side towards the short-circuit point; (2) When $U_{dc}(t)$ decreases to the level of the AC side line voltage, the AC side power supply begins to feed current into the fault point through diodes [7].

According to the substitution theorem, the circuit structure of the converter station can be replaced by a voltage source with a voltage value equal to $U_{dc}(t)$, while maintaining the current in the DC line unchanged. Therefore, the equivalent circuit for the bipolar short-circuit fault can be simplified as shown in Figure 2.

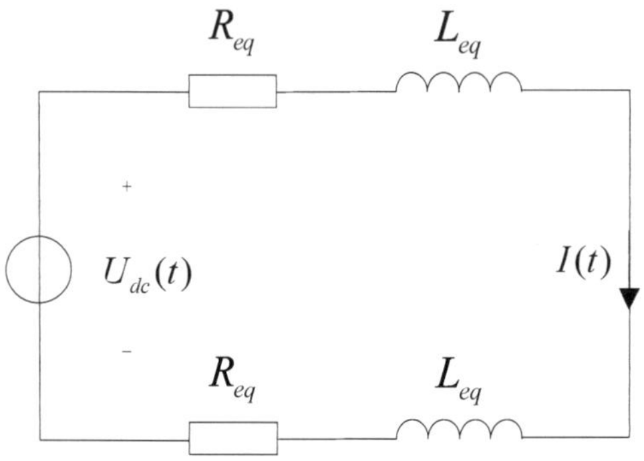

Figure 2. Simplified circuit for bipolar short-circuit fault.

For Figure 2, based on Kirchhoff's Voltage Law (KVL), the following equation can be written:

$$U_{dc}(t) = 2R_{eq}I(t) + 2L_{eq}\frac{dI(t)}{dt} \tag{1}$$

In Equation (1), R_{eq} and L_{eq} represent the equivalent resistance and reactance, respectively, from the DC side outlet to the fault point on the positive and negative lines. Thus, there are:

$$R_{eq} = xR_u \tag{2}$$

$$L_{eq} = xL_u \tag{3}$$

where x is the distance from the DC side outlet to the fault point on the positive or negative line; R_u and L_u are the resistance and inductance per unit length of the line, respectively. Substituting Equations (2) and (3) into Equation (1), we obtain:

$$U_{dc}(t) = 2xR_uI(t) + 2xL_u\frac{dI(t)}{dt} \tag{4}$$

By simplifying Equation (4), the distance from the DC side outlet to the fault point can be obtained as:

$$x = \frac{U_{dc}(t)}{2R_uI(t) + 2L_u\frac{dI(t)}{dt}} \tag{5}$$

2.2. Fault location equation for Monopole Grounding Faults

Taking the positive pole as an example, the equivalent circuit for a single-pole-to-ground fault is shown in Figure 3. In the figure: C_{eq} represents the shunt capacitance on the DC side of the Voltage Source Converter (VSC); $U_1(t)$ and $U_2(t)$ are the measured voltages at the positive pole outlet sides of the converter stations at both ends; $I_1(t)$ and $I_2(t)$ are the measured currents on both sides of the fault point on the line; R_{eq1} and R_{eq2} are the equivalent resistances from the DC side outlets of the converter stations at both ends to the fault point on the positive pole line; L_{eq1} and L_{eq2} are the equivalent reactances from the DC side outlets of the converter stations at both ends to the fault point on the positive pole line; R_f is the value of the transition resistance; R and L are the total equivalent resistance and reactance of the DC line, respectively.

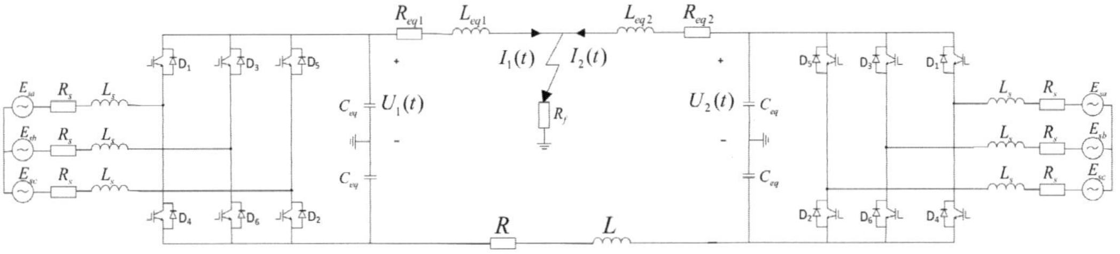

Figure 3. Equivalent circuit for a single-pole-to-ground fault.

According to the substitution theorem, the positive pole circuit structures of the converter stations at both ends can be replaced by voltage sources with voltage values equal to $U_1(t)$ and $U_2(t)$, respectively, while maintaining the current in the DC line unchanged. Therefore, the equivalent circuit for the single-pole-to-ground fault is simplified to the circuit shown in Figure 4.

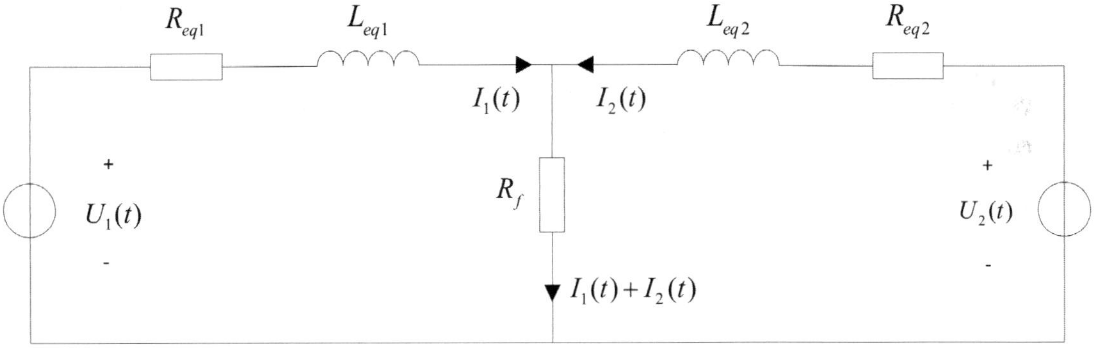

Figure 4. Equivalent circuit for a single-pole-to-ground fault.

For Figure 4, based on Kirchhoff's Voltage Law (KVL), the following equation can be written:

$$U_1(t) = R_{eq1}I_1(t) + L_{eq1}\frac{dI_1(t)}{dt} + R_f I_1(t) + R_f I_2(t) \tag{6}$$

$$U_2(t) = R_{eq2}I_2(t) + L_{eq2}\frac{dI_2(t)}{dt} + R_f I_1(t) + R_f I_2(t) \tag{7}$$

By combining Equations (6) and (7), the influence of the transition resistance on the location equation can be eliminated, resulting in:

$$U_1(t) - U_2(t) = R_{eq1}I_1(t) + L_{eq1}\frac{dI_1(t)}{dt} - R_{eq2}I_2(t) - L_{eq2}\frac{dI_2(t)}{dt} \tag{8}$$

In Equation (8), R_{eq1} and R_{eq2} represent the equivalent resistances, and L_{eq1} and L_{eq2} represent the equivalent reactances, respectively, from the DC side outlets of the converter stations at both ends to the fault point on the positive pole line. Thus, there are:

$$R_{eq1} = xR_u \tag{9}$$

$$R_{eq2} = (l - x)R_u \tag{10}$$

$$L_{eq1} = xL_u \tag{11}$$

$$L_{eq2} = (l - x)L_u \tag{12}$$

In Equations (9)-(12), x represents the distance from the DC side outlet of the rectifier station to the fault point on the positive pole line; l denotes the total length of the line; and R_u and L_u represent the resistance and inductance per unit length of the line, respectively. By substituting Equations (9)-(12) into Equation (8), we can derive:

$$U_1(t) - U_2(t) = xR_uI_1(t) + xL_u\frac{dI_1(t)}{dt} + xR_uI_2(t) + xL_u\frac{dI_2(t)}{dt} - lR_uI_2(t) - lL_u\frac{dI_2(t)}{dt} \tag{13}$$

By simplifying Equation (13), we can obtain the distance from the DC side outlet of the left-end converter station to the fault point on the positive pole line:

$$x = \frac{U_1(t) - U_2(t) + lR_uI_2(t) + lL_u\dfrac{dI_2(t)}{dt}}{R_uI_1(t) + L_u\dfrac{dI_1(t)}{dt} + R_uI_2(t) + L_u\dfrac{dI_2(t)}{dt}} \tag{14}$$

The voltage and current values in Equations (5) and (14) can be directly sampled. However, the derivative of the current cannot be directly obtained, and in practical engineering applications, the difference value is typically used as an approximation for the derivative value, i.e.:

$$\frac{dI(t)}{dt} = \frac{I(t+h) - I(t)}{h} \tag{15}$$

In the equation, h represents the sampling step size.

3. Simulation verification

A simulation model of a two-terminal flexible DC transmission system, as illustrated in Figure 1, is constructed in PSCAD/EMTDC with system parameters listed in Table 1.

Table 1. Simulation system parameters.

Parameter	Parameter Value
DC Voltage (kV)	600
Line Length (km)	600
Shunt Capacitance on DC Side (F)	0.2
Line Resistance (mΩ/km)	14
Line Inductance (mH/km)	0.6

Bipolar short-circuit faults are set at various distances outside the outlet of the rectifier station in the simulation model. Figures 5(a)-(d) show the bipolar voltage waveforms at the outlet side of the rectifier station under fault conditions at distances of 10 km, 50 km, 300 km, and 500 km, respectively.

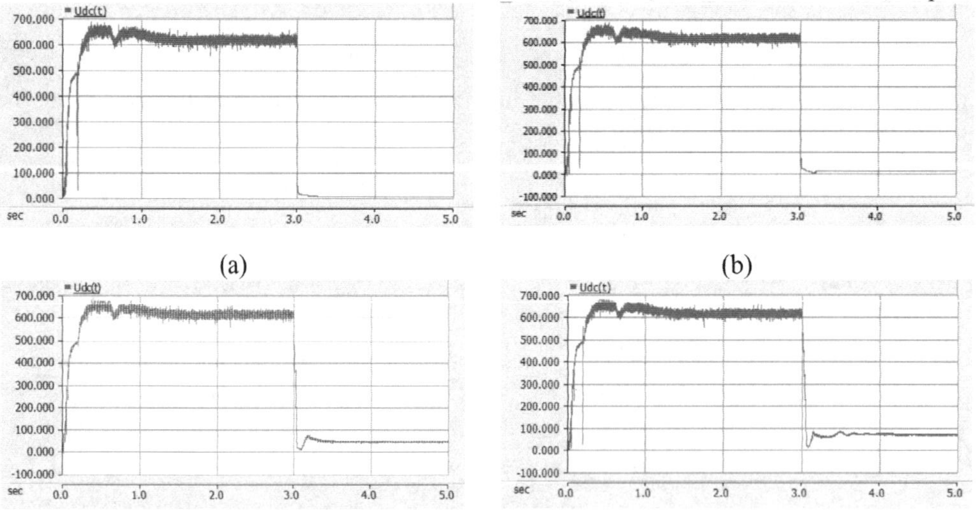

(a) (b)

(c) (d)

Figure 5. Bipolar voltage waveforms at various fault distances.

In Figure 6, (a)-(d) depict the waveform of the line current under fault conditions occurring at distances of 10 km, 50 km, 300 km, and 500 km, respectively.

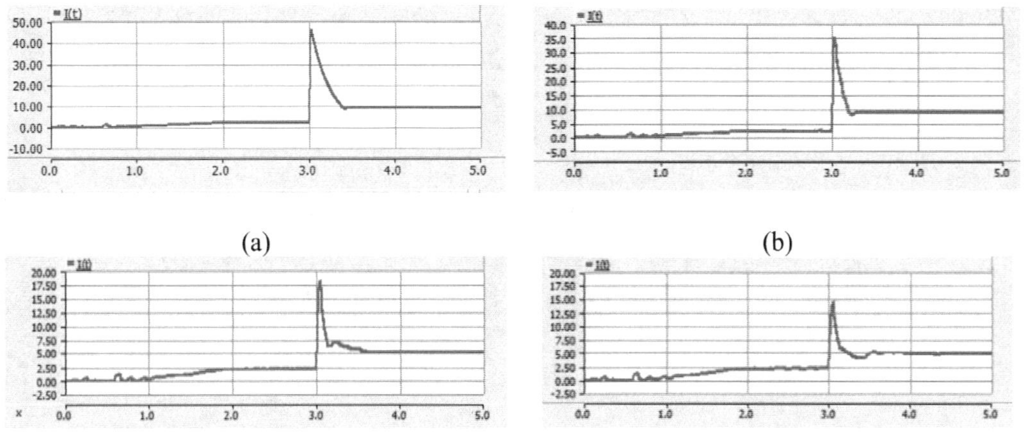

(a) (b)

(c) (d)

Figure 6. Line current waveforms at various fault distances.

The fault initiation is set at the third second. From Figures 5 and 6, it can be observed that upon the occurrence of a bipolar short-circuit fault, the converter station outlet voltage drops sharply, while the line current suddenly rises to the level of 10 kA. Voltage and current data at 1 millisecond after the fault initiation are taken and substituted into Equations (5) and (14). The simulation results obtained are shown in Figure 7.

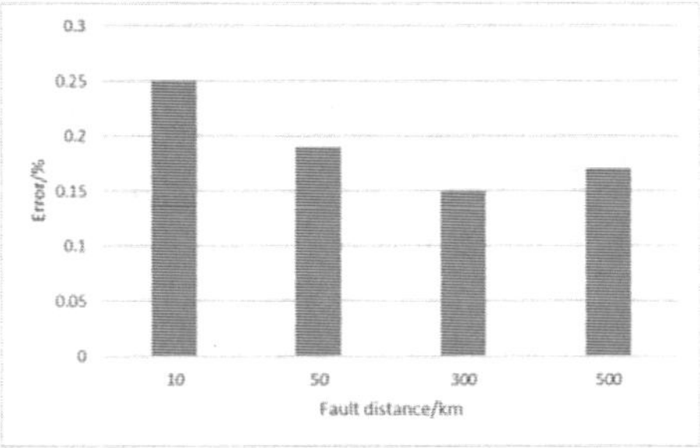

Figure 7. Positioning error for bipolar short-circuit fault.

As shown in Figure 7, when applying the proposed localization method in the case of bipolar short-circuit faults, the positioning errors at various distances can be kept within 0.3%, thus validating the effectiveness of the localization method for bipolar short-circuit faults.

Single-pole grounding faults are set at distances of 10 km, 50 km, 300 km, and 500 km from the rectifier station, with transition resistances configured at 10 Ω, 50 Ω, and 100 Ω, respectively. Following the occurrence of these faults, the trend of changes in the outlet voltage and line current of the faulted pole exhibits similarities to that of a bipolar short-circuit fault. Taking the example of a positive pole grounding fault at a fault distance of 300 km, the positive pole outlet voltage and line current are depicted in Figure 8.

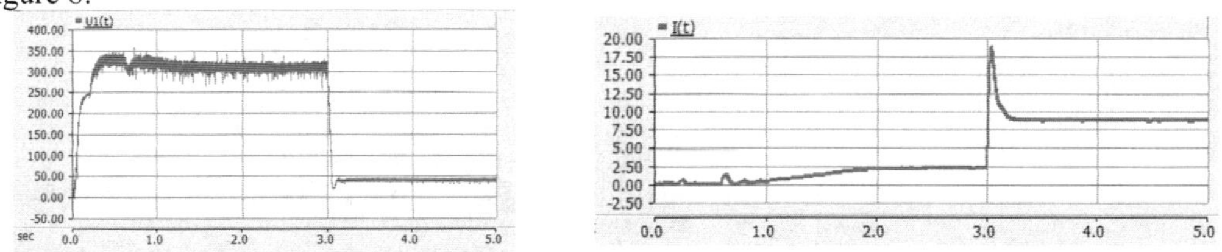

(a) (b)

Figure 8. The fault pole voltage and line current at a fault distance of 300 kilometers.

Similar to the process of a bipolar short-circuit fault, the sampled voltage and current data are substituted into Equations (5) and (14). The resulting simulation calculation outcomes are shown in Figure 9.

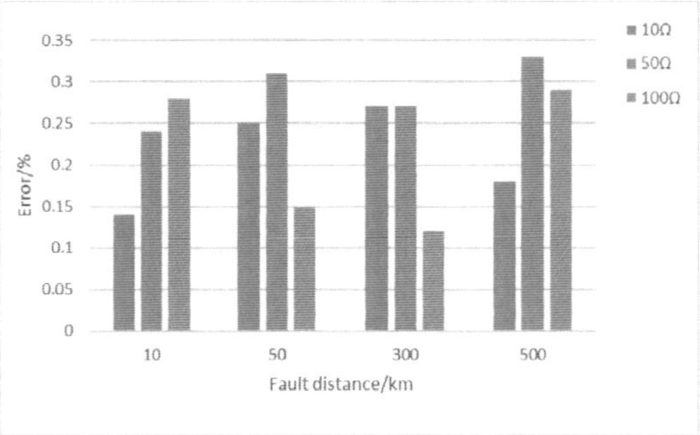

Figure 9. Positioning error for single-pole grounding fault.

As illustrated in Figure 9, when a single-pole grounding fault occurs, the application of the proposed localization method can keep the positioning error within 0.4% for various distances, and the positioning error does not increase with the increase of transition resistance, thereby validating the effectiveness of the localization method for single-pole grounding faults.

4. Conclusion

In this paper, a fault localization method based on the derivative of current values is proposed using the substitution theorem. This method is applicable to two-terminal flexible direct current (DC) transmission systems. The conclusions are as follows:

(1) The flexible direct current (DC) transmission system during a fault is equivalently modeled as a discharge circuit, and Kirchhoff's laws are utilized to establish the localization equation. This method is straightforward and clear.

(2) By utilizing the localization equations from both ends of the converter stations to eliminate the influence of transition resistance, precise localization can be achieved even with a transition resistance of up to 100 Ω.

Acknowledgments

This work was financially supported by the National Natural Science Foundation of China (51567003).

References

[1] Ma W M, Wu S M, Yang Y M, et al. (2014) Analysis on the Status Quo and Application Prospects of Flexible DC Transmission Technology. High Voltage Engineering, 40(08): 2429-2439.
[2] Yang L, Wang B, Dong X Z (2018) Review of Fault Location Techniques for HVDC Transmission Lines. Automation of Electric Power Systems, 42(08): 185-191.
[3] Wang S R, Zhao Y H (2007) Travelling Wave Fault Location for Bipolar HVDC Transmission Lines. High Voltage Engineering, (07): 124-128.
[4] Wei K, Li Z C (2024) Fault Location Method for Unipolar Grounding Faults in Multi-Terminal Flexible DC Transmission Lines. Electrical Switches, 62(02): 65-70+108.
[5] Ding J L, Wang X, Zheng Y H, et al. (2020) An Improved Single-Ended Traveling Wave Fault Location Method Utilizing Current Measurement at the Midpoint of the Line. High Voltage Engineering, 46(05): 1814-1822.
[6] Lan Z, Yuan Y S, He D, et al. (2023) A Fault Location Method for DC Distribution Networks Based on the Initial Value of Current Derivative. Southern Power System Technology, 17(05): 109-116.
[7] Liu W X, Pan X, Zhang X L, et al. (2024) Research on Fault Location in Power Systems Based on Intelligent Algorithms. Information Technology, (06): 160-167.

Two-terminal fault locating method of HVDC transmission

Kunlun Han and Kangtai Zou*

Guangxi Key Laboratory of Power System Optimization and Energy Technology, Nanning, Guangxi, Guangxi University, 530004, China

*Corresponding author's e-mail: 2212392150@st.gxu.edu.cn

Abstract. High-voltage direct current (HVDC) transmission technology has a wide range of application scenarios and development space, but the DC line fault frequency is high, and line faults significantly endanger the stable operation of the system, and fast and accurate fault location can help to isolate the faults quickly to reduce the fault recovery time, which plays a crucial role in ensuring the stable and reliable operation of HVDC systems. Accordingly, an HVDC line fault localization method based on variational mode decomposition (VMD) and Hilbert amplitude is proposed. A sequence of IMF components is obtained by VMD decomposition of the fault traveling wave signals at the rectifier side and inverter side, and the Hilbert transform is performed to obtain the instantaneous amplitude versus time graph to determine the precise arrival time of the initial wave head by selecting the IMF with obvious mutation characteristics. The simulation outcomes demonstrate that the proposed VMD-Hilbert algorithm effectively identifies the fault traveling wave head with minimal impact from transition resistance, ensuring high-accuracy ranging results across various fault locations.

1. Introduction

Owing to the considerable distance of DC transmission lines, across the complex terrain, the operation of the environment is very severe, therefore, the probability of faults is large, the existing research shows that 50 % of HVDC system faults for line faults, the precise fault localization is the key to quickly clear the fault, reduce the workload of patrolling the line to enhance the dependability of the power supply [1]. Fault localization methods for HVDC transmission systems can be primarily categorized into three types: fault analysis method, traveling wave method, and intrinsic frequency method [2]. Among them, the traveling wave method, which has the widest application range, is not easily affected by complex environmental factors and has high-ranging accuracy [3]. The traveling wave method is categorized into single-terminal and two-terminal traveling wave approaches [4]. The key link in the traveling wave method is the identification of the wave head. Currently, the main methods for wave head identification are wavelet transform (WT) [5] and Hilbert-Huang transform (HHT) [6]. WT can carry out effective identification of wavehead after selecting appropriate wavelet bases and wavelet decomposition scales, but the method has poor adaptability under different operating conditions, which makes it difficult to be widely applied [7]; The core link of HHT method is to use empirical mode decomposition (EMD) to decompose the faulty traveling wave signal into several IMF components with fixed center frequencies, and then get the IMF after Hilbert transform (HT). The Hilbert transform (HT) is utilized for acquiring the instantaneous amplitude and frequency of the IMF, which solves the deficiency of the wavelet transform method but introduces the problems of mode aliasing and edge effect, which hinders the precise extraction of the wave head of the faulty traveling wave, and influences the calibration of the arrival time of the wave head [8].

This paper takes long distance LCC-HVDC (Line Commutated Converter High Voltage Direct Current) as the research object, combines VMD and Hilbert transform, and proposes a two-terminal traveling wave localization technique, initially, the fault voltage waveforms at both ends are decomposed by VMD to get a series of IMF components, and then selects the IMF components with obvious mutation characteristics and performs the Hilbert transform to get the instantaneous amplitude diagram, instantaneous amplitude refers to the relationship between amplitude and time in a certain period of time or in a cycle, which can reflect the sudden increase in the amplitude of the signal when the fault occurs, and this sudden change in information can reflect the exact time of the fault travelling wave arriving at the monitoring points at both ends, the arrival moment of the traveling wave is determined by locating the first amplitude mutation point in the instantaneous amplitude diagram and then substituting it into the formula of distance calculation of two-terminal traveling-wave approach for determining the fault distance. The issue of mode aliasing in the EMD has been solved by the VMD-Hilbert. At the same time, because of the iterative method, VMD can choose the number of decomposition layers adaptively, and VMD-Hilbert can accurately detect the singularity of the wave's first arrival, thus obtaining an accurate wave head arrival time, with small localization error and excellent immunity to transition resistance, as verified by establishing an example of high-voltage DC transmission system in PSCAD/ EMTDC.

2. Fault Location Scheme

The traveling wave method is divided into single-terminal and two-terminal methods, Single-terminal ranging method needs to identify the initial wavehead and the reflected wavehead, identifying the initial wavehead is relatively straightforward, whereas accurately detecting the second wavehead poses a significant challenge, because it may be the local traveling wave reflected at the fault point, or the opposite side of the traveling wave reflected at the boundary of the line and come through the fault point. Especially if the fault occurs near the midpoint of the line, the reflected wave and refracted wave can be easily confused, the arrival times of the two waves may be almost the same, thus affecting the judgment. Therefore, the core factor of single-terminal traveling wave ranging is to accurately identify the initial wave head and identify the counter-traveling wave. Compared with single-terminal ranging, two-terminal ranging utilizes the timing gap between the initial arrival of the fault-induced traveling wave at the detection devices on both ends for measurement, thus eliminating the influence of the size and nature of the transition resistor as well as changes in the system operation mode on the ranging accuracy. This makes the ranging results of the two-terminal ranging method better than the single-terminal method, especially in the high-voltage transmission line fault ranging, the two-terminal ranging method has better application prospects. Therefore this paper's VMD-Hilbert ranging scheme uses the two-terminal method and both ranging methods are derived in the next section.

2.1. Principle of Traveling Wave Method

Traveling wave propagation characteristics of the principle shown in Figure 1, m and n are respectively for the two terminals of HVDC, *f* is the fault point, the traveling wave was propagated to the two ends of the first to reach the line m end of the wave head is recorded as $u_m^-(t)$ the moment of arrival is recorded as tm1; to reach the point of the fault and reflect the m end, the time is recorded as tm2. The first to reach the line n end of the wave head is recorded as $u_n^-(t)$ the moment of arrival is recorded as tn1. The wave head that reflects through the n end to form a backward traveling wave and passes through *f* to reach m terminal is denoted as $u_{FR}\left(t\right)$, and the arrival time is recorded as tm3.

From Figure 1., we can get the formula of the single-end ranging method as follows:

$$\begin{cases} \Delta t = t_{m2} - t_{m1} \\ D_{mf} = \dfrac{v}{2}\Delta t \end{cases} \tag{1}$$

Journal of Physics: Conference Series **3011** (2025) 012078 doi:10.1088/1742-6596/3011/1/012078

The formula of the two-terminal ranging method is as follows:

$$\begin{cases} D_{mf} = \dfrac{D_{mn} + v\left(t_{m1} - t_{n1}\right)}{2} \\ D_{nf} = \dfrac{D_{mn} + v\left(t_{n1} - t_{m1}\right)}{2} \end{cases} \tag{2}$$

where D_{mf} is the range from *f* to terminal m, and D_{nf} is the range from *f* to terminal n.

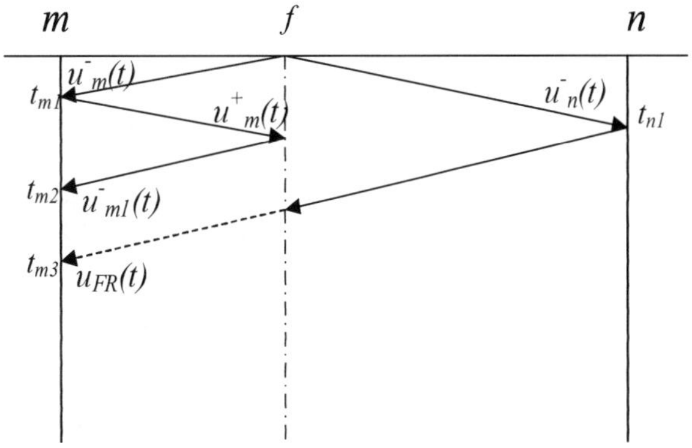

Figure 1. Principle of traveling wave fault location.

2.2. *VMD-Hilbert algorithm*

VMD, a modal variational method, decomposes the primary signal into H IMF components, each characterized by distinct central frequencies and finite bandwidths, and implements the filtering using multiple Wiener filter banks [9-10]. The bandwidth of each IMF is calculated as displayed in the following formula:

$$\begin{cases} \text{s.t.} \sum_{h=1}^{H} \mu_h = f \\ \min\limits_{\{\mu_h\},\{\omega_h\}} \left(\sum_{h=1}^{H} \left\| \partial_t \left(\left(\delta(t) + \dfrac{\mathrm{j}}{\pi t} \right) * \mu_h(t) \right) \mathrm{e}^{-\mathrm{j}\omega_h t} \right\|_2^2 \right) \end{cases} \tag{3}$$

where *f* is the primary signal, μ_h is the h IMF to which the VMD decomposes, and ω_h is the central frequency of the h IMF component. To address the constraint problem of Formula (3), a combination of quadratic penalty term and Lagrange multiplier is used as in Formula (4).

$$L(\{\mu_h\},\{\omega_h\},\lambda) = \alpha \sum\nolimits_h \left\| \partial_t \left[\left(\delta(t) + \frac{j}{\pi t} \right) \mu_h(t) \right] e^{-j\omega_h t} \right\|_2^2 + \left\| f(t) - \sum\nolimits_h \mu_h(t) \right\|_2^2 + \left\langle \lambda(t) + f(t) - \sum\nolimits_h \mu_h(t) \right\rangle \tag{4}$$

where Formula (3). can be reformulated as the saddle point problem in Formula (4). By applying the iterative sub-optimization sequence of the alternating direction method of multipliers (ADMM), the expressions for its modal components and central frequency are derived as in Formula (5):

$$\mu_h^{n+1}(\omega) = \frac{f(\omega) - \sum_{i<h} \mu_i^{n+1}(\omega) - \sum_{i>k} \mu_i^{n+1}(\omega) + \dfrac{\lambda^n(\omega)}{2}}{1 + 2\alpha(\omega - \omega_h^n)^2} \tag{5}$$

$$\omega_k^{n+1} = \frac{\int_0^\infty \omega \left|\mu_k^{n+1}\right|^2 \mathrm{d}\omega}{\int_0^\infty \left|\mu_k^{n+1}\right|^2 \mathrm{d}\omega} \tag{6}$$

The k IMF components can be found by inverting the $\mu_k(\omega)$ Fourier inverse transform of the completed VMD decomposition $\mu_k(t)$, and then applying the Hilbert transform to them.

$$y(t) = \frac{1}{\pi}\int_{-\infty}^{\infty} \frac{\mu_k(\tau)}{t-\tau} d\tau \tag{7}$$

3. Simulation Verification

To validate the effectiveness of the VMD-Hilbert ranging method introduced, a bipolar HVDC system simulation model was constructed in PSCAD/EMTDC, operating at a voltage level of ± 800 kV, as shown in Figure 2. Both the rectifier station and inverter station are 1000 MW, 12-pulse wave converter, transformer with Y/Δ connection and Y/Y connection, and the flat wave reactor is 0.3 H, the line length is 1400 km; Owing to fault ranging requires a sufficiently large number of sampling points, this paper simulation sampling frequency of 200 kHz. In practical applications, the VMD requires the selection of appropriate parameters: the number of modal components K and the quadratic penalty factor α. If K is chosen to be insufficiently large, it may lead to incomplete decomposition, resulting in mode mixing. Conversely, if K is chosen to be too large, it may cause over-decomposition, even generating irrelevant or extraneous components. If α is chosen to be insufficiently large, the bandwidth of the IMF components may become too large, leading to overlapping. On the other hand, if α is chosen to be too large, the bandwidth of the IMF components may become narrow, causing the omission of specific signal features. Each IMF has its characteristics, and the choice of parameters also has a great impact on the selection of IMFs, if the parameters are chosen appropriately, there will be multiple IMFs with obvious singular characteristics, at this time, choosing an appropriate IMF can help to extract the time of arrival. After many experiments on the choice of K and α, the configuration of parameters for VMD were $K = 6$, $\alpha = 2000$. The occurrence of a positive ground fault at 400 km is selected as an example, VMD decomposition of the rectifier-side and inverter-side fault signals is shown in Figure 3(a)and Figure 3(b).

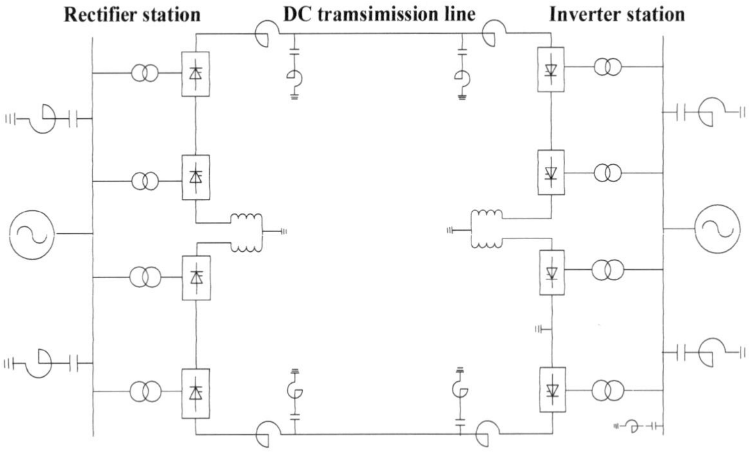

Figure 2. Bipolar HVDC Transmission System.

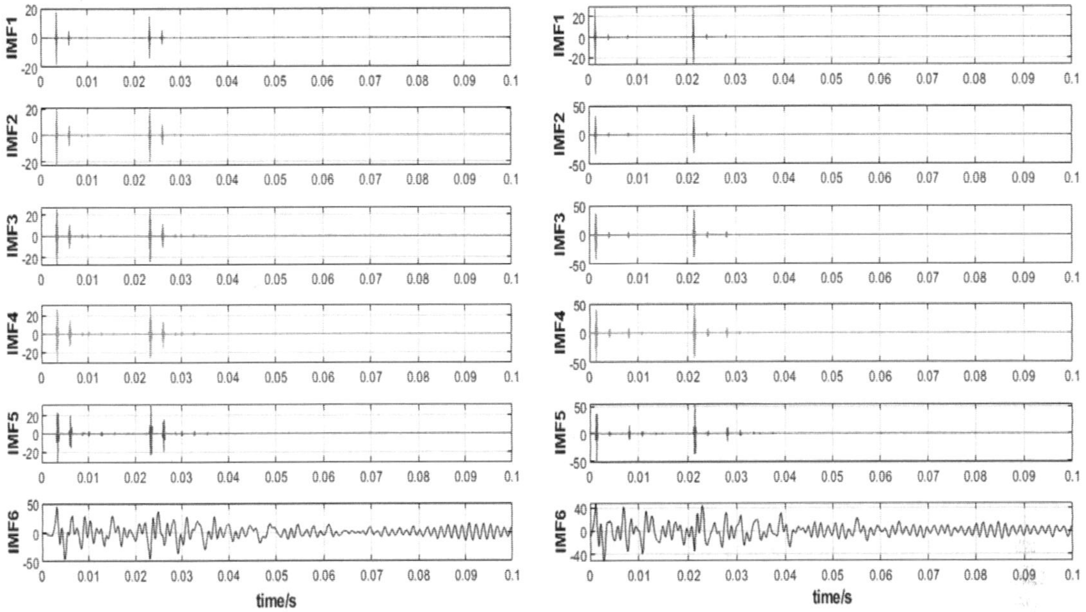

(a) rectifier-side (b) inverter-side

Figure 3. VMD decomposition.

In Figure 3, there are mutation points in IMF1~IMF5 in the VMD decomposition on both rectifier and inverter sides, IMF1 ~IMF5 can all be selected, since the IMF component is more stable in the higher frequency bands, so IMF1 is selected for the Hilbert transform on both the rectifier and inverter sides,
as despalyed in Figure 4(a)and Figure 4(b).

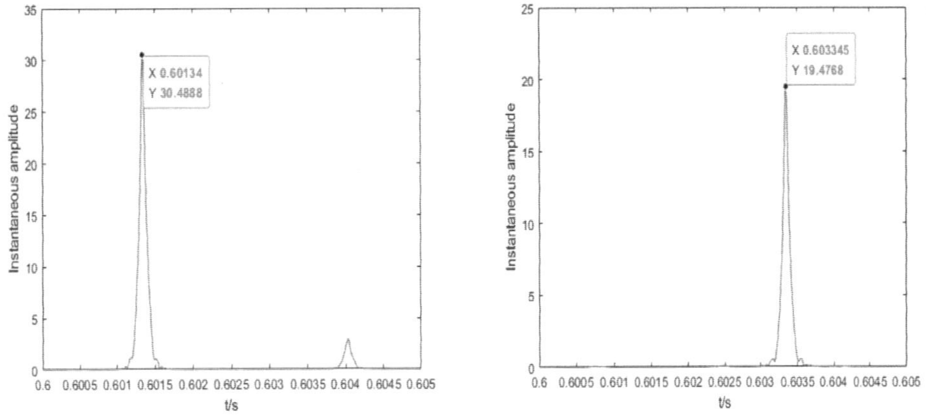

(a) rectifier-side (b) inverter-side

Figure 4. Instantaneous amplitude.

In Figure 4, the first singularity of the instantaneous amplitude plot can be obtained on both the rectifier side and the inverter side. The resulting instantaneous amplitude diagram can reflect the fault signal that suddenly becomes larger. In the traveling wave method, in addition to accurately obtaining the arrival time of the traveling wave, obtaining an accurate velocity is also of great importance. The velocity of the traveling wave is influenced by many factors. Currently, a more dependable approach is to estimate it using multiple datasets. With the fault distance and time pre-set, it can be measured. After multiple measurements, the traveling wave velocity obtained is 298640 km/s. The rectifier side detects the wave in 0.601344s, and the inverter side detects the wave in 0.603345s, according to Formula (2), to obtain the fault distance of 400.6134 km, the error distance of 0.6134k m. Various fault distances, transition resistances, and fault types can all impact the ranking results. The ranging outcomes under various conditions are presented in Table 1, Table 2, and Table 3 below.

Table 1. Distance measurement results in different fault distances.

Fault distances/ km	Ranging results/ km	Error distance/ km	Error percentage/ %
200	200.5246	0.5246	0.2623
400	400.6134	0.6134	0.1534
600	599.9556	0.0444	0.0074
900	900.0888	0.0888	0.0099
1100	1099.431	0.5690	0.0517
1300	1299.5198	0.4802	0.0037

The results show that the ranging error under different fault distances is within 0.3 %, which remains largely unaffected by the factor of fault distance and has high ranging accuracy under different fault distances.

Table 2. Distance measurement results of transition resistances.

Transition resistances /Ω	Ranging results/ km	Error distance/ km	Error percentage/%
10	1299.5198	0.4802	0.0369
300	1298.7732	1.2268	0.0944
500	1298.7732	1.2268	0.0944
700	1307.7324	7.7324	0.5948

The results show that the ranging error under high-resistance ground faults has some elevations, but it is still within 6 % and can be considered to be largely unaffected by the transition resistance.

Table 3. Distance measurement results of different fault types.

Fault distances/ km	Fault types	Ranging results/ km	Error distance/ km	Error percentage /%
200	PG	200.5246	0.5246	0.2623
	NG	201.2712	1.2712	0.6356
	PN	201.2712	1.2712	0.6356
600	PG	599.9556	0.0444	0.0074
	NG	599.9556	0.0444	0.0074
	PN	599.9556	0.0444	0.0074
900	PG	900.0888	0.0888	0.0099
	NG	899.3422	0.6578	0.0731
	PN	899.3422	0.6578	0.0731
1300	PG	1299.5198	0.4802	0.0369
	NG	1298.7732	1.2268	0.0944
	PN	1298.7732	1.2268	0.0944

PG means positive grounding fault, NG means negative grounding fault, and PN means bipolar grounding fault. The results show that the fault errors are within 3 % for different fault types, indicating that the method is largely independent of fault type.

4. Conclusion

In this paper, a VMD-Hilbert-based two-terminal localization method is proposed. VMD-Hilbert can effectively extract wave head features and accurately obtain arrival times. The estimation of wave speed in this localization method is more accurate than the traditional localization method, which improves the ranging accuracy. Simulations are conducted for various fault locations, various fault types, and various transition resistances, and the results show that the ranging accuracy of the VMD-Hilbert algorithm proposed is almost not constrained by the above factors.

Acknowledgments

This work was funded by The National Natural Science Foundation of China (No.51567003).

References

[1] Song, G.; Cai, X.; Gao, S. Survey of fault location research for HVDC transmission lines. Power Syst. Prot. Control 2012, 40, 133–137. (In Chinese)
[2] Wang L, He Y, Li L. A Single-Terminal Fault Location Method for HVDC Transmission Lines Based on a Hybrid Deep Network. Electronics. 2021; 10(3):255.
[3] M. M. Elgamasy, M. A. Izzularab and X. -P. Zhang, "Single-End Based Fault Location Method for VSC-HVDC Transmission Systems," in IEEE Access, vol. 10, pp. 43129-43142, 2022.
[4] Pathirana, A. N. Travelling Wave Based DC Line Fault Location in VSC HVDC Systems. Master's Thesis, University of Manitoba, Winnipeg, MB, Canada, 2013.
[5] Wang, Dafei, et al. "Fault location with high precision of flexible DC distribution system using wavelet transform and convolution neural network." Frontiers in Energy Research 9 (2021): 804405.
[6] J. Wang and Y. Zhang, "Traveling Wave Propagation Characteristic-Based LCC-MMC Hybrid HVDC Transmission Line Fault Location Method," in IEEE Transactions on Power Delivery, vol. 37, no. 1, pp. 208-218, Feb. 2022.
[7] Daniel Marques da Silva, Flavio B. Costa, Vladimiro Miranda, Helder Leite, Wavelet-based analysis and detection of traveling waves due to DC faults in LCC HVDC systems, International Journal of Electrical Power & Energy Systems, Volume 104,2019, Pages 291-300, ISSN 0142-0615.
[8] LI Qiang. Study on fault location of high voltage transmission line based on Hilbert-Huang transform[D]. Nanchang: East China Jiaotong University, 2009.
[9] Wang, D., Sun, Y., Xiao, J., & Lu, J. (2023). Research on pipeline leakage signal denoising using variational mode decomposition and energy value. *Petroleum Science and Technology*, *43*(2), 202–218.
[10] Y. Xu, C. Zhao, S. Xie and M. Lu, "Novel Fault Location for High Permeability Active Distribution Networks Based on Improved VMD and S-transform," in IEEE Access, vol. 9, pp. 17662-17671, 2021.

Journal of Physics: Conference Series **3011** (2025) 012079 doi:10.1088/1742-6596/3011/1/012079

Consideration of the optimization of the regional power supply structure of new energy sources

Shuqiang Zhao, Peiwen Zhou*, Xun Suo

State Key Laboratory of Alternate Electrical Power System With Renewable Energy Sources (North China Electric Power University), Baoding, Hebei,071003, China

*Corresponding author's e-mail: 15284955408@163.com

Abstract. To realize the energy transition, the proportion of new energy in the power supply structure will continue to increase, for the power supply planning, to realize the efficient use of new energy, it is necessary to reasonably configure the installed capacity of different types of power supply, to ensure the new energy consumption. This paper firstly analyzes the power model of the new energy-containing power system and the characteristics of wind-scenery complementarity, and then, to minimize the total cost of power generation and make full use of wind-scenery complementarity, applies the theory of complex adaptive systems to the regional power supply planning, and constructs a power supply planning model based on the theory of complex adaptive systems. Finally, by simulating and analyzing the actual power grid in a region, the power structure of its original power system is re-planned and compared with the genetic algorithm. The results show that the total power generation cost is reduced, the new energy consumption is increased, and the fossil fuel consumption is reduced under the constraints of power and power balance, which verifies the effectiveness and feasibility of the model.

1. Introduction

Global climate change, and fossil energy depletion as the current human survival and development of the key issues that must be resolved, one of the effective measures to solve this problem is the development and utilization of new energy sources such as wind and solar energy Compared with traditional power generation [1], wind energy has a wide distribution, abundant reserves and has the technological advantages of large-scale high-efficiency power generation, and solar energy is a very rich resource, very high security and clean energy. Solar energy is a very abundant and highly secure clean energy source [2]. Due to the stochastic and intermittent characteristics of wind and photovoltaic power generation, it will increase the uncertainty of the system and reduce the quality of power when connected to the grid [3], but the two have strong complementary characteristics in the distribution of time.

At present, there are a lot of research results for the power supply planning research containing new energy at home and abroad. Reference [4] takes the maximum wind power output as the goal and puts forward an optimization method for the power capacity of ultra-high voltage direct current wind and fire joint transmission. Reference [5] based on the basic economic scheduling model, with the maximum amount of wind and light consumption as the goal, to build the integrated scheduling plan model of extra-high voltage direct current outgoing wind and firepower. Reference [6] introduces new energy consumption weights, energy saving, and emission reduction constraints, and combines economic and environmental benefits to construct a multi-objective optimization model for wind-glow-fire joint scheduling. Reference [7] proposes a capacity optimization allocation model of wind-solar hybrid power supply systems based on an improved differential evolution algorithm to minimize the comprehensive power generation cost composed of investment cost, operation, and maintenance cost, and so on.

Most of the above literature applies the cost of power generation and the amount of new energy consumption as the optimization objectives in the power supply planning of power systems with new energy sources, and the application of the complementarity between different power sources is relatively single, which is mainly based on the characteristics of different power sources to carry out a fixed capacity ratio, which is lack of flexibility. To obtain a more reasonable power planning scheme, this study proposes a regional power planning method with new energy based on the theory of complex adaptive systems, which makes full use of the complementary characteristics of wind and solar energy.

2. Complex adaptive systems theory

2.1. Power Adaptation Analysis

Complex adaptive systems theory (CAS) is one of the scientific theories of complex systems based on reductionism and systems theory to analyze and study the system. The theory combines micro-level and macro-level organically [8], The micro-level: defines the research object as the subject, which is adaptive and able to interact with the environment continuously accumulate experience, and adapt to the environmental changes by changing its structure; the macro-level: the subject collection and the environment collection form the system, and the system evolves and develops in the interaction process between the subject and the subject, and between the subject and the environment development [9].

Applying the theory of complex adaptive system to the new energy-containing power system, all kinds of power supplies as adaptive subjects, the subjects adapt and compete with each other, the subjects are not independent, but are interconnected and interact with each other, and change of any one of the subjects will affect the development of the other subjects. As a complex adaptive system with self-adaptive ability, the new energy-containing power system can efficiently adapt to environmental changes, coordinate the energy allocation of various types of power sources, realize the coordinated development of power sources, and ultimately reach dynamic equilibrium.

2.2. Adaptive Subject Model

Subjects described by rules and interacting with each other as well as the external environment constitute complex adaptive systems. Subjects improve their survival ability by accumulating experience and changing their own behavioral rules. The subject's behavioral rules follow the "stimulus-response" model, in which the subject responds by receiving stimulus information. The model of the adaptive subject consists of three parts: the execution system, the credit assignment, and the rule-making.

The executive system is centered on the "stimulus-response" model, in which the detector receives a large amount of environmental information and extracts important information affecting the survival and development of the subject, and then the rule base provides the corresponding response behaviors, which are finally acted on the environment by the effector [10]; the credit assignment reflects the adaptability of the subject, which assigns a probability value to each rule in the rule base, and the size of the probability value reflects the self-interest of each rule for the development of the subject; the rule formulation describes all subjects through a unified grammar, which ensures the consistency among the rules. Each rule in the rule base is assigned a probability value, and the size of the probability value reflects the self-interest of each rule for the development of that subject; rule formulation describes all subjects through a unified syntax, which ensures the consistency among the rules, can cover all the interactions of the subjects, and can be adapted by writing a program to adapt the rules. Figure 1 shows a schematic diagram of the subject model.

As can be seen from Figure 1., the subject through the "stimulus-response" mode, received changing environmental information to change their behavior structure and behavior to gradually adapt to the changing environment, fully embodies the subject of the accumulation of experience, the process of continuous learning.

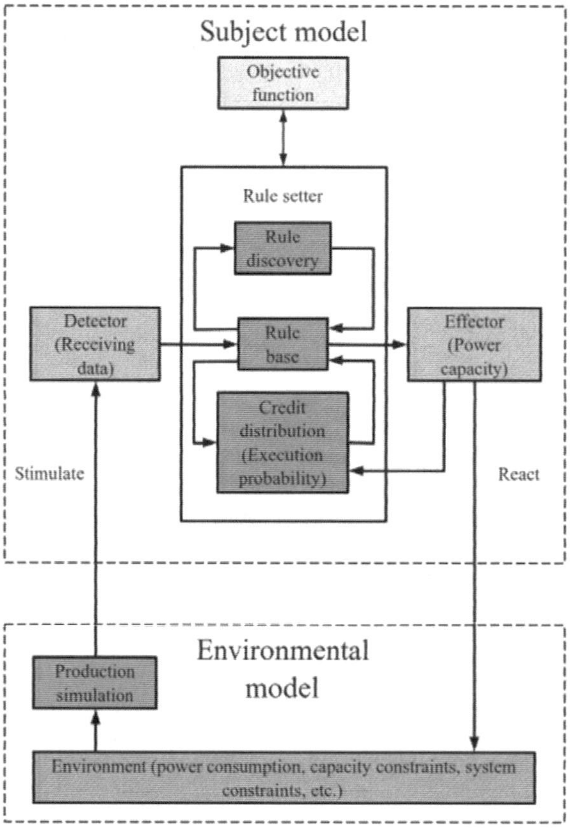

Figure 1. Schematic diagram of the main body model.

3. Regional power planning model

3.1. Power output model

3.1.1. Wind turbine output modeling. The output of a wind turbine is mainly related to the wind speed and is modeled as follows [11]:

$$P_{w,t} = \begin{cases} 0, & v \le v_{ci},\ v > v_{co} \\ \dfrac{v^3 - v_{ci}^{\ 3}}{v_N^{\ 3} - v_{ci}^{\ 3}} * P_{w,N}, & v_{ci} < v \le v_N \\ P_{w,N}, & v_N < v \le v_{co} \end{cases} \tag{1}$$

At present, most scholars use the two-parameter Weibull distribution model to portray wind speed data, which is characterized by high adaptability and accuracy and is easier to calculate [12]. Its probability density equation and wind speed distribution function are expressed as:

$$f(v) = \frac{k}{\lambda}\left(\frac{v}{\lambda}\right)^{k-1} e^{-\left(\frac{v}{\lambda}\right)^k} \tag{2}$$

$$F(v) = 1 - e^{-\left(\frac{v}{\lambda}\right)^k} \tag{3}$$

3.1.2. PV unit output modeling. The output of the PV unit is mainly related to the irradiation intensity, which is modeled as follows [13]:

$$P_{PV,t}=\begin{cases}P_{PV,N}, & \gamma>\gamma_N\\ P_{PV,N}\dfrac{\gamma}{\gamma_N}, & \gamma\le\gamma_N\end{cases} \tag{4}$$

Light resources are similar to wind resources, different weather conditions can affect the solar light intensity. Therefore, a suitable function model is also needed to describe the stochastic distribution of light intensity, which can be described by the Beta distribution model in this paper. Its probability density equation and wind speed distribution function are expressed as:

$$f(g)=\frac{\Gamma(\alpha+\beta)}{\Gamma(\alpha)\Gamma(\beta)}g^{\alpha-1}(1-g)^{\beta-1} \tag{5}$$

$$F(g)=\frac{\Gamma(\alpha+\beta)}{\Gamma(\alpha)\Gamma(\beta)}\int_0^g t^{\alpha-1}(1-t)^{\beta-1}dt \tag{6}$$

3.1.3. Thermal power unit output modeling. Thermal power output $P_{the,t}$ satisfies the following constraints:

$$\begin{cases}P_{the,\min}\le P_{the,t}\le P_{the,\max}\\ P_{the,down}\le P_{the,down\max}\\ P_{the,up}\le P_{the,up\max}\end{cases} \tag{7}$$

3.1.4. Hydropower unit output modeling. The hydroelectric output $P_{hyd,t}$ power satisfies the following constraints:

$$\begin{cases}P_{hyd,\min}\le P_{hyd,t}\le P_{hyd,\max}\\ P_{hyd,down}\le P_{hyd,down\max}\\ P_{hyd,up}\le P_{hyd,up\max}\end{cases} \tag{8}$$

3.2. Modeling regional power planning

The minimum construction unit of different types of power supply is regarded as the "main body of power supply", and the natural resources, unit output characteristics, and source-load balance of the power system are regarded as the "external environment". Through the constant interaction between the "power source body" and the "external environment" as well as between the "power source body" and the "power source body", the power source structure of the power system gradually stabilizes and realizes high efficiency and stability. The power structure of the power system is gradually stabilized to achieve efficient and stable operation.

(1) The power source constitutes its execution system through several stages of "receiving stimuli from the environment - responding to the environment - feedback adaptation - adjusting the rule base - improving the adaptive capacity". In this paper, the amount of abandoned wind, the amount of abandoned light, the climbing limit, the downhill limit, the maximum capacity constraint, and the reliability constraint are selected as the environmental influences, and the six environmental influences are coded as a set of environments: $\mathrm{E}=[e_1,e_2...e_6]$, and the coded values are divided according to the degree of transgression, and the bigger the degree of transgression, the bigger the coded values are.

(2) The credit assignment mechanism of the power source subject reflects its adaptability. In this paper, wind power capacity, photovoltaic capacity, thermal power capacity, and hydropower capacity are taken as controllable factors. Integer coding is performed on the response set composed of four types of power supply capacity: $\mathrm{S}=[s_1,s_2...s_i]$, and the coding value rule is: 1-increase capacity; 0-capacity unchanged; -1-decrease capacity. Set the minimum construction capacity of all four types of power generating units as 1MW, and take the minimum construction capacity as the simulation unit capacity.

(3) Given the environmental impact factors and the reaction set set set set in this paper, construct the power subject behavior rule set R and formulate the modification mechanism of the behavior rule, the power subject behavior rule is shown in Figure 2. If the power subject obtains information from the environment collection and E_k, selects S_{jk} the corresponding behavioral measures, the probability of execution of the behavior is, the target function value can be calculated through the stochastic production simulation program, if the behavior makes the target function value decrease, then increase δ_{jk}, and vice versa, then decrease δ_{jk}. The power supply subject constantly adjusts its own structure and behavioral rules to improve its adaptive ability according to the change in function value.

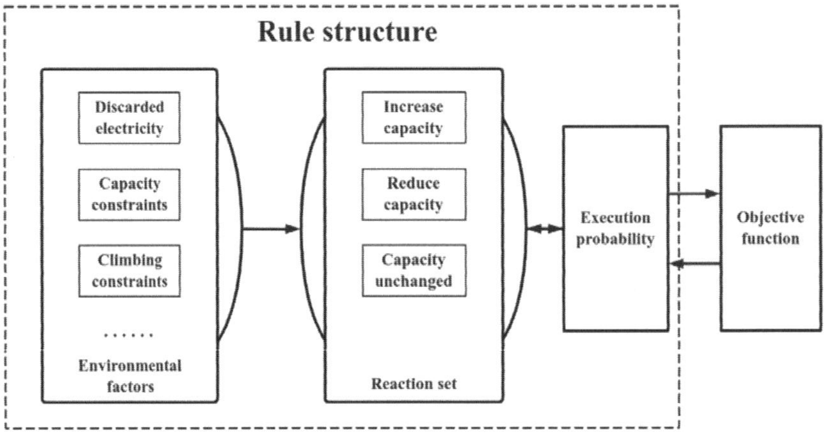

Figure 2. Schematic diagram of behavioral rules for power supply subjects.

(4) Based on the marking mechanism, it realizes the interaction behaviors such as aggregation of power subjects. The marking of subjects is a unified marking customized based on the differentiated characteristics of different subjects, and the interaction behaviors such as identification, selection, and aggregation are carried out between subjects through the marking. For power subjects, the differentiation comes from different timing output characteristics, and different power subjects realize inter-subject cooperation, competition, aggregation, and other behaviors through the marking mechanism.

3.3. *Objective function*

In this paper, we adopt the objective of minimizing the total power generation cost of all power subjects in the planning period, and the total system production cost includes: investment cost C_v, operation and maintenance cost C_m, power shortage cost C_l, fuel cost C_f, and environmental cost C_e. The objective function is shown below:

$$C_{\min} = C_v + C_m + C_l + C_f + C_e \tag{9}$$

(1) Investment costs

$$C_v = \frac{P*V}{L} \tag{10}$$

(2) Operation and maintenance costs

$$C_m = E*m \tag{11}$$

(3) Lack of electricity cost

If the power supply system is insufficient, it is necessary to purchase power from the grid, and the power purchase costs incurred are shared in proportion to the capacity of each type of power source. Take the power purchase costs of wind power as an example:

$$C_l = \frac{P_{win}}{P_{sum}} * E_l * R \tag{12}$$

Journal of Physics: Conference Series **3011** (2025) 012079 doi:10.1088/1742-6596/3011/1/012079

(4) Fuel cost

$$C_f = E * f \tag{13}$$

(5) Environmental costs

$$C_E = E * e \tag{14}$$

3.4. Constraints

(1) Electricity constraints

The total installed capacity of all units in the power system is greater than or equal to the maximum annual demand:

$$\sum_{j=1}^{N_i} P_j \geq P_{li} \tag{15}$$

(2) Power balance constraints

Annual electricity generation from all units in the power system to meet annual electricity demand:

$$\sum_{j=1}^{N_i} E_{ij} \geq E_i \tag{16}$$

(3) Installed capacity constraints

The installed capacity of each type of unit is within certain specified limits:

$$C_{sj.\min.i} \leq \sum C_{ji} \leq C_{sj.\max.i} \quad \text{j=1,2, L ,m} \tag{17}$$

(4) Reliability constraint

Reliability constraints on system operation are imposed using the power shortage probability $LOLP$ and the power shortage expectation $EENS$:

$$\begin{aligned} LOLP_t &\leq LOLP_{t.\max} \\ EENS_t &\leq EENS_{t.\max} \end{aligned} \tag{18}$$

3.5. Model solving

The solution steps are as follows:

(1) Various types of power sources are selected as the research subject, the set of environmental influencing factors and the set of responses are selected, the set of rules is constructed and the initial value is set for the probability of execution. Input the relevant parameters of the power system, set the planning period, and get the typical day scenario through the scenario clustering algorithm.

(2) The timing production simulation is used to obtain the timing output curve of each power subject, and the subject interaction identity is calculated. The power subject accepts the environmental information and makes reaction behavior according to its own behavioral rules, and at the same time, adjusts the execution probability of the corresponding rules according to the changes in the value of the objective function.

(3) When the value of the objective function converges to the optimal value, the optimal planning scheme for the power supply capacity structure is output, and if it does not converge, it returns to step (2) to continue the iterative calculation.

4. Calculus analysis

In this paper, the actual power system of a region in China in 2020 is selected as an example system, and the power structure of the system is optimized and simulated, and the installed capacity of each type of power source is 19,745 MW of wind power, 10,269 MW of photovoltaic power, 5,208,585 MW of thermal power, and 7,344 MW of hydropower, and the main economic and technical parameters are shown in Table 1.

Table 1. Main economic and technical parameters.

Techno-economic parameters	Wind power	Photovoltaic	Thermal power	Hydroelectricity
Unit capacity investment cost (Million yuan/ MW)	5	5.5	3	12
Operating and maintenance costs per unit of power generation (Yuan/ MWh)	0.6	7.8	3.9	1.3
Fuel cost per unit of power generation (Yuan/ MWh)	0	0	57.5	0
Environmental costs per unit of power generation (Yuan/ MWh)	0	0	77.9	0
Service life (Years)	20	20	30	30

Based on the theory of complex adaptive systems, each type of power supply is taken as the main body of the study, and factors such as abandoned power, climbing limit, descending limit, maximum capacity constraint, etc. are set as the environmental influencing factors, and the power supply capacity is taken as the controlling factor. The power supply subject receives the environmental information to make the corresponding reaction behavior, that is, to adjust its capacity. The rules of behavior of the power source body will be modified according to the increase or decrease of power generation cost after the capacity change, to minimize the power generation cost, and finally get the power source capacity planning scheme of the new energy-containing power system. According to the proposed planning method, under the condition of ensuring that the system power generation cost remains unchanged, the power supply capacity of the power system in the region is re-planned to verify the effectiveness of the method proposed in this paper. The schematic diagram of the power supply structure after the planning of the power system in this area is shown in Figure 3.

At the same time, with the objectives of minimizing the generation cost and guaranteeing the amount of new energy consumption, the crossover rate is taken as 0.7 and the variance rate as 0.4, and the genetic algorithm is used to solve the problem and compare it with the CAS scheme. The planning results and production simulation results are shown in Table 2, and the generation cost of each planning scheme is shown in Table 3.

As can be seen from Table 3, after using the genetic algorithm and the method proposed in this paper to re-plan the capacity of new energy power supply, the former increases the installed capacity of wind power and photovoltaic, which makes wind power and photovoltaic consumption improve, and the latter increases the installed capacity of wind power and photovoltaic capacity, and the installed capacity of photovoltaic does not change much; although the wind power consumption is not much improved, the complementary role of wind power and photovoltaic power supply has been strengthened, so that the wind power consumption is obvious. Although the wind power consumption is not much improved, the complementary effect between wind power and PV power is strengthened, which makes the wind power consumption obvious. Compared with the genetic algorithm, the method proposed in this paper makes full use of the complementary characteristics of wind and solar, so the new energy abandonment has been improved to some extent. In addition, the optimized thermal power consumption is reduced, which is conducive to reducing fuel consumption and carbon emissions and reducing environmental pollution. As can be seen from Table 3, the power planning scheme obtained by using a genetic algorithm does not get significant improvement in power generation cost, while the power generation cost is reduced by about 1 % by using CAS scheme, which saves cost investment.

Journal of Physics: Conference Series **3011** (2025) 012079 doi:10.1088/1742-6596/3011/1/012079

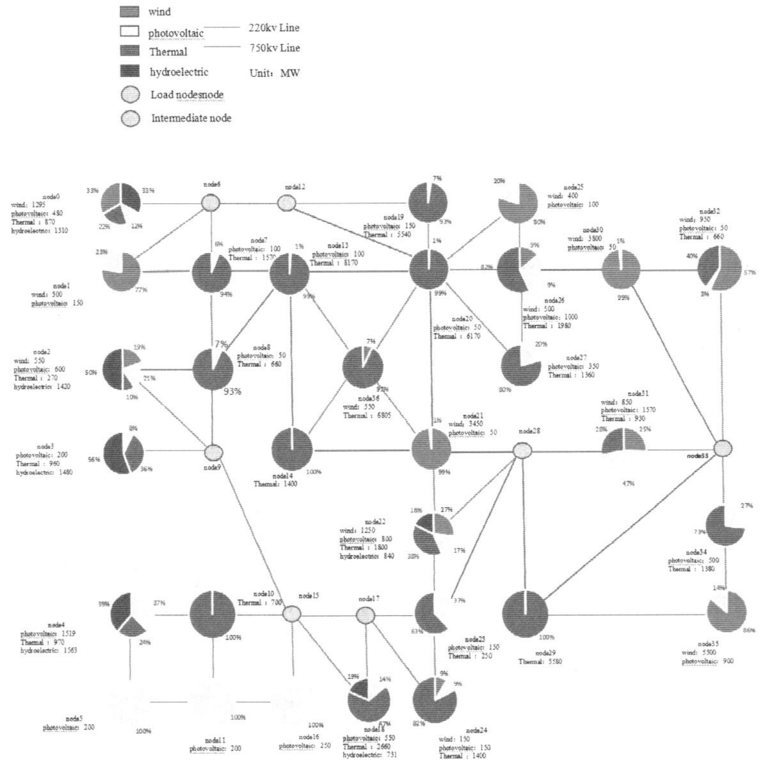

Figure 3. Schematic diagram of power system structure.

Table 2. Power Capacity Planning Program and Consumption.

installed capacity (MW)	Wind power	Photovoltaic	Thermal power	Hydroelectric	Total installed
Current	19745	10269	52085	7344	89443
genetic algorithm	20314	12130	51030	7307	90781
CAS	20750	10350	50365	7148	88613
Consumption (GWh)	Wind power	Photovoltaic	Thermal power	Hydropower	Percentage of new energy
Current	87669.06	14861.89	263052.34	28832.32	26 %
genetic algorithm	89167.14	17496.81	259049.87	28714.06	27.04 %
CAS	92034.74	15079.09	259009.85	28305.97	27.16 %
Abandonment of electricity (GWh)	Wind power abandoned	Amount of light abandoned	Wind abandonment rate	Light abandonment rate	Total new energy power abandonment rate
Current	17224.82	1960.1	16.42 %	11.65 %	15.76 %
genetic algorithm	18861.95	2429.72	17.46 %	12.19 %	16.64 %
CAS	17361.85	1858.36	15.87 %	11 %	15.21 %

Table 3. Power generation costs.

Generation costs (Billion yuan)	wind power	photovoltaic	thermal power	hydroelectricity	total cost
Current	4.989	2.94	41.844	2.975	52.748
genetic algorithm	5.133	3.473	41.189	2.96	52.755
CAS	5.243	2.964	41.117	2.897	52.221

Through the above analysis, the CAS scheme can improve new energy consumption and reduce new energy power abandonment, and at the same time has a good economy, which verifies the effectiveness and practicality of the optimization planning method of power capacity based on the theory of complex adaptive systems.

5. Conclusion

In this paper, the operational characteristics of wind power, photovoltaic power, thermal power, and hydropower are fully considered, the complementary characteristics of photovoltaic are analyzed, and a regional power planning model with new energy sources based on the theory of complex adaptive systems is proposed. The main conclusions are as follows:

(1) Combined with the theory of a complex adaptive system, the output characteristics and complementary characteristics of the new energy-containing power system are analyzed, and the power capacity planning method for the new energy-containing area is proposed.

(2) A power capacity planning model for new energy-containing areas is established to minimize power generation costs, which can effectively improve the consumption of new energy, reduce the economic losses caused by wind and light abandonment, reduce the amount of emissions from thermal power units and improve the pollution of the environment while ensuring the minimization of power generation costs.

(3) It can effectively improve the consumption of new energy, reduce the cost of power generation and the economic loss caused by abandoned wind and light, reduce the amount of emissions from thermal power units, and improve the pollution of the environment.

(4) The power capacity allocation scheme obtained through the model solution can guide the decision maker to determine the final scheme, which provides reference significance to the current power planning in the area containing new energy.

References

[1] Wang J, Huang Y H, Li C Xiang K L and Lin Y (2020) A time series modeling approach for multi-photovoltaic power plants considering spatial correlation and weather type classification. Grid Technology,44 (04): 1376-1384.

[2] Dong Y J (2019) Optimization research on capacity rationing of multi-energy complementary new energy power plant. Nanjing University of Science and Technology, Nanjing.

[3] Hua S (2019) Challenges and responses of new energy grid connection. North China Electric Power University, Bei jing.

[4] Wang Z D (2015) Optimization method of power source scale for joint transmission of wind and thermal power from extra-high voltage direct current. Power Construction,36 (10): 60-66.

[5] Xu D, Wang B, Zhang J L and Li W G (2016) Integrated scheduling plan model of wind-glow-fire power for ultra-high voltage direct current transmission. Power System Automation,40 (06): 25-29.

[6] Tan Q L, Ding H Y, Li Y and Li R (2020) Multi-objective optimization of combined wind-fire dispatch strategy considering economic-environmental balance. Electric Power Construction,41 (08): 129-136.

[7] Liao Y Z, Lu Z X, Zhou Q Y, Han J H and Huang B (2015) Coordinated planning of wind-containing power grid based on reliability balanced optimization. Grid Technology,39 (01): 16-22.

[8] HOLLAND H J (1995) Complex adaptive system. Addison Wesley, Boston.

[9] HOLLAND J (2006) Emergence from chaos to order. Shanghai Science and Technology Press, Shanghai.

[10] Shi Y R (2007) Research on modeling and simulation of socio-economic systems based on complex adaptive systems theory. Huazhong University of Science and Technology, Wuhan.

[11] Zhou W, Sun H, Gu H, Ma Q and Chen X D (2012) Dynamic economic scheduling of wind farm-containing power systems taking into account risky standby constraints. Chinese Journal of Electrical Engineering, 32 (01): 47-55.

[12] Tan T F (2017) Research on key technology of power grid planning considering wind power. Southeast University, Nanjing.
[13] Ankur S R, Fahad I, Anwar S S and Mini S T (2019) Hybrid methodology to analyse reliability and techno-economic evaluation of microgrid configurations. IET Generation, 13 (21): 4778-4787.

Journal of Physics: Conference Series **3011** (2025) 012080 doi:10.1088/1742-6596/3011/1/012080

Impedance modeling of grid-forming converter via decoupling of synchronization control and vector control

Shumin Sun[1,2], Nan Wang[1,2*], Yan Cheng[1,2], Shibo Wang[1,2], Wanzheng Chang[1,2], Yu Yi [1,2], Jingwen Sun [1,2], Yanjun Zou [1,2], Gongrun Wang [1,2], He Cai [1,2], Jiawei Xing [1,2] and Chenglong Wang [1,2]

[1]State Grid Shandong Electric Power Company Electric Power Research Institute, Jinan, Shandong, 250002, China
[2]Shandong Smart Grid Technology Innovation Center, Jinan, Shandong, 250002, China

[*]Corresponding author's e-mail: wangnantju@163.com

Abstract. A novel method for the small-signal stability assessment of grid-connected converters in a grid-forming mode is proposed by adopting a new impedance modeling approach. Firstly, impedance models for both the grid side and the equipment side are established based on the dynamic equations of the inner and outer loops, as well as the synchronous control loop, combined with the dynamic control equations of the power grid. Secondly, the determinant of the impedance model system matrix is solved. Lastly, the stability is examined using the Nyquist stability criterion with the MATLAB simulation tool.

1. Introduction

As the installed capacity of new energy power stations such as photovoltaic in China continues to increase, China is gradually transitioning to a "dual-high" power system characterized by a high proportion of renewable energy and power electronic devices [1-2]. However, due to the lack of frequency and inertia support in the grid connection of new energy, strong oscillation issues, and even instability may occur between power electronic devices and within the power system. To better analyze the stability of new energy power systems, there is a need for simple, accurate, and reliable methods for small-signal stability assessment.

In terms of small-signal stability, grid-forming converters can be categorized into four types of oscillations based on the frequency of potential oscillation phenomena: low-frequency oscillations, sub-synchronous oscillations, power frequency oscillations, and high-frequency oscillations [3] [4] mathematically demonstrated that insufficient damping in virtual synchronous machines leads to power oscillation instability in the system, and proposed that enhancing damping leads to better small-signal stability within the system. In October 2009, sub synchronous oscillation phenomena were observed for the first time in a wind farm in Texas, USA, which were caused by the interaction between the control part of doubly-fed induction generators, leading to the disconnection of many wind turbines and the destruction of crowbar circuits [5] [6] analyzed the interaction between the AC grid and direct-drive permanent magnet synchronous motors using small-signal eigenvalue methods, time-domain simulation methods, and impedance modeling, thereby analyzing the sub-synchronous oscillation phenomena observed in the Hami wind farm in Xinjiang, China [7] modeled the power control loop in a rotating coordinate system, revealing that the frequency shift of grid impedance is the cause of power frequency oscillations in virtual synchronous machine grid-connected systems [8] analyzed grid-forming converters with dual-loop control of voltage and current based on passive theory and proposed a grid current feedforward passive correction method to improve the small-signal stability of grid-forming converter grid-connected systems. The analysis of small-signal stability for grid-forming converters is now relatively comprehensive, gradually expanding to explore the impact of control inner loops on stability based on the analysis of the impact of power outer loops.

At present, it is generally believed that one of the main causes of system oscillation is insufficient damping, and the main methods to study the stability of new power systems that show the damping level are the eigenvalue method [9] and the mode identification method [10]. The eigenvalue method first establishes the small signal model, then solves the eigenvalue of the model, and judges the stability of the system by examining the variation in the eigenvalue decay of the oscillation modes. In addition, the traditional impedance modeling method simulates the synchronous control link and the current inner loop simultaneously, which is more complicated. Therefore, this paper attempts to place the synchronous control loop on the grid side and the inner and outer loops on the equipment side, thereby establishing a new type of impedance model, which is more concise than traditional impedance modeling. At the same time, it abandons the use of eigenvalue analysis and instead analyzes system stability through the Nyquist criterion.

2. Inverter Grid-Tied system modeling

2.1. Grid-Forming Converter Modeling

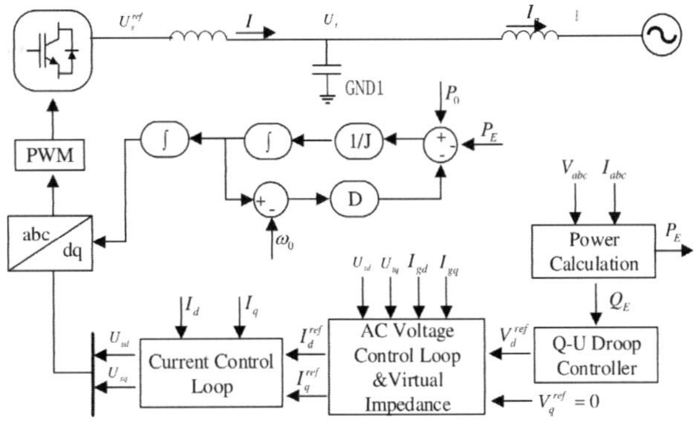

Figure 1. Schematic diagram of the grid-connected system of a typical grid-forming converter.

Figure 1 shows a typical grid-connected system. The control link consists of four parts, which are the power calculation link, network type control link (such as droop control), outer loop voltage control link, and inner loop current control link. Finally, the PWM modulation signal is output by the current loop. The power calculation step is responsible for calculating the active and reactive power that the grid-connected inverter needs to output to meet the needs of the grid and maintain the balance. The control link of the grid structure can realize the active support of the voltage and frequency. The outer loop voltage regulation ensures the sustained stability of the inverter's output voltage, and the inner loop current control link is responsible for controlling the output current of the inverter.

For a single-machine infinite bus power system, the equipment side can be equivalently modeled as impedance and a voltage source, while the grid side can be equivalently modeled as a load. The equivalent circuit can be depicted as shown in Figure 2:

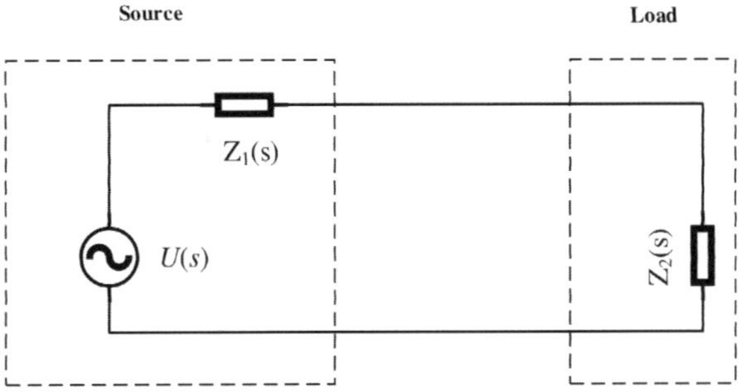

Figure 2. Equivalent Circuit of a Single-Machine Infinite System.

To apply the Nyquist stability criterion for analysis, based on the above figure, the impedance model of the equipment side can be represented as the forward transfer function matrix G(s), and the inverse of the transfer function matrix of the grid side impedance model can be represented as the feedback transfer function H(s), thus forming a complete closed-loop multivariable feedback system. According to the Nyquist stability criterion, it is only necessary to examine whether the Nyquist plot is derived from a characteristic equation.

The dynamic equations of the outer loop and the inner loop are presented below:

$$\begin{cases} I_{dref} = PI_{VC}(s)(U_{tdref} - U_{td}) - U_{tq}\omega_0 C_f + f_{IF}(s)I_{gd} \\ I_{qref} = PI_{VC}(s)(U_{tqref} - U_{tq}) + U_{td}\omega_0 C_f + f_{IF}(s)I_{gq} \\ U_d = PI_{CC}(s)(I_{dref} - I_d) - \omega_0 L_f I_q + f_{VF}(s)U_{td} \\ U_q = PI_{CC}(s)(I_{qref} - I_q) + \omega_0 L_f I_d + f_{VF}(s)U_{tq} \end{cases} \tag{1}$$

where I_{dref} and I_{qref} are VSC current inner loop dq axis component reference value, U_{td} and U_{tq} are PCC bus voltage dq axis component, $PI_{CC}(s)$ and $PI_{VC}(s)$ are the transfer functions of PI regulators in the inner current loop and the outer voltage loop, respectively, $f_{VF}(s)$ and $f_{IF}(s)$ are the transfer functions of the voltage and current feedforward filters, respectively, ω_0 is rated angular frequency.

Filter inductance circuit equation:

$$\begin{cases} U_{sd} - U_d = sL_f I_d - \omega_0 L_f I_q \\ U_{sq} - U_q = sL_f I_q + \omega_0 L_f I_d \end{cases} \tag{2}$$

where U_d and U_q are the dq axis components of the inverter node voltage, respectively.

The network-type converter adopts virtual synchronous control, and its governing equation is presented below:

$$\begin{cases} s\theta = \omega_{JD} \\ \omega_{JD} = \dfrac{f_{PQ}(s)P - P_M + D\omega_0}{sJ + D} \end{cases} \tag{3}$$

where ω_{JD} is the virtual synchronous control angular frequency, J is the inertia parameter and D is the damping parameter.

Ignoring the filter capacitance, the transmission line inductance Lg dynamic equation is:

$$\begin{cases} E_d = U_d + sL_g I_d - \omega_0 L_g I_q \\ E_q = U_q + sL_g I_q + \omega_0 L_g I_d \end{cases} \tag{4}$$

In addition, the component of grid voltage E under the virtual synchronization link is:

$$\begin{cases} E_d = E\cos\theta \\ E_q = -E\sin\theta \end{cases} \tag{5}$$

Linearizing Equations (1)-(5) yields the system's small interference signal model as follows:

$$\begin{cases} \Delta U_{sd} - \Delta U_d = sL_F \Delta I_d - \omega_0 L_F \Delta I_q \\ \Delta U_{sq} - \Delta U_q = sL_F \Delta I_q + \omega_0 L_F \Delta I_d \end{cases} \tag{6}$$

$$\begin{cases} \Delta I_d - \Delta I_{gd} = sC_F \Delta V_d - \omega_0 C_F \Delta V_q \\ \Delta I_q - \Delta I_{gq} = sC_F \Delta V_q + \omega_0 C_F \Delta V_d \end{cases} \tag{7}$$

$$\begin{cases} \Delta U_{td} - \Delta U_d = sL_g \Delta I_d - \omega_0 L_g \Delta I_q \\ \Delta U_{tq} - \Delta U_q = sL_g \Delta I_q + \omega_0 L_g \Delta I_d \end{cases} \tag{8}$$

$$\begin{cases} \Delta U_d = PI_{CC}(s)(\Delta I_d^{ref} - \Delta I_d) - \omega_0 L_F I_q + f_{VF}(s)\Delta V_d \\ \Delta U_q = PI_{CC}(s)(\Delta I_q^{ref} - \Delta I_q) + \omega_0 L_F I_d + f_{VF}(s)\Delta V_q \end{cases} \tag{9}$$

$$\begin{cases}\Delta I_d^{ref} = PI_{\mathrm{VC}}(s)(\Delta V_d^{ref} - \Delta V_d) - \Delta V_q \omega_0 C_F + k_F \Delta I_{gd} \\ \Delta I_q^{ref} = PI_{\mathrm{VC}}(s)(\Delta V_q^{ref} - \Delta V_q) - \Delta V_d \omega_0 C_F + k_F \Delta I_{gq}\end{cases} \tag{10}$$

$$\begin{cases}\Delta P_E = V_{d0}\Delta I_{gd} + I_{gd0}\Delta V_d + V_d I_{gq0} \\ \Delta Q_E = \Delta V_q I_{gd0} - V_{d0}\Delta I_{gq} - I_{q0}\Delta V_d\end{cases} \tag{11}$$

$$\begin{cases}s\Delta\theta = \Delta\omega, \\ Js\Delta\omega = \Delta P_0 - \Delta P_E - D\Delta\omega\end{cases} \tag{12}$$

Establishing impedance models for the inner and outer loops on the equipment side and the synchronous control loop on the grid side, and then linearizing and combining them in parallel can yield the following impedance model:

$$\begin{cases}\begin{bmatrix}\Delta U_d \\ \Delta U_q\end{bmatrix} = \left(Z_{\text{converter}} + Z_{\text{grid}}\right)\begin{bmatrix}\Delta I_d \\ \Delta I_q\end{bmatrix} \\ Z_{\text{converter}} = \begin{bmatrix}Z_1 & 0 \\ 0 & Z_1\end{bmatrix} \\ Z_{grid} = \begin{bmatrix}Z_2 & Z_3 \\ Z_4 & Z_5\end{bmatrix} \\ Z_1 = \dfrac{-[PI_{CC}(s) + sL_f + PI_{CC}(s) f_{IF}(s)]}{1 - f_{VF}(s) + PI_{CC}(s) PI_{VC}(s)} \\ Z_2 = -\dfrac{X_g f_{PQ}(s) + sL_g(s^2 J + sD)}{s^2 J + sD + X_g f_{PQ}(s)} \\ Z_3 = \dfrac{\omega_0 L_g (s^2 J + sD)}{s^2 J + sD + X_g f_{PQ}(s)} \\ Z_4 = -\dfrac{\sqrt{1 - X_g^2} f_{PQ}(s) + \omega_0 L_g (s^2 J + sD)}{s^2 J + sD + \sqrt{1 - X_g^2} f_{PQ}(s)} \\ Z_5 = -\dfrac{sL_g(s^2 J + sD)}{s^2 J + sD + \sqrt{1 - X_g^2} f_{PQ}(s)}\end{cases} \tag{13}$$

$PI_{CC}(s)$ refers to the inner current loop PI link, $PI_{VC}(s)$ refers to the outer voltage loop PI link, L_f is the filter inductance, $f_{IF}(s)$ is the current feedforward filter link, $f_{VF}(s)$ is the voltage feedforward filter link, J is the inertia parameter, D is the damping parameter, L_g is the grid-side inductance, X_g is the grid-side reactance, and ΔU_d, ΔU_q, ΔI_d, ΔI_q are the small-signal voltage and current at the point of common coupling.

Extracting the impedance matrix from the aforementioned Equation (1) and calculating its determinant yields the closed-loop characteristic equation of the system as follows:

$$\det\left(Z_{\text{converter}} + Z_{\text{grid}}\right) = 0 \tag{14}$$

where det () stands for the determinant of a matrix.

2.2. Stability determination

After the impedance determinant is obtained, that is, the impedance modeling is completed, the characteristic equation of the network type control network system can be obtained. The system parameters are set reasonably for simulation, and the Nyquist diagram is drawn by the Nyquist criterion. If the image surrounds the origin, it indicates that the system is stable with small perturbations; otherwise, it indicates that the system is unstable. The steps of modeling with multivariable feedback impedance are as follows:

(a) Obtain the Impedance Determinant: Complete the impedance modeling of the grid-forming converter control network system to derive the impedance determinant.

(b) Formulate the characteristic equation: Utilize the impedance determinant to obtain the characteristic equation of the grid-forming converter control network system.

(c) Set system parameters. Reasonably set the system parameters for simulation purposes to ensure the accuracy of the subsequent analysis.

(d) Draw the Nyquist diagram. Employ the Nyquist criterion to plot the Nyquist diagram of the system. This process entails graphing the frequency response of the open-loop transfer function within the complex plane.

(e) Analyze the Nyquist plot.

(f) Determine system stability.

If the Nyquist plot encircles the origin (0, 0) in the counterclockwise direction, it indicates that the system is stable with small perturbations. Conversely, if it does not encircle the origin or encircles it in a clockwise direction, it suggests that the system is unstable.

The flow diagram of the small disturbance stability assessment method for a grid-connected system of grid-forming converter is shown in Figure 3:

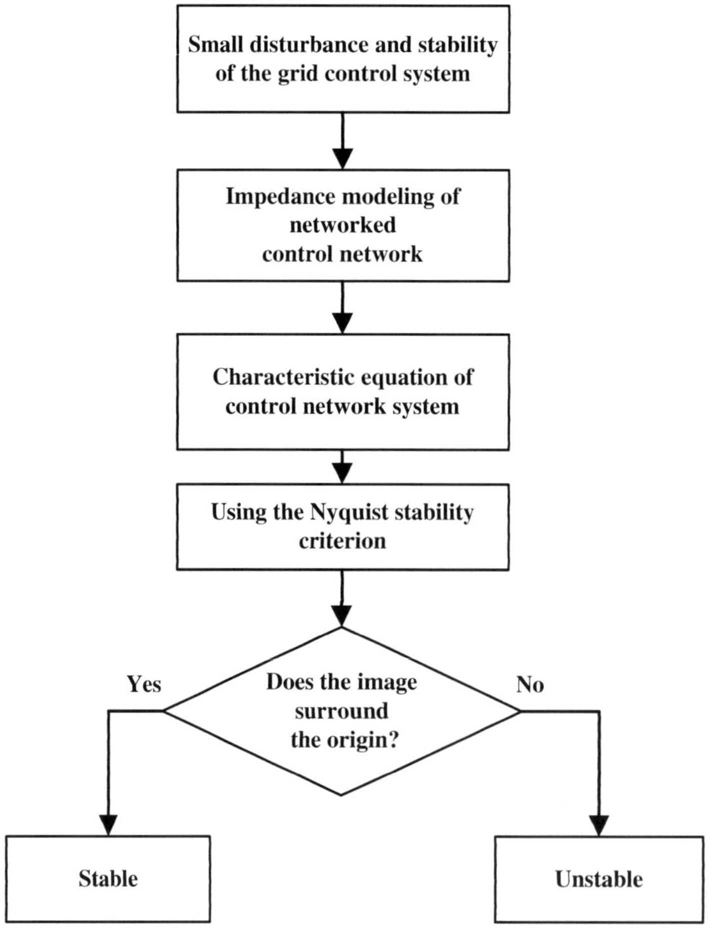

Figure 3. Technical route flow chart of the stability assessment method.

Journal of Physics: Conference Series **3011** (2025) 012080 doi:10.1088/1742-6596/3011/1/012080

3. Stability analysis and simulation

Matlab simulation software is utilized to evaluate the stability of the impedance model and Nyquist stability criterion, and verified by time domain simulation.

Firstly, the parameters set in the simulation experiment are given in Table 1:

Table 1. Simulation parameter setting table.

Symbol	Meaning	Value (p.u.)/ expression
L_f	Filter inductance	0.01
L_g	Net side inductance	0.3 (stable) 0.03 (unstable)
J	Inertia parameter	2
D	Damping parameter	42 (stable) 12 (unstable)
$PI_{CC}(s)$	Current inner loop PI link	2+40/s
$PI_{VC}(s)$	Voltage outer loop PI link	1+10/s
$f_{IF}(s)$	Current feedforward filter	1/(1+0.05 s)
$f_{VF}(s)$	Voltage feedforward filter	1/(1+0.005 s)

By setting the parameters differently, the Nyquist diagram under the two conditions of stability and instability can be obtained as follows.

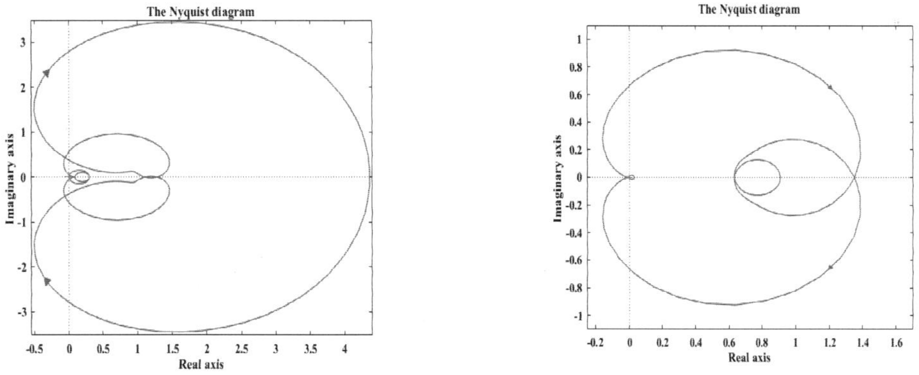

(a) Nyquist diagram under the stable condition (b) Nyquist diagram under the unstable condition

Figure 4. Nyquist plot for two different stable conditions.

Figure 4 shows the Nyquist diagram in two different cases. After solving the closed-loop characteristic equation, according to the above-mentioned theory, the Nyquist diagram is drawn for the characteristic equation for investigating whether it surrounds the origin. Firstly, the grid reactance is set to 0.3 p.u. and the damping parameter D is set to 42 to obtain the corresponding Nyquist diagram as shown in Figure (a). As can be seen from the figure, the Nyquist plot at this time does not surround the origin and is therefore stable. By setting the grid reactance to 0.03 p.u. and adjusting the damping parameter to 12, the corresponding Nyquist diagram can be obtained as shown in Figure (b). As can be seen from the figure, the Nyquist diagram at this time surrounds the origin and is therefore unstable.

From the comparison of the two graphs, it is evident that the system's damping parameters and the power network's robustness exert a notable impact on system stability. Additionally, the validity of the evaluation method has been confirmed.

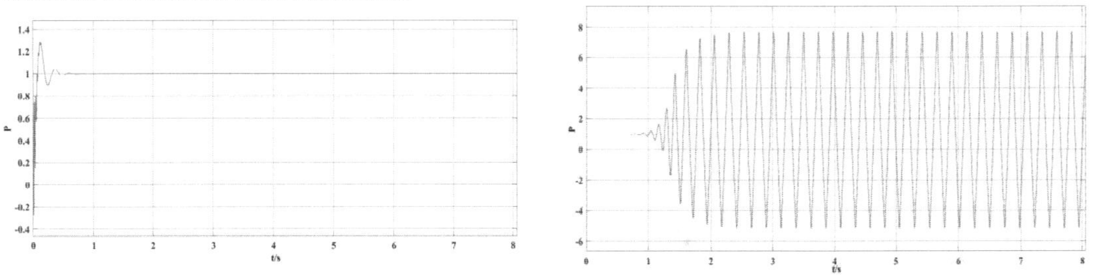

(a) Time-domain simulation waveform under stable conditions

(b) Time domain simulation waveform under unstable conditions

Figure 5. Time domain simulation waveforms under two different stable conditions.

The time domain simulation waveforms made respectively for the above two cases are shown in Figure 5. The values of stability and instability of the system are substituted into the single-machine infinity model, and the corresponding results are obtained. Observing the figure above, it is evident that the waveform achieves convergence. The system is stable when the parameters under stable conditions are adopted. When the parameters under unstable conditions are used, the waveform does not converge and the system is unstable. The simulation results in the time domain align with expectations, thereby further validating the accuracy of the theoretical model.

4. Conclusions

This paper presents a new method for small disturbance stability evaluation of grid-connected converter systems, which is based on an innovative impedance modeling technique. Based on the dynamic equations of the inner and outer loops and synchronous control loops, the impedance models of the network side and the equipment side are constructed. Then, by solving the determinant of the system matrix of the impedance model, the key parameters are provided for the stability analysis of the system. Finally, the system stability is evaluated comprehensively by using the Matlab simulation tool and Nyquist stability criterion. The impedance modeling method proposed in this study provides a new perspective and theoretical basis for small disturbance stability evaluation of grid-connected converter systems.

Acknowledgments

The work is financially supported by the Science and Technology Project of State Grid Shandong Electric Power Company(52062624000N).

References

[1] Shi, Y., Xv Y. and Wu, X. Resilience-enhancing Control Method for Grid-connected Virtual Oscillator. Shandong Electric Power, 2024, 51(08): 10-17.

[2] Wang, Y., Jia, S. and Liu, J. Research on Modeling and Simulation Method of Grid-side Converter Based on Scilab. Shandong Electric Power, 2022, 49(10): 38-43.

[3] Xie, H., Teng X., Zhang Y., Zheng T. and Chen L. Analysis of Economic Influence Factors in Wind-Photovoltaic-Storage Microgrid. Automation of Electric Power Systems, 2019, 43(06): 107-115.

[4] Ren, W., LARSEN, E. A Refined Frequency Scan Approach to Sub-Sy-nchronous Control Interaction (SSCI) Study of Wind Farms. IEEE Transactions on Power Systems: A Publication of the Power Engineering Society, 2016, 31(5): 3904-3912.

[5] Liu H., et al. Subsynchronous Interaction Between Direct-Drive PMSG Based Wind Farms and Weak AC Networks. IEEE Transactions on Power Systems, 2017, 32(6): 4708-4720.

[6] Meng C., Wang Y., et al. Medium-frequency Characteristic Analysis and Oscillation Suppression of Converter with Interaction Between Grid-forming and Grid-following Control. Automation of Electric Power Systems.

[7] Li H., Nian H., et al. Review of Analysis and Suppression Methods for Wide-band Oscillation in Wind Power Grid. Connected Systems, 2023, 1(03): 237-255.

[8] Chen Y., et al. Calculation Method for Unstable Eigenvalue of Power System Based on Exponential Transformation, 2023, 47(03): 161-168.

[9] Zhang C., Qiu B. and Liu J. Power System Low Frequency Oscillation M-odal Identification Based on EFEMD-HT Energy Algorithm, 2022, 16(03): 48-57.

Simulation of secondary overvoltage of voltage divider for wall bushing in flexible direct current transmission system

Honglian Duan[1,*], Pengbo Yin[2], Hong Shen[1], Yifeng Ye[1] and Xin Tong[3]

[1]School of Electric and Electrical Engineering, North China Electric Power University, Changping District, Beijing 102206, China
[2]State Key Laboratory of Power Grid Environmental Protection, Wuhan, Hubei, 430074, China
[3]State Grid Hubei Electric Power Research Institute, Wuhan, Hubei, 430077, China

[*]Correspondingauthor's e-mail: 2677949402@qq.com

Abstract. Wall bushings used in Flexible Direct Current Transmission Systems are key equipment for achieving electrical connections between equipment inside and outside the valve hall, and the ground voltage of the bushings can reflect their operating status. However, there have been cases of damage to the secondary side of the voltage divider in engineering applications using the voltage measurement method of dividing the voltage at the end screen of the wall bushing, which is currently used as a monitoring method. This article first analyzes the structure of the bridge arm capacitive wall bushing and establishes a wideband equivalent circuit model for the wall bushing. Next, we integrate the wall bushing and voltage divider models with the wideband model of the Flexible DC Transmission System. We also simulate the overvoltage on the secondary side of the wall bushing voltage divider under the influence of electromagnetic interference sources during switch operation. The simulation shows that the overvoltage of the wall bushing secondary system is 4p.u, which exceeds the unit value of the voltage borne by the voltage divider.

1. Introduction

Flexible Direct Current Transmission System, as a new generation of direct current transmission technology, is an important component of building a smart grid and a strategic choice to change the development pattern of the power grid. Its operational reliability is crucial. The bridge arm wall bushing located between the bridge arm reactor and the converter valve is mainly used in the Flexible DC Transmission System to guide electrical equipment wires through the wall, so that the equipment inside and outside the valve hall can operate normally. Therefore, monitoring the status of through wall bushing cannot be ignored [1-2].

The capacitive through wall bushing is mainly composed of adhesive impregnated paper insulation core, flange, hollow composite insulator jacket, copper current carrying conduit, and insulation filling paste. The tap from the outermost electrode of the bushing core wire is the terminal screen. Since the grounding of the terminal screen tap connection can serve as a voltage divider, the bridge arm can be drawn through wall bushing using the terminal screen voltage detection method, which can draw on the monitoring experience of transformer bushings. The terminal screen tap is connected to the secondary system for online monitoring and waveform recording of the bushing. In a certain project, the casing monitored using this method malfunctioned immediately after being put into operation for annual maintenance, and there was an abnormal discharge phenomenon at the top of the terminal screen. When the voltage divider box was opened on site, internal components were found to be burned, seriously affecting the normal operation of the system.

The reasons for damage to electrical equipment in the Flexible DC Transmission System are often caused by lightning strikes, switch operations, and short circuit faults. Among the electromagnetic disturbance characteristics experienced by converter stations, the first two have more research results and mature characteristics [3-5]. There are many reports in the academic community on the waveform characteristics of short-circuit faults and overvoltage calculations [6-9]. During the start-up process of the flexible straight system, lightning strikes and short circuit faults are often accidental situations. According to the study of the Flexible DC Transmission System diagram of this project, the electromagnetic transient impact of the isolation switch operation on the through wall bushing voltage divider cannot be ignored, as shown in Figure 1. However, there is currently limited research on the electromagnetic transient characteristics of wall bushings and their secondary systems during the operation of isolation switches. Therefore, it is urgent to conduct a time-domain simulation of the overvoltage on the secondary side of the voltage divider used for the wall bushing in this project to fill the corresponding gap.

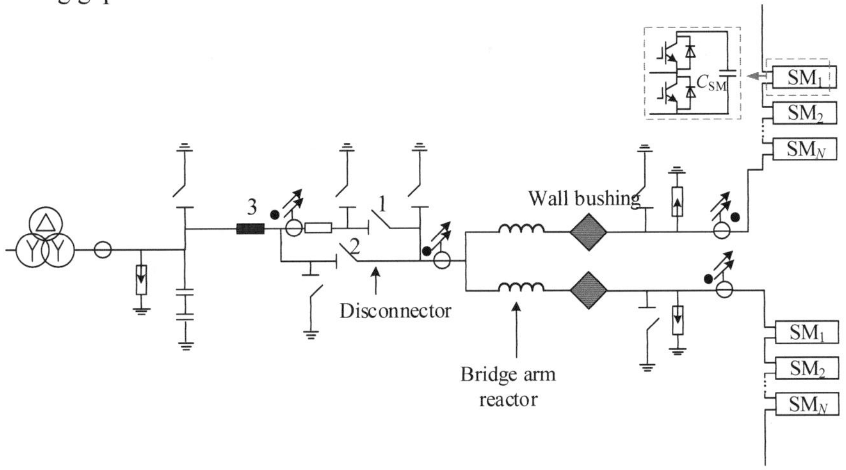

Figure 1. Location diagram of wall bushing and disconnector.

Establishing a broadband model for the main equipment of a converter station is currently an essential means of studying electromagnetic interference in converter stations [10], which can accurately predict the level of electromagnetic interference and is widely used in practical engineering. Due to the small capacitance value of the through wall conduit at the beginning of design, its wideband model is often ignored and only treated as a wire. At present, the capacitance value of through wall bushings has reached the order of hundreds of pF, and its impact on Flexible DC Transmission System cannot be ignored. Therefore, it is necessary to study the wideband model of through wall bushings.

This article first analyzes the capacitive wall bushing structure of the Flexible DC Transmission System, obtains a "T" - shaped equivalent wideband circuit model, and conducts wideband modeling of the main equipment of the Flexible DC Transmission System. Secondly, under the influence of the isolation switch, the broadband model of the established flexible DC transmission system is simulated to calculate the standard value of the overvoltage on the secondary side of the voltage divider used for wall bushing.

2. Broadband modeling of flexible DC transmission system

2.1. Broadband modeling of wall bushing

To study the electromagnetic transient characteristics of wall bushings used in the Flexible DC Transmission System, it is necessary to model the mechanism model of bridge arm wall bushings and consider relevant stray parameters for the model structure. The specific physical image of the capacitive wall bushing is shown in Figure 2. The mechanism model and detailed wideband model of the capacitor core are shown in Figure 3 (a) and (b), respectively.

There is stray inductance in each screen of the wall bushing, and stray capacitance between screens. At frequencies of 100kHz and below, the stray inductance of each screen inside the wall bushing can be ignored, and a stray capacitance C_1 can be used to represent the core wire to the end screen. The stray capacitance C_1 is formed by the series connection of stray capacitance between each screen, so as the number of screens changes, the capacitance value will also change accordingly. When a short circuit occurs in one screen of the wall bushing, the number of stray capacitors in series decreases, and the total capacitance value from the core wire to the end screen increases, resulting in an increase in the leakage current flowing through the wall bushing. In practical engineering, to measure the voltage of the wall bushing, the terminal screen is suspended, and there is still stray capacitance C_2 between the terminal screen and the casing flange. The inductance of each screen inside the wall bushing can be connected in parallel to form an inductance L_1, which is used to flow the corresponding current through the flexible straight system, creating an electrical connection between the equipment inside and outside the valve hall. Thus, a "T" - shaped broadband equivalent circuit for the through wall sleeve is obtained, and the broadband model of the secondary system voltage divider monitoring equipment is considered, as shown in Figure 4.

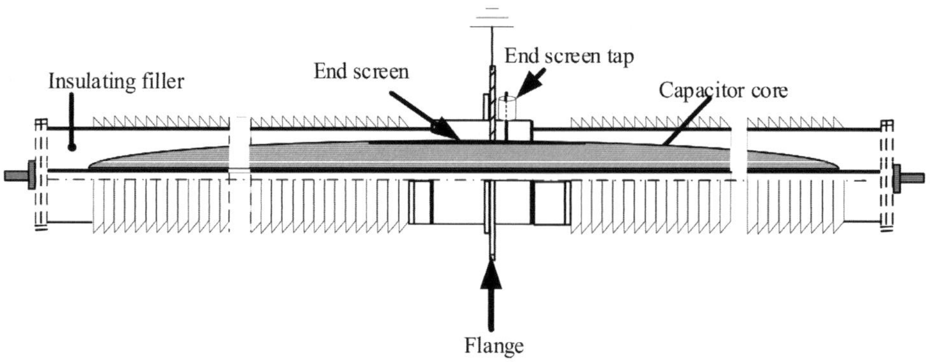

Figure 2. The physical picture of wall bushing.

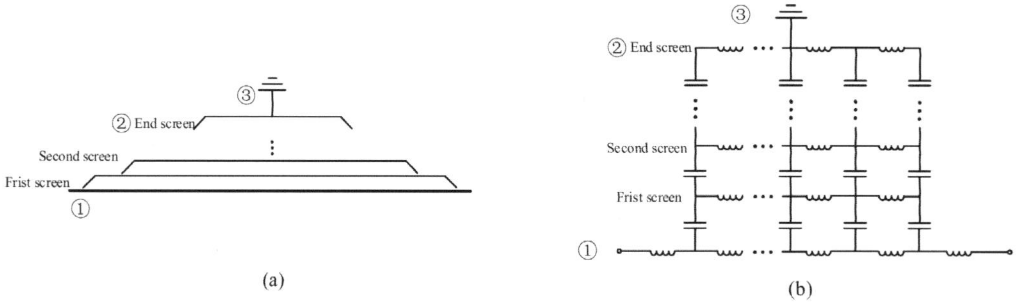

Figure 3. Mechanism model diagram and detailed wideband model of through wall bushing.

Journal of Physics: Conference Series **3011** (2025) 012081 doi:10.1088/1742-6596/3011/1/012081

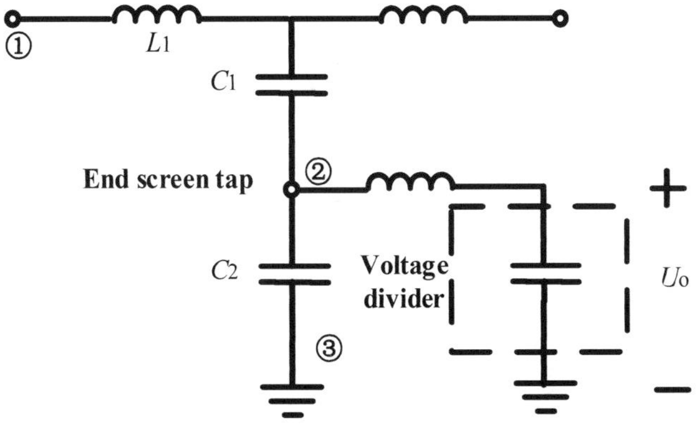

Figure 4. “T” - shaped equivalent wideband model and voltage divider diagram of wall bushing.

2.2. Broadband modeling of other devices in the Flexible DC Transmission System

Wideband modeling of the Flexible DC Transmission System provides a coupling path foundation for disturbance sources. Equipment that needs to be considered for broadband modeling includes converter transformers, voltage transformers, lightning arresters, isolation switches, and bridge arm reactors.

Among them, the wideband model of the converter transformer is shown in Figure 5 (a), where the inductance is at the mH level. At high frequencies, the inductance is considered an open circuit, so the wideband model of the converter transformer can be equivalent to a capacitor.

The bridge arm reactor plays a protective role in the circuit and can effectively prevent high-order harmonics from entering the valve hall and causing damage to the converter in a Flexible DC Transmission System. The wideband circuit diagram is shown in Figure 5 (b). Among them, C_g is the stray capacitance to ground, C_s is the stray capacitance between windings, and the value of stray capacitance is approximately one hundred pF. L_m is the main inductance of the bridge arm reactor, and R_m is the loss resistance. Due to the small value of R_m, it is usually ignored in broadband modeling.

The MMC converter valve bridge arm consists of 1 to 3 converter valve towers. The stray capacitance in the valve tower is mainly distributed as the stray capacitance between the upper and lower layers of the valve module or between two valve modules on the same layer, the stray capacitance to ground of each valve module, the capacitance of the valve module to the shielding plate, and the capacitance to ground of the shielding plate. Due to the presence of bridge arm reactors and large capacitors in submodules in Flexible DC Transmission System, when studying electromagnetic interference in high-frequency systems, the stray capacitance and inductance inside the converter valve can be ignored, and the ground stray capacitance of each module and the ground capacitance of the shielding plate can be regarded as a tower to ground capacitance. In broadband modeling of flexible straight systems, only the influence of the converter valve tower on the ground capacitance C_{Mg} can be considered. The specific model is shown in Figure 5 (c).

Other devices, such as lightning arresters and cables, are mainly considered for their ground stray capacitance in high-frequency oscillation analysis. The model is shown in Figures 5 (d) and (e).

Finally, by analyzing the broadband models of other devices in the Flexible DC Transmission System, a broadband model of the Flexible DC Transmission System was established, which can simulate and analyze the secondary overvoltage of the voltage divider used for wall bushing under the influence of isolation switch operation.

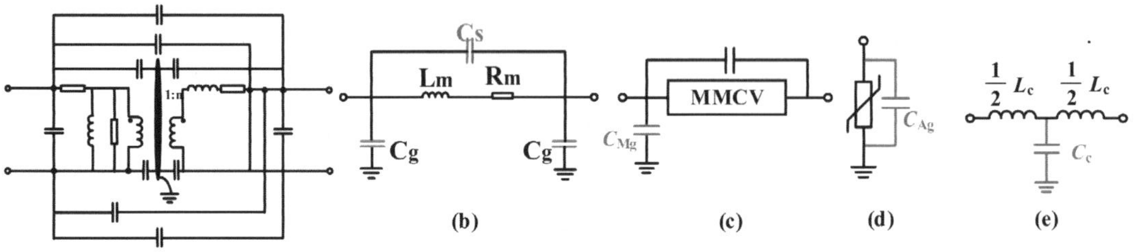

Figure 5. Broadband model of other devices in the Flexible DC Transmission System.

3. System simulation

In traditional engineering, the wall bushing is only used as an inductive element, and its capacitance is ignored. The wideband model of wall bushing proposed in this paper considers the influence of stray capacitance, so it can simulate the overvoltage caused by disturbance source in high frequency. Based on the wideband models of the wall bushing, voltage divider, and flexible straight system established in the previous chapter, this paper simulates the overvoltage on the secondary side of the wall bushing voltage divider in electromagnetic transient simulation software, as shown in Figure 6.

As shown in the figure, at the moment when the isolation switch is activated, the voltage on the secondary side of the voltage divider has reached 4p.u. Far exceeding the overvoltage threshold of 1p. u on the secondary side of the voltage divider used for wall penetrations. After the switch operation, the main frequency of du/dt also reached several hundred kHz to several MHz. Therefore, in traditional model analysis, due to the low system frequency, the overvoltage on the secondary side of the voltage divider cannot be calculated through simulation. At this point, the voltage divider breaks down, which is not conducive to the normal operation of the casing monitoring system in the project.

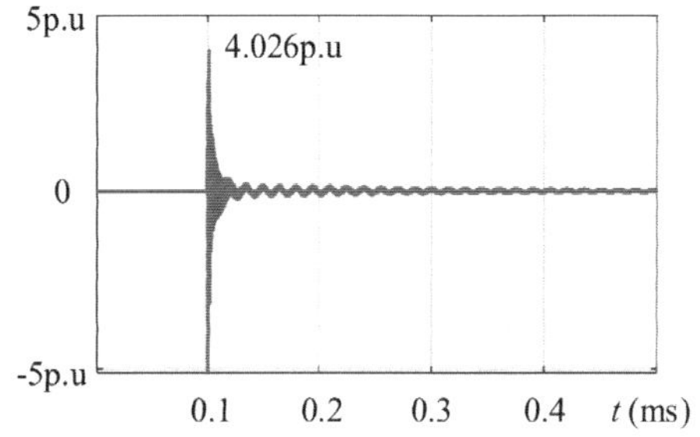

Figure 6. Secondary overvoltage of voltage divider used for wall bushing.

4. Conclusion

This article analyzes the structure of capacitive through wall bushing and establishes a "T" - shaped equivalent wideband model of through wall bushing. This enables simulation of the breakdown voltage of the secondary system voltage divider monitored by the bushing under the influence of isolation switches, and obtains the standard value of the breakdown voltage of the through wall bushing secondary system. The simulation results show that the overvoltage of the secondary system of the through wall bushing is 4p.u, which exceeds the unit value of the voltage borne by the voltage divider by 1p.u. This provides data support for the future protection of the secondary system of the through wall bushing.

In future research, we can focus on the suppression of overvoltage of voltage dividers for through wall bushing. The oscillation circuit where the model is located can be analyzed and studied to better protect the wall bushing and its secondary system in terms of equipment.

Acknowledgments

This work was financially supported by the Science and Technology Project of State Grid Corporation(5108-202218280A-2-364-XG).

References

[1] Wang. Y, Fang. S, Wei. Z and Zhang. X (2021) Thermal Stress Analysis of DC Dry Type Wall Bushing. In: 2021 IEEE 5th Advanced Information Technology, Electronic and Automation Control Conference(IAEAC). Chongqing, China. pp.1969-1973.
[2] Xiao. Y, Hu. W, Xu. Z, Zhou. H and Zhang. J (2018) A Monitoring Scheme for UHV Wall Bushing. In: 2018 International Conference on Power System Technology (POWERCON). Guangzhou, China. pp. 4650-4654.
[3] Shu Y, Chen W and Li Z (2013) Experimental Research on Very-Fast Transient Overvoltage in 1100-kV Gas-Insulated Switchgear. IEEE Transactions on Power Delivery, 28(1):458-466.
[4] Yue G, Liu W and Chen W (2013) Development of Full Frequency Bandwidth Measurement of VFTO in UHV GIS. IEEE Transactions on Power Delivery, 28(4):2550-2557.
[5] Zhou H, Ma G and Wang Y (2019) Optical Sensing in Condition Monitoring of Gas Insulated Apparatus: A Review. High Voltage, 4(4):259-270.
[6] Yang S, Xiang W and Li R (2020)An Improved DC Fault Protection Algorithm for MMC HVDC Grids Based on Modal-Domain Analysis.IEEE Journal of Emerging and Selected Topics in Power Electronics,8(4):4086-4099.
[7] He Z, Hu J and Lin L (2020)Pole-to-ground Fault Analysis for HVDC Grid Based on Common- and Differential-mode Transformation.Journal of Modern Power Systems and Clean Energy,8(3):521-530.
[8] Mitra B, Chowdhury B and Willis A (2020) Protection Coordination for Assembly HVDC Breakers for HVDC Multiterminal Grids Using Wavelet Transform. IEEE Systems Journal, 14(1):1069-1079.
[9] Wakode S, Ballal M and Sheikh A (2021) Oscillation Frequency Component-Based Protection Scheme for DC Microgrid. IEEE Transactions on Industry Applications, 57(6):5747-5757.
[10] Holdyk A, Gustavsen B and Arana I (2014) Wideband Modeling of Power Transformers Using Commercial sFRA Equipment. IEEE Transactions on Power Delivery, 29(3):1446-1453.

Optimal operation of a coal-nuclear-pumped storage hybrid system considering supporting capacity evaluation

Ruijie Chen[1], Bin Chen[2], Xinyu Wang[3], Rong Ye[2], Yichao Zou[2], Suhua Lou[3*]

[1]China-EU Institute for Clean and Renewable Energy, Huazhong University of Science and Technology, Wuhan, 430074, China
[2]Institute of Economic and Technology of State Grid Fujian Electric Power Co., Ltd., Fuzhou 350012, China
[3] State Key Laboratory of Advanced Electromagnetic Engineering and Technology, Huazhong University of Science and Technology, Wuhan, 430074, China

[*]Corresponding author's e-mail: shlou@hust.edu.cn

Abstract. With the increasing penetration of renewable energy and the gradual retirement or transformation of coal-fired power plants, the conflict between decarbonization and power-supply reliability in power systems has become increasingly prominent. To solve the problem, this paper proposes an optimal operation method for a coal-nuclear-pumped storage hybrid system that incorporates an evaluation of power-supporting capacity. The feasibility of partially replacing the functions of coal-fired power plants is analyzed. Firstly, a supporting capacity evaluation framework is developed for supportive power sources, encompassing power and energy support as well as flexibility-regulation capability support. Secondly, the short-term power and flexibility-regulation capability support requirements are assessed. By introducing the concept of ordered clustering, a net load segmentation method based on Fisher optimal segmentation is proposed to analyze the impact of seasonal fluctuations in both renewable energy and load on long-term energy support in power systems. Thirdly, a nuclear-pumped storage joint operation model is constructed for three operation modes: full tracking mode, three-stage tracking mode, and partial tracking mode. Finally, an optimal operation model is established for the hybrid system, aiming to minimize the comprehensive operational cost of the power system while considering supporting capacity evaluation. Case study results validate the effectiveness and rationality of the proposed model and method, and demonstrate the significant potential of nuclear-pumped storage joint operation in replacing the functions of coal-fired power plants.

1. Introduction

With the global energy transition and the intensifying challenge of climate change, decarbonization has become a critical goal for power systems. In this context, renewable energy is gradually replacing traditional coal-fired power plants, emerging as a key component of power systems. However, the inherent intermittency and variability of renewable energy have introduced new challenges to power systems, particularly in ensuring power and energy balance, operational flexibility, and stable system operation. Additionally, the gradual retirement and transformation of conventional coal-fired power plants have further exacerbated the conflict between decarbonization and power supply reliability in power system operations. Achieving decarbonization while ensuring system safety and economic viability is a pressing issue that requires immediate attention.

Nuclear power, as a stable, zero-carbon baseload energy source, is regarded as a promising solution to these challenges. However, nuclear power plants have limited operational flexibility and face difficulties in adapting to rapid load changes. Frequent power adjustments may result in safety risks, reduced fuel utilization, and diminished economic benefits [1–4]. Previous studies have proposed load-following control technologies for nuclear reactors, demonstrating their maturity [5]. The necessity of nuclear power participation in peak load regulation has also been explored [6]. Various fixed-mode peak-shaving strategies, such as "12-3-6-3" and "16-8," have been suggested [7–8], along with a multi-stage flexible peak-shaving model [9]. Despite these advancements, traditional fixed-mode peak-shaving strategies lack flexibility, potentially compromising economic performance and operational safety. As a result, nuclear power plants must integrate with other flexible resources to ensure reliable and adaptable operation.

Pumped storage, as the most widely used and technologically mature energy storage solution, is an economical and efficient peak-shaving resource. However, its energy and flexibility support capabilities are constrained by reservoir capacity and seasonal factors [10]. The integration of nuclear power and pumped storage leverages the long-term energy support capacity of nuclear power with the short-term power and flexibility support capabilities of pumped storage, thereby enhancing the flexibility and stability of power grids. This combined operation not only meets load demand but also partially replaces the regulation functions of coal-fired power plants, achieving both carbon reduction targets and system reliability [11–13].

Current research on nuclear-pumped storage hybrid systems primarily focuses on operational strategies and benefit analyses. Studies have shown that coupling the baseload operation of nuclear power with the peak-shaving capabilities of pumped storage significantly improves nuclear power utilization while reducing pumped storage operating costs [11]. Dynamic joint operation models have validated the economic viability and feasibility of such integration [14]. Furthermore, optimization methods using mixed-integer programming have minimized system costs and maximized renewable energy utilization [15]. Research from the International Hydropower Association (IHA) highlights that nuclear-pumped storage integration significantly increases nuclear plant utilization hours and reduces unit costs for pumped storage plants, with notable advantages in medium- and long-term electricity markets [16].

In summary, considering the lack of comprehensive studies on system support capacity evaluation and the gradual retirement of coal-fired power plants, this paper proposes a support capacity evaluation framework encompassing both power and energy support, as well as flexibility requirements over short- and long-term time scales. Based on three nuclear-pumped storage operation modes, a hybrid operation model is constructed. Simulation results validate the proposed model and methodology, demonstrating their effectiveness in enhancing system support capacity and feasibility in replacing coal-fired power plants.

2. System Supporting Capacity Modelling

2.1. Concept of Supporting Capacity in Power Systems

The concept of support capacity, as referred to in this paper, describes the ability of a dispatch system to coordinate supportive power sources within the system to address power shortages, energy deficiencies, or random power fluctuations over a specific timescale. This primarily focuses on conventional scenarios of renewable energy variability or extreme cases where renewable energy output cannot be guaranteed. It involves power sources capable of providing deterministic support in terms of power, energy, and flexibility, such as coal-fired power plants, gas-fired power plants, pumped storage, and nuclear power. The dispatch system may represent a single energy system or an integrated energy system coupling multiple heterogeneous energy sources.

The support capacity of a power system is characterized by the following three dimensions:

- Dimension 1: Power Supporting Capacity

Power supporting capacity refers to the ability of a power system to ensure a stable power supply at any given moment and under any operating conditions on a short timescale, leveraging supportive power sources. This is achieved while reasonably utilizing renewable energy, which is characterized by variability and intermittency. The primary focus is on meeting load demands at any specific moment. As the proportion of renewable energy generation increases, power support capacity becomes critical, particularly in extreme or special scenarios where renewable energy output remains consistently low. It emphasizes the need for power support at specific moments during such periods.

- Dimension 2: Energy Support Capacity

Energy-supporting capacity refers to the ability of a power system to provide sufficient energy over multi-day, monthly, seasonal, or annual timescales to meet load demands. This is also achieved under the premise of fully utilizing renewable energy with its inherent variability and intermittency. In the context of modern power systems, energy support capacity is crucial in extreme or special scenarios with prolonged periods of low renewable energy output, focusing on ensuring sufficient energy supply during these low-output intervals.

- Dimension 3: Flexibility-Regulation Support Capability

Flexibility-regulation support capacity focuses on short- and medium-timescales and addresses changes in net load between adjacent periods. This includes ramping support on a minute timescale (15 minutes to 4 hours) and peak-shaving support on an hourly timescale (up to 24 hours).

2.2. Representation Methods for Supporting Capacity in Power Systems

For power support capacity, the annual net load curve combined with a net load reserve factor can be used as a constraint. This represents the requirement that the installed capacity of supportive power sources must exceed a certain proportion of the net load peak. The power support capacity constraint is expressed as

$$\sum_{s\in S} P_s^S \geq (1+\rho)\max\left\{P_L(t)-P_N(t)\right\} \tag{1}$$

Where, P_s^S represents the installed capacity of the supportive power sources; and donate the load and renewable generation output at time t respectively; ρ is the load reserve factor of the system, which is taken as 0.1 in this paper.

Energy support capacity primarily reflects the energy balance over multiple days, weeks, or seasonal periods. The issue of long-period energy imbalance in high-penetration renewable energy systems arises from two factors: the sustained low output of renewable energy and the seasonal fluctuation characteristics of both renewable energy and load. Therefore, the energy support capacity constraint needs to incorporate a seasonal energy support capacity constraint in addition to the traditional annual energy constraint.

The annual energy constraint is:

$$\sum_{s\in S} \tau^S P_s^S \geq \sum_{t}^{T}\left(P_L(t)-P_N(t)\right) \tag{2}$$

Where represents the annual utilization hours of the supportive power source units; T represents the corresponding time scale.

To address the seasonal energy support capacity constraint, the temporal characteristics of the net load must be considered. This paper employs ordered clustering based on the Fisher optimization algorithm to segment the net load curve, capturing the system's seasonal supply-demand fluctuations. Ordered clustering divides n ordered samples $X=\{x_1,x_1,\cdots,x_n\}$, to minimize intra-segment differences and maximize inter-segment differences. The variation within each segment can be represented by the variance as follows:

$$d(i,j)=\sum_{a=i}^{j}\left[x_a-\bar{x}(i,j)\right]^2 \tag{3}$$

$$\bar{x}(i,j)=\frac{1}{j-i+1}\sum_{a=i}^{j}x_a \tag{4}$$

Where, $d(i,j)$ is referred to as the sum of squared deviations for the sample sequence (x_i,x_j), which reflects the variation between samples within a segment. A smaller diameter indicates less variation between samples within the segment. $\bar{x}(i,j)$ denotes the centroid of the segment. The Fisher algorithm defines a loss function and uses dynamic programming to determine the optimal segmentation. The definition of the loss function and the recursive formula are as follows:

$$L\left[b(n,k)\right]=\sum_{t=1}^{k}d\left(i_t,i_{t+1}-1\right) \tag{5}$$

$$\begin{cases} L\left[b(n,2)\right]=\min\limits_{2\le j\le n}\left\{d(1,j-1)+d(j,n)\right\} \\ L\left[b(n,k)\right]=\min\limits_{k\le j\le n}\left\{L\left[b(j-1,k-1)\right]+d(j,n)\right\} \end{cases} \tag{6}$$

Where, $L(\cdot)$ represents loss function. $b(n,k)$ represents a way of dividing n ordered samples into k segments. Therefore, the steps of the ordered clustering algorithm based on Fisher optimization are as follows:

Step 1: Input the sample data $X=\{x_1,x_1,\cdots,x_l\}$, and the range of the number of segmentation points, k, satisfies $k\in\left[1,\sqrt{l}\right],k\in Z$.

Step 2: Set the objective function to minimize the sum of squared deviations after segmentation. Select different numbers of segmentation points, k, and use the Fisher optimal segmentation method to obtain the set of segmentation points. Then, divide the net load curve sample data X into k+1 scenario sets.

Step 3: Use the modified silhouette coefficient to compare the clustering performance for different values of k and select the optimal clustering scenario set.

$$\mathrm{MC}=\frac{1}{l}\sum_{i=1}^{l}\frac{b_i-a_i}{\max\left\{a_i,b_i\right\}} \tag{7}$$

Where, MC is the modified silhouette coefficient; l is the sample size; a_i and b_i are the average Euclidean distance and the minimum average Euclidean distance, respectively.

Step 4: Perform K-means clustering on the k+1 centroids of the optimal scenario set, dividing them into several clusters based on the size of the seasonal energy support demand. Then, obtain the duration of each scenario corresponding to the centroids.

Based on the above Fisher optimization ordered clustering algorithm, the average and maximum centroids of the corresponding clusters, q time segments Q_i, can be obtained. Therefore, the seasonal energy support capacity constraint can be expressed as:

$$\sum_{s\in S}\tau_Q^S P_s^S \ge \sum_{i=1}^{q}\sum_{t\in Q_i}\left(P_L(t)-P_N(t)\right) \tag{8}$$

Where, τ_Q^S represents the utilization hours of the supportive power sources within the time series Q.

Taking the net load curve of a power grid in a certain region of China as an example, the seasonal energy support segmentation results are shown in Figure 1. The segmentation point sequence is obtained as {528, 1113, 4137, 6360}, and the average and maximum centroids of the corresponding clusters are the time segments [4137, 6360]. The corresponding seasonal energy support demand is the sum of the net load at each time point within this segment.

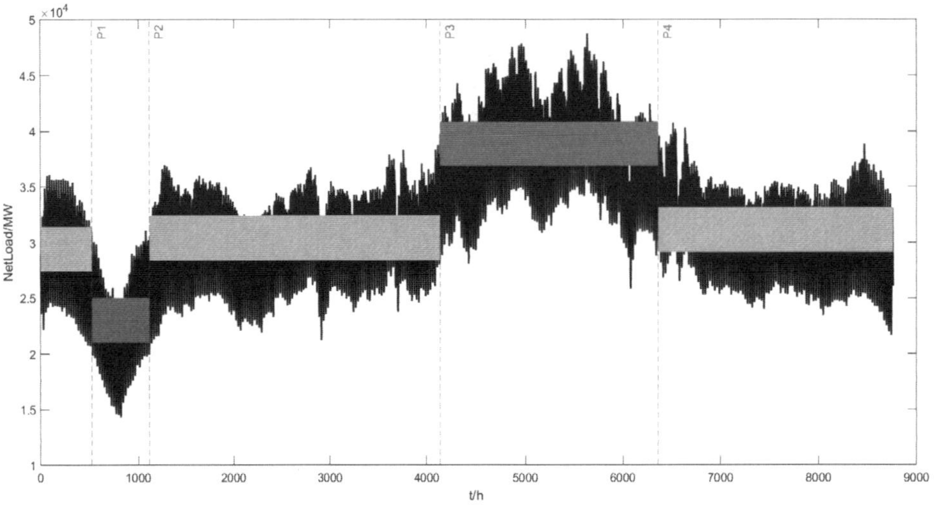

Figure 1. Seasonal Net Load Energy Segmentation Results.

For flexibility-regulation support capacity, it is mainly divided into ramping support and peak-shaving support. Both essentially involve power variations between adjacent periods within the corresponding timescale, with two possible regulation directions: positive and negative [17]. Therefore, the regulation support capacity provided by supportive power sources can be expressed as:

$$P_{R,t}^{\mathrm{up}} = \min[\Delta P_{s,\mathrm{up}}^{S} \Delta Q, P_{s,\max}^{S} - P_{s}^{S}(t)] \tag{9}$$

$$P_{R,t}^{\mathrm{down}} = \min[\Delta P_{s,\mathrm{down}}^{S} \Delta Q, P_{s}^{S}(t) - P_{s,\min}^{S}] \tag{10}$$

Where, $P_{R,t}^{\mathrm{up}}$ and $P_{R,t}^{\mathrm{down}}$ represent the regulation support capacity that can be provided by the supportive power sources at time t; $P_{s,\max}^{S}$ and $P_{s,\min}^{S}$ are the maximum and minimum technical output of the supportive power source units, respectively; $\Delta P_{s,\mathrm{up}}^{S}$ and $\Delta P_{s,\mathrm{down}}^{S}$ are the up and down ramp rates of the units; $P_{s}^{S}(t)$ is the output of the supportive power source at time t; ΔQ is the time scale. Therefore, the regulation support capacity constraint can be expressed as:

$$P_{\mathrm{NL}}(t) = P_{L}(t) - P_{N}(t) \tag{11}$$

$$\begin{cases} P_{R,t}^{\mathrm{up}} \geq P_{\mathrm{NL}}(t+1) - P_{\mathrm{NL}}(t) & \text{if } P_{\mathrm{NL}}(t+1) \geq P_{\mathrm{NL}}(t) \\ P_{R,t}^{\mathrm{down}} \geq P_{\mathrm{NL}}(t) - P_{\mathrm{NL}}(t+1) & \text{if } P_{\mathrm{NL}}(t+1) < P_{\mathrm{NL}}(t) \end{cases} \tag{12}$$

3. Nuclear-Pumped Storage Hybrid Operation Model

3.1Multi-Type Nuclear-Pumped Storage Hybrid Operation Model

The hybrid operation of nuclear power and pumped storage ensures the stable operation of nuclear power units by leveraging the strong regulation support capacity of pumped storage, thereby fully utilizing the energy support capability of nuclear power. At the same time, the combined output of both sources, while ensuring adequate energy support, aims to match load variations as closely as possible and undertake peak-shaving tasks. This approach achieves partial functional replacement of coal-fired power plants. Therefore, this paper proposes that the hybrid operation of nuclear power and pumped storage follows these principles: nuclear power operates as a baseload generator, fully utilizing the regulation capacity of pumped storage, and the combined (nuclear-pumped storage) output curve effectively tracks the daily load factor curve.

Based on the degree of matching between the post-hybrid power curve and the load curve, the nuclear-pumped storage hybrid operation mode can be divided into three categories: Full Tracking Mode, Three-Stage Tracking Mode, and Partial Tracking Mode.

- *Full Tracking Mode*

In the Full Tracking Mode, the combined output of the nuclear-pumped storage unit will fully follow the daily load factor curve of the grid, i.e.:

$$p(i) = \frac{P_{\text{nuc}}(i) + P_{\text{pump}}(i)}{P_{\text{nuc}} + P_{\text{pump,com}}}, \quad i = 1, 2, \cdots, 24 \tag{13}$$

Where, $P_{\text{nuc}}(i)$ represents the output of the nuclear power unit at the time *i*; P_{nuc} is the operational capacity of the nuclear power unit; $P_{\text{pump}}(i)$ represents the output of the pumped storage unit at time *i*, $p(i)$ is the system load factor at time *i*, and $P_{\text{pump,com}}$ is the capacity of the pumped storage unit that operates in conjunction with the nuclear power unit.

- *Three-Stage Tracking Mode*

Three-stage tracking mode is to divide the typical day into three periods: *peak*, *flat*, and *valley* — based on the grid's daily load curve. During each period, the output of the nuclear-pumped storage unit remains constant, and the average output rate of the nuclear-pumped storage hybrid unit is consistent with the average load factor of the grid, i.e.:

$$\begin{cases} p_{\text{f}} = \dfrac{P_{\text{nuc}} + P_{\text{pump,f}}}{P_{\text{nuc}} + P_{\text{pump,com}}} \\ p_{\text{p}} = \dfrac{P_{\text{nuc}} + P_{\text{pump,p}}}{P_{\text{nuc}} + P_{\text{pump,com}}} \\ p_{\text{g}} = \dfrac{P_{\text{nuc}} + P_{\text{pump,g}}}{P_{\text{nuc}} + P_{\text{pump,com}}} \end{cases} \tag{14}$$

$$P_{\text{pump,com}} = \max\left\{P_{\text{pump,f}}, P_{\text{pump,p}}, P_{\text{pump,g}}\right\} \tag{15}$$

Where, p_{f}、p_{p}、p_{g} represent the average load factors for the peak, flat, and valley periods, respectively. $P_{\text{pump,f}}$、$P_{\text{pump,p}}$、$P_{\text{pump,g}}$ represent the pumped storage output during the peak, flat, and valley periods, respectively.

- *Partial Tracking Mode*

In the Partial Tracking Mode, the combined generation's average load factor is ensured to be roughly the same as the daily load factor curve, but without full tracking. In this study, a typical approach is used: the pumped storage unit's pumping capacity and total generation capacity are both considered as its full capacity, following the peak-shaving and valley-filling principle. The water pumping and generation capacities are balanced within the scheduling period. The daily average output rate of the nuclear-pumped storage hybrid unit is consistent with the grid's daily average load factor, i.e.:

$$\gamma = \frac{P_{\text{pump,com}}(t_{\text{f}} - \eta t_{\text{p}}) + 24P_{\text{nuc}}}{24(P_{\text{nuc}} + P_{\text{pump,com}})} \tag{16}$$

$$(P_{\text{pump,com}} + P_{\text{pump,ind}})t_{\text{f}} = \eta P_{\text{pump}} t_{\text{p}} \tag{17}$$

Where, γ is the grid's equivalent daily average load factor, t_{f} and t_{p} represent the daily generation hours and daily pumping hours of the pumped storage unit, respectively, and η is the efficiency of the pumped storage unit.

3.2Joint Optimal Operation Objective Function

In this paper, the objective function for optimizing the joint operation of the nuclear-pumped storage system is to minimize the capacity of the pumped storage unit that operates in conjunction with the nuclear power unit, i.e.,

$$\min P = P_{\text{pump,com}} \tag{18}$$

In this case, the nuclear power unit's power and energy support capabilities, as well as the pumped storage unit's regulation support capabilities, can be fully utilized, while the remaining pumped storage capacity still maintains normal operation.

3.3Constraints

- *Output Constraints within the Pumped Storage Scheduling Period Q*

$$P_{\min} \le P_f(t) \le P_{\max}, t = 1,2,\cdots,Q \tag{19}$$

$$P'_{\min} \le P_c(t) \le P'_{\max}, t = 1,2,\cdots,Q \tag{20}$$

Where, $P_{\min}$ and $P_{\max}$ represent the minimum and maximum generation power of the pumped storage unit, respectively; $P'_{\min}$ and $P'_{\max}$ represent the minimum and maximum pumping power of the pumped storage unit, respectively.

- *Energy Storage Constraints of the Pumped Storage Plant*

$$S(t) = S(t-1) - \eta P(t-1)u(t-1) \tag{21}$$

$$S_{\min} \le S(t) \le S_{\max} \tag{22}$$

Where, $S(t)$ represents the stored energy at time *t*; $u(t)$ is the state variable, where $u(t) = 1$ when the system is in generation mode and $u(t) = -1$ when the system is in pumping mode. And represent the minimum and maximum storage energy of the pumped storage plant, constrained by the maximum capacity of the upper and lower reservoirs.

- *Pumped Storage Operation Duration Constraint*

$$t_f + t_p \le T_e \le Q \tag{23}$$

Where, T_e represents the scheduled operation duration of the pumped storage plant.

3.4Case Study

The installed capacity of a certain nuclear power plant is 4×1250 MW, with the actual grid-connected power output of each unit being 1090 MW. The installed capacity of the pumped storage plant is 4×300 MW, with an overall efficiency of η=0.75. The typical daily load factor of the grid and its corresponding curve are shown in Figure 2 and Table 1.

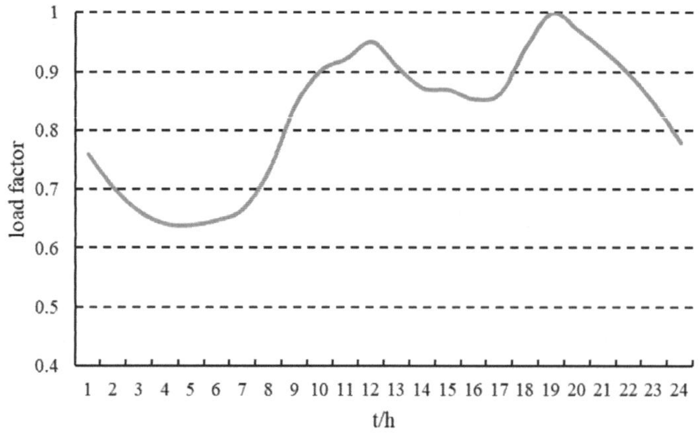

Figure 2. Daily Load Factor Curve of a Typical Power Grid.

Table 1. Daily Load Factor of a Typical Power Grid.

t	1	2	3	4	5	6	7	8	9	10	11	12
Load factor	0.76	0.706	0.66	0.64	0.64	0.65	0.67	0.73	0.84	0.90	0.92	0.95
t	13	14	15	16	17	18	19	20	21	22	23	24
Load factor	0.91	0.87	0.87	0.85	0.86	0.94	1	0.97	0.94	0.90	0.84	0.78

- *Full Tracking Mode*

The combined output curve of the nuclear-pumped storage hybrid system in the Full Tracking Mode is shown in Figure 3. In this mode, the installed capacity of the pumped storage units in a joint operation is 1101 MW, with a daily generation of 7.768 million kWh. The nuclear power units provide 7.011 million kWh of pumping energy daily. As a result, the storage capacity is insufficient, and the remaining 99 MW of pumped storage units operating independently must pump for 34 hours to meet the demand, which requires purchasing electricity from the grid.

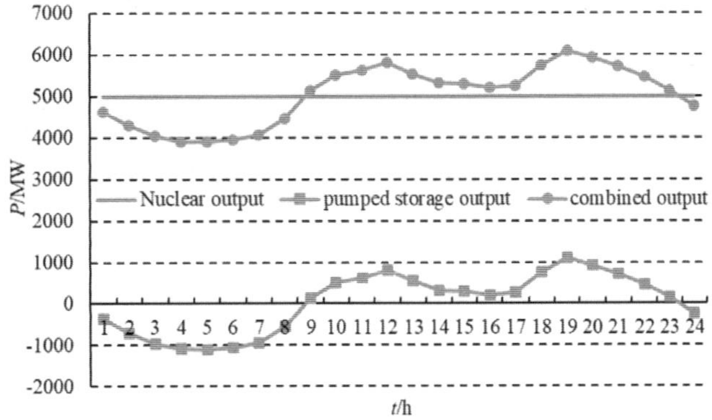

Figure 3. Full Tracking Operation Mode.

- *Three-Stage Tracking Mode*

The combined output curve of the nuclear-pumped storage hybrid system in the Three-Stage Tracking Mode is shown in Figure 4. In this mode, the pumped storage output during the peak, flat, and valley periods is 564.7 MW, 96.0 MW, and -910.5 MW, respectively. The installed capacity of the bundled pumped storage units is 911 MW, with a daily generation of 5.190 million kWh. The nuclear power units provide 8.195 million kWh of pumping energy daily. Therefore, the storage capacity is sufficient, and the pumped storage units operating in conjunction with the nuclear power units can have their pumping energy fully provided by the nuclear power units.

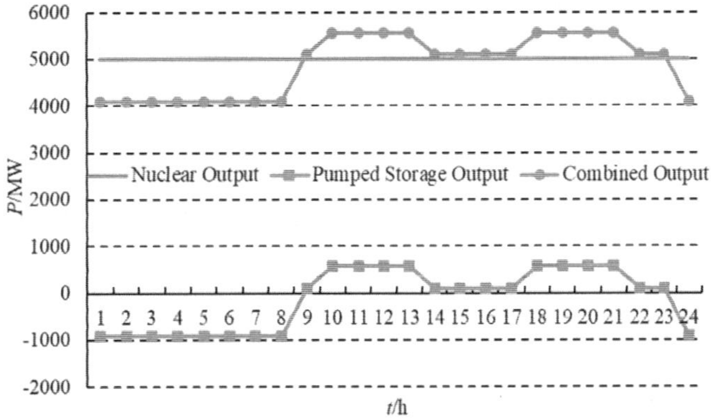

Figure 4. Three-Stage Tracking Operation Mode.

- *Partial Tracking Mode*

The combined output curve of the nuclear-pumped storage hybrid system in the Partial Tracking Mode is shown in Figure 5. In this mode, the pumped storage unit with a capacity of 1200 MW is bundled with a pumped storage generation capacity of 828 MW. The pumping energy required by the pumped storage units is fully supplied by the nuclear power units.

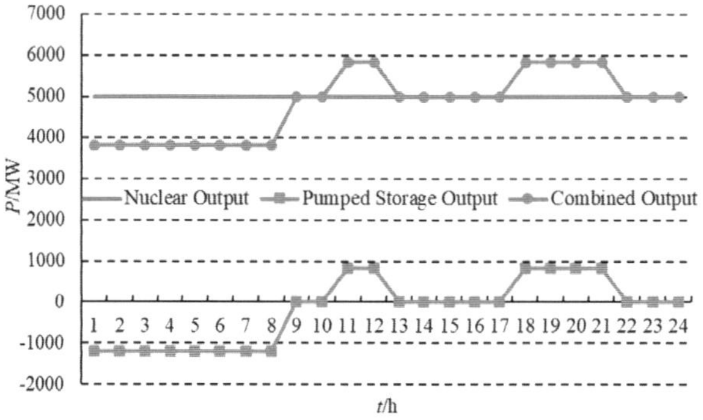

Figure 5. Partial Tracking Operation Mode.

3.5Analysis

The operational conditions of the three modes are shown in Table 2. The table shows that the combined output of the nuclear-pumped storage hybrid system in all three modes effectively tracks the grid load curve and demonstrates good peak-shaving capability. Among them, in the Full Tracking Mode, the nuclear-pumped storage hybrid system has the largest regulation range and the best load-following performance. However, it also requires the largest pumped storage unit capacity and consumes the most energy, resulting in poorer economic efficiency. When considering the constraints of the pumped storage plant, both the Three-Stage Tracking Mode and Partial Tracking Mode provide sufficient pumping energy, fully supplied by the nuclear power units. In the Partial Tracking Mode, the required bundled pumped storage unit capacity is smaller than in the Three-Stage Tracking Mode, but the regulation capacity provided is larger. In the Full Tracking Mode, the daily pumping energy provided by the nuclear power units is insufficient, requiring additional pumping energy to be supplied by the grid.

Table 2. Operating Status of a Nuclear-Pumped Storage Hybrid System.

Mode	$P_{\max}$ /MW	$P'_{\max}$ /MW	Peak-Shaving Capacity/MW	Generation Energy of PS/10^4kWh	Consumption Energy of PS /10^4kWh	Joint Net Generation E_{NS}/10^4kWh
Full Tracking	1100.5	1100.5	2201	776.8	701.1	10539.7
Three-Stage	564.7	910.5	1475.2	519.0	819.5	10163.5
Partial Tracking	828.4	1200	2028.4	497.0	960	10001.0

4. Optimal Operation Model of Coal-Fired Units Considering Nuclear-Pumped Storage Hybrid Operation

4.1Objective

The optimization of coal-fired units considering nuclear-pumped storage aims to minimize the total system operation cost while satisfying the system's power and energy support capacity constraints, as well as the regulation support capacity constraints. The objective function is as follows:

$$\min f = C_{\text{fuel,nuc}} + C_{\text{fuel,fir}} + C_{\text{start}} + C_{\text{loss}} \tag{24}$$

$$\begin{cases} C_{\text{fuel,nuc}} = c_{\text{nuc}} \sum_t P_{\text{nuc}}(t) \\ C_{\text{fuel,fir}} = \sum_i \sum_t \left\{ a[P_{\text{fir},i}(t)]^2 + bP_{\text{fir},i}(t) + c \right\} \\ C_{\text{start}} = \sum_i \sum_t F_{\text{start},i} u_{i,t}^{\text{fir}} (u_{i,t}^{\text{fir}} - u_{i,t-1}^{\text{fir}}) \\ C_{\text{loss}} = \sum_t c_{\text{loss}} \Delta P_t^{\text{Load}} \end{cases} \tag{25}$$

Where, $C_{\text{fuel,nuc}}$, $C_{\text{fuel,fir}}$, C_{start}, and C_{loss} represent the fuel cost of the nuclear power unit, the fuel cost of the coal-fired power unit, the start-up/shut-down cost, and the load shedding penalty, respectively. c_{nuc} is the fuel cost per MWh for the nuclear power unit, taken as 7.4$/MWh in this study. a, b, and c are the cost coefficients for the coal-fired power unit. $P_{\text{fir},i}(t)$ represents the output of the coal-fired power unit i at the t-th time step. $F_{\text{start},i}$ is the start-up/shut-down cost for the coal-fired power unit i. $u_{i,t}^{\text{fir}}$ is a binary variable, which shows the start-up/shut-down state of the coal-fired unit i at the t-th time step, with 1 indicating the start-up state. c_{loss} is the penalty cost for load shedding per unit, and ΔP_t^{Load} is the amount of load shedding at time t.

4.2Constraints

The optimal operation model of coal-fired units considering a nuclear-pumped storage hybrid system must satisfy the nuclear-pumped storage joint operation constraints and system support capacity constraints, in addition to the following constraints:

- *System Power Balance Constraint*

$$\sum_i P_{i,t}^{\text{fir}} + P_{\text{WS}}(t) + P_{\text{nuc}}(t) + P_{\text{pump}}(t) + \Delta P_t^{\text{Load}} = P_{\text{Load}}(t) \tag{26}$$

Where, $P_{\text{WS}}(t)$ is the wind-solar output power at time t; $P_{\text{Load}}(t)$ indicates the load power required of the system at time t.

- *Output Constraint of Coal-Fired Unit*

$$P_{\text{min},i} \le P_{\text{fir},i} \le P_{\text{max},i} \tag{27}$$

$$\Delta P_{\text{fir,ramp,min}} \le P_{\text{fir},i}(t) - P_{\text{fir},i}(t-1) \le \Delta P_{\text{fir,ramp,max}} \tag{28}$$

Where, $P_{\text{min},i}$ and $P_{\text{max},i}$ represent the minimum and maximum output power of coal-fired unit i, respectively; $\Delta P_{\text{fir,ramp,max}}$ and $\Delta P_{\text{fir,ramp,min}}$ are the upper and lower ramping power limits of coal-fired unit i.

- *Minimum Start-Up and Shut-Down Time Constraint of Coal-Fired Unit*

$$\sum_{\tau=t-T_{\text{on},i}^{\text{fir}}}^{t-1} u_{i,\tau}^{\text{fir}} \ge T_{\text{on},i}^{\text{fir}} (u_{i,t-1}^{\text{fir}} - u_{i,t}^{\text{fir}}) \tag{29}$$

$$\sum_{\tau=t-T_{\text{off},i}^{\text{fir}}}^{t-1} (1 - u_{i,\tau}^{\text{fir}}) \ge T_{\text{off},i}^{\text{fir}} (u_{i,t}^{\text{fir}} - u_{i,t-1}^{\text{fir}}) \tag{30}$$

Where, $T_{\text{on},i}^{\text{fir}}$ and $T_{\text{off},i}^{\text{fir}}$ represent the minimum start-up and shut-down time of the coal-fired unit.

5. Case Study

To verify the effectiveness of the proposed operational optimization method for coal-fired units considering nuclear-pumped storage hybrid operation, it is applied to the power grid of a certain region in China for optimization. The maximum daily load is 1500 MW, and the installed capacities of nuclear and pumped storage units correspond to 0.05 times the values presented earlier. The joint operation considers the Three-Stage Tracking Mode and the Partial Tracking Mode. The parameters of the coal-fired units are shown in Table 3.

Table 3. Coal-Fired Unit Parameters.

No.	P_{max}/MW	P_{min}/MW	$T^{fir}_{on,off}$ /h	F_{start}/$	Coefficients of Fuel Cost		
					a	*b*	*c*
1	455	228	8	4500	0.00048	16.19	1000
2	455	228	8	5000	0.00031	17.26	970
3	130	104	24	550	0.002	16.6	700
4	130	104	24	560	0.00211	16.5	680
5	162	130	24	900	0.00398	19.7	450
6	80	64	24	170	0.00712	22.26	370
7	85	68	24	260	0.00079	27.74	480
8	55	44	24	30	0.00413	25.92	660
9	55	44	24	30	0.00222	27.27	665
10	55	44	24	30	0.00173	27.79	670

Based on the aforementioned case study and data, a power and energy balance analysis, as well as a comprehensive cost analysis, is conducted. The operational results are shown in Figure 6 to Figure 8, and the system's comprehensive operational costs are listed in Table 4. Case 1 represents the optimization results without considering the nuclear-pumped storage hybrid operation. Case 2 corresponds to the optimization results under the Three-Stage Tracking Mode, while Case 3 represents the optimization results under the Partial Tracking Mode.

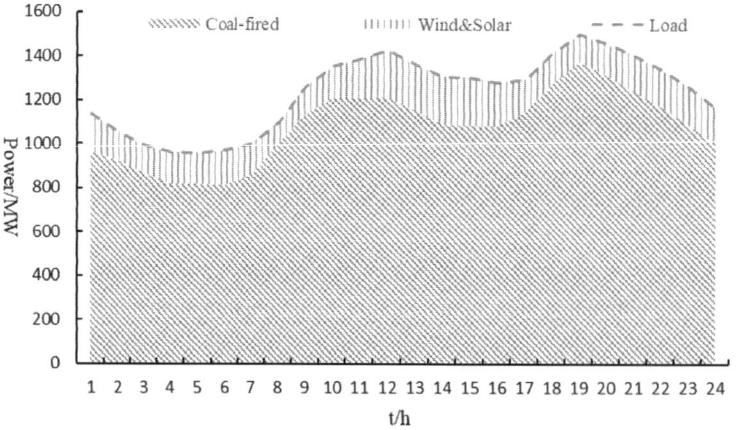

Figure 6. Optimal Operation Result of Case 1.

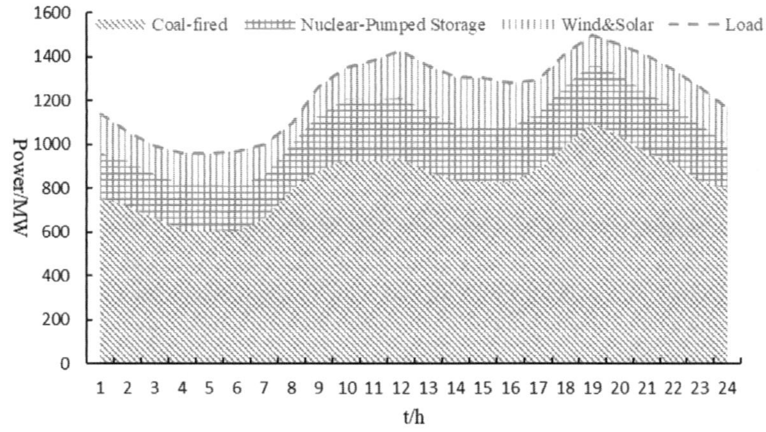

Figure 7. Optimal Operation Result of Case 2.

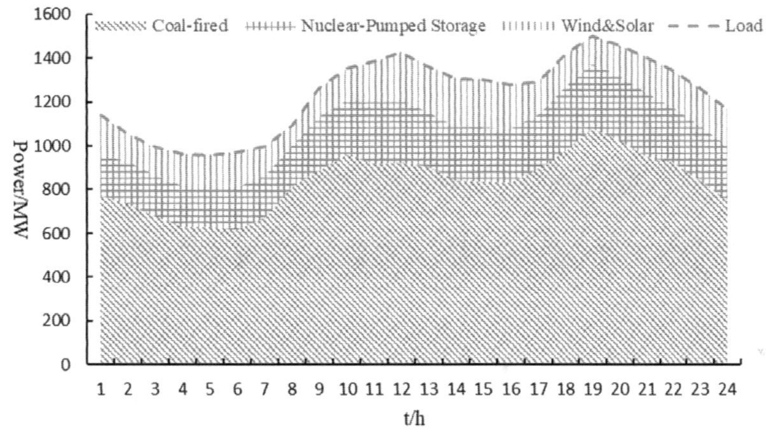

Figure 8. Optimal Operation Result of Case 3.

Table 4. Integrated Operating Cost of System.

Case	Fuel Cost/10^4\$	Start-up/Shut-down Cost/10^4\$	Integrated Operating Cost /10^4\$
1	56.83	0.20	57.03
2	45.51	0.11	45.62
3	44.88	0.20	45.08

From Figure 6 to 8, it can be observed that the nuclear-pumped storage hybrid operation effectively reduces the output of coal-fired units. Combining Table 4 with the output data of each unit, it can be concluded that under the Three-Stage Tracking Mode and the Partial Tracking Mode, the nuclear-pumped storage hybrid units with capacities of 296 MW and 291 MW, respectively, provide stable power and energy support as well as flexibility support to the system. They replace the functionality of 272 MW and 190 MW coal-fired units, respectively, and reduce the comprehensive operational cost of the system by 20.0% and 21.0%, respectively.

6. Conclusion

To address the conflict between low-carbon development and supply reliability in modern power systems, this paper proposes an operational optimization method for nuclear-pumped storage hybrid systems considering power source support capacity evaluation. Based on the case study analysis, the following conclusions can be drawn:

1) The proposed net load segmentation method based on Fisher optimal segmentation effectively captures seasonal fluctuations in net load, providing a foundation for long-term energy support evaluation. The multi-dimensional support capacity evaluation framework also supports subsequent optimization modeling.

2) Among the three nuclear-pumped storage hybrid operation modes, the Three-Stage Tracking Mode and Partial Tracking Mode enable the pumped storage units to be fully powered by nuclear units. The Partial Tracking Mode requires a smaller bundled pumped storage unit capacity compared to the Three-Stage Tracking Mode and achieves better economic performance. In contrast, the Full Tracking Mode provides the best load-following performance but has poorer economic efficiency and requires additional power from the grid.

3) The nuclear-pumped storage hybrid operation modes under the Three-Stage Tracking Mode and Partial Tracking Mode provide stable power and energy support as well as flexibility support to the system. They partially replace the functions of coal-fired units while improving the economic efficiency of system operations.

This study investigates the functional replacement of coal-fired units by the nuclear-pumped storage hybrid system. However, this study does not address coal-fired unit retirement, and future research can explore the power source planning considering nuclear-pumped storage hybrid operation under coal-fired unit retirement cases.

Acknowledgment

This work was financially supported by the Science and Technology Program of State Grid Corporation of China (Key Technologies for Coordinated Planning and Cost Allocation of Supportive Power Sources for Low-Carbon Transformation: 1400-202421300A-1-1-ZN).

References

[1] Veeraya I ,Surachai C . Optimal Variable Renewable Energy Generation Schedules Considering Market Prices and System Operational Constraints[J].Energies,2021,14(17):5320-5320.
[2] Zonggen Y ,Yusheng L ,Tyler W , et al.Deep reinforcement learning-based optimization for a tightly coupled nuclear renewable integrated energy system[J].Applied Energy,2022,328
[3] Mahmoudi M S , Maleki A , Ochbelagh R D . Techno-economic assessment of hydrogen-based energy storage systems in determining the optimal configuration of the nuclear-renewable hybrid energy system[J].Energy Reports,2024,114713-4725.
[4] Mahmud S, Ponkiya B, Katikaneni S, et al. Design and optimization of a modular hydrogen-based integrated energy system to maximize revenue via nuclear-renewable sources[J]. Energy, 2024, 313133763-133763.
[5] Dong Z, Feng J, Huang X . Nonlinear Observer-Based Feedback Dissipation Load-Following Control for Nuclear Reactors[J].IEEE Transactions on Nuclear Science,2009,56(1P2):272-285.
[6] Zhi C, Guofang Z, Dan X, et al.Multi-area joint scheduling model considering peak load regulating of nuclear power[J].Energy Reports,2021,7(S1):1-9.
[7] Cany C ,Mansilla C ,Mathonnière G , et al.Nuclear power supply: Going against the misconceptions. Evidence of nuclear flexibility from the French experience[J].Energy,2018,151289-296.
[8] Zhang J, Leng R, Chen M, et al. The future role of nuclear power in the coal-dominated power system: The case of Shandong[J]. Journal of Cleaner Production,2020,256(prepublish):120744-120744.
[9] Guo Q ,Wu X ,Cai H , et al.Multi-power sources joint optimal scheduling model considering nuclear power peak regulation[J].Energy,2024,293130678-.
[10] Hunt J D , Byers E , Riahi K ,et al.Comparison between seasonal pumped-storage and conventional reservoir dams from the water, energy, and land nexus perspective[J].Energy Conversion and Management, 2018, 166(JUN.):385-401.
[11] Lopes A R ,Junker G R ,Martins J , et al.Characterization and use of energy flexibility in water pumping and storage systems[J].Applied Energy,2020,27.
[12] Bin L, Andrew B, Matthew S, et al.Low-cost, low-emission 100% renewable electricity in Southeast Asia supported by pumped hydro storage[J].Energy,2021,236
[13] Kim W J, Lee Y S, Chun Y H, et al.Reserve-Constrained Unit Commitment Considering Adjustable-Speed Pumped-Storage Hydropower and Its Economic Effect in Korean Power System[J].Energies, 2022, 15.

Journal of Physics: Conference Series **3011** (2025) 012082 doi:10.1088/1742-6596/3011/1/012082

[14] Y. Liu et al., "Coordinated Frequency Regulation Strategy of Pumped Storage Units and Battery Energy Storage System," 2021 IEEE Sustainable Power and Energy Conference (iSPEC), Nanjing, China, 2021, pp. 1195-1200.
[15] Nie X ,Cheng M ,Zuo X , et al.Multi-objective capacity configuration optimization of a nuclear-renewable hybrid system[J].Applied Thermal Engineering,2024,248(PB):123365-.
[16] Hydropower Status Report. International Hydropower Association. IHA. (2022).
[17] Lu Z, Li H, Qiao Y . Flexibility evaluation and supply/demand balance principle of power system with high-penetration renewable electricity[J].Zhongguo Dianji Gongcheng Xuebao/Proceedings of the Chinese Society of Electrical Engineering, 2017, 37(1):9-19.

Journal of Physics: Conference Series **3011** (2025) 012083 doi:10.1088/1742-6596/3011/1/012083

Performance evaluation of PV array configurations under various partial shading conditions

Jhih-Hao Lin and Jian-Hua He*

Intelligent Manufacturing College, Fujian Polytechnic of Information Technology, Fuzhou 350003, China

*Corresponding author's e-mail: hjhnh999@fjpit.edu.cn

Abstract. Solar power is highly sensitive to environmental factors. The shading problem of some photovoltaic (PV) modules may lead to a series of negative effects such as power reduction and hot spots. The PV array configuration is crucial for maximizing power output. By scientifically and rationally adjusting the configurations of the PV system, their efficiency can be significantly improved. In this paper, Matlab/Simulink software is employed to conduct accurate modeling of a 6×6 PV array. This study investigated various performance indicators of six configurations, namely Series (S), Series-Parallel (SP), Honey-Comb (HC), Bridge-Linked (BL), Triple-Tied (TT), and Total-Cross-Tied (TCT), under various partial shading conditions (PSCs). The involved indicators include the global maximum power point (GMPP), shading loss, mismatch loss, fill factor, and efficiency. By comprehensively analyzing the average performance under various PSCs, the advantages and disadvantages of different PV array configurations can be revealed more objectively and accurately. Consequently, the findings of this research offer a reliable reference for the optimized design of PV systems.

1. Introduction

At present, the demand for electricity is on an upward trend. Meanwhile, fossil fuels are being consumed at an alarming rate and have triggered serious environmental issues. This has created ample room for the development of renewable energy sources (RESs). Among the wide-ranging RESs, PV systems are witnessing a substantial growth in popularity. This can be attributed to their outstanding environmental friendliness, diverse application scenarios, modular design, low maintenance requirements, and high reliability [1]. In PV systems, PV modules are precisely interconnected to construct a PV array. This strategic connection is crucial as it enables the achievement of the required output power, ensuring the efficient operation and performance of the PV system. Nevertheless, the performance is significantly affected by environmental conditions, including but not limited to solar irradiance and temperature. When PV modules are under standard test conditions (STC), specifically an irradiance of 1 kW/m^2, a temperature of 25°C, and an air mass spectral distribution of 1.5 AM, their nominal performance can be determined. At this point, a singular GMPP is exhibited on the P-V curve [2].

However, in nature, factors like trees, buildings, overhead lines, bird droppings, clouds, snow, and dust pollution can cause non-uniform irradiance on PV modules, significantly affecting the performance of PV systems [3]. When certain PV modules are under shading, the current generated by the shaded cells is lower compared to that of the unshaded cells. This phenomenon causes the shaded cells to assume the role of loads within the electrical circuit. Consequently, rather than contributing to power generation, these shaded cells consume the output power. This imbalance in power matching will raise the temperature of the shaded modules, thereby increasing the risk of hot spot formation and potential permanent damage. Should hot spots persist for an extended period, they will accelerate the process of module degradation, seriously compromising the reliability of the PV system [4]. To mitigate the adverse effects brought about by partial shading, a bypass diode is commonly connected in reverse parallel at the two ends of the PV module. When some PV modules are in a state of PSCs, the bypass diode can offer an alternative route for the current, thereby preventing the occurrence of hot spots and reducing power losses [5].

Although the bypass diode does, to a certain extent, contribute to alleviating this problem, its conduction gives rise to the appearance of multiple maximum power points (MPPs) on the P-V curve. Only one of these MPPs is denominated as the GMPP, whereas the remaining ones are designated as local maximum power points (LMPPs). The presence of these peaks may potentially misdirect the maximum power point tracking (MPPT) technique, causing it to trace the LMPPs rather than the GMPP. This undoubtedly leads to additional power losses within the PV system [6]. Considering the aforesaid situations, PV modules operating under PSCs or mismatch conditions typically encounter issues of reduced power output and increased power losses. To effectively mitigate mismatch losses, a reasonable configuration of the PV array can be regarded as a viable strategy. Within the domain of PV systems, the design and configuration of the array are crucial in determining its maximum power output. The adjustment of the interconnection approaches of PV modules in a scientific and rational manner is instrumental in effectively enhancing power generation [7].

In [8-10], scholars have conducted a discussion regarding the performance of various configurations to identify the differences among them. However, these references demonstrate a certain degree of inadequacy in terms of comprehensiveness and completeness. To make the relevant research more substantial and complete, this paper employs Matlab/Simulink software to accurately model a 6×6 PV array. Subsequently, a detailed comparative investigation is carried out on multiple performance indicators of six configurations under eight varied PSCs. These indicators include the GMPP, shading loss, mismatch loss, fill factor, and efficiency. In this paper, a comprehensive comparison and discussion are carried out regarding the performance of diverse PV array configurations under varied PSCs. Furthermore, a synthetic analysis is performed on the average values of these performance indicators obtained under each shading condition. In this way, it can more objectively and accurately reflect the advantages and disadvantages of different PV array configurations.

2. Modeling and simulation

Under PSCs, the output power of PV arrays tends to be relatively low. The phenomenon has a significant impact on their performance. For the purpose of analyzing the output characteristics corresponding to the considered array configurations, this study employs MATLAB/Simulink software to conduct modeling and simulation for the 6×6 PV array configuration. The detailed specifications of the modules are presented in Table 1.

Table 1. Parameters of PV modules.

Parameter	Value
Maximum power, P_m	329.96 W
The voltage at the maximum power point, V_m	33.99 V
The current at the maximum power point, I_m	9.71 A
Open circuit voltage, V_{oc}	41.39 V
Short circuit current, I_{sc}	10.36 A
Temperature coefficient of V_{oc}, kv	−0.28 %/°C
Temperature coefficient of I_{sc}, ki	+0.07 %/°C
Ideality factor	1.3
Number of series connected cells in a module	60

Journal of Physics: Conference Series **3011** (2025) 012083 doi:10.1088/1742-6596/3011/1/012083

2.1. PV array configurations

Throughout the simulation, all configurations are maintained in an environment with a constant temperature of 25°C and are analyzed under various PSCs. The PV array configuration types evaluated in this study include Series (S), Series-Parallel (SP), Honey-Comb (HC), Bridge-Linked (BL), Triple-Tied (TT), and Total-Cross-Tied (TCT). Their specific configurations are presented in Figure 1.

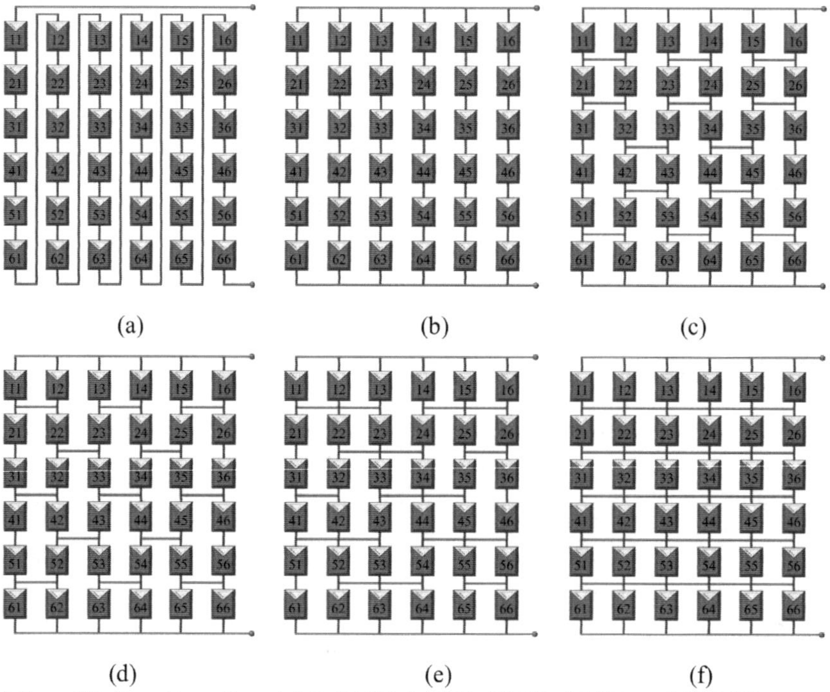

Figure 1. Modeling of 6×6 configurations (a) S, (b) SP, (c) HC, (d) BL, (e) TT, and (f) TCT.

2.2. Various Partial Shading Conditions

Regarding the PSCs considered, ranked by increasing shading intensity, they are diagonal, corner, center, column wise, row wise, frame, staircase, and random shading. The schematic depictions of PSCs are presented in Figure 2.

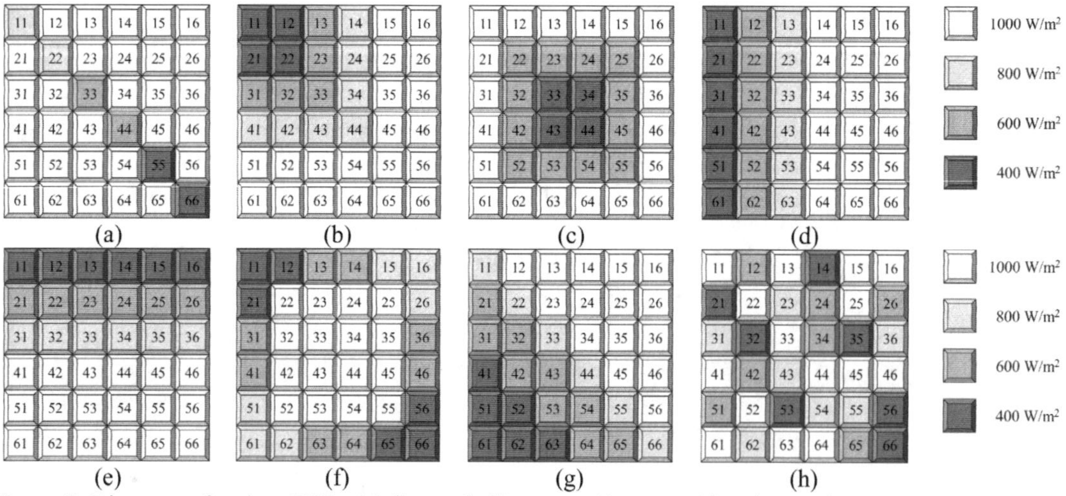

Figure 2. Diagrams of various PSCs. (a) diagonal, (b) corner, (c) center, (d) column wise, (e) row wise, (f) frame, (g) staircase, and (h) random.

3. Performance evaluation

This study is intended to conduct a thorough discussion on the performance of various PV array configurations under different PSCs. Therefore, a performance evaluation is carried out for several crucial factors. The factors involved comprehensively cover aspects such as GMPP, shading loss, mismatch loss, fill factor, and efficiency.

3.1. Shading Loss

The shading loss depends on the irradiance intensity of each PV module, relates closely to actual shading conditions, and is independent of the PV array configuration. It can be calculated by Equation (1) [9].

$$P_{shading\ loss} = P_{max,u} - P_{max,i} \tag{1}$$

where $P_{max,u}$ is the maximum available power of the PV array under uniform irradiation (1000 W/m^2), and $P_{max,i}$ is the sum of the available powers of each module under PSCs.

3.2. Mismatch Loss

Shading loss is an inevitable natural occurrence, and the mismatch loss index can effectively assess the performance of PV array configurations. It can be calculated by Equation (2) [9].

$$P_{mismatch\ loss} = P_{max,i} - P_{GMPP} \tag{2}$$

where P_{GMPP} is the global maximum power under PSCs.

3.3. Fill Factor

The fill factor is a crucial metric for evaluating PV module efficiency. Higher values indicate better power conversion efficiency. It can be calculated by Equation (3) [10].

$$FF = \frac{P_{GMPP}}{V_{oc} \times I_{sc}} \tag{3}$$

where V_{oc} is the open-circuit voltage and I_{sc} is the short-circuit current.

3.4. Efficiency

Efficiency refers to the ratio of the global maximum power generated by the PV array to the input solar power. It can be calculated by Equation (4) [10].

$$\eta\ (\%) = \frac{P_{GMPP}}{L \times A} \times 100 \tag{4}$$

where L is the solar irradiance per square meter of the PV panel (W/m^2), and A is its area (m^2).

4. Comparative study

In this study, the MATLAB/Simulink software was employed to carry out modeling and simulation analyses of six 6×6 PV array configurations. The evaluated configurations encompass S, SP, HC, BL, TT, and TCT. Concurrently, the PSCs taken into account include diagonal, corner, center, column wise, row wise, frame, staircase, and random shading. As shown in Figure 3, research results display the P_{GMPP} under different PSCs. In column-wise scenarios, the S configuration had the lowest P_{GMPP}, while others were similar. For row-wise scenarios, all configurations had similar P_{GMPP}. For other types of shading conditions, the P_{GMPP} are ranked from highest to lowest as follows: TCT, TT, BL, HC, SP, and S. The shading loss remains unaffected by the configuration of the PV array. Instead, it is closely associated with the actual shading conditions, as illustrated in Figure 4. The TCT configuration had the minimum mismatch loss, followed by TT, BL, HC, SP, and S in sequence, as shown in Figure 5. The TCT configuration has the highest Fill factor under frame and random shading. For other shading conditions, the Fill factor of the TCT, TT, and BL configurations are almost the same, as depicted in Figure 6. In terms of the efficiency assessment, across the majority of shading scenarios, the efficiencies are ranked from highest to lowest as follows: TCT, TT, BL, HC, SP, and S, as presented in Figure 7.

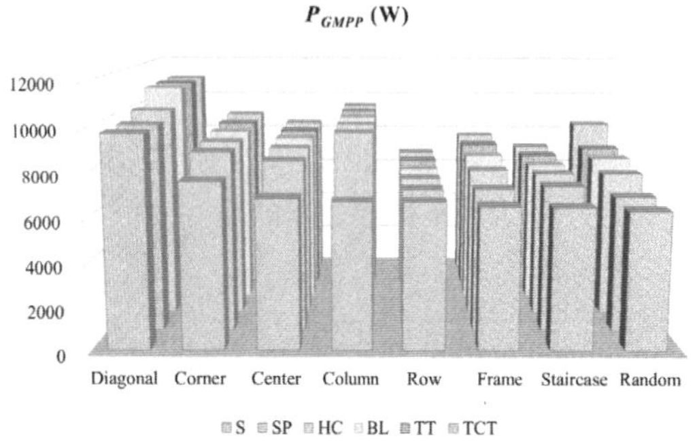

Figure 3. Global maximum power.

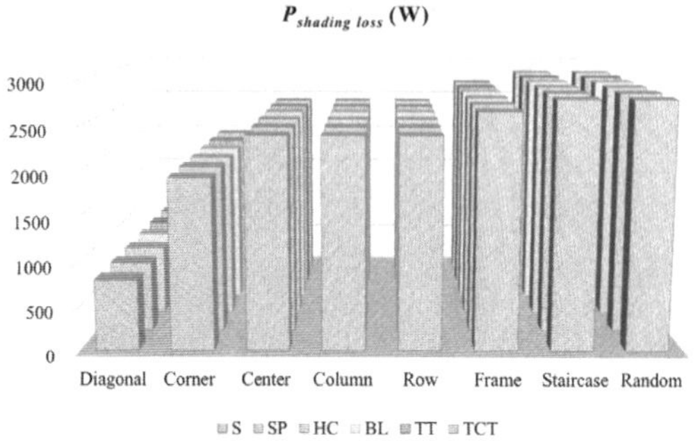

Figure 4. Shading loss.

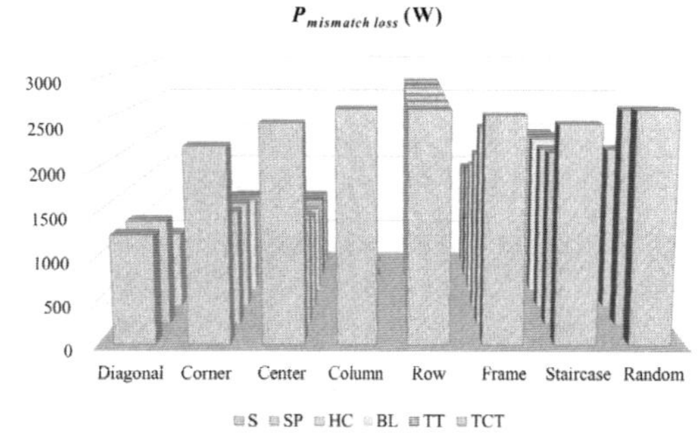

Figure 5. Mismatch loss.

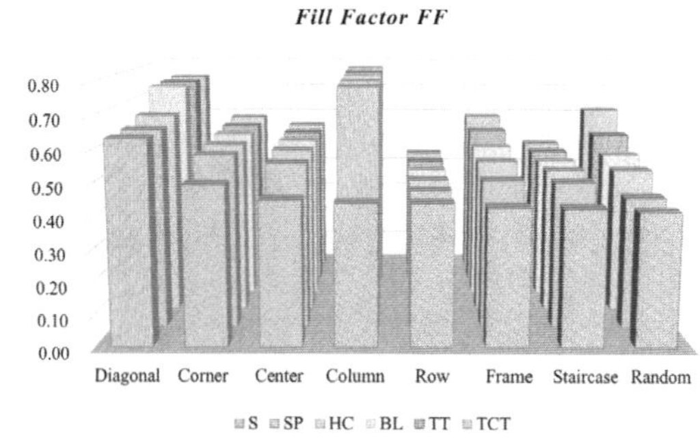

Figure 6. Fill factor.

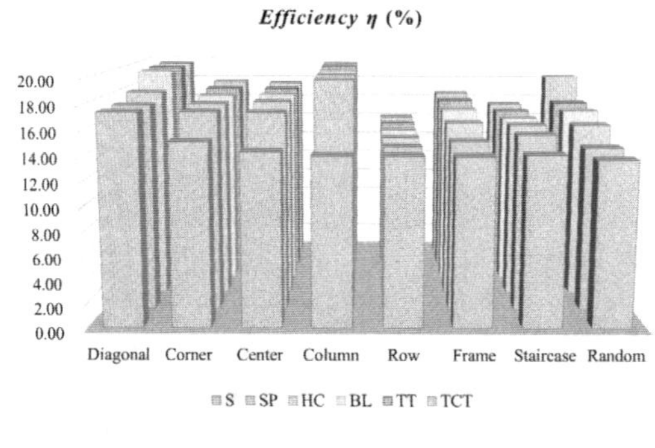

Figure 7. Efficiency.

The specific types of shading and their impact levels are crucial in determining the performance of PV systems. In real-world scenarios, due to the complexity and variability of the environment, it is extremely difficult to predict the shading conditions these systems will encounter. The average performance values derived from various PSCs can more objectively and accurately reflect the advantages and disadvantages of different PV array configurations. For details, they are shown in Table 2. In most cases, the TCT configuration exhibits optimal performance. However, it should be noted that due to the extensive cross-ties required in the TCT configuration, the complexity of wiring is significantly increased. This not only directly raises costs but also gives rise to more wiring loss issues. In comparison, the S configuration, lacking cross-connections, experiences relatively high mismatch losses. Moreover, its GMPP, fill factor, and efficiency are notably lower compared to those of other configurations. For each configuration, the average value of the shading loss under different PSCs is consistent. The BL and TT configurations, on the other hand, present viable compromise alternatives. They exhibit commendable performance across various indicators. Significantly, the performance indicators of the BL configuration are very similar to those of the TT configuration, yet the BL configuration holds an advantage in cost-effectiveness. Regarding the overall performance, the order of precedence from superior to inferior is as follows: TCT, TT, BL, HC, SP, and S.

Table 2. Average performance values under various PSCs.

Configurations	P_{GMPP} (W)	$P_{shading\ loss}$ (W)	$P_{mismatch\ loss}$ (W)	Fill Factor	Efficiency (%)
S	7161.97	2299.16	2417.43	0.47	14.60
SP	7852.39	2299.16	1727.01	0.54	16.03
HC	8057.54	2299.16	1521.86	0.56	16.45
BL	8253.23	2299.16	1326.17	0.57	16.83
TT	8308.13	2299.16	1271.27	0.58	16.95
TCT	8477.51	2299.16	1101.88	0.60	17.31

5. Conclusions

In the context of the escalating global demand for RESs, PV systems have drawn substantial attention for their numerous advantages. Using Matlab/Simulink software, this paper accurately models a 6×6 PV array. It then conducts an analysis of six configurations: S, SP, HC, BL, TT, and TCT. This study focuses on their performance indicators under eight PSCs: diagonal, corner, center, column wise, row wise, frame, staircase, and random. These indicators include the GMPP, shading loss, mismatch loss, fill factor, and efficiency. By comprehensively analyzing the average performance across these shading conditions, this paper objectively evaluates the advantages and disadvantages of each PV array configuration. Research indicates that the shading type and degree are key factors influencing PV array performance. Notably, the TCT configuration generally outperforms others. The S configuration shows significantly lower performance in many indicators. For each configuration, the shading loss average is the same across different PSCs. The BL and TT configurations represent good compromise options. The BL configuration, with performance indicators comparable to the TT configuration, has a cost advantage. Overall, the performance ranking from best to worst is TCT, TT, BL, HC, SP, and S. This research provides a reliable reference for optimizing PV system design.

Acknowledgments

This work was supported by the key scientific research projects of Fujian Polytechnic of Information Technology. The grant numbers are YZDKJ23-05 and YZDKJ24-02.

References

[1] Bonthagorla P K and Mikkili S (2022) Performance investigation of hybrid and conventional PV array configurations for grid-connected/standalone PV systems. *CSEE Journal of Power and Energy Systems*, 8: 682–95.
[2] Smadi M K A and Mahmoud Y (2018) Analysis of photovoltaic systems power losses in partial shading conditions. In: IECON 2018-44th Annual Conference of the IEEE Industrial Electronics Society, Washington, DC, USA.
[3] Murkute S and Kulkarni V A (2023) New high performance PV system architecture for mitigation of partial shading effects. *E-Prime - Advances in Electrical Engineering, Electronics and Energy*, 5: 100189.
[4] Mishra D P, Nanda S, Sethi M and Mohanta T (2023) Performance analysis of photovoltaic cells using partial shading conditions. In: IEEE 3rd International Conference on Smart Technologies for Power, Energy and Control (STPEC), Bhubaneswar, India.
[5] Yadav V K, Yadav R, Singh R, Mishra I, Ganvir I and Manish (2023) Reconfiguration of PV array through recursive addition approach for optimal power extraction under PSC. *Energy Conversion and Management*, 292: 117412.
[6] Krishna G S and Moger T (2019) Improved SuDoKu reconfiguration technique for total-cross-tied PV array to enhance maximum power under partial shading conditions. *Renewable and Sustainable Energy Reviews*, 109: 333–48.
[7] Devakirubakaran S and Bharatiraja C (2023) A novel fused fibonacci-geometric number pattern-based PV array configuration for mitigating the mismatch losses. In: IEEE IAS Global Conference on Renewable Energy and Hydrogen Technologies (GlobConHT), Male, Maldives.

[8] Kumar A S and Reddy V U (2022) Comparative performance evaluation of PV configurations under various shaded conditions. In: International Virtual Conference on Power Engineering Computing and Control: Developments in Electric Vehicles and Energy Sector for Sustainable Future (PECCON), Chennai, India.
[9] Saiprakash C, Mohapatra A, Nayak B, Babu T S and Alhelou H H (2022) A novel Benzene structured array configuration for harnessing maximum power from PV array under partial shading condition with reduced number of cross ties. *IEEE Access*, 10:129712–26.
[10] Desai A A and Mikkili S (2022) Modeling and analysis of PV configurations to extract maximum power under partial shading conditions. *CSEE Journal of Power and Energy Systems*, 8:1670–83.

Parameter analysis of the heat release process of zeolite thermochemical energy storage reactor

Ying Li[1], Shuli Liu[1*], Yihan Wang[1], Tingsen Chen[1], Yongliang Shen[2]

[1] School of Mechanical Engineering, Beijing Institute of Technology, Beijing 100081 China
[2] School of Materials Science and Engineering, Beijing Institute of Technology, Beijing 100081 China

[*]Corresponding author's e-mail: win_li1029@163.com

Abstract. Thermochemical energy storage technology has the advantages of stability and cross-seasonal storage. Adsorption-type TES is a thermal energy storage technology with great development prospects. Zeolite 13X is a cheap and safe material, and the conventional packed reactor bed is a simple and reliable reactor. In this study, a numerical model was established to simulate the output performance of the reactor, such as inlet moist air temperature, inlet moist air flow rate, inlet moist air humidity, and initial adsorption of zeolite on the output performance of the reactor. The sensitivity of the output performance of the reactor to the changes in each parameter is calculated and analyzed, and it is expected to guide the parameter design of the reactor bed. This study found that the output thermal power of the reactor is most sensitive to the change of the initial water absorption of zeolite, and least sensitive to the change of the flow rate of the inlet moist air. Increasing the flow rate of the inlet moist air reduces the output temperature of the reactor while increasing the temperature of the inlet moist air and reducing the initial adsorption amount of zeolite can significantly increase the output temperature of the reactor.

1. Introduction

The carbon emissions from housing construction in China are projected to contribute approximately 38.2 % of the country's total relevant carbon emissions in 2023, with building heating accounting for 16 % of these emissions [1]. Reducing carbon emissions from building heating is urgent. Therefore, the use of renewable energy for building heating has received widespread attention. However, as a widely used renewable energy source, solar energy has two disadvantages that are hard to ignore, namely instability and seasonality [2]. Storing solar energy can solve this problem. The advantage of TES is that the loss is small, and the energy storage density is high, which is conducive to achieving long-term effective energy storage, TES works is the use of solar energy or other energy sources will be absorbed after the drying of the energy storage material to achieve the purpose of storing energy, the need for heat, you can pass the moist air into the material so that it and the adsorption reaction occurs with the water vapor to release the heat.

Zeolite heat storage system designed by Johannes K [3] et al., the maximum output power of the reaction bed under the initial humidity parameters is only 1.5kW.In the study by Gao [4] et al., the energy release process of the cylindrical zeolite reaction bed was calculated. Under the initial parameters, the maximum temperature of the reaction bed energy release was only 36, and the heat release time was only 3 hours. In the study by Scapino [5] et al., zeolite 13X exhibited an energy density of only 0.35 GJ/m^3 in an open reaction system. Hao [6] et al. found that in a multi-module reactor, the output temperature under initial operating parameters was 33.3 % lower than that achieved after optimization. Additionally, the reactor designed by Aydin D [7] et al. demonstrated a heat release power under initial humidity conditions that was only 43 % of the output power after humidity operating parameters were optimized. These findings highlight the significant impact of optimizing operational parameters on enhancing the energy release performance of reactors.

Many scholars have focused on the study of heat and mass transfer in reactors. To optimize these processes, they have designed reactors with embedded heat pipes [8] and fins [9], as well as reactors with honeycomb structures [10]. However, when it comes to studying the various parameters that affect reactor performance, such as inlet moist air flow rate, humidity, temperature, porosity, particle size, and energy storage material size, most researchers have only discussed the influence of some parameters on reactor output performance and have not conducted further sensitivity analysis on the parameters. Han [11] et al. respectively studied the improvement of output performance by hierarchical filling of zeolites, salt hydrates-based composite adsorbents, and the effect of gradient porosity filling on reactor performance. Kant [12] et al. studied the effect of the height of the honeycomb reaction bed on the reaction process. The results indicated that when the reaction bed height was lower, steam was more likely to react with the bed, resulting in higher output temperatures. However, a higher reaction bed height led to a larger amount of filling and a greater cumulative heat release.

Most existing studies focus on the TES systems, and there is little research on the energy release process of the reactor, and few scholars have studied the impact of multiple factors on the energy release performance of the reactor, and the research on parameters is not comprehensive. At the same time, most studies do not involve the sensitivity analysis of the reactor output performance to changes in various parameters.

Therefore, this paper did the following work: (a) the usability of the model was verified based on experimental data; (b) the influence of the parameters (moist air inlet flow rate, moist air inlet temperature, moist air inlet humidity, initial water absorption of zeolite) of the reactor on the energy release process was analyzed; (c) to guide the design of operating parameters, a sensitivity analysis of the parameters was performed.

2. Dimensions of specimens

2.1. Physical model

The size parameters of the TES and the physical properties of zeolite 13X and moist air are shown in Table 1.

To simplify the calculation, the following assumptions were made:

- During the energy release process, the zeolite and moist air are in local thermodynamic equilibrium; that is, the temperature of the dry zeolite, the zeolite after adsorption reaction, and the moist air are the same at all times;
- As a low-temperature TES system, the reactor energy release process is below 100 °C, so only heat conduction and heat diffusion are considered;
- Assuming that the moist air entering the reactor is an ideal gas composed of dry air and water vapor, the thermodynamic properties of moist air are functions of the water vapor content.

2.2. Model Equations

The reaction rate of adsorption processes is represented by Equations (1) and (2).

$$\frac{\partial X}{\partial t} = k\left(X_{eq} - X_0\right) \tag{1}$$

$$k = \frac{15D_{s0}}{d_p^2} e^{\left(-\frac{E}{RT}\right)} \tag{2}$$

where X_0 is the current accumulated adsorption capacity (kg/kg); $\frac{\partial X}{\partial t}$ is the mass of water vapor adsorbed by zeolite 13X per unit time; X_{eq} is the Adsorption capacity at equilibrium (kg/kg); d_p is the particle size (mm).

The continuity equation of zeolite and water vapor in the adsorption reaction can be expressed by Equations (3) and (4).

$$\varepsilon\frac{\partial\rho_a}{\partial t}+\nabla(\rho_a\boldsymbol{u})+S_q=0 \tag{3}$$

$$S_q=(1-\varepsilon)\rho_s\frac{\partial X}{\partial t} \tag{4}$$

where ε is the porosity; u is the flow rate vector (m/s); ρ_s is the density of zeolite 13X (kg/m^3); S_q is the quality source term.

Equation (5) is the momentum equation for the discharge process,

$$\frac{\partial(\rho_a\boldsymbol{u})}{\varepsilon\partial t}+\frac{\nabla(\rho_a\boldsymbol{uu})}{\varepsilon^2}=-\nabla p+\nabla(\mu\nabla\boldsymbol{u})+S_u \tag{5}$$

The source term of the momentum equations can be represented as the Equation (6), (7), and (8)

$$S_u=-\left(\frac{\mu}{\alpha}+\frac{1}{2}C_2\rho_a|\boldsymbol{u}|\right) \tag{6}$$

$$\alpha=\frac{d_p^2\varepsilon^3}{150(1-\varepsilon)^3} \tag{7}$$

$$C_2=\frac{3.5(1-\varepsilon)}{d_p\varepsilon^3} \tag{8}$$

where ε is the porosity; μ is the dynamic viscosity (Pa*s); S_u is the momentum source term; 1/α is the coefficient of viscous resistance; C_2 is the coefficient of inertia resistance; d_p is the particle size (mm).

The energy equation is expressed by Equation (9),

$$\left(\rho C_p\right)_{eff}\frac{\partial T}{\partial t}+\nabla\left(\rho_a C_{pa}T\right)=\nabla[\varepsilon k_a\nabla T+(1-\varepsilon)k_S\nabla T]+S_e \tag{9}$$

$$\left(\rho C_p\right)_{eff}=\varepsilon\rho_a C_{pa}+(1-\varepsilon)\rho_S C_{pS} \tag{10}$$

$$S_q=(1-\varepsilon)\rho_s\frac{\partial X}{\partial t} \tag{11}$$

$$S_e=-S_q\Delta H \tag{12}$$

where $\left(\rho C_p\right)_{eff}$ is per unit effective volume; K_a is the heat conductivity coefficient of moist air (W/(m*K)); Ks is the heat conductivity coefficient of 13X (W/(m*K)).

Equation (13) is the output heat of the reactor energy release process.

$$\begin{aligned}Q_{disch}&=Q_{out}-Q_{in}\\&=C_{pa,out}\rho_{a,in}f_{a,out}\int_0^{t_a}\left(T_{a,out}\right)dt-C_{pa,in}\rho_{a,in}f_{a,in}\int_0^{t_a}\left(T_{a,in}\right)dt\end{aligned} \tag{13}$$

where T is the temperature (K); $\rho_{a,in}$ is the air density of inlet moist air (kg/m^3); $\rho_{a,out}$ is the air density of air at the reactor outlet after the adsorption reaction (kg/m^3); f_a is the airflow rate (m^3/s).

In the adsorption process, the energy release power per hour can be expressed as Equation (14):

$$P_{disch}=\frac{Q_{disch}}{t} \tag{14}$$

where P_{disch} is the energy release power (kW); Q_{disch} is the heat released by the reactor(kJ); and the subscript *disch* represents the discharge process.

Table 1. Parameter settings of the numerical simulation.

Parameter	Value	Parameter	Value
k_s	0.20 W/(m·K)	K_v	0.0196 W/(m·K)
ρ_s	850 kg/m^3	M_w	0.018 kg/mol
C_{ps}	900 J/(kg·K)	X_0	0.33 kg/kg
ρ_d	1.177 kg/m^3	E	1192300
C_{pd}	1004.9 J/(kg·K)	R	8.314 J/(kg·mol)
M_d	0.0289 kg/mol	D_{s0}	$0.02e^{-4}$ m^2/s
K_d	0.262 W/(m·K)	ΔH	29500 J/ mol
C_{pd}	1864 J/(kg·K)		

2.3. Model verification.

By setting up the experimental platform, the data collection of energy release conditions is completed, and the numerical model proposed in this paper is verified by using the experimental data. Zeolite 13X particles with an average particle size of 4mm were used as energy storage materials. It mainly includes fans, humidifiers, plate heat exchangers, electric heaters, thermochemical reactors, etc. At the same time, it also sets up air volume sensors, temperature sensors, humidity sensors, and other remote Settings to achieve the collection of test air volume, temperature, and humidity. The energy release power was calculated, and the energy release process inside the reactor was analyzed by measuring the parameter change of the inlet and outlet air.

It can be concluded that the simulation results of this study are consistent with the experimental results, which proves that the model is feasible.

$$RMSE = \sqrt{\frac{\sum_{i=1}^{n}\left(x_{exp}-x_{sim}\right)^2}{n}} \tag{15}$$

To test the independence of grid division on calculation results, this study conducted grid calculation for three different grid sizes, and the number of grids was 7,1064, 10130, and 179225, respectively. Case 3 in Table 2 is selected as the calculation condition. In Figure 1 (b), the temperature output curves of the three grid numbers almost coincide, indicating that the number of grids does not influence the calculation results. To save calculation time, 71064 mesh numbers were used in the calculation.

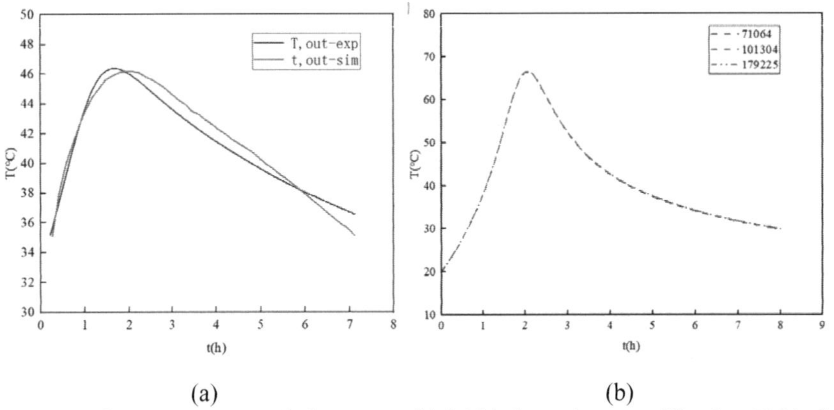

Figure 1. (a)Reactor outlet temperature variation curve; (b)Grid independence verification. Table 2. Parametric analysis of the TES during discharge.

3. Results and discussion

Table 2. Parametric analysis of the TES during discharge.

Case	T(°C)	G(kg/h)	d (%)	X_0(kg/kg)
1	16	80	50	0.17
2	18	80	50	0.17
3	20	80	50	0.17
4	22	80	50	0.17
5	24	80	50	0.17
6	20	80	50	0.17
7	20	100	50	0.17
8	20	120	50	0.17
9	20	140	50	0.17
10	20	160	50	0.17
11	20	80	50	0.15
12	20	80	50	0.16

Journal of Physics: Conference Series **3011** (2025) 012084 doi:10.1088/1742-6596/3011/1/012084

13	20	80	50	0.17
14	20	80	50	0.18
15	20	80	50	0.19
16	20	80	20	0.17
17	20	80	50	0.17
18	20	80	80	0.17

3.1 Effect of inlet moist air temperature

the influence of inlet moist air temperature on the energy release process is studied by analyzing cases 1-5. In Figure 2 (a), the temperature of the inlet moist air has a positive effect on outlet temperature. The peak value of the five groups of working conditions from high to low is 93.0 °C, 80.3 °C, 66.4 °C, 57.3 °C, and 49.2 °C.

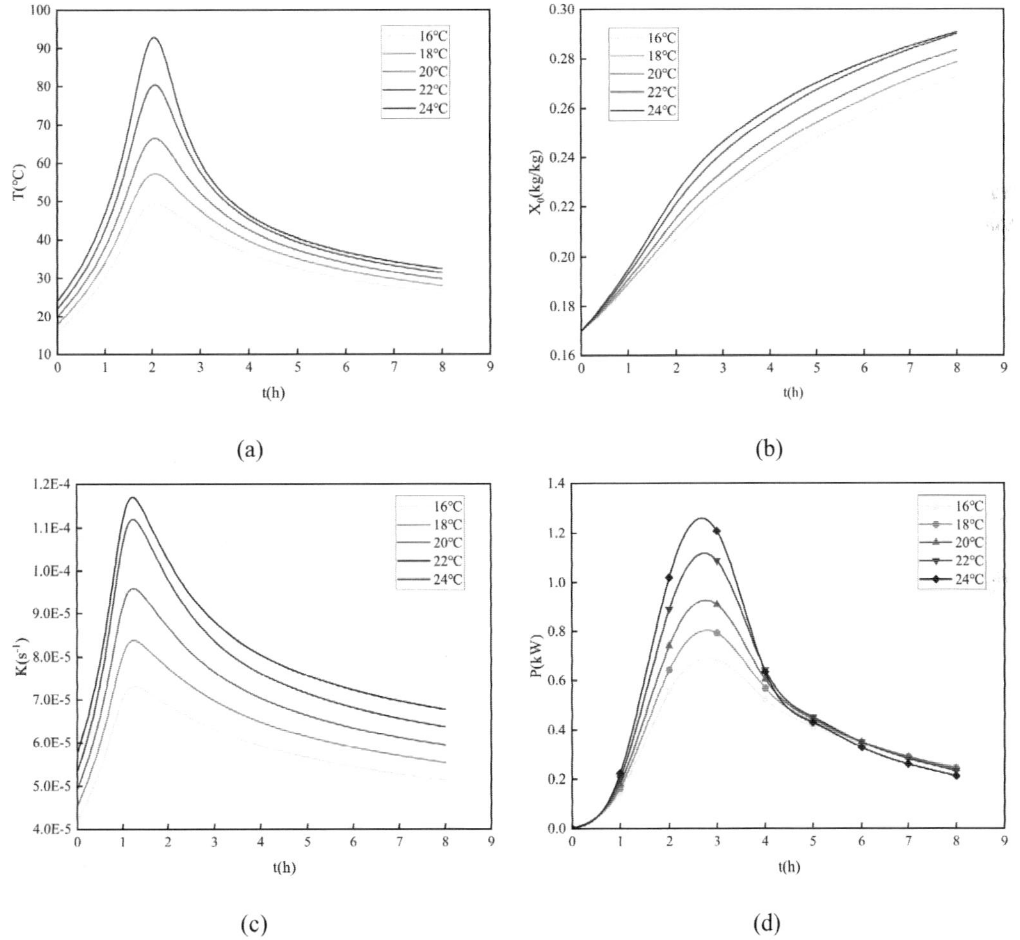

Figure 2. Variation of (a) outlet temperature; (b) zeolite adsorption capacity; (c) reactor reaction rat; (d) reactor output thermal power at different inlet moist air temperature.

Combined with Figure 2, it can be seen that the higher T is, the faster the mass transfer rate in the reactor. With higher T, the equilibrium water absorption of the material increased, and the water absorption increased from 81.98 % to 87.39 %. However, in Figure 2 (b), the T of 22 °C and 24 °C has almost the same water absorption when the energy is released for 8 hours, indicating that the inlet temperature has limited influence on the water absorption of the material. Figure 2 (c) shows that the changing trend of the K is consistent with the changing trend of the T. Figure 2 (d) expresses the change in the reactor thermal power. Its change trend is the same as the change in temperature. As the T increases, the P increases. The average value is 0.4 kW. The T is 18 °C, the peak energy release power is 0.79 kW, and the average is 0.44 kW. When the T increases to 24 °C, the peak energy release power is 1.20 kW, and the average is 0.54 kW.

3.2 Effect of inlet moist air flow rate

the influence of inlet moist air flow rate on the release energy process is studied by analyzing cases 6-10. It can be seen from Figure 3 (a) that the G not only has a significant impact on the outlet temperature of the reactor but also affects the time when the outlet temperature reaches its peak. As G increases, the outlet peak temperature decreases, and the time when the peak temperature reaches the peak is advanced. When G is 160 kg/h, the outlet peak temperature reaches 34.7 °C; when G is 80 kg/h, the outlet temperature reaches a peak of 66.4 °C and reaches the peak in 2 hours.

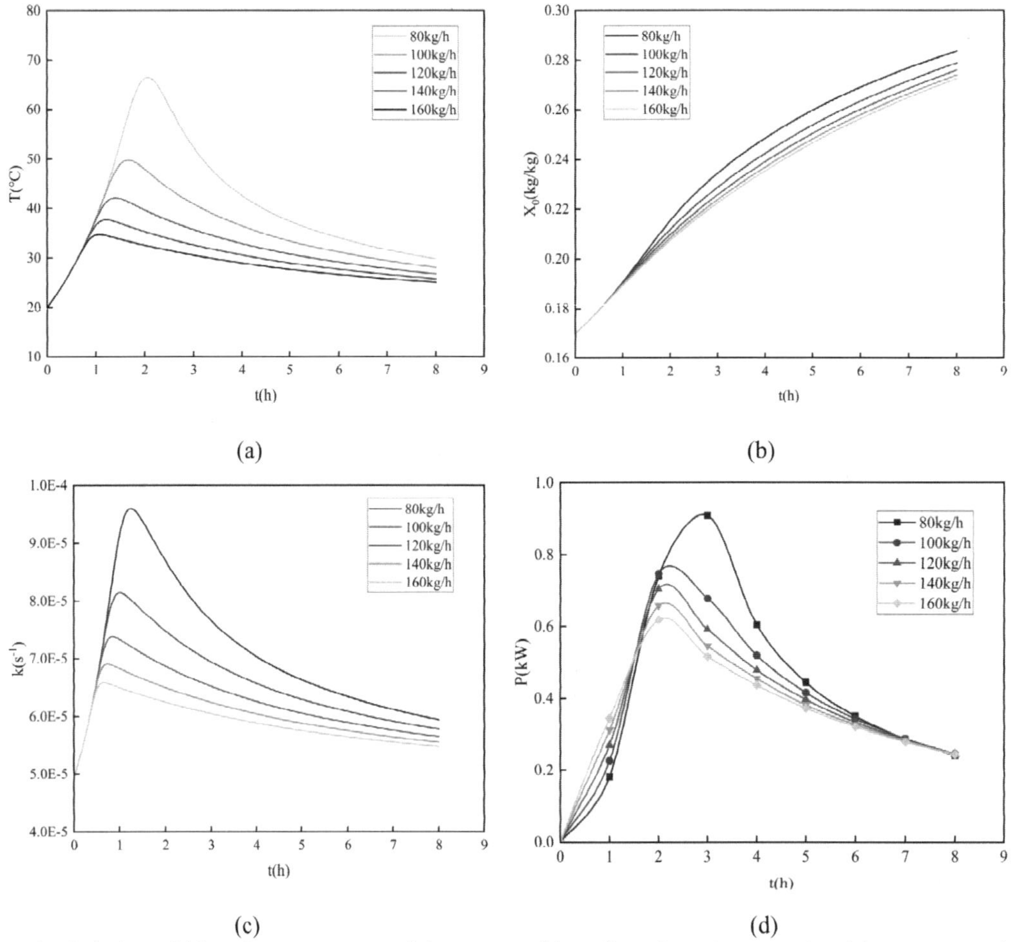

Figure 3. Variation of (a) outlet temperature of the reactor; (b) zeolite adsorption capacity; (c) reactor reaction rat; (d) reactor output thermal power at different inlet moist air flow rate.

Figure 3 (c) shows the influence of different G on the mass transfer coefficient of the adsorbed material. As the G increases, the K decreases. This is because although the higher G carries more water vapor, the faster the flow rate, the more water vapor has already flowed out of the reactor without being adsorbed. The heat generated by the adsorption reaction requires more air to be heated, causing the outlet temperature to decrease as the flow rate increases. Figure 3 (d) shows the variation of the reactor's thermal power with the entrance G. The advantage of increasing the entrance G is to obtain higher PP, but it will reduce the subsequent exit temperature. The peak power release of G at the entrance of 80 kg/h is 0.91 kW, the average is 0.47 kW; the peak P release of G at the entrance of 160 kg/h is 0.62 kW, the average is 0.39 kW. The heat output decreased by 17 % when G at the entrance increased from 80 kg/h to 160 kg/h.

3.3 Effect of initial water absorption of zeolite

This section studies the effect of airflow on the discharge process by analyzing Cases 11-15. As can be seen from Figure 4., the initial water absorption of zeolite has an impact on the outlet temperature of the reactor. The initial value decreases while the outlet temperature increases. The peak values of the five groups of working conditions are 90.7 °C, 76.1 °C, 66.4 °C, 59.3 °C and 53.6 °C successively from high to low. Figure 4 (b) describes the changes in the equilibrium adsorption capacity of the energy storage material under different initial adsorption capacities. It can be seen that with the increase of the initial value, the final equilibrium adsorption capacity increases, but the increment is small. Figure 4 (c) shows the influence of different X_0 on the mass transfer coefficient of the adsorbed material. When the X_0 is increased from 0.15 kg/kg to 0.2 kg/kg, the K of the material is reduced by 25 %. According to the data in Figure 4 (d), when the X_0 is 0.15 kg/kg, the average value is 0.58 kW. When the X_0 is increased to 0.20 kg/kg, the corresponding average value is 0.35 kw, when the X_0 increases from 0.15 kg/kg to 0. 2 kg/kg, the average energy release heat power decreases by 40 %.

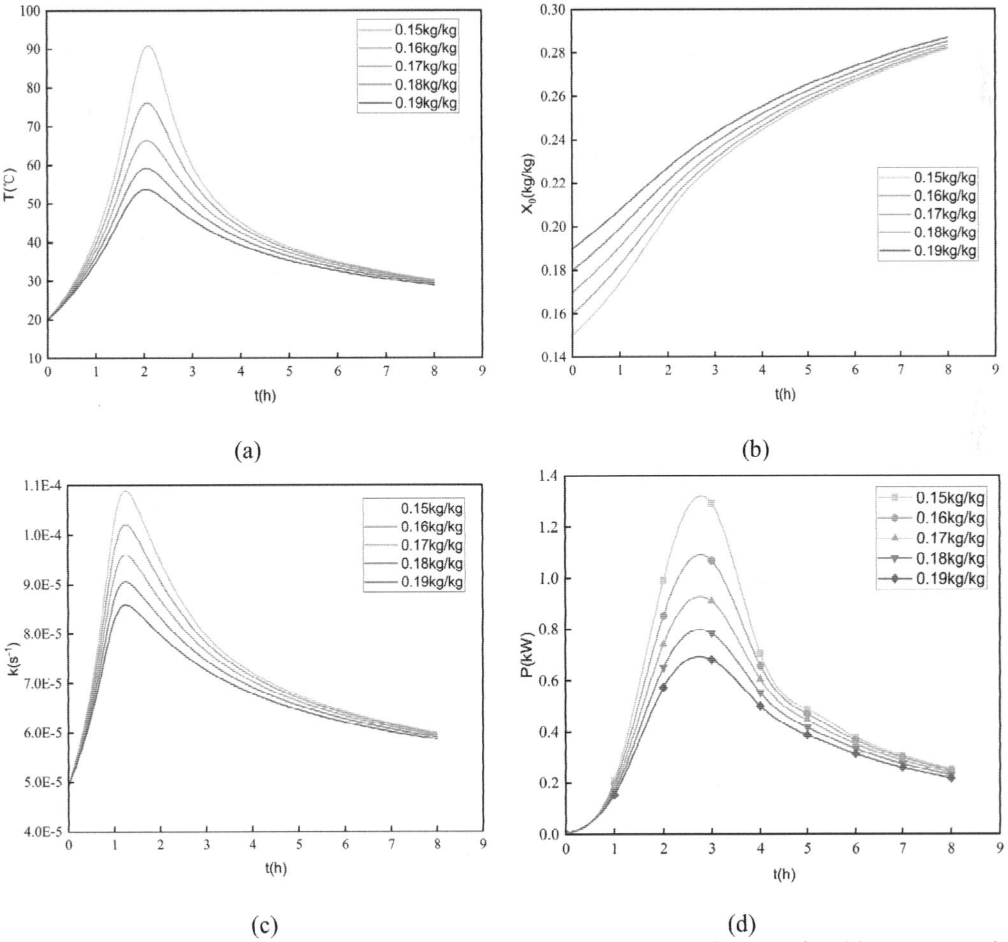

Figure 4. Variation of (a) outlet temperature of the reactor; (b) zeolite adsorption capacity; (c) reactor reaction rat; (d) reactor output thermal power at different initial water absorption of zeolite.

3.4 Effect of inlet moist air humidity

This section examines the thermal effects of the humidity of inlet moist air on the energy release process through cases 16-18. Figure 5 (a) shows the effect of the humidity of moist air on the outlet air temperature of the reactor. In the initial stage, the outlet temperature rises rapidly for all three conditions, reaching a peak after 2 hours. The higher the humidity, the faster the outlet air temperature rises and the higher the peak temperature, which are 55.0 °C, 66.4 °C, and 72.1 °C, respectively. Figure 5 (b) shows that as the d increases, the water absorption rate of the zeolite increases at the end of the energy release process, with absorption rates of 80.78 %, 85.29 %, and 87.09 %, respectively.

Figure 5 (c) indicates that increasing the d raises the temperature of the thermal storage material and increases the mass transfer coefficient. Notably, the mass transfer coefficients for 50 % and 80 % humidity are similar in most periods, except for a significant difference during the peak period. This suggests that within the humidity range of 50 % to 80 %, changes in inlet moist air humidity have a considerable impact on mass transfer during peak periods but a minimal impact on mass transfer during other periods. Figure 5(d) shows that as the d increases, the energy release power increases. When the d is 20 %, the average energy release power is 0.4 kW; at 50 % humidity, it is 0.47 kW; and at 80 % humidity, it is 0.50 kW. The energy release power increases by 25 % as the d from 20 % to 80 %.

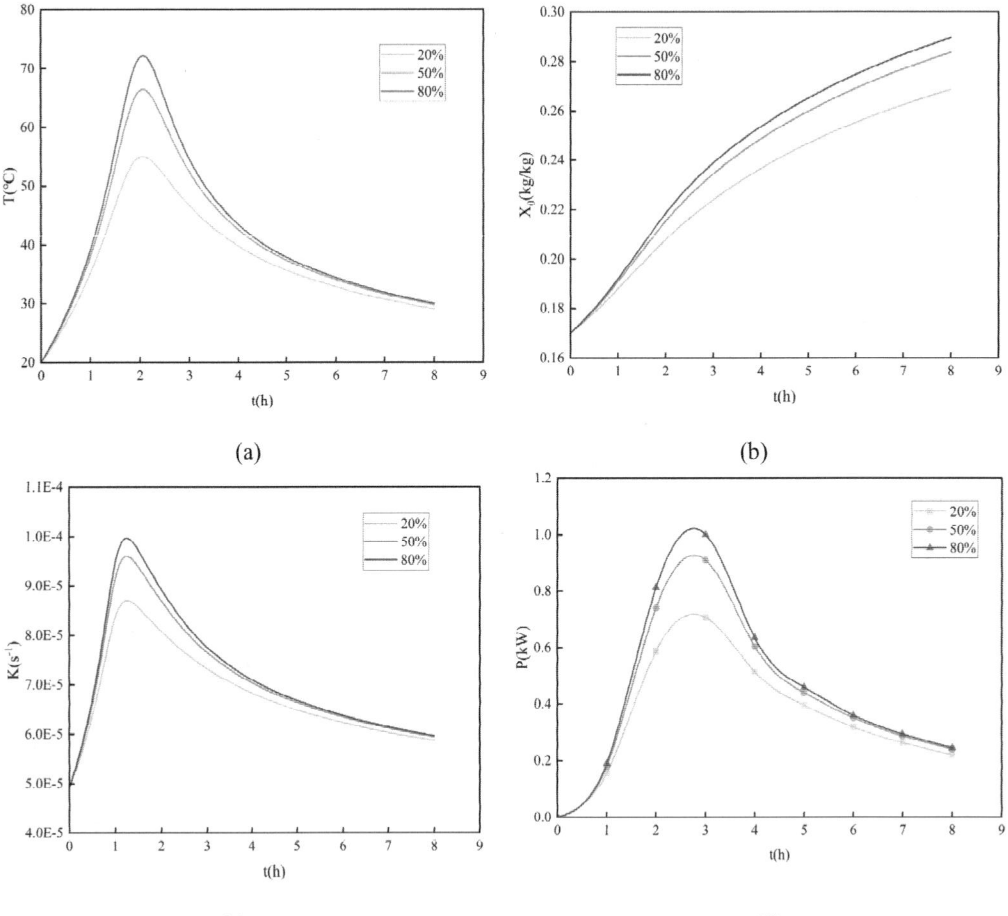

Figure 5. Variation of (a) outlet temperature of the reactor; (b) zeolite adsorption capacity; (c) reactor reaction rat; (d) reactor output thermal power at different humidity.

3.5 Sensitivity analysis

to achieve the purpose of guiding the setting of reactor operating parameters according to the energy release demand, sensitivity analysis is performed on parameters such as the inlet moist air temperature, the inlet moist air flow rate, the inlet moist air humidity, the initial adsorption amount of zeolite 13X to study the extent to which changes in these important parameters affect the energy release performance. Local sensitivity analysis identifies the most important input parameters by applying a one-time change to the nominal value of the parameter to influence the variability of the output. The sensitivity index Si is used to evaluate the sensitivity of a parameter, and the greater the absolute value of Si, the more sensitive the performance is to the parameter. Si is calculated as shown in Equation (16):

$$Si = x\frac{y(x+\Delta x)-y(x)}{\Delta x} \tag{16}$$

thermal power and water absorption rate are two important parameters that reflect the performance of the reactor. The results are shown in Figure 6, which indicates that the sensitivity of water absorption rate in descending order is initial adsorption > > inlet temperature > flow rate, while the sensitivity of thermal power in descending order is initial adsorption >inlet temperature > flow rate.

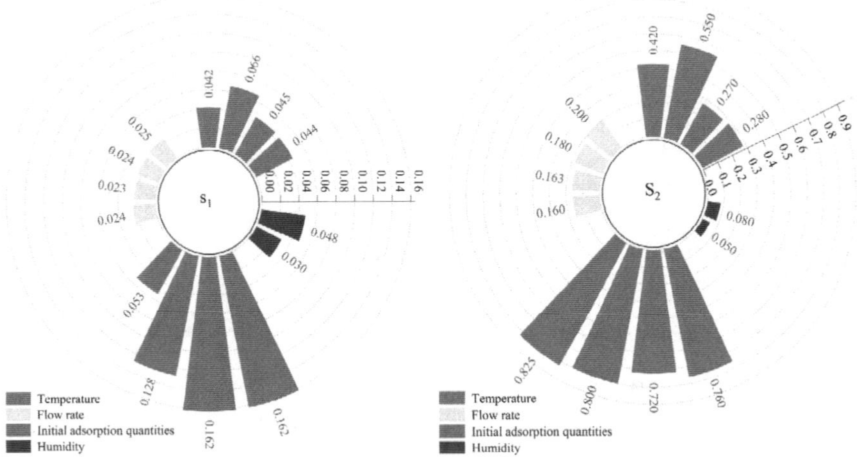

(a) (b)

Figure 6. (a)Sensitivity analysis of water absorption to parameter changes; (b) Sensitivity analysis of power to parameter changes.

4. Conclusion

Given the incomplete research on the parameters of existing thermochemical reactors, this paper studies the influence of the inlet air temperature, humidity, flow rate, initial adsorption amount of zeolite, and other parameters of conventional filled thermochemical reactors on the output performance. The main research findings are as follows:

(1) For conventional filled thermochemical reactors, the thermal power released by the adsorption process is most sensitive to the change of the initial adsorption amount of zeolite and least sensitive to the change of the flow rate of the inlet moist air. Therefore, with the ultimate goal of improving the energy release performance of the reactor, the zeolite should be dried as much as possible during loading; increasing the G does not help much in improving the performance of the reactor, and too high a flow rate will lead to a decrease in the output temperature.

(2) In the process of adjusting the parameters, the increase in the temperature of the reactor inlet moist air and the decrease in the initial adsorption amount of zeolite can make the output temperature higher than 90 °C, and the highest output power is up to 0.54 kW and 0.58 kW respectively.

Acknowledgments

The authors would like to acknowledge the financial support for this study provided by the National Natural Science Foundation of China (No. 52178063) and the Postdoctoral Fellowship Program of CPSF (GZC20233388).

References

[1] Farghali M, Osman A I, Mohamed I M A, Chen Z, Chen L, Ihara I, Yap P-S, and Rooney D W. Strategies to save energy in the context of the energy crisis: A review. *Environmental Chemistry Letters* 21: 2003-39, 2023.
[2] Zhao C, Opolot M, Liu M, Wang J, Bruno F, Mancin S, and Hooman K. Review of analytical studies of melting rate enhancement with fin and/or foam inserts. *Applied Thermal Engineering* 207: 2022.
[3] Johannes K, Kuznik F, Hubert J-L, Durier F, and Obrecht C. Design and characterisation of a high powered energy dense zeolite thermal energy storage system for buildings. *Applied Energy* 159: 80-86, 2015.
[4] Gao S, Wang S, Hu P, Wang J, Sun Y, and Ma Z. Performance of sorption thermal energy storage in zeolite bed reactors: Analytical solution and experiment. *Journal of Energy Storage* 64: 2023.
[5] Scapino L, Zondag H A, Van Bael J, Diriken J, and Rindt C C M. Energy density and storage capacity cost comparison of conceptual solid and liquid sorption seasonal heat storage systems for low-temperature space heating. *Renewable and Sustainable Energy Reviews* 76: 1314-31, 2017.
[6] [Hao C, Feng G, Ma C, Barreneche C, and She X. Performance analysis of a novel multi-module columnar packed bed reactor with salt hydrates for thermochemical heat storage. *Journal of Energy Storage* 86: 2024.
[7] Aydin D, Casey S P, Chen X, and Riffat S. Novel "open-sorption pipe" reactor for solar thermal energy storage. *Energy Conversion and Management* 121: 321-34, 2016.
[8] Liu K, Wu C, Gan H, Liu C, and Zhao J. Latent heat thermal energy storage: Theory and practice in performance enhancement based on heat pipes. *Journal of Energy Storage* 97: 2024.
[9] Nie C, Liu X, Rao Z, and Liu J. Discharging performance evaluation and optimization of a latent heat thermal energy storage unit with helm-shaped fin. *Applied Thermal Engineering* 236: 2024.
[10] Han X C, Xu H J, Xu T, and Zhao C Y. Magnesium-based thermochemical reactor with multiporous structures for medium-temperature solar applications: Transient modelling of discharge capability. *Solar Energy Materials and Solar Cells* 238: 2022.
[11] Han X C, Xu H J, and Zhao C Y. Design and performance evaluation of multi-layered reactor for calcium-based thermochemical heat storage with multi-physics coupling. *Renewable Energy* 195: 1324-40, 2022.
[12] Kant K, Shukla A, Smeulders D M J, and Rindt C C M. Performance analysis of a k2co3-based thermochemical energy storage system using a honeycomb structured heat exchanger. *Journal of Energy Storage* 38: 2021.

Multiphysics simulation analysis of superconducting hybrid energy pipelines

Yizheng Zhao, Qian Li*, Jingdong Chen, Liwen Zhang, and Hailong Liu

School of Electrical Engineering and Information, Southwest Petroleum University, Chengdu, China

* Corresponding author's e-mail: swpuliqian@163.com

Abstract. This article mainly introduces the structure of a superconducting DC energy pipeline and analyzes its transmission principle. A mathematical model of cold energy loss for a superconducting hybrid energy pipeline is established, and the cold energy loss per unit length of the pipeline is calculated to be 11.43 W/m. An electric magnetic thermal current coupling model is established for a 10 kV/1 kA high-temperature superconducting DC energy pipeline through the COMSOL Multiphysics simulation platform. The temperature distribution, current density distribution, flow velocity distribution, and magnetic field distribution of the pipeline are analyzed. Finally, the finite element model is used to verify the mathematical model of cold energy loss. The comparative results show that the structure of the hybrid energy pipeline meets the design requirements of a 10 kV/1 kA high-temperature superconducting DC energy pipeline.

1. Introduction

Due to the uneven distribution of energy in China caused by terrain and environmental factors, electricity, and natural gas are mainly distributed in the western region, while users are mostly distributed in the eastern region. At the same time, with the vigorous development of wind and solar power generation technologies, China has deployed the massive project of West-East Power Transmission and West-East Gas Transmission, among which the energy pipelines and transmission lines used for energy transmission are the key considerations [1-2]. Therefore, it is necessary to research energy gathering and transmission issues.

Superconducting direct current cable is a high-voltage direct current transmission cable made by using superconducting technology, with extremely low electrical resistivity and smaller losses compared to traditional cables when transporting the same power of electrical energy. It has significant energy-saving effects and transmission efficiency [3-5]. Superconducting cables currently commonly use liquid nitrogen cooling, but the production cost of liquid nitrogen cooling systems is relatively high, and the utilization of cold energy is insufficient, which restricts the promotion and application of superconducting direct current transmission [6-7]. Liquefied natural gas (LNG), as a low-temperature liquid mixture formed by deacidification and dehydration of natural gas and cryogenic liquefaction using low-temperature processes, has an energy density about 600 times that of natural gas [8]. Reference proposes combining high-temperature superconducting DC cables with liquefied natural gas transmission pipelines and using LNG as the refrigerant for superconducting systems [9], which can effectively utilize the low-temperature refrigeration properties of LNG and improve the energy transmission efficiency of the system. Bi-2223 tape is used as the conductor material for superconducting DC cables, with an operating temperature range of 85-90 K [10-11]. This can provide effective measures to solve the cooling problem of superconducting cables and provide superior solutions to improve natural gas transportation capacity and efficiency.

This article first establishes a mathematical model for the cold energy loss of superconducting DC energy pipelines. By combining the material parameters of the pipeline, the unit cold energy loss of the pipeline body can be obtained. Then, using the COMSOL simulation platform, an electric magnetic thermal current coupling model is established for the 10 kV/1 kA high-temperature superconducting DC energy pipeline. The temperature distribution, current density distribution, flow velocity distribution, and magnetic field distribution of the pipeline are analyzed. Finally, the correctness of the cold energy loss mathematical model is determined by using a finite element model. This study verifies the feasibility of the superconducting DC energy pipeline and provides a theoretical basis for the demonstration of superconducting DC energy pipelines.

2. Superconducting hybrid energy pipeline structure

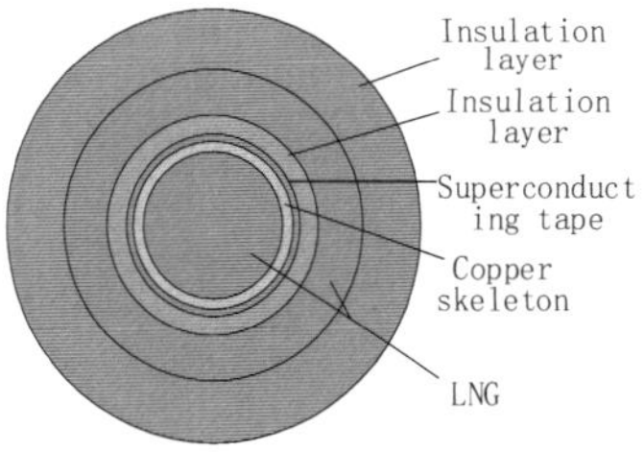

Figure 1. SHEP structure diagram.

The structure of SHEP is shown in Figure 1, and its workflow is as follows: natural gas is cooled and pressurized, liquefied into LNG, and input into a diversion pump. After diversion, it enters SHEP, with some directly entering the outer LNG pipeline for transportation and the other part entering the inner LNG pipeline for transportation through a high-temperature superconducting cable terminal. In the middle, the cold pump station provides the required cooling capacity and transportation pressure to maintain the supercooled state. At the end of the SHEP pipeline, the LNG in the outer LNG pipeline merges with the LNG in the inner LNG pipeline for output. Electricity is transported through high-temperature superconducting cables and output through the terminals of the cables, completing the synchronous transmission of natural gas and electricity. The parameters of the superconducting hybrid energy pipeline transmission system used in this article are shown in Table 1.

Table 1. SHEP system parameters.

Project	Parameter	Numerical value
Rated parameters of superconducting cable	Transmission capacity/MW	10
	Voltage/KV	10
	Electric current/KA	1
	Operation temperature/K	90
	Length/m	1, 000
Cooling and power system	Velocity of flow/m·s^{-1}	1
	Total LNG flow rate/kg·S^{-1}	161
	Number of cold pump stations	2
Insulation layer parameters	Insulation thickness/m	0.5
	Inner diameter of insulation layer/m	0.5
	Thermal conductivity of insulation layer/W/(m·K)	0.011
	Heat transfer coefficient of the inner skin of insulation layer/(m^2·K)	8.14

2.1. Characteristics of superconducting materials

The superconducting cable used by SHEP is essentially a superconductor with three major characteristics: 1) Zero resistance characteristic; 2) Meissner effect; 3) Josephson effect [11]. When SHEP operates normally, the superconductor is superconducting, where the current, temperature, and magnetic field of the superconductor are all kept within critical values, known as the critical current density J_c, critical transition temperature T_c, and critical magnetic field strength H_c, as shown in Figure 2.

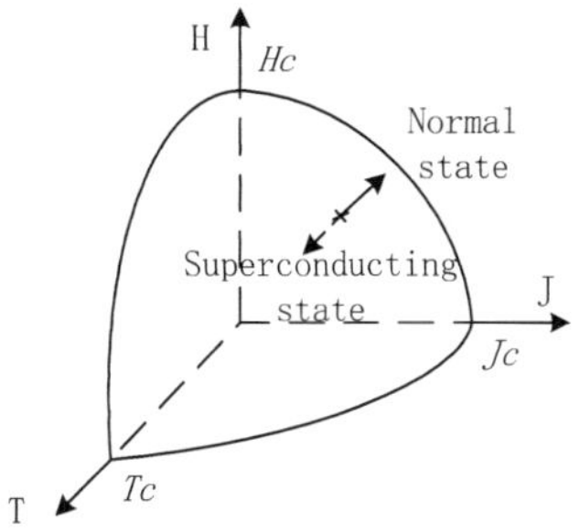

Figure 2. Critical diagram of superconducting state.

2.2. SHEP transmission mathematical model

During the SHEP power transmission process, superconducting cables generate electromagnetic fields, which can cause electromagnetic losses when applied to other metal structures. If the strength of the electric and magnetic fields exceeds the limit, the properties of the superconducting material will change. The relationship between electric field strength E and current density J can be established for the superconducting tape in this article:

$$E = \frac{E_0}{J_c} \cdot \left(\frac{J_{norm}}{J_c} \right)^{n-1} \cdot J \tag{1}$$

When SHEP is powered by direct current and operates in a steady state, the conductor is superconducting, and the magnetic field strength is very small, so electromagnetic heat can be ignored in this model.

2.3. SHEP cold energy loss model

Due to the use of insulation materials in the insulation layer, the convective thermal resistance measured inside the pipeline is almost negligible. Therefore, the temperature of LNG can be approximated as the temperature of the inner surface of the insulation layer, and the calculation formula for the unit length cold energy loss of the pipeline body is:

$$Q_h = (T_{en} - T_{LNG}) \cdot \left(\frac{1}{2\pi\lambda} \ln(1 + \frac{2\delta}{d}) + \frac{1}{\pi\alpha(d + 2\delta)}\right)^{-1} \tag{2}$$

By substituting the parameters shown in Table 1 into the equation, the variation curve of SHEP cooling loss with distance can be obtained, as shown in Figure 3. From the figure, with the determined design parameters of SHEP, the cold energy loss during SHEP transmission is positively correlated with the transmission distance.

When the ambient temperature is 293 K, substituting the SHEP system parameters shown in Table 1 into Equation (2) can calculate that the unit cooling energy loss of the pipeline body is 11.43 W/m. However, after calculation, the maximum cooling energy loss of the pipeline body set in the environment in this article is 11.84 W/m, which is greater than the calculated cooling energy loss of the pipeline body in this article and meets the operating temperature requirements of SHEP.

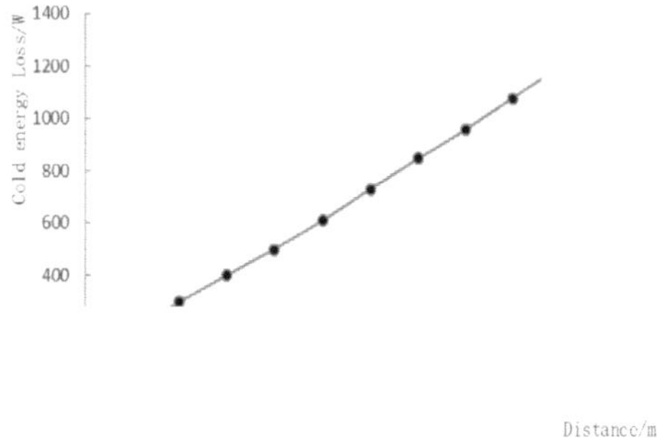

Figure 3. Changes in cold energy loss.

2.4. Geometric model establishment

We establish a two-dimensional axisymmetric geometric model of SHEP by using COMSOL multi-physics simulation software, as shown in Figure 4. The flow velocity of both the inner LNG pipeline and the outer LNG pipeline is 1 m/s. The initial temperature of LNG at the pipeline inlet is set to 90 K, the external ambient temperature is set to 293.15 K, and a current of 1, 000 A is applied to the superconducting tape. Based on this, simulation analysis is conducted on the multi-physics field coupling of SHEP.

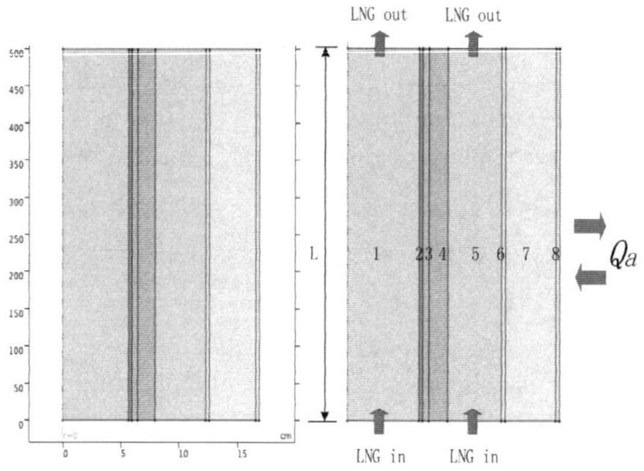

Figure 4. SHEP pipeline two-axis symmetric simulation model.

The SHEP pipeline consists of LNG, copper support layer, superconducting layer, insulation layer from inside to outside LNG, stainless steel support layer, insulation layer, and rigid protective cover. When a current of 1 kA is applied to the superconducting tape and enters the steady state, the magnetic induction intensity distribution cloud map and current density distribution cloud map on the cross-section of SHEP are shown in Figure 5.

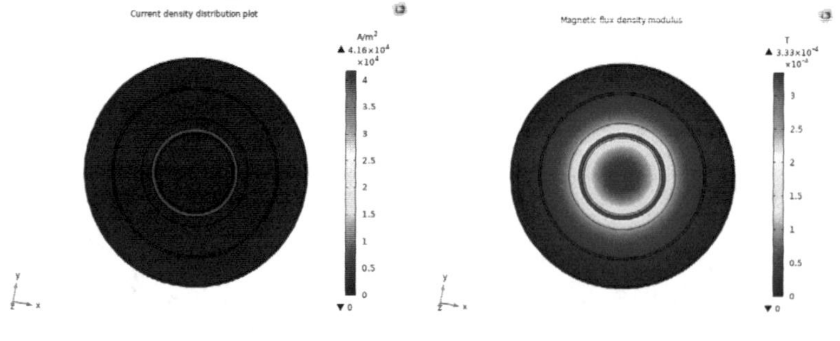

(a) (b)

Figure 5. (a) Current density distribution on SHEP cross-section; (b) Magnetic field intensity distribution.

Comparing Figure 6 (a) and Figure 6 (b), it can be observed that the temperature changes are mainly reflected in the insulation layer, and there is no significant temperature change in the internal area. After a 5 m long transmission, the temperature of the outermost layer of the pipeline increased by about 3.24 mK.

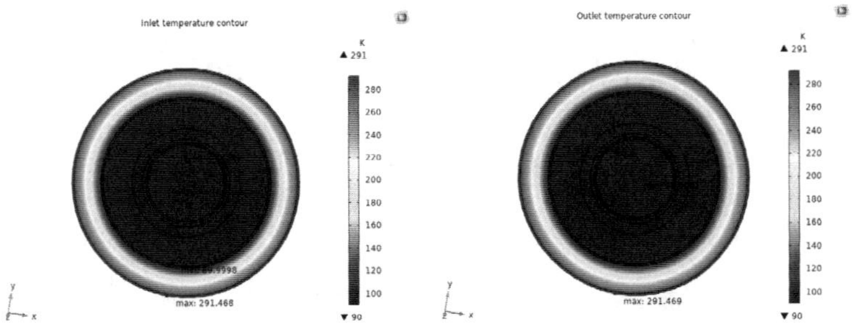

(a) (b)

Figure 6. SHEP cross-sectional inlet and outlet temperature distribution: (a) inlet temperature field cloud map; (b) outlet temperature field cloud map.

At an ambient temperature of 293 K, the unit cooling energy loss of the pipeline body is calculated to be 11.43 W/m, and the unit cooling energy loss obtained through simulation is 10.89 W/m. The error $\delta < 5\%$ is within an acceptable range.

3. Conclusion

This article first establishes a mathematical model for the cold energy loss of a 10 kV/1 kA superconducting DC energy pipeline. After calculation, the unit cold energy loss is obtained as 11.43 W/m. Then, a finite element model of the energy pipeline is established by using COMSOL, and the unit energy loss is obtained as 10.89 W/m, with an error within the allowable range. These works have verified the feasibility of using the SHEP structure for hybrid energy transmission, providing a basis for the feasibility and rationality of regional integrated energy systems based on SHEP energy supply.

References

[1] Yang Shuting, Lan Geng, Wang Pingyu, et al. Research Status and Prospect Analysis of Superconducting Cables [J]. Electric Power and Energy, 2024, 45 (01): 1-7.

[2] Zhu Jinhua, Zhu Hanshan, Wang Enzhe, et al. Research Status and Application Scenario Analysis of High Temperature Superconducting Cables [J]. Zhejiang Electric Power, 2023, 42 (07): 26-37.

[3] Wang Lina. Research on Composite High-Temperature Superconducting Energy Pipeline [D]. Tianjin University, 2022.

[4] Huang Weican, Jiang Xiaohua, Xue Peng, et al. Design of Current Carrying Conductors for Superconducting DC Energy Pipelines [J]. Journal of Tsinghua University (Natural Science Edition), 2022, 62 (10): 1, 715-1, 720.

[5] Wei Kejia. Finite element study on AC losses of superconducting cables composed of high-temperature superconducting tapes with different geometric structures [D]. Kunming University of Science and Technology, 2023.

[6] Zhu Hongliang, Cao Yujun, Xia Fangmin, et al. Simulation and Experimental Study on Decompression Refrigeration of High-Temperature Superconducting Cable [J]. Modern Transmission, 2024, (01): 47-50.

[7] Wang Longbiao, Qian Gang, Lv Yingying, et al. Simulation Model Study on Current Distribution in Shielding Layer of High-Temperature Superconducting Cable [J]. Low Temperature and Superconductivity, 2022, 50 (02): 30-35+100.

[8] Li Jichun, Cao Yujun, Xia Fangmin, et al. Development and Testing of Superconducting DC Energy Pipeline Cable Body [J]. Low Temperature and Superconductivity, 2022, 50 (08): 15-20+27.

[9] Yang Chen, Dong Xiaofeng, Tao Wenbing, et al. Design of Low-Temperature Refrigeration System for High-Temperature Superconducting DC Cable [J]. Low Temperature and Superconductivity, 2023, 51 (09): 59-64.

[10] Liu Yang. Research on the Electromechanical Behavior of Typical High-Temperature Superconducting Wire Strip and Its Cable Structure [D]. Lanzhou University, 2024.

[11] Zong Ming, Chen Yanjun, Duan Ruochen, et al. Numerical Study on Transient Thermal Characteristics of High Temperature Superconducting Cables [J]. Low Temperature and Superconductivity, 2022, 50 (10): 20-25.

Research on lightning-induced voltage of overhead line in the lossy ground

Zhaojian Sun [1*], Tao Yang [1], He Hao [1], Yueqian Ling [1], Huanbo Zhang [1]

[1] Shandong Electric Power Company Yantai Power Supply Company, Yantai, 264001, Shandong, China

[*]Corresponding author: sun_zhaojian@126.com

Abstract. Lightning strikes generate transient high voltage on power lines, which can lead to insulation failure, large-scale power outages, and pose serious threats to the s operation of the power grid. Accurately calculating lightning-induced voltage is crucial in the design, operation, and maintenance of transmission lines. This paper aims to propose and analyze a new model for calculating lightning-induced voltage, considering the effects of lightning channel inclination and soil conductivity under complex geographical and environmental conditions. Techniques based on transient electromagnetic field analysis and the finite element method were employed to study lightning-induced voltage on overhead lines, along with an examination of the thin wire modeling method's applicability. The simulation results indicate that the inclination of the lightning channel significantly affects the induced voltage distribution, with a larger inclination leading to higher induced voltage. Additionally, soil conductivity plays a crucial role, where higher conductivity effectively reduces the induced voltage. The findings suggest that both the inclination of the lightning channel and soil conductivity should be considered critical factors in the design of transmission lines to enhance lightning resistance and operational reliability.

1. Introduction

As the global electricity demand continues to grow, the size and complexity of the power system are expanding rapidly [1-3]. However, as transmission lines spread over a wide area, their vulnerability to exposure to the elements increases. As one of the strongest sources of electromagnetic interference in nature, lightning often poses a serious threat to power transmission lines. The induced high voltage caused by lightning strikes may not only lead to the failure of line insulation but also trigger a series of chain reactions, leading to a wide range of power outage accidents, seriously affecting the stability of the power system and the reliability of the power supply.

In the past few decades, researchers have mainly focused on the phenomenon of high voltage when lightning strikes directly hit the line, and focused on the influence of factors such as the height of the overhead line and the location of the lightning strike point on the induced lightning voltage. However, with further research, there is increasing evidence that the formation mechanism of induced high pressure is more complex than expected. Especially when the lightning strike occurs at a far distance from the line, the geometric shape and location of the lightning channel, soil conductivity, and other environmental parameters will also significantly affect the magnitude and distribution of the induced voltage [4, 5].

Soil conductivity determines the attenuation degree of lightning current during ground propagation, while the inclination angle of the lightning channel affects the path of lightning current in space and the distribution characteristics of the lightning magnetic field. These factors work together to make the induced voltage of transmission lines show complex spatial distribution characteristics. Ignoring these factors may lead to an underestimation of the risk of lightning strikes in transmission line design and protection measures, thus failing to effectively prevent damage caused by lightning strikes [6, 7].

Based on the existing research, a new model for calculating lightning-induced voltage is proposed and analyzed in detail. The model not only considers the traditional overhead line height and lightning strike point location but also introduces the influence of key environmental factors such as the inclination of the lightning channel and soil conductivity [8, 9]. Through the systematic study of these parameters, this paper aims to provide a more comprehensive and accurate calculation method of lightning-induced voltage and provide solid theoretical support for the lightning-resistant design of transmission lines [10, 11]. This improved model can better reflect the actual situation under complex geographical and environmental conditions, thus laying the foundation for improving the safety and reliability of transmission lines.

2. Theoretical model and simulation method

2.1. Theoretical model of lightning-induced voltage

Lightning-induced voltage is the high voltage induced on the line due to the change of electromagnetic field generated by lightning current near the overhead line during a lightning strike. The magnitude and characteristics of this induced voltage depend on many factors, including the amplitude, duration, propagation speed of the lightning current, and the geometry and location of the lightning channel. The calculation of lightning-induced voltage is based on transient electromagnetic field analysis, and the time variation of lightning current is considered. This article uses the RF module of COMSOL Multiphysics software, which provides an interface dedicated to simulating transient physical fields of electromagnetic waves and includes predefined lightning flow models.

The radius of the lightning channel is usually only a few millimeters, so we use the side current signature to describe the path of the lightning current. Since lightning currents often travel at one-third the speed of light and can be further complicated by corona discharge effects around the channel, these factors must be carefully considered in the model. Hin wires are widely found in electrical and electronic systems, and aspect ratios can exceed six orders of magnitude, so it is impractical to mesh wires in volume. There are many methods for modeling thin wires, some of which are based on special assumptions such as the boundary element method (BEM) or the finite Difference Time Domain (FDTD) method, and the main idea for modeling thin wires is usually to place the wires at the edge of a structured grid and manually modify the material properties in the cell containing the wires, as shown in Figure 1.

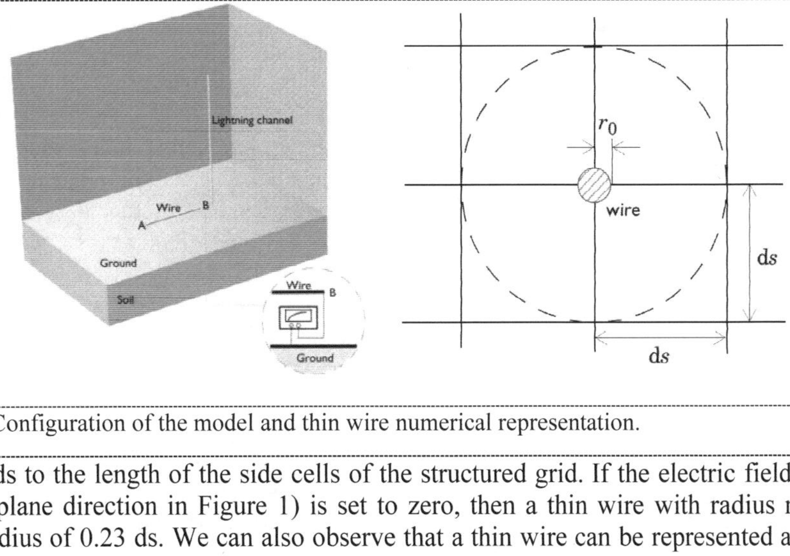

Figure 1. Configuration of the model and thin wire numerical representation.

We set ds to the length of the side cells of the structured grid. If the electric field along the wire axis (out of plane direction in Figure 1) is set to zero, then a thin wire with radius r0 in air has an equivalent radius of 0.23 ds. We can also observe that a thin wire can be represented as equivalent by placing a zero-radius wire in a rectangular prism coaxial with the thin wire, with a cross-sectional area of 2 ds × 2 ds. The corrected relative permittivity and permeability are given by the following formula.

$$\varepsilon_r^* = \frac{\ln(\frac{1}{0.23})}{\ln\frac{ds}{r_0}}, u_r^* = \frac{1}{\varepsilon_r^*} \quad (1)$$

Soil conductivity is influenced by factors like moisture content, temperature, salinity, and soil texture. The electrical conductivity σ (S/m) can be described by the following relation:

$$\sigma = \frac{L}{A} * \frac{I}{V} \quad (2)$$

where: I is the current passing through the soil (in amperes, A), V is the voltage applied across the soil (in volts, V), L is the length of the soil sample (in meters, m), A is the cross-sectional area of the soil through which the current flows (in square meters, m^2). In practice, soil conductivity often depends on factors like soil texture, moisture content, and ion concentration. A more common empirical formula to estimate soil conductivity as a function of temperature and salinity is:

$$\sigma = \sigma_0 * \ (1+bT) \quad (3)$$

where: σ_0 is the conductivity at a reference temperature (typically 25 °C), b is a temperature coefficient (usually around 0.02/°C), and T is the temperature in °C. The inclination of a soil surface can influence the apparent conductivity measured in the field due to the alignment of the electrical current path. The relationship between inclination and conductivity can be expressed in terms of the angle θ\thetaθ of the surface relative to the horizontal. The measured conductivity can be adjusted for inclination with the following relationship:

$$\sigma = \frac{\sigma}{\cos(c)} \quad (4)$$

where: $\sigma_{measured}$ is the conductivity measured along the inclined surface. σ is the conductivity measured on a horizontal surface. *c* is the angle of inclination of the soil surface relative to the horizontal.

2.2. Influence of lightning channel inclination Angle on the induced voltage

The inclination of the lightning channel refers to the deviation angle of lightning flow relative to the vertical direction during the propagation process. The change of channel inclination will directly affect the induced voltage, because the inclination changes the path of lightning current in space, and thus affects the electromagnetic field distribution near the overhead line. In general, a larger inclination will result in a stronger induced voltage because the distance between the lightning current and the overhead line varies more as it travels along the inclined path, which in turn induces a stronger induced effect. Soil electrical conductivity is an important parameter reflecting soil electrical conductivity. High conductivity soils can disperse lightning flow more effectively, reducing the intensity of electromagnetic fields near overhead lines, and thus reducing induced voltage. On the other hand, low-conductivity soil may cause the lightning current to have a shorter conduction path on the ground and concentrate around the overhead line, resulting in a higher induced voltage. Therefore, in the actual transmission line design, it is very necessary to know and measure the conductivity of the soil along the line, and the induced voltage can be reduced by increasing the soil conductivity (for example, increasing soil moisture or applying conductive materials). The calculation of the induced voltage of overhead wire has a high requirement for modeling the precision of thin wire. Because the length of an overhead wire is usually much larger than its diameter, a thin wire has an extremely high aspect ratio. For this situation, the traditional volume grid division method not only consumes a lot of computational resources but also has difficulty in ensuring accuracy. To solve this problem, a simplified modeling technique based on the finite element method is used in this paper, where wires are placed on the edge of a structured grid and the electromagnetic properties of wires are represented by modifying the material properties in the cells containing the wires. The advantage of this method is that it avoids the volume grid division of the wire, which significantly reduces the degree of freedom of the model, and can provide sufficient computational accuracy in engineering applications.

3. Simulation results and discussion

To verify the accuracy of the theoretical model, simulation analysis is carried out in various cases, focusing on the influence of lightning channel inclination and soil conductivity on induced voltage. The simulation results show that the induced voltage at both ends of the overhead line increases significantly with the increase of lightning channel inclination. This result is consistent with theoretical analysis, indicating that the greater the inclination of the lightning channel, the stronger the electromagnetic effect of lightning current near the overhead line, resulting in higher induced voltage. Especially in the case of large inclination Angle, the induced voltage increase is particularly obvious, which indicates that in the lightning-prone area, the transmission line design should avoid choosing the lightning path with large inclination Angle. In addition, the simulation also found that when the lightning channel is close to the vertical, the induced voltage at both ends of the overhead line reaches the minimum value. This is because the electromagnetic field generated by the vertical lightning channel is more evenly distributed in space, and the induction effect on the overhead line is relatively weak. The simulation results of different soil conductivity show that the induced voltage of the overhead line can be significantly reduced by increasing the soil conductivity. Specifically, when the soil conductivity increases from 0.01 S/m to 0.02 S/m, the induced voltage at both ends of the overhead line decreases significantly. This shows that high-conductivity soils can disperse lightning currents more effectively, thereby reducing the intensity of electromagnetic fields near overhead lines and reducing the generation of induced voltages.

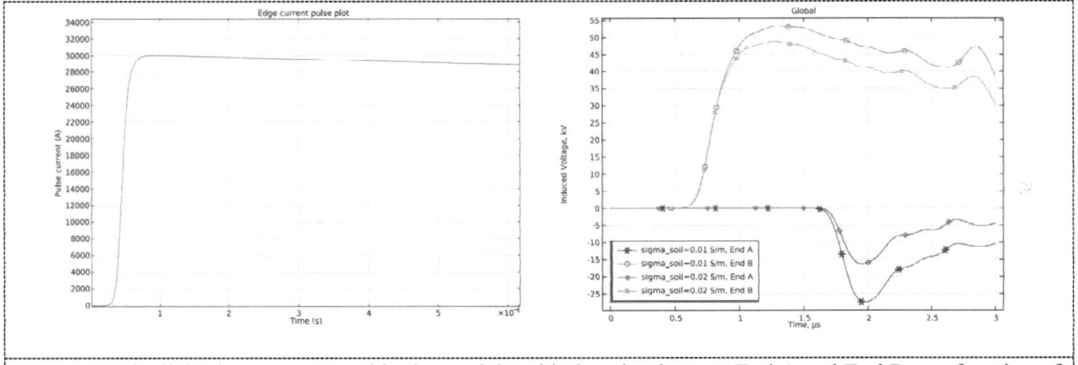

Figure 2. The lightning current used in the model and induced voltage at End A and End B as a function of time under two different soil conductivities.

In the simulation of the thin wire modeling method, the simplified method adopted in this paper not only guarantees the calculation accuracy but also significantly reduces the consumption of computing resources. In particular, the simulation results show that the induced voltage calculated by the simplified modeling method is consistent with the results of the traditional method, which verifies the effectiveness of the method in engineering applications. However, when the wire is too close to other thin wires or objects of study, the simplified method may lead to a decrease in calculation accuracy. To deal with this problem, it is suggested that when using the simplified method, the layout position of the wires should be strictly controlled, and the model should be verified and modified by other numerical methods or experimental data. In addition, future studies can further optimize the method to improve its applicability in complex electromagnetic environments.

4. Conclusion

In this paper, the influence of the inclination Angle of lightning channel and soil conductivity on the induced voltage of overhead line is systematically studied by establishing the calculation model of lightning-induced voltage, and the application of the thin wire modeling method in electromagnetic field simulation is discussed. The simulation results show that the inclination of the lightning channel and soil conductivity are the key factors affecting induced voltage. Improving soil conductivity and choosing the inclination Angle of the lightning channel can significantly reduce the induced voltage of the overhead line, to improve the anti-lightning ability of the transmission line. In addition, the simplified modeling method proposed in this paper can significantly reduce the consumption of computing resources while ensuring calculation accuracy and is suitable for popularization in engineering applications.

Acknowledgment

Thanks to the State Grid Science and Technology Project "Research and Application of Inductive Electricity Multi-parameter Intelligent Sensing and Electric Shock Prevention Technology in Typical Transmission Line Operation Scenarios" No.: 520605240004 for its support.

References

[1] Z. Liu, Y. Chen, X. Yang, and J. Yan, "Power to heat: Opportunity of flexibility services provided by building energy systems," Advances in Applied Energy, vol. 11, p. 100149, 2023/09/01/ 2023, doi: https://doi.org/10.1016/j.adapen.2023.100149.
[2] Z. Liu, Y. Du, C. Song, X. Yang, and J. Yan, "Effect of soil moisture content on thermal performance of ground source heat exchangers: An electromagnetism topology-based analysis," Energy Reports, vol. 10, pp. 3914-3928, 2023/11/01/ 2023, doi: https://doi.org/10.1016/j.egyr.2023.10.061.
[3] Z. Liu et al., "Solar harvest: Enhancing carbon sequestration and energy efficiency in solar greenhouses with PVT and GSHP systems," Renewable Energy, vol. 211, pp. 112-125, 2023/07/01/ 2023, doi: https://doi.org/10.1016/j.renene.2023.04.133.
[4] Z. Liu, X. Yang, H. M. Ali, R. Liu, and J. Yan, "Multi-objective optimizations and multi-criteria assessments for a nanofluid-aided geothermal PV hybrid system," Energy Reports, vol. 9, pp. 96-113, 2023/12/01/ 2023, doi: https://doi.org/10.1016/j.egyr.2022.11.170.
[5] G. Li et al., "Integration of Carbon Dioxide Removal (CDR) Technology and Artificial Intelligence (AI) in Energy System Optimization," Processes, vol. 12, no. 2, doi: 10.3390/pr12020402.
[6] Z. Liu, M. Shi, M. H. Mohammadi, H. Luo, X. Yang, and M. Babaei, "An improved capacitance–resistance model for analysing hydrogen production with geothermal energy utilisation," International Journal of Hydrogen Energy, 2024/08/19/ 2024, doi: https://doi.org/10.1016/j.ijhydene.2024.08.205.
[7] C. Song, Z. Liu, M. Babaei, R. Liu, and G. Hou, "Mitigating low-temperature corrosion in flue-gas heat exchangers for improving thermal storage efficiency in geothermal power plants," Thermal Science and Engineering Progress, vol. 52, p. 102673, 2024/07/01/ 2024, doi: https://doi.org/10.1016/j.tsep.2024.102673.
[8] Sihab M U, Zahid M H, Rahman S F, et al. On the protection of power system: Transmission line fault analysis based on an optimal machine learning approach[J]. Energy Reports,2022,8.
[9] Yaqin Z, Chang X, Yunfeng D, et al. Dual-View Stereovision-Guided Automatic Inspection System for Overhead Transmission Line Corridor[J]. Remote Sensing,2022,14(16).
[10] Yongjie Z, Zhedong H, Qianming W, et al. Multi-Geometric Reasoning Network for Insulator Defect Detection of Electric Transmission Lines[J]. Sensors,2022,22(16).
[11] R. J C, Rosemaira C D, De D T R. Waves and heaviside propagator in transmission lines[J]. Journal of Electromagnetic Waves and Applications,2022,36(13).

AUTHOR INDEX